Advanced Oxidation Processes for Water Treatment

Advanced Oxidation Processes for Water Treatment

Fundamentals and Applications

Edited by
Mihaela I. Stefan

IWA PUBLISHING

Published by **IWA Publishing**
Alliance House
12 Caxton Street
London SW1H 0QS, UK
Telephone: +44 (0)20 7654 5500
Fax: +44 (0)20 7654 5555
Email: publications@iwap.co.uk
Web: www.iwapublishing.com

First published 2018

British Library Cataloguing in Publication Data
A CIP catalogue record for this book is available from the British Library

ISBN: 9781780407180 (Paperback)
ISBN: 9781780407197 (eBook)

Cover images:

TrojanUV system at Orange County Water District, CA, USA. Courtesy of Dr. George Tchobanoglous, UC Davis, CA, USA

RO Membrane filtration system at Orange County Water District, CA, USA. Courtesy of OCWD

WEDECO PDO 1000 ozone generator installed at Sung-Nam water treatment plant, South Korea. Courtesy of WEDECO, a Xylem brand

All other images from istockphoto.com

To all of those who dream big, believe in themselves and work hard to make a difference in the world.

My thoughts go to my parents who taught me the value of perseverance despite humble beginnings and to my family who supported me on this journey.

Mihaela I. Stefan
August 3, 2017

Contents

About the Editor

Dr. Mihaela Stefan received her Ph.D. degree in Photochemistry and Chemical Kinetics from the University of Bucharest, Romania. Dr. Stefan has almost 25 years' experience of academic and industrial research on UV/AOPs for water treatment. She has been affiliated with Western University, London, ON, Canada, and since 2001 with Trojan Technologies as a Senior Research Scientist. In her capacity, Dr. Stefan designed, led, and was directly involved in the execution of various research projects on the fundamentals and application of UV light-based processes, including direct photolysis, UV/H_2O_2 and more recently, UV/chlorine, to water remediation from a variety of environmental micropollutants. Topics included photochemical and radical reactions of chemical contaminants in water sources, degradation kinetics, kinetic modeling and the impact of water quality on UV process performance, reaction mechanisms and byproduct formation, role of post-UV/AOP water treatment steps on the quality of treated water. Dr. Stefan authors book chapters and several articles in peer-reviewed journals, and collaborated on a number of NSERC and Water Research Foundation research projects.

List of Contributors

Tünde Alapi, Ph.D.
Assistant Professor, Environmental Analytical Chemistry, Department of Inorganic and Analytical Chemistry, University of Szeged, Szeged, Hungary
alapi@chem.u-szeged.hu

Eszter Arany, Ph.D.
Research Scientist, Environmental Analytical Chemistry, Department of Inorganic and Analytical Chemistry, University of Szeged, Szeged, Hungary
arany.eszter@chem.u-szeged.hu

Ciara Byrne, B.Sc. (Hons.)
Ph.D. Candidate, Centre for Precision Engineering, Materials & Manufacturing Research; Nanotechnology & Bio-Engineering Division; Department of Environmental Science, Institute of Technology Sligo, Sligo, Ireland
Ciara.Byrne@mail.itsligo.ie

Christophoros Christophoridis, Ph.D.
Researcher, Catalytic-Photocatalytic Processes and Environmental Analysis Laboratory, Institute of Nanoscience and Nanotechnology, NCSR "Demokritos", Athens, Greece
cchrist@chem.auth.gr

Joseph De Laat, Ph.D.
Professor, IC2MP (CNRS 7285), Université de Poitiers, Poitiers, France
joseph.de.laat@univ-poitiers.fr

Dionysios D. Dionysiou, Ph.D.
Professor, Department of Chemical and Environmental Engineering, University of Cincinnati, Cincinnati, OH, USA
dionysios.d.dionysiou@uc.edu

Jörg E. Drewes, Ph.D.
Professor, Chair of Urban Water Systems Engineering, Technical University of Munich, Garching, Germany
jdrewes@tum.de

Polycarpos Falaras, Ph.D.
Research Director, Division of Physical Chemistry, IAMPPNM NCSR "Demokritos", Athens, Greece
papi@chem.demokritos.gr

Alexandra Fishbacher, M.Sc.
Ph.D. Candidate, Instrumental Analytical Chemistry, Faculty of Chemistry, University of Duisburg-Essen, Essen, Germany
alexandra.fischbacher@uni-due.de

Gilbert Galjaard
Chief Technology Officer, PWN Technologies, Velserbroek, The Netherlands
ggaljaard@pwntechnologies.com

Daniel Gerrity, Ph.D.
Assistant Professor, Department of Civil and Environmental Engineering and Construction, University of Nevada – Las Vegas, Las Vegas, NV, USA
daniel.gerrity@unlv.edu

Miguel A. Gracia Pinilla, Ph.D.
Associate Professor, Facultad de Ciencias Físico-Matemáticas, Centro de Investigación en Ciencias Físico Matemáticas (CICFIM), Universidad Autónoma de Nuevo León, San Nicolás de los Garza, N.L., México
miguelchem@gmail.com

Changseok Han, Ph.D.
Post-doctoral Research Associate, Oak Ridge Institute for Science and Education (ORISE), Materials Management Branch, Land and Materials Management Division, NRMRL, ORD, U.S. Environmental Protection Agency, Cincinnati, OH, USA
changseok.han94@gmail.com

Anastasia Hiskia, Ph.D.
Research Director, Catalytic-Photocatalytic Processes and Environmental Analysis Laboratory, Institute of Nanoscience and Nanotechnology, NCSR "Demokritos", Athens, Greece
hiskia@chem.demokritos.gr

Nathalie Karpel Vel Leitner, Ph.D.
Research Director, IC2MP, Université de Poitiers, Poitiers, France
nathalie.karpel@univ-poitiers.fr

Stuart J. Khan, Ph.D.
Associate Professor, School of Civil & Environmental Engineering, University of New South Wales, Sydney, Australia
s.khan@unsw.edu.au

Athanassios G. Kontos, Ph.D.
Senior Research Scientist, Division of Physical Chemistry, IAMPPNM NCSR "Demokritos", Athens, Greece
akontos@chem.demokritos.gr

Erik Koreman, M.Sc.
Scientific Researcher, PWN Technologies, Velserbroek, The Netherlands
ekoreman@pwntechnologies.com

Zsuzsanna Kozmér, M.Sc.
Ph.D. Candidate, Environmental Analytical Chemistry, Department of Inorganic and Analytical Chemistry, University of Szeged, Szeged, Hungary
kozmerzs@chem.u-szeged.hu

Jacob Lalley, M.Sc.
Chemist, Pegasus Technical Services Inc., Cincinnati, OH, USA
jacobmlalley@gmail.com

On-anong Larpparisudthi, Ph.D.
Professor, Department of Environmental Engineering, Chulalongkorn University, Bangkok, Thailand
onny80@gmail.com

Douglas E. Latch, Ph.D.
Associate Professor, Department of Chemistry, Seattle University, Seattle, WA, USA
latchd@seattleu.edu

Bruce R. Locke, Ph.D.
Professor, Department of Chemical and Biomedical Engineering, Florida State University, Tallahassee, FL, USA
locke@eng.fsu.edu

Holger Lutze, Ph.D.
Research Scientist, Instrumental Analytical Chemistry, Faculty of Chemistry, University of Duisburg-Essen, Essen, Germany
holger.lutze@uni-due.de

Ramalinga V. Mangalaraja, Ph.D.
Professor, Department of Materials Engineering, Faculty of Engineering, University of Concepcion, Concepcion, Chile
mangal@udec.cl

Timothy J. Mason, Ph.D.
Professor Emeritus Coventry University; SonoChem Centre Ltd., Kenilworth, Warwickshire
sonochemistry@hotmail.com

Bram Martijn, Ph.D.
Technological Researcher, PWN Technologies, Velserbroek, The Netherlands
bmartijn@pwnt.com

Niall B. McGuinness, Ph.D.
Assistant Lecturer, Centre for Precision Engineering, Materials & Manufacturing Research; Nanotechnology & Bio-Engineering Division; Department of Environmental Science, Institute of Technology Sligo, Sligo, Ireland
McGuinness.Niall@itsligo.ie

Selma Mededovic Thagard, Ph.D.
Associate Professor, Department of Chemical and Biomolecular Engineering, Clarkson University, Potsdam, NY, USA
smededov@clarkson.edu

Christopher J. Miller, Ph.D.
Senior Research Associate, School of Civil & Environmental Engineering, University of New South Wales, Sydney, Australia
c.miller@unsw.edu.au

Mallikarjuna N. Nadagouda, Ph.D.
Physical Scientist, Water Resources Recovery Branch, WSD, NRMRL, ORD, U.S. Environmental Protection Agency, Cincinnati, OH, USA
nadagouda.mallikarjuna@epa.gov

Kevin O'Shea, Ph.D.
Professor, Department of Chemistry and Biochemistry, Florida International University, Miami, FL, USA
osheak@fiu.edu

Larysa Paniwnyk, Ph.D.
Reader, Faculty of Health and Life Sciences, Coventry University, Coventry, UK
apx122@coventry.ac.uk

Suresh C. Pillai, Ph.D.
Professor, Centre for Precision Engineering, Materials & Manufacturing Research; Nanotechnology & Bio-Engineering Division; Department of Environmental Science, Institute of Technology Sligo, Sligo, Ireland
Pillai.Suresh@itsligo.ie

Fernando L. Rosario-Ortiz, Ph.D.
Associate Professor, Department of Civil, Environmental and Architectural Engineering, University of Colorado at Boulder, Boulder, CO, USA
Fernando.Rosario@Colorado.EDU

Torsten C. Schmidt, Ph.D.
Professor, Chair of Instrumental Analytical Chemistry and Centre for Environmental and Water Research, Faculty of Chemistry, University of Duisburg-Essen, Essen, Germany
torsten.schmidt@uni-due.de

Krisztina Schrantz, Ph.D.
Assistant Professor, Environmental Analytical Chemistry, Department of Inorganic and Analytical Chemistry, University of Szeged, Szeged, Hungary
sranc@chem.u-szeged.hu

Virender K. Sharma, Ph.D.
Professor, Program for the Environment and Sustainability, Department of Environmental and Occupational Health, School of Public Health, Texas A&M University, College Station, Texas, USA
vsharma@sph.tamhsc.edu

Holly Shorney-Darby, Ph.D.
Senior Research Engineer, PWN Technologies, Velserbroek, The Netherlands
hshorney@pwntechnologies.com

Benjamin D. Stanford, Ph.D.
Senior Director, Water Research and Development, American Water, Voorhees, NJ, USA
Ben.Stanford@amwater.com

Mihaela I. Stefan, Ph.D.
Senior Research Scientist, Trojan Technologies, London, ON, Canada
mstefan@trojanuv.com

László Szabó, Ph.D.
Researcher, Institute for Energy Security and Environmental Safety, Centre for Energy Research, Hungarian Academy of Sciences, Budapest, Hungary; Department of Organic Chemistry and Technology, Budapest University of Technology and Economics, Budapest, Hungary
szabo.laszlo@energia.mta.hu

Erzsébet Takács, Ph.D.
Professor, Institute for Energy Security and Environmental Safety, Centre for Energy Research, Hungarian Academy of Sciences, Budapest, Hungary; Óbuda University, Sándor Rejtő Faculty of Light Industry and Environmental Engineering, Budapest, Hungary
erzsebet.takacs@energia.mta.hu

Theodoros Triantis, Ph.D.
Senior Researcher, Catalytic-Photocatalytic Processes and Environmental Analysis Laboratory, Institute of Nanoscience and Nanotechnology, NCSR "Demokritos", Athens, Greece
t.triantis@inn.demokritos.gr

Susan Wadley, M.Sc. Eng.
Freelance Writer and Water Treatment Specialist, Sydney, Australia
susan.vanhuyssteen2012@gmail.com

T. David Waite, Ph.D.
Professor, School of Civil & Environmental Engineering, University of New South Wales, Sydney, Australia
d.waite@unsw.edu.au

Troy Walker
Water Reuse Practice Leader, Hazen and Sawyer, Tempe, AZ, USA
twalker@hazenandsawyer.com

Jianlong Wang, Ph.D.
Professor, Laboratory of Environmental Technology, Institute of Nuclear and New Energy Technology, Tsinghua University, Beijing, People's Republic of China
wangjl@mail.tsinghua.edu.cn

Eric C. Wert, Ph.D., P.E.
Project Manager, Southern Nevada Water Authority (SNWA), Las Vegas, NV, USA
eric.wert@snwa.com

László Wojnárovits, Ph.D.
Professor, Institute for Energy Security and Environmental Safety, Centre for Energy Research, Hungarian Academy of Sciences, Budapest, Hungary
wojnarovits.laszlo@energia.mta.hu

Lejin Xu, Ph.D.
Laboratory of Environmental Technology, Institute of Nuclear and New Energy Technology, Tsinghua University, Beijing, People's Republic of China
xulejin@gmail.com

Radek Zboril, Ph.D.
Professor, Regional Centre of Advanced Technologies and Materials, Department of Physical Chemistry and Experimental Physics, Faculty of Science, Palacký University in Olomouc, Olomouc, Czech Republic
radek.zboril@upol.cz

Preface

As a result of climate change, the surface water and groundwater resources are diminishing globally. With decreasing fresh water availability and growing demand for clean water, alternative water sources are used in many parts of the world. At the same time, the increasing environmental pollution with a variety of manmade chemicals with known or unknown effects on the aquatic wildlife and human health is an uncontested reality. With new and more micropollutants in the source waters, the water supply companies are looking for modern, efficient, cost-effective and environmentally friendly technologies for water remediation from both microbial pathogens and chemical contaminants in order to ensure safe drinking water.

Advanced water treatment technologies involve chemical (oxidation), physical (separation) or combination of chemical and physical processes which remove organic and inorganic contaminants of various structures and properties. *Advanced oxidation processes* (AOPs) rely on the *in situ* generation of powerful oxidizing radical species and on their high reactivity toward a wide range of micropollutants. Over the years, extensive studies were dedicated to the theory and application of AOPs, as well as to exploring new, *green* AOP-based technologies. Given their proven efficacy at contaminant removal from water, attractive economics and sustainability, a number of AOPs have been implemented at water facilities around the world for surface water, groundwater and municipal wastewater treatment and water recycling.

This book provides an overview of the most studied AOPs, some of which are largely implemented for water remediation. The fundamental principles, kinetic modeling, water quality impact on process performance, byproduct formation, economics, examples of research and pilot studies, full-scale applications and future research needs are discussed for each advanced oxidation process. In addition to the AOP chapters, the book includes five chapters dedicated to specific topics such as advanced treatment for drinking water production at PWN Water Supply Company North-Holland, AOPs for water reuse, AOPs for municipal and industrial wastewater, photochemistry in the aquatic environment, and green technologies for water treatment. Although not intended to be a comprehensive review on AOPs, the book provides the most recent scientific and technological achievements in this field, and aims to be useful to anyone interested in water treatment processes, including water industry professionals, consulting engineers and scientists, university professors, researchers, and students.

The authors of the book chapters are renowned scientists and water industry professionals. Many of them are leading experts in their fields. I am very grateful to everyone for the outstanding contribution and support for this project.

Mihaela I. Stefan

Chapter 1

A few words about *Water*

Mihaela I. Stefan

The first recorded evidence on "water management" in the history of humanity is considered the macehead of King Scorpion II of the Upper Egypt (*ca.* 2725–2671 B.C.), which also attests the existence of the king. The macehead discovered by the British archeologists James E. Quibell and Frederick W. Green in 1897–1898 illustrates the king holding a hoe, an ancient agricultural hand-tool, interpreted as being used in the ritual involving the pharaoh ceremonially opening the dikes to flood the fields for irrigation. About the same time (*ca.* 2500 B.C.), the most advanced urban settlement of ancient Indus civilization was built – Mohenjo-Daro, now a UNESCO Heritage Site in Larkana District, Pakistan's Sindh Province. Aside from an impressive architectural urban planning, the city of Mohenjo-Daro had a sophisticated water network for providing fresh water to people and for effluent disposal. It was estimated that at least 700 wells were built vertically above or below the ground of wedge-shaped, standard size bricks and engineered to withstand the lateral pressure on 20 m or deeper wells. Most wells were located in private buildings, but one or more public wells were also constructed for each block of buildings, and could be accessed directly from the main streets (Jensen, 1989). The wells were covered to prevent water evaporation and salt crystallization. The wastewater and other sewage of almost every house were channeled into underground cylindrical pipes along the main streets. The archeological site revealed private baths paved with high quality bricks and surrounded by a low brick rim; the effluent was discharged either into a soak pit or to urban sewage drainage system. Jensen (1989) mentions that the Mohenjo-Daro waterworks (inner-urban water supply and effluent disposal system) "were developed to a perfection" which was surpassed only 2000 years later by Romans and "flowering of civil engineering and architecture in classical antiquity".

The above brief note is meant to recognize not only the astonishing achievements of civilizations which existed more than 4500 years ago, but also people's concern to ensure their self-sufficiency and survival through independent water supply and water conservation in a densely populated semi-arid climate.

In our era, water, the natural resource with no substitute, is under unprecedented increasing demand. Although three-quarters of the world is covered by water, over 97% of the planet's water is salt water and less than 3% is fresh water. Approximately 70% of this fresh water is frozen in glaciers and polar ice caps, 29% is stored in underground aquifers and only ~1% of the world's fresh water supply is in rivers, lakes and streams. The impact of climate change on water affects implicitly the Earth's ecosystem, thus our society.

Population growth, urbanization and higher standards of living, industrial expansion and agriculture, and regional imbalances will continue to increase the water demand globally, thus to diminish the fresh water availability. The water demand distribution among these factors varies largely from country to country. UNESCO estimated the cost of adapting to the climate change impact due to a 2°C rise in global average temperature to be from US$70 billion to US$100 billion per year between 2020 and 2050. (http://www.unesco.org/fileadmin/MULTIMEDIA/HQ/SC/pdf/WWDR4%20Background%20Briefing%20Note_ENG.pdf).

As of April 2017, the world's population is believed to have reached 7.5 billion and it is predicted to increase to 10 billion by the year 2056. The population growth is paralleled by increasing global demand for food (*e.g.* expected to go up by 70% by 2050), with the livestock product demand trend ascending rapidly. Production of meat, dairy products and fish are more water (2.9-fold), energy (2.5-fold), fertilizer (13-fold) and pesticides (1.4-fold) demanding than production of vegetables (Angelakis *et al.* 2016).

Through Resolution 64/292 of July 28, 2010, the United Nations General Assembly explicitly recognized the access to water and sanitation as a human right (http://www.un.org/waterforlifedecade/human_right_to_water.shtml). The water should be sufficient for personal and domestic uses, safe (free from microorganisms, chemical substances and radiological hazards that constitute a health threat), acceptable with respect to color, odor and taste, physically accessible *i.e.* within or in the immediate vicinity of the household, educational, health or workplace institution, and affordable. About one billion people do not have access to safe drinking water and as reported in 2010, 2.6 billion people in the world did not have access to adequate sanitation facilities (http://www.unesco.org/fileadmin/MULTIMEDIA/HQ/SC/pdf/WWDR4%20Background%20Briefing%20Note_ENG.pdf). Approximately one-half of the world's population depends on polluted water sources and *ca.* one billion people consume agricultural products originating from lands irrigated with raw or inadequately treated wastewater (Bougnom & Piddock, 2017).

In response to water scarcity trends, engineering solutions to augmentation of the existing supplies were implemented in various geographies around the world. Examples include seawater desalination, wastewater recycling, and bulk water transfer between catchments. Global and regional energy evaluation for water in the context of "water-energy nexus," which links the water demand for energy production with the energy required supplying, treating, and to deliver the water, is the topic of numerous publications (*e.g.* Liu *et al.* 2016; & references therein). Lifecycle assessment (LCA) became a common tool for the evaluation of environmental sustainability of water and wastewater treatment technologies (Friedrich, 2002; Stokes & Horvath, 2006; Lyons *et al.* 2009; Law, 2016).

Currently there are approx. 15,000 water desalination plants in the world (Brookes *et al.* 2014). In the Middle East, desalination satisfies approx. 70% of the water needs in the region. Saudi Arabia is the largest producer of desalinated water, houses the world's largest desalination plant operated with solar photovoltaic energy, and it is predicted that by 2019 all the country's desalination plants will be powered by solar technology (https://www.fromthegrapevine.com/innovation/5-countries-cutting-edge-water-technology#).

Wastewater recycling became an alternative source of water around the world in regions impacted by drought, scarce fresh water resources, and high-water demand. The main purpose of reclaimed water is for direct or indirect potable reuse (groundwater recharge, surface water augmentation), or for non-potable reuse (e.g., landscape, golf course and agriculture irrigation, seawater barrier, industrial and commercial use, natural system restoration, wetlands and wildlife habitat, geothermal energy production). The earliest large water reclamation plant for direct potable reuse was built in Windhoek, Namibia, in 1968. In late 1990s, the plant was no longer technologically up-to-date. A new, larger plant was built and it has been into operation since 2002. The municipal wastewater secondary effluent which was held for 2–4 days into maturation ponds is mixed with surface water (9:1 ratio) then treated following the multiple barrier approach. The multi-step, fully-automated treatment train also includes advanced water treatment consisting of ozonation, biological and granular activated carbon filtration, and ultrafiltration prior to chlorination and distribution (21,000 m^3/day). The water quality is extensively monitored daily along the entire treatment process and its values must adhere to a number of drinking water guidelines and standards, including WHO Guidelines, EU Drinking Water Directive, Rand Water (South Africa) Potable Water Quality Criteria and the Namibian Guidelines (NamWater). Treated water is also used for aquifer recharge (Lahnsteiner & Lempert, 2007).

According to the Australian Bureau of Statistics, in 2010/2011 the major water sources for public consumption were surface waters (92%) and groundwater; a small fraction was provided by water reclamation and desalination plants (Dolnicar *et al.* 2014). In Australia, 13% of households use rainwater from private collection tanks for potable purposes, mostly in the rural areas. Although the risk from consuming rainwater is low in most areas, it is acknowledged that the water from collection tanks is not as well managed and treated as the water from municipal network supplies. The high use of rainwater is linked to the public perception on the quality of water in the distribution network. Dolnicar *et al.*'s study (2014) on the 'water case' in Australia showed how widely the public's perception of different kinds of water – bottled, recycled, desalinated, tap and rainwater – and of their attributes could be, in the context of public acceptance of water from alternative sources.

One common problem to all untreated water sources – natural or alternative – is microbial and chemical pollution. The worldwide occurrence of chemical pollutants at nanogram/L to microgram/L levels in the aquatic environment is well documented in the published literature, and the research on their fate and potential toxicity to the aquatic species expands rapidly. Surface waters and groundwater are impacted by both naturally occurring micropollutants and contaminants originating from human agricultural and industrial activities, and wastewater effluent discharge. Among the naturally occurring water contaminants are cyanotoxins and a wide range of taste-and-odor (T&O)-causing compounds which are released by algal cells (e.g. cyanobacteria and chrysophytes) during the algal bloom seasons which are favored by nutrient levels in the water, high air and water temperatures and sunlight. While most of the T&O compounds are non-toxic, yet impact the water aesthetics, cyanotoxins (e.g. microcystins, saxitoxins, cylindrospermopsin, anatoxin-a, domoic acid – a marine toxin, nodularins) are potent toxins and a potential health threat to humans. Inorganic species such as arsenic, chromium, manganese, iron, vanadium, etc. are among naturally occurring contamination originating from local geology and catchment conditions. Examples of other classes of water pollutants include pesticides (e.g. *s*-triazine pesticides and their metabolites, glyphosate, bromacil, chlortoluron, metaldehyde, linuron, mecoprop), industrial solvents (e.g. trichloroethene, tetrachloroethene, 1,4-dioxane, chlorinated and non-chlorinated aromatic compounds, methyl-*tert*-butylether), human and veterinary drugs, natural and synthetic hormones, ingredients in domestic and personal care products (e.g. caffeine, benzotriazoles, parabens), plasticizers, polyaromatic hydrocarbons, perfluorinated compounds.

Reclaimed water requires advanced treatment due to the presence of a plethora of low molecular weight, neutral molecules such as nitrosamines, 1,4-dioxane, disinfection byproducts, endocrine disruptors, etc. which pass through the membrane filtration steps.

Environmental protection agencies and public health organizations around the world issued quality standards for drinking water and, in many jurisdictions, for the wastewater effluents in order to control and prevent the pollution of receiving waters and to protect the aquatic environment, its habitats and biodiversity. In addition, quality standards specific to the reclaimed water are set and must be met. WHO Guidelines for drinking water quality provide the framework for public health protection and risk management recommendations. Under the Safe Drinking Water Act (SDWA) of 1996, the U.S. EPA enforces the national primary drinking water regulations for 91 contaminants, including microorganisms, disinfectants, disinfection byproducts, radionuclides, and other inorganic and organic compounds. The 1996 SDWA Amendments set the process for further contaminant regulations and standard setting through Contaminant Candidate Lists (CCLs) and subsequent Regulatory Determinations (RDs). Over 100 organic and inorganic compounds are currently listed on the latest CCL (CCL4) to be considered for regulation upon RD. CCL4 includes ~40 pesticides and their degradation products, ~30 industrial solvents, five *N*-nitrosamines, naturally occurring contaminants of which three cyanotoxins and five inorganic compounds, eight hormones or hormone-like compounds, two perfluorinated substances, as well as compounds from other classes such as explosives, ozonation byproducts, pharmaceuticals (human and veterinary drugs), food and cosmetics ingredients. A number of these contaminants already fall under federal or state notification, action or advisory levels in drinking water and reclaimed water for aquifer recharge. In February 2017, the European Commission initiated the process of revision of the Council Directive 98/83/EC on the quality of water intended for human consumption (Drinking Water Directive). Health Canada Drinking Water Quality Guidelines (2014) were established by the Federal-Provincial-Territorial Committee on Drinking Water (CDW) and include five key microbiological parameters, maximum acceptable concentrations (MACs) for approx. 90 organic and inorganic contaminants and six radiological parameters. Australian Drinking Water Guidelines (ADWG) 2011–6 updated in November 2016 provide recommendations for assessing the drinking water quality which ensure safe and good water as well as how that can be achieved from catchment to delivery. The ADWG list health and aesthetics-based guidelines for a wide range of microbial and chemical water pollutants, but those are not mandatory standards. In July 2012, the Chinese Government has established National Standards on Drinking Water Quality with respect to 106 compounds and classes of compounds (e.g. disinfection byproducts) to be monitored in urban water supplies nationwide. The maximum permissible levels in drinking water for many chemical contaminants vary largely among the sets of standards and guidelines mentioned herein.

Maintaining and enhancing the quantity and quality of water, assessment and control of contaminants, development and optimization of sustainable, cost-effective and environmentally-friendly treatment processes and technologies, benchmarking performance indicators, monitoring and modeling the contaminants and their degradation byproducts and managing the impact of treatment process on treated water quality, development of robust bioassays and benchmarking bioactivity of treated water, are just a few examples of trends and challenges in water science, research and technology.

The implementation of water treatment processes will continue to be driven by the water regulations. The suitability of advanced oxidation processes (AOPs) for water and wastewater remediation was demonstrated in numerous research studies. However, only very few AOPs are implemented at water treatment plants around the world to remove micropollutants from either drinking water sources or wastewater effluents for water recycling purposes. This book aims to provide the reader with an overview of the science and application of a number of advanced oxidation processes, as well as to share the authors' views on the future research needs. By comparison to the engineered AOPs, one chapter of this book describes similar processes naturally occurring in the aquatic environment. Three chapters cover specific applications of AOPs, i.e. at a state-of-the art drinking water treatment plant, in water reuse projects around the world, and for wastewater remediation. One last chapter describes briefly the "green" processes for water remediation.

1.1 REFERENCES

Angelakis A. N., Mays L. W., De Feo G., Salgot M., Laureano P. and Drusiani R. (2016). Topics and challenges on water history. In: Global Trends & Challenges in Water Science, Research and Management. 2nd Edition, H. Li (ed.), International Water Association (IWA), London, UK (http://www.iwa-network.org/wp-content/uploads/2016/09/IWA_GlobalTrendReport2016.pdf).

Bougnom B. P. and Piddock L. J. V. (2017). Wastewater for urban agriculture: a significant factor in dissemination of antibiotic resistance. *Environmental Science and Technology*, **51**, 5863–5864.

Brookes J. D., Carey C. C., Hamilton D. P., Ho L., van der Linden L., Renner R. and Rigosi A. (2014). Emerging challenges for the drinking water industry. *Environmental Science and Technology*, **48**, 2099–2101.

Dolnicar S., Hurlimann A. and Grün B. (2014). Branding water. *Water Research*, **57**, 325–338.

Friedrich E. (2002). Life-cycle assessment as an environmental management tool in the production of potable water. *Water Science and Technology*, **46**(9), 29–36.

Jansen M. (1989). Water supply and sewage disposal at Mohenjo-Daro. *World Archaeology: Archaeology of Public Health*, **21**(2), 177–192.

Lahnsteiner J. and Lempert G. (2007). Water management in Windhoek, Namibia. *Water Science and Technology*, **55**(1–2), 441–448.

Law I. B. (2016). An Australian perspective on DPR: technologies, sustainability and community acceptance. *Journal of Water Reuse and Desalination*, **6**(3), 355–361.

Liu Y., Hejazi M., Kyle P., Kim S. H., Davis E., Miralles D. G., Teuling A. J., He Y. and Niyogi D. (2016). Global and regional evaluation of energy for water. *Environmental Science and Technology*, **50**(17), 9736–9745.

Lyons E., Zhang P., Benn T., Sharif F., Li K., Crittenden J., Costanza M. and Chen Y. S. (2009). Life cycle assessment of three water supply systems: importation, reclamation and desalination. *Water Science and Technology: Water Supply*, **9**(4), 439–448.

Stokes J. and Horvath A. (2006). Life cycle assessment of alternative water supply systems. *International Journal of Life Cycle Assessment*, **11**(5), 335–343.

Chapter 2

UV/Hydrogen peroxide process

Mihaela I. Stefan

2.1 INTRODUCTION

Historically, the advanced oxidation processes (AOPs) were described as chemical degradation reactions which involve hydroxyl radical ($^{\bullet}OH$) in the primary step. Glaze *et al.* (1987) defined AOPs as "near ambient temperature and pressure water treatment processes which involve the generation of hydroxyl radicals in sufficient quantity to effect water purification". The first study on the degradation of organic compounds through a process which at that time was not known as proceeding through hydroxyl radical reactions was published by Fenton (Fenton, 1894). Fenton observed that tartaric and racemic acids are degraded in solutions in the presence of trace amounts of ferrous (Fe^{2+}) salts and hydrogen peroxide but he did not investigate the mechanism of this reaction. Haber and Weiss (1932) were the first to propose that Fe^{2+} reduces H_2O_2 with formation of hydroxyl radicals.

Although the photochemical decomposition of H_2O_2 was extensively investigated from the early 1900's through the 1950's, the first study on $^{\bullet}OH$ production *via* H_2O_2 photolysis for the purpose of organic contaminant destruction in aqueous waste streams was reported in 1975 (Koubeck, 1975). The understanding of the process principles, its applicability and engineering design evolved rapidly over the past few decades.

Advanced experimental and analytical techniques, mathematical simulations, computational fluid dynamics and radiation intensity field distribution models are available to predict the reaction mechanisms, chemical and photochemical kinetics and accurate process performance, and to optimize the reactor design and process engineering and controls. The UV/H_2O_2 process is also the most commercially implemented AOP. The first UV system using the UV/H_2O_2 process was installed in 1992 in Gloucester (ON, Canada) to treat 1,4-dioxane (pump & treat with aquifer recharge application). To the best of our knowledge, the UV/H_2O_2 process was first time implemented to water treatment for public consumption in 1998 in Salt Lake City, UT, to remove tetrachloroethene (PCE) from contaminated groundwater. Currently, numerous full-scale UV/H_2O_2-based systems are installed at water utilities around the world to treat micropollutants in contaminated surface waters, groundwater, and wastewater tertiary effluents for water reuse.

This chapter aims at providing an overview on the UV/H_2O_2 AOP with regard to process fundamentals and performance-impacting factors, kinetic modeling and reaction mechanisms, UV reactor design and optimization, case studies and process economics, byproduct formation and mitigation strategies. Since in most UV/H_2O_2 applications the target and/or co-existing contaminants are degraded to a certain extent upon UV photon absorption, the direct UV photolysis process principles and applications will be also briefly covered in this chapter. Although the primary steps in $^{\bullet}OH$-initiated *vs.* photon-initiated degradations are mechanistically different, reactive radicals from chemical bond cleavage are generated in both processes. Often, in the presence of dissolved oxygen, the primary radicals in the two processes follow similar oxidative pathways leading to degradation byproducts of the parent pollutants. Natural water sources and municipal wastewaters contain a variety of inorganic and organic compounds which generate hydroxyl radicals either through light absorption (direct photolysis) or indirectly *via* reactions with other dissolved species. The $^{\bullet}OH$-driven AOPs from these species are not discussed in this chapter.

2.2 ELECTROMAGNETIC RADIATION, PHOTOCHEMISTRY LAWS AND PHOTOCHEMICAL PARAMETERS

2.2.1 Electromagnetic radiation

Light has both wave and particle properties. Maxwell's theory showed that light and sound are characterized by wave-like properties (reflection, refraction, diffraction, interference and polarization), and called them electromagnetic waves. The particle properties of light are associated with the absorption and emission processes. They also explain the photochemical reactions and the photoelectric effect, based on Planck's quantum theory of radiation. *Planck's Law of Radiation* (equation 2.1) expresses the wave/particle duality of light, i.e., light comes in discrete packages of energy E, called photons or quanta, of specific frequency (ν, s^{-1}) and wavelength (λ, m^{-1}).

$$E_\lambda = h\nu = \frac{hc}{\lambda} = hc\bar{\nu} \tag{2.1}$$

where, $h = 6.6260755 \times 10^{-34}$ J · s is Planck's constant, c is the velocity of light in a vacuum (2.99792458×10^8 m s^{-1}), and $\bar{\nu}$ is the wave number (m^{-1}, usually given in cm^{-1}). One mole of photons (quanta), i.e., 6.0221367×10^{23} photons mol^{-1} (Avogadro's number, N_A) is one *einstein* and carries a wavelength-dependent energy (equation 2.2):

$$E_\lambda = N_A\frac{hc}{\lambda} = \frac{0.11962658}{\lambda}\ \text{J einstein}^{-1} \tag{2.2}$$

The UV radiation covers the wavelength range between X-ray and visible regions of the electromagnetic spectrum, i.e., from 4 to 400 nm (Koller, 1965); according to other sources, the UV range is 10–400 nm. Figure 2.1 displays the electromagnetic spectrum from 100 to 1000 nm.

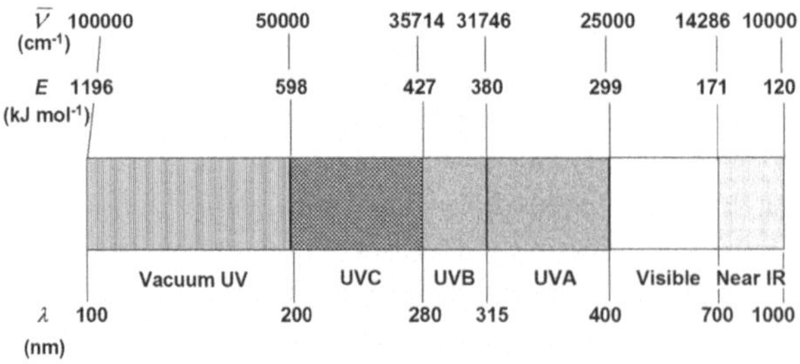

Figure 2.1 Spectrum of the electromagnetic radiation from 100 to 1000 nm (adapted from Phillips, 1983).

The UV spectral range of interest in UV light-based water treatment for environmental contaminant and microbial pathogen removal is the UV-C region (200–280 nm), where many chemical compounds, DNA and hydrogen peroxide absorb the UV radiation. Particular AOP applications use the vacuum UV range (VUV, 100–200 nm) in which water is the main light absorber (see Chapter 5). In principle, any radiation carrying energy equal or higher than the chemical bond dissociation energy (BDE, ΔE°, J mol^{-1}) can break the respective bond. The probability of such process to occur depends on two factors: the strength of light absorption, which is described by the optical properties of chemical compound, and the probability of the excited state generated through the light absorption event to undergo a chemical reaction; the descriptors of these two factors are the molar absorption coefficient and the quantum yield.

2.2.2 Photochemistry laws

The *First Law of Photochemistry*, known as Grotthus (1817) – Draper (1843) law, states that "Only the light which is absorbed by a molecule can be effective in producing photochemical change in the molecule" (Calvert & Pitts, 1966).

When the light beam passes through an absorbing medium, its intensity is attenuated. The quantitative aspects of this process are captured by the Lambert (1760) – Beer (1852) Law. The law expression is given in equation 2.3:

$$P_{\lambda} = P_{\lambda}^{\mathrm{o}} \times 10^{-(\alpha_{\lambda} + \varepsilon_{\lambda,i} C_i) l} \tag{2.3}$$

where P_{λ} and P_{λ}^{o} are the transmitted and the incident spectral radiant power (W m^{-1}, common unit W nm^{-1}) of radiation of wavelength λ, respectively; C_i (mol L^{-1}) is the concentration of compound i in the irradiated solution; l (m, common unit cm) is the pathlength traversed by the radiation, and α_{λ} and $\varepsilon_{\lambda,i}$ are the decadic attenuation coefficient of the medium (if water, commonly referred as the 'water background') (m^{-1}, common unit cm^{-1}) and the decadic molar absorption coefficient (m^2 mol^{-1}, common unit M^{-1} cm^{-1}) of target compound i, at wavelength λ, respectively. In UV light-based applications, the concentrations of chemical pollutants are very low (usually at μg/L levels), thus, their contribution to the overall light absorption by the water matrix is negligible. The term $(\alpha_{\lambda} + \varepsilon_{\lambda,i}\, C_i)$ is the water absorption coefficient a_{λ} (m^{-1}, frequently given in cm^{-1}), and is measured experimentally with a spectrophotometer. The term $(\alpha_{\lambda} + \varepsilon_{\lambda,i}\, C_i)\, l$ is the *absorbance* A_{λ} of a solution, and it is defined at a specific wavelength λ and for a given water layer l. Absorbance is an additive parameter (Beer's Law), i.e., should more than one absorbing species be present in the solution, the total absorbance of the solution is the sum of the individual absorbance values of all constituents within the given water layer l at that specific λ. Absorbance is related to the *transmittance* T_{λ} of a solution through the following expressions:

$$T_{\lambda} = \frac{P_{\lambda}}{P_{\lambda}^{\mathrm{o}}} = 10^{-A_{\lambda}}; \quad A_{\lambda} = -\log\, T_{\lambda} \tag{2.4}$$

Transmittance is commonly expressed as percentage at a given wavelength and for a given water layer thickness (*e.g.*, $T_{254\,\mathrm{nm},1\,\mathrm{cm}} = 95\%$). Both absorbance and transmittance are unitless, and are key characteristics of any water considered for treatment with UV light-based processes.

The Lambert-Beer law assumes that the interactions between the molecules of various solutes in the water matrix are negligible, and, given their low concentrations, chemical pollutants absorb the light independently from one another. Consequently, their degradations are examined independently from one another. Deviations from the law would be observed at very large concentrations of absorbers, when formation of dimers or other molecular aggregates could take place, which case, the linear relationship between absorbance and chemical compound concentration is no longer obeyed.

According to the *First Law of Photochemistry*, the light absorbed by a chemical compound could effect a chemical transformation. The fraction of light absorbed by compound i of concentration C_i in the water exposed to monochromatic radiation of λ is given by equation 2.5, whereas the general expression of light absorbed by i over a wavelength range emitted by a polychromatic light source is given by equation 2.6.

$$F_\lambda^{\text{abs}} = \frac{\varepsilon_{\lambda,i} C_i}{a_\lambda}(1 - 10^{-a_\lambda l}) \tag{2.5}$$

$$E_{p,\lambda} = \int_{\lambda_1}^{\lambda_2} \frac{\varepsilon_{\lambda,i} C_i}{a_\lambda}(1 - 10^{-a_\lambda l}) E_{p,o\lambda}\, d\lambda \tag{2.6}$$

where $E_{p,\lambda}$ and $E_{p,o\lambda}$ are the absorbed and initial photon fluence rate, respectively. The SI unit for photon fluence rate is $m^{-2}\ s^{-1}$. When used on a chemical amount basis, $E_{p,o\lambda}$ is divided by the Avogadro's constant, and the unit becomes mol $m^{-2}\ s^{-1}$ or einstein $m^{-2}\ s^{-1}$. The light absorbed can be also expressed in terms of photons (quanta) absorbed per time unit ($q_{p,\lambda}$, s^{-1}) or, on a moles of photons basis, einstein s^{-1}. According to the IUPAC nomenclature, the symbols for terms referring to moles of photons include "n" in the subscript (e.g., $E_{n,p,o\lambda}$, einstein $m^{-2}\ s^{-1}$; $q_{n,p,\lambda}$, einstein s^{-1}). For simplicity, "n" will not be included in the symbols used in this chapter; einstein-based terms will be properly identified by their units. The reader is referred to the IUPAC Glossary of Terms and Definitions used in Photochemistry (Braslavsky, 2007) for review of the terms and symbols used in this chapter.

Equations 2.5 and 2.6 show that the fraction of light absorbed by a compound is proportional to the molar absorption coefficient and the concentration of chemical compound, and inversely proportional to the water absorption coefficient. The larger the fraction of light absorbed by the chemical pollutant, the more efficient is the direct photolysis process. Similarly, for the $^{\bullet}OH$–initiated oxidations, the larger the fraction of light absorbed by H_2O_2, the higher the yield of $^{\bullet}OH$, thus the more efficient would be the UV/H_2O_2 process (except the H_2O_2 concentration conditions where H_2O_2 becomes a significant competitor for $^{\bullet}OH$).

The *Second Law of Photochemistry*, known as Stark–Bodenstein (1908–1913) and Einstein (1905, 1912) law, states that "The absorption of light by a molecule is a one-quantum process, so that the sum of the primary quantum yields φ must be unity" (Calvert & Pitts, 1966). Powerful light sources (e.g., lasers) can generate very high concentrations of excited states and bi-photonic absorption could occur during the lifetime of the excited state (μs or tens of nanoseconds) originating from the same molecule. Such photo-physical processes do not occur during the water treatment with the light sources used in the UV reactors.

Upon quantum absorption by the molecules in their ground states, unstable, short-lived, high-energy excited states are generated. The electronic transitions with formation of singlet or triplet excited states are of various types, depending on the molecular orbitals involved in the transition. For example, $\pi \rightarrow \pi^*$ transitions occur from unsaturated C=C bonds, aromatic rings, (s)-triazine structures, when one π electron from a bonding orbital is promoted to a higher energy π^* anti-bonding molecular orbital; $n \rightarrow \pi^*$ transitions occur in hetero-atom containing compounds, carbonyl compounds, S—S bonds, etc., when one electron from the non-bonding (n) paired electrons is promoted to an upper π^* energy electronic level. For example, the absorption spectrum of NDMA (Figure 2.2) exhibits two absorption bands, one strong with $\lambda_{max} = 228$ nm ($\varepsilon_{max} = 7378\ M^{-1}\ cm^{-1}$), and one weak, of lower energy, with $\lambda_{max} = 352$ nm ($\varepsilon_{max} = 109\ M^{-1}\ cm^{-1}$). The two bands correspond to a $\pi \rightarrow \pi^*$ intramolecular charge transfer and to a $n \rightarrow \pi^*$ transition, respectively. The electronic excited states can evolve toward radical species as a result of bond dissociation or can deactivate with energy losses *via* radiative (fluorescence or phosphorescence) and/or non-radiative (heat) transitions to the ground state. Schematic representation of photo-physical processes occurring from the ground and excited states of a molecule is given in the Jablonski diagram (Calvert & Pitts, 1966). For in-depth coverage of theoretical and experimental photochemistry, the reader is referred to photochemistry treatises among which Calvert and Pitts (1966); Parker (1968); Wöhrle *et al.* (1998); Turro *et al.* (2010).

2.2.3 Photochemical parameters

Two fundamental photochemical parameters characteristic to chemical compounds absorbing the UV radiation will be discussed in this section: molar absorption coefficient and quantum yield. Both parameters determine the efficiency of direct photolysis process of micropollutants. Hydrogen peroxide photolysis in the UV/H_2O_2 process is also driven by these parameters, as discussed in section 2.4.1.

2.2.3.1 Molar absorption coefficients

The molar absorption coefficient ε_λ (M^{-1} cm^{-1}) was introduced briefly in section 2.2.2. This parameter is a measure of the ability of a molecular structure to absorb the radiation of specific wavelengths from the electromagnetic spectrum. The magnitude of molar absorption coefficients is linked to the energy gap between the highest occupied and the lowest unoccupied molecular orbital i.e., E_{LUMO}–E_{HOMO}. The molar absorption coefficients of a chemical compound are determined experimentally from the linear relationship between absorbance and the concentration of that particular compound at the selected wavelengths. The graphical representation of molar absorption coefficients versus λ is the absorption spectrum of that compound. Given the solute-solvent interactions, the molar absorption coefficients (i.e., absorption spectrum) are solvent-dependent. In some cases, micropollutants which display weak acid/base properties can exist in protonated or deprotonated forms, or the mixture of thereof at the water pH. In general, the absorption spectra of the two forms are different; therefore, the molar absorption coefficients could be pH-dependent. Figures 2.2 give examples of absorption spectra of selected micropollutants and illustrate the pH-dependency of sulfamethoxazole (SMX) molar absorption coefficients.

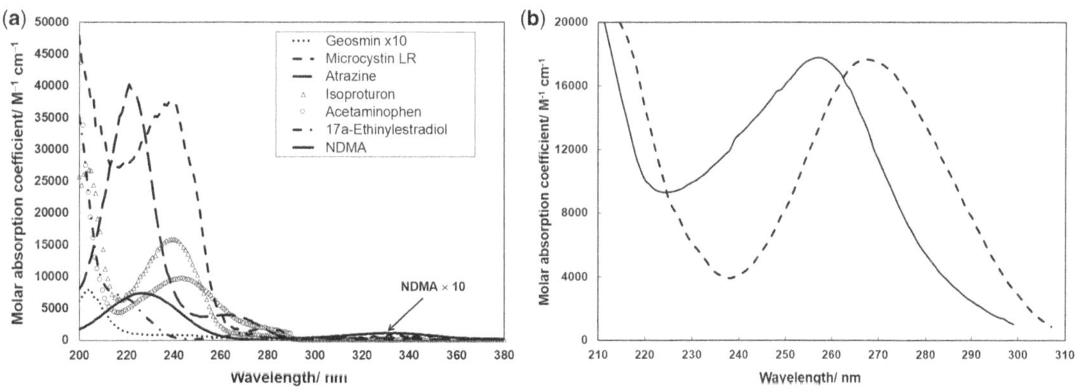

Figure 2.2 Absorption spectra of selected micropollutants (a) and of SMX (b): (- - -) protonated and (–) deprotonated forms of SMX ($pK_{a2} = 5.9$; adapted from Carlson *et al.* 2015).

Despite the large molar absorption coefficients of some micropollutants (e.g., atrazine, microcystin LR, acetaminophen, sulfamethoxazole in Figure 2.2) in the UV range, due to their low concentrations in water sources, the fractions of light absorbed by these compounds are negligible as compared to those absorbed by the water background constituents. Chemical pollutants such as 1,4-dioxane, methyl-*tert*-butyl ether

(MTBE), trichloroethanes, metaldehyde, lack chromophores on their molecular structures, thus, they do not absorb the UV radiation and do not undergo degradation through direct photolysis.

2.2.3.2 Quantum yield

According to the *Second Law of Photochemistry*, the "primary quantum yields" cannot exceed unity. One must distinguish between primary quantum yields (φ) and reaction (overall) quantum yields (Φ). A primary quantum yield φ_i is associated with a single process *i* followed by a fraction of molecules absorbing radiation of λ; such primary processes are collisional deactivation, fluorescence, radiationless transitions, decomposition to primary radicals. Primary quantum yields are of significant theoretical importance, but they are very difficult to measure and require fast kinetics monitoring techniques. The quantum yields usually determined in laboratories and used with the kinetic models for UV treatment performance predictions are the reaction quantum yields (Φ). The IUPAC definition of reaction quantum yield is the number of defined events per photon absorbed by the system. For a photochemical reaction, the following expressions define the quantum yield:

$$\Phi(\lambda) = \frac{\textit{number of moles of reactant transformed or product formed}}{\textit{number of moles of photons of } \lambda \textit{ absorbed}} \tag{2.7}$$

$$\Phi(\lambda) = \frac{-\dfrac{\mathrm{d}[C]}{\mathrm{d}t} \quad (\mathrm{mol\,L^{-1}\,s^{-1}})}{I_a(\lambda) \quad (\mathrm{mol\,L^{-1}\,s^{-1}})} \tag{2.8}$$

where I_a (λ) (mol L^{-1} s^{-1}) is the rate of light absorption by compound *C*, i.e., moles of photons (einsteins) per liter per unit of time. Note that Φ is always determined based on the *absorbed* light. Quantum yield is a dimensionless parameter. Quantum yield is defined only for monochromatic radiation, i.e., at specific λ. In the case of polychromatic light source UV reactors, the quantum yields at radiation wavelengths emitted by the lamp which overlap the absorption spectrum of target compound must be known and used with the kinetic expression of contaminant decay developed over the wavelength range of interest. $\Phi \ll 1.0$ indicates significant deactivation of the excited state *via* processes which did not result in any chemical reaction; $\Phi > 1.0$ indicates secondary reactions involving the original reactant, such as radical reactions and chain reactions.

Quantum yields are measured either directly, using equation 2.8 or indirectly; in the latter case, the direct photolysis kinetics of target compound is compared to that of a 'reference' ('probe') compound added to the solution. The irradiations are performed under 'vanishing absorption' conditions ($A < 0.02$; von Sonntag and Schuchmann, 1992), and the molar absorption coefficients at λ for both target and probe compounds, as well as Φ_λ (probe) must be known. For practical methods for Φ determination, the reader is referred to the published literature; e.g., Johns (1971); Zepp (1978); Leifer (1988); Schwarzenbach *et al.* (1993); Bolton and Stefan (2002); Bolton *et al.* (2015).

The quantum yield could be wavelength-, pH-, temperature-, chemical compound concentration-, solvent-, and dissolved oxygen (DO) level-dependent.

Chemical compounds with chromophoric moieties in their molecular structures absorb the electromagnetic radiation and their absorption spectra display one or more absorption bands. In general, each absorption band corresponds to a specific electronic transition, thus Φ is frequently independent of wavelength in that absorption band (Zepp, 1978; Schwarzenbach *et al.* 1993, and references therein). Examples of organic compounds with wavelength-dependent quantum yields include: isoproturon, $\Phi_{250\,\mathrm{nm}} = \sim 0.045$; $\Phi_{275\,\mathrm{nm}} = \sim 0.0045$; diuron, $\Phi_{253.7\,\mathrm{nm}} = \sim 0.022$; $\Phi_{296\,\mathrm{nm}} = \sim 0.0014$ (Gerecke *et al.* 2001);

diazinon, $\Phi_{253.7\,nm} = \sim 0.082$; $\Phi_{313\,nm} = \sim 0.012$; chlorpyrifos, $\Phi_{253.7\,nm} = 0.016$; $\Phi_{313\,nm} = 0.052$ (Wan *et al.* 1994); nitrite from nitrate, $\Phi_{205\,nm} = \sim 0.215$; $\Phi_{214\,nm} = \sim 0.180$; $\Phi_{220\,nm} = 0.172$; $\Phi_{230\,nm} = 0.149$; $\Phi_{240\,nm} = 0.097$; $\Phi_{253.7\,nm} = 0.065$; $\Phi_{270\,nm} = \sim 0.02$; $\Phi_{300\,nm} = \sim 0.0094$ (Goldstein & Rabani, 2007).

Temperature and pH-dependent quantum yields for selected contaminants are listed in Table 2.1. As mentioned above, depending on their pK_a, some compounds could exist in protonated and non-protonated forms at pH relevant to drinking water sources. Both molar absorption coefficients and quantum yields could be different for the two forms. Since the photolysis first-order rate constants depend on both photochemical parameters, implicitly, they will depend on pH. Figure 2.3 shows the pH effect on photon fluence-based rate constant of sulfamethoxazole photolysis at 253.7 nm (see absorption spectra of SMX in Figure 2.2, and $\Phi_{253.7\,nm,\,pH\text{-}SMX}$ in Table 2.1).

Table 2.1 Examples of pH and temperature-dependent quantum yields.

Compound	$\Phi_{253.7\,nm}$	pH	*t* (°C)	Reference
Linuron	0.0382	5	20	Benitez *et al.* (2006)
	0.0345	7	20	
	0.0367	9	20	
	0.0550	7	40	
	0.0192	7	10	
Chlorotoluron	0.0309	2	20	
	0.0328	9	20	
	0.0533	7	40	
	0.0148	7	10	
Sulfamethoxazole	0.212	<~4		Canonica *et al.* (2008)
	0.046	>7.5		
Triclosan	0.36	<6.2	24	Carlson *et al.* (2015)
	0.20	>10		
N-Nitrosodimethylamine (NDMA)	0.28–0.31	2–8		Lee *et al.* (2005a, b); Stefan *et al.* (2002).
	0.41–0.46		25	Soltermann *et al.* (2013)
	~0.09 → ~0.06	9 → 11		Lee *et al.* (2005a)
	0.27 → 0.16	10 → 12		Stefan *et al.* (2002)

Since the fluence (UV dose)-based rate constants are directly proportional to Φ (Bolton & Stefan, 2002), accurate quantum yield data at the pH and temperature characteristic to the water treated at the water facility must be known for UV equipment sizing and performance prediction. As shown in Figure 2.3, the photon fluence-based rate constant for SMX photolysis varies by up to ~60% over the pH range characteristic to drinking water sources (6.5–8.2). As the apparent molar absorption coefficient (ε^{app}) is *quasi*-constant over that pH range, the variation in $k^{app}_{E^o_p}$ is mostly due to changes of Φ with pH due to the varying ratios of protonated and non-protonated SMX. The fluence required to degrade 90% of SMX in water at pH ~6.5 would be approx. 60% lower than that required to achieve the same treatment level at pH ~ 8.2.

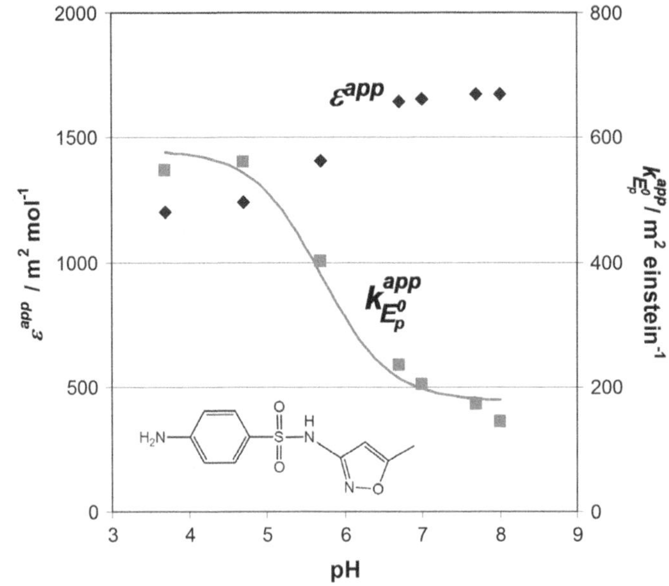

Figure 2.3 Photon fluence-based rate constant of SMX at 253.7 nm as a function of pH; the solid line represents the predicted $k^{app}_{E^0_p}$ (Canonica *et al.* 2008).

The quantum yields of two pesticides, namely, atrazine and metazachlor were found concentration-dependent (Hessler *et al.* 1993); e.g., $\Phi_{253.7\text{ nm, atrazine}}$ was found 0.047 at 3 μM and 0.028 at 160 μM; $\Phi_{253.7\text{ nm,metazachlor}}$ decreased from 0.44 at 4 μM to 0.23 at 360 μM. Lee *et al.* (2005b) determined $\Phi_{253.7\text{ nm,NDMA}} = 0.28$ at pH 7 at various NDMA concentrations (0.001, 0.005, 0.010, and 0.050 M). Sharpless and Linden (2003) reported $\Phi_{\text{NDMA}} = 0.3$ at both 253.7 nm and over the wavelength range emitted by a medium-pressure Hg lamp in synthetic laboratory water of pH ~8 spiked with 1×10^{-6} M NDMA. From these two studies, one concludes that Φ_{NDMA} is both concentration- and wavelength-independent.

The quantum yield for chloride ion formation from chloroacetic acid ($ClCH_2COOH$) photolysis is wavelength-, temperature-, and solvent-dependent (Neumann-Spallart & Getoff, 1979). The wavelength-dependency of Φ (e.g., $\Phi_{213.9\text{ nm},30°\text{C}} = 0.5 \pm 0.1$; $\Phi_{253.7\text{ nm},30°\text{C}} = 0.35 \pm 0.05$) is explained by the cage effects which drive the homolytic scission of C—Cl bond, whereas the solvent influences the polarity of the transition state which is formed during the hydrolysis of the excited state ($\Phi_{253.7\text{ nm},100\%\text{ water}} = 0.35 \pm 0.05$; $\Phi_{253.7\text{ nm},45\%\text{ water}+55\%CH_3OH} = 0.67$).

The $\Phi_{253.7\text{ nm}}$ data in Table 2.1 were determined at ambient DO levels in water. Lee *et al.* (2005a) examined the role of DO on $\Phi_{253.7\text{ nm,NDMA}}$ and measured a $\Phi_{253.7\text{ nm,NDMA}} = {\sim}0.31$ in O_2-saturated conditions. In N_2–saturated solutions, $\Phi_{253.7\text{ nm,NDMA}}$ was ~0.26–0.30 at pH 2–4, and then decreased rapidly from ~0.30 at pH 4 to ~0.14 at pH 8, and to almost 0 at pH 10. Quantum yield of tetrachloroethene (PCE) at 253.7 nm was determined as 0.84 and 0.34 in the presence and in the absence of O_2, respectively (Mertens & von Sonntag, 1995).

The combination of large molar absorption coefficients and quantum yields over the radiation wavelengths emitted by the light source could make direct photolysis an effective process for micropollutant treatment in contaminated waters. Examples of contaminants amenable by UV radiation (253.7 nm or polychromatic 200–400 nm) include NDMA and other aliphatic nitrosamines, dimethylnitramine, trietazine, triclosan, sulfamethoxazole, sulfisoxazole, diclofenac, ketoprofen, RDX (hexahydro-1,3,5-trinitro-1,3,5-triazine).

Photochemistry of organic and inorganic matter occurring in natural waters plays an important role in both aquatic environment and light-based water treatment processes. Reactive species among which, triplet states, singlet oxygen ($^1O_2$), $^•OH$, superoxide radical ($O_2^{•-}$), H_2O_2, are formed as a result of the absorption of UV-Vis radiation by water constituents such as chromophoric dissolved organic matter (CDOM), nitrate, metal complexes. These reactive species may contribute to the degradation of micropollutants whose molecular structures are prone to energy-, electron-, or proton-coupled electron transfer from $^3CDOM^*$, and/or reactions with $^1O_2$, $^•OH$, $O_2^{•-}$, H_2O_2, thus, enhancing the reaction quantum yield determined in pure water. Such 'externally'-induced degradation processes are known as *photosensitized processes*, sometimes incorrectly named 'indirect photolysis'. Given the low concentrations of reactive species generated through water matrix photochemistry, the contribution of these species to the overall degradation of contaminants in the engineered water treatment technologies is rather small. However, these processes are responsible for micropollutant abatement in natural aquatic environment (see Chapter 13).

Photochemical parameters for numerous chemical compounds of environmental and human health concern, including emerging micropollutants, have been determined and are available in the public domain. In addition to the studies discussed above, the following references provide quantum yield data on selected categories of micropollutants, such as: endocrine disrupting compounds (Mazellier & Leverd, 2003; Gmurek *et al.* 2017); parabens (Gmurek *et al.* 2015); nitrosamines (Plumlee & Reinhard, 2007); chlorinated biphenyls (Dulin *et al.* 1986; Langford *et al.* 2011); pesticides (Draper, 1987; Nick *et al.* 1992; Palm & Zetzsch, 1996; Millet *et al.* 1998; Fdil *et al.* 2003; Benitez *et al.* 2006); pharmaceuticals (Boreen *et al.* 2004; Pereira *et al.* 2007; Yuan *et al.* 2009; Rivas *et al.* 2011; Wols *et al.* 2014; Lian *et al.* 2015); algal toxins (Afzal *et al.* 2010; He *et al.* 2012); benzotriazoles and benzothiazoles (Bahnmüller *et al.* 2015); polychlorinated 1,3-butadienes (Lee *et al.* 2017); RDX (explosive, Bose *et al.* 1998); chlorinated dioxins and furans (Dulin *et al.* 1986; Nohr *et al.* 1994); various micropollutants (Baeza & Knappe, 2011; Lester *et al.* 2012; Wols & Hofman-Caris (2012a) and references therein; Benitez *et al.* 2013; Shu *et al.* 2013). One particular class of compounds of both human and environmental health concern is the disinfection byproducts (DBPs). Chuang *et al.* (2016) reported molar absorption coefficients, quantum yields, and fluence-based rate constants at 253.7 nm for twenty five DBPs. Large variations among quantum yields (0.11 to 0.58) and molar absorption coefficients (10 to 1370 M^{-1} cm^{-1}) were observed, which translated into a wide range of fluence-based rate constants (0.05×10^{-4} to 26.5×10^{-4} cm^2 mJ^{-1}).

2.3 UV RADIATION SOURCES

Current light sources generate either continuous or pulsed UV radiation. This section is a brief overview on commercial UV lamps used in the UV systems at water treatment plants and on UV radiation sources under development, currently employed in research studies.

2.3.1 Blackbody radiation

The emission of electromagnetic radiation can be related to the concept of 'blackbody radiator'. A perfectly black body absorbs completely all radiation incident upon it, irrespective of wavelength and angle, and neither reflects nor transmits any of this radiation. A perfect absorber is also a perfect emitter (Kirchhoff's law; Calvert & Pitts, 1966). The total radiant power emitted by a blackbody M (W m^{-2}) depends on the fourth power of temperature T (K) (Stefan-Boltzmann law, 1879), and the maximum of the emittance spectrum (λ_{max}, m) shifts toward lower wavelengths as the temperature increases (Wien displacement law, 1894) (Phillips, 1983):

$$M(T) = 5.67 \times 10^{-8}\, T^4 \;(\sigma \approx 5.67 \times 10^{-8}\ \mathrm{W\,m^{-2}\,K^{-4}} \text{ is the Stefan-Boltzmann constant});$$
$$\lambda_{max} = 2.8978 \times 10^{-3}\, T^{-1} \tag{2.9}$$

Planck's quantum theory (1901) postulates that the energy of the electric oscillators in the atoms in the blackbody walls from which the radiation originated has discrete values, thus, the emission occurs in discrete packages (quanta) of given frequency (energy; see equation 2.1). This theory explained the concept of blackbody radiation, and allowed the derivation of the equation of the energy density distribution in a given wavelength range ($\lambda + \mathrm{d}\lambda$, m). Equation 2.10 gives the Planck's Law on distribution of spectral radiant emittance (Phillips 1983):

$$M_\lambda(T) = \frac{2\pi h c^2}{\lambda^5 (e^{hc/(k_B \lambda T)} - 1)} = \frac{3.742 \times 10^{-25}}{\lambda^5\,(e^{14.39 \times 10^{-3}/\lambda T} - 1)}\ \mathrm{W\,m^{-2}\,nm^{-1}} \tag{2.10}$$

where $k_B = 1.38 \times 10^{-23}$ J K^{-1} is Boltzmann's constant, and the other symbols were defined previously. Equation 2.10 is valid for an *idealized* blackbody radiator; for real radiators, the expression above is multiplied by the *spectral emissivity* factor $\varepsilon(\lambda, T)$ $(0 < \varepsilon(\lambda, T) \leq 1)$, and depends on both wavelength and temperature. Equation 2.10 shows that in order to provide a significant output in the UV-C range, an emitter should operate at very high temperatures. Such very high temperatures are reached in some gas-discharge lamps (e.g., pulsed UV lamps) for very short durations.

2.3.2 Mercury vapor-based UV light sources for water treatment

The lamp technology for water treatment evolved rapidly, particularly over the past two decades. Technological advancements regard enhanced- and application-tailored UV output, wide nominal power range, design, stability and lifetime, robust and long-life power supplies such as high electrical efficiency (>95%)–, variable output power– and high frequency electronic drivers (ballasts), instrumentation and methodology for accurate spectral radiant power measurements, sophisticated models for predicting plasma behavior as a function of various parameters including plasma composition and internal pressure, science of materials used for the lamp components, UV sensors for monitoring fluence rate in UV reactors.

The UV lamps designed for water treatment (disinfection and AOPs) have significant output in the 200–300 nm wavelength range. The Hg vapor-based lamps contain Hg in its physical state (liquid) or in amalgams and a rare gas (argon, neon, xenon, helium), most commonly argon (Ar). Mercury has relatively high vapor pressure and low ionization energy, is *quasi*-chemically inert toward the quartz and electrode materials, and its resonance lines are in the UV range. The lamp assembly has a few components and the lamps come in various lengths and tube diameters.

The artificial light sources emit either monochromatic or polychromatic radiation of wavelengths determined by the nature of the emitting atoms or molecules, concentration and energy of the excited states, and the probability of the radiative transitions. The wavelength, sharpness, and relative intensity of the emitted lines depend on the pressure of both Hg and rare gas. The source of energy for most light sources is electricity, and the excited states are mostly generated *via* collisions with highly energetic electrons, accelerated in the electric field. Grotrian energy-level diagram for mercury atoms in their ground and excited states (singlet and triplet states) illustrates the radiative transitions between Hg energy levels and helps the reader understand the origin of Hg lines in mercury-vapor lamps (Calvert & Pitts, 1966).

The resonance lines of Hg atoms are at 184.9 nm and 253.7 nm. Upper excited states of Hg can be reached and radiative emissions from those states occur depending on the Hg vapor pressure. The fundamentals of photophysics and photochemistry of atoms and molecules are extensively covered in the Photochemistry books. Details on the principles and applications of Hg-atom based lamps are given in Phillips (1983). The characteristics of conventional light sources were described by Gould (1989), Braun *et al.* (1991), Masschelein (2000), Rivera (2001), Schalk *et al.* (2006) and references therein; also available on the websites of UV lamp manufacturing companies.

2.3.2.1 Low-pressure (LP) Hg vapor arc lamps

The standard LP Hg lamp operates in either arc ('hot' cathode) or glow discharge mode ('cold' cathode). The filling is a mixture of mercury and an inert gas, typically argon. The lamp is operated at ~40–42°C and the Hg vapor pressure is 1–10 mbar. The role of argon is to initiate and maintain the discharge, to reduce the required starting voltage for discharge, to act as a buffer protecting the electrodes from sputtering and evaporation, thus extending the lamp lifetime. When the lamp is switched on, the electrons are accelerated in the electric field. The collisions between the energized electrons and atoms are elastic and result in energy transfer with the generation of excited states of the atoms (Hg* and Ar*). The radiative deactivation of Hg* states to the ground state in LP lamps result in the resonance mercury lines of 253.7 and 184.9 nm. Radiative transitions from the upper excited states of the Hg atoms also occur, but their intensities are very low as compared to the 253.7 nm line, given the low populations of those excited states. Figure 2.4a shows the relative radiant emittance from a LP Hg lamp.

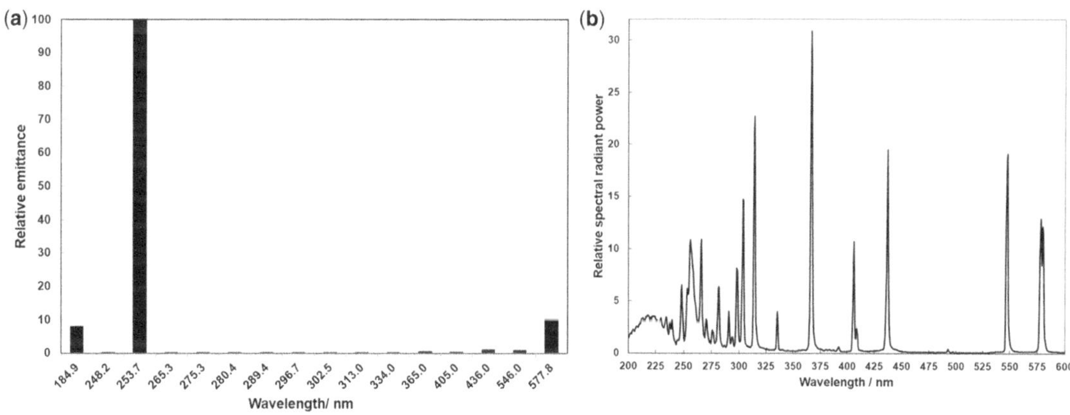

Figure 2.4 Relative spectral distribution of the radiation emitted from (a) LP and (b) MP Hg vapor arc lamps (Stefan, 2004).

The 253.7 nm radiation accounts for most of the lamp radiative emission; therefore, the LP lamps are considered *quasi*-monochromatic light sources. The lamp emits approx. 8–10% in the VUV range (184.9 nm) and approx. 8% in the visible range. The 184.9 nm radiation is filtered by the ordinary quartz, but is transmitted by the high quality quartz; once absorbed by water, hydroxyl radicals are generated and a VUV-AOP is initiated (see Chapter 5).

The radiant power efficiency at 253.7 nm of current commercial LP Hg lamps can be up to ~40%. The output depends on the operating temperature. The concentration of Hg atoms in the discharge must be controlled in order to obtain the maximum UV-C output from the input electrical energy. The optimum pressure is approx. 0.006 torr (Light Sources Inc, 2013).

A better version of standard LP lamp is the LP high-output (LPHO) lamp, whose electrode assembly has a "dead" volume behind the filaments ("cold spot temperature region") which needs to be held at ~40°C in order to achieve optimum efficiency at 253.7 nm. The lamp can be operated at wall temperatures higher than 50°C, and the specific lineal power per arc length (mW/cm) is higher than that of standard LP lamps (Schalk *et al.* 2006).

The impact of ambient temperature on lamp efficiency was overcome by introducing the LP amalgam lamps, in which Hg is introduced as amalgams with bismuth, indium, or gallium – typically with indium. The "long life technology" based on effective coatings developed by Voronov *et al.* (2003) ensures more than 90% of the initial UV-C output over >16,000 h of lamp operation. Table 2.2 compares the key features of various LP lamps.

Table 2.2 Characteristics of commercial LP Hg-vapor arc lamps; adapted from Schalk *et al.* (2006) and Light Sources Inc (2013).

Feature	TiO_2 Doped Doped Quartz, Soft Glass	Natural or Synthetic Fused Quartz		Natural or Synthetic Fused Quartz Amalgam Lamp
		Standard	High Output	
Emitted radiation	253.7 nm	184.9 nm, 253.7 nm	184.9 nm, 253.7 nm	184.9 nm, 253.7 nm
Wall temp. (°C)	30–50	30–50	>50, cold spot 40	90–200
Current (A)	0.2–0.5	0.3–0.4	0.8–1.3	1.2–5.0
Arc length (cm)	7–148	5–155	27–200	27–200
Electrical power (W)	5–80	5–80	10–150	40–1000
Specific electrical power* (W/cm)	0.25–0.3	0.3–0.5	0.5–1.0	1.0–5.0
Specific UV-C radiant power* (mW/cm)	<175	<200	<350	<2000
UV-C (253.7 nm) efficiency (%)	25–35	30–40	25–35	30–38
Lifetime (h) for >90% initial 253.7 nm output	9000 (>85%)	16,000 (>85%)	12,000	16,000

*per unit of arc length.

In addition to the factors mentioned briefly above, the lamp performance depends on the lamp age, ballast operating frequency, frequency of on/off cycles.

Trojan Technologies' Solo Lamp™ (500 or 1000 W) is a powerful amalgam lamp with multiple advantages over the medium pressure Hg arc lamps, among which, high electrical energy efficiency, long life (>15,000 hours), dimmable from 100 to 30% with minimal changes in UV-C output efficiency, low power usage e.g., 1/3 the MP lamp energy usage for the same UV fluence delivered, low carbon footprint, and low environmental impact. These features translate into reduced operation and maintenance costs and low 20 years life cycle assessment, LCA (~1/4 of LCA estimated for MP lamps, based on manufacturing, operation, and disposal) (http://www.trojanuv.com/resources/trojanuv/Products/SoloLamp/SoloLamp_Handout_1000_500W.pdf).

2.3.2.2 Medium-pressure Hg vapor arc lamps

Medium pressure (MP) Hg arc lamps are polychromatic light sources with spectral radiant distribution from VUV to infrared (IR). The emission spectrum (see Figure 2.4b for the significant range) is characterized by multiple lines superimposed on a small amount of continuum radiation (particularly in the short UV range), which is due to the recombination of ions and electrons in the plasma. The MP Hg arc lamps are operated at much higher power density (50–250 W/cm) than the LP lamps. Consequently, all mercury is in the vapor state and the plasma (electrons, Hg vapors, ions, argon) reaches temperatures as high as 5000–7000 K, while the lamp wall temperature reaches 600–950°C. Under those conditions, the Hg vapor pressure is 1–6 bar, i.e., (1–6) × 10^5 Pa (Schalk *et al.* 2006), and plasma constituents reach a "local thermal equilibrium" with very high probability for collisions, thus, a significant amount of Hg* states is generated. The rare gas pressure (argon at a few torr) has a crucial role on the ignition voltage, lamp warm-up characteristics and electrode lifetime, but does not control the steady operation of the lamp unless it is too high (Phillips, 1983). The lamp envelope is made of high purity quartz.

During the warmup time, when the lamp temperature is low, the concentration of Hg vapors is low and the lamp behaves more or less as a LP lamp emitting primarily the 184.9 and 253.7 nm radiation. During operation under steady conditions (very high plasma temperature), the two resonance Hg lines are strongly self-absorbed by the Hg vapors (atoms). Because of the self-absorption of 253.7 nm radiation, the maximum output shifts to a longer wavelength (~256 nm), and the line broadens due to the high pressure. At the same time, collisions between the accelerated electrons and Hg excited states generate upper excited states from which radiation of various wavelengths (dependent on ΔE between the two states involved in transition; see equation 2.1) are emitted. The high pressure and high power density of MP lamps determine the low UV-C efficiency (max. 15%) of these lamps.

The electrical power of MP lamps can be as high as 60 kW, and the arc length within 5–150 cm range; however, the UV-C efficiency per input power is much lower than that of low pressure Hg lamps. The MP lamp lifetime lies within a wide range (1500–10,000) of operating hours.

The spectral radiant power of both LP and MP lamps is measured on 100-h operated lamps. Accurate methods and software-assisted instrumentation are available for that purpose. The lamp ageing over the course of operation in the UV reactors is monitored by "duty" UV sensors installed in the reactors. The accuracy of 'duty' sensor responses are checked periodically against NIST traceable "reference" UV sensors. UV sensors consist of optical and electronic components, and are available commercially in different designs. In all disinfection applications and only recently considered to be introduced in contaminant removal installations, the UV sensor response is used in algorithms computing the UV fluence (UV dose) delivered by the UV reactor. Among recent technological developments on UV sensors is the construction of sensors able to monitor the broadband emission of MP lamps, covering the short wavelength UV radiation ($\lambda < 240$ nm), to which the widely used silicon carbide-based sensors are insensitive. Comprehensive description and investigations into the design and performance guidelines for UV sensors used in UV systems for water treatment can be found in the Water Research Foundation report #91236 (2009). A 360°-view angle micro (0.07 mm^3)-fluorescent silica detector was developed and tested for photon-fluence rate spatial distribution measurements in both LP- and MP-lamp reactors , in waters of various %*T* (Li *et al.* 2017, and references therein).

There are numerous UV systems with either LP- or MP-Hg lamps installed around the world to treat microbial pathogens and/or chemical contaminants in water. The UV equipment (low- or medium-pressure Hg lamps) selection is project-specific.

2.3.2.3 Quartz sleeves

There is a variety of commercially available quartz tubing used to manufacture sleeves for housing the UV lamps in the UV reactors. The quartz selection depends on a number of factors among which the application (disinfection, UV/AOP for micropollutant control, TOC reduction), lamp type, water quality and potential byproduct formation, cost. Figure 2.5 shows the UV transmittance spectra of quartz sleeves most commonly used in UV light-based water treatment. Doped quartz materials are also available if specific spectral regions of the electromagnetic spectrum should be blocked. The TiO_2-doped quartz blocks the high energy, short wavelength radiation. Sometimes, this type of quartz sleeves is used in UV systems treating highly absorbing waters (e.g., nitrate-rich waters) to prevent the formation of regulated or unregulated byproducts from photochemical reactions of water constituents (e.g., nitrite from nitrate). Type 219-quartz or equivalent is required in UV reactors for water disinfection in some jurisdictions to minimize the MP lamp output below 240 nm.

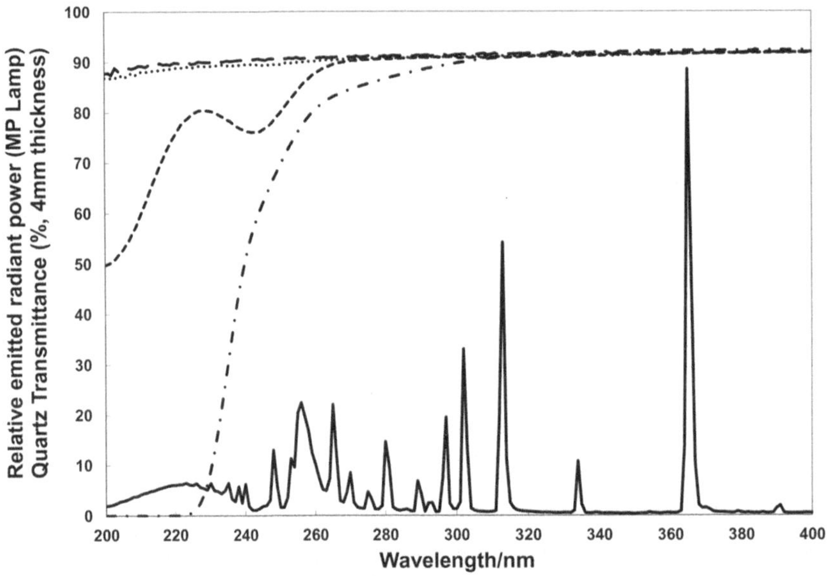

Figure 2.5 Quartz transmittance spectra and typical UV spectral emission from a MP lamp (solid line). Quartz type: suprasil (—–); dry synthetic fused silica (dotted line); standard clear fused silica (- - -); TiO_2-doped quartz (- • - •). Spectra compilation from Stefan (2004) and https://www.heraeus.com/media/media/hqs/doc_hqs/products_ and_solutions_8/tubes/Quartz_lamp_materials_EN.pdf.

The optical, thermal, and mechanical properties of quartz used for the UV lamp sleeves impact both the UV spectral distribution of the radiation entering the water and the sleeve lifetime. Trace chemical elements and/or "frozen network imperfections" in the fused silica matrix affect the optical properties of quartz and its resistance to high energy radiation emitted by the lamps, i.e., its resistance to solarization. Quartz solarization is a known phenomenon and has been investigated and reported in the literature. In short, it consists in gradual transmittance losses due to prolonged exposure of quartz to high intensity radiation, particularly to the VUV radiation but not restricted to that wavelength range. The extent of solarization depends on various factors which include the quartz quality, manufacturing process, radiation wavelength and intensity, duration of exposure, temperature.

Of the fused silica examples shown in Figure 2.5, the standard quartz is the most prone to solarization, particularly when used with MP lamps; suprasil quartz is very resistant to solarization. The transmittance losses of fused quartz due to solarization can be reversed by heat treatment (annealing) at temperatures higher than 500°C (https://www.momentive.com).

The naturally occurring organic matter and inorganic ions (iron and calcium in particular) interact with the UV radiation at the quartz surface, which results in quartz sleeve fouling, thus, transmittance losses (Wait, 2009 and references therein). The UV reactors are equipped with automatic mechanical, chemical or mechanical & chemical sleeve wiping systems which remove the foulants. During long-term operation of the UV system, the inorganic species could alter irreversibly the silica matrix at the sleeve walls through UV radiation-induced local reactions. For a given water background *%T*, in addition to the reactor UV power, the UV sensors respond to both lamp ageing and photochemical and chemical processes altering the quartz sleeve transmittance.

2.3.3 Mercury-free UV lamps

The general trend in science and technology towards the development and implementation of environmentally friendly "green" processes has prompted numerous research studies on Hg-free lamps for water disinfection and contaminant removal. Excimers, Hg-free flash lamps and light emitting diode (LED)-based UV light sources were developed for UV applications. Despite the progress achieved on their technology, these lamps have not reached the overall performance of LP– and MP– Hg vapor arc lamps and are not used in full-scale installations for water treatment.

2.3.3.1 Excilamps

The non-coherent excimer and exciplex lamps are gas-discharge *quasi*-monochromatic light sources. The electromagnetic radiation is produced by either microwave discharges or 'silent' electrical discharges (also called dielectric barrier discharge, DBD) in the same gas or mixture of two gases. The emission spectrum depends on the nature of the fill gas. These light sources operate near room temperature, do not emit IR radiation, and the electric discharge requires high voltages (up to 20 kV) and high frequency ranges (kHz to MHz). The fill gas in these lamps is either a noble gas (Ne, Ar, Kr, Xe) which case, the excited state (e.g., Xe_2^*) emitting the radiation is called *excimer*, or mixtures of a noble gas and a halogen, in which case the excited state (e.g., XeCl*) is called *exciplex*. The excilamps present several advantages over the Hg-vapor lamps, including flexible design geometry, no warmup time, long lifetime (up to 10,000 h depending on design and operating conditions), electrodeless, narrowband emission, high photon flux without absorption of emitted radiation, UV/VUV power efficiency from 1–2 to 60% depending on the fill gas, near room temperature operation, Hg-free, thus, environmentally friendly light sources (Sosnin *et al.* 2006). Spectral emission of excimers and exciplexes, characteristics of excilamps and their limitations, recent advancements in their construction and power supplies, as well as references to the studies in this field are provided in Chapter 5 of this book. Most studies on exciplex lamp applications were in the VUV region and concerned primarily the mineralization of organic compounds. Recently, Matafonova and Batoev (2012) reviewed the research studies on UV radiation from excilamps for contaminant treatment in water by direct photolysis and AOPs, and as a pre-treatment step in combination with biodegradation.

2.3.3.2 Pulsed UV lamps

Pulsed light technology is based on pulsed electrical energy discharges in inert gases (Xe, Ar, Kr). The pulse duration is generally between 50 and 3000 microseconds, and the interval between pulses is in the order of milliseconds. The electrical discharge heats the fill gas (typically Xe) almost instantly to very high temperatures (≥13,000 K) that compresses in a thin 'shell' of plasma (ionized gas, electrons, excited states) on the inner wall of quartz envelope. The heated plasma emits blackbody continuum radiation characteristic to plasma temperature. The shape of spectral emission in the UV range is also plasma-temperature-specific, thus, is determined by the flash lamp operating conditions. Approx. 38–52% of total radiative output can be reached within the 185–400 nm range with an average UV-C efficiency of 15–20% (Haag, 1994). Xenon is the preferred rare gas given its low ionization potential (E^{o} = 12.1 eV; Masschelein, 2000), despite being more expensive than Ar and Kr. The radiation emitted by the plasma is not absorbed by the rare gas. The UV fluence rates from these lamps can be three to four orders of magnitude higher than those from Hg-vapor arc lamps. The characteristics of radiation emitted from pulsed light sources are described in the literature (Phillips, 1983; Marshak, 1984; Haag, 1994; Schaefer *et al.* 2007). An approach to the standardization of fluence determination in bench-scale pulsed light experiments was recently described by Gómez-López and Bolton (2016).

The pulsed UV lamps reach the full-power instantly, and their spectral UV radiation output is driven by the lamp and power supply settings. The excessively high pulse energies (applied voltage as high as 30 kV) increase thermal stresses which shorten drastically the lamp lifetime. Pulsed UV lamps were successfully tested at laboratory scale for both pathogen inactivation and environmental contaminant removal in water. High removal yields in short exposure times were achieved for several water contaminants, among which, NDMA (Liang *et al.* 2003), chlorinated aliphatic compounds (Haag, 1994), pesticides (Baranda *et al.* 2014).

2.3.3.3 Light-emitting diode (LED) lamps

LED-based light sources have been successfully commercialized for more than five decades (since 1962) for indoor and outdoor lighting and are largely used in analytical instrumentation. LEDs convert the electrical energy to radiation *via* electron-hole recombination at the *p-n* junction in the semiconductor chip. The e^--hole (h^+) recombination generates a high energy state which releases a photon upon relaxation; the photon wavelength depends on the semiconductor materials used in LEDs. Electroluminescence in the UV range was reported from doped wide bandgap materials such as diamond (λ = 235 nm; Koizumi *et al.* 2001) and group III element nitrides (e.g., GaN, AlGaN, and AlN), with the shortest wavelength (210 nm) LED (AlN PIN (*p*-type/intrinsic/*n*-type) developed and characterized by Tanyiasu *et al.* (2006). UV radiation covering the 210–365 nm range can be obtained from AlGaN LEDs where AlN and GaN are combined at specific ratios. UV-C LEDs emit in all directions, yet are limited by material and construction to be point source emitters. The optical power density (mW/cm^2) from a 30 mW UV-C LED is approx. 1.8-fold and 50-fold higher than that from a typical MP Hg lamp and a typical LP amalgam lamp, respectively (Pagan & Lawal, 2015). Remarkable advancements were reported in UV LED manufacturing and commercialization, and the output power per single chip of commercial LEDs increased from ~2 mW in 2012 to 30 mW in 2015, with lifetimes in thousands of hours (Pagan & Lawal, 2015).

LEDs can be operated in either pulsed or continuous mode. These light sources present a number of advantages over the Hg-vapor lamps: higher UV power density, are operated of direct current (d.c.), compactness, robustness, instant startup to full power, lower electrical power consumption, longer lifetime, wavelength tenability, flexible geometry and reactor design, environmentally safe. However, currently, the UV-C LEDs are cost-prohibitive, have low wall-plug efficiency and undeveloped product design.

The UV-C LEDs were extensively examined for their capability to inactivate microbial pathogens in water and on surfaces (Song *et al.* 2016). Very limited work was reported on UV-LEDs for micropollutant removal, most of which concerned the UV-A (365 nm) LED/TiO_2 AOP. Autin *et al.* (2013) evaluated Al-In-Ga-nitride LEDs emitting at 255, 310 and 365 nm in combination with either TiO_2 or H_2O_2 for methylene blue and metaldehyde (pesticide) removal in simulated surface water.

In 2012 AquiSense Technologies launched the first UV-C LED product designed for low-flow water disinfection from the PearlAqua platform. In 2017, the company introduced three more UV-C LED compact product lines, one of which is the smallest UV water disinfection system (PearlAqua Micro) in the world.

2.4 UV/H_2O_2 PROCESS FUNDAMENTALS

2.4.1 Photolysis of hydrogen peroxide

Photodecomposition of hydrogen peroxide in aqueous solution upon exposure to UV radiation from a Hg vapor lamp was reported in 1910 (Tian, 1910), and the process was observed to follow a "unimolecular kinetics". Urey *et al.* (1929) determined the molar absorption coefficients of H_2O_2 in water in the 215–375 nm range and discussed the primary dissociation step for the first time. The authors showed that the spectrum is continuous with no structure and, based on thermochemical calculations, they demonstrated that the photon energy for the longest absorbed wavelength would be more than sufficient to cause any of the following three reactions:

$$H_2O_2 + h\nu \rightarrow 2\,^{\bullet}OH \quad (39\ \text{cal}) \tag{2.11}$$

$$H_2O_2 + h\nu \rightarrow H_2O + 1/2\ O_2 \quad (55.8\ \text{cal}) \tag{2.12}$$

$$H_2O_2 + h\nu \rightarrow HO_2^{\bullet} + {}^{\bullet}H \quad (67.3\ \text{cal}) \tag{2.13}$$

Urey *et al.* were the first to postulate the formation of $^{\bullet}OH$ from H_2O_2 photodecomposition and concluded that reaction 2.11 is favored based on the analogy with the primary process occurring in halogens and on their experimental observations during the illumination of low pressure H_2O_2 vapors with zinc spark lines.

Hydrogen peroxide is a weak absorber of UV radiation. The molar absorption coefficients of H_2O_2 and of its conjugate base, hydroperoxide ion HO_2^- ($pK_a = 11.7$) in aqueous solution are available in the published literature. The molar absorption coefficients of HO_2^- are larger than those of H_2O_2; e.g., $\varepsilon(H_2O_2, 210\ nm) = 189 \pm 28\ M^{-1}\ cm^{-1}$, $\varepsilon(HO_2^-, 210\ nm) = 733.7 \pm 63.1\ M^{-1}\ cm^{-1}$; $\varepsilon(H_2O_2, 300\ nm) = 0.96 \pm 0.21\ M^{-1}\ cm^{-1}$; $\varepsilon(HO_2^-, 300\ nm) = 27.3 \pm 1.0\ M^{-1}\ cm^{-1}$ (Morgan *et al.* 1988). At 254 nm, $\varepsilon(H_2O_2)$ is ~19.2 $M^{-1}\ cm^{-1}$. Chu and Anastasio (2005) reported the molar absorption coefficients of H_2O_2 (6.5–26 mM) as a function of wavelength (240–380 nm) and temperature (1–25°C). A steady yet small increase was observed at 25°C relative to 1°C, e.g., from 0.5% at 240 nm to ~6.6% at 280 nm, with 2.2% increase at 254 nm.

Given the poor UV light absorption properties of H_2O_2 relative to those of water matrix constituents, H_2O_2 must be properly dosed in the UV/H_2O_2 applications such that the amount of UV light absorbed by H_2O_2 generates the $^{\bullet}OH$ yield required for meeting the process performance. The optimum H_2O_2 dose is application-specific and depends on several factors among which water quality, lamp type and power, reactor design, contaminant reactivity toward $^{\bullet}OH$, contaminant treatment level, and direct photolysis contribution to the overall treatment. In typical surface water and groundwater, only a small fraction of UV radiation emitted by the light source is utilized to produce $^{\bullet}OH$, and small reductions (up to ~15%) in H_2O_2 concentrations are observed.

Hydrogen peroxide photolysis and subsequent radical reactions in the absence of solutes are described by the following set of reactions:

Light absorption/initiation

$$H_2O_2 + h\nu \rightarrow 2\,{}^{\bullet}OH \quad \phi_{H_2O_2} \cong 0.5;\ \phi_{{}^{\bullet}OH} \cong 1.0 \tag{2.11}$$

$$HO_2^- + h\nu \rightarrow {}^{\bullet}O^- + {}^{\bullet}OH \tag{2.14}$$

Propagation

$$ {}^{\bullet}OH + H_2O_2 \rightarrow H_2O + HO_2^{\bullet} \quad k_{2.15} = 2.7 \times 10^7\ M^{-1}\ s^{-1}\ \text{(avg. of four values; Buxton } et\ al. \text{ 1988)} \tag{2.15}$$

$$k_{2.15} = (3.2 \pm 1.5) \times 10^7\ M^{-1}\ s^{-1}\ \text{(preferred value, Yu, 2004)}$$

$$ {}^{\bullet}OH + HO_2^- \rightarrow OH^- + O_2^{\bullet -} + H^+ \quad k_{2.16} = (7.5 \pm 1.0) \times 10^9\ M^{-1}\ s^{-1}\ \text{(Christensen } et\ al. \text{ 1982)} \tag{2.16}$$

$$ {}^{\bullet}O^- + H_2O_2 \rightarrow OH^- + O_2^{\bullet -} + H^+ \quad k_{2.17} = \leq 5 \times 10^8\ M^{-1}\ s^{-1}\ \text{(Christensen } et\ al. \text{ 1982)} \tag{2.17}$$

$$k_{2.17} = (4.2 \pm 0.2) \times 10^9\ M^{-1}\ s^{-1}\ \text{(preferred value, Yu, 2004)}$$

$$ {}^{\bullet}O^- + HO_2^- \rightarrow OH^- + O_2^{\bullet -} \quad k_{2.18} = 4 \times 10^8\ M^{-1}\ s^{-1}\ \text{(avg. value; Buxton } et\ al. \text{ 1988)} \tag{2.18}$$

$$k_{2.18} = (7.8 \pm 0.8) \times 10^9\ M^{-1}\ s^{-1}\ \text{(preferred value, Yu, 2004)}$$

$$HO_2^{\bullet} + H_2O_2 \rightarrow {}^{\bullet}OH + H_2O + O_2 \quad k_{2.19} = 3.7\ M^{-1}\ s^{-1}\ \text{(Farhataziz \& Ross, 1977)} \tag{2.19}$$

$$O_2^{\bullet -} + H_2O_2 \rightarrow {}^{\bullet}OH + OH^- + O_2 \quad k_{2.20} = (0.13 \pm 0.07)\ M^{-1}\ s^{-1}\ \text{(Weinstein \& Bielski, 1979)} \tag{2.20}$$

Termination

$$HO_2^{\bullet} + HO_2^{\bullet} \rightarrow H_2O_2 + O_2 \quad k_{2.21} = 8.3 \times 10^5\ M^{-1}\ s^{-1}\ \text{(Bielski } et\ al. \text{ 1985)} \tag{2.21}$$

$$HO_2^{\bullet} + O_2^{\bullet -} + H_2O \rightarrow H_2O_2 + O_2 + OH^- \quad k_{2.22} = 9.7 \times 10^7\ M^{-1}\ s^{-1}\ \text{(Bielski } et\ al. \text{ 1985)} \tag{2.22}$$

$$O_2^{\bullet -} + O_2^{\bullet -} + 2\,H_2O \rightarrow H_2O_2 + O_2 + 2\,OH^- \quad k_{2.23} < 0.35\ M^{-1}\ s^{-1}\ \text{(Weinstein \& Bielski, 1979)} \tag{2.23}$$

$$ {}^{\bullet}OH + {}^{\bullet}OH \rightarrow H_2O_2 \quad 2k_{2.24} = 1.1 \times 10^{10}\ M^{-1}\ s^{-1}\ \text{(selected value, Buxton } et\ al. \text{ 1988)} \tag{2.24}$$

$$ {}^{\bullet}OH + HO_2^{\bullet} \rightarrow H_2O + O_2 \quad k_{2.25} = 6 \times 10^9\ M^{-1}\ s^{-1}\ \text{(Buxton } et\ al. \text{ 1988)} \tag{2.25}$$

$$ {}^{\bullet}OH + O_2^{\bullet -} \rightarrow OH^- + O_2 \quad k_{2.26} = 8 \times 10^9\ M^{-1}\ s^{-1}\ \text{(Buxton } et\ al. \text{ 1988)} \tag{2.26}$$

Equilibria of species involved in H_2O_2 photodecomposition mechanism

$$H_2O_2 \rightleftarrows HO_2^- + H^+ \quad pK_a = 11.7 \tag{2.27}$$

$$HO_2^{\bullet} \rightleftarrows O_2^{\bullet -} + H^+ \quad pK_a = 4.7 \tag{2.28}$$

$$ {}^{\bullet}OH \rightleftarrows {}^{\bullet}O^- + H^+ \quad pK_a = 11.9 \tag{2.29}$$

Hydrogen peroxide decomposition proceeds through a primary photolytic step followed by radical-induced reactions. Both H_2O_2 and radical species are in equilibria with their ionized or non-ionized forms, thus, the kinetics and mechanism of H_2O_2 photo-decomposition are pH-dependent. Of all reactions given above, the most important to the overall process at pH values characteristic to natural waters are reactions 2.11, 2.15, 2.21 and 2.22.

The reaction of ${}^{\bullet}OH$ with OH^- ($k = 1.3 \times 10^9\ M^{-1}\ s^{-1}$; Buxton *et al.* 1988) was not included in the above mechanism because the radical formed in this reaction (${}^{\bullet}O^-$) is rapidly protonated to ${}^{\bullet}OH$ at pH values relevant to natural waters or tertiary wastewater effluents.

Reactions 2.15 and 2.19 (and/or 2.20), commonly known as Haber-Weiss cycle, were originally suggested by Haber and Willstätter (1931). Further kinetic investigations of reactions 2.19 and 2.20 led to the conclusion that their probabilities are so low compared to the dismutation reactions 2.21 and 2.22, that those reactions can be ignored (Koppenol (2001) and references therein).

The primary (reaction 2.11) and overall quantum yields (reactions 2.11, 2.15, 2.21 and 2.22) for H_2O_2 photo-decomposition in water in the absence of any ${}^{\bullet}OH$ scavengers were extensively studied and reported. There is a consensus in the published literature that the primary quantum yield is $\phi_{H_2O_2} \cong 0.5$, whereas the overall quantum yield $\Phi_{H_2O_2}$ approaches unity. Studies reporting quantum yields for photodecomposition of H_2O_2 at 253.7 nm, 25–27°C, include Hunt and Taube (1952) ($\phi_{H_2O_2} = 0.49$; $\Phi_{H_2O_2} = 0.98 \pm 0.05$); Weeks and Matheson (1956) ($\phi_{H_2O_2} = 0.490 \pm 0.065$); Baxendale and Wilson (1956) ($\phi_{H_2O_2} = 0.50$; $\Phi_{H_2O_2} = 1.00 \pm 0.02$); Volman and Chen (1959) ($\phi_{H_2O_2} = 0.54 \pm 0.05$; $\Phi_{H_2O_2} = 0.94 \pm 0.05$); Loraine and Glaze (1999) ($\phi_{H_2O_2} = 0.45 \pm 0.06$; $\Phi_{H_2O_2} = 1.1 \pm 0.1$). Quantum yield data diverging from these values were also reported in the literature (see Yu & Barker (2003b) for a critical analysis of those studies). In natural waters, the competition for ${}^{\bullet}OH$ from the matrix constituents reduces the probability of reaction 2.15, and the overall quantum efficiency for photo-decomposition is lower than 1.0 ($0.5 \leq \Phi_{H_2O_2} < 1.0$).

The H_2O_2 overall quantum yield was found independent on wavelength, and on the radiation intensity and H_2O_2 concentration under conditions of high intensity and low concentrations (Baxendale & Wilson, 1956; Volman & Chen, 1959), but it is affected by the temperature, e.g., 0.74 at 2°C and 1.10 at 50°C (Volman & Chen, 1959); 0.76 ± 0.05 at 0°C and 0.98 ± 0.05 at 25°C (Hunt & Taube, 1952); 0.8 at 4°C and 1.00 ± 0.02 at 25°C (Baxendale & Wilson, 1956). Activation energies of 2.7–3.1 kcal mol^{-1} were calculated for the H_2O_2 radiation-induced decomposition. The dependency of the rate of H_2O_2 photolysis on the radiation intensity and H_2O_2 concentration is also discussed in the early literature, e.g., Allmand and Style (1930) and references therein.

In aqueous media, the primary quantum yield for ${}^{\bullet}OH$ formation (reaction 2.11) approaches 1.0, i.e., it is twice the primary quantum yield of H_2O_2. Upon one photon absorption of equal or greater energy than the O—O bond dissociation energy in H_2O_2 (211 kJ mol^{-1}; Calvert & Pitts, 1966), the H_2O_2 molecule reaches an 'activated state' which evolves rapidly into two ${}^{\bullet}OH$ in the solvent (water) cage. The rapid ${}^{\bullet}OH$ recombination within the solvent cage occurs with rates similar to the diffusion into the bulk (within $\sim 10^{-9}$ s), and as a result, only half of the ${}^{\bullet}OH$ generated within the solvent cage "escape" and are involved in chemical reactions in water.

Unlike the early studies which estimated the ${}^{\bullet}OH$ quantum yield based on the primary quantum yield of H_2O_2, the more recent studies determined the ${}^{\bullet}OH$ yield from H_2O_2 photolysis using chemical probes which trapped quantitatively the hydroxyl radicals. Table 2.3 summarizes the $\phi_{{}^{\bullet}OH}$ literature data from indirect quantification of ${}^{\bullet}OH$ formed in the photodissociation step of H_2O_2 and the experimental conditions used in those studies.

The data in Table 2.3 shows that the quantum yield of ${}^{\bullet}OH$ formation from H_2O_2 photolysis is independent on the UV radiation wavelength and solution pH (0–7). A compilation of $\phi_{{}^{\bullet}OH}$ reported from 1949 to 2007 can be found in Herrmann *et al.* (2010). The ${}^{\bullet}OH$ quantum yield is independent on the solution ionic strength (Chu & Anastasio, 2005), but it depends on the temperature. Zellner *et al.* (1990) examined the temperature effect on $\phi_{{}^{\bullet}OH}$ from H_2O_2 photolysis (pH 7) at 308 and 351 nm, over the 5–25°C range. The quantum yield increased at both wavelengths from 0.82 (308 nm) and 0.83 (351 nm) at 5°C to 0.98 (308 nm) and 0.96 (351 nm) at 25°C. The temperature dependence at the two wavelengths corresponded to an activation energy of $E_a = 5.5 \pm 1.6$ kJ mol^{-1}. This value is similar to that reported by Chu and Anastasio (2005) for H_2O_2 photolysis to ${}^{\bullet}OH$, $E_a = 5.5 \pm 0.14$ kJ mol^{-1}, as determined from the Arrhenius equation plot (Fig. 2.6).

Table 2.3 Primary quantum yield of ·OH generation in H_2O_2 photolysis in aqueous solutions.

Wavelength/nm	$\phi_{\cdot OH}$	Temperature/°C	pH	Reference
200	1.12	25		Crowell *et al.* (2004)
205	1.14 ± 0.05	24 ± 1	7.0 ± 0.2	Goldstein *et al.* (2007)
214	1.08 ± 0.03	24 ± 1	7.0 ± 0.2	Goldstein *et al.* (2007)
240	1.12 ± 0.05	24 ± 1	7.0 ± 0.2	Goldstein *et al.* (2007)
248	0.88	25	8.3	Crowell *et al.* (2004)
	1.0 ± 0.1	24 ± 2	0–4	Yu and Barker (2003b)
253.7	1.17 ± 0.09	24 ± 1	7.0 ± 0.2	Goldstein *et al.* (2007)
	1.15 ± 0.05	24 ± 1	7.0 ± 0.2	Goldstein *et al.* (2007)
260	1.11 ± 0.05	24 ± 1	7.0 ± 0.2	Goldstein *et al.* (2007)
270	1.07 ± 0.04	24 ± 1	7.0 ± 0.2	Goldstein *et al.* (2007)
280	1.04 ± 0.04	24 ± 1	7.0 ± 0.2	Goldstein *et al.* (2007)
308	0.8 ± 0.2	24 ± 2	0–4	Yu and Barker (2003b)
	0.98 ± 0.03	25	7	Zellner *et al.* (1990)
313	0.72 ± 0.017	−10	5	Chu and Anastasio (2005)
334	0.69 ± 0.008	−10	5	Chu and Anastasio (2005)
351	0.96 ± 0.04	25	7	Zellner *et al.* (1990)
300–400 nm, max. at 360 nm	0.96 ± 0.09	25		Sun and Bolton (1996)

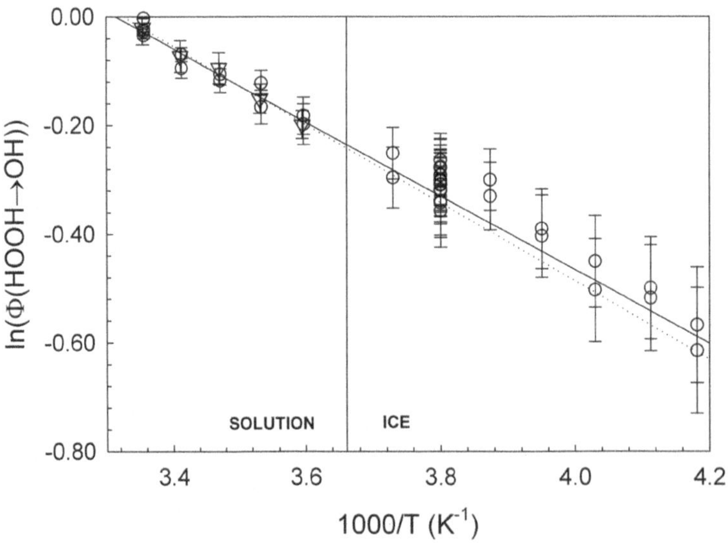

Figure 2.6 Temperature-dependence (239–318 K) of quantum yield of H_2O_2 photolysis to •OH at 313 nm in aqueous and frozen conditions (100 μM H_2O_2, 200 μM benzoic acid, pH 5). The circles and solid (regression) line represent the data from Chu and Anastasio (2005); the inverted triangles and dotted (regression) line represent the data from Zellner *et al.* (1990).

The rate constant for the reaction of $^{\bullet}OH$ with H_2O_2 ($k_{\bullet OH,H_2O_2}$, reaction 2.15) was reported by a number of studies. Yu (2004) compiled and reviewed the thirteen $k_{\bullet OH,H_2O_2}$ values reported from 1962 to 2003. All data were determined experimentally either directly or indirectly through various methods at various pH values, and ranged from 1.2×10^7 M^{-1} s^{-1} (direct measurement, Fricke and Thomas, 1964) to 4.5×10^7 M^{-1} s^{-1} (indirect measurement, Hatada *et al.* 1974). Yu and Barker (2003a) reported a $k_{\bullet OH,H_2O_2} = (4.2 \pm 0.2) \times 10^7$ M^{-1} s^{-1} from direct measurements at pH 2 using laser flash photolysis technique. These authors calculated the unweighted average of the four reported $k_{\bullet OH,H_2O_2}$ values obtained through direct measurements as $k_{\bullet OH,H_2O_2} = (3.5 \pm 1.5) \times 10^7$ M^{-1} s^{-1} and called it as the "preferred value". The commonly accepted and used rate constant is $k_{\bullet OH,H_2O_2} = 2.7 \times 10^7$ M^{-1} s^{-1} (average of four values; Buxton *et al.* 1988).

Christensen *et al.* (1982) measured $k_{\bullet OH,H_2O_2}$ as a function of pH (6.8–13.8) and temperature (14–160°C, pH 7.8) using pulse radiolysis both in N_2O- and in argon-saturated conditions. The rate constant was independent of pH over the 6.8–8.3 range ($k_{\bullet OH,H_2O_2} = (2.7 \pm 0.3) \times 10^7$ M^{-1} s^{-1}) irrespective of the gas sparged through the solution; the "apparent" rate constant (combination of mixed radical and molecular species; see pK_a values above) increased as pH increased to ~12, and then decreased upon further increase of pH to 13.8. The rate constant increased with temperature from 2.1×10^7 M^{-1} s^{-1} at 14°C to 1.5×10^8 M^{-1} s^{-1} at 160°C, with a linear dependence for the Arrhenius plot (ln *k* vs. $1/T$ (K)) over the entire temperature range.

2.4.2 Hydroxyl radical

As described in the chapters of this book, the hydroxyl radicals are generated both in engineered advanced oxidation technologies and in natural waters, through either photolytic or thermal (i.e., dark) processes. Hydroxyl radical is a strong oxidant, thus a very reactive species. Table 2.4 lists the reduction potentials (related to O_2-saturated solutions) of some inorganic radicals of environmental relevance.

Table 2.4 Reduction potentials of selected inorganic radicals (von Sonntag, 2006; p. 93).

Couple	E/V	Couple	E/V
$F^{\bullet}/F^-$	+3.60	O_3, $H^+/HO_3^{\bullet}$ (pH 7)	+1.80
$^{\bullet}OH$, H^+/H_2O	+2.73	$CO_3^{\bullet -}/CO_3^{2-}$	+1.59
$Cl^{\bullet}/Cl^-$	+2.60	$HO_2^{\bullet}$, H^+/H_2O_2 (pH 0)	+1.48
$SO_4^{\bullet -}/SO_4^{2-}$	+2.47	$O_3/O_3^{\bullet -}$ (pH 11–12)	+1.01
$Cl_2^{\bullet -}/2Cl^-$	+2.30	$NO_2^{\bullet}/NO_2^-$	+1.00
$^{\bullet}OH/OH^-$	+1.90	O_2 ($^1\Delta_g$)/$O_2^{\bullet -}$	+0.65

2.4.2.1 Hydroxyl radical properties, detection and quantification in aqueous solutions

Dorfman and Adams (1973) authored a comprehensive monograph on $^{\bullet}OH$ reactivity in aqueous solutions. Recently, a review on $^{\bullet}OH$ generation and detection, properties, kinetics, reaction mechanisms, and environmental implications was published by Gligorovski *et al.* (2015).

Hydroxyl radical is a neutral free radical and one of the most powerful oxidizing radical species. The most important properties of $^{\bullet}OH$ directly related to its chemical reactivity are the reduction potential E^o (pH-dependent; 2.7 V and 1.8 V in acidic and alkaline media, respectively), pK_a (~11.9) and diffusion coefficient (2.3×10^{-5} cm^{-2} s^{-1}) (Dorfman & Adams, 1973). In its reactions with organic compounds, $^{\bullet}OH$

behaves as an electrophile, attacking the electron-rich sites of molecules, whereas its conjugate base $^{\bullet}O^-$ is a nucleophile.

The short lifetime and low concentration (typically $<10^{-12}$ M) of OH radicals due to their high reactivity restrict the $^{\bullet}OH$ detection and monitoring to advanced fast kinetics tools such as laser flash photolysis (LFP) with transient absorption or emission measurements, electron paramagnetic resonance (EPR) or electron spin resonance (ESR), gamma radiolysis. The techniques, methods, data recording, calculation and interpretation, and precautionary steps required in the experimental design to prevent interferences and misleading observations are described in the literature. Thomsen *et al.* (2001) employed LFP with femtosecond pulses at 266 nm to study the $^{\bullet}OH$ and $^{\bullet}Cl$ formation from HOCl photodissociation. Direct formation, kinetics and yield of radical photoproducts were monitored by transient absorption spectroscopy (230–400 nm) on picosecond time-scale. Herrmann (2003) reviewed the LFP techniques used to investigate the $^{\bullet}OH$ kinetics. The direct evidence of $^{\bullet}OH$ was reported for the first time by Smaller *et al.* (1954) from paramagnetic resonance studies on ice exposed to gamma radiation at 77 K. EPR spectra of $^{\bullet}OH$ are very difficult to obtain. Methods using spin-traps, most commonly 1,1- dimethyl-pyrollidine *N*-oxide (DMPO), *N-tert*-butylphenylnitrone (PBN), and 2,2,6,6-tetramethylpiperidine-*N*-oxyl (TEMPO) are used to generate paramagnetic species longer-lived than $^{\bullet}OH$ (Wolfrum *et al.* 1994; Goldstein *et al.* 2004; Clément & Tordo, 2007). The rate constants for the $^{\bullet}OH$ reactions with the spin traps are at diffusion-controlled limits and the resulting OH-spin adducts are characterized by distinct EPR (or ESR) spectra. Direct detection of ionizing radiation-generated $^{\bullet}OH$ based on $^{\bullet}OH$ absorption in the short UV range is rather difficult due to the small absorption coefficients of $^{\bullet}OH$. Accurate detection and quantification of $^{\bullet}OH$ with this technique is commonly achieved with chemical probes, most commonly SCN^-. The reaction of SCN^- with $^{\bullet}OH$ is well characterized (Milosavljevic & LaVerne, 2005) and the resulting radical $(SCN)_2^{-\bullet}$ can be monitored spectroscopically in visible range ($\varepsilon_{475\ nm} = 7600\ M^{-1}\ s^{-1}$) on microsecond time-scale. Chemical probes with well-known rate constants and product yields for the reaction with $^{\bullet}OH$ were used with both pulse radiolysis and steady-state irradiation techniques to quantify $^{\bullet}OH$ yields. Chemical compounds used as probes for $^{\bullet}OH$ quantification include salicylic acid, *p*-hydroxybenzoic acid, methanol, dimethylsulfoxide, terephthalate, dihydroxybenzoate, coumarin-3-carboxylic acid, 3′-*p*-(aminophenyl)fluorescein. The products from these probes can be analyzed by commonly used analytical methods such as HPLC with UV or electrochemical detection, derivatization followed by UV-Vis spectroscopy, fluorescence spectroscopy, GC/MS. The probe selection, $^{\bullet}OH$ generation method, matrix, analytical method, as well as other factors which can influence the accuracy of the experimental outcomes must be carefully considered. The strengths and limitations of the methods and probes employed for the detection and quantification of reactive oxygen species ($^{\bullet}OH$, $HO_2^{\bullet}$, $O_2^{-\bullet}$, $^1O_2$) involved in advanced oxidation processes were reviewed by Fernández-Castro *et al.* (2014).

2.4.2.2 Reactions of hydroxyl radical

As a powerful oxidizing species, $^{\bullet}OH$ attacks the organic and inorganic compounds non-selectively, and most of its reactions approach the diffusion-controlled limits. The high reactivity of $^{\bullet}OH$ explains its very low steady-state concentrations ($10^{-10} - 10^{-12}$ M) in water. The rate constants of $^{\bullet}OH$ reactions with organic compounds of environmental interest range from ~10^6 to $10^{10}\ M^{-1}\ s^{-1}$. The $^{\bullet}OH$ reactions with chemical compounds include hydrogen atom abstraction, addition to electron-rich sites such as unsaturated bonds and aromatic rings, and electron-transfer.

In saturated compounds, *the H-atom abstraction* reactions occur mostly from C—H bonds with formation of a carbon-centered radical ($^{\bullet}R$) (reaction 2.30). This process involves considerable charge separation in the transition state and requires more energy than the other $^{\bullet}OH$ reactions.

$$^{\bullet}OH + R{-}H \rightarrow [R{-}{-}{-}H^{+}{-}{-}{-}O^{-}H^{+}]^{\neq} \rightarrow {}^{\bullet}R + H_2O \quad (2.30)$$

The ${}^{\bullet}OH$ attack to the C—H bond is highly selective, with much higher probability to the tertiary (i.e., $-CHR_2$, where R≠H) than to the primary C (e.g., $-CH_3$). Table 2.5 exemplifies the H-abstraction probability from various C—H bonds in aliphatic alcohols. Note that in the case of H-abstraction from the O—H group, an O-centered radical is formed.

Table 2.5 H-abstraction yields at various C—H positions relative to —OH group (adapted from Asmus *et al.* 1973).

Alcohol	H-abstraction from α-C	H-abstraction from β, γ, δ, etc.-C	H-abstraction from O—H
CH_3OH	93.0		7.0
CH_3-CH_2-OH	84.3	13.2	2.5
CH_3-CH_2-CH_2-OH	53.4	46.0	<0.5
$(CH_3)_2$CH-OH (IPA)	85.5	13.3	1.2
CH_3-CH_2-CH_2-CH_2-OH	41.0	58.5	<0.5
$(CH_3)_3$C-OH (TBA)		95.7	4.3
HO-CH_2-CH_2-OH	100		<0.1
HO-CH_2-CH(OH)-CH_3	79.2	20.7	<0.1

Neighboring e^--donating substituents (e.g., $-CH_3$, $-OR$, $-NR_2$) which will stabilize the resulting C-centered radical will enhance the H-abstraction by ${}^{\bullet}OH$ at that specific C—H bond. On the contrary, neighboring e^--withdrawing substituents will decrease the rate of H-abstraction. For example, although both methanol and *tert*-butanol contain only primary C—H bonds, $k_{{}^{\bullet}OH,CH_3OH}$ (9.7×10^8 M^{-1} s^{-1}) > $k_{{}^{\bullet}OH,\,t\text{-}BuOH}$ (6×10^8 M^{-1} s^{-1}) due to the —OH stabilization effect on the resulting radical (${}^{\bullet}CH_2OH$). Another example is the H-abstraction from the methyl-*tert*-butylether (MTBE) molecule, where the most prone attack occurs at the C—H bond in the methoxy (CH_3O—) group rather than at the C—H bond in one of the three methyl (CH_3—) groups. That was supported by the product studies on the MTBE degradation through the UV/H_2O_2 AOP (Stefan *et al.* 2000). Chlorine atom is a strong e^--withdrawing substituent which affects the rate of H-atom abstraction by ${}^{\bullet}OH$ from chlorinated organics. Maruthamuthu *et al.* (1995) determined the $k_{{}^{\bullet}OH}$ for a series of haloacetates; $k_{{}^{\bullet}OH}$ values decreased as the degree of halogenation increased (e.g., $k_{{}^{\bullet}OH,CH_2ClCOO^-} = 4 \times 10^8$ M^{-1} s^{-1}; $k_{{}^{\bullet}OH,\,CHCl_2COO^-} = 9 \times 10^7$ M^{-1} s^{-1}).

Perhalogenated compounds such as perfluorocarboxylic (PFC) and perfluorosulfonic (PFS) acids are virtually inert to ${}^{\bullet}OH$ reactions. Perfluorooctane sulfonic acid (PFOS) and perflorooctanoic acid (PFOA) are considered micropollutants of concern in drinking water and are listed on the U.S. EPA Contaminant Candidate Lists (CCL3 and draft CCL4, https://www.epa.gov/ccl/chemical-contaminants-ccl-4).

Das and von Sonntag (1986) showed that pH could play an important role on the rates of H-abstraction reactions. For example, protonation of trimethylamine decreases the electron density from the reaction center ($-CH_3$ group) lowering the rate of reaction by a factor of 30.

Hydroxyl radical reacts by H-abstraction with some inorganic species, e.g., NH_3, HSO_4^-, hydroperoxide ion (HO_2^-), H_2O_2, monochloramine (NH_2Cl) with the formation of corresponding radicals and water molecules.

The ${}^{\bullet}OH$ electrophilic addition to unsaturated bonds (e.g., C=C, C≡C, C=N, S=O in sulfoxides) occurs at diffusion-controlled rates. The ${}^{\bullet}OH$ addition to C=O has a very low probability due to the e^--deficiency at the C-atom; ${}^{\bullet}OH$ adds preferentially to the C-atom site in the unsaturated bonds.

The neighboring substituents to the unsaturated bonds as well as at the C-atoms in those bonds reacting with $^{\bullet}OH$ drive the addition site on the bond. For example, the preferred addition site in trichloroethene ($Cl_2C{=}CHCl$) is the less-substituted C-atom, as the other C-atom is e^--deficient due to the e^--withdrawing character of the two Cl-atoms (Köster & Asmus, 1971). All $^{\bullet}OH$ addition reactions to unsaturated bonds result in C- or heteroatom-centered radicals *via* a transient species as exemplified in reaction 2.31.

$$^{\bullet}OH + H_2C{=}CHR \rightarrow [HO\text{---}CH_2\text{---}CH{-}R]^{\neq} \rightarrow HOCH_2{-}^{\bullet}CHR \quad (2.31)$$

The addition of $^{\bullet}OH$ to the aromatic rings proceeds *via* a short-lived π-complex in equilibrium with the original reactants, which evolves into a σ-complex (hydroxycyclohexadienyl radical, reaction sequence 2.32) in which the OH group is bound to the "selected" C-atom (Ashton *et al.* 1995).

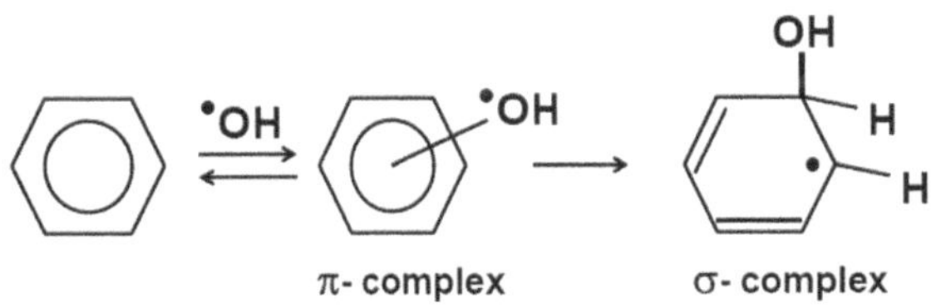

(2.32)

Minakata *et al.* (2015) investigated the addition of $^{\bullet}OH$ to a number of carboxylated and hydroxylated benzenes using pulse radiolysis (PR). This study provides insights into the initial transformation of the aromatic compounds with formation of hydroxylcyclohexadienyl radicals and HO-adducts. The experimental rate constants, thermochemical properties, and transient spectra were used to interpret the effects of carboxylate and hydroxyl functional groups on $^{\bullet}OH$ reactivity towards the investigated compounds.

In general, when aliphatic side chains are available, $^{\bullet}OH$ reacts predominantly with the unsaturated bonds (in aliphatic compounds) or the aromatic ring. In aromatic systems, there is a strong substituent effect on the OH-substitution position on the ring. The e^--donating groups (e.g., $-CH_3$, $-OCH_3$, $-NH_2$, $-NR_2$, $-OH$) direct the substitution to the *ortho-* and *para*-positions, whereas the e^--withdrawing groups (e.g., $-NO_2$, $C{=}O$, $-SO_3H$) direct the substitution to *meta*-position. In case of weak e^--donating and -withdrawing substituents, *quasi*-equal distribution of the hydroxylated products is observed. Addition to the *ipso*-position (i.e., at the substituent C on the ring) is mostly disfavored, particularly in the case of bulky substituents. It was postulated that the regioselectivity of $^{\bullet}OH$ may occur during the transition from the π- to the σ-complex.

The reactions of $^{\bullet}OH$ with heterocyclic compounds follow similar rules, and in such reactions heteroatom centered radicals are also formed. Tauber and von Sonntag (2000) reported the rate constants for $^{\bullet}OH$ reactions with *s*-triazine ($k_{^{\bullet}OH} = 9.6 \times 10^7\ M^{-1}\ s^{-1}$) and with atrazine ($k_{^{\bullet}OH} = 3.0 \times 10^9\ M^{-1}\ s^{-1}$), which is a substituted *s*-triazine. The data suggest that the observed high reactivity of atrazine and other similar *s*-triazines toward $^{\bullet}OH$ is due to fast H-abstraction from the aliphatic chains rather than to the addition to the triazine ring, even though the ring reactivity toward $^{\bullet}OH$ is enhanced by the e^--donating substituents. This is totally different from the substituted benzenes, where the high reactivity of aromatic ring toward $^{\bullet}OH$ makes the ring the preferred reaction site for $^{\bullet}OH$.

The structure-driven selectivity of $^{\bullet}OH$ reactions with complex molecules is illustrated in Song *et al.*'s study (2008a) which reports the $^{\bullet}OH$ rate constants for both three β-lactam antibiotics containing a 6-aminopenicillanic acid (APA) moiety in their molecular structures and APA alone (see figure 2.7).

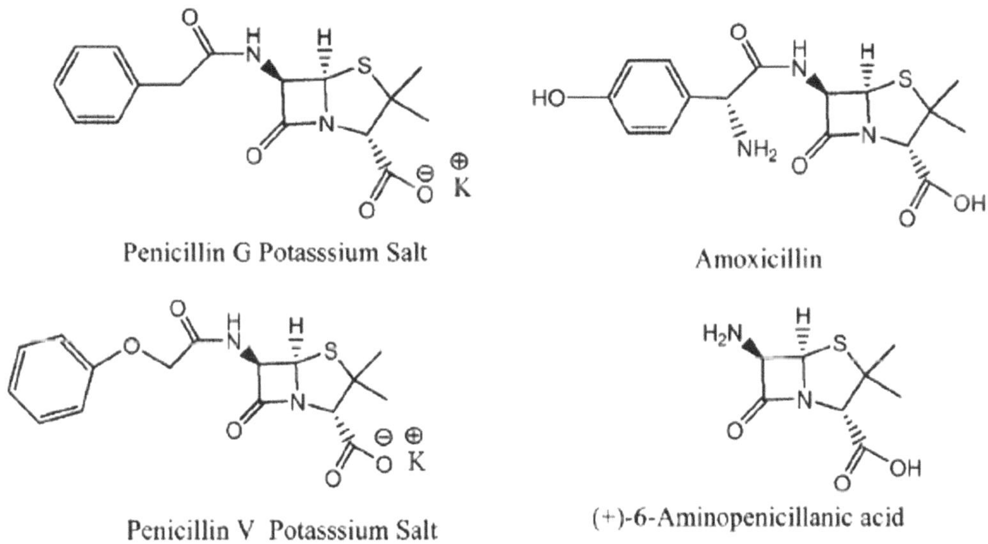

Figure 2.7 Molecular structures of β-lactam antibiotics and of model compound (APA).

The rate constant for APA ($k_{•OH} = 2.4 \times 10^9\ M^{-1}\ s^{-1}$) is ~3 to ~3.7-fold smaller than those of β-lactam antibiotics ($6.94 \times 10^9\ M^{-1}\ s^{-1}$ to $8.76 \times 10^9\ M^{-1}\ s^{-1}$), indicating that only 25 to 30% of the •OH attack the APA moiety in the antibiotic molecules, as the majority of •OH will react with their aromatic ring.

Computational chemistry and molecular modeling are used nowadays to predict both the mechanism of •OH attack on various molecular structures and the product distribution, complementing the experimental information (Nicolaescu *et al.* 2005; Li & Crittenden, 2009; Li *et al.* 2013).

The electron transfer reactions of hydroxyl radical could involve both inorganic and organic compounds. In both cases, the primary step consists in formation of a two-centered three-electron bonded adduct radical followed by the electron transfer. For example, the reactions of •OH with halogen ions (e.g., bromide, chloride) occur at near diffusion-controlled rates ($k_{•OH,X^-}$ ~$10^9 - 10^{10}\ M^{-1}\ s^{-1}$) *via* a three-electron-bonded (weak $\sigma\sigma^*$ bond) intermediate radical ($BrOH^{•-}$ in reaction 2.33), within which the electron transfer takes place with formation of hydroxide ion (OH^-) and a reactive halogen atom (reaction 2.34). The halogen atom reacts rapidly with the halide ion to form dihalide radical anion complex (reaction 2.35). All these reactions are pH-dependent (see Chapter 9 for details).

$$^{•}OH + Br^- \rightleftarrows BrOH^{•-} \quad k_{\rightarrow} = (1.06 \pm 0.08) \times 10^{10}\ M^{-1}\ s^{-1}\ ;\ k_{\leftarrow} = (3.3 \pm 0.4) \times 10^7\ s^{-1} \tag{2.33}$$

(Zchavi & Rabani, 1972)

$$BrOH^{•-} \rightarrow Br^{•} + OH^- \quad k = (4.2 \pm 0.6) \times 10^6\ s^{-1} \text{ (Zehavi \& Rabani, 1972)} \tag{2.34}$$

$$Br^{•} + Br^- \rightleftarrows Br_2^{•-} \quad k_{\rightarrow} = 10^{10}\ M^{-1}\ s^{-1}\ ;\ k_{\leftarrow} = 10^5\ s^{-1} \text{ (Zehavi \& Rabani, 1972)} \tag{2.35}$$

Electron transfer reactions also occur with other inorganic ions, e.g., SCN^-, HCO_3^-, CO_3^{2-}, NO_2^-, AsO_2^-, N_3^-, Fe^{2+}, Ce^{3+}, as well as with some organic compounds (e.g., dimethylsulfide).

The •OH reaction with unsubstituted phenoxide ion ($pK_a = 10$, at 25°C) proceeds through both electron transfer from the oxide group and addition (preferred) to the aromatic ring.

2.4.2.3 Reactions of C-centered radicals, oxyl- and peroxyl radicals

The chemistry and photochemistry of radicals formed *via* •OH reactions are very complex, and drive the byproduct formation in AOP-treated water. In order to be able to predict pollutant degradation mechanisms and byproduct formation and distribution, one must understand the formation and fate of these radicals in the context of the applied treatment process and water quality. Detailed discussion on these topics is far beyond the scope of this chapter.

Generation, detection, properties, reactions, and kinetics of C-, O-, and heteroatom-centered radicals, and oxyl and peroxyl radicals of thereof are extensively covered in the literature (Howard & Scaiano, 1984; Neta *et al.* 1990; von Sonntag & Schuchmann, 1991; Neta *et al.* 1996; Alfassi, 1997; Neta & Grodkowski, 2005; von Sonntag, 2006).

Depending on their structure, concentration, reactivity, and the nature and concentration of water constituents, the radicals formed from the •OH reactions with organic and inorganic species can undergo a variety of reactions. In aerated aqueous solutions, most C-centered radicals react irreversibly with dissolved oxygen with rate constants close to the diffusion-controlled limit ($k > 10^9$ M^{-1} s^{-1}) generating peroxyl radicals (R—O—O•). Resonance-stabilized radicals of indane, fluorene, or similar condensed structures have low reactivity toward O_2 ($k \sim 10^5$–10^7 M^{-1} s^{-1}). Low k values (3×10^8–5×10^8 M^{-1} s^{-1} range) for the reaction with O_2 were observed for the hydroxycyclohexadienyl radicals formed from •OH addition to benzene, toluene, naphthalene and other aromatic hydrocarbons, due to the high stability of the double allylic radical structure. A very low rate constant was determined for the OH-adduct radical of nitrobenzene (2.5×10^6 M^{-1} s^{-1}; Asmus *et al.* 1967) which was interpreted by the effect of high electronegativity of that radical. In general, N-centered radicals react much more slowly with O_2 than C-centered radicals. An example is the reaction of aminyl radical $NH_2^\bullet$ with O_2, for which rate constants more than one order of magnitude apart were reported: 3.8×10^8 M^{-1} s^{-1} (Pagsberg, 1972); 1×10^7 M^{-1} s^{-1} (Neta *et al.* 1978); 1.2×10^8 M^{-1} s^{-1} (pH 11.9; Men'kin *et al.* 1991). Laszlo *et al.* (1998) provide a critical analysis on the experimental methods used in these studies which reported the rate coefficients for peroxylation of $NH_2^\bullet$ and the subsequent reactions of $NH_2O_2^\bullet$. Oxygen-centered radicals are virtually unreactive toward O_2.

Peroxyl radicals decay through unimolecular or bimolecular reactions, whose rate constants depend on the peroxyl radical structure and properties. Unimolecular reactions include $HO_2^\bullet$ or $O_2^{\bullet-}$ elimination, fragmentation, intramolecular rearrangement with addition of —O—O• to a double bond, and intramolecular H-abstraction. Bimolecular reactions include head-to-head self- or cross-termination reactions *via* tetroxide or trioxide intermediate structures, electron-transfer and H-abstraction from stable compounds, etc. Although fairly unreactive in aqueous solutions, hydroperoxyl radical $HO_2^\bullet$ and its conjugate base $O_2^{\bullet-}$ are important in oxyl and peroxyl radical chemistry, as they can reach high concentrations relative to those of the other radicals. Routes to oxyl radicals (R-O•) include peroxyl radical unimolecular or bimolecular reactions and H-abstraction from alcohol or phenol group (-OH). Of note, organic radicals including oxyl and peroxyl radicals are formed not only from micropollutants, but also from the •OH-induced oxidation of natural organic matter constituents present in the water. The reactivity of oxygen-centered radicals generally follows the order •OH ≫ alkoxyl > peroxyl > phenoxyl (Howard & Scaiano, 1984). Radical kinetic studies employ fast kinetics monitoring techniques such as pulse radiolysis and nanosecond and microsecond flash photolysis.

2.4.3 Rate constants of •OH reactions with organic and inorganic compounds

As a strong electrofile, •OH reacts with a variety of organic and inorganic compounds with specific rate coefficients ranging over 4–5 orders of magnitude. Only very few water contaminants can be treated by direct photolysis, thus, the vast majority of UV/H_2O_2 applications rely on the effectiveness of •OH-initiated degradation of micropollutants. Compounds with $^{\bullet}OH$ rate constants approaching the diffusion-controlled limits ($k_{\bullet OH} > 10^9\ M^{-1}\ s^{-1}$) are well treated with the UV/H_2O_2 AOP. The rate constants for $^{\bullet}OH$ reactions with chemical compounds have been determined experimentally or predicted with methods based on statistical relationships between their reactivity and the chemical structures and chemical and thermodynamic properties, among which quantitative structure-property/activity relationship (QSPR/QSAR), linear free energy relationship (LFER), group contribution method (GCM). Water quality parameters e.g., pH and temperature affect the $^{\bullet}OH$ rate constants.

2.4.3.1 Brief review on $k_{\bullet OH}$ literature data

Dorfman and Adams (1973), Farhataziz and Ross (1977) and Buxton *et al.* (1988) compiled the experimental rate constants for the reactions of hydrated electron (e^-_{aq}), hydrogen atom ($H^{\bullet}$), hydroxyl radical ($^{\bullet}OH$), and oxide anion ($O^{\bullet -}$) with organic and inorganic compounds, recommended either an average or a selected value, and provided references to the original studies. NIST website (http://kinetics.nist.gov/solution/) provides an easy access to the rate constants for radical reactions with a variety of chemical compounds. The NIST database covers literature data published before mid-1990s. Over the past two decades, numerous micropollutant treatability studies which include chemical and photochemical kinetic parameters relevant to the direct photolysis and UV/H_2O_2 process have been reported.

Wols and Hofman-Caris (2012) tabulated the $k_{\bullet OH}$ and the quantum yields and molar absorption coefficients at 253.7 nm for over 100 organic compounds, mostly emerging contaminants, and for $^{\bullet}OH$ scavengers relevant to water treatment, including inorganic species and DOM of various origins. In another work, Wols *et al.* (2013) investigated the kinetics of 40 pharmaceuticals in milliQ water, tap water, and pre-treated surface water exposed to LP- and MP UV radiation with and without H_2O_2 addition. The authors determined the $k_{\bullet OH}$ data and LP- and MP-fluence-based rate constants for the investigated pharmaceuticals except tramadol, terbutaline, and penicillin V which were rapidly removed through dark reaction with H_2O_2. Yan and Song (2014) summarized the studies published from 2003 to 2013 on solar and solar-simulated photodegradation kinetics (quantum yields, $k_{\bullet OH}$, and k_{1O_2}) of a wide range of pharmaceutically active compounds. Jin *et al.* (2012) determined $k_{\bullet OH}$ and k_{O_3} for 24 micropollutants including endocrine disruptors, pharmaceuticals, and personal care products, using bench-scale reactors. WaterReuse Foundation Project report (2010) provides rate constants for the reactions of $^{\bullet}OH$ and e^-_{aq} with 51 emerging contaminants (algal toxins, pharmaceuticals, X-ray contrast agents, bisphenol A, atrazine and its primary degradation products) and nine organic matter isolates. O_3 and $^{\bullet}OH$ rate constants for selected compounds from the U.S. EPA CCL3 can be found in Mestankova *et al.* (2016). Herrmann *et al.* (2010) compiled literature kinetic data ($k_{\bullet OH}$ and Arrhenius parameters) for a large number of organic compounds and a few inorganic compounds. Armbrust (2000) determined the $^{\bullet}OH$ rate coefficients for 14 pesticides frequently identified in water sources using a competition kinetics method with acetophenone as a probe. The studies on the degradation of pesticides *via* UV/H_2O_2 and Fenton-type AOPs were reviewed by Ikehata and Gamal El-Din (2006). Along with experimental conditions and postulated degradation pathways, the authors compiled the $k_{\bullet OH}$'s for selected pesticides. Peter and von Gunten (2007) reported the rate constants for the $^{\bullet}OH$ and O_3 reactions with 13 taste-and-odor (T&O) causing compounds with odor threshold ranging from 0.03 ng/L (2,4,6-trichloroanisole) to 70,000 ng/L (cis-3-hexen-1-ol). The $^{\bullet}OH$ rate constants for the reactions with selected benzothiazoles and benzotriazoles are reported by Bahnmüller *et al.* (2015). Fang *et al.* (2000) discussed the mechanism of $^{\bullet}OH$ reactions with polyhalophenolates and determined their second-order kinetics rate constants. Chuang *et al.* (2016) measured experimentally the $^{\bullet}OH$ rate constants for 42 DBPs and compared them with the predicted values calculated with the group contribution method (GCM). Examples of rate constants for $^{\bullet}OH$ reactions with selected organic compounds are given in Table 2.6.

Table 2.6 Rate constants for •OH reactions with selected compounds.

Class	Compound	$k_{\cdot OH} \times 10^9\ M^{-1}\ s^{-1}$	Reference
Industrial solvents	1,2,3-Trichloropropane	<0.5	Mestankova *et al.* (2016)
	Tetrachloroethene (PCE)	2.3 – 2.8	Buxton *et al.* (1988)
	Trichloroethene (TCE)	4.0 – 4.3	Buxton *et al.* (1988)
	1,4-Dioxane	2.5 – 3.1	Buxton *et al.* (1988)
	N-Methyl-2-pyrrolidone	3.1 – 4.9	Gligorovski *et al.* (2009)
DBPs	Bromodichloromethane	0.064	Chuang *et al.* (2016)
	Trichloromethane	0.054	Chuang *et al.* (2016)
	Triiodomethane	7.2	Chuang *et al.* (2016)
	Trichloronitromethane	0.08	Chuang *et al.* (2016)
	Dichloroacetic acid	0.13	Chuang *et al.* (2016)
	Dichloroacetonitrile	0.025	Chuang *et al.* (2016)
Gasoline additives	Methyl-*tert*-butyl-ether (MTBE)	1.6 – 3.9	Cooper *et al.* (2009)
		1.0	Acero *et al.* (2001)
	Hexane	6.6	Buxton *et al.* (1988)
Cyanotoxins	Anatoxin-a	3.0	Onstad *et al.* (2007)
		5.3	Afzal *et al.* (2010)
	Cylindrospermopsin	5.5	Onstad *et al.* (2007)
		5.1	He *et al.* (2013)
	Microcystin-LR	11	Onstad *et al.* (2007)
		23	Song *et al.* (2009)
		11.3	He *et al.* (2015)
	Microcystin-RR	14.5	He *et al.* (2015)
	Microcystin-LA	11	He *et al.* (2015)
	Microcystin-YR	16.3	He *et al.* (2015)
N-Nitrosamines	*N*-Nitrosodimethylamine (NDMA)	0.43	Landsman *et al.* (2007)
		0.45	Lee *et al.* (2007)
	N-Nitrosomethylethylamine (NMEA)	0.495	Landsman *et al.* (2007)
	N-Nitrosodiethylamine (NDEA)	0.7	Landsman *et al.* (2007)
	N-Nitrosodi-n-propylamine (NDPA)	2.3	Landsman *et al.* (2007)
	N-Nitrosoethylbutylamine (NEBA)	3.1	Landsman *et al.* (2007)
	N-Nitrosopyrrolidine (NPYR)	1.75	Landsman *et al.* (2007)
	N-Nitrosodiphenylamine (NDPhA)	6.5	Mestankova *et al.* (2014)
Food additives	Butylated hydroxyanisole	7.4	Jin *et al.* (2012)

(*continued*)

Table 2.6 Rate constants for $^{\bullet}OH$ reactions with selected compounds (*continued*).

Class	Compound	$k_{^{\bullet}OH} \times 10^9\ M^{-1}\ s^{-1}$	Reference
Pesticides	Dicamba	3.5	Jin *et al.* (2012)
	Trifluralin	1.3	Jin *et al.* (2012)
	Methoxychlor	3.9	Jin *et al.* (2012)
	Acephate	7.1	Mestankova *et al.* (2016)
	Dicrotophos	5.5	Mestankova *et al.* (2016)
	Methamidophos	8.2	Mestankova *et al.* (2016)
	Fenamiphos	6.2	Mestankova *et al.* (2016)
	Metolachlor	6.1	De Laat *et al.* (1996)
	Metaldehyde	1.3	Autin *et al.* (2012)
	Glyphosate	0.18	Haag and Yao (1992)
		2.8	Balci *et al.* (2009)
Explosives	Hexahydro-1,3,5-trinitro-1,3,5-triazine (RDX)	0.34 (UV/H_2O_2)	Bose *et al.* (1998)
		1.1 (UV/H_2O_2)	Elovitz *et al.* (2008)
		5.2 (PR)	Elovitz *et al.* (2008)
		0.25 (O_3/H_2O_2)	Chen *et al.* (2008)
	Trinitrotoluene (TNT)	0.43	Bose *et al.* (1998)
Taste-and-odor causing compounds	2-Methyl-*iso*-borneol (MIB)	5.09	Peter and von Gunten (2007)
	Geosmin	7.8	Peter and von Gunten (2007)
	β-Cyclocitral	7.42	Peter and von Gunten (2007)
	2,4,6-Tribromoanisole	3.74	Peter and von Gunten (2007)
	2,4,6-Trichloroanisole	5.1	Peter and von Gunten (2007)
	2,6-nonadienal	10.49 (UV/H_2O_2)	Peter and von Gunten (2007)
		8.95 (γ-radiolysis)	Peter and von Gunten (2007)

Rate constants for $^{\bullet}OH$ reactions with inorganic species relevant to water treatment are discussed briefly in sections 2.4.2.2 and 2.6, and extensive compilations were mentioned above. A number of inorganic micropollutants are oxidized by $^{\bullet}OH$ at rates near diffusion-controlled limits; e.g., $k_{^{\bullet}OH,CN^-} = 7.6 \times 10^9\ M^{-1}\ s^{-1}$; $k_{^{\bullet}OH,Fe(CN)_6^{4-}} = 1.05 \times 10^{10}\ M^{-1}\ s^{-1}$; $k_{^{\bullet}OH,SeO_3^{2-}} = 3.5 \times 10^9\ M^{-1}\ s^{-1}$; $k_{^{\bullet}OH,AsO_2^-} = 9.0 \times 10^9\ M^{-1}\ s^{-1}$ (Buxton *et al.* 1988).

2.4.3.2 Temperature-dependence of $^{\bullet}OH$ reactions

Herrmann *et al.* (2010) compiled the Arrhenius parameters available for the $^{\bullet}OH$ reactions of various classes of aliphatic compounds including alcohols, carbonyl and carboxyl derivatives, ethers, and halogenated compounds. The typical temperature range in those kinetic studies was ~283 K to ~328 K. The $^{\bullet}OH$ reaction activation energies E_a and the pre-exponential factors A were determined as 5 to 20 kJ mol^{-1} and 10^8 to $10^{11}\ M^{-1}\ s^{-1}$, respectively. Minakata *et al.* (2011) reported the $^{\bullet}OH$ reaction rate constants for a series of haloacetates of environmental relevance in the range of $(0.60 \sim 4.77) \times 10^8\ M^{-1}\ s^{-1}$ over the 22–50°C, and determined the Arrhenius factors A and the reaction activation energies E_a (13.3 to 33.5 kJ mol^{-1}). Their experimental observations indicated that an increase of 10°C in the reaction temperature conditions enhanced the $k_{^{\bullet}OH}$ by a factor of 1.05 ~ 1.2. Beltran-Heredia *et al.* (1996) studied the quantum yield and $k_{\cdot OH}$ temperature-dependence for pesticide bentazone. The authors reported the $^{\cdot}OH$ rate constants of 2.66×10^9 M^{-1} s^{-1} and 4.14×10^9 M^{-1} s^{-1}, at 10 °C and 40 °C, respectively, and developed the $k_{\cdot OH}$ – Arrhenius correlation for the $^{\cdot}OH$ – bentazone reaction over the 10–40°C range. Zhu *et al.* (2003) examined the temperature rate coefficients ($k_{\cdot OH}$) for the reaction of $^{\cdot}OH$ with the oxidation byproducts of dimethylsulfide, i.e., dimethylsulfoxide, dimethylsulfone, and methanesulfonate in aqueous solutions within the temperature range of 2–37 °C. Other studies reporting temperature-dependency of $^{\cdot}OH$ rate constants for reactions with organic compounds include Chin *et al.* (1994); Ashton *et al.* (1995); Herrmann (2003); Gligorovski *et al.* (2009).

McKay *et al.* (2011) used pulse radiolysis to determine the temperature-dependent $^{\cdot}OH$ rate coefficients for three NOM isolates (Elliot Soil humic acid, Pony Lake fulvic acid, and Suwannee River fulvic acid) and three bulk effluent organic matter (EfOM) samples over the temperature range of ~9 to 43 °C. The $^{\cdot}OH$ reaction rate constants of NOM isolates at 20.5–24.4 °C were found within the 1.21 ~ 6.9×10^8 $M(C)^{-1}s^{-1}$ range, whereas the corresponding Arrhenius activation energy varied from 15.18 to 29.93 kJ mol^{-1}

2.4.3.3 Experimental and theoretical methods for $k_{\cdot OH}$ determination

The experimental $k_{\cdot OH}$ data were obtained either as absolute values from the kinetics of formation of substrate OH-adducts or from competition kinetics methods relative to a chemical probe (reference) with known $^{\cdot}OH$ rate constant. The OH radicals can be generated *via* various processes, such as ionizing radiation, O_3/H_2O_2, UV/O_3, UV/H_2O_2, UV/TiO_2, Fenton or Fenton-like reaction. The following are a few examples of studies using experimental methods for $k_{\cdot OH}$ determination.

Song *et al.* (2008a) used pulse and γ-radiolysis to measure $k_{\cdot OH}$ for β-lactam antibiotics by both absolute and competition kinetics. The absolute $k_{\cdot OH}$ data were determined from the kinetics of OH-adduct transient specific to the investigated substrate over selected μs time-scale, as a function of substrate concentration (Figure 2.8a). The transient kinetics was monitored spectroscopically at its absorption maximum. The slope of *pseudo*-first order rate constant *vs.* β-lactam antibiotic concentration corresponds to $k_{\cdot OH}$ for that specific compound (Figure 2.8b).

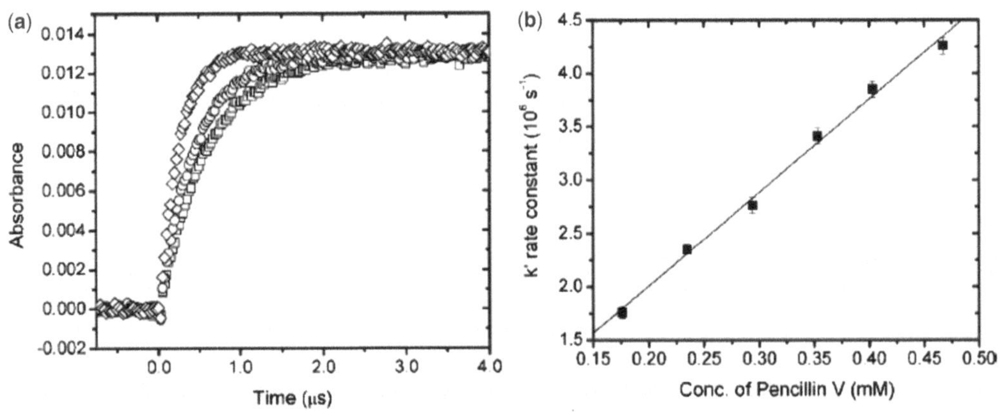

Figure 2.8 Transient absorption at 320 nm in pulse radiolysis of penicillin V solution at pH 7 as a function of penicillin concentration (a), and first-order rate constants ($k \times 10^6/s^{-1}$) determined from (a) at various penicillin V concentrations (b). The slope represents $k_{\cdot OH,penicillin\ V} = (8.76 \pm 0.28) \times 10^9$ M^{-1} s^{-1}. From Song *et al.* (2008a).

The competition kinetics method used thiocyanate ion (SCN^-) as a probe to determine $k_{\bullet OH}$ for 6-aminopenicillanic acid (APA). The competition for $^\bullet OH$ between APA and SCN^- (fixed concentration) was analyzed from the decrease of $(SCN)_2^{\bullet -}$ concentration as a function of increasing APA concentration ([APA]). $(SCN)_2^{\bullet -}$ is the transient formed in the $^\bullet OH$ reaction with SCN^-, with a well-characterized absorption spectrum, and it was monitored at 475 nm ($\varepsilon = 7600$ M^{-1} cm^{-1}). The $k_{\bullet OH,APA}$ was calculated from the slope of the plot of $[(SCN)_2^{\bullet -}]_0/[(SCN)_2^{\bullet -}]$ *vs.* [APA]/[SCN^-], where $[(SCN)_2^{\bullet -}]_0$ is transient concentration in the absence of APA and $k_{\bullet OH,SCN^-} = 1.05 \times 10^{10}$ M^{-1} s^{-1} (equation 2.36).

$$\frac{[(SCN)_2^{\bullet -}]_0}{[(SCN)_2^{\bullet -}]} = 1 + \frac{k_{\bullet OH,APA}}{k_{\bullet OH,SCN^-}} \frac{[APA]}{[SCN^-]} \tag{2.36}$$

Kochany and Bolton (1992) implemented an elegant competition kinetics method for $k_{\bullet OH}$ determination for reactions of benzene and six halogenated benzenes with $^\bullet OH$, using the spin-trapping technique with EPR detection of the OH-adduct. Hydroxyl radicals were generated from H_2O_2 photolysis in the EPR quartz cell. DMPO was used as a spin trap and probe compound ($k_{\bullet OH,DMPO} = 4.3 \times 10^9$ M^{-1} s^{-1}). The initial rate of DMPO-OH adduct formation determined from the EPR signal amplitude changed with the substrate (halobenzene) concentration.

Chemical probes commonly used in competition kinetics methods include *p*-chlorobenzoic acid (*p*-CBA), disodium terephthalate, 3′-*p*-aminophenyl-fluorescein, coumarin-3-carboxylic acid, hydroxybenzoic acid, benzoic acid. *p*-CBA is extensively used as a probe compound, as it can be easily quantified by HPLC, has a small k_{O_3}, thus can be used with O_3/H_2O_2 process, and low direct photolysis yield at 253.7 nm when used with the UV/H_2O_2 process. However, corrections to *p*-CBA kinetics for either O_3 reactions or UV photolysis must be considered. A basic competition kinetics equation where *p*-CBA and $UV_{253.7\,nm}/H_2O_2$ are used as a chemical probe and $^\bullet OH$ generation process, respectively, is shown below. In equation 2.37, $k_{1,UV/H_2O_2}$ is the first-order rate constant for either the probe or test compound decay in the UV/H_2O_2 process, $k_{1,UV}$ (*p*-CBA) is the first-order rate constant for *p*-CBA direct photolysis, and $k_{\bullet OH}$(*p*-CBA) and $k_{\bullet OH}$(C) are the second-order rate constants for $^\bullet OH$ reactions with *p*-CBA (5×10^9 M^{-1} s^{-1}) and test compound, respectively. Herein it is assumed that the test compound does not undergo direct photolysis. The rate constants in the left term in equation 2.37 can be either time- or fluence-based.

$$\frac{k_{1,UV/H_2O_2}^{p\text{-CBA}} - k_{1,UV}^{p\text{-CBA}}}{k_{1,UV/H_2O_2}^{C}} = \frac{k_{\bullet OH}^{p\text{-CBA}}}{k_{\bullet OH}^{C}} \tag{2.37}$$

Alternative to the above method is to firstly determine the $^\bullet OH$ exposure for given water quality, oxidant concentration and UV dose conditions using a well-characterized chemical probe, then to use this information and identical conditions to determine the kinetic decays of one or more target compounds and to solve for $k_{\bullet OH,C}$ from the individual first-order kinetics expression (Gerrity *et al.* 2016).

Accurate $k_{\bullet OH}$ data reporting is paramount given that this parameter is a key input parameter of kinetic models used for AOP performance predictions and equipment sizing, and the error in $k_{\bullet OH}$ is carried into the kinetic model output data (see sections 2.5 and 2.7).

Theoretical methods to predict $k_{\bullet OH}$ values for a wide range of molecular structures have been developed and the predictions were compared to the available experimental $k_{\bullet OH}$ data as a means of model validation. Extensive databases on molecular parameters and properties of chemical compounds are available, which are used to build computational predictive tools on chemical, physical, biochemical, and toxicological reactivity of micropollutants. Quantitative structure- activity/property relationship (QSAR/QSPR) models provide valuable information on the response of chemical compounds to specific treatment technologies, thus, aiding with selection and performance evaluation of those processes (e.g., Sudhakaran *et al.* 2012; Sudhakaran *et al.* 2013), reaction mechanisms and degradation pathways, substituting the experimental approaches which are time-consuming, expensive, and frequently difficult to implement due to analytical method and instrumentation limitations.

QSAR models are predictive tools based on chemical structure parameters and their relationship with chemical reactivity of that specific molecule. Among molecular structural descriptors used to develop QSAR models for the purpose of quantification of micropollutant reactivity toward hydroxyl radical are average molecular polarizability, dipole moment, standard heat of formation, total and electronic energy, E_{HOMO} and E_{LUMO}, positive and negative charge distribution on specific atoms, molecular weight, molecular volume, molecular surface area, molecular atom distribution and compactness, Hammett/Taft constants, hydration equilibrium constants. Selection of molecular descriptors depends on their relevance to the reaction of respective molecule with $^{\bullet}OH$, which is further assessed through extensive statistical analysis. Wang *et al.* (2009) developed a QSAR model to predict $k_{\bullet OH}$ values for reactions in aqueous phase of a number of substituted phenols and benzenes, alcohols, and alkanes. It was found that the quantum chemical descriptors governing $k_{\bullet OH}$ for these compounds are E_{HOMO}, average net atomic charges on hydrogen atoms, molecular surface area, and the dipole moment.

Sudhakaran and Amy (2013) built QSAR models for micropollutant removal from selected surface water sources *via* ozonation and advanced oxidation ($^{\bullet}OH$), and determined the $^{\bullet}OH$ rate coefficients using specific molecular descriptors. The approach is described by the following steps: evaluation of dataset, correlation analysis between the descriptors and the rate constants, principal component analysis (PCA) for variable reduction, multiple regression analysis (MLR) as a final step for QSAR model building, and model validation. Water micropollutants from various classes with known rate constants were used to validate the model and to define the role of descriptors. The $k_{\bullet OH}$ QSAR model applicability domain was identified for pH 5–8 and $k_{\bullet OH}$ values of 0.04×10^9–18×10^9 M^{-1} s^{-1}. Wols and Vries (2012) developed a QSAR model to predict micropollutant degradation with the UV/H_2O_2 process and to estimate $k_{\bullet OH}$ data. The approach to the model development followed the OECD (Organization for Economic Cooperation and Development) principles for reliable QSAR models. Kušic *et al.* (2009) reported a QSAR study for predicting $k_{\bullet OH}$ for the degradation of aromatic compounds in water. The authors investigated 78 compounds with various functional groups. The primary objective of the work was to find a mechanistic structure-degradation relationship and then, acceptable quantitative models. E_{HOMO} was the best descriptor among all selected. The authors concluded that the 4-variable model can be used to predict degradation of micropollutants and their persistence in natural waters. A comprehensive approach to QSAR model development was taken by Borhani *et al.* (2016). The authors used QSPR method to model the $^{\bullet}OH$ rate constants for 457 water contaminants from 27 chemical classes. The models built with different approaches were validated using various techniques, and the data analysis indicated that the log $k_{\bullet OH}$ for a large number of water contaminants can be predicted with less than 4% absolute relative error. Borhani *et al.* compared their models with other recently developed models and discussed the role of structural characteristics of chemical compounds in AOP efficiency. Recently, Luo *et al.* (2017) elaborated a QSAR model with larger and wider applicability domain as compared with the previous models. Critical reviews on QSAR models are also available in the literature, among which, Canonica and Tratnyek (2008); Wojnárovits and Takács (2013); Doussin and Monod (2013).

Minakata *et al.* (2009) developed a comprehensive group contribution method (GCM) to predict $k_{\bullet OH}$ for water micropollutants reacting with $^{\bullet}OH$ through the following mechanisms: H-atom abstraction, addition to C=C bonds in alkenes, addition to the aromatic ring, and $^{\bullet}OH$ reactions with hetero-atom (N, S, P)-containing organic compounds. In this method, the $k_{\bullet OH}$ is computed based on the type of reaction characterized by the "group rate constants" (e.g., H-abstraction reaction and the $k_{\bullet OH,H\text{-}abstr}$ "group" from all potential sites in the molecule such as primary-, secondary-, tertiary-C, -OH, -COOH) and on the effect of neighboring functional groups, i.e., "group contribution factors". The GCM developed in this study to predict $k_{\bullet OH}$ in water, integrated the reaction mechanisms in the aqueous phase and the "essential features" of GCM developed by Atkinson (1987) which is currently implemented in the U.S. EPA EPI Suite^{TM} software (2012a) for the gas-phase. Minakata *et al.*'s GCM includes 66 group rate constants for the abovementioned $^{\bullet}OH$ mechanisms and 80 group contribution factors. The GCM was calibrated using the experimental $k_{\bullet OH}$ data reported for 310 compounds. The calibrated group rate constants and group contribution factors were used to predict the $k_{\bullet OH}$ values for 124 compounds from various chemical classes and bearing various functional groups. Seventy-seven predicted $k_{\bullet OH}$ data (i.e., 62% of all predicted $k_{\bullet OH}$ data) were within a factor of 0.5–2.0 from the experimental rate constants. The GCM software is available as a MS Excel spreadsheet and compiled Fortran program in the Supplementary Information of Minakata *et al.*'s (2009) publication. Lee and von Gunten (2012) used the GCM to predict $k_{\bullet OH}$ values for 16 micropollutants including BPA, four steroid estrogens, eight pharmaceuticals, triclosan, and two cyanotoxins (MC-LR and anatoxin-a). The predicted to the experimental $k_{\bullet OH}$ ratios for 11 of the 16 compounds were ≤1.5; the largest factor (3.3) was calculated for anatoxin-a and its protonated form. The accuracy of GCMs developed for predicting rate coefficients for $^{\bullet}OH$-initiated micropollutant degradation in aqueous phase is affected by the limited thermodynamic and kinetic data for many functional groups, by the approach of averaging the effect of functional groups, and by certain assumptions among which the rate constant additivity, role of hydrogen bonding, steric and solvation effects, electronic intra- and intermolecular effects (Minakata & Crittenden, 2011).

Using *ab initio* quantum mechanics calculations for the gas-phase $^{\bullet}OH$ reactions (H-atom abstraction and addition to double bonds in alkenes) and a method of solvation to account for the water effect, Minakata and Crittenden (2011) calculated the standard free energies of activation ($\Delta^{\ddagger}G_{rxn,aq}$) for these elementary reactions for a number of organic compounds (alkanes, oxygenated and halogenated structures) with single functional groups, and correlated the $\Delta^{\ddagger}G_{rxn,aq}$ data with the experimental $k_{\bullet OH}$ data reported in the literature for the respective selected compounds. The calculated $\Delta^{\ddagger}G_{rxn,aq}$ data were in good agreement (±3 kcal/mol) with the experimental $\Delta^{\ddagger}G_{rxn,aq}$ reported in the literature, which validated the theoretical approach. The developed LFERs between $\Delta^{\ddagger}G_{rxn,aq}$ and log $k_{\bullet OH}$ (experimental) were used to predict $k_{\bullet OH}$ for a series of H-atom abstraction and addition reactions with oxygenated and halogenated compounds with multiple functional groups, for which the authors calculated the standard free energies of activation. The predicted $k_{\bullet OH}$ data were within a factor of 5 from the experimental data reported in the literature. The authors noted that according to the transition state theory, a 1 kcal/mol difference in $\Delta^{\ddagger}G_{rxn,aq}$ causes a 5.4-fold difference in the estimated $k_{\bullet OH}$.

2.5 KINETIC MODELING OF UV/H_2O_2 PROCESS

For a given, well-characterized light source, the development of a kinetic model for predicting the UV/H_2O_2 process performance requires good understanding of direct photolysis, radical generation and radical reaction mechanisms involving target compound(s), H_2O_2, and at least the key water matrix constituents namely the dissolved organic matter, nitrate, and carbonate species, as well as the pH and water absorption impact. As not all radical reactions occurring in a UV/H_2O_2 –based water treatment contribute significantly to the process performance, usually, sensitivity analyses are performed and the kinetic model is constructed based on relevant reactions and equations. Then, the ordinary differential equations (ODEs) describing photolysis, radical reactions and relevant mono- and bimolecular reactions of stable compounds are solved with numerical methods. Given the low concentrations of micropollutants in water, their degradation follows *pseudo*-first order kinetics.

The kinetic models predicting the UV/H_2O_2 process performance account for the nominal power and absolute UV spectral power distribution of the light source, spectral transmittance of the quartz sleeve, average reactor hydraulic efficiency determined from CFD modeling, average light absorption efficiency inside the reactor determined from radiation models, and a number of safety factors, which describe the lamp ageing and sleeve quartz ageing and fouling. The more sophisticated approaches to UV/H_2O_2 performance evaluation employ models where the flow patterns (CFD model), radiation distribution field (radiation model), and chemical kinetics (kinetic model) are integrated and solved together spatially and temporally in the reactor chamber. This section summarizes the information available in the published literature on the mostly used models for predicting the contaminant decay and UV/H_2O_2 process performance.

2.5.1 *Pseudo*-steady-state approximation and dynamic kinetic models

In principle, the steady-state approximation models assume that given the high reactivity of radical species, those will be consumed in reactions with water constituents as soon as they are generated, i.e., the rate of formation of each radical equals the rate of disappearance of that radical. Under this assumption, *the change* in the concentrations of radicals considered in the reaction mechanism is negligible in time, i.e., the radicals reach *pseudo*-steady state concentrations. Equation 2.38 gives the general expression of *pseudo*-steady-state concentration of hydroxyl radical for the UV/H_2O_2 process.

$$[\mathrm{OH}]_{\mathrm{ss}}\big|_{\lambda} = \frac{\Phi_{\bullet\mathrm{OH}}\, f_{\mathrm{abs},\lambda}^{\mathrm{H_2O_2}}\, q_{p,\lambda}(1-\mathrm{e}^{-A_\lambda})}{k_{\mathrm{OH,H_2O_2}}[\mathrm{H_2O_2}] + \sum_i k_{\mathrm{OH},S_i}[S_i]} \tag{2.38}$$

In equation 2.38, the numerator and the denominator terms represent the rate of formation and the rate of decay of $^\bullet$OH, respectively, and $q_{p,\lambda}$ and $f_{\mathrm{abs},\lambda}(H_2O_2)$ are the photon flow (einstein $L^{-1}\ s^{-1}$) entering the water and the fraction of light absorbed by H_2O_2, respectively, at λ. The denominator accounts for $^\bullet$OH reactions with both H_2O_2 and all scavengers (S_i) present in the water, including micropollutant(s), DOM, carbonate species, etc. The simplified version of *pseudo*-steady-state approximation kinetic model assumes the H_2O_2 and S_i concentrations constant and equal to their initial concentrations. The term $\Sigma_i\ k_{\mathrm{OH},S_i}\ [S_i]$ is commonly called the $^\bullet$OH water background demand or simply, the scavenging term. In the kinetic models reported in the literature, this parameter is either calculated from water quality parameters or determined experimentally. This parameter is further discussed in section 2.5.1.2. Therefore, the rate expression for contaminant decay by both direct photolysis and $^\bullet$OH-induced reactions at λ, under the *pseudo*-steady state approximation approach becomes:

$$-\frac{\mathrm{d}C}{\mathrm{d}t}\bigg|_{\lambda} = \Phi_\lambda^C\, f_{\mathrm{abs},\lambda}^C\, q_{p,\lambda}(1-\mathrm{e}^{-A_\lambda}) + k_{\mathrm{OH},C}[^\bullet\mathrm{OH}]_{\mathrm{ss},\lambda}[C] \tag{2.39}$$

$$-\frac{\mathrm{d}C}{\mathrm{d}t}\bigg|_{\lambda} = \Phi_\lambda^C\, f_{\mathrm{abs},\lambda}^C\, q_{p,\lambda}(1-\mathrm{e}^{-A_\lambda}) + k_{\mathrm{OH},C}\frac{\Phi_{\bullet\mathrm{OH}}\, f_{\mathrm{abs},\lambda}^{\mathrm{H_2O_2}}\, q_{p,\lambda}(1-\mathrm{e}^{-A_\lambda})}{k_{\mathrm{OH,H_2O_2}}[\mathrm{H_2O_2}] + \sum_i k_{\mathrm{OH},S_i}[S_i]}[C] \tag{2.40}$$

Glaze *et al.* (1995) developed a kinetic model based on radical *pseudo*-steady state approximation to predict the oxidation of 1,2-dibromo-3-chloropropane (DBCP) in simulated and real groundwater. The experimental work was performed in a batch UV (253.7 nm) reactor equipped with 4 lamps. Various DBCP treatment conditions were explored, with various light intensity-, H_2O_2 and alkalinity concentration-, and pH-combinations. The kinetic model included ${}^{\bullet}OH$ reactions with DBCP ($k_{\bullet OH,DBCP}$ estimated in that work), H_2O_2 and carbonate species at their pH-dependent concentrations, and also a series of radical-radical reactions involving ${}^{\bullet}OH$, $CO_3^{-\bullet}$, $HO_2^{\bullet}$ and $O_2^{-\bullet}$. DBCP photolysis was investigated but not considered in the kinetic model since it was found insignificant. The radical steady-state concentrations were explicitly derived and used to describe the decay of molecular species such as DBCP, H_2O_2 and HCO_3^-. The agreement between experimental and predicted values was very good over the 0.26–3.00 mM H_2O_2 range and low alkalinity, but the DBCP destruction was under-predicted at high alkalinity. H_2O_2 decay was predicted reasonably well. In natural groundwater, the model over-predicted DBCP degradation, presumably because the model did not account for DOM contribution to both UV light intensity attenuation and ${}^{\bullet}OH$ scavenging.

Liao and Gurol (1995) published a UV/H_2O_2 kinetic model for chlorobutane (BuCl) degradation in a continuous flow-stirred tank reactor (CSTR) at various H_2O_2 doses, light intensity, and carbonate/bicarbonate concentrations. The authors' kinetic model included the role of DOM on BuCl degradation rate. Standard humic acid (HA) was used for that purpose, and the $k_{\bullet OH,HA} = 1.6 \times 10^4$ L^{-1} mgC s^{-1} was estimated from modeling exercises. The steady-state approximation was used for radical concentrations and the non-linear ordinary equations were solved numerically in the MATLAB software. The model predicted well both BuCl and H_2O_2 patterns under various experimental conditions and pH range of 3–9.

Unlike the above models, the 'dynamic' kinetic models do not use the steady-state approximation for the description of kinetics of free radicals in the irradiated system. Crittenden *et al.* (1999) undertook a critical evaluation of several UV/H_2O_2 kinetic models and indicated that in some cases, the *pseudo*-steady-state approximation could result in errors. The authors outlined the improvements of the advanced oxidation kinetic model (AdOx™) developed in their work, among which: all known photochemical pathways are considered; the rate expression is not simplified by the *pseudo*-steady state assumption; changes in solution pH and consequently, in the species chemistry, as well as the reactions of humic substances and other water constituents with ${}^{\bullet}OH$ and the impact of these reactions on UV/H_2O_2 performance are accounted for; the light absorption by humic substances and its impact on H_2O_2 and contaminant photolysis efficiency are considered; the ODEs can be solved in a powerful solver. The AdOx™ model was developed for a completely mixed batch reactor (CMBR) and the rate expressions for each species considered in the reaction mechanism were included in the respective governing mass-balance simultaneous ODEs, which were solved using a backward differentiation formula method (Gear's method). The model was successfully validated against the large experimental data acquired by Glaze *et al.* (1995) and was used to predict the electrical energy per order (E_{EO}) for DBCP degradation under various experimental conditions. The AdOx™ software is commercially available, it is applicable to single or tank-in series (TIS) reactors, and has a built-in tracer study analysis.

Song *et al.*'s kinetic model (2008b) improved upon Crittenden *et al.*'s model by including the DOM inner filter effect, DOM photolysis, and its ${}^{\bullet}OH$ scavenging capacity, using the UV absorbance at 310 nm (UVA_{310}) as a surrogate. Both Song *et al.*'s and Crittenden *et al.*'s models predicted well the DBCP experimental data from Glaze *et al.*'s work. Song *et al.* used alachlor (a pesticide regulated in drinking water) as a probe compound for testing their model. Excellent agreement was observed between the model predictions and the experimental data for alachlor under multiple variable combinations. The study gives the model-predicted time-dependent concentrations of the radicals formed in the system in the presence of DOM, the dynamics of ${}^{\bullet}OH$ scavenging capacity of DOM ($k_{\bullet OH,DOM}[{}^{\bullet}OH][DOM]$) and carbonate species, and the time-dependent f_{abs,H_2O_2} and $f_{abs,DOM}$.

Li *et al.* (2004) examined trichloroethene (TCE) photolysis, byproduct formation, and their degradation to complete mineralization of the original organic carbon in TCE. The study was performed in laboratory grade water, using a 1 kW MP lamp batch reactor in a recirculation mode. Based on the product study, a comprehensive reaction mechanism was developed, which was further used to generate the kinetic model equations (32 ODEs and 8 acid-base equilibria were considered). The kinetic model accounted for the photon flow spectral distribution, photochemistry of TCE and its degradation products across the polychromatic radiation emitted by the lamp, $Cl^{\bullet}$, $CO_2^{\bullet-}$, C-centered and oxyl radical reactions, pH change and its impact on carboxylic acid speciation. The model predicted well the experimental data for TCE degradation, formation and decay of intermediate products, chloride ion formation and pH pattern. In a subsequent study, Li *et al.* (2007) modeled TCE degradation and the product formation and decay for the UV/H_2O_2 process using the AdOx™ model-frame. In addition to direct photolysis and $^{\bullet}OH$ reactions for all involved species, the model accounted for $Cl^{\bullet}$, $Cl_2^{\bullet-}$, $HO_2^{\bullet}$, $CO_2^{\bullet-}$, C-centered and oxyl radical reactions, dark ("thermal") reactions between molecular species present in the water, and acid-base equilibria of species impacted by the water pH. The model predicted well the experimental data and the dynamics of UV photolysis-, $^{\bullet}OH$-, $Cl^{\bullet}$-, and $Cl_2^{\bullet-}$ radical contributions to the TCE degradation is described graphically in the study.

Hofman-Caris *et al.* (2012) considered two kinetic models together with the CFD model for predicting the UV/H_2O_2-based removal of nine priority pollutants (atrazine, NDMA, ibuprofen, bromacil, carbamazepine, diclofenac, metoprolol, phenazone and sulfamethoxazole) at pilot-scale. Reactors equipped with MP-, LP-, and DBD-UV lamps were installed at three sites in The Netherlands, and MP- and LP- reactors were used at one site in the USA. The UVPerox I kinetic model was based on $^{\bullet}OH$ *pseudo*-steady state assumption, and the $^{\bullet}OH$ water background demand was calculated from the experimentally measured water quality parameters. UVPerox II model was developed based on the experimentally measured fluence-based rate constants for each of the nine contaminants in the respective water qualities. Both approaches were integrated with the CFD model for various lamp power and flow operating conditions. Both models predicted reasonably well the experimental data, except for the MP-reactor data from one site in Netherlands and UVPerox II model, in which case the discrepancy was attributed to uncertainties in the MP lamp emittance spectrum used in the model. The authors concluded that while the (theoretical) UVPerox I model can be used only if all kinetic and water quality data are known for the specific application, the UVPerox II model (UV dose-based model) can be applied to any reactor and any lamp type, if the water absorption properties are known.

Wols *et al.* (2014) developed a non-steady-state radical concentration approximation kinetic model for the degradation of 36 micropollutants *via* direct photolysis and $UV_{253.7\ nm}/H_2O_2$ process in laboratory water and two pre-treated surface waters. The ODEs and acid-base equilibria were set and solved in MATLAB. The model considered nitrate photolysis and subsequent N_xO_y species reactions, as well as $CO_3^{\bullet}$ reactions given the high selectivity and, in some cases, large rate constants of this radical. Quantum yields, $^{\bullet}OH$, and $CO_3^{\bullet}$ reaction rate constants were obtained as fitted parameters, and compared well with the literature reported data. Model sensitivity to each incorporated reaction showed that nitrate photolysis could become important at low H_2O_2 and DOC levels, and $CO_3^{\bullet}$ reactions should be considered in high alkalinity waters. The model predicted well the experimental data for 31 of the 36 compounds. This kinetic model was used with CFD model to predict the degradation of 35 pharmaceutical compounds at (LP-UV reactor) pilot-scale in tap water (Wols *et al.* 2015a). The temperature effect on the degradation rates was examined in bench-scale experiments and included in the kinetic model for pilot-scale predictions. In this study, the kinetics was determined within the radiation field and flow patterns generated in the CFD model. As in the previous model, the ODEs were solved in MATLAB for each particle track, and the average photon fluence rate along the particle track was calculated from the fluence and residence time in the reactor. Absolute differences of 5–10% were determined between the predictions and observed degradation yields.

Wols *et al.* (2015b) described a kinetic model for polychromatic radiation. Similar approaches to the study and same water matrices as in Wols *et al.*'s (2014) study were used to investigate the degradation of 31 pharmaceuticals *via* direct photolysis and UV/H_2O_2 process with a MP lamp. Nitrate photolysis with

MP radiation generated nitrite at much higher yields than in LP lamp-reactor. In such cases, where strong •OH scavengers are generated *in situ*, the *pseudo*-steady state approximation-based models could generate erroneous predictions.

As discussed in this chapter, the most important degradation pathways in the UV/H_2O_2 process are direct photolysis and hydroxyl radical reactions. Bolton and Stefan (2002) derived the fluence-based rate constant ($k'_{1,\lambda}$, $m^2\ J^{-1}$) for direct photolysis and showed that it depends only on the fundamentals parameters Φ_λ and ε_λ ($M^{-1}\ cm^{-1}$):

$$k'_{1,\lambda} = \frac{\Phi_\lambda^C\ \varepsilon_\lambda^C\ \ln(10)}{10 \times U_\lambda} \tag{2.41}$$

Expression 2.41 can be converted to photon fluence-based rate constant ($k'_{1,po,\lambda}$, m^2 einstein^{-1}) by multiplying the terms on both sides of the equation by U_λ. The integrated rate expression for first-order kinetics decay of compound C *via* combined processes as a function of fluence (H', J m^{-2}) becomes:

$$\ln\frac{[\mathrm{C}]_0}{[\mathrm{C}]_{H'}}\bigg|_\lambda = \left[\frac{\ln(10)}{U_\lambda}\left(\Phi_\lambda^C\ \varepsilon_\lambda^C + \frac{k_{\mathrm{OH,C}}\ \varepsilon_\lambda^{\mathrm{H_2O_2}}\ \Phi_{\bullet\mathrm{OH}}[\mathrm{H_2O_2}]}{k_{\mathrm{OH,H_2O_2}}[\mathrm{H_2O_2}] + \sum_i k_{\mathrm{OH,S}_i}[\mathrm{S}_i]}\right)\right] H'_\lambda \tag{2.42}$$

A similar expression (2.43) was derived by Wols and Hofman-Caris (2012a) under the *pseudo*-steady state approximation for •OH concentration and used to predict the removal of ~90 micropollutants in surface water in a LP-lamp collimated apparatus at 400 mJ cm^{-2} and 10 mg/L H_2O_2. Note that for consistency, in equation 2.43, the symbols used in Wols and Hofman-Caris' paper were replaced with those used in this chapter.

$$\ln\frac{\overline{[\mathrm{C}]_t}}{\overline{[\mathrm{C}]_0}}\bigg|_\lambda = -E'_{p,0,\lambda}\frac{1-10^{-al}}{al}t\left(\Phi_\lambda^C\ \varepsilon_\lambda^C + 2\phi_{\mathrm{H_2O_2}}\ \varepsilon_\lambda^{\mathrm{H_2O_2}}\ \overline{[\mathrm{H_2O_2}]}\frac{k_{\mathrm{OH,C}}}{\sum_i k_{\mathrm{OH,S}_i}[\mathrm{S}_i]}\right) \tag{2.43}$$

In equation 2.43, $\overline{[\mathrm{C}]}, \overline{[\mathrm{H_2O_2}]}$ represent the volume averaged concentrations of C and H_2O_2, $E'_{p,0,\lambda}$ is photon fluence rate (einstein $m^{-2}\ s^{-1}$) and all the other symbols were defined previously. Under steady-state approximation, the H_2O_2 and S_i concentrations are set constant and equal to the original values. In equation 2.43, the •OH scavenging term includes the contribution of H_2O_2. The term $E'_{p,o,\lambda}\ (1\text{–}10^{-al})t/al$ multiplied by the energy of one mole of photons (U_λ, J einstein^{-1}) gives the fluence (H'_λ). Figure 2.9 shows the removal of selected micropollutants under the treatment conditions specified above, calculated with equation 2.43.

Lee *et al.* (2016) used a *pseudo*-steady state kinetic model with an equation similar to (2.42) to estimate the abatement of 16 micropollutants, including pharmaceuticals, BPA, pesticides and NDMA in treated wastewater effluents and to compare the electrical energy requirements for various AOPs (O_3, O_3 followed by UV, UV/H_2O_2, O_3/H_2O_2) and their combinations for potential water reuse applications. The authors estimated the •OH water demand based on DOM, carbonate species, nitrite and bromide as •OH scavengers. Of note, the $k_{\bullet\mathrm{OH,DOM}}$ values used with the model were determined experimentally, using ozone as a •OH source, according to a previously reported method (Lee *et al.* 2013). The contributions of •OH-initiated reactions to the overall degradation rate were determined and discussed.

Chuang *et al.* (2016) determined the •OH rate coefficients for the reactions with a large number of common and emerging disinfection byproducts. Further, the authors used the quantum yields, molar absorption coefficients and $k_{\bullet\mathrm{OH}}$ of these compounds to model the removal of halogenated DBPs with the UV/H_2O_2 process under water reuse treatment conditions (pH ~ 5, 3.4 mg/L H_2O_2). They used 170 reactions with the chemical kinetics model Kintecus 4.55, and observed a good agreement between the experimental loss of DBPs and predictions as a function of UV fluence.

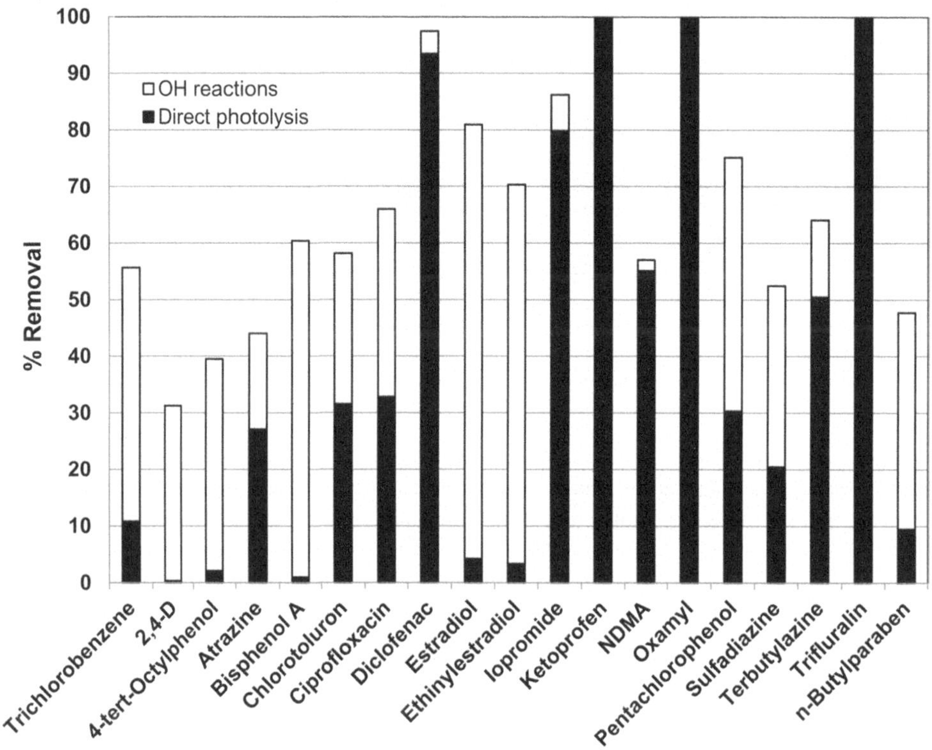

Figure 2.9 Removal yields of micropollutants in surface water with UV/H_2O_2 at 400 mJ cm^{-2} (adapted from Wols & Hofman-Caris, 2012).

The fluence rate distribution in the UV reactors is predicted by computational radiation models. Approaches to the experimental validation of model-predicted fluence rate distribution in bench- and pilot-scale reactors are described in the literature (Bohrerova *et al.* 2005; Rahn *et al.* 2006; Zhao *et al.* 2009; Shen *et al.* 2009; Gandhi *et al.* 2012; Solari *et al.* 2015). The UV/H_2O_2 process modeling with computational models (CFD) which integrate the hydraulics, radiation and kinetic models is briefly described in section 2.5.2.

2.5.1.1 Modeling the UV/H_2O_2 process with the $R_{OH,UV}$ parameter

The $R_{OH,UV}$ concept was introduced by Rosenfeldt and Linden (2005) to characterize and to model the UV/H_2O_2 process, by analogy to the R_{ct} concept introduced by Elovitz and von Gunten (1999) for ozone applications. The $R_{OH,UV}$ metric was further described and used in subsequent publications (Rosenfeldt *et al.* 2006, Rosenfeldt & Linden, 2007). $R_{OH,UV}$ is defined as the $^{\bullet}$OH exposure ($\int[^{\bullet}OH]dt$, M s) per UV fluence (H'_{λ}, J m^{-2}) for given water matrix, initial H_2O_2 concentration, and relative spectral distribution of the radiation emitted by the light source entering the water. $R_{OH,UV}$ is an experimentally determined parameter based on the kinetics of a probe compound, e.g., *p*-CBA. $R_{OH,UV}$ (M s m^2 J^{-1}, common unit M s cm^2 mJ^{-1}) is calculated with equation 2.44, which for simplicity, is given for monochromatic radiation.

$$R_{OH,UV} = \frac{\int_0^t [^{\bullet}OH]\,dt}{H'_{\lambda}} = \frac{k'_{1,UV/H_2O_2} - k'_{1,UV}}{k_{OH,pCBA}} \tag{2.44}$$

$R_{OH,UV}$ is wavelength-dependent and is determined under controlled fluence rate conditions (e.g., collimated beam apparata). The UV fluences used in the calculation of *p*-CBA fluence-based rate constants $k'_{1,UV/H_2O_2}$ (UV/H_2O_2 process) and $k'_{1,UV}$ (direct photolysis) are determined as described in Bolton and Linden (2003).

$R_{OH,UV}$ can be used to determine experimentally the $^{\bullet}OH$ water matrix demand which can be further used in the model kinetic equations to predict the contaminant removal with the UV/H_2O_2 process. Using the *pseudo*-steady state approximation and *p*-CBA as a chemical probe in a given water matrix exposed to 253.7 nm in presence of various H_2O_2 concentrations, the following equation can be derived (Rosenfeldt & Linden, 2007):

$$\frac{E'_{avg}}{[^{\bullet}OH]_{ss}} = \frac{U_{253.7nm}\left(\sum_i k_{OH,S_i}[S]_i + k_{OH,pCBA}[pCBA]\right)}{\phi_{OH}\,\varepsilon^{H_2O_2}_{253.7nm}} \times \frac{1}{[H_2O_2]} + \frac{U_{253.7nm} \times k_{OH,H_2O_2}}{\phi_{OH}\,\varepsilon^{H_2O_2}_{253.7nm}} \tag{2.45}$$

where E'_{avg} is the average fluence rate (W m^{-2}, common unit mW cm^{-2}) through the water, and the $E'_{avg}/[^{\bullet}OH]_{ss}$ is $1/R_{OH,UV}$. A plot of $1/R_{OH,UV}$ data *vs.* $1/[H_2O_2]$ at various initial H_2O_2 concentrations allows the calculation of $^{\bullet}OH$ demand of that specific water matrix.

The $R_{OH,UV}$ parameter was used to examine the effect of water quality on the $^{\bullet}OH$ exposure for the UV/H_2O_2 process (Rosenfeldt & Linden, 2007) and to compare the $^{\bullet}OH$ exposure among various AOPs (O_3, O_3/H_2O_2, UV/H_2O_2) and for two UV lamps (MP and LP) as a function of fluence and selected H_2O_2 concentrations (Rosenfeldt *et al.* 2006).

Gerrity *et al.* (2016) modeled the degradation of 17 micropollutants in wastewater effluents ($T_{254\,nm,1\,cm}$ <80%) collected from ten water facilities in the United States, Switzerland, and Australia. The $^{\bullet}OH$ exposure was determined from *p*-CBA degradation kinetics and then was used to estimate $k_{^{\bullet}OH}$ for reactions with selected contaminants. The authors developed a semi-empirical model which uses the $R_{OH,UV}$ values to estimate the $^{\bullet}OH$ exposure and to correlate this parameter with the $UV_{253.7\,nm}$ fluence/DOC ratio. This can be used as a semi-empirical strategy to account for the effects of variable $^{\bullet}OH$ water matrix demand and to predict the removal of micropollutants with the UV/H_2O_2 process.

2.5.1.2 Experimental determination of $^{\bullet}OH$ water matrix background demand

Accurate determination of $\Sigma k_{^{\bullet}OH,S_i}[S_i]$ is required for each UV/H_2O_2 application. Section 2.5.1.1 described the $R_{OH,UV}$ method. Nöthe *et al.* (2009) described a method for measurement of the $^{\bullet}OH$ scavenging capacity of wastewater samples in the ozonation process. The method is based on the competition between the water matrix and *tert*-butanol (TBA) for a given O_3 dose and exposure. A similar approach can be used for the UV/H_2O_2 process, where the UV fluence and H_2O_2 concentration are kept constant in a set of exposures where TBA concentration was varied. The competition is calculated based on formaldehyde yield from reaction of $^{\bullet}OH$ with TBA at various TBA initial concentrations.

Lee and von Gunten (2010) used the method developed by Zepp *et al.* (1987) to calculate the rate of formation and consumption of hydroxyl radicals in wastewater effluents. The method determines the kinetics of a •OH probe compound (e.g., *p*-CBA, $k_{1,pCBA}$) in the presence of known concentrations of an added •OH scavenger (e.g., TBA), where the OH radicals are generated *via* H_2O_2 photolysis at 253.7 nm. The equation allowing the calculation of •OH scavenging rate was derived as:

$$\frac{1}{k_{1,pCBA}} = \frac{\sum_i k_{OH,S_i}[S_i]}{v\,k_{OH,pCBA}} = \frac{k_{S_i^0}}{v\,k_{OH,pCBA}} + \frac{k_{OH,TBA}[TBA]}{v\,k_{OH,pCBA}} \tag{2.46}$$

where k_{Sio} (s^{-1}) is the •OH water matrix demand upon H_2O_2 and *p*-CBA addition, v [$(\Delta^{\bullet}OH)_t/t$, M s^{-1}] is the rate of •OH formation, and $k_{OH,pCBA} = 5 \times 10^9$ M^{-1} s^{-1} and $k_{OH,TBA} = 6.0 \times 10^8$ M^{-1} s^{-1}. The k_{Sio} value is calculated from the ratio of the slope and the intercept of the plot of the first term of equation *vs.* [TBA], and then, upon subtraction of H_2O_2 and *p*-CBA contributions, the •OH water matrix demand is calculated. This approach was developed further by Katsoyiannis *et al.* (2011) to determine the •OH rate constants for the reaction with DOM in lake water samples.

Rosenfeldt (2010) reported a rapid method for measuring •OH water background demand based on methylene blue (MB) degradation rate monitored spectrophotometrically in the visible range of MB absorption spectrum. The spectrophotometric method based on both MB and fluorescein as probes was used in combination with Lee and von Gunten's (2010) approach to determine the $k_{\bullet OH,NOM}$ values for NOM isolates, and further to correlate these data with NOM characteristics (Donham *et al.* 2014). Keen *et al.* (2014a) combined Rosenfeldt's MB method with Glaze *et al.*'s (1995) model to determine the total •OH scavenging demand of 28 filtered wastewater effluents collected from 8 different water utilities. The $k_{\bullet OH,EfOM}$ values were also calculated. Kwon *et al.* (2014) used the $R_{OH,UV}$ parameter with rhodamine B as a rapid and convenient method to determine the scavenging term. Rhodamine B decay was monitored spectrophotometrically at 554 nm.

2.5.2 Computational fluid dynamics models for the UV/H_2O_2 process

CFD model equations describe the spatial and temporal fluid dynamics, accounting for the conservation of mass, momentum, and energy. Hydrodynamics and turbulence modeling include also the velocity and pressure profiles within the reactor. The CFD models for UV reactors incorporate the spatial distribution of fluence rate. Among others, the radiation sub-model accounts for the lamp arrangement (number, geometry, and spacing) and spectral power density, quartz optics, reflection and refraction at the interfaces, fluid and reactor components optical properties. Examples of radiation models include multiple point summation (MPSS) model, line source integration (LSI) model, multi-segment source summation (MSSS) model, modified LSI model (RAD-LSI), and radiative transfer equation using discrete ordinate (DO) method. Liu *et al.* (2004) evaluated various fluence rate models used with CFD models to predict the radiation field distribution inside the reactor by comparing the predictions with the experimental measurements of fluence rate at selected points in the reactor. Chemical changes due to photochemical and chemical reactions are described by the kinetic model equations which are integrated with the combined fluid hydrodynamics and radiation models to describe the micropollutant treatment with UV alone or UV/H_2O_2 processes. The UV AOP models are described in the CFD models either in the Eulerian or the Lagrangian frameworks. Within the Eulerian framework, the concentrations of chemical compounds and radical species are resolved within the computational grid, whereas in the Lagrangian framework, "particles" are introduced in the flow. The particle trajectories are determined from the hydraulic model and chemical concentrations are calculated along the particle tracks in the radiation field (Wols & Hofman-Caris, 2012b). Elyasi and

Taghipour (2010) described a step-by-step approach to the development of a CFD model for simulating the •OH-initiated oxidation of pollutants in UV reactors, using the Eulerian framework. The authors measured the concentration profiles of the probe (rhodamine WT) inside a LP-lamp UV reactor using the planar laser-induced fluorescence technique. This technique allowed the authors to evaluate the hydrodynamics and radiation sub-models relative to the experimental profiles, thus, the approach can be used to simulate and optimize UV reactors of various geometries and operating conditions. Wols and Hofman-Caris (2012b) discussed a number of modeling approaches with different levels of complexity. The authors evaluated the Eulerian and the Lagrangian frameworks for the transport of chemical species. Degradation of atrazine with the UV/H_2O_2 process (5 and 10 mg/L H_2O_2) in a LP-collimated beam apparatus and a cross-flow four LP lamp flow-through reactor was simulated with the CFD models and the data were compared to the experimental results. The kinetic model included direct photolysis and •OH reactions, as well as bicarbonate and DOC reactions. The complex hydraulics in the flow-through reactor was defined with a 3D CFD model which solved the Reynolds-averaged Navier-Stokes (RANS) equations in combination with a k-ε turbulence model implemented in the COMSOL 4.2 software package. The authors observed good agreement between the simulations and experimental data for all conditions used. Moreover, there was a good agreement between the results obtained with the two frameworks. The authors indicated that the main advantage of the Lagrangian method over the Eulerian method is that once the particle tracks are known, the micropollutant kinetics can be solved by ODEs, reducing the high computational capacity and time required by the Eulerian method. Later, Wols *et al.* (2015a) used the Lagrangian method in combination with a validated UV/H_2O_2 kinetic model to predict the degradation of a large group of pharmaceuticals at pilot-scale. A good agreement was observed between the CFD simulations and experimental data for most compounds. Santoro *et al.* (2010) developed a dynamic kinetic model adapted from AdOx™ software for the $UV_{253.7\,nm}/H_2O_2$ degradation of tri(2-chloroethyl) phosphate (TCEP) and tributylphosphate (TBP), and incorporated it in the CFD model using the Eulerian framework. An annular reactor and a cross-flow reactor were modeled. Fluence rate distribution was calculated by solving the radiative transport equations (RTE) using the DO model, whereas the flow distribution was determined by solving the RANS equations in the CFD model. The model predicted the •OH concentration contours using the particle tracking approach, thus, assessing the contaminant exposure to •OH. The authors found that TCEP and TBP degradation yields were consistently higher in the annular reactor than in the cross-flow reactor and explained these findings by the absence of recirculation zones in the annular reactor. Alpert *et al.* (2010) evaluated the performance of combined CFD/radiation/UV/H_2O_2 models at predicting the UV/H_2O_2 degradation of an indicator (methylene blue, MB) in an annular single LP-lamp pilot reactor. The CFD model prediction sensitivity to turbulence and radiation models as well as to the water quality and $k_{\bullet OH,MB}$ was examined. Three turbulence closure sub-models (k-ε, RNG k-ε, and k-ω) and two fluence rate sub-models (MSSS and RAD-LSI) were tested with varying flowrate (5–30 gpm) and water quality parameters. The CFD model tended to under-predict the MB removal yields. The MSSS radiation sub-model predicted higher degradation than the RAD-LSI sub-model, while the turbulence sub-models had negligible impact on the yields. The model was sensitive to $k_{\bullet OH,MB}$ and to the •OH scavenging effect of DOC.

2.6 WATER QUALITY IMPACT ON UV/H_2O_2 PROCESS PERFORMANCE

Water quality plays a crucial role in both selecting the suitable AOP for a given application and predicting the performance of the selected AOP with respect to contaminant treatment. There are numerous publications describing the impact of water characteristics on the UV/H_2O_2 process performance. Discussion of those studies is outside the scope of this chapter. Instead, a brief description of the role of key water quality parameters – pH, temperature, and composition – is presented herein.

2.6.1 pH

The water pH could play an important role both in direct photolysis and •OH-initiated oxidation performance. Chemical micropollutants with pK_a values within ±1 unit of water pH would be present in both ionized and non-ionized forms whose photochemical (Φ, ε) and chemical ($k_{\bullet OH}$) parameters relevant to the UV/H_2O_2 process, could be largely different. The same is valid for water matrix constituents (DOM, carbonate species). In general, •OH rate constants for the reactions of dissociated species are larger than those of the respective non-dissociated forms (see Buxton *et al.* 1988).

Alkalinity is one of the key water quality parameters. Carbonate ion is almost two orders of magnitude more reactive with •OH than bicarbonate ion ($k_{\bullet OH,carbonate} = 3.9 \times 10^8$ M^{-1} s^{-1}; $k_{\bullet OH,bicarbonate} = 8.5 \times 10^6$ M^{-1} s^{-1}; Buxton *et al.* 1988). That is, even in slightly alkaline waters, carbonate ion would impact the UV/H_2O_2 process efficiency.

The pH effect on direct photolysis, i.e., on Φ_λ and ε_λ, was briefly discussed in section 2.2.3.

2.6.2 Temperature

As discussed in sections 2.2.3.2 and 2.4.3.2, both Φ_λ and $k_{\bullet OH}$ could be temperature-dependent. Changes in the reaction kinetics with temperature are difficult to predict given the extremely limited information in the literature on the thermodynamic parameters for key photochemical and chemical reactions occurring during the water treatment.

The water temperature could also affect the UV output of Hg lamps, thus, the photon fluence rate inside the reactor; consequently, the photochemical rates will be affected. The UV output as a function of temperature is determined routinely, particularly for LP-Hg vapor arc lamps, thus, the temperature effect can be accounted for in the radiation models.

2.6.3 Water matrix composition

Speciation, photochemistry and chemistry of inorganic and organic compounds present in the water affect the degradation kinetics of micropollutants in various ways. Water constituents could be strong absorbers over the UV-C or the entire UV wavelength range and/or could compete efficiently for hydroxyl radicals. They can also generate reactive oxygen species which could enhance the contaminant removal yields. Water turbidity caused by inorganic and/or organic particles interfere with the treatment process through light absorption and light scattering, and often decompose H_2O_2. In general, the processes implemented at water utilities upstream to UV/H_2O_2 reduce the water turbidity to levels which do not impact the advanced oxidation performance (<5 NTU). The role played by the most relevant inorganic and organic water constituents in the UV/H_2O_2 treatment performance is briefly discussed below.

2.6.3.1 Inorganic compounds

Nitrate is regulated in drinking water by both the U.S. EPA (MCL of 45 mg/L) and the EU Drinking Water Directive (MCL of 50 mg/L). Nitrate ion absorption spectrum displays a strong band in the 200–230 nm range and a very weak band in the UV-A range (Armstrong, 1963; Sharpless & Linden, 2001), thus it partly overlaps the spectral emission of MP-Hg lamps; NO_3^- has a small molar absorption coefficient at 254 nm (4 M^{-1} cm^{-1}, Mark *et al.* 1996). In nitrate-rich waters, the fraction of light absorbed by H_2O_2, thus the yield of •OH, are strongly impacted by nitrate in MP-UV/H_2O_2 applications. Nitrate photolysis generates nitrite with wavelength-, pH-, concentration-, and temperature-dependent yields (Sharpless and Linden, 2001; Goldstein and Rabani, 2007, and references therein; Beem Bendict *et al.* 2017). Nitrite is regulated in drinking water to 0.1 mg/L NO_2^- (EU) and 3 mg/L NO_2^- (North America). DOM interferes with nitrate photochemistry, as it reacts with both •OH and reactive nitrogen oxide radicals. Nitrite is a strong •OH scavenger ($k_{\bullet OH,NO_2^-} = 1.1 \times 10^{10}$ M^{-1} s^{-1}, Buxton *et al.* 1988), such that even at very low levels (e.g., μg/L), it will increase *in situ* the •OH water demand. Noticeable impact of nitrate photolysis *via* nitrite formation with LP Hg lamp-based reactors is observed only in high-transmittance (>~98 $\%T_{254\,nm,1\,cm}$) waters and large nitrate levels (>10–15 mg/L).

Bicarbonate and carbonate ions are commonly represented by water 'alkalinity'. Their speciation at water pH is important, as their reactivity toward •OH is different, and such would be their contribution to the overall •OH water background demand.

Metal ions such as iron (II) and manganese (II) are often present in groundwater due to the anoxic conditions. These ions and their complexes with DOM constituents absorb the UV radiation, thus, depending on their concentrations and absorption properties, they would interfere with H_2O_2 light absorption. They can also compete for •OH with the target micropollutant(s); e.g., $k_{\bullet OH,Fe(II)} = 3.5 \times 10^8$ M^{-1} s^{-1}; $k_{\bullet OH,FeEDTA2-} = 5.0 \times 10^9$ M^{-1} s^{-1}; $k_{\bullet OH,Mn(II)} = 3.0 \times 10^7$ M^{-1} s^{-1}; $k_{\bullet OH,MnEDTA2-} = 1.5 \times 10^9$ M^{-1} s^{-1}.

Chloride ion (Cl^-) concentrations are typically around 4 ~ 5 mg/L in surface waters and 10 ~ 20 mg/L in secondary effluents from wastewater treatment. The electron transfer reaction between Cl^- and •OH and subsequent reactions is pH-driven and becomes relevant only in acidic media.

Bromide ion reaction with •OH approaches the diffusion-controlled limits (reaction 2.33) and initiates the mechanism of bromate (BrO_3^-) formation in advanced oxidation of bromide-containing waters (von Gunten and Oliveras, 1998). Bromide levels up to 2 mg/L were detected in inland surface waters and groundwater around the world (Magazinovic *et al.* 2004). Disproportionation of dibromine radical anion ($Br_2^{\bullet -}$) formed in reaction 2.35 generates molecular bromine (Br_2), which dissolves in water yielding hypobromous acid (HOBr). HOBr and its conjugate base (OBr^-) are requisite intermediates in the bromate formation since these species are the precursors of bromoxide radical ($BrO^\bullet$) whose further reactions lead to bromate (von Gunten & Oliveras, 1998).

Although the initiation reactions 2.33–2.35 occur in the UV/H_2O_2 process and HOBr is generated, the reactions leading to bromate are suppressed by the fast reactions of H_2O_2 with HOBr and its conjugate base OBr^-, converting the bromine species back to Br^-. Therefore, unlike in ozone-based processes, bromate is not formed in UV/H_2O_2 treatment of bromide-rich waters.

$$H_2O_2 + HOBr \rightarrow Br^- + H^+ + H_2O + O_2 \quad k = 1.5 \times 10^4\ M^{-1}\ s^{-1}\ \text{(Taube, 1942)} \tag{2.47}$$

$$H_2O_2 + OBr^- \rightarrow Br^- + H_2O + O_2 \quad k = 1.3 \times 10^6\ M^{-1}\ s^{-1}\ \text{(von Gunten \& Oliveras, 1997)} \tag{2.48}$$

Chloramines are formed during the water and wastewater chlorination. In water reuse applications, chlorine (as hypochlorite) is dosed prior to the reverse osmosis (RO) membrane filtration of wastewater effluent to prevent the membrane fouling. At this step, chloramines are formed through the reaction of chlorine with free ammonia. As chloramines are not well rejected by RO membranes, they will be present in the RO permeate (ROP) at concentrations ranging from ~2 to 4 mg/L. UV/H_2O_2 treatment of ROP is required for NDMA and 1,4-dioxane reduction in both direct and indirect potable reuse water (see Chapter 14). Chloramines could be also present at low levels in the UV reactor influent in surface water treatment when chlorination is practiced by the plant at the water intake for *Zebra mussel* control. Unlike free chlorine ($k_{H_2O_2\text{-}OCl^-,25°C} = 1.7 \times 10^5$ M^{-1} s^{-1}, Connick, 1947), chloramines react very slowly with H_2O_2, ($k_{H_2O_2\text{-}NH_2Cl,\,24.6°C} = {\sim}2.76 \times 10^{-2}$ M^{-1} s^{-1} and $k_{H_2O_2\text{-}NHCl_2,35°C} = 1.14 \times 10^{-5}$ M^{-1} s^{-1}, McKay *et al.* 2013). Therefore, H_2O_2 and chloramines could coexist in the UV influent. Mono- and dichloramine absorb the UV radiation much stronger than H_2O_2, thus, will compete for the UV light affecting the •OH yield. Chloramines react with •OH (e.g., $k_{\bullet OH,NH_2Cl,22°C} = (2.8 \pm 0.2) \times 10^9$ M^{-1} s^{-1}, Johnson *et al.* 2002; $k_{\bullet OH,NH_2Cl,22°C} = 5.2 \times 10^8$ M^{-1} s^{-1}, Poskrebyshev *et al.* 2003) thus they contribute to the •OH water background demand.

2.6.3.2 Dissolved organic Matter (DOM)

Natural Organic Matter (NOM) referred herein as *dissolved organic matter* (DOM) is a key water quality parameter considered in the design and operation of water treatment processes. DOM is a complex mixture of hydrophobic and hydrophilic components of various structures, properties, and molecular weights. DOM constituents contain various functional groups, many of which have different dissociation constants; consequently, the photochemical and chemical properties of DOM will be pH-dependent. DOM composition shows geographical and seasonal variations.

DOM interferes with the UV/H_2O_2 process in multiple ways. It absorbs the UV radiation limiting the fraction of light absorbed by H_2O_2, therefore, the •OH yield. DOM is the major •OH scavenger in the water (see $k_{\cdot OH}$ examples in Table 2.7), competing for •OH with the target micropollutants.

Table 2.7 Experimental $k_{\cdot OH,DOM}$ values. Water quality conditions are reported in the citations.

DOM origin	Bulk/Isolated	$k_{\cdot OH,DOM}$ (M (C)$^{-1}$ s^{-1})	Reference
SRFA	Isolated	2.06×10^8	McKay *et al.* (2011)
ESHA	Isolated	1.21×10^8	McKay *et al.* (2011)
EfOM	Bulk	$(6.79–9.37) \times 10^8$	McKay *et al.* (2011)
Lake waters	Bulk	$(2.4–3.2) \times 10^8$	Katsoyiannis *et al.* (2011)
EfOM	Bulk	4.2×10^8	Katsoyiannis *et al.* (2011)
EfOM	Bulk	$(6.3–14.1) \times 10^8$	Dong *et al.* (2010)
EfOM	Bulk	3.6×10^8	Nöthe *et al.* (2009)
EfOM	Bulk	$(2.7–12.1) \times 10^8$	Rosario-Ortiz *et al.* (2008)
SRFA	Isolated	$(1.4–1.9) \times 10^8$	Westerhoff *et al.* (2007)
Saguaro Lake	Isolated	$(1.5–2.2) \times 10^8$	Westerhoff *et al.* (2007)
Nogales EfOM	Isolated	$(1.7–4.5) \times 10^8$	Westerhoff *et al.* (2007)
SRFA	Isolated	3.2×10^8	Goldstone *et al.* (2002)
SRHA	Isolated	2.3×10^8	Goldstone *et al.* (2002)
17 surface waters	Isolated	$(2.6–8.1) \times 10^8$	Westerhoff *et al.* (1999)
Lake waters	Bulk	$(1.8–3.7) \times 10^8$	Brezonik and Fulkerson-Brekken (1998)

EfOM: Effluent organic matter; SRFA: Suwanee River fulvic acid; SRHA: Suwanee River humic acid; ESHA: Elliot Soil humic acid.

During the UV/H_2O_2 process, the hydrophobic constituents of DOM are partly converted to hydrophilic fractions which may have different reactivity toward •OH than the parent structures. Also, depending on the H_2O_2 dose and the fluence applied, fragmentation and partial oxidation of DOM can occur. Consequently, the optical properties and •OH scavenging capacity of water being treated would be different than those of the UV influent water. These changes in water characteristics during the UV/H_2O_2 treatment are difficult to predict, but their net cumulative effect would be reflected in the target contaminant removal yield.

2.7 PERFORMANCE METRICS FOR UV LIGHT-BASED AOPs

2.7.1 Electrical energy per order

Electrical Energy per Order (E_{EO}) was introduced by Bolton *et al.* (1996) as a figure-of-merit for technical development and application of advanced oxidation processes. A few years later, the E_{EO} concept was covered by the authors in an IUPAC Technical Report, and its applicability was extended to solar-driven systems (Bolton *et al.* 2001).

The E_{EO} is defined as the electrical energy in kilowatt hours (kWh) required to reduce the concentration of a chemical compound by one order of magnitude (90%) in a unit volume [e.g., 1 m^3 (1000 L) or, frequently in North America, 1000 US gallons (1 US gal = 3.785 L)]. The E_{EO} values are calculated from experimental data with the following equations:

$$E_{EO}\,(\text{kWh/m}^3/\text{order}) = \frac{P(\text{kW}) \times t(\text{h})}{V(\text{m}^3) \times \log\left(\frac{c_i}{c_f}\right)} \quad (\text{batch operation}) \tag{2.49}$$

$$E_{EO}\,(\text{kWh/m}^3/\text{order}) = \frac{P(\text{kW})}{F(\text{m}^3/\text{h}) \times \log\left(\frac{c_i}{c_f}\right)} \quad (\text{flow-through operation}) \tag{2.50}$$

In the above expressions, P, V, t, c_i, and c_f, and F are electrical power at the wall, volume of water treated in the time t, initial and final concentrations (mol L^{-1}) of contaminant, and flowrate, respectively. The E_{EO} figure-of-merit is applicable to the conditions where the contaminant degradation follows *pseudo*-first order kinetics and relates to the rate constant k_1 through equation 2.51, where V and k_1 have units of L and min^{-1}, respectively.

$$E_{EO} = \frac{38.4\,P}{V\,k_1} \tag{2.51}$$

The E_{EO} expresses the electrical energy efficiency of the equipment using a given process to treat micropollutants in a given water quality. E_{EO} is used in sizing the UV equipment required to meet specific treatment goals and as a system performance parameter. This metric can be used to compare the water treatment performance of various AOPs and to assess the UV process efficacy for treatment of various contaminants in a given water source. A low E_{EO} results in low power costs through the project lifetime. The E_{EO} is related exclusively to the electrical energy cost incurred by the UV treatment, which is only a fraction of the overall costs associated with the implementation, operation and maintenance of the UV system, and oxidant-related costs.

The E_{EO} depends on several factors, thus, this metric must be properly understood and used. Among others, E_{EO} is dependent on: reactor design (geometry; lamp spacing and orientation; "effective" pathlength); lamp type, spectral power distribution and efficiency; quality and optical properties of quartz material used for lamp sleeves; hydraulic and optical (absorption) efficiencies of the UV reactor; power supply (ballast) efficiency; process conditions (direct photolysis and/or advanced oxidation); chemical and photochemical kinetic parameters of target contaminants in the given water matrix; oxidant (e.g., H_2O_2) concentration, absorption characteristics, chemistry and photochemistry; water quality, including spectral absorption properties, $^{\bullet}OH$ background demand, and photochemistry and chemistry of water constituents. Essentially, all these "factors" impact the rate constant for contaminant decay (k_1), which determines the treatment performance. Nowadays, E_{EO} is computed with kinetic models which account for all factors mentioned above, some of them determined experimentally, others generated by integrated radiation and CFD models.

In theory, E_{EO} should not depend on F; however, it has been observed that E_{EO} decreases when F increases over a wide range of flowrates, due to a narrowing UV dose distribution i.e., better hydraulic efficiency at higher flow rates. E_{EO} is sensitive to the effective radiation pathlength i.e., the better absorption of light inside the reactor and the lower UV radiation losses at the reactor walls, the lower the E_{EO} is. Therefore, optimization of UV reactor performance with regard to the combined effects of pollutant chemistry and photochemistry, hydraulics, and reactor optics is paramount.

Over the years, E_{EO} was proven a simple, reliable and practical metric to evaluate the UV-AOP system performance at water treatment facilities. Long-term treatment performance data for a wide range of chemical contaminants confirmed that E_{EO} can be used with robust kinetic models as a design parameter for UV system scale-up and as a parameter in the UV reactor control algorithms. Figure 2.10 exemplifies the robustness of the E_{EO} metric as a process performance monitoring and control parameter.

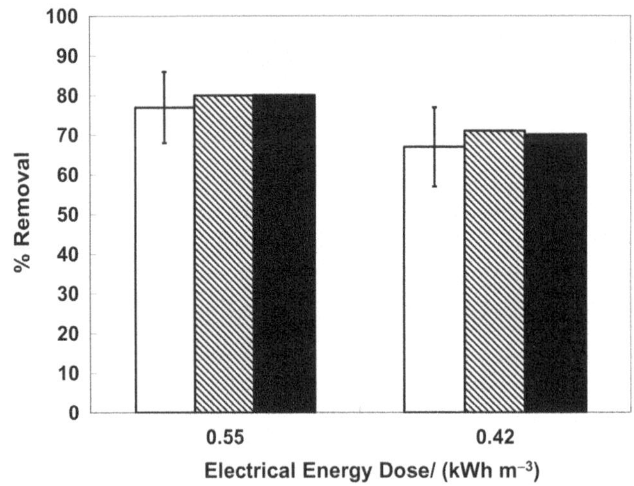

Figure 2.10 Atrazine removal with the UV (MP)/H_2O_2 process during the full-scale performance testing in Andijk, Netherlands: measured (empty bars); reactor control algorithm (grey bars); kinetic model (black bars). Adapted from Kruithof and Martijn (2013).

2.7.2 UV Fluence (UV dose)

UV Fluence (UV dose) has been recently advanced as a parameter for UV AOP equipment sizing and control, and for contaminant removal performance monitoring. Essentially, this approach relies on the experimental determination or kinetic model computation of fluence-based rate constant for contaminant decay which is further used with the calculated UV dose ("reduction equivalent dose", RED) required to achieve the contaminant treatment level (log reduction) in that specific application. The calculated dose is then input as a set-point in the UV reactor controls algorithm.

Aflaki *et al.* (2015) describes the approach to the UV equipment scale-up, implementation, and control for 1,4-dioxane treatment at Terminal Island Water Reclamation Plant (TIWRP) Advanced Water Purification Facility (AWPF). Bench- and pilot-scale testing data were used to determine the UV dose and oxidant (chlorine) concentration required to achieve 0.5-log reduction in 1,4-dioxane concentration in the RO permeate water, which was further used to scale-up the UV system. The UV dose delivered by the UV system was computed with radiation (point source summation, PSS) and CFD models and set as a critical alarm point in the UV reactor controls. The approach to UV dose-based equipment sizing and UV/H_2O_2 process performance control for MIB treatment at the municipal drinking water plant in Siheung (Korea) is described by Scheideler *et al.* (2016). The scale-up to the required UV dose for a given H_2O_2 concentration (6 mg/L) was done based on the pilot RED. The real-time UV sensor data are input parameters in the UV dose algorithm. The authors indicated that controlling the UV AOP system based on the UV dose approach ensures both quick response to the water quality changes and significant electrical energy savings.

Theoretical approaches to UV-AOP system scale-up based on fluence-based rate constants of contaminant decay determined from bench-scale experiments were advanced by Sharpless (2005) and Bircher (2015). Sharpless (2005) expanded Bolton and Stefan's analysis (2002) to the UV/H_2O_2 process where the contaminant decays by both photolysis and •OH-driven oxidation. In the derivation of kinetic equation, the author used the [•OH] steady-state approximation and the R_{OH} concept, which specifies the •OH concentration obtained for a given E'_{avg}. For polychromatic radiation, the contaminant-weighted and H_2O_2-weighted average fluence-based rate constant and average fluence-normalized R_{OH} values, respectively, would be calculated. The weighing factors for direct photolysis and •OH-induced oxidation rate constants would be the molar absorption coefficients and quantum yields of contaminant, and molar absorption coefficients of H_2O_2, respectively (Sharpless, 2005; Hofman-Caris *et al.* 2015a). The fluence-based expression of first-order kinetics decay of a chemical compound was discussed in section 2.5.1, and is shown below, where H' was substituted by $E'_{avg} \times t$.

$$\ln\left.\frac{[C]_0}{[C]}\right|_{\lambda} = \left[\frac{\ln(10)}{U_\lambda}\left(\Phi_\lambda^C\,\varepsilon_\lambda^C + \frac{k_{OH,C}\,\varepsilon_\lambda^{H_2O_2}\,\Phi_{\bullet OH}[H_2O_2]}{k_{OH,H_2O_2}[H_2O_2] + \sum_i k_{OH,S_i}[S_i]}\right)\right]E'_{avg,\lambda}t \qquad (2.42)$$

In equation 2.42 the term in the large bracket is the *pseudo*-first order fluence-based rate constant $k'_{1,UV/H_2O_2}$ ($cm^2\ mJ^{-1}$), which can be determined experimentally or computed if all parameters defining it are known. The UV fluence rate $E'_{avg,\lambda}$ is UV reactor design-, water quality- and flow conditions- specific, and is computed using optical and fluid hydraulics models.

The first-order kinetics described by equation 2.42 can be also expressed as a function of $R_{OH,UV}$ (M s $cm^2\ mJ^{-1}$, Gerrity *et al.* 2016), which is determined experimentally (section 2.5.1.1). The $R_{OH,UV}$-based contaminant kinetics is described by equation 2.52, where $k'_{1,\lambda}$ and $H'_{avg,\lambda}$ are the *pseudo*-first order fluence-based rate constant for direct photolysis and the average UV dose delivered in the reactor at λ, respectively:

$$\ln\left.\frac{[C]_0}{[C]}\right|_{\lambda} = (k'_{1,\lambda} + k_{OH,C} \times R_{OH,UV}) \times H'_{avg,\lambda} \qquad (2.52)$$

Bircher (2015) suggested the use of UV dose per log contaminant reduction (D_L) with CFD modeling to size full-scale UV reactors, then to test the reactor performance. The author stresses that although many parameters which define the E_{EO} can be easily scaled-up, CFD is needed to determine hydraulic and mixing efficiency of a flow-through reactor; as indicated above, hydraulic efficiency does impact E_{EO}. Therefore, an empirical method which uses the bench-scale-determined D_L in CFD modeling to determine the contaminant destruction at each point in the UV reactor would allow proper determination of UV dose in the reactor, accounting for hydraulic efficiency, thus, the UV reactor performance. The general expression of the UV dose delivered in a flow-through reactor equipped with 253.7 nm lamps was given as:

$$D = \sum_{d_1}^{d_n} \frac{P_e F_i (1 - 10^{-a_{254} d_i})}{Q \times a_{254} \ln(10)} \qquad (2.53)$$

Equation 2.53 is reproduced from Bircher (2015), where D, P_e, F_i, Q and d_i represent the UV dose, total radiant power at 253.7 nm entering the water, fraction of photons traversing the distance d_i before being absorbed, flowrate, and the pathlength traversed by the photons inside the UV reactor, respectively. A method to calculate the pathlength distribution from the reactor geometry is suggested, and the UV dose trends as a function of pathlength and water transmittance (85 to 100%) were calculated and reported. The findings showed that UV dose is relatively independent of pathlength longer than ~20 cm for water of $\%T_{1\,cm,254\,nm} < 90\%$ due to total absorption, but large diameter reactors should be designed for highly transmitting waters (>97%). These calculations do not account for reactor hydraulic efficiency, thus, proper UV dose calculations should be based on CFD modeling. For polychromatic radiation light sources, the UV dose should be weighted by the H_2O_2 absorption spectrum.

The E_{EO} and UV dose metrics for either UV equipment sizing or treatment performance control should not be regarded as two independent entities. Bolton and Stefan (2002) showed that the fluence-based rate constant (k_1', $m^2\,J^{-1}$) for contaminant decay by direct photolysis process depends only on the fundamental photochemical parameters (equation 2.41) and derived the relationship between E_{EO} and average fluence rate (E'_{avg}, W m^{-2}) in a batch UV reactor operated with a *quasi*-monochromatic radiation UV lamp (equation 2.54). Therefore, E_{EO} is linked directly to the fundamental photochemical parameters of treated contaminant (equation 2.55).

$$E_{EO} = \frac{0.6396\,P}{V\,k_1'\,E'_{avg}} \tag{2.54}$$

$$E_{EO} = \frac{6.396 P U_\lambda}{V \Phi_C \varepsilon_C \ln(10) E'_{avg}} \tag{2.55}$$

In the above equations, the units for ε_C and V are $M^{-1}\,cm^{-1}$ and L, respectively; all parameters except V and P are wavelength-dependent. All parameters were defined previously. Equation 2.55 was used to determine the average fluence rate in a semi-batch recirculating LP reactor, using atrazine as a probe compound. This approach was used in Sharpless and Linden's (2005) study, with both LP- and MP-UV reactor, and atrazine and NDMA as chemical probes.

By analogy to equation 2.54, the E_{EO} (kWh/m^3/order) for a flow-through reactor can be expressed as a function of the UV dose delivered in the reactor (H'_{avg}, mJ cm^{-2}) as shown in equation 2.56, using the overall fluence-based rate constant for the UV/H_2O_2 AOP from equations (2.42) or (2.52). The kinetic models for predicting the UV/H_2O_2 performance in a given application, as well as for sizing the UV equipment, are based on the relationship between E_{EO} and UV fluence.

$$E_{EO} = \frac{38.38\,P}{F \times k'_{1,UV/H_2O_2} \times H'_{avg}} \tag{2.56}$$

In equation 2.56, F is given in L min^{-1}. The UV dose delivered to the water is determined with combined CFD and UV radiation models which account for the flow patterns and the UV radiation distribution and absorption inside the reactor at a given flow rate. Clearly, E_{EO} and UV dose concepts used or advanced for UV system sizing, control, and treatment performance monitoring are inter-related and ultimately depend on the same factors outlined above for E_{EO}, given that both rely on the same key performance indicator: contaminant removal yield (i.e., log reduction). The implementation of either of the two metrics requires robust kinetic models and UV system control algorithms. While the changes in water transmittance are intercepted and the UV system operational conditions are modified accordingly in both E_{EO}- and UV dose-controlled contaminant treatment, neither metric responds to potential real-time changes in $^{\bullet}OH$ water matrix demand.

2.8 UV/H_2O_2 AOP EQUIPMENT DESIGN AND IMPLEMENTATION

A wide variety of bench-scale UV systems are used for the purpose of studying micropollutant photochemistry in laboratory settings, among which are collimated beam apparata, merry-go-round reactors, single-lamp batch reactors, and single-lamp semi-batch with recirculation loop reactors. These reactors are not discussed in this section. Instead, basic information is provided on full-scale UV reactor design considerations and components, sizing full-scale UV systems from bench- and pilot-scale data, and UV system integration in the treatment train at water utilities that have implemented UV light-based processes for micropollutant control.

2.8.1 UV Reactor design concepts

The basic objective of UV reactor design is ensuring uniform photon fluence rate distribution and high absorption efficiency of UV radiation within the water volume (i.e., not wasted at the reactor walls), good mixing and well-distributed water flow patterns within the radiation field (i.e., high hydraulic efficiency). The reactor optics and hydraulics are optimized and integrated together with the chemical and photochemical kinetics of target micropollutant(s) in order to achieve the treatment goal(s) sustainably and cost-efficiently over the entire duration of the project. Full-scale UV systems are designed for defined water transmittance and flowrate ranges, which also drive the UV reactor selection for specific applications. The optimization of reactor design, which includes the reactor geometry, is achieved with CFD modeling.

UV reactors can be either pressurized (closed vessels) or unpressurized (open-channel) reactors and are equipped with either monochromatic or polychromatic Hg vapor-arc lamps. For the purpose of contaminant removal by UV/AOP, mainly pressurized, closed-vessel UV reactors are used. Figures 2.11 exemplify closed vessel and open channel UV reactors.

Figures 2.11 Full-scale closed vessel and open-channel UV reactors.

Inlet and outlet water flow conditions, sometimes in combination with mixing elements within the reactor, are design parameters that can affect the reactor efficiency. Such elements are intended to ensure uniform distribution of water and H_2O_2 throughout the radiation field inside the reactor.

The key components of the UV system are the reactor housing, UV lamps, quartz sleeves and sleeve cleaning devices, sensors, lamp power supplies (drivers or "ballasts"), and the reactor controls.

The design of reactor geometry considers the lamp power, configuration and orientation, the number of lamps, flowrate and water transmittance ranges. In response to small footprint or space constraints at water treatment facility sites, modern full-scale UV reactors have flexible modular designs.

UV Lamps. Lamp positioning relative to the flow (perpendicular, parallel, or angled) and the distance between lamps (lamp spacing) inside the UV reactor are critical design parameters. A parallel lamp configuration is often used in order to minimize the head loss and is efficient if the water velocities are high enough in order to achieve good radial mixing along the length of the lamp. Perpendicular lamp configurations are often used to distribute light uniformly with respect to the direction of fluid flow and to enhance mixing, while angled configurations attempt to achieve the advantages of both parallel and normal configurations. Considered together with the fluid flow conditions and water and H_2O_2 optical properties over the spectral power distribution of the lamp, there exists an optimum lamp spacing that will maximize the radiation absorption efficiency inside the reactor

The lamp type (LP or MP) is also considered in the reactor design, as the lamp selection is highly project-specific. Given their high UV-C efficiency and suitability to most water qualities as compared to MP lamps, the LPHO UV lamps are used predominantly in modern UV light-based applications. Side-by-side comparison of LP- and MP-lamps for treatment of a number of micropollutants by both direct photolysis and UV/H_2O_2 processes in collimated beam apparata and bench-scale reactors showed 30–50% lower E_{EO} values for the LP- than for the MP-lamp based process (IJpelaar *et al.* 2010). Similar observations were reported from pilot tests on UV/H_2O_2 treatment of atrazine, bromacil, ibuprofen and NDMA in pre-treated surface water at Dunea (The Netherlands) treatment plant (Lekkerkerker-Teunissen *et al.* 2013). Therefore, although both monochromatic (LP lamp) and polychromatic radiation (MP lamp) are effective at contaminant treatment in water, the lamp UV-C efficiency impacts the electrical energy costs. However, cost evaluations demonstrated that for specific applications such as seasonal taste-and-odor control in drinking water, MP-lamp reactors operated in dual mode (UV/H_2O_2 during the T&O events and UV disinfection only most of the year) are the preferred solutions.

Quartz Sleeves. Selection of quartz quality used for the lamp housing depends on the type of lamp and water quality. For example, in applications based on short-wavelength UV radiation (e.g., 200–240 nm range), sleeves of high-purity fused silica which is highly resistant to solarization are used in MP-lamp reactors in order to benefit from the light energy emitted in that spectral region. Cyclic mechanical and/or chemical cleaning of quartz surface is programmed in the UV reactor controls to ensure constant spectral power delivered to the water through the sleeves. The cleaning mechanism is also part of the reactor design process.

Sensors. UV reactor systems use fluid level sensors, pressure sensors, temperature sensors, irradiance sensors and flow meters to control the water treatment conditions. The number of UV sensors and their positions inside the UV reactor are properly defined at the reactor design stage and also should be in compliance with the U.S. EPA Disinfection Guidance Manual (2006). In addition to the sensor and lamp power supply data, water transmittance at 254 nm is continuously monitored online by calibrated UV optical meters.

Lamp Power Supplies. The UV lamp drivers are installed outside the UV reactors, often in designated spaces with ambient humidity and temperature control. The lamp drivers control the power into the UV lamps. Highly efficient power supplies are used with the UV reactors. The power from the lamp driver into the lamp can be adjusted on the reactor control interface.

Reactor Controls. The UV system operation and control are possible *via* Programmable Logic Controller (PLC) algorithms. In principle, the controls algorithms are based on the relationship between the E_{EO} and logarithmic contaminant reduction, UV system power (real-time input of number of lamps in operation, lamp power), lamp and quartz sleeve age (i.e., number of hours in operation), real-time input of water temperature, flowrate, water $\%T_{254\,nm}$, and H_2O_2 dose. As shown in section 2.7.2, there is a direct relationship between the E_{EO} and UV fluence delivered inside the reactor; that relationship is built into the controls algorithm and the UV fluence is displayed on the UV reactor monitoring screen along with other operational parameters. The UV reactor performance controls using the E_{EO} metric ensure compliance with the project design specifications and also allows energy and H_2O_2 savings based on the real-time site conditions, e.g., water quality and flowrate. Moreover, as often the contaminant concentration varies in the water source, the plant operator is able to set the required contaminant log reduction and/or the influent/effluent contaminant concentrations in the water, on the reactor control panel (touch-screen human machine interface, HMI). In response, the controls algorithm calculates and optimizes the required combination of power and H_2O_2 dose for the selected treatment level with minimum energy and oxidant costs, and drives the UV/H_2O_2 treatment process according to the new settings. The real-time information on the UV system and flow conditions is logged into SCADA (supervisory control and data acquisition). Alarms based on sensor input are also built into the reactor controls and transferred to SCADA.

Other reactor control philosophies that are based on bench- and pilot-scale empirical testing prior to the full-scale installation of the UV system have been recently promoted and implemented. Scheideler *et al.* (2016) described a UV reactor control philosophy based on the UV fluence delivered inside the reactor, which was implemented for the UV/H_2O_2 treatment of 2-MIB at Siheung WTP, South Korea. The UV dose is calculated with the PSS method by the PLC algorithm using the UV sensor input, flowrate, and %*T*, and the power is adjusted to maintain the UV fluence set point determined and confirmed from collimated beam and pilot tests.

Sophisticated and computationally-intensive models which integrate the fluid flow characteristics and species transport for selected piping configuration and inside the reactor, fluence rate distribution, and contaminant kinetics, are used to optimize the UV reactor design for given applications. These models as well as the UV system control algorithms are specific to the various UV equipment suppliers. Design aspects of UV reactors for UV/H_2O_2 applications are addressed in numerous publications. Wols *et al.* (2015c) describe a systematic investigation of the UV reactor design using an analytical model. Critical analysis of the impact of each design parameter on reactor performance is discussed. Among selected parameters were the velocity profiles relative to fluence rate profiles, water layer, quartz sleeve diameter, H_2O_2 distribution inside the reactor. New UV/H_2O_2 reactor designs were developed with CFD simulations and tested experimentally for the degradation of 34 pharmaceuticals. Up to 30% increased degradation (on a log basis) was observed with the improved reactor design. A systematic approach to the CFD-based reactor design was described in a previous publication of the group (Wols *et al.* 2011). Coenen *et al.* (2013) modeled and optimized the geometry of single- and multiple-lamp photochemical reactors. The evaluation of CFD for modeling the UV/H_2O_2-based process and reactor design, including a literature review and operational cost optimization were investigated within the scope of Water Research Foundation project #3176 (WRF 2011). Wols' (2012) treatise on CFD for drinking water treatment covers the key topics on CFD modeling, including the use of this powerful tool to optimize the UV system design and to predict their performance.

2.8.2 Sizing full-scale UV equipment from bench- and pilot-scale

The UV equipment sizing models process the input information and select the most economical operational conditions which meet the treatment objectives set by the customer. The widely used metric for process economics and equipment sizing purposes is the E_{EO}. The cost associated with H_2O_2 is also considered when optimizing the process conditions that meet the project design specifications.

Successful implementation of direct photolysis and/or UV/H_2O_2 process requires correct understanding of chemical and photochemical kinetics of target micropollutant(s), H_2O_2, and water constituents in the site-specific water matrix, under specific water flowrate and radiation field conditions. Bench-scale testing is necessary to determine the micropollutant chemical and photochemical parameters and the water quality data among which the absorption spectrum, pH, turbidity (NTU), alkalinity, nitrate and TOC concentrations, total suspended solids (TSS), •OH water background demand. Preliminary information on treatment feasibility as a function of H_2O_2 concentration and lamp type can be also acquired from bench-scale studies using site-specific water. Photo-reactor scale-up methodologies from laboratory- or pilot-scale to full-scale UV systems are described in the published literature, e.g., Ghafoori *et al.* (2014) and references therein.

Pilot scale tests involve small UV reactors with similar lamp spectral characteristics and similar conceptual design as the full-scale system, set up at the water treatment facilities where the full-scale installations will be implemented. Although the pilot tests are great opportunities to validate the kinetic models, scaling up to full-size treatment systems from such testing is not trivial. It has been extensively demonstrated that:

- Pilot testing often leads to higher E_{EO} values when compared to full-scale results.
- Hydraulic efficiencies often increase in full-scale systems from mixing effects due to increased turbulence at higher flow rates.
- Optical efficiencies (absorption efficiencies) are often higher in full-scale systems due to larger reactors and therefore longer pathlengths traversed by the light emitted from the lamp, thus absorbed by the water. This is particularly valid for high-transmittance waters, e.g., >80% $T_{254\,nm,1\,cm}$. An example is given in figure 2.12 which illustrates the E_{EO} (NDMA) trend as a function of the optical pathlength and water transmittance for an annular reactor equipped with a LP lamp (Bolton & Stefan, 2002).
- Increased optical reactor efficiency translates into better use of H_2O_2, i.e., higher •OH yield and contaminant treatment are observed in full-scale than in pilot reactors for the same H_2O_2 dose.
- The small reactors used in pilot testing may be equipped with UV lamps with lower UV-C efficiency and different UV-C efficiency *vs.* temperature and/or ballast power level (BPL) patterns than the powerful lamps used in the full-scale UV reactors.

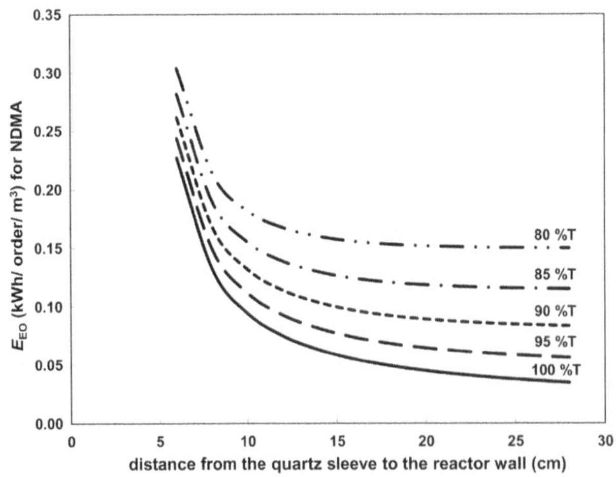

Figure 2.12 Reactor efficiency trend with $\%T_{254\,nm,1\,cm}$ and optical pathlength (Bolton & Stefan, 2002).

The scale-up process is *reactor-specific* and must consider the inter-dependence of contaminant photo-degradation kinetic parameters (ε, Φ, $k_{\bullet OH}$), H_2O_2 concentration, water quality parameters and photochemistry and chemistry of light-sensitive water constituents and their byproducts, UV (LP- or MP-lamp) reactor hydraulic and optical efficiencies. The robust UV equipment sizing models combine all these key factors and determine their optimum combination to meet the treatment goals with the minimum costs.

The implementation of the UV system at the water treatment plant is followed by full-scale performance evaluation as a contractual obligation. The experience with full-scale UV/H_2O_2 process installation for treatment of various contaminants in various water qualities and conditions and the extensive validation of the sizing models result in very good agreement between the full-scale performance data and the predicted data. In such cases, where the accuracy of the sizing tool was extensively demonstrated, there is no need for performing on-site pilot tests and the UV systems can be safely sized with the mechanistic UV/H_2O_2 model.

2.8.3 Incorporating the UV light-based processes into water treatment trains

Locating the UV system in a treatment train for removal of chemical contaminants is project specific. In general, the UV technologies work very well with other treatment technologies, such as activated carbon filtration, biodegradation, ion exchange (IX), and membrane technologies. The multi-barrier approach to microbial and chemical contaminants is taken by many water treatment utilities, with UV or UV/H_2O_2 process being one of the implemented "barriers".

In surface water treatment for drinking water production, processes upstream of the UV reactor that improve the water quality typically include coagulation, sedimentation, and filtration. Ion exchange prior to MP-UV/H_2O_2 process in high-nitrate water reduces the nitrate levels, which both increases the water transmittance in the short wavelength range of radiation emitted by the MP-lamps and prevents the increase of $^{\bullet}OH$ demand due to nitrite formed from nitrate photolysis. Results from extensive research conducted by PWN Technologies (PWNT) at their WTP in Andijk indicated significant energy savings upon replacing the coagulation-rapid sand filtration (CSF) pre-treatment steps with IX-membrane microfiltration (MF). Pilot tests using NDMA and 1,4-dioxane as chemical probes showed that the E_{EO} values for the UV/H_2O_2 treatment of these compounds in IX-UF pretreated water were up to 50% lower than those determined in CSF-pretreated water (Martijn *et al.* 2010; see also Chapter 15). In 2015, the Andijk WTP was retrofitted with PWNT processes SIX® (ion-exchange) and CeraMac® (microfiltration) prior to the UV/H_2O_2 treatment.

Groundwater can be impacted by iron and manganese which reduce the water transmittance and cause quartz sleeve fouling. Pre-chlorination for Fe and Mn oxidation followed by filtration are used prior to the UV/H_2O_2 treatment. Remediation of groundwater heavily impacted by VOCs, may require air stripping, carbon adsorption, ion exchange, or biological process or combination of thereof prior to the UV/H_2O_2 treatment.

The widely implemented advanced treatment train for direct or indirect potable water reuse involves microfiltration (MF), ultrafiltration (UF) and reverse osmosis (RO) prior to UV/H_2O_2 treatment step. The UV influent has high $\%T_{254\,nm,1\,cm}$ (≥95%) and very low $^{\bullet}OH$ water background demand. See Chapter 14 for details on water reuse applications.

GAC or BAC filtration is commonly implemented as a post-UV/H_2O_2 treatment step. Short empty bed contact time (EBCT ~ 6 min) is required to remove the H_2O_2 residual, whereas EBCTs up to ~20–30 min are needed to remove byproducts formed either before or after UV/H_2O_2 treatment. GAC and

BAC represent additional barriers to micropollutants in water (Tabrizi & Mehrvar, 2004; Kruithof *et al.* 2007).

Post-UV/H_2O_2 chlorination is also practiced at water treatment plants to remove the H_2O_2 residual and to ensure residual disinfectant in the water distribution network. If chloramine is the disinfectant of choice, ammonia is added to the chlorinated water. The chlorine – H_2O_2 reaction (1:1 on molar basis; 2.09:1 on mass basis) occurs rapidly (within a few seconds) and no harmful byproducts are thought to be formed. Sodium sulfite and sodium thiosulfate were explored as potential H_2O_2 quenching agents, but are not recommended given their slow, complex, and pH-dependent reactions with H_2O_2 (Keen *et al.* 2013).

Table 2.8 lists treatment trains at selected WTPs which implemented the UV/H_2O_2 process for chemical contaminant removal.

Table 2.8 Examples of UV/H_2O_2 process location in water treatment trains.

WTP	Water Source	Treatment Train
Cornwall, ON, Canada	River water	Chlorination–coagulation–media filtration–UV/H_2O_2–chlorination
West Elgin, ON, Canada	Lake water	Chlorine dioxide–ultrafiltration–UV/H_2O_2–chlorination
Andijk, The Netherlands	Lake water	IX–MF–UV/H_2O_2–BAC–chlorine dioxide
Patoka Lake, TX, USA	Surface water	Chlorination–coagulation–filtration–UV/H_2O_2–chlorine/ammonia
San Gabriel Valley, CA	Groundwater	IX for perchlorate removal–UV photolysis–air stripping–chlorination
Tucson, AZ, USA	Groundwater	Air stripping–UV/H_2O_2–GAC–chlorination
Hall (Lincoln), UK	River water	GAC-ultrafiltration–UV/H_2O_2–BAC–UV disinfection–chlorination
Greenbrook, ON, Canada	Groundwater	Chlorination for Fe/Mn oxidation–media filtration– UV/H_2O_2– GAC–chloramination

2.9 UV/H_2O_2 AOP FOR MICROPOLLUTANT TREATMENT IN WATER

UV/H_2O_2 AOP is the most extensively investigated process for water remediation from a wide range of organic and inorganic micropollutants. Many literature studies were and continue to be driven by the water regulations, whereas several other studies demonstrate proactive research towards viable solutions against currently unregulated compounds whose adverse human or ecological health effects have been demonstrated or are anticipated. More recently, several classes of "emerging contaminants" have caught researchers' attention. Examples of emerging contaminants include pesticides, pharmaceuticals, personal care products, flame and fire retardants, new industrial additives and byproducts, surfactants, synthetic hormones, algal toxins, nanomaterials, water treatment byproducts. Reports on occurrence of priority pollutants, legislative requirements, and remediation strategies are available in the literature. Examples include Water Research Foundation report (2010); Shi *et al.* (2012); Merel *et al.* (2013); Stone *et al.* (2014); Ribeiro *et al.* (2015); Barbosa *et al.* (2016); Bui *et al.* (2016). Over the years, several reviews covered the literature on the environmental contaminant removal *via* AOPs, including the UV/H_2O_2 AOP, among which Peyton (1990); Legrini *et al.* (1993); Zhou and Smith (2002); Oturan and Aaron (2014); Yang *et al.* (2014).

The following sub-sections provide examples of laboratory- and pilot-scale studies on micropollutant treatment with UV light in the presence and in the absence of H_2O_2, and describe selected full-scale UV/H_2O_2 installations at water utilities for drinking water production. The last sub-section gives a high-level perspective on UV/H_2O_2 process economics as relayed in the published literature.

2.9.1 Laboratory-scale research studies

The research studies conducted in laboratory-scale setups concerned primarily with the understanding of UV photolysis or UV/H_2O_2 process effectiveness at the removal of micropollutants from water. The experiments were conducted under a variety of water qualities, often in modified laboratory-grade water, with monochromatic or polychromatic light sources.

Stefan and Williamson (2004) reviewed the literature studies on UV photolysis of organic compounds identified as priority pollutants of drinking water sources, including halogenated aliphatic VOCs, polycyclic aromatic hydrocarbons, phenol and derivatives, chlorinated benzenes, chlorinated biphenyls, dioxins and furans, nitroaromatic compounds, various classes of pesticides, nitrosamines. It is beyond the scope of this chapter to provide a comprehensive literature review on micropollutant treatment by UV light-based processes. Instead, only a few examples will be selected from a few classes of water pollutants. Where applicable, both direct photoysis and $^{\bullet}OH$-induced processes will be briefly discussed.

2.9.1.1 N-Nitrosamines

Aliphatic and aromatic *N*-nitrosamines were identified as priority pollutants and classified as hepatotoxic and potent carcinogens. The U.S. EPA's Integrated Risk Information System (IRIS) database shows that at concentrations as low as nanogram per liter (ng/L) in drinking water, *N*-nitrosamines are associated with a lifetime cancer risk of 10^{-6} (http://www.epa.gov/iris/). The U.S. EPA IRIS database lists eight nitrosamines (NDMA, NDEA, NMEA, NDPA, NEBA, NPYR, NDPhA, and *N*-nitrosopiperidine NPIP), five of which are included on the CCL3 and CCL4 (NDMA, NDEA, NDPA, NPYR, and NDPhA). Health Canada's (2014) drinking water standard for NDMA is 40 ng/L, whereas WHO guideline (2011) for NDMA is 100 ng/L. In 2009, California Department of Public Health (CDPH) established a notification level of 10 ng/L for three nitrosamines (NDMA, NDEA, and NDPA).

The absorption spectra of aliphatic and heterocyclic *N*-nitrosamines are characterized by two distinct bands, one in the UV-C range (200–280 nm, $\pi \rightarrow \pi^*$ transition, maximum at ~220–240 nm) and a much weaker band in the UV-B – UV-A range (290–400 nm, $n \rightarrow \pi^*$ transition, maximum at ~330–350 nm) (Plumlee & Reinhard, 2007). The NDPhA spectrum shows a band from 200 to ~250 nm with a shoulder at ~230 nm, and a second band extended into UV-B range with a maximum at ~295 nm. The absorption spectrum of NDMA is given in Stefan and Bolton (2002). Plumlee and Reinhard (2007) determined the quantum yields for photolysis of seven of the eight nitrosamines listed in the EPA IRIS database, under simulated sunlight conditions ($290 < \lambda < 800$ nm) at pH 6. The wavelength-averaged quantum yields varied from 0.41 ± 0.02 for NDMA to 0.61 ± 0.03 for NMEA. Despite the large number of publications on NDMA photolysis, the reported quantum yields at 253.7 nm vary from ~0.30 to ~0.43. Section 2.2.3.2 provides additional quantum yield data for NDMA and discusses the pH, concentration and wavelength-dependency.

Lee *et al.* (2005a, b) proposed the most detailed degradation mechanism for NDMA photolysis to-date (Figure 2.13). The authors suggested three major pathways for the evolution of NDMA excited state, all of which involving protonation of NDMA* followed by N—NO bond dissociation. In Figure 2.13, these pathways are as follows: homolytic splitting (1), heterolytic cleavage (2), and oxidation (3). Product studies on NDMA photolysis (Lee *et al.* 2005a, b; Stefan & Bolton, 2002) showed the formation of the following compounds: dimethylamine, monomethylamine, formaldehyde, *N*-methylformamide, nitrite, nitrate, formic acid.

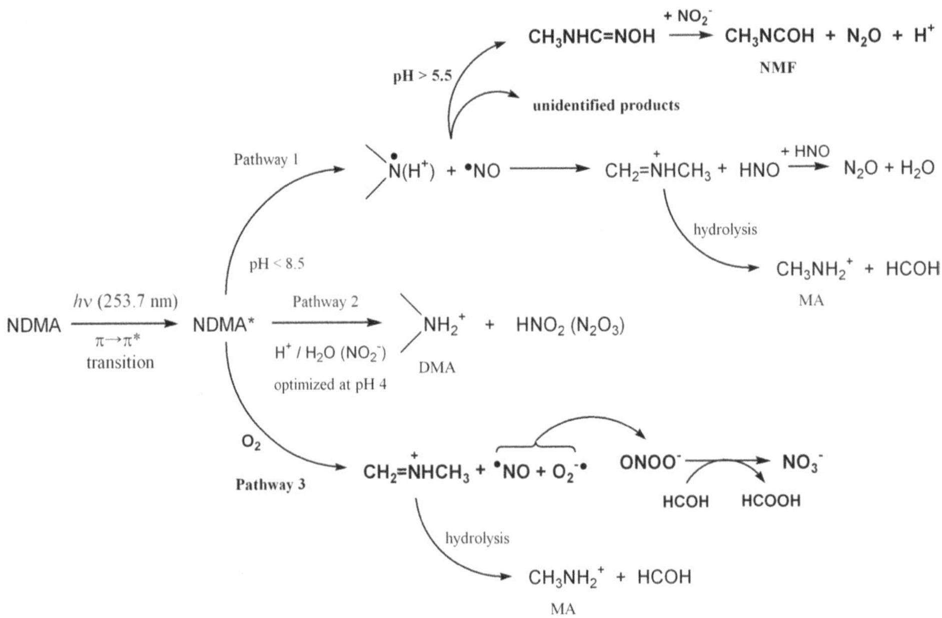

Figure 2.13 Proposed photolytic pathways for NDMA and subsequent reaction in aqueous solution (Lee *et al.* 2005b). MA – methylamine; DMA – dimethylamine; NMF – *N*-methylformamide.

Xu *et al.* (2009) reported on the photodegradation of NPYR and NPIP under irradiation at 253.7 nm in a bench reactor. The pH (3–10.5) effect and product study were examined in laboratory-grade water, whereas the impact of water quality on the removal yields was examined in drinking water ($a_{254\,nm} = 0.08$ cm^{-1}) and river water ($a_{254\,nm} = 0.14$ cm^{-1}) matrices. No pH effect was observed on NPYR photolysis, whereas the ratios of *pseudo*-first order rate constants obtained for NPIP at pH 6.0, 7.1, and 10.5 relative to that at pH 3.1 were 0.96, 0.80, and 0.05, respectively.

The •OH rate constants for the reactions of *N*-nitrosamines considered herein vary over more than one order of magnitude (Table 2.6). Photochemical characteristics of *N*-nitrosamines make these compounds amenable by direct photolysis, thus, in the UV/H_2O_2 process, the contribution of the •OH-induced reactions to the overall degradation yields is expected to be relatively small, except for NDPhA whose aromatic ring is prone to •OH attack. Lee *et al.* (2007) studied the kinetics of NDMA degradation with O_3 and O_3/H_2O_2 processes and the role of pH, bicarbonate, and H_2O_2 on thereof. The bimolecular rate constants were reported as $k_{O_3,NDMA} = 0.052 \pm 0.0016$ M^{-1} s^{-1} and $k_{\bullet OH,NDMA} = (4.5 \pm 0.21) \times 10^8$ M^{-1} s^{-1}. NDMA was not well removed by O_3 alone, but its degradation increased to 50% at pH 8 in deionized water due to reactions with •OH which were formed from O_3 reaction with OH^-. Zhou *et al.* (2012) investigated the degradation of the eight *N*-nitrosamines listed above and of *N*-nitrosodibutylamine (NDBA) in their mixture by direct photolysis (253.7 nm), H_2O_2 only, and UV/H_2O_2 process. The removal yields by direct photolysis follow the order NDBA≅ NPIP≅ NDPA≅ NMOR> NDEA> NPYR≅ NMEA> NDMA≫ NDPhA. The UV/H_2O_2 (25 µM H_2O_2) process increased slightly the removals of NDBA, NDPA, NMOR, NPIP and NDEA; moderate increase was observed for NMEA and NPYR, whereas approx. 50% and 100% increased degradation was reported for NDMA and NDPhA, respectively. For more information on *N*-nitrosamine treatment, particularly in water recycling projects see Chapter 14.

2.9.1.2 Pesticides

The widespread contamination of water and soil with pesticides, their persistence in the environment and the link between the neurodevelopmental and neurodegenerative disorders, alterations in immune response, diabetes, predisposition to different types of cancers and pesticide exposure have prompted regulatory standards in drinking water for pesticides. Pesticides are controlled in drinking water under the EU Directive to a maximum contaminant level (MCL) of 0.1 µg/L for the individual pesticide but not more than 0.5 µg/L in total, including pesticide metabolites. More relaxed drinking water standards are set by the U.S. EPA, with fewer regulated pesticides and higher MCLs than in Directive 2013/39/EU. The U.S. EPA CCLs 3&4 include approx. 40 pesticides and their environmental degradates. Burrows *et al.* (2002) reviewed the photodegradation of pesticides, with particular emphasis on studies that describe the reaction mechanisms, including direct photolysis, $^1O_2$- and $^{\bullet}OH$-induced reactions. This review references approx. 300 studies. Ikehata and Gamal El-Din's review (2006) covers the degradation of pesticides in aqueous solutions with UV/H_2O_2 and Fenton-type AOPs. The groups of pesticides included in the review are: aniline derivatives, carbamates, chlorophenoxy compounds, chlorinated organics, organophosphate, pyridine and pyrimidine derivatives, triazines and triazinones, substituted urea compounds, and a few miscellaneous pesticides. The authors summarized the basic properties of pesticides, reaction conditions and kinetics for the AOP, quantum yield and $k_{^{\bullet}OH}$ data, reaction byproducts. Over 150 citations are listed in this review. Stefan and Williamson (2004) reviewed the literature on UV photolysis of triazine-, urea-, carbamate-, organophosphorous-, chloracetamide-, dinitroaniline-, and chlorophenol pesticides. Ribeiro *et al.* (2015) provided an overview on the AOPs described in the literature for treating priority pollutants, including five major classes of pesticides, listed on Directive 2013/39/EU.

s-Triazine pesticides. Several studies concern photodegradation of *s*-triazine pesticides with and without H_2O_2 addition. Photolysis of atrazine (ATZ), simazine (SIM), propazine (PPZ) and terbuthylazine (TBZ) at 253.7 nm was investigated by Nick *et al.* (1992). Quantum yields and molar absorption coefficients were reported for the four pesticides, i.e., ATZ (3860 M^{-1} cm^{-1}; $\Phi = 0.05$); SIM (3330 M^{-1} cm^{-1}; $\Phi = 0.083$); PPZ (3370 M^{-1} cm^{-1}; $\Phi = 0.099$); TBZ (3830 M^{-1} cm^{-1}; $\Phi = 0.094$) and some of their degradation byproducts. Photohydrolysis of chlorine atom was the major reaction pathway, followed by dealkylation. Other values for photochemical parameters of atrazine were reported in the literature; e.g., $\varepsilon_{254\,nm} = 3683$ M^{-1} cm^{-1}; $\Phi_{254\,nm} = 0.033$ (Bolton & Stefan, 2002); $\Phi_{\lambda<250\,nm} = 0.038$; $\Phi_{\lambda>250\,nm} \cong 0.06$; Sharpless and Linden (2005). Acero *et al.* (2000) identified atrazine primary and subsequent degradation byproducts *via* $^{\bullet}OH$ reactions, proposed the degradation mechanism, and ascertained the contribution of each pathway. The authors determined the rate constants for $^{\bullet}OH$ reactions with ATZ (3×10^9 M^{-1} s^{-1}; $t = 20°C$) and with its primary byproducts (1.2×10^9 M^{-1} s^{-1}–1.9×10^9 M^{-1} s^{-1}; $t = 20°C$). Álvarez *et al.* (2016) reported the removal of terbuthylazine from water with various methods including adsorption onto activated carbon and multiwalled carbon nanotubes, UV(253.7 nm) photolysis, $UV_{253.7\,nm}/H_2O_2$, O_3, O_3/H_2O_2, catalytic ozonation and solar-driven processes. The UV/H_2O_2 process was found the best process, due to the measured large $k_{^{\bullet}OH,TBZ} = 3.3 \times 10^9$ M^{-1} s^{-1}. The UV fluence required for 90% destruction of TBZ by UV alone was calculated as 2074 mJ cm^{-2}, based on the measured $\varepsilon_{254\,nm} = 3615$ M^{-1} cm^{-1} and $\Phi_{253.7\,nm} = 0.064$; in the presence of 5 mg/L H_2O_2, a fluence of 1404 mJ cm^{-2} per log TBZ removal was determined. A reaction mechanism is proposed for both $^{\bullet}OH$ and O_3-initiated degradation. Sorlini *et al.* (2014) observed similar photolysis yields for TBZ degradation in groundwater (1.9 mg/L DOC, 0.03 mg/L Fe, 0.035 mg/L Mn, 242 mg/L alkalinity as $CaCO_3$), i.e., 85% removal at 2000 mJ cm^{-2}. In the presence of 5 mg/L H_2O_2, ~75% TBZ removal was achieved at a fluence of 1200 mJ cm^{-2}.

Ijpelaar *et al.* (2010) compared the LP- and MP-UV/H_2O_2 process efficacy at the removal of a suite of chemical pollutants in pre-treated surface water samples collected from water treatment plants. Atrazine, cyanazine, simazine, desethylatrazine and desisopropylatrazine were the studied triazine pesticides. Other pesticides included in the tests were metazachlor and alachlor. Five estrogens, ten pharmaceuticals and MTBE were also investigated. Bench-scale reactor data showed $E_{EO} < 1.0$ kWh/m^3/order for all compounds except MTBE and cyanazine. The energy required to achieve 90% removal of ATZ with LP-UV/H_2O_2 was found approx. 40% lower than with MP-UV/H_2O_2.

Urea pesticides. Photochemistry of phenylurea herbicides depends on the nature and position of substituents on the aromatic ring. Photo-rearrangement and photohydrolysis are mostly observed processes. Amine-Khodja *et al.* (2004) reviewed the photochemical properties of the following phenylurea herbicides: non-halogenated (fenuron, isoproturon); monohalogenated (chlorotoluron, metobromuron, metoxuron, monolinuron, monuron); dihalogenated (chlorbromuron, diuron, linuron). These compounds display three absorption bands, with the middle one in the 225–275 nm range (ε up to ~20,000 M^{-1} cm^{-1}) and a much weaker band extending into the UV-A range. The quantum yields at 253.7 nm vary from 0.002 to 0.13, depending on the pesticide structure. Reaction mechanisms and detailed interpretation based on the structural characteristics of pesticides are given for direct photolysis, photosensitized and radical reactions. The review contains 53 references. Nitrate-sensitized degradation of micropollutants including a wide variety of pesticides in MP-UV-irradiated aqueous solutions is well represented in the published literature; e.g., Nélieu *et al.* (2009) and references therein.

The impact of pH (2–9) and temperature (10–40 °C) on the direct photolysis $\Phi_{253.7\,nm}$ of linuron, chlorotoluron, diuron, and isoproturon in milliQ water and the $^{\bullet}OH$ rate coefficients for these pesticides were investigated by Benitez *et al.* (2006). The reported $\Phi_{253.7\,nm}$ data over the experimental conditions used are as follows: isoproturon (0.0035–0.0066), diuron (0.0057–0.0238), chlorotoluron (0.0148–0.0533), linuron (0.0192–0.055). The $k_{^{\bullet}OH}$ values were similar for all herbicides, ranging from 4.3×10^9 M^{-1} s^{-1} (linuron) to 5.2×10^9 M^{-1} s^{-1} (isoproturon). The removal of these herbicides in their mixtures in ultra-pure water, mineral water, groundwater, and lake water was modeled successfully.

Chlorophenoxy-carboxylic acid pesticides. Meunier and Boule (2000) studied the effect of pH, dissolved O_2, and wavelength on direct photolysis of mecoprop [2-(4-chloro-2-methylphenoxy)-propionic acid] (MCPP) in aqueous solution, identified the early degradation products and proposed the phototransformation mechanisms. Molecular and anionic forms of mecoprop ($pK_a = 3.78$) have similar spectra, with a band centered at ~280 nm ($\varepsilon = 1390$ and 1604 M^{-1} cm^{-1}, molecular and anionic forms, respectively). The quantum yields for molecular form were determined at 280 nm as 0.34 and 0.23 in air-saturated and deoxygenated solutions, respectively, and 0.75 (independent on O_2) for the ionized form. The products and their yields were found dependent on the irradiation wavelength(s). Direct photolysis and UV/H_2O_2 oxidation of two pesticides (MCPP and 2,4-dichlorophenoxyacetic acid, 2,4-D) and six pharmaceuticals (naproxen, carbamazepine, diclofenac, gemfibrozil, ibuprofen, and caffeine) in pH 7 aqueous solution under exposure to polychromatic radiation from a MP lamp were reported by Shu *et al.* (2013). Wavelength-averaged (200–300 nm) quantum yields were reported as 0.0866 ± 0.0034 and 0.0036 ± 0.0003 for MCPP and 2,4-D, respectively. The Φ_{MCPP} is almost one order of magnitude smaller than that determined by Meunier and Boule (2000), whereas $\Phi_{2,4\text{-}D}$ is very low and similar to that reported by Benitez *et al.* (2004) at 253.7 nm (0.008). The $k_{^{\bullet}OH,MCPP}$ data were reported as 1.9×10^9 M^{-1} s^{-1} (Beltran *et al.* 1994), 3.2×10^9 M^{-1} s^{-1} (Fdil *et al.* 2003), and 2.5×10^9 M^{-1} s^{-1} (Armbrust, 2000).

Benitez *et al.* (2004) examined the direct photolysis of 2,4-D and 4-chloro-2-methylphenoxy-acetic acid (MCPA) as a function of pH at 253.7 nm and their treatment with UV/H_2O_2 process. The authors concluded that $\Phi_{2,4\text{-D}}$ is independent on pH (~0.008, pH 3–9), while Φ_{MCPA} was found 0.004 (pH 3) and 0.15 (pH 5–9); the •OH rate constants were reported as $k_{\bullet OH,MCPA} = 6.6 \times 10^9$ M^{-1} s^{-1} and $k_{\bullet OH,2,4\text{-D}} = 5.1 \times 10^9$ M^{-1} s^{-1}.

Fdil *et al.* (2003) compared the degradation kinetics of selected chlorophenoxyalcanoic herbicides in aqueous solutions through three processes: $UV_{253.7\ nm}$, $UV_{253.7\ nm}/H_2O_2$, and photo-Fenton reaction ($UV_{253.7\ nm}/H_2O_2/Fe(III)$). The investigated herbicides and their kinetic parameters were reported as follows: $\Phi = 0.38$, $k_{\bullet OH} = 3.2 \times 10^9$ M^{-1} s^{-1} (MCPP); $\Phi = 0.41$, $k_{\bullet OH} = 3.6 \times 10^9$ M^{-1} s^{-1} (MCPA); $\Phi = 0.20$ (2,4-D); $k_{\bullet OH} = 1.6 \times 10^9$ M^{-1} s^{-1} [2-(2,4-dichlorophenoxy)propionic acid] (2,4-DP); $\Phi = 0.10$, $k_{\bullet OH} = 1.5 \times 10^9$ M^{-1} s^{-1} [2,4,5-trichlorphenoxyacetic acid] (2,4,5-T). The •OH rate constants were determined by competition kinetics relative to $k_{\bullet OH,2,4\text{-D}} = 1.5 \times 10^9$ $M^{-1}s^{-1}$. Reaction mechanisms and complete mineralization kinetics are discussed. Rate constants for •OH reactions with the above and other pesticides were reported by Armbrust (2000).

Miscellaneous pesticides. Metaldehyde (2,4,6,8-tetramethyl-1,3,5,7-tetraoxocane) is a molluscicide used to control snails and slugs on turfs, ornamentals, vegetables, citrus, berries. Metaldehyde is a cyclic acetaldehyde tetramer, with no chromophores, thus, no specific absorption properties in the UV-Vis range. Metaldehyde hydrolyzes to acetaldehyde in acidic conditions, is poorly adsorbed onto GAC, and is not treated by O_3.

Only very few peer-reviewed publications on metaldehyde treatment in water are available in the literature. Autin *et al.* (2012) compared the UV/H_2O_2 and UV/TiO_2 process efficiencies for metaldehyde removal in synthetic water and pre-treated surface water collected from a water treatment plant. Less than 2% removal was observed at a UV (253.7 nm) dose of 1750 mJ cm^{-2} in the absence of H_2O_2. A $k_{\bullet OH,Metaldehyde} = 1.3 \times 10^9$ M^{-1} s^{-1} was determined with a competition kinetics method using *p*-CBA as a probe. This rate constant is much smaller than $k_{\bullet OH}$ values reported for the vast majority of pesticides. The •OH attack is expected to occur at the tertiary-C on the molecule, followed by sequential ring opening with formation of acetaldehyde as a byproduct. Semitsoglou-Tsiapou *et al.* (2016b) examined the degradation of metaldehyde, clopyralid, and mecoprop by direct photolysis and $UV_{253.7\ nm}/H_2O_2$ process. In the presence of 5 mg/L H_2O_2, the *pseudo*-first order rate constants followed the order mecoprop ≫ metaldehyde > clopyralid.

Clopyralid (3,6-dichloro-2-pyridine-carboxylic acid; $pK_{a1} = 1.4$; $pK_{a2} = 4.4$) is a herbicide used to control broadleaf weeds and could damage some crops including peas, tomatoes, spinach. Only a few reports were published on UV light-induced degradation of clopyralid. Clopyralid absorption spectrum consists of two distinct bands, one in the ~200–250 nm range and a weaker band in the 250–300 nm region (Tizaoui *et al.* 2011). Tizaoui *et al.* compared UV/TiO_2, UV/H_2O_2, and O_3 processes for the degradation of clopyralid in water at pH 5.6, with UV light from a MP lamp. Removals were observed in all processes, yet no comparison can be made given that the treatment conditions were not optimized for each process. Xu *et al.* (2013) published a study on clopyralid degradation in the absence and the presence of H_2O_2. Direct photolysis was found insignificant as compared to the •OH-induced oxidation. Short-chain carboxylic acids and chloride ion were identified as degradation products. Orellana-García *et al.* (2014) evaluated the effectiveness of UV(253.7 nm) radiation on the degradation of clopyralid and three other pesticides (amitrole, fluoxypyr, and diuron) as a function of pH and pesticide concentration, in ultrapure water, tap water, and wastewater. Quantum yields and molar absorption coefficients were determined as 0.0677 and 845 M^{-1} cm^{-1} (clopyralid); 0.0478 and 151 M^{-1} cm^{-1} (amitrole); 0.0539 and 2142 M^{-1} cm^{-1} (fluoxypyr); 0.0127 and 9837 M^{-1} cm^{-1} (diuron). The authors observed a significant impact of pH on the degradation rates, with higher reactivity of the anionic forms than molecular forms of pesticides. Autin (2012) calculated a $k_{\bullet OH,clopyralid} = 1.15 \times 10^7$ M^{-1} s^{-1} using Minakata *et al.*'s GCM (2009). Semitsoglou-Tsiapou *et al.* (2016b) reported three values for $k_{\bullet OH,clopyralid}$ (5.0×10^8 $M^{-1}s^{-1}$, 7.5×10^8 $M^{-1}s^{-1}$, and 2.0×10^8

$M^{-1} s^{-1}$) as determined by competition kinetics with *p*-CBA as a probe and from kinetic modeling. The large discrepancies between the $k_{\bullet OH,clopyralid}$ reported data prompt the need for accurate determination of this rate constant.

Glyphosate (*N*-phosphonomethyl glycine) is a broad-spectrum organophosphorous herbicide, marketed as Roundup, one of the most extensively used pesticides in the world. It is rapidly biodegraded in soil to aminometyl-phosphonic acid (AMPA) and glyoxylate. Although glyphosate is regulated in drinking water in NA, EU, and other jurisdictions around the world, its photodegradation is insufficiently studied and inconsistent data on its reactivity toward $^{\bullet}OH$ are reported in the literature. Manassero *et al.* (2010) examined the degradation of glyphosate with the $UV_{253.7\ nm}/H_2O_2$ process. No significant direct photolysis was observed. The experimental $k_{\bullet OH,glyphosate}$ values reported in the literature are given in Table 2.6. A model-estimated value of $k_{\bullet OH,glyphosate} = 3.37 \times 10^7\ M^{-1}\ s^{-1}$ was considered in the study on glyphosate degradation with UV/H_2O_2 (Vidal *et al.* 2015), which makes the reported values for this rate constant span over approx. three orders of magnitude.

2.9.1.3 Cyanotoxins

During the seasonal algae growth due to favorable environmental conditions (temperature, light, nutrients), the blue-green algae (cyanobacteria) may constitute the dominant community of phytoplankton in freshwater (lakes, ponds, reservoirs) and even in coastal marine waters. Certain cyanobacteria produce and release toxins (cyanotoxins, also known as algal toxins) which impair the water quality and pose serious threats to human and ecological health. Cyanotoxins are powerful neurotoxins, hepatotoxins, and dermatotoxins. Microcystins (MCs) LR and RR, anatoxin-a (ATX-a), cylindrospermopsin (CYN) and saxitoxins (STXs) are the most extensively studied cyanotoxins. Excellent reviews and reports with hundreds of references are available in the literature on topics regarding cyanobacteria and cyanotoxins; e.g., WRF (2010); Hitzfeld *et al.* (2000); Svrcek and Smith (2004); Westrick *et al.* (2010); Merel *et al.* (2013a, b); Pantelić *et al.* (2013); USGS (2016); Tarrah *et al.* (2017). Drinking water standards are in effect in many countries, e.g., 1.5 µg/L MCs (Canada); 1.3 µg/L MC-LR and 1 µg/L CYN (Australia); 1.0 µg/L MC-LR and 1 µg/L CYN (New Zealand), and health advisories were issued, e.g., 1 µg/L MC-LR (WHO, 1998); 0.3 and 1.6 µg/L MC-LR and 0.7 and 3 µg/L CYN (age-dependent, U.S. EPA 2015). No drinking water regulations are set for algal toxins in the EU and the U.S.; cyanotoxins, as a group including ATX-a, CYN, microcystins and saxitoxin but not limited to, are on the U.S. EPA CCLs 3&4.

The WRF report (2010) documents that of the 243 samples collected from utility source waters located in the U.S. and Canada during the algae bloom seasons, 181 (75%) samples were tested positive for T&O compounds and 148 (82%) samples were tested positive for microcystins. T&O compounds and cyanotoxins co-occurred in 91% of the algal blooms, with microcystins co-occurring with both MIB and geosmin and anatoxin-a co-occurring primarily with geosmin. The general approach at removing cyanotoxins from the impacted waters consists in their removal prior to release into the water from cyanobacteria (i.e., intracellular stage) through conventional processes (coagulation, sedimentation, filtration) or, if released into the water (i.e., extracellular stage) through typical water treatment processes such as GAC filtration, ozonation, chlorination, advanced oxidation, as applicable.

The molecular structures of most cyanotoxins contain chromophores which enable them to absorb strongly the UV radiation emitted by the LP- and MP-lamps. In general, algal toxins are also highly reactive toward $^{\bullet}OH$ (Table 2.6), thus, are effectively treated with $^{\bullet}OH$-based AOPs, including the UV/H_2O_2 process.

Microcystins treatment in water by conventional and advanced oxidation processes was reviewed by Sharma *et al.* (2012). The authors discussed process efficacy, reaction kinetics, intermediate byproducts and their toxicity, and potential degradation pathways. De Freitas *et al.* (2013) compared MC-LR degradation by solar photo-Fenton, UV-A/photo-Fenton, and UV-C (MP lamp)/H_2O_2 AOPs. High removal yields were determined with UV-C/H_2O_2. The WRF report (2010) showed that only 300 mJ cm^{-2} (253.7 nm) was needed to degrade 58% (0.37-log) MC-LR in pre-treated Lake Washington water containing 5.1 mg/L DOC (~82 $\%T_{254\ nm,1\ cm}$). This result can be explained through both direct photolysis and photosensitized degradation of MC-LR. At a UV fluence of 990 mJ cm^{-2} in the presence of 2 mg/L H_2O_2, 1.3-log MC-LR removal was measured. He *et al.* (2012) examined the direct photolysis and UV/H_2O_2-based degradation of MC-LR at 253.7 nm. A fluence-based rate constant of $k_1' \sim 3.6 \times 10^{-3}$ cm^2 mJ^{-1} was determined for direct photolysis, which is more than 50% larger than that of NDMA at 253.7 nm. The quantum yield was calculated as 0.06. Therefore, MC-LR is efficiently treated with UV alone. Given the complexity of MC-LR molecular structure, the fluence-based rate constant ($k'_{1\,UV/H_2O_2}$) for MC-LR degradation with UV/H_2O_2 process was found pH dependent, with $k'_{1\,UV/H_2O_2}$ at pH 6 twice that measured in the pH range of 7–9. In a later study, Xe *et al.* (2015) determined the $k_{\cdot OH}$ values for four microcystins i.e., MC–LR, –RR, –LA, and –YR (Table 2.6), and their degradation products for both direct photolysis and •OH reactions. Degradation mechanisms were postulated for the two processes based on the byproduct patterns. Zong *et al.* (2013) selected MC-LR as a model compound for microcystins and conducted a thorough product study for the UV/H_2O_2-based degradation. Seven types of oxidation products were identified by MS/MS, isolated and purified (>90% purity). The authors postulated the •OH attack sites on MC-LR molecule and proposed a degradation mechanism. Molecular toxicity assays on MC-LR and its oxidation products indicated that one of byproducts was more potent than MC-LR. Overall, toxicity decreased during the treatment, but the mixtures still exhibited residual toxicity.

Cylindrospermopsin occurrence, detection, toxicity and means of degradation were reviewed by De La Cruz *et al.* (2013). He *et al.* (2013) examined the degradation of CYN at 253.7 nm by direct photolysis and UV/H_2O_2, UV/$S_2O_8^{2-}$, and UV/HSO_5^- processes and determined the rate constants for reactions of CYN with the respective radicals. Unlike MC-LR, no degradation was observed with UV alone. In a later publication, He *et al.* (2014) reported a product study based on the intermediates detected from CYN degradation with UV/H_2O_2 process, and proposed a mechanism. Song *et al.* (2012) determined $k_{\cdot OH,CYN} = (5.09 \pm 0.16) \times 10^9$ M^{-1} s^{-1} which is in very good agreement with other reported data (Table 2.6) and were the first to publish a degradation mechanism for CYN by •OH reactions.

Anatoxin-a degradation in various water matrices exposed to MP-lamp radiation or to the VUV (Xe_2*, 172 nm) radiation was studied by Afzal *et al.* (2010). At an integrated polychromatic fluence of 1285 mJ cm^{-2}, 88% and 50% removal were reported for direct photolysis of 0.6 and 1.8 mg/L ATX-a, respectively. Wavelength-averaged quantum yields were calculated as 0.15 and 0.05, for initial concentrations of ATX-a of 0.6 and 1.8 mg/L, respectively. Verma and Sillanpää (2015) used $UV_{260\ nm}$-LEDs to investigate the kinetics of ATX-a photolysis and •OH-induced oxidation in milliQ water (with and without the addition of humic acid and/or bicarbonate) and in pre-treated surface water. Toxicity decreased in treated as compared to untreated samples. The quantum yield for ATX-a photolysis at 260 nm was found as 0.26. Vlad *et al.* (2014) reviewed the removal of ATX-a through the practiced drinking water processes. The literature review indicates that chlorine-based processes are ineffective, whereas ozonation, advanced oxidation and permanganate treatments are successful at treating ATX-a.

Saxitoxins occur in various carbamate-based structures and are the most toxic cyanotoxins among those discussed herein. The most studied processes for extracellular STX removal are chlorination and powdered activated carbon filtration (Nicholson *et al.* 2003; Lo *et al.* 2009; Zamyadi *et al.* 2010). Orr *et al.* (2004) investigated the removal of extracellular STXs in O_3-treated (for O_3-demand) raw water from an Australian drinking water plant with O_3 (up to 15 mg/L), BAC filtration (15 min EBCT) with or without prior O_3 treatment, and O_3/H_2O_2 (10:1, w:w ratio). Ozonation and peroxone processes did not remove STX, whereas 100% removal was determined for BAC filtration. Water toxicity decreased from 28.5 to 1 μg (STX-equivalents)/L through BAC filtration.

2.9.1.4 Taste-and-odor (T&O) causing compounds

In general, T&O compounds have seasonal occurrence as most of them are natural products released during the algal blooms. Other T&O causing compounds are associated with water treatment processes, such as chlorination and chloramination. Unlike the naturally occurring T&O compounds which impact primarily drinking water aesthetics, the halogenated products raised public health concerns. Of all T&O-causing compounds, MIB and geosmin are frequently detected during T&O-occurring events, and several water treatment utilities have implemented AOPs for their reduction in drinking water. Several reviews are available in the literature. Antonopoulou *et al.* (2014) covered extensively the literature on the nature and origin of various categories of T&O compounds and the AOPs used at laboratory scale and to some extent in pilot trials for T&O treatment. Among the T&O compounds discussed are benzothiazoles, geosmin and MIB, mercaptans and sulfides, aromatics, odorous aliphatic aldehydes, ketones, and alcohols, halogenated anisoles. Key experimental conditions and kinetic data are discussed, and to a limited extent the degradation mechanisms. Srinivasan and Sorial (2011) reviewed the occurrence and removal of geosmin and MIB *via* conventional and advanced treatment processes.

Glaze *et al.* (1990) pioneered the research on AOPs for removal of T&O causing compounds from surface water (Colorado River) under conditions similar to those practiced at water treatment plants. Six T&O compounds (MIB, geosmin, 1-hexanal, 1-heptanal, 2,4-decadienal, and dimethyltrisulfide) were examined using typical drinking water processes and three AOPs: O_3/UV, O_3/H_2O_2, and UV/H_2O_2, with the UV radiation generated from LP-lamps. The authors were the first to estimate the $^{\bullet}OH$ rate constants from the *pseudo*-steady state approximation kinetic model used with the experimental data. These rate constants were reported as $k_{^{\bullet}OH,MIB} = (8.2 \pm 0.4) \times 10^9\ M^{-1}\ s^{-1}$ and $k_{^{\bullet}OH,geosmin} = (1.4 \pm 0.3) \times 10^{10}\ M^{-1}\ s^{-1}$. Later on, more accurate determination of rate constants for eleven T&O causing compounds, including MIB and geosmin, using an experimental competition kinetics method, was performed and reported by Peter and von Gunten (2007) (see Table 2.6). The ozonation rate constants were also measured. Rosenfeldt *et al.* (2005) examined the oxidation of MIB and geosmin in raw blend and clear well water samples collected from a drinking water plant, using LP- and MP-lamp collimated beam apparata and UV/H_2O_2 (0 – 7.2 mg/L) process, with UV doses up to 5000 mJ cm^{-2}. Molar absorption coefficients for MIB and geosmin are reported. Degradation *via* direct photolysis of both MIB and geosmin was observed with polychromatic radiation. Agus *et al.* (2011) studied the levels of odorous compounds in municipal wastewater effluents and their removal in advanced treatment trains at potable reuse facilities. Twelve of 15 odorous compounds were detected in secondary effluents at levels of ~100 ng/L. RO membranes removed most of odor-causing compounds, but some of them were still above their threshold levels in the RO permeate. Ozone-BAC-UV/H_2O_2 and RO-UV/H_2O_2 advanced treatment trains eliminated the odorous compounds, but no individual process could achieve treatment levels below the odor thresholds. Simultaneous treatment of DBPs, geosmin and MIB with the $UV_{253.7\,nm}/H_2O_2$ process in deionized water was demonstrated at bench-scale by Jo *et al.* (2011). At 1200 mJ cm^{-2} with 6 mg/L H_2O_2, 90% and 60% removals of geosmin and MIB, respectively, were observed.

2.9.1.5 Volatile organic compounds (VOCs)

Organic compounds from various classes are classified as VOCs, although some of them are largely miscible with water and they cannot be air-stripped at the levels encountered in the drinking water sources. Several

VOCs are monitored and regulated in drinking water. In UV/H_2O_2 applications, some VOCs are well treated by a combination of direct photolysis and •OH-induced oxidation, others are degraded exclusively through reactions initiated by •OH. Trichloroethene (TCE) and tetrachloroethene (PCE) were extensively studied over the years. A few examples include Sundstrom *et al.* (1986); Mertens and von Sonntag (1994); Mertens and von Sonntag (1995); Hirvonen *et al.* (1996); Hirvonen *et al.* (1998); Li *et al.* (2004); Li *et al.* (2007); Ikehata *et al.* (2016). See also Stefan and Williamson (2004) for photolysis of TCE, PCE, and other chlorinated VOCs. Both TCE and PCE absorb the UV-C radiation and photolyze with large quantum yields; e.g., $\Phi_{\mathrm{TCE,200-300\,nm}} = 0.46$ (Li *et al.* 2004); $\Phi_{\mathrm{PCE,253.7\,nm,O_2}} = 0.84$ and $\Phi_{\mathrm{PCE,253.7\,nm,-O_2}} = 0.34$ (Mertens & von Sonntag, 1995). Direct photolysis product studies were undertaken for both TCE (Li *et al.* 2004) and PCE (Mertens & von Sonntag, 1995). Based on the identified and quantified TCE intermediate byproducts, Li *et al.* (2004) proposed the degradation mechanism and postulated that the primary photolytic processes consisted in homolytic cleavage of C—Cl bond, loss of molecular chlorine, photohydrolysis, and loss of HCl. In a later study, Li *et al.* (2007) studied TCE degradation to complete mineralization of the initial organic carbon with the UV/H_2O_2 process using a bench-scale MP-lamp reactor. A mechanistic kinetic model was developed which predicted well the experimental results. The model-predicted contribution of direct photolysis to the overall TCE degradation decreased from ~12% to ~8% over the exposure time, whereas the Cl atom and dichlorine radical ($Cl_2^{-\bullet}$) reactions accounted for approx. 1% and ~4% of the total degradation reactions of TCE, respectively. The •OH-induced degradation represented 86–88% of the overall mechanism of TCE destruction. Figure 2.14 is the schematic representation of TCE degradation with UV/H_2O_2 process. Detailed molecular and radical reaction mechanism and modeling of this system are available in the published work.

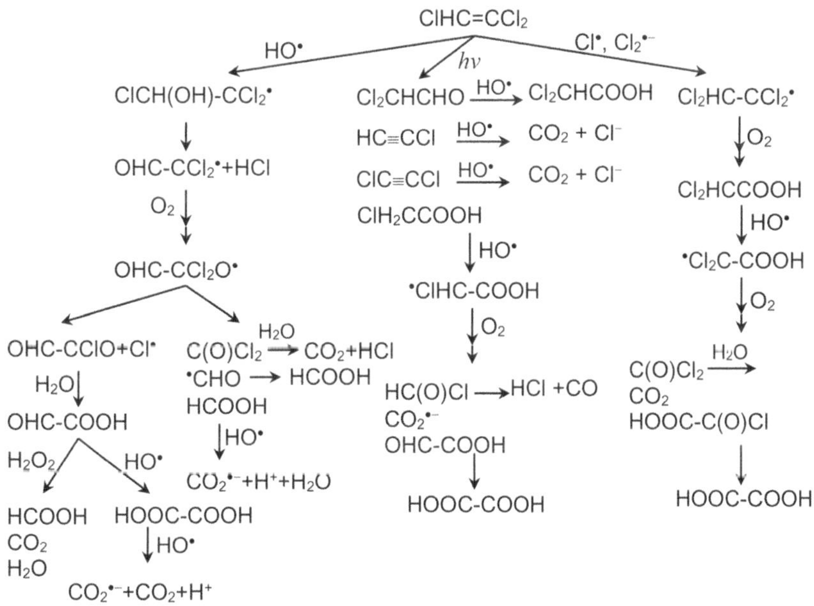

Figure 2.14 Simplified reaction mechanisms for degradation of TCE and its byproducts with UV/H_2O_2 process (Li *et al.* 2007).

Guo *et al.* (2014) developed for the first time a computer-based first-principles kinetic model to predict the degradation mechanisms of chemical compounds in aqueous phase through the UV/H_2O_2 process. The model contains a rule-based pathway generator, a reaction rate constant estimator, a mechanistic reduction module, an ODE generator and a solver which calculates and generates the real-time data of all species. The model also includes a toxicity estimator to predict the toxicity of major species and the toxicity trend during the degradation process. The computational model was validated for acetone and TCE degradation with UV/H_2O_2 AOP using the experimental product study data published by Stefan *et al.* (1996) and Stefan and Bolton (1999) on acetone, and Li *et al.* (2007) on TCE. In a later publication, Guo *et al.* (2015) described their on-the-fly kinetic Monte Carlo (KMC) model for predicting the formation and decay of reaction intermediates from micropollutants treated with UV/H_2O_2 AOP. The novelty of the approach taken in KMC model is that the pathways are developed as the degradation proceeds rather than being developed prior to using the KMC to solve the ODEs.

Ikehata *et al.* (2016) conducted a study on the effectiveness, costs, and byproduct formation during the treatment of contaminated groundwater at a Superfund site in Simpsonville, South Carolina, using four AOPs: O_3/H_2O_2, O_3/UV, $O_3/H_2O_2/UV$, and UV/H_2O_2. A LP-collimated beam apparatus was used for the UV/H_2O_2 process. The raw water contamination included chlorinated ethanes and ethenes, 1,4-dioxane, and tetrahydrofuran (THF); up to 360 µg/L bromide was measured in the samples. Air-stripping is practiced at the Superfund site, such that VOCs were greatly reduced relative to the raw water in the samples used in AOP tests. At a UV dose of 1000 mJ cm^{-2} and 20 mg/L H_2O_2, TCE, 1,1-dichloroethene, and 1,4-dioxane were reduced from 200, 110, and 10 µg/L, respectively, to below their respective required treatment levels. With UV alone, PCE was removed by ~25%, whereas only ~5% removal was observed for TCE. Bromate was formed in all O_3-based AOPs.

1,4-Dioxane is considered a semi-volatile organic compound. Currently, there is no U.S. EPA drinking water standard for 1,4-dioxane (proposed health advisory of 3 µg/L), but it is included on CCL3 & 4. Local advisory and notification levels across the USA are in effect. 1,4-Dioxane is a cyclic aliphatic ether with no chromophore, thus, no direct photolysis is expected. The rate constant for 1,4-dioxane reaction with $^{\bullet}OH$ is ~3×10^9 M^{-1} s^{-1} (Table 2.6). 1,4-Dioxane treatment with UV/H_2O_2 process was well studied at laboratory- and pilot-scale, and demonstrated during full-scale performance testing at water utilities. Examples of laboratory studies include Stefan and Bolton (1998); Quen and Chidambara Raj (2006); Coleman *et al.* (2007); Martijn *et al.* (2010); Chitra *et al.* (2012). Stefan and Bolton (1998) undertook a comprehensive product study of 1,4-dioxane removal with $MP-UV/H_2O_2$ process. 1,2-Ethanediol mono- and diformate esters, formic and methoxyacetic acids were identified as primary degradation intermediates, and glycolic, acetic and oxalic acids as byproducts quantified upon longer treatment. Complete mineralization was achieved and good carbon mass balance was obtained. A complete degradation mechanism was proposed.

1,4-Dioxane and THF treatment for the purpose of increasing industrial wastewater effluent biodegradability was investigated by Quen and Chidambara Raj (2006). The authors found that UV/O_3 treatment resulted in higher biodegradability of 1,4-dioxane than the UV/H_2O_2 process, whereas the latter process was better than UV/O_3 process at biodegrading THF. Martijn *et al.* (2010) examined the impact of two pretreatment processes, i.e., coagulation & sand filtration (CSF) and ion exchange & ultrafiltration (IX-UF) on the performance of UV/H_2O_2 process for NDMA and 1,4-dioxane removal from surface water used for drinking water production. The studies were conducted at both bench- (MP-collimated beam apparatus) and pilot- (MP-UV system) scale. A decrease in the fraction of light absorbed by the matrix was observed upon pretreatment (89% raw water, 85% CSF-treated and 72% IX-UF treated-water), while the fraction of light absorbed by 0.5 mg/L NDMA (7% raw water, 9% CSF-treated and 17% IX-UF treated-water) and 6 mg/L H_2O_2 (4% raw water, 6% CSF-treated and 11% IX-UF treated-water) increased. The patterns were explained by the DOC removal by CSF and DOC and nitrate removal by IX-UF. As a result, both NDMA photolysis and 1,4-dioxane oxidation yields increased. The E_{EO} (NDMA) decreased from 1.24 kWh/m^3/order in CSF- to 0.41 kWh/m^3/order in IX-UF-treated water, and it was independent of the H_2O_2 concentration (0–10 mg/L range). The E_{EO} (1,4-dioxane with 5 mg/L H_2O_2) for IX-UF pre-treated water (1.41 kWh/m^3/order) was more than 50% lower than that calculated for CSF pre-treated water (3.0 kWh/m^3/order). The E_{EO} (1,4-dioxane) decreased as the H_2O_2 dose was increased.

Examples of studies on UV/H_2O_2-based removal of other ethers of environmental concern include: Li *et al.* (1995) on bis(2-chloroethyl)ether, BCEE, which reports the kinetics and degradation mechanism; Christensen *et al.* (2009) on BCEE, demonstrating the degradation feasibility, dechlorination yields and byproduct formation, and biodegradability of treated water; Schuchmann and von Sonntag (1982), pulse radiolysis study on diethyl-ether reaction with $^{\bullet}OH$ and byproducts; Chang and Young (2000) on methyl-*tert*-butylether (MTBE) kinetics; Stefan and Bolton (2000) and Cooper *et al.*'s review (2009) on degradation products of and full mechanism for MTBE; Cater *et al.* (2000) which reports E_{EO} data for MTBE treatment in a bench-scale MP-reactor; Acero *et al.* (2001) which examined MTBE removal and byproduct formation with O_3 and O_3/H_2O_2 processes and reported $k_{\cdot OH}$ for MTBE and a few byproducts.

Numerous aromatic compounds are volatile or semi-volatile, and their treatment in drinking water sources and wastewater by AOPs, including UV light-based processes is extensively covered in the literature. The following are just a few examples: Guittonneau *et al.* (1988); Shen *et al.* (1995); Masten *et al.* (1996); Hirvonen *et al.* (2000); Daifullah and Mohamed (2004); Czaplicka (2006).

2.9.1.6 Endocrine disrupting compounds (EDCs)

Endocrine disrupting compounds are environmental micropollutants that interfere at extremely low levels (ng/L) with the normal functions of endocrine system in living organisms by acting as hormones, perturbing the normal reproduction and development in wildlife; potential threats to human health *via* insufficiently treated drinking water are not excluded. These compounds do not occur naturally in the environment but originate from human use and discharge in the environment *via* treated wastewater effluents. EDCs are not restricted to natural or synthetic estrogens e.g., 17α-ethinylestradiol (EE2), 17β-estradiol (E2), estrone (E1), estriol, progesterone, testosterone, androstenedione, as similar effects are triggered by compounds beloging to other classes, including pesticides (DDT, lindane, glyphosate, atrazine), alkylphenols and their ethoxylates (4-*tert*-octylphenol, OP; nonylphenol, NP; bisphenol A, BPA; nonylphenol ethoxylate, NPE), phtalates (dibutylphtalate, di(2-ethylhexyl) phtalate), preservatives, brominated flame retardant decabromodiphenyl ether, polychlorinated biphenyls, dioxins and furans, heavy metals (cadmium, mercury, arsenic), and so on. Some chemicals displaying endocrine disrupting activity are regulated in drinking water, others fall under the European Water Framework Directive which sets maximum levels in the aquatic environment (e.g., 0.035 ng/L for EE2, 0.4 ng/L for E2) or other environmental policies around the world, or are considered for regulation in the future, e.g., EE2, E2, estriol, equilin, equilenin, norethindrone which are on the U.S. EPA CCLs 3&4. Examples of reviews on EDC treatment include Silva *et al.* (2012); Luo *et al.* (2014); Sornalingam *et al.* (2016); Gmurek *et al.* (2017).

Rosenfeldt and Linden (2004) determined the photochemical and chemical parameters pertaining direct photolysis and $^{\bullet}OH$ reaction for EE2, E2, and BPA, using LP- and MP-collimated beam apparata. The absorption spectra of all three compounds consist of three bands within the 200–300 nm range. Quantum yields were determined as 0.026, 0.043, and 0.0085 at 253.7 nm, and as 0.061, 0.10, and 0.019 over the 200–300 nm range, for EE2, E2, and BPA, respectively. The $k_{\cdot OH}$ values ($1.02–1.41 \times 10^{10}$ M^{-1} s^{-1}) indicate very fast, diffusion-controlled reactions with $^{\bullet}OH$. The kinetic model developed using these parameters fit well the experimental data for EE2, E2, and BPA treatment with UV/H_2O_2 process in milliQ water, simulated natural water, and 0.45 μm-filtered river water. Mazellier *et al.* (2008) studied the photodegradation of EE2 and E2 at 253.7 nm and under polychromatic exposure ($\lambda > 290$ nm). The molar absorption coefficients and quantum yields at 253.7 nm were found as $\varepsilon = 420 \pm 20$ M^{-1} cm^{-1}, $\Phi = 0.067 \pm 0.007$, and $\varepsilon = 440 \pm 35$ M^{-1} cm^{-1}, $\Phi = 0.062 \pm 0.007$, for E2 and EE2, respectively. The quantum yields determined in the long-wavelength absorption band are close to those calculated at 253.7 nm, i.e., ~0.07–0.08, which indicates that the quantum yield is independent of wavelength. Zhang *et al.* (2010) examined photodegradation of EE2 in the absence and in the presence of H_2O_2 and, based on yeast estrogen screen (YES) assay performed at different stages of EE2 removal, concluded that EE2 accounted for ~95% of sample estrogenicity, whereas its byproducts did not exhibit biological activity.

Carlson *et al.* (2015) determined the pK_a values, quantum yields (253.7 nm), molar absorption coefficients and fluence-based rate constants for ten pharmaceuticals and five EDCs (EE2, E2, E1, NP, and BPA). The quantum yield of NP was found concentration-dependent, decreasing from 0.32 ± 0.08 (23 μg/L) to 0.092 ± 0.006 (230 μg/L), whereas photochemical parameters of triclosan (antimicrobial) ($pK_a = 8.2$) are pH-dependent, i.e., specific to triclosan forms: $\varepsilon = 1440$ M^{-1} cm^{-1}, $\Phi = 0.34 \pm 0.02$ (pH 4.2, protonated form); $\varepsilon = 4590$ M^{-1} cm^{-1}, $\Phi = 0.25 \pm 0.02$ (pH 10.8, deprotonated form). Mazellier and Leverd (2003) determined the quantum yield $\Phi_{253.7\ nm} = 0.058$, $k_{\cdot OH} = (6.4 \pm 0.5) \times 10^9$ M^{-1} s^{-1} and photolysis and UV/H_2O_2 degradation products of OP. Degradation mechanisms and kinetic modeling are discussed in this work. Larger $k_{\cdot OH}$ values were reported for NP (1.1×10^{10} M^{-1} s^{-1}) and OP (1.4×10^{10} M^{-1} s^{-1}) by Ning *et al.* (2007) in a study on ozonation of these compounds.

An early study on NP and NP-monoethoxylate photolysis and their sensitized degradation in lake water matrix under simulated sunlight ($\lambda > 280$ nm) was published by Ahel *et al.* (1994). The authors concluded that in natural waters, the degradation of alkylphenols and their ethoxylates occurs through both direct photolysis and ^{3}DOM-sensitized processes, with negligible contribution from $^1O_2$. Ikehata and Gamal El-Din (2004) covered the published literature on the removal of linear alkylbenzene sulfonates, alkylphenol ethoxylates (APE), and quaternary ammonium surfactants by O_3 and AOPs. The average number of ethoxylate unit in APEs could range from 7 or less up to 20 or more, and the alkyl chain is frequently highly branched. The review discusses the AOPs, including UV/H_2O_2, able to degrade the three groups of contaminants to more biodegradable products.

Karci *et al.* (2014) determined an $E_{EO} = 1.2$ kWh/m^3/order for NP polyethoxylate degradation in surface water with the UV/H_2O_2 process. Toxicity and genotoxicity assays were performed and their patterns were contoured as a function of treatment time. Photodegradation of NP under simulated sunlight and the effects of nitrate, ferric ions and bicarbonate were explored by Peng *et al.* (2016). Chen *et al.* (2007) determined the kinetics of photodegradation of a mix of NP-ethoxylates, with an average of 10 ethoxylate units, and identified the photolysis products.

Phthalic acid esters (phthalates) were found mutagenic, carcinogenic, and potent endocrine disruptors. Wang *et al.* (2016) studied the degradation of plasticizer dibutylphthalate (DBP) with $UV_{253.7\ nm}/H_2O_2$ process. A 77% removal yield of 1 μM DBP was obtained in the presence of 1 mM H_2O_2 for an applied UV dose of 160 mJ cm^{-2} at pH 7.6. Lau *et al.* (2005) showed that dibutylphthalate photolysis and the nature of degradation intermediates are pH-dependent. Acid-catalyzed photohydrolysis predominates at pH 3, whereas within the pH range of 3–5, hydrolysis and oxidation/reduction at the butyl moiety co-occur. The fastest degradation was observed in alkaline conditions (pH > 7) when base-catalyzed ester hydrolysis takes place. The degradation kinetics and the pH-dependent mechanisms are discussed.

An excellent source of information on the removal of a wide range of EDCs, pharmaceuticals, and personal care products from surface water, drinking water, and reuse water at environmentally relevant levels is the AwwaRF report on the project "Removal of EDCs and pharmaceuticals in drinking water and reuse treatment processes" (AwwaRF, 2007).

2.9.1.7 Pharmaceuticals

The term "Pharmaceuticals" covers several classes of human and veterinary medicinal compounds, of various molecular structures and modes of action. Pharmaceuticals enter the aquatic environment primarily *via* wastewater effluent discharge into surface waters. Improper disposal as domestic waste and the use of manure for farming purposes result in soil contamination with pharmaceuticals which leach out further into the groundwater wells. The impact of pharmaceutical residues (ng/L levels) on the wildlife and aquatic species is documented in the literature. The potential impact on human health upon chronic exposure to low levels of pharmaceuticals in their mixtures *via* drinking water consumption is unknown (Khetan & Collins, 2007).

The UV light-based processes were reported as viable treatment options for numerous pharmaceuticals, from various classes, among which hormones and steroids, antimicrobials and antibiotics, analgesics and anti-inflammatory, psychoactive, heart medication, antidiabetic medication, X-ray contrast agents, veterinary drugs. Several reviews with hundreds of citations summarize laboratory studies and their outcomes, including photochemical and chemical parameters relevant to direct photolysis and $^{\bullet}OH$-induced oxidation of pharmaceutical compounds; e.g., Boreen *et al.* (2003); Ikehata *et al.* (2006); Klavarioti *et al.* (2009); Fatta-Kassinos *et al.* (2011); Homem and Santos (2011); Rivera-Utrilla *et al.* (2013); Yan and Song (2014); Keen *et al.* (2014b); Postigo and Richardson (2014); Trawinski and Skibinksi (2017). Wols and Hofman-Caris (2012a) reviewed the published literature reporting quantum yields, molar absorption coefficients and $k_{^{\bullet}OH}$ data for a large number of micropollutants which include pharmaceuticals, EDCs and personal care products, tabulated these parameters and provided the corresponding references.

Wols *et al.* (2014 and 2015b) determined kinetic parameters relevant to photolysis and UV/H_2O_2 process for a group of 35 pharmaceuticals and compared the experimental and predicted data for their degradation with LP- and MP-radiation in laboratory water and two surface water matrices. Temperature-dependency of $^{\bullet}OH$ rate constants for the investigated micropollutants is also illustrated in this work. Keen and Linden (2013) examined the degradation of six antibiotics, including erythromycin A, in milliQ water and two secondary wastewater effluents, using LP- and MP-collimated beam apparata. Quantum yields, $k_{^{\bullet}OH}$ values, and fluence-based rate constants are reported ($k_{^{\bullet}OH,erythromycin} = 3.7 \times 10^9$ M^{-1} s^{-1}). Biological activity of treated samples decreased in four of the six antibiotics; erythromycin and doxycycline appeared to form biologically active intermediates upon treatment with UV/H_2O_2 process up to 500 mJ cm^{-2}, but those were further removed at higher UV doses. Photodegradation and toxicity changes of oxytetracycline, doxycycline and ciprofloxacin by direct photolysis and UV/H_2O_2 process in ultrapure water, surface water and treated waters from local drinking water and wastewater treatment plants were studied by Yuan *et al.* (2011). UV photolysis increased toxicity in *Vibrio fisheri* assay, whereas during the UV/H_2O_2 process toxicity increased initially then was completely removed. Yao *et al.* (2013) examined the degradation of four ionophore antibiotics (anti-parasitic veterinary pharmaceuticals) by direct photolysis and $UV_{253.7\,nm}/H_2O_2$ process in milliQ water, pre-treated surface water from a drinking water treatment plant and secondary wastewater effluent using a bench-top reactor. The $^{\bullet}OH$ rate constants were determined to be within a narrow range of 3.49×10^9 to 4.00×10^9 M^{-1} s^{-1}. E_{EO} data are reported and the patterns of formation and decay of primary intermediates are presented. Using a kinetic model, the authors predicted the E_{EO} variation with the H_2O_2 concentration for the UV/H_2O_2 process in surface water and secondary effluent, and illustrated the individual contribution of direct photolysis and $^{\bullet}OH$-reactions to the overall E_{EO}.

Jeong *et al.* (2010) determined $k_{^{\bullet}OH}$ for four iodinated X-ray contrast agents (iohexol, iopromide, iopamidol, iomeprol), identified the $^{\bullet}OH$-reaction intermediates and proposed degradation mechanisms.

Kim *et al.* (2008) assessed the degradability of 29 pharmaceuticals and N,N-diethyl-m-toluamide (DEET) with O_3, UV alone, UV/O_3, UV/H_2O_2, and O_3/H_2O_2 in a high quality water (TOC < 50 μg/L) using a LP bench-scale reactor. Good photolysis yields were measured for many compounds, among which ketoprofen, diclofenac, sulfamethoxazole, antipyrine, fenoprofen; other pharmaceuticals were removed by combined UV and •OH reactions in the UV/H_2O_2 process with 4.9 mg/L H_2O_2, e.g., naproxen, metoprolol, clarithromycin, carbamazepine. Kim *et al.* (2009) investigated the removal of 41 pharmaceuticals detected in a biologically treated secondary wastewater effluent with UV alone and $UV_{253.7\,nm}/H_2O_2$ process. Three 3 × 65 W LP-lamp reactors in series were used at various flowrates in order to deliver various UV doses; 7.8 mg/L H_2O_2 was added to the water for the UV/H_2O_2 process. Variable removal yields were measured in direct photolysis process, depending on pharmaceutical structures. More than 90% removal efficiency was observed at 5 min hydraulic retention time with the UV/H_2O_2 process for all pharmaceuticals except norfloxacin and caffeine. The impact of water quality is discussed.

Several studies on UV light-induced degradation of individual pharmaceutically active compounds are available in the literature.

2.9.1.8 Miscellaneous micropollutants

Military Explosives such as 1,3,5-trinitrotriazacyclohexane (RDX, acronym for Research Department eXplosive), octahydro-1,3,5,7-tetranitro-1,3,5,5-tetrazocine (HMX, acronym for various names including High-velocity Military eXplosive, High Melting eXplosive), 2,4,6-trinitrotoluene (TNT), 1,3-dinitrooxypropan-2-yl nitrate (nitroglycerin, NG), 3-nitrooxy-2,2-bis(nitroxymethyl)propyl nitrate (PETN), *N*-methyl-*N*,2,4,6-tetranitroaniline (tetryl) and the caged structured polycyclic saturated nitramine CL-20 are the most studied compounds given their widespread contamination of the environment. An excellent review on the occurrence and fate of these pollutants in the environment and on the state-of-the art remediation processes was initiated by IUPAC and published by Kalderis *et al.* (2011). The review provides more than 400 references. Soil contamination could occur through unexploded ordnance from military operations, wastewaters from manufacturing processes, and disposal. Rainfall-driven dissolution of these energetic materials starts the process of diffusion into the aquifer, thus groundwater contamination. Biodegradation of nitramine explosives leads to byproducts such as hydrazine and substituted hydrazines which are known as *N*-nitrosamine precursors, thus, a pathway to contamination of groundwater with nitrosamines and nitramines. Environmental protection agencies in various jurisdictions issued guidelines/regulations and/or health advisories for explosives in drinking water. There is no drinking water standard for these compounds in the USA; however, there are health advisories for RDX, HMX, TNT, 2,4- and 2,6-dinitrotoluene issued by the U.S. EPA Office (2012b).

Early research studies on UV light-based processes for the removal of ordnance explosives were funded by the US Army in late 1970s – early 1980s (e.g., Kubose & Hoffsommer, 1977; Noss & Chyrek, 1984; Burrows & Brueggemann, 1986). RDX, HMX and TNT absorb the UV-C radiation with maxima around 230–240 nm; the absorption spectrum of RDX extends into UV-A, which explains the slow attenuation of this compound under daylight in contaminated lagoons. Peyton *et al.* (1999) reviewed the prior art and studied the RDX photochemistry at 253.7 nm. The author generated a comprehensive mechanism involving photochemical and radical reactions and developed a kinetic model for RDX photolysis. Photodegradation of a number of energetic explosives, namely nitromethane, dimethylnitramine (DMNA), 1,3,5-triamino-2,4,6-trinitrobenzene (TATB), RDX, HMX and CL-20 and their degradation mechanisms was reviewed by Sun *et al.* (2012).

Bose *et al.* (1998) were the first to report a quantum yield for RDX at 253.7 nm ($\Phi = 0.13$), calculated from the experimental *pseudo*-first order rate constant of direct photolysis and the measured $\varepsilon_{254\,nm} = 5000\ M^{-1}\ cm^{-1}$. In the same study, RDX degradation with the UV/H_2O_2 process was modeled using a previously reported $k_{\cdot OH,RDX} = 3.4 \times 10^8\ M^{-1}\ s^{-1}$ and keeping the quantum yield as an adjustable parameter.

The good fit of experimental data was obtained for $\Phi = 0.34$ which is approx. 2.5-fold larger than that determined from the direct photolysis experiment. The authors attributed this large discrepancy to the competition for light from RDX photolysis byproducts which was not accounted for in the calculations. This explanation is supported by Gares *et al.*'s study (2015) which reported $\Phi_{RDX,229\,nm}$ as ~0.35 and 0.32 from RDX photolysis experiments, in which RDX was quantified by HPLC and from the decay of the intensity of Raman band at 1580 cm^{-1}, respectively. Gares *et al.* calculated a quantum yield of 0.19 from the decay of Raman band intensity at 920 cm^{-1}, where interferences from byproducts are expected. Since both wavelengths (229 and 253.7 nm) are in the same absorption band of RDX spectrum, it is conceivable to assume that Φ_{RDX} is ~0.34 in the strong absorption band. A quantum yield of $\Phi_{RDX,302\,nm}$ ~0.16 ($\varepsilon_{302\,nm} = 113\ M^{-1}\ cm^{-1}$) is cited by Bordeleau *et al.* (2013), and further used by the authors to estimate the quantum yield for nitroglycerin photolysis at 302 nm ($\Phi_{NG,302\,nm}$ ~0.23; $\varepsilon_{NG,302\,nm} = 5\ M^{-1}\ cm^{-1}$). Gares *et al.* (2015) identified the RDX byproducts and correlated their formation sequence with the spectral changes. Chen *et al.* (2008) determined $k_{\bullet OH}$ values for a number of contaminants on CCL4, including RDX. Competition kinetics and two •OH source processes were used. Reported $k_{\bullet OH,RDX}$ data are scattered over almost one order of magnitude: $(0.25 \pm 0.01) \times 10^9\ M^{-1}\ s^{-1}$ (O_3/H_2O_2 process, bromobenzene as a probe); $(1.6 \pm 0.2) \times 10^9\ M^{-1}\ s^{-1}$ (UV/H_2O_2 process, nitrobenzene as a probe). Other $k_{\bullet OH,RDX}$ are listed in Table 2.6.

Liou *et al.* (2003) compared the oxidation of seven explosives with Fenton and photo (253.7 nm)-Fenton processes. The following compounds were studied: 2,4,6-trinitrophenol (TNP), ammonium picronitrate (AP), DNT, tetryl, TNT, RDX, and HMX. Direct photolysis data were inconclusive, but the •OH-induced degradation rate with Fenton process (pH 2.8) followed the order DNT> TNP> AP> TNT> Tetryl> RDX> HMX, which is consistent with the low reactivity of RDX towards •OH observed by Bose *et al.* (1998).

Ayoub *et al.* (2010) reviewed the studies published from 1990 through 2009 on TNT degradation by various AOPs, including direct photolysis and UV/H_2O_2 process. Gares *et al.* (2014) studied the degradation of TNT upon irradiation at 229 nm. Raman spectroscopy and LC/MS techniques were used to identify the intermediates and to explain the degradation mechanism. A quantum yield of ~0.015 was calculated. A literature review on homogeneous and heterogeneous UV light-based processes for degradation of cyclic nitramine and nitroaromatic explosives was published recently by Mahbub and Nesterenko (2016). Another interesting review on computational chemistry as a powerful predicting tool for the environmental impact, fate, and toxicity of nitroaromatic and nitramine explosives, based on the relationships between their structural characteristics and reactivity was published by Qasim *et al.* (2007).

Disinfection Byproducts (*DBPs*) belong to a stand-alone class of compounds which continues to expand rapidly from the traditional (chlorinated and brominated) THMs and HAAs regulated in drinking water to the new generation of DBPs among which, iodinated THMs and HAAs, halocarbonyl compounds, halonitromethanes, haloacetonitriles and haloamides, nitrosamines, and other C-DBPs and N-DBPs, as well as byproducts formed from disinfection processes other than chlorine/chloramines/chlorine dioxide, e.g., ozonation. Jones and Carpenter (2005) determined the degradation rates of low concentrations of CH_2I_2, CH_2IBr and CH_2ICl and the formation of photoproducts upon exposure to natural light in water, salt- and seawater matrices. Fang *et al.* (2013) studied the photodegradation of halonitromethanes (HNMs) at 253.7 nm as a function of pH (3–9), determined the kinetics and proposed the photolysis mechanisms. The rate constants increased with increasing pH, except for trichloronitromethane (TCNM), whose k_{app} was found *quasi*-independent of pH. The quantum yields were found pH-independent and determined as 0.114, 0.409, 0.327, and 0.46 for bromonitromethane, dibromonitromethane, dichloronitromethane, and TCNM, respectively. Xiao *et al.* (2014) published a detailed study on photolysis and UV/H_2O_2 – based degradation of a number of THMs, including $CHCl_2I$ (DCIM), CHClBrI (CBIM), $CHClI_2$ (CDIM), $CHBr_2I$ (DBIM), $CHBrI_2$ (BDIM), CHI_3 (TIM), $CHBr_3$ (TBM), $CHClBr_2$ (CDBM) and $CHCl_2Br$ (DCBM). Quantum yields and fluence-based rate constants were determined for all compounds. The quantum yields varied from 0.02 for DCBM to (0.58 ± 0.07) for DBIM. The impact of NOM, alkalinity and nitrate on the degradation yields is discussed. At a fluence of 540 mJ cm^{-2} in the presence of 6 mg/L H_2O_2, removal yields >97% were observed in single component solutions, and >84% were determined for the individual DBPs in their mixture. A QSAR model was developed based on the correlation of DBP reactivity with specific molecular descriptors and a good agreement between the predicted and experimental log *k* data was observed. More recently, Chuang *et al.* (2016) reported comprehensive experimental and theoretical studies on various classes of DBPs and determined the photochemical parameters at 254 nm for 26 DBPs. Experimental $k_{\bullet OH}$ were determined for 42 DBPs.

Personal Care Products (*PCPs*) represent a category of compounds which belong to various chemical classes and are commonly used as ingredients in products such as cosmetics, shampoos, hand creams, sunscreens, nutritional supplements, household products (detergents, fragrances, deodorizers, disinfectants), etc. There is evidence on the interference of these compounds with the reproduction and development systems of the aquatic species. *Triclosan*, a widely used antibacterial, is known to undergo photolysis with high yields across the UV radiation spectrum. Its photochemical properties were discussed briefly in other sections of this chapter. The formation of harmful photoproducts from triclosan photolysis under environmentally relevant conditions was mentioned in this chapter and is reviewed in Chapter 13. In engineered UV technologies, triclosan is well removed by direct photolysis, with a pH-dependent quantum yield. Triclosan reacts with $^\bullet OH$ at diffusion-controlled regimes; $k_{\bullet OH,TCS} = (5.4 \pm 0.3) \times 10^9$ M^{-1} s^{-1} (Latch *et al.* 2005). *N,N-Diethyl-m-toluamide* (*DEET*) is an active compound in insect repellents for protection against insect bites. Benitez *et al.* (2013) investigated the photo-oxidation of DEET and *Chlorophene* (*CF*), an antimicrobial used in hospitals and households for cleaning and disinfection purposes, using a bench-scale LP-lamp reactor. Ultrapure water, surface water, and two secondary wastewater effluents were used as matrices. The molar absorption coefficients at 254 nm of DEET and CF were determined as 1642 and 1350 M^{-1} cm^{-1}, respectively. The quantum yields and $k_{\bullet OH}$ were found as $\Phi_{DEET} = 0.0012$, $\Phi_{CF} = 0.429$, $k_{\bullet OH,DEET} = (7.51 \pm 0.07) \times 10^9$ M^{-1} s^{-1} and $k_{\bullet OH,CF} = (8.47 \pm 0.19) \times 10^9$ M^{-1} s^{-1}. In the presence of 0.1 µM H_2O_2, 90% CF and less than 5% DEET were removed in natural water at the same treatment time, which suggests that the major degradation route of CF is direct photolysis; similar removals were observed in the secondary effluent containing 11 mg/L TOC, in the presence of 0.5 µM H_2O_2. *Parabens* (*PBs*) are esters of *p*-hydroxybenzoic acid added as preservatives in cosmetics, PCPs, pharmaceuticals, and food products. The endocrine disrupting potential of these compounds at very low levels was demonstrated. Gmurek *et al.* (2015) showed that methyl-, ethyl-, propyl-, butyl- and benzyl-parabens absorb the UV-C radiation with the absorption band centered at ~240 – 260 nm and with molar absorption coefficients at 254 nm ranging from 8741 M^{-1} cm^{-1} (benzyl-PB) to 16,866 M^{-1} cm^{-1} (propyl-PB). Direct photolysis is not an option for treating these compounds given their very low quantum yields (~10^{-3}). The $k_{\bullet OH}$ values vary from $(3.76 \pm 0.2) \times 10^9$ M^{-1} s^{-1} (methyl-PB) to $(1.33 \pm 0.3) \times 10^{10}$ M^{-1} s^{-1} (benzyl-PB), and compare well with those reported in other studies. The authors determined the UV/H_2O_2 efficiency at the removal of these PBs from an actual wastewater effluent and attempted cost and energy consumption calculations. Gao *et al.* (2016) used a computational approach to investigate the $^\bullet OH$-initiated degradation of four PBs, kinetics and aquatic toxicity of treated water using a green algae (*Daphnia*) and fish as test species.

2.9.2 Pilot-scale tests

Pilot-scale tests employing UV reactors are reported in the literature on a variety of micropollutants and treatment conditions. In general, these tests are conducted either for research purposes following bench-scale tests or in preparation for the implementation of UV light-based processes at the water utilities. This section provides examples of pilot-scale tests described in the literature, with a particular emphasis on those conducted at water treatment plants for further implementation of UV processes or to expand knowledge on the ability of an existing UV system to act as a barrier for 'emerging' micropollutants.

PWN Water Supply Company North Holland implemented the UV/H_2O_2 process at two water treatment plants, namely, Andijk (2004) and Heemskerk (2008), to reduce the pesticide levels in surface water used for drinking water production. Prior to full-scale implementation, extensive research at bench- and pilot-scale was conducted in collaboration with Trojan Technologies (Canada). Twelve pesticides from various classes were investigated at the bench-scale: atrazine, simazine, bromacil, pyrazone, methabenzthiazuron, dicamba, isoproturon, diuron, 2,4-D, bentazone, trichloroacetic acid and triclopyr. Trichloroacetic acid was not considered in the pilot tests. The surface water is impacted by the agricultural activities, with seasonal variations in TOC (1.7 to 5 mg/L) and nitrate (2 to 15 mg/L); consequently, the water $\%T_{254\,nm,1\,cm}$ varies from ~87–89% (summer) to ~82% (winter). Information on bench-scale studies and pilot results is given in Stefan *et al.* (2005) and Kruithof *et al.* (2007). A TrojanUVSwift™ 4L12 reactor equipped with 4 × 3 kW MP-lamps operated at variable power level was used in the pilot testing. Flowrate ranged from 20 to 60 m^3/h. The removal target set for atrazine (80%) was achieved for an electrical energy dose (EED) of 0.56 kWh/m^3 (~540 mJ cm^{-2}) and 6 mg/L H_2O_2. Martijn *et al.* (2006) reported byproduct formation data from the same pilot setup operated for one-year under AOP conditions similar to the full-scale treatment in Andijk. Byproduct removal through BAC filtration is also discussed. The authors showed that NO_2^- was formed up to 250 μg/L and was not removed by filtration through virgin carbon; upon becoming biological, the carbon filtration at 15 min EBCT reduced NO_2^- below the EU drinking water standard, and low water temperature did not inhibit the nitrification process. Similarly, AOC level increased from ~20 to ~85 μg/L upon AOP treatment, but it was reduced to ~30 μg/L upon BAC filtration at an EBCT of 30 min. H_2O_2 residual (~5.5 mg/L) was near non-detect at 7.5 min EBCT. Information on THM trend and speciation, AOC, nitrite and contaminant removal in the full-scale installation is also provided. In another study, Martijn (2014) investigated the removal of selected herbicides, pharmaceuticals, and perfluorinated compounds with the UV/H_2O_2 process at EED = 0.54 kWh/m^3 and 6 mg/L H_2O_2 with and without subsequent GAC filtration (Norit ROW 0.8; 20 min EBCT). In 'summer' water, the herbicides were degraded by ~80% to ~92%, whereas removal yields of 60% to 80% were reported in treated 'winter' water. Almost 100% removal of herbicides was observed upon GAC filtration of the UV/H_2O_2-treated 'winter' water. Pharmaceuticals (carbamazepine, diclofenac, metoprolol, pentoxifylline, sotalol, metformin) were reduced from ~60 to 100% with the UV/H_2O_2 process, except metformin, which was degraded by only 20%. Subsequent GAC filtration resulted in overall removal yields higher than 90%. Perfluorinated compounds, i.e., perfluorobutane sulfonate (PFBS), perfluorohexane sulfonate (PFHS), perfluorooctane sulfonate (PFOS), perfluorobutanoic acid (PFBA) perfluorohexanoic acid (PFHA), perfluorooctanoic acid (PFOA) and perfluorononanoic acid (PFNA) were very poorly treated with the UV/H_2O_2 process (~1 – ~15% removal yields). Varying adsorption onto GAC was observed for these pollutants, with PFHA, PFOS, PFOA and PFNA removed by > 95%, whereas PFBS, PFHA, and PFBA were reduced by 80%, ~70%, and ~25%, respectively.

Anglian Water Services (AWS) Company (UK) supplies drinking water to approx. 4.5 million customers. Roughly half of the water comes from surface water sources and the other half is extracted from aquifers *via* ~450 boreholes. The major contaminants in the well water are pesticides, iron/manganese, nitrate and microbial pathogens. AWS performed pilot testing at Riddlesworth WTP for groundwater treatment with UV and UV/H_2O_2 processes to reduce the levels of two pesticides (trietazine and bentazone) below the EU standard of 0.1 μg/L (Holden & Richardson, 2009). The treatment performance with TrojanUVPhox™ 8AL20 (LP lamps) and TrojanUVSwift™ 4L12 (MP lamps) reactors were compared for this application. Borehole #3 groundwater could contain up to ~80 mg/L nitrate (EU regulation set at 50 mg/L) due to farming in the area, 240 mg/L alkalinity and ~400 hardness as $CaCO_3$; $T_{254\,nm,1\,cm}$~ 96–97%. Flowrate was varied up to 17 m^3/h, with H_2O_2 dose ranging from 0 to 15 mg/L. At an EED = 0.66 kWh/m^3, 85% and 56% trietazine reduction was observed with the LP- and MP-reactor, respectively. In the presence of 15 mg/L H_2O_2, >95% and ~71% trietazine removals were obtained with the LP- and MP-reactor, respectively. Nitrate (54 mg/L at the time of testing) impacted the MP-lamp reactor performance. Moreover, approx. 0.24 mg/L NO_2^- was formed with MP- *vs.* ~0.07 mg/L with the LP-reactor, which increased the $^{\bullet}OH$ water background demand in the MP-UV/H_2O_2 process. In the process selection, the MP-UV process option was rejected. Cost comparison on a 15 year-net present value (NPV)-basis between GAC (GBP2495k) and UV technology (GBP1040k) led to the implementation of two TrojanUVPhox™ 72AL75 UV reactors in a parallel configuration alternating on duty/standby every 24 h. No H_2O_2 dosing was required to meet the project specifications. The UV system achieves >0.5 log reduction (from 0.2 µg/L to 0.06 µg/L) of trietazine, while removing bentazone to below 0.01 µg/L upon final chlorination. The Borehole #3-treated water is blended with water from two other boreholes then passed through nitrate removal plant in order to meet the NO_3^- EU standard, prior to chlorination and distribution.

Dunea Water Company, The Netherlands, provides drinking water to over 1.3 million customers, with an average production of ~9000 m^3/h (Knol, 2017). The water source is River Meuse, which is treated *via* multiple steps, including coagulation and rapid sand filtration, managed aquifer recharge (MAR) by dune passage, softening, activated carbon filtration, rapid sand filtration and slow sand filtration. Although the majority of organic micropollutants (OMPs) were either biodegraded or adsorbed at the MAR stage, some compounds (MTBE, diglyme, bentazone, 1,4-dioxane) were quantified in the water after MAR (Lekkerkerker-Teunissen *et al.* 2012). The effectiveness of AOPs at the removal of OMPs from pre-treated river water, prior to MAR was evaluated at pilot-scale research (2009–2012) performed at Dunea location in Bergambacht (pilot plant capacity of 5 m^3/h). Lekkerkerker-Teunissen *et al.* (2012) reported the results from over 6 month-testing of individual and serial application of O_3/H_2O_2 and $UV_{253.7\,nm}/H_2O_2$ AOPs for reduction of fourteen OMPs (9 pharmaceuticals, 4 pesticides, and diglyme) from water at various O_3:H_2O_2 ratios (6 and 10 mg/L H_2O_2) and UV dose between 700 and 950 mJ cm^{-2}. Both ozone generator and LP-reactor (LBX 10) were provided by Wedeco (a Xylem brand, Germany). Dunea treatment goal for a full-scale installation was set as 80% removal of atrazine, which was one of the tested pesticides. Over 90% removals were observed for bromacil, bentazone and 5 pharmaceuticals with 6 mg/L H_2O_2 and 1.5 mg/L O_3. Atrazine was degraded between ~35 and 60%, whereas metformin was the least degraded pollutant (<30%). The serial treatment O_3 (1.5 mg/L)/H_2O_2 (6 mg/L)/LP-UV resulted in additional degradation of the six OMPs. An $E_{EO\ atrazine}$ = 0.55 kWh/m^3/order was determined in the serial AOP, which was lower than that observed for LP-UV/H_2O_2 in another test (0.73 kWh/m^3/order). A significantly lower E_{EO} was predicted for a full-scale system than that observed in the pilot work. Bromate did not exceed 0.5 µg/L (Dunea standard). Additional data from the pilot work conducted at Dunea's Bergambacht WTP were provided and discussed by Scheideler *et al.* (2011), Hofman-Caris *et al.* (2012), Lekkerkerker-Teunissen *et al.* (2013). Based on the extensive pilot testing, Dunea board approved the installation of the serial AOPs at Bergambacht with a flow capacity of 2200 m^3/h. Wedeco was awarded the project and the serial AOPs are expected to be operational in October 2017. The full new treatment system would comprise O_3 reactors designed by Dunea and Wedeco, 2 × K-216 Wedeco UV reactors in series, and three GAC (ROW 0.8 cat, 3 min EBCT) contactors for H_2O_2 residual quenching. The treatment target is 80% removal of micropollutants with an energy consumption of 0.15 kWh/m^3 (Knol, 2017).

Extensive pilot-scale research (2005–2012) was performed at Greater Cincinnati Water Works (GCWW)' Richard Miller WTP (OH, USA) on the effectiveness of UV-AOP with various light sources (LP, MP, DBD) at the removal of a wide range of contaminants including pesticides, T&O-causing compounds, pharmaceuticals. The optimum location of UV process in the treatment train was assessed based on the treatment performance, but also in relation to other criteria including the quality of treated water. Detailed information on the experimental approach, data analysis, results, and process economics were disseminated in several publications, e.g., Dotson *et al.* 2010; IJpelaar *et al.* 2010; Metz *et al.* (2011); Heringa *et al.* 2011; Hofman-Caris *et al.* 2012; Metz *et al.* 2012. An 1820 m^2-, \$30 million state-of-the-art UV disinfection facility was constructed at GCWW. 240 MGD of GAC-filtered water is treated with 8 × Sentinel® 48″ Chevron reactors (Calgon Carbon Corporation, Pittsburgh, PA) installed in a parallel configuration. Each reactor has 5 × 20 kW MP-lamps. Two additional reactors are in the plans for further plant expansion. Currently, the UV system is operated in the disinfection mode but flexibility is built into the system to operate in the UV/AOP (UV/Chlorine) mode for seasonal T&O treatment. The UV disinfection project involved installation of 160 solar panels on the roof of UV facility which can offset as much as 7% of the UV process power requirements and reduced the WTP carbon footprint.

City of Kennewick (WA)'s Water Filtration Plant treats 15 MGD water from the Columbia River using conventional treatment followed by membrane filtration. Pilot testing at the water facility was conducted comparing $KMnO_4$, O_3, and $UV_{253.7\ nm}/H_2O_2$ on membrane-filtered water (>97% $T_{254\ nm}$, ~1 mg/L TOC) for MIB and geosmin removal. $KMnO_4$ up to 2 mg/L had virtually no effect on T&O compounds, while O_3 at ~0.6 mg/L could reduce geosmin by 60% but increased the threshold odor number (TON). Pilot data using Trojan Technologies' UV reactor and UV/H_2O_2 process showed the best performance for both MIB and geosmin removal. More than 90% removal of geosmin was observed with 2 mg/L H_2O_2. Increasing H_2O_2 to 6 mg/L changed the odor profile and increased TON (Chang, 2010).

Wang *et al.* (2015a) compared the UV/H_2O_2 and UV/Chlorine AOP performance for MIB and geosmin removal from pre-treated St. Lawrence River water at the full-scale UV/H_2O_2 treatment plant of City of Cornwall, ON, Canada. The WTP operates four TrojanUV SwiftECT™ 8L24 MP-lamp reactors for year-round UV disinfection and T&O control during the T&O events. The reactor power was 83.5 kW, H_2O_2 dose was varied from 1 to 4.8 mg/L, and pH was controlled at 6.5, 7.5 and 8.5. MIB, geosmin and caffeine were dosed to the UV influent. The $E_{EO}s$ were found pH-dependent at all H_2O_2 concentrations and for all compounds tested. The lowest $E_{EO}s$ were observed for pH 6.5 and 4.8 mg/L H_2O_2, and found as 0.23, 0.31, and 0.36 kWh/m^3/order for geosmin, MIB and caffeine, respectively. Byproduct formation in the two AOPs was reported in Wang *et al.* (2015b). Detailed description of UV/Chlorine *vs.* UV/H_2O_2 data as reported in Wang *et al.* (2015 a, b) is given Chapter 9.

A two-year (2011–2012) AOP pilot test was conducted at Choa Chu Kang Waterworks (CCKWW, Singapore) to assess process performance and associated costs of O_3, O_3/H_2O_2 and $UV_{253.7\ nm}/H_2O_2$ AOPs for treatment of a suite of 19 micropollutants including T&O compounds, PPCPs, EDCs, artificial sweeteners, and one industrial solvent (1,4-dioxane) in sand-filtered water. The pilot plant was constructed and supplied by Wedeco (Xylem, Germany) and was designed for a flowrate of 5–20 m^3/h and hydraulic retention time (HRT) up to 10 min. The pilot results were reported by Wang *et al.* (2015c). Six of the ten PPCPs and all four EDCs were very well removed by O_3 at 2 mg/L and 5 min HRT; of the four artificial sweeteners, cyclamate and acesulfame were removed by >80 % at ≥3 mg/L O_3 or 2 mg/L O_3/(unspecified mg/L) H_2O_2 processes at 5 min HRT. Geosmin was removed by ≥80 % by both O_3 and peroxone processes under all tested conditions; ≥80 % MIB removal was obtained only with peroxone process. 1,4-Dioxane was the least treatable compound with O_3-based processes. Bromate was controlled to below 5 μg/L. The UV AOP was examined at UV doses ranging from ~510 to ~875 mJ cm^{-2} with or without the addition of either 5 or 10 mg/L H_2O_2. According to Wang *et al.*'s study, removal of MIB, DEET, and ibuprofen by ≥80 % required minimum 820 mJ cm^{-2} and 10 mg/L H_2O_2; geosmin removal yields >80 % were reported for UV doses >510 mJ cm^{-2} in the presence of 10 mg/L H_2O_2. The pilot E_{EO} values for MIB, geosmin, and 1,4-dioxane treated with the O_3/H_2O_2 process were found lower than those determined for the UV/H_2O_2 process.

Chu *et al.* (2016) reported the data from another UV/H_2O_2 pilot testing performed at the same site (CCKWW, Singapore) as in the work described above. The pilot plant consisted of Trojan UVSwift™ 4L12 reactor and external H_2O_2 and chemical dosing tanks as well as a downstream activated carbon unit for removal of H_2O_2 residual and any oxidation byproducts. Prior to UV, the water was treated by coagulation, flocculation and ultrafiltration. Eight micropollutants were spiked and treated, namely, BPA, E2, diclofenac, PFOA, PFOS, NDMA, MIB and geosmin, at three flowrates (40, 50, and 60 m³/h) and 0, 5, and 10 mg/L H_2O_2. In the absence of H_2O_2, all compounds except PFOA and PFOS, showed $E_{EO} < 0.5$ kWh/m³/order, and the E_{EO} values decreased markedly for the UV/H_2O_2 process. For example, the E_{EO} (MIB) values were calculated as 0.32, 0.23, and 0.15 kWh/m³/order at 0, 5, and 10 mg/L H_2O_2, respectively. The MIB and geosmin E_{EO}s were found dependent on the initial concentrations of these compounds, i.e., approx. 50% lower at 300 ng/L than at 150 ng/L; this trend was observed also in a study cited by the authors, and could be attributed to the contribution of direct photolysis to the overall degradation kinetics of MIB and geosmin treated with polychromatic radiation.

MTBE treatment with the LPHO- and MP-UV/H_2O_2 process at pilot scale was investigated by Kommineni *et al.* (2008). Trojan Technologies' UVPhox™ 8AL20 and Swift™ 4L12 reactors were used in the pilot work. The tests were performed on Santa Monica, CA (USA) groundwater and Lake Huron, ON (Canada) surface water, and the key objectives were to evaluate the water quality impact on UV-AOP performance, to determine the levels of MTBE byproducts formed, and to estimate the economics for a full-scale WTP treating MTBE in Santa Monica groundwater. The groundwater had high alkalinity and nitrate, and low TOC, whereas the surface water had high TOC, low alkalinity and low nitrate levels. Under all pilot operating conditions, MP-UV/H_2O_2 tests resulted in much higher $E_{EO,MTBE}$ values than the LP-UV/H_2O_2 tests in both water sources. The E_{EO} data for surface water were lower than those for groundwater tests. The authors developed a kinetic model and estimated $E_{EO,\,MTBE}$ as a function of $^{\bullet}OH$ water matrix demand and H_2O_2 concentration, for both pilot- and full-scale conditions, recognizing the higher optical and hydraulic efficiency of large full-scale reactors as compared to pilot-scale reactors. The identified byproducts were also reported in a comprehensive product study on MTBE degradation with the UV/H_2O_2 process (Stefan *et al.* 2000).

Removal of emerging contaminants such as pharmaceuticals and EDCs from drinking water sources at UV-AOP pilot scale is well represented in the literature. Merel *et al.* (2015) published an interesting study where the authors employed three different analytical methods to monitor the efficacy of UV and UV/H_2O_2 processes at removal of trace organic contaminants (TOrCs) from secondary wastewater effluent from a 7500 m³/day capacity wastewater treatment facility located in southern Arizona, USA. The purpose of the work was to assess the suitability of the analytical methods to be used or to be developed as online sensors or as robust procedures to assess the presence and fate of a wide range of (known or unknown) micropollutants and their byproducts in water sources. A stream from the secondary effluent was diverted to the pilot plant comprising a LBX 20 LP-UV-AOP system (Wedeco, Xylem, Germany) operated at max. 6 m³/h. Three UV fluences with or without H_2O_2 addition were applied. The pilot plant was operated for 30 months and the occurrence, concentrations, and treatment of 29 micropollutants with different photochemical and chemical characteristics relevant to UV-AOP were monitored through eight sampling campaigns. The authors selected 16 TOrCs as indicators of the UV-AOP performance based on 75% detection frequency and min. 25 ng/L levels in the secondary effluent. The UV-AOP efficacy at the removal of the selected indicators was summarized by the authors as shown in Figure 2.15.

Zhang *et al.* (2016) investigated the synergistic contributions of various treatment processes practiced at water utilities, including dissolved air flotation, pre- and intermediate ozonation with or without H_2O_2, intermediate chlorination, dual media filtration, GAC, and UV/H_2O_2, to the overall removal of pharmaceuticals, steroids, and pesticides. Kim *et al.* (2009) reported experimental data for UV photolysis and $UV_{253.7\,nm}$/(7.8 mg/L) H_2O_2 treatment of 41 pharmaceuticals from various classes in biologically-treated, sand-filtered- wastewater. Three 65W LP-reactors in series were used, and the water was aerated prior to UV treatment. Of the 41 pharmaceuticals, 29 were not well removed by UV alone up to an UV dose of ~2800 mJ cm^{-2}; 39 pharmaceuticals were removed by at least 90% at an UV dose of 923 mJ cm^{-2} and 7.8 mg/L H_2O_2.

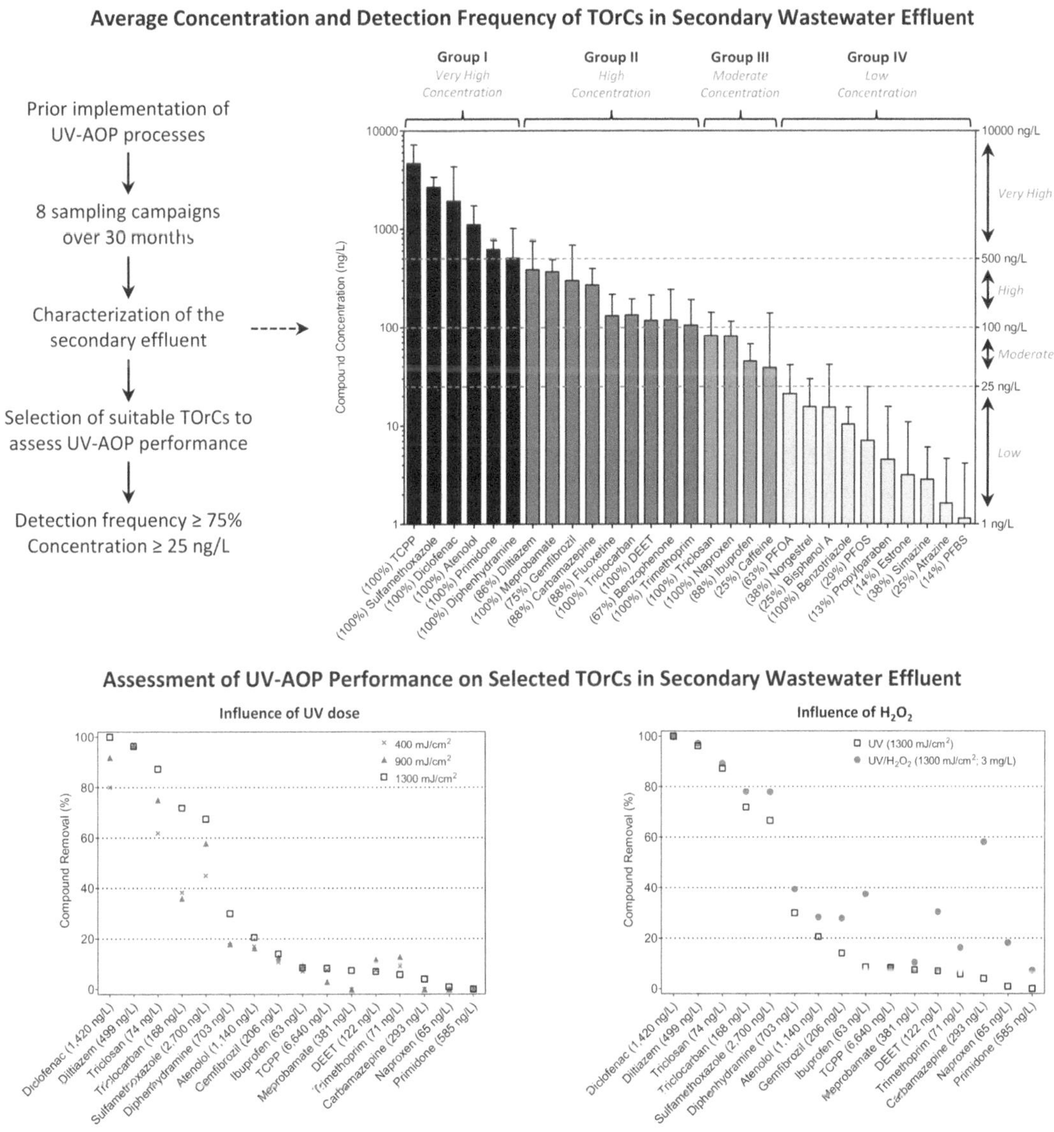

Figure 2.15 Frequency and removal of trace organic indicators in secondary wastewater effluent (Merel *et al.* 2015).

Appleman *et al.* (2014) undertook a thorough investigation on the occurrence of poly- and perfluoroalkyl substances (PFASs) in 18 raw drinking water sources and 2 treated wastewater effluents and their removal at 15 water utilities across the USA, which use different processes in their water treatment trains.

Perfluorocarboxylic acids (PFCAs) and perfluorosulfonic acids (PFSAs) were detected in more than 70% of the sampled raw waters. In 2016 the U.S. EPA established health advisory levels of 70 parts per trillion (0.070 μg/L) for PFOA and PFOS. If both compounds are found in drinking water, the combined concentrations of PFOA and PFOS should be compared to the 0.070 μg/L health advisory level. Of the 15 WTPs included in this study, three utilities use the UV/H_2O_2 AOP in their treatment trains; e.g., microfiltration (MF)/ RO/ UV-H_2O_2/ chlorination; river bank filtration (RBF)/ aquifer recharge and recovery/ softening/ UV-H_2O_2/ BAC/ GAC; MF/ ultrafiltration (UF)/ RO/ UV-H_2O_2/ chlorination. Water treatment processes such as coagulation, MF, UF, permanganate, ozonation, chlorination, chloramination, and UV/H_2O_2 were found mostly ineffective at removing PFASs. Anion exchange and GAC filtration removed the long chain PFASs and PFSAs, while RO membranes were very effective at removing all perfluorinated compounds.

As discussed elsewhere in this chapter, pilot-scale studies are useful sources of information on the ability of the investigated process to degrade environmental contaminants. However, caution should be exercised when attempting to compare data originating from various process conditions, including UV systems, operating settings, water source and characteristics. Frequently, the reported data are not for optimized process conditions. Moreover, scaling up pilot data to full-scale treatment conditions could be both challenging and risky.

2.9.3 Full-scale UV/H_2O_2 AOP installations

Several UV light-based systems using the UV/H_2O_2 process are currently installed at water utilities around the world to treat micropollutants in drinking water sources with simultaneous disinfection. Full-scale UV systems are also installed at water recycling facilities treating the wastewater secondary effluents for indirect or direct potable reuse (see Chapter 14). Fewer systems were commissioned to treat industrial wastes prior to discharge or reuse. Examples of full-scale UV/H_2O_2-based systems are given below. Trojan Technologies Case studies presented below are mostly taken from the company website: http://www.trojanuv.com/.

Andijk, The Netherlands. In 2004, Trojan Technologies installed a MP-lamp based UV system at PWN Water Supply Company North Holland treatment plant in Andijk. Three parallel trains, with 4 × TrojanUV SwiftECT™ 16L30 reactors per train are operated with the UV/H_2O_2 process. The water originates from IJssel Lake, and at that time it was treated by coagulation and rapid sand filtration prior to entering the UV plant. The UV system was designed to achieve min. 80% reduction in atrazine concentration at an EED = 0.56 kWh/m^3 with 6 mg/L H_2O_2. As described in this chapter and in Chapter 15, over the years, PWN explored new technologies for improving the water quality entering the UV system while achieving significant energy savings and high quality drinking water production. In 2015, PWN retrofitted the Andijk WTP, now Andijk III (http://pwntechnologies.com/portfolio-item/andijk-iii/). The current surface water treatment train consists of the following processes: softening, ion-exchange (SIX® technology), microfiltration (CeraMac® technology), UV/H_2O_2 AOP, GAC/BAC filtration (Norit ROW 0.8, 20 min EBCT), clearwell with microscreens for removal of any trace of microbial growth or carbon particles from GAC, and final disinfection with chlorine dioxide. This state-of-the-art WTP operates at max. 5000 m^3/h (120,000 m^3/day) and provides safe drinking water to people through the multi-barrier micropollutant treatment and disinfection philosophy implemented by PWN in Andijk. Figure 2.16 shows one of the UV-AOP trains at Andijk III.

In 2008, four parallel trains with 5 × TrojanUV SwiftECT™ 16L30 reactors per train operated with the UV/H_2O_2 process were installed at Heemskerk WTP, The Netherlands. The UV system was designed to treat pesticides based on 80% and 70% reduction of atrazine and bromacil, respectively, at a flow per train of 1500 m^3/h with max. 20 mg/L H_2O_2. The UV/H_2O_2-treated water is filtered through activated carbon at 9 min EBCT, and then pumped into the dunes.

Figure 2.16 TrojanUV SwiftECT™ 16L30 reactors in Andijk III, Netherlands.

Hall Water Treatment Works (WTW), Lincoln, UK. Anglian Water Services (AWS) is geographically the largest Water Supply Company in the UK, with ~140 WTPs delivering drinking water at approx. 1,200,000 m^3/day. In 2008, new pesticides were detected in surface waters, among which metaldehyde, clopyralid, and chlormequat, which are not removed by O_3, poorly removed with O_3/H_2O_2, and very poorly reduced through adsorption onto activated carbon. Additionally, high bromide levels (up to ~450 μg/L) in River Trent water, the sole drinking water source for Hall WTW, raised concerns about bromate formation with ozone-based processes. Research collaboration between AWS and Trojan Technologies at bench- and pilot-scale investigating the UV/H_2O_2 process with both LP- and MP-lamp based reactors indicated that LP-UV/H_2O_2 is a suitable and cost-effective technology for Hall WTW project. High nitrate levels in source water and low transmittance of the UV influent impacted the UV/H_2O_2 process performance with MP-lamp based reactors. AWS decided to install a Trojan UV system consisting of 4 × TrojanUVTorrent™ reactors in a parallel configuration. The UV system was designed for 80 $\%T_{254\ nm,1\ cm}$ water at a flowrate of 20,000 m^3/day, to reduce metaldehyde, atrazine and clopyralid by 65%, 80%, and 45%, respectively, in the presence of up to 50 mg/L H_2O_2. Each UV Torrent reactor uses 96 × 1.0 kW LPHO TrojanUV Solo lamps. The schematic of Hall WTW processes is shown in Figure 2.17. Of note, although the system design parameters required only 0.7-log reduction of atrazine to ensure compliance with the EU drinking water standard of 0.1 μg/L-individual pesticide, the UV system installed at Hall WTW achieves simultaneous removal of 1.12-log atrazine and 0.45-log metaldehyde. Two sets of GAC filtration vessels are in the treatment train, which control the THM precursors. The post-UV/AOP GAC vessels are used for quenching H_2O_2 residual and removal of potential byproducts formed during the oxidation process. The UV disinfection system at 25 mJ cm^{-2} ensures the inactivation of any biomass which could potentially leach off the GAC contactors. The construction of Hall WTW started in 2012 and the WTW were fully operational in March 2015.

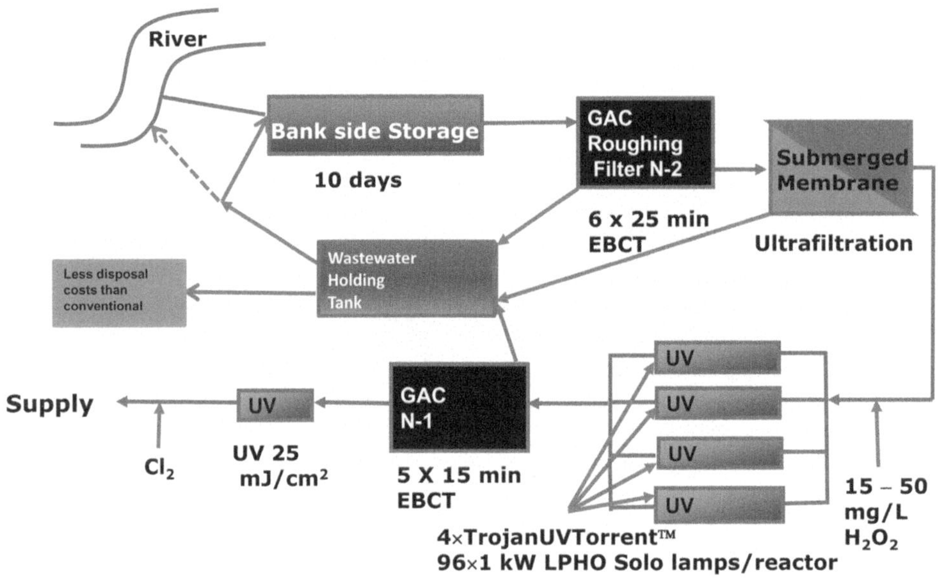

Figure 2.17 Graphical description of water treatment at Hall WTW (courtesy of AWS).

Cornwall, Ontario, Canada. City of Cornwall WTP draws water from St. Lawrence River for drinking water production. The agricultural activities in the region make tributary waters carry high loads of phosphorus into the nearshore area of Lake St. Francis, a fluvial lake along the St. Lawrence River near Cornwall. Consequently, algae blooms are favored during the high temperature seasons, and the release of T&O causing compounds impacts the water aesthetics. In 2004, City of Cornwall (ON) selected the TrojanUVSwift™ECT technology for treatment of T&O compounds in the drinking water source during the T&O events and to ensure year-round water disinfection. Four parallel trains, one Swift™ECT 8L24 UV reactor per train, were installed to treat 100,000 m^3 water/day (peak flow). The UV system was fully operational in 2006. The river water is chlorinated at the plant intake for *Zebra mussel* control, followed by coagulation, sedimentation and GAC filtration prior to UV treatment. Chlorine is dosed downstream to the UV process to remove the H_2O_2 residual and to ensure residual disinfection capacity in the drinking water distribution system. The UV system is operated in dual mode, i.e., in the disinfection mode at min. 40 mJ cm^{-2} providing >1-log *Giardia* inactivation year-round and in the UV-AOP mode, during the T&O events. In the AOP mode, the UV reactor controls algorithm adjusts the UV reactor power by powering more lamps than used for water disinfection and initiates and controls the H_2O_2 dosing for the UV/H_2O_2 process. The UV/H_2O_2 system was designed to reduce geosmin concentration by 90%. Based on the literature $k_{\cdot OH,geosmin}/k_{\cdot OH,MIB} \cong 1.5$ (Table 2.6) and algal toxin kinetics during UV/H_2O_2 process, it is predicted that while removing 90% geosmin from water (96% $T_{254\,nm}$; 3 mg/L H_2O_2) the UV system would simultaneously achieve 78%, 88%, 99.5% and 78% reduction in MIB, anatoxin-a, MC-LR and cylindrospermopsin, respectively.

Ilsan, South Korea. The Ilsan multi-regional water supply system provides domestic water to Ilsan new town and Giyang city since 1992 (http://english.kwater.or.kr/eng). Ilsan WTP draws water from the Han River and has a daily water supply capacity of 150,000 m^3/day with future expansion to 250,000 m^3/day. Seasonal algae blooms are associated with T&O events which impact the quality of drinking water. Korea

Water Resources Corporation (K-Water) awarded the UV/H_2O_2 AOP project for T&O control to Trojan Technologies. In 2016, two TrojanUVTorrent™ 96SL48 reactors with 1 kW LPHO Solo lamps were installed in parallel (Figure 2.18). The UV system removes 70% MIB while ensuring inactivation of >3-log *Cryptosporidium parvum* from water of ~92.5 to ~97% $T_{254\,nm,1\,cm}$ at a total flowrate of 150,000 m^3/day. Prior to UV treatment, the water goes through coagulation, flocculation and sedimentation processes. H_2O_2 is dosed in the settled water, which is pumped to the point where it splits among the two UV reactors. The UV/H_2O_2-treated water is filtered through GAC and chlorinated prior to distribution. The UV system is operated in the AOP-mode during the T&O events and in the UV-disinfection mode (low energy) for the rest of the year.

Figure 2.18 UV system installed at Ilsan WTP, South Korea.

Additional examples of UV/H_2O_2-based full scale UV systems installed in North America for T&O treatment include: *Lorne Park WTP* (Mississauga, ON, Canada). Lorne Park WTP, Region of Peel, ON, is the largest UV/H_2O_2 installation for T&O treatment in the world. Site description: eight TrojanUVSwift™ECT 16L30 reactors installed post-membrane filtration in parallel trains; 83% geosmin reduction at peak flow of 390,000 m^3/day, and 95% geosmin and 90% MIB reduction at 200,000 m^3/day with a maximum of 10 mg/L H_2O_2; residual H_2O_2 is quenched *via* GAC filtration; in operation since 2011. *Neshaminy Falls WTP* (Neshaminy, PA, USA). Two parallel trains, 1 × TrojanUVSwift™ECT 16L30 reactor per train; 90% geosmin reduction at average flow of 45,000 m^3/day or 80% at the peak flow of 57,000 m^3/day;

residual H_2O_2 is quenched with free chlorine; operational since 2010. *Patoka Lake WTP* (Patoka Lake, IN, USA). Two parallel trains with 3 × TrojanUVSwift™ECT 16L30 reactors per train; 97% MIB reduction at flowrate of ~38,000 m^3/day; residual H_2O_2 is removed with chlorine; operational since 2013. Currently, Trojan Technologies has over 20 UV systems installed in the world which are operated in a dual mode, i.e., treating T&O causing compounds during the T&O events and disinfecting the water year-round.

Siheung, South Korea. In 2014, Wedeco, a Xylem brand, implemented the UV/H_2O_2 process at Siheung WTP for disinfection and micropollutant control in drinking water, as the first phase of the infrastructure upgrade at the plant. The UV system consists of three Wedeco K143 LPHO-lamp UV reactors mounted in parallel, with the forth reactor planned to be installed upon plant expansion. The Siheung project requires 3-log removal of *Cryptosporidium parvum* and when operated in AOP-mode 0.5 log reduction of MIB (from 50 to 15 ng/L) with ≤ 10 mg/L H_2O_2 in addition to disinfection, for a total flow of 4419 m^3/h (1473 m^3/h per reactor). The water quality is specified as 92.7–97.3% $T_{254\,nm}$ and 1.13–1.85 mg/L TOC. The UV AOP system performance and control is based on UV dose, which is calculated, among other parameters, from the UV sensor input (Scheideler, 2016). Figure 2.19 is a photo of the UV system installed at Siheung WTP.

Figure 2.19 UV system installed at Siheung WTP, South Korea.

Bench- and pilot-scale tests were conducted prior to full-scale design and installation. The exprimental data, process economics as well as the approach to the UV dose-based controls algorithm for full-scale UV system performance are discussed in Scheideler *et al.* (2015), Scheideler *et al.* (2016), and Scheideler (2016).

Tucson, Arizona, USA. The Tucson International Airport Area was listed as a Superfund Site in 1983 given the extensive groundwater contamination with elevated levels of VOCs (TCE and PCE). Cleanup activities started in late 1980s using technologies adequate to TCE removal. The cleanup project is known as Tucson Airport Remediation Project (TARP). In 2002, 1,4-dioxane was discovered throughout the TARP area. 1,4-Dioxane was a stabilizer in industrial solvents used in aircraft manufacturing companies in the Tucson airport area from 1940s to 1970s. Tucson Water Company decided on construction of an AOP facility adjacent to the TARP plant, which was completed in 2014. In 2010, pilot studies were performed on $UV_{253.7\,nm}/H_2O_2$ effectiveness at 1,4-dioxane removal, byproduct formation, and residual H_2O_2 quenching. Tests at flowrates of 11 and 22 m^3/h with 15 mg/L H_2O_2 showed 1.55-log 1,4-dioxane & 1.71-log TCE, and 0.97-log 1,4-dioxane & 1.69-log TCE removal, respectively. Six TrojanUVPhox™ D72AL75 reactors were installed in three parallel trains with each train containing 2 × D72AL75 units (Figure 2.20). Each unit has 144 LPHO lamps. The system design parameters included 1.6-log 1,4-dioxane removal at a peak flow capacity of 1317 m^3/h with H_2O_2 dosed from an advanced H_2O_2-dosing system. Secondary contaminants TCE, PCE and 1,1-DCE are also removed.

Figure 2.20 UV system installed at TARP water purification facility, Tucson, AZ.

Greenbrook, Ontario, Canada. In July 2004, 1,4-dioxane was detected up to 285 μg/L in wells used as drinking water supplies for Greenbrook, Region of Waterloo, ON. As a precautionary measure the WTP usually running at 150 L/s was shut down. Air-stripping and GAC filtration tests were found ineffective at 1,4-dioxane removal to the target level of <10 μg/L. Bench- an pilot-scale tests were conducted for 1,4-dioxane removal with three AOPs: UV/H_2O_2 (Trojan Technologies, Canada), O_3/H_2O_2 (Applied Process Technology, APT, CA, USA), and UV/TiO_2 (Purifics, London, ON, Canada). Trojan's process was selected given its superior performance relative to the other technologies. Additionally, the small UV system footprint suited the existing facility. Under tested conditions, UV/H_2O_2 process achieved 1.28-log and 1.39-log 1,4-dioxane reduction in groundwater, at flowrates of 75 and 50 L/s, respectively. Three TrojanUVSwift™ 16L30 UV reactors installed in three parallel trains remove >1.3-log 1,4-dioxane with the UV/H_2O_2 process while achieving simultaneous disinfection of local groundwater at 12,940 m^3/day. New iron and manganese filters were installed before the UV process, and GAC filtration removes residual H_2O_2 and undesirable byproducts. The UV-AOP system was commissioned in 2008.

Llangollen, Delaware, USA. Llangollen wellfield operated by the Artesian Water Company Inc. consists of four public supply wells and one aquifer storage and recovery well and began operation in the 1950s, with a water supply withdrawal of 3.8 MGD (~14,400 m^3/day) in the late 1960s (Civardi *et al.* 2014). Upon groundwater contamination from two nearby Superfund Sites with bis(2-chlorocthyl) ether (BCEE), and trace levels of benzene, 1,2-dichlorethane, radium and chromium, Artesian Water ceased operation of Llangollen wellfield. Delaware Department of Public Health (DDPH) set an advisory level of 0.96 μg/L for BCEE in October 2000. GAC was installed, but two years later, breakthroughs were experienced, and the cost associated with GAC replacement increased significantly. In 2013, upon DDPH's request for 1,4-dioxane testing, this contaminant was found at elevated levels in the largest well (G-3R) of the field, and as a result, the well was removed from service by Artesian Water. Two AOPs were examined for 1,4-dioxane treatment in the wells: UV/H_2O_2 with LPHO- and MP-lamps and O_3/H_2O_2. Economic and non-economic criteria were used to evaluate the three options based on select design scenarios (see 2.9.4 for economic analysis). Upon careful evaluation and based on the scoring chart, Artesian Water selected Trojan Technologies' UV (LPHO)/ H_2O_2 process. Bench-scale tests were conducted with multiple objectives. Detailed description of the water quality, experimental design and data are given in Civardi *et al.* (2014). The full-scale UV/H_2O_2 system design criteria were set as follows: 2 trains with provisions for a third future train; 1 × TrojanUVPhox™ D72AL75 reactor (144 LPHO lamps) per train; average flow per reactor of 1.1 MGD (4163 m^3/day); min. 95% $T_{254\,nm}$; 2-log reduction of 1,4-dioxane (from max. 300 μg/L); 1.3-log reduction of BCEE (from max. 8 μgL); maximum H_2O_2 dose of 18.0 mg/L. One 7.8 kgal (29.5 m^3) double walled high-density polyethylene H_2O_2 storage tank with dosing pumps were among the equipment installed at Well G-3R site. GAC filtration follows the UV/H_2O_2 process. The UV system has been in operation since September 2014.

Trojan Technologies has installed over 40 UV systems with the UV/H_2O_2 process to remove 1,4-dioxane from contaminated water (flowrate range from 380,000 to 80 m^3/day), in water reuse and groundwater remediation projects around the world.

Calgon Carbon UV Technologies pioneered the UV light-based advanced oxidation technology for water remediation over 30 years ago. Of their many installations, over 20 systems are installed for treatment of 1,4-dioxane in groundwater and drinking water applications. Calgon Carbon UV commercializes powerful MP-lamp based UV reactors in combination with H_2O_2, offering small footprint UV-AOP treatment options. The 30 kW Rayox® reactors are implemented for low flow applications whereas the Sentinel™ series 24-inch and 48-inch reactors can be used for high flowrates, e.g., up to 52 MGD (196,820 m^3/day) per reactor (http://www.calgoncarbon.com/wp-content/uploads/2015/02/Calgon_Carbon_DioxaneFactSheet-Final2.pdf). No case studies on AOPs for contaminant destruction are available on Calgon Carbon UV Technologies website, thus, no examples are included in this chapter.

Hastings, Nebraska, USA. The Hastings Ground Water Contamination Site is located in and around Hastings, NE, and is one of the U.S. EPA's largest and most complex groundwater cleanup projects. U.S. EPA placed the site on the Superfund program's National Priorities List in 1986. Public and private water supplies were contaminated with chemicals of concern including VOCs (TCE, PCE, vinyl chloride, 1,1-DCE, 1,1,1-trichloroethane, polyaromatic hydrocarbons), explosives (e.g., RDX and TNT) and metals (https://semspub.epa.gov/work/07/30303295.pdf). The groundwater monitoring program started in 2001. UV photolysis and air-stripping were found viable technologies for groundwater remediation from RDX and VOCs, respectively. The cleanup goals for RDX and TCE are 2 μg/L and 5 μg/L, respectively. Two TrojanUVPhox™ 18AL50 reactors equipped with LPHO lamps were installed in parallel trains at Hastings, NE, site. The UV system is designed to remove 1.42-log (96% from max. 33 μg/L) RDX in 88% $T_{254\,nm}$ water at a flowrate of up to 241.5 gpm (~55 m^3/h) per train. The UV system would be in operation for 20 of the 30-year groundwater extraction and treatment program. Two Aquionics MP-lamp reactors (10 kW/reactor) are installed at two separate locations at the same site. The UV-treated water is pumped to an air-stripper equalization tank, then to air-stripper tanks for VOC removal. Treated groundwater is either reused for irrigation or discharged to a nearby creek.

2.9.4 Process economics, sustainability and life-cycle assessment

Advanced treatment processes, among which UV/H_2O_2-AOP, are able to remove organic and inorganic micropollutants from increasingly contaminated waters. It is worth mentioning again Sudhakaran *et al.*'s study (2013) which outlines the multi-criteria analysis (MCA) approach used in the selection of the best advanced treatment option for a given application.

The UV/H_2O_2 process presents several advantages over other AOPs which require chemical addition, among which, no mass transfer limitation, no off-gas treatment is required, no bromate formation, but it could be high energy-demanding due to the rather low conversion of the lamp electrical power to UV radiation power and to the water quality impact on process performance. A high-level comparison of advantages and disadvantages of advanced treatment technologies is given in Bui *et al.* (2016).

Katsoyiannis *et al.* (2011) compared the energy consumptions of conventional ozonation, O_3/H_2O_2 and $UV_{253.7\,nm}/H_2O_2$ for treatment of atrazine, sulfamethoxazole and NDMA in surface (lake) waters and a wastewater matrix. For those specific water qualities, the UV/H_2O_2 process was found more energy intensive than O_3-based processes, except for NDMA reduction in Lake Zürich water, where the EEDs for 90% removal were determined as 0.44 and 0.50 kWh/m^3 for UV/H_2O_2 and ozone, respectively. Lee *et al.* (2016) used a validated UV/H_2O_2 kinetic model to calculate the electrical energies required to degrade 16 organic contaminants in a secondary wastewater effluent (6 mg/L DOC, 73 $\%T_{254\,nm}$) through various advanced treatment scenarios involving ozone, $UV_{253.7\,nm}$, and H_2O_2, and the $O_3/H_2O_2/UV_{253.7\,nm}$ combined AOPs viewed as a potential treatment option in water recycling projects. The energy required for O_3 and H_2O_2 production were included in calculations, and an optimum pathlength was used for UV and UV/H_2O_2 processes. The results showed that UV/H_2O_2 required 4–25 more energy (0.20–1.06 kWh/m^3) than O_3 (0.02–0.15 kWh/m^3) to achieve 90% removal yields except for NDMA, where UV/H_2O_2 required 5.4 times less energy than O_3 (0.25 kWh/m^3 *vs.* 1.38 kWh/m^3); in the absence of H_2O_2, the required energy for 1-log NDMA reduction was only 0.15 kWh/m^3, which makes the UV process the preferred treatment when NDMA is the target. Figure 2.21 summarizes the micropollutant removal yields and the corresponding energy requirements with the four AOPs. The electrical energy consumption in the sequential AOPs was reduced by utilizing the H_2O_2 residual from the peroxone process in the downstream UV/H_2O_2 process.

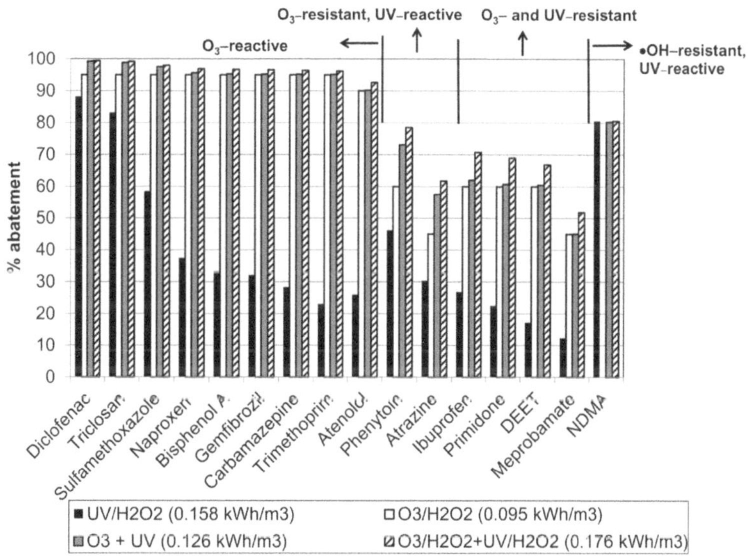

Figure 2.21 Energy requirements for micropollutant abatement in a secondary wastewater effluent with various AOPs (adapted from Lee *et al.* 2016).

There are several full-scale UV systems installed for contaminant treatment using the UV/H_2O_2 process but very few reports are available in the public domain on life cycle sustainability analysis (LCSA), which broadened the traditional environmental life cycle assessment (LCA) based mainly on the energy usage. LCSA reflects the impact of AOPs on global warming potential (GWP), greenhouse gas (GHG) emissions, and other sustainability indicators described by life cycle inventory analysis (LCIA) e.g., flows into and out of system including raw resources, chemicals and other materials, types of energy, emissions to air, water and land, etc. Over the past decade or so, LCSAs were published almost exclusively on water recycling projects and their outcomes are worth reviewing e.g., Lyons *et al.* (2009); Law (2016). Some of the examples given below on full-scale UV-AOP economics included life cycle analysis.

Prior to installing the UV/H_2O_2 process at Neshaminy Falls WTP, Aqua Pennsylvania, which owns and operates the plant, explored the option of using powdered activated carbon (PAC) filtration in combination with the existing coagulation treatment step for T&O control (Civardi & Luca, 2012). For the 15MGD plant, the implementation of PAC would have required large carbon vessels and twice the existing solids production, handling, and disposal. The costs associated with PAC option were reported as \$2.2 MM (capital), \$310,000/year (operating costs for 90 days per year at 12 MGD, without including the solids disposal costs), and \$475,000 (equivalent uniform annual cost at 4% discount over 20 years). By comparison, the costs for the UV/H_2O_2 treatment option, which achieved 90% geosmin removal, were reported as \$2.5 MM (capital), \$200,000 (operating costs which include energy, lamp replacement, H_2O_2 consumption, chlorine for quenching H_2O_2 residual), \$384,000 (equivalent uniform annual cost at 4% discount over 20 years) (Civardi and Luca, 2012). The 20-year net present value (NPV) analysis of the two technologies showed costs of ~\$5,280 MM and ~\$4,780 MM (USD) for PAC and UV/H_2O_2 technology, respectively. The analysis was based on a 90-day T&O event and 4% discount factor, and the costs include capital, construction, operation and maintenance (including solids handling and disposal associated with PAC) (http://www.trojanuv.com/resources/trojanuv/casestudies/ECT/ECT_CaseStudy_Neshaminy_FINAL.pdf). Trojan Technologies in collaboration with Western University, London, ON, conducted a LCA of PAC *vs.* UV/H_2O_2 process installed at Neshaminy Falls, PA, for seasonal operation for T&O control in drinking water. The data showed that the UV process would release 74% less CO_2-equivalent (CO_2-e) than PAC. The environmental 20-year total carbon footprint estimated for the T&O control (90-day event) at Neshaminy Falls WTP by PAC (30 mg/L) and TrojanUVSwift™ ECT system were determined as 32,000 and 8,000 tons of CO_2-e emissions, respectively.

Swaim *et al.*'s study (2011) provides a thorough comparison of three technologies (ozone, PAC, and UV/H_2O_2) for T&O control in drinking water in terms of capital cost, O&M cost, and GHG footprint as a measure of sustainability. The authors recognized that cost comparisons are very site-specific; their analysis was based on a typical drinking water plant of 114,000 m^3/day and an average flow of 76,000 m^3/day. The estimates were based on three seasonal T&O event durations, i.e., 4, 16, and 52 weeks per year. BAC filtration and chlorination, in addition to the T&O removal processes, were included in the designed treatment trains. Treatment goal was 90% reduction of geosmin. Detailed description of each process, key design parameters, cost tools including the life-cycle cost tool, GHG module and assumptions are discussed in the paper. Table 2.9 exemplifies capital and NPV costs for the T&O treatment trains considered in the analysis. Table 2.9 shows that between the two technologies which provide disinfection and T&O control (O_3 and UV/H_2O_2), the UV-AOP implemented post-BAC filtration has a lower cost than ozone, for the 4- and 16-week/year T&O events. Year-round T&O treatment is more costly for UV/H_2O_2 than ozone (data not shown). The need for booster pumping can increase significantly the costs. PAC has the lowest cost of all scenarios, but does not provide any disinfection. PAC has also the largest carbon footprint among the three technologies due to carbon losses, solids handling and disposal, and energy intensive-reactivation. GHG emissions associated with construction and 20-yrs, 4-week/yr O&M, for BAC filtration-UV/H_2O_2 combined processes (option 5) for T&O control are the lowest among the six treatment options in Table 2.9 (~7,000 tons-CO_2-e), and comparable to O_3/BAC for 16-week T&O treatment per year (~15,000 tons-CO_2-e).

Table 2.9 Cost comparisons for the T&O control technologies examined by Swaim *et al.* (2011).

Process	Capital	4-Week NPV	16-Week NPV
(1) O_3–BAC	\$16,268,000	\$25,869,000	\$26,089,000
(2) PAC with solids handling	\$4,819,000	\$10,340,000	\$17,739,000
(3) UV/H_2O_2–BAC	\$18,067,000	\$29,699,000	\$33,008,000
(4) UV/H_2O_2–BAC with booster pumping	\$21,265,000	\$35,378,000	\$38,687,000
(5) BAC–UV/H_2O_2–Cl_2	\$12,056,000	\$18,500,000	\$22,942,000
(6) BAC–UV/H_2O_2–Cl_2 with booster pumping	\$15,255,000	\$24,179,000	\$28,621,000

MacNab *et al.* (2015) took a LCA approach to evaluate the GWP and GHG emissions associated with MIB and geosmin treatment in drinking water at three hypothetical WTPs using different treatment technologies: (*i*) PAC (bituminous coal, 35 mg/L); (*ii*) ozone (generated from ambient air and dosed into the water at 10 mg/L); (*iii*) UV/H_2O_2 AOP. For the purpose of GWP and GHG emission estimates for the three hypothetical WTPs, it was assumed that all three facilities were designed to remove 0.7-log (63%) MIB and 0.9-log (81%) geosmin at a design flow of 37,800 m^3/day for a T&O event duration of two months. In this exercise, H_2O_2 residual was considered to be removed by BAC filtration at an EBCT of 2 min as observed in bench-scale studies. The GWPs were expressed in CO_2-e emissions, with methane and nitrous oxide also expressed as CO_2-e based on the generally adopted conversion factors. The estimated GWPs included the contributions from all individual processes, such as extraction of raw material and its further conversion into usable supplies and equipment, transportation, energy consumption, etc. The highest environmental impact was determined for PAC (1.07×10^7 kg CO_2-e), as compared to O_3 (3.07×10^6 kg CO_2-e) and UV/H_2O_2 (3.10×10^6 kg CO_2-e) technologies. Production of raw material (activation) contributed the most to PAC carbon footprint, whereas the O_3 and UV/H_2O_2 GWPs were heavily weighted by the electrical energy production. The authors provide the GWP breakdown for UV/H_2O_2 process as 80% electrical energy production, 12% H_2O_2 production, ~5% other equipment, and ~3% infrastructure and transportation. Replacement of conventional energy source with renewable sources would reduce the UV/H_2O_2 and O_3 carbon footprint.

In September 2014, an LPHO-lamp based UV system was installed at Artesian Water Company's Llangollen wellfield in Delaware, USA, to treat 1,4-dioxane and BCEE in the largest contaminated well, using the UV/H_2O_2 process. Civardi *et al.* (2014) described the project and decision making process on the technology selection. At the conceptual stage, economic and non-economic criteria were used to evaluate three treatment options, namely, Trojan LPHO-UV/H_2O_2, Calgon MP-UV/H_2O_2, and ATP O_3/H_2O_2, based on two select design scenarios targeting the same contaminant treatment goal of ≤3 μg/L 1,4-dioxane: (*i*) 1.2-log 1,4-dioxane reduction in two AOP trains at a total flow of 341 m^3/h given the existing 1,4-dioxane levels; (*ii*) up to 2.0-log reduction, should 1,4-dioxane level increase to 300 μg/L due to the nearby Superfund sites. GAC filtration was considered after AOP treatment. Table 2.10 summarizes the estimated costs and the economic and non-economic scores which led to the treatment option ranking; cost breakdown is provided in the publication. The total maximum possible score was 100, divided as 60 and 40 for economic and non-economic scores, respectively.

Table 2.10 Ranking of 1,4-dioxane treatment options at Llangollen wellfield (adapted from Civardi *et al.* 2014).

Criteria	LPHO-UV/H_2O_2	MP-UV/H_2O_2	O_3/H_2O_2
Economic			
Construction	\$3,100,000	\$2,901,000	\$3,787,000
O&M (annual)	\$121,000	\$224,000	\$77,000
Lifecycle (present worth)	\$4,290,000	\$5,100,000	\$4,540,000
Total Economic Score	60	50	57
Total Non-Economic Score	38	32	19
Total Score	98	82	76

Operating costs were estimated for continuous operation i.e., 24 h/day, 365 days/year, at an average flow of 1500 gpm (341 m^3/h). For the lifecycle costs, the capital and O&M costs associated with each technology were converted to a present worth lifecycle cost for 20 years at an interest rate of 8%. Among the non-economic factors scored were supplier's experience, equipment complexity, level of permitting, simultaneous BCEE removal level, ease of integration in the existing facilities, and operator training requirements.

Capital and operational costs at conceptual stage were estimated for full-scale UV/H_2O_2 treatment of MTBE in water quality comparable to Santa Monica, CA, groundwater by Kommineni *et al.* (2008). The design considerations and treatment costs were based on the E_{EO} values estimated for full-scale reactors using a kinetic model validated at pilot scale, and optical and hydraulic performances of full-scale UV systems. The costs for the UV/H_2O_2 systems were developed for 1, 3, and 10MGD (3785; 11,355; and 37,850 m^3/day) using an $E_{EO,MTBE} = 0.35$ kWh/m^3/order for MP-UV/H_2O_2 and an $E_{EO,MTBE} = 0.13$ kWh/m^3/order for LP-UV/H_2O_2, both with 15 mg/L H_2O_2. The capital, O&M, and total costs as well as the assumptions considered in the cost estimates are extensively presented and discussed in the publication. For a 10MGD plant, the authors estimated a cost of \$0.41 and \$0.32 per 3.8 m^3 water treated with MP- and LP-UV/H_2O_2, respectively. An additional cost of \$0.20/3.8 m^3 was estimated for 20 min-EBCT BAC filtration of AOP-treated water for removal of H_2O_2 residual and MTBE byproducts (details in the AwwaRF project referenced in the work).

Pilot testing precluded the installation of full-scale O_3/H_2O_2 – UV/H_2O_2 combined AOPs at Dunea's Bergambacht WTP for treatment of pesticides and other micropollutants. One set of tests were conducted to estimate the efficiency of individual process and combined processes for 80% removal of four probe pollutants (atrazine, bromacil, NDMA, and ibuprofen), with bromate formation control. The test conditions and data were reported by Scheideler *et al.* (2011). The treatment costs for a 240,000 m^3/day plant were estimated for the three AOP scenarios. Capital cost (20-years depreciation time at 4% discount rate/year) and O&M cost per m^3 of water treated, as well as O&M breakdown (maintenance, energy, H_2O_2, oxygen for O_3 generation) were calculated and compared. Peroxone process (2 g O_3/5 g H_2O_2)/m^3 had the lowest cost (~0.01 €/m^3), but only 58% and <10% removals were obtained for atrazine and NDMA, respectively. The UV/H_2O_2 at 0.26 kWh/m^3 and 6 g/m^3 H_2O_2 was found to remove >80% NDMA, but only 53% atrazine, 59% ibuprofen and 50% bromacil. The associated cost was approx. 0.037 €/m^3. Increasing the EED to 0.42 kWh/m^3 would achieve >80% removal of all four contaminants, but with >60% cost increase. The combined AOP case 2 g/m^3 O_3/ 6 g/m^3 H_2O_2/ 0.26 kWh/m^3 EED_{UV} ensured the ≥80% treatment goal for all four compounds at a cost of 0.044 €/m^3, which is only 20% higher than that estimated for the UV/H_2O_2 scenario. The cost breakdown showed that capital represented ~20–25% of the total costs in each case. For the selected conditions, O&M costs for O_3/H_2O_2 process were mostly for H_2O_2 (~25%), O_2 for O_3

generation (~25%) and energy (~20%), whereas for the UV/H_2O_2 and O_3/H_2O_2/UV AOPs the energy was found as the major cost contributor (~40%), followed by maintenance (~25%); H_2O_2-related costs were less than 15% in the UV-based processes.

Costs associated with full-scale advanced treatment in water reclamation projects were extensively analyzed at a conceptual level by Plumlee *et al.* (2014). The authors developed capital and O&M cost curves for selected advanced treatment unit processes to allow cost estimates per unit flow under various treatment train scenarios over a range of plant flow capacities. Unit processes included MF/UF membranes, NF or RO membranes, ozone with and without H_2O_2, UV/H_2O_2, and BAC. The costs for advanced treatment of a wide range of trace organic contaminants as a function of targeted effluent water quality were briefly discussed. The authors estimated the increase in energy consumption and CO_2-e emissions as well as the related financial costs passed on to the public.

2.10 BYPRODUCT FORMATION AND MITIGATION STRATEGIES

Any chemical or photochemical water treatment process generates chemical byproducts, whose nature and yields are often unknown and difficult to predict. During UV/H_2O_2 water treatment, byproducts can be formed from the target micropollutants, DOM, nitrate and/or other water constituents, or from subsequent reactions of generated byproducts with water matrix and/or chemicals added to the water (e.g., chlorine, chloramines, chlorine dioxide) during the post-UV/H_2O_2 treatment steps. Byproduct formation and speciation are driven by the water quality, upstream and downstream to UV/AOP treatment steps, UV/AOP process operating conditions, light-source characteristics, and post-UV/AOP disinfection practices – i.e., are highly application-specific.

Drinking water and direct and indirect potable water reuse regulations in jurisdictions around the world set permissible levels of a wide range of individual chemical compounds and specific classes of byproducts (e.g., DBPs) in the finished water. Action and notification levels are also in effect, whereas monitoring requirements are enforced for a number of unregulated chemicals considered for future regulation.

The degradation intermediates of chemical micropollutants are identified during laboratory product studies. Frequently, for given water quality, LP- and MP-lamp based processes generate similar byproducts, yet, their relative yields depend on the concentration and photochemical properties of micropollutants. Byproducts particularly related to polychromatic radiation-based treatment of nitrate-rich waters have been reported in the literature, and postulated or demonstrated as nitroso- and/or nitro-derivatives. Byproduct formation during the oxidation of pharmaceuticals with ozone, UV, UV/H_2O_2, chlorine, chloramine, and ClO_2, with a particular emphasis on those byproducts which belong to the DBP group, was reviewed by Postigo and Richardson (2014).

Directive 2013/39/EU framework policy includes several pesticides as priority pollutants. Photodegradation and $^{\bullet}OH$-initiated oxidation of *s*-triazines was extensively studied (Nick *et al.* 1992: Hessler *et al.* 1993; Beltran *et al.* 1996; Müller *et al.* 1997; Acero *et al.* 2000; Evgenidou & Fytianos, 2002; Álvarez *et al.* 2016). Dealkylated and hydroxy-*s*-triazine structures are among the photolysis and $^{\bullet}OH$-oxidation byproducts, and were identified also as metabolites.

Chlorinated ethenes (1,2-dichloroethene DCE, TCE, PCE) both photolyze and are prone to $^{\bullet}OH$ reactions. Mono- and dichloroacetic acids were identified among the TCE degradation products (Haag 1994; Hirvonen *et al.* 1996; Li *et al.* 2004, 2007), whereas di- and trichloroacetic acids were found as PCE degradation byproducts (Mertens & von Sonntag, 1995; Hirvonen *et al.* 1998). These acids are among the haloacetic acids (HAAs) regulated and monitored in drinking water as DBPs. Chloroacetaldehyde is also a PCE degradation byproduct and dichloroacetic acid precursor. Given the very low concentrations of chlorinated ethenes in water, the HAA levels formed through their degradation are expected to be insignificant.

Triclosan is an antibacterial compound widely used in personal care products. It was found in surface waters impacted by wastewater effluent discharge. 2,8-Dichlorodibenzo-*p*-dioxin (2,8-DCDD) was identified as a photochemical byproduct of triclosan in water exposed to natural or simulated sunlight (Tixier *et al.* 2002; Latch *et al.* 2003; Latch *et al.* 2005; Kliegman *et al.* 2013). Polychlorinated dioxins are known powerful toxins and carcinogens, with toxicological potency increasing with the degree of chlorination. The 2,8-DCDD levels in irradiated triclosan solutions depended on the experimental conditions with yields ranging from 1.2 to 12% (Latch *et al.* 2003). Although chlorinated dioxins and phenols could be formed from triclosan and its chlorinated derivatives during UV light-based water treatment, these compounds are not found in finished water due to their effective degradation under typical AOP conditions (Gao *et al.* 2014).

N-Nitrosodimethylamine formation from various compounds either *via* 'thermal' (dark) reactions or as a result of photochemical or chemical oxidative treatment is documented in the published literature. Shen and Andrews (2011) demonstrated the NDMA formation potential (FP) from 20 pharmaceuticals and personal care products upon disinfection with chloramine. All selected compounds contained a tertiary amine, i.e., a *N,N′*-dimethylamine moiety, in their chemical structures. On a molar basis, ranitidine showed the highest NDMA-FP (>80%), whereas metformin and tramadol had the lowest NDMA-FP (~0.1%). Sgroi *et al.* (2015) examined the NDMA formation upon chloramination of UV/H_2O_2-treated RO permeate at a full-scale indirect potable reuse water facility. NDMA formation was observed in all studied scenarios (low and high UV dose, with and without 3 mg/L H_2O_2). The UV/H_2O_2 process at half of the usual UV dose applied in full-scale NDMA treatment conditions enhanced the NDMA formation relative to the level observed in the UV only process. At a practical UV dose of ~1000 mJ cm^{-2}, reduced NDMA formation during post-chloramination and water stabilization was reported. The effect of pre-ozonation for membrane fouling control (O_3/MF/RO) on NDMA formation was studied at pilot and bench-scale. High levels of NDMA (117–227 ng/L) were measured. Ozonation followed by chloramination increased the NDMA yield. Formation of NDMA precursors during advanced treatment was considered as a possible explanation for the observed data.

In a recent pilot study, Li *et al.* (2017) reported NDMA formation upon exposure of wastewater effluents treated with either O_3 or UV/H_2O_2 to biological activated carbon filtration. NDMA formation was observed in the BAC columns fed with O_3-treated water and to a lesser extent in the BAC columns used with the UV/H_2O_2-treated water. The EBCT played a crucial role in NDMA formation; no NDMA was observed at EBCTs shorter than 12–16 min, but NDMA was consistently formed at longer EBCTs (e.g., 20 min). The authors examined the microbial populations along the BAC columns and explained their observations on the NDMA trends through the nitrite oxidation at short EBCTs with increasing nitrate levels during the BAC transit, and the interplay between nitrogen cycling *via* microbial speciation and activity along the BAC columns which favored nitrosation reactions at long EBCTs.

Zeng *et al.* (2016) examined the occurrence and formation of nine *N*-nitrosamines and 32 halogenated DBPs from various classes (Cl- and Br-THMs, I-THMs, HAAs, haloketones, haloacetonitriles, haloacetaldehydes, haloacetamides, and chloropicrin) along the potable reuse treatment trains (post-MF, post-RO, and post-AOP) and in the distribution system at five full-scale advanced water treatment plants in the U.S. Chloramination for membrane fouling control increased slightly some of the DPBs in the post-MF waters. NDMA, *N*-nitrosopyrrolidine, and *N*-nitrosomorpholine were quantified within the 4.5 –1020, <2–17, and 2.7–39 ng/L range, respectively. The RO membranes removed most DBPs below their method reporting limits, and NDMA levels dropped to 2–490 ng/L. The UV/H_2O_2 process reduced the concentrations of all nitrosamines below 2 ng/L, and of most DBPs which passed through RO membrane; chloroform was the least treated DBP, with levels measured as 4–8.7 μg/L.

The impact of UV/H_2O_2 AOP on DOM composition and properties as related to *DBP formation* upon post-UV/AOP chlorination and to the increased biodegradability was extensively investigated and reported in the literature. Research indicated that in general, DBPs are formed from the hydrophobic, non-polar natural organic matter (NOM) fractions. New evidence into the correlation between the DBP formation potential and molecular composition of NOM was provided recently (Wang *et al.* 2017). The DBP drinking water standards could vary largely from one jurisdiction to another.

Sarathy and Mohseni (2010) examined the impact of UV/H_2O_2 on NOM aromaticity and hydrophobicity, and on the DBP-FP changes in water due to AOP-induced transformations in NOM. Raw surface water (2.53 mg/L TOC; 81 $\%T_{254\text{ nm},1\text{ cm}}$) as-is and passed through a DAX-8 resin (1.16 mg/L TOC; 95 $\%T_{254\text{ nm},1\text{ cm}}$) were exposed to UV fluences from 500 to 2000 mJ cm^{-2} in the presence of 15 mg/L H_2O_2. No increase in DBP-FP was observed under all conditions; less aromatic and more hydrophilic fractions were formed from NOM. In a UV/H_2O_2 – BAC filtration pilot-scale study equipped with both LP- and MP-lamp UV reactors, Sarathy *et al.* (2011) investigated the changes in TOC, $\%T$, THM-FP, formaldehyde and acetaldehyde concentrations, and BDOC in surface water, as a function of UV dose and lamp type. The initial H_2O_2 concentration was kept constant at 10 mg/L. Significant reductions in CDOM and aromatic substitution were observed; marginal increase in THM-FP (~14 µg/L) and up to 4-fold increase in BDOC fraction were determined at the maximum UV dose applied (880 mJ cm^{-2}). BAC filtration (20 min EBCT) removed the residual H_2O_2 and BDOC, and reduced the THM-FP and HAA-FP by up to 60% and 75%, respectively.

Dotson *et al.* (2010) investigated the formation of THMs, HAAs, and total organic halide (TOX) in water samples collected from Greater Cincinnati Water Works (GCWW) full-scale water treatment plant. The water samples were amended with either 5 or 10 mg/L H_2O_2 and exposed to 500 and 1000 mJ cm^{-2} from either a LP- or MP-lamp collimated beam apparatus. The data showed a total THM increase by up to ~30 µg/mg-C and ~20 µg/mg-C at 1000 mJ cm^{-2} and 10 mg/L H_2O_2, upon exposure to LP- and MP-lamp, respectively. Under the same conditions, HAA9 increased by ~12–17 µg/mg-C under LP-lamp irradiation; no change was observed for the MP-lamp case. Large increase in TOX was reported for LP-lamp experiments, with no or minimal increase under MP-lamp exposure. Metz *et al.* (2011) conducted a one-year pilot study at GCWW treatment plant to evaluate the DBP formation in two plant-process water sources (conventional and GAC treatments) with different DOM composition and TOC content treated with the UV/H_2O_2 process in MP- and LP-lamp reactors; 3-day SDS (simulated distribution system) tests were used to determine the DBP formation. A 10 mg/L H_2O_2 was targeted and the pilot units were operated to achieve 80% reduction in atrazine concentration. Large increases in THM-FPs (20–118%) through the UV/H_2O_2 reactors were reported, with lower values when GAC-filtered water was the UV-influent.

Unlike the above reports, Toor and Mohseni's study (2007) indicated no statistically significant difference between THM-FP data for untreated and UV/H_2O_2-treated surface water in both LP-collimated beam and flow-through annular photoreactors at UV doses up to 3350 mJ cm^{-2} and 4 to 23 mg/L H_2O_2. Marginal increase in HAA-FP was observed with 23 mg/L H_2O_2 at UV doses up to ~1500 mJ cm^{-2}; HAA-FP decreased below the level measured in the untreated water when higher UV fluences were applied. Moreover, BAC filtration of UV/H_2O_2-treated water resulted in great reductions in DBP-FP, TOC and $\%T_{254\text{ nm}}$.

Guo *et al.* (2016) examined the effect of UV dose (0–~1700 mJ cm^{-2}), pH, nitrate and bromide on DBP-FP in water samples collected from a drinking water treatment plant of Shanghai upon subsequent chloramination. Both LP- and MP-lamp irradiations were performed and compared. Detailed information is provided in the published work, and the overall conclusion was that MP-UV irradiation resulted in more DBPs than the LP-UV, especially N-DBPs (HANs and HNMs). On the other hand, Chu *et al.* (2014) showed that UV (253.7 nm) alone (19.5–585 mJ cm^{-2}) and H_2O_2 alone (2–20 mg/L) did not significantly change the total haloacetamides (HAMs), whereas the UV/H_2O_2 at a UV dose of 585 mJ cm^{-2} with 10 mg/L H_2O_2 controlled the formation of HAMs and several other N-DBPs in drinking water.

The impact of disinfectant on the type of byproducts and their concentrations from MP-UV-treated surface water samples collected from a water treatment plant and the correlation between DOM fluorescence and DBP formation were described by Lyon *et al.* (2014). Approx. 37% increase in THM4 and almost 4-fold increased chloral hydrate were observed at 1000 mJ cm^{-2} both in ambient water and in water amended with 10 mg/L nitrate, upon chlorination; no change in THMs was observed upon chloramination. Cyanogen chloride was measured at very low levels (0.8–1.9 µg/L) upon chloramination of samples exposed to 1000 mJ cm^{-2}.

The increase in NOM biodegradability could result in bacterial growth in the distribution system lacking the required residual disinfectant. Bazri *et al.* (2012) studied the impact of $UV_{253.7\,nm}$/(10 mg/L) H_2O_2 process on molecular weight distribution and water biostability. Both *assimilable organic carbon (AOC)* and BDOC increased in the two natural waters (5 and 10 mg/L TOC) as the UV dose increased up to 2000 mJ cm^{-2}. An increase in the amount of biodegradable organic compounds, particularly in formaldehyde and acetaldehyde, upon $UV_{253.7\,nm}$/(20 mg/L) H_2O_2 treatment of a raw surface water and a filtered (ultrafiltration) surface water was observed by Sarathy and Mohseni (2009). At a UV fluence of 1500 mJ cm^{-2}, 15% and 27% mineralization of NOM was determined in the raw and filtered water, respectively. Toor and Mohseni (2007) examined the biodegradability of NOM (~1.9 mg/L TOC) in surface water used as drinking water source for Vancouver, BC, Canada, upon $UV_{253.7\,nm}/H_2O_2$ treatment. The AOP with 20 mg/L H_2O_2 increased the AOC from 62 µg/L (source water) to 100 µg/L (treated water); post-AOP BAC filtration at 8.2 min EBCT decreased the AOC to the source water level. BDOC followed a similar trend, increasing from ~7% of TOC in untreated water to ~45% of TOC in AOP-treated sample, then decreased to ~22% of TOC upon BAC filtration.

Chlorination for H_2O_2 residual removal from water treated with the UV/H_2O_2 process is practiced at many water utilities, most of which have implemented the UV-AOP for controlling the seasonal T&O causing compounds. Keen *et al.* (2013) examined various H_2O_2 quenching agents and concluded that insignificant increase in DBP formation would be expected as a result of H_2O_2 residual removal with chlorine. Batterman *et al.* (2000) observed lower amounts of DBPs upon H_2O_2 quenching with chlorine than in the chlorination of the same water without H_2O_2.

Wang *et al.* (2015b) compared the UV/chlorine and UV/H_2O_2 process performance for geosmin and 2-MIB removal at Cornwall (ON, Canada) Water Purification Plant. DBPs from various classes were also monitored: THMs, HAA9, HANs, HKs, chloropicrin, adsorbable organic halide (AOX), inorganic DBPs. H_2O_2 dose and pH were varied, whereas the UV dose was constant and estimated as 1800 mJ cm^{-2}; short (up to 1 min)- and 24 h-contact time with (rather high) 6.5 mg/L Cl_2 residual were tested. Virtually no increase in any of the monitored DBPs was observed at short chlorination time at any of the tested UV/H_2O_2 operational conditions. Upon 24 h-incubation time, THMs and HAAs increased in treated samples relative to those in untreated water, but did not exceed the U.S. EPA THM standard of 80 µg/L, while slightly exceeding the HAA standard of 60 µg/L. HANs increased by ~8 µg/L in the 24 h-chlorinated samples, whereas marginal change in AOX was observed.

Full-scale T&O treatability tests at Neshaminy Falls (PA, USA) WTP showed no statistically significant difference in the immediate formation of HAAs and THMs between the UV/H_2O_2 treated water and the traditional chemical treatment (dual media filtration/chlorination (gas)/chloramination) (Civardi and Lucca, 2012). The results of a large-scale field study on MIB and geosmin treatment using TrojanUVSwift™ 16L30 reactors are discussed by Royce and Stefan (2005). Flowrates up to 6 MGD (22,710 m^3/day) and H_2O_2 doses from 0 to 11 mg/L were tested. Several sample sets were analyzed for DBPs, AOC and bromate before and after UV/H_2O_2 treatment. The UV-AOP reduced the THM and HAA levels by an average of 25% and 12%, respectively, with no bromate formation, whereas no statistically different AOC data were reported by the commercial laboratory between the untreated and treated samples. The observed discrepancies between the byproduct data acquired under bench- and small pilot-scale conditions and full-scale AOP treatment conditions as implemented at the water utilities raise awareness on the need for careful interpretation and cautious extrapolation of the data across different experimental settings.

The U.S. EPA drinking water standard for *nitrite ion* (NO_2^-) is 1 mg/L as N (or 3.0 mg/L as NO_2^-), whereas the EU standard is 0.03 mg/L as N (0.1 mg/L as NO_2^-). Some surface waters are impacted by high nitrate (NO_3^-) levels, thus, nitrite can be generated as a nitrate photolysis byproduct in UV and UV/H_2O_2-based water treatment. Therefore, along with the nitrite impact on UV/H_2O_2 process performance discussed earlier in this chapter, there is the regulatory aspect, particularly in the EU given the low NO_2^- standard. Nitrate and nitrite photolysis generates intermediate radicals which may react with the organic compounds present in the water matrix forming nitroso- and nitro-derivatives, some of which are potential mutagenic and genotoxic compounds. If formed in the UV or UV/H_2O_2 processes, nitrite is rapidly removed upon chlorination or BAC filtration.

Sharpless and Linden (2001) studied nitrate photolysis with nitrite formation as a function of DOC and alkalinity levels, pH, and UV dose delivered by either a LP- or MP-lamp. The authors concluded that in only one case (5 mg/L DOC, pH 8, ~300 mJ cm^{-2}) the U.S. EPA standard for NO_2^- was exceeded by ~10%. Semitsoglou-Tsiapou *et al.* (2016a) determined the nitrite formation from 50 mg/L nitrate-containing synthetic water samples exposed to 253.7 nm radiation in a collimated beam apparatus. The role of NOM-source and concentration, UV dose (0–2100 mJ cm^{-2}), and H_2O_2 concentration (0, 10, 25, and 50 mg /L) on nitrite yield was studied. The authors noted that the EU NO_2^- standard may be exceeded upon LP-UV/H_2O_2 treatment of 50 mg/L- NO_3^- containing water with UV doses >1500 mJ cm^{-2} and H_2O_2 doses >10 mg/L.

As expected from nitrate photochemistry, significantly higher nitrite yields were observed upon MP-UV/H_2O_2 treatment of nitrate-rich waters than upon the LP-UV/H_2O_2 treatment (Lekkerkerker-Teunissen *et al.* 2013; Hofman-Caris *et al.* 2015b; Kruithof *et al.* 2007). Nitrite yields exceeding the EU standard were measured after the UV/H_2O_2 process. Nitrite was removed from the water upon BAC filtration; nitrite removal during water filtration through virgin GAC was problematic, particularly at low water temperature (Kruithof *et al.* 2007; Martijn & Kruithof, 2012).

Photochemistry of nitrate in the context of water treatment raised concerns over potential reactions between N-containing reactive species and water matrix constituents leading to harmful byproducts. Nitric oxide (NO) and nitrogen dioxide (NO_2) are among the reactive species and nitrosation and nitration reactions involving these species are known and reviewed in the literature (e.g., Williams, 2004). Nitration and nitrosation of probe aromatic compounds (phenol, catechol, naphthalene) in nitrate- or nitrite-containing water upon exposure to UV were investigated and reported (e.g., Vione *et al.* 2005; Machado & Boule, 1995). Susceptibility of soil and aquatic NOM to nitrosation is also known.

Heringa *et al.* (2011) evaluated the *genotoxic activity* of surface water after UV/H_2O_2 treatment (MP-lamp reactors) and post- GAC/BAC filtration. Water samples were collected from either pilot or full-scale tests at three different locations, namely, Dunea's Bergambacht WTP (The Netherlands), Greater Cincinnati Water Works (OH, USA), and Andijk WTP (The Netherlands). Detailed description of water source and quality parameters, and test conditions is provided in the publication. No genotoxic response was observed in any of the samples in the Comet assay. Ames II assay was positive in *Salmonella typhimurium* TA98 strains for all UV/H_2O_2-treated samples at all locations. Mutagenic activity was completely removed upon BAC filtration. Following this work, a number of studies focused on the impact of water quality and AOP conditions on mutagenic activity of treated water and emphasized the effectiveness of BAC filtration at removing that activity. Lekkerkerkek-Teunissen *et al.* (2013) compared the genotoxicity (Comet assay) and mutagenicity (Ames II test) of surface water treated with the UV/H_2O_2 process and three different lamps i.e., MP, LP, and DBD (emission around 240 nm), at pilot scale (Dunea WTP, The Netherlands). No responses in Comet assay were recorded, but positive responses were measured in the Ames II test in all samples treated with UV/H_2O_2 in MP-, LP- and DBD-lamp reactors. The water contained ~15 mg/L nitrate and ~4 mg/L TOC. Fewer positive wells were counted for LP- and DBD-lamp reactors than for the MP-lamp reactor for similar UV doses, yet, calculated over different spectral power distributions.

Two full-scale water treatment plants owned and operated by PWN Water Supply Company North Holland, Andijk WTP and Heemskerk WTP, using the MP-UV/H_2O_2 process followed by BAC filtration were described briefly in section 2.9.3. PWN undertook thorough research on the processes implemented at their WTPs, including on the MP-UV/H_2O_2 AOP. The outcomes were disseminated in scientific events and journal publications. Martijn and Kruithof's study (2012) presents data on nitrite formation and responses in Ames II test for samples collected from the two full-scale WTPs. Nitrite yield increased upon increasing the EED and it was higher when high purity quartz housed the MP lamps than in the case of natural fused silica (214 quartz). Nitrate was measured as 9 mg/L. A marginal increase in Ames II (TA98 strains with and without metabolic activation, i.e., +S9 and –S9) response was observed after conventional coagulation/sand filtration, but a significant response was measured after MP-UV/H_2O_2 treatment (~540 mJ cm^{-2}, Andijk WTP) in strain TA98-S9. A much lower response was observed in strain TA98 + S9. The mutagenicity levels in the BAC (30 min EBCT)-filtered samples were comparable to the negative control response. Samples collected from Heemskerk WTP showed a gradual increase in Ames II test response in samples collected from first through the fifth MP-reactor, then a reduction in strain TA98-S9 response in BAC (9 min EBCT)-filtered samples; very little response was measured in the dune infiltration effluent. A linear correlation was observed between the number of positive wells and nitrite yield. Preliminary Ames II assays on NOM-irradiated solutions with and without nitrate showed no response in the absence of NO_3^-, and significant responses in strain TA98 ± S9 from samples containing NO_3^-. The authors stressed that no responses were observed in MP-UV/H_2O_2-treated samples when the classic Ames test was used. Further investigations (Martijn *et al.* 2014) supported the earlier findings and revealed that solutions of well characterized NOM (IHSS Pony Lake NOM) showed marginal if any response in Ames II test (TA98-S9) upon MP-UV (with or without H_2O_2) in the absence of NO_3^- relative to mineral water, whereas large numbers of positive wells were counted in the same solutions in the presence of NO_3^-. Experiments conducted in the presence of phenol showed the formation of 2- and 4-nitrophenol and 4-nitrocatechol through the reaction of N_xO_y reactive species generated through nitrate photolysis with the aromatic compounds. More in-depth studies using stable isotope ^{15}N-labeled nitrate in synthetic water samples containing IHSS Pony Lake NOM in combination with high-resolution mass-spectrometry allowed the identification of 84 distinct N-containing byproducts which were quantified relative to bentazone-d_6 and atrazine-d_5 depending on the detection ionization mode (Kolkman *et al.* 2015). Of those 84 compounds, 22 were also detected in the full-scale MP-UV/H_2O_2 treated water, among which 4-nitrophenol, 4-nitrocatechol, and 2-methoxy-4,6-dinitrophenol. An approach to enable a preliminary risk assessment based on a known carcinogen (4-nitroquinoline oxide, 4-NQO) was used to express the genotoxic effect of water treated with the UV/H_2O_2 process (Martijn *et al.* 2015; Martijn *et al.* 2016). A 4-NQO equivalent concentration of 107 ng/L was determined in MP-UV/H_2O_2 treated water in the full-scale installation, which, in the authors' opinion, is a safety concern if the water is provided to the consumers without further post-treatment. Surface water pre-treatment with IX-MF, which reduces the organic content and nitrate level, lowered significantly the response in Ames II test and the 4-NQO-equivalent concentrations in MP-UV and MP-UV/H_2O_2-treated water samples (Martijn *et al.* 2015). The key messages from all byproduct (including mutagenicity) studies published by the group are that the UV/H_2O_2 process is a reliable barrier for primary disinfection and chemical contaminants and that an integrated treatment approach which involves pre- and post-UV processes should be taken in order to mitigate the byproduct formation.

Several studies examined the biological activity (toxicity, mutagenicity and/or genotoxicity, estrogenicity, etc.) of individual micropollutants, including chemical compounds regulated in drinking water and emerging contaminants, under treatment with various processes, in various water matrices. Most investigations were conducted at concentrations much larger than those detected in water sources, e.g., mg/L as compared to low μg/L and ng/L, and often under process conditions irrelevant to water treatment, e.g., $[H_2O_2] \gg 10$ mg/L and extremely high UV doses.

N-Nitrosamines are considered carcinogenic environmental contaminants. Genotoxicity of five nitrosamines in *Salmonella* strain YG7108 and in mammalian cells was compared by Wagner *et al.* (2012). Nitrosamines are effectively removed from water *via* direct photolysis, and some nitrosamines (e.g., NDPA, NPYR, NDPhA) react with •OH at near diffusion-controlled rates. Mestankova *et al.* (2014) used Ames test with three *S. typhimurium* strains (TA98, TAMix, and YG7108), of which YG7108 is sensitive to nitrosamines, to monitor the mutagenicity trends during direct photolysis and •OH degradation of these compounds. Direct photolysis of NDMA, NDEA, NDPA and NPYR resulted in removal of their specific mutagenic activity in strain YG7108. Yet, during the •OH-initiated degradation of NDPA and NPYR increasing mutagenicity was observed in both specific strain YG7108 and TAMix, in the absence of metabolic activation (S9). NDPhA was not found mutagenic. However, its photolysis generated mutagenic byproducts causing positive response in strain TA98. In another study, Mestankova *et al.* (2016) examined the mutagenicity and estrogenicity of 23 organic compounds from the U.S. EPA CCL3 with $k_{\cdot OH} > 5 \times 10^8$ M^{-1} s^{-1}, and of their degradation byproducts formed from •OH and O_3 reactions ($UV_{253.7\,nm}$ photolysis was found relevant only for a few compounds in the context of their approach). Mutagenicity was assessed with Ames test (*Salmonella* strains TA98 and TAMix, ±S9), whereas estrogenicity was measured with the recombinant yeast estrogen screen (YES) test. Biological activity decreased upon oxidation of most examined compounds without any form of mutagenicity and/or estrogenicity detected in the treated samples. Increased mutagenic activity was found upon oxidation of the following contaminants listed on CCL3: nitrobenzene (•OH, photolysis), aniline and NDPhA (photolysis), quinoline (•OH, O_3), methamidophos, NPYR, and NDPA (•OH). Increased estrogenic activity was observed only for quinoline (•OH).

Widespread contamination of the aquatic environment is an uncontested reality and the ecotoxicity of micropollutants has been demonstrated. The presence of mutagens in surface waters was also highlighted in numerous publications and reviewed (Ohe *et al.* 2004). The occurrence of regulated and emerging DBPs and their genotoxicity and carcinogenicity were comprehensively reviewed, and gaps and future research needs were identified (Richardson *et al.* 2007, 2008). Undoubtedly, studies on the impact of water treatment technologies on the quality of drinking water are invaluable sources of information, and every possible measure should be taken to safeguard drinking water quality and to protect the environment. However, the interpretation of genotoxicity and mutagenicity data acquired on concentrated water samples for the purpose of risk assessment of real water matrix quality in its integrity and complexity was found to be extremely challenging and is still being explored. Single bioassays could not represent the environmental sample bioactivities in their mixtures, thus, batteries of *in vitro* bioassays with various end points representative to various modes of actions were developed (Escher *et al.* 2008; Escher *et al.* 2012; Jia *et al.* 2015). The first study to integrate quantitative *in vitro* toxicological data (chronic cytotoxicity and acute genotoxicity) in mammalian cells (Chinese hamster ovary cells, CHO cells) with analytical chemistry data and the available human epidemiological records for drinking water DBPs was published by Jeong *et al.* (2012). The study covered eleven drinking water treatment plants across five European countries, and very good correlation was found between the chronic mammalian cell cytotoxicity and the number and concentrations of the identified DBPs. The genotoxic responses did not correlate as well with the analytical data, which was explained by possible unresolved interferences of unidentified DBPs or other toxic water contaminants.

2.11 FUTURE RESEARCH NEEDS

Since 1910, when the first *large* UV unit at that time was installed to disinfect water in a suburb of Rouen, tremendous scientific and technological advancements were made in both disinfection and contaminant treatment with UV light-based processes, through the knowledge acquired on pathogen, chemical micropollutant and water matrix photochemistry and chemistry, UV lamp and power supply development, UV reactor design, engineering and implementation of sophisticated process control algorithms, development of powerful theoretical models to predict the hydrodynamics, radiation distribution and the treatment performance in UV systems, better understanding of UV technology integration in the water treatment trains, and so on.

The implementation of UV/H_2O_2 process at water utilities is expected to increase given its proven efficacy at the removal of a wide range of water pollutants. Research opportunities on this process are also expected to grow and to diversify in various areas, among which on-line sensors for contaminant or contaminant surrogate "fingerprint" monitoring, as a means of process performance control; real-time determination of $^{\bullet}OH$ water background demand and its input into the process control algorithm; understanding the impact of water constituents' photochemistry and chemistry on UV/H_2O_2 process performance; '*in-situ*' or on-site H_2O_2 generation; product studies on micropollutants regulated in drinking water; byproduct formation and economical, reliable, and environmentally friendly mitigation strategies; computational tools for reliable prediction of $k_{^{\bullet}OH}$, degradation mechanisms and byproduct formation; development of sensitive biochemical tools/assays to evaluate water toxicity/mutagenicity/genotoxicity as a whole; synergies between UV/H_2O_2 process and upstream/downstream processes and optimization of integrated processes; life cycle sustainability analysis.

2.12 ACKNOWLEDGMENTS

The author of this Chapter is very grateful to Dr. Silvio Canonica for providing the original drawing for Fig. 2.3, to Dr. Cort Anastasio for providing the original chart for Fig. 2.6, to Dr. Wols and Dr. Hofman-Caris for providing the kinetic model data for Fig. 2.9, to Dr. Changha Lee for providing the original drawing for Fig. 2.13, to Dr. Shane Snyder for providing the file for Fig. 2.15, to Mr. Jens Scheideler for providing the high-resolution image for Fig. 2.19, to Dr. Yunho Lee for providing the original data for Fig. 2.21, and to Ms. Sarah Brown for providing the high-resolution images of Trojan Technologies' UV systems inserted in this chapter.

2.13 REFERENCES

Acero J. L., Stemmler K. and von Gunten U. (2000). Degradation kinetics of atrazine and its degradation products with ozone and OH radicals: a predictive tool for drinking water treatment. *Environmental Science and Technology*, **34**, 591–597.

Acero J. L., Haderlein S. B., Schmidt T. C., Suter M. J. F. and von Gunten U. (2001). MTBE oxidation by conventional ozonation and the combination ozone/hydrogen peroxide: efficiency of the processes and bromate formation. *Environmental Science and Technology*, **35**(21), 4252–4259.

Aflaki R., Hammond S., Tag Oh S., Hokanson D., Trussell S. and Bazzi A. (2015). Scaling-up step-by-step and AOP investment decision. *Proceedings of the Water Environment Federation's Technical Exhibition and Conference (WEFTEC)*, September 26–30, Chicago, IL, USA.

Afzal A., Oppenländer T., Bolton J. R. and Gamal El-Din M. (2010). Anatoxin-a degradation by advanced oxidation processes: vaccum-UV at 172 nm, photolysis using medium pressure UV and UV/H_2O_2. *Water Research*, **44**, 278–286.

Agus E., Lim M. H., Zhang L. and Sedlak D. (2011). Odorous compounds in municipal wastewater effluent and potable water reuse systems. *Environmental Science and Technology*, **45**, 9347–9355.

Ahel M., Scully F. E. Jr., Hoigné J. and Giger W. (1994). Photochemical degradation of nonylphenol and nonlyphenol polyethoxylates in natural waters. *Chemosphere*, **28**, 1361–1368.

Alfassi Z. B. (ed.) (1997). The Chemistry of Free Radicals: Peroxyl Radicals. John Wiley & Sons, New York, NY, USA, pp. 1–18; 19–26; 27–48; 173–234; 407–438; 439–456; 483–506.

Allmand A. J. and Style D. W. G. (1930). The photolysis of aqueous hydrogen peroxide solutions. Part I. Experimental methods. *Journal of Chemical Society*, 596–606; Part II. Experimental results. *ibid.*, 606–626.

Alpert S. M., Knappe D. R. U. and Ducoste J. J. (2010). Modeling the UV/Hydrogen peroxide advanced oxidation process using computational fluid dynamics. *Water Research*, **44**, 1797–1808.

Álvarez P. M., Quinõnes D. H., Terrones I., Rey A. and Beltrán F. J. (2016). Insights into the removal of terbutylazine from aqueous solution by several treatment methods. *Water Research*, **98**, 334–343.

Amine-Khodja A., Boulkamh A. and Boule P. (2004). Photochemical behavior of phenylurea herbicides. *Photochemical and Photobiological Sciences*, **3**, 145–156.

Appleman T. D., Higgins C. P., Quinones O., Vanderford B. J., Kolstad C., Zeigler-Holady J. C. and Dickenson E. R. V. (2014). Treatment of poly- and perfluoroalkyl substances in U.S. full-scale water treatment systems. *Water Research*, **51**, 246–255.

Armbrust K. L. (2000). Pesticide hydroxyl radical rate constants: measurements and estimates of their importance in aquatic environment. *Environmental Technology and Chemistry*, **19**(9), 2175–2180.

Armstrong F. A. J. (1963). Determination of nitrate in water by ultraviolet spectrophotometry. *Analytical Chemistry*, **35**(9), 1292–1294.

Ashton, L., Buxton G. V. and Stuart C. R. (1995). Temperature dependence of the rate of reaction of OH with some aromatic compounds in aqueous solution. *Journal of Chemical Society Faraday Transactions*, **91**(11), 1631–1633.

Asmus K.-D., Cercek B., Ebert M., Henglein A. and Wigger A. (1967). Pulse radiolysis of nitrobenzene solutions. *Transactions of Faraday Society*, **63**, 2435–2441.

Asmus K.-D., Möckel H. and Henglein A. (1973). Pulse radiolytic study of the site of OH radical attack on aliphatic alcohols in aqueous solution. *Journal of Physical Chemistry*, **77**(10), 1218–1221.

Atkinson R. (1987). A structure-activity relationship for the estimation of rate constants for the gas-phase reactions of OH radicals with organic compounds. *International Journal of Chemical Kinetics*, **19**, 799–828.

Autin O. (2012). Micropollutant Removal by Advanced Oxidation Processes. PhD thesis, Cranfield University, Cranfield, UK.

Autin O., Hart J., Jarvis P., MacAdam J., Parsons S. A. and Jefferson B. (2012). Comparison of UV/H_2O_2 and UV/TiO_2 for the degradation of metaldehyde: kinetics and the impact of background organics. *Water Research*, **46**(17), 5655–5662.

Autin O., Romelot C., Rust L., Hart J., Jarvis P., MacAdam J., Parsons S. A. and Jefferson B. (2013). Evaluation of a UV-light emitting diodes unit for the removal of micropollutants in water for low energy advanced oxidation processes. *Chemosphere*, **92**, 745–751.

AwwaRF (2007). Removal of EDCs and Pharmaceuticals in Drinking and Reuse Treatment Processes. Project #2758. Awwa Research Foundation, Denver, CO, USA.

Ayoub K., van Hullebusch E. D., Cassir M. and Bermond A. (2010). Application of advanced oxidation processes for TNT removal: a review. *Journal of Hazardous Materials*, **178**, 10–28.

Baeza C. and Knappe D. R. (2011). Transformation kinetics of biochemically active compounds in low-pressure UV photolysis and UV/H_2O_2 advanced oxidation processes. *Water Research*, **45**, 4531–4542.

Bahnmüller S., Loi C. H., Linge K. L., von Gunten U. and Canonica S. (2015). Degradation rates of benzotriazoles and benzothiazoles under UV-C irradiation and the advanced oxidation process UV/H_2O_2. *Water Research*, **74**, 143–154.

Balci B., Oturan M. A., Oturan N. and Sirés I. (2009). Decontamination of aqueous glyphosate, (aminomethyl) phosphonic acid, and glufosinate solutions by electro-Fenton-like process with Mn^{2+} as the catalyst. *Journal of Agricultural and Food Chemistry*, **57**, 4888–4894.

Baranda A. B., Fundazuri O. and Martínez de la Maranón I. (2014). Photodegradation of several triazinic and organophosphoric pesticides in water by pulsed light technology. *Journal of Photochemistry and Photobiology A: Chemistry*, **286**, 29–39.

Barbosa M. O., Moreira N. F. F., Ribeiro A. R., Pereira M. F. R. and Silva A. M. T. (2016). Occurrence and removal of organic micropollutants: an overview of watch list of EU Decision 2015/495. *Water Research*, **94**, 257.

Baxendale J. H. and Wilson J. A. (1957). The photolysis of hydrogen peroxide at high light intensities. *Transactions of the Faraday Society*, **53**, 344–356.

Bazri M. M., Barbeau B. and Mohseni M. (2012). Impact of UV/H_2O_2 advanced oxidation treatment on molecular weight distribution of NOM and biostability of water. *Water Research*, **46**, 5297–5304.

Beem Benedict K., McFall A. S. and Anastasio C. (2017). Quantum yield of nitrite from aqueous nitrate above 300 nm. *Environmental Science and Technology*, **51**(8), 4387–4395.

Beer A. (1852). Bestimmung der Absorption des rothen Lichts in farbigen Flüssigkeiten (Determination of the absorption of red light in colored liquids). *Annalen der Physik und Chemie*, **86**, 78–88.

Beltran F. J., Gonzales M., Rivas J. and Marin M. (1994). Oxidation of mecoprop in water with ozone and ozone combined with hydrogen peroxide. *Industrial and Engineering Chemistry Research*, **33**, 125–136.

Beltran F. J., Gonzales M., Rivas J. F. and Alvarez P. (1996). Aqueous UV radiation and UV/H_2O_2 oxidation of atrazine first degradation products: deethylatrazine and deisopropylatrazine. *Environmental and Toxicological Chemistry*, **15**(6), 868–872.

Beltran-Heredia J., Benitez F. J., Gonzalez T., Acero J. L. and Rodriguez B. (1996). Photolytic decomposition of bentazone. *Journal of Chemical Technology and Biotechnology*, **66**, 206–212.

Benitez F. J., Acero J. A., Real F. J. and Roman S. (2004). Oxidation of MCPA and 2,4-D by UV radiation, ozone, and the combinations UV/H_2O_2 and O_3/H_2O_2. *Journal of Environmental Science and Health. Part B-Pesticides, Food Contaminants, and Agricultural Wastes*, **B39**, 393–409.

Benitez F. J., Real F. J., Acero J. L. and Garcia C. (2006). Photochemical oxidation processes for the elimination of phenyl-urea herbicides in waters. *Journal of Hazardous Materials*, **B138**, 278–287.

Benitez F. J., Acero J. L., Real F. J., Roldan G. and Rodriguez E. (2013a). Photolysis of model emerging contaminants in ultra-pure water: kinetics, by-product formation and degradation pathways. *Water Research*, **47**, 870.

Benitez F. J., Acero J. L., Real F. J. and Roldan G. (2013b). Modeling the photodegradation of emerging contaminants in waters by UV radiation and UV/H_2O_2 system. *Journal of Environmental Science and Health. Part A*, **48**, 120–128.

Bielski B. H. J., Cabelli D. E., Arudi R. L. and Ross A. B. (1985). Reactivity of HO_2/O_2^- radicals in aqueous solution. *Journal of Physical Chemistry Reference Data*, **14**(4), 1041–1100.

Bircher K. G. (2015). Calculating UV dose for UV/AOP reactors using dose/log as a water quality metric. *IUVA News*, **17**(2), 19–24.

Bohrerova Z., Bohrer G., Mohanraj S. M., Ducoste J. and Linden K. G. (2005). Experimental measurements of fluence rate distribution in a UV reactor using fluorescent microspheres. *Environmental Science and Technology*, **39**, 8925–8930.

Bolton J. R. and Linden K. G. (2003). Standardization of methods for fluence (UV dose) determination in bench-scale UV experiments. *Journal of Environmental Engineering*, **129**(3), 209–215.

Bolton J. R. and Stefan M. I. (2002). Fundamental photochemical approach to the concepts of fluence (UV dose) and electrical energy efficiency in photochemical degradation reactions. *Research on Chemical Intermediates*, **28**(7–9), 857–870.

Bolton J. R., Bircher K. G., Tumas W. and Tolman C. A. (1996). Figures-of-merit for the technical development and application of advanced oxidation processes. *Journal of Advanced Oxidation Processes*, **1**(1), 13–17.

Bolton J. R., Bircher K. G., Tumas W. and Tolman C. A. (2001). Figures-of-merit for the technical development and application of advanced oxidation technologies for both electric- and solar-driven systems. *Pure and Applied Chemistry*, **73**(4), 627–637.

Bolton J. R., Mayor-Smith J. and Linden K. G. (2015). Rethinking the concepts of fluence (UV dose) and fluence rate: the importance of photon-based units – a systemic review. *Photochemistry and Photobiology*, **91**(6), 1252–1262.

Bordeleau G., Martel R., Ampleman G. and Thiboutot S. (2013). Photolysis of RDX and nitroglycerin in the context of military training ranges. *Chemosphere*, **93**, 14–19.

Boreen A. L., Arnold W. A. and McNeill K. (2003). Photodegradation of pharmaceuticals in the aquatic environment. A review. *Aquatic Sciences*, **65**, 320–421.

Boreen A. L., Arnold W. A. and McNeill K. (2004). Photochemical fate of sulfa drugs in the aquatic environment: sulfa drugs containing five-membered heterocyclic groups. *Environmental Science and Technology*, **38**(14), 3933–3940.

Borhani T. N. G., Saniedanesh M., Bagheri M. and Lim J. S. (2016). QSPR prediction of the hydroxyl radical rate constants of water contaminants. *Water Research*, **98**, 344–353.

Bose P., Glaze W. H. and Maddox D. S. (1998). Degradation of RDX by various advanced oxidation processes: I. Reaction rates. *Water Research*, **32**(4), 997–1004.

Braslavsky S. E. (2007). International union of pure and applied chemistry (IUPAC): glossary of terms used in photochemistry. *Pure and Applied Chemistry*, **79**(3), 293–465.

Braun A. M., Maurette M.-T. and Oliveros E. (1991). Photochemical Technology. Wiley, Chichester, UK.

Brezonik P. L. and Fulkerson-Brekken J. (1998). Nitrate-induced photolysis in natural waters: controls on concentrations of hydroxyl radical photo-intermediates by natural scavenging agents. *Environmental Science and Technology*, **32**(19), 3004–3010.

Bui X. T., Vo T. P. T., Ngo H. H., Guo W. S. and Nguyen T. T. (2016). Multicriteria assessment of advanced treatment technologies for micropollutant removal at large-scale applications. *Science of the Total Environment*, 563–564, 1050–1067.

Burrows W. D. and Brueggemann E. E. (1986). Tertiary treatment of effluent from Holston AAP industrial liquid waste treatment facility. V. Degradation of nitramines in Holston AAP wastewaters by ultraviolet radiation. Technical Report 8602, ADA176195; US Army Toxic and Hazardous Materials Agency, Aberdeen Proving Gound, MD.

Burrows H. D., Canle L. M., Santaballa J. A. and Steenken S. (2002). Reaction pathways and mechanisms for photodegradation of pesticides. *Journal of Photochemistry and Photobiology B: Biology*, **67**, 71–108.

Buxton G. V., Greenstock C. L., Helman W. P. and Ross A. B. (1988). Critical review of rate constants for reactions of hydrated electrons, hydrogen atoms and hydroxyl radicals ($^{\bullet}HO/^{\bullet}O^-$) in aqueous solution. *Journal of Physical and Chemical Reference Data*, **17**(2), 513–886.

Calvert J. G. and Pitts J. N. Jr. (1966). Photochemistry. Wiley, New York.

Canonica S. and Tratnyek P. G. (2003). Quantitative structure-activity relationships for oxidation reactions of organic chemicals in water. *Environmental Toxicology and Chemistry*, **22**(8), 1743–1754.

Canonica S., Meunier L. and von Gunten U. (2008). Phototransformation of selected pharmaceuticals during UV treatment of drinking water. *Water Research*, **42**, 121–128.

Carlson J. C., Stefan M. I., Parnis M. J. and Metcalfe, C. D. (2015). Direct UV photolysis of selected pharmaceuticals, personal care products, and endocrine disruptors in aqueous solution. *Water Research*, **84**, 350–361.

Cater S. R., Stefan M. I., Bolton J. R. and Safarzadeh-Amiri A. (2000). UV/H_2O_2 treatment of methyl *tert*-butyl ether in contaminated waters. *Environmental Science and Technology*, **34**, 659–662.

Chang Y. (2010). UV/Peroxide oxidation process for taste & odor control. *IUVA News*, **12**(2), 14–16.

Chang P. B. and Young T. M. (2000). Kinetics of methyl *tert*-butyl ether degradation and by-product formation during UV/hydrogen peroxide water treatment. *Water Research*, **34**, 2233–2240.

Chen L., Zhou H. J. and Deng Q. Y. (2007). Photolysis of nonylphenol ethoxylates: the determination of the degradation kinetics and intermediate products. *Chemosphere*, **68**, 354–359.

Chen W. R., Wu C., Elovitz M. S., Linden K. G. and Suffet I. H. (2008). Reactions of thiocarbamate, triazine and urea herbicides, RDX and benzenes on EPA Contaminant Candidate List with ozone and with hydroxyl radicals. *Water Research*, **42**, 137–144.

Chin M. and Wine P. H. (1994). A temperature-dependent competitive kinetics study of the aqueous phase reactions of OH radicals with formate, formic acid, acetate, acetic acid, and hydrated formaldehyde. In: Aquatic and Surface Photochemistry, G. R. Helz, R. G. Zepp and D. G. Crosby (eds), Lewis Publishers, CRC Press, Boca Raton, FL, U.S.A.

Chitra S., Paramasivan K., Cheralathan M. and Sinha P. K. (2012). Degradation of 1,4-dioxane using advanced oxidation processes. *Environmental Science & Pollution Research*, **19**, 871–878.

Christensen H., Sehested K. and Corftizen H. (1982). Reactions of hydroxyl radicals with hydrogen peroxide at ambient and elevated temperatures. *Journal of Physical Chemistry*, **86**(9), 1588–1590.

Christensen A., Gurol M. D. and Garoma T. (2009). Treatment of persistent organic compounds by integrated advanced oxidation processes and sequential batch reactor. *Water Research*, **43**, 3910–3921.

Chu L. and Anastasio C. (2005). Formation of hydroxyl radical from the photolysis of frozen hydrogen peroxide. *Journal of Physical Chemistry A*, **109**(28), 6264–6271.

Chu W., Gao N., Yin D., Krasner S. W. and Mitch W. A. (2014). Impact of UV/H_2O_2 pre-oxidation on the formation of haloacetamides and other nitrogenous disinfection byproducts during chlorination. *Environmental Science and Technology*, **48**(20), 12190–12198.

Chu X., Xiao Y., Quek E., Xie R., Pang T. and Xing Y. (2016). Pilot-scale UV/H_2O_2 study for emerging contaminants decomposition. *Reviews on Environmental Health*, **31**(1), 71–74.

Chuang Y.-H., Parker K. M. and Mitch W. A. (2016). Development of predictive models for the degradation of halogenated disinfection byproducts during the UV/H_2O_2 advanced oxidation process. *Environmental Science and Technology*, **50**, 11209–11217.

Civardi J. and Lucca M. A. (2012). Aqua PA's experience with UV hydrogen peroxide for reduction of taste and odors at two water treatment plants. *Proceedings of the IUVA Conference "Moving Forward: Sustainable Solutions to Meet Evolving Regulatory Standards"*, Washington DC, August 12–14, 2012.

Civardi J., Prosser A., Gray J. M., Marie J. and Greising K. (2014). Treatment of 1,4-dioxane using advanced oxidation at Artesian Water Company's Llangollen wellfield. *IUVA Americas Regional Conference*, White Plains, NY, USA, October 26–29.

Clément J. L. and Tordo P. (2007). Advances in spin trapping. *Electron Paramagnetic Resonance*, **20**, 29–49.

Coenen T., Van de Moortel W., Logist F., Luyten J., Van Impe J. F. M. and Degrève J. (2013). Modeling and geormetry optimization of photochemical reactors: single- and multi-lamp reactors for UV-H_2O_2 AOP systems. *Chemical Engineering Science*, **96**, 174–189.

Coleman H. M., Vimonses V., Leslie G. and Amal R. (2007). Degradation of 1,4-dioxane in water using TiO_2 based photocatalytic and H_2O_2/UV processes. *Journal of Hazardous Materials*, **146**, 496–501.

Connick R. E. (1947). The interaction of hydrogen peroxide and hypochlorous acid in acidic solutions containing chloride ion. *Journal of the American Chemical Society*, **69**, 1509–1514.

Cooper W. J., Cramer C. J., Martin N. H., Mezyk S. P., O'Shea K. E. and von Sonntag C. (2009). Free radical mechanisms for the treatment of methyl-*tert*-butyl ether (MTBE) *via* advanced oxidation/reductive processes in aqueous solutions. *Chemical Reviews*, **109**, 1302–1345.

Crittenden J. C., Hu S., Hand D. W. and Green S. A. (1999). A kinetic model for H_2O_2/UV process in a completely mixed batch reactor. *Water Research*, **33**(10), 2315–2328.

Crowell R. A., Lian R., Sauer M. C. Jr., Oulianov D. A. and Shkrob I. A. (2004). Geminate recombination of hydroxyl radicals generated in 200 nm photodissociation of aqueous hydrogen peroxide. *Chemical Physics Letters*, **383**, 481–485.

Czaplicka M. (2006). Photo-degradation of chlorophenols in the aqueous solution (a review). *Journal of Hazardous Materials B*, **134**, 45–59.

Daifullah A. H. and Mohamed M. M. (2004). Degradation of benzene, toluene, ethylbenzene and *p*-xylene (BTEX) in aqueous solutions using UV/H_2O_2 system. *Journal of Chemical Technology and Biotechnology*, **79**, 468–474.

Das S. and von Sonntag, C. (1986). Oxidation of trimethylamine by OH radicals in aqueous solution as studied by pulse radiolysis, ESR and product analysis. The reactions of the alkylamine radical cation, the aminoalkyl radical and the protonated aminoalkyl radical. *Zeitschrift für Naturforschung*, **41b**, 505–513.

De Freitas A. M., Sirtori C., Lenz C. A. and Zamora P. G. P. (2013). Microcystin-LR degradation by solar photo-Fenton and UV-C/H_2O_2: a comparative study. *Photochemical and Photobiological Sciences*, **12**, 696–702.

De La Cruz A. A., Hiskia A., Kaloudis T., Chernoff N., Hill D., Antoniou M. G., He X., Loftin K., O'Shea K., Zhao C., Pelaez M., Han C, Lynch T. J. and Dionysiou D. D. (2013). A review on cylindrospermopsin: the global occurrence, detection, toxicity and degradation of a potent cyanotoxin. *Environmental Sciences Processes & Impacts*, **15**, 1979–2003.

De Laat J., Doré M. and Suty H. (1995). Oxydation de s-triazines par les procédés d'oxydation radicalaire. Sous-produits de réaction et constants cinétiques de réaction [Oxidation of s-triazines by advanced oxidation processes. By-products and kinetic rate constants]. *Revue des Sciences de l'Eau*, **8**, 23–42.

De Laat J., Maouala-Makata P. and Doré M. (1996). Rate constants for reactions of ozone and hydroxyl radicals with several phenyl-ureas and acetamides. *Environmental Technology*, **17**(7), 707–716.

Dong M., Mezyk S. P. and Rosario-Ortiz F. L. (2010). Reactivity of effluent organic matter (EfOM) with hydroxyl radical as a function of molecular weight. *Environmental Science and Technology*, **44**(15), 5714–5720.

Donham J. E., Rosenfeldt E. J. and Wigginton K. R. (2014). Photometric hydroxyl radical scavenging analysis of standard natural organic matter isolates. *Environmental Sciences Processes & Impacts*, **16**, 764.

Dorfman L. M. and Adams G. E. (1973). Reactivity of the hydroxyl radical in aqueous solutions. *National Standards and Reference Data Series*, **46**, 1–59. U.S. National Bureau of Standards, Washington DC, USA.

Dotson A. D., Keen V. O. S. and Linden K. G. (2010). UV/H_2O_2 treatment of drinking water increases post-chlorination DBP formation. *Water Research*, **44**(12), 3703–3713.

Doussin J. F. and Monod A. (2013). Structure-activity relationship for the estimation of OH-oxidation rate constants of carbonyl compounds in the aqueous phase. *Atmospheric Chemistry and Physics*, **13**, 11625–11641.

Draper W. M. (1987). Measurement of quantum yields in polychromatic light: dinitroaniline herbicides. *ACS Symposium Series*, **327**, 268–281.

Dulin D., Drossman H. and Mill T. (1986). Products and quantum yields for photolysis of chloroaromatics in water. *Environmental Science and Technology*, **20**, 72–77.

Elovitz M. and von Gunten U. (1999). Hydroxyl radical ozone ratios during ozonation processes. I – The R-ct concept. *Ozone: Science & Engineering*, **21**, 239–260.

Elovitz M. S., Shemer H., Peller J. R., Vinodgopal K., Sivaganesan M. and Linden K. G. (2008). Hydroxyl radical rate constants: comparing UV/H_2O_2 and pulse radiolysis for environmental pollutants. *Journal of Water Supply: Research and Technology-AQUA*, **57**(6), 391–401.

Elyasi S. and Taghipour F. (2010). Simulation of UV photoreactor for degradation of chemical contaminants: model development and evaluation. *Environmental Science and Technology*, **44**, 2056–2063.

Escher B. I. and Leusch F. (2012). Bioanaytical Tools in Water Quality Assessment. IWA Publishing, London, UK.

Escher B. I., Bramaz N., Mueller J. F., Quayle P., Rutishauser S. and Vermeirssen E. L. M. (2008). Toxic equivalent concentrations (TEQs) for baseline tocxicity and specific modes of action as a tool to improve interpretation of ecotoxicity testing in environmental samples. *Journal of Environmental Monitoring*, **10**, 612–621.

Evgenidou E. and Fytianos K. (2002). Photodegradation of triazine herbicides in aqueous solutions and natural waters. *Journal of Agricultural and Food Chemistry*, **50**, 6423–6427.

Fang X., Schuchmann H.-P. and von Sonntag, C. (2000). The reaction of the OH radical with pentafluoro-, pentachloro- and 2,4,6-triiodophenol in water: electron transfer vs. addition to the ring. *Journal of Chemical Society, Perkin Transactions*, **2**, 1391–1398.

Fang J. Y., Ling L. and Shang C. (2013). Kinetics and mechanisms of pH-dependent degradation of halonitromethanes by UV photolysis. *Water Research*, **47**, 1257–1266.

Farhataziz P. C. and Ross A. B. (1977). Selected Specific Rates of Reactions of Transients from Water in Aqueous Solution. III. Hydroxyl Radical and Perhydroxyl Radical and their Radical Ions. *National Standards and Reference Data Series*, 59. U.S. National Bureau of Standards, Washington DC, USA.

Fatta-Kassinos D., Vasquez M. I. and Kümmerer K. (2011). Transformation products of pharmaceuticals in surface waters and wastewater formed during photolysis and advanced oxidation processes – degradation, elucidation of byproducts and assessment of their biological potency. *Chemosphere*, **85**, 693–709.

Fdil F., Aaron J.-J., Oturan N., Chaouch A. and Oturan M. A. (2003). Photochemical degradation of chlorophenoxyalcanoic herbicides in aqueous media. *Revue des Sciences de l'Eau*, **16**, 123–142.

Fenton H. J. H. (1894). LXXIII-Oxidation of tartaric acid in presence of iron. *Journal of Chemical Society Transaction (London)*, **65**, 899–910.

Fernández-Castro P., Vallejo M., Fresnedo San Román M. and Ortiz I. (2015). Insight on the fundamentals of advanced oxidation processes. Role and review of the determination methods of reactive oxygen species. *Journal of Chemical Technology and Biotechnology*, **90**, 796–820.

Fricke H. and Thomas J. K. (1964). Basic mechanisms in the radiation chemistry of aqueous media. *Radiation Research Supplement*, **4**, 35–53.

Gandhi V. N., Roberts P. J. W. and Kim J.-H. (2012). Visualizing and quantifying dose distribution in a UV reactor using three-dimensional laser-induced fluorescence. *Environmental Science and Technology*, **46**, 13220–13226.

Gao Y., Ji Y., Li G. and An T. (2014). Mechanism, kinetics and toxicity assessment of OH-initiated transformation of triclosan in aquatic environment. *Water Research*, **49**, 360–370.

Gao Y., Ji Y., Li G. and An T. (2016). Theoretical investigation on the kinetics and mechanisms of hydroxyl radical-induced transformation of parabens and its consequences for toxicity: influence of alkyl-chain length. *Water Research*, **91**, 77–85.

Gares K. L., Bykov S. V., Godugu B. and Asher S. A. (2014). Solution and solid trinitrotoluene (TNT) photochemistry: persistence of TNT-like ultraviolet (UV) resonance Raman bands. *Applied Spectroscopy*, **68**, 49–56.

Gares K. L., Bykov S. V., Brinzer T. and Asher S. A. (2015). Solution and solid hexahydro-1,3,5-trinitro-1,3,5-triazine (RDX) ultraviolet (UV) 229 nm photochemistry. *Applied Spectroscopy*, **69**, 545–554.

Gerecke A. C., Canonica S., Müller S. R., Schärer M. and Schwarzenbach R. P. (2001). Quantification of dissolved natural organic matter (DOM) mediated phototransformation of phenylurea herbicides in lakes. *Environmental Science and Technology*, **35**(19), 3915–3923.

Gerrity D., Lee Y., Gamage S., Lee M., Pisarenko A. N., Trenholm R. A., von Gunten U. and Snyder S. A. (2016). Emerging investigators series: prediction of trace organic contaminant abatement with UV/H_2O_2: development and validation of semi-empirical models for municipal wastewater effluents. *Environmental Science Water Research and Technology*, **2**, 460–473.

Ghafoori S., Mehvar M. and Chan P. K. (2014). Photoreactor scale-up for degradation of aqueous poly(vinyl alcohol) using UV/H_2O_2 process. *Chemical Engineering Journal*, **245**, 133–142.

Glaze W. H., Kang J. W. and Chapin D. H. (1987). The chemistry of water treatment processes involving ozone, hydrogen peroxide and UV-radiation. *Ozone: Science & Engineering*, **9**, 335–352.

Glaze W. H., Schep R., Chauncey W., Ruth E. C., Zarnoch J. J., Aieta E. M., Tate C. H. and McGuire M. J. (1990). Evaluating oxidants for the removal of model taste and odor compounds from a municipal water supply. *Journal of the American Water Works Association*, **82**(5), 79–84.

Glaze W. H., Lay Y. and Kang J.-W. (1995). Advanced oxidation processes. A kinetic model for the oxidation of 1,2-dibromo-3-chloropropane in water by the combination of hydrogen peroxide and UV radiation. *Industrial and Engineering Chemistry Research*, **34**, 2314–2323.

Gligorovski S., Rousse D., George C. H. and Herrmann H. (2009). Rate constants for the OH reactions with oxygenated organic compounds in aqueous solution. *International Journal of Chemical Kinetics*, **41**(5), 309–326.

Gligorovski S., Strekowski R., Barbati S. and Vione D. (2015). Environmental implications of hydroxyl radicals (•OH). *Chemical Reviews*, **115**, 13051–13092.

Gmurek M., Rossi A. F., Martins R. C., Quinta-Ferreira R. M. and Ledakowicz S. (2015). Photodegradation of single and mixture of parabens – Kinetic, by-product identification and cost-efficiency analysis. *Chemical Engineering Journal*, **276**, 303–314.

Gmurek M., Olak-Kucharczyk M. and Ledakowicz S. (2017). Photochemical decomposition of endocrine disrupting compounds – a review. *Chemical Engineering Journal*, **310**(2), 437–456.

Goldstein S. and Rabani J. (2007). Mechanism of nitrite formation by nitrate photolysis in aqueous solutions: the role of peroxynitrite, nitrogen dioxide, and hydroxyl radical. *Journal of the American Chemical Society*, **129**, 10597–10601.

Goldstein S., Rosen G. M., Russo A. and Samuni A. (2004). Kinetics of spin trapping superoxide, hydroxyl and aliphatic radicals by cyclic nitrones. *Journal of Physical Chemistry A*, **108**(32), 6679–6685.

Goldstone J. V., Pullin M. J., Bertilsson S. and Voelker B. M. (2002). Reactions of hydroxyl radical with humic substances: bleaching, mineralization, and production of bioavailable carbon substrates. *Environmental Science and Technology*, **36**(3), 364–372.

Goméz-Lopéz V. M. and Bolton J. R. (2016). An approach to standardize methods for fluence determination in bench-scale pulsed light experiments. *Food and Bioprocess Technology*, **9**, 1040–1048.

Gould I. R. (1989). Conventional light sources. Chapter 5 In: Handbook of Organic Photochemistry, J. C. Scaiano (ed.), CRC Press, Boca Raton, FL, Vol. 1.

Guittonneau S., De Laat J., Doré M., Duguet J. P. and Bonnel C. (1988). Comparative study of the photodegradation of aromatic compounds in water by UV and H_2O_2/UV. *Environmental Technology Letters*, **9**, 1115.

Guo X., Minakata D., Niu J. and Crittenden J. C. (2014). Computer-based first-principles kinetic modeling of degradation pathways and byproduct fates in aqueous-phase advanced oxidation processes. *Environmental Science and Technology*, **48**, 5718–5725.

Guo X., Minakata D. and Crittenden J. C. (2015). On-the-fly kinetic Monte Carlo simulation of aqueous phase advanced oxidation processes. *Environmental Science and Technology*, **49**, 9230–9236.

Guo Z. B., Lin Y. L., Xu B., Hu C. Y., Huang H., Zhang T. Y., Chu W. H. and Gao N. Y. (2016). Factors affecting THM, HAN and HNM formation during UV-chlor(am)ination of drinking water. *Chemical Engineering Journal*, **306**, 1180–1188.

Haag W. R. (1994). Photooxidation of organic compounds in water and air using low-wavelength pulsed xenon lamps. In: Aquatic and Surface Photochemistry, G. R. Helz, R. G. Zepp and D. G. Crosby (eds), Lewis Publishers, Boca Raton, FL, pp. 517–530.

Haag W. R. and Yao C. C. D. (1992). Rate constants for reaction of hydroxyl radicals with several drinking water contaminants. *Environmental Science and Technology*, **26**, 1005–1013.

Haber F. and Weiss J. (1932). Über the Einwirkung von Hydroperoxyd auf Ferrosalz. *Naturwissenschaft*, **51**, 948–950.

Haber F. and Willstätter R. (1931). Unpaarigheheit und Radikalketten im Reaktion-Mechanismus organischer unde enzymatischer Vorgänge. *Chemische Berichte*, **64**, 2844 2856.

Hatada M., Kraljic I., El Samahy A. and Trumbore C. N. (1974). Radiolysis and photolysis of the hydrogen peroxide -*p*-nitrosodimethylaniline -oxygen system. *Journal of Physical Chemistry*, **78**(9), 888–891.

He X., Pelaez M., Westrick J. A., O'Shea K. E., Hiskia A., Triantis T., Kaloudis T., Stefan M. I., de la Cruz A. A. and Dionysiou D. D. (2012). Efficient removal of microcystin-LR by UV-C/H_2O_2 in synthetic and natural water samples. *Water Research*, **46**, 1501–1510.

He X., de la Cruz A. A. and Dionysiou D. D. (2013). Destruction of cyanobacterial cylindrospermopsin by hydroxyl radicals and sulfate radicals using UV-254 nm activation of hydrogen peroxide, persulfate and peroxomonosulfate. *Journal of Photochemistry and Photobiology A: Chemistry*, **251**, 160–166.

He X., Zhang G., de la Criz A. A., O'Shea K. E. and Dionysiou D. D. (2014). Degradation mechanism of cyanobacterial toxin cylindrospermopsin by hydroxyl radicals in homogeneous UV/H_2O_2 process. *Environmental Science and Technology*, **48**, 4495–4504.

He X., de la Cruz A. A., Hiskia A., Kaloudis T., O'Shea K. E. and Dionysiou D. D. (2015). Destruction of microcystins (cyanotoxins) by UV-254nm-based direct photolysis and advanced oxidation processes (AOPs): influence of variable aminoacids on the degradation kinetics and reaction mechanisms. *Water Research*, **74**, 227–238.

Heringa M. B., Harmsen D. J. H., Beerendonk E. F., Reus A. A., Krul C. A. M., Metz D. H. and Ijpelaar G. F. (2011). Formation and removal of genotoxic activity during UV/H_2O_2-GAC treatment of drinking water. *Water Research*, **45**, 366–374.

Herrmann H. (2003). Kinetics of aqueous phase reactions relevant for atmospheric chemistry. *Chemical Reviews*, **103**, 4691–4716.

Herrmann H., Hoffmann D., Schaefer T., Bräuer P. and Tilgner A. (2010). Tropospheric aqueous-phase free-radical chemistry: radical sources, spectra, reaction kinetics and prediction tools. *ChemPhysChem*, **11**, 3796–3822.

Hessler D. P., Gorenflo V. and Frimmel F. H. (1993). Degradation of aqueous atrazine and metazachlor solutions by UV/H_2O_2 – Influence of pH and herbicide concentration. *Acta Hydrochimica and Hydrobiologica*, **21**, 209–214.

Hirvonen A., Tuhkanen T. and Kalliokoski P. (1996). Removal of chlorinated ethylenes in contaminated ground water by hydrogen peroxide mediated oxidation processes. *Environmental Technology*, **17**, 263–272.

Hirvonen A., Tuhkanen T., Ettala M., Korhonen S. and Kalliokoski P. (1998). Evaluation of a field-scale UV/H_2O_2-oxidation system for the purification of groundwater contaminated with PCE. *Environmental Technology*, **19**, 821–828.

Hirvonen A., Trapido M., Hentunen J. and Tarhanen J. (2000). Formation of hydroxylated and dimeric intermediates during oxidation of chlorinated phenols in aqueous solution. *Chemosphere*, **41**, 1211–1218.

Hitzfeld C. C., Höger S. J. and Dietrich D. R. (2000). Cyanobacterial toxins: removal during drinking water treatment and human risk assessment. *Environmental Health Perspectives*, **108**(Suppl. 1), 113–122.

Ho L., Tanis-Plant P., Kayal N., Slyman N. and Newcombe G. (2009). Optimising water treatment practices for the removal of *Anabaena circinalis* and its associated metabolites, geosmin and saxitoxins. *Journal of Water and Health*, **7**, 544–556.

Hofman-Caris C. H. M., Harmsen D. J. H., Beerendonk E. F., Knol T. H., Houtman C. J., Metz D. H. and Wols B. A. (2012). Prediction of advanced oxidation performance in various pilot UV/H_2O_2 reactor systems with MP- and LP- and DBD-UV lamps. *Chemical Engineering Journal*, **210**, 520–528.

Hofman-Caris C. H. M., Harmsen D. J. H., Wols B. A., Beerendonk E. F. and Keltjens L. L. M. (2015a). Determination of reaction rate constants in a collimated beam setup: the effect of water quality and water depth. *Ozone: Science & Engineering*, **37**(2), 134–142.

Hofman-Caris R. C. H. M., Harmsen D. J. H., Puijker L., Baken K. A., Wols B. A., Beerendonk E. F. and Keltjens L. L. M. (2015b). Influence of process conditions and water quality on the formation of mutagenic byproducts in UV/H_2O_2 process. *Water Research*, **74**, 191–202.

Holden B. and Richardson A. (2009). Reduction of trietazine from groundwater by UV. *Proceedings of the 5th International Congress on Ultraviolet Technologies*, Amsterdam, The Netherlands, September 21–23.

Homem V. and Santos L. (2011). Degradation and removal methods of antibiotics from aqueous matrices – A review. *Journal of Environmental Management*, **92**, 2304–2347.

Howard J. A. and Scaiano J. C. (1984). Radical reaction rates in liquids: Oxyl, peroxyl and related radicals. In: Landolt-Börnstein: Numerical Data and Functional Relationships in Science and Technology. Molecules and Radicals, H. Fisher (ed.), Springer-Verlag, Berlin-Heidelberg, Germany, Vol. 13D.

Hunt J. P. and Taube H. (1952). The photochemical decomposition of hydrogen peroxide. Quantum yields, tracer and fractionation effects. *Journal of the American Chemical Society*, **74**, 5999–6002.

IJpelaar G. F., Harmsen D. J. H., Beerendonk E. F., van Leerdam R. C., Metz D. H., Knol A. H., Fulmer A. and Krijnen S. (2010). Comparison of low pressure and medium pressure UV lamps for UV/H_2O_2 treatment of natural waters containing micropollutants. *Ozone: Science & Engineering*, **32**, 329–337.

Ikehata K. and Gamal El-Din M. (2004). Degradation of recalcitrant surfactants in wastewater by ozonation and advanced oxidation processes. A review. *Ozone: Science & Engineering*, **26**, 327–343.

Ikehata K. and Gamal El-Din M. (2006). Aqueous pesticide degradation by hydrogen peroxide/ultraviolet irradiation and Fenton-type advanced oxidation processes: a review. *Journal of Environmental Engineering and Science*, **5**, 81–135.

Ikehata K., Naghashkar N. J. and Gamal El-Din M. (2006). Degradation of aqueous pharmaceuticals by ozonation and advanced oxidation processes. A review. *Ozone: Science & Engineering*, **28**, 353–414.

Ikehata K., Wang-Stanley L., Qu X. and Li Y. (2016). Treatment of groundwater contaminated with 1,4-dioxane, tetrahydrofuran, and chlorinated volatile organic compounds using advanced oxidation processes. *Ozone: Science & Engineering*, **38**, 413–424.

Jeong J., Jung J., Cooper W. J. and Song W. (2010). Degradation mechanisms and kinetic studies for the treatment of X-ray contrast media compounds by advanced oxidation/reduction processes. *Water Research*, **44**, 4391–4398.

Jeong C. H., Wagner E. D., Siebert V. R., Anduri S., Richardson S. D., Daiber E. J., McKague A. B., Kogevinas M., Villaneuva C. M., Goslan E. H., Luo W., Lorne M. I., Pankow J. F., Grazuleviciene R., Cordier S., Edwards S. C., Righi E., Nieuwenhuijsen M. J. and Plewa M. J. (2012). Occurrence and toxicity of disinfection byproducts in European drinking waters in relation with the HIWATE epidemiology study. *Environmental Science and Technology*, **46**, 12120–12128.

Jia A., Escher B. I., Leusch F. D. L., Tang J. Y. M., Prochazka E., Dong B., Snyder E. M. and Snyder S. A. (2015). *In vitro* bioassays to evaluate complex chemical mixtures in recycled water. *Water Research*, **80**, 1–11.

Jin X., Peldszus S. and Huck P. M. (2012). Reaction kinetics of selected micropollutants in ozonation and advanced oxidation processes. *Water Research*, **46**, 6519–6530.

Jo C. H., Dietrich A. M. and Tanko J. M. (2011). Simultaneous degradation of disinfection byproducts and earthy-musty odorants by the UV/H_2O_2 advanced oxidation process. *Water Research*, **45**, 2507–2516.

Johns H. E. (1971). Quantum yields and kinetics of photochemical reactions in solution. Chapter 3 In: Creation and Detection of the Excited States, A. A. Lamola (ed.), Marcel Dekker, New York, Vol. 1 Part A, pp. 123–171.

Johnson H. D., Cooper W. J., Mezyk S. P. and Bartels D. M. (2002). Free radical reactions of monochloramine and hydroxylamine in aqueous solution. *Radiation Physics and Chemistry*, **65**, 317–326.

Jones C. E. and Carpenter L. J. (2005). Solar photolysis of CH_2I_2, CH_2ICl and CH_2IBr in water, saltwater and seawater. *Environmental Science and Technology*, **39**, 6130–6137.

Kalderis D., Juhasz A. L., Boopathy R. and Comfort S. (2011). Soil contaminated with explosives: environmental fate and evaluation of state-of-the-art remediation processes (IUPAC Technical Report). *Pure and Applied Chemistry*, **83**(7), 1407–1484.

Karci A., Arslan-Alaton I., Bekbolet M., Ozhan G. and Alpertunga B. (2014). H_2O_2/UV-C and photo-Fenton treatment of a nonlyphenol polyethoxylate in synthetic freshwater: follow-up of degradation products, acute toxicity and genotoxicity. *Chemical Engineering Journal*, **241**, 43–51.

Katsoyiannis I. A., Canonica S. and von Gunten U. (2011). Efficiency and energy requirements for the transformation of organic micropollutants by ozone, O_3/H_2O_2 and UV/H_2O_2. *Water Research*, **45**(13), 3811–3822.

Keen O. S. and Linden K. G. (2013). Degradation of antibiotic activity during UV/H_2O_2 advanced oxidation and photolysis in wastewater effluent. *Environmental Science and Technology*, **47**, 13020–13030.

Keen O. S., Dotson A. and Linden K. G. (2013). Evaluation of hydrogen peroxide chemical quenching agents following an advanced oxidation process. *Journal of Environmental Engineering*, **1**, 137–140.

Keen O. S., McKay G., Mezyk S. P., Linden K. G. and Rosario-Ortiz F. L. (2014a). Identifying the factors that influence the reactivity of effluent organic matter with hydroxyl radicals. *Water Research*, **50**, 408–419.

Keen O. S., Bell K. Y., Cherchi C., Finnegan B. J., Mauter M. S., Parker A. M., Rosenblum J. S. and Stretz H. A. (2014b). Emerging pollutants Part II: treatment. *Water Environment Research*, **86**, 2036–2096.

Khetan S. K. and Collins T. J. (2007). Human pharmaceuticals in the aquatic environment: a challenge to green chemistry. *Chemical Reviews*, **107**, 2319–2364.

Kim I. H., Tanaka H., Iwasaki T., Takubo T., Morioka T. and Kato Y. (2008). Classification of the degradability of 30 pharmaceuticals in water with ozone, UV and H_2O_2. *Water Science and Technology*, **57**, 195–200.

Kim I., Yamashita N. and Tanaka H. (2009). Performance of UV and UV/H_2O_2 processes for the removal of pharmaceuticals detected in secondary effluent of a sewage treatment plant. *Journal of Hazardous Materials*, **166**, 1134–1140.

Klavaroti M., Mantzavinos D. and Kassinos D. (2009). Removal of residual pharmaceuticals from aqueous systems by advanced oxidation processes. *Environment International*, **35**, 402–417.

Kliegman S., Eustis S. N., Arnold W. A. and McNeill K. (2013). Experimental and theoretical insights into the involvement of radicals in triclosan phototransformation. *Environmental Science and Technology*, **47**, 6756–6763.

Knol A. H. (2017). Effective OMP control in drinking water treatment by serial AOP with O_3/H_2O_2 and UV/H_2O_2 in front of MAR. *Wasser Berlin International; International Ozone Symposium*, Berlin, Germany, March 28–31.

Kochany J. and Bolton J. R. (1992). Mechanism of photodegradation of aqueous organic pollutants. 2. Measurement of the primary rate constants for reaction of •OH radicals with benzene and some halobenzenes using an EPR spin-trapping method following the photolysis of H_2O_2. *Environmental Science and Technology*, **26**(2), 1992.

Koizumi S., Watanabe K., Hasegawa M. and Kanda H. (2001). Ultraviolet emission from a diamond p-n junction. *Science*, **292**, 1899–1901.

Kolkman A., Martijn B. J., Vughs D., Baken K. A. and van Wezel, A. P. (2015). Tracing nitrogenous disinfection byproducts after medium pressure UV water treatment by stable isotope labeling and high resolution mass spectrometry. *Environmental Science and Technology*, **49**(7), 4458–4465.

Koller L. R. (1965). Ultraviolet Radiation. 2nd edn, Wiley, New York, pp. 1–312.

Kommineni S., Chowdhury Z., Kavanaugh M., Mishra D. and Croue J. P. (2008). Advanced oxidation of methyl-tertiary-butyl ether: pilot findings and full-scale implications. *Journal of Water Supply: Research and Technology – AQUA*, **57**(6), 403–419.

Koppenol W. H. (2001). The Haber-Weiss cycle – 70 years later. *Redox Report*, **6**(4), 229–234.

Köster R. and Asmus K.-D. (1971). Die Reaktionen chlorierter Äthylene mit hydratisirten Elektronen und OH-Radikalen in wäßriger Lösung. *Zeitschrift für Naturforschung*, **26b**, 1108–1116.

Koubek E. (1975). Photochemically induced oxidation of refractory organics with hydrogen peroxide. *Industrial Engineering Chemistry Process Design and Development*, **14**(3), 348–350.

Kruithof J. C. and Martijn B. J. (2013). UV/H_2O_2 treatment: an essential process in a multi-barrier approach against trace chemical contaminants. *Water Science and Technology: Water Supply*, **13**(1), 130–137.

Kruithof J. C., Kamp P. C. and Martijn B. J. (2007). UV/H_2O_2 Treatment: a practical solution for organic contaminant control and primary disinfection. *Ozone: Science & Engineering*, **29**, 273–280.

Kubose D. A. and Hoffsommer J. C. (1977). Photolysis of RDX in Aqueous Solution. Initial Studies. Report No. NSWC/WOL/TR-77–20, ADA042199; Naval Surface Weapons Center, White Oak Laboratory, White Oak, MD.

Kušic H., Rasulev B., Leszczynska D., Leszczynski J. and Koprivanac N. (2009). Prediction of rate constants for radical degradation of aromatic pollutants in water matrix. *Chemosphere*, **75**, 1128–1134.

Kwon M., Kim S., Yoon Y., Jung Y., Hwang T-M and Kang J-W. (2014). Prediction of the removal efficiency of pharmaceuticals by a rapid spectrophotometric method using Rhodamine B in the UV/H_2O_2 process. *Chemical Engineering Journal*, **236**, 438–447.

Lambert J. H. (1760). Photometria sive de mensura et gradibus luminis, colorum et umbrae [Photometry or on the Measure and Gradations of Light, Colors, and Shade]. Eberhardt Klett, Augsburg, Germany.

Landsman N. A., Swancutt K. L., Bradford C. N., Cox C. R., Kiddle J. J. and Mezyk S. P. (2007). Free radical chemistry of advanced oxidation process removal of nitrosamines in water. *Environmental Science and Technology*, **41**, 5818–5823.

Langford C. H., Achari G. and Izadifard M. (2011). Wavelength dependence of luminescence and quantum yield in dechlorination of PCBs. *Journal of Photochemistry and Photobiology A: Chemistry*, **222**, 40–46.

Laszlo B., Alfassi Z. B., Neta P. and Huie R. E. (1998). Kinetics and mechanism of the reaction of $^{\bullet}NH_2$ with O_2 in aqueous solutions. *Journal of Physical Chemistry A*, **102**, 8498–8504.

Latch D. E., Packer J. L., Arnold W. A. and McNeill K. (2003). Photochemical conversion of triclosan to 2,8-dichlorodibenzo-p-dioxin in aqueous solution. *Journal of Photochemistry and Photobiology A: Chemistry*, **158**, 63–66.

Latch D. E., Packer J. L., Stender B. L., Van Overbecke J., Arnold W. A. and McNeill K. (2005). Aqueous photochemistry of triclosan: formation of 2,4-dichlorophenol, 2,8-dichlorodibenzo-p-dioxin, and oligomerization products. *Environmental Toxicology and Chemistry*, **24**(3), 517–525.

Lau T. K., Chu W. and Graham N. (2005). The degradation of endocrine disruptor di-*n*-butyl phthalate by UV irradiation: a photolysis and product study. *Chemosphere*, **60**, 1045–1053.

Law I. B. (2016). An Australian perspective on DPR: technologies, sustainability and community acceptance. *Journal of Water Reuse and Desalination*, **6**(3), 355–361.

Lee Y. and von Gunten U. (2010). Oxidative transformation of micropollutants during municipal wastewater treatment: comparison of kinetic aspects of selective (chlorine, chlorine dioxide, ferrate^{VI}, and ozone) and non-selective oxidants (hydroxyl radical). *Water Research*, **44**, 555–566.

Lee Y. and von Gunten U. (2012). Quantitative structure-activity relationships (QSARs) for the transformation of organic micropollutants during oxidative water treatment. *Water Research*, **46**, 6177–6195.

Lee C., Choi W. and Yoon J. (2005a). UV Photolytic mechanism of *N*-nitrosodimethylamine in water: roles of dissolved oxygen and solution pH. *Environmental Science and Technology*, **39**, 9702–9709.

Lee C., Choi W., Kim Y. G. and Yoon J. (2005b). UV Photolytic mechanism of *N*-nitrosodimethylamine in water: dual pathways to methylamine versus dimethylamine. *Environmental Science and Technology*, **39**, 2101–2106.

Lee C., Yoon J. and von Gunten U. (2007). Oxidative degradation of *N*-nitrosodimethylamine by conventional ozonation and the advanced oxidation process ozone/hydrogen peroxide. *Water Research*, **41**, 581–590.

Lee Y., Gerrity D., Lee M., Bogeat A. E., Salhi E., Gamage S., Treholm R. A., Wert E. C., Snyder S. A. and von Gunten U. (2013). Prediction of micropollutant elimination during ozonation of municipal wastewater effluents: use of kinetic and water specific information. *Environmental Science and Technology*, **47**, 5872–5881.

Lee Y., Gerrity D., Lee M., Gamage S., Pisarenko A. N., Trenholm R. A., Canonica S., Snyder S. A. and von Gunten U. (2016). Organic contaminant abatement in reclaimed water by UV/H_2O_2 and a combined process consisting of O_3/H_2O_2 followed by UV/H_2O_2: prediction of abatement efficiency, energy consumption, and by-product formation. *Environmental Science and Technology*, **50**(7), 3809–3819.

Lee M., Merle T., Rentsch D., Canonica S. and von Gunten U. (2017). Abatement of polychloro-1,3-butadienes in aqueous solution by ozone, UV photolysis, and advanced oxidation processes (O_3/H_2O_2 and UV/H_2O_2). *Environmental Science and Technology*, **51**(1), 497–505.

Legrini O., Oliveros E. and Braun A. M. (1993). Photochemical processes for water treatment. *Chemical Reviews*, **93**, 671–698.

Leifer A. (1988). The Kinetics of Environmental Aquatic Photochemistry: Theory and Practice. ACS Professional Reference Book. American Chemical Society, Washington DC.

Lekkerkerker-Teunissen K., Knol A. H., van Altena L. P., Houtman C. J., Verberk J. Q. J. and van Dijk J. C. (2012). Serial ozone/ peroxide/ low pressure UV treatment for synergetic and effective organic micropollutant conversion. *Separation and Purification Technology*, **100**, 22–29.

Lekkerkerker-Teunissen K., Knol A. H., Derks J. G., Heringa M. B., Houtman C. J., Hofman-Caris R. C. H. M., Beerendonk E. F., Reus A., Verberk J. Q. J. C. and van Dijk J. C. (2013). Pilot plant results with three different types of UV lamps for advanced oxidation. *Ozone: Science & Engineering*, **35**, 38–48.

Lester Y., Mamane H. and Dror A. (2012). Enhanced removal of micropollutants from groundwater using pH modification coupled with photolysis. *Water, Air and Soil Pollution*, **223**, 1639.

Li K. and Crittenden J. (2009). Computerized pathway elucidation for hydroxyl radical-induced chain reaction mechanisms in aqueous phase advanced oxidation processes. *Environmental Science and Technology*, **43**, 2831–2837.

Li K. Y., Liu C. C., Ni Q., Liu Z. F., Huang F. Y. C. and Colapret J. A. (1995). Kinetic study of UV peroxidation of bis(2-chloroethyl) ether in aqueous solution. *Industrial and Engineering Chemistry Research*, **34**, 1960.

Li K., Stefan M. I. and Crittenden J. C. (2004). UV Photolysis of trichloroethylene: product study and kinetic modeling. *Environmental Science and Technology*, **38**, 6685–6693.

Li K., Stefan M. I. and Crittenden J. C. (2007). Trichloroethene degradation by UV/H_2O_2 advanced oxidation process: product study and kinetic modeling. *Environmental Science and Technology*, **41**, 1696–1703.

Li X., Zeng F. and Li K. (2013). Computer assisted pathway generation for atrazine degradation in advanced oxidation processes. *Journal of Environmental Protection*, **4**, 62–69.

Li D., Stanford B., Dickenson E., Khunjar W. O., Homme C. L., Rosenfeldt E. J. and Sharp J. O. (2017a). Effect of advanced oxidation on N-nitrosodimethylamine (NDMA) formation and microbial ecology during pilot-scale biological activated carbon filtration. *Water Research*, **113**, 160–170.

Li M., Qiang Z., Bolton J. R. and Blatchley E. R. (2017b). Experimental assessment of photon fluence rate distributions in a medium-pressure UV photoreactor. *Environmental Science and Technology*, **51**(6), 3453–3460.

Lian J., Qiang Z., Li M., Bolton J. R. and Qu J. (2015). UV photolysis of sulfonamides in aqueous solution based on optimized fluence quantification. *Water Research*, **75**, 43–50.

Liang S., Min J. H., Davis M. K., Green J. F. and Remer D. S. (2003). Use of pulsed-UV processes to destroy NDMA. *Journal of the American Water Works Association*, **95**(9), 121–131.

Liao C. H. and Gurol M. D. (1995). Chemical oxidation by photolytic decomposition of hydrogen peroxide. *Environmental Science and Technology*, **29**, 3007–3014.

Light Sources Inc. (2013). Germicidal Lamps Basics. http://www.light-sources.com/wp-content/uploads/2015/05/Germicidal_Lamp_Basics_-_2013.pdf.

Liou M. J., Lu M. C. and Chen J. N. (2003). Oxidation of explosives by Fenton and photo-Fenton processes. *Water Research*, **37**, 3172–3179.

Liu D., Ducoste J., Jin S. and Linden K. G. (2004). Evaluation of alternative fluence rate distribution models. *Journal of Water Supply: Research and Technology – AQUA*, **53**, 391–408.

Loraine G. A. and Glaze W. H. (1999). The ultraviolet photolysis of aqueous solutions of 1,1,1-trichloroethane and hydrogen peroxide at 222 nm. *Journal of Advanced Oxidation Technologies*, **4**(4), 424–433.

Luo Y., Guo W., Ngo H., Nghiem L. D., Hai F. I., Zhang J., Liang S. and Wang X. C. (2014). A review on the occurrence of micropollutants in the aquatic environment and their fate and removal during wastewater treatment. *Science of the Total Environment*, 473–474, 619–641.

Luo X., Yang X., Qiao X., Wang Y., Chen J., Wei X. and Peijnenburg W. J. G. M. (2017). Development of a QSAR model for predicting aqueous reaction rate constants for organic chemicals with hydroxyl radicals. *Environmental Science Processes & Impacts*, **19**, 350–356.

Lyon B. A., Cory R. M. and Weinberg H. S. (2015). Changes in dissolved organic matter fluorescence and disinfection byproduct formation from UV and subsequent chlorination/chloramination. *Journal of Hazardous Materials*, **264**, 411–419.

Lyons E., Zhang P., Benn T., Sharif F., Li K., Crittenden J., Costanza M. and Chen Y. S. (2009). Life cycle assessment of three water supply systems: importation, reclamation and desalination. *Water Science and Technology: Water Supply*, **9**(4), 439–448.

Machado F. and Boule P. (1995). Photonitration and photonitrosation of phenolic derivatives induced in aqueous solution by excitation of nitrite and nitrate ions. *Journal of Photochemistry and Photobiology A: Chemistry*, **86**, 73–80.

MacNab A., Bindner S. and Festger A. (2015). Effectiveness and carbon footprint of UV-oxidation for the treatment of taste and odor-causing compounds in drinking water. *Proceedings of Water New Zealand Annual Conference and Exhibition*, Hamilton, New Zealand, September 16–18.

Magazinovic R. S., Nicholson B. C., Mulcahy D. E. and Davey D. E. (2004). Bromide levels in natural waters: its relationship to levels of both chloride and total dissolved solids and the implications for water treatment. *Chemosphere*, **57**, 329–335.

Mahbub P. and Nesterenko P. N. (2016). Application of photo-degradation for remediation of cyclic nitramine and nitroaromatic explosives. *RSC Advances*, **6**, 77603–77621.

Manassero A., Passalia C., Negro A. C., Cassano A. E. and Zalazar C. S. (2010). Glyphosate degradation in water employing the H_2O_2/UVC process. *Water Research*, **44**, 3875–3882.

Mark G., Korth H.-G., Schuchmann H.-P. and von Sonntag C. (1996). The photochemistry of aqueous nitrate revisited. *Journal of Photochemistry and Photobiology A: Chemistry*, **101**, 80–103.

Marshak I. S. (1984). Radiation Characteristics of Flashlamps. Chapter 5 in Pulsed light sources. Consultants Bureau, New York, NY.

Martijn A. J. (2014). How drugs in wastewater enforce advanced drinking water treatment. In "Wastewater and Biosolids Treatment and Reuse: Bridging Modeling and Experimental Studies"; *Engineering Conferences International Symposium Series*. http://dc.engconfintl.org.

Martijn A. J. and Kruithof J. C. (2012). UV and UV/H_2O_2 treatment: the silver bullet for by-product and genotoxicity formation in water production. *Ozone: Science & Engineering*, **34**, 92–100.

Martijn B. J., Kruithof J. C. and Welling M. (2006). UV/H_2O_2 treatment: the ultimate solution for organic contaminant control and primary disinfection. *Proceedings of the AWWA Water Quality and Technology Conference*, Denver, CO.

Martijn B. J., Fuller A. L., Malley J. P. and Kruithof J. C. (2010). Impact of IX-UF pretreatment on the feasibility of UV/H_2O_2 treatment for degradation of NDMA and 1,4-dioxane. *Ozone: Science & Engineering*, **32**(6), 383–390.

Martijn A. J., Boersma M. G., Vervoorf J. M., Rietjens I. M. C. M. and Kruithof, J. C. (2014). Formation of genotoxic compounds by medium pressure ultraviolet treatment of nitrate-rich water. *Desalination and Water Treatment*, **52**, 6275–6281.

Martijn B. J., Kruithof J. C., Hughes R. M., Mastan R. A., Van Rompay A. R. and Maller J. P. Jr. (2015). Induced genotoxicity in nitrate-rich water treated with medium-pressure ultraviolet processes. *Journal of American Water Works Association*, **107**(6), 301–312.

Martijn B. J., Van Rompay A. R., Penders E. J. M., Alharbi Y., Baggelaar P. K., Kruithof J. C. and Rietjens I. M. C. M. (2016). Development of a 4-NQO toxic equivalency factor (TEF) approach to enable a preliminary risk assessment of unknown genotoxic compounds detected by Ames II test in UV/H_2O_2 water treatment samples. *Chemosphere*, **144**, 338–345.

Maruthamuthu P., Padmaja S. and Huie R. E. (1995). Rate constants for some reactions of free radicals with haloacetates in aqueous solution. *International Journal of Chemical Kinetics*, **27**, 605–612.

Masschelein W. J. (2000). Utilisation des U.V. dans le traiement des eaux. *Tribune de l'eau*, **53**(4–5), 7–107.

Masten S. J., Shu M., Galbraith M. J. and Davies S. H. R. (1998). Oxidation of chlorinated benzenes using advanced oxidation processes. *Hazardous Waste & Hazardous Materials*, **13**, 265–282.

Matafonova G. and Batoev V. (2012). Recent progress on application of UV excilamps for degradation of organic pollutants and microbial inactivation. *Chemosphere*, **89**, 637–647.

Mazellier P. and Leverd J. (2003). Transformation of 4-*tert*-octylphenol by UV irradiation and by an H_2O_2/UV process in aqueous solution. *Photochemical and Photobiological Sciences*, **2**, 946–953.

Mazellier P., Meite L. and De Laat J. (2008). Photodegradation of the steroid hormones 17 beta-estradiol (E2) and 17 alpha-ethinylestradiol (EE2) in dilute aqueous solution. *Chemosphere*, **73**, 1216–1223.

McKay G., Dong M. M., Kleinman J. L., Mezyk S. P. and Rosario-Ortiz F. L. (2011). Temperature dependence of the reaction between the hydroxyl radical and organic matter. *Environmental Science and Technology*, **45**, 6932–6937.

McKay G., Sjelin B., Chagnon M., Ishida K. I. and Mezyk S. P. (2013). Kinetic study of the reactions between chloramine disinfectants and hydrogen peroxide: temperature dependence and reaction mechanism. *Chemosphere*, **92**, 1417–1422.

Men'kin V. B., Makarov I. E. and Pikaev A. K. (1991). Mechanism for formation of nitrite in radiolysis of aqueous solutions of ammonia in the presence of oxygen. *High Energy Chemistry*, **25**, 48–52.

Merel S., Walker D., Chicana R., Snyder S., Baurès E. and Thomas O. (2013a). State of knowledge and concerns on cyanobacterial blooms and cyanotoxins. *Environment International*, **59**, 303–327.

Merel S., Villarin M. C., Chung K. and Snyder S. (2013b). Spatial and thematic distribution of research on cyanotoxins. *Toxicon*, **76**, 118–131.

Merel S., Anumol T., Park M. and Snyder S. A. (2015). Application of surrogates, indicators, and high-resolution mass-spectrometry to evaluate the efficacy of UV processes for attenuation of emerging contaminants in water. *Journal of Hazardous Materials*, **282**, 75–85.

Mertens R. and von Sonntag C. (1994). Reaction of OH radical with tetrachloroethene and trichloroacetaldehyde (hydrate) in oxygen-free aqueous solutions. *Journal of Chemical Society Perkin Transactions*, **2**, 2175–2180.

Mertens R. and von Sonntag C. (1995). Photolysis ($\lambda = 254$ nm) of tetrachloroethene in aqueous solutions. *Journal of Photochemistry and Photobiology A: Chemistry*, **85**, 1–9.

Mestankova H., Schirmer K., Canonica S. and von Gunten U. (2014). Development of mutagenicity during degradation of N-nitrosamines by advanced oxidation processes. *Water Research*, **66**, 399–410.

Mestankova H., Parker A. M., Bramaz N., Canonica S., Schirmer K., von Gunten, U. and Linden K. G. (2016). Transformation of Contaminant Candidate List (CCL3) compounds during ozonation and advanced oxidation processes in drinking water: assessment of biological effects. *Water Research*, **93**, 110–120.

Metz D. H., Meyer M., Dotson A., Beerendonk E. and Dionysiou D. D. (2011). The effect of UV/H_2O_2 treatment on disinfection by-product formation potential under simulated distribution system conditions. *Water Research*, **45**, 3969–3980.

Metz D. H., Meyer M., Vala B., Beerendonk E. F. and Dionysiou D. D. (2012). Natural organic matter: effect on contaminant destruction by UV/H_2O_2. *Journal of American Water Works Association*, **104**(12), 31–32.

Meunier L. and Boule P. (2000). Direct and induced phototransformation of mecoprop [2-(4-chloro-2-methylphenoxy)-propionic acid] in aqueous solution. *Pest Management Science*, **56**, 1077–1085.

Millet M., Palm W. U. and Zetzsch C. (1998). Investigation of the photochemistry of urea herbicides (chlorotoluron and isoproturon) and quantum yields using polychromatic irradiation. *Environmental Toxicology and Chemistry*, **17**(2), 258–264.

Milosavljevic B. H. and LaVerne J. A. (2005). Pulse radiolysis of aqueous thiocyanate. *Journal of Physical Chemistry A*, **109**, 165–168.

Minakata D. and Crittenden J. (2011). Linear free energy relationships between the aqueous phase hydroxyl radical ($HO^•$) reaction rate constants and the free energy of activation. *Environmental Science and Technology*, **45**, 3479–3486.

Minakata D., Li K., Westerhoff P. and Crittenden J. (2009). Development of a Group Contribution Method to predict aqueous phase hydroxyl radical ($HO^•$) reaction rate constants. *Environmental Science and Technology*, **43**, 6220–6227.

Minakata D., Song W., Mezyk S. P. and Cooper W. J. (2015). Experimental and theoretical studies on aqueous-phase reactivity of hydroxyl radicals with multiple carboxylated and hydroxylated benzene compounds. *Physical Chemistry Chemical Physics*, **17**, 11796–11812.

Morgan M. S., Van Trieste P. F., Garlick S. M., Mahon M. J. and Smith A. L. (1988). Ultraviolet molar absorptivities of aqueous hydrogen peroxide and hydroperoxyl ion. *Analytica Chimica Acta*, **215**, 325–329.

Müller S. R., Berg M., Ulrich M. M. and Schwarzenbach R. P. (1997). Atrazine and its primary metabolites in Swiss lakes: input characteristics and long-term behavior in the water column. *Environmental Science and Technology*, **31**, 2104–2113.

Nélieu S., Perreau F., Bonnemoy F., Ollitrault M., Azam D., Lagadic L., Bohatier J. and Einhorn J. (2009). Sunlight nitrate-induced photodegradation of chlorotoluron: evidence of the process in aquatic mesocosms. *Environmental Science and Technology*, **43**, 3148–3154.

Neta P. and Grodkowski J. (2005). Rate constants for reactions of phenoxyl radicals in solution. *Journal of Physical Chemistry Reference Data*, **34**(1), 109–199.

Neta P., Maruthamuthu P., Carton P. M. and Fessenden R. W. (1978). Formation and reactivity of the amino radical. *Journal of Physical Chemistry*, **82**, 1875–1878.

Neta P., Huie R. E. and Ross A. B. (1990). Rate constants for reactions of peroxyl radicals in fluid solutions. *Journal of Physical Chemistry Reference Data*, **19**(2), 413–513.

Neta P., Grodkowski J. and Ross A. B. (1996). Rate costants for reactions of aliphatic carbon-centered radicals in aqueous solution. *Journal of Physical Chemistry Reference Data*, **25**(3), 709–937.

Neumann-Spallart, M. and Getoff, N. (1979). Photolysis and radiolysis of monochloroacetic acid in aqueous solution. *Radiation Physical Chemistry*, **13**, 101–105.

Nicholson B. C., Shaw G. R., Morrall J., Senogles P. J., Woods T. A., Papageorgiou J., Kapralos C., Wickramasinghe W., Davis B. C., Eaglesham G. K. and Moore M. R. (2003). Chlorination for degrading saxitoxins (paralytic shellfish poisons) in water. *Environmental Technology*, **24**, 1341–1348.

Nick K., Schoeler H. F., Mark G., Söylemez T., Akhlaq M. S., Schuchmann H.-P. and von Sonntag C. (1992). Degradation of some triazine herbicides by UV radiation such as in the UV disinfection of drinking water. *Journal of Water Supply: Research and Technology-Aqua*, **41**(2), 82–87.

Nicolaescu A. R., Wiest O. and Kamat P. V. (2005). Mechanistic pathways of hydroxyl radical reactions of quinoline. 2. Computational analysis of hydroxyl radical attack at C atoms. *Journal of Physical Chemistry A*, **109**, 2829–2835.

Ning B., Graham N. J. D. and Zhang Y. (2007). Degradation of octylphenol and nonylphenol by ozone – Part II: indirect reaction. *Chemosphere*, **68**, 1173–1179.

Nohr R. S., MacDonald J. G., Kogelschatz U., Mark G., Schuchmann H.-P. and von Sonntag C. (1994). Application of excimer incoherent-UV sources as a new tool in photochemistry: photodegradation of chlorinated dibenzodioxins in solution and adsorbed on aqueous pulp sludge. *Journal of Photochemistry and Photobiology A: Chemistry*, **79**, 141–149.

Noss C. I. and Chyrek R. H. (1984). Tertiary Treatment of Effluent from Holston AAP Industrial Liquid Waste Treatment Facility. IV. Ultraviolet Radiation and Hydrogen Peroxide Studies: TNT, RDX, HMX, TASX, and SEX. Technical Report 8308; US Army Medical Bioengineering Research & Development Laboratory, Fort Detrick, Frederick, Maryland.

Nöthe T., Fahlenkamp H. and von Sonntag C. (2009). Ozonation of wastewater: rate of ozone consumption and hydroxyl radical yield. *Environmental Science and Technology*, **43**(15), 5990–5995.

Ohe T., Watanabe T. and Wakabayashi K. (2004). Mutagens in surface waters: a review. *Mutation Research*, **567**, 109–149.

Onstad G. D., Strauch S., Meriluoto J., Codd G. A. and von Gunten U. (2007). Selective oxidation of key functional groups in cyanotoxins during drinking water ozonation. *Environmental Science and Technology*, **41**, 43976–4404.

Orellana-García F., Álvarez M. A., López-Ramón V., Rivera-Utrilla J., Sánchez-Polo M. and Mota A. J. (2014). Photodegradation of herbicides with different chemical natures in aqueous solution by ultraviolet radiation. Effects of operational variables and solution chemistry. *Chemical Engineering Journal*, **255**, 307.

Orr P. I., Jones G. J. and Hamilton G. R. (2004). Removal of saxitoxins from drinking water by granular activated carbon, ozone and hydrogen peroxide – implications for compliance with the Australian drinking water guidelines. *Water Research*, **38**, 4455–4461.

Oturan M. A. and Aaron J.-J. (2014). Advanced oxidation processes in water/wastewater treatment: principles and applications. A review. *Critical Reviews in Environmental Science and Technology*, **44**, 2577–2641.

Pagan J. and Lawal O. (2015). Coming of age – UV-C LED technology update. *IUVA News*, **17**(1), 21–24.

Pagsberg P. B. (1972). Investigation of the NH_2 radical produced by pulse radiolysis of ammonia in aqueous solution. *Risø* **256**, 209–221.

Palm W. U. and Zetzsch C. (1996). Investigation of the photochemistry and quantum yields of triazines using polychromatic irradiation and UV spectroscopy as an analytical tool. *International Journal of Environmental and Analytical Chemistry*, **65**(4), 313–329.

Pantelić D., Svirčev Z., Simeunovic J., Vidović M. and Trajković I. (2013). Cyanotoxins: characteristics, production and degradation routes in drinking water treatment with reference to the situation in Serbia. *Chemosphere*, **91**, 421–441.

Parker C. A. (1968). Photoluminiscence of Solutions: With Applications to Photochemistry and Analytical Chemistry. Elsevier Publishing Co., New York.
Peng J., Wang G., Zhang D., Zhang D. and Li X. (2016). Photodegradation of nonlylphenol in aqueous solution by simulated solar irradiation: the comprehensive effect of nitrate, ferric ion and bicarbonate ions. *Journal of Photochemistry and Photobiology A: Chemistry*, **326**, 9–15.
Pereira V. J., Weinberg H. S., Linden K. G. and Singer P. C. (2007). UV degradation kinetics and modelling of pharmaceutical compounds in laboratory grade and surface water via direct and indirect photolysis at 254 nm. *Environmental Science and Technology*, **41**, 1682–1688.
Peter A. and von Gunten U. (2007). Oxidation kinetics of selected taste and odor compounds during ozonation of drinking water. *Environmental Science and Technology*, **41**, 626–631.
Peyton G. (1990). Oxidative treatment methods for removal of organic compounds from drinking water supplies. In: Significance and Treatment of Volatile Organic Compounds in Water Supplies, N. M. Ram, R. F. Christman and K. P. Cantor (eds), Lewis, Chelsea, MI (USA), pp. 313–362.
Peyton G., LeFaivre M. H. and Maloney S. W. (1999). Verification of RDX Photolysis Mechanism. Technical Report 99/93. US Army Construction Engineering Research laboratory, Champaign, IL.
Phillips R. (1983). Sources and Applications of Ultraviolet Radiation. Academic Press Inc., New York.
Plumlee M. H. and Reinhard M. (2007). Photochemical attenuation of *N*-nitrosodimethylamine (NDMA) and other nitrosamines in surface water. *Environmental Science and Technology*, **41**, 6170–6176.
Plumlee M. H., Stanford B. D., Debroux J. F., Hopkins D. C. and Snyder S. A. (2014). Costs of advanced treatment in water reclamation. *Ozone: Science & Engineering*, **36**(5), 485–495.
Poskrebyshev G. A., Huie R. E. and Neta P. (2003). Radiolytic reactions of monochloramine in aqueous solutions. *Journal of Physical Chemistry A*, **107**, 7423–7428.
Postigo C. and Richardson S. D. (2014). Transformation of pharmaceuticals during oxidation/disinfection processes in drinking water treatment. *Journal of Hazardous Materials*, **279**, 461–475.
Qasim M. M., Moore B., Taylor L., Honea P., Gorb L. and Leszczynski J. (2007). Structural characteristics and reactivity relationships of nitroaromatic and nitramine explosives – A review of our computational chemistry and spectroscopic research. *International Journal of Molecular Sciences*, **8**, 1234–1264.
Quen H. L. and Chidambara Raj C. B. (2006). Evaluation of UV/O_3 and UV/H_2O_2 processes for nonbiodegradable compounds: implications for integration with biological processes for effluent treatment. *Chemical Engineering Communications*, **193**, 1263–1276.
Rahn R. O., Bolton J. R. and Stefan M. I. (2006). The iodate/iodate actinometer in UV disinfection: determination of fluence rate distribution in UV reactors. *Photochemistry and Photobiology*, **82**, 611–615.
Ribeiro A. R., Nunes O. C., Pereira M. F. R. and Silva A. M. T. (2015). An overview of the advanced oxidation processes applied for the treatment of water pollutants defined in the recently launched Directive 2013/39/EU. *Environment International*, **75**, 33–51.
Richardson S. D., Plewa M. J., Wagner E. D., Schoeny R. and DeMarini D. M. (2007). Occurrence, genotoxicity, and carcinogenicity of regulated and emerging disinfection by-products in drinking water: a review and roadmap for research. *Mutation Research*, **636**, 178–242.
Richardson S. D., Thruston A. D., Krasner S. W., Weinberg H. S., Miltner R. J., Schenck K. M., Narotsky M. G., McKague A. B. and Simmons J. E. (2008). Integrated disinfection by-products mixtures research: comprehensive characterization of water concentrates prepared from chlorinated and ozonated/post-chlorinated drinking water. *Journal of Toxicology and Environmental Health Part A*, **71**(17), 1165.
Rivas J., Gimeno O., Borralho T. and Sagasti J. (2011). UV-C and UV-C/peroxide elimination of selected pharmaceuticals in secondary effluents. *Desalination*, **279**, 115–120.
Rivera R. L. (2001). Mercury amalgam lamps: Characteristics, design parameters and comparison to other UV technologies. *Proceedings of First International Congress on Ultraviolet Technologies*, IUVA, June 14–16, Washington DC.
Rivera-Utrilla J., Sánchez-Polo M., Ferro-García M. A., Prados-Joya G. and Ocampo-Pérez R. (2013). Pharmaceuticals as emerging contaminants and their removal from water. A review. *Chemosphere*, **93**, 1268–1287.

Rosario-Ortiz F. L., Mezyk S. P., Doud D. F. R. and Snyder S. A. (2008). Quantitative correlation of absolute hydroxyl radical rate constants with non-isolated effluent organic matter bulk properties in water. *Environmental Science and Technology*, **42**(16), 5924–5930.

Rosenfeldt E. J. (2010). Rapid method for measuring background hydroxyl radical scavenging for impacts on design of advanced oxidation processes. *Proceedings of AWWA Water Quality and Technology Conference and Exhibition*, November 14–18, Savannah, GA, USA.

Rosenfeldt E. J. and Linden K. G. (2004). Degradation of endocrine disrupting chemicals bisphenol A, ethinylestradiol, and estradiol during UV photolysis and advanced oxidation processes. *Environmental Science and Technology*, **38**, 5476–5483.

Rosenfeldt E. J. and Linden K. G. (2005). The $R_{OH,UV}$ concept to characterize and model UV/H_2O_2 processes in natural waters. *Proceedings of the 3-rd International Congress on Ultraviolet Technologies*, IUVA, Whistler, BC, Canada, May 24–27.

Rosenfeldt E. J. and Linden K. G. (2007). The $R_{OH,UV}$ concept to characterize and to model UV/H_2O_2 process in natural waters. *Environmental Science and Technology*, **41**, 2548–2553.

Rosenfeldt E. J., Melcher B. and Linden K. G. (2005). UV and UV/H_2O_2 treatment of methylisoborneol (MIB) and geosmin in water. *Journal of Water Supply: Research and Technology-AQUA*, **54**, 423–435.

Rosenfeldt E. J., Linden K. G., Canonica S. and von Gunten U. (2006). Comparison of the efficiency of OH radical formation during ozonation and the advanced oxidation processes O_3/H_2O_2 and UV/H_2O_2. *Water Research*, **40**, 3695–3704.

Royce A. and Stefan M. I. (2005). Application of UV in drinking water treatment for simultaneous disinfection and removal of taste and odor compounds. *Proceedings of AWWA Water Quality and Technology Conference*, Quebec City, QC, Canada, November 6–10.

Santoro D., Raisee M., Moghaddami M., Ducoste J., Sasges M., Liberti L. and Notarnicola M. (2010). Modeling hydroxyl radical distribution and trialkyl phosphates oxidation in $UV-H_2O_2$ photoreactors using computational fluid dynamics. *Environmental Science and Technology*, **44**, 6233–6241.

Sarathy S. and Mohseni M. (2009). The fate of natural organic matter during UV/H_2O_2 advanced oxidation of drinking water. *Canadian Journal of Civil Engineering*, **36**, 140–169.

Sarathy S. and Mohseni M. (2010). Effects of UV/H_2O_2 advanced oxidation on chemical characteristics and chlorine reactivity of surface water natural organic matter. *Water Research*, **44**, 4087–4096.

Sarathy S., Stefan M. I., Royce A. R. and Mohseni M. (2011). Pilot-scale UV/H_2O_2 advanced oxidation process for surface water treatment: effects on natural organic matter characteristics, DBP formation potential, and downstream biological treatment. *Environmental Technology*, **33**(11–16), 1709–1718.

Schaefer R. B., Grapperhaus M., Schaefer I. and Linden K. G. (2007). Pulsed UV lamp performance and comparison with UV mercury lamps. *Journal of Environmental Engineering and Science*, **6**, 303–310.

Schalk S., Adam V., Arnold E., Brieden K., Voronov A. and Witzke H.-D. (2006). UV lamps for disinfection and advanced oxidation: lamp types, technologies, and applications. *IUVA News*, **8**(1), 32–37.

Scheideler J. (2016). Development of a novel control philosophy for UV AOPs. *Proceedings of the IUVA Congress*, January 31–February 3, Vancouver, BC, Canada.

Scheideler J., Lekkerkerker-Teunissen K., Knol T., Ried A., Verbeck J. and van Dijk H. (2011). Combination of O_3/H_2O_2 and UV for multiple barrier micropollutant treatment and bromate formation control – an economic attractive option. *Water Practice & Technology*, **6**(4); DOI:10.2166/wpt.2011.0063.

Scheideler J., Lee K. H., Raichle P., Choi T. and Dong H. S. (2015). UV-advanced oxidation process for taste and odor removal – comparing low pressure and medium pressure UV for a full-scale installation in Korea. *Water Practice & Technology*, **10**(1), 66–72.

Scheideler J., Lee K.-H., Gebhardt J. and Fassbender M. (2016). Development of a novel control philosophy for UV AOPs in S. Korea. *IUVA News*, **18**(3), 17–18.

Schuchmann M. N. and von Sonntag C. (1982). Hydroxyl radical-induced oxidation of diethyl ether in oxygenated aqueous solution. A product and pulse radiolysis study. *Journal of Physical Chemistry*, **86**, 1995–2000.

Schwarzenbach R. P., Gschwend P. M. and Imboden D. M. (1993). Photochemical transformation reactions. In: Environmental Organic Chemistry, John Wiley & Sons, Inc., New York, pp. 436–484.

Semitsoglou-Tsiapou S., Mous A., Templeton M. R., Graham N. J. D., Hernández-Leal L. and Kruithof J. C. (2016a). The role of natural organic matter in nitrite formation by LP-UV/H_2O_2 treatment of nitrate-rich water. *Water Research*, **106**, 312–319.

Semitsoglou-Tsiapou S., Templeton M. R., Graham N. J. D., Hernández-Leal L., Martijn B. J., Royce A. and Kruithof J. C. (2016b). Low pressure UV/H_2O_2 treatment for the destruction of pesticides metaldehyde, clopyralid and mecoprop – Kinetics and reaction product formation. *Water Research*, **91**, 285–294.

Sgroi M., Roccaro P., Oelker G. L. and Snyder S. A. (2015). N-nitrosodimethylamine (NDMA) formation at an indirect potable reuse facility. *Water Research*, **70**, 174–183.

Sharma V. K. (2012). Kinetics and mechanism of formation and destruction of N-nitrosodimethylamine in water – A review. *Separation and Purification Technology*, **88**, 1–10.

Sharma V. K., Triantis T. M., Antoniou M. G., He X., Pelaez M., Han C., Song W., O'Shea K. E., De La Cruz A. A., Kaloudis T., Hiskia A. and Dionysiou D. D. (2012). Destruction of microcystins by conventional and advanced oxidation processes. A review. *Separation and Purification Technology*, **91**, 3–17.

Sharpless C. M. (2005). Weighted fluence-based parameters for assessing UV and UV/H_2O_2 performance and transferring bench-scale results to full-scale water treatment reactor models. *IUVA News*, **7**(3), 13–19.

Sharpless C. M. and Linden K. G. (2001). UV photolysis of nitrate: effects of natural organic matter and dissolved inorganic carbon and implications for water disinfection. *Environmental Science and Technology*, **35**, 2949–2955.

Sharpless C. M. and Linden K. G. (2003). Experimental and model comparisons of low- and medium-pressure Hg lamps for the direct and H_2O_2 assisted UV photodegradation of *N*-nitrosodimethylamine in simulated drinking water. *Environmental Science and Technology*, **37**, 1933–1940.

Sharpless C. M. and Linden K. G. (2005). Interpreting collimated beam ultraviolet photolysis rate data in terms of electrical efficiency of treatment. *Journal of Environmental Engineering and Science*, **4**, S19–S26.

Shen R. and Andrews S. A. (2011). Demonstration of 20 pharmaceuticals and personal care products (PPCPs) as nitrosamine precursors during chloramine disinfection. *Water Research*, **45**, 944–952.

Shen Y. S., Ku Y. and Lee K. C. (1995). The effect of light absorbance on the decomposition of chlorophenols by ultraviolet radiation and UV/H_2O_2 processes. *Water Research*, **29**, 907–914.

Shen C., Scheible O. K., Chan P., Mofidi A., Yun T. I., Lee C. C. and Blatchley E. R. III (2009). Validation of medium pressure UV disinfection reactors by Lagrangian actinometry using dyed microspheres. *Water Research*, **43**, 1370–1380.

Shi H., Cheng X., Wu Q., Mu R. and Ma Y. (2012). Assessment and removal of emerging water contaminants. *Journal of Environmental and Analytical Toxicology*, **S2**, 1–14.

Shu Z., Bolton J. R., Belosevic M. and Gamal El Din M. (2013). Photodegradation of emerging micropollutants using the medium-pressure UV/H_2O_2 advanced oxidation process. *Water Research*, **47**(8), 2881–2889.

Silva C. P., Otero M. and Esteves V. (2012). Processes for the elimination of estrogenic steroid hormones from water: a review. *Environmental Pollution*, **165**, 38–58.

Smaller B., Matheson M. S. and Yasaitis E. L. (1954). Paramagnetic resonance in irradiated ice. *Physical Review*, **94**(1), 202.

Solari F., Girolimetti G., Montanari R. and Vignali G. (2015). A new method for the validation of ultraviolet reactors by means of photochromic materials. *Food and Bioprocess Technology*, **8**(11), 2192–2211.

Soltermann F., Lee M., Canonica S. and von Gunten U. (2013). Enhanced N-nitrosamine formation in pool water by UV irradiation of chlorinated secondary amines in the presence of monochloramine. *Water Research*, **47**, 79–90.

Song W., Chen W., Cooper W. J., Greaves J. and Miller G. E. (2008a). Free-radical destruction of β-lactam antibiotics in aqueous solution. *Journal of Physical Chemistry A*, **112**, 7411–7417.

Song W., Ravindran V. and Pirbazari M. (2008b). Process optimization using a kinetic model for the ultraviolet radiation-hydrogen peroxide decomposition of natural and synthetic organic compounds in groundwater. *Chemical Engineering Science*, **63**, 3249–3270.

Song W., Xu T., Cooper W. J., Dionysiou D. D., de la Cruz A. A. and O'Shea K. E. (2009). Radiolysis studies on the destruction of microcystin-LR in aqueous solution by hydroxyl radicals. *Environmental Science and Technology*, **43**(5), 1487–1492.

Song W., Yan S., Cooper W. J., Dionysiou D. D. and O'Shea K. E. (2012). Hydroxyl radical oxidation of cylindrospermopsin (cyanobacterial toxin) and its role in the photochemical transformation. *Environmental Science and Technology*, **46**, 12608–12615.

Song K., Mohseni M. and Taghipour F. (2016). Application of ultraviolet light-emitting diodes (UV-LEDs) for water disinfection: a review. *Water Research*, **94**, 341–349.

Sorlini S., Gialdini F. and Stefan M. (2014). UV/H_2O_2 oxidation of arsenic and terbuthylazine in drinking water. *Environmental Monitoring and Assessment*, **186**, 1311–1316.

Sornalingam K., McDonagh A. and Zhou J. L. (2016). Photodegradation of estrogenic endocrine disrupting steroidal hormones in aqueous systems: progress and future challenges. *Science of the Total Environment*, **550**, 209–224.

Sosnin E. A., Oppenländer T. and Tarasenko V. F. (2006). Applications of capacitive and barrier discharge excilamps in photoscience. *Journal of Photochemistry and Photobiology C: Photochemistry Reviews*, **7**, 145–163.

Srinivasan R. and Sorial G. A. (2011). Treatment of taste and odor compounds 2-methylisoborneol and geosmin in drinking water: a critical review. *Journal of Environmental Sciences*, **23**, 1–13.

Stefan M. I. (2004). UV photolysis: Background. In: Advanced Oxidation Processes for Water and Wastewater Treatment, S. Parsons (ed.), IWA Publishing, London, UK, pp. 7–48.

Stefan M. I. and Bolton J. R. (1998). Mechanism of the degradation of 1,4-dioxane in dilute aqueous solution using the UV/hydrogen peroxide process. *Environmental Science and Technology*, **32**, 1588–1595.

Stefan M. I. and Bolton J. R. (1999). Reinvestigation of acetone degradation mechanism in dilute aqueous solution by the UV/H_2O_2 process. *Environmental Science and Technology*, **33**, 870–873.

Stefan M. I. and Bolton J. R. (2002). UV direct photolysis of N-nitrosodimethylamine (NDMA): kinetic and product study. *Helvetica Chimica Acta*, **85**, 1416–1426.

Stefan M. I. and Williamson C. T. (2004). UV light-based applications. In: Advanced Oxidation Processes for Water and Wastewater Treatment, S. Parsons (ed.), IWA Publishing, London, UK, pp. 49–85.

Stefan M. I., Hoy A. R. and Bolton J. R. (1996). Kinetics and mechanism of the degradation and mineralization of acetone in dilute aqueous solutions sensitized by the UV photolysis of hydrogen peroxide. *Environmental Science and Technology*, **30**, 2382–2390.

Stefan M. I., Mack J. and Bolton J. R. (2000). Degradation pathways during the treatment of methyl-*tert*-butylether by the UV/H_2O_2 process. *Environmental Science and Technology*, **34**(4), 650–658.

Stefan M. I., Atasi K., Linden K. G. and Siddiqui M. (2002). Impact of water quality on the kinetic parameters of NDMA photodegradation. *Proceedings AWWA Water Quality and Technology Conference*, Seattle, WA.

Stefan M. I., Kruithof J. C. and Kamp P. C. (2005). Advanced oxidation treatment of herbicides: From bench-scale studies to full-scale installation. *Proceedings of the Third International Congress on Ultraviolet Technologies*, Whistler, BC, Canada, May 24–27.

Stone W. W., Gilliom R. J. and Ryberg K. R. (2014). Pesticides in U.S. streams and rivers: occurrence and trends during 1992–2011. *Environmental Science and Technology*, **48**, 11025–11030.

Sudhakaran S. and Amy G. L. (2013). QSAR models for oxidation of organic micropollutants in water based on ozone and hydroxyl radical rate constants and their chemical classification. *Water Research*, **47**, 1111–1122.

Sudhakaran S., Calvin J. and Amy G. (2012). QSAR models for the removal of organic micropollutants in four different river water matrices. *Chemosphere*, **87**, 144–150.

Sudhakaran S., Latterman S. and Amy G. (2013). Appropriate drinking water treatment processes for organic micropollutants removal based on experimental and model studies – A multi-criteria analysis study. *Science of the Total Environment*, **442**, 478–488.

Sun L. and Bolton J. R. (1996). Determination of the quantum yield for the photochemical generation of hydroxyl radicals in TiO_2 suspensions. *Journal of Physical Chemistry*, **100**(10), 4127–4134.

Sun Y., Shu Y., Xu T., Shui M., Zhao Z., Gu Y. and Wang X. (2012). Review of the photodecomposition of some important energetic materials. *Central European Journal of Energetic Materials*, **9**(4), 411–423.

Sundstrom D. W., Klei H. E., Nalette T. A., Reidy D. J. and Weir B. A. (1986). Destruction of halogenated aliphatics by ultraviolet catalyzed oxidation with hydrogen peroxide. *Hazardous Waste & Hazardous Materials*, **3**, 101.

Svrcek C. and Smith D. W. (2004). Cyanobacteria toxins and current state of knowledge on water treatment options: a review. *Journal of Environmental Engineering Science*, **3**, 155–185.

Swaim P. D., Ridens M., Festger A. and Royce A. (2011). UV Advanced oxidation for taste and odor control: Understanding life-cycle cost and sustainability. *Proceedings of the 2nd North American Conference on Ozone, Ultraviolet and Advanced Oxidation Technologies*, Toronto, ON, Canada, September 18–21.

Tabrizi G. B. and Mehrvar M. (2004). Integration of advanced oxidation technologies and biological processes: recent developments, trends, and advances. *Journal of Environmental Science and Health. Part A-Toxic/Hazardous Substances & Environmental Engineering*, **A39**(11–12), 3029–3081.

Taniyasu Y., Kasu M. and Makimoto T. (2006). An aluminum nitride light-emitting diode with a wavelength of 210 nanometres. *Nature*, **441**, 325–328.

Tarrah H., Plummer S. and Roberton A. (2017). Occurrence and state approaches for addressing cyanotoxins in US drinking water. *Journal of American Water Works Association*, **109**, 40–47.

Taube H. (1942). Reactions of solutions containing ozone, H_2O_2, H^+ and Br^-. *Journal of the American Chemical Society*, **64**, 2468.

Tauber A. and von Sonntag C. (2000). Products and kinetics of the OH radical-induced dealkylation of atrazine. *Acta hydrochimica et hydrobiologica*, **28**(1), 15–23.

Thomsen C. L., Madsen D., Poulsen J. Aa., Thøgersen J. and Knak-Jensen S. J. (2001). Femtosecond photolysis of aqueous HOCl. *Journal of Chemical Physics*, **115**(20), 9361–9369.

Tian M. (1910). Sur la nature de la décomposition de l'eau oxygénée produite par la lumière. *Comptes rendus*, **151**, 1040.

Tixier C., Singer H. P., Canonica S. and Müller S. R. (2002). Phototransformation of triclosan in surface waters: a relevant elimination process for this widely used biocide – laboratory studies, field measurements, and modeling. *Environmental Science and Technology*, **36**, 3482–3489.

Tizaoui C., Mezughi K. and Bickley R. (2011). Heterogeneous photocatalytic removal of the herbicide clopyralid and its comparison with UV/H_2O_2 and ozone oxidation techniques. *Desalination*, **273**, 197–204.

Toor R. and Mohseni M. (2007). UV-H_2O_2 based AOP and its integration with biological activated carbon treatment for DBP reduction in drinking water. *Chemosphere*, **66**, 2087–2095.

Trawinski J. and Skibinski R. (2017). Studies on photodegradation processes of psychotropic drugs: a review. *Environmental Science and Pollution Research*, **24**, 1152–1199.

Turro N. J., Ramamurthy V. and Scaiano J. C. (2010). Modern molecular photochemistry of organic molecules. J. Stiefel (ed.), University Science Books, Mill-Valley, CA, USA.

U.S. EPA (2006). UV Disinfection Guidance Manual. Office of Water, U.S. Environmental Protection Agency, Washington DC.

U.S. EPA (2012a). Estimation Programs Interface (EPI) Suite™ for Microsoft® Windows v.4.11. AOPWIN™ Program. United States Environmental Protection Agency, Washington DC, USA.

U.S. EPA (2012b). 2012 Edition of the Drinking Water Standards and Health Advisories. Office of Water, U.S. EPA, Washington DC, April 2012.

Urey H. C., Dawsey L. H. and Rice F. O. (1929). The absorption spectrum and decomposition of hydrogen peroxide by light. *Journal of the American Chemical Society*, **51**, 1371–1383.

USGS (2016). Total cylindrospermopsins, microcystins/nodularins, and saxitoxins data for the 2007 United States Environmental Protection Agency National Lake Assessement. US Geological Survey Report, Data Series 929, Reston, VA, USA.

Verma S. and Sillanpää M. (2015). Degradation of anatoxin-a by UV-C LED and UV-C LED/H_2O_2 advanced oxidation processes. *Chemical Engineering Journal*, **274**, 274–281.

Vidal E., Negro A., Cassano A. and Zalazar C. (2015). Simplified reaction kinetics, models and experiments for glyphosate degradation in water by the UV/H_2O_2 process. *Photochemical and Photobiologocal Sciences*, **14**, 366–377.

Vione D., Maurino V., Minero C. and Pelizzetti E. (2005). Nitration and photonitration of naphthalene in aqueous system. *Environmental Science and Technology*, **39**, 1101–1110.

Vlad S., Abderson W. B., Peldszus S. and Huck P. M. (2014). Removal of the cyanotoxin anatoxin-a by drinking water treatment processes: a review. *Journal of Water and Health*, **12**(4), 601–617.

Volman D. H. and Chen J. C. (1959). The photochemical decomposition of hydrogen peroxide in aqueous solutions of allyl alcohol at 2537 Å. *Journal of the American Chemical Society*, **81**, 4141–4144.

von Gunten U. and Oliveras Y. (1997). Kinetics of the reaction between hydrogen peroxide and hypobromous acid: implication on water treatment and natural systems. *Water Research*, **31**(4), 900–906.

von Gunten U. and Oliveras Y. (1998). Advanced oxidation of bromide-containing waters: bromate formation mechanisms. *Environmental Science and Technology*, **32**, 63–70.

von Sonntag C. (2006). Free-radical-induced DNA damage and its repair. In: A Chemical Perspective, S. Schreck (ed.), Springer-Verlag Heidelberg, Germany.

von Sonntag C. and Schuchmann H.-P. (1991). The elucidation of peroxyl radical reactions in aqueous solution with the help of radiation-chemical methods. *Angewandte Chemie International Edition English*, **30**, 1229–1253.

von Sonntag C. and Schuchmann H.-P. (1992). UV Disinfection of drinking water and by-product formation - some basic considerations. *Journal of Water Supply, Research and Technology-Aqua*, **41**(2), 67–74.

Voronov A., Arnold E. and Roth E. (2003). Long-life technology of high power amalgam lamps. *Proceedings of the 2nd International Congress of IUVA*, Vienna, Austria, July 9–11.

Wagner E. D., Hsu K. M., Lagunas A., Mitch W. A. and Plewa M. J. (2012). Comparative genotoxicity of nitrosamine drinking water disinfection byproducts in *Salmonella* and mammalian cells. *Mutation Research*, **741**, 109.

Wait I. (2009). Lamp Sleeve Fouling in Ultraviolet Disinfection Reactors: the Accumulation of Inorganic Foulants on Potable Water UV Reactor Lamp Sleeves: Composition, Rate, Effects, and Modelling. VDM Verlag, Dr. Müller GmbH & Co.; Saarbrücken, Germany. ISBN-10: 363912023X.

Wan H. B., Wong M. K. and Mok C. Y. (1994). Comparative study on the quantum yields of direct photolysis of organophosphorus pesticides in aqueous solution. *Journal of Agricultural Food and Chemistry*, **42**(11), 2625–2630.

Wang Y.-N., Chen J., Li X., Zhang S. and Qiao X. (2009). Estimation of aqueous phase reaction rate constants of hydroxyl radical with phenols, alkanes, and alcohols. *QSAR & Combinatorial Science*, **28**, 1309–1316.

Wang D., Bolton J. R., Andrews S. A. and Hofmann R. (2015a). UV/chlorine control of drinking water taste and odour at pilot and full-scale. *Chemosphere*, **136**, 239–244.

Wang D., Bolton J. R., Andrews S. A. and Hofmann R. (2015b). Formation of disinfection by-products in the ultraviolet/chlorine advanced oxidation process. *Science of the Total Environment*, 518–519, 49–57.

Wang J., Ried A., Stapel H., Zhang H., Zhang Y., Chen M., Ang W. S., Xie R., Duarah A., Zhang L. and Lim M. H. (2015c). A pilot-scale investigation of ozonation and advanced oxidation processes at Choa Chu Kang Waterworks. *Water Practice & Technology*, **10**(1), 43–49.

Wang D., Duan X., He X. and Dionysiou D. D. (2016). Degradation of dibutyl phthalate (DBP) by UV-254 nm/H_2O_2 photochemical oxidation: kinetics and influence of various process parameters. *Environmental Science and Pollution Research*, **23**, 23772–23780.

Wang X., Zhang H., Zhang Y., Shi Q., Wang J., Yu J. and Yang M. (2017). New insights into trihalomethane and haloacetic acid formation potentials: correlation with the molecular composition of natural organic matter in source water. *Environmental Science and Technology*, **51**(4), 2015–2021.

Water Research Foundation (2009). Design and Performance Guidelines for UV Sensor Systems. Project #91236. Published by Water Research Foundation, Denver, CO, USA.

Water Research Foundation (2010). Treating Algal Toxins Using Oxidation, Adsorption, and Membrane Technologies. Project #2839. Published by Water Research Foundation, Denver, CO, USA.

Water Research Foundation (2011). Evaluation of Computational Fluid Dynamics for Modelling the UV-initiated Advanced Oxidation Processes. Project #3176. Published by Water Research Foundation, Denver, CO, USA.

WateReuse Foundation (2010). Reaction Rates and Mechanisms of Advanced Oxidation Processes (AOPs) for Water Reuse. Project number: WRF-04-017, Published by WateReuse Foundation, Alexandria, VA, USA.

Weeks J. L. and Matheson M. S. (1956). The primary quantum yield of hydrogen peroxide decomposition. *Journal of the American Chemical Society*, **78**, 1273–1279.

Weinstein J. and Bielski B. H. J. (1979). Kinetics of the interaction of HO_2 and O_2^- radicals with hydrogen peroxide. The Haber-Weiss reaction. *Journal of the American Chemical Society*, **101**(1), 58–62.

Westerhoff P., Aiken G., Amy G. and Debroux J. (1999). Relationships between the structure of natural organic matter and its reactivity towards molecular ozone and hydroxyl radicals. *Water Research*, **33**, 2265–2276.

Westerhoff P., Mezyk S. P., Cooper W. J. and Minakata D. (2007). Electron pulse radiolysis determination of hydroxyl radical rate constants with Suwannee river fulvic acid and other dissolved organic matter isolates. *Environmental Science and Technology*, **41**, 4640–4646.

Westrick J., Szlag D. C., Southwell B. J. and Sinclair J. (2010). A review of cyanobacteria and cyanotoxin removal/inactivation in drinking water treatment. *Analytical and Bioanalytical Chemistry*, **397**(5), 1705–1714.

Williams D. L. H. (2004). Nitrosation Reactions and the Chemistry of Nitric Oxide. Elsevier, Amsterdam.

Wöhrle D., Tausch M. W. and Stohrer W. D. (1998). Photochemie, Konzepte, Methoden, Experimente. Wiley-VCH, Weinheim, Germany.

Wojnárovits L. and Takács E. (2013). Structure dependence of the rate coefficients of hydroxyl radical + aromatic molecule reaction. *Radiation Physics and Chemistry*, **87**, 82–87.

Wolfrum E. J., Ollis D. F. and Lim P. K. (1994). The UV-H_2O_2 process: quantitative EPR determination of radical concentrations. *Journal of Photochemistry and Photobiology, A: Chemistry*, **78**, 259–265.

Wols B. A. (2012). Computational Fluid Dynamics in Drinking Water Treatment. B. A. Wols ed. IWA Publishing, London, UK.

Wols B. A. and Hofman-Caris C. H. M. (2012a). Review of photochemical reaction constants of organic micropollutants required for UV advanced oxidation processes in water. *Water Research*, **46**, 2815–2827.

Wols B. A. and Hofman-Caris C. H. M. (2012b). Modelling micropollutant degradation in UV/H_2O_2 systems: Lagrangian versus Eulerian method. *Chemical Engineering Journal*, **210**, 289–297.

Wols B. and Vries D. (2012). On a QSAR approach for the prediction of priority compound degradation by water treatment processes. *Water Science & Technology*, **66**(7), 1446–1453.

Wols B. A., Hofman J. A. M. H., Beerendonk E. F., Uijttewaal W. S. J. and van Dijk J. C. (2011). A systematic approach for the design of UV reactors using computational fluid dynamics. *AIChE Journal*, **57**(1), 193–207.

Wols B. A., Hofman-Caris C. H. M., Harmsen D. J. H. and Beerendonk E. F. (2013). Degradation of 40 selected pharmaceuticals by UV/H_2O_2. *Water Research*, **47**(15), 5876–5888.

Wols B. A., Harmsen D. J. H., Beerendonk E. R. and Hofman-Caris C. H. M. (2014). Predicting pharmaceutical degradation by UV (LP)/H_2O_2 processes: a kinetic model. *Chemical Engineering Journal*, **255**, 334–343.

Wols B. A., Harmsen D. J. H., Wanders-Dijk J., Beerendonk E. F. and Hofman-Caris C. H. M. (2015a). Degradation of pharmaceuticals in UV (LP)/H_2O_2 reactors simulated by means of kinetic modeling and computational fluid dynamics. *Water Research*, **75**, 11–24.

Wols B. A., Harmsen D. J. H., Beerendonk E. R. and Hofman-Caris C. H. M. (2015b). Predicting pharmaceutical degradation by UV (MP)/H_2O_2 processes: a kinetic model. *Chemical Engineering Journal*, **263**, 336–345.

Wols B. A., Harmsen D. J. H., van Remmen T., Beerendonk E. F. and Hofman-Caris C. H. M. (2015c). Design aspects of UV/H_2O_2 reactors. *Chemical Engineering Science*, **137**, 712–721.

Xiao Y., Fan R., Zhang L., Yue J., Webster R. D. and Lim T. T. (2014). Photodegradation of iodinated trihalomethanes in aqueous solution by UV 254 irradiation. *Water Research*, **49**, 275–285.

Xu B., Chen Z., Qi F., Ma J. and Wu, F. (2009). Rapid degradation of new disinfection by-products in drinking water by UV irradiation: *N*-Nitrosopyrrolidine and *N*-nitrosopiperidine. *Separation and Purification Technology*, **69**, 126–133.

Xu G., Liu N., Wu M., Bu T. and Zheng M. (2013). The photodegradation of clopyralid in aqueous solutions: effects of light sources and water constituents. *Industrial and Engineering Chemistry Research*, **52**, 9770–9774.

Yan S. and Song W. (2014). Photo-transformation of pharmaceutically active compounds in the aqueous environment: a review. *Environmental Science: Processes & Impacts*, **16**, 697–720.

Yang W., Zhou H. and Cicek N. (2014). Treatment of organic micropollutants in water and wastewater by UV-based processes: a literature review. *Critical Reviews in Environmental Science and Technology*, **44**, 1443.

Yao H., Sun P., Minakata D., Crittenden J. C. and Huang C. H. (2013). Kinetics and modeling of degradation of ionophore antibiotics by UV and UV/H_2O_2. *Environmental Science and Technology*, **47**, 4581–4589.

Yu X.-Y. (2004). Critical evaluation of rate constants and equilibrium constants of hydrogen peroxide photolysis in acidic aqueous solutions containing chloride ions. *Journal of Physical and Chemical Reference Data*, **33**(3), 747–763.

Yu X.-Y. and Barker J. R. (2003a). Hydrogen peroxide photolysis in acidic aqueous solutions containing chloride ions. I. Chemical mechanism. *Journal of Physical Chemistry A*, **107**, 1313–1324.

Yu X.-Y. and Barker J. R. (2003b). Hydrogen peroxide photolysis in acidic aqueous solutions containing chloride ions. II. Quantum yield of $HO^{\bullet}$ (aq.) radicals. *Journal of Physical Chemistry A*, **107**, 1325–1332.

Yuan F., Hu C., Hu X., Qu J. and Yang M. (2009). Degradation of selected pharmaceuticals in aqueous solution with UV and UV/H_2O_2. *Water Research*, **43**, 1766–1774.

Yuan F., Hu C., Wei D., Chen Y. and Qu J. (2011). Photodegradation and toxicity changes of antibiotics in UV and UV/H_2O_2 process. *Journal of Hazardous Materials*, **185**, 1256–1263.

Zamyadi A., Ho L., Newcombe G., Daly R. I., Burch M., Baker P. and Prévost M. (2009). Release and oxidation of cell-bound saxitoxins during chlorination of *Anabaena circinalis* cells. *Environmental Science and Technology*, **44**, 9055–9061.

Zehavi D. and Rabani J. (1972). The oxidation of aqueous bromide ions by hydroxyl radicals. A pulse radiolysis investigation. *Journal of Physical Chemistry*, **76**(3), 312–319.

Zellner R., Exner M. and Herrmann H. (1990). Absolute OH quantum yields in the laser photolysis of nitrate, nitrite and dissolved H_2O_2 at 308 and 351 nm in the temperature range 278–353 K. *Journal of Atmospheric Chemistry*, **10**, 411–425.

Zeng T., Plewa M. J. and Mitch W. A. (2016). *N*-Nitrosamines and halogenated disinfection byproducts in U.S. full advanced treatment trains for potable reuse. *Water Research*, **101**, 176–186.

Zepp R. G. (1978). Quantum yields for reaction of pollutants in dilute aqueous solution. *Environmental Science and Technology*, **12**(3), 327–329.

Zepp R. G., Hoigné J. and Bader H. (1987). Nitrate-induced photooxidation of trace organic chemicals in water. *Environmental Science and Technology*, **21**, 443–450.

Zhang Z., Feng Y., Liu Y., Sun Q., Gao P. and Ren N. (2010). Kinetic degradation model and estrogenicity changes of EE2 (17α ethinylestradiol) in aqueous solution by UV and UV/H_2O_2 technology. *Journal of Hazardous Materials*, **181**, 1127–1133.

Zhang S., Gitungo S., Axe L., Dyksen J. E. and Raczko R. F. (2016). A pilot plant study using conventional and advanced water treatment processes: evaluating removal efficiency of indicator compounds representative of pharmaceuticals and personal care products. *Water Research*, **105**, 85–96.

Zhao X., Alpert S. M. and Ducoste J. J. (2009). Assessing the impact of upstream hydraulics on the dose distribution of ultraviolet reactors using fluorescence microspheres and computational fluid dynamics. *Environmental Engineering Science*, **26**(15), 947–959.

Zhou H. and Smith D. W. (2002). Advanced technologies in water and wastewater treatment. *Journal of Environmental Engineering Science*, **1**, 247–264.

Zhou C., Gao N., Deng Y., Chu W., Rong W. and Zhou S. (2012). Factors affecting ultraviolet irradiation/ hydrogen peroxide (UV/H_2O_2) degradation of mixed N-nitrosamines in water. *Journal of Hazardous Materials*, 231–232, 43–48.

Zhu L., Nicovich J. M. and Wine P. (2003). Temperature-dependent kinetics studies of aqueous phase reactions of hydroxyl radicals with dimethylsulfoxide, dimethylsulfone, and methanesulfonate. *Aquatic Sciences*, **65**, 425–435.

Zong W., Sun F. and Sun X. (2013). Oxidation by-products formation of microcystin-LR exposed to UV/H_2O_2: toward the generative mechanism and biological toxicity. *Water Research*, **47**, 3211–3219.

Chapter 3

Application of ozone in water and wastewater treatment

Daniel Gerrity, Fernando L. Rosario-Ortiz, and Eric C. Wert

3.1 INTRODUCTION

The use of ozone for water treatment dates back to the early part of the twentieth century, with the first implementation of ozone in a drinking water facility for disinfection purposes in 1906 (Nice, France) (Rice *et al.* 1981). Since then, ozone has been used in drinking water applications to meet multiple objectives including disinfection, minimizing taste and odor, reducing the formation of disinfection byproducts, and oxidizing organic contaminants. Over the past decade, the popularity of ozone for wastewater/water reuse treatment and industrial applications has increased because of its ability to provide inactivation of pathogens and oxidation of organic contaminants. Today, advancements in ozone generation technology have promoted the production of ozone at high percent weight (10–12%) using less electrical power per pound of ozone produced (Thompson & Drago, 2015). The oxidative power of ozone combined with more cost efficient generation methodologies set the stage for future growth in municipal and industrial applications.

There is a wealth of literature on the history, chemistry and applications of ozone for water and wastewater treatment (see recent summary by (von Sonntag & von Gunten, 2012) and other reviews in the area (Hubner *et al.* 2015; von Gunten, 2003a, b). This chapter is not intended to offer a comprehensive discussion of all of the aspects related to the application of ozone for different treatment technologies, but rather to offer the reader an overview of recent developments on the use of ozone for water and wastewater treatment (including applications for disinfection and organic contaminant oxidation), which represent a significant fraction of existing ozone applications and projected growth.

3.2 PROPERTIES OF OZONE

The ozone molecule is composed of three oxygen atoms (Figure 3.1). The electron density and distribution across the ozone molecule causes it to be highly unstable and reactive toward a wide variety of water constituents. Ozone has a solubility in water of approximately 0.01 M and a Henry's law constant of 100 atm/M at 20°C (von Sonntag & von Gunten, 2012).

Figure 3.1 Structure of the ozone molecule.

3.3 DECOMPOSITION OF OZONE IN WATER

The mechanism by which ozone decomposes in water has been of great interest to the scientific community, starting with the pioneering work by Hoigné and others (Buhler *et al.* 1984; Hoigné & Bader, 1975; Staehelin *et al.* 1984; Staehelin & Hoigné, 1982). The initial work on the decomposition of ozone examined the reaction between this species and the hydroxide ion (Eq. 3.1). The reaction rate constant for this initial step has been reported as 70 M^{-1} s^{-1} (Staehelin *et al.* 1984). However, this process will be of importance only at high pH values (pH > 10), where the observed half-life of ozone would be expected to be in the range of minutes.

$$OH^- + O_3 \rightarrow \text{Products} \quad k_1 = 70\ M^{-1}s^{-1} \tag{3.1}$$

In this mechanism, it was initially suggested that the reaction occurred via transfer of an oxygen atom to form HO_2^- and molecular oxygen (Eq. 3.2) (Staehelin & Hoigné, 1982; Tomiyasu *et al.* 1985). The reaction would then yield $^{\bullet}OH$ via a series of additional reactions.

$$OH^- + O_3 \rightarrow HO_2^- + O_2 \tag{3.2}$$

However, recent work suggests that the process occurs via the formation of an adduct, which decomposes to radical species leading eventually to $^{\bullet}OH$ (Reactions 3.3–3.7) (Merenyi *et al.* 2010).

$$OH^- + O_3 \rightarrow HO_4^- \tag{3.3}$$

$$HO_4^- \rightleftarrows HO_2^{\bullet} + O_2^{\bullet -} \tag{3.4}$$

In the presence of ozone, reactions 3.5–3.7 occur, with the formation of $^{\bullet}OH$.

$$O_2^{\bullet -} + O_3 \rightarrow O_2 + O_3^{\bullet -} \tag{3.5}$$

$$O_3^{\bullet -} \rightarrow O_2 + O^{\bullet -} \tag{3.6}$$

$$O^{\bullet -} + H_2O \rightarrow {}^{\bullet}OH + OH^- \tag{3.7}$$

In the absence of ozone, $HO_2^{\bullet}$ and its conjugate base, $O_2^{\bullet -}$, disproportionate to HO_2^- (which protonates rapidly to H_2O_2; $pK_a(H_2O_2 = 11.8)$) and O_2, with a rate constant of $k = 9.7 \times 10^7$ M^{-1} s^{-1} (Bielski *et al.* 1985) which is approximately 50 times slower than the reverse of reaction 3.4. The formation of $^{\bullet}OH$ is an important aspect of the chemistry of ozone, as this species is highly unselective and reacts at rates close to the diffusion-controlled limit with many organic contaminants (Buxton *et al.* 1988). For more details regarding the series of reactions regarding the decomposition of ozone, please refer to the recent publication by von Sonntag and von Gunten (2012).

In natural waters (including surface water) and also in wastewaters, the decomposition of ozone occurs via reactions with reactive groups within the dissolved organic matter (DOM) (Buffle *et al.* 2006b; Buffle & von Gunten, 2006; Nöthe *et al.* 2009). The decay of ozone in natural waters has been characterized as multi-phased, with 2–3 different kinetics regimes observed (von Sonntag & von Gunten, 2012). These regimes were determined using stopped-flow techniques, which allow the experimentalist to monitor fast kinetics reactions, i.e., within milliseconds of reaction time (Buffle *et al.* 2006b). In this discussion, the observed kinetics regimes will be described as an initial phase and a second phase.

In the initial phase (less than 20 seconds), ozone decays following a sequence of fast reactions with the DOM (Buffle *et al.* 2006b; Buffle & von Gunten, 2006). Ozone has a high reactivity with specific DOM moieties such as secondary and tertiary amines, and phenols (Buffle & von Gunten, 2006). In these reactions, ozone is reduced by electron-donating moieties, forming the ozonide radical (Eq. 3.8).

$$DOM + O_3 \rightarrow DOM^{\bullet+} + O_3^{\bullet-} \tag{3.8}$$

The formed ozonide radical then decomposes into $^{\bullet}OH$ following reactions 3.6 and 3.7. An important observation regarding this pathway is that reactive groups within DOM are reformed upon other reactions between ozone and DOM (Nöthe *et al.* 2009). These reactions are expected to continue the formation of phenolic-type structures, which can transfer electrons to ozone molecules. However, it has been recently reported that ozone reduces the electron-donating capacity of the DOM to a limited extent (Wenk *et al.* 2013).

The slow decomposition phase (t > ~20 seconds) has a slower ozone decay and is modeled as pseudo first-order (Buffle *et al.* 2006a; Buffle *et al.* 2006b; Buffle & von Gunten, 2006). In this range, the pseudo-first order degradation rate of ozone has been shown to be in the order of 0.001 to 0.020 s^{-1} for DOM isolates (von Sonntag & von Gunten, 2012), to 0.01–0.05 s^{-1} for wastewater samples and wastewater-derived organic matter fractions (Gonzales *et al.* 2012; Wert *et al.* 2009a).

An important consideration regarding the reactions involving DOM is the scavenging of the formed $^{\bullet}OH$. The reaction rate constant between DOM and $^{\bullet}OH$ has been measured with various methods (pulse radiolysis, competition kinetics, etc.) and reported in the literature by different research groups (Dong *et al.* 2010; Hoigné & Bader, 1979; Keen *et al.* 2014; Lee *et al.* 2013; Rosario-Ortiz *et al.* 2008; Westerhoff *et al.* 2007). The reported values range from 1×10^4 to 9.5×10^4 (mg C/L)$^{-1}$ s^{-1} (Keen *et al.* 2014; Lee *et al.* 2013; Rosario-Ortiz *et al.* 2008; Westerhoff *et al.* 2007), although recent studies suggest that the range is smaller (Lee *et al.* 2013; Keen *et al.* 2014). In the published work, greater variability has been observed for values measured using pulse radiolysis when compared to competitive kinetics using alternative sources of $^{\bullet}OH$. It is not clear whether the differences in observed values are artifacts of the different methods used or whether they are explained by intrinsic differences in the reactivity of the DOM, which could depend on the specific sample collected. However, a recent study compared kinetic values using both competitive kinetics and pulse radiolysis and found overall good agreement between both methods (Keen *et al.* 2014), suggesting that the differences observed in different samples are probably due to intrinsic differences between the set of samples analyzed.

Given the importance of the formation of $^{\bullet}OH$ in the application of ozone for water and wastewater treatment, there is a need to quantify $^{\bullet}OH$. The concentration of ozone can be easily quantified using different methods, including a colorimetric method and with the use of modern electrochemical sensors

(Bader & Hoigne, 1982; Kaiser *et al.* 2013; Rakness *et al.* 2010). However, there are no methods to quantify $^{\bullet}OH$ at the concentrations expected during the ozonation of water. Instead, the use of probe molecules that can be monitored has been the preferred approach. The most widely used method involves the quantification of *para*-chlorobenzoic acid (pCBA) (Elovitz & von Gunten, 1999), although in theory any organic compound with limited reactivity towards ozone could be used. The reaction rate constant between pCBA and $^{\bullet}OH$ is 5×10^9 M^{-1} s^{-1}, whereas the reaction rate constant with ozone is ≤ 0.15 M^{-1} s^{-1} (Elovitz & von Gunten, 1999). The decay of pCBA can be monitored, and from equation 3.9 the overall $^{\bullet}OH$ exposure can be calculated.

$$\ln\left(\frac{[\mathrm{pCBA}]_{t=t}}{[\mathrm{pCBA}]_{t=0}}\right) = -k_{\mathrm{pCBA}} \int_{t=0}^{t=t} \mathrm{OH}\, dt \tag{3.9}$$

Work by von Gunten has shown that the ratio of the $^{\bullet}OH$ to ozone exposure, known as the R_{ct}, is constant during the second phase of ozonation (Elovitz & von Gunten, 1999). Therefore, by knowing the concentration of ozone, an estimate of the concentration of $^{\bullet}OH$ can be obtained. The values for the R_{ct} are on the order of 10^{-8} for a wide variety of water and wastewater samples (Elovitz & von Gunten, 1999; Gonzales *et al.* 2012; Hollender, 2009). This method has been widely used to model the decay of organic contaminants during the ozonation process.

3.4 OZONATION FOR CONTAMINANT REMOVAL

3.4.1 Overview

The application of ozone for contaminant oxidation has gained widespread interest (Acero *et al.* 2001; Acero & von Gunten, 2001; Dodd *et al.* 2006; Gerrity *et al.* 2011; Gerrity & Snyder, 2011; Huber *et al.* 2003; Huber *et al.* 2005; Lee *et al.* 2013; Rivas *et al.* 2009; Roth & Sullivan, 1981; Wert *et al.* 2009a). Recent advances in the quantification of organic compounds has led to the widespread detection of contaminants of emerging concern (CECs) such as endocrine disrupting compounds (EDCs), pharmaceuticals and personal care products (PPCPs) and pesticides in the part-per-trillion (ppt) to part-per-billion (ppb) level in many different fresh water sources (Snyder *et al.* 2003a; Snyder *et al.* 2003b). A majority of these chemicals are consumed by humans for the prevention, control and cure of diseases or used for agricultural purposes. Ingested compounds are generally not fully metabolized and end up in human waste and ultimately the influent of a wastewater treatment plant. Because the traditional wastewater treatment processes do not fully remove these trace organics, the more recalcitrant compounds are discharged into the aquatic environment and find their way into downstream drinking water intakes (Snyder *et al.* 2003b). In industrialized countries, more than 90% of wastewater is treated in centralized wastewater treatment plants, making them major point sources for trace organic contaminants as well as the most efficient location for mitigation (Hollender, 2009). Because of concerns with regards to the occurrence of organic contaminants in surface waters, there has been a lot of interest in the use of ozone to degrade these species, both in wastewater and drinking water systems.

3.4.2 Direct reactions with ozone

There are three main mechanisms for the reaction of ozone with organic compounds. The first is a cycloaddition reaction where ozone targets the carbon–carbon double bond (C=C) such as olefinic compounds (Beltran, 2003). The most well-known cycloaddition mechanism is the Criegee mechanism (Figure 3.2). The Criegee mechanism occurs when ozone forms an unstable five-member ring or ozonide.

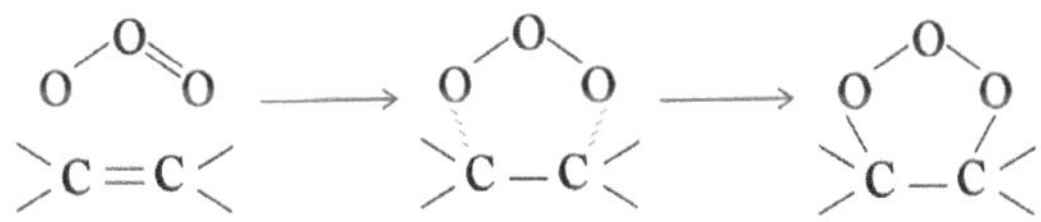

Figure 3.2 Ozonide ring formed by the Criegee mechanism.

In aqueous solutions, the completion of the reaction results in the formation of ketones, aldehydes, or acids. Ozone could also react with aromatic compounds through cycloaddition reactions resulting in the breakup of the aromatic ring (Beltran, 2003; Hubner *et al.* 2015). The second mechanism is electrophilic substitution reactions. Electrophilic substitution is when ozone attacks one nucleophilic position on the organic compound resulting in the substitution of one part of the molecule. Aromatic compounds are likely to undergo electrophilic substitution due to the stability of the aromatic ring structure. A third mechanism is *via* direct oxidation of a reducing species by ozone, e.g., ozone reactions with compounds containing sulfur in reduced states. For a more in-depth review of the mechanisms for ozone reactivity, including examples of the reactivity of different contaminant groups, refer to excellent reviews published recently (Hubner *et al.* 2015, von Sonntag & von Gunten, 2012).

Generally, the reactions of organic compounds with ozone are slower than those with •OH. Table 3.1 presents second-order reaction rate constants for the reactions of ozone and •OH with selected organic compounds. The limited reactivity of some organic compounds with ozone will limit the efficacy of the removal of these contaminants within the initial stages of ozone application, when the concentrations of ozone are highest. However these contaminants will still be removed *via* direct reactions with •OH, close to the diffusion controlled limit (i.e., 10^{10} M^{-1} s^{-1}) (Buxton *et al.* 1988).

The reactivity of ozone toward a wide range of trace organic compounds and the second-order rate constants for the respective reactions have been extensively studied and reported in the literature. Various experimental techniques (e.g., competition kinetics, stopped-flow fast kinetics) and sophisticated theoretical models have been employed over the years in order to determine kinetic rate constants. Canonica and Tratnyek (2003) reviewed the quantitative structure-activity relationships (QSAR) for the oxidation of various organic compounds with a number of commonly used oxidants, including ozone, and emphasized the interrelation between traditional empirical models, as well as the potential for development of QSARs based on theoretical models. Li *et al.* (2013) have developed QSAR models to determine the bimolecular rate constants of ozone as a function of temperature. The model descriptors were validated for various compounds, including alkenes, cycloalkenes, haloalkenes, alkynes, oxygen- and nitrogen-containing compounds, and aromatic structures. Sudhakaram and Amy (2013) developed QSAR models for O_3 and •OH-based processes and predicted k_{O_3} and $k_{\bullet OH}$ data for a number of chemical compounds with various structural characteristics. The k_{O_3} rate constants varied from 5×10^{-4} to 10^5 M^{-1} s^{-1}, whereas the $k_{\bullet OH}$ rate constants ranged from 0.04×10^9 to 18×10^9 M^{-1} s^{-1}. The authors identified several molecular descriptors that determined the reactivity toward ozone or •OH, including double bond equivalency, ionization potential, electron affinity, and weakly-polar component of solvent accessible surface area. The models were validated through specific methods. Figure 3.3 exemplifies the correlation between predicted and experimental data.

Table 3.1 Ozone and ·OH radical rate constants with selected organic contaminants.

Compound Name	Rate Constant		Reference
	k_{O_3} ($M^{-1}s^{-1}$)	$k_{\cdot OH} \times 10^{-9}$ ($M^{-1}s^{-1}$)	
Carbamazepine	3×10^5	8.8	Huber *et al.* (2003)
Diazepam	0.75	7.2	Huber *et al.* (2003)
Diclofenac	1×10^6	7.5	Huber *et al.* (2003)
Ibuprofen	9.6	7.4	Huber *et al.* (2003)
Iopromide	<0.8	3.3	Huber *et al.* (2003)
Naproxen	2×10^5	9.6	Huber *et al.* (2003)
Sulfamethoxazole	5.7×10^5	5.5	Huber *et al.* (2003)
2-methylisoborneol (MIB)	0.35	5.1	Peter and von Gunten (2007)
Geosmin	0.10	7.8	Peter and von Gunten (2007)
Microcystin-LR	4.1×10^5	11	Rodriguez *et al.* (2007)
Cylindrospermopsin (pH 8)	3.4×10^5	5.5	Rodriguez *et al.* (2007)
Anatoxin-a (pH 8)	6.4×10^5	3.0	Rodriguez *et al.* (2007)
NDMA	0.052 ± 0.0016	0.45 ± 0.021	Lee *et al.* (2007)
1,4-Dioxane	<1	3	Bowman *et al.* (2001)
Atrazine	6	3	Acero *et al.* (2000)
Bisphenol A (BPA)	7×10^5	10	Deborde *et al.* (2005) Rosenfeldt and Linden (2004)
Methoxychlor	250 ± 24	3.9 ± 0.9	Jin *et al.* (2012)
Dicamba	<0.1	3.5 ± 0.1	Jin *et al.* (2012)
Hexachlorobenzene	<0.01	0.24 ± 0.12	Jin *et al.* (2012)
Tris(2-chloroethyl) phosphate (TCEP)	0.8 ± 0.2	0.56 ± 0.02	Jin *et al.* (2012); Watts and Linden (2009)

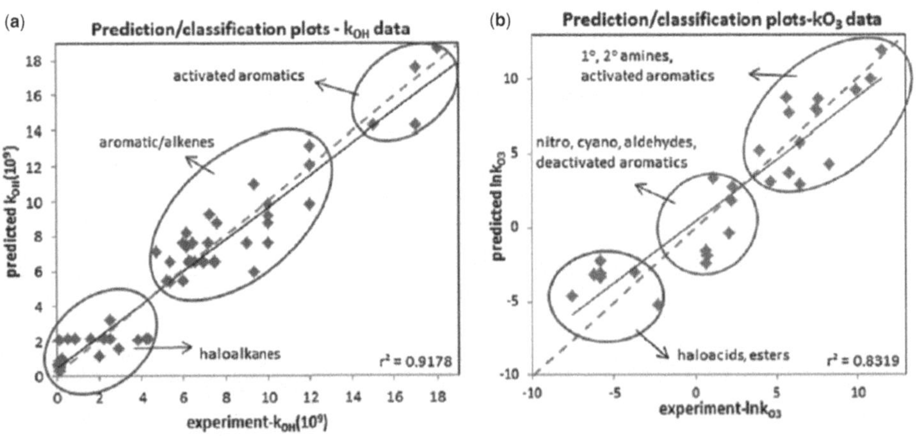

Figure 3.3 Predicted (QSAR) versus experimental k_{O_3} and k_{OH} data. Adapted from Sudhakaran and Amy (2013), with permission.

Lee and von Gunten (2012) developed 18 QSARs to describe the oxidation of organic and inorganic compounds by chlorine, chlorine dioxide, ferrate, and ozone, and used the models to predict the rate constants for the reactions of these oxidants with contaminants of emerging concern. The authors discussed how the uncertainty in QSAR model-predicted rate constants may affect the accuracy of kinetic model predictions of trace organic compound removal during water treatment.

Recently, Lee *et al.* (2015) reported theoretical models to predict k_{O_3} for a variety of organic compounds (aromatics, olefins, and amines), based on quantum chemical molecular orbital calculations using *ab initio* Hartree-Fock and density functional theory methods. The authors found good correlations ($r^2 = 0.77 - 0.96$) between the rate constants predicted with QSAR models based on Hammett and Taft constants and those predicted with the developed quantum mechanics models using molecular orbital descriptors.

3.4.3 Impact of water quality on process performance

Process performance in ozone systems is highly dependent on water quality, specifically the quantity and composition of both organic and inorganic constituents. Because the performance of ozonation for the elimination of a particular compound depends on contributions from both ozone and $^{\bullet}OH$ oxidation (Eq. 3.10), it is important to describe how process efficiency is affected by the competing reactions of various water constituents (scavengers) for the oxidizing species (O_3 and $^{\bullet}OH$).

$$-\ln\left(\frac{[c]_{t=t}}{[c]_{t=0}}\right) = k_{O_3}\int_{t=0}^{t=t}[O_3]\,dt + k_{^{\bullet}OH}\int_{t=0}^{t=t}[^{\bullet}OH]\,dt \tag{3.10}$$

For ozone, nitrite and DOM are the principal components affecting the initial ozone demand phase. The ozone process is sometimes supplemented with hydrogen peroxide to expedite the decomposition of ozone into $^{\bullet}OH$, but this will be discussed in the next chapter. The rapid reaction between ozone and nitrite is shown in Eq. 3.11 (Crittenden *et al.* 2012). Because the molecular weights of NO_2^- (46 g/mole) and O_3 (48 g/mole) are approximately equal, the reaction requires a ~1:1 mass ratio in order to satisfy the ozone demand caused by nitrite.

$$O_3 + NO_2^- \rightarrow \text{Products} \quad k_{O_3,NO_2^-} = (3.7 \pm 0.5) \times 10^5\ M^{-1}\ s^{-1} \tag{3.11}$$

With respect to DOM, the O_3/DOC (or O_3/TOC) (DOC stands for dissolved organic carbon and TOC for total organic carbon) ratio has been shown to be effective in accounting for the effects of varying concentrations of DOM in different water matrices and for predicting the elimination of trace organic compounds in wastewater treatment applications (Lee *et al.* 2013). Many of the concepts discussed later are presented in the context of the O_3/DOC ratio. This ratio is simply the ozone dose divided by the DOC (i.e., O_3/DOC) or TOC (i.e., O_3/TOC). For full-scale applications, O_3/DOC ratios may range from 0.25–1.5, but the design ratio will ultimately depend on a variety of factors, including water quality, intended use of the treated effluent, treatment objectives, and potential for disinfection byproduct formation. For example, the O_3/DOC ratio for a potable reuse application might be limited by the formation of bromate, which has a 10-μg/L maximum contaminant level in the United States. When treating a wastewater with a high concentration of organic matter, there might be a practical limit on the O_3/DOC ratio, particularly in a high-flow system, because of capacity limitations on ozone generators.

When using this ratio for ozone dosing, the applied ozone dose can be calculated according to Eq. 3.12. The relationship between O_3/DOC ratio and ozone exposure (or ozone CT) is illustrated in Figure 3.4 for ten different secondary effluents. For this figure, the ozone exposure was based on the entire demand-decay period for each dosing condition.

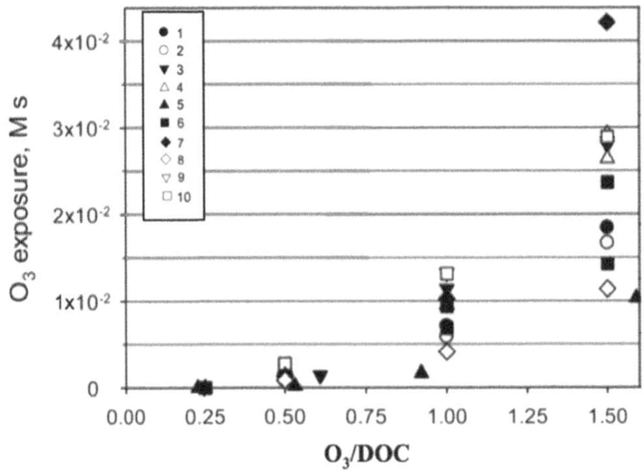

Figure 3.4 Dissolved ozone exposure as a function of O_3/DOC ratio in ten different secondary effluents. Adapted from Lee *et al.* (2013).

$$\text{Applied } O_3\left(\frac{mg}{L}\right) = \frac{O_3}{\text{DOC}} \times \text{DOC}\left(\frac{mg_C}{L}\right) + [NO_2^-]\left(\frac{mg_{NO_2}}{L}\right) \tag{3.12}$$

Figure 3.4 indicates that there is a 'lag' in dissolved ozone exposure when plotted against O_3/DOC ratio. In effect, ozone must be dosed at O_3/DOC ratios greater than ~0.25 in order to achieve a measurable dissolved ozone residual after 30 seconds. The 30-sec threshold is sometimes used to define the upper boundary of the initial ozone demand phase; the justification for this threshold will be described later in the context of quantifying dissolved ozone concentrations. In this context, the ozone exposure (or ozone CT) is effectively zero for low O_3/DOC ratios because the applied ozone dose is less than the initial ozone demand. However, some treatment objectives, including the elimination of some trace organic compounds, can still be achieved in these low-dose scenarios, typically for compounds that are fast reacting with ozone (Wert *et al.* 2009a, b). In other words, ozone CT is not always the most appropriate measure of ozone process efficiency, particularly in the context of advanced oxidation.

After the nitrite demand is satisfied, ozone rapidly oxidizes the reactive moieties (e.g., amines, phenols, and alkoxylated aromatics) of DOM in source water or effluent organic matter (EfOM) in wastewater. The initial demand reactions between ozone and EfOM typically carry over into the secondary decay phase because of the abundance of these reactive moieties and the expected reformation of reactive sites (Nöthe *et al.* 2009). This explains the greater variability in dissolved ozone exposure for different wastewaters at higher O_3/DOC ratios (Figure 3.4). The secondary ozone decay phase is further expedited by an increase in pH ($k_{OH^-,O_3} = 70\ M^{-1}\ s^{-1}$; (Crittenden *et al.* 2012)), as shown in Figure 3.5 and described above.

Some of these reactions, particularly between ozone and DOM, lead to the formation of $^{\bullet}OH$ as a primary, highly reactive intermediate (Buffle & von Gunten, 2006; Gonzales *et al.* 2012; Nöthe *et al.* 2009; Pablo Pocostales *et al.* 2010). As described earlier, this can be beneficial for trace organic compound elimination but can adversely impact disinfection. Many trace organic contaminants are susceptible to oxidation by both ozone and $^{\bullet}OH$ (Lee *et al.* 2013; von Gunten, 2003a; Wert *et al.* 2009a), but due to the complex structure of some microbes, particularly spore-forming microbes, extended exposure to dissolved ozone is typically required to achieve disinfection targets (Gamage *et al.* 2013). Therefore, predicting the performance of ozone, particularly for disinfection applications, requires an estimate of ozone exposure, or ozone CT, which is a function of the initial ozone dose and the reactivity between ozone and the various water matrix constituents.

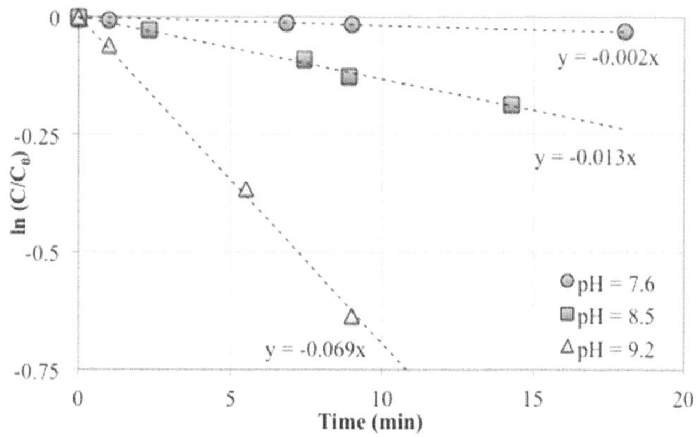

Figure 3.5 Decay of dissolved ozone as a function of pH in laboratory-grade water. The pseudo first order rate constant describing ozone decay increases from 0.002–0.069 min^{-1} as the pH increases from 7.6 to 9.2. Adapted from Crittenden *et al.* (2012).

Figure 3.6 illustrates these relationships for three different secondary treated wastewater effluents. This type of 'demand-decay curve' can be generated by spiking water matrices with aliquots of ozone from a stock solution with a known concentration. The dissolved ozone concentration in the sample is then measured and recorded throughout the decay phase. In these bench-scale systems, dissolved ozone can be measured using a gravimetric method based on oxidation of an indigo solution (Rakness *et al.* 2010). In full scale applications, other types of processes can be used to measured ozone, including electrochemical sensors (Kaiser *et al.* 2013). The instantaneous ozone demand (IOD) (i.e., the initial demand phase) for each dosing condition can be interpreted multiple ways. Due to practical limitations with the bench-scale demand-decay test, it is difficult to collect a representative sample prior to 30 seconds. Therefore, the IOD is sometimes defined as the difference between the applied ozone dose and the dissolved ozone residual after 30 seconds. More advanced quench-flow systems (Lee *et al.* 2013) are capable of measuring the dissolved ozone residual almost immediately after ozone addition, which allows for more accurate characterizations of the initial demand phase, where significant ozone consumption and $^{\bullet}OH$ formation has been observed (Nöthe *et al.* 2009; Buffle & von Gunten, 2006). When using a quench-flow system and an O_3/DOC ratio of 0.25 in a secondary effluent, (Lee *et al.* 2013) observed nearly complete depletion of the dissolved ozone residual within two seconds, and (Buffle *et al.* 2006b) was able to quantify ozone exposures as low as 0.01 mg-min/L for O_3/DOC ratios less than 0.25.

With respect to Figure 3.6, the corresponding TOC concentrations, IODs, and pseudo first-order rate constants for the secondary ozone decay phase (k_{O_3}) are summarized in Table 3.2. These values reflect their dependence on water quality in that they vary between matrices, and they also reflect the complexity of reactions between ozone and organic matter in that they vary as a function of applied ozone dose (or O_3/DOC ratio) within the same matrix. As the applied ozone dose or the concentration of DOM increases, the rate of change in the dissolved ozone concentration, including the IOD, will also increase in magnitude.

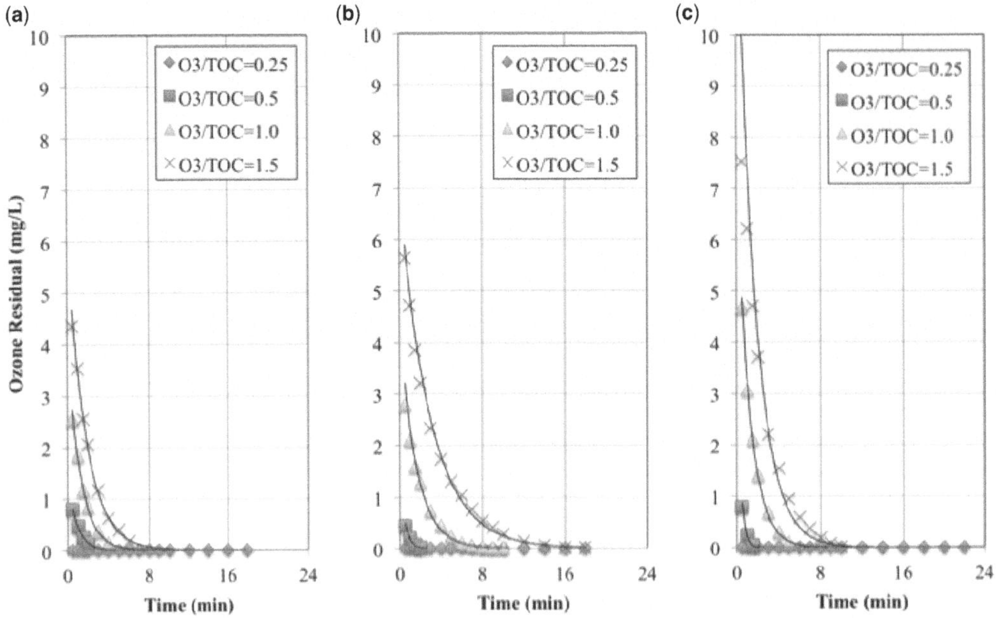

Figure 3.6 Ozone decay curves for three different secondary treated wastewater effluents. The wastewaters are listed in order of increasing dissolved organic carbon concentration (Table 3.2). Adapted from Snyder *et al.* (2015).

Table 3.2 Summary of DOC, IOD, and pseudo first order rate constants for ozone decay for the wastewaters in Figure 3.6. Adapted from Snyder *et al.* (2015).

	DOC (mg-C/L)	IOD (mg/L)			k_{O_3} (min^{-1})			O_3 CT (mg-min/L)			O_3 Exposure (10^{-3} M-s)		
O_3/DOC	–	0.5	1.0	1.5	0.5	1.0	1.5	0.5	1.0	1.5	0.5	1.0	1.5
A	6.1	1.6	2.0	2.9	1.2	0.8	0.6	1.3	5.0	11	1.6	6.2	14
B	6.8	2.0	2.4	3.2	2.1	0.6	0.3	0.7	7.0	21	0.8	8.7	26
C	13.6	3.0	6.2	6.8	3.0	0.8	0.6	1.3	8.9	23	1.6	11	29

In addition to characterizing the reactivity of water matrices, demand-decay curves can be used to calculate the accumulated ozone exposure at any time. One approach is to use a simple numerical method, such as the trapezoidal rule, to calculate the area under the demand-decay curve, which equates to the ozone exposure or ozone CT (the slight differences in the values in Table 3.2 are due to unit conversions). Ozone exposure is typically reported in units of M-s to estimate ozone performance based on kinetics or in units of mg-min/L to estimate ozone performance in disinfection applications. An analytical approach can also be used if the demand-decay curves are fitted to an exponential decay model, as shown in Eq. 3.13 (dissolved ozone residual as a function of time; exponential regression equation) and Eq. 3.14 (ozone exposure or ozone CT as a function of time) (assumes scavenging from nitrite is insignificant). In this approach, the IOD is calculated as the difference between the applied ozone doses and the constants of the exponential regression equations. The model parameters and the corresponding ozone CTs (or exposures) for the curves in Figure 3.6 were summarized previously in Table 3.2.

$$C_{O_3}(t) = (O_3/DOC \times DOC - IOD) \times e^{-k_{O_3}t} \tag{3.13}$$

$$\int_0^t [O_3]\, dt = \frac{(O_3/DOC \times DOC - IOD)}{k_{O_3}} \times (1 - e^{-k_{O_3}t}) \tag{3.14}$$

where

$C_{O_3}(t)$ = dissolved ozone concentration at time t (mg/L)

$\int_0^t [O_3]\, dt$ = ozone exposure or CT from time 0 to t (mg-min/L).

As noted earlier for the ozone exposures in Figure 3.4, another issue that is apparent in Figure 3.6 is the effect of dosing below the IOD (e.g., $O_3/DOC < 0.25$), which provides no measurable ozone residual with a conventional bench-scale setup. Despite the very low ozone CT in these scenarios, recent studies indicate that low-dose ozonation can still achieve oxidation of fast-reacting compounds (Lee *et al.* 2013; Wert *et al.* 2009a, b) and limited inactivation of viruses and non-spore-forming bacteria (Gamage *et al.* 2013). However, higher ozone doses capable of achieving a measurable CT are needed for other treatment applications – sometimes for the extended exposure to dissolved ozone (i.e., for disinfection) or simply for the higher ozone doses (i.e., for the elimination of ozone-resistant trace organic compounds). (Gamage *et al.* 2013) reported a threshold O_3/TOC of 0.33 for achieving measurable CT values based on the 30-second limitation of the conventional bench-scale setup.

The previous discussion provides a general overview of ozone decay and the methodology for calculating CT. However, the required procedure for calculating exposure times and log inactivation credits may differ in practice. The reader should refer to relevant regulatory guidelines (e.g., U.S. EPA guidance manuals (USEPA, 2010)) for specific instructions on how to calculate exposure times and log inactivation credits in full-scale (i.e., non-ideal) reactors for compliance purposes.

Although greater ozone exposure is generally advantageous for treatment efficacy, large contactors may be needed to allow for complete ozone decay and maximum treatment. In wastewater applications, ozone decay in applications with high-applied ozone doses (e.g., $O_3/TOC > 1.0$) may require 10–30 minutes of contact time (see Figure 3.6), which translates to greater capital costs and larger structural footprints. More stable water matrices, such as reverse osmosis permeate, groundwater, or surface water, may require even longer contact times for similar ozone doses.

As will be described in the next chapter, the O_3/H_2O_2 process can be used to mitigate this issue because it expedites the decomposition of ozone into fast-reacting $^{\bullet}OH$ [$k_{O_3,H_2O_2} = 1.4 \times 10^3\, M^{-1}s^{-1}$ (Crittenden *et al.* 2012)], which significantly reduces the necessary contact time. However, ozone can also react with other water quality constituents, including DOM, to generate $^{\bullet}OH$, as was described earlier in this chapter. Therefore, it is important to consider how water quality can impact the efficacy of $^{\bullet}OH$ oxidation.

$^{\bullet}OH$ exposure is sometimes described as the product of the 'steady state' concentration and exposure time (i.e., $[^{\bullet}OH]_{ss} \times t$). The $[^{\bullet}OH]_{ss}$ parameter can defined as the ratio of the $^{\bullet}OH$ generation rate (with units of $M\, s^{-1}$), which is due to reactions between ozone and various water constituents, and the $^{\bullet}OH$ scavenging rate constant ($k_{i,^{\bullet}OH}$; with units of s^{-1}), which is defined in Eq. 3.15. The principal $k_{i,^{\bullet}OH}$ values to be used in Eq. 3.15 are as follows: $k_{DOC,^{\bullet}OH}$ = matrix-specific $(mg\, C/L)^{-1}\, s^{-1}$ (Lee *et al.* 2013), $k_{HCO_3^-,^{\bullet}OH} = 8.5 \times 10^6\, M^{-1}s^{-1}$ (Buxton *et al.* 1988), $k_{CO_3^{-2},^{\bullet}OH} = 3.9 \times 10^8$ (Buxton *et al.* 1988), $k_{NH_3,^{\bullet}OH} = 9.0 \times 10^7\, M^{-1}s^{-1}$ (Buxton *et al.* 1988), $k_{Br^-,^{\bullet}OH} = 1.1 \times 10^9 M^{-1}s^{-1}$ (Buxton *et al.* 1988), and $k_{NO_2^-,^{\bullet}OH} = 1 \times 10^{10} M^{-1}s^{-1}$ (Coddington *et al.* 1999). The matrix-specific rate constant describing the reactivity between DOM and $^{\bullet}$OH can be determined using different methods (as described above), but average values have also been reported in the literature. For example, the average $k_{DOC,^{\bullet}OH}$ value based on experiments in 10 different secondary treated wastewater effluents was determined to be $(2.1 \pm 0.6) \times 10^4$ (mg C/L)$^{-1}$ s^{-1} (Lee *et al.* 2013), similar to other values recently reported (Keen *et al.* 2014).

$$k_{i,^{\bullet}OH}(s^{-1}) = k_{DOC,^{\bullet}OH} \times [DOC] + k_{HCO_3^-,^{\bullet}OH} \times [HCO_3^-] + k_{CO_3^{-2},^{\bullet}OH} \times [CO_3^{-2}] + k_{NH_3,^{\bullet}OH)} \times [NH_3] + k_{Br^-,^{\bullet}OH} \times [Br^-] + k_{NO_2^-,^{\bullet}OH} \times [NO_2^-] \quad (3.15)$$

where

$k_{i,^{\bullet}OH}$ = pseudo first order $^{\bullet}$OH scavenging rate constant (s^{-1})
$k_{DOC,^{\bullet}OH}$ = second order rate constant between DOC and $^{\bullet}$OH ((mg-C/L)$^{-1}$ s^{-1})
DOC = dissolved organic carbon concentration (mg-C/L)
$k_{i,^{\bullet}OH}$ = second order rate constants between the major scavengers and $^{\bullet}$OH

The $^{\bullet}$OH scavenging contribution from nitrite can generally be omitted in ozone applications because low concentrations of nitrite are quickly converted to nitrate by ozone and $^{\bullet}$OH. Lee *et al.* (2013) described the relative contribution of each major constituent included in Eq. 3.15 to $^{\bullet}$OH scavenging in 10 secondary treated wastewater effluents (Lee *et al.* 2013). The authors used a competition kinetics method based on the conversion of *tert*-butanol to formaldehyde (Nöthe *et al.* 2009; von Sonntag & von Gunten, 2012) to quantify the second order rate constant for the reaction between the effluent organic matter (EfOM) characteristic to those particular water matrices and $^{\bullet}$OH, and the overall pseudo first-order $^{\bullet}$OH scavenging rate constant in the respective matrices (see Eq. 3.15). The water quality data for the secondary effluents in that study are summarized in Table 3.3, and Figure 3.7 illustrates the overall $^{\bullet}$OH scavenging rate constants and the relative contribution from each scavenger.

Table 3.3 Summary of water quality data for 10 secondary treated wastewater effluents used in characterizing $^{\bullet}$OH scavenging potential. Adapted from Lee *et al.* (2013).

WW	SRT[a] (days)	pH	DOC[c] (mg-C/L)	Alkalinity (as mM HCO_3^-)	NH_4^+ (mg-N/L)	NO_2^- (mg-N/L)	Br^- (μg/L)	$k_{DOC,^{\bullet}OH}$ (mg/L)$^{-1}$ s^{-1} (10^{-4})	$k_{^{\bullet}OH}$ (s^{-1}) (10^{-5})
1	11	7.0	4.7	2.9	0.023	0.07	37	2.00	1.20
2	17	7.2	4.7	4.4	0.025	<0.05	40	2.05	1.36
3	3	7.2	6.0	1.3	0.11	0.16	940	1.95	1.41
4	7	6.9	7.1	2.5	0.09	0.06	174	1.80	1.52
5	11	7.1	7.0	2.1	0.09	0.05	330	2.00	1.63
6	7	7.6	5.7	2.7	0.07	<0.05	93	2.65	1.77
7	12	7.3	7.0	4.1	0.02	<0.05	730	2.20	2.00
8	11	7.3	6.3	3.4	5.8	0.30	31	3.40	2.45
9	N/A[b]	7.3	26	5.9	46.8	0.45	140	1.05	3.35
10	1.5	7.3	15	6.6	46.9	0.17	409	2.15	3.90

[a]SRT = solids retention time, [b]N/A = Not applicable; Trickling filter, [c]DOC = dissolved organic carbon.

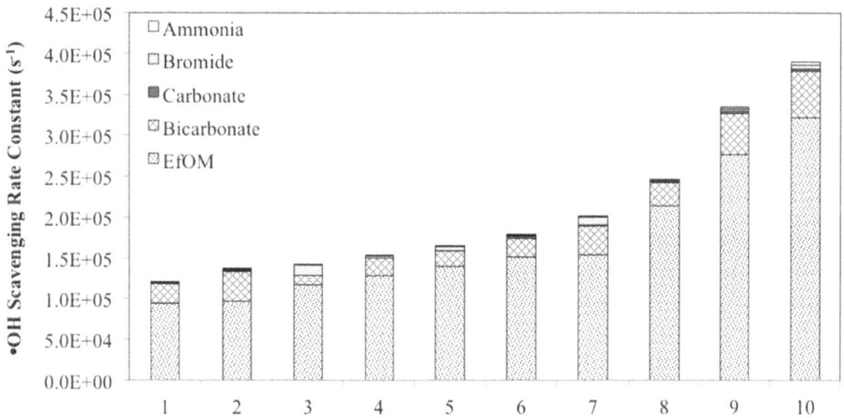

Figure 3.7 Contribution of major •OH scavengers to the overall pseudo first order •OH scavenging rate constant for 10 secondary treated wastewater effluents (see Table 3.3). Adapted from Lee *et al.* (2013).

As shown in Figure 3.7, EfOM is the principal constituent contributing to •OH scavenging during ozonation of wastewater effluents (Keen *et al.* 2014). Bicarbonate also contributes a significant amount, but the remaining constituents generally have minimal impact on the overall scavenging rate. These observations further support the utility of the O_3/DOC ratio – this time in the context of •OH exposure. Figure 3.8 illustrates the linear relationship between O_3/DOC ratio and •OH exposure in secondary treated wastewater effluents, and the corresponding regression equation is described in Eq. 3.16. Similar to the ozone exposures described earlier, Figure 3.8 and Eq. 3.16 indicate that there is a minimum O_3/DOC value (i.e., horizontal intercept) that must be reached before a measurable •OH exposure is achieved. The O_3/DOC "threshold" value ranged from 0.06–0.24 (or 0.10 based on Eq. 3.16) for the 10 secondary effluents in Figure 3.8 (Lee *et al.* 2013).

$$\int_0^t [{}^\bullet\mathrm{OH}]\,dt\ (\mathrm{M-s}) = 3.8\times10^{-10}\times \mathrm{O_3/DOC} - 3.7\times10^{-11}(\mathrm{M-s})\,R^2 = 0.89 \tag{3.16}$$

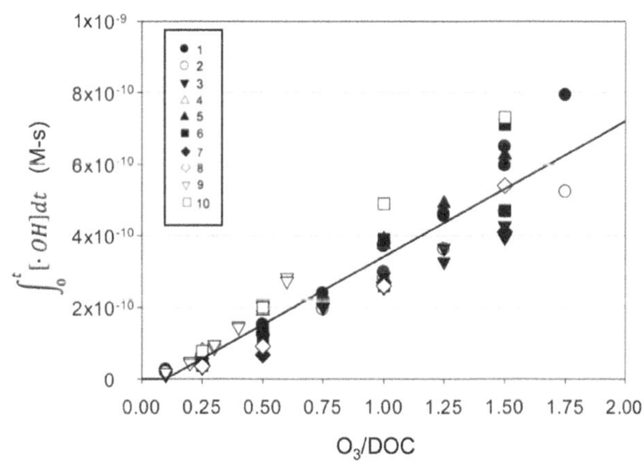

Figure 3.8 Linear relationship between •OH exposure and O_3/DOC for ozonation of 10 secondary treated wastewater effluents (see Table 3.3). Adapted from Lee *et al.* (2013).

As shown earlier in Figure 3.4 and Figure 3.8, the O_3/DOC ratio is a useful parameter for estimating ozone and •OH exposure, respectively. Lee *et al.* (2013) further describes why the O_3/DOC ratio is useful for predicting trace organic compound elimination in a variety of wastewater matrices. First, oxidation via •OH exposure is often the more significant reaction mechanism when considering relative contribution to overall elimination. As shown in Eq. 3.10, the overall elimination of a compound will depend on its exposure to both dissolved ozone and •OH and the second order rate constants describing the target compound's reactivity with each oxidant. As shown in Table 3.1, trace organic compounds are orders of magnitude more reactive with •OH than with dissolved ozone. Despite the limited •OH exposure in typical ozone applications, •OH oxidation still serves as the dominant reaction mechanism for ozone-resistant compounds (i.e., compounds with $k_{s,O_3} < 10^3 - 10^4\,M^{-1}s^{-1}$) because of the significantly slower kinetics with dissolved ozone (Figure 3.9). As shown in Figure 3.8, •OH exposure in wastewater treatment applications typically ranges from $<1 \times 10^{-10}\,M^{-1}\,s^{-1}$ to $8 \times 10^{-10}\,M^{-1}\,s^{-1}$ for O_3/DOC ratios ranging from 0.25–1.5, respectively (Lee *et al.* 2013).

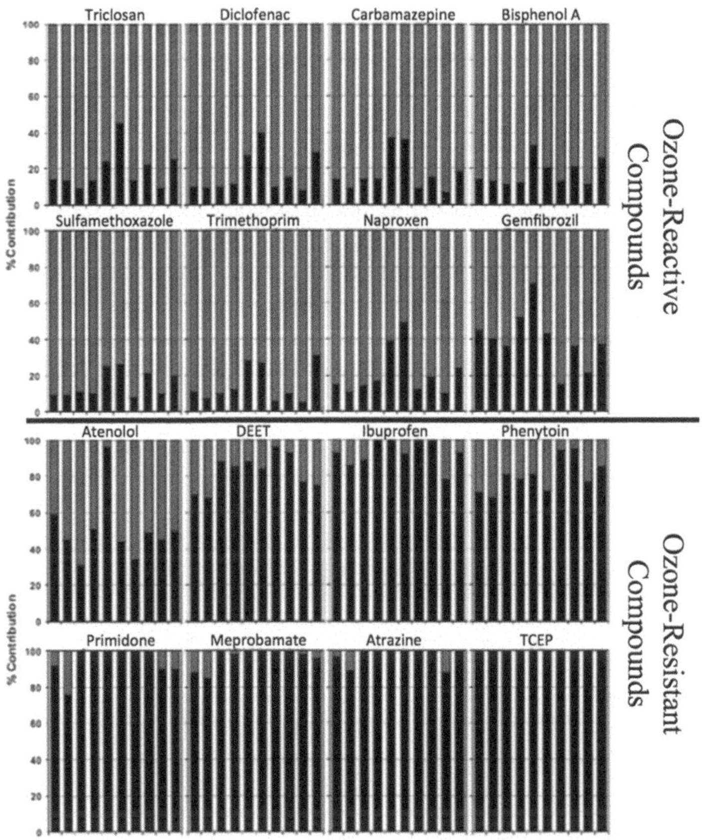

Figure 3.9 Relative contribution to overall elimination from ozone (light gray) and •OH (black) during ozonation of 10 secondary treated wastewater effluents. Compounds are grouped based on whether they are reactive ($k_{i,O_3} > 10^4\,M^{-1}s^{-1}$) or resistant ($k_{i,O_3} < 10^4\,M^{-1}s^{-1}$). Atenolol is moderately reactive with ozone ($k_{i,O_3} = 1.7 \times 10^3\,M^{-1}s^{-1}$), which explains the approximately equal contributions from ozone and •OH. Adapted from Lee *et al.* (2013).

Second, •OH exposure yields a strong linear correlation when described as a function of O_3/DOC ratio (Figure 3.8 and Eq. 3.16), and the relationship is more consistent than for dissolved ozone exposure (Figure 3.4). In fact, •OH exposure differs by less than a factor of two for the same O_3/DOC ratio, while dissolved ozone exposure differs by more than a factor of four at higher O_3/DOC ratios. As a result, one might expect the elimination of ozone-reactive compounds at high O_3/DOC ratios to be highly wastewater-specific. However, ozone-reactive compounds (i.e., $k_{s,O_3} > 10^3 - 10^4 M^{-1}s^{-1}$) experience consistent and nearly complete elimination even with low O_3/DOC ratios (i.e., $O_3/DOC < 0.5$). In this dosing range, the differences in dissolved ozone exposures are less significant.

The reason for the relatively strong linear correlation between •OH exposure and O_3/DOC ratio in different secondary effluents is twofold: (1) DOM is the principal •OH scavenger during wastewater ozonation, accounting for majority scavenging (Keen *et al.* 2014; Lee *et al.* 2013), and (2) the second-order rate constant between DOM and •OH are expected to be relatively constant between different samples (Keen *et al.* 2014; Lee *et al.* 2013). Therefore, the contribution to overall elimination from ozone is only significant for ozone-reactive compounds, which have $k_{s,O_3} > 10^3 - 10^4 M^{-1}s^{-1}$, and those compounds experience nearly complete elimination even at low O_3/DOC ratios. Many other trace organic compounds are primarily susceptible to •OH oxidation, and •OH exposure is primarily controlled by the ozone dose and the concentration of DOC, which can vary from <5 mg C/L to >20 mg C/L depending on the secondary effluent in question. Collectively, these observations explain the utility of the O_3/DOC ratio in predicting trace organic compound elimination during wastewater ozonation, as illustrated in Figure 3.10. Table 3.4 summarizes the expected percent eliminations for a variety of trace organic compounds at O_3/DOC ratios ranging from 0.25–1.5. The first order rate constants describing the elimination of each compound with respect to O_3/DOC ratio are also included in the table.

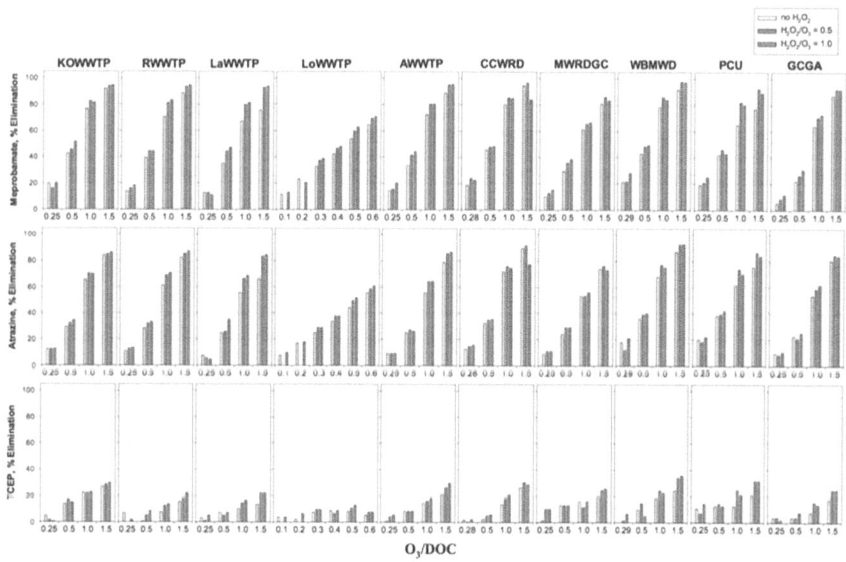

Figure 3.10 Relative elimination of three ozone-resistant trace organic compounds as a function of O_3/DOC and H_2O_2/O_3 ratios (described in the next chapter) in 10 different secondary effluents. Note that the O_3/DOC scale for "LoWWTP" is different from the other wastewaters. The target compounds include the anti-anxiety drug meprobamate (top), the herbicide atrazine (middle), and the flame retardant tris (2-chloroethyl) phosphate (TCEP) (bottom). Adapted from Lee *et al.* (2013).

Table 3.4 Summary of first order rate constants describing the elimination of each compound with respect to O_3/DOC ratio ($k_{s,O_3/DOC}$) and expected percent eliminations for each compound as a function of O_3/DOC ratio. The values represent average percent eliminations based on experiments in 10 different secondary treated wastewater effluents. The compounds are grouped according to their reactivity with ozone and •OH (see footnotes), and a representative 'indicator' compound is also listed for each group. Adapted from Snyder *et al.* (2015).

Group	Compound	($k_{s,O_3/DOC}$)	O_3/DOC = 0.25	O_3/DOC = 0.5	O_3/DOC = 1.0	O_3/DOC = 1.5
1[a]	Triclosan	8.96	93 ± 9	97 ± 1	97 ± 1	97 ± 1
	Naproxen	8.12	90 ± 16	98 ± 0	98 ± 0	98 ± 0
	Sulfamethoxazole	7.96	84 ± 13	98 ± 0	99 ± 1	99 ± 1
	Carbamazepine	7.85	92 ± 15	99 ± 0	99 ± 0	99 ± 0
	Diclofenac	7.34	91 ± 13	98 ± 1	98 ± 1	98 ± 1
	Trimethoprim	6.84	92 ± 15	99 ± 0	99 ± 0	99 ± 0
	Bisphenol A	5.90	91 ± 14	98 ± 1	98 ± 1	98 ± 1
	Indicator	7.57	90 ± 14	98 ± 0	98 ± 0	98 ± 0
2[b]	Gemfibrozil	6.19	81 ± 18	99 ± 0	99 ± 0	99 ± 0
	Atenolol	4.20	47 ± 8	97 ± 1	98 ± 1	98 ± 1
	Indicator	5.20	64 ± 13	98 ± 1	99 ± 1	99 ± 1
3[c]	Phenytoin	2.84	34 ± 15	67 ± 13	94 ± 4	98 ± 1
	Ibuprofen	2.62	38 ± 10	69 ± 7	94 ± 4	98 ± 1
	Primidone	2.55	30 ± 9	60 ± 8	91 ± 5	97 ± 2
	DEET	2.35	26 ± 9	57 ± 9	88 ± 6	97 ± 3
	Indicator	2.59	32 ± 10	63 ± 9	92 ± 5	98 ± 2
4[d]	Meprobamate	1.40	18 ± 5	40 ± 8	71 ± 9	86 ± 8
	Atrazine	1.10	15 ± 5	33 ± 6	64 ± 8	81 ± 8
	Indicator	1.25	17 ± 5	37 ± 6	68 ± 8	84 ± 8
5[e]	TCEP	0.18	−1 ± 13	9 ± 5	15 ± 3	23 ± 3

*Shading represents >80% oxidation.
[a]Group 1: $k_{s,O_3} \geq 1 \times 10^5 M^{-1}s^{-1}$
[b]Group 2: $10\ M^{-1}s^{-1} \leq k_{s,O_3} < 1 \times 10^5 M^{-1}s^{-1}$
[c]Group 3: $k_{s,O_3} < 10\ M^{-1}s^{-1}$ and $k_{s,^{\bullet}OH} \geq 5 \times 10^9 M^{-1}s^{-1}$
[d]Group 4: $k_{s,O_3} < 10\ M^{-1}s^{-1}$ and $1 \times 10^9 M^{-1}s^{-1} \leq k_{s,^{\bullet}OH} < 5 \times 10^9 M^{-1}s^{-1}$
[e]Group 5: $k_{s,O_3} < 10\ M^{-1}s^{-1}$ and $k_{s,^{\bullet}OH} < 1 \times 10^9 M^{-1}s^{-1}$

3.4.4 Summary

As described by Eq. 3.10, predicting the performance of ozone systems requires knowledge of (1) the rate constants for the target contaminant(s) with the reactive species, (2) the overall ozone exposure, and (3) the overall •OH exposure. Ozone demand and decay often vary considerably between matrices because of rapid reactions with nitrite (when present) and varying reaction rates with organic matter present in the water. Therefore, estimates of overall ozone exposure are often developed through bench-scale or pilot-scale ozone demand-decay testing with multiple dosing conditions. •OH exposure is also a function of ozone dose and the concentrations of various scavengers, but matrix-independent empirical models have been developed to aid in predicting •OH exposure. Because of the relatively similar reactivity between •OH and organic matter in different waters, the O_3/DOC (or O_3/TOC) framework serves as a useful strategy to standardize ozone dosing. For more matrix-specific estimates, the relative elimination of a probe compound, for which the second order rate constant describing its reactivity with •OH is known, can be used to estimate the •OH exposure for a given ozone dosing condition.

In a recently published study, Schindler Wildhaber *et al.* (2015) propose a laboratory-tiered methodology to evaluate the micropollutant treatment feasibility through ozonation at wastewater treatment plants, based on chemical and biological measurements. The methodology comprises four distinct modules that allow the experimentalist to draw valuable conclusions on the treatability of wastewater effluents with ozone. The key considerations include (1) water matrix effects on ozone stability; (2) efficacy of ozonation for trace organic compound removal; (3) formation of potentially toxic byproducts such as NDMA and bromate, as an indicator of ozonation impact on treated water quality; and (4) *in vitro* and *in vivo* toxicity assays on ozonated wastewater before and after biological treatment relative to untreated water. A similar methodology can be applied to evaluate ozonation for drinking water production.

3.5 FORMATION OF BYPRODUCTS

Water ozonation results in a variety of inorganic and organic byproducts (von Gunten, 2003a). Ozonation of bromide-containing waters may be at risk of forming bromate (Haag & Hoigne, 1983). Bromate is a known carcinogen and is regulated in drinking water at a concentration of 10 µg/L by the United States Environmental Protection Agency (USEPA, 1998). Formation results from the oxidation of bromide by ozone and •OH through several potential pathways (Fischbacher *et al.* 2015; von Gunten, 2003b; von Gunten & Hoigne, 1994; von Gunten & Oliveras, 1998). During the direct pathway, ozone oxidizes hypobromite (OBr^-) to bromite (BrO_2^-), which is then oxidized to bromate (BrO_3^-). During the direct-indirect and indirect-indirect pathways, •OH and ozone participate in intermediate oxidation reactions resulting in bromate formation. The HOBr/OBr- equilibrium ($pK_a = 8.8$) is a key factor affecting each of these bromate formation pathways. Ammonia addition, pH depression, and chloramine-based strategies are the most practical techniques to minimize bromate formation in practice (Buffle *et al.* 2004; von Gunten, 2003a; Wert *et al.* 2007).

Ozone oxidation of waters containing DOM from natural or anthropogenic sources can produce a variety of organic byproducts (von Gunten, 2003a). As larger complex organic molecules are oxidized, smaller byproducts are formed including aldehydes, carboxylic acids, and ketones. The formation of these products can also be assessed through assimilable organic carbon (AOC) measurements. Due to increases in AOC during ozonation, utilities incorporating ozone are at increased risk of coliform regrowth in the distribution system. AOC concentrations less than 100 µg/L have been recommended to minimize this risk (LeChevallier *et al.* 1992). As a result, the ozone process is commonly followed by biofiltration processes to remove these organic byproducts (i.e., AOC) formed during ozonation and increase the stability of the water.

During wastewater applications, ozonation can result in increased toxicity due to the production of organic byproducts (Stalter *et al.* 2010b). Subsequent studies have shown that aldehyde mixtures resulting from ozonation can result in deformed fish larvae (Yan *et al.* 2014). When the ozone process was followed by activated carbon treatment, the toxicity was reduced by removing these organic byproducts (Stalter *et al.* 2010a; Stalter *et al.* 2011). Ozone-biofiltration process performance has since been the focus of several research projects investigating efficacy for water reuse (Gerrity *et al.* 2011; Gerrity & Snyder, 2011; Reungoat *et al.* 2012; Reungoat *et al.* 2011; Reungoat *et al.* 2010).

Consideration must also be given to transformation products from ozonation of trace organic compounds in drinking water, wastewater or industrial waters (Hubner *et al.* 2015). The identification and toxicity of these products is quite complex and continues to be an active research area. Nitrosamine formation (most notably N-nitrosodimethylamine (NDMA)) has also been shown to occur through direct reactions with ozone in wastewater (Gerrity *et al.* 2015; Pisarenko *et al.* 2012; Schmidt & Brauch, 2008). Polymers added during wastewater treatment (Padhye *et al.* 2011), dyes from textile processing or clothes laundering (Zeng & Mitch, 2015), and some trace organic compounds found in wastewater are considered to be the primary sources of precursor material that react with ozone to form nitrosamines (Marti *et al.* 2015; Schmidt & Brauch, 2008). Marti *et al.* (2015) studied a number of model compounds to evaluate their role as NDMA precursors and found that some hydrazones, semicarbazides, and carbamates had relatively high NDMA yields. That same study indicated that NDMA precursors identified during ozonation did not necessarily yield NDMA during chloramination (i.e., precursor groups are somewhat distinct) and that bromide was a significant catalyst for NDMA formation for some precursor compounds. The bromide catalysis effect was also reported during ozonation of dimethylsulfamide (DMS), which is an NDMA precursor and bacterial metabolite of the fungicide tolylfluanid (von Gunten *et al.* 2010). Multiple studies concluded that molecular ozone, rather than $^{\bullet}OH$, was primarily responsible for NDMA formation during ozonation (Marti *et al.* 2015). To date, few mitigation measures other than eliminating the precursors or possibly minimizing the ozone dose have been identified. The formation of nitrosamines during ozone processes is still not very well understood and remains an active research area.

3.6 MICROBIOLOGICAL APPLICATIONS

The initial drivers for ozone in water treatment were control of iron and manganese, its ability to oxidize compounds contributing to taste and odor problems, and its disinfection efficacy (USEPA, 1999a; von Gunten, 2003b; Zuma *et al.* 2009; Zuma & Jonnalagadda, 2010), particularly for *Cryptosporidium* oocysts and *Giardia* cysts (von Gunten *et al.* 2001). Interestingly, drinking water and wastewater treatment facilities still rely on its disinfection efficacy in many applications, but in a seeming contradiction, an increasing number of facilities are also leveraging ozone's ability to promote biological growth. In fact, some facilities even employ ozone for both objectives: (1) to transform bulk organic matter to promote biological growth and improve biodegradation in downstream biofiltration processes (such as biological filters) and (2) as a final disinfection step. The following sections address each of these applications. Although there is some literature related to ozone disinfection in wastewater, studies have generally focused on drinking water treatment applications. Therefore, the following discussion is adapted largely from (Gamage *et al.* 2013), which provides a broad overview of ozone disinfection in wastewater applications.

3.6.1 Disinfection in drinking water and wastewater applications

Operational control of ozone systems has evolved over time in conjunction with technological advances and shifting treatment goals. In the U.S. EPA's 1986 disinfection guidance manual (USEPA, 1986), off-gas concentration and the product of off-gas concentration and time were identified as tools to estimate microbial inactivation. For a given applied ozone dose, lower off-gas concentrations correspond with higher ozone transfer efficiencies and, in effect, greater ozone exposures and improved disinfection. In recent years, the 'CT' value has become the standard dosing parameter for disinfection (Rakness *et al.* 2005; von Gunten, 2003b).

Ozone reacts rapidly with EfOM in wastewater applications so it is not always possible to measure dissolved ozone residuals and determine the corresponding CT values (Buffle *et al.* 2006a; Lee *et al.* 2013; Rosario-Ortiz *et al.* 2008). This is particularly true for the ozone/H_2O_2 advanced oxidation process (AOP) in which ozone reacts with H_2O_2 to generate $^{\bullet}OH$ with minimal dissolved ozone exposure (Lee *et al.* 2013).

Even without H_2O_2 addition, the dissolved ozone residual can be depleted within seconds when low ozone doses fail to meet the instantaneous ozone demand exerted by high EfOM concentrations (i.e., apparent CT of 0 mg-min/L). Therefore, alternative process control strategies and measures of disinfection efficacy, such as O_3/TOC or differential UV absorbance, have recently been proposed. A study by (von Gunten *et al.* 2001) even proposed a method to assess disinfection efficacy in drinking water by monitoring disinfection byproduct formation. They discussed the relationship between ozone CT and microbial inactivation (*Escherichia coli*, *G. lamblia* cysts, *Bacillus subtilis* spores, *C. parvum* oocysts) and also correlated the data with bromate formation, which is a common disinfection byproduct in ozone systems. Based on these correlations, they reported a first order kinetic relationship between *B. subtilis* spore inactivation and bromate formation.

Although ozone is a powerful disinfectant, one of its major limitations is its relative instability in water. In contrast, free chlorine and chloramine are highly stable in many water matrices, particularly in drinking water applications, so these oxidants can provide long-term residuals to reduce biological growth in distribution systems. Ozone decays at a much faster rate so it cannot be used as a residual disinfectant. On the other hand, ozone is much more effective than chlorine and chloramine when targeting spore-, oocyst-, or cyst-forming microorganisms. This became critically important after the cryptosporidiosis outbreaks in Milwaukee in 1993 (Mackenzie *et al.* 1994) and in Las Vegas in 1994 (Goldstein *et al.* 1996; USEPA, 2001). Therefore, ozone and chlorine are often used in tandem to inactivate a wide range of microorganisms and also provide a stable disinfectant residual.

3.6.2 Microbial surrogates and indicators

The primary goal of disinfection in water and wastewater treatment is to inactivate *pathogenic* microorganisms capable of causing disease. However, these microbes are generally more difficult and costly to quantify in full-scale systems and even in laboratory settings. Therefore, microbial *indicators* are often used as alternatives to assess disinfection efficacy and characterize the risk of disease. Ideal microbial indicators are microbes that are exclusively linked to fecal contamination, present in water in higher numbers than the target pathogens, and considered more resistant to disinfection than the target pathogens. Based on these criteria, the presence of indicator organisms in a water matrix equates to a higher level of fecal contamination and risk of pathogen exposure and subsequent waterborne disease. Because ideal indicators are more resistant to disinfection, their absence from water following treatment suggests that pathogenic microorganisms have also been removed. Unfortunately, a true *ideal* indicator has not been identified so the water industry is forced to use non-ideal indicators, such as *E. coli*, that satisfy these criteria to varying degrees.

In addition to the microbial indicator framework, microbial *surrogates* are also used in experimental research and, more recently, in validation testing for full-scale disinfection technologies. Microbial surrogates are microbes, particles, dyes, or any other additive that behave similarly to pathogens in various water and wastewater treatment processes. These surrogates are not necessarily linked to fecal contamination, but they are generally easy to assay and provide an estimate of how a pathogen would respond during treatment.

For ozonation, common microbial indicators/surrogates include *E. coli*, MS2 bacteriophage (a virus that infects bacteria), and *B. subtilis* spores. *E. coli* is often used as a surrogate for the inactivation of vegetative bacterial pathogens, and *E. coli* is also considered an indicator of fecal contamination. *E. coli* is also a member of the fecal and total coliform groups. Regulatory frameworks sometimes focus on *E. coli*, fecal coliforms, total coliforms, or some combination of the three. MS2 infects only bacteria, but it is often detected in environmental samples in conjunction with indicator bacteria, such as *E. coli*, and human pathogens. Because of its link to fecal contamination, it has been proposed as an indicator microorganism.

However, MS2 is more commonly used as a surrogate for the inactivation of pathogenic viruses (e.g., poliovirus, coxsackievirus, and echovirus) in disinfection processes. *B. subtilis* is a useful surrogate for protozoan parasites (e.g., *Cryptosporidium* and *Giardia*) because it forms disinfectant-resistant spores, they are significantly easier to assay, and they are not pathogenic to humans or animals.

With respect to UV disinfection, some microbes exhibit significantly different dose response curves than other microbes within the same group (e.g., UV-resistant adenovirus vs. UV-susceptible poliovirus). In contrast, ozone efficacy is relatively consistent within each microbial group, but as with other disinfectants, significant variability is observed between groups. Therefore, ozone disinfection efficacy will be described in the context of these major microbial groups in the sections below.

3.6.3 Ozone dosing frameworks for disinfection

Ozone CT is generally considered the industry standard for ozone process control in disinfection applications, particularly for drinking water. However, due to its versatility in describing ozone efficacy for trace organic contaminant oxidation (Lee *et al.* 2013), O_3/TOC has also been proposed as a potential framework for predicting disinfection efficacy. Based on Figure 3.11, which illustrates the relationship between O_3/TOC ratio and CT for five secondary effluents, there appears to be a relatively strong correlation for O_3/TOC values less than 1.3, but the correlation weakens considerably for O_3/TOC values greater than 1.3. Based on this observation, Eq. 3.17 was developed to predict ozone CT in wastewater disinfection applications when the O_3/TOC ratio is less than 1.3 (Gamage *et al.* 2013).

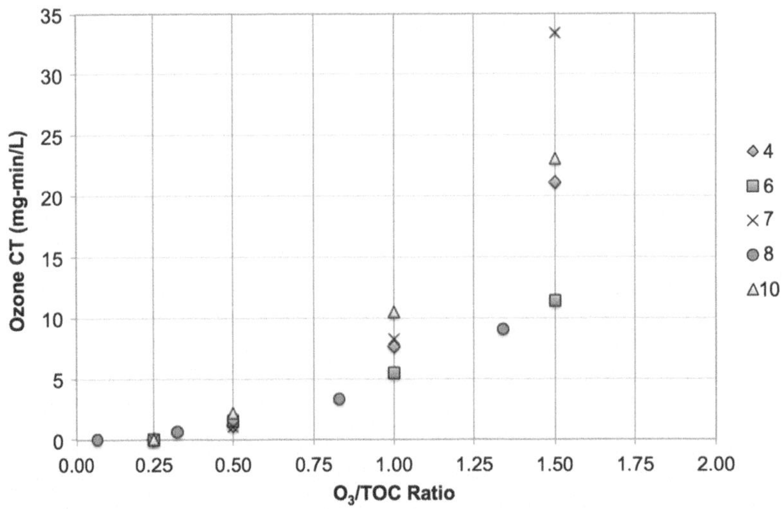

Figure 3.11 Relationship between ozone CT and O_3/TOC ratio in five secondary effluents. The labels correspond with the wastewaters in Table 3.3. Adapted from Gamage *et al.* (2013).

$$\text{Ozone CT (mg-min/L)} = 10 \times O_3/TOC - 3.3 \quad R^2 = 0.86 \tag{3.17}$$

As mentioned in previous sections, the pseudo first-order ozone decay rate constant (k_{O_3}) is wastewater-specific and inversely correlated with O_3/TOC ratio. Therefore, higher O_3/TOC ratios lead to greater stability of dissolved ozone, and the effect is particularly pronounced in some wastewaters, such as wastewater 7 in Figure 3.11.

Multiple kinetic models have been proposed for ozone disinfection, including the Chick (Eq. 3.18), Chick-Watson (Eq. 3.19), and delayed Chick-Watson models (Eq. 3.20) (Crittenden *et al.* 2012). Furthermore, (Crittenden *et al.* 2012) summarizes activation energies for the inactivation kinetics of a number of microorganisms (Eq. 3.21). These activation energies can be used in conjunction with the Arrhenius equation to adjust the kinetic parameters in Eqs. 3.18–3.20 for changes in temperature. A subset of the activation energies is provided in Table 3.5, and a modification to the Arrhenius equation is provided in Eq. 3.21 (Crittenden *et al.* 2012).

Table 3.5 Activation energies that can be used to correct k_C (Eq. 3.18) or Δ_{CW} (Eqs. 3.19 and 3.20) for changes in temperature according to the modified Arrhenius equation (Eq. 3.21). Adapted from Crittenden *et al.* (2012).

Microorganism	E_a (kJ/mole)
Escherichia coli	37.1
Bacillus subtilis	46.8
Cryptosporidium parvum	76.0
Cryptosporidium muris	92.8
Giardia lamblia	39.2
Giardia muris	70.0
Naegleria gruberi	31.4

$$\ln\left(\frac{N_t}{N_0}\right) = -kct \tag{3.18}$$

$$\ln\left(\frac{N_t}{N_0}\right) = -\Delta cwCT \tag{3.19}$$

$$\ln\left(\frac{N_t}{N_0}\right) = -\Delta cw(CT - b) \quad (\text{for } CT > b) \tag{3.20}$$

$$\frac{k_{T_1}}{k_{T_2}} = (e^{E_a/RT_1T_2})^{(T_1-T_2)} \tag{3.21}$$

where

N_t = number of microorganisms at time t
N_0 = number of microorganisms at time 0
k_C = pseudo first order Chick's law rate constant (min^{-1})
Δ_{CW} = coefficient of specific lethality $(mg\text{-}min/L)^{-1}$
CT = ozone exposure (mg-min/L)
b = lag coefficient (mg-min/L)

$k_{T_1} = k_C$ or Δ_{CW} at T_1 (min^{-1} or $(mg\text{-}min/L)^{-1}$)
$k_{T_2} = k_C$ or Δ_{CW} at T_2 (min^{-1} or $(mg\text{-}min/L)^{-1}$)
E_a = activation energy (kJ/mole)
R = ideal gas constant = 8.314×10^{-3} (kJ/mole-K)
T_1 = Temperature 1 (K)
T_2 = Temperature 2 (K)

As will be discussed in later sections, differential UV absorbance is another useful surrogate for verifying ozone dosing and predicting ozone process performance. As such, this dosing framework has also been evaluated for its ability to predict disinfection efficacy. Therefore, disinfection efficacy will be presented in the following sections in the context of kinetic models, ozone CT, O_3/TOC, and differential UV_{254} absorbance.

3.6.4 Vegetative bacteria

Multiple studies have indicated that ozone inactivation of vegetative bacteria, such as *E. coli*, can be highly variable (Blatchley *et al.* 2012; Gamage *et al.* 2013; Ishida *et al.* 2008). This is problematic because regulatory frameworks often target specific concentration thresholds (e.g., <2.2 most probable number/100 mL or non-detect) for indicator bacteria. In contrast, viruses and protozoan parasites are often regulated based on log inactivation credits. Therefore, disinfection variability during ozonation, which could be caused by particle shielding or highly resistant subpopulations, is more likely to result in regulatory non-compliance for indicator bacteria.

This variability is reflected in the error bars in Figure 3.12, which illustrates *E. coli* inactivation in ozonated secondary effluent. In addition, Figure 3.12 illustrates the tailing effect observed for some disinfection processes. This tailing effect has been described as an artifact of the detection limits of the corresponding assays, and it has also been attributed to particle shielding and highly resistant subpopulations. Figure 3.12b also highlights a potential issue with the CT framework, particularly in wastewater disinfection applications, in which the CT is difficult to measure because of low ozone doses and/or high reactivity of the EfOM. In these cases, the apparent ozone CT is ~0 mg-min/L, although significant microbial inactivation may still be observed.

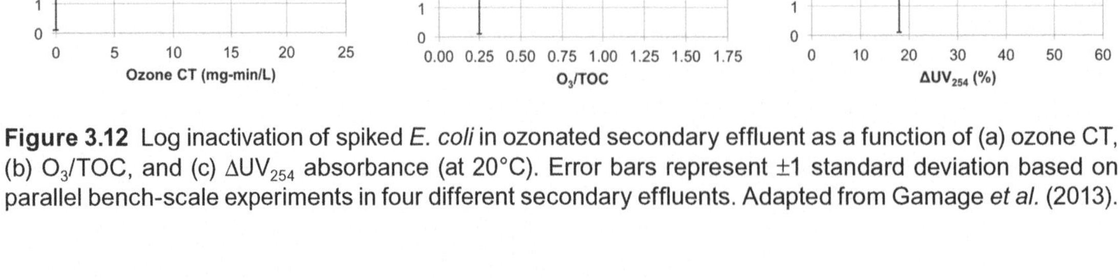

Figure 3.12 Log inactivation of spiked *E. coli* in ozonated secondary effluent as a function of (a) ozone CT, (b) O_3/TOC, and (c) ΔUV_{254} absorbance (at 20°C). Error bars represent ±1 standard deviation based on parallel bench-scale experiments in four different secondary effluents. Adapted from Gamage *et al.* (2013).

According to the data in Gamage *et al.* (2013), ozone CT values greater than 2 mg-min/L generally achieve at least 5-log inactivation of spiked *E. coli* (Gamage *et al.* 2013). Similar results were observed for a pilot-scale ozone reactor (Ishida *et al.* 2008). That study also indicated that an ozone CT of at least 1 mg-min/L was generally sufficient to reduce the native total coliform concentration to <1 CFU/100 mL.

However, experiments performed in drinking water matrices have resulted in slightly different results. With respect to the Chick-Watson equation (Eq. 3.19), previous studies reported a Λ_{CW} of 130 (mg-min/L)$^{-1}$ for reactions between ozone and *E. coli* (Hunt & Marinas, 1997; von Gunten, 2003b). Based on this value, $\ln(10^{-5})$ (i.e., 5-log inactivation) would correspond to an ozone CT of 0.09 mg-min/L, which is significantly lower than the wastewater values. Therefore, the greater complexity of wastewater disinfection may necessitate separate CT targets.

Multiple studies also indicate that the inactivation of *E. coli* (and other microorganisms) is primarily attributable to ozone oxidation (Gamage *et al.* 2013; von Gunten, 2003b; Zuma *et al.* 2009). In fact, von Gunten (2003) suggests that microbial inactivation is primarily attributable to DNA damage, which would be less likely with $^{\bullet}OH$ due to radical consumption by the bacterial cell wall (von Gunten, 2003b). Figure 3.13, which illustrates cell-bound versus free adenosine triphosphate (ATP) concentration as a function of O_3/TOC, supports this theory. ATP is used by organisms as a source of cellular energy, and it is increasingly being used as a surrogate for bacterial presence and bioactivity in water and wastewater applications. In the context of Figure 3.13, cell-bound ATP is indicative of the presence of intact bacterial cells, while free ATP is indicative of the presence of bacterial cells that have been severely damaged or lysed. According to Figure 3.12b, an O_3/TOC ratio of 0.25 results in *E. coli* inactivation ranging from 0 to 4 logs. However, that same dosing condition yields minimal release of ATP according to Figure 3.13, which suggests that inactivation is occurring despite minimal damage to bacterial cell walls. Therefore, ozone is presumably diffusing through the cell walls and membranes and subsequently attacking the genetic material to achieve the reported levels of inactivation. Higher ozone doses eventually compromise the integrity of the bacterial cell walls, but cell-bound ATP can still be observed even at an O_3/TOC ratio of 1.5 in some wastewaters.

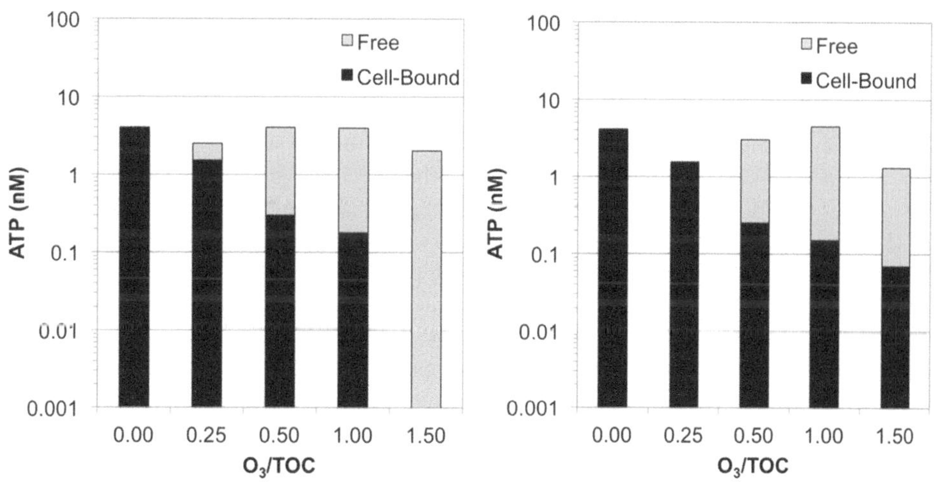

Figure 3.13 Comparison of cell-bound and free adenosine triphosphate concentration as a function of O_3/TOC in two different secondary effluents. Adapted from Snyder *et al.* (2014).

3.6.5 Viruses

Figure 3.14 illustrates the inactivation of MS2 bacteriophage in ozonated secondary effluent. In comparison with Figure 3.12, MS2 inactivation appears to be greater and more consistent (i.e., smaller standard deviations) than that of *E. coli*. An O_3/TOC ratio of 0.25 achieves 1-log to 4-log inactivation, and O_3/TOC ratios greater than 0.25 consistently achieve greater than 5-log inactivation. At least 5-log MS2 inactivation can generally be achieved with measurable CTs (based on bench-scale ozone demand-decay testing), and CT values greater than 1.0 mg-min/L generally achieve at least 6-log MS2 inactivation. Figure 3.12 and Figure 3.14 also indicate that *E. coli* and MS2 exhibit similar inactivation profiles. Therefore, it may be plausible to use *E. coli* inactivation as a conservative surrogate for MS2 (or viral) inactivation in ozone applications. In some instances this strategy will underestimate viral inactivation, but it provides a useful alternative to implementing MS2 assays in wastewater treatment plant laboratories.

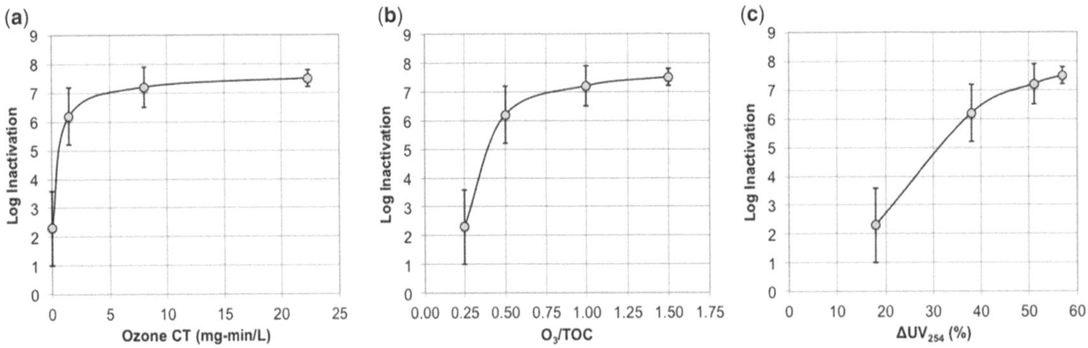

Figure 3.14 Log inactivation of spiked MS2 bacteriophage in ozonated secondary effluent as a function of (a) ozone CT, (b) O_3/TOC, and (c) ΔUV_{254} absorbance (at 20°C). Error bars represent ±one standard deviation based on parallel bench-scale experiments in four different secondary effluents. Adapted from Gamage *et al.* (2013).

Some regulatory frameworks require manufacturers to demonstrate specified levels of virus inactivation with a pilot-scale validation study before their technologies can be implemented at full-scale. For example, California's "Title 22" certification requires a disinfection process to achieve at least 5-log virus (historically poliovirus) inactivation before it can be used in water reuse applications. In this context, Ishida *et al.* (2008) first demonstrated that 5-log poliovirus inactivation equated to 6.5-log MS2 inactivation during ozonation (Ishida *et al.* 2008). Subsequently, that study determined that an ozone CT of only 0.2 mg-min/L was sufficient to achieve the 6.5-log MS2 inactivation target. This value is also relevant to potable reuse applications because recent risk assessments indicate that a total of 12-log virus removal/inactivation should be achieved in any potable reuse system (Trussel, 2012; Trussell *et al.* 2013), with no more than 6 logs awarded to any unit process (CDPH, 2014).

With respect to drinking water applications, the U.S. EPA published ozone CT values for virus inactivation to aid municipalities in complying with the Surface Water Treatment Rule, which requires public water systems to accumulate 4 logs of virus inactivation credits. Figure 3.15 illustrates the corresponding ozone CT values for virus inactivation as a function of temperature.

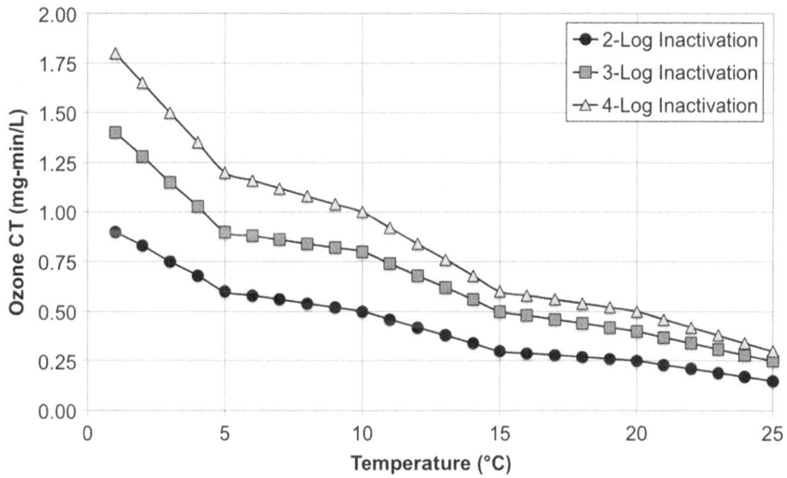

Figure 3.15 Ozone CT values for virus inactivation as a function of temperature. Adapted from USEPA (1999b).

3.6.6 Spore-forming microbes

Figure 3.16 illustrates the inactivation of *B. subtilis* spores in ozonated secondary effluent. Due to their highly resistant nature, extended exposure to dissolved ozone is necessary to achieve any level of spore inactivation. Specifically, minimal inactivation is achieved with CT values less than 8 mg-min/L or O_3/TOC ratios less than 1.0. Beyond those thresholds, the level of inactivation is characterized by significant variability between secondary effluents.

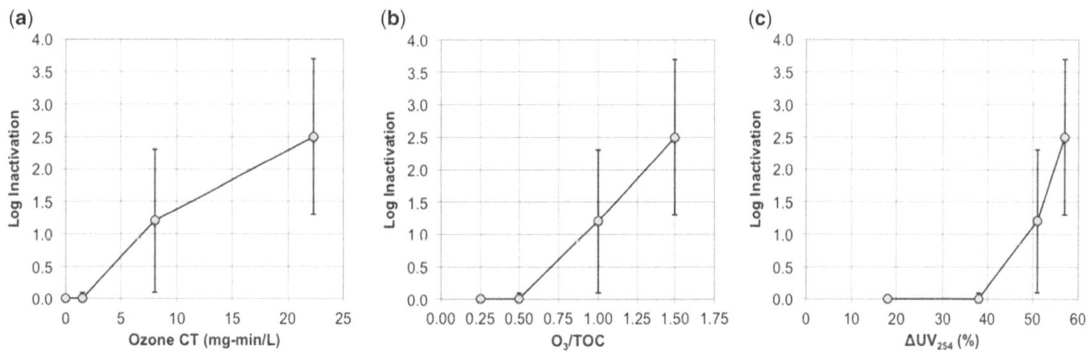

Figure 3.16 Log inactivation of spiked *B. subtilis* spores in ozonated secondary effluent as a function of (a) ozone CT, (b) O_3/TOC, and (c) ΔUV_{254} absorbance (at 20°C). Error bars represent ±1 standard deviation based on parallel bench-scale experiments in four different secondary effluents. Adapted from Gamage *et al.* (2013).

With respect to the delayed Chick-Watson model (Eq. 3.20), the literature suggests a CT lag of 2.9 mg-min/L and a Δ_{CW} of 2.9 $(\text{mg-min/L})^{-1}$ for *B. subtilis* spores (Driedger *et al.* 2001; von Gunten,

2003b). Another study proposed a CT lag of 4.91 mg-min/L and a Λ_{CW} of 2.12 (mg-min/L)$^{-1}$ (Crittenden *et al.* 2012; Larson & Marinas, 2003). These kinetic models are consistent with the dose response curve in Figure 3.16a.

As discussed earlier, *B. subtilis* spores are often used as surrogates for disinfectant-resistant *Cryptosporidium* oocysts and *Giardia* cysts, which are the true target microbes in water and wastewater disinfection. The U.S. EPA Surface Water Treatment Rule requires 3-log removal/inactivation of *Giardia* cysts and up to 3-log inactivation/removal of *Cryptosporidium* oocysts. With respect to potable reuse, recent risk assessments recommend at least 10-log removal/inactivation of *Giardia* cysts (CDPH, 2014) and *Cryptosporidium* oocysts (Trussell, 2012; CDPH, 2014; Trussell *et al.* 2013). The corresponding published CT values for the inactivation of *Giardia* cysts with ozonation are illustrated in Figure 3.17. CT estimates for 2-log inactivation of *Cryptosporidium spp.* range from 2.4–7.8 mg-min/L at 20–25°C (USEPA, 1999a). Kinetic parameters for *Bacillus*, *Giardia*, and *Cryptosporidium* are summarized in Table 3.6.

Table 3.6 Summary of Chick-Watson (Eq. 3.18) and delayed Chick-Watson (Eq. 3.19) kinetic parameters for the inactivation of disinfectant-resistant microorganisms with ozone (at 20–25°C). Adapted from Crittenden *et al.* (2012).

Microbe	Λ_{CW} (mg-min/L)$^{-1}$	b (mg-min/L)	2-Log CT (mg-min/L)	3-Log CT (mg-min/L)	Reference
Bacillus	2.12	4.91	7.1	8.2	Larson and Marinas (2003)
Bacillus	2.9	2.9	4.5	5.3	Driedger *et al.* (2001)
Giardia	1.9	–	2.4	3.6	Wallis *et al.* (1989)
Cryptosporidium	0.84	0.83	6.3	9.1	Driedger *et al.* (2001)
Cryptosporidium	1.7	0.22	2.9	4.3	Oppenheimer *et al.* (2000)

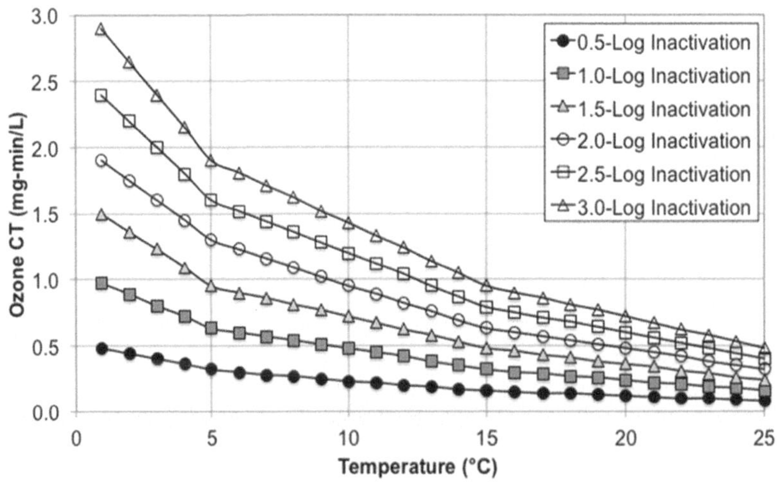

Figure 3.17 Ozone CT values for *Giardia* cyst inactivation as a function of temperature. Adapted from USEPA (1999b).

3.7 IMPLEMENTATION AT FULL SCALE FACILITIES

3.7.1 Ozone systems

The four primary components of an ozone system include the feed gas supply, ozone generator, ozone contactor, and off-gas ozone destruct (Rakness, 2005). The feed gas supply often utilizes air, liquid oxygen (LOX), or vacuum pressure swing adsorption to produce oxygen for the ozone generator. The ozone generator then convert this oxygen supply into ozone feed gas. The feed gas is transferred from gaseous phase to aqueous phase ozone during mass transfer, which can occur via fine bubble diffusion or side-stream injection. The ozone contactor provides time for the aqueous ozone residual to react with the specific contaminant or microbe of interest. The off-gas system destroys any gaseous phase ozone that is not transferred into the water.

3.7.2 Ozone contactor

The method of ozone dissolution defines the two types of ozone contactors: fine bubble diffusers and sidestream injection (Rakness, 2005). Many ozone facilities rely upon fine bubble diffusion to transfer ozone into the water using porous stone diffusers (Figure 3.18). Following dissolution, the ozonated water passes through a baffled contact chamber to complete oxidation and disinfection reactions. In recent years, sidestream injection has emerged as an alternative to fine bubble diffusion for ozone dissolution (Figure 3.19). Sidestream injection systems dissolve ozone into a sidestream flow via an injection process or into the full process flow stream. The sidestream flow is mixed with full flow to achieve the targeted ozone dose. In some cases, a degas vessel may eliminate the remaining ozone feed gas prior to blending with the full process flow.

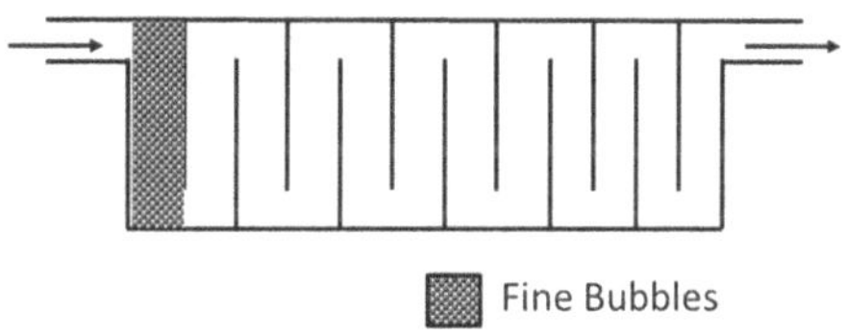

Figure 3.18 Typical contactor schematic with fine bubble diffusion.

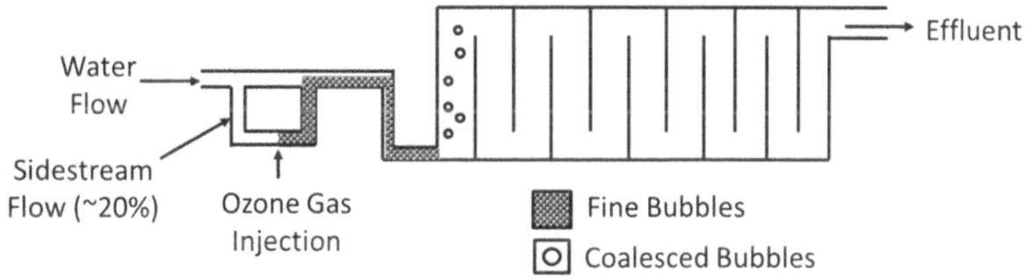

Figure 3.19 Typical contactor schematic with sidestream ozone injection.

3.7.3 Mass transfer efficiency

Both ozone dissolution systems are capable of achieving transfer efficiencies of >85%. In bubble diffusion systems, the water column depth above the stone diffuser must be between 18 and 22 ft. (Rakness, 2005). The transfer efficiency for air-fed ozone systems (1% wt.) is approximately 85%, while oxygen-fed ozone systems (8–10% wt.) can achieve 95% transfer efficiency (Rakness, 2005). In sidestream systems, the gas to liquid (G/L) ratio is a key variable in achieving good transfer efficiency. When the G/L ratio is around 0.1, a transfer efficiency of 90% can be achieved. The downstream pressure at the point of mixing should be maintained between 5–10 psig in order to achieve transfer efficiency >90%. Sidestream injection systems also provide maintenance advantages since all equipment is easily assessable and diffuser maintenance is eliminated (Rakness, 2007).

3.7.4 Cost estimates

In some applications, ozone is considered a costly, energy intensive treatment option, but in more advanced applications (e.g., potable reuse), ozone might be considered a more sustainable, cost effective alternative. Therefore, it is important to consider the context of the application and the suitability of alternative treatment options. The following discussion of cost and energy consumption, which is adapted from Gerrity *et al.* (2014), Plumlee *et al.* (2014), and Snyder *et al.* (2014), addresses this goal.

It is important to note that the numbers provided below are based on an Association for the Advancement of Cost Engineering (AACE) Class 4 cost estimate. Therefore, the costs are conceptual in nature and are typically used in planning phases (≤1% of design completed) when estimates ranging from −30% and +50% of actual costs are appropriate. It is also important to note that the costs (adjusted to 2011 U.S. dollars) and energy estimates are based on vendor- and facility-reported data for full-scale installations from the early 2000s. With technological advancements and more widespread implementation, many of these values will change over time. However, these conceptual level figures still provide a useful basis for comparison. The reader is referred to the aforementioned references for a detailed description of the cost estimate methodology.

The following unit cost curves can be used to estimate the capital (Eq. 3.22) and annual operations and maintenance (O&M) (Eq. 3.23) costs for ozone systems at flow rates ranging from 1 to 500 million gallons per day (MGD). Note that when using the costs curves, the unit cost is calculated in millions of dollars (2011 U.S. dollars) per MGD so the flow rate must be applied twice to calculate the final cost in $M. Unit costs are useful because they explicitly illustrate economy of scale in that the unit costs decrease as the design capacity increases. The unit cost curves were developed using a baseline applied ozone dose of 3 mg/L. The differential capital (Eq. 3.24) and differential annual O&M (Eq. 3.25) equations can be used to correct the baseline cost curves for different ozone doses.

$$\text{Capital Cost (\$M/MGD)} = 2.26 \times Q^{-0.54} \tag{3.22}$$

$$\text{Annual O\&M Cost (\$M/MGD)} = 0.0068 \times Q^{-0.051} \tag{3.23}$$

$$\Delta\ \text{Capital Cost (\$M)} = 0.0156 \times Q \times \left(\frac{D}{3} - 1\right) \tag{3.24}$$

$$\Delta\ \text{Annual O\&M Cost (\$M)} = 0.005 \times Q \times \left(\frac{D}{3} - 1\right) \tag{3.25}$$

where

Q = design flow rate (MGD)
D = design applied ozone dose (mg/L)

The cost curves above were used to create summaries of the total capital costs, annual O&M costs, and estimated energy consumption for ozone as a function of design flow rate and ozone dose (Table 3.7). Total capital and annual O&M costs associated with some of the most common treatment trains in potable reuse applications are summarized in Table 3.8 and Table 3.9, respectively. These treatment trains are discussed in greater detail in the next section. It is important to consider these estimated costs in discussions of alternative treatment trains because of the significant savings that can be realized with ozone-based treatment, assuming the treatment train is still able to achieve the desired water quality.

Table 3.7 Estimated capital costs (2011 U.S. dollars), annual O&M costs (2011 U.S. dollars), and energy consumption for ozone as a function of design flow rate and applied ozone dose. Adapted from Snyder *et al.* (2014).

Design Flow (MGD)	Cost and Energy[a]	Applied Ozone Dose (mg/L)			
		1.5	**3**	**6**	**9**
1	Capital ($M)	2.3	2.3	2.3	2.3
	Annual O&M ($M)	0.01	0.01	0.01	0.01
	Energy Consumption (MWh)	25	50	99	149
5	Capital ($M)	4.7	4.7	4.8	4.9
	Annual O&M ($M)	0.02	0.03	0.06	0.08
	Energy Consumption (MWh)	124	249	497	746
10	Capital ($M)	6.4	6.5	6.7	6.8
	Annual O&M ($M)	0.04	0.06	0.11	0.16
	Energy Consumption (MWh)	249	497	995	1492
25	Capital ($M)	9.7	9.9	10.3	10.7
	Annual O&M ($M)	0.08	0.14	0.27	0.39
	Energy Consumption (MWh)	622	1243	2487	3730
50	Capital ($M)	13.3	13.7	14.4	15.2
	Annual O&M ($M)	0.15	0.28	0.53	0.78
	Energy Consumption (MWh)	1243	2487	4973	7460

[a]Assumes a specific energy consumption of 0.012 kWh/g O_3 for ozone generation.

Table 3.8 Total capital costs (2011 U.S. dollars) for advanced treatment trains in water reuse applications. Adapted from Snyder *et al.* (2014).

Capacity (MGD)	Process Trains and Capital Costs ($M)				
	O_3-BAC	**MF-O_3-BAC**	**MF-RO**	**MF-RO-UV/H_2O_2**	**MF-O_3-RO (O_3-MF-RO)**
1	$5.2	$9.0	$11	$11	$13
5	$11	$24	$38	$40	$42
10	$16	$38	$65	$69	$71
25	$31	$75	$132	$142	$142
50	$50	$126	$226	$245	$240
80	$71	$180	$327	$356	$344

Table 3.9 Total annual O&M costs (2011 U.S. dollars) for advanced treatment trains in water reuse applications. Adapted from Snyder *et al.* (2014).

	Process Trains and Capital Costs ($M)				
Capacity (MGD)	**O_3-BAC**	**MF-O_3-BAC**	**MF-RO**	**MF-RO-UV/H_2O_2**	**MF-O_3-RO (O_3-MF-RO)**
1	$0.1	$0.4	$0.5	$0.6	$0.5
5	$0.3	$1.4	$2.6	$2.7	$2.6
10	$0.6	$2.4	$4.8	$5.1	$4.8
25	$1.4	$5.1	$11	$11	$11
50	$2.8	$9.1	$19	$21	$19
80	$4.0	$13	$29	$31	$29

Finally, Table 3.10 provides a summary of costs for a hypothetical 50-mgd treatment train composed of ozone and biological activated carbon (BAC) treating wastewater with a TOC of 6 mg/L. The costs are presented as a function of ozone dose and expected level of trace organic contaminant elimination. Cost comparisons of ozone and various advanced oxidation processes (e.g., O_3/H_2O_2 and UV/H_2O_2) are also available in the literature (Katsoyiannis *et al.* 2011).

Table 3.10 Total capital and annual O&M costs (2011 U.S. dollars) for a hypothetical 50-mgd ozone BAC treatment train treating a wastewater with 6 mg/L of TOC. Compound groups (and example compounds) were defined earlier in Table 3.4. Adapted from Snyder *et al.* (2014).

O_3 Dose **O_3/TOC**	**1.5 mg/L** **0.25**	**3 mg/L** **0.5**	**6 mg/L** **1.0**	**9 mg/L** **1.5**
Conceptual Level Cost Estimate				
Capital Costs	$49 M	$50 M	$52 M	$53 M
Annual O&M	$2.7 M	$2.8 M	$3.1 M	$3.3 M
Average percent elimination of trace organic compounds				
Group 1[a]	>90%	>90%	>90%	>90%
Group 2[b]	>60%	>90%	>90%	>90%
Group 3[c]	>30%	>60%	>90%	>90%
Group 4[d]	>15%	>30%	>60%	>80%
Group 5[e]	<5%	>5%	>15%	>20%

[a]Group 1: $k_{s,O_3} \geq 1\times10^5 M^{-1}s^{-1}$
[b]Group 2: $10\ M^{-1}s^{-1} \leq k_{s,O_3} < 1\times10^5 M^{-1}s^{-1}$
[c]Group 3: $k_{s,O_3} < 10\ M^{-1}s^{-1}$ and $k_{s,\bullet OH} \geq 5\times10^9 M^{-1}s^{-1}$
[d]Group 4: $k_{s,O_3} < 10\ M^{-1}s^{-1}$ and $1\times10^9 M^{-1}s^{-1} \leq k_{s,\bullet OH} < 5\times10^9 M^{-1}s^{-1}$
[e]Group 5: $k_{s,O_3} < 10\ M^{-1}s^{-1}$ and $k_{s,\bullet OH} < 1\times10^9 M^{-1}s^{-1}$

3.7.5 Process control

During drinking water applications, process control is determined from either an ozone dose setpoint or ozone CT to meet a desired disinfection target. When targeting a CT value, online dissolved ozone analyzers are used to provide real-time concentration measurements. When using a minimum of ozone residuals, the first order decay rate can be calculated and the area under the decay curve integrated to determine the ozone CT (Rakness *et al.* 2005). In other applications, direct ozone decay measurements are also used to control the ozone dose, including for contaminant oxidation (Kaiser *et al.* 2013). At some water utilities treating groundwater for hydrogen sulfide removal, water oxidation-reduction potential (ORP) has been applied successfully for process control purposes.

During wastewater applications, process control can be achieved through use of the O_3/DOC ratio and the rate constants for the selected trace organic compounds of interest (Lee *et al.* 2013). In some instances, sufficient micropollutant elimination may be achieved without generating a measurable dissolved ozone residual. In these cases, process performance may be assessed in real time through the use of online UV absorbance or fluorescence measurements (Bahr *et al.* 2007; Gerrity *et al.* 2012; Wert *et al.* 2009b). A framework for implementing online UV absorbance measurements for process control and verification is illustrated in Figure 3.20, and example correlations between differential UV_{254} absorbance and percent reduction in phenytoin in bench- and pilot-scale ozone systems are illustrated in Figure 3.21.

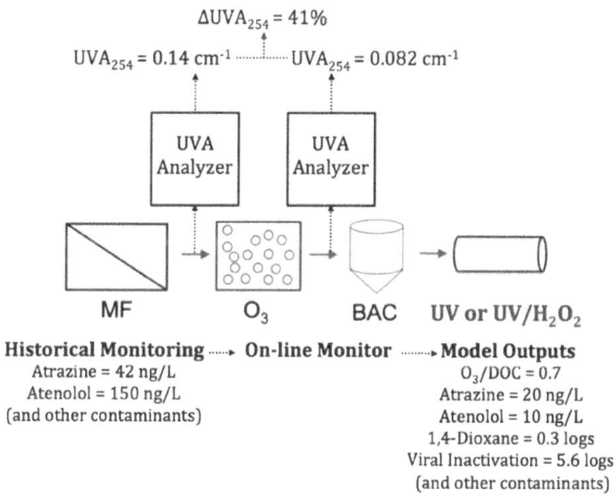

Figure 3.20 Example framework for implementing on-line UV absorbance analyzers to monitor and predict ozone process performance in a hypothetical potable reuse treatment train. Model parameters are adapted from Gerrity *et al.* (2012).

3.8 CASE STUDIES AND REGULATORY DRIVERS

3.8.1 Drinking water applications

The use of ozone is growing in drinking water due to the multiple benefits of this oxidant, including disinfection, oxidation of inorganic and organic contaminants, destruction of DBP precursors, and flocculation benefits. Ozone is recognized as a suitable treatment technology for the inactivation of *Cryptosporidium* according to the Long Term 2 Enhanced Surface Water Treatment Rule (LT2ESWTR) (USEPA, 2006). However, utilities treating bromide-rich water must consider bromate formation, which is regulated at 10 μg/L by the USEPA (USEPA, 1998).

Other drivers for ozone implementation in drinking water treatment plants may include the oxidation of cyanotoxins. Ozone has proved to be very effective in oxidizing multiple cyanotoxins, including microcystin, cylindrospermopsin, and anatoxin-a (Rodriguez *et al.* 2007). Multiple taste and odor compounds can also be removed through the ozone process (Peter & von Gunten, 2007). However, the two most common taste and odor compounds, MIB and geosmin, are removed via $^{\bullet}OH$ oxidation rather than reactions with molecular ozone (Lalezary *et al.* 1986; Westerhoff *et al.* 2006). The formation of chlorinated disinfection byproducts may also be reduced following ozonation. In one study using three NOM isolates, trihalomethane formation potential was reduced by 24–46% after ozonation (Galapate *et al.* 2001). These examples further illustrate the benefit of ozone during drinking water treatment.

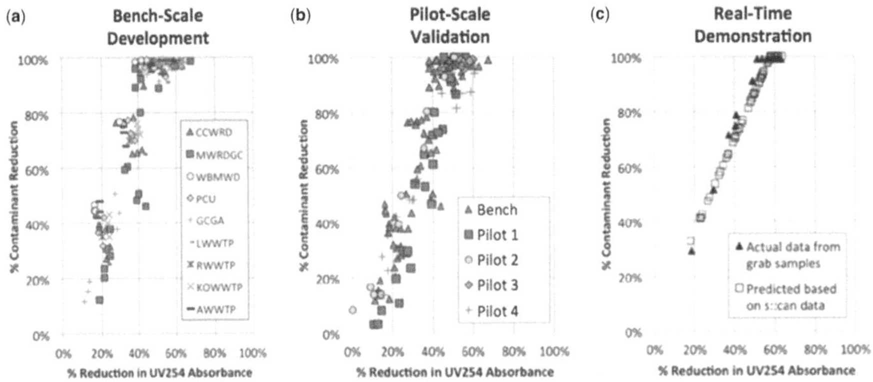

Figure 3.21 Percent reduction in phenytoin concentrations as a function of differential UV_{254} absorbance in bench-scale and pilot-scale ozone systems. The correlations were first developed in nine different secondary treated wastewater effluents (left), then validated in independent pilot-scale ozone systems (middle), and finally implemented in conjunction with on-line UV_{254} analyzers in a pilot-scale ozone system. Model development for other trace organic compounds and microbial surrogates is described in Gerrity *et al.* (2012).

The Southern Nevada Water Authority implemented ozone-based water treatment at the Alfred Merritt Smith and River Mountains Water Treatment Facilities in 2003. The ozone facilities were designed to achieve a 2.0-log *Cryptosporidium* inactivation in response to an outbreak in the 1990s (Roefer *et al.* 1996). Source water is supplied by the Colorado River via Lake Mead and contains around 100 µg/L of bromide. Therefore, a bromate mitigation strategy was needed. Alternatives such as pH depression and ammonia addition were considered; however, subsequent addition of chlorine and ammonia provided enhanced bromate control during pilot-scale and full-scale studies (Wert *et al.* 2007). SNWA also utilizes free chlorine for secondary disinfection in the distribution system. Upon conversion to ozone, the trihalomethane concentration decreased by approximately 10 µg/L, and haloacetic acids decreased by 5 µg/L (Wert & Rosario-Ortiz, 2011). In addition to the overall decrease in concentration, there was a shift in speciation from di- and tri-chlorinated species to di- and tri-brominated species following conversion to ozone.

3.8.2 Wastewater and potable reuse applications

In many areas, wastewater is now perceived as a resource – rich in water, nutrients, and energy – instead of a waste that must be discharged. Technological advancements related to bioassays and analytical methods are also providing researchers, water utilities, and regulators with a more comprehensive understanding of the unintended consequences of wastewater discharge. For example, the yeast estrogen screen (YES) assay provides a quantitative estimate of the estrogenicity of a wastewater. In environmental discharge applications, this estrogenicity has been linked to the feminization of aquatic species. In addition, wastewater often contains sufficient nutrients to induce algal blooms in surface waters – even after advanced chemical and biological treatment. Therefore, applications in which poorly treated wastewaters are discharged to the ocean or other bodies of water are becoming increasingly rare. Instead, many wastewater utilities are implementing advanced treatment, and ozone is becoming an increasingly popular option as a single barrier or as part of a multi-barrier advanced treatment train. These upgrades are being driven by necessity, new regulatory frameworks, or simply a proactive response to emerging water quality problems, such as estrogenicity and trace organic contaminants. Several benchmark ozone facilities are shown in Figure 3.22 and described in the text below. Specific details related to plant capacity, ozone dose, etc. are available in the literature (Gerrity *et al.* 2013; Oneby *et al.* 2010; Audenaert *et al.* 2014).

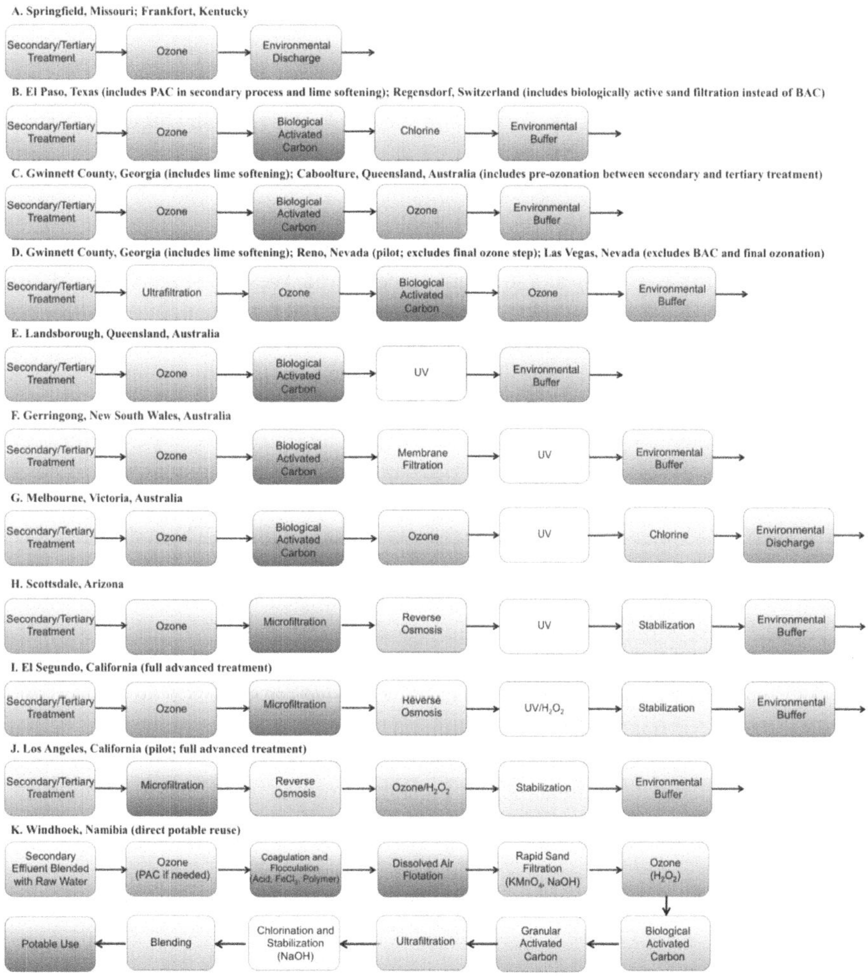

Figure 3.22 Schematics of benchmark wastewater and water reuse treatment trains employing ozonation. Adapted from Gerrity *et al.* (2013).

Prior to 2000, potable reuse systems employing ozone and biological activated carbon were constructed in Texas (Fred Hervey Water Reclamation Plant; Figure 3.22b) and Georgia (F. Wayne Hill Water Resources Center; Figure 3.22c and d) to augment local water resource portfolios. However, with improved technology and the success of benchmark systems in Orange County and Singapore, utilities started turning to reverse osmosis in the early 2000s for their potable reuse needs (e.g., the Western Corridor Recycled Water Project in Australia). Despite their excellent track record, treatment trains employing reverse osmosis proved to be costly and energy intensive. As a result, several utilities in Australia (Figure 3.22c, e, and f) recently opted for ozone-BAC to achieve comparable water quality (except for total dissolved solids) with reduced costs and energy consumption. Many of the Australian facilities have since been decommissioned as the 'Millennium drought' subsided and conventional supplies became more readily available.

On the regulatory side, the California Division of Drinking Water's Groundwater Replenishment Reuse Regulations (CDPH 2014) and the European Union (EU) Water Framework Directive (2000/60/EC) are now having significant impacts on treatment train selection. In California, utilities that directly inject recycled water into an aquifer (and likely into surface water by 2017) are required to use full advanced treatment consisting of reverse osmosis and an AOP; microfiltration is typically used as a pretreatment step as well. In this type of application, ozone has been used to reduce organic fouling (Stanford *et al.* 2011) on the microfiltration membranes at the West Basin Municipal Water District (Figure 3.22i) and to reduce NDMA formation during downstream chloramination at the City of Scottsdale's Water Campus (Figure 3.22h). It is important to note that reducing NDMA formation potential with ozone is not practical in all systems. Some wastewaters contain high concentrations of ozone-specific NDMA precursors (Gerrity *et al.* 2015) so upstream ozonation can actually lead to net increases in NDMA in a combined ozone-chloramine application. This has been a significant issue in the West Basin system. At full advanced treatment facilities where NDMA can be controlled without UV/H_2O_2 as the final treatment process, there is also potential to implement ozone (or ozone/H_2O_2) as the AOP – a concept that was piloted in California (Figure 3.22j) (Tiwari *et al.* 2012).

Ozone is also applicable for recycled water spreading applications in California, any potable reuse project outside of California, and any nonpotable reuse project targeting (1) bulk organic matter transformation, (2) trace organic contaminant oxidation, and/or (3) disinfection. In these applications, ozone is sometimes dosed at multiple locations within the process train to achieve these three major treatment objectives. For example, the F. Wayne Hill Water Resources Center in Georgia (Figure 3.22c and d) doses ozone before its BAC process and for final disinfection, and the South Caboolture Water Reclamation Plant in Australia (Figure 3.22c) doses three different times. With respect to the EU Water Framework Directive, ozone is expected to become increasingly common in Europe due to impending regulations on trace organic contaminants in water supplies (Audenaert *et al.* 2014). Even without regulatory drivers, some utilities, such as the Clark County Water Reclamation District in Nevada (Figure 3.22d), are proactively upgrading their facilities with ozone for trace organic contaminant mitigation and to reduce the potential adverse impacts of wastewater discharges on aquatic species. These relatively new applications supplement the historical use of ozone for disinfection in numerous locations throughout the world (Audenaert *et al.* 2014; Oneby *et al.* 2010).

3.9 REFERENCES

Acero J. L. and von Gunten U. (2001). Characterization of oxidation processes: ozonation and the AOP O-3/H2O2. *Journal of American Water Works Association*, **93**(10), 90–100.

Acero J. L., Stemmler K. and von Gunten U. (2000). Degradation kinetics of atrazine and its degradation products with ozone and OH radicals. *Environmental Science and Technology*, **34**, 591–597.

Acero J. L., Haderlein S. B., Schmidt T. C., Suter M. J. F. and von Gunten U. (2001). MTBE oxidation by conventional ozonation and the combination ozone/hydrogen peroxide: efficiency of the processes and bromate formation. *Environmental Science and Technology*, **35**(21), 4252–4259.

Audenaert W., Chys M. and Van Hulle S. (2014). (Future) regulation of trace organic compounds in WWTP effluents as a driver for advanced wastewater treatment, IOA-PAG Annual Conference, August 24–27, 2014, Montreal, Canada.

Bader H. and Hoigne J. (1982). Determination of ozone in water by the indigo method: a submitted standard method. *Ozone Science and Engineering*, **4**(4), 169–176.

Bahr C., Schumacher J., Ernst M., Luck F., Heinzmann B. and Jekel M. (2007). SUVA as control parameter for the effective ozonation of organic pollutants in secondary effluent. *Water Science & Technology*, **55**(12), 267–274.

Beltran F. J. (2003). Ozone Reaction Kinetics for Water and Wastewater Systems. CRC Press, Boca Raton, Florida.

Bielski B. H. J., Cabelli D. E., Arudi R. L. and Ross A. B. (1985). Reactivity of HO_2/O_2^- radicals in aqueous solution. *Journal of Physical Chemistry Reference Data*, **14**, 1041–1100.

Blatchley E. R., III, Weng S., Afifi M. Z., Chiu H.-H., Reichlin D. B., Jousset S. and Erhardt R. S. (2012). Ozone and UV254 radiation for municipal wastewater disinfection. *Water Environment Research*, **84**(11), 2017–2027.

Bowman R. H., Miller P., Purchase M. and Schoellerman R. (2001). Ozone-peroxide Advanced Oxidation Water Treatment System for Treatment of Chlorinated Solvents and 1,4-Dioxane. American Chemical Society Meeting, April 1–5, 2001, San Diego, CA.

Buffle M. O. and von Gunten U. (2006). Phenols and amine induced HO center dot generation during the initial phase of natural water ozonation. *Environmental Science and Technology*, **40**(9), 3057–3063.

Buffle M.-O., Galli S. and von Gunten U. (2004). Enhanced bromate control during ozonation: the chlorine-ammonia process. *Environmental Science and Technology*, **38**(19), 5187–5195.

Buffle M. O., Schumacher J., Meylan S., Jekel M. and von Gunten U. (2006a). Ozonation and advanced oxidation of wastewater: effect of O_3 dose, pH, DOM and HO-scavengers on ozone decomposition and HO generation. *Ozone Science and Engineering*, **28**(4), 247–259.

Buffle M. O., Schumacher J., Salhi E., Jekel M. and von Gunten U. (2006b). Measurement of the initial phase of ozone decomposition in water and wastewater by means of a continuous quench-flow system: Application to disinfection and pharmaceutical oxidation. *Water Research*, **40**(9), 1884–1894.

Buhler R. E., Staehelin J. and Hoigne J. (1984). Ozone decomposition in water studied by pulse-radiolysis .1. Ho2/O2-and Ho3/O3- as intermediates. *Journal of Physical Chemistry*, **88**(12), 2560–2564.

Buxton G. V., Greenstock C. L., Helman W. P. and Ross A. B. (1988). Critical review of rate constants for reactions of hydrated electrons, hydrogen atom and hydroxyl radicals (OH/•O–) in aqueous solutions. *Journal of Physical and Chemical Reference Data*, **17**, 513–886.

Canonica S. and Tratnyek P. G. (2003). Quantitative structure-activity relationships for oxidation reactions of organic chemicals in water. *Environmental Toxicology and Chemistry*, **22**(8), 1743–1754.

California Department of Public Health (CDPH) (2014). Groundwater replenishment using recycled water. Title 22 California code of regulations. DPH-14-003E.

Coddington J. W., Hurst J. K. and Lymar S. V. (1999). Hydroxyl radical formation during peroxynitrous acid decomposition. *Journal of the American Chemical Society*, **121**(11), 2438–2443.

Crittenden J. C., Trussell R. R., Hand D. W., Howe K. J. and Tchobanoglous G. (2012). Water Treatment: Principles and Design. 3rd edn, John Wiley & Sons Inc., Hoboken, New Jersey, USA.

Deborde M., Rabouan S., Duguet J. and Legube B. (2005). Kinetics of aqueous ozone-induced oxidation of some endocrine disruptors. *Environmental Science and Technology*, **39**(16), 6086–6092.

Dodd M. C., Buffle M. O. and von Gunten U. (2006). Oxidation of antibacterial molecules by aqueous ozone: Moiety-specific reaction kinetics and application to ozone-based wastewater treatment. *Environmental Science and Technology*, **40**(6), 1969–1977.

Dong M. M., Mezyk S. P. and Rosario-Ortiz F. L. (2010). Reactivity of effluent organic matter (EfOM) with hydroxyl radical as a function of molecular weight. *Environmental Science and Technology*, **44**(15), 5714–5720.

Driedger A., Staub E., Pinkernell U., Marinas B., Koster W. and von Gunten U. (2001). Inactivation of Bacillus subtilis spores and formation of bromate during ozonation. *Water Research*, **35**(12), 2950–2960.

Elovitz M. S. and von Gunten U. (1999). Hydroxyl radical ozone ratios during ozonation processes. I-The R-ct concept. *Ozone Science and Engineering*, **21**(3), 239–260.

Fischbacher A., Loppenberg K., Von Sonntag C. and Schmidt T. C. (2015). A New Reaction Pathway for Bromite to Bromate in the Ozonation of Bromide. *Environmental Science and Technology*, **49**(19), 11714–11720.

Galapate R. P., Baes A. U. and Okada M. (2001). Transformation of dissolved organic matter during ozonation: effects on trihalomethane potential. *Water Research*, **35**(9), 2201–2206.

Gamage S., Gerrity D., Pisarenko A. N., Wert E. C. and Snyder S. A. (2013). Evaluation of process control alternatives for the inactivation of escherichia coli, MS2 bacteriophage, and Bacillus subtilis spores during wastewater ozonation. *Ozone Science and Engineering*, **35**(6), 501–513.

Gerrity D. and Snyder S. (2011). Review of ozone for water reuse applications: toxicity, regulations, and trace organic contaminant oxidation. *Ozone Science and Engineering*, **33**(4), 253–266.

Gerrity D., Gamage S., Holady J. C., Mawhinney D. B., Quinones O., Trenholm R. A. and Snyder S. A. (2011). Pilot-scale evaluation of ozone and biological activated carbon for trace organic contaminant mitigation and disinfection. *Water Research*, **45**(5), 2155–2165.

Gerrity D., Gamage S., Jones D., Korshin G. V., Lee Y., Pisarenko A., Trenholm R. A., von Gunten U., Wert E. C. and Snyder S. A. (2012). Development of surrogate correlation models to predict trace organic contaminant oxidation and microbial inactivation during ozonation. *Water Research*, **46**(19), 6257–6272.

Gerrity D., Holady J. C., Mawhinney D. B., Quinones O., Trenholm R. A. and Snyder S. A. (2013). The effects of solids retention time in full-scale activated sludge basins on trace organic contaminant concentrations. *Water Environment Research*, **85**(8), 715–724.

Gerrity D., Owens-Bennett E., Venezia T., Stanford B. D., Plumlee M. H., Debroux J. and Trussell R. S. (2014). Applicability of ozone and biological activated carbon for potable reuse. *Ozone Science and Engineering*, **36**(2), 123–137.

Gerrity D., Pisarenko A. N., Marti E., Trenholm R. A., Gerringer F., Reungoat J. and Dickenson E. (2015). Nitrosamines in pilot-scale and full-scale wastewater treatment plants with ozonation. *Water Research*, **72**, 251–261.

Goldstein S. T., Juranek D. D., Ravenholt O., Hightower A. W., Martin D. G., Mesnik J. L., Griffiths S. D., Bryant A. J., Reich R. R. and Herwaldt B. L. (1996). Cryptosporidiosis: an outbreak associated with drinking water despite state-of-the-art water treatment. *Annals of Internal Medicine*, **124**(5), 459–468.

Gonzales S., Pena A. and Rosario-Ortiz F. L. (2012). Examining the role of effluent organic matter components on the decomposition of ozone and formation of hydroxyl radicals in wastewater. *Ozone Science and Engineering*, **34**(1), 42–48.

Haag W. R. and Hoigne J. (1983). Ozonation of bromide-containing waters: kinetics of formation of hypobromous acid and bromate. *Environmental Science and Technology*, **17**(5), 261–267.

Hoigné J. and Bader H. (1975). Ozonation of water-Role of hydroxyl radicals as oxidizing intermediates. *Science*, **190**(4216), 782–784.

Hoigné J. and Bader H. (1979). Ozonation of water-Oxidation-competition values of different types of waters used in Switzerland. *Ozone Science and Engineering*, **1**(4), 357–372.

Hollender J., Zimmermann, S. G., Koepke, S., Krauss, M., McArdell, C. S., Ort, C., Singer, H., von Gunten, U. and Siegrist, H. (2009). Elimination of organic micropollutants in a municipal wastewater treatment plant upgraded with a full-scale post-ozonation followed by sand filtration. *Environmental Science and Technology*, **43**(20), 7862–7869.

Huber M. M., Canonica S., Park G. Y. and von Gunten U. (2003). Oxidation of pharmaceuticals during ozonation and advanced oxidation processes. *Environmental Science and Technology*, **37**(5), 1016–1024.

Huber M. M., Gobel A., Joss A., Hermann N., Loffler D., McArdell C. S., Ried A., Siegrist H., Ternes T. A. and von Gunten U. (2005). Oxidation of pharmaceuticals during ozonation of municipal wastewater effluents: a pilot study. *Environmental Science and Technology*, **39**(11), 4290–4299.

Hubner U., von Gunten U. and Jekel M. (2015). Evaluation of the persistence of transformation products from ozonation of trace organic compounds – a critical review. *Water Research*, **68**, 150–170.

Hunt N. K. and Marinas B. J. (1997). Kinetics of Escherichia coli inactivation with ozone. *Water Research*, **31**(6), 1355–1362.

Ishida C., Salveson A., Robinson K., Bowman R. and Snyder S. (2008). Ozone disinfection with the HiPOX (TM) reactor: streamlining an "old technology" for wastewater reuse. *Water Science and Technology*, **58**(9), 1765–1773.

Jin X., Peldszus S. and Huck P. M. (2012). Reaction kinetics of selected micropollutants in ozonation and advanced oxidation processes. *Water Research*, **46**(19), 6519–6530.

Kaiser H.-P., Koester O., Gresch M., Perisset P. M. J., Jaeggi P., Salhi E. and von Gunten U. (2013). Process control for ozonation systems: a novel real-time approach. *Ozone Science and Engineering*, **35**(3), 168–185.

Katsoyiannis I. A., Canonica S. and von Gunten U. (2011). Efficiency and energy requirements for the transformation of organic micropollutants by ozone, O3/H2O2 and UV/H2O2. *Water Research*, **45**(13), 3811–3822.

Keen O. S., McKay G., Mezyk S. P., Linden K. G. and Rosario-Ortiz F. L. (2014). Identifying the factors that influence the reactivity of effluent organic matter with hydroxyl radicals. *Water Research*, **50**, 408–419.

Lalezary S., Pirbazari M. and McGuire M. J. (1986). Oxidation of five earthy-musty taste and odor compounds. *Journal of the American Water Works Association*, **78**(3), 62–69.

Larson M. A. and Marinas B. J. (2003). Inactivation of Bacillus subtilis spores with ozone and monochloramine. *Water Research*, **37**(4), 833–844.

LeChevallier M. W., Becker W. C., Schorr P. and Lee R. G. (1992). Evaluating the performance of biologically active rapid filters. *Journal of American Water Works Association*, **84**(4), 136–140.

Lee C., Yoon J. and von Gunten U. (2007). Oxidative degradation of N-nitrosodimethylamine by conventional ozonation and the advanced oxidation process ozone/hydrogen peroxide. Water Research, **41**(19), 581–590.

Lee M., Zimmermann-Steffens S. G., Arey J. S., Fenner K. and von Gunten U. (2015). Development of prediction models for the reactivity of organic compounds with ozone in aqueous solution by quantum chemical calculations: the role of delocalized and localized molecular orbitals. *Environmental Science and Technology*, **49**(16), 9925–9935.

Lee Y. and von Gunten U. (2012). Quantitative structure-activity relationships (QSARs) for the transformation of organic micropollutants during oxidative water treatment. *Water Research*, **46**(19), 6177–6195.

Lee Y., Gerrity D., Lee M., Bogeat A. E., Salhi E., Gamage S., Trenholm R. A., Wert E. C., Snyder S. A. and von Gunten U. (2013). Prediction of micropollutant elimination during ozonation of municipal wastewater effluents: use of kinetic and water specific Iiformation. *Environmental Science and Technology*, **47**(11), 5872–5881.

Li X., Zhao W., Li J., Jiang J., Chen J. and Chen J. (2013). Development of a model for predicting reaction rate constants of organic chemicals with ozone at different temperatures. *Chemosphere*, **92**(8), 1029–1034.

Mackenzie W. R., Hoxie N. J., Proctor M. E., Gradus M. S., Blair K. A., Peterson D. E., Kazmierczak J. J., Addiss D. G., Fox K. R., Rose J. B. and Davis J. P. (1994). A massive outbreak in Milwaukee of Cryptosporidium infection transmitted through the public water supply. *New England Journal of Medicine*, **331**(3), 161–167.

Marti E. J., Pisarenko A. N., Peller J. R. and Dickenson E. R. V. (2015). N-nitrosodimethylamine (NDMA) formation from the ozonation of model compounds. *Water Research*, **72**, 262–270.

Merenyi G., Lind J., Naumov S. and von Sonntag C. (2010). The reaction of ozone with the hydroxide ion: mechanistic considerations based on thermokinetic and quantum chemical calculations and the role of HO4- in superoxide dismutation. *Chemistry-A European Journal*, **16**(4), 1372–1377.

Nöthe T., Fahlenkamp H. and von Sonntag C. (2009). Ozonation of wastewater: rate of ozone consumption and hydroxyl radical yield. *Environmental Science and Technology*, **43**(15), 5990–5995.

Oneby M. A., Bromley C. O., Borchardt J. H. and Harrison D. S. (2010). Ozone treatment of secondary effluent at US municipal wastewater treatment plants. *Ozone Science and Engineering*, **32**(1), 43–55.

Oppenheimer J. A., Aieta E. M., Trussell R. R., Jacangelo J. G. and Najm I. N. (2000). Evaluation of Cryptosporidium Inactivation in Natural Waters. Water Research Foundation, Denver, CO.

Pablo Pocostales J., Sein M. M., Knolle W., von Sonntag C. and Schmidt T. C. (2010). Degradation of ozone-refractory organic phosphates in wastewater by ozone and ozone/hydrogen peroxide (peroxone): the role of ozone consumption by dissolved organic matter. *Environmental Science and Technology*, **44**(21), 8248–8253.

Padhye L., Luzinova Y., Cho M., Mizaikoff B., Kim J.-H. and Huang C.-H. (2011). PolyDADMAC and dimethylamine as precursors of N-nitrosodimethylamine during ozonation: reaction kinetics and mechanisms. *Environmental Science and Technology*, **45**(10), 4353–4359.

Peter A. and von Gunten U. (2007). Oxidation kinetics of selected taste and odor compounds during ozonation of drinking water. *Environmental Science and Technology*, **41**(2), 626–631.

Pisarenko A. N., Stanford B. D., Yan D. X., Gerrity D. and Snyder S. A. (2012). Effects of ozone and ozone/peroxide on trace organic contaminants and NDMA in drinking water and water reuse applications. *Water Research*, **46**(2), 316–326.

Plumlee M. H., Stanford B. D., Debroux J.-F., Hopkins D. C. and Snyder S. A. (2014). Costs of advanced treatment in water reclamation. *Ozone Science and Engineering*, **36**(5), 485–495.

Rakness K. L. (2005). Ozone in Drinking Water Treatment: Process Design, Operation, and Optimization. American Water Works Association, Denver, CO.

Rakness K. L. (2007). Ozone side-stream design options and operating considerations. *Ozone Science and Engineering*, **29**(4), 231–244.

Rakness K. L., Najm I., Elovitz M., Rexing D. and Via S. (2005). Cryptosporidium log-inactivation with ozone using effluent CT10, geometric mean CT10, extended integrated CT10 and extended CSTR calculations. *Ozone Science and Engineering*, **27**(5), 335–350.

Rakness K. L., Wert E. C., Elovitz M. and Mahoney S. (2010). Operator-friendly technique and quality control considerations for indigo colorimetric measurement of ozone residual. *Ozone Science and Engineering*, **32**(1), 33–42.

Reungoat J., Macova M., Escher B. I., Carswell S., Mueller J. F. and Keller J. (2010). Removal of micropollutants and reduction of biological activity in a full scale reclamation plant using ozonation and activated carbon filtration. *Water Research*, **44**(2), 625–637.

Reungoat J., Escher B. I., Macova M. and Keller J. (2011). Biofiltration of wastewater treatment plant effluent: effective removal of pharmaceuticals and personal care products. *Water Research*, **45**(9), 2751–2762.

Reungoat J., Escher B. I., Macova M., Argaud F. X., Gernjak W. and Keller J. (2012). Ozonation and biological activated carbon filtration of wastewater treatment plant effluents. *Water Research*, **46**(3), 863–872.

Rice R. G., Robson C. M., Miller G. W. and Hill A. G. (1981). Uses of ozone in drinking water treatment. *Journal of the American Water Works Association*, **73**(1), 44–57.

Rivas J., Gimeno O., de la Calle R. G. and Beltran F. J. (2009). Ozone treatment of PAH contaminated soils: Operating variables effect. *Journal of Hazardous Materials*, **169**(1–3), 509–515.

Rodriguez E., Onstad G. D., Kull T. P. J., Metcalf J. S., Acero J. L. and von Gunten U. (2007). Oxidative elimination of cyanotoxins: comparison of ozone, chlorine, chlorine dioxide and permanganate. *Water Research*, **41**(15), 3381–3393.

Roefer P. A., Monscvitz J. T. and Rexing D. J. (1996). The Las Vegas cryptosporidiosis outbreak. *Journal of the American Water Works Association*, **88**(9), 95–106.

Rosario-Ortiz F. L., Mezyk S. P., Doud D. F. R. and Snyder S. A. (2008). Quantitative correlation of absolute hydroxyl radical rate constants with non-isolated effluent organic matter bulk properties in water. *Environmental Science and Technology*, **42**(16), 5924–5930.

Rosenfeldt E. J. and Inden K. G. (2004). Degradation of endocrine disrupting chemicals bisphenol A, ethinyl estradiol, and estradiol during UV photolysis and advanced oxidation processes. *Environmental Science and Technology*, **38**(20), 5476–5483.

Roth J. A. and Sullivan D. E. (1981). Solubility of ozone in water. *Industrial & Engineering Chemistry Fundamentals*, **20**(2), 137–140.

Schindler Wildhaber Y., Mestankova H., Schärer M., Schirmer K., Salhi E. and von Gunten U. (2015). Novel test procedure to evaluate the treatability of wastewater with ozone. *Water Research*, **75**, 324–335.

Schmidt C. K. and Brauch H.-J. (2008). N,N-Dimethylsulfamide as Precursor for N-Nitrosodimethylamine (NDMA) formation upon Ozonation and its fate during drinking water treatment. *Environmeantal Science and Technology*, **42**(17), 6340–6346.

Snyder S., Vanderford B., Pearson R., Quinones O. and Yoon Y. (2003a). Analytical methods used to measure endocrine disrupting compounds in water. *Practice Periodical of Hazardous, Toxic, and Radioactive Waste Management*, **7**(4), 224–234.

Snyder S., Westerhoff P., Yoon Y. and Sedlak D. L. (2003b). Pharmaceuticals, personal care products, and endocrine disruptors in water: implications for the water industry. *Environment Engineering Science*, **20**(5), 449–469.

Snyder S. A., von Gunten U., Amy G., Debroux J. and Gerrity D. (2014). Use of Ozone in Water Reclamation for Contaminant Oxidation. WateReuse Foundation, Alexandria, VA.

Staehelin J. and Hoigné J. (1982). Decomposition of ozone in water – rate of initiation by hydroxide ions and hydrogen-peroxide. *Environmental Science and Technology*, **16**(10), 676–681.

Staehelin J., Buhler R. E. and Hoigné J. (1984). Ozone decomposition in water studied by pulse-radiolysis .2. Oh and Ho4 as chain intermediates. *Journal of Physical Chemistry*, **88**(24), 5999–6004.

Stalter D., Magdeburg A. and Oehlmann J. (2010a). Comparative toxicity assessment of ozone and activated carbon treated sewage effluents using an in vivo test battery. *Water Research*, **44**(8), 2610–2620.

Stalter D., Magdeburg A., Weil M., Knacker T. and Oehlmann J. (2010b). Toxication or detoxication? In vivo toxicity assessment of ozonation as advanced wastewater treatment with rainbow trout. Water Research, **44**(2), 439–448.

Stalter D., Magdeburg A., Wagner M. and Oehlmann J. (2011). Ozonation and activated carbon treatment of sewage effluents: removal of endocrine activity and cytotoxicity. *Water Research*, **45**(3), 1015–1024.

Stanford B. D., Pisarenko A. N., Holbrook R. D. and Snyder S. A. (2011). Preozonation effects on the reduction of reverse osmosis membrane fouling in water reuse. *Ozone Science and Engineering*, **33**(5), 379–388.

Sudhakaran S. and Amy G. L. (2013). QSAR models for oxidation of organic micropollutants in water based on ozone and hydroxyl radical rate constants and their chemical classification. *Water Research*, **47**(3), 1111–1122.

Thompson C. M. and Drago J. A. (2015). North American installed water treatment ozone systems. *Journal of the American Water Works Association*, **107**(10), 45–55.

Tiwari S., Hokanson D., Stanczak G. and Trussell R. R. (2012). Pilot-scale UV/H2O2 and O3/H2O2 AOP performance comparison for groundwater recharge, WateReuse California Annual Conference, Sacramento, CA.

Tomiyasu H., Fukutomi H. and Gordon G. (1985). Kinetics and mechanism of ozone decomposition in basic aqueous-solution. *Inorganic Chemistry*, **24**(19), 2962–2966.

Trussell R. (2012). Examining the Criteria for Direct Potable Reuse. WateReuse Foundation, Arlington, VA.

Trussell R. R., Salveson A., Snyder S. A., Trussell R. S., Gerrity D. and Pecson B. M. (2013). Potable Reuse: State of the Science Report and Equivalency Criteria for Treatment Trains. WateReuse Foundation, Arlington, VA.

USEPA (1986). Design Manual: Municipal Wastewater Disinfection. EPA/625/1-86/021. Office of Research and Development, Water Engineering Research Laboratory, Center for Environmental Research Information, Cincinnati, OH, USA.

USEPA (1998). National primary drinking water regulations: disinfectants and disinfection byproducts final rule. *Federal Register*, **63**(241), 69390.

USEPA (1999a). Alternative Disinfectants and Oxidants Guidance Manual, EPA 815-R-99-014. Washington, DC.

USEPA (1999b). Disinfection Profiling and Benchmarking Guidance Manual. EPA 815-R-99-013. Washington, DC.

USEPA (2001). Cryptosporidium: Human Health Criteria Document. EPA 822-K–94-001. Washington, DC.

USEPA (2006). National Primary Drinking Water Regulations: Long Term 2 Enhanced Surface Water Treatment Rule; Final Rule. *Federal Register*, **71**(3), 654.

USEPA (2010). Long Term 2 Enhanced Surface Water Treatment Rule Toolbox Guidance Manual. EPA 815-R-09-016. Washington, DC.

von Gunten U. (2003a). Ozonation of drinking water: Part I. Oxidation kinetics and product formation. *Water Research*, **37**(7), 1443–1467.

von Gunten U. (2003b). Ozonation of drinking water: Part II. Disinfection and by-product formation in presence of bromide, iodide or chlorine. *Water Research*, **37**(7), 1469–1487.

von Gunten U. and Hoigne J. (1994). Bromate formation during ozonation of bromide-containing water: interaction of ozone and hydroxyl radical reactions. *Environmental Science and Technology*, **28**(7), 1234–1242.

von Gunten U. and Oliveras Y. (1998). Advanced Oxidation of Bromide-Containing Waters: Bromate Formation Mechanisms. *Environmental Science and Technology*, **32**(1), 63–70.

von Gunten U., Driedger A., Gallard H. and Salhi E. (2001). By-products formation during drinking water disinfection: A tool to assess disinfection efficiency? Water Research, **35**(8), 2095–2099.

von Gunten U., Salhi E., Schmidt C. K. and Arnold W. A. (2010). Kinetics and mechanisms of N-nitrosodimethylamine formation upon ozonation of N,N-dimethylsulfamide-containing waters: Bromide catalysis. *Environmental Science and Technology*, **44**(15), 5762–5768.

von Sonntag C. and von Gunten U. (2012). Chemistry of Ozone in Water and Wastewater Treatment: From Basic Principles to Applications. IWA Publishing, London.

Wallis P., van Roodselaar A., Neurwirth M., Roach P., Buchanan-Mappin J. and Mack H. (1989). Inactivation of Giardia Cysts in a Pilot Plant using Chlorine Dioxide and Ozone. American Water Works Association, Philadephia, PA, 695–708.

Watts M. J. and Linden K. G. (2009). Advanced oxidation kinetics of aqueous trialkyl phosphate flame retardants and plasticizers. *Environmental Science and Technology*, **43**(8), 2937–2942.

Wenk J., Aeschbacher M., Salhi E., Canonica S., von Gunten U. and Sander M. (2013). Chemical oxidation of dissolved organic matter by chlorine dioxide, chlorine, and ozone: effects on Its optical and antioxidant properties. *Environmental Science and Technology*, **47**(19), 11147–11156.

Wert E. C. and Rosario-Ortiz F. L. (2011). Effect of ozonation on trihalomethane and haloacetic acid formation and speciation in a full-scale distribution system. *Ozone Science and Engineering*, **33**(1), 14–22.

Wert E. C., Neemann J. J., Johnson D., Rexing D. and Zegers R. (2007). Pilot-scale and full-scale evaluation of the chlorine-ammonia process for bromate control during ozonation. *Ozone Science and Engineering*, **29**(5), 363–372.

Wert E. C., Rosario-Ortiz F. L. and Snyder S. A. (2009a). Effect of ozone exposure on the oxidation of trace organic contaminants in wastewater. *Water Research*, **43**(4), 1005–1014.

Wert E. C., Rosario-Ortiz F. L. and Snyder S. A. (2009b). Using ultraviolet absorbance and color to assess pharmaceutical oxidation during ozonation of wastewater. *Environmental Science and Technology*, **43**(13), 4858–4863.

Westerhoff P., Nalinakumari B. and Pei P. (2006). Kinetics of MIB and geosmin oxidation during ozonation. *Ozone Science and Engineering*, **28**(5), 277–286.

Westerhoff P., Mezyk S. P., Cooper W. J. and Minakata D. (2007). Electron pulse radiolysis determination of hydroxyl radical rate constants with Suwannee River Fulvic Acid and other dissolved organic matter isolates. *Environmental Science and Technology*, **41**(13), 4640–4646.

Yan Z., Zhang Y., Yuan H., Tian Z. and Yang M. (2014). Fish larval deformity caused by aldehydes and unknown byproducts in zonated effluents from municipal wastewater treatment systems. *Water Research*, **66**, 423–429.

Zeng T. and Mitch W. A. (2015). Contribution of N-nitrosamines and their precursors to domestic sewage by greywaters and blackwaters. *Environmental Science and Technology*, **49**(22), 13158–13167.

Zuma F., Lin J. and Jonnalagadda S. B. (2009). Ozone-initiated disinfection kinetics of Escherichia coli in water. *Journal of Environmental Science and Health Part A-Toxic/Hazardous Substances & Environmental Engineering*, **44**(1), 48–56.

Zuma F. N. and Jonnalagadda S. B. (2010). Studies on the O-3-initiated disinfection from Gram-positive bacteria Bacillus subtilis in aquatic systems. *Journal of Environmental Science and Health Part A-Toxic/Hazardous Substances & Environmental Engineering*, **45**(2), 224–232.

Chapter 4

Ozone/H_2O_2 and ozone/UV processes

Alexandra Fischbacher, Holger V. Lutze and Torsten C. Schmidt

4.1 INTRODUCTION

Water treatment processes that lead to highly reactive hydroxyl radicals ($^{\bullet}OH$) are referred to as Advanced Oxidation Processes (AOPs). These radicals are capable of degrading pollutants that survive other oxidation processes such as conventional ozonation, thus, the implementation of OH radical-based water treatment processes represent a viable option in order to meet the low drinking water standards for regulated micropollutants.

The two common ozone-based AOPs rely on the reaction of ozone with hydrogen peroxide (peroxone process) and the photolysis of ozone with UV light.

4.2 O_3/H_2O_2 (PEROXONE) PROCESS FUNDAMENTALS

4.2.1 Mechanism of hydroxyl radical generation

The peroxone process describes the generation of $^{\bullet}OH$ from the reaction of ozone (O_3) with hydrogen peroxide (H_2O_2). The reaction of O_3 with H_2O_2 is very slow ($k < 0.01$ M^{-1} s^{-1}), but that with its anion, HO_2^-, is several orders of magnitude faster ($k = 5.5 \times 10^6$ M^{-1} s^{-1}). Hence, only the latter species is relevant for the peroxone process resulting in a strong pH dependency of the apparent reaction rate constant k_{obs} described by equation (4.1).

$$k_{obs} = k(HO_2^- + O_3) \times 10^{(pH - pK_a)} \tag{4.1}$$

Due to the high pK_a (H_2O_2) of 11.8, the reaction is fast only at elevated pH values. The k_{obs}–values for pH 6, 7 and 8 were calculated to be 8.72 M^{-1} s^{-1}, 87.2 M^{-1} s^{-1} and 872 M^{-1} s^{-1}, respectively.

In the early 1980s Staehelin and Hoigné (1982) suggested the mechanism for the peroxone process as shown in the reactions 4.2–4.7. A summarized overall reaction of the elementary steps

($H_2O_2 + 2O_3 \rightarrow 2\,{}^{\bullet}OH + 3O_2$) gives the yield of ${}^{\bullet}OH$ generated per consumed O_3. For several decades, it was assumed that this yield was unity on a molar basis with respect to O_3 consumption.

$$H_2O_2 \rightleftharpoons H^+ + HO_2^- \tag{4.2}$$

$$HO_2^- + O_3 \rightarrow HO_2^{\bullet} + O_3^{\bullet-} \tag{4.3}$$

$$HO_2^{\bullet} \rightleftharpoons O_2^{\bullet-} + H^+ \tag{4.4}$$

$$O_2^{\bullet-} + O_3 \rightarrow O_2 + O_3^{\bullet-} \tag{4.5}$$

$$O_3^{\bullet-} + H^+ \rightleftharpoons HO_3^{\bullet} \tag{4.6}$$

$$HO_3^{\bullet} \rightarrow {}^{\bullet}OH + O_2 \tag{4.7}$$

Recently, it was stated that the electron transfer reaction of HO_2^- with O_3 proceeds *via* adduct formation (HO_5^-) (reaction 4.8) and the reaction of $O_3^{\bullet-}$ en route to ${}^{\bullet}OH$, formerly described by reactions 4.6 and 4.7, rather proceeds by reactions 4.10 and 4.11. (Sein *et al.* 2007; Merenyi *et al.* 2010)

$$HO_2^- + O_3 \rightleftharpoons HO_5^- \tag{4.8}$$

$$HO_5^- \rightarrow HO_2^{\bullet} + O_3^{\bullet-} \tag{4.9}$$

$$O_3^{\bullet-} \rightleftharpoons O_2 + O^{\bullet-} \tag{4.10}$$

$$O^{\bullet-} + H_2O \rightleftharpoons {}^{\bullet}OH + OH^- \tag{4.11}$$

Merenyi *et al.* (2010) stated that the adduct formed in reaction 4.8 could also decay into O_2 in its ground state and OH^- (reaction 4.12). This decay occurs in competition to the reaction initiating ${}^{\bullet}OH$ formation (reaction 4.9). As only one of both reactions yields ${}^{\bullet}OH$, the ${}^{\bullet}OH$ yield in the peroxone process must be below unity.

$$HO_5^- \rightarrow 2\,{}^3O_2 + OH^- \tag{4.12}$$

This *new* assumption was experimentally confirmed in a recent work of Fischbacher *et al.* (2013). To that end, three different approaches were applied to verify this assumption. The first evidence was derived from competition kinetics experiments. Here, the consumption of the ozone refractory compounds *p*-chlorobenzoic acid (*p*CBA), *p*-nitrobenzoic acid (*p*NBA) and atrazine was monitored as a function of the ozone concentration at a constant H_2O_2 concentration and in the presence of *tert*-butanol (*t*BuOH) as competitor for ${}^{\bullet}OH$. Further evidence was given by product analysis assays. Thereby, the generated ${}^{\bullet}OH$ were quantitatively scavenged by *t*BuOH and dimethylsulfoxide (DMSO) and these reactions give products with known yields per radical attack [i.e., 50% formaldehyde (*t*BuOH assay) and 92% methanesulfinic acid and methanesulfonic acid (DMSO assay)]. Quantification of these products in turn gives ${}^{\bullet}OH$ yields of the studied reaction. The approaches were tested by comparing the results of competition experiments and the product analysis assays (*t*BuOH) with analogous experiments using γ-radiolysis as ${}^{\bullet}OH$ source with a defined ${}^{\bullet}OH$ concentration. The ${}^{\bullet}OH$ yields in the peroxone process obtained by different approaches are summarized in Table 4.1.

Table 4.1 •OH yields per O_3 in the peroxone process determined by different approaches (Fischbacher *et al.* 2013).

Approach	•OH Yield
Competition kinetics (*p*CBA as a probe)	56%
Competition kinetics (*p*NBA as a probe)	49%
Competition kinetics (Atrazine as a probe)	60%
Product analysis (*t*BuOH as •OH scavenger)	50.9%
Product analysis (DMSO as •OH scavenger)	49.5%

Thus, it was concluded, that the •OH yield in the peroxone process is around 50% of the consumed ozone concentration (on a molar basis) corroborating the postulated mechanism of Merenyi *et al.* (2010), (reactions 4.8–4.12).

4.2.2 O_3 and •OH exposures: the R_{ct} concept

The R_{ct} concept is a tool to characterize the oxidation of micropollutants in water by ozone and •OH (Elovitz & von Gunten, 1999). It can be used for the prediction/modeling of oxidation processes in ozone-based water treatment. For a characterization of these oxidation processes the concentration or exposure (integral of concentration over time) of ozone and •OH have to be known. The concentration of ozone can be easily measured by calorimetric, spectrometric and electrochemical methods. However, the determination of the •OH concentration is much more difficult, as •OH are highly reactive and their steady-state concentration is too low to be measured directly. An indirect method to determine the •OH exposure is measuring the decay of a probe compound. The prerequisites for a probe compound are a low reactivity with O_3 and a high reactivity with •OH. A typically used probe compound is *p*CBA, as it reacts very slowly with O_3 ($k_{O_3+pCBA} \leq 0.15\ M^{-1}\ s^{-1}$) (Neta *et al.* 1988), but readily with •OH ($k_{\bullet OH+pCBA} = 5 \times 10^9\ M^{-1}\ s^{-1}$) (Buxton *et al.* 1988).

The reaction of the probe compound can be described as a second order reaction. Since the probe compound does not react with O_3, just the reaction with •OH is taken into account.

$$\frac{-d[\text{probe}]}{dt} = k_{\bullet\text{OH+probe}} \times [\text{probe}][\bullet\text{OH}] \tag{4.13}$$

Integration and rearrangement of equation 4.13 leads to equation 4.14

$$-\ln\left(\frac{[\text{probe}]}{[\text{probe}]_0}\right) = k_{\bullet\text{OH+probe}} \times \int[\bullet\text{OH}]dt \tag{4.14}$$

R_{ct} describes the ratio of the exposures of •OH and O_3.

$$R_{ct} = \frac{\int[\bullet\text{OH}]dt}{\int[\text{O}_3]dt} \tag{4.15}$$

Insertion of equation 4.15 in equation 4.14 results in equation 4.16

$$-\ln\left(\frac{[\text{probe}]}{[\text{probe}]_0}\right) = k_{\bullet\text{OH+probe}} \times R_{ct}\int[\text{O}_3]dt \tag{4.16}$$

By plotting the decay of the probe compound ($\ln([\text{probe}]/[\text{probe}]_0)$) over ozone exposure ($\int[O_3]dt$) the R_{ct}-value can be calculated from the slope, as $k_{\cdot OH+probe}$ is known. Once a certain water is "calibrated" by determination of the R_{ct}-value, the •OH-exposure can be calculated *via* equation 4.15 if the O_3 exposure is known. Since O_3 can be determined on-line over time by suitable electrodes, the •OH-exposure is available as real-time value. This enables a continuous prediction of pollutant degradation (equation 4.14), which can be used for process control.

The R_{ct}-value in ozonation depends on the water quality, e.g., dissolved organic carbon (DOC) concentration, pH, alkalinity and temperature and thus, largely varies for different kinds of water sources. Typically, waters with a high ozone stability yield low R_{ct}-values. Since in conventional ozonation, O_3 is mainly scavenged by organic matter, low-DOC groundwaters have lower R_{ct}-values than rich-DOC surface waters (Elovitz *et al.* 2000a). Furthermore, ozone decay may reveal different kinetic stages in conventional ozonation. Figure 4.1 shows an example of the first order decay of ozone in ozone-based processes. In conventional ozonation, typically two kinetic regimes of ozone decay can be observed, i.e., a first rapid phase and a second slower phase (Figure 4.1). In the first phase fast-reacting electron rich moieties of DOC result in a rapid O_3 consumption (ca. 5–30% in drinking water ozonation). These moieties are consumed within the first seconds of reaction time resulting in slower ozone decay kinetics in the second phase. This is also reflected by the R_{ct}-values which are higher in the first phase of ozone decay compared to the second phase. In the second phase of ozone decomposition, the R_{ct}-value is fairly stable over a longer period of time (often throughout the whole hydraulic retention time of full scale reactors) and thus, describes the ratio of •OH to ozone consumption at any time of the reaction (Elovitz & von Gunten, 1999; Acero & von Gunten, 2001b). However, in some cases even three or more phases can be observed.

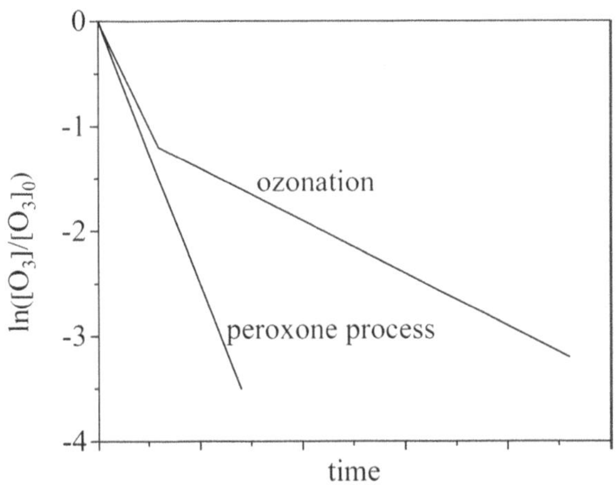

Figure 4.1 Generic representation of ozone decay in ozonation and in the peroxone process. Kinetics of ozone decay in the peroxone process is constant over time, whereas it displays two phases in ozonation only, with a first rapid phase followed by a slower phase.

At high DOC concentrations (e.g., wastewater) the ozone decay and thus the R_{ct} resembles the values in the peroxone process. In the peroxone process, such stages of ozone decay kinetics are typically not observed (Figure 4.1). This is due to the fact that the chemical nature of the initiator (HO_2^-) does not change during the reaction.

During drinking water treatment a high R_{ct}-value indicates a good degradation of ozone refractive micropollutants. However, since a high R_{ct}-value is often accompanied by a low ozone exposure disinfection efficiency is weak (see section 4.6 Disinfection). Typical values for R_{ct} in ozonation are in the range of 10^{-9}–10^{-8}, while in the peroxone process the values are up to two orders of magnitude higher.

4.2.3 Reaction kinetics and modeling

The degradation of micropollutants reacting with ozone and $^{\bullet}$OH can be predicted for batch or plug-flow reactors based on their second-order reaction kinetics. The degradation of a micropollutant (MP) can be expressed as follows:

$$-\frac{d[\mathrm{MP}]}{dt} = k_{^\bullet\mathrm{OH+MP}} \times [^\bullet\mathrm{OH}] \times [\mathrm{MP}] + k_{\mathrm{O_3+MP}} \times [\mathrm{O_3}] \times [\mathrm{MP}] \quad (4.17)$$

Replacing [$^{\bullet}$OH] in equation 4.17 by $R_{ct} \times [O_3]$ according to equation 4.15 gives:

$$-\frac{d[\mathrm{MP}]}{dt} = k_{^\bullet\mathrm{OH+MP}} \times R_{\mathrm{ct}} \times [\mathrm{O_3}] \times [\mathrm{MP}] + k_{\mathrm{O_3+MP}} \times [\mathrm{O_3}] \times [\mathrm{MP}] \quad (4.18)$$

Integrating equation 4.18 results in:

$$\ln\frac{[\mathrm{MP}]_t}{[\mathrm{MP}]_0} = -\left(\int_0^t [\mathrm{O_3}]dt\right)\left(k_{^\bullet\mathrm{OH+MP}} \times R_{\mathrm{ct}} + k_{\mathrm{O_3+MP}}\right) \quad (4.19)$$

After determination of R_{ct} (cf. section 4.2.2) and with the second-order rate constants of micropollutants (k_{O_3+MP} and $k_{^\bullet OH+MP}$) at hand, it is possible to predict the oxidation of a micropollutant as a function of ozone exposure using Eq. 4.19. Applying different H_2O_2/O_3 ratios in raw water and determining ozone exposures and R_{ct}-values at different H_2O_2/O_3 ratios thus, allows a rough assessment of the degradation of several micropollutants without the need to measure their degradation at each H_2O_2/O_3 ratio. This is a useful tool to evaluate the optimal H_2O_2/O_3 ratio with just few experiments (Acero & von Gunten, 2001b). It has to be noted that the error in the reaction rate k may bias such a prediction of the degree of pollutant degradation.

It must be considered that in waters with high O_3 stability, the addition of H_2O_2 accelerates the degradation of micropollutants, but may not necessarily change the overall oxidation efficiency significantly (von Gunten, 2003b).

The fraction of $^{\bullet}$OH reacting with a pollutant is typically very small, since most of $^{\bullet}$OH are consumed by water matrix constituents. The reaction rate constants of such matrix constituents are compiled in Table 4.2. With that data at hand it is possible to calculate the actual percentage of radical reactions targeting a micropollutant (here *p*CBA) by applying equation (4.20). Eq. (4.20) includes all relevant water matrix components and added chemicals (H_2O_2) that scavenge $^{\bullet}$OH. Thus, it gives the ratio of the rate of *p*CBA oxidation by $^{\bullet}$OH and the sum of the rates of the reactions of $^{\bullet}$OH with all $^{\bullet}$OH scavengers (S) present in the water.

$$\%^\bullet\mathrm{OH}_{p\mathrm{CBA}} = 100\frac{k_{^\bullet\mathrm{OH}+p\mathrm{CBA}}[p\mathrm{CBA}]}{\left\{k_{^\bullet\mathrm{OH}+p\mathrm{CBA}}[p\mathrm{CBA}] + \sum k_{^\bullet\mathrm{OH+S}}[\mathrm{S}]\right\}} \quad (4.20)$$

Matrix compounds present in natural waters that need to be considered as possible •OH scavengers are listed in Table 4.2. At typical water treatment conditions (1 mg/L DOC, 1 mM HCO_3^-, 10 μg/L Br^-) only a fraction of ca. 6.69% •OH is reacting with *p*CBA (0.5 μM) and most radicals are consumed by DOC (66.89%) followed by HCO_3^- (22.74%). Such calculations allow assessing which matrix constituents are the most important factors affecting the pollutant degradation efficiency.

Table 4.2 Reaction rate constants of matrix compounds in natural waters and H_2O_2 with •OH.

Matrix Compound	Reaction Rate Constant with •OH	Reference
Dissolved organic carbon	2.5×10^4 L mg^{-1} s^{-1}	(Larson & Zepp, 1988)
HCO_3^-	8.5×10^6 M^{-1} s^{-1}	(Hoigné & Bader, 1983a)
CO_3^{2-}	3.9×10^8 M^{-1} s^{-1}	(Buxton *et al.* 1988)
NO_2^-	8×10^9 M^{-1} s^{-1}	(Buxton *et al.* 1988)
Br^-	1.1×10^{10} M^{-1} s^{-1}	(Zehavi & Rabani, 1972)
H_2O_2	2.7×10^7 M^{-1} s^{-1}	(Buxton *et al.* 1988)

A possible scavenger for •OH may also be O_3. However, Rosenfeldt *et al.* (2006) tested the approach including and excluding the reaction of O_3 with •OH and observed a minimal difference in the •OH fraction reacting with *p*CBA ($\Delta\%^\bullet OH_{pCBA}$) of 0.5% between the two test scenarios. Thus, the •OH reaction with ozone can be neglected.

It must be considered that the calculated fraction is just valid for the initial conditions of a reaction and that errors in measuring the concentrations of the matrix compounds and discrepancies in published reaction rates can significantly change the result of the •OH fraction reacting with the target compound.

Reaction rate constants *k* are useful to evaluate the reaction of a micropollutant with an oxidant and obtain half-lives of micropollutants. Many rate constants are compiled for ozone (Hoigné & Bader, 1983a; Hoigné & Bader, 1983b; von Sonntag & von Gunten, 2012) and •OH (Buxton *et al.* 1988; Neta *et al.* 1988). If reaction rate constants cannot be found in literature, they can be experimentally determined by different methods (Huber, 2004) or be predicted. Quantitative structure-activity relationships (QSARs) are tools to predict rate constants for reactions of oxidants with micropollutants.

Lee and von Gunten (2012) developed QSARs for the reactions of O_3 with phenols, phenolates, benzene derivatives, olefins and amines. These QSARs are based on the correlation between Hammett σ (σ, σ^+ and σ^-) or Taft σ^* constants and reaction rate constants (some values are compiled by Hansch *et al.* (1991)). The authors also developed cross-correlations between *k*-values for the reaction of O_3 vs. ClO_2 and O_3 vs. HOCl. Two multiple linear regression-based standardized QSAR model equations for $k_{\bullet OH}$ and k_{O_3} were introduced by Sudhakaran and Amy (2013). A plot of the predicted $k_{\bullet OH}$ vs. experimentally obtained $k_{\bullet OH}$ for haloalkanes, aromatics, alkanes and activated aromatics shows a good correlation ($r^2 = 0.9178$). The models can be applied for reactions with •OH in a pH range of 5–8 and $k_{\bullet OH}$-values between 4×10^7 and 1.8×10^{10} M^{-1} s^{-1} and for O_3 reactions in a pH range of 5–8 and k_{O_3}-values between 5×10^{-4} and 10^5 M^{-1} s^{-1}.

Li *et al.* (2013) developed a QSAR model to predict *k*-values for the reactions of O_3 with organic chemicals at different temperatures. The equation includes 13 descriptors and is suitable for the prediction of *k*-values for the reaction of O_3 with structurally diverse compounds with various functional groups (e.g., >C=C<, −C≡C−, −OH, −CHO, −O−, >C=O, −COOH, −C≡N, $-NO_2$ and −X(Cl, F).

4.2.4 Water quality impact on process performance: O_3 and H_2O_2 dose selection criteria

In analogy to O_3 based processes the water matrix influences both the efficiency of pollutant degradation and by-product formation. In that regard the concentration of other solutes (i.e., DOC, CO_3^{2-}/HCO_3^-, and Br^-) and the pH value play a role (Figure 4.2). The effect of pH on the reaction rate of O_3 with H_2O_2 (reaction 4.21) that initiates $^•OH$ formation was already explained in section (4.2.1). At a low $k_{obs}(O_3 + H_2O_2)$ the fraction of O_3 reactions with DOC (reaction 4.22) increases, turning the peroxone process into conventional ozonation. This can be compensated by increasing the H_2O_2 dosage. However, with increasing concentrations of H_2O_2 the reaction of $^•OH$ with H_2O_2 becomes important (reaction 4.24). $^•OH$ react with H_2O_2 *via* H-abstraction yielding $HO_2^•$ (k ($^•OH + H_2O_2$) = 2.7×10^7 M^{-1} s^{-1} (Buxton *et al.* 1988)) giving rise to superoxide ($O_2^{•-}$) (Christensen *et al.* 1982) upon $HO_2^•$ deprotonation ($pK_a = 4.8$ (Bielski *et al.* 1985)). $O_2^{•-}$ reacts fast with O_3 yielding $^•OH$, resulting in a catalyzed O_3 decomposition. This deteriorates both the disinfection strength and the pollutant degradation. However, the reaction of O_3 with electron rich moieties of DOC (i.e., phenolic moieties and deprotonated amines) also initiates $^•OH$ formation (von Sonntag & von Gunten, 2012) with yields comparable to the peroxone process in DOC rich waters (von Gunten, 2003a). DOC, carbonate and bicarbonate consume $^•OH$ (reaction 4.23 and 4.25) ($k(^•OH + DOC) = 2.5 \times 10^4$ L mg^{-1} s^{-1} (Larsson & Zepp, 1988), $k(^•OH + HCO_3^-) = 8.5 \times 10^6$ M^{-1} s^{-1}, $k(^•OH + CO_3^{2-}) = 3.9 \times 10^8$ M^{-1} s^{-1}) (Buxton *et al.* 1988). This leads to a loss of oxidation strength in the peroxone process which in particular targets the degradation of O_3 recalcitrant pollutants *via* the $^•OH$ pathway. The reaction of $^•OH$ with organic compounds typically yields carbon centered radicals ($R^•$), which rapidly add oxygen ($k(O_2 + R^•) \approx 10^9$ M^{-1} s^{-1} (von Sonntag *et al.* 1997)) (von Sonntag & von Gunten, 2012). The ensuing peroxyl radical ($ROO^•$) may cleave superoxide ($O_2^{•-}$) (reaction 4.26) which contributes in the turnover of O_3 into $^•OH$ (reaction 4.27) and thus, increase the rate of pollutant degradation. Reactions accelerating $^•OH$ formation are called propagation reactions (von Gunten, 2003a).

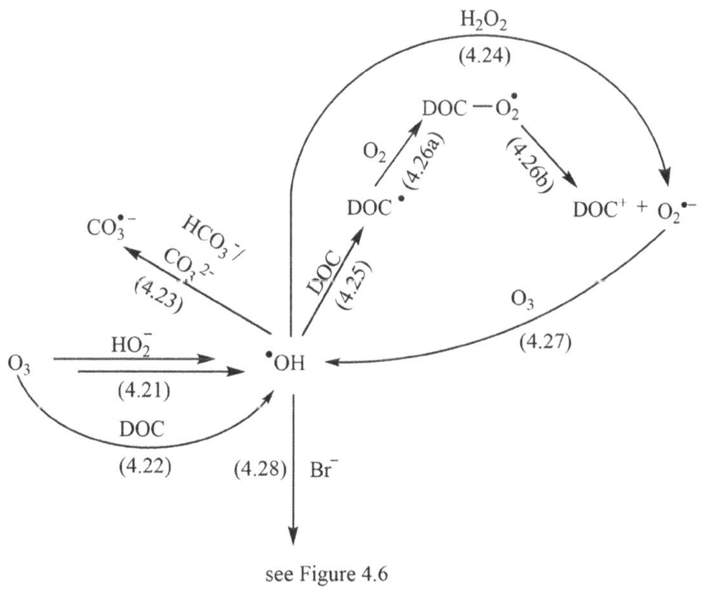

Figure 4.2 Relevant reactions of $^•OH$ with water matrix constituents influencing the peroxone process.

The reaction of $^{\bullet}OH$ with HCO_3^- and CO_3^{2-} (reaction 4.23) yields carbonate radicals ($CO_3^{\bullet -}$), which do neither significantly contribute in pollutant degradation nor react with O_3. Such kind of reactions which exclusively consume $^{\bullet}OH$ are called scavenging reactions (von Gunten, 2003a). Waters with a high alkalinity and a low DOC concentration (e.g., ground waters) typically have a high scavenging and a low radical initiation/propagation capacity. This dramatically lowers the degradation efficiency for O_3 recalcitrant pollutants in ozonation and the degradation of such pollutants can substantially be improved by addition of the initiator H_2O_2 (von Gunten, 2003a).

It is noteworthy that H_2O_2 may affect the oxidation of reduced manganese species. Colloidal MnO_2 is rapidly reduced by H_2O_2, which may hamper the O_3 based removal of reduced Mn species. The oxidation of reduced Mn and Fe species in conventional ozonation is mainly driven by O_3, since the reaction is fast ($k(O_3 + Fe^{2+}) \approx 8 \times 10^5\ M^{-1}\ s^{-1}$; $k(O_3 + Mn^{2+}) \approx 10^3\ M^{-1}\ s^{-1}$) (von Sonntag & von Gunten, 2012) and O_3 concentrations are typically several orders of magnitude higher than $^{\bullet}OH$ concentration (Elovitz *et al.*, 2000a). Hence, destabilization of O_3 by addition of H_2O_2 is counterproductive regarding the removal of these metal species. Indeed, all treatment targets, which rely on reactions of O_3, may be hampered by addition of H_2O_2. Hence, the peroxone process is a tradeoff of bromate control and degradation of O_3 recalcitrant compounds with a loss of most other treatment aims of conventional ozonation such as micro-flocculation, disinfection, enhanced biodegradation and removal of reduced metal species (see above).

4.3 O_3/H_2O_2 AOP FOR MICROPOLLUTANT REMOVAL

4.3.1 Bench-scale research studies

There are a few reviews on micropollutant degradation in the peroxone process. Ikehata and Gamal El-Din published two reviews on the degradation of aqueous pesticides by O_3 and O_3-based AOPs (Ikehata & Gamal El-Din, 2005a; Ikehata & Gamal El-Din, 2005b). They come to the conclusion that most of the pesticides are considerably reactive towards O_3. In general, O_3-based AOPs are even more effective than ozonation. The same group also reviewed the literature on the degradation of aqueous pharmaceuticals by O_3 and AOPs (Ikehata *et al.* 2006). Many pharmaceuticals e.g., carbamazepine, diclofenac, 17β-estradiol and some antibiotics are very reactive towards O_3. For these compounds ozonation is sufficient for micropollutant abatement but AOPs are more effective than ozonation alone for pharmaceuticals which react slow with O_3 (e.g., triiodinated X-ray contrast agents). In wastewater treatment, the main goal is the improvement of biodegradability of pharmaceuticals that are resistant to biodegradation (e.g., antibiotics, cytostatic agents, hormones, X-ray contrast agents, and some acidic drugs). However, in most studies this is just the case for some antibiotics and the reduction in toxicity by O_3 and $^{\bullet}OH$ has been observed only in few cases.

Yargeau and Leclair (2008) published a review on the impact of operating conditions such as pH, temperature, H_2O_2 and O_3 dosage, reactor setup and wastewater composition on the degradation of antibiotics during ozonation and the peroxone process.

The review published by Ning *et al.* (2007) is dealing with the degradation of endocrine disrupting chemicals (EDCs) by ozone and several AOPs. The authors concluded that hormones and BPA can be degraded by ozone alone while phthalate compounds have lower reactivity towards ozone, but good reactivity towards $^{\bullet}OH$. In general, they suggested applying O_3-based AOPs for the treatment of EDCs.

Beside these reviews there are many important publications dealing with the removal of micropollutants by the peroxone process at lab-scale. Selected peroxone-based water treatment studies are described below.

Methyl-*tert*-butyl ether (MTBE) is an important gasoline component and due to leakage from the underground gasoline storage tanks, it is frequently detected in surface water and groundwater. Acero *et al.* (2001a) investigated MTBE oxidation by O_3/H_2O_2 in water samples from two lakes and an underground well. Water quality parameters for both lake waters were 1.4 and 2.7 mg/L DOC, and 2.5 and 3.8 mM HCO_3^-, respectively, whereas the well water contained 0.8 mg/L DOC and 5.0 mM HCO_3^-. These waters represented three different scenarios, (i) low DOC content and low alkalinity, (ii) high DOC content and high alkalinity, and (iii) low DOC content and high alkalinity. MTBE removal of 34–46% was achieved without exceeding the drinking water standard for bromate (10 μg/L) at an initial bromide concentration of 50 μg/L. Best results were obtained for scenario (i), where 46% MTBE was degraded with formation of only 8.8 μg/L bromate. The applied molar H_2O_2:O_3 ratio was 0.5.

Due to the large-scale application of pesticides and their persistence in the environment, their degradation is of great importance. The oxidation of pesticides (bromoxynil and trifluralin) in natural waters by the peroxone process and ozonation was compared by Chelme-Ayala *et al.* (2011). The authors investigated river water (DOC: 2.8 mg/L, alkalinity as $CaCO_3$: 136 mg/L) and irrigation return flow water (DOC: 20.6 mg/L, alkalinity as $CaCO_3$: 230 mg/L). The addition of H_2O_2 enhanced the degradation of both pesticides in natural waters by about 10–15%. The optimal molar H_2O_2:O_3 ratio was 0.5. At an ozone dose of 2×10^{-5} M and a H_2O_2 dose of 1×10^{-5} M, bromoxynil (3.6×10^{-6} M) was degraded in ultrapure water by 99%, in river water by 70% and in irrigation return flow water by 49%. Similar results were achieved for trifluralin (3.0×10^{-6} M) under similar treatment conditions. The degradation in ultrapure water was of 91%, 54% in river water, and 47% in irrigation return flow water. Higher DOC content resulted in an O_3 consumption and high alkalinity in scavenging of •OH. Both cases lead to a decrease in degradation as less O_3 ($k_{Bromoxynil+O_3} = 5.2 \times 10^2$ M^{-1} s^{-1}, $k_{Trifluralin+O_3} = 1 \times 10^2$ M^{-1} s^{-1}) and •OH ($k_{Bromoxynil+^\bullet OH} = 2 \times 10^{10}$ M^{-1} s^{-1}, $k_{Trifluralin+^\bullet OH} = 7.1 \times 10^9$ M^{-1} s^{-1}) were available for the reaction with pesticides.

The presence of many pharmaceuticals in surface waters has been reported in the past (Ternes, 1998; Kümmerer, 2001; Al-Rifai *et al.* 2007; Lin *et al.* 2016). Many of these pharmaceuticals are persistent in conventional wastewater treatment (Carballa *et al.* 2004). Zwiener and Frimmel (2000) investigated the oxidation of environmentally relevant pharmaceuticals (clofibric acid, ibuprofen and diclofenac) by the peroxone process in river Ruhr water (3.7 mg/L DOC and 122 mg/L HCO_3^-). At concentrations of 3.7 mg/L O_3 and 1.4 mg/L H_2O_2 90% degradation could be achieved and at 5.0 mg/L O_3 and 1.8 mg/L H_2O_2 even up to 98%.

The removal of 32 pharmaceuticals in wastewater by ozone and the peroxone process was studied by Rosal *et al.* (2008). The experiments were conducted in urban and domestic wastewater in a 5-L reactor. The produced ozone was introduced to the sample by bubbling it into the reactor. It was observed that the oxidation yields of nearly all pharmaceuticals were similar for ozonation and the peroxone process. Within 5 minutes, > 99% removal was achieved for most compounds with and without the addition of hydrogen peroxide, but the addition of hydrogen peroxide increased the total organic carbon (TOC) removal significantly. TOC was used as a measure of the extent of mineralization. In ozonation a TOC decrease of about 15% was achieved after 1 h of operation. The addition of hydrogen peroxide (injecting pulses of 0.15 mL H_2O_2 30% w/v every 5 min) resulted in a TOC decrease of about 90%. However, such treatment conditions are not applicable for water treatment due to restrictions with regard to energy demand in water treatment by peroxone process. It was shown that O_3 and H_2O_2 doses of 3.456 kg/m^3 and 108 g/m^3, respectively, are required to degrade 90% TOC. This results in an energy demand of 52.92 kWh/m^3. Comparing the demand with typical demands in water treatment (cf. Section 4.3.5), it is obvious that this dosage is not economically feasible. Therefore, the mineralization of micropollutants is not achieved in the peroxone process under economically feasible conditions, neither is it targeted in surface water, groundwater, or water reuse treatment for public consumption.

Pocostales *et al.* (2010) investigated the degradation of organic phosphates (tri-n-butyl phosphate and tris-2-chloroisopropyl phosphate) by O_3/H_2O_2 in wastewater (DOC: 11 mg/L, 3.5 mM HCO_3^-, pH 8). O_3 was introduced to the wastewater as stock solution in distilled water. This resulted in a dilution of the wastewater by a factor of two. At low O_3 concentration (i.e., 42 µM O_3) the addition of 21 µM H_2O_2 had only a little effect, as the reaction of O_3 with DOC from wastewater nearly outcompetes the reaction with H_2O_2. But at higher O_3 concentrations (i.e., ≥ 84 µM O_3) and a consistent molar H_2O_2:O_3 ratio of 0.5 the addition of H_2O_2 became more important leading to an •OH yield increase by a factor of 1.5. This is probably due to the fact that more of the reactive electron rich moieties of DOC are oxidized in case of the higher ozone dose, thus, reducing the reaction rate of DOC with O_3.

1,4-Dioxane is a solvent used in the manufacture of paper, textile products and cosmetics. As it is classified as possibly carcinogenic, the World Health Organisation (WHO) published a drinking water quality standard of 50 µg/L. Takahashi *et al.* (2013) studied the degradation of 1,4-dioxane in buffered deionized water by O_3/H_2O_2 and O_3/UV. A total degradation of 150 mg/L 1,4-dioxane was achieved within 60 min by O_3/H_2O_2 and within 120 min by O_3/UV.

Algal toxins are released into water through the cyanobacteria cell lysis and were found in surface waters around the globe during the algal blooming seasons. As these toxins are on the Candidate Chemical List (e.g., US EPA CCL3) research is of great interest. Al Momani (2007) studied the use of ozone and peroxone for the degradation of anatoxin-a in buffered pure water. The apparent rate constants for the degradation of 1 mg/L anatoxin-a were $87 \pm 2\ s^{-1}$ in the peroxone process (2 mg/L O_3 and 0.001 mg/L H_2O_2) and $5.1 \pm 0.2\ s^{-1}$ in ozonation (2 mg/L O_3). The corresponding half-life times were 21 s in the peroxone process and 180 s in ozonation. Orr *et al.* (2004) investigated the removal of saxitoxins (algal toxins (including saxitoxin, neosaxitoxin, C toxins and gonyautoxins) produced by freshwater cyanobacteria *A. circinalis*) from raw drinking water obtained from a drinking water reservoir by O_3/H_2O_2. When applying the peroxone process in batch experiments (H_2O_2:O_3 ratio of 1:10 (w:w) at an O_3 concentration of 1 mg/L; reaction time: 10 min) it was not possible to reduce the initial toxicity (defined as µg saxitoxin (STX) toxicity-equivalents/L) of 30 µg (STX equivalents)/L below the suggested Australian drinking water guideline concentration of 3 µg (STX equivalents)/L. The toxicity could just be reduced to 28.7 µg (STX equivalents)/L.

Further algal-derived surface water contaminants are 2-methylisoborneol (2-MIB) and geosmin. These compounds are classified as taste and odor (T&O) compounds, as their taste and smell deteriorates the organoleptic quality of drinking water. Mizuno *et al.* (2011) studied the application of the peroxone process and ozonation with regard to the removal of these T&O compounds and bromate formation in river water. With an initial Br^- concentration of 50 µg/L in ozonation (2 mg/L O_3) bromate formation was controlled to less than 10 µg/L. However, the concentration of 2-MIB was still above the standard for drinking water quality in Japan of 10 ng/L (at 100 ng/L initial concentration). In the peroxone process (2 mg/L O_3; 0.5–1.7 mg/L H_2O_2) a degradation of 2-MIB to below 10 ng/L was achieved, while keeping the bromate concentration below the drinking water standard of 10 µg/L, even though higher initial Br^- (up to 500 µg/L) and higher 2-MIB (up to 500 ng/L) concentrations were present.

Westerhoff *et al.* (2006) investigated the degradation of T&Os in several surface waters with different water quality parameters (DOC, alkalinity and Br^-) by O_3/H_2O_2 and O_3. In all waters ≥ 98% of 2-MIB was degraded by O_3/H_2O_2 (molar H_2O_2:O_3 ratio: 0.07), while in ozonation at same conditions the degradation varies from 66–86%. In both processes •OH play a major role, since with addition of *t*BuOH as •OH scavenger the degree of degradation dropped to 6–20%. This is in accordance with the reaction rates determined by Peter and von Gunten (2007) for several T&O compounds, which revealed that 2-MIB hardly reacts with ozone ($k = 0.35\ M^{-1}\ s^{-1}$) but rapidly with •OH ($k = 5.09 \times 10^9\ M^{-1}\ s^{-1}$).

4.3.2 Pilot-scale studies

As the results on the oxidation of micropollutants in water obtained from lab-scale studies cannot be often transferred to full-scale applications, there is a necessity for pilot testing. In a pilot study for the

Metropolitan Water District of Southern California, Ferguson *et al.* (1990) compared the peroxone process with ozonation for the degradation of taste and odor-causing compounds in two different waters: the Colorado River water (moderate DOC and alkalinity, low Br^- level (0.03–0.07 mg/L)) and state project water (moderate DOC and alkalinity, high Br^- level (0.33–0.48 mg/L)). The pilot plant included four ozone contactors with a capacity of 2.73 m^3/h and simulated a full-scale treatment train including pre-oxidation (peroxone or ozone) with a contact time of 12 min, followed by coagulation-flocculation, sedimentation and filtration. The oxidation of 2-MIB and geosmin in Colorado River water was more effective with the peroxone process than with ozone alone. For the degradation of 90% 2-MIB by ozone alone 4 mg/L ozone was required, while only 2 mg/L ozone was necessary to degrade 90% 2-MIB in the peroxone process (molar H_2O_2:O_3 ratio 0.3). In state project water ~80% 2-MIB degradation was achieved at an O_3 dose of 4 mg/L in ozonation and an O_3 dose of 1 mg/L in the peroxone process. It was concluded that the optimal H_2O_2:O_3 ratio for the oxidation of taste and odor compounds is 0.15 to 0.3 (state project water with lower alkalinity) and 0.4 or higher (Colorado River water with higher alkalinity). The different optimal ratios for both waters were caused by the difference in water quality.

Duguet *et al.* (1990) investigated the application of O_3/H_2O_2 for the removal of chloronitrobenzenes from groundwater (pH 6.7, DOC: 1.9 mg/L, alkalinity: 225 mg/L $CaCO_3$). The main contaminant of the groundwater was o-chloronitrobenzene (up to 1828 μg/L). The pilot plant was operated at 450 L/h, and was equipped with two ozone contactors, an ozone generator (up to 10 g/h), a dosage system for hydrogen peroxide and two parallel-running filters (sand and granular activated carbon (GAC)). Under optimized conditions, removal of 99% of aromatic compounds was achieved. The applied O_3 and H_2O_2 concentrations were 8 and 3 mg/L, respectively, with a 0.5 H_2O_2:O_3 molar ratio, and 20 min contact time. Ozonation alone with 16 mg/L O_3 and 40 min contact time resulted in 88% degradation of chloronitrobenzene (k_{pCNB+O_3} = 1.6 M^{-1} s^{-1}, (Shen *et al.* 2008)).

Wu and Englehardt (2015) applied O_3/H_2O_2 for the reduction of chemical oxygen demand (COD) in wastewater. The collected wastewater with an initial COD of 10–16.5 mg/L was treated biologically followed by an iron mediated aeration, flocculation, low-pressure vacuum ultrafiltration and O_3/H_2O_2. The pilot-scale experiments were conducted in a 3785 L tank. 2300–3030 L wastewater was recirculated at a flow of 151 L/min. To reduce 90% COD 60 h of treatment were required at 2.3 mg/L/h O_3 (O_3:H_2O_2 mass ratio: 3.4) and less than 5 h at an O_3 dose of 18.83 mg/L/h (O_3:H_2O_2 mass ratio: 3.61). The specific O_3 consumption per COD was determined to be between 7.4 and 13.15 mg O_3/mg COD. The energy demand was calculated to be 1.73–2.49 kWh/m^3/order. The authors show that O_3/H_2O_2 is an attractive method for the mineralization of compounds contributing to COD, as required for water reuse and reclamation.

Lake Huron is one of the Great Lakes and is used as a drinking water source in Canada and USA. It was shown that this lake water is contaminated with pharmaceuticals and personal care products (PPCPs) and EDCs. Thus, a pilot-scale treatment plant operated in two trains at a total flow of 12.4 L/min was installed at the Walkerton Clean Water Centre, Canada (Uslu *et al.* 2012). Both trains include a coagulation-flocculation chamber (hydraulic retention time: 90 min), a clarifier (hydraulic retention time: 20 min) followed by dual media filtration. While train 1 consists of just these units, in train 2 O_3/H_2O_2 is additionally applied prior to coagulation. O_3 is dosed at a concentration of 2–2.3 mg/L and H_2O_2 at 0.2 mg/L. Raw water concentrations for alkalinity ($CaCO_3$) and DOC range between 80–106 mg/L and 1.5–1.9 mg/L, respectively. The raw water was spiked with nine micropollutants (PPCPs and EDCs). Samples were taken after the entire processes at the end of train 1, train 2 and from train 2 after passing the AOP reactor (before coagulation). The micropollutant removal in train 1 was for all compounds less than 15%. The results for the samples of train 2 and after the AOP reactor did not significantly differ from each other. A complete removal of five (bisphenol A, naproxen, gemfibrozil, diclofenac and atrovastatin) of nine target compounds was achieved after the AOP reactor, and thus at the end of train 2 as the AOP reactor is a part of train 2. O_3-refractory compounds (e.g., atrazine and ibuprofen) were removed insufficiently. The authors concluded that this could be due to the low molar H_2O_2:O_3 ratio (0.14).

Dunea Duin & Water Company, The Netherlands, uses Meuse River water as a source for drinking water production for The Hague and surrounding areas. Organic micropollutants, among which pesticides, EDCs, PPCPs, were found in surface waters at levels ranging from ng/L to a few µg/L. Dunea provides a multi-barrier treatment of River Meuse water, through a complex infrastructure and various treatment steps. After collection in a "dead end side stream", the river water goes through conventional treatment such as coagulation, sedimentation, and micro-sieve filtration in combination with aeration. After pre-treatment, the water is pumped over 27 km to Bergambacht for additional pre-treatment (dual media filtration), prior to transportation and infiltration into the dunes (managed aquifer recharge, MAR) via two pipelines. After an average residence time of ~120 days, the dune-filtered water is pumped to three different locations for multiple step-treatment (softening, activated carbon filtration, cascade aeration, rapid sand filtration, and slow-sand filtration) prior to be distributed to the consumers (Lekkerkerker *et al.* 2009). Dunea Water Company conducted extensive research in collaboration with WEDECO, a Xylem Inc. brand company, on advanced oxidation processes for organic micropollutant (OMP) removal, as part of Dunea's strategy to preventing any undesired compounds from being present in their treated water. A WEDECO AOP pilot system was installed onsite at the pre-treatment location Bergambacht, and used to examine the effectiveness of O_3/H_2O_2, UV/H_2O_2, and their combination for the removal of fourteen selected OMP (pesticides, pharmaceuticals, x-ray contrast agents) under various treatment conditions. Given the concern over bromate formation from bromide present in the water, bromate was a key parameter monitored in the treated water. The pilot plant research outcomes are available in the published literature (Lekkerkerker *et al.* 2009; Scheideler *et al.* 2011; Lekkerkerker-Teunissen *et al.* 2012; Lekkerkerker-Teunissen *et al.* 2013; Knol *et al.* 2015). Lekkerkerker-Teunissen *et al.* (2012) reported data for both peroxone process alone and serial AOP (O_3/H_2O_2/LP-UV/H_2O_2) under selected process conditions, i.e., 5 m^3/h flow rate, 0.7–2 g/m^3 O_3, 0, 6 and 10 mg/L H_2O_2, and 700 and 950 mJ/cm^2 UV (253.7 nm) dose. The pre-treated water influent to either O_3/H_2O_2 or serial AOP has a relatively high DOC concentration (3–5 mg/L), low transmittance (73–83% $T_{254\,nm,\,1\,cm}$), and an average bromide concentration of 120 µg/L. At 1.5 mg/L O_3 and 6 mg/L H_2O_2, the O_3/H_2O_2 AOP achieved more than 90% removal of bromacil, phenazone, bentazone, carbamazepine, diclofenac, furosemide, isoproturon, and trimethoprim, while keeping the bromate level below the reporting limit of 0.5 µg/L. Lower removal yields were observed for metoprolol and ibuprofen (70–90%) and diglyme, iopromide, atrazine, and metformin (20–70%). The treatment yields were about 10% lower in wintertime than in summertime for most compounds. As the treatment goal set by Dunea of 80% atrazine removal was not met by the peroxone process, a serial AOP was investigated. Sequential treatment by O_3(1.5 mg/L)/H_2O_2(6 mg/L) – LP UV(0.73 kWh/m^3)/H_2O_2(5.5 mg/L residual from peroxone process) achieved 80% removal of atrazine, and substantially increased conversion yields of all other contaminants, and demonstrated a synergistic effect relative to treatments observed in the two individual AOPs. Metformin, a refractory pollutant, was removed by 37% in the serial AOP treatment. Concomitantly, the water transmittance increased by up to 5%, and DOC was removed by approx. 0.2 mg/L. The authors indicated that distributing the UV dose across three reactors rather than one-time UV dose in a single reactor would increase the treatment efficiency of serial AOP (5–15% higher removal yields) due to the increase of water %*T* across the three reactors. In 2014, Dunea Duin & Water Company decided to implement the serial AOP full-scale system at their water facility of Bergambacht. Following the tender process open to all AOP technology providers, in 2015 Dunea awarded the full-scale system implementation to WEDECO, Xylem Inc.

In a recently published study based on the extensive pilot work at Dunea, Knol *et al.* (2015) described the efficacy of O_3/H_2O_2 process at the removal of the 14 model compounds mentioned above, and emphasized the role of water quality parameters, i.e., bicarbonate, DOC, and temperature on the removal yields. The bromate formation from bromide contained in the pilot water (104–136 μg/L) during the peroxone treatment was monitored from August 2011 through March 2012. The ozone-loop reactor (WEDECO, Xylem Inc.) with sequential ozone dosing injection ports and reaction/degassing chamber, and the source and water quality for the pilot plant were similar to those used in the study described above. The micropollutants (1.8–31.5 μg/L) and H_2O_2 were dosed to the influent water (flowrate of 5 m^3/h), whereas ozone was applied at 1m-spaced four dosing points followed by static mixers (Figure 4.3). The authors reported at least 70% removal of micropollutants when dosing 6 mg/L H_2O_2 and 1.5 mg/L O_3 (5.65 H_2O_2:O_3 molar ratio), while keeping bromate well below the set limit of 0.5 μg/L. The degradation rate of micropollutants increased with increasing water temperature and decreasing DOC and HCO_3^- concentrations, except for metformin which was the most poorly treated compound (~20% irrespective the water quality conditions). Varying the H_2O_2 concentration had minimal effect on the micropollutant degradation yields, but limited the bromate formation. Bromate formation yield increased with the temperature and bromide concentration, and decreased with increasing alkalinity and pH. From these studies, Knol *et al.* (2015) concluded that the degradation of micropollutants and the bromate formation can be controlled by adjusting the ozone and H_2O_2 doses to the water temperature, and the DOC and bicarbonate concentrations. Thus, in wintertime the degradation of micropollutants can be increased by increasing the ozone dose without exceeding the bromate limit. In summertime, the bromate formation can be controlled by increasing the H_2O_2 concentration, which does not influence noticeably the degradation of micropollutants.

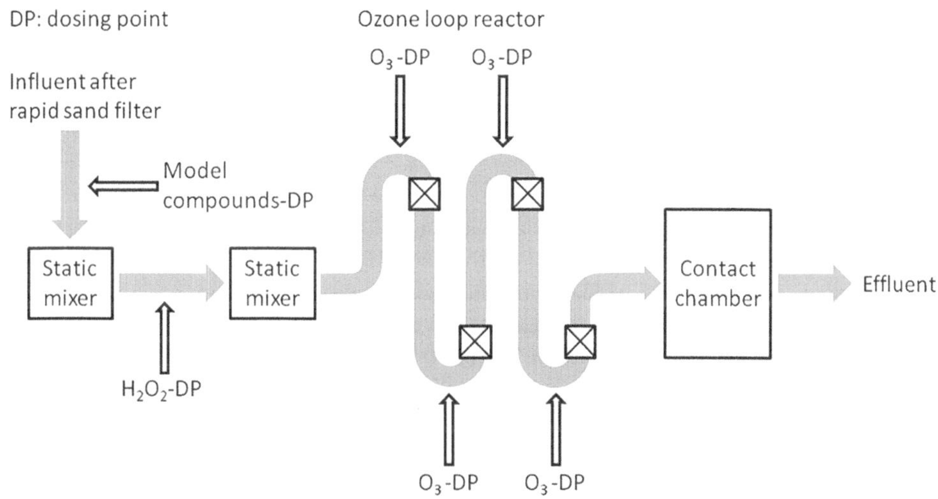

Figure 4.3 Schematic view of the pilot plant in Bergambracht, Netherlands. Adapted from Knol *et al.* (2015).

Metaldehyde is commonly used as a molluscicide in horticulture and agriculture. It contaminates groundwater and surface waters, and its maximum permissible level in drinking water in the EU countries is set to 0.1 μg/L. Metaldehyde is refractory to conventional treatment processes, does not photolyze, and is not removed by ozone. It reacts with the •OH with a reaction rate constant of $k = 1.3 \times 10^9$ M^{-1} s^{-1} (Autin *et al.* 2012). Scheideler and Bosmith (2014) reported the metaldehyde treatment data for peroxone, UV (253.7 nm)/H_2O_2, and their combination obtained during a 8-month pilot testing at a water utility operated by Anglian Water Services (UK). The pre-treated surface water quality used in the pilot plant was characterized by ~80 μg/L bromide, <5 μg/L bromate, ~5 mg/L DOC, ~207 mg/L hardness as $CaCO_3$, pH 8, and 88% $T_{254\,nm,1\,cm}$. The main components of the O_3-AOP equipment (WEDECO's PRO_3MIX) consisted of ozone generator (from liquid O_2), H_2O_2 dosing, reaction chamber, the degassing and catalytic destruction of residual O_3, measuring devices. The O_3-loop reactor had five O_3 dosing ports followed by static mixers. The water flow rate varied from 30 to 40 m^3/h, H_2O_2 (up to 24 g/m^3) was dosed at the inlet, whereas O_3 (up to 16 g/m^3) was incrementally dosed across the loop reactor to minimize the bromate formation. Metaldeyde treatment goal set by Anglian Water was 0.5 log removal. The O_3/H_2O_2 study showed that 0.4–0.5 log metaldehyde reduction can be achieved with 8 g/m^3 O_3 and a O_3/H_2O_2 mass ratio of 2–2.5, while bromate level was kept below 3 μg/L. In order to achieve a consistent 0.5 log metaldehyde removal, one needs to increase the O_3 dose. Lower peroxide doses achieved better degradation yields, given the lower $^{\bullet}OH$ consumption by H_2O_2. The UV-AOP pilot (3 × 80 W low-pressure lamp reactor) data showed that in order to obtain 0.5 log removal of metaldehyde, a UV dose of 1300 mJ/cm^2 and 20 mg/L H_2O_2 would be required.

The combined AOPs indicated that for the H_2O_2 residual from the peroxone process, the yield of metaldehyde destruction increased with increasing UV dose. The pilot data provided by Scheideler and Bosmith (2014) indicates that 0.5 log metaldehyde removal can be achieved with 8 mg/L O_3/(16–22 mg/L H_2O_2) followed by less than 500 mJ/cm^2 UV dose, which is much lower than the UV dose required for the UV-AOP alone. This observation suggests a synergistic effect of the combined AOPs. Should 90% (1.0 log) removal of metaldehyde be targeted, a post-O_3-AOP UV dose of 1200 mJ/cm^2 would suffice (as compared to 2600 mJ/cm^2 for the UV-AOP alone). Bromate standard was not exceeded under any of the pilot conditions used in this work.

4.3.3 Full-scale applications

Sung-Nam Water Treatment Plant (WTP), South Korea: (http://www.xylem.com/treatment/nz/case-studies). The Sung-Nam Metro City WTP is operated by K-Water and produces tap drinking water for more than 3 million people and clean water for a beverage industry filling around 45,000 drinking water bottles per day. The treated water originates from the Han-River. During the seasonal algae blooming, T&O compounds (e.g., geosmin and 2-MIB) are released into the water, affecting the aesthetics of drinking water delivered to the consumers. These compounds are effectively treated with hydroxyl radical-based AOPs. In order to provide high quality water to the customers, K-Water selected the peroxone advanced oxidation process combined with activated carbon filtration. The T&O compounds are removed using the WEDECO MiPRO™ eco_3 AOP system. The AOP system is designed to achieve 0.5 log (70%) removal of 2-MIB. It consists of three ozone generators (WEDECO PDO 1000; Xylem) with a total capacity of 51 kg O_3/h. Ozone is generated from liquid oxygen, and is injected into the water at a dose of 2 g/m^3. The ozone feed is combined with H_2O_2 (0.5 mg/L) at a H_2O_2:O_3 molar ratio of 0.35, to generate efficiently hydroxyl radicals for the oxidation of the T&O compounds. The 34,390 m^3 water/h flows through two ozone injection lines, then into two separate concrete contact tanks for reaction and degassing. The equipment includes also the hydrogen peroxide storage and the H_2O_2 delivery system, which ensures an efficient mixing of the two oxidants and their uniform distribution into the water streams. The O_3-AOP is in use at Sung-Nam WTP since 2012.

Moosbrunn Waterworks, Austria: Vienna Waterworks supplies its customers with drinking water from pristine sources. The demand of 400,000 m^3/d is covered by 32 springs allocated to two water supply mains. These were installed in the 1870s and have been operated since then without any reconstruction. In case of reconstruction of one of the two mains or increased water demand during hot and dry summer time, the Vienna Waterworks decided to provide a further water source as back-up. The new source, two horizontal groundwater wells (Moosbrunn Waterworks), has been in operation since 2006 (Werderitsch, 2013). This groundwater is contaminated with tetrachloroethene (PCE) and trichloroethene (TCE). The contaminant concentrations were monitored over several years (1989–2005). A rising contamination trend has been observed in both wells and in 2005 the concentrations of the chlorinated hydrocarbons in well 1 and 2 were in the range of 25–30 µg/L and 7–10 µg/L, respectively. The total chlorinated hydrocarbon (TCE and PCE) drinking water standard of 10 µg/L was exceeded. After a half-year pilot testing (864 m^3/d), the Vienna Waterworks decided to install an AOP treatment step. H_2O_2 is injected to the raw water followed by O_3 injection. The water is mixed in a static mixer, and then pumped into the O_3/H_2O_2-reaction chamber. The treated water is stored in a reservoir and disinfected with ClO_2 before being discharged. Optimal process conditions were defined during the pilot plant work as 2.5 g/m^3 O_3, 1.75 g/m^3 H_2O_2 (H_2O_2:O_3 molar ratio of 0.98), and a minimum 10 min-hydraulic residence time in the reaction tank in order to achieve at least 90% degradation of both TCE and PCE, and to mitigate the bromate formation. However, due to increased chlorine dioxide demand caused by residual H_2O_2, the H_2O_2 dose was decreased; therefore, the H_2O_2:O_3 molar ratio decreased from 1 to 0.7 without significantly influencing the degradation rate of the chlorinated hydrocarbons. The throughput of the full-scale treatment plant is 64,000 m^3/d. Bromate formation was mitigated by the optimized dosing of oxidants and a short hydraulic retention time (≤10 min).

Andalucia, Spain: (http://www.mcilvainecompany.com/Decision_Tree/subscriber/articles/ITT_Water_and_Wastewater_Case_Studies.pdf). In order to comply with the EU drinking water regulations on trihalomethanes (100 µg/L) and pesticides (0.1 µg/L each pesticide, maximum level of total pesticides and their metabolites of 0.5 µg/L), GIAHSA, one of the water suppliers in Andalucia, Spain, decided to implement pre-ozonation, intermediate ozonation coupled with H_2O_2 (O_3-AOP), and GAC filtration at the drinking water treatment plants of Aljaraque, Tinto, and Lepe. The WTPs are equipped with ozonation stations which provide O_3 gas for pre- and intermediate ozonation steps. The station has two WEDECO SMO 700 O_3 generators with automatic O_3 dose control and O_3 analyzer, gas flowmeters, gas diffusion systems, catalytic O_3 destruction, and SCADA. Each O_3 unit has a max. capacity of 9300 g/h at 10% (by weight) from liquid O_2 and a water cooling temperature of 20°C. The O_3 design dose for each ozonation step is 1–4 mg/L, and the minimum and maximum water flow rate at the plants are 2160 and 4320 m^3/h, respectively. The project was funded by The Public Administration of Andalusia, and the O_3 units were installed in 2007.

ATPwater (Sacramento, CA, USA) systems are used for remediation of contaminated water. There are two different operating systems offered, namely, in situ chemical oxidation (PulseOx®) and a continuous in-line system (HiPOx®). The PulseOx® process is based on the peroxone process for the remediation of contaminated soil and groundwater. A sequence of pressurized pulses of ozone, hydrogen peroxide, oxygen and air are injected directly into the subsurface. The oxidation occurs in situ. The PulseOx® process was/is operated among others at following sites:

- Active Gas Station, Bridgeport, CA, USA (http://www.wateronline.com/doc/pulseox-in-well-reactor-iwr-case-study-0002)

  A release at an active gasoline service station in 1993 caused a contamination of soil and groundwater with BTEX, MTBE and TPH-g. In 2007 the concentration of benzene, BTEX, MTBE and TPH-g were 56 µg/L, 243 µg/L, 9 µg/L and 660 mg/L, respectively. Hence, in 2008 two in-well reactors (IWR) were installed in two existing monitoring wells. The IWRs were operated for six weeks. Ozone and hydrogen peroxide were injected at a dose of 0.45 kg/d and 3.8 L/d in each IWR, respectively at a recirculation rate of 3.8–19 L/min. After six weeks of operation the contaminant concentrations were reduced below regulatory detection limits.

- Cooper Drum Superfund Site, CA, USA (http://www.wateronline.com/doc/pulseox-case-study-cooper-drum-superfund-0001)

  Since 1941, the site (an area of about 15,000 m^2) was used by Cooper Drum Co. and several other companies to recondition used steel drums. The drum process waste was collected in open concrete trenches. This waste collection resulted in leaching of contaminants (e.g., 1,4-dioxane and volatile organic compounds (VOCs)) into the soil and groundwater beneath the site. From 1996 to 2001 the United States Environmental Protection Agency (US EPA) conducted remedial investigation experiments on this site and added it to the National Priority List of hazardous waste sites requiring remedial action. The contaminated site is located close to drinking water wells (the nearest well is located within less than 1 km of the site) supplying 335,000 people. The contaminant plume was 250 m long, 75 m wide and 12–25 m below the ground. Ozone and hydrogen peroxide doses were 0.9 kg/d and 90 L/d, respectively. The pilot plant operation period was from July 2005 to June 2006. 1,4-Dioxane was degraded by 88% in one of the test points and ≥49% in all other points. After successful piloting, US EPA indicate that it will suggest full-scale implementation of PulseOx process for *in situ* removal of 1,4-dioxane at Cooper Drum Company site.

HiPOx is a continuous in-line system using the peroxone process for water treatment with patented injection and mixing techniques. The HiPOx® process was/is operated at following sites:

- US Air Force Plant 44, AZ, USA (http://www.wateronline.com/doc/hipox-case-study-us-air-force-plant-44-0002)

  The US Air Force Plant 44 facility is located on the Tucson International Airport superfund site. Since the 1950's it was used for manufacturing missile and aircraft components. The process waste was disposed of in open ponds. A leaching into soil and groundwater results in contamination with 1,4-dioxane and TCE. The contaminant plume impacted the drinking water supply for the City of Tucson. In the 1980's a remediation of groundwater and soil started. The water was treated by air-stripping followed by activated carbon adsorption and the soil by a soil vapor extraction system. In 2002 1,4-dioxane was identified in influent and effluent of the treatment system. As it is not very volatile and does not adsorb on the activated carbon, it was passing the treatment system. To eliminate 1,4-dioxane besides TCE the HiPOx® AOP system was installed in 2009 and replaced the 20+ year equipment at the US Air Force site. The plant has a capacity of 13.1 million L/d and the applied ozone dosage is 27 kg/d. Groundwater is pumped to the plant and re-injected after treatment at the site. The new system successfully degrades TCE and 1,4-dioxane and reduces water losses caused by evaporation by over 121 million L per year.
- Polyester Manufacturing Facility, South Korea (http://aptwater.com/hipox/case-studies/hipox-case-study-polyester-aop-water-treatment-south-korea/)

  A polyester manufacturer in South Korea produces wastewater with high concentrations of 1,4–dioxane (up to 65 mg/L) and other surfactants resulting in a COD of about 200 mg/L. 1,4-Dioxane passed through the treatment plant (anaerobic treatment process) and was discharged to a river. To improve the river water quality industries are required by the Ministry of the Environment to remove contaminants, e.g., 1,4-dioxane from their wastewater prior to discharging. The HiPOx® system was installed in 2005. The system is operated with a flow of 0.5–0.75 m^3/min and an ozone dosing capacity of 90 kg/d. The reduction of 1,4-dioxane to below the regulatory limit of 10 mg/L could be achieved. The plant was operated for over 48,000 hours and treated more than 34 million L of water with only minimal maintenance.
- Keefe Environmental Services Superfund Site, NH, USA (http://www.wateronline.com/doc/hipox-case-study-keefe-environmental-0001)

In 2003, the US EPA detected 1,4-dioxane in the groundwater in Epping at the Keefe Environmental Services superfund site. The site was operated as hazardous waste storage from 1978–1981. 1,4-dioxane was detected in the water treatment (metal removal and air-stripping remediation systems) effluent. A bench testing in October 2004 showed a good degradation of 1,4-dioxane from 270 μg/L to below the project regulatory limit of 3 μg/L. Based on these results a full-scale plant installation was completed in 2005. The HiPOx® system is operated at a flow of 100 L/min and an ozone dosage of 1.6 kg/d.

- Nebraska Ordnance Plant, Mead, NE, USA (http://www.wateronline.com/doc/hipox-case-study-nebraska-ordnance-plant-0002)

During the World War II and the Korean War bombs, shells and rockets were assembled at the Nebraska Ordnance Plant site. These site activities resulted in a contamination of the groundwater with chlorinated solvents, e.g., TCE. In 2002, the US Army Corps of Engineers started a remediation project at the site. The treatment facility consisted of air-stripping and vapor-phase carbon adsorption. Due to very high TCE contaminations of 5.8 mg/L, the treatment facility could not economically meet remediation goals. In 2008, a HiPOx® system was installed operated at a flow of 2.3 m^3/min and dosages for O_3 and H_2O_2 of 45 kg/d and 57 L/d, respectively. A contaminant removal of about >99.8% was achieved and thus the treated water meets the public health requirements with a TCE concentration of less than 10 μg/L. In the first six months 540,000 m^3 water was treated and 770 kg of TCE degraded. The treatment costs decreased by 71% ($ 164,000/yr) compared to the existing air-stripping system.

- South Lake Tahoe, CA, USA (http://aptwater.com/hipox/case-studies/hipox-case-study-in-south-lake-tahoe-ca/)

In 1998, the City of South Lake Tahoe discovered a contamination in 13 of 34 drinking water supply wells with MTBE and its degradation product *t*BuOH. These contaminated wells had to be closed. In order to restore the wells, the City of South Lake Tahoe installed one HiPOx® systems in 2002 (flow: 3 m^3/min) and a second one in 2004 (flow: 5.7 m^3/min) at a different well. The MTBE and *t*BuOH concentrations in the second well were 6 and 8 μg/L, respectively. A degradation of both contaminants below the detection limits could be achieved (for MTBE <0.2 μg/L). The ozone capacity of both systems is 27 kg/day.

- Wichita Aquifer Storage Recovery (ASR), KS, USA (http://aptwater.com/hipox/case-studies/wichita-asr-project-ks-bromate-control/)

This application regards seasonal (spring and fall) treatment of storm water from Little Arkansas River for groundwater recharge. The initial design of the 1.1×10^5 m^3/d surface water treatment for aquifer recharge consisted of river water intake, sedimentation, ultrafiltration, and chlorination. Concerns over contamination with atrazine from the agricultural runoff, pathogens including viruses, high TOC levels, and turbidity led City of Wichita to examine potential treatment processes for storm water. Powdered or granular activated carbon required additional chemicals, O&M- and on-site waste handling high costs, UV technology is not applicable in high-turbidity water, and O_3 alone is not effective at atrazine removal and generates bromate. In 2008 and 2009, a series of bench- and pilot-scale tests using the ATPwater HiPOx® system were conducted under a variety of experimental conditions. The pilot test results demonstrated that HiPOx® technology could achieve atrazine removal below the drinking water MCL of 3 μg/L, control the bromate formation below the regulatory limit of 10 μg/L, while providing 4 to 5 log inactivation of poliovirus. The City of Wichita and CDM recommended implementation of HiPOx® technology for storm water treatment for ASR project at the 1.1×10^5 m^3/d plant, approved in September 2009 by the Kansas Department of Health and Environment, and installed in 2011.

4.3.4 Process economics and limitations

One important factor for comparing and assessing the efficiency of water treatment processes is the electrical energy demand for achieving a certain pollutant treatment goal. Bolton *et al.* (2001) introduced two figures of merit for electrical energy-driven-systems i.e., electrical energy per mass (E_{EM}) and electrical energy per order (E_{EO}).

E_{EM} is defined as the electric energy in kWh required to degrade a contaminant by a mass unit (e.g., 1 kg). It is used when the contaminant concentration is high, as its degradation rate is directly proportional to the electric energy rate (zero-order kinetics). As such conditions were not found in water treatment, since the concentration of pollutants is typically small, E_{EM} is not further explained here.

The second approach is E_{EO}. It is defined as the electrical energy in kWh required to degrade a contaminant by one order of magnitude (90%) in 1 m^3 water. E_{EO} is used for situations where the contaminant concentration is low and the required energy is independent of the contaminant concentration (first-order kinetics). This means that it takes the same amount of electric energy to degrade 10 mg/L to 1 mg/L of a contaminant as it would take to degrade 10 μg/L to 1 μg/L. The energy demand can be calculated for batch reactors (eq. 4.29) and flow-through operation (eq. 4.30).

$$E_{EO} = \frac{P \times t \times 1000}{V \times \log\left(\frac{c_i}{c_f}\right)} \tag{4.29}$$

$$E_{EO} = \frac{P}{F \times \log\left(\frac{c_i}{c_f}\right)} \tag{4.30}$$

P is the rated power in kW of the system, t the treatment time in h, V the treated volume of water in L, F the flow rate of water in m^3/h, M the molar mass of the contaminant in g/mol, c_i and c_f are the initial or influent and final or effluent concentration of the contaminant in mol/L, respectively.

As a rule of thumb the required energy to produce 1 kg of ozone is assumed to be around 15 kWh. The production of H_2O_2 is estimated to be 10 kWh/kg (Rosenfeldt *et al.* 2006).

For processes using a UV-light source the energy of the lamp can be calculated using the following equation:

$$ER_{lamp} = \frac{H_\lambda(90\%)}{l^{avg} \times \eta_{UV}} \times \frac{kWh}{3.6 \times 10^6 J}, \tag{4.31}$$

where ER_{lamp} is the specific energy consumption (kWh/m^3), H_λ (90%) is the calculated fluence for 90% conversion of a compound, l^{avg} the average optical path length and η_{UV} the efficiency of the UV lamp. Katsoyiannis *et al.* (2011) considered an efficiency of the UV lamp of 0.3 ($\eta_{UV} = 0.3$).

Katsoyiannis *et al.* (2011) compared the required energy to degrade *p*CBA by O_3, O_3/H_2O_2 and UV/H_2O_2 in a batch reactor in various waters, Lake Zurich (DOC: 1.3 mg/L, carbonate-alkalinity: 2.6 mM), Lake Jonsvatnet (DOC: 3 mg/L, carbonate-alkalinity: 0.4 mM), Lake Greifensee (DOC: 3.1 mg/L, carbonate-alkalinity: 4.0 mM) and wastewater Dübendorf (DOC: 3.9 mg/L, carbonate-alkalinity: 6.5 mM). It could be shown that the peroxone process has a slightly higher energy demand than ozonation, and the UV/H_2O_2 process requires around one order of magnitude more energy for the lake waters and around four times more energy for the *p*CBA removal in wastewater (Table 4.3).

Table 4.3 Energy demand for the removal of 90% *p*CBA in various water matrices by O_3/H_2O_2 (H_2O_2:O_3 molar ratio of 0.5), O_3, and UV/H_2O_2 (5 cm path length and 0.2 mM H_2O_2) (Katsoyiannis *et al.* 2011).

Process	Energy (kWh/m³)			
	Lake Zurich	Lake Jonsvatnet	Lake Greifensee	Wastewater Dübendorf
O_3/H_2O_2	0.043	~0.043*	~0.08*	~0.25*
O_3	0.035	0.035	0.065	0.2
UV/H_2O_2	0.23	0.45	0.61	0.82

** Estimated values*

Lekkerkerker Teunissen *et al.* (2012) calculated the E_{EO} data for 14 micropollutants treated with peroxone process and the serial AOPs, as investigated in the pilot work at Dunea WTP (see Section 4.3.2). The 14 contaminants have different structural characteristics, such that some of them react moderately to well with O_3, others are prone to •OH degradation (moderate to high reactivity), and one (metformin) reacts poorly with both O_3 and •OH. The E_{EO} values reflect such reactivities in both peroxone and serial AOP treatment. For example, the E_{EO} for bromacil treatment with peroxone was within 0.01–0.02 kWh/m³/order for all used oxidant ratios, and increased for the serial AOP treatment to 0.13–0.17 kWh/m³/order. This trend reflects the high reactivity towards O_3, with very little benefit from additional •OH-based treatment via the UV/H_2O_2 process. Similar examples are diglyme, ibuprofen, carbamazepine, trimethoprim, and diclofenac. On the contrary, atrazine is not well treated with O_3, has a moderate rate constant for •OH reaction, and exhibits a moderate direct photolysis at 253.7 nm. As a result, the E_{EO} for atrazine treatment with peroxone process ranged from 0.11 to 0.18 kWh/m³/order, depending on the oxidant ratios, and increased for serial AOP treatment to 0.43–0.79 kWh/m³/order, depending on the O_3/H_2O_2 ratio and applied UV dose. Metformin would require the higher electrical energy for 90% removal with both O_3/H_2O_2 (0.21–0.39 kWh/m³) and serial AOP (0.90–2.45 kWh/m³). The pilot data showed that 8 out of 14 compounds were removed by >90% by O_3 (1.5 mg/L)/H_2O_2 (6 mg/L) process with $E_{EO} \leq 0.027$ kWh/m³/order. Atrazine treatment goal set by Dunea was of 80%. The 80% removal yield was achieved with an electrical energy consumption of 0.52 kWh/m³ in the serial AOP treatment, which is much lower than that required by the LP UV/H_2O_2 process alone (0.73 kWh/m³).

Scheideler and Bosmith (2014) provided cost analysis for metaldehyde treatment by O_3-AOP, UV-AOP, and their sequential application based on the pilot testing conducted at one of the Anglian Water (UK) water treatment plants (see Section 4.3.2). The costs included were for chemicals (H_2O_2 and liquid oxygen), energy, maintenance, and CAPEX, and assumed certain prices. Investment costs and costs for parts including spare UV lamps and electronic ballasts driving the lamps were not included. Under the assumed flow rate of 800 m³/h and a 10-year depreciation period, the costs associated with the treatment for 0.5 log metaldehyde removal by peroxone and UV-AOP individual processes were approx. 0.048 and 0.064 €/m³, respectively. The treatment costs to achieve 90% removal of metaldehyde were estimated at 0.091 and 0.085 €/m³, for the UV-AOP and the O_3-AOP followed by UV/AOP, respectively. The authors concluded that the serial AOP is the treatment option when high reduction in metaldehyde concentration is needed (e.g., 1-log), as significant cost savings can be realized (~77,000 €/yr). On the other hand, the combined AOPs offer a robust barrier against micropollutants with various molecular structures, with different reactivity towards O_3, •OH, and UV radiation.

Lee *et al.* (2016) presented calculated data on the abatement of several micropollutants by four different AOPs (O_3 followed by UV, O_3/H_2O_2 followed by UV/H_2O_2, UV/H_2O_2 and O_3/H_2O_2,) in wastewater (Table 4.4). Although the calculated energy demand for O_3/H_2O_2 is lower than for all other processes, O_3/H_2O_2

is not sufficient to degrade *N*-Nitrosodimethylamine (NDMA) or its precursors from wastewaters. The highest degrees of micropollutant degradation are predicted to be achieved by the combination of O_3/H_2O_2 followed by UV/H_2O_2. The high energy demand for this combined process can be reduced by optimized conditions, which makes the process combination more economically feasible.

Table 4.4 Calculated abatement of different micropollutants (bisphenol A, atrazine, ibuprofen and NDMA) by different AOPs (UV/H_2O_2, O_3/H_2O_2, O_3 followed by UV and O_3/H_2O_2 followed by UV/H_2O_2) and the required energy (Lee *et al.* 2016).

Micropollutant	Micropollutant Abatement [%] at Required Energy			
	UV/H_2O_2 0.158 kWh/m³	O_3/H_2O_2 0.095 kWh/m³	O_3 + UV 0.126 kWh/m³	O_3/H_2O_2 + UV/H_2O_2 0.176 kWh/m³
Bisphenol A	33	94	94	96
Atrazine	30	44	57	62
Ibuprofen	27	60	62	71
NDMA	80	0	80	80

4.4 O_3/UV PROCESS

4.4.1 Process fundamentals

The photolysis of ozone in water is known as an AOP, as •OH are generated. Since the molar absorption coefficient of ozone ($\varepsilon = 3300\ M^{-1}\ cm^{-1}$) is much higher at relevant wavelengths ($\lambda = 253.7$ nm) than that of H_2O_2 ($\varepsilon = 19.6\ M^{-1}\ cm^{-1}$) (Legrini *et al.* 1993), the photolysis of ozone should be much more efficient than the photolysis of H_2O_2. As indicated by Figure 4.4, O_3 has an absorption maximum at 260 nm. A low-pressure mercury-lamp (LP Hg-lamp) is suitable for the photolysis of O_3 as it emits light at 253.7 nm. Light sources emitting light at wavelength ≥300 nm are not applicable for the photolysis of O_3.

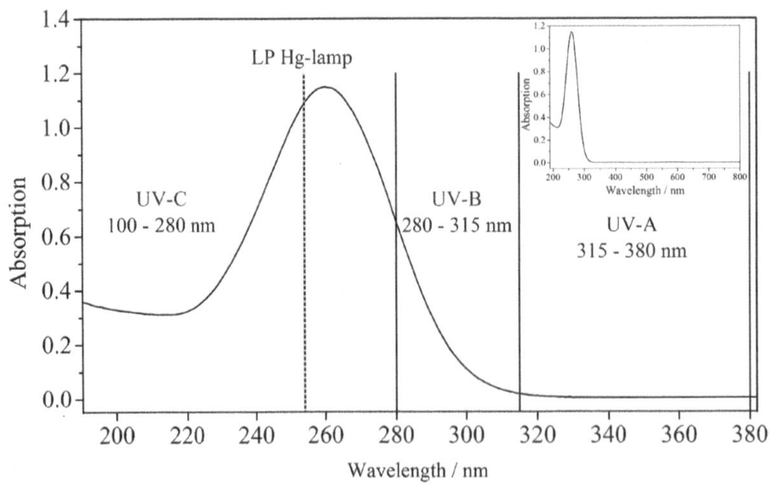

Figure 4.4 UV-spectrum of O_3 in water with indicated UV-A, UV-B and UV-C ranges. Inset: Full spectrum (190–800 nm) of O_3.

The photolysis of ozone yields oxygen atoms and molecular oxygen, in their ground and/or excited states (reactions 4.32a–d) in the solvent cage. The species are $O(^1D)$ (oxygen atom in its excited state), $O(^3P)$ (oxygen atom in its ground state), singlet oxygen ($O_2(^1\Delta_g)$) as well as oxygen in its ground state ($O_2(^3\Sigma_g^-)$). These species are involved both in recombination reactions within the "cage", and, if escaped from the solvent cage, in reactions with water molecules and/or other compounds present in the solution.

$$O_3 + h\nu \rightarrow O(^1D) + O_2(^1\Delta_g) \tag{4.32a}$$

$$O_3 + h\nu \rightarrow O(^1D) + O_2(^3\Sigma_g^-) \tag{4.32b}$$

$$O_3 + h\nu \rightarrow O(^3P) + O_2(^1\Delta_g) \tag{4.32c}$$

$$O_3 + h\nu \rightarrow O(^3P) + O_2(^3\Sigma_g^-) \tag{4.32d}$$

$O(^1D)$ is highly energetic, more electrophilic and prone to undergo bond-forming addition reactions than the triplet ground state ^3P, thus, it readily reacts with water to generate "hot" H_2O_2 (reaction 4.33). This can thermalize (reaction 4.34) or break into two $^\bullet OH$ (reaction 4.35). If $^\bullet OH$ are formed, the solvent cage effect favors the recombination of $^\bullet OH$ to H_2O_2 (reaction 4.36), thus, limiting the radical diffusion into the bulk solution. Therefore, most of the generated $O(^1D)$ yield H_2O_2 (Peyton & Glaze, 1988; von Sonntag, 2008; von Sonntag & von Gunten, 2012).

$$O(^1D) + H_2O \rightarrow H_2O_2(\text{hot}) \quad k = 1.8 \times 10^{10}\ M^{-1}\ s^{-1} \tag{4.33}$$

$$H_2O_2(\text{hot}) \rightarrow H_2O_2 \tag{4.34}$$

$$H_2O_2(\text{hot}) \rightarrow 2\,^\bullet OH \tag{4.35}$$

$$[^\bullet OH + {}^\bullet OH]_{\text{cage}} \rightarrow H_2O_2 \tag{4.36}$$

The formed H_2O_2 can be photolyzed (reaction 4.37) if the radiation is absorbed, or can react with ozone and initiate the peroxone process (see above).

$$H_2O_2 + h\nu \rightarrow 2\,^\bullet OH \tag{4.37}$$

However, as in case of ozone photolysis, cage recombination of $^\bullet OH$ (reaction 4.36) also occurs in H_2O_2 photolysis (H_2O_2 photolysis primary quantum yield = 0.5). In competition to reaction (4.33), $O(^1D)$ can react with $O_2(^1\Delta_g)$ to form ozone (4.38).

$$O(^1D) + O_2(^1\Delta_g) \rightarrow O_3 \tag{4.38}$$

Ozone can be also formed by the reversed reaction 4.32d of the oxygen atom in its ground state with ground state oxygen (reaction 4.39).

$$O(^3P) + O_2(^3\Sigma_g^-) \rightarrow O_3 \tag{4.39}$$

The apparent quantum yield of direct ozone photolysis is $\Phi = 0.5$. By adding *t*BuOH as OH radical scavenger to the system Reisz *et al.* (2003) could show that the quantum yield for •OH generation is considerably lower ($\Phi = 0.1$).

4.4.2 Research studies and applications

Although the photolysis of ozone is not often applied at pilot- or full-scale, there are a number of lab-scale studies in different water matrices. Several publications deal with the oxidation of pharmaceuticals by O_3/UV in water. Lee *et al.* (2011) showed that the combination of O_3 and UV may be a very effective option for the degradation of the antibiotic chlorotetracycline in livestock wastewater (DOC: 2.5 g/L, alkalinity as $CaCO_3$: 4.8 g/L, pH 8.5). The authors investigated the chlorotetracycline degradation by O_3, O_3/UV, O_3/H_2O_2 and $O_3/UV/H_2O_2$. All experiments were conducted in a cylindrical 1.2-L reactor. Gaseous O_3 was continuously bubbled into the reactor filled with 1 L water at a flow rate of 0.7 L/min and a concentration of 7 g/m^3. This resulted in a dissolved O_3 concentration of 0.6 mg/L. H_2O_2 was added in the range of 0–200 mg/L and a LP Hg-lamp emitting light at 253.7 nm at 4.9 W UV output was used. A complete removal of chlorotetracycline from wastewater was achieved by $O_3/UV/H_2O_2$ within 15 min and by O_3/UV within 20 min. This result could not be reached applying O_3/H_2O_2 (65% removal in 40 min) or ozone alone (30% removal in 40 min).

The removal of ketoprofen (anti-inflammatory painkiller) by O_3 and O_3/UV was observed by Illés *et al.* (2014). The experiments were conducted in pure water. A LP Hg-lamp emitting light at 185 nm and 253.7 nm was used as a light source. The UV output was 2.7 W and the emitted photon flux was 5.7×10^{-6} einstein/s. Ketoprofen degradation was significantly faster (~56 times) using O_3/UV than O_3 alone and a high mineralization yield (95%) could be achieved with O_3/UV within 1 h, while with O_3 just 65% was mineralized.

The *pseudo*-first order rate constants for the removal of 30 PPCPs by O_3/UV were determined to be higher than those for the removal by UV/H_2O_2 and O_3 (Kim *et al.* 2008). In a subsequent study by the same group, Kim and Tanaka (2010) investigated the oxidation of 40 pharmaceuticals in sand filtered wastewater (DOC: 3.1–3.8) by O_3/UV, O_3/H_2O_2 and O_3. The experimental setup consisted of three reactors connected in series. The volume of each reactor was 35 L and the hydraulic retention time of each reactor was 5 min. O_3 was introduced to the water by continuously bubbling gaseous O_3 (42 mg/L) at a flow rate of 1.0 L/min into reactor 1 and 2 resulting in an O_3 concentration of 6 mg/L. Reactor 1 and 2 were equipped with three 65 W LP Hg-lamps (fluence rate per lamp: 1.025 mW/cm^2) per reactor. The UV fluence for these three UV lamps was 1845 mJ/cm^2 at a contact time of 10 min. A >90% degradation of nearly all pharmaceuticals was achieved in all investigated processes. There was a great difference for the bromate formation though. No bromate was formed in the treatment with O_3/H_2O_2, while ozone alone yielded the highest bromate concentrations (up to 6 µg/L). When O_3/UV was applied bromate formation could not be completely avoided, but was significantly lower than in the sole use of ozone. The costs for the treatment with ozone and O_3/UV were compared. Although the energy requirements in O_3/UV were much higher than in the other two processes, it should be taken into account that treating water with O_3/UV reduces the bromate formation *and* largely increases the disinfection efficiency due to UVC-exposure (Kim & Tanaka, 2011).

Xu *et al.* (2010) compared UV alone and O_3/UV for the degradation of *N*-nitrosodiethylamine (NDEA). NDEA is of great interest as its carcinogenicity is much greater than that of NDMA. The used UV light source was a LP Hg-lamp (8 W, emission at 253.7 nm) with a photon flow of 3.3×10^{-6} einstein/(L × s). The experiments were conducted at an O_3 concentration of 6.6 mg/L and an UV intensity of 1 mW/cm^2. The degradation efficiency was similar for both processes, but the degradation products and pathways were different. O_3/UV had the advantage that the concentrations of the two main products diethylamine (DEA) and NO_2^- were reduced and thus the reformation of NDEA was inhibited.

EDCs are of great concern, as they interfere with the action of natural hormones in the body responsible for homeostatis, reproduction, development and behavior (US EPA, 1997). Lau *et al.* (2007) investigated the degradation of the EDC carbofuran (insecticide) in pure water by UV irradiation, ozonation and O_3/UV. Direct photolysis experiments were conducted in a 1-L quartz beaker surrounded by eight LP Hg-lamps emitting at 253.7 nm (1.5×10^{-6} einstein/(L × s)). O_3 was introduced by continuously bubbling resulting in a concentration of 18 μM. O_3/UV was proposed to be an effective treatment for the degradation of carbofuran. A 24% removal of TOC, as measure of mineralization was observed within 30 min reaction time. Irmak *et al.* (2005) investigated the oxidation of the EDCs 17*β*-estradiol and bisphenol A (BPA) by O_3/UV. A 15 W LP Hg UV-lamp was used. 17*β*-estradiol was completely removed at an O_3/17*β*-estradiol ratio of 6.6 in the O_3/UV process, and at a ratio of 8.9 by ozone alone. BPA was completely removed at an O_3/BPA ratio of 15 in ozonation and at an O_3/BPA ratio of 14 in the application of O_3/UV.

4.5 BYPRODUCT FORMATION AND MITIGATION STRATEGIES

4.5.1 O_3/H_2O_2 process

Bromate is a by-product formed in ozonation processes. As it is considered to be carcinogenic, the drinking water standard is 10 μg/L. Bromate formation is a multi-step process (Haag & Hoigné, 1983; von Gunten & Hoigné, 1994), that was recently revised in the last step ($BrO_2^- + O_3$) (reaction 4.52) (Fischbacher *et al.* 2015). The formation of bromate and possible mitigation strategies are presented in Figure 4.5.

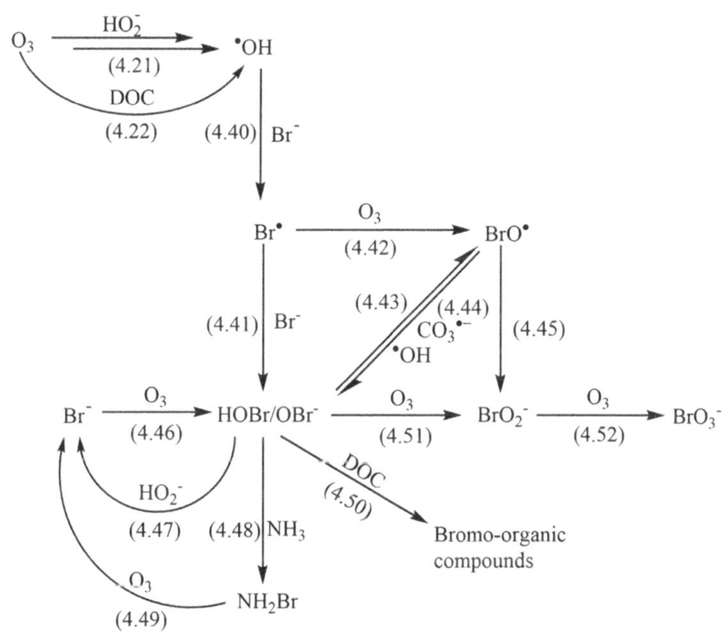

Figure 4.5 Bromate formation in ozone based processes and mitigation strategies.

Mitigation of bromate formation is an important aim in the peroxone process. The bromate minimizing effect is based on interception of an intermediate in bromate formation (i.e., hypobromous acid ($HOBr/OBr^-$), formed in reactions 4.41 and 4.46) by H_2O_2 (reaction 4.47). This reaction reveals a distinctive pH-dependency (von Gunten & Oliveras, 1997). The rate of this reaction increases with increasing pH up to a maximum at pH 10.2, which corresponds to the average of the pK_a values of H_2O_2 (≈11.8) and HOBr (≈8.5) (von Gunten & Oliveras, 1997). This behavior was explained by a reaction of a free acid and an anion and it was proposed that the mechanism is a nucleophilic substitution (von Gunten & Oliveras, 1997). Since HO_2^- is a stronger nucleophile than H_2O_2 and HOBr a stronger electrophile than OBr^- reaction 4.53 was proposed to happen (von Gunten & Oliveras, 1997).

$$HO_2^- + HOBr \rightarrow Br^- + H_2O + O_2 \quad k = 7.1 \times 10^5 \text{ M}^{-1}\text{ s}^{-1} \tag{4.53}$$

The pH-dependency of reaction 4.53 is very important regarding the efficiency of H_2O_2 to control bromate formation. At pH ≤ 7 the reaction becomes so slow ($k_{app} \leq 25$ M^{-1} s^{-1}) (von Gunten & Oliveras, 1997), that H_2O_2 may have to be added in excess over O_3 to compete against oxidation of HOBr (reactions 4.44 and 4.51). This however, can result in the ozone decomposing loop of reactions 4.24 and 4.27 and thus, largely lower the efficiency of pollutant degradation and disinfection.

The presence of NH_3/NH_4^+ and DOC also contributes to scavenging the intermediate $HOBr/OBr^-$ (reactions 4.48 and 4.49). HOBr readily brominates NH_3/NH_4^+ and the ensuing bromamines are oxidized by O_3 yielding nitrate and bromide (Pinkernell & von Gunten, 2001; von Sonntag & von Gunten, 2012). In case of DOC as $HOBr/OBr^-$ scavenger, electron rich phenolic moieties of DOC react with HOBr. Remarkably, incorporation of bromine into DOC *via* electrophilic substitution is of minor importance in this reaction (≈20–30% yield with respect to HOBr consumed) (Criquet *et al.* 2015). Instead, a large yield of bromide was observed which was explained with an electron transfer reaction (Criquet *et al.* 2015). The presence of DOC and ammonia principally lower the yield of bromate, which may reduce the required H_2O_2 doses for bromate control. However, HCO_3/CO_3^{2-} possibly enhance formation of bromate, since $CO_3^{\bullet -}$ from reaction 4.23 may contribute in oxidation of HOBr (reaction 4.44) (von Sonntag & von Gunten, 2012).

The interception of $Br^\bullet$ in reaction 4.42 opens a pathway of bromate formation without $HOBr/OBr^-$ as intermediate via reaction 4.4 (dismutation of $BrO^\bullet$). Hence, bromate formation cannot completely be controlled by addition of H_2O_2 albeit reaction 4.42 is disfavored in the peroxone process due to destabilization of O_3.

Acero *et al.* (2001a) could show that the addition of H_2O_2 to ozone leads to a reduction of the bromate formation by approximately a factor of two. Three different natural waters were investigated at pH 7 and 8. All water samples were spiked with bromide to achieve an initial concentration of 50 µg/L. The molar ratio of H_2O_2 and O_3 was 0.5 (0.34 mg H_2O_2 per mg O_3). The bromate concentrations in all water samples exceeded the drinking water standard (10 µg/L) when treated with ozone alone ($[BrO_3^-] = 12$ and 20.7 µg/L). The concentration of bromate formed in the peroxone process varied between 4.2 and 12.5 µg/L at an ozone dose of 2 mg/L.

A further important by-product formed in drinking water treatment is the assimilable organic carbon (AOC). AOC is defined as the amount of substrate (organic carbon) utilized by microorganisms. As most bacteria in drinking water distribution systems are heterotrophic, carbon is essential for their growth. AOC is a small fraction of total organic carbon that tends to be composed of small molecular weight compounds and thus, can readily be degraded by microorganisms. Water is defined as biologically stable, if AOC is present in concentrations below 50 µg/L (Kaplan *et al.* 1993). In contrast to bromate AOC can be decreased by applying biological filters after the oxidation step. Hacker *et al.* (1994) investigated the AOC formation in ozonation and the peroxone process. They could show that the addition of hydrogen peroxide resulted in a higher AOC formation. However, they could not show a clear relationship between higher H_2O_2/O_3 ratios and higher AOC formation. Filtration of ozone- and peroxone process-treated water by a dual-media filtration (biologically active anthracite/sand filters) resulted in an AOC reduction by 75–90% which corresponded to an AOC concentration that was equal or even lower than the AOC concentration in the raw waters.

NDMA is a human carcinogen and there have been growing concerns about NDMA contamination, as trace levels were found in surface and drinking waters (WHO, 2006). It is a secondary pollutant produced in water from nitrogeneous precursors. It was found to be formed during water disinfection with chloramines (Najm & Trussell, 2001; Mitch & Sedlak, 2002). If dimethylamine is present in water it reacts with monochloramine to 1,1-dimethylhydrazine, that is oxidized to NDMA. Recently, it was shown that NDMA can also be formed in ozonation, if N,N-dimethylsulfamide (DMS), a degradation product of the fungicide tolyfluanide is present (Lee *et al.* 2007). Von Gunten *et al.* (2010) found that the direct reactions of DMS with O_3 and $^{\bullet}OH$ do not lead to NDMA. NDMA is only formed, if Br^- is present. Br^- is oxidized by O_3 and $^{\bullet}OH$ to HOBr which reacts with the primary amine of DMS yielding a Br-DMS species. This Br-DMS species reacts then with O_3 to NDMA.

Lee *et al.* (2007) studied the NDMA degradation in conventional ozonation and O_3/H_2O_2 in two different lake waters (Lake Zurich: DOC = 1.2 mg/L, HCO_3^- = 2.6 mM, pH 7.9; Lake Greifensee: DOC = 3.9 mg/L, HCO_3^- = 4.1 mM, pH 8.1). They showed that applying O_3/H_2O_2 results in a higher degradation of NDMA than in conventional ozonation. It was degraded in conventional ozonation (160 μM O_3) by 25% in Lake Zurich water, and by 9% in Lake Greifensee water. During the peroxone process at same O_3 doses and a H_2O_2:O_3 ratio of 0.5, NDMA was oxidized by 55% in Lake Zurich and 28% in Lake Greifensee water. In Lake Greifensee water more $^{\bullet}OH$ were scavenged by HCO_3^- which resulted in a lower NDMA degradation in both processes. As in conventional ozonation O_3 was not completely consumed, the NDMA oxidation per mol of consumed O_3 was the same in ozonation and peroxone process. Higher O_3 doses would result in an increase of NDMA degradation, but also in an increase of bromate formation. Thus, bromate formation may be the limiting factor for NDMA oxidation.

Lee *et al.* (2016) investigated the abatement of NDMA by several AOPs and their combinations in wastewaters. As the reaction rate constant for the reaction of NDMA with $^{\bullet}OH$ is low ($k_{^{\bullet}OH+NDMA} = 4 \times 10^8$ M^{-1} s^{-1}), O_3/H_2O_2 is not suitable to degrade NDMA in wastewater. However, direct photolysis contributes almost 100% to the abatement of NDMA and a UV-based process should be applied, if NDMA degradation is the treatment goal. In addition the formation of NDMA in ozonation and O_3/H_2O_2 from the reaction of NDMA precursors with O_3 was studied. In three of seven investigated wastewaters significant concentrations of NDMA were formed in ozonation. The application of O_3/H_2O_2 in comparison led to an only slight reduction in NDMA formation (–4%–14%). As the reaction of NDMA precursors with O_3 is fast it is difficult to avoid NDMA formation in ozonation and the peroxone process. The authors could show that O_3/H_2O_2 is a suitable treatment process for micropollutant degradation in wastewater, as a high abatement of several micropollutants, except NDMA, could be achieved and ambient and formed NDMA is mainly attenuated by UV-based processes.

4.5.2 O_3/UV process

As bromate is formed in the reaction of bromide and ozone and/or $^{\bullet}OH$ (see Section 4.5.1.), bromate formation must be controlled in the O_3/UV process.

In the peroxone process BrO^- can be reduced by hydrogen peroxide to bromide. In the O_3/UV process even bromate can be reduced by UV to bromite, then to hypobromous acid and finally to bromide (reaction 4.54–4.56) (Siddiqui *et al.* 1996).

$$2\,BrO_3^- + h\nu \rightarrow 2\,BrO_2^- + O_2 \qquad (4.54)$$

$$2\,BrO_2^- + h\nu \rightarrow 2\,BrO^- + O_2 \qquad (4.55)$$

$$2\,BrO^- + h\nu \rightarrow 2\,Br^- + O_2 \qquad (4.56)$$

Siddiqui *et al.* (1996) examined bromate photolysis under different UV exposure and water quality conditions. Four different UV light sources were used, namely VUV (184.9 nm)/UV(253.7 nm) LP-, UV(253.7 nm) LP-, and (200–400 nm) MP Hg-arc lamps, and a pulse electric arc discharge source (VUV-UV range). Of note, both LP lamps emitted both the 184.9 nm and 253.7 nm radiation, but the operating power and the quartz sleeves were different for the two LP lamps, which allowed the isolation of 253.7 nm radiation from one of the lamps. As bromate has an absorption maximum at 195 nm, the VUV (184.9 nm)/UV(253.7 nm) LP-Hg was the most effective at bromate conversion to bromide among the investigated UV radiation sources. Dissolved organic carbon and nitrate affected the removal yield at 184.9 nm due to the competition for photons with BrO_3^-, but did not change the overall performance of VUV light source relative to the other UV light sources. The decrease in temperature lowered the removal yields, whereas pH did not have any measurable effect. The study provides UV dose data for specific bromate removal yields, but it is unclear how the UV radiation field was monitored at the reactor wall and whether the fluid dynamics were incorporated in the UV dose calculations.

Zhao *et al.* (2013) investigated the bromate formation and humic acid removal in water treatment by O_3/UV. The used LP Hg-lamp emitted light at 254 nm and 185 nm, but due to experimental setup light at 185 nm was absorbed by the quartz sleeves and the oxygen in the quartz sleeves. Ozone dose was 2 mg/L. At 100 μg/L bromide, a reaction time of 60 minutes and without adding humic acids to the water, bromate concentrations were higher in O_3/UV (17.1–77.6 μg/L) than for ozone alone (8–33.8 μg/L). The authors explained this high bromate formation with the fact that there was no DOC present to scavenge O_3 and •OH. After a reaction time of 60 minutes (first stage) UV irradiation was applied for 30 min without further addition of ozone (second stage). In this stage bromate was reduced by UV light by about 20%. When humic acid was added to the water bromate formation in the O_3/UV process was lower than in ozonation. However, the bromate concentration in the O_3/UV process exceeded the 10 μg/L drinking water standard at high ozone dose and low humic acid concentrations.

4.6 DISINFECTION

Ozonation is widely practiced for the effective inactivation of viruses, bacteria and their spores, and to a lower extent for inactivation of protozoa. Ozone targets a broad spectrum of pathogens, which makes it more effective than other common disinfectants such as chlorine dioxide, free chlorine and chloramines (von Gunten, 2003b). The kinetics for the inactivation of bacteria, spores and protozoa by ozone are analogous to the kinetics for their inactivation by UV radiation. Typically the kinetic starts with a lag which is caused by repair mechanisms of the microorganisms, followed by a first order decay. However, in many cases the lag phase is very short and can be neglected (von Sonntag & von Gunten, 2012). In the inactivation process by O_3 and UV radiation, the DNA is the main target. The mode of action of O_3 is different than that of UV and O_3 is more effective. A single lesion of the DNA is not leading to inactivation, since the cells have repair mechanisms for their DNA. Ozone causes single base lesions, while double-strand breaks and crosslinks are of minor importance (von Sonntag & von Gunten, 2012).

As it is not feasible to monitor all possible pathogens, indicator microorganisms are monitored instead. A typical indicator microorganism in water is *E. coli* that indicates a fecal contamination of the water. This approach works well only if the inactivation of other pathogens is as efficient as the inactivation of the indicator microorganism or even better. A much more reliable approach to assess the disinfection efficiency is the ct-concept. It can be described by Chick and Watson equation:

$$\ln\left(\frac{N}{N_0}\right) = -k \times c \times t, \tag{4.57}$$

where N and N_0 are the numbers of microorganisms after a certain exposure time t (N) and the initial number of microorganisms N_0, k is the inactivation rate constant for a certain microorganism, c is the disinfectant concentration and, t is the disinfectant exposure time. The exposure (or ct-value) is calculated by integrating the disinfectant concentration over time.

The oxidant exposure during the lag phase on the inactivation curve (ct_{lag}) is subtracted from the ct-value. The assessment of the disinfection is based on published ct values (see Table 4.5) obtained from the measurement of the inactivation of a specific microorganism. If a ct-value is not available, it can be obtained from the determination of O_3 exposure, which can practically and theoretically be determined by different methods (Roustan *et al.* 1993; von Gunten *et al.* 1999; Do-Quang *et al.* 2000; Rakness *et al.* 2005).

Table 4.5 Kinetics of the inactivation of selected microorganisms (bacteria, spores, virus and protozoa) by ozone.

Microorganism	k_{ozone} (L mg^{-1} min^{-1})	ct_{lag} (mg min L^{-1})	ct (mg min L^{-1}) for Inactivation of 4-log	Temperature (°C)	Reference
E. coli	7800	–	0.001	20	(Hunt & Mariñas, 1997)
B. subtilis spores	4.91	2.12	4	20	(Larson & Mariñas, 2003)
	2.9	2.9	6.1	20	(Driedger *et al.* 2001)
Rotavirus	76	–	0.12	5	(von Sonntag & von Gunten, 2012)
G. lamblia cysts	29	–	0.32	25	(Wickramanayake *et al.* 1984a)
G. muris cysts	2.4	–	3.9	25	(Wickramanayake *et al.* 1984b)
C. parvum oocysts	0.84	0.83	11.8	20	(Driedger *et al.* 2001)

The combination of O_3 with UV improves the water disinfection, given the complementary modes of action of the two disinfection agents. In the O_3/UV AOP, the UV-C radiation inactivates the microorganisms, as it is absorbed by DNA and RNA.

Lee *et al.* (2011) describe the bacterial disinfection in chlorotetracyline (antibiotic) containing wastewater. The disinfection was investigated by UV and ozone alone and by their combination (O_3/UV). While ozone alone lead to a $<$2-log inactivation within 1 h reaction time, UV resulted in a 4-log reduction within 20 min and a 5-log reduction within 40 min reaction time. The best results though were achieved by the combination of ozone and UV, which showed a synergistic effect. Here, the disinfection was enhanced, showing more than a 5-log reduction within 20 min.

Wolfe *et al.* (1989) investigated the inactivation of *E. coli*, two coliphages, (MS2 and f2) and *G. muris* by ozone and peroxone process in surface water. The applied molar H_2O_2:O_3 ratios were between 0.07 and 0.4. No significant difference between the inactivation of *E. coli* and the two coliphages in ozonation and in the peroxone process was observed. The inactivation of *G. muris* was slightly higher in the peroxone process than in ozonation alone.

Addition of H_2O_2 leads to a faster production of •OH and a faster consumption of ozone. Thus, less ozone is available for disinfection. The role of •OH in the disinfection of drinking water is not clear. At R_{ct}-values (see section 4.2.2) of 10^{-8} typically found in ozonation, •OH can be neglected as disinfectant. This is different from AOPs where the R_{ct}-values are about 10^{-6} and •OH may play a role in the inactivation of microorganisms with small k_{ozone}, e.g., *B. subtilis* spores, *G. muris* cysts and *C. parvum* oocysts (Table 4.5). One also has to consider that the main target for the inactivation of a microorganism is its DNA and that •OH are scavenged by the cell wall or other non-critical cell constituents before reaching the DNA (von Gunten, 2003b).

4.7 REFERENCES

Acero J. L. and von Gunten U. (2001b). Characterization of oxidation processes: ozonation and the AOP O_3/H_2O_2. *Journal of the American Water Works Association*, **93**(10), 90–100.

Acero J. L., Haderlein S. B., Schmidt T. C., Suter M. J. F. and von Gunten U. (2001a). MTBE oxidation by conventional ozonation and the combination ozone/hydrogen peroxide: efficiency of the processes and bromate formation. *Environmental Science & Technology*, **35**(21), 4252–4259.

Al-Rifai J. H., Gabefish C. L. and Schaefer A. I. (2007). Occurrence of pharmaceutically active and non-steroidal estrogenic compounds in three different wastewater recycling schemes in Australia. *Chemosphere*, **69**(5), 803–815.

Al Momani F. (2007). Degradation of cyanobacteria anatoxin-a by advanced oxidation processes. *Separation and Purification Technology*, **57**(1), 85–93.

Autin O., Hart J., Jarvis P., MacAdam J., Parsons S. A. and Jefferson B. (2012). Comparison of UV/H_2O_2 and UV/TiO_2 for the degradation of metaldehyde: kinetics and the impact of background organics. *Water Research*, **46**(17), 5655–5662.

Bielski B. H. J., Cabelli D. E., Arudi R. L. and Ross, A. B. (1985). Reactivity of HO_2/O_2^- radicals in aqueous solution. *Journal of Physical and Chemical Reference Data*, **14**(4), 1041–1100.

Bolton J. R., Bircher K. G., Tumas W. and Tolman C. A. (2001). Figures-of-merit for the technical development and application of advanced oxidation technologies for both electric- and solar-driven systems – (IUPAC technical report). *Pure and Applied Chemistry*, **73**(4), 627–637.

Buxton G. V., Greenstock C. L., Helman W. P. and Ross A. B. (1988). Critical-review of rate constants for reactions of hydrated electrons, hydrogen-atoms and hydroxyl radicals (•OH/•O⁻) in aqueous-solution. *Journal of Physical and Chemical Reference Data*, **17**(2), 513–886.

Carballa M., Omil F., Lema J. M., Llompart M., Garcia-Jares C., Rodriguez I., Gomez M. and Ternes T. (2004). Behavior of pharmaceuticals, cosmetics and hormones in a sewage treatment plant. *Water Research*, **38**(12), 2918–2926.

Chelme-Ayala P., Gamal El-Din M., Smith D. W. and Adams C. D. (2011). Oxidation kinetics of two pesticides in natural waters by ozonation and ozone combined with hydrogen peroxide. *Water Research*, **45**(8), 2517–2526.

Christensen H., Sehested K. and Corfitzen H. (1982). Reactions of hydroxyl radicals with hydrogen peroxide at ambient and elevated temperatures. *Journal of Physical Chemistry*, **86**(9), 1588–1590.

Criquet J., Rodriguez E. M., Allard S., Wellauer S., Salhi E., Joll C. A. and von Gunten U. (2015). Reaction of bromine and chlorine with phenolic compounds and natural organic matter extracts – electrophilic aromatic substitution and oxidation. *Water Research*, **85**, 476–486.

Do-Quang Z., Roustan M. and Duguet J. P. (2000). Mathematical modeling of theoretical cryptosporidium inactivation in full-scale ozonation reactors. *Ozone Science & Engineering*, **22**(1), 99–111.

Driedger A., Staub E., Pinkernell U., Mariñas B., Köster W. and von Gunten, U. (2001). Inactivation of bacillus subtilis spores and formation of bromate during ozonation. *Water Research*, **35**(12), 2950–2960.

Duguet J. P., Anselme C., Mazounie P. and Mallevialle J. (1990). Application of combined ozone-hydrogen peroxide for the removal of aromatic compounds from a groundwater. *Ozone: Science & Engineering*, **12**(3), 281–294.

Elovitz M. S. and von Gunten U. (1999). Hydroxyl radical/ozone ratios during ozonation processes. I. The R_{ct} concept. *Ozone: Science & Engineering*, **21**(3), 239–260.

Elovitz M. S., von Gunten U. and Kaiser H. P. (2000b). Hydroxyl radical/ozone ratios during ozonation processes. II. The effect of temperature, pH, alkalinity, and DOM properties. *Ozone: Science and Engineering*, **22**(2), 123–150.

Elovitz M. S., von Gunten U. and Kaiser H. P. (2000a). The influence of dissolved organic matter character on ozone decomposition rates and R_{ct}. In: Natural Organic Matter and Disinfection By-Products, S. E. Barrett, S. W. Krasner and G. L. Amy (eds), 761, American Chemical Society, pp. 248–269.

Ferguson D. W., McGuire M. J., Koch B., Wolfe R. L. and Aieta E. M. (1990). Comparing peroxone and ozone for controlling taste and odor compounds, disinfection by-products, and microorganisms. *Journal of the American Water Works Association*, **82**(4), 181–191.

Fischbacher A., von Sonntag J., von Sonntag C. and Schmidt T. C. (2013). The •OH radical yield in the $H_2O_2 + O_3$ (peroxone) reaction. *Environmental Science & Technology*, **47**(17), 9959–9964.

Fischbacher A., Löppenberg K., von Sonntag C. and Schmidt T. C. (2015). A new reaction pathway for bromite to bromate in the ozonation of bromide. *Environmental Science & Technology*, **49**(19), 11714–11720.

Haag W. R. and Hoigné J. (1983). Ozonation of bromide-containing waters: kinetics of formation of hypobromous acid and bromate. *Environmental Science & Technology*, **17**(5), 261–267.

Hacker P. A., Paszko-Kolva C., Stewart M. H., Wolfe R. L. and Means E. G. (1994). Production and removal of assimilable organic carbon under pilot-plant conditions through the use of ozone and peroxone. *Ozone: Science & Engineering*, **16**(3), 197–212.

Hansch C., Leo A. and Taft R. W. (1991). A survey of hammett substituent constants and resonance and field parameters. *Chemical Reviews*, **91**(2), 165–195.

Hoigné J. and Bader H. (1983a). Rate constants of reactions of ozone with organic and inorganic compounds in water.1. Non-dissociating organic compounds. *Water Research*, **17**(2), 173–183.

Hoigné J. and Bader H. (1983b). Rate constants of reactions of ozone with organic and inorganic compounds in water.2. Dissociating organic compounds. *Water Research*, **17**(2), 185–194.

Huber M. M. (2004). Elimination of pharmaceuticals during oxidative treatment of drinking water and wastewater: application of ozone and chlorine dioxide. PhD thesis, ETH Zurich, Switzerland.

Hunt N. K. and Mariñas B. J. (1997). Kinetics of Escherichia coli inactivation with ozone. *Water Research*, **31**(6), 1355–1362.

Ikehata K. and Gamal El-Din M. (2005a). Aqueous pesticide degradation by ozonation and ozone-based advanced oxidation processes: a review (Part I). *Ozone-Science & Engineering*, **27**(2), 83–114.

Ikehata K. and Gamal El-Din M. (2005b). Aqueous pesticide degradation by ozonation and ozone-based advanced oxidation processes: a review (Part II). *Ozone-Science & Engineering*, **27**(3), 173–202.

Ikehata K., Jodeiri Naghashkar N. and Gamal El-Din M. (2006). Degradation of aqueous pharmaceuticals by ozonation and advanced oxidation processes: a review. *Ozone: Science & Engineering*, **28**(6), 353–414.

Illés E., Szabó E., Takács E., Wojnárovits L., Dombi A. and Gajda-Schrantz K. (2014). Ketoprofen removal by O_3 and O_3/UV processes: kinetics, transformation products and ecotoxicity. *Science of the Total Environment*, **472**, 178–184.

Irmak S., Erbatur O. and Akgerman A. (2005). Degradation of 17 β-estradiol and bisphenol A in aqueous medium by using ozone and ozone/UV techniques. *Journal of Hazardous Materials*, **126**(1–3), 54–62.

Kaplan L. A., Bott T. L. and Reasoner D. J. (1993). Evaluation and simplification of the assimilable organic carbon nutrient bioassay for bacteria growth in drinking water. *Applied and Environmental Microbiology*, **59**(5), 1532–1539.

Katsoyiannis I. A., Canonica S. and von Gunten U. (2011). Efficiency and energy requirements for the transformation of organic micropollutants by ozone, O_3/H_2O_2 and UV/H_2O_2. *Water Research*, **45**(13), 3811–3822.

Kim I. and Tanaka H. (2010). Use of ozone-based processes for the removal of pharmaceuticals detected in a wastewater treatment plant. *Water Environment Research*, **82**(4), 294–301.

Kim I. and Tanaka H. (2011). Energy consumption for PPCPs removal by O_3 and O_3/UV. *Ozone-Science & Engineering*, **33**(2), 150–157.

Kim I. H., Tanaka H., Iwasaki T., Takubo T., Morioka T. and Kato Y. (2008). Classification of the degradability of 30 pharmaceuticals in water with ozone, UV and H_2O_2. *Water Science and Technology*, **57**(2), 195–200.

Knol A. H., Lekkerkerker-Teunissen K., Houtman C. J., Scheideler J., Ried A. and van Dijk J. C. (2015). Conversion of organic micropollutants with limited bromate formation during the peroxone process in drinking water treatment. *Drinking Water Engineering and Science*, **8**(2), 25–34.

Kümmerer K. (2001). Drugs in the environment: emission of drugs, diagnostic aids and disinfectants into wastewater by hospitals in relation to other sources – a review. *Chemosphere*, **45**(6–7), 957–969.

Larson M. A. and Mariñas B. J. (2003). Inactivation of Bacillus subtilis spores with ozone and monochloramine. *Water Research*, **37**(4), 833–844.

Larson R. A. and Zepp R. G. (1988). Environmental chemistry. Reactivity of the carbonate radical with aniline derivatives. *Environmental Toxicology and Chemistry*, **7**(4), 265–274.

Lau T. K., Chu W. and Graham N. (2007). Degradation of the endocrine disruptor carbofuran by UV, O_3 and O_3/UV. *Water Science and Technology*, **55**(12), 275–280.

Lee C., Yoon J. and von Gunten U. (2007). Oxidative degradation of N-nitrosodimethylamine by conventional ozonation and the advanced oxidation process ozone/hydrogen peroxide. *Water Research*, **41**(3), 581–590.

Lee H., Lee E., Lee C. H. and Lee K. (2011). Degradation of chlorotetracycline and bacterial disinfection in livestock wastewater by ozone-based advanced oxidation. *Journal of Industrial and Engineering Chemistry*, **17**(3), 468–473.

Lee Y. and von Gunten U. (2012). Quantitative structure–activity relationships (QSARs) for the transformation of organic micropollutants during oxidative water treatment. *Water Research*, **46**(19), 6177–6195.

Lee Y., Gerrity D., Lee M., Gamage S., Pisarenko A., Trenholm R. A., Canonica S., Snyder S. A. and von Gunten U. (2016). Organic contaminant abatement in reclaimed water by UV/H_2O_2 and a combined process consisting of O_3/H_2O_2 followed by UV/H_2O_2: prediction of abatement efficiency, energy consumption, and byproduct formation. *Environmental Science & Technology*, **50**(7), 3809–3819.

Legrini O., Oliveros E. and Braun A. M. (1993). Photochemical processes for water treatment. *Chemical Reviews*, **93**(2), 671–698.

Lekkerkerker K., Scheideler J., Maeng S. K., Ried A., Verberk J. Q. J. C., Knol A. H., Amy G. and van Dijk J. C. (2009). Advanced oxidation and artificial recharge: a synergistic hybrid system for removal of organic micropollutants. *Water Science and Technology: Water Supply*, **9**(6), 643–651.

Lekkerkerker-Teunissen K., Knol A. H., van Altena L. P., Houtman C. J., Verberk J. Q. J. C. and van Dijk J. C. (2012). Serial ozone/peroxide/low pressure UV treatment for synergistic and effective organic micropollutant conversion. *Separation and Purification Technology*, **100**, 22–29.

Lekkerkerker-Teunissen K., Knol A. H., Derks J. G., Heringa M. B., Houtman C. J., Hofman-Caris C. H. M., Beerendonk E. F., Reus A., Verberk J. Q. J. C. and van Dijk J. C. (2013). Pilot plant results with three different types of UV lamps for advanced oxidation. *Ozone-Science & Engineering*, **35**(1), 38–48.

Li X., Zhao W., Li J., Jiang J., Chen J. and Chen J. (2013). Development of a model for predicting reaction rate constants of organic chemicals with ozone at different temperatures. *Chemosphere*, **92**(8), 1029–1034.

Lin T., Yu S. and Chen W. (2016). Occurrence, removal and risk assessment of pharmaceutical and personal care products (PPCPs) in an advanced drinking water treatment plant (ADWTP) around Taihu Lake in China. *Chemosphere*, **152**, 1–9.

Merenyi G., Lind J., Naumov S. and von Sonntag C. (2010). Reaction of ozone with hydrogen peroxide (peroxone process): a revision of current mechanistic concepts based on thermokinetic and quantum-chemical considerations. *Environmental Science & Technology*, **44**(9), 3505–3507.

Mitch W. A. and Sedlak D. L. (2002). Formation of N-nitrosodimethylamine (NDMA) from dimethylamine during chlorination. *Environmental Science and Technology*, **36**(4), 588–595.

Mizuno T., Ohara S., Nishimura F. and Tsuno H. (2011). O_3/H_2O_2 process for both removal of odorous algal-derived compounds and control of bromate ion formation. *Ozone: Science & Engineering*, **33**(2), 121–135.

Najm I. and Trussell R. R. (2001). NDMA formation in water and wastewater. *Journal of the American Water Works Association*, **93**(2), 92–99.

Neta P., Huie R. E. and Ross A. B. (1988). Rate constants for reactions of inorganic radicals in aqueous solution. *Journal of Physical and Chemical Reference Data*, **17**(3), 1027–1284.

Ning B., Graham N., Zhang Y., Nakonechny M. and Gamal El-Din M. (2007). Degradation of endocrine disrupting chemicals by ozone/AOPs. *Ozone: Science & Engineering*, **29**(3), 153–176.

Orr P. T., Jones G. J. and Hamilton G. R. (2004). Removal of saxitoxins from drinking water by granular activated carbon, ozone and hydrogen peroxide – implications for compliance with the Australian drinking water guidelines. *Water Research*, **38**(20), 4455–4461.

Peter A. and von Gunten U. (2007). Oxidation kinetics of selected taste and odor compounds during ozonation of drinking water. *Environmental Science and Technology*, **41**(2), 626–631.

Peyton G. R. and Glaze W. H. (1988). Destruction of pollutants in water with ozone in combination with ultraviolet radiation. 3. Photolysis of aqueous ozone. *Environmental Science & Technology*, **22**(7), 761–767.

Pinkernell U. and von Gunten U. (2001). Bromate minimization during ozonation: mechanistic considerations. *Environmental Science and Technology*, **35**(12), 2525–2531.

Pocostales J. P., Sein M. M., Knolle W., von Sonntag C. and Schmidt T. C. (2010). Degradation of ozone-refractory organic phosphates in wastewater by ozone and ozone/hydrogen peroxide (peroxone): the role of ozone consumption by dissolved organic matter. *Environmental Science & Technology*, **44**(21), 8248–8253.

Rakness K. L., Najm I., Elovitz M., Rexing D. and Via S. (2005). Cryptosporidium log-inactivation with ozone using effluent CT_{10}, geometric mean CT_{10}, extended integrated CT_{10} and extended CSTR calculations. *Ozone: Science & Engineering*, **27**(5), 335–350.

Reisz E., Schmidt W., Schuchmann H. P. and von Sonntag C. (2003). Photolysis of ozone in aqueous solutions in the presence of tertiary butanol. *Environmental Science & Technology*, **37**(9), 1941–1948.

Rosal R., Rodríguez A., Perdigón-Melón J. A., Mezcua M., Hernando M. D., Letón P., García-Calvo E., Agüera A. and Fernández-Alba A. R. (2008). Removal of pharmaceuticals and kinetics of mineralization by O_3/H_2O_2 in a biotreated municipal wastewater. *Water Research*, **42**(14), 3719–3728.

Rosenfeldt E. J., Linden K. G., Canonica S. and von Gunten U. (2006). Comparison of the efficiency of OH radical formation during ozonation and the advanced oxidation processes O_3/H_2O_2 and UV/H_2O_2. *Water Research*, **40**(20), 3695–3704.

Roustan M., Beck C., Wable O., Duguet J. P. and Mallevialle J. (1993). Modeling hydraulics of ozone contactors. *Ozone: Science & Engineering*, **15**(3), 213–226.

Scheideler J. and Bosmith A. (2014). AOP für den Abbau von Metaldehyd: Weitergehende Oxidationprozesse gegen Pestizid im Oberflächenwasser. *Aqua & Gas*, **94**(3), 52–57.

Scheideler J., Lekkerkerker-Teunissen K., Knol T., Ried A., Verberk J. and van Dijk H. (2011). Combination of O_3/H_2O_2 and UV for multiple barrier micropollutant treatment and bromate formation control – an economic attractive option. *Water Practice and Technology*, **6**(4), 1–8.

Sein M. M., Golloch A., Schmidt T. C. and von Sonntag C. (2007). No marked kinetic isotope effect in the peroxone ($H_2O_2/D_2O_2 + O_3$) reaction: mechanistic consequences. *ChemPhys Chem.*, **8**(14), 2065–2067.

Shen J. M., Chen Z. L., Xu Z. Z., Li X. Y., Xu B. B. and Qi F. (2008). Kinetics and mechanism of degradation of p-chloronitrobenzene in water by ozonation. *Journal of Hazardous Materials*, **152**(3), 1325–1331.

Siddiqui M. S., Amy G. L. and McCollum L. J. (1996). Bromate destruction by UV irradiation and electric arc discharge. *Ozone: Science and Engineering*, **18**(3), 271–290.

Staehelin J. and Hoigné J. (1982). Decomposition of ozone in water: rate of initiation by hydroxide ions and hydrogen peroxide. *Environmental Science & Technology*, **16**(10), 676–681.

Sudhakaran S. and Amy G. L. (2013). QSAR models for oxidation of organic micropollutants in water based on ozone and hydroxyl radical rate constants and their chemical classification. *Water Research*, **47**(3), 1111–1122.

Takahashi N., Hibino T., Torii H., Shibata S., Tasaka S., Yoneya J., Matsuda M., Ogasawara H., Sugimoto K. and Fujioka T. (2013). Evaluation of O_3/UV and O_3/H_2O_2 as practical advanced oxidation processes for degradation of 1,4-dioxane. *Ozone: Science & Engineering*, **35**(5), 331–337.

Ternes T. A. (1998). Occurrence of drugs in german sewage treatment plants and rivers. *Water Research*, **32**(11), 3245–3260.

US EPA (1997). Special report on environmental endocrine disruption: an effects assessment and analysis, Report EPA/630/R-96/012, Risk Assessment Forum, U.S. EPA, Washington, D.C., USA.

Uslu M. O., Rahman M. F., Jasim S. Y., Yanful E. K. and Biswas N. (2012). Evaluation of the reactivity of organic pollutants during O_3/H_2O_2 process. *Water, Air & Soil Pollution*, **223**(6), 3173–3180.

von Gunten U. (2003a). Ozonation of drinking water: part I. Oxidation kinetics and product formation. *Water Research*, **37**(7), 1443–1467.

von Gunten U. (2003b). Ozonation of drinking water: part II. Disinfection and by-product formation in presence of bromide, iodide or chlorine. *Water Research*, **37**(7), 1469–1487.

von Gunten U. and Hoigné J. (1994). Bromate formation during ozonation of bromide-containing waters: interaction of ozone and hydroxyl radical reactions. *Environmental Science & Technology*, **28**(7), 1234–1242.

von Gunten U. and Oliveras Y. (1997). Kinetics of the reaction between hydrogen peroxide and hypobromous acid: implication on water treatment and natural systems. *Water Research*, **31**(4), 900–906.

von Gunten U., Elovitz M. and Kaiser H. P. (1999). Calibration of full-scale ozonation systems with conservative and reactive tracers. *Journal of Water Supply Research and Technology-Aqua*, **48**(6), 250–256.

von Gunten U., Salhi E., Schmidt C. K. and Arnold W. A. (2010). Kinetics and mechanisms of N-nitrosodimethylamine formation upon ozonation of N,N-dimethylsulfamide-containing waters: bromide catalysis. *Environmental Science & Technology*, **44**(15), 5762–5768.

von Sonntag C. (2008). Advanced oxidation processes: mechanistic aspects. *Water Science and Technology*, **58**(5), 1015–1021.

von Sonntag C. and Schuchmann H. P. (1997). Peroxyl radicals in aqueous solutions. In: Peroxyl Radicals, Z. B. Alfassi (ed.), John Wiley & Sons Ltd., Chichester, England, pp. 173–234.

von Sonntag C. and von Gunten U. (2012). Chemistry of Ozone in Water and Wastewater Treatment – from Basic Principles to Applications. IWA Publishing, London.

Werderitsch M. (2013). Planning, construction and operation of a full scale advanced oxidation treatment plant for drinking water. AOT-Conference, San Diego.

Westerhoff P., Nalinakumari B. and Pei P. (2006). Kinetics of MIB and geosmin oxidation during ozonation. *Ozone: Science & Engineering*, **28**(5), 277–286.

WHO (2006). N-Nitrosodimethylamine in drinking-water. Background document for development of WHO Guidelines for Drinking-water Quality, Report, WHO, Geneva, Switzerland.

Wickramanayake G. B., Rubin A. J. and Sproul O. J. (1984a). Inactivation of Giardia lamblia cysts with ozone. *Applied and Environmental Microbiology*, **48**(3), 671–672.

Wickramanayake G. B., Rubin A. J. and Sproul O. J. (1984b). Inactivation of *Naegleria* and *Giardia* cysts in water by ozonation. *Journal of Water Pollution Control Federation*, **56**(8), 983–988.

Wolfe R. L., Stewart M. H., Scott K. N. and McGuire M. J. (1989). Inactivation of Giardia muris and indicator organisms seeded in surface water supplies by peroxone and ozone. *Environmental Science & Technology*, **23**(6), 744–745.

Wu T. and Englehardt J. D. (2015). Peroxone mineralization of chemical oxygen demand for direct potable water reuse: kinetics and process control. *Water Research*, **73**, 362–372.

Xu B. B., Chen Z. L., Qi F., Ma J. and Wu F. C. (2010). Comparison of N-nitrosodiethylamine degradation in water by UV irradiation and UV/O_3: efficiency, product and mechanism. *Journal of Hazardous Materials*, **179**(1–3), 976–982.

Yargeau V. and Leclair C. (2008). Impact of operating conditions on decomposition of antibiotics during ozonation: a review. *Ozone: Science & Engineering*, **30**(3), 175–188.

Zehavi D. and Rabani J. (1972). Oxidation of aqueous bromide ions by hydroxyl radicals. A pulse radiolytic investigation. *Journal of Physical Chemistry*, **76**(3), 312–319.

Zhao G., Lu X., Zhou Y. and Gu Q. (2013). Simultaneous humic acid removal and bromate control by O_3 and UV/O_3 processes. *Chemical Engineering Journal*, **232**, 74–80.

Zwiener C. and Frimmel F. H. (2000). Oxidative treatment of pharmaceuticals in water. *Water Research*, **34**(6), 1881–1885.

Chapter 5

Vacuum UV radiation-driven processes

Tünde Alapi, Krisztina Schrantz, Eszter Arany and Zsuzsanna Kozmér

5.1 FUNDAMENTAL PRINCIPLES OF VACUUM UV PROCESSES

Vacuum ultraviolet (VUV) photolysis is one of the Advanced Oxidation Processes (AOPs) for the elimination of pollutants from water and air. The ultraviolet (UV) radiation below 200 nm was named VUV in 1893 by the German physicist Victor Schumann, because it is strongly absorbed by air. The electromagnetic spectrum of UV radiation is defined most broadly as 10–400 nm and is subdivided into a number of spectral ranges (http://spacewx.com/solar_spectrum.html). The VUV radiation covers the spectral range from 100 to 200 nm. The photochemistry in the VUV spectral domain focuses on the region between 140–200 nm.

5.1.1 VUV radiation sources for water treatment

In photo-initiated AOP applications, such as VUV photolysis, the lamp type determines the process effectiveness. There are two types of light sources commonly used in VUV photolysis: the low-pressure mercury vapor (LP) lamps and the excimer lamps.

The spectral radiation from low-pressure mercury plasma (wherein the optimum pressure of mercury is approximately 1 Pa) is dominated by the two Hg resonance lines at 253.7 nm and 184.9 nm. The 253.7 nm line represents 85% of the total emitted UV intensity. The intensity of 184.9 nm line relative to 253.7 nm radiation (quoted as 100%) decreases to 8% through the quartz sleeve for a conventional LP lamp (Masschelein, 2002) (Figure 5.1). These lamps are often called "ozone (O_3) producing UV lamps" *via* absorption of 184.9 nm radiation by O_2. The photo-initiated decomposition of the generated O_3 takes place ($\varepsilon_{O_3,\,254\,nm} = 2952\ M^{-1}\ cm^{-1}$, $\Phi_{230-280\,nm}(^{\bullet}O^{\bullet}) = 1.0$ (Atkinson *et al.* 2004)) simultaneously.

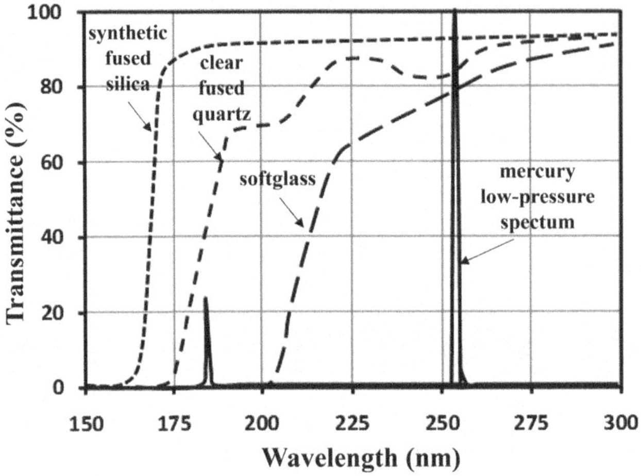

Figure 5.1 Transmittance of different quartz glass types and relative spectral emittance from a low-pressure mercury lamp. Reprinted with permission from Schalk *et al.* (2005).

The optimum operating temperature of these lamps is 40 °C. Lower temperatures result in partial condensation of mercury vapors on the lamp envelope. At temperatures above 40 °C, the self-absorption by the mercury vapors increases. Therefore, the UV output of the lamp is temperature-dependent. The lineal power density (electrical power per unit arc length) is typically between 0.3 and 0.6 W cm^{-1} and the total UV output is in the 0.2 to 0.3 W(UV) cm^{-1} range, which means that the UV efficiency (the ratio of UV output to the electrical input) is between 25% and 35%. The energy losses are mainly in the form of heat (Masschelein, 2002; Schalk *et al.* 2005).

Typically, the LP amalgam lamps use the mercury/indium amalgam, which reaches the optimum mercury vapor pressure at wall temperatures close to 100 °C. Thus, a higher fraction of electrical power is used, and their UV output is less dependent on the ambient temperature (Van der Pol & Krijnen, 2005).

LP lamps require warm-up time in order to reach 90% of their full UV output after the start-up. The solarization of lamp quartz envelope results in loss of transmission at 184.9 nm, which could be as much as 50% after 700 h (Masschelein, 2002). Another reason for the aging is the UV absorbing mercury oxide layer which forms on the inner wall surface of the lamp. This layer stems from a reaction of mercury ions with O_2 in the quartz glass, resulting in a decline of UV output over the long course of operation. Al_2O_3-based protective coatings are commonly used to diminish this effect and to extend the lamp lifetime up to 16000 h (Voronov *et al.* 2003). By combining the high transmittance quartz material, the amalgam lamp, and the long life technology, an optimized lamp with high specific output at 184.9 nm can be achieved.

The LP lamps are available in both "O_3-free" and "O_3 producing" versions depending on the quartz quality of lamp envelope. Ordinary quartz, which is produced by fusing natural quartz, has low transmittance below 200 nm and contains metallic impurities such as Al and Ti. Synthetic fused silica can be made from Si-rich chemical precursor, usually through a continuous process which involves flame oxidation of volatile Si compounds to SiO_2 and thermal fusion of the resulting dust. The transmittance of natural fused quartz is only ~50% at 184.9 nm as compared to ~90% transmittance of synthetic fused silica (Schalk *et al.* 2005; Witzke, 2001) (Figure 5.1).

LP lamps are usually cylindrical, but flat lamp technology is also marketed by Hereaus. These light sources are readily available at comparatively low cost, and their emission spectra are well established and quantified. Several light source manufacturers produce and distribute a large variety of "O_3 producing" or VUV/UV LP lamps; e.g., SEN, Japan (http://www.senlights.com/), Jelight, USA (http://www.jelight.com/), LightTech, USA (http://www.light-sources.com/).

LP lamps are used worldwide to disinfect, sanitize, and oxidize water, as well as to reduce total organic carbon (TOC) and chlorine/chloramines in water for specific applications. The combination of 184.9 nm and 253.7 nm radiation is necessary for photooxidation of organic compounds and simultaneous sterilization of water. The LP lamps with high purity silica sleeves are used in O_3 generators (http://www.jelight.com/ozone-generator.html). The same lamps are used in the UVOX system (http://www.uvox.com/en/choice-of-right-uvox-system.html), which combines the disinfecting effect of UV light with the oxidizing effect of O_3 and hydroxyl radicals (HO•) in one single system for water treatment. Besides water treatment and disinfection, the VUV/UV LP lamps are used in several other applications, e.g., surface cleaning and modification, photo chemical vapor deposition, ionization, deletion of IC memory, light source of measuring instruments, etc.

Excimer and exciplex light sources based on the formation of noble gas and halogen excimers, or rare gas/halogen exciplex represent a relatively novel lamp generation. They have been developed in the last few decades and became the most important incoherent sources, which can operate over a wide range of wavelengths both in the UV and the VUV spectral regions (Figure 5.2). Their operation is based on the formation of excited dimers.

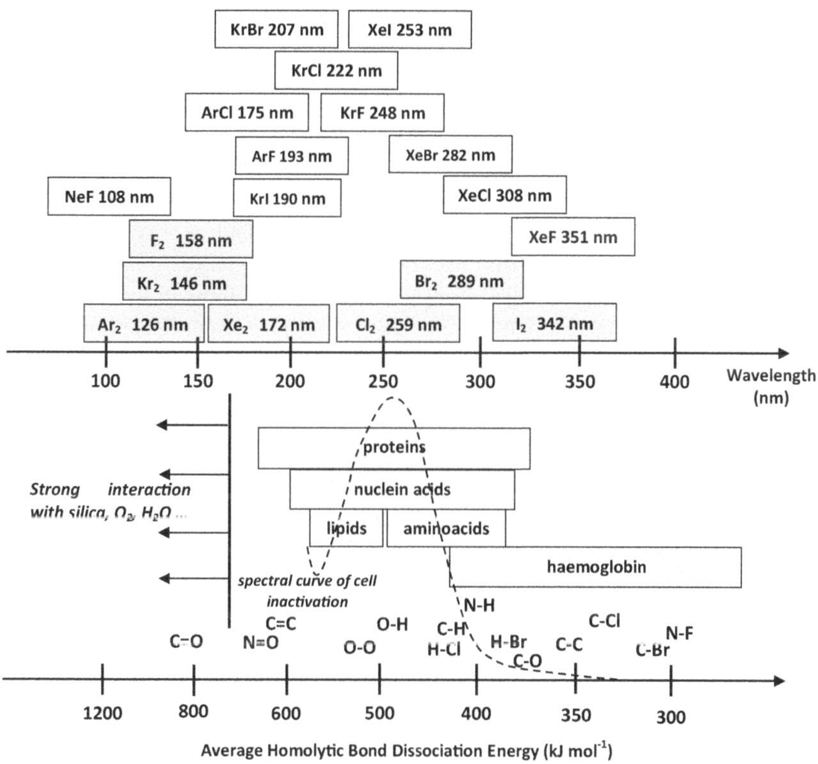

Figure 5.2 Emission wavelengths of common excimer and noble gas/halogen exciplex lamps, absorption bands of biological molecules, microbial pathogen inactivation spectral range, and chemical bond dissociation energies.

The term excimer (excited state dimer) is, strictly speaking, limited to cases in which both components of the dimer are the same atom or molecule. The term exciplex refers to the heterodimeric case; however, common usage expands excimer to cover exciplex. The spectrum of rare gas and halogen gas excimer lamp radiation is characterized by an intense narrow emission band (Eliasson & Kogelschatz, 1988; Gellert & Kogelschatz, 1991; Oppenländer, 2003), therefore these lamps are usually called "*quasi*-monochromatic light sources". The half-width of the spectral emission bands of these light sources depends mainly on the type of gas and excitation conditions.

A noble gas excimer VUV source based on electron beam excitation was first described by Wieser *et al.* (1997). Recently, several types of excimer lamps are built using various types of excitation (Kitamura *et al.* 2004; Lomaev *et al.* 2006b; Sosnin *et al.* 2002; Sosnin *et al.* 2011b), among which the capacitive-discharge (CD) and dielectric-barrier discharge (DBD) are the most attractive ones. Information on excimer lamp technology and potential use of excimers in large scale industrial applications are extensively covered in the literature (Kogelschatz, 1990, 2003, 2012; Kogelschatz *et al.* 2000; Kogelschatz *et al.* 1997; Lomaev *et al.* 2003; Lomaev *et al.* 2006b; Sosnin *et al.* 2006).

In general, the fixed geometry of most of the available lamps limits the flexibility in the reactor design. However, the novel incoherent excimer sources present the advantage that their design is independent of the electrode configuration (Kogelschatz, 1990, 2003; Oppenländer & Sosnin, 2005). Several publications are available on the detailed optimization of working parameters (optimal pressure, gas composition, excitation method and various parameters, e.g., voltage pulse shape, excitation pulse repetition rate and power supplies) (Kogelschatz, 1990, 2004; Kogelschatz *et al.* 2000; Lomaev *et al.* 2003; Lomaev *et al.* 2012; Lomaev *et al.* 2006b; Oppenländer, 2007; Oppenländer & Sosnin, 2005; Oppenländer *et al.* 2005; Sosnin *et al.* 2006; Tarasenko *et al.* 1999).

Xenon excimer (Xe_2*) lamps are the most studied VUV light sources and are widely used in research related to water purification. In order to generate excimer molecules, high energy electrons are required to generate excited state noble gas atoms. Noble gases can be exposed to extremely high electrical power densities, when energized electrons collide with noble gas atoms, and as a result, ionized species and excited states are formed. The short-lived (ns) excimer molecule (e.g., Xe_2*) is formed in a three-body reaction involving the excited state of rare gas atom (e.g., Xe*), and atoms in the ground state (e.g., Xe), where the third collision partner takes away the energy excess of the excimer. The photons are emitted from the excited dimer (excimer) with the formation of two noble gas atoms in the ground state (Figure 5.3). Since the excimer molecule structure is different from that of noble gas atoms and there is no excimer ground state, there is no self-absorption of the emitted radiation. The energy excess dissipated by ions is lost energy as that is not sufficient to generate excimers (Zvereva & Gerasimov, 2001). The Xe_2* decays with emission of radiation peaking at 172 nm, with the half-width of the spectral emission band of 14 nm (Braslavsky, 2007). About 70–80% of the radiation power of an excimer lamp is concentrated in this single emission band (Eliasson & Kogelschatz, 1991).

Theoretically, under optimal conditions 80% of the discharge power can be converted to VUV radiation of Xe_2* (Kogelschatz, 2003, 2012). In practice, the typical radiant power efficiency of Xe_2* lamps is about 40% (Avdeev *et al.* 2008; Gerasimov *et al.* 2006; Lomaev *et al.* 2003; Lomaev *et al.* 2006a; Lomaev *et al.* 2012; Lomaev *et al.* 2006b; Tarasenko *et al.* 1999; Zhang & Boyd, 2000). Pulsed Xe_2* lamps may have up to 40% efficiency (Carman & Mildren, 2003). Beleznai *et al.* (2008) performed theoretical and experimental studies on a dielectric barrier Xe discharge lamp and achieved an overall VUV output efficiency of 66.8%, of which 47.2% was at 172 nm. The other emitted radiation were at 147 nm (2.4%) and 150 nm (17.2%). In addition to the Xe_2* lamps, Lomaev *et al.* (2006b) investigated the windowless Kr_2* and Xe_2* excimer lamps, and reported radiation efficiencies of 25% for Kr_2* (146 nm) and 45% for Xe_2* (172 nm), respectively. Ar_2* lamps were also investigated by Elsner *et al.* (2006) and Baricholo *et al.* (2011).

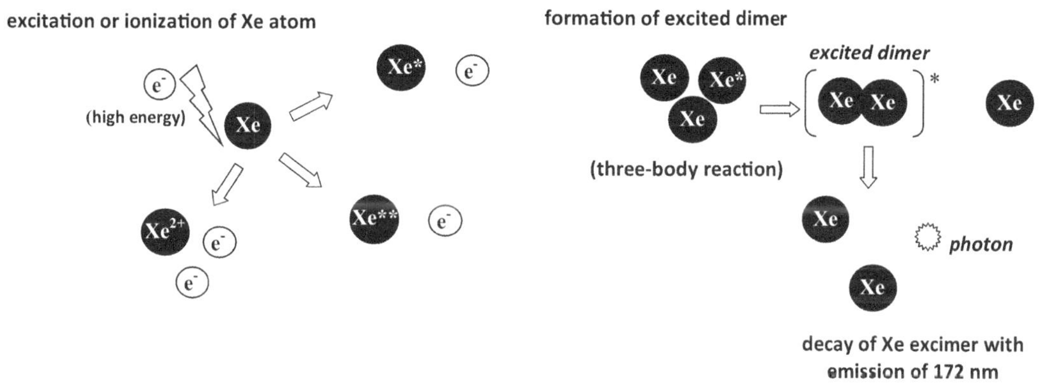

Figure 5.3 Main processes for the formation and decay of Xe_2*.

The average output power density of excimer lamps usually does not exceed 50 mW cm^{-2}. The excitation power and the radiation power density can be enhanced by using two-barrier excilamps (Erofeev *et al.* 2010; Lomaev *et al.* 2008). The average radiation power densities for one- and two-barrier excilamps were found to be 20–30 and ~40 mW cm^{-2}, respectively (Lomaev *et al.* 2008). A high-power sealed-off DBD Xe_2* lamp having output power density of 120 mW cm^{-2} was designed, constructed, and tested by Lomaev *et al.* (2003).

Volkova and Gerasimov (1997) and Gerasimov *et al.* (2000, 2002) studied the emission spectra of a mixture of Kr and Xe and showed that small additions of Xe led to excitation and simultaneous deactivation of molecules of the basic Kr gas. The redistribution of emission energy was accompanied by a nonlinear amplification of radiation near 147 nm. Further investigations were made with binary Xe–X and Kr–Y mixtures (X is He, Ne, Ar, or Kr; Y is He, Ne, or Ar). The emission bands investigated are related to electronic transitions in heteronuclear dimers (Gerasimov *et al.* 2003).

The extremely narrow radiation line at 121.6 nm is due to the transition of a H atom (Lyman-line) and can be obtained in He or Ne gas with traces (less than 0.1%) of H_2 with an energy efficiency of about 10% (Yan *et al.* 2002; Yan & Gupta, 2003). The spectrum, optical power, stability and efficiency of this light source were also investigated in Yan *et al.*'s works. Traces of water (0.02%) in Ar-filled lamp led to an efficient and selective excitation of the O—H band. Shuaibov *et al.* (2012) investigated the energy transfer from Ar to H_2O. The emission maxima are at 297.6 and 308.9 nm in the UV region, and at 156.0, 180.3, and 186.0 nm in the VUV region.

A wide variety of both excimer and exciplex lamps are available (Figure 5.2). Noble gas- and halogen excimer lamps (Avdeev *et al.* 2008) and noble gas/halogen exciplex lamps were also extensively investigated (Kogelschatz, 2012; Lomaev *et al.* 2012; Lomaev *et al.* 2007; Tarasenko *et al.* 1999). Among all, the most commonly investigated are KrCl* (Erofeev *et al.* 2010; Shuaibov *et al.* 2013; Sosnin *et al.* 2011a; Sosnin *et al.* 2015a; Sosnin *et al.* 2015b; Zhuang *et al.* 2010), XeCl* (Avtaeva *et al.* 2013; Baadj *et al.* 2013), and XeBr* (Lomaev *et al.* 2012) lamps with emission at 222, 308, and 282 nm, respectively. Typically, 5–18% of radiant efficiency is reached. KrCl* is formed in Kr gas in the presence of 1% chlorine donor (Cl_2, HCl or CCl_4 (Shuaibov *et al.* 2013)). Zhuang *et al.* (2010) investigated the effect of gas composition and pressure, and the maximum efficiency of the 222 nm radiation was found to be 9.2% for a KrCl* filled with Kr (198 mbar) and Cl_2 (2 mbar). Other excimer light sources have been investigated in laboratory studies, e.g., excimers radiating in the visible range, multi-wavelength excimer sources, etc. Internal phosphor coatings were used to convert the 172 nm radiation of the Xe_2* to near UV or visible one (Beleznai *et al.* 2006; Beleznai *et al.* 2008; Malinin, 2006). This is the basis of mercury-free fluorescent lamps and of flat plasma-display panels with a large screen.

Zhang and Boyd (2000) compared the lifetime of 172, 222, and 308 nm excimer lamps, as well as the overall efficiency, stability and output fluctuations. The Xe_2* lamp caused the formation of "color centers" in the Suprasil quartz, which reduced the intensity transmitted through quartz by 22% during the first 60 hours of operation. In contrast, 100% of the original UV intensity output through the quartz of the 222 and 308 nm lamps was still maintained after up to 4000 h operating time. The fluctuation of the average radiation power of these air-cooled excimer lamps were found to be only 2–5% (Lomaev *et al.* 2003; Zhang & Boyd, 2000).

Efficient lamp operation requires an envelope material with a high transmittance in the VUV spectral range and resistance to high energy radiation. The photo-induced generation of "color centers" in quartz glass is due to various defects. Extrinsic defects are mainly trace impurities. Intrinsic defects are always generated as a result of thermal effect during the production process and are present as equilibrium or frozen-in defects. These are network imperfections, such as two-fold and three-fold coordinated Si atoms, Si to Si bonds, O deficiency centers, non-bridging O atoms, interstitial O atoms and interstitial O_2. Additionally, there are technology-related defects. For example, in quartz glass fused in an atmosphere containing H_2, the typical defects are hydride (SiH), hydroxyl (OH) and free hydrogen. These defects are also present in synthetic fused silica. Other technology-related defects are halogen atoms, e.g., Cl and F, which are often used for OH removal in order to produce dry synthetic fused silica. In dry fused silica, Si—Cl or Si—F bonds, as well as interstitial Cl and F, may be present. Schreiber *et al.* (2005) investigated the colorization and radiation resistance of quartz glass for VUV lamps. Synthetic fused silica containing 250 ppm OH and quartz glass fused from cultured crystals were identified as the best materials for 172 nm VUV applications.

The short wavelength of Ar_2* radiation (126 nm) requires special windows (e.g., CaF_2, MgF_2, LiF) which are expensive and available only in small sizes. To circumvent this problem, open discharge configurations, called as "windowless excimer lamps" were designed by Kogelschatz (1992). An open windowless excilamp capable of operating on Ar_2* (126 nm), Kr_2* (146 nm), and Xe_2* (172 nm) was described by Lomaev *et al.* (2006a).

There are several benefits of excimer lamps, such as high average specific power of either VUV or UV radiation, high energy of emitting photons, quasi-monochromatic radiation, high spectral power density, absence of visible and IR radiation, low heating of radiating surface (cold lamps), no fixed geometry, no warm up time, etc. The availability of multiple-wavelength UV radiation by simultaneous excitation of several kinds of working excimer molecules is also possible. Finally, excimer lamps based on noble gases are non-hazardous and are much more environmentally friendly than mercury vapor lamps.

As shown in Figure 5.2, the VUV radiation can break most of the chemical bonds. VUV excimer lamps emit at short wavelengths, and their high energy radiation is generally used for large scale surface modifications, including low-temperature oxidation, deposition of metal patterns on heat sensitive substrates, photochemical polymerization and cleaning (dry cleaning). Ar_2* lamps are mainly used in photolithography. These VUV excimer lamps also have great potential as light or ionization sources in analytical instrumentation. Xe-based fluorescent excimer lamps used as mercury-free image processing lamps in scanners and in copy machines are manufactured by different companies, such as Ushio and Osram (Kogelschatz, 2012). VUV and UV excimer lamps are an alternative to conventional light sources used for UV disinfection, as the spectral curve of cell inactivation shows in the Figure 5.2. Excimer lamps have great potential in medical research and applications, e.g., phototherapy (psoriasis), wavelength-selective drug phototoxicity testing (Oppenländer, 1994, 1996). Detailed information on excimer lamps and their applications can be found in the published literature (Kogelschatz, 2012; Lomaev *et al.* 2012; Oppenländer & Schwarzwälder, 2002; Oppenländer & Sosnin, 2005; Sosnin *et al.* 2006).

A wide variety of excimer (Ar_2*, Kr_2* and Xe_2*) and exciplex (KrCl*, XeCl*, XeBr*) lamps are already commercially available. Osram (http://www.osram.com/osram_com/), Ushio (http://www.ushio.com/) and Hamamatsu (http://www.hamamatsu.com/) are the major manufacturers. The mostly applied light source at lab-scale is the XERADEX Xe_2* lamp from Osram. This light source does not require cooling in normal operation, has no startup time, the switching cycle is unlimited and the estimated lifetime is 2500 h. The VUV output at 172 nm is 40 mW cm^{-2} and can be enhanced with active cooling to 80 mW cm^{-2}. The lamp wattage is 60–300 W and the nominal VUV efficiency at 172 nm is 40% (http://www.osram.com/osram_com/).

5.1.2 VUV irradiation of water

5.1.2.1 VUV photolysis of pure water

The VUV photolysis is mainly used and investigated for the elimination and mineralization of various pollutants in aqueous solutions. Organic and inorganic molecules or ions have relatively high absorption coefficients in the VUV region. However in aqueous solutions, the VUV radiation is absorbed almost exclusively by water when its concentration (55.5 mol L^{-1}) substantially exceeds those of the dissolved compounds. Figure 5.4 is a compilation of water absorption coefficients determined at 25 °C (Barrett & Baxendale, 1960; Barrett & Mansell, 1960; Halmann & Platzner, 1965; Kröckel & Schmidt, 2014; Querry *et al.* 1978; Segelstein, 1981; Weeks *et al.* 1963). The absorption coefficients and the photolysis quantum yield of water were found dependent on the radiation wavelength in the 140–200 nm region.

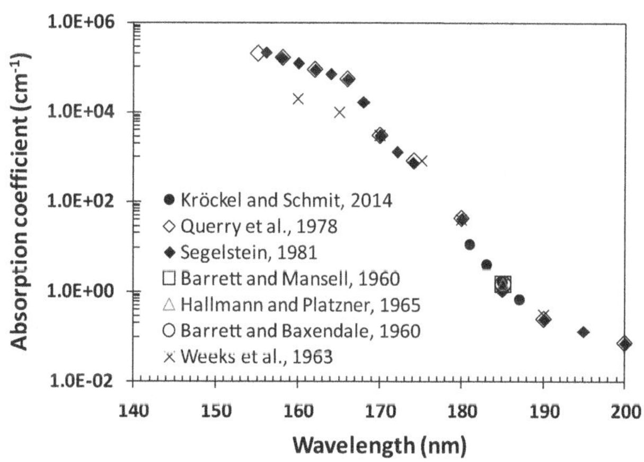

Figure 5.4 Absorption coefficients of pure water determined at 25°C.

The absorption coefficient of water at 184.9 nm was found to be 1.46–1.80 cm^{-1} (Barrett & Mansell, 1960; Halmann & Platzner, 1965; Kröckel & Schmidt, 2014; Weeks *et al.* 1963) at 25 °C. The latest value reported by Kröckel and Schmidt (2014) was 1.62 cm^{-1}. In this work, a linear dependence of the absorption coefficient at 187 nm on the temperature (10 to 30 °C) was observed (e.g., 0.45 cm^{-1} at 10 °C to 0.67 cm^{-1} at 30 °C). The absorption coefficient of water at 172 nm is 550 cm^{-1}. Consequently, the penetration depth of VUV radiation through the water layer is within a few millimeters for 184.9 nm, or a fraction of a millimeter for 172 nm (Heit & Braun, 1996, 1997). Absorption of the VUV radiation results in the homolysis and photochemical ionization of water molecules (5.1 and 5.2).

$$H_2O + h\nu(<190\,nm) \rightarrow H^\bullet + HO^\bullet \tag{5.1}$$

$$H_2O + h\nu(<200\,nm) \rightarrow [e^-, H_2O^+] + H_2O \rightarrow [e^-, H_2O^+] + (H_2O) \rightarrow e_{aq}^- + HO^\bullet + H_3O^+ \tag{5.2}$$

As shown in Figure 5.5, the quantum yield of water homolysis is wavelength-dependent.

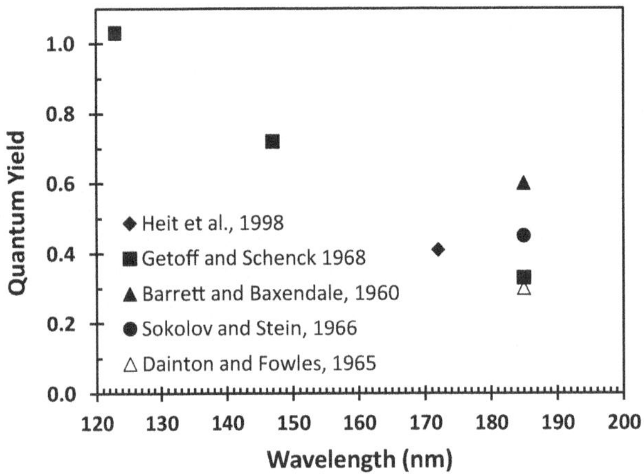

Figure 5.5 Quantum yield of pure water photolysis.

Table 5.1 compiles the literature data on water photolysis and the corresponding HO•, H atom (H•), and hydrated electron (e_{aq}^-) quantum yields at 172 and 184.9 nm. The threshold energy for the water homolysis is between 6.41 and 6.71 eV (Nikogosyan & Görner, 1992).

The VUV irradiation of water leads primarily to H• and HO•. Since e_{aq}^- concentrations are very low, in general, the contribution of e_{aq}^- reactions to overall mechanisms in VUV systems is rather small. $\Phi(e_{aq}^-)$ depends slightly on pH and it was observed to increase in alkaline solutions of pH > 9 due to the contribution of the reaction of H• with HO$^-$ (5.3) (Gonzalez *et al.* 2004).

$$HO^- + H^\bullet \rightarrow H_2O + e_{aq}^- \tag{5.3}$$

Inversely, $\Phi(e_{aq}^-)$ decreases drastically in solutions of pH < 4, as e_{aq}^- is scavenged by H_3O^+ (5.4) (Gonzalez *et al.* 2004).

$$H^+ + e_{aq}^- \rightarrow H^\bullet \qquad k = 2.2 \times 10^{10}\ \text{L mol}^{-1}\ \text{s}^{-1}\ \text{(Gonzalez et al. 2004)} \tag{5.4}$$

Table 5.1 Quantum yields of homolysis and ionization of water determined at 172 nm using $Xe_2{}^*$ lamp and 184.9 nm using LP lamp.

	Quantum Yield $H_2O + h\nu(<190\ nm) \rightarrow H^\bullet + HO^\bullet$	Quantum Yield $H_2O + h\nu(<200\ nm) \rightarrow e_{aq}{}^- + HO^\bullet + H_3O^+$	Reference
172 nm	0.42		Heit *et al.* (1998)
184.9 nm	0.33	0.045	Getoff and Schenck (1968)
	0.6		Barrett and Baxendale (1960)
	0.45		Sokolov and Stein (1966)
	0.3		Dainton and Fowles (1965)

Moreover, the reaction of $e_{aq}{}^-$ with $HO^\bullet$ results in a much more stable HO^-, increasing the water pH (5.5).

$$e_{aq}{}^- + HO^\bullet \rightarrow HO^- \quad k = 3 \times 10^{10}\ L\ mol^{-1}\ s^{-1} \text{ (Gonzalez et al. 2004)} \tag{5.5}$$

In strong alkaline solution (pH > 12.7), the reaction of the conjugated base of $HO^\bullet$ ($O^{\bullet-}$) with dissolved O_2 (DO) yields significant amounts of ozonide radical anion ($O_3{}^{\bullet-}$) (5.6 and 5.7), which is a source of $HO^\bullet$ (reactions 5.8 and 5.9) (Gonzalez *et al.* 2004; Hayon & McGarvey, 1967; Martire & Gonzalez, 2001).

$$HO^\bullet + HO^- \rightarrow H_2O + O^{\bullet-} \quad pK_a = 11.9 \text{ (Gonzalez et al. 2004)} \tag{5.6}$$

$$O^{\bullet-} + O_2 \rightarrow O_3{}^{\bullet-} \quad k = 3.5 \times 10^3\ L\ mol^{-1}\ s^{-1} \text{ (Gonzalez et al. 2004)} \tag{5.7}$$

$$O_3{}^{\bullet-} + H^+ \rightarrow H_2O + HO_3{}^\bullet \quad pK_a = 8.2 \text{ (Gonzalez et al. 2004)} \tag{5.8}$$

$$HO_3{}^\bullet \rightarrow O_2 + HO^\bullet \tag{5.9}$$

The primary radicals ($H^\bullet$ and $HO^\bullet$) are formed in a solvent cage (Thomsen *et al.* 1999), where the recombination is favored (László *et al.* 1998). That explains the quantum yield lower than unity. $H^\bullet$ and $HO^\bullet$ that escape from the "cage" are able to initiate a series of diffusion controlled reactions. Their self-recombination can result in H_2O_2, H_2 and H_2O (5.10–5.12).

$$2HO^\bullet \rightarrow H_2O_2 \quad k = 4 \times 10^9 - 2.0 \times 10^{10}\ L\ mol^{-1}\ s^{-1} \text{ (Buxton et al. 1988; Gonzalez et al. 2004)} \tag{5.10}$$

$$2H^\bullet \rightarrow H_2 \quad k = 1.0 \times 10^{10}\ L\ mol^{-1}\ s^{-1} \text{ (Buxton et al. 1988; Gonzalez et al. 2004)} \tag{5.11}$$

$$HO^\bullet + H^\bullet \rightarrow H_2O \quad k = 2.4 \times 10^9\ L\ mol^{-1}\ s^{-1} \text{ (Buxton et al. 1988; Gonzalez et al. 2004)} \tag{5.12}$$

The DO changes the radical reactions. The reaction of O_2 with $H^\bullet$ (5.14) reduces greatly the probability of recombination of $H^\bullet$ and $HO^\bullet$, and as a consequence, the concentration of $HO^\bullet$ increases while the concentration of $H^\bullet$ is drastically reduced. László and Dombi (2002) studied the oxidizing and reducing properties of the $H^\bullet$ and $HO^\bullet$ in aqueous solutions of $[Fe(CN)_6]^{4-}$ and $[Fe(CN)_6]^{3-}$ exposed to 172 nm radiation. In O_2-free solutions, there was no significant difference between the rate of oxidation of $[Fe(CN)_6]^{4-}$ and reduction of $[Fe(CN)_6]^{3-}$. However, in the presence of DO, the rate of oxidation of $[Fe(CN)_6]^{4-}$ increased considerably and the rate of $[Fe(CN)_6]^{3-}$ reduction was significantly suppressed, probably because O_2 scavenges e_{aq}^- and $H^\bullet$ to yield $O_2^{\bullet-}$ and its conjugated acid $HO_2^\bullet$, respectively (5.13–5.15).

$$O_2 + e_{aq}^- \rightarrow O_2^{\bullet-} \qquad k = 2\times10^{10}\ \mathrm{L\ mol^{-1}\ s^{-1}}\ \text{(Buxton } et\ al. \text{ 1988; Gonzalez } et\ al. \text{ 2004)} \tag{5.13}$$

$$H^\bullet + O_2 \rightarrow HO_2^\bullet \qquad k = 2.1\times10^{10}\ \mathrm{L\ mol^{-1}\ s^{-1}}\ \text{(Buxton } et\ al. \text{ 1988)} \tag{5.14}$$

$$HO_2^\bullet + H_2O \rightleftharpoons H_3O^+ + O_2^{\bullet-} \qquad pK_a = 4.8\ \text{(Bielski } et\ al. \text{ 1985)} \tag{5.15}$$

In the presence of DO, H_2O_2 is mostly formed through the disproportionation reaction of $HO_2^\bullet$/$O_2^{\bullet-}$ (5.16).

$$O_2^{\bullet-} + HO_2^\bullet + H_2O \rightarrow O_2 + H_2O_2 + HO^- \qquad k = 9.7\times10^{7}\ \mathrm{L\ mol^{-1}\ s^{-1}}\ \text{(Buxton } et\ al. \text{ 1988)} \tag{5.16}$$

The rate constant of reaction 5.16 is pH-dependent (Bielski *et al.* 1985), with a maximum value at $pH \approx pK_a$.

During the VUV photolysis of O_2-free pure water, negligible amount of H_2O_2 is formed. In the presence of DO, the H_2O_2 concentration increases, mostly through the disproportionation reaction 5.16. Organic compounds, such as oxalic acid, aromatic compounds, formic acid and methanol (Arany *et al.* 2012; Azrague *et al.* 2005; Imoberdorf & Mohseni, 2011; Robl *et al.* 2012) substantially increase the concentration of H_2O_2 in the presence of DO. The results reported by Azrague *et al.* (2005) show a steep increase of the concentration of H_2O_2, up to a maximum corresponding to approx. 25% conversion of oxalic acid. With diminishing the substrate concentration, the concentration of H_2O_2 decreased to the concentration determined in pure water. Imoberdorf and Mohseni (2011) used formic acid as the model compound, and showed that H_2O_2 concentration increased until approx. 75% conversion of the organic substrate was reached. Robl *et al.* (2012) confirmed the enhanced production of H_2O_2 in aqueous solutions of methanol. During the oxidative transformation of organic pollutants, the formation and accumulation of $HO_2^\bullet$ is possible through the reactions of organic peroxyl radicals. Consequently, the increase of H_2O_2 concentration observed in these studies (Alapi & Dombi, 2007; Azrague *et al.* 2005; Robl *et al.* 2012) is explained by the enhanced concentration of the $HO_2^\bullet/O_2^{\bullet-}$ formed during the oxidative degradation mechanisms of the organic compounds.

5.1.2.2 Heterogeneity of the VUV-irradiated aqueous solutions

The inhomogeneity of VUV-irradiated systems is mainly due to the low penetration depth of the VUV radiation. Water homolysis with the formation of primary reactive species ($HO^\bullet$, $H^\bullet$, e_{aq}^-) takes place within a thin liquid layer. The short-lived (<10 μs, (Hoigné, 1998)) primary radicals cannot diffuse far outside the irradiated volume, thus, all their bimolecular reactions take place within this photoreaction zone (Figure 5.6).

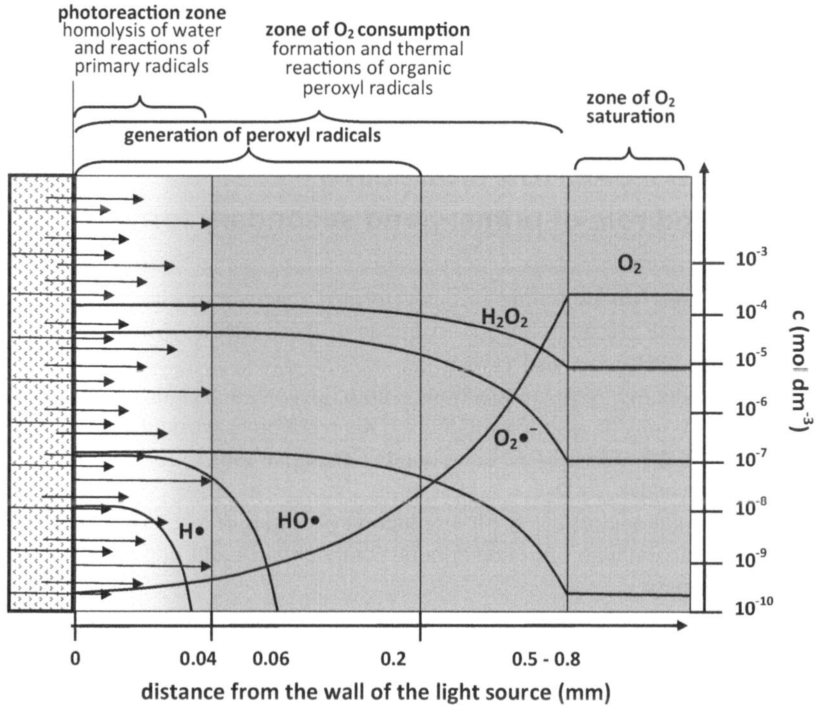

Figure 5.6 Heterogeneity of the VUV-irradiated aqueous solutions (the figure is based on data and figures presented by Heit and Braun (1996, 1997), László (2001), and Oppenländer (2003)).

In a solution containing DO, $H^{\bullet}$ and e_{aq}^{-} are trapped by DO, consequently the transformation of the organic substances is initiated mainly by the reactions with $HO^{\bullet}$. The formed carbon-centered radicals are also trapped by DO, which results in DO depletion in the photoreaction zone. The DO depletion was verified in a well-designed experiment by Heit and Braun (1996) where the DO concentration profile was determined as a function of the radial distance from the Xe_2^* lamp surface. The results confirmed the existence of a thin (0.036 mm) photoreaction zone, which is characterized by diffusion controlled reactions of the short-lived primary radicals. The addition of DO to the $H^{\bullet}$ and carbon-centered radicals creates the zone of O_2 consumption, which is outside the irradiated zone. The degree of DO depletion depended on the concentration of the organic substrate and photon flux, and slightly depended on the flow rate of aqueous solution. Within the zone of O_2 consumption the reactions of peroxyl radicals dominate. In contrast to the short lived primary radicals, $HO_2^{\bullet}/O_2^{\bullet -}$ and organic peroxyl radicals have longer lifetimes, which allows their diffusion into the "dark" zone, up to a maximum distance of about 0.5 mm from the surface of the Xe_2^* lamp.

The heterogeneity of VUV-irradiated solutions can be mitigated by using high flow rates which generate radial mixing and turbulent flow conditions (Dobrovic *et al.* 2007) enhancing the mass transfer to the photoreaction zone, thus increasing the oxidation and mineralization rates of contaminants. The combination of VUV photolysis and electrolysis with DO generation at the anode close to the irradiated zone enhanced the rate of oxidation and mineralization of benzoic acid (Wörner *et al.* 2003). Tasaki *et al.* (2009) investigated the effect of O_2 microbubbles both in UV (253.7 nm) and VUV/UV (184.9/253.7 nm)-irradiated solutions of methyl orange. The rate of mineralization is controlled by the diffusion of DO from the non-irradiated bulk to the photoreaction zone. The use of ceramic gassing units or the injection of O_2 or air through a porous glass plate prevented the DO depletion in the VUV-irradiated zone (Han *et al.* 2004; Oppenländer *et al.* 2005).

5.2 KINETICS AND REACTION MODELING

5.2.1 Reactions and role of primary and secondary formed reactive species

In VUV-irradiated aqueous solutions, the primary formed reactive species are $HO^\bullet$, $H^\bullet$ and e_{aq}^-. An excellent compilation of rate constants for the reactions of these species with a wide range of chemical compounds can be found in Buxton *et al.* (1988).

The e_{aq}^- is a strong reducing agent. In reactions with halogenated organic compounds, the e_{aq}^- acts as a nucleophile, with halide ions as reaction products. This reaction is particularly relevant to the removal of perhalogenated saturated hydrocarbons, compounds unreactive toward the $HO^\bullet$ (Gonzalez *et al.* 2004; Oppenländer & Schwarzwälder, 2002).

The $H^\bullet$ is the conjugated acid of e_{aq}^- with a reduction potential of –2.3 V (Buxton *et al.* 1988). This species reacts by H-abstraction with saturated organic compounds, or addition to double bonds in reactions with unsaturated compounds.

The $HO^\bullet$ is by far the most important reactive species in the VUV-irradiated aqueous solutions, particularly in the presence of DO. The rate constants of $HO^\bullet$ reactions with organic and inorganic compounds extend over several orders of magnitude, most of which approach the diffusion controlled limits ($10^9 - 10^{10}$ L mol^{-1} s^{-1}).

Despite the fact that $HO^\bullet$ is considered a highly reactive and unselective oxidant, one observes a significant selectivity in the H-abstraction reactions with the structural properties of the substrate. Hydroxyl radical reacts with unsaturated compounds by electrophilic addition to π-electron-rich moieties, such as the unsaturated double or triple bonds. $HO^\bullet$ is also involved in e^- transfer reactions, typically with the inorganic compounds. The most common reaction of the aromatic structures with $HO^\bullet$ is the addition to the aromatic ring, while e^- transfer and H-abstraction from the side chain generally occur less frequently. In general, the rate constants for the $HO^\bullet$ addition reactions are larger than those for H-abstraction. $HO^\bullet$ is a strong electrophile and the addition occurs preferentially to the negatively polarized sites. Most of $HO^\bullet$ reactions with the organic compounds result in carbon-centered radicals.

In DO-free solutions the carbon-centered radicals are involved in a series of reactions including combination and disproportionation reactions. In the presence of DO, the carbon-centered radicals are converted to peroxyl radicals *via* diffusion-controlled reactions (von Sonntag & Schuchmann, 1991). Detailed information on the formation and fate of peroxyl radicals in aqueous solutions can be found in von Sonntag and Schuchmann (1997).

The reactivity of $HO_2^\bullet$ and $O_2^{\bullet-}$ is much lower than that of $HO^\bullet$ and $H^\bullet$ (Bielski *et al.* 1985); the reaction rate constants for $HO_2^\bullet$ and $O_2^{\bullet-}$ radicals with phenol were reported as 2.7×10^3 L mol^{-1} s^{-1} (Kozmér *et al.* 2014) and 5.8×10^2 L mol^{-1} s^{-1} (Tsujimoto *et al.* 1993), respectively. However, if present at high concentrations, these species may contribute to the degradation of organic contaminants (Alapi & Dombi, 2007; Arany *et al.* 2015; Kozmér *et al.* 2014). One typical reaction of $HO_2^\bullet$ and $O_2^{\bullet-}$ is the addition to aromatic rings (Getoff, 1996). The $O_2^{\bullet-}$ can also participate in e^- transfer reactions, which can be very fast (e.g., reduction of quinones). During the mineralization process of target compounds, the organic peroxyl radical reactions (e.g., Russell mechanism (von Sonntag & Schuchmann, 1997)) yield $HO_2^\bullet$ and $O_2^{\bullet-}$. Consequently, the concentration of $HO_2^\bullet$ and $O_2^{\bullet-}$ is highly enhanced in VUV-irradiated DO-containing solutions of organic compounds.

In 172 nm- or 184.9/253.7 nm -irradiated systems, DO could have either an accelerating or an inhibitory effect on chemical degradation rates. Studies have reported increased transformation rates (Gonzalez & Braun, 1996; Han *et al.* 2004; Quici *et al.* 2008; Szabó *et al.* 2011), and efficient mineralization of the organic carbon (Arany *et al.* 2015; Gonzalez & Braun, 1996; Han *et al.* 2004; Oppenländer *et al.* 2005) in the presence of DO. Gonzales *et al.*'s studies (1994, 1995) showed that the photodegradation rates (172 nm) of atrazine and 3-amino-5-methylisoxazole did not depend on DO concentration, and higher mineralization yields were obtained in O_2-free than in oxygenated conditions.

5.2.2 Kinetics and mechanistic modeling of VUV AOP

The VUV transformation of contaminants is initiated by second-order kinetics reactions with the radicals generated during the VUV photolysis of water (5.1 and 5.2). At a constant photon flux, the reactive radicals approach *quasi*-steady state concentrations; consequently, the degradation rate is usually described by a *pseudo*-first-order kinetics.

The modeling of the VUV process is complicated mainly because of the heterogeneity of the medium caused by the short penetration depth of the VUV radiation in water (see in Section 5.1.2.2). Furthermore, some of the species are formed *in situ* as a result of light-dependent reactions, while others are formed by subsequent light-independent reactions. Moreover, the reactivity and lifetime of the reactive species are very different.

Gonzalez and Braun (1995) developed a kinetic model that considers 12 different species and a total of 28 elementary reactions. Later, Gonzalez *et al.* (2004) used methanol as a model compound and investigated its oxidation mechanism taking into account 27 different species and 54 elementary reactions.

László (2001) modeled the concentrations of radicals and formation of H_2O_2 in 172 nm irradiated water. The concentration profiles of $H^{\bullet}$, $HO^{\bullet}$, $HO_2^{\bullet}$, $O_2^{\bullet-}$ and H_2O_2, as a function of the radial distance from the Xe_2^* lamp surface were computed, and were found in agreement with those reported by Heit and Braun (1997). $H^{\bullet}$ and $HO^{\bullet}$ were found to be present only within the 0.03 mm and 0.07 mm water layer, respectively, while the concentrations of the $HO_2^{\bullet}$, $O_2^{\bullet-}$ and H_2O_2 reached steady-state conditions at 0.5 mm from the lamp surface. The author demonstrated a good agreement between the model predictions and experimental results on H_2O_2 formation in a mixed batch reactor.

Zvereva (2010) modeled the concentrations of the products of liquid water decomposition under 172 nm VUV radiation. The author considered a continuous irradiation, and modeled the kinetics over 10^{-2} s time intervals; the diffusion of primary formed reactive species was neglected. The concentrations of $H^{\bullet}$, $HO^{\bullet}$, and e_{aq}^- were calculated based on the quantum yields of their formation and local (spatial) light intensity. The model was used to estimate the rate of decomposition of polychlorinated biphenyls through the reaction with $HO^{\bullet}$.

Imoberdorf and Mohseni (2011) examined the VUV AOP for environmental contaminant destruction in laboratory-scale reactors using chemical probes. Formic acid was selected as a simple molecule to develop and to validate a kinetic model for the VUV (184.9 nm)/UV (253.7 nm) AOP in a single lamp annular flow-through reactor operated in a batch mode. The kinetic model included the water photochemistry, photolysis of formic acid and of H_2O_2 which was formed during the VUV irradiation, and the radical reactions occurring in the solution. The kinetic model was combined with the radiation model solved using the Monte Carlo method, which allowed the calculation of local (spatial) rate of photon absorption based on the photon flow propagation through the reactor volume and optical characteristics of the solution. The experimental and modeling data showed zero-order kinetics for VUV-radiation driven degradation of formic acid, which indicated limited availability of $HO^{\bullet}$ in the reaction system. The model was extended to H_2O_2/VUV and H_2O_2/UV AOPs, and good agreement was observed between the experimental and simulated data (Imoberdorf & Mohseni, 2011).

In another study, Imoberdorf and Mohseni (2012) reported the kinetics of VUV/UV degradation of pesticide 2,4-D (2,4-dichlorophenoxyacetic acid) in ultrapure water and in raw surface water samples collected from three different sites in British Columbia, Canada. The model equations, mass balances, and radial radiation profiles for the 184.9 nm and 253.7 nm radiation are provided, along with the experimental patterns of 2,4-D and of the identified degradation by-products. In a study dedicated to the degradation of natural organic matter (NOM), Imoberdorf and Mohseni (2011) determined an overall quantum efficiency of 0.10 for NOM degradation by VUV photolysis at 50% TOC reduction.

Recently, Bagheri and Mohseni (2015b) developed and validated experimentally a comprehensive computational fluid dynamics (CFD) model, incorporating flow hydrodynamics, 184.9 nm and 253.7 nm radiation propagation, and a comprehensive kinetic scheme. The authors monitored the phototransformation of atrazine and found that the VUV/UV process performance depends strongly on the flow characteristics inside the photoreactor. Similar observations were reported in a study on the impact of turbulence and mixing on the performance of VUV/UV-AOPs (Bagheri & Mohseni, 2015a). *para*-Chlorobenzoic acid was used as a probe compound, and the electrical energy-per-order (E_{EO}) as the process performance metric. Baffles were inserted in order to enhance the mixing inside the UV reactor. The treatment efficacy of the VUV/UV process displays much stronger correlation with the extent of mixing than in the case of UV/H_2O_2 process. The enhanced mixing and vortexes ("circulation zones"), due to the presence of baffles, resulted in up to 50% reduction in the energy cost associated with the VUV/UV treatment, whereas no significant impact on the E_{EO} was observed for the UV/H_2O_2 (5 mg L^{-1}) process.

A novel mechanistic model that describes the VUV photolysis in an annular photoreactor with either 172 nm or 184.9 nm (in combination with 253.7 nm, with and without added H_2O_2) was published by Crapulli *et al.* (2014). The study showed that, depending on the reactor characteristics and operating conditions, the reaction zone during the 172 nm-VUV process could be more than one order of magnitude deeper than the photon penetration layer. The model confirmed that short-lived $HO^•$ were present at a radial distance far beyond the radiation penetration depth. The kinetic simulations showed that the presence of $HO^•$ at unexpected long radial distances from the lamp surface is the effect of non-linear behavior of the complex reaction kinetics.

5.3 VACUUM UV RADIATION FOR WATER REMEDIATION

5.3.1 VUV for removal of specific compounds

Over the past two decades, the VUV treatment of contaminated water has been investigated extensively. These studies regarded transformation yields of specific compounds, mineralization of chemical pollutants, by-product formation, and prediction of mechanistic pathways.

5.3.1.1 Aliphatic and chlorinated volatile organic compounds

In DO-containing aqueous solutions, the transformation of non-halogenated aliphatic compounds is initiated primarily by $HO^•$ *via* H-abstraction or addition to unsaturated bonds. The VUV (172 nm) and VUV (184.9 nm)/UV (253.7 nm) treatment of methanol in water was investigated in detail by Heit *et al.* (1998) and Gonzalez *et al.* (2004) (Figure 5.7). The attack of $HO^•$ ($k_{methanol,\ HO^•} = 9.7 \times 10^8$ L mol^{-1} s^{-1} (Heit *et al.* 1998)) or $H^•$ ($k_{methanol,\ H^•} = 2.6 \times 10^6$ L mol^{-1} s^{-1} (Heit *et al.* 1998)) leads almost solely to the formation of hydroxymethyl radicals ($HOCH_2^•$) (Gonzalez *et al.* 2004). The reaction mechanism involving these radicals is determined by the presence or the absence of O_2.

Oppenländer and Gliese (2000) investigated and compared the rates of the 172 nm radiation-initiated transformation and mineralization of twenty organic micropollutants of various structures, including C_1 to C_8 alcohols; the initial TOC content (40–50 ppm) was similar for all compounds. The mineralization rates of aromatic C_6 compounds were faster than that for the C_6 saturated alcohol, and Cl-substitution on the aromatic ring had an activating effect. The efficiency of TOC removal decreased with the increase of the number of C atoms in the alcohol series.

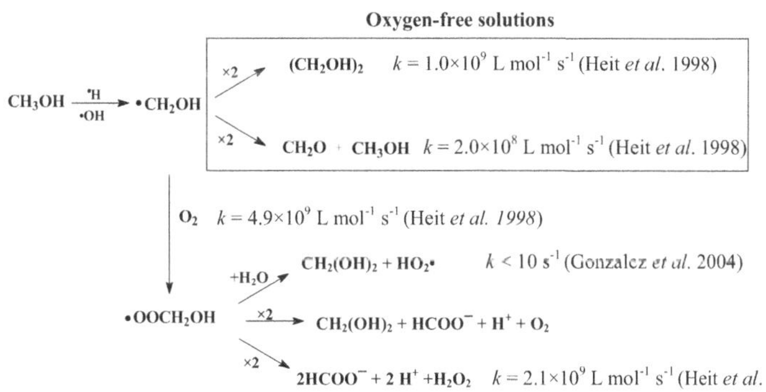

(5.17)
(5.18)
(5.19)
(5.20)
(5.21)
(5.22)

Figure 5.7 Simplified VUV-induced degradation mechanism of methanol in aqueous solution. Rate constants taken from Heit *et al.* (1998) and Gonzalez *et al.* (2004).

Chlorinated methanes, such as CCl_4 ($k_{CCl_4,{}^\bullet OH} < 10^5$ L mol^{-1} s^{-1}) and $CHCl_3$ react slowly with HO$^\bullet$ and H$^\bullet$, whereas their reaction with e_{aq}^- is almost diffusion controlled. The degradation of highly halogenated compounds is slowed down in the presence of O_2 (see reaction 5.13).

$$CCl_4 + e_{aq}^- \rightarrow {}^\bullet CCl_3 + Cl^- \quad k = 3\times10^{10}\ \text{L mol}^{-1}\ \text{s}^{-1}\ \text{(Gonzalez \textit{et al.} 2004)} \tag{5.23}$$

$$CHCl_3 + HO^\bullet \rightarrow {}^\bullet CCl_3 + H_2O \quad k = 5\times10^{7}\ \text{L mol}^{-1}\ \text{s}^{-1}\ \text{(Gonzalez \textit{et al.} 2004)} \tag{5.24}$$

$$CHCl_3 + H^\bullet \rightarrow {}^\bullet CCl_3 + H_2 \quad k = 1.1\times10^{7}\ \text{L mol}^{-1}\ \text{s}^{-1}\ \text{(Gonzalez \textit{et al.} 2004)} \tag{5.25}$$

$$CHCl_3 + e_{aq}^- \rightarrow {}^\bullet CHCl_2 + Cl^- \quad k = 3\times10^{10}\ \text{L mol}^{-1}\ \text{s}^{-1}\ \text{(Gonzalez \textit{et al.} 2004)} \tag{5.26}$$

More than 93% of the organic chlorine from $CHCl_3$ is released as Cl^- upon 172 nm exposure in the presence of DO, indicating that the ${}^\bullet CCl_3$ reactions lead to the formation of HCl and CO_2 (Gonzalez *et al.* 2004).

Trichloroethene (TCE), tetrachloroethene (PCE), and 1,2-dichloroethene (DCE) are volatile organic compounds (VOCs), which, along with their metabolites such as vinyl chloride or trichloroacetic acid are toxic and regulated in drinking water (Weissflog *et al.* 2004). Groundwater contaminated with TCE, DCE and PCE can be treated with the 172 nm radiation (Baum & Oppenländer, 1995). The addition of HO$^\bullet$, H$^\bullet$ and e_{aq}^- to the unsaturated bonds initiates the degradation process (5.27–5.29).

$$Cl_2C{=}CCl_2 + HO^\bullet \rightarrow Cl_2(OH)C{-}C^\bullet Cl_2 \quad k = 4.9\times10^{8}\ \text{L mol}^{-1}\ \text{s}^{-1}\ \text{(Getoff, 1990)} \tag{5.27}$$

$$Cl_2C{=}CCl_2 + H^\bullet \rightarrow Cl_2HC{-}C^\bullet Cl_2 \quad k = 5\times10^{9}\ \text{L mol}^{-1}\ \text{s}^{-1}\ \text{(Getoff, 1990)} \tag{5.28}$$

$$Cl_2C{=}CCl_2 + e_{aq}^- \rightarrow Cl_2C_5C^\bullet Cl + Cl^- \quad k = 4.2\times10^{10}\ \text{L mol}^{-1}\ \text{s}^{-1}\ \text{(Getoff, 1990)} \tag{5.29}$$

The carbon-centered radical $Cl_2(OH)C{-}C^{\bullet}Cl_2$ reacts with DO at diffusion-controlled rates, and undergoes further decomposition to aldehydes and carboxylic acids. 1,1,2-trichloroethane (1,1,2-TCA) was found as a by-product of the DCE degradation (Baum & Oppenländer, 1995; Gonzalez *et al.* 2004). Direct photolysis of PCE, TCE, DCE, 1,1,1-TCA, 1,1,2-TCA, $CHCl_3$ and CCl_4 in VUV (184.9 nm)/UV (253.7 nm)-irradiated aqueous solution was discussed by Shirayama *et al.* (2001). The degradation rates of these chlorinated hydrocarbons increased in the absence of DO. The photolysis at 253.7 nm was not efficient in the case of chlorinated methanes, whereas TCE and PCE were easily degraded. The absorption of 184.9 nm radiation initiated the transformation of chlorinated ethanes and methanes *via* dissociation of C–Cl bonds. The negative impact of DO was attributed to the DO competition for the 184.9 nm photons with the chlorinated compounds (Shirayama *et al.* 2001). Wang *et al.* (2014) stated that the competition between DO and organics for VUV photons is not significant, and that the radiation scattering by the O_2 bubbles is negligible in the reactor. Most likely, the negative impact of DO on the degradation rates of the chlorinated compounds is related to the e_{aq}^- scavenging by DO. Gu *et al.* (2013) did not detect any organic intermediates during the VUV/UV treatment of TCE, PCE, and TCA aqueous solutions.

5.3.1.2 Perfluorinated organic compounds

The perfluorinated compounds (PFCs) are persistent organo-fluorine environmental contaminants given their resistance toward biodegradation and oxidation. Perfluorooctanoic acid (PFOA), perfluorodecanoic acid (PFDeA) and perfluorooctanesulfonic acid (PFOS) were quantified in human and marine biota plasma and liver tissue, in water sources, sediments, domestic sludge, and even in the remote Arctic region. Data from animal studies indicate that PFOA can cause several types of tumors and neonatal death, and may affect the immune and endocrine systems (Steenland *et al.* 2010). Because of their high stability and virtually no reactivity towards $HO^{\bullet}$, PFC elimination from water sources with AOPs is not possible.

Studies concerning VUV (172 nm) and VUV (184.9 nm)/UV (253.7 nm) treatment of perfluorinated compounds suggest that the reaction with e_{aq}^- (reduction process) and, to some extent, VUV photolysis represent the mechanistic pathways for the degradation of these compounds. PFOA, PFDeA and PFOS can be degraded by the 184.9 nm radiation, but cannot be removed by the 253.7 nm radiation (Cao *et al.* 2010; Chen *et al.* 2007; Giri *et al.* 2011a; Giri *et al.* 2012; Jin & Zhang, 2015; Wang *et al.* 2010a; Wang & Zhang, 2014). Chen *et al.* (2007) hypothesized that the direct VUV photolysis of PFCs results in photo-Kolbe decarboxylation. Hydrolysis of generated radical ($C_7F_{15}^{\bullet}$) occurs with formation of a shorter chain PFC, i.e., prefluoroheptanoic acid (Chen *et al.* 2007; Wang & Zhang, 2014). Shorter-chain perfluorocarboxylic acids are formed consecutively with a stepwise loss of CF_2 units (Chen *et al.* 2007).

Wang and Zhang (2014) studied the effect of pH and DO on PFOA decomposition under 184.9/253.7 nm irradiation. In the presence of DO, the efficiency of PFOA degradation decreased as the pH was increased from 4.5 to 12.0. In DO-free solutions, the increase of pH enhanced the degradation rate. The authors concluded that in the absence of DO, in acidic and neutral solutions, PFOA degrades primarily *via* direct VUV photolysis, whereas in alkaline conditions the reaction with e_{aq}^- is the major degradation pathway. The role of e_{aq}^--initiated transformation of PFCs in DO-free slightly alkaline solutions was also postulated by Jin and Zhang (2015). Other studies (Cao *et al.* 2010; Wang *et al.* 2010a) confirmed the role of e_{aq}^- in PFC degradation in the VUV process. Detailed investigation of the PFOA decomposition by the VUV (184.9 nm)/UV (253.7 nm) process was reported in a series of publications by (Giri *et al.* 2011a; Giri *et al.* 2011b; Giri *et al.* 2012). The impact of pH, initial concentration, and water quality (Giri *et al.* 2012) and the matrix effect (Giri *et al.* 2011b) on PFOA removal are discussed. Defluorination efficiency in river water and wastewater treatment plant effluent was negligible, due to the strong e_{aq}^- scavenging capacity of water constituents other than PFOA.

5.3.1.3 Aromatic compounds

Phenol is often used as a model compound for the investigation of the efficiency of various AOPs. The VUV degradation of phenol is initiated by $HO^{\bullet}$ and $H^{\bullet}$ reactions at diffusion controlled rates ($k_{phenol,\ HO^{\bullet}} = 8.4 \times 10^9$ L mol^{-1} s^{-1} (Bonin *et al.* 2007), $k_{phenol,\ H^{\bullet}} = 1.7 \times 10^9$ L mol^{-1} s^{-1} (Buxton *et al.* 1988)) yielding dihydroxy-cyclohexadienyl (DHCD$^{\bullet}$) and hydroxyl-cyclohexadienyl (HCD$^{\bullet}$) radicals, respectively. Dimerization and dismutation reactions of these radicals result in bicyclohexadienes, dihydroxybenzenes/cyclohexadiens and phenol (Gonzalez *et al.* 2004). Further transformations of the aromatic radicals can result ring-opening products, and finally CO_2 and H_2O (Gonzalez *et al.* 2004). Huang *et al.* (2013) studied the transformation of 4-*tert*-octylphenol by the 184.9/253.7 nm radiation. The degradation occurs *via* both direct photolysis (C—C bond cleavage) yielding phenol, and $HO^{\bullet}$-based oxidation with the formation of 4-*tert*-octylcatechol.

Chlorophenols represent another environmentally relevant group of pollutants. 2,4-Dichlorophenol is an industrial intermediate product of chlorophenoxy herbicides (Wild *et al.* 1993). Efficient photomineralization of chlorophenols was achieved with a Xe_2^* lamp, converting organically bound Cl atoms to Cl^- (Baum & Oppenländer, 1995; Jakob *et al.* 1993). The Cl-substitution on the aromatic ring exerts an activating effect towards $HO^{\bullet}$ addition, accelerating the rate of mineralization, which was found to be lower of 4-chlorophenol and 2,4-dichlorophenol than that of phenol (Oppenländer & Gliese, 2000).

The treatment of 4-chloro-3,5-dinitrobenzoic acid (CDNBA) with 172 nm or 184.9/253.7 nm radiation released Cl^- and NO_2^- at a rate similar to that of substrate depletion. NO_2^- was efficiently oxidized to NO_3^-, which was further reduced to NH_4^+ in the presence of organic matter *via* $H^{\bullet}/e_{aq}^-$ initiated reactions (Lopez *et al.* 2000). No aromatic intermediates were detected, which led to the conclusion that the ring opening followed the reactions of carbon-centered radicals with DO. The rates of transformation and mineralization of CDNBA were slightly slower under the VUV/UV process using a LP lamp than using the 172 nm VUV treatment, most probably due to the difference in the $HO^{\bullet}$ yield at the two wavelengths (Lopez *et al.* 2000).

5.3.1.4 Pesticides

The stability and mobility of various pesticides result in the pollution of soil, surface waters and groundwater, which induce long-term effects on living organisms. Many pesticides have been proven endocrine disruptors in fish, amphibians, birds, reptiles, laboratory rats and even humans. Over twenty pesticides are regulated in drinking water by the US EPA with maximum contaminant levels (MCLs) anywhere from 0.2 μg L^{-1} (lindane) to 0.5 mg L^{-1} (picrolam), whereas under the European Union regulations, the total concentration of pesticides and their metabolites in drinking water should not exceed 0.5 μg L^{-1}, while the concentration of any one pesticide may not exceed 0.1 μg L^{-1}.

Atrazine (2-chloro-4-ethylamino-6-isopropylamino-*s*-triazine) is one of the most widely used *s*-triazine herbicides. Pulse radiolysis studies indicated that in DO-free solutions, atrazine degradation occurred mostly *via* reaction with e_{aq}^-, with other mechanistic pathways involving $HO^{\bullet}$ and $H^{\bullet}$ (Khan *et al.* 2015). Gonzales *et al.* (1994) examined the mineralization of atrazine in aqueous solutions exposed to the 172 nm radiation, and observed two different stages in TOC decay. During the first stage, TOC was removed rapidly and the kinetics was independent on the oxygen level (same rates in argon-, air-, and O_2-saturated solutions). Approximately two-thirds into the mineralization process, the TOC removal reached a plateau under all studied conditions indicating that non-degradable products were formed; the plateau was reached faster in the presence of oxygen than in argon-saturated solutions. Atrazine was degraded to various products, among which the stable, non-degradable cyanuric acid. The conversion yields of atrazine to cyanuric acid were found oxygen dependent, and reported as 10%, 30%, and 50%, in argon-, air-, and O_2-saturated conditions, respectively. This study showed that mineralization of atrazine to CO_2 and inorganic nitrogen-products *via* 172 nm radiation does not necessarily require the presence of DO and other mechanistic routes involving reducing species are involved.

A similar effect was observed in the 172 nm initiated degradation of 3-amino-5-methylisoxazole (Gonzalez *et al.* 1995). Cyanide was a degradation product of 3-amino-5-methylisoxazole and it was completely removed *via* $HO^•$ reaction. Using a theoretical approach, Liu (2014) confirmed the resistance to oxidative treatment of cyanuric acid, which was attributed to the electron deficiency of the triazine ring. Dechlorination and hydroxylation were reported as the main reactions involving the *s*-triazine ring, and the degradation intermediates were assessed as less toxic than the parent pollutants (Gonzalez *et al.* 1994; Khan *et al.* 2014; Khan *et al.* 2015).

2,4-Dichlorophenol and H_2O_2 were detected as by-products of VUV degradation of herbicide 2,4-dichlorophenoxyacetic acid (Imoberdorf & Mohseni, 2012). Moussavi *et al.* (2014) evaluated the performance of UV (253.7 nm) and VUV (184.9 nm)/UV (253.7 nm) processes for the degradation of diazinon. At the optimum pH (5 for UV, 9 for VUV/UV), the transformation rate was two orders of magnitude higher in the presence of the 184.9 nm radiation.

5.3.1.5 Pharmaceuticals

Pharmaceuticals and personal care products (PPCPs) belong to the group of emerging contaminants detected in surface waters, groundwater and even in drinking waters, and were proven to cause adverse effects on the aquatic species, and could impact human health.

Kim and Tanaka (2009) studied the 253.7 nm and 184.9/253.7 nm degradation of 30 PPCPs commonly found in surface water. At a dose of ~230 mJ cm^{-2} approx. 3% (theophylline) to 100% (diclofenac), and ~15% (clarithromycin) to 100% (diclofenac) removals were observed under UV and VUV/UV irradiation, respectively. The PPCP degradation rates were on average 1.4 times faster for the VUV/UV than UV process. It was concluded that some PPCPs containing amide bonds (e.g., cyclophosphamide) were highly resistant to photodegradation at 253.7 nm, whereas at 184.9 nm the removal rates increased considerably. The results proved that the 184.9 nm radiation enhances the oxidative transformation, but the contribution of the $HO^•$-initiated degradation to the overall rate depends on the initial concentrations of PPCPs, and their molar absorption coefficients and quantum yields at 253.7 nm (Szabó *et al.* 2011). Direct UV photolysis was found to dominate during the VUV/UV decomposition of ketoprofen and diclofenac, whereas 184.9 nm radiation accelerated the transformation rate of ibuprofen. DO enhanced the rate of decay of ibuprofen, but had no effect on ketoprofen removal yield, in either UV or VUV/UV-irradiated solutions (Szabó *et al.* 2011). However, DO enhanced the rate of mineralization. Arany *et al.* (2013) have reported that DO has no significant effect on the transformation rate of naproxen at 253.7 nm, while it decreases the removal efficiency for the 172 nm or 184.9 nm/253.7 nm processes. Only the 172 nm irradiation was found to be able to completely mineralize naproxen. Degradation of ibuprofen, ketoprofen, naproxen, and diclofenac under VUV (172 nm) exposure is initiated by H-abstraction, $HO^•/H^•$-addition and decarboxylation reactions. DO enhances significantly the TOC removal rate for each pollutant (Arany *et al.* 2015).

5.3.1.6 Other water contaminants

Alkylphenols are endocrine disrupting chemicals often detected in surface waters in the concentration range of 1.0×10^{-3}–1.0 µg L^{-1}. The transformation rate of 4-*tert*-octylphenol (presumably the most powerful estrogenic compound among alkylphenols (Routledge & Sumpter, 1997)) was much larger upon exposure to 184.9 nm/253.7 nm combined radiation than that to 253.7 nm radiation alone. The C—C bond cleavage *via* absorption at 184.9 nm was proposed to occur, although direct photolysis at 184.9 nm is likely of a minor importance relative to the $HO^•$-initiated reactions (Huang *et al.* 2013).

Taste and odor-causing compounds (e.g., geosmin (*trans*-1,10-dimethyl-*trans*-9-decalol) and 2-methylisoborneol (MIB)), although non-toxic, impact the aesthetics of drinking water. Currently, there are several full-scale installations in the world using the UV/H_2O_2 process to treat these compounds in drinking water sources impacted by the seasonal algal blooming. The combination of 184.9 nm and 253.7 nm radiation (with or without simultaneous generation of O_3) led to efficient removal of these compounds (Kutschera *et al.* 2009; Zoschke *et al.* 2012).

1,4-Dioxane is a persistent, potentially carcinogenic chemical, widely used as a stabilizing agent and solvent, and is reported as an impurity of surfactants in many household detergents. 1,4-Dioxane, geosmin and MIB were efficiently degraded using VUV/UV treatment, photocatalysis, and the combination of these processes. The degradation rates followed the order geosmin > MIB > 1,4-dioxane, which is consistent with their reactivity toward the $HO^•$.

Anatoxin-a is a potent neurotoxin produced by freshwater cyanobacteria during the algal blooming seasons. The degradation rate of anatoxin-a at 172 nm was significantly slower in natural water containing 3.94 mg L^{-1} dissolved organic carbon (DOC) and in synthetic water with 6.63 mg L^{-1} DOC than in pure water, given the strong competition for the $HO^•$ from the water matrix components (Afzal *et al.* 2010).

The VUV/UV radiation can transform non-biodegradable textile dyes into biodegradable oxidized intermediates, thus enhancing their elimination in a VUV-biofiltration combined treatment (Al-Momani *et al.* 2002).

Arsenic (As) is a carcinogenic water pollutant released from soil into the groundwater during weathering of arsenic-containing minerals. The maximum contaminant level of arsenic in drinking water is regulated at 10 μg L^{-1} by the US EPA (https://www3.epa.gov/). Arsenic removal techniques are based on adsorption, oxidation, and coagulation/precipitation. As(V) is better removed by these methods than As(III). As(III) was oxidized more efficiently in the VUV/UV (184.9/253.7 nm) treatment than by the UV/H_2O_2, photo-Fenton, or photocatalytic AOPs (Yoon *et al.* 2008). In all these processes, $HO^•$ is the oxidizing species of As(III), with a minor contribution of molecular oxidation by H_2O_2 generated *in situ* in alkaline solutions (pH > 9). The addition of Fe(III) or H_2O_2 increased the efficiency of As(III) oxidation due to the additionally generated $HO^•$ and Fe(IV) species in the presence of 253.7 nm radiation. The presence of humic acids (7.5–15 mg L^{-1}) did not affect significantly the As(III) oxidation efficiency. The combination of the VUV/UV process with adsorption onto activated alumina or with coagulation/precipitation using $FeCl_3$ were found to be effective strategies for As removal (Yoon *et al.* 2008).

5.3.2 VUV in combination with other treatment technologies

5.3.2.1 VUV and VUV/UV in combination with H_2O_2

The molar absorption coefficients of H_2O_2 at 172 nm and 184.9 nm ($\varepsilon_{172\,nm} = 782$ L mol^{-1} cm^{-1} (Schürgers & Welge, 1968) and $\varepsilon_{184.9\,nm} = 297$ L mol^{-1} cm^{-1} (Weeks *et al.* 1963)) exceed those of water ($\varepsilon_{172\,nm} = 9.9$ L mol^{-1} cm^{-1} (Heit & Braun, 1996; Heit *et al.* 1998) and $\varepsilon_{184.9\,nm} = 0.04$ L mol^{-1} cm^{-1} (Weeks *et al.* 1963)). However, water is the major VUV light absorbing compound in aqueous solutions of H_2O_2. Upon absorption of VUV radiation, H_2O_2 undergoes photolysis with $HO^•$ formation (5.30).

$$H_2O_2 + h\nu(<300\text{ nm}) \rightarrow 2HO^• \tag{5.30}$$

In aqueous solution, the quantum yield of $HO^•$ formation (reaction 5.30) is 1.11 ± 0.07 (Goldstein *et al.* (2007), which also exceeds that from water (0.42 (172 nm) and 0.33 (184.9 nm)). Although H_2O_2 is a weak absorber of the 253.7 nm radiation, the addition of H_2O_2 to water exposed to VUV/UV radiation increases the overall HO• yield.

The VUV (184.9 nm)/UV (253.7 nm)/H_2O_2 process was found more effective for clofibric acid degradation than UV (253.7 nm), VUV/UV, and UV (253.7 nm)/H_2O_2 processes (Li *et al.* 2011). Simonsen *et al.* (2013) demonstrated a synergistic effect of the VUV (184.9 nm)/UV (253.7 nm)/H_2O_2 process in degradation of *para*-nitrosodimethylaniline.

The efficiency of VUV/UV process for 1,4-dioxane removal increased upon addition of increasing H_2O_2 concentration. The VUV/UV/H_2O_2 process in the presence of 1 mg/L H_2O_2 was found to be more effective than the UV/H_2O_2 process. Increasing the H_2O_2 concentration to 5.0 mg L^{-1} reduced the difference between the transformation rates determined for VUV/UV/H_2O_2 and UV/H_2O_2 processes (Matsushita *et al.* 2015). The degradation rates of nitrobenzene upon exposure to 172 nm radiation increased upon addition of H_2O_2 up to an optimal H_2O_2:nitrobenzene initial concentration ratio of 7:1; above this ratio, the excess of H_2O_2 competed with nitrobenzene for the HO• decreasing the transformation rate (Li *et al.* 2006).

Reverse osmosis concentrate (ROC) of a municipal wastewater secondary effluent was effectively treated with the UV/H_2O_2 process and, with even a better performance, with the VUV/UV/H_2O_2 process. The highly conjugated compounds of ROC characterized by low biodegradability were efficiently fragmented and oxidized by HO•, thus increasing their biodegradability and mineralization (Liu *et al.* 2011).

5.3.2.2 VUV and VUV/UV in combination with photocatalysis

Titanium dioxide (TiO_2) is the commonly used photocatalyst, either in suspension or immobilized on solid surfaces. Because the VUV light is absorbed in a very thin water layer, the TiO_2 particles can be excited by the VUV radiation only within this layer. In VUV/UV process using LP lamp, the excitation of photocatalyst can occur also over a longer path length through the suspension *via* the absorption of 253.7 nm radiation. However, in a 1.00 g L^{-1} TiO_2 containing suspension, the 253.7 nm radiation is absorbed within less than 2 mm.

The VUV (184.9 nm)/UV (253.7 nm) or VUV (172 nm) radiation were found more effective for the degradation of organic pollutants than the UV (253.7 nm)/TiO_2 process (Dombi *et al.* 2002; Han *et al.* 2004; Matsushita *et al.* 2015; Wang *et al.* 2014). The addition of TiO_2 to the VUV/UV-irradiated solutions slightly enhanced the transformation rates of various pollutants and accelerated the rate of mineralization. No synergism was observed between the VUV/UV process and photocatalysis when the distance between the sheet with immobilized TiO_2 and the lamp was large, e.g., 3.5 cm (Matsushita *et al.* 2015) or 1.5 cm (Han *et al.* 2004).

The VUV/UV/TiO_2 process efficiency can be increased upon addition of H_2O_2. Although the performance of this combination was better than those of the VUV/UV or VUV/UV/TiO_2 processes, it was worse than that of the VUV/UV/H_2O_2 process, most likely because the TiO_2 in suspension competes strongly with H_2O_2 for the photons, and has a lower HO• yield (~0.05) than H_2O_2 photolysis (~1.0) (Shen & Liao, 2007).

The VUV (172 nm, 7.2 mW cm^{-2})/TiO_2 process was also found to be more effective in transformation of phenol than the UV (253.7 nm, 10 mW cm^{-2})/TiO_2, UV (310–380 nm, 4 mW cm^{-2})/TiO_2 or VUV (172 nm) processes (Ochiai *et al.* 2013a; Ochiai *et al.* 2013b).

5.3.2.3 VUV and VUV/UV in combination with ozone

Ozone is unstable in water, and its degradation rate depends on various factors, including pH and water composition. O_3 absorbs the UV radiation with large absorption coefficient at 253.7 nm (3300 M^{-1} cm^{-1}, (Reisz *et al.*)). HO• is one of reactive species formed through O_3 photolysis. VUV photolysis of O_2 in gas phase enables *in situ* photochemical generation of O_3. Consequently, the reactor configuration gives possibility for the combination of VUV, VUV/UV or UV photolysis and ozonation without external ozone generator.

The combination of 172 nm or 184.9 nm/253.7 nm radiation with ozonation was studied using one light source for the photochemical generation of O_3 from O_2 in the gas phase. Hashem *et al.* (1997) used this technique for VUV (172 nm)/O_3 treatment of 4-chlorophenol in laboratory water. Zoschke *et al.* (2014) used the *in situ*-generated O_3 in combination with the VUV (184.9 nm)/UV (253.7 nm) radiation to study the degradation of geosmin and MIB.

The combination of 184.9 nm/253.7 nm radiation with ozonation was more efficient for the NOM mineralization than the individual VUV/UV, UV/O_3, O_3 or direct UV (253.7 nm) processes (Ratpukdi *et al.* 2010). Synergistic effects were observed at pH 7 and 9, but not at pH 11; increased biodegradability of treated water was also observed. Synergistic was also observed during the degradation of organophosphoric acid triesters (OPEs) by the VUV (184.9 nm)/UV (253.7 nm)/O_3 at pH 7.5 as compared to the sequential application of the individual processes (Echigo *et al.* 1996). The efficiency of this combination was reported to increase with the increasing concentration of O_3. This combined process was also more effective than the O_3/H_2O_2 process in distilled water, but only at low OPE concentrations. Treatment of 20 mg L^{-1} OPE exhibited similar efficiencies for the two processes in pure water, whereas the treatment of a solid waste landfill effluent indicated a slightly higher efficiency for the O_3/H_2O_2 process than for the VUV/UV/O_3 combination. This was explained in terms of the negative impact of NO_3^- on the performance of VUV/UV process.

5.4 WATER QUALITY IMPACT ON VACUUM UV PROCESS PERFORMANCE AND BY-PRODUCT FORMATION

5.4.1 The effect of inorganic ions

The inorganic ions are known to impact the VUV process performance. Thus, when designing a VUV process for water remediation, it is important to know the water matrix composition, including the concentrations of ions such as NO_2^-, NO_3^-, Cl^-, CO_3^{2-}, HCO_3^-, SO_4^{2-}, HPO_4^{2-} and $H_2PO_4^-$. Some of these ions, e.g., NO_2^- and CO_3^{2-} are highly reactive toward the $HO^•$. The rate constants for the reactions of many inorganic ions with $HO^•$ are available in the published literature (Buxton *et al.* 1988). The majority of inorganic ions present in natural waters are orders of magnitude less reactive than the organic pollutants towards the $HO^•$, such that the impact of those ions on the contaminant treatment performance becomes significant only at large concentrations.

At low pH (< 3), Cl^- becomes a significant $HO^•$ scavenger, reducing the efficiency of the VUV process. The reaction of Cl^- with $HO^•$ results in the formation of $Cl^•$ *via* the decomposition of $HOCl^{•-}$, an unstable intermediate. $Cl^•$ is highly reactive toward organic compounds and could increase the adsorbable organic halogen (AOX) level of water (Oppenländer, 2003). An inhibitory effect of Cl^- (1 mmol L^{-1} and 100 mmol L^{-1}) on the transformation of 1,1,1-trichloroethane in VUV/UV-irradiated solution was observed by Gu *et al.* (2013).

The HCO_3^- or CO_3^{2-} can compete for $HO^•$ especially in high alkalinity waters. The reactions of HCO_3^- and CO_3^{2-} with $HO^•$ are electron transfer reactions. The $HO^•$ scavenging by CO_3^{2-} becomes significant mainly at pH > 10.5, where CO_3^{2-} is the predominant carbonate species ($k_{HO^•,\,carbonate} = 3.9 \times 10^8$ L mol^{-1} s^{-1} (Buxton *et al.* 1988)).

Although the photolysis at 172 nm radiation of SO_4^{2-}, HPO_4^{2-} and $H_2PO_4^-$ yields highly reactive and selective radicals, i.e., $SO_4^{•-}$, $HPO_4^{•-}$ and $H_2PO_4^•$, respectively, this process is insignificant relative to water photolysis, due to the high molar absorption coefficient of water at $\lambda < 175$ nm. In aqueous solutions containing less than 10^{-2} mol L^{-1} of dissolved organic compounds with molar absorption coefficients of ~6×10^3 mol L^{-1} cm^{-1}, the water photolysis is the dominant process in the VUV-based treatment ($\lambda < 200$ nm) (Gonzalez *et al.* 2004).

The VUV (172 nm) water photolysis in the presence of nitrate (NO_3^-) and nitrite ions (NO_2^-) is a very complex reaction system. It involves 16 N-containing species and 64 reactions, where NO_2^- is generated from NO_3^- and *vice versa*, yielding N_2O and N_2 as final products. Due to the small rate constant of the reaction of NO_3^- with $HO^•$ (see 5.31), NO_3^- reacts preferentially with reducing species (reactions 5.32 and 5.33) and it competes efficiently for e_{aq}^- with DO. Nitrite is an intermediate product of NO_3^- photolysis which reacts at diffusion controlled rates with all primary radicals of water VUV photolysis (Gonzalez & Braun, 1995) (reactions 5.34–5.36).

$$NO_3^- + HO^• \rightarrow NO_3^• + HO^- \quad k < 1\times10^5\,\text{L mol}^{-1}\,\text{s}^{-1}\ \text{(Gonzalez \& Braun, 1995)} \tag{5.31}$$

$$NO_3^- + H^• \rightarrow NO_3H^{•-} \quad k = 2.4\times10^7\,\text{L mol}^{-1}\,\text{s}^{-1}\ \text{(Gonzalez \& Braun, 1995)} \tag{5.32}$$

$$NO_3^- + e_{aq}^- \rightarrow NO_3^{•2-} \quad k = 1\times10^{10}\,\text{L mol}^{-1}\,\text{s}^{-1}\ \text{(Gonzalez \& Braun, 1995)} \tag{5.33}$$

$$NO_2^- + HO^• \rightarrow NO_3H^{•-} \quad k = 2.5\times10^9\,\text{L mol}^{-1}\,\text{s}^{-1}\ \text{(Gonzalez \& Braun, 1995)} \tag{5.34}$$

$$NO_2^- + e_{aq}^- \rightarrow NO_2^{•2-} \quad k = 6\times10^9\,\text{L mol}^{-1}\,\text{s}^{-1}\ \text{(Gonzalez \& Braun, 1995)} \tag{5.35}$$

$$NO_2^- + H^• \rightarrow NO_2H^{•-} \quad k = 1\times10^9\,\text{L mol}^{-1}\,\text{s}^{-1}\ \text{(Gonzalez \& Braun, 1995)} \tag{5.36}$$

It was found that both in the presence and the absence of DO, the N-containing species that form during the 172 nm irradiation of NO_3^--containing solutions ($c_0 = 2.6 \times 10^{-4} - 2.3 \times 10^{-3}$ mol L^{-1}) promote the transformation and mineralization of phenol (Gonzalez & Braun, 1996). An excellent description on the mechanistic pathways and dynamics of inorganic N-species in VUV-irradiated systems as well as their impact on contaminant treatment efficiency are given in Gonzales *et al.*'s review (2004).

NO_2^- is not a common water constituent but it could be generated *in situ via* NO_3^- photolysis particularly with medium-pressure lamps and 184.9 nm radiation. Therefore, its contribution to the overall $HO^•$ water demand should be anticipated and properly accounted for in VUV- and medium-pressure lamp-based AOP applications. In VUV (172 nm)/O_3 and/or VUV (184.9 nm)/UV (253.7 nm)/O_3 combined processes, NO_2^- is rapidly removed through the reaction with O_3 (Kutschera *et al.* 2009; Zoschke *et al.* 2012).

5.4.2 The effect of dissolved natural organic matter (NOM)

Studies conducted on the potential application of VUV (184.9 nm)/UV (253.7 nm) radiation as a pretreatment step to biological filtration of drinking water proved that the VUV/UV process increases the biodegradability and mineralizes NOM to a greater extent than the UV (253.7 nm) radiation alone (Buchanan *et al.* 2006; Buchanan *et al.* 2005; Ratpukdi *et al.* 2010; Thomson *et al.* 2002). The enhanced biodegradability of VUV/UV-irradiated NOM in water samples was attributed to the sequential breakdown of high molecular weight, conjugated compounds in the hydrophobic acidic NOM fractions to generate biodegradable, charged (hydrophilic) and neutral (hydrophilic) structures. Low molecular weight, hydrophilic neutral fractions of NOM however, were refractory to photo-oxidation but more amenable by biological filtration than the hydrophobic fractions (Buchanan *et al.* 2005).

Xing *et al.* (2015) investigated UV (253.7 nm) and VUV (184.9 nm)/UV (253.7 nm) photolysis as pre- and post-treatment of coking wastewater and compared the effect of processes on TOC, COD, BOI, NH_4^+ and NO_2^- concentration, total-nitrogen content, biodegradability and toxicity of the treated water samples.

It was demonstrated that the combination of UV or VUV radiation with O_3 was more effective at the decomposition of high molecular weight compounds than ozonation alone. The combined processes are effective in the mineralization of hydrophobic neutral and acidic fractions of NOM. The oxidized residue mainly consisted of hydrophilic neutral NOM, which is more biodegradable than the original NOM (Ratpukdi *et al.* 2010). Thus, VUV/UV photooxidation or its combination with other oxidants followed by biological treatment appears to be a promising approach to NOM removal (Matilainen & Sillanpää, 2010; Zoschke *et al.* 2014).

The NOM removal during the pre-treatment steps of water is desirable particularly in water treatment scenarios which require contaminant removal *via* $HO^•$-based processes. It is well known that humic and fulvic acids, as well as other NOM constituents are both $HO^•$ scavengers and UV light absorbing species. The negative impact of water matrix constituents on the efficiency of VUV/UV process is well represented in the published literature (Imoberdorf & Mohseni, 2012; Kutschera *et al.* 2009; Li *et al.* 2009; Li *et al.* 2011).

5.4.3 Effect of pH

The effect of pH on the yield of primary radicals ($HO^•$, $H^•$ and e_{aq}^-) and on their subsequent reactions was discussed briefly in Section 5.1.2. The reaction rate constants of the protonated and deprotonated forms of the target compounds as well as of pH-dependent structures of NOM constituents with $HO^•$ and other reactive species could be different. The pH also affects the HCO_3^-/CO_3^{2-}concentration ratio in water, which consequently affects the contribution of these species to the $HO^•$ water background demand. Therefore, the effect of pH on the VUV-irradiated aqueous solutions is complex.

Quici *et al.* (2008) investigated the pH effect on the 172 nm-irradiated solutions of citric acid and gallic acid. The removal of citric acid at pH 3.4 was faster than at pH 11, which was explained as due to a lower steady-state concentration of $HO^•$ at pH 11 ($pK_{a(HO^•)} = 11.9$ (Gonzalez *et al.* 2004)) than at pH 3.4, and to the lower reactivity of citrate ion toward $O^{•-}$ than toward $HO^•$. Additionally, the coulombic repulsion between the ionized species and $O^{•-}$ at high pH decreases the degradation rate of the substrate. Gallic acid (pK_a values of 4.44, 8.45, and 10.05) degradation rate was not significantly affected by pH within the 2.5–7.5 pH range, despite the difference between the reported $HO^•$ rate constants for the fully protonated (6.4×10^9 L mol^{-1} s^{-1} at pH 0) and monoprotonated form (1.1×10^{10} L mol^{-1} s^{-1} at pH 6.8) of gallic acid (Dwibedy *et al.* 1999).

The degradation efficiency of 4-*tert*-octylphenol (4-OP) with the 253.7 nm radiation was significantly higher in alkaline media than under acidic conditions (pH 3 or 6) (Huang *et al.* 2013), which could be due to a larger molar absorption coefficient at 254 nm of 4-OP ionized form than that of non-ionized form. The $HO^•/O^{•-}$ speciation as a function of pH, as well as the pH-dependent oxidation potential of $HO^•$ are factors to be considered during the assessment of the VUV/UV for water contaminant treatment (Huang *et al.* 2013). The oxidation potential of $HO^•$ was reported as 2.59 V at pH 0 and 2.18 V at pH 7, while that of $O^{•-}$ is 1.64 V at pH 14 (Koppenol & Liebman, 1984).

The degradation rates of 1,4-dioxane, geosmin, and MIB in buffered dechlorinated tap water treated with VUV/UV and VUV/UV/TiO_2 processes were reported to decrease with the increase of pH in the 5.5–8.0 range (Matsushita *et al.* 2015), effect attributed to the pH-dependent speciation of carbonate ions and their reactivity with $^•OH$. Formation and accumulation of H_2O_2 in VUV light irradiated DO containing solutions is mainly due to the pH-dependent disproportionation reactions of $HO_2^•/O_2^{•-}$ (Arany *et al.* 2012;

Azrague *et al.* 2005; Robl *et al.* 2012). Given that $pK_a(HO_2^{\bullet}) = 4.8$, higher H_2O_2 concentrations formed through the $HO_2^{\bullet}$ disproportion reaction are expected at pH < 6 than at neutral and alkaline pH.

5.4.4 By-product formation during the VUV process and their removal through biological activated carbon filtration

5.4.4.1 Chlorination disinfection by-products (DBPs)

Buchanan *et al.* (2006) and Matsushita *et al.* (2015) have studied the THM formation potential as a result of NOM and 1,4-dioxane treatment with VUV/UV radiation, respectively. The applied UV doses were several times larger than those used in full-scale AOP applications. The authors observed that VUV/UV doses lower than 10 J cm^{-2} decreased the THM formation potential, particularly for chloroform, dibromochloro- and bromodichloromethanes (Matsushita *et al.* 2015). At doses of 30–40 J cm^{-2}, the THM formation potential increased, which was attributed to the halogenation of low molecular weight compounds generated from the high molecular weight NOM constituents. Although the total THM formation potential decreased upon VUV/UV treatment, a linear increase in bromoform formation was observed. Bromoform formation was associated only with the treatment of the hydrophilic fraction of NOM. The major species accounting for the THM formation potential were characterized as chlorinated hydrophobic compounds. The superiority of the VUV/UV treatment over UV photolysis on the reduction of THM formation potential was demonstrated (Buchanan *et al.* 2006; Matsushita *et al.* 2015).

The nine HAAs (i.e., monochloroacetic, dichloroacetic, trichloroacetic, monobromoacetic, dibromoacetic, tribromoacetic, bromochloroacetic, dibromochloroacetic and bromodichloroacetic acids) are potentially genotoxic and carcinogenic. The HAA formation potential on VUV/UV-treated 1,4-dioxane and NOM was found nearly constant at doses lower than 32 J cm^{-2} and decreased upon exposure to doses ≥ 48 J cm^{-2}. In contrast, the UV photolysis caused no significant change in the HAA formation potential. Most of the HAAs were chlorinated compounds, but significant concentrations of brominated HAAs were also observed (Buchanan *et al.* 2006; Matsushita *et al.* 2015).

Biological activated carbon (BAC) filtration is a well-established, widely implemented process in water treatment. BAC removes effectively the biodegradable fraction of DOC, a vast range of water micropollutants, inorganic species, and toxic as well as genotoxic compounds.

The combination of VUV/UV treatment with BAC or granular activated carbon (GAC) filtration reduced significantly the THM and HAA formation potentials, which was explained by the ability of BAC to remove the hydrophilic biodegradable compounds, including carbonyl and carboxylic compounds, generated through VUV-induced degradation of hydrophobic fraction of NOM (Buchanan *et al.* 2008; Matsushita *et al.* 2015).

5.4.4.2 Aldehydes, nitrite and H_2O_2

The low-molecular weight compounds such as aldehydes and carboxylic acids are assimilable organic carbon and are responsible for the biofilm growth in the water distribution network and on filtration membranes used in water treatment. The VUV/UV treatment of NO_3^--containing waters generates NO_2^-. Exposure to NO_2^- may cause methaemoglobinaemia in infants and mutations in mammalian cells. H_2O_2 formed during the VUV water photolysis in the presence of DO should be removed from the treated water prior to use or discharge. H_2O_2 reacts with chlorine and consequently increases the chlorine demand in the water disinfection process. The concentration of H_2O_2 formed in solutions treated with VUV/UV process exceeds that formed during the UV (253.7 nm) irradiation (Buchanan *et al.* 2006; Matsushita *et al.* 2015).

Matsushita *et al.* (2015) showed that the chlorination and VUV/UV treatment of raw water from Lake Hakucho and raw groundwater generated formaldehyde, glyoxal, and formaldehyde-forming compounds from NOM. These compounds were efficiently removed by the subsequent BAC or GAC treatment. Imoberdorf and Mohseni (2014) reported that VUV/UV doses in the range of 1–3 J cm^{-2} increased the aldehyde levels, which were reduced upon extended treatment (> 4 J cm^{-2}).

The VUV/UV/BAC or VUV/UV/GAC processes removed NO_2^- and H_2O_2 from the VUV/UV treated waters (Buchanan *et al.* 2006, 2008; Matsushita *et al.* 2015). The NO_2^- concentration in VUV/O_3 or VUV/UV/O_3 treated water was below the detection limit (Kutschera *et al.* 2009; Zoschke *et al.* 2012).

5.4.4.3 Bromate

Although the efficiency of the VUV or VUV/UV processes could be significantly increased in the presence of O_3, the application of the combined processes raises concerns over bromate (BrO_3^-) formation. In the VUV/UV/O_3 process, the oxidation of Br^- by O_3 *via* $HO^\bullet$-based mechanism yields BrO_3^-. Bromate absorbs the VUV radiation and is converted back to Br^-. When the UV (253.7 nm) or VUV (184.9 nm)/UV (253.7 nm) photolysis and the ozonation of bromide-containing water were performed in the same reactor and sufficient time was allowed for BrO_3^- reduction by the VUV/UV light, the measured BrO_3^- concentrations in the treated water were 6 times lower for the VUV/UV/O_3 process than those observed for the UV/O_3 process at pH 9 (Ratpukdi *et al.* 2011). An opposite outcome was found in a system configuration where ozonation and photolysis were performed separately, in two reactors, with a short VUV/UV treatment time (Collivignarelli & Sorlini, 2004). Since the increase of the lamp electrical power, O_3 concentration, and pH results in both the increase of BrO_3^- concentration and DOC removal, the optimization of these factors is required to control the BrO_3^- formation and to maximize the pollutant removal efficiency (Ratpukdi *et al.* 2011).

5.5 WATER DISINFECTION

An advantage of UV-based oxidation processes is the simultaneous disinfection. The 253.7 nm radiation emitted by the LP lamp causes the inactivation of microorganisms *via* absorption-induced changes in the DNA structure. The disinfection mechanism of the VUV radiation is different than that of UV radiation. In the VUV process, the disinfection is the effect of the radical species which induce surface oxidation at the cell wall, resulting in the cell membrane damage (Wang *et al.* 2010b). Diffusion of $HO^\bullet$ through the damaged membrane creates conditions for radical-induced alteration of biologically active cell constituents (Mamane *et al.* 2007).

Wang *et al.* (2010b) determined the fluence-based rate constants of inactivation of *Bacillus subtilis* spores at 172, 222 and 253.7 nm as 0.0023, 0.122 and 0.069 cm^2 mJ^{-1}, respectively. These results indicated quantitatively that VUV light alone is not satisfactory for water and wastewater disinfection, but the combination of the VUV (184.9 nm) and UV (253.7 nm) radiation emitted by the LP lamps would be effective both at micropollutant removal (AOP process) and microorganism inactivation.

The efficiency of the VUV process depends on the $HO^\bullet$ concentration, which is reduced in the presence of radical scavengers, such as NOM. Liu and Ogden (2010) reported a faster inactivation of *Xanthomonas* species in pure water under VUV (184.9 nm)/UV (253.7 nm) than under UV (253.7 nm) irradiation. Zoschke *et al.* (2012) examined the inactivation of coliforms and *Escherichia coli* under VUV/UV and UV irradiation in drinking water, and observed no significant difference in the results. Similar conclusions were drawn on the inactivation of *Bacillus subtilis* spores. The authors explained the results as an impact of the $HO^\bullet$ scavengers, which makes the contribution of the 184.9 nm radiation to the microorganism inactivation negligible, such that the disinfection is achieved mostly by the 253.7 nm light. Moreover, the regrowth potential after VUV/UV exposure was somewhat higher than after the UV alone due to the formation of lower molecular weight compounds during the VUV-radical-initiated oxidation of the organics in the water matrix (Zoschke *et al.* 2012).

Ochiai *et al.* (2013a) compared the efficacy of the combination of two AOPs, namely, UV (253.7 nm)/TiO_2-mesh filter/O_3 and VUV (172 nm)/TiO_2-mesh filter at the contaminant removal and waterborne pathogen inactivation. The authors concluded that the VUV/TiO_2 process was more effective at pollutant removal than the UV (253.7 nm)/TiO_2-mesh filter/O_3 but less efficient for *Escherichia coli* inactivation both in laboratory water and in sewage effluents.

5.6 REACTOR/EQUIPMENT DESIGN AND ECONOMIC CONSIDERATIONS

5.6.1 Actinometry for VUV photon flow measurements

In order to design a photoreactor, the incident photon flow must be known. A chemical system that undergoes a light-induced reaction at a certain wavelength, for which the quantum yield is known and the measuring of the reaction rate allows the calculation of the absorbed photon flux, can be used as an actinometer. It has to be simple to use, easy to reproduce and should not require special equipment. The quantum yield of the actinometer reaction should be independent of the reactant concentration and light intensity, and, if any, its temperature dependence should be known.

Gas- and liquid-phase chemical actinometers are used to quantify the incident photon flows of VUV radiation. László *et al.* (1998) determined the radiation power of a Xe_2* lamp (172 nm) by indirect O_2 actinometry, following the O_3 formation from O_2. Model calculations led to a simple correlation between the quantum yield of O_3 formation and the concentrations of O_3 and O_2 at room temperature ($\Phi = 2–1.10\ [O_3]/[O_2]^{1.94}$). This method was also used by Vicente *et al.* (2009) in a cascade reactor system.

Three liquid-phase chemical actinometers are described for 184.9 nm and 172 nm VUV light (Kuhn *et al.* (2004)). Ethanol/water actinometer (Farkas & Hirshberg, 1937), known as Farkas actinometer, is based on gas chromatography (GC) analysis of H_2 formed during ethanol photolysis (184.9 nm) in water. Because of the $\Phi(H_2)$ dependence on ethanol concentration, this method is not used routinely. Adam and Oppenländer (1984) used a reproducible method based on *cis-trans* photo-isomerization of *cis*-cyclooctene to quantify the 184.9 nm radiation. The concentration of *cis*-cyclooctene was followed by GC analysis. Due to the formation of minor degradation products after ~20 min of irradiation, this actinometer was described as suitable only for short irradiation times.

Nowadays methanol is the most commonly used actinometer for VUV radiation. Heit *et al.* (1998) described for the first time the actinometer based on methanol degradation, using a Xe_2* lamp (172 nm). The authors determined the quantum yield for water photolysis from the experimental rate of methanol degradation and the rate of $HO^•$ production. The overall quantum yield of water photolysis ($\Phi(H_2O)$) at 172 nm was found as 0.42 ± 0.04 (Heit *et al.* 1998). This methanol actinometer was calibrated against the *cis*-cyclooctene actinometer. Oppenländer and Schwarzwälder (2002) determined the photon flow emitted from a Xe_2* lamp (172 nm) using methanol photolysis in competition kinetics with water photolysis. The degradation kinetics of methanol was found to depend on its initial concentration. At low concentrations ($< 1 \times 10^{-3}$ mol L^{-1}), methanol degradation follows *pseudo*-first-order kinetics, whereas at high concentrations (0.075–0.250 mol L^{-1}) zero-order kinetics is obeyed ($k_0^{obs.} = 4.9 \pm 0.2 \times 10^{-4}$ mol L^{-1} min^{-1}). Under the latter experimental conditions, the VUV photolysis of methanol occurs concomitantly with water VUV photolysis and the $HO^•$-initiated degradation of methanol increases the $k_0^{obs.}$ values (Heit *et al.* 1998; Oppenländer & Schwarzwälder, 2002). The actinometry must be performed in the same reactor which is used for the photodegradation experiments.

In recent years, alternatives to the methanol actinometry were also developed. For example, diamond-based sensors have been developed and tested over a wide spectral range, from the extreme UV (20 nm) to the near IR region (2400 nm) (Balducci *et al.* 2005), and showed VUV to UV sensitivity ratios larger than 400 (Bergonzo *et al.* 2000), indicating their solar blindness. This is a consequence of the band-gap energy of diamond (5.5 eV), which corresponds to the 225 nm radiation energy, thus the sensor is essentially sensitive only to radiation carrying higher energies, such as 184.9 nm radiation from LP lamps and 172 nm radiation of Xe_2* lamp. Hayashi *et al.* (2005) demonstrated that the output signals from the diamond sensors are reproducible and stable for more than 500 h continuous irradiation from the Xe_2* lamp. Suzuki *et al.* (2006) reported direct measurement data for 184.9 nm radiation from low-pressure mercury lamps used in VUV/O_3 surface dry cleaning process monitored with diamond-based VUV sensor.

Small photoionization lamps operating in the short-VUV spectral region are used in analytical engineering, particularly in gas analysis devices with ionization detectors. These sources are generally Kr_2* or Xe_2* lamps. Budovich and Il'in (2014) described a method to measure the light intensities of Kr_2* (116.5 nm and 123.6 nm) and Xe_2* (129.6 nm and 147.0 nm) photoionization lamps using a flow ionization total absorption chamber operating at atmospheric pressure. The chamber has no window and radiation is directed into the mixture of vapors of volatile organics (e.g., methane, hexane, and/or isobutylene), O_2 and/or N_2, with known individual concentrations. The photon flux is determined indirectly from the measured ionization current, ionization quantum yield, and the fraction of radiation absorbed by the organics. The method proved to be applicable in the wavelength range of 105–147 nm and photon fluxes of $10^5 - 10^{15}$ photon s^{-1}.

5.6.2 Reactor design

Photochemical reactors can be designed in batch, semi-batch or flow-through configurations. Most often the semi-batch configuration (Figure 5.8) is used to study VUV processes, with the lamp placed at the center of a water-cooled, tubular reactor. The VUV lamps are housed by high purity quartz sleeves, to allow high transmittance of the VUV radiation and to be resistant to solarization.

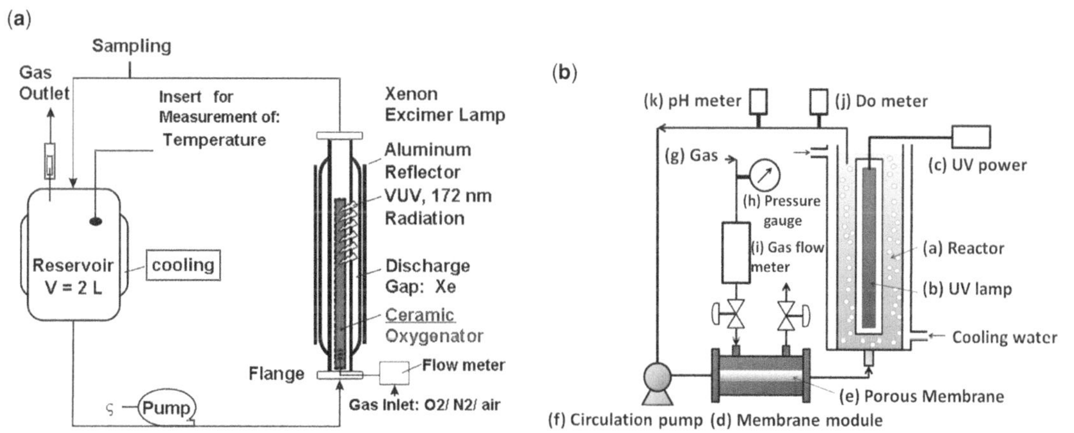

Figure 5.8 (a) Scheme of 100 W photochemical apparatus containing a ceramic gassing unit; (b) Photochemical apparatus containing Shirasu porous glass (SPG) membrane module for microbubble formation. Reprinted with permission from Oppenländer *et al.* (2005) (Fig. 5.8a) and Tasaki *et al.* (2009) (Fig. 5.8b). Copyright Elsevier 2016.

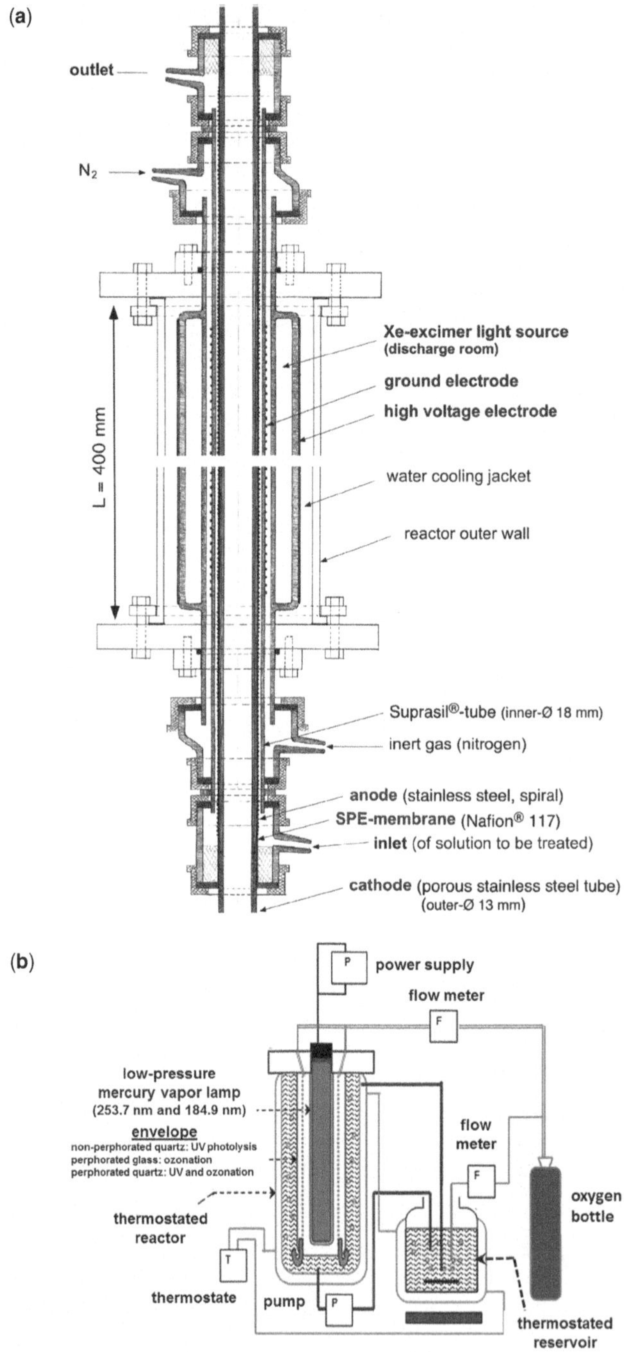

Figure 5.9 (a) Schematic representation of the photochemical reactor combined with electrochemical generation of O_2. Reprinted with permission from Gonzalez *et al.* (2004). Copyright Elsevier 2016. (b) Schematic representation of VUV/UV photochemical reactor with O_3 generation; adapted from Alapi *et al.* (2013).

Completely mixed batch cylindrical vessels are also popular reactor configurations. A typical batch operation requires a chemical storage reservoir and the UV reactor equipped with an O_2 injection port. The reactor chamber is usually surrounded by a cooling jacket.

Given the temperature dependency of the quantum yield of water homolysis, the studies are usually conducted in thermostated reactors. Oppenländer and Xu (2008) demonstrated that the rates of VUV (172 nm)-initiated oxidation and mineralization of methanol in water in a flow-through photoreactor are independent of the water temperature in the 20–50 °C range.

It was shown that the presence of DO has a significant role on the transformation rates of pollutants (Oppenländer *et al.* 2005). Moreover, DO concentration is reduced near to the lamp wall. To reduce the negative effect of the DO demand layer, Han *et al.* (2004) injected O_2 or air through a porous glass plate. This procedure drastically increased the rate of TOC removal. Similarly, Oppenländer *et al.* (2005) used a ceramic oxygenator mounted axially within the Xe_2* lamp, which facilitated the transfer of O_2 directly into the irradiation zone (Figure 5.8a).

Tasaki *et al.* (2009) used uniform-size O_2 microbubbles (average diameter of 5.79 nm) in irradiated solutions of methyl orange (model compound). It was found that the decolorization reaction rate constant in the microbubble system was 2.1 times higher than that in conventional large bubble systems. Photodegradation experiments were conducted with LP lamps (253.7 nm and 184.9/253.7 nm) (Figure 5.8b). The O_2 microbubbles accelerated methyl orange decolorization and TOC reduction rates with 184.9/253.7 nm radiation, but no significant effect was observed on the kinetics for the 253.7 nm radiation.

In situ electrochemical generation of O_2 during the VUV irradiation was experimented by Braun *et al.* (2004) and Wörner *et al.* (2003). This method resulted in fast mineralization of the studied organic compounds. In this case, the mass transfer limitations were overcome *via* electrochemical generation of O_2 within or close to the irradiated volume and with optimum reactor design (Figure 5.9a).

As mentioned briefly earlier in this chapter, VUV radiation emitted from the LP lamps can be used for O_3 generation from O_2 in gas phase. Figure 5.9b shows a reactor configuration where ozone is generated in the air/oxygen atmosphere within the quartz sleeve housing the VUV/UV lamp. The O_3 containing gas is passed through the irradiated solution, where the photolysis of O_3 at 253.7 nm radiation results in $HO^•$ formation, such that the conditions for UV/O_3 treatment are created. This reactor configuration was used by Zhao *et al.* (2013) and Alapi *et al.* (2013).

Oppenländer *et al.* (1995) used the combination of 172 nm (VUV) and 222 nm (UV) radiation in a VUV/UV-excimer flow through photoreactor for wastewater treatment. The VUV/UV-excimer "double irradiation" unit consisted of coupled Xe_2* and KrCl* light sources. The results showed more effective photomineralization of organics in this "double radiation" than in single-radiation configuration. The 172 nm irradiation of pure water saturated with either air or O_2, led to *in situ* generation of both H_2O_2 and O_3. Therefore, the VUV/UV-oxidation process in excimer flow-through photoreactor could combine the advantages of multiple photo-oxidation processes (H_2O_2/O_3/UV) without the addition of chemical oxidizing agents.

Alekseev *et al.* (2006) constructed a flow-through reactor suitable for irradiation of gases and solutions at pressures up to 40 atm. They used a sealed single barrier Xe_2* lamp with about 700 cm^2 radiation surface. The mean VUV radiation power was found to be inversely proportional with the lamp temperature.

Dobrovic *et al.* (2007) investigated the effect of Reynolds number on the rate of NOM transformation in a flow-through VUV/UV reactor. In the presence of 184.9 nm radiation (VUV/UV process), the overall degradation rate increased ten-fold and seventeen-fold, at lower and at higher Reynolds numbers, respectively, as compared to the UV (253.7 nm) process under the same flow conditions. This proved that the hydrodynamic characteristics of the photoreactor are important for the overall energy efficiency of the VUV treatment. Al-Gharabli *et al.* (2016) compared the effect of flow rate on the efficiency of Rhodamine B transformation using either low-pressure mercury lamp or Xe_2* lamp. No flowrate effect was found in the 184.9/253.7 nm tests, whereas the hydrodynamics played a significant role in 172 nm-based tests, which indicates an enhanced mass transfer into the photoreaction zone with increasing flow rate, particularly required for very narrow photoreaction zones as in the case of 172 nm radiation.

Bagheri and Mohseni (2015b) stressed the role of flowrate on the VUV AOP performance, based on the profiles and distribution of $HO^•$ in the VUV (184.9 nm)/UV (253.7 nm) reactor. The computed volume-weighted average $HO^•$ concentration was two times higher at 0.5 L min^{-1} than at 6.7 L min^{-1}. Higher $HO^•$ concentrations are expected at higher UV doses (i.e., lower flowrates), but the flowrate should be optimized such that extensive mixing is achieved. Degradation rates of micropollutants in the VUV/UV photoreactors are limited by mass transfer in the VUV-irradiated photoreaction zone. The extent of mixing and circulation zones was found to be the key parameter controlling the treatment economics and energy-efficiency (Bagheri & Mohseni, 2015a, 2015b).

5.6.3 Economics considerations

The treatment of organic pollutants is an energy-intensive process, thus the use of an appropriate "figure-of-merit" based on the energy input is justified. The largest part of the treatment costs are the electrical energy costs. Bolton *et al.* (2001) proposed two figures-of-merit, i.e., Electric energy per mass (E_{EM}, kWh kg^{-1}), to evaluate process treatment performance for waters containing high concentrations of pollutants (in which case the zero-order kinetics is obeyed) and the Electric energy per order (E_{EO}, kWh m^{-3} $order^{-1}$) to assess process treatment performance for water containing low micropollutant concentrations (in this case *pseudo*-first order kinetics is followed).

Thomson *et al.* (2004) concluded that the removal of NOM from highly colored surface water by the VUV/UV process is not economically feasible, as an electrical energy dose of 290 kWh m^{-3} was required to reduce the DOC by one order of magnitude. Puspita *et al.* (2011) compared the E_{EO} values for the removal of humic acids from wastewater through the UV (253.7 nm) and VUV/UV (184.9/253.7 nm) processes, with and without the addition of H_2O_2. Under the studied water quality and operating conditions, the determined E_{EO} values (kWh m^{-3} $order^{-1}$) followed the order: 29 (UV) > 15 (VUV/UV) > 5 (UV/H_2O_2 (16 mg L^{-1})) = 5 (VUV/UV/H_2O_2 (16 mg L^{-1})) > 2.5 (UV/H_2O_2 (32 mg L^{-1})).

Andreozzi *et al.* (1999) stated that an E_{EO} of 2.5 kWh m^{-3} $order^{-1}$ would be acceptable for practical treatment of organic micropollutants in water. Matsushita *et al.* (2015) investigated the efficacy of a series of AOPs for the degradation of 1,4-dioxane, geosmin, and MIB in water and found that under the experimental conditions used in that study the process efficiency followed the order: UV/TiO_2 < VUV/UV < VUV/UV/TiO_2. All these processes were assessed as economically feasible, with the best performance determined for the VUV/UV and VUV/UV/TiO_2 AOPs, whose E_{EO} values were lower than 1 kWh m^{-3} $order^{-1}$.

Zoschke *et al.* (2012) calculated the E_{EO} values for the degradation of geosmin and MIB under VUV (184.9 nm)/UV (253.7 nm) irradiation, and compared them with those obtained for the UV/O_3 and UV/H_2O_2 processes. In ultrapure water, all treatments yielded E_{EO} values close or below 1 kWh m^{-3} $order^{-1}$ for both pollutants. In raw water from drinking water reservoir (Saxony, Germany) both pollutants could be removed at ~1.5, ~0.5 and ~3.5 kWh m^{-3} $order^{-1}$ for the VUV/UV, UV/O_3 and UV/H_2O_2 processes, respectively. The O_3 dose in these experiments was 0.9 mg min^{-1}, and O_3 was generated by silent electric discharge. It was also shown, that the E_{EO} values decreased with increasing O_3 dose, and leveled off below 1 kWh m^{-3} $order^{-1}$ at O_3 dosages above 1 mg min^{-1}.

Afzal *et al.* (2010) reported the degradation of the algal toxin anatoxin-a (as fumarate salt) in pure, natural, and synthetic waters using the VUV (172 nm) and UV (253.7 nm)/H_2O_2 treatments. The results showed that an UV dose of ~200 mJ cm^{-2} was sufficient to achieve more than 70% degradation of anatoxin-a with $HO^•$-based AOPs (172 nm VUV and UV (253.7 nm)/H_2O_2 (30 mg L^{-1})), whereas an UV (medium-pressure lamp, 200–300 nm range) dose of 1285 mJ cm^{-2} was needed to degrade 88% of 0.6 mg L^{-1} anatoxin-a by direct photolysis.

Bagheri and Mohseni (2015b) used a validated computer model to optimize the energy-efficiency of the VUV/UV process for a variety of reactor configurations. E_{EO}, $HO^•$ concentration, and delivered UV and VUV dose distributions were included in the model. The mixing and circulation zones were recognized as key parameters in controlling the treatment economics and energy-efficiency of the VUV/UV process. The authors used a computer-aided reactor design with baffle insertion to control good mixing and circulation zones inside the reactor, which led to up to 72% reduction in the total electrical energy requirement for atrazine degradation *via* VUV/UV treatment. The theoretical predictions were validated experimentally for the optimized reactor/baffle configuration.

5.7 APPLICATIONS OF VACUUM UV LIGHT SOURCES

5.7.1 Applications in instrumental chemical analysis

The instrumentation for TOC analysis in water samples commonly uses the VUV radiation for water photolysis as a means of $HO^•$ generation. The process mineralizes the organic carbon present in the water to CO_2 (http://www.lfe.de/en/process-water-analysis/toc-810). Satou *et al.* (2013) used a home-made lamp-pass-through photoreactor with a built-in narrow (2 mm inner diameter) reaction tube passing through a 40 W VUV (184.9 nm)/UV (253.7 nm) LP lamp, for the mineralization of DOM in river water. This "reagent-free" photoreactor was also suitable for TOC analysis of river-water samples. The detection limit was determined as 6.2 μg C · L^{-1} using potassium hydrogen phthalate as a standard compound.

Liquid chromatographs with organic carbon – organic nitrogen detection (LC-OCD-OND) used for the identification and quantification of NOM fractions are equipped with VUV (184.9 nm)/UV (253.7 nm) lamps. The VUV/UV lamp is placed in the so-called Graentzel thin-film reactor which has a light shielded (for inorganic carbon, IC) and a light exposed (for organic carbon, OC) section. For detailed description on the process principle the reader is referred to the manufacturer website (http://www.doc-labor.de/OCD.html). The LC-OCD/OND instruments are widely employed in drinking water, wastewater, and marine water characterization, as well as in steam condensate analysis in power plants. They are also routinely used in analytical laboratories in chemical, pharmaceutical, and semiconductor industry.

Incoherent VUV excimer light is also used for single photon ionization (SPI) in time-of-flight mass spectrometers (TOF-MS). Most common light sources in these instruments are based either on brilliant (focused) lasers, which are expensive, or on conventional VUV deuterium lamps, which have low output. The rare gas excimer VUV lamps do not emit brilliant radiation, therefore they are not suitable for this application without further modification. Muhlberger *et al.* (2002) generated compact brilliant VUV light using electron beam excitation of dense rare gases through a thin ceramic foil window. This novel VUV light source was then coupled to a compact and mobile TOF-MS. This device can be adapted for analytical application or monitoring purposes. Chen *et al.* (2014) designed a *quasi*-trapping chemical ionization (QT-CI) source with a commercial VUV 10.6 eV Kr_2^* for time-of-flight mass spectrometry. The use of Kr_2^* lamp improved the sensitivity and extended the range of ionizable molecules with ionization potential higher than 10.6 eV.

VUV Analytics Inc. (http://www.vuvanalytics.com/) developed a new VUV/UV absorption detector for gas chromatography, which acquires data from ~120 nm to 240 nm. Chemical compounds eluting from the gas chromatograph column enter a heated transfer line. At the end of the transfer line, a makeup flow of carrier gas is introduced, which carries the analyte into the flow cell irradiated by a deuterium lamp. The eluted components absorb light resulting in reduced transmission and a detectable signal. This detector (Vapor Generation Accessory, VGA-100) has claimed the distinguished "Best New Analytical Instrument" award at the Gulf Coast Conference in 2014, emerging from the competitive market of the World's newest scientific instruments.

An application of a KrCl* lamp (222 nm) and a conventional LP lamp (184.9 nm/253.7 nm) was described by Matter *et al.* (1994). They examined how the ionization of the combustion aerosols depends on the electric power of the lamps and on the irradiation time, and showed that particles from different sources (petrol engine, diesel engine, cigarette smoke) can be differentiated by measuring the photocurrent. The advantage of the exciplex lamp is the possibility to rapidly switching off and on, eliminating the time delay in the measurements. This system could be made more compact by the miniaturization of the exciplex lamp (http://www.epa.gov/ordntrnt/ORD/NRMRL/archive-etv/pubs/01_vs_ecochem_pas2000.pdf). These instruments can be used in the field of pollution detection like the analysis of indoor and outdoor air quality or quantification of combustion aerosols, and in air conditioning or fire alarm systems.

Ellipsometers are used for investigating the dielectric properties (complex refractive index or dielectric function) of thin films. This technique has found applications in many different fields, from semiconductor physics to microelectronics and biology, from basic research to industrial applications. The VUV-VASE® spectroscopic ellipsometer is the standard in optical characterization of materials used in lithography applications. Equipped with deuterium, Xe_2* and tungsten halogen lamps, it covers the spectral range from 140 nm up to 1700 nm (http://www.jawoollam.com/vuv_home.html).

VUV Resonance Fluorescence CO Instrument (VUV-CO) was developed as a commercial version of the instrument described by Gerbig *et al.* (1999). It quantitatively determines airborne CO using resonant fluorescent (RF) discharge gas lamp emitting in the VUV range. The narrow emission band centered at 151 nm is achieved using an optical filter and CO fluorescence is detected by photon counting. The instrument was integrated into the HAIS O_3 instrument used within the NASA Airborne Science Program (https://airbornescience.nasa.gov/instrument/VUV-CO).

A new method for explosives detection based on the field asymmetric ion mobility spectrometry (FAIMS) and ionization by an Ar_2* (126 nm) emitter has been developed jointly with a portable detector by Chistyakov *et al.* (2014). Vapor of explosives are introduced into the ionization region containing the Ar_2* lamp. The ions are separated based on their relative mobility. A varying compensation voltage (CV) is applied generating spectra of ion current as a function of CV.

5.7.2 Ultrapure water production

Ultrapure water (UPW) production is one of the most important industrial applications of the VUV process. UPW with strict quality specifications is used in semiconductor, solar photovoltaic, pharmaceutical, and power generation industries, as well as in research laboratories. Point of use (POU) treatment is often applied in critical tool applications such as immersion lithography. Typically wafer manufacture uses ~5500 m^3 city water per day to produce UPW, which is only used once before becoming waste. Recycling rinse water from the semiconductor manufacturing processes has been encouraged. Also, given the increasing water and wastewater treatment costs, the recovery and reuse of spent UPW is currently being investigated (Yazdani, 2016).

UPW is treated to the highest purity from all types of contamination, including organic and inorganic compounds, dissolved and particulate matter, volatiles and non-volatiles, reactive and/or inert species, dissolved gases and bacteria. Treatment steps include gross filtration for large particulate removal, carbon filtration, water softening, reverse osmosis (RO), VUV/UV process for TOC reduction and disinfection, ion exchange or electrodeionization, heat exchange and finally, filtration or ultrafiltration. Depending on the required water quality, UPW treatment plants often also feature degassing, microfiltration, and ultrafiltration (Figure 5.10).

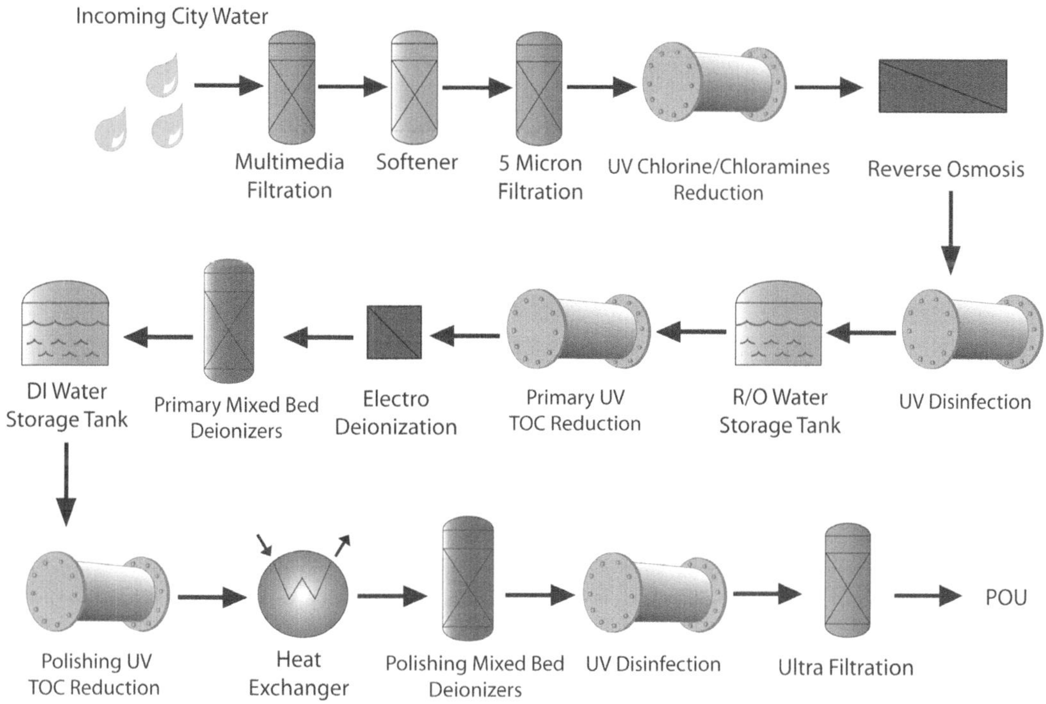

Figure 5.10 Microelectronics water treatment process (courtesy of Aquafine Corporation, Valencia, CA).

The biofilm potentially formed onto RO membranes causes operational problems such as decrease in the membrane water flux, and increased operating pressure. Bohus *et al.* (2010) conducted microbiological investigations at an industrial UPW treatment plant and found that most of the identified bacteria could contribute to microbial-induced corrosion of semiconductor industry products.

Particles, which can cause defects in photolithography products, are controlled through filtration and ultrafiltration. TOC contributes to bacterial proliferation by providing nutrients and, in severe cases, leaves unwanted residues on products and product lines. Water entering the UPW purification system can contain TOC levels between 0.7–15 mg/L (ppm). Removal of TOC requires activated carbon filtration, RO, intermediate oxidation processes, and finally the VUV/UV process to decrease it typically below 1 μg/L (ppb), preferably below 0.5 ppb. Elimination of metallic and anionic (mainly silica) contaminants by ion exchange or electrodeionization and dissolved gases (O_2, O_3 and CO_2) is also required.

Typically, city water feed containing several unwanted contaminants and chlorine as a disinfectant is taken through a series of purification steps depending on the UPW specifications. UPW systems generally consist of three subsystems, i.e., pretreatment, primary treatment, and polishing. Pretreatment produces purified water by using multimedia filtration and ultrafiltration to remove suspended solids and to reduce turbidity, as well as activated carbon for chlorine removal and organic loading reduction. Commercially available activated carbon can effectively remove large non-polar molecules, but shows low removal efficiency for low molecular weight neutral compounds, such as isopropyl alcohol (IPA) and acetone; acetone is the IPA oxidation by-product. Among the most difficult-to-treat compounds in UPW production are urea, trihalomethanes, and low molecular weight neutral compounds, such as acetone. The elimination of chloroform and bromoform is important in minimizing the TOC level. The efficiency of RO followed by UV disinfection can be enhanced by increasing the pH, which results in improved rejection rate of both organics and boron.

The primary treatment (see Figure 5.10) for residual TOC reduction is based on the VUV/UV process, with or without oxidant addition (e.g., sodium persulfate or H_2O_2). Persulfate use is limited as both persulfate residual and sulfate are UPW ionic contaminants. The VUV/UV process efficiency is impacted by suspended solids (light shadowing effect), the presence of Fe^{2+} and Mn^{2+}, which are limited to 0.3 ppm and 0.05 ppm, respectively, to avoid discoloration of quartz housing the VUV/UV lamps, turbidity, and lamp sleeve fouling. Still, the advantages of VUV/UV primary treatment far outweigh these limitations (http://www.prudentialtechgy.com/data/TOC-reduction-article-JD.pdf). Gareth and Avijit (2003) discussed the removal of THMs, particularly of chloroform which was found in RO permeate, and showed the role of 184.9 nm radiation in $CHCl_3$ removal. Another option is the use of heat exchangers, which enable the gas transfer membranes to extract the volatile organics formed, thus reducing residual TOC level. During the primary treatment step, IPA is very efficiently degraded to acetone.

Polishing system established for the final treatment steps (VUV/UV photolysis for TOC reduction, heat exchange, deionization, final disinfection by UV light and ultrafiltration) includes a recirculation and distribution system in order to maintain stable and high quality UPW. High quality UPW is characterized by high resistivity, typically ~18 MΩ cm. When the TOC is less than 1 ppm and the water is saturated with O_2, the VUV/UV process is effective at oxidizing the organics to CO_2. TOC level below 0.5 ppb can be reached using VUV/UV lamps having high output at 184.9 nm (Baas, 2003). Mineralization of TOC results in resistivity drop and ionization of some organics, therefore a deionizer is also required in the polishing system. The oxidative degradation of organics generates CO_2, which is eliminated by degasifier.

Today, the semiconductor manufacturing processes demand typically < 1 ppb (preferably 0.5 ppb) TOC. Aquafine Corporation, a Trojan Technologies business, manufactures VUV (184.9 nm)/UV (253.7 nm)-lamp based reactors which provide an enhanced, synergistic effect toward the reduction of trace organics and microbial contamination for UPW production. These systems are used in applications for TOC reduction, disinfection, ozone (commonly used in the pre-treatment step, as well as for sanitizing process and re-circulating systems) destruction, and chlorine/chloramines removal. Aquafine Corporation also offers UV/H_2O_2-based advanced oxidation solutions for remediation and reuse of spent rinse water, which translates into significant cost savings to utilities and reduces the waste disposal.

There are increasing requirements for TOC reduction (<100 ppb) in industrial water used at nuclear power plants in order to prevent the formation of corroding acids in the steam used to drive the power generating turbines. The system developed and installed by Aquafine for Frosmark nuclear power plant (Sweden) is operated at 2.7 kW and at flow rate of 58 m^3 h^{-1} (http://www.aquafineuv.com).

Evoqua introduced a high flow solution specifically for TOC reduction. The Vanox AOP System can consistently reduce TOC to 0.5 ppb and treat seasonal TOC variations in feed water (http://www.evoqua.com). Species, such as urea, THMs, 1,4-dioxane and IPA are efficiently eliminated with this system. The UPW produced by Vanox AOP Systems contains low levels of particles (<100 units/L, 0.05 μm average diameter) and low residual metal levels (<1.0 ng/L (ppt)).

UPW is also extensively used in analytical chemistry and fundamental research. EMD Millipore (Darmstadt, Germany) manufactures a wide range of Milli-Q® water systems producing high quality UPW (Milli-Q water, resistivity > 18 MΩ, < 4 µg/L TOC) using tap water as feed water. The system contains pretreatment units including RO, electrodeionization and UV irradiation for disinfection, and a polishing system. The polishing system contains two VUV (184.9 nm)/UV (253.7 nm) LP lamps. The first is designed for TOC reduction, and the second one is part of the in-line TOC analyzer. The TOC is determined from the difference between the resistivity of the water before and after the VUV/UV photooxidation.

5.8 VACUUM UV AOP – GENERAL CONCLUSIONS

VUV photolysis is a simple and clean method to produce high concentrations of $HO^•$ in water without the addition of oxidizing agents. The overall yield of $HO^•$ in VUV photolysis of water is much higher than in any other AOPs, because water is the main absorber. However, its application to water treatment is limited, mainly because of the short penetration depth of VUV radiation and inhomogeneity of aqueous VUV systems. Various studies have shown that the irradiation of water by VUV light results in a fast degradation of organic micropollutants and a sequential decomposition of larger NOM structures. Moreover, VUV light sources offer the possibility for the combination of the VUV photolysis with ozonation, when O_3 is photochemically generated from O_2 in the gas phase. Thus, application of external ozone generator can be avoided. Although the VUV photolysis enhances the formation of NO_2^- which is an inorganic by-product of concern in drinking water, the combination of VUV irradiation and O_3 generated by the same lamp enhances the oxidation and disinfection efficiencies of the system and minimizes the formation of NO_2^-.

Currently, the LP lamps with high quality quartz envelopes are used as VUV light sources. Thus, disinfection due to the germicidal effect of UV light and mineralization of organic substances *via* generation of $HO^•$ due to the 184.9 nm light can be simultaneously achieved. Moreover, addition of H_2O_2 enhances the efficiency of the VUV/UV process due to the UV photolysis of H_2O_2 to $HO^•$.

The VUV excimer lamps offer new opportunities for expanding the VUV AOP applications and develop versatile reactor designs. Excimer lamps have several advantages over mercury vapor-based lamps, among which, they are mercury-free, have high photon flux, sharp emission lines, instant on-off switch without UV power losses, and their geometry can be adapted to various reactor configurations. Further development of VUV excimer lamps opens up new dimensions in water treatment.

Despite current limitations, the VUV AOP could be a viable solution for niche applications, such as on-site treatment of industrial wastewater for biodegradability enhancement, and in small residential equipment for concomitant chemical micropollutant destruction and disinfection, in combination with post-VUV activated carbon filtration.

However, at first, the VUV AOP has to evolve from lab-scale experimentation to full-scale implementation. Among others, research is needed on VUV lamp efficiency enhancement, optimization of photoreactor concepts and design to counter-balance the short VUV radiation penetration depth by optimized hydrodynamics and mass transfer into the photoreaction zone, development of reliable kinetic models for equipment sizing and performance guarantee.

5.9 ACKNOWLEDGEMENTS

The authors would like to thank Prof. Dr. Thomas Oppenländer and Dr. Mihaela Stefan for their valuable contributions to this chapter. Special thanks to Prof. Dr. André Braun and Dr. Masato Kukizaki for providing the high quality diagrams of the VUV reactors used in their works (reproduced with permission from Elsevier as Figure 5.9a and 5.9b in this chapter).

5.10 REFERENCES

Adam W. and Oppenländer T. (1984). Absolute intensity method for the determination of quantum yields in 185 nm photochemistry: application to the azoalkane 2,3-diazabicyclo[2.2.1]hept-2-ene. *Photochemistry and Photobiology*, **39**(6), 719–723.

Afzal A., Oppenländer T., Bolton J. R. and El-Din M. G. (2010). Anatoxin-a degradation by advanced oxidation processes: vacuum-UV at 172 nm, photolysis using medium pressure UV and UV/H_2O_2. *Water Research*, **44**(1), 278–286.

Alapi T. and Dombi A. (2007). Comparative study of the UV and UV/VUV-induced photolysis of phenol in aqueous solution. *Journal of Photochemistry and Photobiology A: Chemistry*, **188**(2–3), 409–418.

Alapi T., Berecz L. and Arany E. (2013). Comparison of the UV-induced photolysis, ozonation, and their combination at the same energy input using a self-devised experimental apparatus. *Ozone Science and Engineering*, **35**(5), 350–358.

Alekseev S. B., Kuvshinov V. A., Lisenko A. A., Lomaev M. I., Orlovskii V. M., Panarin V. A., Rozhdestvenskii E. A., Skakun V. S. and Tarasenko V. F. (2006). A photoreactor on the basis of a Xe_2 excilamp. *Instruments and Experimental Techniques*, **49**(1), 132–134.

Al-Gharabli S., Engesser P., Gera D., Klein S. and Oppenländer T. (2016). Engineering of a highly efficient Xe_2*-excilamp (xenon excimer lamp, λ_{max} = 172 nm, η = 40%) and qualitative comparison to a low-pressure mercury lamp (LP-Hg, λ = 185/254 nm) for water purification. *Chemosphere*, **144**, 811–815.

Al-Momani F., Touraud E., Degorce-Dumas J. R., Roussy J. and Thomas O. (2002). Biodegradability enhancement of textile dyes and textile wastewater by VUV photolysis. *Journal of Photochemistry and Photobiology A: Chemistry*, **153**(1–3), 191–197.

Andreozzi R., Caprio V., Insola A. and Marotta R. (1999). Advanced oxidation processes (AOP) for water purification and recovery. *Catalysis Today*, **53**, 51–59.

Arany E., Oppenländer T., Gajda-Schrantz K. and Dombi A. (2012). Influence of H_2O_2 formed in situ on the photodegradation of ibuprofen and ketoprofen. *Current Physical Chemistry*, **2**(3), 286–293.

Arany E., Szabó R. K., Apáti L., Alapi T., Ilisz I., Mazellier P., Dombi A. and Gajda-Schrantz K. (2013). Degradation of naproxen by UV, VUV photolysis and their combination. *Journal of Hazardous Materials*, **262**, 151–157.

Arany E., Alapi T. and Schrantz K. (2015). Reactive Species against Selected Nonsteroidal Anti-inflammatory Drugs. Lap Lambert Academic Publishing, Saarbrücken, Germany.

Atkinson R., Baulch D. L., Cox R. A., Crowley J. N., Hampson R. F., Hynes R. G., Jenkin M. E., Rossi M. J. and Troe J. (2004). Evaluated kinetic and photochemical data for atmospheric chemistry: volume I gas phase reactions of O_x, HO_x, NO_x and SO_x species. *Atmospheric Chemistry and Physics*, **4**, 1461–1738.

Avdeev S. M., Sosnin É. A., Skakun V. S., Tarasenko V. F. and Schitz D. V. (2008). Two-band emission source based on a three-barrier KrCl-XeBr excilamp. *Technical Physics Letters*, **34**(9), 725–727.

Avtaeva S. V., Sosnin E. A., Saghi B., Panarin V. A. and Rahmani B. (2013). Influence of the chlorine concentration on the radiation efficiency of a XeCl exciplex lamp. *Plasma Physics Reports*, **39**(9), 768–778.

Azrague K., Bonnefille E., Pradines V., Pimienta V., Oliveros E., Maurette M. T. and Benoit-Marquie F. (2005). Hydrogen peroxide evolution during V-UV photolysis of water. *Photochemical & Photobiological Sciences*, **4**(5), 406–408.

Baadj S., Harrache Z. and Belasri A. (2013). Electrical and chemical properties of XeCl* (308 nm) exciplex lamp created by a dielectric barrier discharge. *Plasma Physics Reports*, **39**(12), 1043–1054.

Baas M. (2003). Enhanced 185 nm UV-source to achieve a TOC reduction below 1 ppb. *Ultrapure Water*, **20**, 26–30.

Bagheri M. and Mohseni M. (2015a). Impact of hydrodynamics on pollutant degradation and energy efficiency of VUV/UV and H_2O_2/UV oxidation processes. *Journal of Environmental Management*, **164**, 114–120.

Bagheri M. and Mohseni M. (2015b). A study of enhanced performance of VUV/UV process for the degradation of micropollutants from contaminated water. *Journal of Hazardous Materials*, **294**, 1–8.

Balducci A., Marinellia M., Milani E., Morgada M. E., Tucciarone A. and Verona-Rinati G. (2005). Extreme ultraviolet single-crystal diamond detectors by chemical vapor deposition. *Applied Physics Letters*, **86**(19), 193509.

Baricholo P., Hlatywayo D. J., Von Bergmann H. M., Stehmann T., Rohwer E. G. and Collier M. (2011). Influence of gas discharge parameters on emissions from a dielectric barrier discharge excited argon excimer lamp. *The South African Journal of Science*, **107**(11–12), 46–52.

Barrett J. and Baxendale J. H. (1960). The photolysis of liquid water. *Transactions of the Faraday Society*, **56**, 37–43.

Barrett J. and Mansell A. L. (1960). Ultraviolet absorption spectra of the molecules H_2O, HDO and D_2O. *Nature*, **187**, 138–139.

Baum G. and Oppenländer T. (1995). Vacuum-UV-oxidation of chloroorganic compounds in an excimer flow-through photoreactor. *Chemosphere*, **30**(9), 1781–1790.

Beleznai S., Mihajlik G., Agod A., Maros I., Juhász R., Németh Z., Jakab L. and Richter P. (2006). High-efficiency dielectric barrier Xe discharge lamp: theoretical and experimental investigations. *Journal of Physics D: Applied Physics*, **39**(17), 3777–3787.

Beleznai S., Mihajlik G., Maros I., Balázs L. and Richter P. (2008). Improving the efficiency of a fluorescent Xe dielectric barrier light source using short pulse excitation. *Journal of Physics D: Applied Physics*, **41**(11), 115202.

Bergonzo P., Brambilla A., Tromson D., Mer C., Guizard B. and Foulon F. (2000). Diamond devices as characterisation tools for novel photon sources. *Applied Surface Science*, **154–155**, 179–185.

Bielski B. H. J., Cabelli D. E., Arudi R. L. and Ross A. B. (1985). Reactivity of HO_2/O_2^- radicals in aqueous-solution. *Journal of Physical and Chemical Reference Data*, **14**(4), 1041–1100.

Bolton J. R., Bircher K. G., Tuman W. and Tolman C. A. (2001). Figure-of-merit for the technical developement and application of advanced oxidation technologies for both electric- and solar-driven systems. *Pure and Applied Chemistry*, **73**(4), 627–637.

Bonin J., Janik I., Janik D. and Bartels D. M. (2007). Reaction of the hydroxyl radical with phenol in water up to supercritical conditions. *The Journal of Physical Chemistry A*, **111**(10), 1869–1878.

Braslavsky S. E. (2007). Glossary of terms used in photochemistry, 3rd edition (IUPAC Recommendations 2006). *Pure and Applied Chemistry*, **79**(3), 293–465.

Braun A. M., Pintori I. G., Popp H. P., Wakahata Y. and Worner M. (2004). Technical development of UV-C- and VUV-photochemically induced oxidative degradation processes. *Water Science and Technology*, **49**(4), 235–240.

Buchanan W., Roddick F., Porter N. and Drikas M. (2005). Fractionation of UV and VUV pretreated natural organic matter from drinking water. *Environmental Science & Technology*, **39**(12), 4647–4654.

Buchanan W., Roddick F. and Porter N. (2006). Formation of hazardous by-products resulting from the irradiation of natural organic matter: comparison between UV and VUV irradiation. *Chemosphere*, **63**(7), 1130–1141.

Buchanan W., Roddick F. and Porter N. (2008). Removal of VUV pre-treated natural organic matter by biologically activated carbon columns. *Water Research*, **42**(13), 3335–3342.

Budovich V. L. and Il'in V. P. (2014). Measurements of the intensities of vacuum ultraviolet radiation sources using the flow ionization chamber. *Instruments and Experimental Techniques*, **57**(2), 195–200.

Buxton G. V., Greenstock C. L., Helman W. P. and Ross A. B. (1988). Critical-review of rate constants for reactions of hydrated electrons, hydrogen-atoms and hydroxyl radicals (•OH/•O⁻) in aqueous-solution. *Journal of Physical and Chemical Reference Data*, **17**(2), 513–886.

Cao M. H., Wang B. B., Yu H. S., Wang L. L., Yuan S. H. and Chen J. (2010). Photochemical decomposition of perfluorooctanoic acid in aqueous periodate with VUV and UV light irradiation. *Journal of Hazardous Materials*, **179**(1–3), 1143–1146.

Carman R. J. and Mildren R. P. (2003). Computer modelling of a short-pulse excited dielectric barrier discharge xenon excimer lamp ($\lambda \sim 172$ nm). *Journal of Physics D: Applied Physics*, **36**(1), 19–33.

Chen J., Zhang P. and Liu J. (2007). Photodegradation of perfluorooctanoic acid by 185 nm vacuum ultraviolet light. *Journal of Environmental Sciences*, **19**(4), 387–390.

Chen P., Hou K., Hua L., Xie Y., Zhao W., Chen W., Chen C. and Li H. (2014). Quasi-trapping chemical ionization source based on a commercial VUV lamp for time-of-flight mass spectrometry. *Analytical Chemistry*, **86**(3), 1332–1336.

Chistyakov A. A., Kotkovskii G. E., Sychev A. V. and Budovich V. L. (2014). An excimer-based FAIMS detector for detection of ultra-low concentration of explosives. Conference: SPIE Defense + Security. Chromdet/Analytical Instruments, Russia.

Collivignarelli C. and Sorlini S. (2004). AOPs with ozone and UV radiation in drinking water: contaminants removal and effects on disinfection byproducts formation. *Water Science and Technology*, **49**(4), 51–56.

Crapulli F., Santoro D., Sasges M. R. and Ray A. K. (2014). Mechanistic modeling of vacuum UV advanced oxidation process in an annular photoreactor. *Water Research*, **64**, 209–225.

Dainton F. S. and Fowles P. (1965). The photolysis of aqueous systems at 1849 Å. I. Solutions containing nitrous oxide. *Proceedings of Royal Society*, **A287**, 295–311.

Dobrovic S., Juretic H. and Ruzinski N. (2007). Photodegradation of natural organic matter in water with UV irradiation at 185 and 254 nm: importance of hydrodynamic conditions on the decomposition rate. *Separation Science and Technology*, **42**(7), 1421–1432.

Dombi A., Ilisz I., László Z. and Wittmann G. (2002). Comparison of ozone-based and other (VUV and TiO_2/UV) radical generation methods in phenol decomposition. *Ozone Science and Engineering*, **24**(1), 49–54.

Dwibedy P., Dey G. R., Naik D. B., Kishore K. and Moorthy P. N. (1999). Pulse radiolysis studies on redox reactions of gallic acid: one electron oxidation of gallic acid by gallic acid OH adduct. *Physical Chemistry Chemical Physics*, **1**(8), 1915–1918.

Echigo S., Yamada H., Matsui S., Kawanishi S. and Shishida K. (1996). Comparison between O_3/VUV, O_3/H_2O_2, VUV and O_3 processes for the decomposition of organophosphoric acid triesters. *Water Science and Technology*, **34**(9), 81–88.

Eliasson B. and Kogelschatz U. (1988). UV excimer radiation from dielectric-barrier discharges. *Applied Physics B-Photo*, **46**(4), 299–303.

Eliasson B. and Kogelschatz U. (1991). Modeling and applications of silent discharge plasmas. *IEEE Transactions on Plasma Science*, **19**(2), 309–323.

Elsner C., Lenk M., Prager L. and Mehnert R. (2006). Windowless argon excimer source for surface modification. *Applied Surface Science*, **252**(10), 3616–3624.

Erofeev M. V., Schitz D. V., Skakun V. S., Sosnin E. A. and Tarasenko V. F. (2010). Compact dielectric barrier discharge excilamps. *Physica Scripta*, **82**(4), 045403.

Farkas L. and Hirshberg Y. (1937). The photochemical decomposition of aliphatic alcohols in aqueous solution. *Journal of the American Chemical Society*, **59**(11), 2450–2453.

Gareth T. and Avijit D. (2003). Removal of trihalomethanes from RO product water using UV 185 nm technology. *Ultrapure Water*, **20**, 18–22.

Gellert B. and Kogelschatz U. (1991). Generation of excimer emission in dielectric barrier discharges. *Applied Physics B-Photo*, **52**(1), 14–21.

Gerasimov G. N., Volkova G. A., Hallin R., Zvereva G. N. and Heikensheld F. (2000). VUV spectrum of the barrier discharge in a krypton-xenon mixture. *Optics and Spectroscopy*, **88**(6), 814–818.

Gerasimov G. N., Krylov B. E., Hallin R., Morozov A. O., Arnesen A. and Heijkenskjold F. (2002). Stimulated emission of inert gas mixtures in the VUV range. *Optics and Spectroscopy*, **92**(2), 290–297.

Gerasimov G. N., Krylov B. E., Hallin R., Morozov A. O., Arnesen A. and Heijkenskjold F. (2003). Vacuum ultraviolet spectra of heteronuclear dimers of inert gases in a direct-current discharge. *Optics and Spectroscopy*, **94**(3), 374–383.

Gerasimov G. N., Krylov B. E., Hallin R. and Arnesen A. (2006). Parameters of VUV radiation from a DC capillary discharge in a mixture of krypton with xenon. *Optics and Spectroscopy*, **100**(6), 825–829.

Gerbig C., Schmitgen S., Kley D., Volz-Thomas A., Dewey K. and Haaks D. (1999). An improved fast-response vacuum-UV resonance fluorescence CO instrument. *Journal of Geophysical Research*, **104**(D1), 1699–1704.

Getoff N. (1990). Decomposition of biological resistant pollutants in water by irradiation. *Radiation Physics and Chemistry*, **35**, 432–439.

Getoff N. (1996). Radiation-induced degradation of water pollutants – state of the art. *Radiation Physics and Chemistry*, **47**(4), 581–593.

Getoff N. and Schenck G. O. (1968). Primary products of liquid water photolysis at 1236 Å, 1470 Å and 1849 Å. *Journal of Photochemistry and Photobiology A: Chemistry*, **8**, 167–178.

Giri R. R., Ozaki H., Morigaki T., Taniguchi S. and Takanami R. (2011a). UV photolysis of perfluorooctanoic acid (PFOA) in dilute aqueous solution. *Water Science and Technology*, **9**, 276–282.

Giri R. R., Ozaki H., Okada T., Takikita S., Taniguchi S. and Takanami R. (2011b). Water matrix effect on UV photodegradation of perfluorooctanoic acid. *Water Science and Technology*, **64**(10), 1980–1986.

Giri R. R., Ozaki H., Okada T., Taniguchi S. and Takanami R. (2012). Factors influencing UV photodecomposition of perfluorooctanoic acid in water. *Chemical Engineering Journal*, **180**, 197–203.

Goldstein S., Aschengrau D., Diamant Y. and Rabani J. (2007). Photolysis of aqueous H_2O_2: quantum yield and applications for polychromatic UV actinometry in photoreactors. *Environmental Science & Technology*, **41**(21), 7486–7490.

Gonzalez M. C. and Braun A. M. (1995). VUV photolysis of aqueous solutions of nitrate and nitrite. *Research on Chemical Intermediates*, **21**, 837–859.

Gonzalez M. C. and Braun A. M. (1996). Vacuum-UV photolysis of aqueous solutions of nitrate: effect of organic matter I. Phenol. *Journal of Photochemistry and Photobiology A: Chemistry*, **93**, 7–19.

Gonzalez M. C., Braun A. M., Prevot A. B. and Pelizzetti E. (1994). Vacuum-ultraviolet (VUV) photolysis of water - mineralization of atrazine. *Chemosphere*, **28**(12), 2121–2127.

Gonzalez M. C., Hashem T. M., Jakob L. and Braun A. M. (1995). Oxidative-degradation of nitrogen-containing organic-compounds – Vacuum-ultraviolet (VUV) photolysis of aqueous-solutions of 3-amino 5-methylisoxazole. *Fresenius J. Anal. Chem.*, **351**(1), 92–97.

Gonzalez M. C., Oliveros E., Worner M. and Braun A. (2004). Vacuum-ultraviolet photolysis of aqueous reaction systems. *Journal of Photochemistry and Photobiology C: Photochemistry Reviews*, **5**(3), 225–246.

Gu X., Lu S., Qiu Z., Sui Q., Banks C. J., Imai T., Lin K. and Luo Q. (2013). Photodegradation performance of 1,1,1-trichloroethane in aqueous solution: in the presence and absence of persulfate. *Chemical Engineering Journal*, **215–216**, 29–35.

Halmann M. and Platzner I. (1965). The photochemistry of phosphorus compounds. Part III. Photolysis of ethyl dihydrogen phosphate in aqueous solution. *Journal of the Chemical Society*, **1440**, 5380–5385.

Han W., Zhang P., Zhu W., Yin J. and Li L. (2004). Photocatalysis of p-chlorobenzoic acid in aqueous solution under irradiation of 254 nm and 185 nm UV light. *Water Research*, **38**(19), 4197–4203.

Hashem T. M., Zirlewagen M. and Braum A. M. (1997). Simultaneous photochemical generation of ozone in the gas phase and photolysis of aqueous reaction systems using one VUV light source. *Water Science and Technology*, **35**(4), 41–48.

Hayashi K., Tachibana T., Kawakami N., Yokota Y., Kobashi K., Ishihara H., Uchida K., Nippashi K. and Matsuoka M. (2005). Diamond sensors durable for continuously monitoring intense vacuum ultraviolet radiation. *The Japanese Journal of Applied Physics*, **44**(10), 7301–7304.

Hayon E. and McGarvey J. J. (1967). Flash photolysis in the vacuum ultraviolet region of $S_2O_4^{2-}$, CO_3^{2-} and OH^- ions in aqueous solutions. *Journal of Physical Chemistry*, **71**(5), 1472–1477.

Heit G. and Braun A. M. (1996). Spatial resolution of oxygen measurements during VUV-photolysis of aqueous systems. *Journal of Information Recording*, **22**(5–6), 543–546.

Heit G. and Braun A. M. (1997). VUV-photolysis of aqueous systems: spatial differentiation between volumes of primary and secondary reactions. *Water Science and Technology*, **35**(4), 25–30.

Heit G., Neuner A., Saugy P. Y. and Braun A. M. (1998). Vacuum-UV (172 nm) actinometry. The quantum yield of the photolysis of water. *The Journal of Physical Chemistry A*, **102**(28), 5551–5561.

Hoigné J. (1998). Chemistry of aqueous ozone and transformation of pollutants by ozonation and advanced oxidation processes, Part C Quality and treatment of drinking water II. In: The Handbook of Environmental Chemistry, J. Hrubec (ed.), Springer, Berlin.

Huang L., Jing H., Cheng Z. and Dong W. (2013). Different photodegradation behavior of 4-tert-octylphenol under UV and VUV irradiation in aqueous solution. *Journal of Photochemistry and Photobiology A: Chemistry*, **251**, 69–77.

Imoberdorf G. and Mohseni M. (2011). Modeling and experimental evaluation of vacuum-UV photoreactors for water treatment. *Chemical Engineering Science*, **66**(6), 1159–1167.

Imoberdorf G. and Mohseni M. (2012). Kinetic study and modeling of the vacuum-UV photoinduced degradation of 2,4-D. *Chemical Engineering Science*, **187**, 114–122.

Imoberdorf G. and Mohseni M. (2014). Comparative study of the effect of vacuum-ultraviolet irradiation on natural organic matter of different sources. *Journal of Environmental Engineering*, **140**(3), 04013016.

Jakob L., Hashem T. M., Bürki S., Guindy N. M. and Braun A. M. (1993). Vacuum-ultraviolet (VUV) photolysis of water: oxidative degradation of 4-chlorophenol. *Journal of Photochemistry and Photobiology A: Chemistry*, **75**(2), 97–103.

Jin L. and Zhang P. (2015). Photochemical decomposition of perfluorooctane sulfonate (PFOS) in an anoxic alkaline solution by 185 nm vacuum ultraviolet. *Chemical Engineering Journal*, **280**, 241–247.

Khan J. A., He X., Shah N. S., Khan H. M., Hapeshi E., Fatta-Kassinos D. and Dionysiou D. D. (2014). Kinetic and mechanism investigation on the photochemical degradation of atrazine with activated H_2O_2, $S_2O_8^{2-}$ and HSO_5^-. *Chemical Engineering Journal*, **252**(0), 393–403.

Khan J. A., Shah N. S., Nawaz S., Ismail M., Rehman F. and Khan H. M. (2015). Role of e_{aq}^-, •OH and H• in radiolytic degradation of atrazine: a kinetic and mechanistic approach. *Journal of Hazardous Materials*, **288**, 147–157.

Kim I. and Tanaka H. (2009). Photodegradation characteristics of PPCPs in water with UV treatment. *Environment International*, **35**(5), 793–802.

Kitamura M., Mitsuka K. and Sato M. A. (2004). Practical high-power excimer lamp excited by a microwave discharge. *Applied Surface Science*, **79–80**, 507–513.

Kogelschatz U. (1990). Silent discharges for the generation of ultraviolet and vacuum ultraviolet excimer radiation. *Pure and Applied Chemistry*, **62**(9), 1667–1674.

Kogelschatz U. (1992). Silent-discharge driven excimer UV sources and their applications. *Applied Surface Science*, **54**, 410–423.

Kogelschatz U. (2003). Dielectric-barrier discharges: their history, discharge physics, and industrial applications. *Plasma Chemistry and Plasma Processing*, **23**(1), 1–46.

Kogelschatz U. (2004). Excimer lamps: history, discharge physics, and industrial applications. Proceedings of the SPIE 5483, Atomic and Molecular Pulsed Lasers V. Bellingham, WA.

Kogelschatz U. (2012). Ultraviolet excimer radiation from nonequilibrium gas discharges and its application in photophysics, photochemistry and photobiology. *Journal of Optical Technology*, **79**(8), 484–493.

Kogelschatz U., Esrom H., Zhang J. Y. and Boyd I. W. (2000). High-intensity sources of incoherent UV and VUV excimer radiation for low-temperature materials processing. *Applied Surface Science*, **168**, 29–36.

Kogelschatz U., Eliasson B. and Egli W. (1997). Dielectric-barrier discharges. Principle and applications. *J. Phys. IV France*, **7**(C4), 47–66.

Koppenol W. H. and Liebman J. F. (1984). The oxidizing nature of the hydroxyl radical – a comparison with the ferryl ion (FeO_2^+). *Journal of Physical Chemistry*, **88**(1), 99–101.

Kozmér Z., Arany E., Alapi T., Takács E., Wojnárovits L. and Dombi A. (2014). Determination of the rate constant of hydroperoxyl radical reaction with phenol. *Radiation Physics and Chemistry*, **102**, 135–138.

Kröckel L. and Schmidt M. A. (2014). Extinction properties of ultrapure water down to deep ultraviolet wavelengths. *Optical Materials Express*, **4**(9), 1932–1942.

Kuhn H. J., Braslavsky S. E. and Schmidt R. (2004). International Union of pure and applied chemistry organic and biomolekular chemistry division subcommittee on photochemistry, Chemical actinometry (IUPAC Technical Report). *Pure and Applied Chemistry*, **76**(12), 2105–2146.

Kutschera K., Bornick H. and Worch E. (2009). Photoinitiated oxidation of geosmin and 2-methylisoborneol by irradiation with 254 nm and 185 nm UV light. *Water Research*, **43**(8), 2224–2232.

László Z. (2001). Application of the VUV Photolysis in the Mineralization of Pollutants of Water. PhD thesis, University of Szeged, Szeged, Hungary.

László Z. and Dombi A. (2002). Oxidation of $[Fe(CN)_6]^{4-}$ and reduction of $[Fe(CN)_6]^{3-}$ in VUV irradiated aqueous solutions. *Chemosphere*, **46**, 491–494.

László Z., Ilisz I., Peintler G. and Dombi A. (1998). VUV intensity measurement of a 172 nm Xe excimer lamp by means of oxygen actinometry. *Ozone Science and Engineering*, **20**(5), 421–432.

Li Q. R., Gu C. Z., Di Y., Yin H. and Zhang J. Y. (2006). Photodegradation of nitrobenzene using 172 nm excimer UV lamp. *Journal of Hazardous Materials*, **133**(1–3), 68–74.

Li W., Lu S., Chen N., Gu X., Qiu Z., Fan J. and Lin K. (2009). Photo-degradation of clofibric acid by ultraviolet light irradiation at 185 nm. *Water Science and Technology*, **60**(11), 2983–2989.

Li W., Lu S., Qiu Z. and Lin K. (2011). UV and VUV photolysis vs. UV/H_2O_2 and VUV/H_2O_2, treatment for removal of clofibric acid from aqueous solution. *Environmental Technology*, **32**(9–10), 1063–1071.

Liu G. Y. (2014). Recalcitrance of cyanuric acid to oxidative degradation by OH radical: theoretical investigation. *RSC Advances*, **4**(70), 37359–37364.

Liu K., Roddick F. A. and Fan L. (2011). Potential of UV/H_2O_2 oxidation for enhancing the biodegradability of municipal reverse osmosis concentrates. *Water Science and Technology*, **63**(11), 2605–2611.

Liu Y. and Ogden K. (2010). Benefits of high energy UV 185 nm light to inactivate bacteria. *Water Science and Technology*, **62**(12), 2776–2782.

Lomaev M. I., Skakun V. S., Sosnin E. A., Tarasenko V. F., Shitts D. V. and Erofeev M. V. (2003). Excilamps: efficient sources of spontaneous UV and VUV radiation. *Physics-Uspekhi*, **46**(2), 193–209.

Lomaev M. I., Skakun V. S., Tarasenko V. F., Shitts D. V. and Lisenko A. A. (2006a). A windowless VUV excilamp. *Technical Physics Letters*, **32**(7), 590–592.

Lomaev M. I., Sosnin E. A., Tarasenko V. F., Shits D. V., Skakun V. S., Erofeev M. V. and Lisenko A. A. (2006b). Capacitive and barrier discharge excilamps and their applications (Review). *Instruments and Experimental Techniques*, **49**(5), 595–616.

Lomaev M. I., Tarasenko V. F. and Schitz D. V. (2007). On the formation of a barrier discharge in excilamps. *Technical Physics*, **52**(8), 1046–1052.

Lomaev M. I., Skakun V. S., Tarasenko V. F. and Schitz D. V. (2008). One- and two-barrier excilamps on xenon dimers operating in the VUV range. *Technical Physics*, **53**(2), 244–248.

Lomaev M. I., Sosnin E. A. and Tarasenko V. F. (2012). Excilamps and their applications (Review). *Quantum Electronics*, **36**(1), 51–97.

Lopez J. L., Einschlag F. S. G., Gonzalez M. C., Capparelli A. L., Oliveros E., Hashem T. M. and Braun A. M. (2000). Hydroxyl radical initiated photodegradation of 4-chloro-3,5-dinitrobenzoic acid in aqueous solution. *Journal of Photochemistry and Photobiology A: Chemistry*, **137**(2–3), 177–184.

Malinin A. N. (2006). An excimer source of visible light. *Instruments and Experimental Techniques*, **49**(1), 96–100.

Mamane H., Shemer H. and Linden K. G. (2007). Inactivation of *E. coli*, *B. subtilis* spores, and MS2, T4, and T7 phage using UV/H_2O_2 advanced oxidation. *Journal of Hazardous Materials*, **146**, 479–486.

Martire D. O. and Gonzalez M. C. (2001). Aqueous phase kinetic studies involving intermediates of environmental interest: phosphate radicals and their reactions with substituted benzenes. *Progress in Reaction Kinetics and Mechanism*, **26**(2–3), 201–218.

Masschelein W. J. (2002). Ultraviolet Light in Water and Wastewater Sanitation. Lewis Publishers. Boca Raton, FL.

Matilainen A. and Sillanpää M. (2010). Removal of natural organic matter from drinking water by advanced oxidation processes - review. *Chemosphere*, **80**, 351–365.

Matsushita T., Hirai S., Ishikawa T., Matsui Y. and Shirasaki N. (2015). Decomposition of 1,4-dioxane by vacuum ultraviolet irradiation: study of economic feasibility and by-product formation. *Process Safety and Environmental Protection*, **94**, 528–541.

Matter D., Burtscher H., Kogelschatz U. and Scherrer L. (1994). Photoemission of combustion aerosols using an excimer UV radiation source. *Staub-Reinhaltung der Luft*, **54**, 163–166.

Moussavi G., Hossaini H., Jafari S. J. and Farokhi M. (2014). Comparing the efficacy of UVC, UVC/ZnO and VUV processes for oxidation of organophosphate pesticides in water. *Journal of Photochemistry and Photobiology A: Chemistry*, **290**, 86–93.

Muhlberger F., Wieser J., Ulrich A. and Zimmermann R. (2002). Single photon ionization (SPI) via incoherent VUV-excimer light: robust and compact time-of-flight mass spectrometer for on-line, real-time process gas analysis. *Analytical Chemistry*, **74**(15), 3790–3801.

Nikogosyan D. N. and Görner H. (1992). Photolysis of aromatic amino acids in aqueous solution by nanosecond 248 and 193 nm laser light. *Journal of Photochemistry and Photobiology B: Biology*, **13**(3–4), 219–234.

Ochiai T., Masuko K., Tago S., Nakano R., Nakata K., Hara M., Nojima Y., Suzuki T., Ikekita M., Morito Y. and Fujishima A. (2013a). Synergistic water-treatment reactors using a TiO_2-modified Ti-mesh filter. *Water*, **5**(3), 1101–1115.

Ochiai T., Masuko K., Tago S., Nakano R., Niitsu Y., Kobayashi G., Horio K., Nakata K., Murakami T., Hara M., Nojima Y., Kurano M., Serizawa I., Suzuki T., Ikekita M., Morito Y. and Fujishima A. (2013b). Development of a hybrid environmental purification unit by using of excimer VUV lamps with TiO_2 coated titanium mesh filter. *Chemical Engineering Journal*, **218**, 327–332.

Oppenländer T. (1994). Novel incoherent excimer UV irradiation units for the application in photochemistry, photobiology/-medicine and for waste water treatment. *European Photochemistry Association Newsletter*, **50**, 2–8.

Oppenländer T. (1996). The contribution of organic photochemistry to investigations of phototoxicity. In: The Photostability of Drugs and Drug Formulations, H. Tonnesen (ed.), Taylor & Francis, London, pp. 217–265.

Oppenländer T. (2003). Photochemical Purification of Water and Air. Wiley-VCH, Weinheim.

Oppenländer T. (2007). Mercury-free sources of VUV/UV radiation: application of modern excimer lamps (excilamps) for water and air treatment. *Journal of Environmental Engineering and Science*, **6**(3), 253–264.

Oppenländer T. and Gliese S. (2000). Mineralization of organic micropollutants (homologous alcohols and phenols) in water by vacuum-UV-oxidation (H_2O-VUV) with an incoherent xenon-excimer lamp at 172 nm. *Chemosphere*, **40**, 15–21.

Oppenländer T. and Schwarzwälder R. (2002). Vacuum-UV oxidation (H_2O-VUV) with a xenon excimer flow-through lamp at 172 nm: use of methanol as actinometer for VUV intensity measurement and as reference compound for OH-radical competition kinetics in aqueous systems. *Journal of Advanced Oxidation Technologies*, **5**(2), 155–163.

Oppenländer T. and Sosnin E. (2005). Mercury-free vacuum-(VUV) and UV excilamps: lamps of the future? *IUVA News*, **7**(4), 16–20.

Oppenländer T. and Xu F. (2008). Temperature effects on the vacuum-UV (VUV)-initiated oxidation and mineralization of organic compounds in aqueous solution using a xenon excimer flow-through photoreactor at 172 nm. *Ozone Science and Engineering*, **30**(1), 99–104.

Oppenländer T., Baum G., Egle W. and Hennig T. (1995). Novel vacuum-UV-(VUV) and UV-excimer flow-through photoreactors for waste water treatment and for wavelength-selective photochemistry. *Proceedings of the Indian Academy of Sciences, Chemical Sciences*, **107**(6), 621–636.

Oppenländer T., Walddorfer C., Burgbacher J., Kiermeier M., Lachner K. and Weinschrott H. (2005). Improved vacuum-UV (VUV)-initiated photomineralization of organic compounds in water with a xenon excimer flow-through photoreactor (Xe_2* lamp, 172 nm) containing an axially centered ceramic oxygenator. *Chemosphere*, **60**(3), 302–309.

Puspita P., Roddick F. A. and Porter N. A. (2011). Decolourisation of secondary effluent by UV-mediated processes. *Chemical Engineering Journal*, **171**(2), 464–473.

Querry M. R., Cary P. G. and Waring R. C. (1978). Split-pulse laser method for measuring attenuation coefficients of transparent liquids: application to deionized filtered water in the visible region. *Applied Optics*, **17**(22), 3587–3592.

Quici N., Litter M. I., Braun A. M. and Oliveros E. (2008). Vacuum-UV-photolysis of aqueous solutions of citric and gallic acids. *Journal of Photochemistry and Photobiology A: Chemistry*, **197**(2–3), 306–312.

Ratpukdi T., Siripattanakul S. and Khan E. (2010). Mineralization and biodegradability enhancement of natural organic matter by ozone-VUV in comparison with ozone, VUV, ozone-UV, and UV: effects of pH and ozone dose. *Water Research*, **44**(11), 3531–3543.

Ratpukdi T., Casey F., DeSutter T. and Khan E. (2011). Bromate formation by ozone-VUV in comparison with ozone and ozone-UV: effects of pH, ozone dose, and VUV power. *Journal of Environmental Engineering*, **137**(3), 187–195.

Reisz E., Schmidt W., Schochmann H.-P. and von Sonntag C. (2003). Photolysis of ozone in aqueous solution in the presence if tertiary butanol. *Environmental Science & Technology*, **37**(9), 1941–1948.

Robl S., Worner M., Maier D. and Braun A. M. (2012). Formation of hydrogen peroxide by VUV-photolysis of water and aqueous solutions with methanol. *Photochemical & Photobiological Sciences*, **11**(6), 1041–1050.

Routledge E. J. and Sumpter J. P. (1997). Structural features of alkylphenolic chemicals associated with estrogenic activity. *The Journal of Biological Chemistry*, **272**(6), 3280–3288.

Satou T., Nakazato T. and Tao H. (2013). Online TOC analysis based on reagent-free oxidation of dissolved organic matter using a mercury lamp-pass-through photoreactor. *Analytical Sciences*, **29**, 233–238.

Schalk S., Adam V., Arnold E., Brieden K., Voronov A. and Witzke H.-D. (2005). UV-lamps for disinfection and advanced oxidation – lamp types, technologies and applications. *IUVA News*, **8**(1), 32–37.

Schreiber A., Kühn B., Arnold E., Schilling F. J. and Witzke H. D. (2005). Radiation resistance of quartz glass for VUV discharge lamps. *Journal of Physics D: Applied Physics*, **38**(17), 3242–3250.

Schürgers M. and Welge K. H. (1968). Absorptionskoeffizient von H_2O_2 und N_2H_4 zwischen 1200 und 2000 Å. *Z. Naturforsch.*, **23a**, 1508–1510.

Segelstein D. J. (1981). The Complex Refractive Index of Water. M.S. thesis, University of Missouri, Kansas City, USA.

Shen Y. S. and Liao B. H. (2007). Study on the treatment of Acid Red 4 wastewaters by a laminar-falling-film-slurry-type VUV photolytic process. *Water Science and Technology*, **55**(12), 13–18.

Shirayama H., Tohezo Y. and Taguchi S. (2001). Photodegradation of chlorinated hydrocarbons in the presence and absence of dissolved oxygen in water. *Water Research*, **35**(8), 1941–1950.

Shuaibov A. K., Minya A. I., Malinin A. N., Homoki Z. T. and Hrytsak R. V. (2012). VUV lamp based on mixtures of inert gases with water molecules pumped by a pulsed-periodic capacitive discharge. *Journal of Applied Spectroscopy*, **78**(6), 867–872.

Shuaibov A. K., Gomoki Z. T., Minya A. I. and Shevera I. V. (2013). Emission characteristics of an ultraviolet emitter based on mixtures of krypton with low-aggressive halogen carriers pumped by a barrier discharge. *Optics and Spectroscopy*, **114**(2), 189–192.

Simonsen M. E., Jensen C. V. and Sogaard E. G. (2013). Comparison of different UV-activated AOP methods. *Journal of Advanced Oxidation Technologies*, **16**(1), 179–187.

Sokolov U. and Stein G. (1996). Photolysis of liquid water at 1849 Å. *Journal of Chemical Physics*, **44**(9), 3329–3337.

Sosnin E. A., Erofeev M. V., Tarasenko V. F. and Shitz D. V. (2002). Capacitive discharge excilamps. *Instruments and Experimental Techniques*, **45**(6), 838–839.

Sosnin E. A., Oppenländer T. and Tarasenko V. F. (2006). Applications of capacitive and barrier discharge excilamps in photoscience. *Journal of Photochemistry and Photobiology C: Photochemistry Reviews*, **7**(4), 145–163.

Sosnin E. A., Avdeev S. M., Panarin V. A., Tarasenko V. F., Pikulev A. A. and Tsvetkov V. V. (2011a). The radiative and thermodynamic processes in DBD driven XeBr and KrBr exciplex lamps. *The European Physical Journal D*, **62**(3), 405–411.

Sosnin E. A., Pikulev A. A. and Tarasenko V. F. (2011b). Optical characteristics of cylindrical exciplex and excimer lamps excited by microwave radiation. *Technical Physics*, **56**(4), 526–530.

Sosnin E. A., Avdeev S. M., Tarasenko V. F., Skakun V. S. and Schitz D. V. (2015a). KrCl barrier-discharge excilamps: energy characteristics and applications (Review). *Instruments and Experimental Techniques*, **58**(3), 309–318.

Sosnin E. A., Korzenev A. N., Avdeev S. M., Volkind D. K., Novakovskii G. S. and Tarasenko V. F. (2015b). Numerical simulation and experimental study of thermal and gas-dynamic processes in barrier-discharge coaxial excilamps. *High Temperature*, **53**(4), 558–563.

Steenland K., Fletcher T. and Savitz D. (2010). Epidemiologic evidence on the health effects of perfluorooctanoic acid (PFOA). *Environmental Health Perspectives*, **118**(8), 1100–1108.

Suzuki F., Ono K., Sakai K. and Hayashi K. (2006). Direct measurement of 185 nm radiation from low-pressure mercury lamps using diamond-based vacuum ultraviolet sensors. *The Japanese Journal of Applied Physics*, **45**(8A), 6484–6485.

Szabó R. K., Megyeri C., Illés E., Gajda-Schrantz K., Mazellier P. and Dombi A. (2011). Phototransformation of ibuprofen and ketoprofen in aqueous solutions. *Chemosphere*, **84**(11), 1658–1663.

Tarasenko V. F., Chernov E. B., Erofeev M. V., Lomaev M. I., Panchenko A. N., Skakun V. S., Sosnin E. A. and Shitz D. V. (1999). UV and VUV excilamps excited by glow, barrier and capacitive discharges. *Applied Physics A*, **69**, 327–329.

Tasaki T., Wada T., Fujimoto K., Kai S., Ohe K., Oshima T., Baba Y. and Kukizaki M. (2009). Degradation of methyl orange using short-wavelength UV irradiation with oxygen microbubbles. *Journal of Hazardous Materials*, **162**(2–3), 1103–1110.

Thomsen C. L., Madsen D., Keiding S. R., Thøgersen J. and Christiansen O. (1999). Two-photon dissociation and ionization of liquid water studied by femtosecond transient absorption spectroscopy. *Journal of Chemical Physics*, **110**(7), 3453–3462.

Thomson J., Roddick F. A. and Drikas M. (2002). Natural organic matter removal by enhanced photo-oxidation using low pressure mercury vapour lamps. *Water Science and Technology*, **2**(5–6), 435–443.

Thomson J., Roddick F. A. and Drikas M. (2004). Vacuum ultraviolet irradiation for naturai organic matter removal. *Journal of Water Supply: Research & Technology*, **53**(4), 196–206.

Tsujimoto Y., Hashizume H. and Yamazaki M. (1993). Superoxide radical scavenging activity of phenolic compounds. *International Journal of Biochemistry*, **25**(4), 491–494.

Van der Pol A. J. H. P. and Krijnen S. (2005). Optimal UV output in different application of low-pressure UV-C lamps. Proceedings of the Third International Congress on Ultraviolet Technologies, International Ultraviolet Association. Whistler, BC, Canada.

Vicente J. S., Gejo J. L., Rothenbacher S., Sarojiniamma S., Gogritchiani E., Worner M., Kasper G. and Braun A. M. (2009). Oxidation of polystyrene aerosols by VUV-photolysis and/or ozone. *Photochemical & Photobiological Sciences*, **8**(7), 944–952.

Volkova G. A. and Gerasimov G. N. (1997). Amplification of $\lambda = 147$ nm radiation from a barrier discharge in a mixture of krypton with xenon. *Quantum Electronics*, **27**(3), 213–216.

von Sonntag C. and Schuchmann H. P. (1991). The elucidation of peroxyl radical reactions in aqueous solution with the help of radiation-chemical methods. *Angewandte Chemie International Edition (England)*, **30**, 1229–1253.

von Sonntag C. and Schuchmann H. P. (1997). Peroxyl Radicals in Aqueous Solutions. John Wiley and Sons, Chichester.

Voronov A., Arnold E. and Roth E. (2003) Long life technology of high power amalgam lamps. Proceedings of the Second International Conference on Ultraviolet Technologies, International Ultraviolet Association. Vienna, Austria.

Wang B. B., Cao M. H., Tan Z. J., Wang L. L., Yuan S. H. and Chen J. (2010a). Photochemical decomposition of perfluorodecanoic acid in aqueous solution with VUV light irradiation. *Journal of Hazardous Materials*, **181**(1–3), 187–192.

Wang D., Oppenländer T., El-Din M. G. and Bolton J. R. (2010b). Comparison of the disinfection effects of vacuum-UV (VUV) and UV light on Bacillus subtilis spores in aqueous suspensions at 172, 222 and 254 nm. *Photochemistry and Photobiology*, **86**(1), 176–181.

Wang J., Yang C., Wang C., Han W. and Zhu W. (2014). Photolytic and photocatalytic degradation of micro pollutants in a tubular reactor and the reaction kinetic models. *Separation and Purification Technology*, **122**, 105–111.

Wang Y. and Zhang P. (2014). Effects of pH on photochemical decomposition of perfluorooctanoic acid in different atmospheres by 185 nm vacuum ultraviolet. *Journal of Environmental Sciences (China)*, **26**(11), 2207–2214.

Weeks J. L., Meaburn G. M. A. C. and Gordon S. (1963). Absorption coefficients of liquid water and aqueous solutions in the far ultraviolet. *Radiation Research*, **19**(3), 559–567.

Weissflog L., Elansky N., Putz E., Krueger G., Lange C. A., Lisitzina L. and Pfennigsdorff A. (2004). Trichloroacetic acid in the vegetation of polluted and remote areas of both hemispheres - Part II: salt lakes as novel sources of natural chlorohydrocarbons. *Atmospheric Environment*, **38**(25), 4197–4204.

Wieser J., Murnick D. E., Ulrich A., Huggins H. A., Liddle A. and Brown W. L. (1997). Vacuum ultraviolet rare gas excimer light source. *Review of Scientific Instruments*, **68**(3), 1360–1364.

Wild S. R., Harrad S. J. and Jones K. C. (1993). Chorophenols in digested UK sewage sludges. *Water Research*, **27**(10), 1527–1534.

Witzke H.-D. (2001) Recent studies on fused quartz and synthetic fused silica for light sources. Proceedings of 9th International Symposium on the Science and Technology of Light Sources. Ithaca, NY, USA.

Wörner M., Eggers J., Nunes M., Schnabel C., Rudolph S., Zegenhagen f., Workman A. and Baum A. M. (2003). Combination of VUV (Vacuum-Ultraviolet)-photolysis and electrolysis for the accelerated mineralization of organic pollutant in aqueous systems. 3rd International Conference on Oxidation Technologies for Water and Wastewater – Special Topic: AOP's for Recycling and Reuse. Clausthal.

Xing R., Zheng Z. and Wen D. (2015). Comparison between UV and VUV photolysis for the pre- and post-treatment of coking wastewater. *Journal of Environmental Sciences (China)*, **29**, 45–50.

Yan J. and Gupta M. C. (2003). High power 121.6 nm radiation source. *Journal of Vacuum Science & Technology B*, **21**(6), 2839–2842.

Yan J., El-Dakrouri A., Laroussi M. and Gupta M. C. (2002). 121.6 nm radiation source for advanced lithography. *Journal of Vacuum Science & Technology B*, **20**(6), 2574–2577.

Yazdani A. (2016). Is POU water recovery and reuse practical in microelectronics fabs? *Ultrapure Water*, 1–3. https://www.ultrapurewater.com/articles/micro/is-pou-water-recovery-and-reuse-practical-in-microelectronics-fabs

Yoon S. H., Lee J. H., Oh S. and Yang J. E. (2008). Photochemical oxidation of As(III) by vacuum-UV lamp irradiation. *Water Research*, **42**(13), 3455–3463.

Zhang J.-W. and Boyd I. W. (2000). Lifetime investigation of excimer UV sources. *Applied Surface Science*, **168**, 296–299.

Zhao G., Lu X., Zhou Y. and Gu Q. (2013). Simultaneous humic acid removal and bromate control by O_3 and UV/O_3 processes. *Chemical Engineering Journal*, **232**, 74–80.

Zhuang X., Han Q., Zhang H., Feng X., Roth M., Rosier O., Zhu S. and Zhang S. (2010). The efficiency of coaxial KrCl* excilamps. *Journal of Physics D: Applied Physics*, **43**(20), 205202.

Zoschke K., Dietrich N., Bornick H. and Worch E. (2012). UV-based advanced oxidation processes for the treatment of odour compounds: efficiency and by-product formation. *Water Research*, **46**(16), 5365–5373.

Zoschke K., Bornick H. and Worch E. (2014). Vacuum-UV radiation at 185 nm in water treatment – A review. *Water Research*, **52**, 131–145.

Zvereva G. N. (2010). Investigation of water decomposition by vacuum ultraviolet radiation. *Optics and Spectroscopy*, **108**(6), 915–922.

Zvereva G. N. and Gerasimov G. N. (2001). Numerical simulation of a barrier discharge in Xe. *Optics and Spectroscopy*, **90**(3), 321–328.

Chapter 6

Gamma-ray and electron beam-based AOPs

L. Wojnárovits, E. Takács and L. Szabó

6.1 INTRODUCTION

The field of radiation chemistry has been intensively investigated for more than 100 years. However, there is a basic change in the approach. Until the 1950's and 1960's the main aim was to know more about the basic processes of absorption of radiation energy and about the chemical consequences of irradiation. In recent years radiation chemistry evolved more and more as a reliable tool for fundamental studies, such as reaction kinetics by which important kinetic parameters can be measured and reaction mechanisms can be clarified. Irradiation techniques are often applied as essential parts of technology, e.g., in polymer processing, food processing and sterilization (Woods & Pikaev, 1994). Over the past 20 years there has been an increasing interest in using radiation chemistry in the field of environmental protection, i.e., air and water treatment (IAEA, 2007).

When they penetrate matter, the high-energy photons or electrons interact with the molecules of the medium. As a consequence of the inelastic collisions the particles lose energy with concomitant formation of energy-rich excited or ionized molecules. The primary energy deposition is unselective; it is basically determined by the number of electrons in the molecules. This is a basic difference from the traditional photochemistry, where the chromophores determine the energy absorption characteristics.

The sources of ionizing radiation used for research or industrial purposes can be divided into two groups: sources containing radionuclides emitting γ-rays, such as ^{60}Co or ^{137}Cs, and electron accelerators. In laboratory-scale experiments, the ionizing-radiation sources used are mostly radionuclides. The use of radionuclide sources for large scale water treatment involves the application of rather robust equipment, which allows batch type operation only. For treating several thousand cubic meters of water daily in continuous operation the much more versatile electron accelerators are needed (IAEA, 2007). Electron beam (EB) sources provide much higher absorbed dose rates than γ-irradiators, which are required in full-scale installations. Such electron-beam accelerators have good reliability, high efficiency (the electric energy can be converted to the kinetic energy of accelerated electrons with 60–80% efficiency), and they are available to an installed power up to 400 kW.

This chapter introduces the radiation chemistry of water and of aqueous solutions, and gives an up-to-date overview on the status of the research and the application of radiation technology in water treatment.

6.2 RADIOLYSIS AS A UNIVERSAL TOOL TO INVESTIGATE RADICAL REACTIONS AND AS A PROCESS FOR LARGE SCALE INDUSTRIAL TECHNOLOGY

6.2.1 Techniques in radiation chemistry for establishing reaction mechanisms

Two main, basically different, experimental approaches are used to investigate radiation-induced chemical reactions. One of them (steady-state radiolysis) is applied to detect, qualitatively or quantitatively determine intermediates (e.g., ions or free radicals "frozen" in solid matrices) and final products that are stable and do not change during the time of measurement. The other approach (pulse radiolysis) uses time-dependent measurements in order to observe intermediates or the buildup of final products (Tabata, 1991a, 1991b).

The *final products* are analyzed by various analytical techniques, among which spectrophotometry, gas-chromatography, and liquid-chromatography are the most common ones. Product identification is relatively easy when the system is equipped with a mass-spectrometer detector (GC-MS/MS, LC-MS/MS).

In radiolysis reactions excited molecules, cations, free electrons, anions and radicals are the main *intermediates*. The intermediates can be studied directly by the *pulse radiolysis* method. In this technique short, ps-μs pulses of electrons accelerated to several MeV energies are directed into a cell containing the sample to be irradiated (Baxendale & Busi, 1982; Tabata, 1991b; Wishart, 2008). The buildup and decay of intermediate species and stable products are monitored with time-resolved optical detection. This approach, also called kinetic spectrophotometry or time-resolved spectrophotometry, differs from the conventional spectrophotometric technique since it has very short response time. The kinetic spectrophotometer operates with a single path: the reference level is obtained as the light level passing the sample prior to irradiation. This is the basis for the high sensitivity of kinetic spectrophotometry, where changes less than 1% in the transmitted light are measured accurately. Averaging of a large number of traces is possible with a computerized data acquisition system.

A schematic illustration of a pulse radiolysis setup is shown in Figure 6.1. In the usual arrangements the measuring system is divided into two parts. The first part, including the light source for the sample illumination, the lens system, the shutter preventing the heating and photolysis of the sample, light filters, and the measuring cell itself, is housed in the irradiation room. The other parts of the setup are outside the radiation shielding concrete wall and the Faraday cage protecting against electromagnetic interference. These include the monochromator, the photodetector, the oscilloscope/digitizer and the computer system for data acquisition and processing. Light is usually transmitted from the irradiation room to the measuring room through a hole in the shielding by means of mirrors and lenses. Less frequently light guide cable is used.

In the spectrophotometric method for kinetic measurements, the time-dependent absorbance of the observed species is measured and the intermediate concentrations are calculated using the Beer-Lambert law. In most of experiments the changes in absorbance over time are monitored at a given wavelength. After obtaining the time dependence at many selected wavelengths, the spectrum of the intermediates can be composed by taking the absorbance at a given time after the pulse. The identification of intermediates is based on the absorption spectra. In radiolysis of aqueous solution of organic molecules, the transient species are mostly carbon-centered or oxygen-centered radicals. *Simple alkyl radicals* (like pentyl, cyclopentyl and cyclohexyl) have a wide absorption band in the 220–300 nm wavelength range with maximum at around 250 nm and a molar absorption coefficient of ~1000 $mol^{-1}\ dm^3\ cm^{-1}$ (Swallow, 1982). In the case of *α-hydroxyalkyl radicals* (e.g., $^{\bullet}CH_2OH$, $CH_3{}^{\bullet}CHOH$, $CH_3{}^{\bullet}C(OH)CH_3$) λ_{max} is shifted to shorter wavelengths (≤220 nm) as compared to that of the simple alkyl radicals. However, when a carboxyl or an amide functional group is attached in the α-position to the radical site (e.g., $CH_3{}^{\bullet}CHCOOH$, $CH_3{}^{\bullet}CHCONH_2$)

the absorption band is located at longer wavelengths. In the case of radicals formed from acrylic acids and acrylic acid esters the characteristic UV band is between 280 and 330 nm, ε_{max} = 300–600 mol^{-1} dm^3 cm^{-1}. For acrylamides λ_{max} = 360–480 nm and ε_{max} = 600–1000 mol^{-1} dm^3 cm^{-1} (Wojnárovits *et al.* 2001).

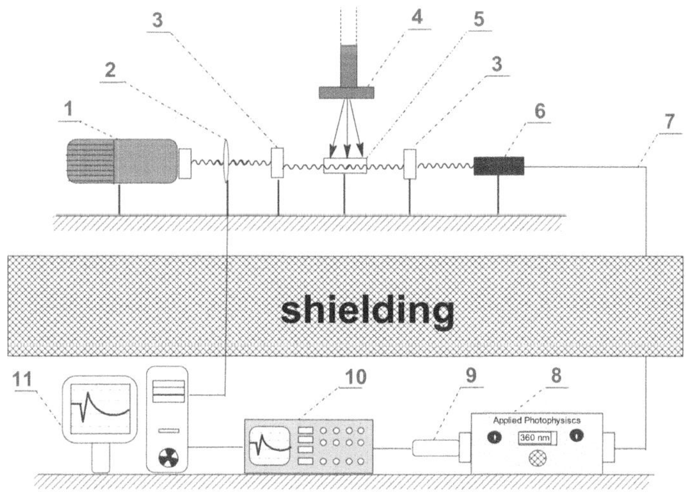

Figure 6.1 Schematic representation of a pulse radiolysis setup. 1 – xenon lamp, 2 – shutter, 3 – lenses, 4 – electron accelerator window, 5 – sample, 6 – entrance of the light guide cable, 7 – light guide cable, 8 – monochromator, 9 – photomultiplier, 10 – digital oscilloscope, 11 – computer.

In the radicals mentioned above the radical site is localized mostly onto a single carbon atom. Many radicals, however, have delocalized spin, and in such cases ε_{max} is usually larger. For example, in alkyl-substituted aromatic molecules where a hydrogen atom was abstracted from α-position of the side chain (e.g., $C_6H_5CH_2^{\bullet}$, *benzyl radical*) the spin is distributed over the aromatic ring and the α-carbon atom. Delocalization causes a red shift in the spectrum (from $\lambda_{max} \approx 250$ nm to $\lambda_{max} = 318$ nm) and an increase in the molar absorption coefficient (e.g., for benzyl radical ε_{max} = (19,200 ± 3840) mol^{-1} dm^3 cm^{-1}) (Hodgkins & Megarity, 1965).

When a hydrogen atom or a hydroxyl radical attacks an aromatic molecule *cyclohexadienyl type radical* is obtained. These radicals have unstructured absorption bands between 300 and 400 nm, λ_{max} around 340 nm and ε_{max} in the range of several hundred to several thousand mol^{-1} dm^3 cm^{-1} (Wojnárovits *et al.* 2002; Albarran & Schuler, 2007). The absorption above 400 nm is very weak. Figure 6.2 gives the absorption spectrum of dihydroxylcyclohexadienyl radical formed upon addition of $^{\bullet}OH$ to phenol as a mixture of *ortho-*, *meta-* and *para* $^{\bullet}OH$-adducts, in comparison with the spectrum of the phenoxyl radical. The dihydroxycyclohexadienyl radical is a C-centered radical, under certain conditions it transforms to an O-centered phenoxyl radical.

When the H-atom from the OH group of phenol is eliminated phenoxyl radical is obtained. This radical has considerable spin densities at the oxygen atom and also at carbons in *ortho* and *para* position. The absorption spectrum of the phenoxyl radical is rather complex. It has a wide and unstructured absorption band with maximum between 280 and 350 nm, a structured absorption band with $\lambda_{max} \approx 400$ nm and $\varepsilon_{max} \approx 3000$ mol^{-1} dm^3 cm^{-1}, and a wide unstructured band with well-defined maximum around 600 nm (Johnston *et al.* 1993; Wojnárovits *et al.* 2002; Steenken & Neta, 2003).

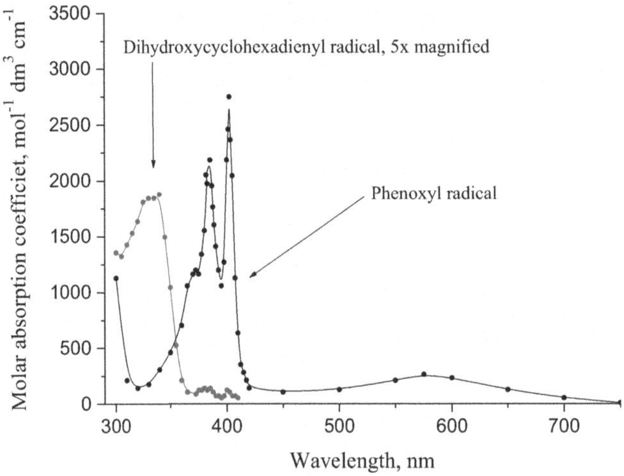

Figure 6.2 Absorption spectra of intermediates of $^{\bullet}OH$ reaction with phenol (Johnston *et al.* 1993; Albarran & Schuler, 2007).

6.2.2 Sources of ionizing radiation in water treatment

In order to test the large scale applicability of radiation technology for water treatment, before industrial implementation, several flow-through laboratory and pilot-plant facilities were designed and constructed. Some of these laboratory test reactors used ^{60}Co gamma rays for irradiation (e.g., Lee & Yoo, 2004). However, under industrial settings, very large quantities of wastewater should be treated, and the electron accelerators with small footprint requirement are better alternatives. Electron accelerators are easy to implement to an existing water treatment train at wastewater treatment plants. Figure 6.3 exemplifies high-power electron accelerators used in radiation processing.

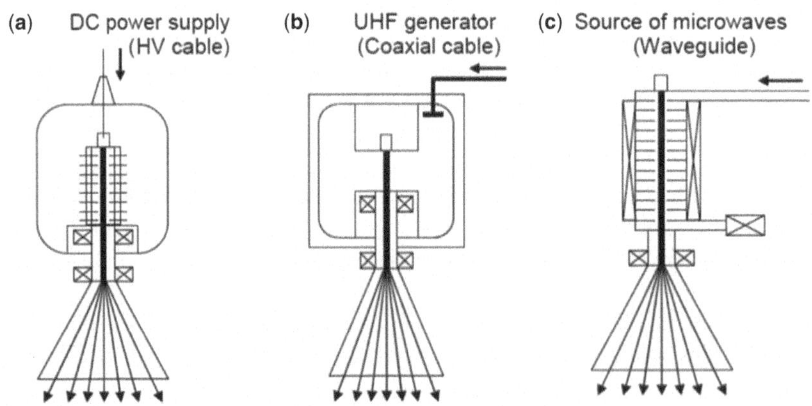

Figure 6.3 Electron accelerators applied in radiation processing: (a) direct high voltage accelerator; (b) single cavity radiofrequency accelerator; (c) linear microwave accelerator (Zimek & Bulka, 2007).

For cost-effective technologies, accelerators with highly efficient conversion of electric power to accelerated electron energy are needed. Among various types of accelerators, the conversion efficiencies are the highest for the so-called direct high-voltage accelerators, where the electrons are accelerated by a static electric field (Figure 6.3a). These accelerators apply high voltage for electron acceleration in contrast to the other accelerators that use radiofrequency or microwave for that purpose. The electron energy is limited to a few MeV for the direct accelerators. For water treatment 1–5 MeV electron accelerators are recommended at powers up to 400 kW (IAEA, 2007). Such energies provide adequate penetration of accelerated electrons into the water in admissible hydrodynamic regimes of water flow. The dose-depth profiles of accelerated electrons after an initial build-up period go through a maximum and then decrease nearly linearly with the distance (Figure 6.4). The single cavity radiofrequency accelerators (Figure 6.3b) and the linear microwave accelerators (Figure 6.3c) have lower conversion efficiencies and higher electron energies than direct accelerators. In general, such accelerators are used for radiation-based sterilization or polymer crosslinking.

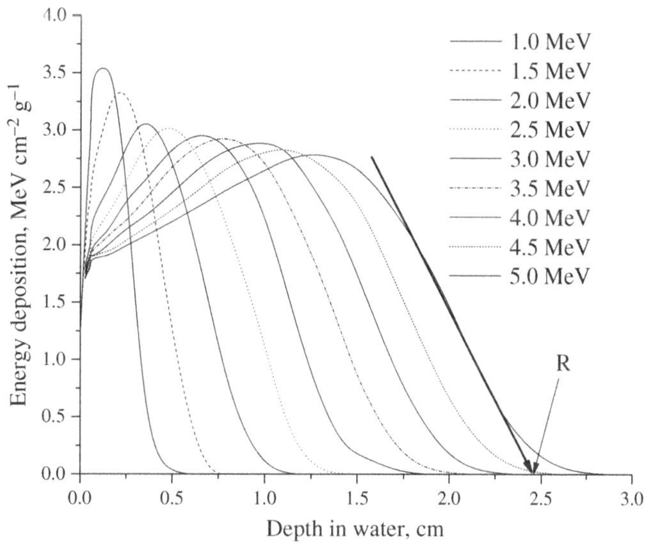

Figure 6.4 Dose-depth profiles of 1–5 MeV electrons in water. On the Y-axis the rate of energy loss per gram per square centimeter, MeV g^{-1} cm^{-2} (called mass stopping power) is shown. The figure also shows the determination of the penetration depth (*R*) for 5.0 MeV electrons (Strydom *et al.* 2005).

6.2.3 *G*-value, dosimetric quantities, penetration depth

The radiation chemical yield, $G(\mathrm{X})$, is defined as the quotient of the amount $n(\mathrm{X})$ of a substance of a specified entity, X, produced, destroyed or changed by radiation, by the mean energy imparted, E, to the irradiated matter:

$$G(\mathrm{X}) = \frac{n(\mathrm{X})}{E} \tag{6.1}$$

Hence, the SI unit of the *G*-value is mol J^{-1}. In most of the literature up to the 1990's, the *G*-value used to be expressed as the number of molecules produced, destroyed or changed per 100 eV of energy absorbed. The conversion to SI units is as follows (Spinks & Woods, 1990):

$$1\ \mu\text{mol J}^{-1} = 9.65\ \text{molecules}\ (100\ \text{eV})^{-1} \tag{6.2}$$

$$1\ \text{molecule}\ (100\ \text{eV})^{-1} = 0.1036\ \mu\text{mol J}^{-1} \tag{6.3}$$

Since only the absorbed fraction of the incident radiation flux can induce physical or chemical changes, the absorbed energy, in other words the absorbed dose is an important quantity. The absorbed dose *D* is defined as the mean energy, d*E*, delivered to an incremental quantity of matter, divided by the mass of that matter, d*m*:

$$D = \frac{\mathrm{d}E}{\mathrm{d}m} \tag{6.4}$$

The SI unit of the absorbed dose is the Gray (Gy), which is defined as the energy (Joule) per unit of mass (kilogram) of the medium, i.e., J kg^{-1}. The energy absorbed by the unit mass in unit time is called the absorbed dose rate; it is expressed in Gy s^{-1}. At a constant dose rate ($D^{\bullet}$), the absorbed dose is directly proportional to the time of irradiation. In γ-radiolysis experiments the dose rates are typically between 1 and 10 Gy s^{-1}. In industrial water treatment in flow-through system using electron accelerators the dose rates are in the 1000 Gy s^{-1} range, and the "contact time" is a fraction of a second (McLaughlin *et al.* 1989).

The penetration depth (*R*) is defined as the depth at which extrapolation of the tail of the dose-depth curve meets the X axis (see Figure 6.4). (For more precise definition see Strydom *et al.* 2005.) The penetration depth of an electron beam is directly proportional to the energy of the incident electrons and inversely proportional to the density of the fluid. In water the *R* of electrons with 1 MeV energy is approximately 0.4 cm.

6.3 WATER RADIOLYSIS

6.3.1 Process fundamentals, yields and reactions of reactive intermediates

In the radiolysis of water hydrated electron (e_{aq}^-), hydroxyl radical ($^{\bullet}OH$) and hydrogen atom ($H^{\bullet}$) reactive intermediates are produced in the primary processes (Buxton, 1982, 1987, 2001), a general scheme is shown below, in brackets the radiation chemical yields (*G*-values) are given in µmol J^{-1} units.

$$H_2O \rightsquigarrow [0.28]^{\bullet}OH + [0.27]e_{aq}^- + [0.06]H^{\bullet} + [0.07]H_2O_2 + [0.27]H_3O^+ + [0.05]H_2 \tag{6.5}$$

The initial radiolysis intermediates are generated in isolated volume elements called spurs. As the spurs expand through diffusion a fraction of the species reacts together, while the others escape into the bulk solution. This process is finished in about 10^{-7} s. In water, 10^{-7} s is the lifetime of a radical reacting in a spur at the diffusion-limited rate constant with a solute S, whose concentration is ~10^{-3} mol dm^{-3}, i.e. $k[S] = 10^7\ s^{-1}$ where *k* is the bimolecular rate constant ($k \approx 10^{10}$ $mol^{-1}\ dm^3\ s^{-1}$) and [S] is the solute concentration. (The *k*[S] product is called scavenging capacity.) The yields of e_{aq}^-, $^{\bullet}OH$ and $H^{\bullet}$ intermediates between pH 3 and 11 are 0.27, 0.28 and 0.06 µmol J^{-1}, respectively. These yields are practically independent of the energy of fast electrons or gamma-rays.

$^{\bullet}OH$ is a strongly oxidizing species with a standard reduction potential *vs.* normal hydrogen electrode (NHE) of $E^\circ(^{\bullet}OH/OH^-) = 1.9$ V (Table 6.1). The hydrated electron has strong reducing character with E° of −2.9 V. $H^{\bullet}$ is slightly less powerful reducing agent than e_{aq}^- ($E^\circ = -2.42$ V).

Table 6.1 One-electron reduction potentials *vs.* normal hydrogen electrode (NHE) of some radicals used in radiation chemical studies (Steenken, 1985; Wardman, 1989; Parker, 1992).

Pair	$E°$, V
aq/e_{aq}^-	−2.9
$H_{aq}^+/H^•$	−2.42
$CO_2/CO_2^{•-}$	−1.90
$(CH_3)_2CO$, $H^+/(CH_3)_2{}^•COH$	−1.39
$O_2/O_2^{•-}$	−0.33
$C_6H_5O^•/C_6H_5O^-$	0.79
$ClO_2^•/ClO_2^-$	0.94
$NO_2^•/NO_2^-$	1.03
$(SCN)_2^{•-}/2SCN^-$	1.33
$BrO_2^•/BrO_2^-$	1.33
$N_3^•/N_3^-$	1.33
$^•OH/OH^-$	1.90
$SO_4^{•-}/SO_4^{2-}$	2.43

In addition to these reactive radical intermediates H_2O_2 is also produced in the early radiation chemical processes with a yield of 0.07 μmol J^{-1} (Spinks & Woods, 1990).

In aerated solutions, e_{aq}^- and $H^•$ mostly transform to the $O_2^{•-}/HO_2^•$ pair. Superoxide radical anion ($O_2^{•-}$) has reducing properties with $E°(O_2/O_2^{•-}) = -0.33$ V (Table 6.1). This reduction potential relates, by definition, to O_2-saturated solution. For comparison with other values that are based on molarities, a value of −0.179 V should be considered (von Sonntag, 2006). $O_2^{•-}$ is the deprotonated form of hydroperoxyl radical ($HO_2^•$) pK_a ($HO_2^•$) = 4.8 (Bielski *et al.* 1985).

$$e_{aq}^- + O_2 \rightarrow O_2^{•-} \quad k_{6.6} = 1.9 \times 10^{10}\ \text{mol}^{-1}\ \text{dm}^3\ \text{s}^{-1} \tag{6.6}$$

$$H^• + O_2 \rightarrow HO_2^• \quad k_{6.7} = 1.2 \times 10^{10}\ \text{mol}^{-1}\ \text{dm}^3\ \text{s}^{-1} \tag{6.7}$$

$H^•$ and e_{aq}^- react with the dissolved O_2 with rate constants of 1.2×10^{10} and 1.9×10^{10} mol^{-1} dm^3 s^{-1}, respectively (Buxton *et al.* 1988). At 20°C the water absorbs 2.8×10^{-4} mol dm^{-3} O_2 from the air. Due to the high rate constants, the e_{aq}^- and $H^•$ scavenging capacities (k[S], where S represents O_2 concentration) are high, 5.3×10^6 s^{-1} and 3.4×10^6 s^{-1}, respectively. Any impurity can compete with the $e_{aq}^- + O_2$ and $H^• + O_2$ reactions if the scavenging capacity calculated for the dissolved compound is in this range. With a diffusion controlled rate constant of ~10^{10} mol^{-1} dm^3 s^{-1} it requires an impurity concentration of ~10^{-4} mol dm^{-3}. Since this is a high concentration for environmental water samples, e_{aq}^- and $H^•$ reactions rarely play determining role in the micropollutant degradation. However, at relatively high solute concentrations, and when at very high dose rates (electron beam irradiation) considerable O_2 depletion occurs, the e_{aq}^- reactions may not be negligible.

Unlike $^{\bullet}OH$, e_{aq}^{-} and $H^{\bullet}$, the $O_2^{\bullet-}$ and $HO_2^{\bullet}$ are rather un-reactive with most of organic solutes (Bielski *et al.* 1985). In the absence of suitable reaction partners, $O_2^{\bullet-}$ and $HO_2^{\bullet}$ disappear from the solution in slow radical-radical reactions giving H_2O_2; in neutral solutions the cross-reaction (6.10) dominates:

$$2\,HO_2^{\bullet} \rightarrow H_2O_2 + O_2 \qquad k_{6.8} = 8.3 \times 10^5\ \text{mol}^{-1}\ \text{dm}^3\ \text{s}^{-1} \tag{6.8}$$

$$2\,O_2^{\bullet-} + 2\,H_2O \rightarrow H_2O_2 + O_2 + 2\,OH^- \qquad k_{6.9} < 3 \times 10^1\ \text{mol}^{-1}\ \text{dm}^3\ \text{s}^{-1} \tag{6.9}$$

$$HO_2^{\bullet} + O_2^{\bullet-} + H_2O \rightarrow H_2O_2 + O_2 + OH^- \qquad k_{6.10} = 9.7 \times 10^7\ \text{mol}^{-1}\ \text{dm}^3\ \text{s}^{-1} \tag{6.10}$$

Therefore, in water radiolysis hydrogen peroxide is produced in the early processes (reaction 6.5, $G = 0.07\ \mu\text{mol J}^{-1}$), and also in the reactions of $O_2^{\bullet-}/HO_2^{\bullet}$ pair (reactions 6.8–6.10). Under certain conditions, H_2O_2 can be involved in reactions with the reactive intermediates generating species which can influence considerably the degradation of chemical pollutants. However, its effect in the degradation mechanism is usually neglected and only few publications mention the possible role of H_2O_2 in the degradation (e.g., Szabó *et al.* 2014; Liu *et al.* 2014). In laboratory gamma radiolysis experiments conducted in laboratory grade water in the presence of dissolved O_2 the formation of H_2O_2 was observed at a relatively high yield ($G \approx 0.3\ \mu\text{mol J}^{-1}$). At doses >1 kGy the H_2O_2 concentration–dose curve reaches a plateau due to the competition between the formation and degradation reactions.

In real water treatment scenarios, in the presence of traces of transition metal ions H_2O_2 is expected to disappear in Fenton-like reactions.

6.3.1.1 *Hydroxyl radical*

In ionizing radiation-induced wastewater treatment the hydroxyl radical is considered to be the main reactive intermediate, which initiates the degradation of the solute. In alkaline solution $^{\bullet}OH$ converts to $O^{\bullet-}$ with a pK_a of 11.9:

$$^{\bullet}OH + OH^- \rightleftarrows O^{\bullet-} + H_2O \qquad k_{6.11} = 1.3 \times 10^{10}\ \text{mol}^{-1}\ \text{dm}^3\ \text{s}^{-1} \tag{6.11}$$

$$K_{6.-11} = 1.8 \times 10^6\ \text{mol}^{-1}\ \text{dm}^3\ \text{s}^{-1} \tag{6.-11}$$

The hydroxyl radicals in reaction with dissolved organic molecules may react in three different ways: addition to the double bonds, abstraction of hydrogen atoms from C−H bonds, and direct electron transfer.

Due to the electrophilic character, $^{\bullet}OH$ readily reacts with C=C and C=N bonds but not with C=O bonds which are electron-deficient at the carbon atom. Although $^{\bullet}OH$ reacts with C=C bonds at near diffusion-controlled rates, the reaction is highly regioselective largely due to the electrophilic nature of hydroxyl radical reactions (von Sonntag, 2006).

In the case of aromatic structures the rate constant of the $^{\bullet}OH$ reaction varies within the narrow range of 2×10^9 to $1.0 \times 10^{10}\ \text{mol}^{-1}\ \text{dm}^3\ \text{s}^{-1}$ (Wojnárovits & Takács, 2013), i.e., approaching diffusion controlled reaction rate. For example, the published average rate constant for the reaction of $^{\bullet}OH$ with benzene is $k_{OH} = 7.8 \times 10^9\ \text{mol}^{-1}\ \text{dm}^3\ \text{s}^{-1}$ (Buxton *et al.* 1988). Electron-withdrawing substituent, such as $-Cl$, $-NO_3$, or $-COOH$, decreases the $^{\bullet}OH$ rate constants with the respective aromatic compounds, whereas the electron-donating substituent such as $-NH_2$, $-OH$, or $-CH_3$, increases the rate constants, and the k_{OH} approaches the diffusion-limited values (Table 6.2).

Table 6.2 Rate constants of selected $^{\cdot}OH$, $H^{\cdot}$ and e_{aq}^{-} reactions.

Compound	$^{\cdot}OH$	(k, $mol^{-1}\ dm^3\ s^{-1}$) $H^{\cdot}$	e_{aq}^{-}
Simple inorganic compounds or ions			
O_2		1.2×10^{10}[a]	1.9×10^{10}[a]
H_2O_2	2.7×10^{7}[a]	9×10^{7}[a]	1.1×10^{10}[a]
H_3O^+			2.3×10^{10}[a]
OH^-	1.3×10^{10}[a]		
Simple organic compounds			
Benzene	7.8×10^{9}[b]	1.1×10^{9}[c]	1×10^{7}[a]
Phenol	8.4×10^{9}[b]	1.7×10^{9}[d]	2.0×10^{7}[a]
p-Cresol	9.2×10^{9}[b]	4.1×10^{9}[e]	
Aniline	8.6×10^{9}[b]	1.9×10^{9}[f]	3×10^{7}[f]
Chlorobenzene	5.6×10^{9}[b]	1.4×10^{9}[a]	5×10^{8}[g]
Nitrobenzene	3.5×10^{9}[b]	1×10^{9}[a]	3.7×10^{10}[a]
Benzoic acid	1.9×10^{9}[b]	9.2×10^{8}[a]	7.1×10^{9}[a]
Benzoate ion	5.9×10^{9}[b]	1.1×10^{9}[a]	3.2×10^{9}[a]
Endocrine disrupters			
17β-estradiol	5.3×10^{9}[h]		2.5×10^{8}[i]
p-Nonylphenol	1.1×10^{10}[j]		
Bisphenol	6.9×10^{9}[k]		
Dimethyl-phthalate	3.4×10^{9}[l]		1.6×10^{10}[l]
Pesticides			
2,4-Dichlorophenol	6.0×10^{9}[m]		5.0×10^{8}[n]
2,4-D	5.5×10^{9}[m]	1.4×10^{9}[o]	2.5×10^{9}[o]
Atrazine	2.4×10^{9}[m]		4.8×10^{9}[p]
Simazine	2.6×10^{9}[m]		
Prometon	2.8×10^{9}[m]		
Fenuron	8.3×10^{9}[m]		1×10^{9}[r]
Monuron	7.3×10^{9}[m]		
Diuron	6.0×10^{9}[m]		1×10^{10}[s]
Pharmaceuticals			
Chloramphenicol	1.8×10^{9}[t]	1.0×10^{9}[t]	2.3×10^{10}[t]
Sulfacetamide	5.3×10^{9}[u]		
Sulfamethoxazole	6.5×10^{9}[v]		1.0×10^{10}[w]
Sulfamethazine	6.7×10^{9}[v]		2.4×10^{10}[w]
Penicillin G	8.4×10^{9}[v]		2.7×10^{9}[x]
Amoxicillin	6.4×10^{9}[v]		5.2×10^{9}[y]
Ampicillin	6.7×10^{9}[v]		5.7×10^{9}[x]
Cloxacillin	7.1×10^{9}[v]		7.5×10^{9}[x]

(Continued)

Table 6.2 Rate constants of selected $^{\bullet}OH$, $H^{\bullet}$ and e_{aq}^- reactions (*Continued*).

Compound	$^{\bullet}OH$	(*k*, $mol^{-1}\,dm^3\,s^{-1}$) $H^{\bullet}$	e_{aq}^-
Salicylic acid	$1.07 \times 10^{10,y}$	$2.3 \times 10^{9,d}$	$9 \times 10^{9,z}$
Paracetamol	$5.6 \times 10^{9,aa}$		$5 \times 10^{8,aa}$
Diclofenac	$8.12 \times 10^{9,bb}$		$1.7 \times 10^{9,bb}$
Ketoprofen	$4.6 \times 10^{9,bb}$		$2.6 \times 10^{10,bb}$
Ibuprofen	$6.1 \times 10^{9,bb}$	$4.0 \times 10^{9,cc}$	$8.9 \times 10^{9,bb}$
Organic dyes			
Acid Red 265	$9.3 \times 10^{9,dd}$		
Acid Blue 62	$1 \times 10^{10,ee}$	$3.8 \times 10^{9,ee}$	$3 \times 10^{10,ee}$
Isobutylnaphthalene sulfonate	$1.8 \times 10^{10,ff}$ (overestimated)	$5.0 \times 10^{9,ff}$	$2.5 \times 10^{10,ff}$

a) Buxton *et al.* 1988; b) Wojnárovits and Takács, 2013; c) Roduner and Bartels, 1992; d) Madden and Mezyk, 2011; e) Wojnárovits *et al.* 2002; f) Solar *et al.* 1986; g) Lichtscheidl and Getoff, 1979; h) Kosaka *et al.* 2003; i) Mezyk *et al.* 2013; j) Ning *et al.* 2007; k) Peller *et al.* 2009; l) Wu *et al.* 2011; m) Wojnárovits and Takács, 2014; n) Hoy and Bolton, 1994; o) Zona *et al.* 2002a; p) Varghese *et al.* 2006; r) Kovács *et al.* 2014; s) Kovács *et al.* 2015; t) Kapoor and Varshney, 1997; u) Sági *et al.* 2015; v) Average of several mesurements; w) Mezyk *et al.* 2007; x) Philips *et al.* 1973a; y) Szabó *et al.* 2016a; z) Ayatollahi *et al.* 2013; aa) Szabó *et al.* 2012; bb) Jones, 2007; cc) Illés *et al.* 2013; dd) Wojnárovits and Takács, 2008; ee) Perkowski *et al.* 1989; ff) Gogolev *et al.* 1992; Pikaev, 2001.

In the reaction between $^{\bullet}OH$ and phenols and phenolates in the first step cyclohexadienyl type radical forms. The intermediate dihydroxycyclohexadienyl radical (reactions 6.12 and 6.13), in the neutral pH range – under the usual experimental conditions – terminates in radical-radical reactions. At low and high pH the intermediate decomposes in acid/base catalyzed H_2O/OH^- elimination to phenoxyl radical (Steenken, 1987; Ashton *et al.* 1995; Roder *et al.* 1999). In aerated solutions these reactions are in competition with peroxyl radical formation.

$$C_6H_5OH + {}^{\bullet}OH \rightarrow [{}^{\bullet}C_6H_5OHOH] \rightarrow C_6H_5O^{\bullet} + H_2O \quad (6.12)$$

$$C_6H_5O^- + {}^{\bullet}OH \rightarrow [{}^{\bullet}C_6H_5OOH^-] \rightarrow C_6H_5O^{\bullet} + OH^- \quad (6.13)$$

Figure 6.2 shows the absorption spectra of the dihydroxycyclohexadienyl and phenoxyl radical intermediates: the transformation of dihydroxycyclohexadienyl radical to phenoxyl radical below pH 3 and above pH 9 is completed within a few μs (Wojnárovits *et al.* 2002).

The HO−H bond dissociation energy is 499 kJ mol^{-1}; in saturated hydrocarbons the C−H bonds are much weaker. In alkanes the energies of primary ($-CH_3$), secondary ($-CH_2-$) and tertiary (>CH–) C−H bonds are 423, 412 and 403 kJ mol^{-1}, respectively. The C−H bond energy in a substituent at α-position to a C=C bond (e.g., $CH_2{=}CH{-}CH_2{-}H$, $C_6H_5{-}H_2C{-}H$) is around 365 kJ mol^{-1}. Thus, there is a considerable driving force for the H-abstraction in these cases. Despite this driving force, there is some remarkable selectivity. Hydrogen abstraction from a primary C is less likely to occur than from secondary- and tertiary-C (von Sonntag, 2006). Reactions of $^{\bullet}OH$ with saturated organic compounds usually proceed *via* H-abstraction (e.g., reactions 6.14 and 6.15, with *tert*-butanol and 2-propanol, respectively) (Alam *et al.* 2001). The rate constants of H-abstractions are one to two orders of magnitude smaller than those corresponding to the diffusion-controlled reactions.

$$^{\bullet}OH + (CH_3)_3COH \rightarrow H_2O + {}^{\bullet}CH_2C(CH_3)_2OH \qquad k_{6.14} = 6 \times 10^8 \text{ mol}^{-1} \text{ dm}^3 \text{ s}^{-1} \qquad (6.14)$$

$$^{\bullet}OH + (CH_3)_2CHOH \rightarrow H_2O + (CH_3)_2C^{\bullet}OH \qquad k_{6.15} = 1.9 \times 10^9 \text{ mol}^{-1} \text{ dm}^3 \text{ s}^{-1} \qquad (6.15)$$

In the case of isopropanol (reaction 6.15) the H-abstraction occurs at the tertiary-C and the α-hydroxyisopropyl radical formed in the reaction exhibits strongly reducing character (Table 6.1). The radical formed in reaction with *tert*-butanol (6.14) has no reducing character; it is rather inert in most reaction systems.

Although the reduction potential of the $^{\bullet}OH/OH^-$ couple is very high, direct electron transfer is rarely observed in $^{\bullet}OH$ reactions with organic molecules, and where it is observed intermediate complexes are likely to be involved (Wojnárovits, 2011). Electron transfer requires a considerable rearrangement of the water molecules around the reaction center; this rearrangement disfavors the reaction (high entropy change). As an example of electron transfer is the oxidation of metal M^{n+} ions to $M^{(n+1)+}$ ions (e.g., $Fe^{2+} \rightarrow Fe^{3+}$). However, this reaction may also take place through an intermediate ($M^{n+\bullet}OH$) complex (addition-elimination mechanism).

The electron transfer from phenolates is highly exothermic, and although the electron transfer is thermodynamically favored to the addition reaction, the mechanism of phenolate ion oxidation by $^{\bullet}OH$ involves primarily the radical addition to the aromatic ring (reaction 6.13). In some cases, however, addition and electron transfer reactions compete, e.g., when the preferred position of addition is blocked by a bulky substituent.

In laboratory experiments the $^{\bullet}OH$ reactions are usually studied in N_2O-saturated solutions. In such solutions the conversion of hydrated electrons to hydroxyl radicals is a two-step process with the formation of intermediate $O^{\bullet-}$ ions (reactions 6.16 and 6.17):

$$N_2O + e_{aq}^- \rightarrow N_2 + O^{\bullet-} \quad k_{6.16} = 9.1 \times 10^9 \text{ mol}^{-1} \text{ dm}^3 \text{ s}^{-1} \qquad (6.16)$$

$$O^{\bullet-} + H_2O \rightarrow OH^- + {}^{\bullet}OH \quad k_{6.17} = 1.8 \times 10^6 \text{ mol}^{-1} \text{ dm}^3 \text{ s}^{-1} \qquad (6.17)$$

In N_2O-saturated solutions the yield of $^{\bullet}OH$ is 0.56 μmol J^{-1}. Under such conditions, the $H^{\bullet}$ reactions also contribute to the chemical transformations with a yield of 0.06 μmol J^{-1} (Swallow, 1982).

6.3.1.2 Hydrated electron

In reactions with organic molecules e_{aq}^- reacts as a nucleophile. Hydrated electron reacts preferentially with low-lying molecular orbital structures, such as aromatic hydrocarbons, conjugated olefins, carboxyl compounds, halogenated hydrocarbons (Swallow, 1982; Buxton, 1982, 1987, 2001). The e_{aq}^- is characterized by a strong light absorption band centered at 720 nm and a maximum molar absorption coefficient of ~20,000 mol^{-1} dm^3 cm^{-1}. This band is used for the determination of e_{aq}^- reaction rate constants by pulse radiolysis.

Formerly we mentioned that e_{aq}^- reacts with high rate constants with dissolved O_2 (reaction 6.6). In its reaction with hydrogen peroxide hydroxyl radical forms.

$$e_{aq}^- + H_2O_2 \rightarrow {}^{\bullet}OH + OH^- \quad k_{6.18} = 1.1 \times 10^{10} \text{ mol}^{-1} \text{ dm}^3 \text{ s}^{-1} \qquad (6.18)$$

The e_{aq}^- reacts with many compounds which are capable of releasing an anion by dissociative electron capture. In a number of reactions that are written as dissociative electron attachment, short-lived radical anions are intermediates. Dissociative electron capture can occur only when single bonds are involved in the process (von Sonntag, 2006). Typical reactions are with halogenated compounds, where the electron "scavenging" at the halogen site is followed by the elimination of the halide ion. One example is the reaction of hydrated electron with chloroacetic acid (reaction 6.19), resulting in Cl^- and $^{\bullet}CH_2CO_2H$ (Swallow, 1973):

$$e_{aq}^- + ClCH_2CO_2H \rightarrow Cl^- + {}^{\bullet}CH_2CO_2H \qquad (6.19)$$

The reaction with chlorobenzene is assumed to be a two-step process *via* an intermediate radical anion ($C_6H_5Cl^{\bullet -}$), which dissociates to phenyl radical and chloride ion (reaction 6.20):

$$e_{aq}^- + C_6H_5Cl \rightarrow [C_6H_5Cl^{\bullet -}] \rightarrow C_6H_5^{\bullet} + Cl^- \quad k_{6.20} = 5 \times 10^8\ mol^{-1}\ dm^3\ s^{-1} \qquad (6.20)$$

Compounds with high electron affinity (e.g., nitro- and cyano-derivatives) react with e_{aq}^- with diffusion-controlled rate constants of ~3×10^{10} $mol^{-1}\ dm^3\ s^{-1}$ with formation of the corresponding radical anions. The rate constant with aldehydes and ketones is ~4×10^9 $mol^{-1}\ dm^3\ s^{-1}$. The electron is accommodated on the carbonyl carbon. The e_{aq}^- rate constants for the reactions with carboxylic acids, esters, and amides are in the order of ~10^7 $mol^{-1}\ dm^3\ s^{-1}$.

Simple olefins do not react with e_{aq}^- at appreciable rates, but compounds with extended π-electron delocalization react rapidly with the hydrated electron. The rate constants k_{eaq-} of maleates and fumarates, which contain conjugated double bonds (e.g., O=CH−HC=CH−HC=O) are in the range of diffusion controlled-limit (Takács *et al.* 2000; Wojnárovits *et al.* 2003). The radical anion formed in reaction (6.21) protonates instantaneously (Grotthuss mechanism). The protonation reaction results in a carbon-centered radical at the carbonyl (reaction 6.22). (In the reactions below R = alkyl group, H-atom).

$$e_{aq}^- + ROOC{-}CH{=}CH{-}COOR \rightarrow RO^-O^{\bullet}C{-}CH{=}CH{-}COOR \qquad (6.21)$$
$$k_{6.21} = (2-3) \times 10^{10}\ mol^{-1}\ dm^3\ s^{-1}$$

$$H_3O^+ + RO^-O^{\bullet}C{-}CH{=}CH{-}COOR \rightarrow RO(HO)^{\bullet}C{-}CH{=}CH{-}COOR + H_2O$$
$$K_{6.22} = (5-6) \times 10^{10}\ mol^{-1}\ dm^3\ s^{-1} \qquad (6.22)$$

In reactions with aromatic molecules, e_{aq}^- adds to the ring with formation of a radical anion. The rate constant is small (≤10^7 $mol^{-1}\ dm^3\ s^{-1}$) in the case of benzene and alkylbenzenes, and the process is reversible; e_{aq}^- addition may be followed by a thermally-activated dissociation (von Sonntag, 2006). This dissociation is in competition with the anion protonation, in which cyclohexadienyl radical ($C_6H_7^{\bullet}$) is produced (reaction 6.23).

$$e_{aq}^- + C_6H_6 \rightleftarrows C_6H_6^{\bullet -} \xrightarrow[-OH^-]{+H_2O} C_6H_7^{\bullet} \quad k_{6.23} = 1 \times 10^7\ mol^{-1}\ dm^3\ s^{-1} \qquad (6.23)$$

k_{eaq-} is higher ($10^8 - 10^9$ $mol^{-1}\ dm^3\ s^{-1}$) when an electron-withdrawing substituent, such as –COOH, –Cl, or $-NO_2$ group, is attached to the ring (Table 6.2). The rate constant of e_{aq}^- reaction with benzoic acid, chlorobenzene (electron addition followed by Cl^- elimination (reaction 6.20), and nitrobenzene are 7.1×10^9, 5×10^8, and 3.7×10^{10} $mol^{-1}\ dm^3\ s^{-1}$, respectively.

6.3.1.3 Hydrogen atom

The yield of H-atoms in neutral or alkaline solutions is low, $G = 0.06$ μmol J^{-1}. In acidic solutions due to the $e_{aq}^- \rightarrow H^{\bullet}$ conversion (reaction 6.24) the yield can be as high as $G = 0.33$ μmol J^{-1}.

$$e_{aq}^- + H_3O^+ \rightarrow H^{\bullet} + H_2O \quad k_{6.24} = 2.3 \times 10^{10}\ mol^{-1}\ dm^3\ s^{-1} \qquad (6.24)$$

Therefore, $H^{\bullet}$ reactions are mostly studied in acidic *tert*-butanol-containing solutions. As it was mentioned earlier, this additive is used to convert the $^{\bullet}OH$ to the less reactive *tert*-butanol radical (reaction 6.14); the reactivity of $H^{\bullet}$ with *tert*-butanol is low (1×10^6 mol^{-1} dm^3 s^{-1}, Wojnárovits *et al.* 2004). $H^{\bullet}$ and $^{\bullet}OH$ do not absorb the light in the usual wavelength range (250–800 nm) of pulse radiolysis studies. In transient measurements the product of $H^{\bullet}$ (or $^{\bullet}OH$) reaction or competitive techniques are used to determine the rate constants.

Direct reduction is observed in reactions with several metal ions, e.g., $H^{\bullet}$ reduces Cu^{2+} to Cu^{+} with k_H of 9.1×10^7 mol^{-1} dm^3 s^{-1} (Buxton *et al.* 1988). In reaction with organic molecules $H^{\bullet}$ – similarly to $^{\bullet}OH$ – abstracts an H-atom from the aliphatic chain, or adds to the double bonds of molecules. The rate constants of H-atom abstraction reactions by $H^{\bullet}$ are generally smaller by 1–3 orders of magnitude than those of $^{\bullet}OH$. In the addition reactions $H^{\bullet}$ behaves similarly to $^{\bullet}OH$, i.e., as electrophiles (Wojnárovits *et al.* 2003; von Sonntag, 2006). The rate constants for addition reactions of the two radicals are similar, although the $^{\bullet}OH$ is more reactive than the $H^{\bullet}$. The rate constant of $H^{\bullet}$ reaction with benzene is 1.1×10^9 mol^{-1} dm^3 s^{-1}. When electron-releasing –OH (in phenol), or $-NH_2$ (in aniline) group is attached to the benzene ring, k_H increases to 1.7×10^9, and 2.4×10^9 mol^{-1} dm^3 s^{-1}, respectively (Buxton *et al.* 1988; Madden & Mezyk, 2011).

6.3.2 Reactions of primary species with common inorganic ions

The highly reactive $^{\bullet}OH$, $H^{\bullet}$ and e_{aq}^- intermediates of water radiolysis react with inorganic species (ions) generating usually less reactive radical intermediates (Neta *et al.* 1988). Therefore, water radiolysis in the presence of various inorganic solutes is complex as a variety of reactive radical species can be formed, which, depending on their reduction/oxidation potentials can be involved in subsequent reactions with other water constituents. Examples of reactions of primary species formed in water radiolysis with some common inorganic ions are given below:

$$^{\bullet}OH + NO_2^- \rightarrow OH^- + NO_2^{\bullet} \quad k_{6.25} = 1 \times 10^{10}\ mol^{-1}\ dm^3\ s^{-1} \tag{6.25}$$

$$^{\bullet}OH + ClO_2^- \rightarrow OH^- + ClO_2^{\bullet} \quad k_{6.26} = 6.6 \times 10^{9}\ mol^{-1}\ dm^3\ s^{-1} \tag{6.26}$$

$$e_{aq}^- + BrO_3^- + 2H_3O^+ \rightarrow BrO_2^{\bullet} + 3H_2O \quad k_{6.27} = 2.6 \times 10^{9}\ mol^{-1}\ dm^3\ s^{-1} \tag{6.27}$$

$$e_{aq}^- + S_2O_8^{2-} \rightarrow SO_4^{\bullet-} + SO_4^{2-} \quad k_{6.28} = 1.2 \times 10^{10}\ mol^{-1}\ dm^3\ s^{-1} \tag{6.28}$$

Such reactions are often examined and used in advanced oxidation process research to investigate the degradation mechanisms of chemical pollutants or for technological application purposes (e.g. sulfate radical anion formation, reaction 6.28). In reactions of $^{\bullet}OH$ with NO_2^- or ClO_2^- ions the $NO_2^{\bullet}$ and $ClO_2^{\bullet}$ oxidizing radicals are formed; the reduction potentials of these radicals are 1.03 and 0.94 V, respectively (Table 6.1). e_{aq}^- reaction with bromate (BrO_3^-) or persulfate ($S_2O_8^{2-}$) ions converts the strongly reductive hydrated electron to strongly oxidizing $BrO_2^{\bullet}$ and $SO_4^{\bullet-}$ with reduction potentials of 1.33 and 2.43 V, respectively.

6.3.2.1 Reactions of carbonate radical anion

In practice, the wastewater always contains carbonate and hydrogen carbonate ions. In reactions of CO_3^{2-} and HCO_3^- with $^{\bullet}OH$, carbonate radical anion $CO_3^{\bullet-}$, a strong one-electron oxidant, $E^{\circ} = 1.78$ V (*vs.* NHE at pH 7) is generated *via* electron transfer reactions (Huie *et al.* 1991).

$$^{\bullet}OH + CO_3^{2-} \rightarrow CO_3^{\bullet-} + OH^- \quad k_{6.29} = 3.9 \times 10^8 \text{ mol}^{-1} \text{ dm}^3 \text{ s}^{-1} \tag{6.29}$$

$$^{\bullet}OH + HCO_3^- \rightarrow H_2O + CO_3^{\bullet-} \quad k_{6.30} = 8.5 \times 10^6 \text{ mol}^{-1} \text{ dm}^3 \text{ s}^{-1} \tag{6.30}$$

$CO_3^{\bullet-}$ reacts selectively with organic molecules (Table 6.3). The rate constants are in the 10^2–10^9 mol^{-1} dm^3 s^{-1} range. $CO_3^{\bullet-}$ may react by electron transfer (aromatic amines, thiols, metal ions, inorganic anions), or by hydrogen abstraction (alcohols, primary amines, thiols, phenols) (Neta *et al.* 1988).

Table 6.3 Rate constants of selected inorganic radicals with organic molecules.

Compound	$O_2^{\bullet-}$	$HO_2^{\bullet}$	(k, mol^{-1} dm^3 s^{-1}) $CO_3^{\bullet-}$	$Cl_2^{\bullet-}$	$SO_4^{\bullet-}$
Benzene			3×10^3 [c]		
Aniline			5.0×10^8 [c]		
Phenol	5.8×10^2 [a]	2.7×10^3 [b]	2.2×10^7 [c]	2.5×10^8 [c]	2.1×10^8 [d]
Benzoate ion				2×10^6 [c]	1.2×10^9 [c]

a) Tsujimoto et al. 1993; b) Kozmér et al. 2014; c) Neta et al. 1988; d) Das, 2005.

6.3.2.2 Reactions of dichloride radical anion

In $^{\bullet}OH$ induced oxidation of Cl^- and rapid complexation of $Cl^{\bullet}$ with chloride ion, the dichloride radical anion ($Cl_2^{\bullet-}$) is produced. This species is a strong oxidant with the standard one-electron reduction potential of 2.1 V *vs*. NHE (Hasegawa & Neta, 1978).

$$^{\bullet}OH + Cl^- \rightarrow Cl^{\bullet} + OH^- \quad k_{6.31} = 4.3 \times 10^9 \text{ mol}^{-1} \text{ dm}^3 \text{ s}^{-1} \tag{6.31}$$

$$Cl^{\bullet} + Cl^- \rightarrow Cl_2^{\bullet-} \quad k_{6.32} = 8.5 \times 10^9 \text{ mol}^{-1} \text{ dm}^3 \text{ s}^{-1} \tag{6.32}$$

$Cl_2^{\bullet-}$ can abstract an H-atom from aliphatic compounds with rate constants between 10^3–10^6 mol^{-1} dm^3 s^{-1} (Neta *et al.* 1988). The reaction with olefins occurs with $Cl^{\bullet}$ addition and the k_{Cl2-}'s are in the 10^6–10^9 mol^{-1} dm^3 s^{-1} range. Addition to the aromatic ring may also take place with k_{Cl2-} of $\leq 10^7$ mol^{-1} dm^3 s^{-1}, but the direct oxidation of the aromatic ring by electron transfer seems to be the predominant pathway. The oxidation reaction occurs only when electron-donating functional groups such as –OH, $-OCH_3$ or $-NH_2$ are on the aromatic ring.

6.3.2.3 Reactions of sulfate radical anion

In a large number of publications $SO_4^{\bullet-}$ ($[E°(SO_4^{\bullet-}/SO_4^{2-})] = 2.43$ V *vs*. NHE) is considered as an oxidant in water treatment (see reaction 6.28) (Lutze, 2013; Paul (Guin) *et al.* 2014a, 2014b; Criquet & Karpel Vel Leitner, 2015). $SO_4^{\bullet-}$ has some unique features, such as being a very strong electron acceptor enabling the degradation of recalcitrant pollutants refractory to $^{\bullet}OH$ treatment. One example is perfluorinated carboxylic acids (PFCAs), which can, albeit slowly, be degraded by $SO_4^{\bullet-}$, but persist in other AOPs.

When persulfate is added to the solution the e_{aq}^- reaction with dissolved O_2 (6.6) is replaced by the persulfate + e_{aq}^- reaction (6.28). $SO_4^{\bullet-}$ may also form in persulfate reaction with $O_2^{\bullet-}$ (6.33) (Fang *et al.* 2013).

$$O_2^{\bullet -} + S_2O_8^{2-} \rightarrow O_2 + SO_4^{\bullet -} + SO_4^{2-} \tag{6.33}$$

It was shown that in the irradiation/persulfate AOP the sulfate radical anion enhanced the degradation of organic compounds, e.g., carboxylic acids or aromatic molecules (Criquet & Karpel Vel Leitner, 2011, 2012; Paul (Guin) *et al.* 2014a, 2014b). $SO_4^{\bullet -}$ reacts with the aromatic ring by electron transfer with the formation of a radical cation. This is in contrast to ${}^{\bullet}OH$ reactions, where addition to the ring dominates. Therefore, different products are expected in ${}^{\bullet}OH$ and $SO_4^{\bullet -}$ reactions. The reaction of sulfate radical anion with an aromatic compound is exemplified below for ibuprofen (Scheme 6.1) (Paul (Guin) *et al.* 2014a); $k_{SO4-} = 3.8 \times 10^9$ mol^{-1} dm^3 s^{-1}.

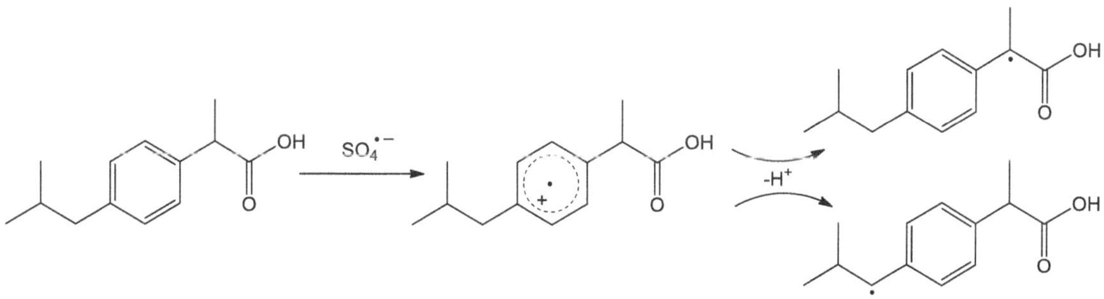

Scheme 6.1 $SO_4^{\bullet -}$ reaction with ibuprofen.

$SO_4^{\bullet -}$-based water treatment processes are strongly affected by the presence of water constituents (Lutze, 2013), such as Cl^- and HCO_3^-; $k_{SO4-,Cl-} = 2 \times 10^8$ mol^{-1} dm^3 s^{-1}; $k_{SO4-,HCO3-} = 9.1 \times 10^6$ mol^{-1} dm^3 s^{-1} (Neta *et al.* 1988).

6.3.2.4 Reactions in the presence of ozone

In the presence of ozone/oxygen the primary water radiolysis intermediates, $H^{\bullet}$ and e_{aq}^- react quickly with ozone with rate constants in the range of 10^{10} mol^{-1} dm^3 s^{-1}. The rate constant of the (${}^{\bullet}OH + O_3$) reaction is 1.1×10^8 mol^{-1} dm^3 s^{-1} (Buxton *et al.* 1988). The reactions involving the dissolved molecular ozone result in reactive intermediates, among which the highly oxidizing ${}^{\bullet}OH$ (von Gunten, 2003; Drzewicz *et al.* 2004; Gehringer *et al.* 2008). The injection of O_3 before or during irradiation converts the reducing species of water radiolysis into the key intermediate, i.e., the ozonide radical anion ($O_3^{\bullet -}$), responsible for enhancing the ${}^{\bullet}OH$ yield in this system. Here we show a simplified reaction mechanism based on the works of Merényi *et al.* (2010) and Fischbacher *et al.* (2013):

$$H^{\bullet} + O_3 \rightarrow HO_3^{\bullet} \quad k_{6.34} = 2.2 \times 10^{10} \text{ mol}^{-1} \text{ dm}^3 \text{ s}^{-1} \tag{6.34}$$

$$HO_3^{\bullet} \rightarrow O_3^{\bullet -} + H^+ \tag{6.35}$$

$$O_3^{\bullet -} \rightleftarrows O^{\bullet -} + O_2 \xrightarrow{+H_2O} + {}^{\bullet}OH + OH^- + O_2 \tag{6.36}$$

$$e_{aq}^- + O_3 \rightarrow O_3^{\bullet -} \quad k_{6.37} = 3.6 \times 10^{10} \text{ mol}^{-1} \text{ dm}^3 \text{ s}^{-1} \tag{6.37}$$

$$^{\bullet}OH + O_3 \rightarrow HO_2^{\bullet} + O_2 \quad k_{6.38} = 1.1 \times 10^8 \text{ mol}^{-1} \text{ dm}^3 \text{ s}^{-1} \tag{6.38}$$

$$O_2^{\bullet-} + O_3 \rightarrow O_3^{\bullet-} + O_2 \quad k_{6.39} = 1.5 \times 10^9 \text{ mol}^{-1} \text{ dm}^3 \text{ s}^{-1} \tag{6.39}$$

Literature studies on the combination of ozone and electron-beam radiation are rather inconsistent with regard to a potential synergy between the two processes. For example, Gehringer *et al.* (2008) reported that applying the irradiation/O_3 combined method decreased the dose required to degrade 1,5-NDSA (naphthalene-1,5-disulfonate) in tap water to half as compared to a single irradiation treatment. With regard to mineralization of 1,5-NDSA, unexpectedly, the combined treatment was less efficient than the single electron beam irradiation process: after total 1,5-NDSA decomposition, in the combined process the TOC reduction was only ~60%, whereas in the single treatment it reached 83%.

A more detailed description on the combined ozonation/electron-beam radiation technology is given by Cooper *et al.* (2004).

6.3.3 Kinetics and modeling of ionizing radiation-induced processes

In water treatment processes the time dependences of solute concentrations can be described by using stationary kinetics: the rate of formation of reactive intermediates (r_X) is equal to the rate of decay of intermediates (Wojnárovits & Takács, 2008). The formation rates of the three reactive intermediates (r_{OH}, r_{eaq-} and r_H) are directly proportional to the dose rate ($D^{\bullet}$, J kg^{-1} s^{-1}) and the density (ρ, for water 1 kg dm^{-3}):

$$r_{OH} = 2.8 \times 10^{-7} D^{\bullet} \rho(\text{mol dm}^{-3} \text{ s}^{-1}); \; r_{eaq-} = 2.7 \times 10^{-7} D^{\bullet} \rho(\text{mol dm}^{-3} \text{ s}^{-1}); \; r_H = 0.6 \times 10^{-7} D^{\bullet} \rho(\text{mol dm}^{-3} \text{ s}^{-1})$$

The kinetics of transformation of R solute to P stable products in reaction with X intermediate (X = $^{\bullet}$OH, e_{aq}^- or H$^{\bullet}$) depends on the reactivity of P with X. When the rate constant of X in reaction with R is much higher than with product P, linear time dependence of R concentration is obtained:

$$R + X \rightarrow\rightarrow P \text{ (stable product)} \tag{6.40}$$

$$\frac{d[X]}{dt} = r_X - k_X [R][X] \approx 0 \quad [X] = \frac{r_X}{k_X[R]} \tag{6.41}$$

$$\frac{d[R]}{dt} = -k_X [R][X] \approx k_X[R]\frac{r_X}{k_X[R]} \tag{6.42}$$

$$[R] = [R]_0 - r_X t = [R]_0\left(1 - \frac{r_X}{[R]_0}t\right) = [R]_0(1 - k_{obsd}t) \tag{6.43}$$

In practice linear time dependence of the R concentration is very rarely observed. Figure 6.5 shows an example, namely, the e_{aq}^- reaction with the azo dye Apollofix Red (AR-28). The hydrated electron reacts with the –N=N– azo bond and this reaction destroys the extended conjugation through this connection. This is shown in panel c, the absorbance in the visible spectral range decreases monotonously with the absorbed dose. Panel d shows the decrease of dye concentration (see also 6.4.5.1).

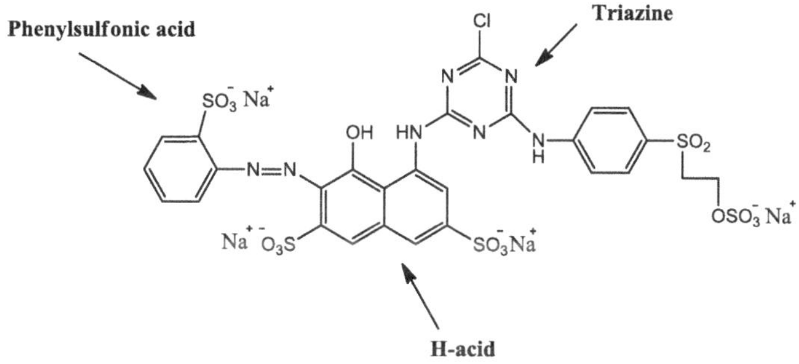

Apollofix Red (AR-28)

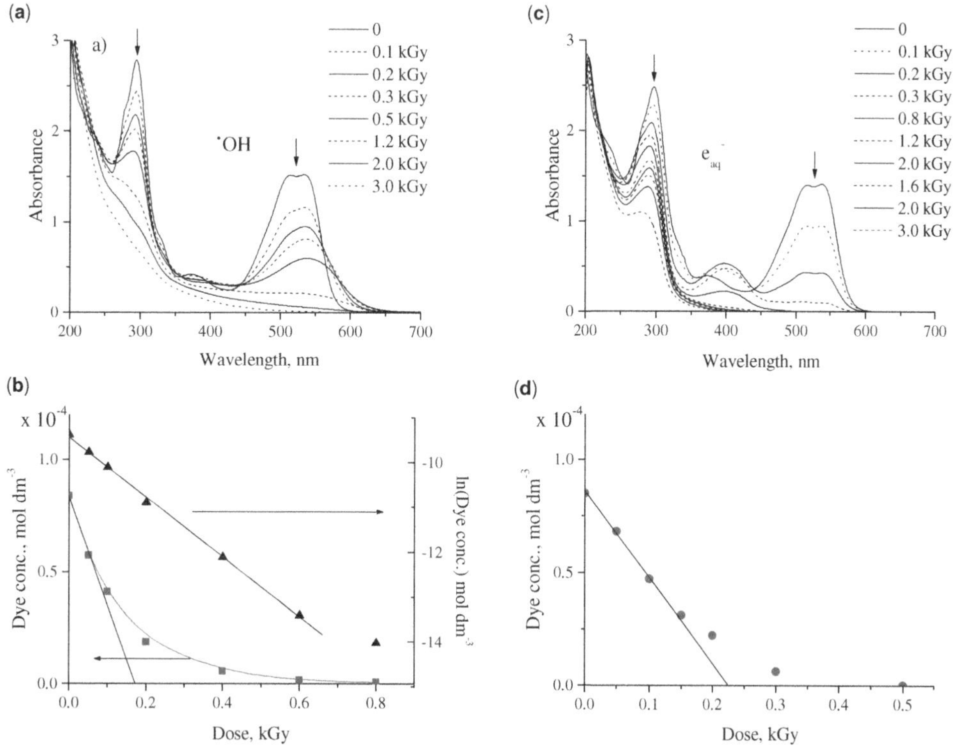

Figure 6.5 Degradation of AR-28 during gamma radiolysis through reaction with $^{\bullet}OH$ (a, b), and e_{aq}^- (c, d) at 8.5×10^{-5} mol dm^{-3} initial dye concentration. Panels a and c show the absorption spectra of the irradiated solutions; panels b and d give the dye concentration – absorbed dose kinetics (Wojnárovits & Takács, 2008).

When the X reactive intermediate reacts with high (practically the same) rate constant with both the starting R molecules and with the product P ($k_X \approx k_{X,P}$), assuming also that $[R] + [P] = [R]_0$, after integration of the respective equations exponential time dependence is obtained:

$$\frac{d[X]}{dt} = r_X - k_X\,[R][X] - k_{X,P}\,[P][X] \approx r_X - k_X\,[R]_0[X] \approx 0 \qquad [X] = \frac{r_X}{k_X[R]_0} \tag{6.44}$$

$$\frac{d[R]}{dt} = -k_X[R][X] \approx -k_X[R]\frac{r_X}{k_X[R]_0} \tag{6.45}$$

$$[R] = [R]_0 \exp\left(-\frac{r_X}{[R]_0}t\right) = [R]_0 \exp(-k_{obsd}t) \tag{6.46}$$

In radiolytic reactions of hydroxyl radicals carried out under constant dose rate conditions, it is quite general that the concentration of the starting molecules decreases exponentially with the time (dose). Figure 6.5 panel b shows this exponential dependence for the $^{\bullet}OH$ + AR-28 reaction, more details are given in 6.4.5.1.

With the progress of degradation more and more products are accumulated in the reaction mixture and the kinetics can be described by changes of such bulk parameters as COD or TOC. Figure 6.6 shows the decrease of the chemical oxygen demand (COD) values in aerated solutions of phenol (2×10^{-3} mol dm^{-3}) and three other compounds as a function of absorbed dose (gamma irradiation). Under fixed-dose rate irradiation, the reactive intermediates form in the solution with a constant rate, therefore, it is not surprising that during the initial period COD decreases linearly with the dose (Homlok *et al.* 2013; Csay *et al.* 2014): i.e. the oxidation proceeds with a constant rate.

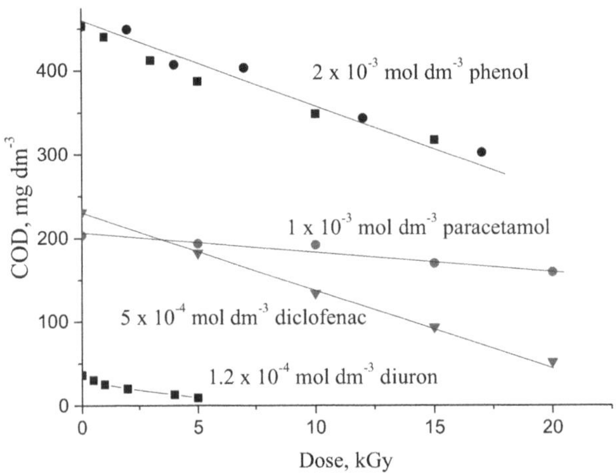

Figure 6.6 COD decay in phenol, paracetamol, diclofenac and diuron air-saturated-solutions. Data were collected from the works of Csay *et al.* (2014), Homlok *et al.* (2013) and Kovács *et al.* (2015).

6.3.4 Toxicity of ionizing radiation-treated water

A large number of toxicity tests are used in practice for water quality control. Among those, the Microtox test is one of the simplest. The bioassay uses *Vibrio fischeri*, a luminescent marine bacterium. The bacteria are exposed to the toxic sample, and the amount of light emitted is measured by a photometer relative to a non-toxic sample. Healthy bacteria emit a greater amount of light than those affected by the quality of sample. The amount of light measured indicates the metabolic state of the bacteria, thus the toxicity of the sample (Ruiz *et al.* 1997; Gunatilleka & Poole, 1999). The bacteria are provided by the manufacturers in lyophilized form, and can be stored for several months and used on demand.

Radiolysis experiments on the degradation of individual compounds showed that in general, the toxicity first increases and then decreases with the absorbed dose. Such pattern observed in gamma and electron-beam radiolysis may have two reasons: (*i*) some of the organic degradation products are more toxic than the starting compound; (*ii*) hydrogen peroxide is formed in radiation-induced processes at levels toxic to the bacteria. For example, the irradiation of phenol solutions results in hydroxylated products, one of which, hydroquinone, is about 20 times more toxic than phenol. Zona and Solar (2003) showed that introducing catalase enzyme in the samples to destroy H_2O_2 prior to conducting the toxicity bioassay would provide an accurate assessment on the byproduct toxicity in radiation-induced experiments. Figure 6.7 exemplifies the *Vibrio fischeri* luminescence trend in the presence and the absence of catalase.

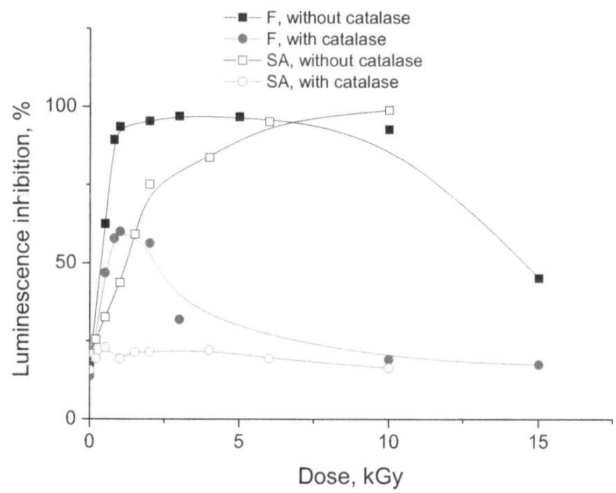

Figure 6.7 Luminescence patterns in Microtox bioassay on irradiated solutions of 2×10^{-4} mol dm^{-3} fenuron (F) and 5×10^{-4} mol dm^{-3} salicylic acid (SA) in pure water, with and without catalase addition (Kovács *et al.* 2014; Szabó *et al.* 2014).

6.4 RESEARCH STUDIES ON WATER RADIOLYSIS-MEDIATED DEGRADATION OF ORGANIC POLLUTANTS

The ionizing radiation-induced degradation of many classes of organic pollutants was investigated by numerous research groups, ranging from simple structures such as small chlorinated alkanes to large molecules such as dyes containing conjugated or unconjugated aromatic rings. This section provides selected examples of research studies reported mainly over the past 15 years on ionizing radiation-induced degradation of water micropollutants, with a particular emphasis on endocrine disrupting compounds, pesticides, pharmaceuticals, dyes, and emulsifiers.

6.4.1 Aromatic compounds

The radiolytic degradation of phenol in aqueous solution was reported in a large number of publications (e.g., Sato *et al.* 1978; Hashimoto *et al.* 1980; Mvula *et al.* 2001; Miyazaki *et al.* 2006). The degradation products may be classified in three categories: (*i*) hydroxylation products (dihydroxybenzenes); (*ii*) dimers (*o,o′*-biphenol), and (*iii*) ring degradation products (e.g., dihydroxymucondialdehyde, muconic acid derivatives, maleic acid).

Although the •OH addition may take place to any of the carbon atoms on the ring, in substituted rings there is a considerable selectivity due to the directing effect of the substituent. The •OH attack site depends on the charge distribution on the ring, and is preferentially selected at the position with high electron density (Schuler & Albarran, 2002; Albarran & Schuler, 2007). The distribution of radicals produced provides information on the charge distribution on the ring. Electron-donating substituents direct the addition preferentially to *ortho-* and *para* positions. In the case of phenol (Scheme 6.2) *ipso-*, *ortho-*, *meta-* and *para*-additions take place with probabilities of 0.08, 0.50, 0.08 and 0.34, respectively.

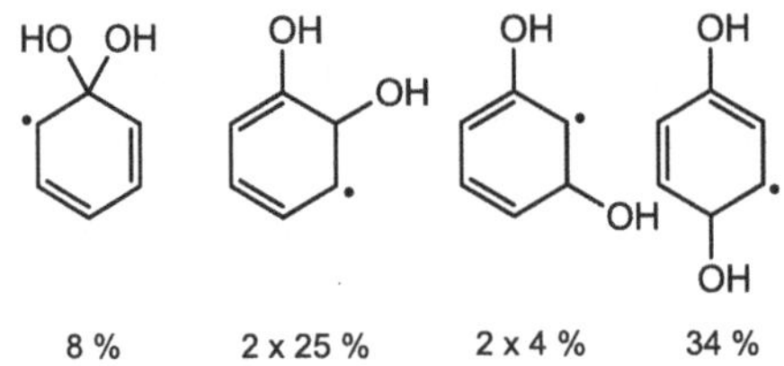

Scheme 6.2 Dihydroxycyclohexadienyl radical isomers formed in the •OH + phenol reaction (Albarran & Schuler, 2007).

In aerated solutions, the dihydroxycyclohexadienyl radicals react with O_2 forming peroxyl radicals in equilibrium with the original species (Fang *et al.* 1995). The O_2 addition is fast and the adduct is relatively stable when there is an electron-donating substituent on the ring; the opposite is true for electron-withdrawing substituent. It is generally assumed that the ring-opening takes place from a peroxyl radical structure (Getoff, 1997; von Sonntag & Schuchmann, 1997, 2001).

Based on the works of Getoff's and von Sonntag's groups, in Scheme 6.3 we suggest a possible reaction mechanism for the initial steps of •OH induced degradation of phenol. The Scheme shows •OH attack at *para* position.

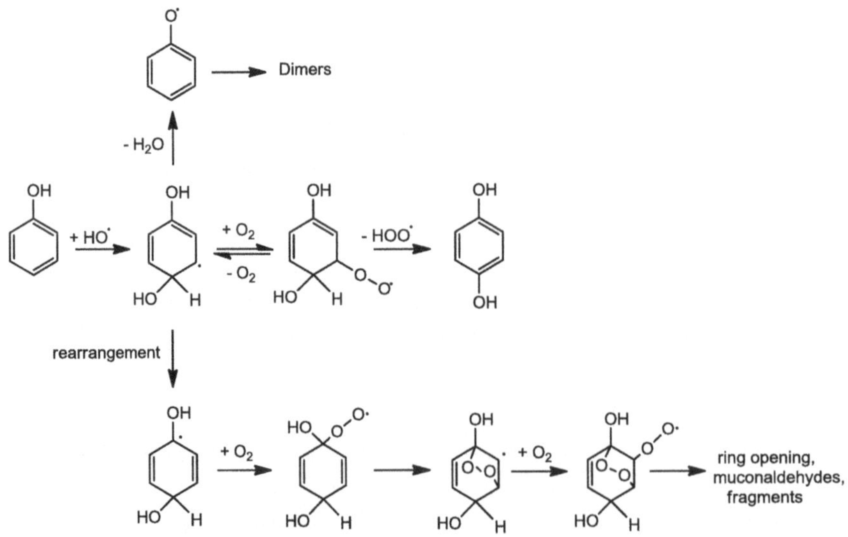

Scheme 6.3 •OH induced degradation of phenol.

The initial rate of phenol oxidation (ΔCOD/kGy) is 8.8 ± 0.6 mg dm^{-3} kGy^{-1}, i.e. 2.75×10^{-4} mol dm^{-3} kGy^{-1} (see Figure 6.6). With 1 kGy dose 2.8×10^{-4} mol dm^{-3} $^{\bullet}OH$ forms in the solution. The ratio of the last two values ($2.75 \times 10^{-4}/2.8 \times 10^{-4}$) shows that 1 $^{\bullet}OH$ induces the incorporation of 0.98 O_2 molecule into phenol. When an O_2 molecule is incorporated into the product, four-electron oxidation occurs. Therefore, the one-electron oxidant, $^{\bullet}OH$, on average induces 3.92 electron oxidations. Upon the $^{\bullet}OH$ attack on organic molecules, organic radicals are formed. The carbon-centered radicals react with dissolved O_2 forming peroxyl radicals (second oxidation step). Further, uni- or bimolecular reactions of the peroxyl radicals may induce additional oxidations as it is shown for phenol in Scheme 6.3.

The degradation of chlorinated and brominated aromatic molecules is a frequent subject of radiation chemical studies (Al-Sheikhly *et al.* 1997; Poster *et al.* 2003; Zhang *et al.* 2007; Tang *et al.* 2010). Often the water soluble phenol type chlorinated and brominated aromatics served as model substances for more complex molecules, e.g., pesticides. In the degradation studies of badly water soluble polychlorinated biphenyls or benzenes water-organic solvents (like isopropanol, acetone or methanol) were used to increase the solubility. Polychlorinated biphenyls were widely used in the past (in many countries till the end of 1970s) in transformers, capacitors, heat transfer and hydraulic fluids, etc. A large fraction of these chemicals was released to the environment and - due to the low-solubility in water - accumulated in the sediment of rivers and lakes. Several papers were also devoted to the radiolytic degradation of these compounds in the sediment.

The basic goal of irradiation of the polychlorinated molecules is the dechlorination. According to the general experiences both $^{\bullet}OH$ and e_{aq}^- reaction can lead to dechlorination. When $^{\bullet}OH$ adds to a Cl-substituted carbon atom in the ring this *ipso* adduct undergoes very rapid HCl elimination to form phenoxyl type radical:

$$^{\bullet}OH + ArCl \rightarrow {}^{\bullet}ArCl(OH) \rightarrow ArO^{\bullet} + HCl \tag{6.47}$$

The C-Cl bond may dissociate rapidly through a dissociative electron capture reaction:

$$e_{aq}^- + ArCl \rightarrow {}^{\bullet}Ar + Cl^- \tag{6.48}$$

In the usual reaction systems this reaction is in competition with the $e_{aq}^- + O_2$ reaction.

In alkaline methanol or isopropanol solutions chain decomposition of polychlorinated aromatics was observed (Singh *et al.* 1985; Al-Sheikhly *et al.* 1997). This chain reaction is due to intermediate alcohol radical anions which serve as chain carriers. For example in the case of methanol the following reactions are suggested:

$$^{\bullet}OH + CH_3OH \rightarrow H_2O + {}^{\bullet}CH_2OH \tag{6.49}$$

$$^{\bullet}CH_2OH + OH^- \rightleftarrows {}^{\bullet}CH_2O^- + H_2O \tag{6.50}$$

$$^{\bullet}CH_2O^- + ArCl \rightarrow CH_2O + Cl^- + Ar^{\bullet} \tag{6.51}$$

$$Ar^{\bullet} + CH_3OH \rightarrow ArH + {}^{\bullet}CH_2OH \tag{6.52}$$

Dissolved O_2 by reacting with the radicals inhibits the chain reaction. For obtaining good efficiency careful optimization of reaction conditions is needed. The doses required to degrade polyhalogenated compounds are generally much higher than the doses required for the degradation of the other most common organic pollutants. Yet, irradiation technology may be a competitive alternative to other technologies which also suffer from several drawbacks.

6.4.2 Endocrine disrupting compounds

Endocrine disrupting compounds (EDCs) belong to a variety of chemical classes, including natural hormones and a large variety of industrial products. Most of these compounds are produced in large quantities some of which are released into the environment (Hu & Aizawa, 2003; Gültekin & Ince, 2007). Endocrine disruptors alter hormonal functions in vertebrates. Among the chemical compounds largely used in industrial processes and identified as EDCs, which have received scientific and public attention over the concerns regarding their highly potent bioactivity, are alkylphenols, bisphenols, and phthalates.

6.4.2.1 Alkylphenols

The compounds with estrogen receptor binding activity usually have very low concentrations in the treated wastewater (e.g., ng dm^{-3}). Frequently, the estrogen activity of samples containing micropollutants was selected as "the treatment criterion" rather than the micropollutant concentration. Water sample estrogenicity was often assessed using the yeast two-hybrid assay. 17*β*-estradiol (E2) is commonly used standard in estrogenicity bioassays. E2 is a steroid hormone; it is released into the environment from humans and domestic animals.

17*β*-Estradiol reacts with $^{\bullet}OH$ with a rate constant of 5.3×10^9 mol^{-1} dm^3 s^{-1} (Kosaka *et al.* 2003). Ionizing radiation experiments on E2 in laboratory water indicated that at the beginning of irradiation the estrogen activity increases, and upon further exposure it decreases (Kimura *et al.* 2004, 2006, 2007). This pattern shows that during the primary oxidation steps E2 is degraded to byproducts bearing the core structure of the molecule responsible for the endocrine disrupting property. Such an intermediate byproduct was identified as 2-hydroxyestradiol (Wheeler & Montalvo, 1967; Kimura *et al.* 2004). In these studies, it was shown that when starting from an initial concentration of 1.8×10^{-9} mol dm^{-3} E2 in pure water, a dose of 10 Gy is required to degrade the estrogen, whereas the estrogenic activity can be removed only upon application of a dose of 30 Gy. Elimination of estrogen activity of real wastewater is considered to be interfered by the other impurities (Kimura *et al.* 2007). In a model wastewater with 1.8×10^{-9} mol dm^{-3} E2 and 1 mg dm^{-3} TOC ~100 Gy would suffice to remove the estrogen activity.

Due to the structural similarities to E2 some alkylphenols, especially those with longer alkyl chains, have also been identified as endocrine disruptors. Nonylphenol (NP) exhibits one of the highest estrogen activities among artificial chemicals, having about 0.1% cross reactivity with the estrogen receptor. This activity is observed above the concentration level of about 1×10^{-6} mol dm^{-3}. The *p*-nonylphenol isomer shown below (4-(1,2,4-trimethylhexyl)phenol) resembles the estradiol structure.

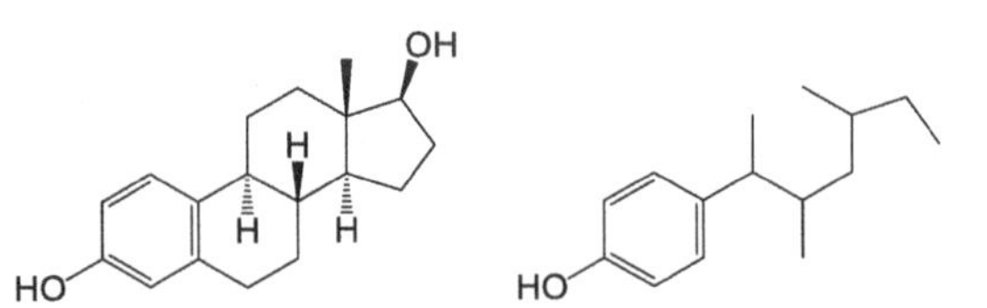

Estradiol 4-((1,2,4-Trimethylhexyl)phenol)

In radiolysis experiments, NPs react with $^{\bullet}OH$ to produce two intermediate byproducts, namely, *p*-nonylcatechol and 1-(*p*-hydroxyphenyl)-1-nonanol (Kimura *et al.* 2006, 2007). In gamma radiolysis/H_2O_2 reaction systems great increase of the degradation rate of nonylphenol ethoxylates was reported as compared to gamma radiolysis alone. H_2O_2 promoted the formation of acidic degradation products (Iqbal & Bhatti, 2015). Petrovic *et al.* (2007) conducted detailed investigations on surfactants (mainly nonyl- and octylphenol and derivatives) decomposition in a sewage treatment plant effluent using a pilot plant electron beam irradiation system. The initial concentration of alkyl phenol type compounds exceeded 265 μg dm^{-3}. With a dose of 3 kGy about one order of magnitude decrease was observed. Irradiation treatment strongly decreased the toxicity of the solutions.

6.4.2.2 Bisphenols

The representative compound of this class is bisphenol A, which is largely used in the synthesis of plasticizers, polycarbonate, epoxy and unsaturated polyester resins, flame retardants, fungicides, antioxidants, and rubber chemicals.

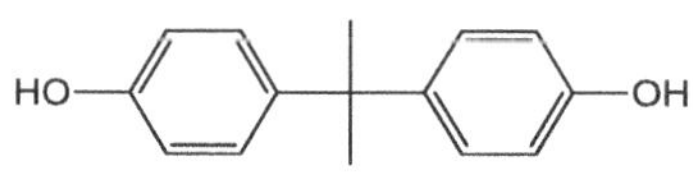

Bisphenol A

The rate constant for the $^{\bullet}OH$ addition to bisphenol A was determined in a pulse radiolysis study as 6.9×10^9 mol^{-1} dm^3 s^{-1} (Peller *et al.* 2009). The hydroxycyclohexadienyl radical type transient is completely transformed to hydroxylated byproducts within 0.5 ms. These products show some degree of resistance to subsequent $^{\bullet}OH$ reactions, since they accumulate as stable, detectable products during the oxidation of bisphenol A. The *G*-value of the hydroxylated products in N_2O-saturated solution is 0.45 ± 0.02 μmol J^{-1}, corresponding to an overall degradation efficiency of 76% in the reactions of $^{\bullet}OH$.

When the experiments were carried out in pre-treated wastewater the hydroxylated product formation was inhibited by dissolved substances in the water. There is a competition for $^{\bullet}OH$ between bisphenol A and other contaminants. Similar competition was also reported in the radiolytic degradation of low concentrations of E2 and *p*-nonylphenol in treated wastewater (Kimura *et al.* 2007). The degradation efficiency was found higher under oxidative than under reductive conditions, and increased with the dose but decreased with the increasing initial concentration (Xu *et al.* 2011).

6.4.2.3 Phthalates

Phthalate esters are widely used as plasticizers for polyvinyl chloride (PVC) resins and cellulose film coatings; they are also used in cosmetics, insect repellents, and solid rocket propellants. Of all, di-(2-ethylhexyl)-phthalate is produced at the largest scale (Gültekin & Ince, 2007).

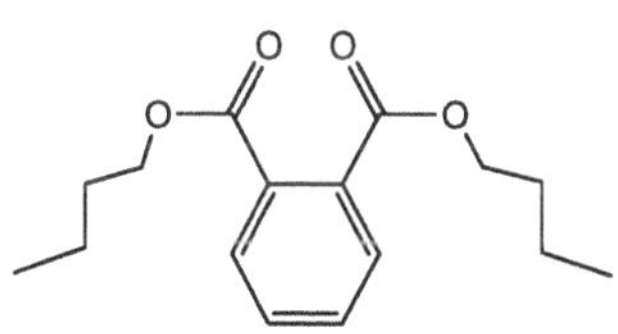

Di-*n*-butyl phthalate

The ionizing radiation effects on aqueous solutions of phthalic acid, di-*n*-butyl phthalate, and dimethyl phthalate were reported in a few literature studies (Tezuka *et al.* 1978; Yoshida *et al.* 2003; Wu *et al.* 2011). The rate constants of the e_{aq}^- reactions with phthalate di-anion and mono-ion were determined as 1.9×10^9 and 1.1×10^{10} mol^{-1} dm^3 s^{-1}, respectively, whereas the rate constants for the $^{\bullet}OH$ reaction with di-anion form is reported as 5.9×10^9 mol^{-1} dm^3 s^{-1} (Buxton *et al.* 1988). Tezuka *et al.* (1978) studied the ionizing radiation-induced degradation of di-*n*-butyl phthalate and identified mono-*n*-butyl phthalate as the first stable reaction product. Yoshida *et al.* (2003) showed that upon irradiation of this compound (17 mg dm^{-3}) in laboratory water for a total dose of 750 Gy, the complete mineralization was achieved.

6.4.3 Pesticides

6.4.3.1 Chlorophenoxy pesticides

The chlorophenoxy herbicide, 2,4-dichlorophenoxyacetic acid (2,4-D) is considered to be a moderately toxic chemical (Zona & Solar, 2003). Nevertheless, 2,4-D is also suggested to be an endocrine disruptor and a possible carcinogen. The fresh waters often contain measurable levels of this compound and its degradation products (Peller *et al.* 2004).

The radiolytic decomposition of 2,4-D in dilute aqueous solutions was investigated by several research groups (Peller *et al.* 2001, 2003, 2004; Zona & Solar, 2003; Drzewicz *et al.* 2004; Jankowska *et al.* 2004; Peller & Kamat, 2005). Peller *et al.* (2004) compared the degradation of 2,4-D and its derivatives, 2,4-dichlorophenol, 2-(2,4-dichlorophenoxy)propionic acid and 2,4,6-trichlorophenol, with three advanced oxidation processes, i.e., TiO_2-photocatalysis, sonolysis and radiolysis showing that in all these processes, the $^{\bullet}OH$ is the reactive species which initiates the oxidation of pesticide.

Radiolysis of 2×10^{-4} mol dm^{-3} 2,4-D solution resulted in approx. 50% destruction of 2,4-D and its aromatic degradation products at a dose of 1 kGy, irrespective of solution pH in the 4–8 range. This comparative degradation study has also indicated that in the $^{\bullet}OH$-mediated transformation of 2,4-D, the predominant byproduct formed in all three AOPs was 2,4-dichlorophenol (2,4-DCP). In addition to this major byproduct, Zona *et al.* (2002a, 2002b) and Zona and Solar (2003) reported some minor hydroxylation byproducts. The mechanism postulated on 2,4-D degradation to 2,4-DCP involves the $^{\bullet}OH$ addition to *ipso* position on the aromatic ring followed by the elimination of glycolic acid from the adduct (Scheme 6.4).

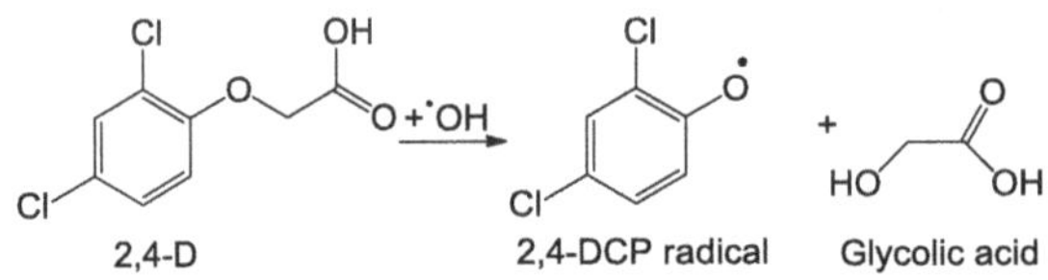

Scheme 6.4 Degradation of 2,4-D (Zona *et. al.* 2002a).

In the above reactions, the aromatic ring is not destroyed. Ring destruction should occur in the subsequent processes. The ring destruction yield ($G \approx 0.1$ μmol J^{-1}) is much smaller than the $^{\bullet}OH$ yield ($G = 0.56$ μmol J^{-1}), and smaller than the overall yield of phenolic products determined by Zona *et al.* (2002a), i.e. $G = 0.179$ μmol J^{-1}. In gamma radiolysis studies the oxygen concentration did not play an important role in 2,4-D degradation (Drzewicz *et al.* 2004). Hydrogen peroxide does not react with 2,4-D; however, the addition of H_2O_2 as a supplementary source of oxygen and hydroxyl radicals enhanced substantially the degradation of 2,4-D and of its intermediate products.

2,4-DCP is more toxic than 2,4-D; toxicity increased during the first stage of irradiation, then decreased as 2,4-DCP was oxidized (Zona & Solar, 2003).

In the radiolysis of 4-chloro-2-methylphenoxyacetic acid (MCPA) (in the 2,4-D structure the *ortho* Cl is replaced by CH_3) similar $^{\bullet}OH$ induced degradation was reported as shown in Scheme 6.4 for 2,4-D (Bojanowska-Czajka *et al.* 2006). In purified water more effective mineralization was observed at alkaline than at acidic pH. This pH dependence was attributed to bicarbonate and especially to carbonate that form during mineralization and compete for $^{\bullet}OH$ scavenging. The authors showed that one should be careful with H_2O_2 addition to the solutions before irradiation in order to increase the degradation rate, because at higher concentration the $^{\bullet}OH + H_2O_2$ reaction also decreases the degradation rate. In the experiments carried out on wastes from the industrial production process of MCPA (concentration not reported) a dose of 5 kGy was sufficient to degrade MCPA, at 10 kGy most of the organic products were also decomposed. 1.3 g dm^{-3} H_2O_2 added to the liquid before irradiation considerably increased the degradation rate. However, when the irradiation was carried out in the presence of H_2O_2, the level of toxicity was much higher than in the absence of hydrogen peroxide in the treated waste sample.

6.4.3.2 Triazine pesticides

The *s*-triazine core structure is widely used in many herbicides. These compounds are characterized by high chemical stability. Atrazine (2-chloro-4-(ethylamine)-6-(isopropylamine)-*s*-triazine) is a representative of the group. The degradation of *s*-triazines by various processes, including gamma-radiation, is well represented in the literature (Bucholtz & Lavy, 1977; Karpel Vel Leitner *et al.* 1999; Angelini *et al.* 2000; Basfar *et al.* 2009; Mohamed *et al.* 2009). The rate constants for the reactions with e_{aq}^- are in the ~(1–5) × 10^9 mol^{-1} dm^3 s^{-1} range (Varghese *et al.* 2006). The recommended k_{OH} of atrazine is 2.4 × 10^9 mol^{-1} dm^3 s^{-1} and the rate constants of other triazines are not much different from this value (Wojnárovits & Takács, 2014). Both k_{OH} and k_{eaq-} are smaller by a factor of 5–10 than the diffusion controlled limit.

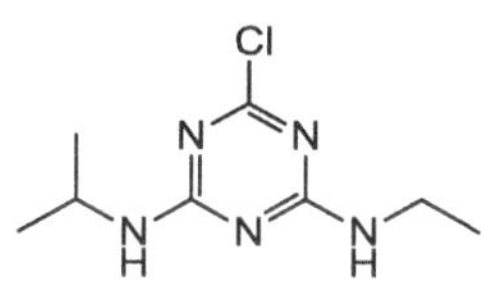

Atrazine

Hydrated electron adds to one of the electron-deficient nitrogen atoms in the heterocyclic ring with formation of a radical anion, which is rapidly protonated in reaction with water. The intermediate radicals formed in e_{aq}^- reactions are N-protonated electron adducts, with an unpaired electron at the vicinal carbon atom (Varghese *et al.* 2006). e_{aq}^- reactions are relevant in the absence of dissolved oxygen; in practical applications of radiation-induced AOP, such reactions have very low probability.

Hydroxyl radical can react with the triazine molecules by either addition to the ring or H-abstraction from the C-H bonds in the side chain (Karpel Vel Leitner *et al.* 1999; Khan *et al.* 2015). The major primary degradation products of atrazine were identified as desisopropyl-atrazine (DIA), desethyl-atrazine (DEA), and desethyl-desisopropyl-atrazine (DEDIA), which result from H-atom abstraction by $^{\bullet}OH$ from the secondary and tertiary C-atoms in the side chains.

In gamma irradiation experiments, higher degradation rates of atrazine were found in deoxygenated than in oxygenated solutions (Khan *et al.* 2015). This finding shows the e_{aq}^- contribution to atrazine degradation. Experiments carried out in the presence of ozone showed considerable enhancement in the degradation yield, i.e. a synergistic effect. Atrazine degradation experiments carried out in different groundwater matrixes showed large dependence of the degradation rate on the type of the groundwater. Especially, the carbonate, bicarbonate and nitrate anion concentrations influenced strongly the degradation rate due to their competing reactions with reactive intermediates (Mohamed *et al.* 2009).

Acetochlor is used to control weeds in corn, it is particularly useful as a replacement for atrazine in the case of some important weeds. Its rate of radiolytic degradation was found higher in alkaline than in acidic solutions and in γ-radiolytic experiments the rate increased with the increasing dose rate. In aqueous solutions the compound was shown to decompose easily under both oxidative and reductive conditions (Liu *et al.* 2004, 2005).

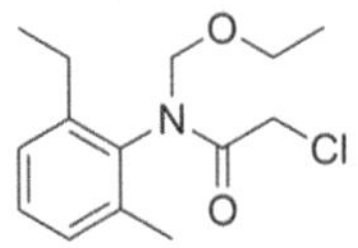

Acetochlor

6.4.3.3 Phenylurea herbicides

Phenylurea herbicides are photosynthesis inhibitors, and are applied to kill broadleaf weeds. Many phenylurea herbicides are *N*-dimethyl derivatives bearing various substituents on the aromatic ring. These compounds are highly persistent in the environment, with half-lives of several months in soil. Fenuron, monuron and diuron are among the most used phenylurea pesticides.

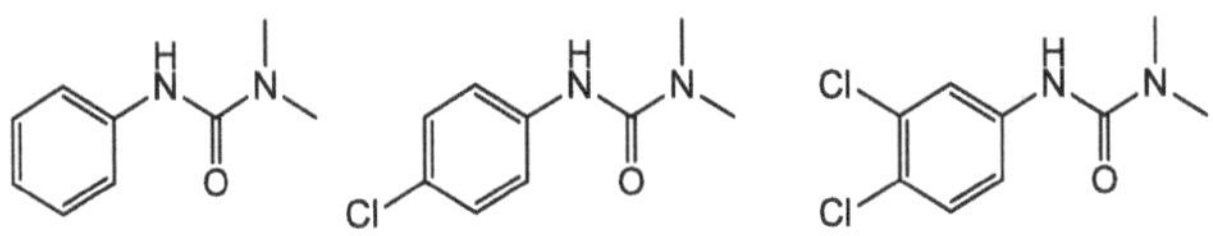

Fenuron Monuron Diuron

The hydroxyl radical rate constants for the reactions with fenuron, monuron and diuron are 8.3×10^9, 7.3×10^9, and 6.0×10^9 mol^{-1} dm^3 s^{-1}, respectively (Wojnárovits & Takács, 2014; Kovács *et al.* 2014, 2015). The decrease of k_{OH} with the number of chlorine atoms on the ring is attributed to the electron-withdrawing effect of the Cl atoms. $^{\bullet}$OH reacts with the aromatic ring resulting in hydroxycyclohexadienyl radicals, which transform further to dechlorinated and hydroxylated, or hydroxylated products (Zhang *et al.* 2008; Kovács *et al.* 2014, 2015). $^{\bullet}$OH attack on the methyl groups was found less important than reaction with the aromatic ring, in agreement with the results of most of other AOP investigations (e.g., Oturan *et al.* 2010). However, Mazellier and Sulzberger (2001) observed only one product in heterogenous photo-Fenton degradation of diuron (3-(3,4-dichlorophenyl)-1-formyl-1-methylurea), which is formed in the $^{\bullet}$OH attack on a methyl group. Dissolved O_2 enhances the rate of dechlorination and also the overall rate of degradation. The rate of degradation is much higher for diuron than for monuron or fenuron (Kovács *et al.* 2014, 2015). This is due to the highly labile intermediates that form in radical reactions of diuron. The intermediates easily undergo dechlorination, or in the presence of dissolved O_2 ring degradation.

6.4.4 Pharmaceutical compounds

Due to the extensive use of pharmaceuticals, a large variety of these compounds are found in the municipal wastewaters mostly originating from the households and hospitals; a large fraction (up to ~90%) of the medicines after intake is excreted either as original compounds or as their metabolites. Due to their large scale use and, in some cases, low biodegradability, pharmaceuticals are regularly detected in surface waters. Their long-term effects on the aquatic species are not elucidated, but concerns extend over pharmaceuticals presence in the water consumed by humans, when the contaminated surface waters are used for drinking water production. As an example is Lake Geneva, Switzerland, the "receiver" of treated wastewater effluents, which serves as a drinking water production source. From the micropollutants detected in the treated wastewater effluents at the Lausanne wastewater treatment plant, 40% were also detected in the drinking water. The majority of these micropollutants were pharmaceutical compounds (De la Cruz *et al.* 2012).

This section presents the radiation chemistry of selected antibiotics and non-steroidal anti-inflammatory drugs.

6.4.4.1 Antibiotics

In the past decade, the removal of antibiotics by various AOPs has caught significant attention in the scientific community (Michael *et al.* 2013). A major concern is associated with the increasing resistance of bacteria to the antibiotics identified in the environment. Urban wastewater treatment plants offer ideal conditions for the spread of bacterium resistance given the coexistence of antibiotic contaminants and diverse bacterial populations during the biological treatment processes. There is a huge pool of genes encoding resistance mechanism (Wright, 2007) which is also available for pathogenic bacteria, undermining the achievements of the modern medicine (Rizzo *et al.* 2013).

6.4.4.1.1 Chloramphenicol

Chloramphenicol is the first broad-spectrum antibiotic used in treating microbial infections. It acts by interfering with bacterial protein synthesis and is mainly bacteriostatic.

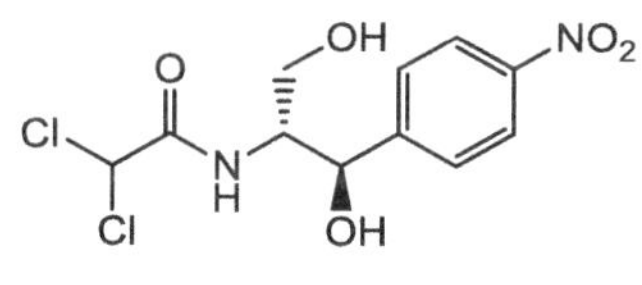

Chloramphenicol

The fundamental radiation chemistry of this compound in aqueous solution was studied by Kapoor and Varshney (1997) using pulse radiolysis technique, whereas the degradation mechanism and product identification were reported by Varshney and Patel (1994), and more recently by Zhou *et al.* (2009) and Csay *et al.* (2012).

Hydrated electrons and hydrogen atoms react with chloramphenicol with rate constants of 2.3×10^{10} and 1×10^9 mol^{-1} dm^3 s^{-1}, respectively (Kapoor & Varshney, 1997). The rate constant of $^{\bullet}$OH reaction was measured as 1.8×10^9 mol^{-1} dm^3 s^{-1} at pH 7. The low k_{OH} is due to the electron-withdrawing nitro group on the aromatic ring disfavoring the electrophilic $^{\bullet}$OH attack. The low k_H may have the same reason. On the contrary, the rate constant of the nucleophilic e_{aq}^- attack is high; it results in the formation of radical anion. At the beginning of irradiation the ring and side chain hydroxylated compounds dominate the product spectrum. Complete removal of 1×10^{-4} mol dm^{-3} chloramphenicol was observed at 2.5 kGy dose (Csay *et al.* 2012).

6.4.4.1.2 Sulfonamides

Sulfa drugs (sulfonamides) are derived from sulfanilamide. Many pharmaceuticals belong to this class; either aliphatic group or a 5-membered- or 6-membered- heterocyclic ring is attached to the sulfanilamide unit. Three representative structures are shown below. These antibiotics block the folic acid metabolism in bacteria inhibiting the synthesis of one of nucleotides.

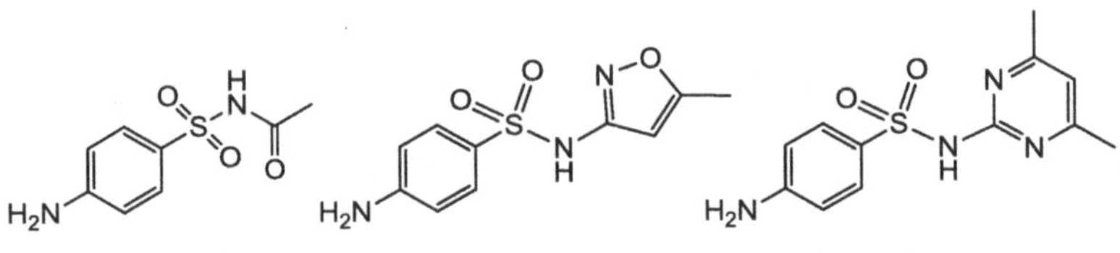

Sulfacetamide Sulfamethoxazole Sulfamethazine

Sulfonamides have two acid-base dissociation equilibria. At pH ~2 the free NH_2 group protonates/deprotonates, while at pH 5 –7 the protolytic dissociation takes place at the sulfonamide ($-SO_2-NH-R$) group. Most radiolysis experiments (Sabharwal *et al.* 1994; Mezyk *et al.* 2007; Kim *et al.* 2009, 2012; Liu *et al.* 2013, 2014; Sági *et al.* 2014, 2015) were carried out in un-buffered solutions (pH ~ 5 – 6).

The rate constants of $^{\bullet}OH$ reactions with the sulfa drugs were found within the $4 \times 10^9 - 8 \times 10^9$ mol^{-1} dm^3 s^{-1} range, which is typical for aromatic compounds. $^{\bullet}OH$ reacts primarily with the benzene ring; given the electron delocalization across the heterocycle ring, if such ring is present in the molecule, the $^{\bullet}OH$ attack at that site also occurs.

In experiments carried out in aerated solutions of a sulfonamide at $\sim 1 \times 10^{-4}$ mol dm^{-3} concentration, the yields of hydroxylated products reach maxima at an absorbed dose of ~0.6 kGy. Phenol formation, ring opening and fragmentation occur simultaneously. That is indicated by the high amount of O_2 incorporated in the products (~0.75) upon one $^{\bullet}OH$ attack, and also by the strong decrease of pH and the appearance of low molecular mass organic acids even at low doses. Sulfate (SO_4^{2-}) and ammonium (NH_4^+) ions are also observed at low doses, however, their concentration-dose curves are shifted in time as compared to the curves of the intact sulfonamide molecules. Phenols disappear from the solution shortly after the decay of sulfonamide molecules; above 1.0 kGy hardly any organic product was detectable. The SO_4^{2-} and NH_4^+ concentration-dose dependences are similar; these ions may form simultaneously with ring opening. The nitrate (NO_3^-) ion mainly forms toward the end of system radiolysis, and can originate from the amide-N or from the N-atoms in the heterocyclic ring. The mechanistic patterns, i.e. hydroxylation followed by degradation are in agreement with the observed faster decrease of COD than of TOC at low doses. Inorganic molecules/ions (mineralization) form in multistep processes that could take place either simultaneously or consecutively.

In experimental studies where H_2O_2 was added to the solutions before irradiation, a significant increase in the degradation efficiency was observed. The presence of carbonate ions in the solutions lowered slightly the degradation rate (Guo *et al.* 2012; Liu & Wang, 2013).

6.4.4.1.3 β-Lactam antibiotics

The penicillin and cephalosporin classes of β-lactam antibiotics are largely used to treat bacterial infections, being the most prescribed antibiotics in community and hospital settings (ECDC, 2014). The β-lactam moiety is a cyclic amide with an inherent strain, and gives the antibacterial property to the molecule. Penicillins share the same core structure, a four-membered β-lactam ring fused to a thiazolidine ring;

cephalosporins have a similar four-membered β-lactam moiety, but condensed with a dihydrothiazine ring (Figure 6.8). β-Lactams exert their antimicrobial action by hampering the biosynthesis pathway of the cell wall in bacteria *via* targeting transpeptidases by mimicking their natural D-Ala-D-Ala substrate (Walsh, 2003). The β-lactam structure is an efficient acylating agent, and binds covalently to the serine residue of the enzyme, generating an inactive acyl-enzyme complex. In the absence of the four-membered lactam structure, the molecules lose the antibacterial activity. Targeting a remote side chain group on the structure would result in reduced process efficiency. The primary attack sites of the reactive species in various advanced oxidation processes are shown in Figure 6.8 (Dodd *et al.* 2006, 2010; Rickman & Mezyk, 2010; Karlesa *et al.* 2014; Szabó *et al.* 2016b, 2016c).

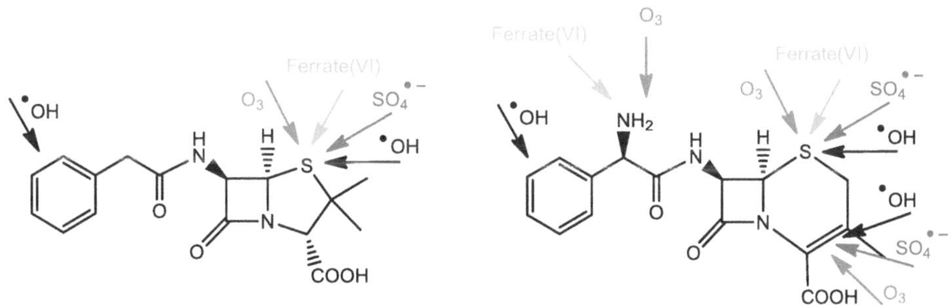

Penicillin G

Cephalexin

Figure 6.8 Validated targets of reactive species under advanced oxidation of β-lactams (Dodd *et al.* 2006, 2010; Rickman & Mezyk, 2010; Karlesa *et al.* 2014; Szabó *et al.* 2016b, 2016c).

The radiation chemistry of β-lactams has been extensively studied (Song *et al.* 2008; Yu *et al.*; 2008; Chung *et al.* 2009; Dail & Mezyk, 2010; Zhang *et al.* 2011; Mezyk & Otto, 2013; Otto *et al.* 2015; Szabó *et al.* 2016b). The rate constants of •OH reactions with β-lactams (Table 6.2) are close to the diffusion-controlled limit (Dail & Mezyk, 2010). Penicillin structures possess many sites prone to •OH oxidation, fact substantiated by the large variety of products formed under steady state γ-radiolysis of amoxicillin (the compound differs from penicillin G that it has an OH group in *para* position of the benzene ring and a primary amino group similar to cephalexin). Since •OH exhibits prominent reactivity towards aromatics and organic sulfides, with rate constants in the $0.2 \times 10^{10} - 1 \times 10^{10}$ mol^{-1} dm^3 s^{-1} range (Hiller *et al.* 1981; Wojnárovits & Takács, 2013), a competition is expected to occur between the remote aromatic ring and thioether site in penicillins. In cephalosporins, the competition involves the thioether site, the aromatic/heterocyclic side chain, and the double bond in the dihydrothiazine ring.

The partitioning of •OH between the different parts of the molecules drives the efficiency of the process, since targeting the aromatic moiety reduces the possibility of inactivation of the antibiotic, oxidation of the sulfur adjacent to the β-lactam ring can lead to ring-opening, i.e., the destruction of the original molecular structure. Scheme 6.5 shows the •OH oxidation mechanism of penicillin G, as postulated by Pogocki and Bobrowski (2014) and Szabó *et al.* (2016b) for structurally similar derivatives. •OH attack on the aromatic ring generates the corresponding hydroxycyclohexadienyl radical **(a)**, which can react with dissolved O_2 to form the stable OH-substituted product **(b)**. Oxidation of the sulfur of the heterocycle generally starts with an adduct formation **(c)**, which converts into the sulfur radical cation **(d)** (Asmus *et al.* 1977). This intermediate is unstable, and rearranges to an intramolecular structure **(e)**.

Three electron bonded species of this type could also form with a compound containing a heteroatom with free *p*-electron pair, e.g., Cl^- addition to the S-center (Bonifacic & Asmus, 1980). Deprotonation of the α-carbon adjacent to sulfur in **(d)** leads to the α-(alkylthio)alkyl radical **(f)**, which exists in two mesomeric forms **(f)** and **(g)**. An intramolecular electron transfer *via* intermediate **(h)**, with subsequent decarboxylation (pseudo-Kolbe mechanism) proceeds with considerable efficiency in the thiazolidine ring, yielding the α-amino radicals **(i)**. These species undergo β-fragmentation to form thiyl radicals **(j)** in equilibrium. It is assumed that the opening of the β-lactam ring with antibiotic inactivation starts from the α-(alkylthio)alkyl and thiyl radicals.

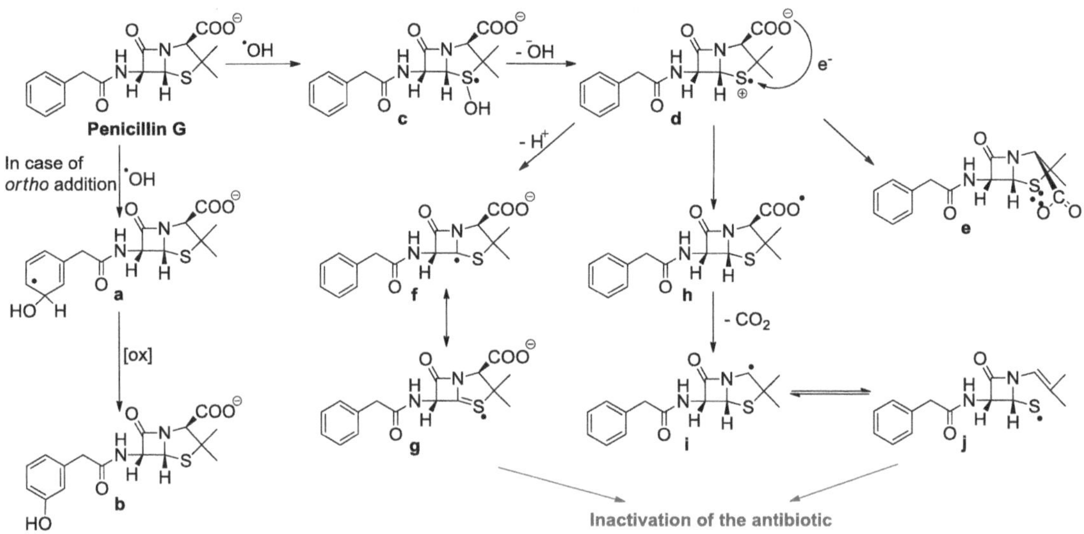

Scheme 6.5 •OH induced degradation pathway proposed for penicillin G.

The fraction of •OH that initiates the molecule destruction reaction which results in the loss of antimicrobial activity can be determined based on end-product experiments. Philips *et al.* (1973b) in the degradation of penicillin G observed a yield of ~0.05 μmol J^{-1} for the hydroxylated derivatives, which corresponds to the consumption of ~18% of the initially available •OH.

The destruction of the β-lactam ring can be followed by a quantitative IR method (Szabó *et al.* 2016c), revealing a yield of ~0.1 μmol J^{-1} for the opening of the β-lactam ring of amoxicillin, which is 36% of the •OH yield. This would suggest a stoichiometry of ~3:1 (•OH:amoxicillin) for the OH reaction with amoxicillin resulting in antibacterial inactivation. The same ratio was found in microbial tests (Otto *et al.* 2015) in highly alkaline water.

The following gives a basic calculation of fraction of •OH required for removal of the activity of an antibacterial drug using the ionizing radiation process of wastewater effluent right after the activated sludge process before the disinfection takes place. For the purpose of such calculation, the following conditions were selected from literature: Effluent Organic Matter (EfOM) TOC – 7 mg dm^{-3} (average value of 28 samples from Keen *et al.* (2014)); •OH scavenging capacity of EfOM – ~1.6×10^5 s^{-1} (Hoigné, 1997; von Gunten, 2003); alkalinity – 105 mg dm^{-3} (as $CaCO_3$, average value from Keen *et al.* (2014)); amoxicillin concentration – 120 ng dm^{-3} (Andreozzi *et al.* 2004) for a treated wastewater effluent sample from Italy.

The calculated percentage of $^{\bullet}OH$ used for elimination of the antibacterial activity under these conditions is ~0.00025%, which is a very low value. The dose requirement can be calculated using the equation:

$$\%(^{\bullet}OH) \times 10^{-2} \times G(^{\bullet}OH)\ [\text{mol J}^{-1}] \times D\ [\text{Gy}] \times \rho\ [\text{kg dm}^{-3}] = [\text{Pollutant}]\ [\text{mol dm}^{-3}] \quad (6.53)$$

which yields (with ρ density of 1 kg dm^{-3}) a realistic dose requirement of ~500 Gy. This model is not sensitive to pollutant concentration, e.g., 100-fold increase in AMX concentration would not affect significantly the dose.

6.4.4.2 Non-steroidal anti-inflammatory drugs

6.4.4.2.1 Aspirin

Aspirin (2-acetoxybenzoic acid, ASA) relieves minor aches and pains, reduces fever and decreases inflammation. At ambient temperature ASA undergoes a slow hydrolysis to salicylic acid (2-hydroxybenzoic acid, SA) and acetic acid. In wastewater samples SA and not ASA is detected. In SA there is strong hydrogen-bonding between the phenolic OH and the carboxyl group in *ortho* position.

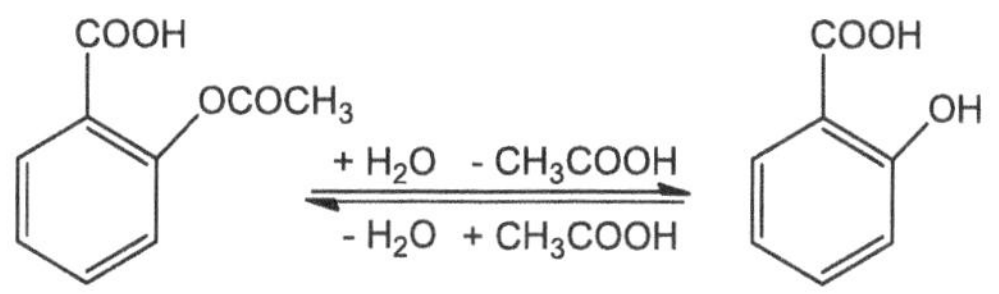

The degradation of SA is a frequent topic of radiation chemical studies (Amphlett *et al.* 1968; Albarran & Schuler, 2003; Kishore & Mukherjee, 2006; Ayatollahi *et al.* 2013); the degradation of ASA/SA hydrolysis mixture is detailed only in one publication (Szabó *et al.* 2014). In the reaction with phenol, $^{\bullet}OH$ adds preferentially to *ortho* and *para* positions of the OH-substituent. In salicylic acid the *ortho* position is favored rather than the *para* position (rate constant 1.07×10^{10} mol^{-1} dm^3 s^{-1}). This difference suggests that the electron density at the *ortho* position in the salicylate anion is enhanced as a result of the hydrogen bonding. The major degradation products of SA were identified as 2,3-dihydroxybenzoic acid, 2,5-dihydroxybenzoic acid and catechol, whereas 2,4-dihydroxybenzoic acid was a minor product (Albarran & Schuler, 2003). Under prolonged irradiation trihydroxylated derivatives were also detected (Szabó *et al.* 2014). The hydrated electron reacts with SA with k_{eaq-} of 9×10^9 mol^{-1} dm^3 s^{-1} (Ayatollahi *et al.* 2013). The electron addition is immediately followed by protonation with formation of a cyclohexadienyl-type radical.

Salicylates are moderately toxic compounds. However, upon irradiation of 5×10^{-4} mol dm^{-3} SA aerated solutions, the luminescence was increasingly inhibited in the *Vibrio fischeri* test up to an absorbed dose of 10 kGy, then gradually decreased (Figure 6.7), indicating that the degradation products were more toxic than the parent compound (Szabó *et al.* 2014). Introduction of catalase enzyme into the irradiated sample before the toxicity test resulted in a huge decrease in the fluorescence inhibition, showing that the toxicity is mainly due to H_2O_2. The increase in toxicity upon irradiation was smaller when the experiments were carried out in tap water instead of purified water. Tap water always contains traces of transition metal ions which catalyze the degradation of H_2O_2 in Fenton-like process.

6.4.4.2.2 Paracetamol

Paracetamol (*N*-(4-hydroxyphenyl)acetamide) is used as an analgesic and antipyretic drug. The rate of paracetamol degradation is slow at the beginning of irradiation (Figure 6.6), which was explained by the rapid conversion of hydroxycyclohexadienyl radical to a low reactivity phenoxyl-type radical (Szabó *et al.* 2012). This radical is also called semi-iminoquinone radical, because the unpaired electron can be either on the oxygen or on the nitrogen atom (Bisby & Tabassum, 1988).

The initial slope of the COD – dose relation indicates that ~0.2 O_2 molecules were incorporated into the products upon one $^{\bullet}OH$ attack (one-electron oxidation). This value is smaller than found for other phenolic compounds. However, once the paracetamol structure is degraded to organic byproducts, the products decompose fast. The first products of $^{\bullet}OH$ induced reactions are *N*-(3,4-dihydroxyphenyl)acetamide, *N*-(2,4-dihydroxyphenyl)acetamide, acetamide and hydroquinone. Due to the latter two products and also to hydrogen peroxide, during the first part of irradiation, high increase in sample toxicity was determined using the Microtox test (Szabó *et al.* 2012). When starting from 1×10^{-4} mol dm^{-3} paracetamol solutions, the maximum toxicity was observed at 0.5 kGy dose.

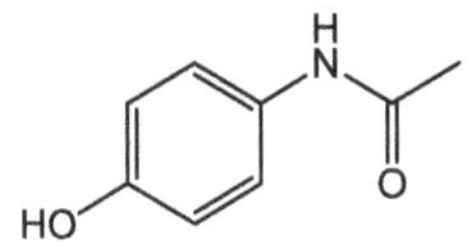

Paracetamol

6.4.4.2.3 Diclofenac

Diclofenac (2-(2,6-dichlorophenylamino)phenylacetic acid, DCF) is one of the most popular non-steroidal anti-inflammatory drugs used to treat inflammatory conditions of rheumatic and non-rheumatic origin. Although it is susceptible to photodegradation under solar light, diclofenac is still one of the most frequently detected pharmaceuticals in surface waters, up to μg dm^{-3} levels. The ecotoxicity of DCF is relatively low; it has been demonstrated, however, that in combination with other drugs present in water the toxic effect increases considerably. It has been reported that the drug released into the sewage is barely removed during the conventional wastewater treatment processes (Jones, 2007).

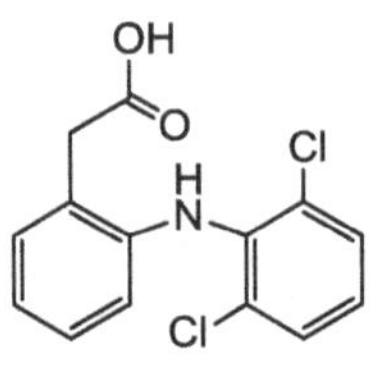

Diclofenac

DCF is removed efficiently *via* $^{\bullet}OH$-induced reactions (Sein *et al.* 2008; Liu *et al.* 2011; Trojanowicz *et al.* 2012). At an absorbed dose of 1 kGy, 1×10^{-4} mol dm^{-3} diclofenac and most of its degradation products are removed, while one order of magnitude higher dose is needed to achieve a complete mineralization of the initial organic carbon (Homlok *et al.* 2011). The high rate constant of $^{\bullet}OH$ + DCF reaction (8.12×10^9 mol^{-1} dm^3 s^{-1}, Jones (2007)) suggests that the $^{\bullet}OH$ attack occurs at the activated sites of the ring without chlorine atoms. In the reaction various hydroxycyclohexadienyl type radicals may form (Sein *et al.* 2008). Hydroxylated products were also detected in several other AOPs (TiO_2 photocatalysis, ozonation), where $^{\bullet}OH$ plays an important role in degradation (e.g., Sein *et al.* 2008).

The degradation mainly starts with $^{\bullet}OH$ addition to the $-CH_2COOH$ bearing ring at *para* position to the –NH– group. The hydroxycyclohexadienyl type radical thus formed may be stabilized in the form of

5-hydroxydiclofenac, which is one of the major degradation products. The yield of this product is greatly enhanced in the presence of O_2. $^{\bullet}OH$ addition to *ipso* position to the amino group is also reported in the literature (Sein *et al.* 2008). *Ipso* adduct is suggested to undergo rapid decomposition by releasing the substituent yielding 2,6-dichloroaniline. This compound observed in radiolysis was also detected in other AOPs. $^{\bullet}OH$ attack may also take place on the aromatic ring at the carbon-chlorine bond. As a result, fast HCl elimination and conjugated cyclohexadienyl type radical formation are expected to occur. These radicals may stabilize as quinoid-type compounds. Such products are also expected to form from other OH-adducts.

6.4.4.2.4 Ketoprofen and ibuprofen

Due to the stereocenter in the α-position of the propionate moiety both ketoprofen ((*RS*)-2-(3-benzoylphenyl)propionic acid) and ibuprofen ((*RS*)-2-(4-(2-methylpropyl)phenyl)propanoic acid) exist as two stereoisomers. Ketoprofen is frequently used as a photosensitizer in biological systems, both *in vivo* and *in vitro*.

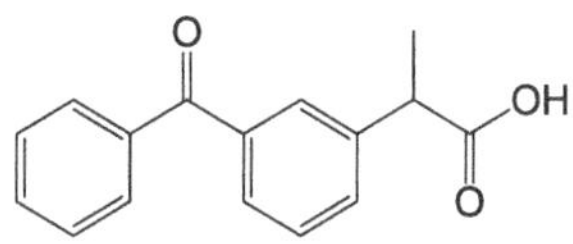

Ketoprofen

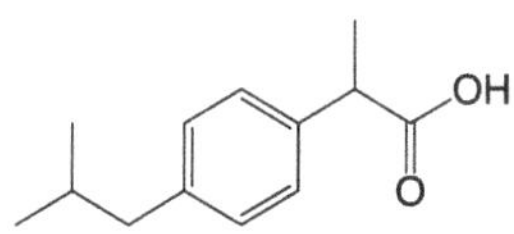

Ibuprofen

Ketoprofen is a substituted benzophenone ($C_6H_5{-}CO{-}C_6H_5$). The photophysical and photochemical processes of benzophenone were thoroughly investigated; its photochemical behavior is due to its low-lying triplet excited state. The photochemical degradation of ketoprofen is dominated by decarboxylation, with formation of (3-benzoylphenyl)ethane as a final product (Borsarelli *et al.* 2000; Musa *et al.* 2007).

In radiolysis of aqueous solutions, e_{aq}^- reacts with ketoprofen with a rate constant of 2.6×10^{10} mol^{-1} dm^3 s^{-1} (Jones, 2007). The high rate constant suggests electron attachment to the carbon atom of the carbonyl group. By analogy with benzophenone reaction (Brede *et al.* 1975); the hydroxyl radical is suggested to add to one of the aromatic rings, $k_{OH} = 4.6 \times 10^9$ mol^{-1} dm^3 s^{-1} (Jones, 2007). The transient absorption spectrum obtained in N_2O-saturated solution of ketoprofen is similar to the spectrum formed in the $^{\bullet}OH$ reaction with benzophenone (Brede *et al.* 1975; Sharma *et al.* 1997; Illés *et al.* 2012). Based on the spectral and kinetics similarities, it is assumed that the transient absorption spectrum belongs to hydroxycyclohexadienyl radical. The degradation products of benzophenone radiolysis through the OH radical-initiated process are reported as 2-, 3-, and 4-hydroxybenzophenones (Sharma *et al.* 1997). Similar hydroxylated products are expected to form from ketoprofen radiolysis, however, H-abstraction from the tertiary position of the side chain should also contribute to the degradation.

The $^{\bullet}OH$ reaction rate constant with *ibuprofen* is published as 6.1×10^9 mol^{-1} dm^3 s^{-1}. The value for e_{aq}^- reaction is 8.9×10^9 mol^{-1} dm^3 s^{-1} (Jones, 2007). The transient absorption spectra indicate that the $^{\bullet}OH$ reacts *via* two processes, through addition to the ring forming hydroxycyclohexadienyl radicals and H-abstraction at the tertiary carbon positions of the side chains giving carbon-centered radicals. In radiolysis studies besides the ring hydroxylated products, products hydroxylated at the tertiary positions were also observed in early stages of the degradation (Zheng *et al.* 2011; Illés *et al.* 2013; Gajda-Schrantz *et al.* 2013). At 1×10^{-4} mol dm^{-3} initial ibuprofen concentration, complete degradation was observed in pure water with 1 kGy dose. The presence of CO_3^{2-} and NO_3^- ions in the solution reduced the degradation rate due to their $^{\bullet}OH$ scavenging capacity. When potassium persulfate was added to ibuprofen solutions the degradation rate increased due to the transformation of e_{aq}^- to the strongly oxidizing $SO_4^{\bullet-}$ (Paul (Guin) *et al.* 2014a). $SO_4^{\bullet-}$, unlike $^{\bullet}OH$, produces preferentially benzyl type radicals *via* benzene radical cations (see Scheme 6.1).

6.4.5 Organic dyes

Organic dyes used for dyeing fabrics are expected to be adherent, long lasting, and resistant to sunshine and chemical processes. The dyes should be removed from industrial effluents because even at very low concentration ($\leq 10^{-5}$ mol dm^{-3}) they change the water quality and esthetics. The dyes are recalcitrant to biodegradation, and their chemical/photochemical degradation products (e.g., aromatic amines, aldehydes) are often highly toxic (Janos, 2009).

6.4.5.1 Azo dyes

Azo dyes have the common structure of R−N=N−R′ (R and R′ are aryl groups). Because of the electron delocalization across the two aromatic rings and the azo (N=N) group, these compounds have vivid colors. The most commonly used are the H-acid containing dyes which belong to the class of "Reactive" dyes. Herein, Apollofix Red (AR-28) was selected as a model compound. This is a triazine and H-acid containing azo dye with intensive red color (λ_{max} = 514 and 532 nm, $\varepsilon_{max,514 \text{ and } 532 \text{ nm}}$ = 31,400 mol^{-1} dm^3 cm^{-1}). The degradation of AR-28 (Figure 6.5) and some of the 1-arylazo-2-naphthol type dyes (e.g., Acid Orange 7, AO7) was extensively investigated by both pulse- and gamma radiolysis (Wojnárovits & Takács, 2008).

During the dye degradation through $^{\bullet}OH$ reactions, the absorbance decreases with the absorbed dose, both in visible and UV regions (Figure 6.5a). The spectrum in the visible region gradually shifts to longer wavelengths. However, there is no clear isobestic point, which indicates that the original compound does not transform quantitatively to a single product, also colored, but with weaker absorption properties in the visible range than AR-28. Chemical analysis showed a sharp decrease in concentration of the starting compound with the dose (Figure 6.5b). However, the spectral characteristics and elution times of the new products are similar to that of the starting compound, indicating that these are substituted versions of AR-28. These observations can be related to the complex structure of AR-28, which contains several potential sites for the $^{\bullet}OH$ attack. Most probably, the "preferred" radical attack site is not the –N=N– bond as was suggested from theoretical calculations by Hihara *et al.* (2006), but different positions on the aromatic rings. The organic radical intermediates in the subsequent reactions end-up as ring hydroxylated versions of AR-28. These products, similarly to AR-28 also have the extended conjugated system and large molar absorption coefficient. However, due to the extra OH group on the aromatic ring their absorption spectra are somewhat shifted to longer wavelength as that of AR-28 (Wojnárovits *et al.* 2005; Pálfi *et al.* 2007, 2011).

e_{aq}^- (and $H^{\bullet}$) react with the azo bond yielding hydrazyl-type radicals in the first step. The reaction at the azo bond destroys the extended delocalization of electrons and results in discoloration. There is no shift in the visible absorption band which may indicate that the OH radical reaction products are different than the e_{aq}^- reaction products (Figure 6.5c).

6.4.5.2 Anthraquinone dyes

Reactive anthraquinone dyes represent the second largest class of textile dyes after azo dyes used in the textile industry. Anthraquinones (derivatives of 9,10-anthracenedione) are widely distributed molecules in living organisms; they are present in bacteria, fungi, lichens and several families of higher plants (Orbán *et al.* 2009). Natural derivatives of 9,10-anthracenedione are well known and widely used products to dye textiles in many regions of the world since ancient times. A wide range of synthetic dyes are also available for dying purposes. These dyes are less soluble (or they are insoluble, e.g., Disperse Red 60) in water than the azo dyes and they are often present in water in disperse form (Pikaev *et al.* 1997a; Bagyo *et al.* 1998). In order to increase their solubility usually $-SO_3^-$ or $-NH_3^+$ groups are incorporated into the molecules.

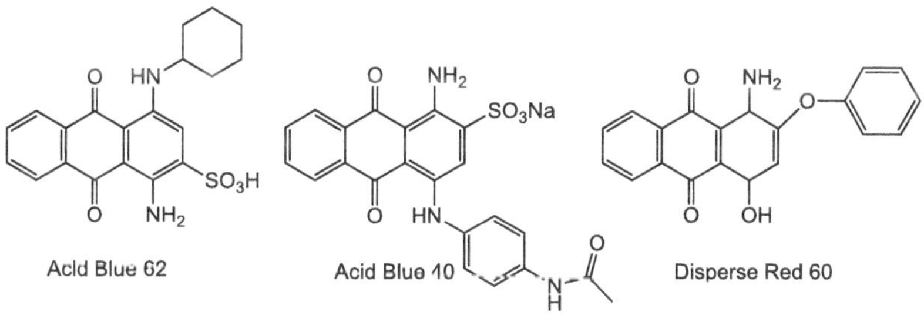

Acid Blue 62 Acid Blue 40 Disperse Red 60

The degradation of anthraquinone dyes was investigated by various research groups; one example is Acid Blue 62 (ABSBR). It reacts with $^{\bullet}OH$ at diffusion-controlled rate with k_{OH} of 1×10^{10} mol^{-1} dm^3 s^{-1} (Perkowski *et al.* 1989). The addition reaction is assumed to produce a single intermediate (Perkowski *et al.* 1989; Perkowski & Mayer, 1989). The radiolysis experiments showed that the discoloration is due to $^{\bullet}OH$ reactions. Similar conclusion was drawn from the degradation of Acid Blue 40 (Nagai & Suzuki, 1976). Concomitantly with the discoloration associated with the decrease in absorption at ~600 nm upon irradiation of both ABSBR and Acid Blue 40 in O_2-free solutions, a new absorption band appeared in the 400–500 nm range. The discoloration is not accompanied by destruction of the carbon skeleton. The dissolved O_2 affects the secondary reactions of OH-adducts, and in those experiments no new absorption band in the 400–500 nm range was observed (Nagai & Suzuki, 1976). The role of dissolved O_2 in carbon skeleton destruction upon irradiation was also suggested for other anthraquinone dyes (Vysotskaya *et al.* 1986; Hashimoto *et al.* 1979).

When the dyes are present in water in disperse form, due to the low accessibility, the radiolysis is an inefficient process at their degradation (Pikaev *et al.* 1997a; Bagyo *et al.* 1998). In case of Dispersive Red 60 complete purification of the wastewater containing this dye was achieved by the coagulation and flocculation method (Pikaev *et al.* 1997a). By this treatment both the water-soluble and insoluble dyes can be removed.

6.4.6 Naphthalene sulfonic acid derivatives

Due to their wide application in various chemical processes (e.g., surfactant and dye production), the sulfonated aromatic compounds have been produced in large quantities. Of the naphthalene sulfonates, naphthalene-1,5-disulfonate (1,5-NDSA) is the most persistent and regularly found in surface waters (Knepper *et al.* 2004; Gehringer *et al.* 2006, 2008). 1,5-NDSA is not readily biodegradable, thus it is not removed in conventional biological treatment plants, and it also has a low reactivity toward ozone ($k = 41$ mol^{-1} dm^3 s^{-1}).

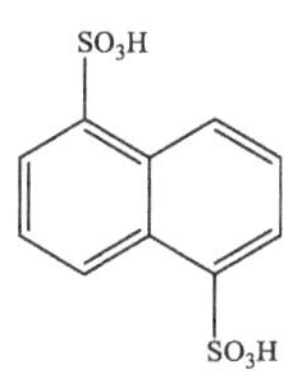

Naphthalene-1,5-disulfonate (1,5-NDSA)

Gehringer *et al.* (2006, 2008) performed comprehensive studies on the degradation of 30 mg dm^{-3} 1,5-NDSA in tap water through radiolysis reactions, by ozone, and by radiolysis/O_3 combined treatment. Under the given experimental conditions single electron beam (EB) irradiation was found more efficient than the other two processes. Ozonation was the least efficient in mineralization of 1,5-NDSA. EB irradiation decomposed all 1,5-NDSA and its metabolites at a low absorbed dose of 2 kGy, and the treatment was considered efficient given the complex matrix of wastewater effluent.

The radiation chemistry of aqueous solutions of isobutylnaphthalene sulfonates (also named Nekal) of general formula $C_{10}H_6(R)SO_3Na$, where R is $(CH_3)_3C$–, $(CH_3)_2CHCH_2$– or $CH_3CH_2CH(CH_3)$–, was studied in details by Pikaev's group (Gogolev *et al.* 1992; Pikaev *et al.* 1998; Pikaev, 2001). In these studies, the authors showed that all three primary reactive species of water radiolysis were involved in the degradation of isobutylnaphthalene sulfonate (IBNS) molecules. $^{\bullet}OH$ and e_{aq}^- react with these molecules with diffusion controlled rate constant of 1.8×10^{10} mol^{-1} dm^3 s^{-1} (overestimated) and 2.5×10^{10} mol^{-1} dm^3 s^{-1}, respectively; the rate constant for H-atom reaction with IBNS is smaller than the other two, and determined as $k_H = 5.0 \times 10^9$ mol^{-1} dm^3 s^{-1} (Gogolev *et al.* 1992). In $^{\bullet}OH$ reaction an intermediate with λ_{max} 430 nm was formed, which is different than the intermediate produced in e_{aq}^- (or $O_2^{\bullet-}$) reactions. Although IBNS degradation is assumed to occur through both $^{\bullet}OH$ and e_{aq}^- reactions, the $^{\bullet}OH$ reactions prevail (Pikaev, 2001). The yield of Nekal degradation was found to be 0.20 μmol J^{-1} independently of the initial concentration in the 45–180 mg dm^{-3} range. The removal of either the alkyl or sulfonate group would suffice to make the compounds biodegradable.

6.5 IONIZING RADIATION FOR WATER TREATMENT: PILOT- AND INDUSTRIAL SCALE APPLICATIONS

First studies on gamma irradiation for treatment of wastes were carried out in the 1950s, and were focused primarily on disinfection (IAEA, 2007). In the 1960s the topics of such studies were extended to the purification of water and wastewater (e.g., Amphlett *et al.* 1968). Laboratory research on industrial wastewater and polluted groundwater started in the 1970s, and since then, there has been an increasing interest in developing the radiation chemistry AOP for full-scale applications (Woods & Pikaev, 1994). Based on laboratory studies, the first pilot plants were designed in the 1990s and later on industrial-scale water treatment installations were established.

6.5.1 General considerations

Radiation technology (γ-ray and electron beam) without using additives provides the most efficient way to generate radical species for advanced oxidation processes, laboratory and pilot studies demonstrated that radiation processing of wastewater can be a competitive technology (Han, 2009). In principle, the water streams that can be treated with ionizing radiation processes fall into two categories: (*i*) groundwater and drinking water, and (*ii*) industrial and municipal wastewater.

The main differences between the two groups are the concentration of the pollutants and the level of microbial contamination, both of which are higher in the second category. In general, the treatment goal for groundwater and drinking water is microbial disinfection, whereas for wastewater that concerns the increase of biodegradability (Cooper *et al.* 1998; IAEA, 2007).

The quality parameters of groundwater and drinking water sources are in general well-characterized, and their variation over the course of time is either known or predictable. On the contrary, most of the time, wastewater streams have variable composition, quantitatively and qualitatively, and because of that, bulk parameters, such as chemical oxygen demand (COD), biological oxygen demand (BOD), and total organic carbon (TOC) are used to characterize those streams.

A municipal wastewater treatment train comprises several steps, among which are mechanical screening, sedimentation and biological treatment. The electron beam irradiation process can be implemented either prior to biological treatment to improve the biodegradability of the organic content or on the treated wastewater effluent to degrade recalcitrant micropollutants and/or for disinfection. Depending on the treatment purpose, either high absorbed doses (~20 kGy) or low absorbed doses (<2 kGy) are required. Installing the ionizing radiation AOP before the biological process presents a series of advantages from the point of view of fighting against antimicrobial resistance, such as (*i*) addresses the concern of increasing bacterial resistance through the degradation of antibiotics, (*ii*) limits or eliminates the exposure of high density microbial population in the activated sludge to antibiotic residues, and gives an additional profit with reducing the solids retention time of the biological treatment (Yamazaki *et al.* 1983; Kim *et al.* 2007; Han *et al.* 2012).

The industrial wastes contain high concentrations of organics which may not be amenable to degradation by the ionizing radiation alone. In such cases, combined processes can be applied.

6.5.2 Ionizing radiation reactors for water treatment

Electron-beam accelerators can treat several hundred thousand m^3 of wastewater per day at the dose level of 1 kGy. Key requirements for consistent treatment performance are: (*i*) continuous operation with constant beam current, (*ii*) uniform dose distribution; (*iii*) constant flow rate, independent of the temperature and pressure variation. Basically, there are three main types of ionizing radiation reactor designs: upflow reactor, spray type reactor, and nozzle reactor.

In the *upflow reactor* the water flows upwards into a reaction vessel and overflows the rim of the vessel (Figure 6.9). The reactor is placed in a bigger vessel in which the treated water is collected. In some cases air bubbling is used to ensure a good mixing. The gas is injected from below by special bubblers. The reactor chamber is usually vertically divided into several sections. Several versions of the upflow reactor were constructed and have been used at a pilot plant in Brazil in tests to treat real effluents from industries and municipal wastewater treatment plants (Rela *et al.* 2000, 2008).

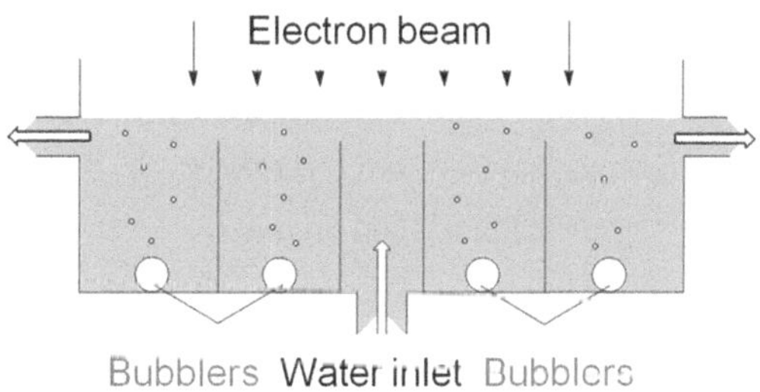

Figure 6.9 Wastewater irradiation in upflow arrangement (IAEA, 2007).

In the *spraying type* reactor the wastewater is pumped through a sprayer forming fine water droplets (Figure 6.10). This reactor ensures a very good mixing with air which enhances the oxidation rate of pollutants. This design was tested in combination with ozone in a pilot plant in Raduzhnyi, Russia (Podzorova *et al.* 1998; Pikaev *et al.* 1997b; Pikaev, 1998).

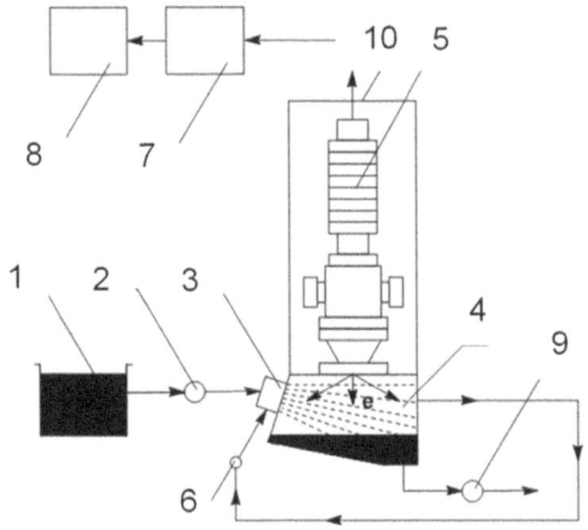

Figure 6.10 Wastewater irradiation in a sprayer arrangement: (1) reservoir of wastewater intake, (2) pump for wastewater, (3) sprayer, (4) irradiation chamber, (5) electron accelerator, (6) turboblower, (7) power supply, (8) control desk, (9) pump for purified water removal, and (10) biological shielding (Pikaev *et al.* 2001).

The *nozzle reactor* treats the wastewater dispersed as a wide continuous water jet (Figure 6.11). The thickness of the water jet is adjusted according to the penetration depth of accelerated electrons. The width of the water jet corresponds to the width of the electron beam. This jet is irradiated by a transverse electron beam and the water is collected below in a channel (Han *et al.* 2002, 2005a, b, 2012; Makarov *et al.* 2004). Nozzle reactor is used at an industrial scale installation in Daegu, Republic of Korea, to treat textile dyeing wastewater.

Laboratory, 50 m^3 day^{-1} Pilot Plant, 1000 m^3 day^{-1} Industrial Plant, 10,000 m^3 day^{-1}

Figure 6.11 Nozzle-type injectors used in textile dyeing wastewater treatment. Courtesy of Dr. Bumsoo Han, EB-Tech Co., Daejon, Republic of Korea.

Mobile water irradiation systems were also constructed where the electron-beam accelerator and the water handling, controlling and safety systems are placed on a trailer, and functions as a ready-to-use, self-shielded and self-powered sustainable unit for treatment services where effluents are generated (Nickelsen *et al.* 1998; Han *et al.* 2009). The mobile accelerators facilitate the logistics and promote the use of EB technology for municipal and industrial wastewater treatment; they are also used for demonstration purposes.

6.5.3 Ionizing radiation for water treatment: pilot studies

6.5.3.1 The Miami (USA) electron beam research facility (EBRF)

The Miami Electron Beam Research Facility (EBRF), operating since 1988, has performed process-scale irradiation of many solutes in batch or continuous mode, including mixtures of solutes in natural waters (Cooper *et al.* 2004). The accelerator is a vertically arranged, 75 kW, 1.5 MeV, 50 mA insulated core transformer type. The beam may be scanned to provide uniform, continuous irradiation of an area of 122 × 7.6 cm, and is directed onto a stream of water falling over a weir. At the design the flow rate is 460 $dm^3\ min^{-1}$, the thickness of the irradiated falling water stream is 0.38 cm. This pilot plant installation was used for studying the technical details of degradation of large number of molecules, for instance different small molecular mass aromatic hydrocarbons or methyl *tert*-butyl ether (MTBE) (an additive used in gasoline to increase octane number). At an initial MTBE concentration of 2.3 mg dm^{-3} ~6 kGy dose reduced the concentration below the detectable level (Cooper & Tornatore, 1999; Cooper *et al.* 2009). Other studies with this system involve, e.g., testing of trichloroethylene (TCE) or perchloroethylene (PCE) removal in a flow system (Nickelsen *et al.* 2002).

6.5.3.2 Removal of organic and petrochemical pollutants in Brazil

This pilot plant was set up to treat wastewater and industrial effluents in the electron beam facility of IPEN (Instituto de Pesquisa em Energia Nuclear, Brazil). It operates with a Dynamitron type electron beam accelerator with accelerating voltage varying from 500 keV to 1.5 MeV and beam current from 1 mA to 25 mA. The beam with a frequency of 100 Hz can scan an area of 60 cm length and 2 cm width (Rela *et al.* 2000, 2008; Sampa *et al.* 2004). The pilot plant can process water with up to 70 $dm^3\ min^{-1}$ at an average dose of 5 kGy. Two tanks with 1200 dm^3 capacity are used for storage and collect the liquid and two pumps are used to homogenize and pump the liquid through the irradiation device. In a series of experiments different upflow reactors were tested in order to find an optimal solution.

This installation was used in experiments with drinking water, municipal and industrial wastewater effluents. The drinking water in a big industrial city contained 78 μg dm^{-3} $CHCl_3$, 12 μg dm^{-3} $CHBrCl_2$, and 168 μg dm^{-3} $CHBr_2Cl$. With 1.3 $m^3\ h^{-1}$ flow rate and 2 kGy dose 87% of chloroform was removed, the concentrations of other halogenated compounds decreased below the detection limit (Sampa *et al.* 1995).

6.5.3.3 Austrian drinking water treatment plant using e-beam combined with ozone

The Austrian pilot plant was equipped with an ICT-500 High Voltage (USA) electron beam accelerator (500 keV, 25 mA, 1.2 m scan width) with constant water flow up to 4 $m^3\ h^{-1}$. This system was especially designed for irradiation/O_3 combined treatments (Gehringer *et al.* 2003, 2008). The concentration of O_3 in the gases leaving the ozone generator is much lower than the concentration of O_2. Due to the different solubility of O_2 and O_3 in water, the gas that leaves the ozonizator first is injected into purified water under pressure, and a part of the excess oxygen is removed, and then the ozone/water is mixed with groundwater selected for pilot plant tests. The advantages of the irradiation/O_3 combined treatments were discussed previously in 6.3.2.4. Here one more advantage will be mentioned. When the water contains higher concentration of nitrate, the hydrated electrons may reduce NO_3^- to hazardous NO_2^-. O_3 removes some part of e_{aq}^- in fast reaction; moreover O_3 oxidizes NO_2^- to NO_3^- (Gehringer *et al.* 1993).

6.5.3.4 Irradiation of wastewater aerosols in Russia

The low penetration depth of low energy electrons may be a serious disadvantage in several applications. To overcome this problem a pilot plant was established and tested to treat municipal wastewater in Raduzhnyi,

Russia. In this system, liquid wastewater is sprayed using injection nozzles and the aerosol is irradiated. The density of the aerosol is more than an order of magnitude smaller than that of the liquid water and therefore the penetration depth is more than an order of magnitude higher. The accelerator produces electrons of 300 keV energy at maximum power of 15 kW. The plant treated 500 m^3 municipal wastewater per day with doses less than 4–5 kGy. The results indicate that the concentration of pollutants in irradiated wastewater decreased by 2–3 orders of magnitude (Pikaev *et al.* 1997b; Podzorova *et al.* 1998).

6.5.3.5 Pilot plant installation in China to remove HCN dissolved in water

The pilot plant operating with a self-shielded electron beam accelerator has energy of 0.5 ~ 1.0 MeV and beam current of 10 ~ 15 mA. Aqueous streams of wastewater containing cyanide are exposed to a scanned beam in a water film from two nozzles. The size of each nozzle is designed to be 10 cm wide and 0.2 cm thick according to the penetration depth of electrons in water. The average speed of the water under the beam is approximately 1.37 m s^{-1}, the residence time is approximately 0.05 s at a flow rate of about 1.82 m^3 h^{-1}. This pilot plant is used to optimize the conditions for HCN removal from waste gases of a carbon fiber factory. The gases leaving the fiber producing technological line pass through two spray towers in turn to decrease the HCN concentration in the gas flow. Then the remaining HCN is entrapped in absorption solution containing sodium hypochlorite. The residual aqueous cyanide solution is introduced into a process vessel and treated with the electron beam/ozonation technique. Influent CN^- concentration is controlled at 15 ± 2 mg dm^{-3} with a water regulating tank. This treatment allows CN^- to remain under the regulatory limit (of 0.5 mg dm^{-3}) for safe industrial wastewater discharge with an absorbed dose of 12 kGy. The obtained results show that the combined process is effective for removing HCN from the waste gas (Ye *et al.* 2013). A full-scale water irradiation plant is under construction in China.

6.5.4 Industrial scale installations using radiation-based AOP

6.5.4.1 Voronezh (Russia) electron beam-biological filtration wastewater facility

The first full scale application treating 2000 m^3 effluent daily was reported for the purification of wastewater generated at the Voronezh rubber plant in Russia (Pikaev, 2001; IAEA, 2007). The compounds, named Nekal are used as emulsifiers in the production of synthetic rubber. In the vicinity of rubber factories the groundwater was heavily contaminated with this surfactant. It was found that application of a combined electron beam/biological method could be successful for purification of wastewater from Nekal. This plant had two lines equipped with 50 kW accelerators to convert the non-biodegradable emulsifier present in the wastewater to more biodegradable form. The dose required to mineralize Nekal at concentration of 1×10^{-3} mol dm^{-3} was 300 kGy. However, it was only necessary to remove alkyl or sulfonate groups to render the molecule more easily biodegradable.

6.5.4.2 Daegu (Republic of Korea) electron beam – biological filtration wastewater facility

Other full scale application is a combination of irradiation and biological treatments of textile dying wastewater. Initial laboratory investigations indicated that electron beam treatment of textile dying wastewater was prospective means for purification. These investigations resulted in discoloration and destruction of organic pollutant with a dose of 1–2 kGy. After laboratory scale experiments in a flow-through system (Figure 6.11), a pilot plant (output 1000 m^3 day^{-1}) was established for combined electron-beam and biological treatment of industrial textile dyeing wastewater in Daegu Dyeing Industrial Complex

(Daegu, Republic of Korea) in 1998 (Han *et al.* 2002, 2005a, b, 2012; Makarov *et al.* 2004). This plant has shown considerable reduction of chemicals used in the chemical wastewater treatment step, and also a reduction in retention time in the biological treatment, with an increase in removal efficiencies of COD and BOD. On the basis of data obtained from pilot plant operation, an industrial scale plant was constructed in 2005. The plant is located on the area of an existing wastewater treatment facility and its capacity is 10,000 m^3 of wastewater per day. The facility is operating with 1 MeV, 400 kW accelerator and the radiation pre-treatment is combined with the existing bio-treatment facility. The operation of this facility provided additional data on reliability and for a detailed economic evaluation. The total construction cost for this plant was USD 4 M and the operation cost was not more than USD 1 M per year and about USD 0.3 per m^3 of wastewater (Han *et al.* 2012).

6.5.5 Economics

The economics of the EB-based water treatment depends on many factors, and prior to the industrial implementation of this technology, a detailed assessment is necessary. Technology cost evaluations are available in the literature (Gehringer, 2004; IAEA, 2007; Emmi & Takács, 2008; Han, 2009). The total cost is determined by accelerator characteristics, type and size, voltage and power, the cost of electricity. The energy utilization efficiency is directly influenced by process characteristics, accelerator construction and electron beam application. It can be optimized by achieving the proper relation between electron energy, beam current and irradiated material properties. On the other hand the cost strongly depends on the volume and characteristics of wastewater (microbiological contamination, chemical composition, COD, BOD, toxicity), the aim of the treatment (disinfection, discoloration, mineralization) and also on the other steps of the treatment technology (combination with biological, chemical methods). All of these parameters have an impact on dose requirement.

As mentioned in section 6.2.2, electron energies in the range of 1–5 MeV are required for industrial water treatment applications. In pilot plant and full scale operations, the absorbed dose varies from 0.2 to 2 kGy. Table 6.4 shows the estimated installation and maintenance costs for two 1 MeV (power 100 and 400 kW) and one 2.5 MeV (power 100 kW) accelerators.

Table 6.4 Cost assessment for EB accelerators (Courtesy of Dr. Bumsoo Han, EB-Tech Co., Daejon, Republic of Korea).

					Remarks
Energy (max), MeV		1.0	1.0	2.5	
Power (max), kW		100	400	100	
Capital cost, kUSD	Accelerator	880	2000	960	
	Shield vault	400	700	400	Local cost
	Handling system	200	300	200	Reactor
	Licensing, etc.	100	100	100	
	Total	**1580**	**3100**	**1660**	
Maintenance cost, kUSD	E-gun, Ti foil	2.5	3.0	2.5	1–2 years
	Ion pumps	2.5 × 2	2.5 × 6	2.5 × 2	2–3 years
	Other	2.0	2.0	2.0	Per year
	Annually	**9.5**	**20.0**	**9.5**	

The above capital costs do not include the land- and building-associated costs. In the maintenance cost the most expensive part is the electron gun and the Ti foil (replacement costs are 2500–3000 USD annually, or every two years). Ti foil is used as accelerator window, where the electrons exit from the high vacuum to the air. Examples of estimated water treatment costs are shown in Table 6.5.

Table 6.5 Treatment cost assessment (Courtesy of Dr. Bumsoo Han, EB-Tech Co., Daejon, Republic of Korea).

				Remarks
Energy (max), MeV	1.0	1.0	2.5	
Power (max), kW	100	400	100	
Fixed cost, kUSD	150	160	160	
Variable cost, kUSD (8000 h/year)	125	345	125	
Theoretical throughput, m^3 day^{-1}	3,600	15,000	3,600	at 1 kGy
	360	1,500	360	at 10 kGy
	150	600	150	at 25 kGy
Treatment cost (variable cost), USD m^{-3}	0.1	0.07	0.1	at 1 kGy
	1.0	0.72	1.0	at 10 kGy
	2.6	1.8	2.6	at 25 kGy
Treatment cost (total), USD m^{-3}	0.23	0.11	0.24	at 1 kGy
	2.3	1.1	2.4	at 10 kGy
	5.7	2.6	5.9	at 25 kGy

The fixed costs are not dependent on the facility throughput in primary terms. They are related to investment cost and administrative overhead. The equipment cost including accelerator, renting or buying the land are the most common examples of fixed expenses. Additional fixed components are employee health care, liability insurance, environmental costs, research and development, taxes, etc.

The variable expenditures incurred through the facility throughput and such expenses as material, utilities and labor costs should also be included. Utility costs in electron beam operation are mostly electricity. Water consumption is relatively low due to commonly applied close loop in water cooling systems of accelerators. It should be noted that labor cost and cost of electricity vary widely depending on the country. Energy consumption is becoming one of the significant parts of any cost analysis.

The costs are a function of the power (Figure 6.12), but they are more dependent on the energy as shown in Tables 6.4 and 6.5. Both the capital and the maintenance costs are about two times higher for a 1 MeV 400 kW accelerator than that for a 1 MeV 100 kW accelerator, thus there is a moderate increase in cost with the power. The treatment cost in USD m^{-3} is the lowest for the 400 kW accelerator. The variable expenses calculated per unit amount of irradiated water are not constant for different throughputs. For industrial scale treatment (large volume), the high power accelerator seems to be the most economic, the cost being as cheap as 0.11 USD m^{-3} for 1 kGy dose with a throughput of 15,000 m^3 per day. Since a lower dose would allow increased flow and thus a lower treatment cost, the indicated costs per m^3 are the maximum values to be expected.

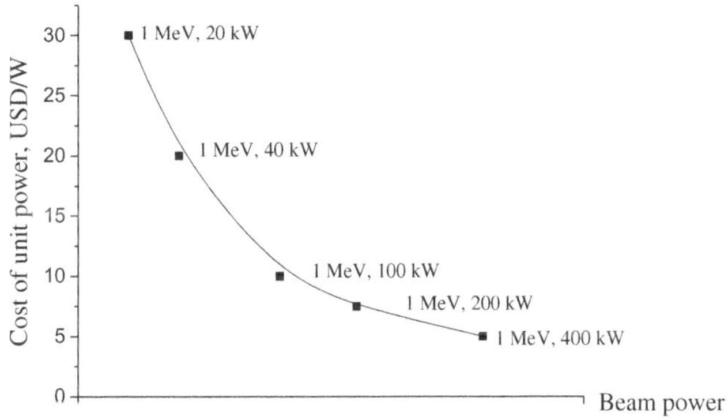

Figure 6.12 Cost of unit power for 1 MeV electron accelerators with different power. Courtesy of Dr. Bumsoo Han, EB-Tech Co., Daejon, Republic of Korea.

6.6 CONCLUSIONS

Radiation processing has been one of the promising processes for environmental protection, such as water/wastewater treatment since the 1970s. Accelerators of several hundred kilowatt power are already available in the market, and some of them have proved their reliability in long term operation in flue gas or wastewater treatment, in polymer processing and disinfection. In some cases, the radiation technology is the only and unique solution for treatment.

In dilute aqueous solutions the radiation energy is mainly absorbed by the solvent, i.e., water, and the water radiolysis reactive intermediates (hydroxyl radical, hydrated electron, hydrogen atom) induce the degradation of the solute. The direct effect is negligible. The yield of the reducing intermediates (e_{aq}^-, $H^\bullet$) is nearly equal to the yield of the oxidizing intermediates [$^\bullet OH$, hydrogen peroxide (minor product)] in pure deoxygenated water. In air-saturated water, if the solute concentration is lower than 10^{-4} mol dm^{-3}, the e_{aq}^- and $H^\bullet$ species are mainly converted to the rather unreactive $O_2^{\bullet-}/HO_2^\bullet$ pair in reaction with dissolved O_2. In water treatment with ionizing radiation, the chemistry is primarily initiated by the hydroxyl radicals. However, at high solute concentrations, and when very high dose rates are applied in electron beam irradiations, considerable O_2 depletion occurs, in which case, the e_{aq}^- reactions may contribute significantly to the overall contaminant degradation.

The radiolytic degradation of many harmful compounds in aqueous solutions has been intensively studied during the last decades. The basic radiation chemistry and in some cases the details of chemical reactions are now well understood. The investigations refer to the decay of the starting compounds, although often the scientists follow the fate of the studied compounds until complete transformation to H_2O, CO_2 and other inorganic substances (mineralization). Many studies are supplemented by toxicity measurements as well.

In real wastewater there is always a competition between the reactions of different constituents with the reactive water radiolysis intermediates. The dose requirement for complete mineralization is 5–10 times higher than that needed for discoloration. Very often, however, the discoloration increases the biodegradability, so the irradiation is needed just for a small modification (e.g., dechlorination) of the chemical structure to improve the biodegradability. In bioassay experiments the toxicities of the irradiated solutions are sometimes higher at the beginning of the degradation than the toxicities of the starting solutions. This is due to the higher toxicity of the degradation products than the initial molecules. In laboratory experiments carried out using aerated purified water formation of hydrogen peroxide also contributes to the increase of toxicity. In real wastewaters H_2O_2 contributes to disinfection and it is expected to decompose quickly in Fenton-like reactions.

In the case of endocrine disruptors, many pharmaceuticals, pesticides, organic dyes and disulfonic acids, the principal radiation chemical process is the addition of $^{\bullet}OH$ to the aromatic ring. When dissolved O_2 is present the $^{\bullet}OH$-adduct hydroxycyclohexadienyl radical often reacts in fast reaction with the dissolved O_2 molecules forming peroxyl radical. The ring opening reaction and thereby the start of mineralization may take place through the reactions of the peroxyl radicals. Dissolved O_2 generally enhances both the rate of discoloration and that of the mineralization. The rate of oxidation is low, when the organic radicals do not react with dissolved O_2.

After many laboratory experiments, technologies were worked out on pilot plant and full scale level. Pilot plants were established for the treatment of water contaminated with naphthalene sulfonic acids (Nekal), chlorinated hydrocarbons, endocrine disruptors, cyanides or azo dyes. Industrial installations were made for the treatment of wastewater containing Nekal and for dying wastewater.

Among the advantages of irradiation treatment we mention that the process does not need additives (catalyst), heat and there is no secondary waste generation. The energy absorption is independent of the turbidity, temperature and state of aggregation. The energy conversion efficiency from electric energy to energy of accelerated electrons is excellent; it may go up to 80%. Very large quantities can be treated, up to hundred thousand m^3 per day. Radiation technology can be adapted to existing technological lines; it is easy to make automated systems. Pilot plant and industrial applications also demonstrated its economic competitiveness to other AOP techniques. Applying irradiation treatment as end-of-pipe technology in wastewater treatment plants removal of the remaining harmful organic pollutants and disinfection occur simultaneously. The cost of irradiation treatment with a dose requirement of 1 ~kGy is estimated to be 0.1–0.2 USD m^{-3}.

6.7 ACKNOWLEDGEMENT

The authors express their thanks to Professor Stephen Mezyk for his helpful advice. The financial supports of IAEA (Contract No. 16485) and OTKA (Hungarian Science Foundation, NK 105802) are also acknowledged.

6.8 REFERENCES

Alam M. S., Rao B. S. M. and Janata E. (2001). A pulse radiolysis study of H atom reactions with aliphatic alcohols: evaluation of kinetics by direct optical absorption measurement. *Physical Chemistry Chemical Physics*, **3**, 2622–2624.

Albarran G. and Schuler R. H. (2003). Concerted effects in the reaction of $^{\bullet}OH$ radicals with aromatics: radiolytic oxidation of salicylic acid. *Radiation Physics and Chemistry*, **67**, 279–285.

Albarran G. and Schuler R. H. (2007). Hydroxyl radical as a probe of the charge distribution in aromatics: phenol. *The Journal of Physical Chemistry A*, **111**, 2507–2510.

Al-Sheikhly M., Silverman J., Neta P. and Karam L. (1997). Mechanisms of ionizing radiation-induced destruction of 2,6-dichlorobiphenyl in aqueous solutions. *Environmental Science and Technology*, **31**, 2473–2477.

Amphlett C. B., Adams G. E. and Michael B. D. (1968). Pulse radiolysis studies of deaerated aqueous salicylate solutions. In: Radiation Chemistry, Advances in Chemistry, E. J. Hart (ed.) American Chemical Society, Washington, DC, USA, pp. 231–250.

Andreozzi R., Caprio V., Ciniglia C., De Champdoré M., Guidice R. L., Marotta R. and Zuccato E. (2004). Antibiotics in the environment: occurrence in Italian STPs, fate, and preliminary assessment on algal toxicity of amoxicillin. *Environmental Science and Technology*, **38**, 6832–6838.

Angelini G., Bucci R., Carnevaletti F. and Colosimo M. (2000). Radiolytic decomposition of aqueous atrazine. *Radiation Physics and Chemistry*, **59**, 303–307.

Ashton L., Buxton G. V. and Stuart C. R. (1995). Temperature dependence of the rate of reaction of OH with some aromatic compounds in aqueous solution. Evidence for the formation of a π-complex intermediate? *Journal of the Chemical Society, Faraday Transactions*, **91**, 1631–1633.

Asmus K.-D., Bahnemann D., Bonifacic M. and Gillis H. A. (1977). Free radical oxidation of organic sulphur compounds in aqueous solution. *Faraday Discussions of the Chemical Society*, **63**, 213–225.

Ayatollahi S., Kalnina D., Song W., Turks M. and Cooper W. J. (2013). Radiation chemistry of salicylic and methyl substituted salicylic acids: models for the radiation chemistry of pharmaceutical compounds. *Radiation Physics and Chemistry*, **92**, 93–98.

Bagyo A. N. M., Andayani W. and Surtipanti S. (1998). Radiation-induced degradation and decoloration of disperse dyes in water. In: Environmental Applications of Ionizing Radiation, W. J. Cooper, R. D. Curry and K. E. O'Shea (eds), John Wiley & Sons, New York, USA, pp. 507–520.

Basfar A. A., Mohamed K. A., Al-Abduly A. J. and Al-Shahrani A. A. (2009). Radiolytic degradation of atrazine aqueous solution containing humic substances. *Ecotoxicology and Environmental Safety*, **72**, 948–953.

Baxendale J. H. and Busi F. (eds) (1982). The Study of Fast Processes and Transient Species by Electron Pulse Radiolysis, NATO Advanced Study Institutes Series. D. Reidel Publishing Company, Dordrecht, Netherlands.

Bielski B. H., Cabelli D. E., Arudi R. L. and Ross A. B. (1985). Reactivity of HO_2/O_2^- radicals in aqueous solution. *Journal of Physical and Chemical Reference Data*, **14**, 1041–1100.

Bisby R. H. and Tabassum N. (1988). Properties of the radicals formed by one-electron oxidation of acetaminophen – a pulse radiolysis study. *Biochemical Pharmacology*, **37**, 2731–2738.

Bojanowska-Czajka A., Drzewicz P., Kozyra C., Nałęcz-Jawecki G., Sawicki J., Szostek B. and Trojanowicz M. (2006). Radiolytic degradation of herbicide 4-chloro-2-methyl phenoxyacetic acid (MCPA) by γ-radiation for environmental protection. *Ecotoxicology and Environmental Safety*, **65**, 265–277.

Bonifacic M. and Asmus K.-D. (1980). Stabilization of oxidized sulphur centres by halide ions. Formation and properties of $R_2S\therefore X$ radicals in aqueous solutions. *Journal of the Chemical Society, Perkin Transaction* 2, 758–762.

Borsarelli C. D., Braslavsky S. E., Sortino S., Marconi G. and Monti S. (2000). Photodecarboxylation of ketoprofen in aqueous solution. A time resolved laser-induced optoacoustic study. *Photochemistry and Photobiology*, **72**, 163–171.

Brede O., Helmstreit W. and Mehnert R. (1975). Pulsradiolyse von Benzophenon in wäßriger Lösung (Pulse radiolysis of benzophenone in aqueous solution). *Zeitschrift für Physikalische Chemie*, **256**, 513–521.

Bucholtz D. L. and Lavy T. L. (1977). Effects of ^{60}Co radiation on herbicides in aqueous solution. *Weed Science*, **25**, 200–202.

Buxton G. V. (1982). Basic radiation chemistry of liquid water. In: The Study of Fast Processes and Transient Species by Electron Pulse Radiolysis, NATO Advanced Study Institutes Series, J. H. Baxendale and F. Busi (eds), D. Reidel Publishing Company, Dordrecht, Netherlands, pp. 241–266.

Buxton G. V. (1987). Radiation chemistry of the liquid state: (1) water and homogeneous aqueous solutions. In: Radiation Chemistry, Principles and Applications, Farhataziz and M. A. J. Rodgers (eds), VHC Publishers, New York, USA, pp. 321–349.

Buxton G. V. (2001). High temperature water radiolysis. In: Radiation Chemistry: Present Status and Future Trends, Studies in Physical and Theoretical Chemistry 87, C. D. Jonah and B. S. M. Rao (eds), Elsevier, Amsterdam, Netherlands, pp. 145–162.

Buxton G. V., Greenstock C. L., Helman W. P. and Ross A. B. (1988). Critical review of rate constants for reactions of hydrated electrons, hydrogen atoms and hydroxyl radicals ($^{\bullet}OH/^{\bullet}O^-$) in aqueous solution. *Journal of Physical and Chemical Reference Data*, **17**, 513–886. An extended database is available on the internet: http://kinetics.nist.gov/solution/.

Chung B. Y., Kim J.-S., Lee M. H., Lee K. S., Hwang S. A. and Cho J. Y. (2009). Degradation of ampicillin in pig manure slurry and an aqueous ampicillin solution using electron beam irradiation. *Radiation Physics and Chemistry*, **78**, 711–713.

Cooper W. J. and Tornatore P. M. (1999). Destruction of MTBE in Contaminated Groundwater Using Electron Beam Injection. ACS Proceeding, Anaheim, California.

Cooper W. J., Curry R. D. and O'Shea K. E. (eds) (1998). Environmental Applications of Ionizing Radiation. John Wiley & Sons, New York, USA.

Cooper W. J., Gehringer P., Pikaev A. K., Kurucz C. N. and Mincher B. J. (2004). Radiation processes. In: Advanced Oxidation Processes for Water and Wastewater Treatment, S. Parsons (ed), IWA Publishing, London, UK, pp. 209–245.

Cooper W. J., Cramer C. J., Martin N. H., Mezyk S. P., O'Shea K. E. and Sonntag C. (2009). Free radical mechanisms for the treatment of methyl tert-butyl ether (MTBE) via advanced oxidation/reductive processes in aqueous solutions. *Chemical Reviews*, **109**, 1302–1345.

Criquet J. and Karpel Vel Leitner N. (2011). Electron beam irradiation of aqueous solution of persulfate ions. *Chemical Engineering Journal*, **169**, 258–262.

Criquet J. and Karpel Vel Leitner N. (2012). Electron beam irradiation of citric acid aqueous solutions containing persulfate. *Separation and Purification Technology*, **88**, 168–173.

Criquet J. and Karpel Vel Leitner N. (2015). Reaction pathway of the degradation of the *p*-hydroxybenzoic acid by sulfate radical generated by ionizing radiations. *Radiation Physics and Chemistry*, **106**, 307–314.

Csay T., Rácz G., Takács E. and Wojnárovits L. (2012). Radiation induced degradation of pharmaceutical residues in water: chloramphenicol. *Radiation Physics and Chemistry*, **81**, 1489–1494.

Csay T., Homlok R., Illés E., Takács E. and Wojnárovits L. (2014). The chemical background of advanced oxidation processes. *Israel Journal of Chemistry*, **54**, 233–241.

Dail M. K. and Mezyk S. P. (2010). Hydroxyl-radical-induced degradative oxidation of β-lactam antibiotics in water: absolute rate constant measurements. *The Journal of Physical Chemistry A*, **114**, 8391–8395.

Das T. N. (2005). Oxidation of phenol in aqueous acid: characterization and reactions of radical cations vis-à-vis the phenoxyl radical. *The Journal of Physical Chemistry A*, **109**, 3344–3351.

De la Cruz N., Giménez J., Esplugas S., Grandjean D., de Alencastro L.F. and Pulgarin C. (2012). Degradation of 32 emergent contaminants by UV and neutral photo-Fenton in domestic wastewater effluent previously treated by activated sludge. *Water Research*, **46**, 1947–1957.

Dodd M. C., Buffle M.-O. and von Gunten U. (2006). Oxidation of antibacterial molecules by aqueous ozone: moiety-specific reaction kinetics and application to ozone-based wastewater treatment. *Environmental Science and Technology*, **40**, 1969–1977.

Dodd M. C., Rentsch D., Singer H. P., Kohler H.-P. E. and von Gunten U. (2010). Transformation of β-lactam antibacterial agents during aqueous ozonation: reaction pathways and quantitative bioassay of biologically-active oxidation products. *Environmental Science and Technology*, **44**, 5940–5948.

Drzewicz P., Trojanowicz M., Zona R., Solar S. and Gehringer P. (2004). Decomposition of 2,4-dichlorophenoxyacetic acid by ozonation, ionizing radiation as well as ozonation combined with ionizing radiation. *Radiation Physics and Chemistry*, **69**, 281–287.

European Centre for Disease Prevention and Control (2014). Surveillance of Antimicrobial Consumption in Europe 2012. ECDC, Stockholm, Sweden.

Emmi S. S. and Takács E. (2008). Water remediation by the electron-beam treatment. In: Radiation Chemistry – From Basics to Applications in Material and Life Sciences, M. Spotheim-Maurizot, M. Mostafavi, T. Douki and J. Belloni (eds), EDP Sciences, Paris, France, pp. 79–95.

Fang G.-D., Dionysiou D. D., Al-Abed S. R., Zhou D.-M. (2013). Superoxide radical driving the activation of persulfate by magnetite nanoparticles: implications for the degradation of PCBs. *Applied Catalysis B: Environmental*, **129**, 325–332.

Fang X., Pan X., Rahmann A., Schuchmann H.-P. and von Sonntag C. (1995). Reversibility in the reaction of cyclohexadienyl radicals with oxygen in aqueous solution. *Chemistry - A European Journal*, **1**, 423–429.

Fischbacher A., von Sonntag J., von Sonntag C. and Schmidt T. C. (2013). The $^{\bullet}OH$ radical yield in the $H_2O_2 + O_3$ (peroxone) reaction. *Environmental Science and Technology*, **47**, 9959–9964.

Gajda-Schrantz K., Arany E., Illés E., Szabó E., Pap Zs., Takács E. and Wojnárovits L. (2013). Advanced oxidation processes for ibuprofen removal and ecotoxicological risk assessment of degradation intermediates. In: Ibuprofen: Clinical Pharmacology, Medical Uses and Adverse Effects, W. C. Carter and B. R. Brown (eds), Nova Science Publishers, New York, USA, pp. 152–232.

Gehringer G., Eschweiler H., Szinovatz W., Fiedler H., Steiner R. and Sonneck G. (1993). Radiation-induced OH radical generation and its use for groundwater remediation. *Radiation Physics and Chemistry*, **42**, 711–714.

Gehringer P. (2004). Technical and economical aspects of radiation technology for wastewater treatment applications in industrial scale. In: IAEA-TECDOC-1407, Status of Industrial Scale Radiation Treatment of Wastewater and its Future, International Atomic Energy Agency, Vienna, Austria, 2003, pp. 19–27.

Gehringer P., Eschweiler H., Leth H., Pribil W., Pfleger S., Cabaj A., Haider T. and Sommer R. (2003). Bacteriophages as viral indicators for radiation processing of water: a chemical approach. *Applied Radiation and Isotopes*, **58**, 651–656.

Gehringer P., Eschweiler H., Weiss S. and Reemtsma T. (2006). Decomposition of aqueous naphthalene-1,5-disulfonic acid by means of oxidation processes. *Ozone: Science and Engineering*, **28**, 437–443.

Gehringer P., Eschweiler H., Weiss S. and Reemtsma T. (2008). Effluent polishing by means of advanced oxidation. In: IAEA-TECDOC-1598, Irradiation Treatment of Polluted Water and Wastewater, International Atomic Energy Agency, Vienna, Austria, 2006, pp. 15–26.

Getoff N. (1997). Peroxyl radicals in the treatment of waste solutions. In: Peroxyl Radicals, Z.B. Alfassi (ed.), John Wiley & Sons, New York, USA, pp. 483–506.

Gogolev A. V., Kabakchi S. A. and Pikaev A. K. (1992). Pulse radiolysis of aqueous solutions of Nekal. *Khimiya Vysokikh Energii*, **25**, 531–535.

Gunatilleka A. D. and Poole C. F. (1999). Models for estimating the non-specific aquatic toxicity of organic compounds. *Analytical Communications*, **36**, 235–242.

Guo Z., Zhou F., Zhao Y., Zhang C., Liu F., Bao C. and Lin M. (2012). Gamma irradiation-induced sulfadiazine degradation and its removal mechanisms. *Chemical Engineering Journal*, **191**, 256–262.

Gültekin I. and Ince N. H. (2007). Synthetic endocrine disruptors in the environment and water remediation by advanced oxidation processes. *Journal of Environmental Management*, **85**, 816–832.

Han B. (2009). Electron beam for environmental conversion. International topical meeting on nuclear research application and utilization of accelerators. Vienna, Austria, 4–8 May 2009. http://www-pub.iaea.org/MTCD/publications/PDF/P1433_CD/datasets/presentations/SM-EB-23.pdf

Han B., Ko J., Kim J., Kim Y., Chung W., Makarov I. E., Ponomarev A. V. and Pikaev A. K. (2002). Combined electron-beam and biological treatment of dyeing complex wastewater. Pilot plant experiments. *Radiation Physics and Chemistry*, **64**, 53–59.

Han B., Kim J. K., Kim Y. R., Makarov I. E. and Ponomarev A. V. (2005a). Electron beam treatment of textile dying wastewater: operation of pilot plant and industrial plant construction. In: IAEA-TECDOC-1473, Radiation Treatment of Gaseous and Liquid Effluents for Contaminant Removal, International Atomic Energy Agency, Vienna, Austria, pp. 101–110.

Han B., Kim J., Kim Y., Choi J. S. and Makarov I. E. (2005b). Electron beam treatment of textile dyeing wastewater: operation of pilot plant and industrial plant construction. *Water Science and Technology*, **52**, 317–324.

Han B., Kim J. K., Kim Y., Choi J. S. and Jeong K. Y. (2012). Operation of industrial-scale electron beam wastewater treatment plant. *Radiation Physics and Chemistry*, **81**, 1475–1478.

Hasegawa K. and Neta P. (1978). Rate constants and mechanism of reaction of $Cl_2^{\bullet-}$ radicals. *The Journal of Physical Chemistry*, **82**, 854–857.

Hashimoto S., Miyata T., Suzuki N. and Kawakami W. (1979). Decoloration and degradation of an anthraquinone dye aqueous solution in flow system using an electron accelerator. *Radiation Physics and Chemistry*, **13**, 107–113.

Hashimoto S., Miyata T. and Kawakami W. (1980). Radiation-induced decomposition of phenol in flow system. *Radiation Physics and Chemistry*, **16**, 59–65.

Hihara T., Okada Y. and Morita Z. (2006). Photo-oxidation of pyrazolinylazo dyes and analysis of reactivity as azo and hydrazone tautomers using semiempirical molecular orbital PM5 method. *Dyes and Pigments*, **69**, 151–176.

Hiller K.-O., Masloch B., Göbl M. and Asmus K.-D. (1981). Mechanism of the •OH radical induced oxidation of methionine in aqueous solution. *Journal of the American Chemical Society*, **103**, 2734–2743.

Hodgkins J. E. and Megarity E. D. (1965). Study of the benzyl free radical and substituted benzyl free radicals. *The Journal of Physical Chemistry*, **87**, 5322–5326.

Hoigné J. (1997). Inter-calibration of OH radical sources and water quality parameters. *Water Science and Technology*, **35**, 1–8.

Homlok R., Takács E. and Wojnárovits L. (2011). Elimination of diclofenac from water using irradiation technology. *Chemosphere*, **85**, 603–608.

Homlok R., Takács E. and Wojnárovits L. (2013). Degradation of organic molecules in advanced oxidation processes: relation between chemical structure and degradability. *Chemosphere*, **91**, 383–389.

Hoy A. R. and Bolton J. R. (1994). Determination of rate constants for reactive intermediates in the aqueous photodegradation of pollutants using a spin-trap/EPR method. In: Aquatic and Surface Photochemistry, G. R. Helz, R. G. Zepp and D. G. Crosby (eds), CRC Press, Boca Raton, USA, pp. 491–498.

Hu J.-Y. and Aizawa T. (2003). Quantitative structure-activity relationships for estrogen receptor binding affinity of phenolic chemicals. *Water Research*, **37**, 1213–1222.

Huie R. E., Clifton C. L. and Neta P. (1991). Electron transfer reaction rates and equilibria of the carbonate and sulfate radical anions. *Radiation Physics and Chemistry*, **38**, 477–481.

IAEA (2007). Radiation Processing, Environmental Applications (2007). International Atomic Energy Agency, Vienna, Austria.

Iqbal M. and Bhatti I.A. (2015). Gamma radiation/H_2O_2 treatment of a nonylphenol ethoxylates: degradation, cytotoxicity, and mutagenicity evaluation. *Journal of Hazardous Materials*, **299**, 351–360.

Illés E., Takács E., Dombi A., Gajda-Schrantz K., Gonter K. and Wojnárovits L. (2012). Radiation induced degradation of ketoprofen in dilute aqueous solution. *Radiation Physics and Chemistry*, **81**, 1479–1483.

Illés E., Takács E., Dombi A., Gajda-Schrantz K., Rácz G., Gonter K. and Wojnárovits L. (2013). Hydroxyl radical induced degradation of ibuprofen. *Science of the Total Environment*, **447**, 286–292.

Jankowska A., Biesaga M., Drzewicz P., Trojanowicz M. and Pyrzynska K. (2004). Chromatographic separation of chlorophenoxy acid herbicides and their radiolytic degradation products in water samples. *Water Research*, **38**, 3259–3264.

Janos P. (2009). Non-conventional sorbents for the dye removal from waters: mechanism and selected applications. In: Dyes and Pigments: New Research, A. R. Lang (ed.), Nova Science Publishers, New York, USA, pp. 201–224.

Johnston L. J., Mathivanan N., Negri F., Siebrand W. and Zerbetto F. (1993). Assignment and vibrational analysis of the 600 nm absorption band in the phenoxyl radical and some of its derivatives. *Canadian Journal of Chemistry*, **71**, 1655–1662.

Jones K. G. (2007). Applications of radiation chemistry to understand the fate and transport of emerging pollutants of concern in coastal waters. PhD thesis, North Caroline State University, Raleigh, North Carolina, USA.

Kapoor S. and Varshney L. (1997). Redox reactions of chloramphenicol and some aryl peroxyl radicals in aqueous solutions: a pulse radiolytic study. *The Journal of Physical Chemistry A*, **101**, 7778–7782.

Karlesa A., De Vera G. A. D., Dodd M. C., Park J., Espino M. P. B. and Lee Y. (2014). Ferrate(VI) oxidation of β-lactam antibiotics: reaction kinetics, antibacterial activity changes, and transformation products. *Environmental Science and Technology*, **48**, 10380–10389.

Karpel Vel Leitner N., Berger P. and Gehringer P. (1999). γ-irradiation for the removal of atrazine in aqueous solution containing humic substances. *Radiation Physics and Chemistry*, **55**, 317–322.

Keen O. S., McKay G., Mezyk S. P., Linden K. G. and Rosario-Ortiz F. L. (2014). Identifying the factors that influence the reactivity of effluent organic matter with hydroxyl radicals. *Water Research*, **50**, 408–419.

Khan J. K., Shah N. S., Nawaz S., Ismail M., Rehman, F. and Khan H. M. (2015). Role of e_{aq}^-, $^{\bullet}OH$ and $H^{\bullet}$ in radiolytic degradation of atrazine: a kinetic and mechanistic approach. *Journal of Hazardous Materials*, **288**, 147–157.

Kim H. Y., Yu S. H., Lee M. J., Kim T.-H. and Kim S. D. (2009). Radiolysis of selected antibiotics and their toxic effects on various aquatic organisms. *Radiation Physics and Chemistry*, **78**, 267–272.

Kim T.-H., Lee J.-K. and Lee M.-J. (2007). Biodegradability enhancement of textile wastewater by electron beam irradiation. *Radiation Physics and Chemistry*, **76**, 1037–1041.

Kim T.-H., Kim S. D., Kim H. Y., Lim S. J., Lee M. and Yu S. (2012). Degradation and toxicity assessment of sulfamethoxazole and chlortetracycline using electron beam, ozone and UV. *Journal of Hazardous Materials*, **227–228**, 237–242.

Kimura A., Taguchi M., Arai H., Hiratsuka H., Namba H. and Kojima T. (2004). Radiation-induced decomposition of trace amounts of 17β-estradiol in water. *Radiation Physics and Chemistry*, **69**, 295–301.

Kimura A., Taguchi M., Ohtani Y., Takigami M., Shimada Y., Kojima T., Hiratsuka H. and Namba H. (2006). Decomposition of *p*-nonylphenols in water and elimination of their estrogen activities by ^{60}Co γ-ray irradiation. *Radiation Physics and Chemistry*, **75**, 61–69.

Kimura A., Taguchi M., Ohtani Y., Shimada Y., Hiratsuka H. and Kojima T. (2007). Treatment of wastewater having estrogen activity by ionizing radiation. *Radiation Physics and Chemistry*, **76**, 699–706.

Kimura A., Osawa M. and Taguchi M. (2012). Decomposition of persistent pharmaceuticals in wastewater by ionizing radiation. *Radiation Physics and Chemistry*, **81**, 1508–1512.

Kishore K. and Mukherjee T. (2006). A pulse radiolysis study of salicylic acid and 5-sulpho-salicylic acid in aqueous solutions. *Radiation Physics and Chemistry*, **75**, 14–19.

Knepper T. P., Barcelo D., Lindner K., Seel P., Reemtsma T., Ventura F., De Wever H., Van der Voet E., Gehringer P. and Schönerklee M. (2004). Removal of persistent polar pollutants through improved treatment of wastewater effluents (P-THREE). *Water Science and Technology*, **50**, 195–202.

Kosaka K., Yamada H., Tsuno H., Shimizu Y. and Matsumi S. (2003). Reaction rate constants of di-*n*-butyl phthalate and 17β-estradiol with ozone and hydroxyl radical. *Journal of Japan Society on Water Environment*, **26**, 215–221.

Kovács K., Mile V., Csay T., Takács E. and Wojnárovits L. (2014). Hydroxyl radical-induced degradation of fenuron in pulse and gamma radiolysis: kinetics and product analysis. *Environmental Science and Pollution Research*, **21**, 12693–12700.

Kovács K., He S., Mile V., Csay T., Takács E. and Wojnárovits L. (2015). Ionizing radiation induced degradation of diuron in dilute aqueous solution. *Chemistry Central Journal*, **9**, doi: 10.1186/s13065-015-0097-0, Open access.

Kozmér Z., Arany E., Alapi T., Takács E., Wojnárovits L. and Dombi A. (2014). Determination of the rate constant of hydroperoxyl radical reaction with phenol. *Radiation Physics and Chemistry*, **102**, 135–138.

Lee M.-J. and Yoo D.-H. (2004). Gamma rays treatment of groundwater polluted by TCE and PCE. In: IAEA-TECDOC-1407, Status of Industrial Scale Radiation Treatment of Wastewater and its Future, International Atomic Energy Agency, Vienna, Austria, 2003, pp. 81–86.

Lichtscheidl J. and Getoff N. (1979). Pulsradiolytische Untersuchungen der Reaktion des e_{aq}^- mit halogenierten aromatischen Verbindungen (Pulse radiolysis investigations on the reaction of e_{aq}^- with halogenated aromatic compounds). *Monatshefte für Chemie*, **110**, 1357–1375.

Liu S.-Y., Chen Y.-P. and Yu H.-Q. (2004). Degradation pathways of acetochlor by γ-radiolysis. *Chemistry Letters*, **33**, 1164–1165.

Liu S.-Y., Chen Y.-P., Yu H.-Q. and Zhang S.-J. (2005). Kinetics and mechanism of radiation-induced degradation of acetochlor. *Chemosphere*, **59**, 13–19.

Liu Q., Luo X., Zheng Z., Zheng B., Zhang J., Zhao Y., Yang X., Wang J. and Wang L. (2011). Factors that have effect on degradation of diclofenac in aqueous solutions by gamma ray irradiation. *Environmental Science and Pollution Research*, **18**, 1243–1252.

Liu Y. and Wang J. (2013). Degradation of sulfamethazine by gamma irradiation in the presence of hydrogen peroxide. *Journal of Hazardous Materials*, **250–251**, 99–105.

Liu Y., Hu J. and Wang J. (2014). Fe^{2+} enhancing sulfamethazine degradation in aqueous solution by gamma irradiation. *Radiation Physics and Chemistry*, **96**, 81–87.

Lutze H. (2013). Sulfate radical based oxidation in water treatment. Dissertation Dr. rer. nat., Universität Duisburg, Essen, Germany.

Madden K. P. and Mezyk S. P. (2011). Critical review of aqueous solution reaction rate constants for hydrogen atoms. *Journal of Physical and Chemical Reference Data*, **40**, 1–43.

Makarov I. E., Ponomarev A. V. and Han B. (2004). Demonstration plant for electron-beam treatment of Taegu dye industrial complex wastewater. In: IAEA-TECDOC-1386, Emerging Applications of Radiation Processing, International Atomic Energy Agency, Vienna, Austria, pp. 138–152.

Mazellier P. and Sulzberger B. (2001). Diuron degradation in irradiated, heterogeneous iron/oxalate systems: the rate determining step. *Environmetnal Science and Technology*, **35**, 3314–3320.

McLaughlin W. L., Boyd A. W., Chadwick K. H., McDonald J. C. and Miller A. (1989). Dosimetry for Radiation Processing. Taylor and Francis, London, UK.

Merenyi G., Lind J., Naumov S. and von Sonntag C. (2010). Reaction of ozone with hydrogen peroxide (peroxone process): a revision of current mechanistic concepts based on thermokinetic and quantum-chemical considerations. *Environmental Science and Technology*, **44**, 3505–3507.

Mezyk S. P. and Otto S. C. (2013). Quantitative removal of β-lactam antibiotic activity by hydroxyl radical reaction in water: how much oxidation is enough? *Journal of Advanced Oxidation Technologies*, **16**, 117–122.

Mezyk S. P., Neubauer T. J., Cooper W. J. and Peller J. R. (2007). Free-radical-induced oxidative and reductive degradation of sulfa drugs in water: absolute kinetics and efficiencies of hydroxyl radical and hydrated electron reactions. *The Journal of Physical Chemistry A*, **111**, 9019–9024.

Mezyk S. P., Rickman K. A., Hirsch C. M., Dail M. K., Scheeler J. and Foust T. (2013). Advanced oxidation and reduction process radical generation in the laboratory and on a large scale: an overview. In: Monitoring Water Quality: Pollution Assessment, Analysis, and Remediation, S. Ahuja (ed.), Elsevier, Waltham, USA, pp. 227–248.

Michael I., Rizzo L., McArdell C. S., Manaia C. M., Merlin C., Schwartz T., Dagot C. and Fatta-Kassinos D. (2013). Urban wastewater treatment plants as hotspots for the release of antibiotics in the environment: a review. *Water Research*, **47**, 957–995.

Miyazaki T., Katsumura Y., Lin M., Muroya Y., Kudo H., Taguchi M., Asano M. and Yoshida M. (2006). Radiolysis of phenol in aqueous solution at elevated temperatures. *Radiation Physics and Chemistry*, **75**, 408–415.

Mohamed K. A., Basfar A. A. and Al-Shahrani A. A. (2009). Gamma-ray induced degradation of diazinon and atrazine in natural groundwaters. *Journal of Hazardous Materials*, **166**, 810–814.

Musa K. A. K., Matxain J. M. and Eriksson L. A. (2007). Mechanism of photoinduced decomposition of ketoprofen. *Journal of Medicinal Chemistry*, **50**, 1735–1743.

Mvula E., Schuchmann M. N. and von Sonntag C. (2001). Reactions of phenol-OH-adduct radicals. Phenoxyl radical formation by water elimination vs. oxidation by dioxygen. *Journal of the Chemical Society, Perkin Transactions* 2, 264–268.

Nagai T. and Suzuki N. (1976). The radiation-induced degradation of anthraquinone dyes in aqueous solution. *The International Journal of Applied Radiation and Isotopes*, **27**, 699–705.

Neta P., Huie R. E. and Ross A. B. (1988). Rate constants for reactions of inorganic radicals in aqueous solution. *Journal of Physical and Chemical Reference Data*, **17**, 1027–1284.

Nickelsen M. G., Kajdi D. C., Cooper W. J., Kurucz C. N., Waite T. D., Gensel F., Lorenzl H. and Sparka U. (1998). Field application of mobile 20-kW electron-beam treatment system on contaminated groundwater and industrial wastes. In: Environmental Applications of Ionizing Radiation, W. J. Cooper, R. D. Curry and K. E. O'Shea (eds), John Wiley & Sons, New York, USA, pp. 451–466.

Nickelsen M. G., Cooper W. J., Secker D. A., Rosoch L. A., Kurucz C. N. and Waite T. D. (2002). Kinetic modeling and simulation of PCE and TCE removal in aqueous solutions by electron-beam irradiation. *Radiation Physics and Chemistry*, **65**, 579–587.

Ning B., Graham N. J. D. and Zhang Y. (2007). Degradation of octylphenol and nonylphenol by ozone - Part II: Indirect reaction. *Chemosphere*, **68**, 1173–1179.

Orbán N., Boldizsár I. and Bóka K. (2009). Enhanced anthraquinone dye production in plant cell cultures of *Rubiaceae* species: emerging role of signalling pathways. In: Dyes and Pigments: New Research, A. R. Lang (ed.), Nova Science Publishers, New York, USA, pp. 403–420.

Otto S. C., Mezyk S. P. and Zimmerman K. D. (2015). Complete β-lactam antibiotic activity removal from wastewaters: hydroxyl radical-mediated oxidation efficiencies. In: Food, Energy, and Water: The Chemistry Connection, S. Ahuja (ed.), Elsevier, Amsterdam, Netherlands, pp. 113–128.

Oturan M. A., Edelahi M. C., Oturan N., El Kacemi K. and Aaron J.-J. (2010). Kinetics of oxidative degradation/mineralization pathways of the phenylurea herbicides diuron, monuron and fenuron in water during application of the electro-Fenton process. *Applied Catalysis B: Environmental*, **97**, 82–89.

Pálfi T., Takács E. and Wojnárovits L. (2007). Degradation of H-acid and its derivative in aqueous solution by ionising radiation. *Water Research*, **41**, 2533–2540.

Pálfi T., Wojnárovits L. and Takács E. (2011). Mechanism of azo dye degradation in Advanced Oxidation Processes: degradation of sulfanilic acid azochromotrop and its parent compounds in aqueous solution by ionizing radiation. *Radiation Physics and Chemistry*, **80**, 462–470.

Parker, V. D. (1992). Homolytic bond (H-A) dissociation free energies in solution. Application of the standard potential of the ($H^+/H^•$) couple. *Journal of the American Chemical Society*, **114**, 7458–7462.

Paul (Guin) J., Naik D. B., Bhardwaj Y. K. and Varshney L. (2014a). Studies on oxidative radiolysis of ibuprofen in presence of potassium persulfate. *Radiation Physics and Chemistry*, **100**, 38–44.

Paul (Guin) J., Naik D. B., Bhardwaj Y. K. and Varshney L. (2014b). An insight into the effective advanced oxidation process for treatment of simulated textile dye waste water. *RSC Advances*, **4**, 39941–39947.

Peller J. and Kamat P. V. (2005). Radiolytic transformations of chlorinated phenols and chlorinated phenoxyacetic acids. *The Journal of Physical Chemistry A*, **109**, 9528–9535.

Peller J., Wiest O. and Kamat P. V. (2001). Sonolysis of 2,4-dichlorophenoxyacetic acid in aqueous solutions. Evidence for •OH-radical-mediated degradation. *The Journal of Physical Chemistry A*, **105**, 3176–3181.

Peller J., Wiest O. and Kamat P. V. (2003). Mechanism of hydroxyl radical-induced breakdown of the herbicide 2,4-dichlorophenoxyacetic acid (2,4-D). *Chemistry - A European Journal*, **9**, 5379–5387.

Peller J., Wiest O. and Kamat P. V. (2004). Hydroxyl radical's role in the remediation of a common herbicide, 2,4-dichlorophenoxyacetic acid (2,4-D). *The Journal of Physical Chemistry A*, **108**, 10925–10933.

Peller J. R., Mezyk S. P. and Cooper W. J. (2009). Bisphenol A reactions with hydroxyl radicals: diverse pathways determined between deionized water and tertiary treated wastewater solutions. *Research on Chemical Intermediates*, **35**, 21–34.

Perkowski J. and Mayer J. (1989). Gamma radiolysis of anthraquinone dye aqueous solution. *Journal of Radioanalytical and Nuclear Chemistry*, **132**, 269–280.

Perkowski J., Gebicki J. L., Lubis R. and Mayer J. (1989). Pulse radiolysis of anthraquinone dye aqueous solution. *Radiation Physics and Chemistry*, **33**, 103–108.

Petrovic M., Gehringer P., Eschweiler H. and Barceló D. (2007). Radiolytic decomposition of multi-class surfactants and their biotransformation products in sewage treatment plant effluents. *Chemosphere*, **66**, 114–122.

Philips G. O., Power D. M., Robinson C. and Davies J. V. (1973a). Interactions of bovine serum albumin with penicillins and cephalosporins studied by pulse radiolysis. *Biochemica et Biophysica Acta*, **295**, 8–17.

Philips G. O., Power D. M. and Robinson C. (1973b). Chemical changes following γ-irradiation of benzylpenicillin in aqueous solution. *Journal of the Chemical Society, Perkin Transaction 2*, 575–582.

Pikaev A. K. (1998). Electron-beam purification of water and wastewater. In: Environmental Applications of Ionizing Radiation, W. J. Cooper, R. D. Curry and K. E. O'Shea (eds), John Wiley & Sons, New York, USA, pp. 495-506.

Pikaev A. K., Makarov I. E., Ponomarev A. V., Kim Y., Han B. and Kang H. J. (1997a). A combined electron-beam and coagulation method of purification of water from dyes. *Mendeleev Communications*, **7**, 176–177.

Pikaev A. K., Podzorova E. A. and Bakhtin O. M. (1997b). Combined electron-beam and ozone treatment of wastewater in the aerosol flow. *Radiation Physics and Chemistry*, **49**, 155–157.

Pikaev A. K. (2001). Application of pulse radiolysis and computer simulation for the study of the mechanism of radiation purification of polluted water. *Research on Chemical Intermediates*, **27**, 775–786.

Pikaev A. K., Podzorova E. A., Bakhtin O. M., Lysenko S. L. and Belyshev V. A. (2001). Electron beam technology for purification of municipal wastewater in the aerosol flow. In: IAEA-TECDOC-1225, Use of irradiation for chemical and microbial decontamination of water, wastewater and sludge, International Atomic Energy Agency, Vienna, Austria, 1999, pp. 45–55.

Podzorova E. A., Pikaev A. K., Belyshev V. A. and Lysenko S. L. (1998). New data on electron-beam treatment of municipal wastewater in aerosol flow. *Radiation Physics and Chemistry*, **52**, 361–364.

Pogocki D. and Bobrowski K. (2014). Oxidative degradation of thiaproline derivatives in aqueous solutions induced by •OH radicals. *Israel Journal of Chemistry*, **54**, 321–332.

Poster D. L., Chaychian M., Neta P., Huie R. E., Silverman J. and Al-Sheikhly M. (2003). Degradation of PCBs in a marine sediment treated with ionizing and UV radiation. *Environmental Science and Technology*, **37**, 3808–3815.

Rela P. R., Sampa M. H. O., Duarte C. L., Costa F. E. D. and Sciani V. (2000). Development of an up-flow irradiation device for electron beam wastewater treatment. *Radiation Physics and Chemistry*, **57**, 657–660.

Rela P. R., Sampa M. H. O., Duarte C. L., Costa F. E., Sciani V., Borrely S. I., Mori M. N., Somessari E. S. R. and Silveira C. G. (2008). The status of radiation process to treat industrial effluents in Brazil. In: IAEA-TECDOC-1598, Irradiation treatment of polluted water and wastewater, International Atomic Energy Agency, Vienna, Austria, pp. 27–41.

Rickman K. A. and Mezyk S. P. (2010). Kinetics and mechanisms of sulfate radical oxidation of β-lactam antibiotics in water. *Chemosphere*, **81**, 359–365.

Rizzo L., Manaia C., Merlin C., Schwartz T., Dagot C., Ploy M. C., Michael I. and Fatta Kassinos D. (2013). Urban wastewater treatment plants as hotspots for antibiotic resistant bacteria and genes spread into the environment: a review. *Science of the Total Environment*, **447**, 345–360.

Roder M., Wojnárovits L., Földiák G., Emmi S. S., Beggiato G. and D'Angelantonio M. (1999). Addition and elimination kinetics in OH radical induced oxidation of phenol and cresols in acidic and alkaline solutions. *Radiation Physics and Chemistry*, **54**, 475–479.

Roduner E. and Bartels D. M. (1992). Solvent and isotope effects on addition of atomic hydrogen to benzene in aqueous solution. *Berichte der Bunsengesellschaft für Physikalische Chemie*, **96**, 1037–1042.

Ruiz M. J., López-Jaramillo L., Redondo M. J. and Font G. (1997). Toxicity assessment of pesticides using the Microtox test: application to environmental samples. *Bulletin of Environmental Contamination and Toxicology*, **59**, 619–625.

Sabharwal S., Kishore K. and Moorthy P. N. (1994). Pulse radiolysis study of oxidation reactions of sulphacetamide in aqueous solutions. *Radiation Physics and Chemistry*, **44**, 499–506.

Sági Gy., Csay T., Pátzay Gy., Csonka E., Wojnárovits L. and Takács E. (2014). Oxidative and reductive degradation of sulfamethoxazole in aqueous solutions: decomposition efficiency and toxicity assessment. *Journal of Radioanalytical and Nuclear Chemistry*, **301**, 475–482.

Sági Gy., Csay T., Szabó L., Pátzay G., Csonka E., Takács E. and Wojnárovits L. (2015). Analytical approaches to the OH radical induced degradation of sulfonamide antibiotics in dilute aqueous solutions. *Journal of Pharmaceutical and Biomedical Analysis*, **106**, 52–60.

Sampa M. H. O., Borrely S. I., Silva B. L., Vieira, J. M., Rela, P. R., Calvo W. A. P., Nieto, R. C., Duarte C. L., Perez H. E. B., Somessari E. S. and Lugão A. B. (1995). The use of electron beam accelerator for the treatment of drinking water and wastewater in Brazil. *Radiation Physics and Chemistry*, **46**, 1143–1146.

Sampa M. H. O., Rela P. R., Las Casas A., Mori M. N. and Duarte C. L. (2004). Treatment of industrial effluents using electron beam accelerator and adsorption with activated carbon: a comparative study. *Radiation Physics and Chemistry*, **71**, 459–462.

Sato K., Takimoto K. and Tsuda S. (1978). Degradation of aqueous phenol solution by gamma irradiation. *Environmental Science and Technology*, **12**, 1043–1046.

Schuler R. H. and Albarran G. (2002). The rate constants for reaction of •OH radicals with benzene and toluene. *Radiation Physics and Chemistry*, **64**, 189–195.

Sein M. M., Zedda M., Tuerk J., Schmidt T. C., Golloch A. and von Sonntag C. (2008). Oxidation of diclofenac with ozone in aqueous solution. *Environmental Science and Technology*, **42**, 6656–6662.

Sharma S. B., Mudaliar M., Rao B. S. M., Mohan H. and Mittal J. P. (1997). Radiation chemical oxidation of benzaldehyde, acetophenone, and benzophenone. *The Journal of Physical Chemistry A*, **101**, 8402–8408.

Singh A., Kremers W., Smalley P. and Bennett G.S. (1985). Radiolytic dechlorination of polychlorinated biphenyls. *Radiation Physics and Chemistry*, **25**, 11–19.

Solar S., Solar W. and Getoff N. (1986). Resolved multisite OH-attack on aqueous aniline studied by pulse radiolysis. *Radiation Physics and Chemistry*, **28**, 229–234.

Song W., Chen W., Cooper W. J., Greaves J. and Miller G. E. (2008). Free-radical destruction of β-lactam antibiotics in aqueous solution. *The Journal of Physical Chemistry A*, **112**, 7411–7417.

Spinks J. W. T. and Woods R. J. (1990). An Introduction to Radiation Chemistry. 3rd edn, Wiley-Interscience, New York, USA.

Steenken S. (1985). Electron transfer equilibria involving radicals in aqueous solution. In: Landolt-Börnstein, Neue Serie, Gruppe II, Vol. 13e, E. Fischer (ed.), Springer Verlag, Heidelberg, Germany, pp. 147–293.

Steenken S. (1987). Addition-elimination paths in electron-transfer reactions between radicals and molecules. Oxidation of organic molecules by OH radical. *Journal of the Chemical Society, Faraday Transactions I*, **83**, 113–124.

Steenken S. and Neta P. (2003). Transient phenoxyl radicals. Formation and properties in aqueous solution. In: Chemistry of Phenols, Part 2, Z. Rappoport (ed.), John Wiley & Sons, Chichester, UK, pp. 1107–1152.

Strydom W., Parker W. and Olivares M. (2005) Electron beams: physical and clinical aspects. In: Radiation Oncology Physics: a Handbook for Teachers and Students, E.B. Podgorsak (techn. ed.), International Atomic Energy Agency, Vienna, Austria. pp. 273–299.

Szabó L., Tóth T., Homlok R., Takács E. and Wojnárovits L. (2012). Radiolysis of paracetamol in dilute aqueous solution. *Radiation Physics and Chemistry*, **81**, 1503–1507.

Szabó L., Tóth T., Homlok R., Rácz G., Takács E. and Wojnárovits L. (2014). Hydroxyl radical induced degradation of salicylates in aerated aqueous solution. *Radiation Physics and Chemistry*, **97**, 239–245.

Szabó L., Tóth T., Rácz G., Takács E. and Wojnárovits L. (2016a). •OH and e_{aq}^- are yet good candidates for demolishing the β-lactam system of a penicillin eliminating the antimicrobial activity. *Radiation Physics and Chemistry*, **124**, 84–90.

Szabó L., Tóth T., Rácz G., Takács E. and Wojnárovits L. (2016b). Drugs with susceptible sites for free radical induced oxidative transformations: the case of a penicillin. *Free Radical Research*, **50**, 26–38.

Szabó L., Tóth T., Rácz G., Takács E. and Wojnárovits L. (2016c). Change in hydrophilicity of penicillins during advanced oxidation by radiolytically generated •OH compromises the elimination of selective pressure on bacterial strains. *Science of the Total Environment*, **551–552**, 393–403.

Swallow A. J. (1973). Radiation Chemistry: An Introduction. Longman, London, UK.

Swallow A. J. (1982). Application of pulse radiolysis to study of aqueous organic systems. In: The Study of Fast Processes and Transient Species by Electron Pulse Radiolysis, NATO Advanced Study Institutes Series, J. H. Baxendale and F. Busi (eds), D. Reidel Publishing Company, Dordrecht, Netherlands, pp. 289–315.

Tabata Y. (ed.) (1991a). CRC Handbook of Radiation Chemistry. CRC Press, Boca Raton, USA.

Tabata Y. (ed.) (1991b). Pulse Radiolysis. CRC Press, Boca Raton, USA.

Tang L., Xu G., Wu W., Shi W., Liu N., Bai Y. and Wu M. (2010). Radiolytic decomposition of 4-bromodiphenyl ether. *Nuclear Science and Techniques*, **21**, 72–75.

Takács E., Dajka K., Wojnárovits L. and Emmi S. S. (2000). Protonation kinetics of acrylate anions. *Physical Chemistry Chemical Physics*, **2**, 1431–1433.

Tezuka M., Okada S. and Tamemasa O. (1978). Radiolytic decontamination of di-*n*-butyl phthalate from water (Article in Japanese). *Radioisotopes*, **27**, 306–310.

Trojanowicz M., Bojanowska-Czajka A., Kciuk G., Bobrowski K., Gumiela M., Koc A., Nalecz-Jawecki G., Torun M. and Ozbay D.S. (2012). Application of ionizing radiation in decomposition of selected pollutants in water. *European Water*, **39**, 15–26.

Tsujimoto Y., Hashizume H. and Yamazaki M. (1993). Superoxide radical scavenging activity of phenolic compounds. *International Journal of Biochemistry*, **25**, 491–494.

Varghese R., Mohan H., Manoj P., Manoj V. M., Aravind U. K., Vandana K. and Aravindakumar C. T. (2006). Reactions of hydrated electrons with triazine derivatives in aqueous medium. *Journal of Agricultural and Food Chemistry*, **54**, 8171–8176.

Varshney L. and Patel K. M. (1994). Effects of ionizing radiations on a pharmaceutical compound, chloramphenicol. *Radiation Physics and Chemistry*, **43**, 471–480.

von Gunten U. (2003). Ozonation of drinking water: part I. Oxidation kinetics and product formation. *Water Research*, **37**, 1443–1467.

von Sonntag C. (2006). Free-Radical-Induced DNA Damage and its Repair. A Chemical Perspective. Springer, Heidelberg, Germany.

von Sonntag C. and Schuchmann H.-P. (1997). Peroxyl radicals in aqueous solution. In: Peroxyl Radicals, Z.B. Alfassi (ed.), John Wiley & Sons, New York, USA, pp. 173–274.

von Sonntag C. and Schuchmann H.-P. (2001). The chemistry behind the application of ionizing radiation in water-pollution abatement. In: Radiation Chemistry: Present Status and Future Trends, Studies in Physical and Theoretical Chemistry 87, C. D. Jonah and B. S. M. Rao (eds), Elsevier, Amsterdam, Netherlands, pp. 657–670.

Vysotskaya N. A., Bortun L. N., Ogurtsov N. A., Migdalovich E. A., Revina A. A. and Voldko V. V. (1986). Radiolysis of anthraquinone dyes in aqueous solutions. *Radiation Physics and Chemistry*, **28**, 469–472.

Walsh C. (2003). Antibiotics: Actions, Origins, Resistance. ASM Press, Washington, DC, USA.

Wardman P. (1989). Reduction potentials of one-electron couples involving free radicals in aqueous solution. *Journal of Physical and Chemical Reference Data*, **18**, 1637–1755.

Wheeler O. H. and Montalvo R. (1967). Irradiation of estrone in aqueous solutions. *The International Journal of Applied Radiation and Isototopes*, **18**, 127–131.

Wishart J. F. (2008). Tools for radiolysis studies. In: Radiation Chemistry – From Basics to Applications in Material and Life Sciences, M. Spotheim-Maurizot, M. Mostafavi, T. Douki and J. Belloni (eds), EDP Sciences, Paris, France, pp. 17–33.

Wojnárovits L. (2011). Radiation chemistry. In: Handbook of Nuclear Chemistry, A. Vértes, S. Nagy, Z. Klencsár, R. Lovas and F. Rösch (eds), Springer US, New York, USA, pp. 1267–1331.

Wojnárovits L. and Takács E. (2008). Irradiation treatment of azo dye containing wastewater: an overview. *Radiation Physics and Chemistry*, **77**, 225–244.

Wojnárovits L. and Takács E. (2013). Structure dependence of the rate coefficients of hydroxyl radical + aromatic molecule reaction. *Radiation Physics and Chemistry*, **87**, 82–87.

Wojnárovits L. and Takács E. (2014). Rate coefficients of hydroxyl radical reactions with pesticide molecules and related compounds: a review. *Radiation Physics and Chemistry*, **96**, 120–134.

Wojnárovits L., Takács E., Dajka K., D'Angelantonio M. and Emmi S. S. (2001). Pulse radiolysis of acrylamide derivatives in dilute aqueous solution. *Radiation Physics and Chemistry*, **60**, 337–434.

Wojnárovits L., Földiák G., D'Angelantonio M. and Emmi S. S. (2002). Mechanism of OH radical-induced oxidation of *p*-cresol to *p*-methylphenoxyl radical. *Research on Chemical Intermediates*, **28**, 373–386.

Wojnárovits L., Takács E., Dajka K., Emmi S. S., Russo M. and D'Angelantonio M. (2003). Rate coefficient of H atom addition with acrylate monomers in aqueous solution. *Tetrahedron*, **59**, 8353–8358.

Wojnárovits L., Takács E., Dajka K., Emmi S. S., Russo M. and D'Angelantonio M. (2004). Re-evaluation of the rate constant for the H atom reaction with *tert*-butanol in aqueous solution. *Radiation Physics and Chemistry*, **69**, 217–219.

Wojnárovits L., Pálfi T. and Takács E. (2005). Radiolysis of azo dyes in aqueous solution: Apollofix Red. *Research on Chemical Intermediates*, **31**, 679–690.

Woods R. J. and Pikaev A. K. (1994). Applied Radiation Chemistry: Radiation Processing. Wiley, New York, USA.

Wright G. D. (2007). The antibiotic resistome: the nexus of chemical and genetic diversity. *Nature Reviews Microbiology*, **5**, 175–186.

Wu M.-H., Liu N., Xu G., Ma J., Tang L., Wang L. and Fu H.-Y. (2011). Kinetics and mechanisms study on dimethyl phthalate degradation in aqueous solutions by pulse radiolysis and electron beam radiolysis. *Radiation Physics and Chemistry*, **80**, 420–425.

Xu G., Ren H., Wu M.-H., Liu N., Yuan Q., Liang T. and Liang W. (2011). Electron-beam induced degradation of bisphenol A. *Nuclear Science and Techniques*, **22**, 277–281.

Yamazaki M., Sawai T., Sawai T., Yamazaki K. and Kawaguchi S. (1983). Combined γ-ray irradiation-activated sludge treatment of humic acid solution from landfill leachate. *Water Research*, **17**, 1811–1814.

Ye L., He S., Yang C., Wang J. and Yu J. (2013). A comparison of pilot scale electron beam and bench scale gamma irradiations of cyanide aqueous in solution. *Nuclear Science and Techniques*, **24**, 1–9.

Yoshida T., Tanabe T., Chen A., Miyashita Y., Yoshida H., Hattori T. and Sawasaki T. (2003). Method for the degradation of dibutyl phthalate in water by gamma-ray irradiation. *Journal of Radioanalytical and Nuclear Chemistry*, **255**, 265–269.

Yu S., Lee B., Lee M., Cho I.-H. and Chang S.-W. (2008). Decomposition and mineralization of cefaclor by ionizing radiation: kinetics and effects of the radical scavengers. *Chemosphere*, **71**, 2106–2112.

Zhang J., Zheng Z., Luan J., Yang G., Song W., Zhong Y. and Xie Z. (2007). Degradation of hexachlorobenzene by electron beam irradiation. *Journal of Hazardous Materials*, **142**, 431–436.

Zhang J., Zheng Z., Zhao T., Zhao Y., Wang L., Zhong Y. and Xu Y. (2008). Radiation-induced reduction of diuron by gamma-ray irradiation. *Journal of Hazardous Materials*, **151**, 465–472.

Zhang X. H., Cao D. M., Zhao S. Y., Gong P., Hei D. Q. and Zhang H. Q. (2011). Gamma radiolysis of ceftriaxone sodium for water treatment: assessments of the activity. *Water Science and Technology*, **63**, 2767–2774.

Zheng B. G., Zheng Z., Zhang J. B., Luo X. Z., Wang J. Q., Liu Q. and Wang L. H. (2011). Degradation of the emerging contaminant ibuprofen in aqueous solution by gamma irradiation. *Desalination*, **276**, 379–385.

Zhou J., Wu M., Xu G., Liu N. and Zhou Q. (2009). Irradiation degradation of chloramphenicol, thiamphenicol and florfenicol with electron beam. IEEE Xplore, International Conferece on Information and Automation, Proceedings of the 2009 IEEE, Zhuhai, Macau, China, 2009, pp. 659–664. doi: 10.1109/ICINFA.2009.5205004.

Zimek Z. and Bulka S. (2007). Electron accelerator facilities. Regional Training Course on Validation and Process Control for Electron Beam Radiation Processing, Technical Cooperation Project RER/8/010 'Quality Control Methods and Procedures for Radiation Technology', 2007, Warsaw, Poland.

Zona R. and Solar S. (2003). Oxidation of 2,4-dichlorophenoxyacatic acid by ionizing radiation: degradation, detoxification and mineralization. *Radiation Physics and Chemistry*, **66**, 137–143.

Zona R., Solar S., Sehested K., Holcman J. and Mezyk S. P. (2002a). OH-Radical induced oxidation of phenoxyacetic acid and 2,4-dichlorophenoxyacetic acid. Primary radical steps and products. *The Journal of Physical Chemistry A*, **106**, 6743–6749.

Zona R., Solar S. and Gehringer P. (2002b). Degradation of 2,4-dichlorophenoxyacetic acid by ionizing radiation: influence of oxygen concentration. *Water Research*, **36**, 1369–1374.

Chapter 7

Fenton, photo-Fenton and Fenton-like processes

Christopher J. Miller, Susan Wadley, and T. David Waite

7.1 INTRODUCTION

The Fenton process, first described by Fenton (1894), involves the addition of aqueous solutions of Fe^{2+} ions and hydrogen peroxide (H_2O_2) to produce a highly reactive nonselective oxidant, which was proposed by Haber and Weiss (1934) to be the hydroxyl radical ($^{\bullet}OH$). More recent research has shown that the reactive intermediate may not be the hydroxyl radical itself, but is perhaps the ferryl ion, $Fe^{IV}_{(aq)}$ (Bossmann *et al.* 1998; Barbusiński, 2009). These reactive intermediates (i.e., either the hydroxyl radical or the ferryl ion) are capable of unselectively oxidizing most organic compounds. As discussed in Section 7.3, several other radicals and reactive intermediates are also involved in a complex system of chain reactions, in which iron acts as a catalyst by cycling between various oxidation states.

The classical Fenton process takes place in a homogeneous system. Processes that involve reactions between iron in an insoluble form, either in the unsupported form as a mineral particle or supported on a substrate (which could be a particle or a membrane), and H_2O_2 are referred to as "heterogeneous Fenton processes". The iron can initially be in the Fe(II) or Fe(III) form, because the reaction of Fe(III) with H_2O_2 regenerates Fe(II).

The effectiveness of the Fenton process can be improved by using external non-chemical energy sources such as ultraviolet light, electrical current, and ultrasound. These processes will be referred to as "extended Fenton processes".

The term "Fenton-like process" is sometimes used to refer to any process that has similar features to the Fenton process but is not the classical process, and would include heterogeneous processes and processes that use Fe initially in the Fe(III) form. We recommend including all these processes in the category of "Fenton type processes" provided they involve Fe (in any form) as the catalyst and use H_2O_2 as the oxidant regardless of whether it is added directly as a reagent or formed in situ, such as by electrochemical reactions and by the photolysis of Fe(III) complexes. The term "Fenton-like processes" should be reserved for all processes that have similar chemistry to the Fenton process, but involve metallic catalysts other than iron and/or oxidants that are not H_2O_2. More details are provided in Section 7.2 on the various types of Fenton process, as well as some Fenton-like processes.

Persistent organic pollutants (POPs) such as chlorinated pesticides, other chlorinated compounds, and combustion by-products are of major concern in the area of global water quality (EPA, 2015). Although many POPs are no longer produced in the US and several other countries, they still have a long lasting effect on the environment due to their low degradation rates and tendency to bioaccumulate in the food chain. These toxic organic compounds are resistant to both microbiological degradation and direct photo-degradation by sunlight. In addition, POPs cannot be adequately removed from the water by physicochemical separation processes such as adsorption, flocculation and sedimentation, dissolved air flotation, and pressure-driven membrane processes. The wastes that arise from these separation processes still contain the POPs (usually at higher concentrations than in the original wastewater), and although the volume is significantly reduced they remain difficult to dispose. Fenton processes can be used to increase the biodegradability of the organic compounds in wastewater and in some cases complete mineralisation can be achieved.

In addition to POPs, the United States EPA (EPA, 2014a) has also identified several contaminants of emerging concern (CEC). These CECs include pharmaceuticals and personal care products (PPCPs), perfluorinated compounds (PFCs), and polybrominated diphenyl ethers (PBDEs). For European Union countries, the current Directive on Environmental Quality Standards lists several priority substances (European Commission, 2015), some of which are identified as priority hazardous substances, as well as other pollutants. The priority substances include agricultural pesticides, biocides, metals, and polyaromatic hydrocarbons, while the other pollutants are mainly certain organochlorine insecticides that are no longer manufactured or used in the European Union or the United States, and organochlorine solvents.

The US EPA's effluent guidelines currently address 56 categories of industries, which either discharge wastewater into the natural waterways or into wastewater treatment plants and publicly owned treatment works (EPA, 2014b). Typically, the guidelines for each category specify the maximum allowable levels of pollutants that can be discharged, and are based on the performance of certain available treatment technologies. Industries can also use any alternative treatment process that can meet the specified pollutants limits. The regulations also vary depending on whether the industry is a newly constructed or existing facility. Since standards for industrial wastewater discharge are becoming increasingly strict, although existing treatment processes can still be used to meet current standards, development of improved wastewater treatment processes are required to exceed the current standards and to provide treatment processes at a lower cost. Fenton type processes have been shown to provide cost effective solutions for treating a broad range of wastewaters (see Section 7.4).

7.2 TYPES OF FENTON PROCESSES

Fenton processes can be either homogeneous (all reactions take place in the aqueous phase) or heterogeneous (the reactions take place in both the aqueous phase and the solid phase). It is the iron catalyst that is in the solid phase and this can either be supported on a substrate (such as a zeolite or a polymeric membrane) or exist in an unsupported form.

7.2.1 Fenton processes

7.2.1.1 Homogeneous Fenton processes

Homogeneous Fenton processes make use of $Fe^{II}_{(aq)}$ or $Fe^{III}_{(aq)}$ as the catalyst. The distinction between these systems is somewhat arbitrary, because $Fe^{III}_{(aq)}$ also forms in the $Fe^{II}_{(aq)}/H_2O_2$ system, with the same reactions occurring but with differing significance. When these processes occur in the absence of light of suitable wavelength to cause photo-reduction of the Fe(III) complexes, they are called dark Fenton processes or thermal Fenton processes; unless otherwise specified, it is assumed that the processes are dark processes.

$Fe^{II}_{(aq)}/H_2O_2$

The origins of this process date back to the late 19th century when Henry John Horstman Fenton first noted that Fe^{2+} and H_2O_2 could degrade tartaric acid (Fenton, 1876; Fenton, 1894). The formation of $^{\bullet}OH$ as the oxidative intermediate was demonstrated by the work of Haber and Willstätter (1931), and by Haber and Weiss (1932, 1934), with the now generally accepted radical chain mechanism advanced by Barb *et al.* (1951a, b). There has been dispute about the nature of the oxidative intermediate since $^{\bullet}OH$ was first proposed, with early workers also advancing the notion that a high-valent state of iron, Fe(IV), could be the actual oxidant (Bray & Gorin, 1932). There is still ongoing debate regarding the nature of the oxidant (Dunford, 2002), with new evidence supporting intermediates different to $^{\bullet}OH$ still being uncovered (Kremer, 1999, 2003). It is, however, generally accepted that under the acidic conditions typically relevant to application of the $Fe^{II}_{(aq)}/H_2O_2$ system in AOPs that $^{\bullet}OH$ is the most likely oxidant formed (Lee *et al.* 2013; Bataineh, 2012; Keenan & Sedlak, 2008), with the principal scheme of free radical reactions shown below.

$$Fe^{II} + H_2O_2 \rightarrow Fe^{III} + {}^{\bullet}OH + OH^- \quad (7.1)$$

$$Fe^{II} + {}^{\bullet}OH \rightarrow Fe^{III} + OH^- \quad (7.2)$$

$$^{\bullet}OH + H_2O_2 \rightarrow HO_2^{\bullet} + H_2O \quad (7.3)$$

$$HO_2^{\bullet} \rightleftharpoons H^+ + O_2^{\bullet -} \quad (7.4)$$

$$Fe^{III} + O_2^{\bullet -} \rightarrow Fe^{II} + O_2 \quad (7.5)$$

$$Fe^{II} + HO_2^{\bullet}/O_2^{\bullet -} \rightarrow Fe^{III} + H_2O_2 \quad (7.6)$$

$$Fe^{III} + H_2O_2 \rightarrow Fe^{II} + HO_2^{\bullet} + H^+ \quad (7.7)$$

$$Fe^{II} + O_2 \rightarrow Fe^{III} + O_2^{\bullet -} \quad (7.8)$$

$$RH + {}^{\bullet}OH \rightarrow R^{\bullet} + H_2O \quad (7.9)$$

In this scheme RH is a representative contaminant, shown here to react with $^{\bullet}OH$ by H-abstraction, although other routes are also possible, with addition of $^{\bullet}OH$ to aromatic rings to form cyclohexadienyl radicals being another important possibility (von Sonntag, 2008). The relative importance of each reaction will depend strongly on the nature of the system under investigation, as the reactivity of Fe^{II} and Fe^{III} in each reaction will depend upon their speciation (including various hydrolysis products formed as the pH is varied), e.g., the oxidation of Fe^{II} by O_2 is only important at circumneutral pH and can be neglected at the acidic pHs normally employed.

$Fe^{III}_{(aq)}/H_2O_2$

As H_2O_2 is able to reduce $Fe^{III}_{(aq)}$ as well as oxidize $Fe^{II}_{(aq)}$, it is possible to use $Fe^{III}_{(aq)}$ as the iron-source (De Laat & Gallard, 1999; De Laat *et al.* 2004; De Laat & Le, 2005). In these systems H_2O_2 must first reduce Fe^{III} to Fe^{II}, before the Fe^{II} thus formed reacts with H_2O_2 to yield $^{\bullet}OH$, as for the $Fe^{II}_{(aq)}/H_2O_2$

system. As the reaction between $Fe^{III}_{(aq)}$ and H_2O_2 is comparatively slow, this system will typically have a lower rate of oxidant production than a comparable $Fe^{II}_{(aq)}/H_2O_2$ process, which can be advantageous if mixing rates are low, or dispersion of the reagents over a wide area before they are consumed is desirable. In this system, it is critical that the $Fe^{III}_{(aq)}$ remains in solution, so acidic conditions are mandated to prevent precipitation of oxyhydroxides at any appreciable $Fe^{III}_{(aq)}$ concentration (Stefánsson, 2007). The reduction of Fe(III) to Fe(II) can also be greatly accelerated by addition of reducing agents such as hydroxylamine to the process, which if added in excess can also maintain cycling of Fe for longer periods (Chen *et al.* 2011).

Fe^{II} – $ligand_{(aq)}/H_2O_2$

The inclusion of a ligand can alter the reactivity of both Fe^{II} and Fe^{III}, which has the two key advantages of ensuring the formation of $^{\bullet}OH$ at circumneutral pH, subject to appropriate ligand selection (Miller *et al.* 2016; Remucal & Sedlak, 2011), and extending the suitable pH-range of this process by preventing the precipitation of Fe^{III}. Provided that the Fe^{III}-$ligand_{(aq)}$ complex is readily reduced by $O_2^{\bullet-}$ (or H_2O_2), a qualitatively similar Fe^{II}-Fe^{III} catalytic cycle can be achieved as for the ligand-free system.

The ligand itself needs to be carefully selected so as to not only perform the correct chemistry, but to also not be a potential contaminant in its own right; e.g., although ethylenediamine tetraacetate (EDTA) performs well in this sense and is well-researched, it is generally considered to be persistent in natural systems and its discharge should be avoided (Bucheli-Witschel & Egli, 2001). However, the structurally similar ethylenediamine-*N,N′*-disuccinate (EDDS) is a readily-biodegradable substitute for EDTA which shows promise in these systems (Huang *et al.* 2013). The concentration of any ligand should also be carefully controlled to minimize the extent to which the ligand itself can act as a sink for $^{\bullet}OH$.

Inorganic ligands can also be employed to both solubilize Fe and to enhance desired reactivity. Tetrapolyphosphate has been shown to increase the amount of $^{\bullet}OH$ formed when Fe(II) is oxidized by H_2O_2 at circumneutral pH, albeit only marginally and still such that it is likely to be formed in a yield of less than 1% (Biaglow & Kachur, 1997). Nonetheless, the addition of tetrapolyphosphate to a range of Fenton type processes has shown substantial improvements comparable to or exceeding that obtained by addition of EDTA (Wang *et al.* 2013, 2014, 2015).

Fe^{III} – $ligand_{(aq)}/H_2O_2$

Fe^{III}-ligand/H_2O_2 systems rely upon the reduction of Fe^{III}-ligand to Fe^{II}-ligand to undergo reaction with an additional molecule of H_2O_2 to form $^{\bullet}OH$. The reduction of the Fe^{III}-ligand to the active Fe^{II}-ligand form is typically the rate determining step in this process, with the feasibility of this reaction critical for the attainment of a catalytic cycle. For example, although Fe^{II}-DTPA is an effective Fenton catalyst, the ferric complex is only negligibly reduced by both H_2O_2 and $O_2^{\bullet-}$ (Gutteridge, 1990; Aruoma *et al.* 1989; Buettner *et al.* 1983), rendering it ineffective in an Fe^{III}-ligand/H_2O_2 system without the addition of an appropriate reductant. The use of Fe^{III}-EDTA in this system has been well-studied, including the significant influence of pH and solution conditions on its reactivity (Walling *et al.* 1975). As for the Fe^{II}-ligand/H_2O_2 system, the use of EDDS as a ligand is also promising, with qualitatively similar performance to EDTA observed (Huang *et al.* 2013). As well as these systems, there are also additional chemistries that employ Fe^{III}-ligand/H_2O_2 systems that do not follow Fenton-based chemistry. For example, the ferric tetraamido macrocyclic ligand (Fe^{III}-TAML) family of H_2O_2-activators are capable of oxidizing substrates (Kundu *et al.* 2013; Shappell, 2008), but do so following a pathway that is purported to involve Fe^{IV} and/or Fe^{V} intermediates as the oxidants, although the precise mechanism is still under investigation (Ryabov *et al.* 2013). Although this is a promising treatment technology, it is not a Fenton type system and will thus not be considered further here.

7.2.1.2 Heterogeneous Fenton processes

Heterogeneous Fenton processes make use of Fe^{II}_s or Fe^{III}_s as the catalyst. The use of iron that is initially in the solid form can reduce the amount of sludge that is formed. As outlined below, the iron can be present in the unsupported form as catalytic iron-containing mineral or metal particles, or it can be adsorbed on a solid material that provides support but does not participate as a catalyst.

The use of iron oxide minerals and clays as Fenton type catalysts are of interest because they are inexpensive, abundant, and environmentally friendly materials. These minerals have the advantage that, in some cases, they retain their activity over multiple treatments and are easy to separate after each treatment. They can be used in a wide pH range, including close to neutral, and a wide temperature range (Garrido-Ramírez *et al.* 2010; Pereira *et al.* 2012).

Most of the types of unsupported heterogeneous catalysts are microparticulate and include $\mathbf{Fe^{III}oxide_{(s)}}$, e.g., hematite ($\alpha$-$Fe_2O_3$), goethite ($\alpha$-$Fe^{III}OOH$), lepidocrocite ($\gamma$-FeOOH), and two-line ferrihydrite ($Fe_2(OH)_6$) (Matta *et al.* 2007), as well as, mixed silica/iron(III) oxides (Hanna *et al.* 2008); $\mathbf{Fe^{II,III}oxide_{(s)}}$, e.g., magnetite ($Fe_3O_4$) (Matta *et al.* 2007); and $\mathbf{Fe^{II}sulphide_{(s)}}$, e.g., pyrite ($FeS_2$) (Matta *et al.* 2007). In addition, synthetic materials such as nanoparticulate $\mathbf{Fe^0_{(s)}}$, i.e., zerovalent iron (nZVI) (Keenan & Sedlak, 2008; Xu & Wang, 2011) can also be used, with $\mathbf{Fe^{III}OCl}$ being particularly reactive, orders of magnitude more so than goethite (Yang *et al.* 2013, 2015).

Although in all cases the Fe is added as a solid form, the solid phase is not always the catalytically active site, e.g., although pyrite has been shown to be highly effective, this is due to its ability to constantly leach Fe(II), as well as the rapid reduction of Fe(III) by pyrite, maintaining a reasonably constant Fe(II) concentration driven by continual pyrite oxidation (Choi *et al.* 2014). For other materials, both homogeneous and heterogeneous processes are important, with the balance being dependent upon factors such as pH (Fang *et al.* 2013), crystallinity, and surface area as demonstrated for magnetite (Prucek *et al.* 2009). The presence of an iron oxide surface can also catalyse the homogeneous oxidation of Fe(II) (Jones *et al.* 2014), with adsorbed Fe(II) also being able to transform the mineral phase (Boland *et al.* 2014), further complicating interpretation of these systems. These processes are also operative in the surface-associated Fenton reaction that occurs during corrosion of nano- and micro-ZVI in oxic solutions (He *et al.* 2016; Ma *et al.* 2016). The importance of the surface can also be seen from studies where silica is included, which retards surface-catalyzed Fe(II) oxidation by both O_2 and H_2O_2 (Kinsela *et al.* 2016; Pham *et al.* 2012), but has no influence on efficiency of conversion of H_2O_2 to an oxidant (Pham *et al.* 2012).

In many Fe minerals the reactivity has been found to correlate with the Fe(II)-content, where the order is approximately Fe^{2+} > pyrite > green rust > magnetite ≈ hematite > goethite (Lee *et al.* 2006; Matta *et al.* 2008). It is also generally observed that the efficiency of conversion of H_2O_2 to an oxidant is inversely proportional to the H_2O_2 reaction rate (Pham *et al.* 2012). The role of the Fe^{II} sites in magnetite has been demonstrated by isomorphous substitution of the Fe^{II} with Co^{2+}, Mn^{2+}, and Cr^{3+}, leading to more effective catalysts, or with Ni^{2+} (which is redox inert), leading to an inert material (Costa *et al.* 2006; Magalhães *et al.* 2008); the inclusion of Cr^{3+} can also alter the kinetics of the process by changing the degree of adsorption (Liang *et al.* 2011). The presence of the Fe^{II} surface sites on magnetite (and not the homogeneous chemistry) are integral to formation of $^{\bullet}OH$, with oxidation of magnetite eventually leading to hematite (Ardo *et al.* 2015; He *et al.* 2014); Fe^0 can be incorporated into the catalyst to reductively regenerate magnetite and maintain the surface reactivity for longer (Costa *et al.* 2008).

The supported heterogeneous catalysts, $\mathbf{Fe^{III}_{(ads)}}$, consist of Fe(III) adsorbed onto microparticulate materials such as Fe-ZSM-5 (a synthetic zeolite), Fe/Montmorillonite, pillared interlayered clays (containing aluminium oxide/iron oxide pillars), and γ-Fe_2O_3/Y zeolite (Garrido-Ramírez *et al.* 2010). The ability for the support to strongly adsorb and hence concentrate contaminants near the active site can also enhance the process, and has been effectively employed using iron-impregnated zeolites to adsorb and then oxidize methyl *tert*-butyl ether (Gonzales-Olmos *et al.* 2013). In addition, nZVI can be used in the form $\mathbf{Fe^0}_{\mathbf{(ads)}}$, e.g., laponite clay-based Fe nanocomposites (Feng *et al.* 2006) and kaolin-supported nZVI (Liu *et al.* 2014).

Nanocatalysts have high surface areas and low diffusional resistance; hence they are more efficient than conventional heterogeneous catalysts made from pure microparticulate minerals (Garrido-Ramírez *et al.* 2010).

7.2.2 Extended Fenton processes

Several types of external non-chemical energy sources have been used in conjunction with Fenton reagents. These include electromagnetic radiation sources (such as ultraviolet (UV) and near UV light, microwave (Yang *et al.* 2009), and gamma radiation), ultrasonic radiation (Torres *et al.* 2007), and pulsed electron beams. In addition, some or all of the Fenton reagents can be produced in situ by electrochemical reactions. The types of extended Fenton processes in which the external energy source is UV-visible light (photo-Fenton), electrical energy (electro-Fenton), and ultrasound (sono-Fenton) are described below. Combinations of the extended Fenton processes, known as hybrid Fenton processes, are also possible, e.g., the sono-electro-Fenton process, sono-photo-Fenton process, and photo-electro-Fenton process (Nidheesh *et al.* 2013).

7.2.2.1 Homogeneous photo-Fenton processes

$\mathbf{Fe^{II}}_{\mathbf{(aq)}}\mathbf{/H_2O_2/UV}$

In these systems the standard chemistry observed in $Fe^{II}_{(aq)}/H_2O_2$ systems is operative, however, the presence of UV light enhances the process in two main ways. Firstly, UV can photolyse $Fe^{III}_{(aq)}$, leading to reduction of this species to $Fe^{II}_{(aq)}$ as well as the concomitant production of $^{\bullet}OH$ directly; the reactivity of the $Fe(OH)^{2+}$ form is highest, leading to a distinct pH-effect on this process.

$$Fe^{3+} \rightsquigarrow Fe^{2+} + {}^{\bullet}OH \tag{7.10}$$

$$Fe(OH)^{2+} \rightsquigarrow Fe^{2+} + {}^{\bullet}OH \tag{7.11}$$

The quantum yield of Fe^{II} production in reaction 7.11 has been estimated by Faust and Hoigné (1990) to be $\Phi(Fe^{II}) = 0.14$ at 313 nm. H_2O_2 is also homolysed directly to two $^{\bullet}OH$ with $\Phi(^{\bullet}OH) = 1$ (i.e., ~50% efficiency) (Yu, 2004).

$$H_2O_2 \rightsquigarrow 2\,{}^{\bullet}OH \tag{7.12}$$

It is also possible that UV light can directly degrade some contaminants with active photochemistries.

$\mathbf{Fe^{III}}$ – $\mathbf{ligand_{(aq)}/H_2O_2/h\nu}$

The addition of a ligand to the photo-Fenton system can both enhance the solubility of Fe^{III} at higher pH and increase the range of light which is photochemically active, allowing more efficient use of visible light to drive the process. The photoreduction quantum yield can also increase by an order of magnitude when Fe^{III} is complexed with a carboxylic acid anion such as oxalate. For example, $\Phi_{Fe(II)} = 1.24$ at 313 nm, with 6 mM ferrioxalate (Murov *et al.* 1993). The ferrioxalate complex $[Fe(C_2O_4)_3]^{3-}$ is highly photoreactive with the reduction of Fe(III) to Fe(II) extending into the visible spectrum (about 550 nm) (Hislop & Bolton, 1999); this compound is sufficiently photosensitive to be employed as a chemical actinometer for low light-fluxes (Hatchard & Parker, 1956).

The reaction sequence of ferrioxalate is representative of that of many ligands. It is initiated by the absorption of light leading to excitation via a ligand-to-metal charge-transfer, resulting in an excited charge-transfer state (reaction 7.13), followed by decomposition of this excited complex, producing Fe(II) and an oxalate radical (reaction 7.14) (Balmer & Sulzberger 1999).

$$Fe^{III}\left(C_2O_4\right)_n^{(3-2n)} \rightsquigarrow Fe^{II}\left(C_2O_4^{\bullet}\right)\left(C_2O_4^{\bullet}\right)_{n-1}^{(3-2n)*} \tag{7.13}$$

$$Fe^{II}\left(C_2O_4^{\bullet}\right)\left(C_2O_4\right)_{n-1}^{(3-2n)*} \rightarrow Fe^{II} + \left(n-1\right)C_2O_4^{2-} + C_2O_4^{\bullet-} \tag{7.14}$$

The oxalate radical can either decompose into a carbon dioxide radical ($CO_2^{\bullet-}$) and CO_2 (reaction 7.15), or it can react with molecular oxygen to produce $O_2^{\bullet-}$ and CO_2 (reaction 7.16). Under oxygenated conditions the $CO_2^{\bullet-}$ will rapidly react with O_2 yielding $O_2^{\bullet-}$ and CO_2 (i.e., same result as for reaction 7.16). $CO_2^{\bullet-}$ may also reduce further Fe^{III} (reaction 7.17).

$$C_2O_4^{\bullet-} \rightarrow CO_2^{\bullet-} + CO_2 \tag{7.15}$$

$$C_2O_4^{\bullet-} + O_2 \rightarrow O_2^{\bullet-} + 2CO_2 \tag{7.16}$$

$$CO_2^{\bullet-} + Fe^{III}\left(C_2O_4\right)_3^{3-} \rightarrow Fe^{II} + CO_2 + 3\,C_2O_4^{2-} \tag{7.17}$$

The superoxide can disproportionate to hydrogen peroxide, oxidize Fe^{II} (reaction 7.18), or reduce Fe^{III} (reactions 7.18 and 7.20) (Balmer & Sulzberger, 1999; Hislop & Bolton, 1999).

$$2\,HO_2^{\bullet}/O_2^{\bullet-} \rightarrow H_2O_2 + O_2 \tag{7.18}$$

$$Fe^{II} + HO_2^{\bullet}/\,O_2^{\bullet-} \rightarrow Fe^{III} + H_2O_2 \tag{7.19}$$

$$Fe^{III}\left(C_2O_4\right)_3^{3-} + O_2^{\bullet-} \rightarrow Fe^{II} + O_2 + 3\,C_2O_4^{2-} \tag{7.20}$$

The greater than one quantum yield for Fe^{II} formation is due to the possibility of the $C_2O_4^{\bullet-}$ radical (or its subsequent intermediate transformation products) being able to further reduce Fe^{III} to Fe^{II}. In the presence of sufficient H_2O_2, this may translate to a quantum yield for $HO^{\bullet}$ formation also greater than one if oxidation by H_2O_2 is the main sink for Fe^{II}. In the presence of excess oxalate, the Fe^{III} formed from oxidation of Fe^{II} will reform ferrioxalate, leading to a catalytic cycle.

The inherent production of a reduced ligand radical can be beneficial for compounds that are resistant to oxidation, such as perchloroalkanes. It is possible to sensitise the removal of the halogen atoms by reduction mediated by the $CO_2^{\bullet-}$ (in the absence of H_2O_2). The reaction product may then be oxidized by adding H_2O_2 to generate $^{\bullet}OH$ radicals (Huston & Pignatello, 1996).

The initial oxidation of organic compounds usually produces oxygenated intermediates with carboxylic functional groups. These can form complexes with Fe(III) that are photoactive, producing CO_2, organic radicals, and Fe(II) on irradiation. Hence, the organic compounds can be degraded without the participation of $^{\bullet}OH$ in some cases (Safarzadeh-Amiri *et al.* 1996).

Oxalate is only one of many ligands employed in this process, with the type of ligand used influencing the reactivity and efficiency of the process. The photoreactivity of Fe^{III} EDTA has been studied in detail by Kocot *et al.* (2007), and shown to proceed in the presence of an oxidizable substrate D through reactions 7.21 and 7.22.

$$\left[\mathrm{Fe^{III}(EDTA)(H_2O)}\right]^{-} \rightsquigarrow \left[\mathrm{Fe^{II}(EDTA^{\bullet})(H_2O)}\right]^{-*} \tag{7.21}$$

$$\left[\mathrm{Fe^{II}(EDTA^{\bullet})(H_2O)}\right]^{-*} + \mathrm{D} \rightarrow \left[\mathrm{Fe^{III}(EDTA)(H_2O)}\right]^{-} + \mathrm{D^{\bullet+}} \tag{7.22}$$

There is also a competing pathway involving reaction of the excited state with additional $Fe^{III}EDTA$, leading to breakdown of the ligand (to ethylenediaminetriacetic acid, ED3A) as well as formation of $Fe^{II}EDTA$, which can undergo reaction with H_2O_2 forming an Fe^{II}-ligand$_{(aq)}$/H_2O_2 system:

$$\left[\mathrm{Fe^{II}(EDTA^{\bullet})(H_2O)}\right]^{-*} + \left[\mathrm{Fe^{III}(EDTA)(H_2O)}\right]^{-} \rightarrow \left[\mathrm{(EDTA^{\bullet})Fe^{II}(\mu - OH_{\mathit{x}})Fe^{III}(EDTA)}\right]^{x-4} \tag{7.23}$$

$$\left[\mathrm{(EDTA^{\bullet})Fe^{II}(\mu - OH_{\mathit{x}})Fe^{III}(EDTA)}\right]^{x-4} \rightarrow \left[\mathrm{Fe^{II}(EDTA)(H_2O)}\right]^{2-} + \mathrm{Fe^{II}} + \mathrm{ED3A} + \mathrm{CO_2} + \mathrm{H_2CO} \tag{7.24}$$

In many applications the oxidation of the chelating ligand may be considered an advantage provided it keeps the iron dissolved and catalytically active long enough for the desired extent of degradation of the contaminant to be achieved. This is because it removes the potential requirement for the ligand to be removed by other processes (Sun & Pignatello, 1992). The use of similar ligands that are inherently biodegradable has also been developed successfully, with EDDS being a suitable material (Wu *et al.* 2014).

Fe^{III}-ligand$_{(aq)}$/hν

Although addition of H_2O_2 substantially accelerates the degradation of contaminants during the photochemical reaction of Fe^{III}-ligand, as $O_2^{\bullet-}$ is formed in many of these processes, eventually sufficient H_2O_2 may build up to levels sufficient to perform this role without need for external additions.

7.2.2.2 Heterogeneous photo-Fenton processes

Heterogeneous photo-Fenton process can usually be carried out using iron catalysts in the supported form only, as unsupported solid iron catalysts tend to block the light path.

The supported heterogeneous catalysts that can be used in the presence of light include adsorbed Fe(III) ions and Fe(III) oxides. For **$Fe^{III}_{(ads)}$**, the ferric ions are immobilized by ion-exchange onto perfluorinated membranes (Fernandez *et al.* 1999) or perfluorinated membranes/glass fibers (Dhananjeyan *et al.* 2001a), both of which make use of expensive Nafion® membranes due to their resistance to degradation by hydroxyl radicals. The ferric ions can also be immobilized by ion-exchange onto structured silica fabrics (Bozzi *et al.* 2003). In the case of **$Fe^{III}oxide_{(ads)}$**, Fe_2O_3 can be supported on polyethylene copolymers (Dhananjeyan *et al.* 2001b), which is a less expensive polymer than Nafion®, but is also less resistant to degradation by hydroxyl radicals.

7.2.2.3 Electro-Fenton processes

Two types of Fenton process involve the use of in-situ electrochemically produced reagents, differing mainly in the form in which the iron enters the system. In cathodic Fenton processes, the iron is added as an Fe(II) or Fe(III) salt, whereas in anodic Fenton processes, the source of the iron is a sacrificial iron (Fe(0)) anode.

Cathodic Fenton processes: In cathodic processes, one or both of the reagents Fe(II) and H_2O_2 are produced in situ. The source of Fe(II) may be direct Fe(II) addition (via addition of a ferrous salt or, as recently suggested by Ammar *et al.* (2015), a heterogeneous source such as pyrite) or reduction of Fe(III) at the cathode (Eq. 7.25). The source of H_2O_2 may be either direct H_2O_2 addition or reduction of dioxygen at the cathode (Eq. 7.26) (Brillas *et al.* 1996; Ventura *et al.* 2002).

$$Fe^{III} + e^- \rightarrow Fe^{II} \tag{7.25}$$

$$O_2 + 2H^+ + 2e^- \rightarrow H_2O_2 \tag{7.26}$$

Simultaneous Fe(III) reduction and O_2 reduction can take place at the cathode with rates dependent on the concentration of Fe(III) present and the supply of O_2 to the cathode, with models developed to successfully describe the O_2 binding and reaction by Liu *et al.* (2007a). Since Fe(II) and H_2O_2 are continuously produced at controlled rates, a more efficient and more complete degradation of the contaminant can be achieved than with classic Fenton systems, where all the reagents are typically added at the beginning of the reaction. This is because competitive reactions, which consume the reagents without producing hydroxyl radicals, are less significant in the electrochemical process (Ventura *et al.* 2002; Duesterberg *et al.* 2006). Additionally, the rate of heterogeneous reduction of Fe(III) at the electrode surface is more rapid than the solution-phase H_2O_2-mediated reduction of Fe(III) in the homogeneous Fenton process (Qiu *et al.* 2015) (see Figure 7.1).

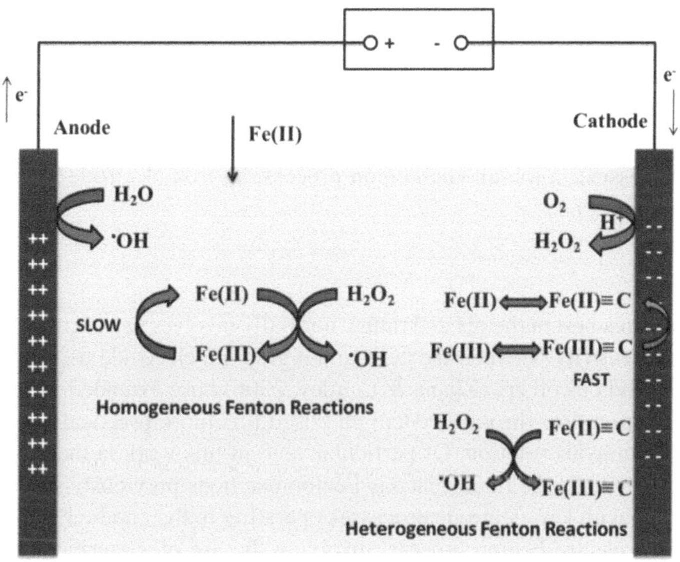

Figure 7.1 Schematic showing the homogeneous and heterogeneous (electrode located) Fenton reactions responsible for contaminant degradation. Note that Fe(III) ≡ C represents an Fe(III) species attached to the carbonaceous cathode surface (adapted from Qiu *et al.* 2015).

The electrochemical cell in most implementations is undivided and typically has an anode made from inert material such as platinum or platinised titanium. Anodes covered with a boron doped diamond (BDD) film have also been used as the extent of oxidation of contaminants may be enhanced either by hydroxyl radicals produced directly at the BDD surface as a result of water oxidation (Eq. 7.27) or by an increase in H_2O_2 generation rate as a result of hydroxyl radical dimerization at the BDD surface (Eq. 7.28) (Brillas, 2011), i.e.,

$$BDD + H_2O \rightarrow BDD\,(^{\bullet}OH) + H^+ + e^- \tag{7.27}$$

$$2BDD\,(^{\bullet}OH) \rightarrow BDD + H_2O_2 \tag{7.28}$$

The cathode is made from carbon-containing materials such as graphite, carbon felt (Oturan, 2000) or carbon-polytetrafluoroethylene (Brillas *et al.* 2000) and is supplied with oxygen. Recent developments in this area include use of a granular activated carbon packed bed electrode with diffusion of oxygen through the packed bed (Banuelos *et al.* 2014).

The nature of H^+ consumption or production in the electrode reactions can also be exploited to control the pH in a multi-compartment process. Liu *et al.* (2007b) have demonstrated that if the main anode and cathode are separated by an agar salt bridge, H^+ production in an initial anodic chamber can drop the pH of neutral solutions to ~3.5, with transfer to a second combined-chamber with both cathode and anode for the main treatment stage. The solution is finally transferred to the compartment containing the cathode coupled to the first compartment, with H^+ consumption raising the pH to circumneutral levels, appropriate for discharge or further treatment processes, with the added advantage of precipitating the Fe for easy removal or recycling back to the main combined reactor.

The electro-Fenton systems may also be exposed to irradiation by UV/visible light, leading to photoelectro-Fenton processes (Boye *et al.* 2002). This approach has been shown to be particularly advantageous where either the original contaminant (such as dyes) or the degradation products of the original contaminant (such as Fe(III) complexes with carboxylic acids) are chromophores (Brillas and Martinez-Huitle, 2015; Olvera-Vargas *et al.* 2015).

Anodic Fenton processes: In the anodic Fenton process, an iron electrode is used as the anode and becomes the source of Fe(II), i.e.,

$$Fe^0 \rightarrow Fe^{2+} + 2e^- \tag{7.29}$$

The process, which operates best in the pH 2–3 range, normally involves use of two cells, with a salt bridge providing electrical connectivity between the cells and a graphite electrode used as a cathode (Wang & Lemley, 2001). Lemley and coworkers (Zhang & Lemley, 2006) have extended their earlier batch anodic Fenton treatment (AFT) to a flow-through system and used this more practical approach to degradation of various pesticides in aqueous solution. Of particular note in this work is the development and use of a dynamic model that combines the homogeneous Fenton reactions previously successfully modelled by Duesterberg *et al.* (2005) with key electrode processes operating in the anodic Fenton system.

A modification of the electro-Fenton process, involving the use of a sacrificial iron anode, has been developed by Brillas and Casado (2002). This process makes use of an undivided electrochemical cell. Since the solution cannot be maintained in the acidic pH range, this process is characterised by precipitation of iron hydroxides.

7.2.2.4 Sono-Fenton processes

H_2O_2 is produced during the standard sonochemical treatment process, hence the addition of Fe(II) can further enhance this process by converting this H_2O_2 to $^{\bullet}OH$, thereby accelerating contaminant destruction (Jiang & Waite, 2006; Torres *et al.* 2007).

Ma (2012) has reviewed the use of the sono-Fenton process for water treatment. The ultrasound frequency has a major effect on the success of the process. Increasing the ultrasonic frequency increases the rate of formation of active cavitation bubbles, leading to the generation of more $^{\bullet}OH$, however it also increases the rate of collapse of these cavitation bubbles. This can lead to the presence of a higher concentration of $^{\bullet}OH$ at lower ultrasonic frequencies. The optimum pH range is 2–4, with inhibition of $^{\bullet}OH$ formation taking place at lower pH values and loss of iron from the system due to precipitation of Fe(III) hydroxides taking place at higher pH values. This is the same optimum pH range as that for the classical Fenton process; hence the use of ultrasound does not extend the pH range. The main advantage of using the sono-Fenton process over ultrasound on its own is that the cost of sono-Fenton process is much lower due to shorter treatment time required by the sono-Fenton process.

7.2.3 Fenton-like processes

7.2.3.1 Fenton-like homogeneous reagents

A wide variety of other trace metals can also react with H_2O_2 to yield useful reactive oxidants (which may or may not be $^{\bullet}OH$); it is also possible to generate reactive oxidants employing hypochlorite (HOCl), peroxydisulfate ($S_2O_8^{2-}$), or even organic hydroperoxides (ROOH, of particular relevance to atmospheric chemistry) in the place of H_2O_2 (Chevallier *et al.* 2004; Lister, 1956; Walling & Camaioni, 1978). Commonly used trace metals include Cu^{I}, Co^{II}, Mn^{II}, and Ti^{III} (Armstrong, 1969; Bandala *et al.* 2007; Watts *et al.* 2005). The reactive intermediate of the Cu^{I}-H_2O_2 system is likely to be a Cu^{III} species and not the $^{\bullet}OH$ radical under circumneutral conditions (Lee *et al.* 2013; Pham *et al.* 2013).

7.2.3.2 Fenton-like heterogeneous reagents

As well as the reactions of Fe-containing minerals and H_2O_2, qualitatively similar processes occur with other transition metal minerals and/or with other oxidants, such as peroxydisulfate. For example, Ruan *et al.* (2015) have successfully employed a magnetite/$S_2O_8^{2-}$ system to degrade trichloroethylene and Yip *et al.* (2005) have developed a Cu-loaded clay catalyst and demonstrated its effectiveness for the degradation of the azo-dye acid black 1 with UV and H_2O_2 addition. Similar to Fe^{II}, Ag^{0} nanoparticles can also be oxidized by H_2O_2 to produce $^{\bullet}OH$, although only at acidic pH (He *et al.* 2014).

7.3 REACTION KINETICS AND PROCESS MODELLING

The set of plausible reactions and the rate at which they occur determines a system's performance. In almost all Fenton type systems a range of reaction pathways are possible, some desired and some deleterious to performance; the key to designing a successful process is understanding these reactions and being able to predict their importance under any given condition. Kinetic modelling can provide the tools to perform these assessments. In early studies on these mechanisms extensive numerical work and intuitive assumptions were required for all but the simplest reaction schemes (e.g., the work of Barb *et al.* (1951a, 1951b) and Merz & Waters (1947)), nonetheless, quantitative evaluation of a proposed reaction sequence was an invaluable tool to probe the veracity of a proposed scheme. These same approaches are also useful when applying these derived reaction schemes to design a water treatment solution, with modern computing resources greatly simplifying the implementation.

The main goal of Fenton processes is typically to produce $^{\bullet}OH$ that will then react with some contaminant, RH, in such a way that will mineralize it or at least degrade it to other products such that its problematic properties are eliminated. The immediate product of any such reaction is always a radical species itself, as shown in Eq. 7.30.

$$^{\bullet}OH + RH \rightarrow H_2O + R^{\bullet} \quad (7.30)$$

Although this initial reaction is an important step that often controls the kinetics and viability of removal, what subsequently happens to this short-lived organic radical on its pathway to eventually becoming a diamagnetic species, although often neglected, can also have significant impact on the process. There are excellent resources for compliations of rate constants for the initial reaction of $^{\bullet}OH$ with a large variety of substrates, primarily derived from radiation chemistry, such as the early compilation of Buxton *et al.* (1988) and the updated compilation from Notre Dame Radiation Laboratory now available from the National Institute of Standards and Technology (http://kinetics.nist.gov/solution/). The reactivity of the radicals that are formed is, however, often not as well known, although the reactivity of major classes of compounds is reasonably well-defined (von Sonntag *et al.* 1997; Pignatello *et al.* 2006).

Compilations of the reactions of carbon-centred radicals and peroxyl radicals (the typical product from the reaction of carbon-centred radicals with O_2) have also been published by Neta *et al.* (1990, 1996) and are instructive for determining the likely viability of reactions. Unfortunately, similar information is much less readily available for the reaction of most of these radical species with Fe^{II} or Fe^{III} (complexed or otherwise), including those interactions that could possibly impact the process performance under some conditions.

Determination of the degradation pathway of even a simple compound can be very challenging. As an illustrative example, the reaction mechanism for 2-methylpropan-2-ol (TBA) will be examined. As shown in Figure 7.2, even though there are only two types of hydrogen atom available to react with $^{\bullet}OH$, with only one of these being signifcant, the resulting mechanism is quite complex. After the formation of the initial carbon-centred 2-hydroxy-2-methylpropyl radical, three reaction pathways are possible: dimerization to a diol, reduction by Fe(II), or oxidation by O_2 to a peroxyl radical. Depending upon the redox potential of the initially formed carbon-centred radical, its oxidation by Fe(III) may be more important than reduction of this initial radical by Fe(II). The relative importance of each process depends upon both the rate constant and concentration of the reactant; dimerization requires a high radical concentration, typically limiting it to situations such as pulse radiolysis, not typical treatment processes. As the rate constant for the reaction with O_2 to form a peroxyl radical is ~10^3 times as large as that for the reaction with Fe(II)L, formation of the peroxyl radical will dominate for Fe(II)L concentrations below ~0.25 M, i.e., nearly all relevant conditions.

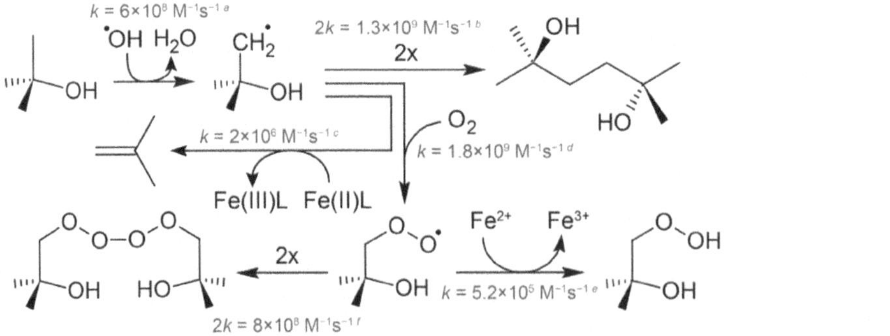

Figure 7.2 Oxidation of TBA by $^{\bullet}OH$ and subsequent reactions. [a]Buxton *et al.* (1988), [b]Simic *et al.* (1969), [c]rate constant for Fe(II)EDTA from Croft *et al.* (1992), nature of product from Eberhardt (1984), [d]Reisz *et al.* (2003), [e]Khaikin *et al.* (1996), [f]Schuchmann and von Sonntag (1979).

It is seen that the 2-hydroxy-2-methylpropyl radical is able to oxidize Fe(II), which, as this prevents Fe(II) from reacting with H_2O_2 to form more $^\bullet OH$, is a chain-terminating process. Alternatively, reducing radicals such as $^\bullet CH_2OH$ are able to reduce Fe(III) to Fe(II) (Buxton & Green, 1978), thus regenerating Fe(II), which is then able to form more $^\bullet OH$ to propagate the chain. The presence of chain propagating or terminating species can greatly influence the efficiency of a Fenton process (Merz & Waters, 1949; Duesterberg & Waite, 2006).

The fate of the 2-hydroxy-2-methylpropylperoxyl radical will depend upon both the steady-state concentration of the species, influenced by its production rate (which is governed by the production rate of the $^\bullet OH$ radical), as well as the Fe^{2+} concentration. Because the concentration of the peroxyl radical will generally be orders of magnitude lower than that of Fe^{2+}, both processes can still compete under many relevant conditions even though the coupling rate constant is ~10^3 times greater. The product of the coupling reaction is a short-lived tetroxide species (see Figure 7.2), which can decompose by four main pathways (Figure 7.3). Under the Russell-type pathway O_2 is liberated as well as an aldehyde and an alcohol; under the Bennett-type reacion H_2O_2 and two aldehydes are formed; O_2 can be released to form a dialkyl-peroxide, or, the tetroxide can homolytically break down to O_2, formaldehyde, and the 2-hydroxypropyl radical. Although alkyl hydroperoxides have been shown to be capable of promoting a Fenton-like reaction (Chevallier *et al.* 2004), dialkyl peroxides are unreactive (Sawyer *et al.* 1996).

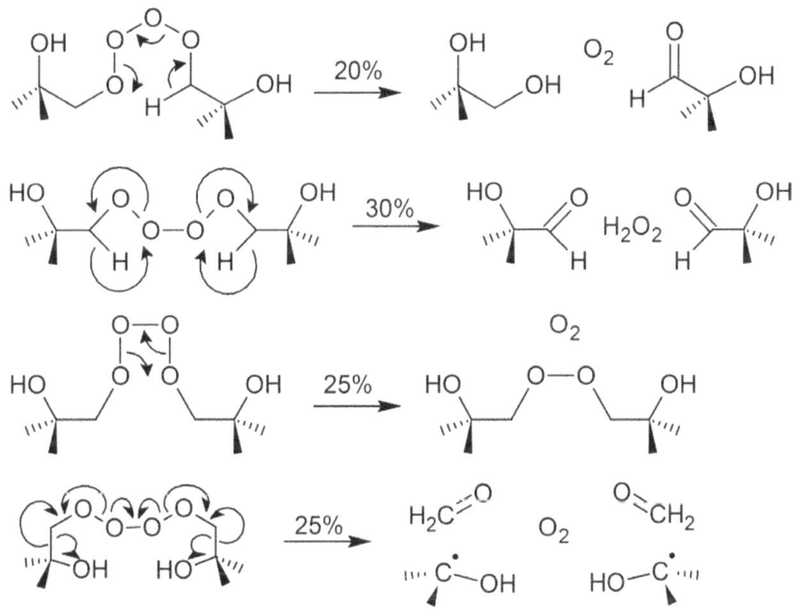

Figure 7.3 Degradation of tetroxide formed from dimerization of 2-hydroxy-2-methylpropylperoxyl radical (Reisz *et al.* 2003).

The reactions of the 2-hydroxypropyl radical are shown in Figure 7.4; unlike the 2-hydroxy-2-methylpropyl radical, this species is better able to reduce Fe^{3+} than to oxidize Fe^{2+}, so it is therefore a chain propagating radical. However, under oxic conditions, the concentration of Fe^{3+} needs to be at least 10 times greater than that of O_2 for reactions with Fe to become significant, because the reaction with

O_2 forms an α-hydroxyperoxyl radical, which then rapidly decomposes to acetone and $O_2^{\bullet-}$. Since $O_2^{\bullet-}$ is itself a good Fe^{3+} reductant (Rush & Bielski, 1985), the overall impact is still of a chain-propagating nature.

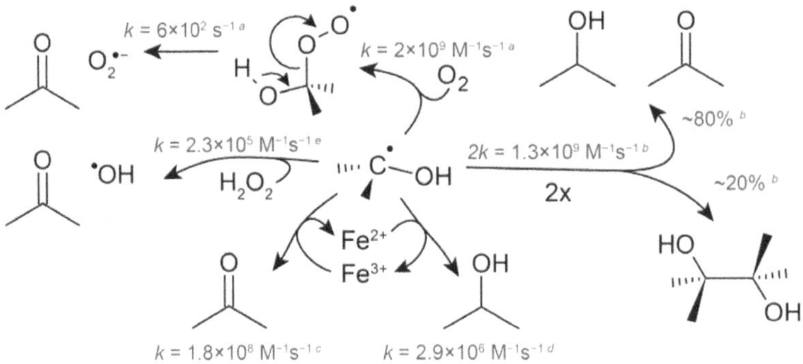

Figure 7.4 Typical reactions undertaken by the isopropanol radical. [a]Reisz *et al.* (2003), [b]Lehni and Fischer (1983), [c]Berdnikov *et al.* (1977), [d]Pribush *et al.* (1975), [e]Kishore *et al.* (1987).

Under a scenario where reactions with O_2 dominates the fate of all radical species, this yields the final products shown in Figure 7.5. As the process progresses, each of these products will then also start to become important sinks for $^{\bullet}OH$ as their concentrations increase.

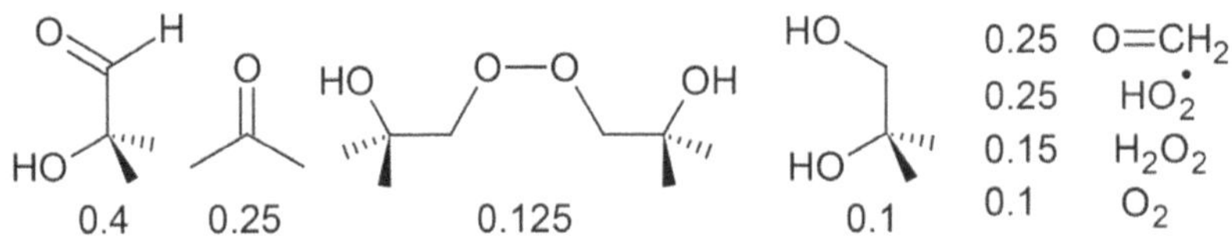

Figure 7.5 Final products of reaction of 1 mole of 2-methylpropan-2-ol with $^{\bullet}OH$ and sufficient O_2. The values given are the apparent stoichiometric coefficients per mole of 2-methylpropan-2-ol reacted derived from the mechanism presented above.

The ability to precisely determine (and model) the breakdown mechanism of any arbitrarily-complex contaminant, especially when present as part of a more complex mixture, is clearly unattainable in the foreseeable future. Nonetheless, useful insight can still be obtained from consideration of the potential, general, pathways. The reactions of aromatic compounds are commonly of interest in Fenton oxidation processes, and are generally more complicated than the aliphatic case presented above. A common pathway for aromatic oxidations involves addition of $^{\bullet}OH$ to the aromatic ring to form a hydroxycyclohexadienyl radical, which can then reversibly add O_2 to form a peroxycyclohexadienyl radical (Dorfman *et al.* 1962; Fang *et al.* 1995). This process, shown in Figure 7.6, is generally applicable for many aromatic compounds, with the rate constants for each step of the process being influenced by the nature of the ring substituents, and a Hammett correlation being found suitable to describe the reactivity for simple compounds (Neta & Dorfman, 1968).

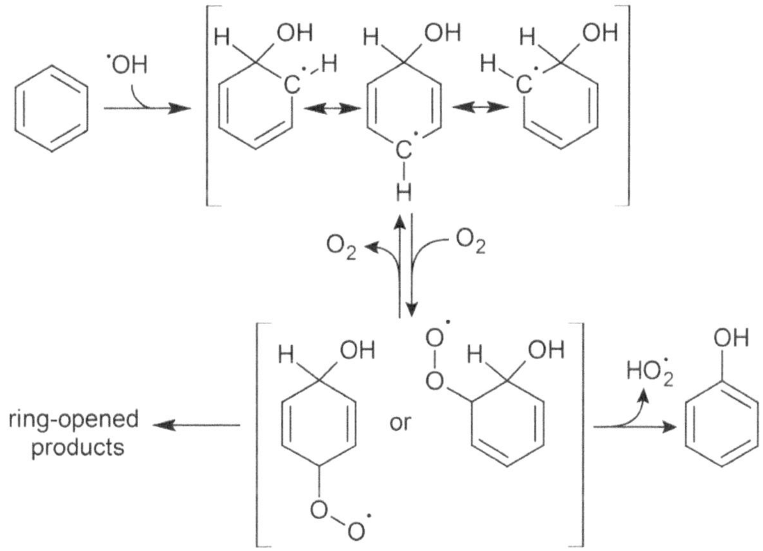

Figure 7.6 Initial steps in the oxidation of benzene by •OH including the formation of the hydroxycyclohexadienyl radical and its subsequent oxidation by oxygen (via a peroxyl radical intermediate) to yield either phenol or ring-opened products.

Although the mechanistic details of contaminant destruction are quite complex and influenced by a myriad of operating parameters, application to real world systems is both attainable and of value in system optimisation. Duesterberg *et al.* (2005) examined the $Fe^{II}_{(aq)}/H_2O_2$ system under acidic conditions for the degradation of formic acid, developing a kinetic model explicitly including the reactivity of all intermediates formed. It was demonstrated that the nature of the radical species (in this system $CO_2^{\bullet-}$) formed after oxidation of the organic contaminant and the presence or absence of O_2 (which competes with Fe^{III} for $CO_2^{\bullet-}$) determined whether this intermediate radical was able to promote redox-cycling of Fe. Further examination of the importance of pH and the presence of chain-propagating and -terminating organic radicals was undertaken by Duesterberg and Waite (2006), with the substantial decrease in performance in the presence of TBA quantitatively accounted for by consideration of the formation of the relatively inert (at least towards Fe) 2-hydroxy-2-methylpropyl radical.

Duesterberg and Waite (2007) extended this work by considering the full mechanistic pathway for degradation of *p*-hydroxybenzoic acid and developed a kinetic model containing 49 reactions (see the scheme in Figure 7.7) that was able to adequately model the observed loss of this material and appearance of degradation intermediates. The modelling approach also allowed for the important role of the quinone and quinone-like intermediates in driving the redox-cycling of Fe to be identified. This depth of understanding allows the system to be readily optimized for contaminant destruction. Extending this approach to progressively more complex compounds is possible through determination of the breakdown of a material to this particular compound (as well as dealing with the reactivity of any other unique fragments that form along this pathway). Future work advancing this mechanistic whole-of-system approach towards Fenton processes to progressively more complex and realistic target contaminants would seem of value to both aid process optimization and allow for better system rectification when performance is not as expected.

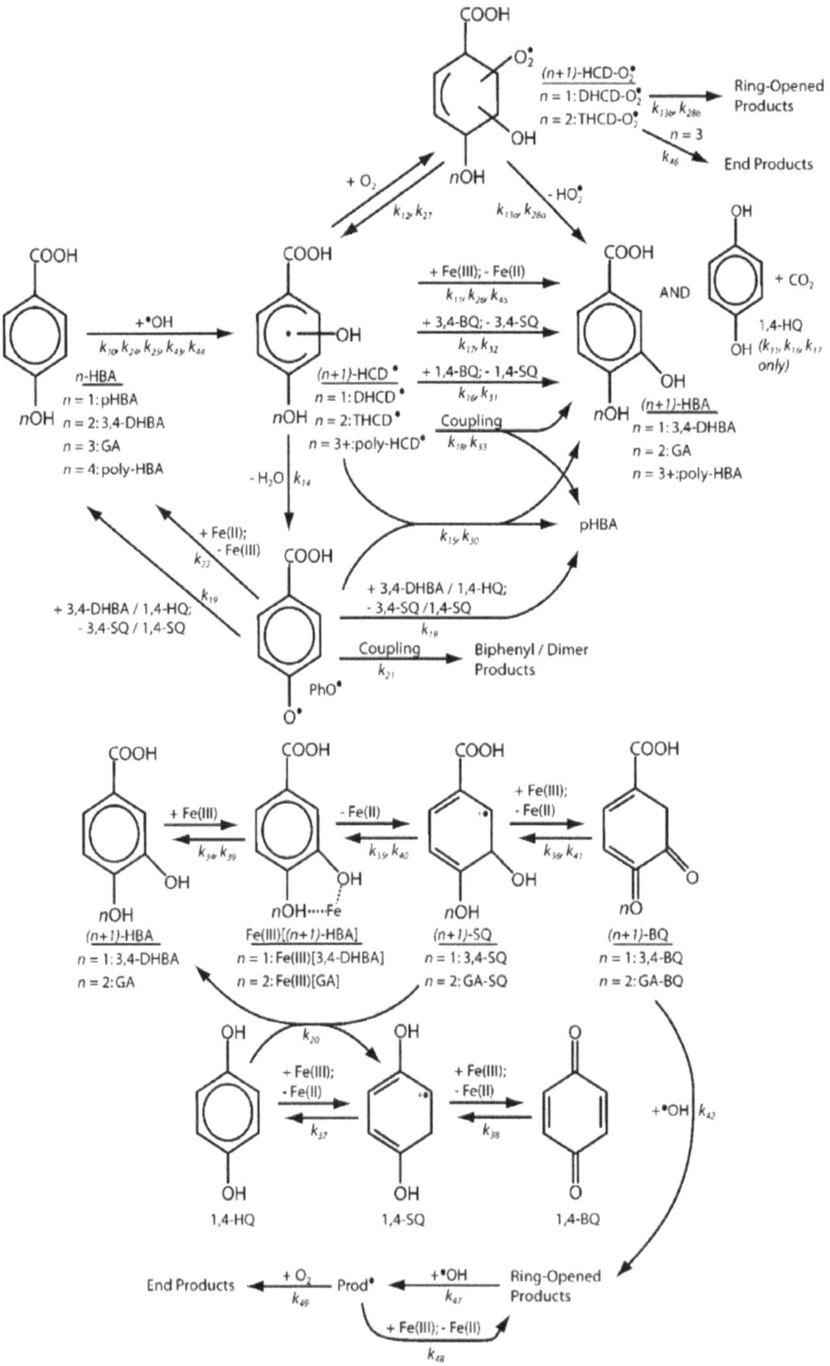

Figure 7.7 Reaction scheme for modelling the Fenton oxidation of *p*-hydroxybenzoic acid. Reprinted from Duesterberg and Waite (2007) with permission. Copyright 2007 American Chemical Society.

7.4 APPLICATIONS AND IMPLICATIONS

The Fenton and photo-Fenton processes (and modifications thereof) are the most well-known applications of Fenton type chemistry. The photo-Fenton process is more suited to industrial applications than the dark Fenton process because the light significantly increases the rate of reaction, so it can be complete within a few minutes rather than several hours. (Photo-)Fenton reactions also play a major role in the photochemical and dark degradation of organic matter (both natural and synthetic) in water droplets in the atmosphere, and in sunlit surface freshwater and seawater (Southworth & Voelker, 2003). As well as acting as a sink for $^{\bullet}OH$, thus mineralizing natural organic matter (NOM) in the process, the complexation of Fe^{II} by NOM (or other organic ligands) is also necessary for production of $^{\bullet}OH$ by the Fe^{II}/H_2O_2 under circumneutral conditions, with uncomplexed Fe^{II} forming an oxidant that is not $^{\bullet}OH$ (Miller *et al.* 2013); the presence of NOM can also facilitate extensive cycling between Fe(II) and Fe(III) under natural conditions, increasing $HO^{\bullet}$ production (Burns *et al.* 2010). Fenton type reactions are also believed to play important roles in human health and disease pathology (Halliwell & Gutteridge, 1990). The reaction rates tend to be considerably lower in natural systems than in industrial applications due to lower concentrations of H_2O_2 and Fe^{2+}, the lower intensity of natural sunlight, obscuring of light by natural features of the system, the presence of competing ions, non-optimal pH conditions, and inadequate mixing. In natural system the success of the photo-Fenton process depends on having large surface areas open to the sunlight and the availability of long residence times. Nonetheless, photo-Fenton processes still play important roles in natural systems, and have been implicated as a major pathway for oxidation of methylmercury to less-toxic products (Hammerschmidt & Fitzgerald, 2010).

7.4.1 Treatment objectives

A Fenton type process can have one or more of the following roles in an overall wastewater treatment system:

1. Degrade the target organic compounds into other organic compounds
2. Mineralize the target organic compounds
3. Decrease the toxicity of the wastewater
4. Improve the biodegradability of the organic load of the wastewater
5. Decrease the biological oxygen demand (BOD) or chemical oxygen demand (COD) of the wastewater
6. Decrease the total organic carbon (TOC)
7. Decolourise the wastewater
8. Deodorise the wastewater

When considering how a Fenton type process would fit into an overall water treatment system, some of the following additional processes may be required:

- *Pre-treatment processes, including:*
  - Filtration, dissolved air floatation, or coagulation and settling to remove suspended solids,
  - Treatment using other degradation processes or separation processes,
  - pH adjustment and decarbonation, and
  - Combination with a different type of waste stream.
- *Post-treatment processes, including:*
  - pH adjustment,
  - Sludge separation, and
  - Further treatment process.

Depending on its composition, the final treated water may be suitable for:

- Reuse in an industrial process (either the original process or a different process),
- Reuse in agricultural irrigation,
- Disposal to an industrial or municipal sewerage works,
- Disposal into natural water bodies, or
- Disposal into lined evaporation ponds.

7.4.2 Types of compounds suited to treatment

A wide range of organic contaminants have been degraded by Fenton type processes. Some of the most commonly studied compounds, many of which are classified as persistent organic pollutants (POPs), contaminants of emerging concern CECs, and priority substances, are as follows:

- Textile dyes (Kiwi *et al.* 2000)
- Leather tanning chemicals and by-products (Kurt *et al.* 2007)
- Formaldehyde (Guimaraes *et al.* 2012)
- Petroleum refinery by-products, including BTEX (benzene, toluene, ethylbenzene, and xylenes) (Da Silva *et al.* 2012)
- Phenols and phenolic compounds from petrochemical units, oil refineries, and the polymer and pharmaceutical industries, including the following priority substances: alkyl phenols (APs), alkylphenol ethoxylates (APEOs), and bisphenol A (BPA) (Gözmen *et al.* 2003; Molkenthin *et al.* 2013)
- Natural phenolic compounds from the wood pulping and food industry (Gernjak *et al.* 2003)
- Some chlorinated organic compounds, e.g., carbon tetrachloride (Che & Lee, 2011)
- Agrochemicals, including pesticides (insecticides, herbicides, biocides, fungicides) (Ballesteros Martín *et al.* 2008; Lapertot *et al.* 2006)
- Pharmaceuticals and personal care products (PPCPs) (Fatta-Kassinos *et al.* 2011)
- Surfactants (Lin *et al.* 1999)
- Aromatic amines (Brillas & Casado, 2002)
- Nitroaromatic explosives in groundwater and soil (Matta *et al.* 2007)
- Contaminants in landfill leachate (Deng & Englehardt, 2006)
- Contaminants in treated municipal wastewater (De la Cruz *et al.* 2012)

7.4.3 Process advantages

Fenton type processes have several advantages over most other advanced oxidation processes including higher reaction rates, lower capital and operating costs (Cañizares *et al.* 2007), and, in the case of homogeneous Fenton processes, the absence of mass transfer limitations caused by the presence of solids and/or gases in heterogeneous reactions. Iron is highly abundant and may be naturally occurring in the system being treated. Iron based catalysts are generally non-toxic and inexpensive, whereas many other potentially suitable metallic catalysts tend to be toxic and are more expensive than iron based catalysts. Hydrogen peroxide is relatively easy to transport and handle, as well as being environmentally benign in diluted form.

The light source for the photo-Fenton process can often be the sun (solar photo-Fenton). At ground level, the sunlight has a significant photon flux above 300 nm, while most photoactive iron species can absorb light up to around 400 nm (Zuo & Hoigne, 1992). Hence the wavelength range 300 to 400 nm can be used for the photo-Fenton process. Other advanced oxidation processes require ultraviolet light of lower wavelength, for example, the direct photolysis of contaminants in water usually takes place in the wavelength range 200 to 280 nm; hence sunlight is not suitable as a light source for direct photolysis.

The electro-Fenton process and sono-Fenton process require the use of more specialised and expensive equipment and have higher energy costs than the Fenton process or the solar photo-Fenton process.

7.4.4 Process limitations

The homogeneous Fenton process (in the absence of suitable complexing agents) is efficient only in the pH range 2 to 4. Hence, the process is inefficient in the pH range of most natural waters (pH 5 to 9) and biologically treated waters (pH 6 to 7). This is mainly due to precipitation of ferric oxyhydroxide, which has a low catalytic activity, at pH > 3 to 4. Hence pH adjustment is required before the Fenton process is started. The acidification step is also required to remove carbonate (usually by rapid stirring) because carbonates are hydroxyl radical scavengers.

After the Fenton process is complete, pH adjustment is usually required again before disposal or reuse of the treated water. This is an additional expense and leads to the generation of large volumes of iron sludge (Andreozzi *et al.* 1999). It has been found that when certain organic ligands that form complexes with Fe(III) are used the process can be carried out at a higher pH (Balmer & Sulzberger, 1999). This occurs both because the complex formation limits the loss of Fe(III) by ferric oxyhydroxide precipitation, and (in the case of the photo-Fenton process) because organically complexed Fe(III) is more efficiently photolysed than hydrated or inorganically bound Fe(III) (Andreozzi *et al.* 1999). The higher the concentration of the organic ligand, the greater the extent of Fe(III) complexation. However, since most organic compounds also react with radicals and other oxidants, increasing the concentration of these organic ligands can also lead to a decrease in the rate of oxidation of the target contaminant.

The use of organic compounds that can complex Fe(III) is an important modification of the process because the addition of large amounts of acid to achieve the optimum process efficiency, followed by base to neutralise the water after the oxidation is complete, makes the process unsuitable for many applications, due to the increase in salinity, as well as presenting operational problems such as the need to store and handle these potentially hazardous acids/bases on-site. Although the organic ligands that can be used are usually at least partially consumed in the process, they should also be biodegradable. Suitable ligands should bind Fe(III) strongly but allow transformation of Fe(III) to Fe(II) due to reaction with reductants (such as superoxide) produced in the process. If the ferric complex is photosensitive, light will enhance the formation of Fe(II) as a result of ligand-to-metal charge transfer-induced reduction of the metal centre. Complexing agents that have been used for this purpose in the pH range 3 to 9 include naturally occurring polycarboxylates such as oxalic acid and citric acid, and synthetic aminoplycarboxylic acids such as EDTA and EDDS. EDTA is not biodegradable and cannot be removed by chemical wastewater treatment processes. One of the three stereoisomers of EDDS, [S,S]-EDDS, is biodegradable and gives the best performace of the four complexing agents mentioned above (Huang *et al.* 2012).

The use of complexing agents does not solve the problem of iron remaining in solution after treatment and their use may prevent Fe precipitation by alkali or coagulant addition (if the ligands have not been degraded by the Fenton process). The amount of dissolved iron can be significantly decreased by the use of heterogeneous Fenton processes where the iron remains substantially in the solid phase, either as a mineral or as an adsorbed ion. Most of these heterogeneous processes have the added advantage that they can be operated in the close to neutral pH range.

The use of Fenton processes can lead to the complete mineralisation of some organic compounds, converting them to CO_2, H_2O, and inorganic ions. However, this would usually involve a large excess of chemicals, often preventing the process from being cost effective, hence only partial degradation usually occurs (Huston & Pignatello, 1999). Partial degradation usually reduces the toxicity of the contaminants and hence increases the biodegradability of the residue. However it is sometimes possible for the process to generate products with the same or higher toxicity than the parent compound (Fernández-Alba *et al.* 2002).

7.4.5 Laboratory and pilot plant scale studies

A large range of applications of Fenton type processes have been tested at the laboratory or pilot plant scale, particularly in the last 25 years. Pignatello *et al.* (2006) have published a comprehensive review of the processes that involve Fenton type chemistry. Most of the processes described here are ex-situ processes used to treat wastewater that can be stored in tanks prior to treatment. Some in-situ Fenton processes used to treat groundwater or soil are also described.

7.4.5.1 Homogeneous dark (thermal) Fenton processes

Pesticides: Atrazine (2-chloro-4-ethylamino-6-isopropylamino-*s*-triazine) is a commonly used herbicide that is toxic to animals and has a low biodegradability. In a study at pH 2 using various ratios of Fe^{II} to H_2O_2, Chan & Chu (2005) found that up to 98% of the atrazine (initially 10 μM) could be degraded in 60 min, with optimal conditions occurring for $[Fe^{II}] = 200$ μM and $[H_2O_2] = 600$ μM. A complex degradation pathway was found, which involved 10 organic intermediates in which atrazine had been fully or partially dealkylated and in some cases dechlorinated, but ring cleavage did not occur. The final degradation product in the degradation pathway is ammeline, which is more biodegradable than atrazine. Further treatment may be required to remove residual chloro-*s*-triazines before releasing the treated water into the environment (Arnold *et al.* 1995).

Textile dyes: Textile dyehouse effluents often contain high concentrations of azo dyes, which are strongly coloured and highly resistant to biodegradation, as well as a large range of other contaminants. Hydroxyl radicals tend to react with dyes very rapidly, for example for the reaction with the azo dye Orange II, the rate constant is 6.0×10^9 $M^{-1} \cdot s^{-1}$ (Kiwi *et al.* 2000). This reaction destroys the dye chromophore and forms uncoloured hydroxylated products; hence this water can often be reused for dyeing. Like most unmodified Fenton processes, the study by Kiwi *et al.* (2000) was carried out at $pH < 3$, however textile dyehouse effluents can range in pH from 4 to 12. Studies using Fe^{III} complexing agents have been carried out to extend the pH range of the Fenton process (Aplin, 2001). Several common and inexpensive Fe^{III} complexing agents can be used to allow operation of the Fenton process at pH 6 (Sun & Pignatello, 1993). Some more complex and expensive ligands have been found which allow operation at very high pH, for example, Fe^{III} complexes of tetra amido macrocyclic ligands (Fe^{III}–TAML) can be used to degrade Orange II when combined with H_2O_2 at pH 9 to 11 to produce non-toxic products (Chahbane *et al.* 2007).

7.4.5.2 Heterogeneous dark (thermal) Fenton processes

The main advantage of using heterogeneous Fenton processes is the avoidance of the large amount of Fe-containing sludge that is produced as a result of homogeneous Fenton processes, particularly the dark Fenton process. In some cases the solid phase acts as a source of Fe^{2+} ions, e.g., by the dissolution of ferrihydrite (Barreiro *et al.* 2007) or the desorption of Fe^{2+} ions from a resin. In other cases a true heterogeneously catalysed process may occur, in which the iron remains in the solid phase.

Heterocyclic aromatic compounds: Valentine & Wang (1998) found that goethite had the highest catalytic activity towards quinoline oxidation when using a heterogeneous Fenton process patented under the name IROX by Gurol and Lin (1998). The process makes use of a microparticulate mineral oxide catalyst (10 μm to 5 mm), usually goethite (α-FeOOH) and an oxidant, usually hydrogen peroxide. Alternative iron(III) oxide catalysts include lepidocrocite (γ-FeOOH), hematite (α-Fe_2O_3), limonite ($FeO(OH) \cdot nH_2O$), magnetite (Fe_3O_4) as well as other ferric oxides and hydroxy oxides. The catalytic efficiency increases with increasing surface area of the mineral. Goethite is almost insoluble in aqueous media over a wide pH range; hence it is able to function as a heterogeneous catalyst without losing a substantial amount of its mass.

7.4.5.3 Homogeneous photo-Fenton processes

Natural phenolic compounds: Naturally-occurring phenolic compounds from the food processing and wood pulping industry are not highly toxic, however, because they are present in high concentrations, they can kill the bacterial populations in biological wastewater treatment plants. Gernjak *et al.* (2003) found that these substances (1 mM) could be completely mineralized by the photo-Fenton process (with a medium pressure Hg lamp as the light source) within 1–3 h using a 12–24 times stoichiometric excess of H_2O_2 and 100 mM Fe(II). No non-degradable intermediates were formed.

Chlorinated aromatic compounds: Chlorinated aromatic compounds are found in many wastewaters and groundwaters. They tend to be highly toxic and only partially biodegradable. Benitez *et al.* (2001) found that use of the photo-Fenton process enabled partial degradation of a range of chlorophenols. The extent of degradation during the experiments decreased with increasing chlorine substitution of the aromatic ring.

Textile dyes: In a study of the degradation of Reactive Red 235 by the photo-Fenton process (using a mercury arc lamp as the light source) with Fe(III) complexation using oxalate, citrate, and glucarate, Aplin (2001) found that the effectiveness of the ligand depended on the operating pH. At pH 3, dye degradation was the most rapid for oxalate (and occurred at the same rate as in the absence of any ligand). At pH 4.5, dye degradation was also the most rapid for oxalate, however at pH 6 citrate gave the highest rate. When glucarate was used, the dye degradation rate was relatively low due to the low photoactivity of Fe(III)-glucarate complexes.

Pesticides: Balmer & Sultzberger (1999) studied atrazine degradation by the ferrioxalate process, which is similar to the photo-Fenton process except that hydrogen peroxide is not added as a reagent but is produced in situ due to photolysis of Fe^{III}oxalate complexes. They found that 0.47 μM atrazine could be degraded completely within 125 min at pH 3.2 when using 6 μM $Fe_{(aq)}$ and 18 μM oxalate when the light source was a xenon arc lamp (simulated sunlight). The reaction time could be reduced to about 20 min by increasing the oxalate concentration by a factor of 10. No degradation was observed when the pH was above 7. In the absence of oxalate, atrazine degradation was considerably slower and ceased to occur when the pH was above 4.

7.4.5.4 Heterogeneous photo-Fenton processes

Chlorophenols: The degradation of chlorophenols by heterogeneous Fenton and photo-Fenton processes using iron ions immobilised on Nafion® perfluorinated membranes has been studied by Sabhi and Kiwi (2001) and Maletzky *et al.* (1999). Nafion® membranes (DuPont™ N117) are resistant to attack by hydroxyl radicals. These processes can take place in pH neutral solutions with close to the same efficiency as in acidic solutions, which removes the need for costly pH adjustments when subsequent biological degradation is to be used (Balanosky *et al.* 1999). The reactions of Fe(II) and Fe(III) ions on the membrane surface are equivalent to the reactions of dissolved Fe(II) and Fe(III) ions. The membrane could be reused through many cycles, without leaching out of a significant amount of Fe(III) ions from the membrane, which would cause loss of efficiency (Sabhi & Kiwi, 2001).

Textile dyes: Fernandez *et al.* (2000) carried out a study on Fenton-enhanced decolourisation/degradation of the azo dye Orange II using an encapsulated Fe catalyst. The iron was supported on alginate gel beads by complexation with carboxylate groups. The Fe-alginate beads were about 2 mm in diameter and contained highly dispersed Fe species of about 0.5 nm in size. Using energy-dispersive X-ray microanalysis it was shown that most of the iron was on the bead surface. In the presence of simulated solar light and H_2O_2, the rate of degradation of Orange II when using this catalyst was only slightly higher at pH 5.6 than at pH 7.8, however the degradation was much slower at pH 10.

Nitrophenols: Pulgarin *et al.* (1995) used a heterogeneous photo-Fenton process involving a zeolite supported iron catalyst (Fe-ZSM-5) to degrade 4-nitrophenol. A solution containing 3.6 mM 4-nitrophenol was photodegraded in the presence of 1.5% Fe-zeolite and 13 mM H_2O_2 in about 8 h at 35°C; however it required more than 12 h for complete mineralisation to be achieved.

7.4.5.5 Electro-Fenton processes

The range of currently available electro-Fenton processes and their applications has been reviewed by Brillas *et al.* (2009).

Pesticides: In a study on the electro-Fenton and photoelectro-Fenton process (with in-situ H_2O_2 production) for degrading 2,4-D at pH 3, Brillas *et al.* (2000) found that a 230 ppm solution of 2,4-D with low salt content could be completely mineralized at low current by the photoelectro-Fenton process, while the electro-Fenton process lead to about 90% mineralisation. The rate of loss of 2,4-D itself was the same by both methods, however the higher degradation power of the photo-enhanced form of the process is attributed to the fast photodecomposition of some of the intermediates. These intermediates included chlorinated phenols, such as 2,4-dichlorophenol, 4,6-dichlororesorcinol, chlorohydroquinone and chlorobenzoquinone, as well as short-chain acids, such as glycolic, glyoxylic, maleic, fumaric, and oxalic acids.

Aromatic amines: Brillas & Casado (2002) have carried out pilot-scale studies on degradation of aniline by the Electro-Fenton® and Peroxi-coagulation processes at constant currents up to 20 A. They used 1000 ppm solutions of aniline in 50 mM Na_2SO_4/H_2SO_4 solution at pH 3 and 40°C. The reactor consisted of a filter-press cell operated in recirculation mode and contained an anode and an oxygen diffusion cathode (made of carbon-PTFE), each of area 0.010 m^2. For the Electro-Fenton Process®, the anode was a platinised titanium mesh or a DSA® plate. For the Peroxi-coagulation process, the anode was an iron plate. For the Electro-Fenton process (using 1 mM Fe(II)), 61% degradation of aniline was achieved after 2 h at 20 A. The Peroxi-coagulation process appeared to have a higher degradation power, enabling removal of more 95% of dissolved aniline at 20 A, because some intermediates coagulate with the ferric hydroxide precipitate that forms. Both these processes have moderate energy costs, which increase with increasing electrolysis time and applied current.

7.4.5.6 Fenton-like processes

Fuels: In a study on the degradation of 1,1-dimethylhydrazine (UDMH), Pestunova *et al.* (2002) compared a range of Cu- and Fe-hydroxide containing catalysts, supported on oxide carriers and containing less than 2% by weight active metal. UDMH is the main propellant for space rockets and missiles and is a highly toxic substance, with a maximum allowable concentration in water of 0.01 ppm. They found that UDMH could be efficiently oxidized in aqueous solutions by both hydrogen peroxide and oxygen in the presence of metal hydroxide catalysts. The pH of the solution affects the mechanism of the process. Under neutral conditions, the main product was found to be the non-toxic 1,1,5,5-tetramethylformazanium cation (at up to 30% yield), which further decomposes to other products with low toxicity. However in alkaline medium the reaction is not selective, and results in the formation of toxic and stable products, such as 1,1-dimethylnitrosamine (DMNA) and dimethylformamide (DMFA) at yields of up to 15%. The Fe-containing catalysts were found to be more resistant to loss of catalytic metal ions than the Cu-containing catalysts, as 0.5% and 30% of the active metal was leached from the catalyst at pH 7, respectively.

Inorganically contaminated groundwater: A process that is similar to the photo-Fenton process has been developed by Hug *et al.* (2001) to treat groundwater contaminated with arsenic. In this process As(III) is oxidised to As(IV) and then to As(V). The process can take place at a pH close to that of the groundwater, usually 6.5 to 8. In this process, citric acid (in the form of lemon juice) is added to the water, which is then exposed to sunlight. Iron is not usually added, because Fe(II) is naturally occurring in As-containing groundwater, however Fe(II) or Fe(III) salts can be added, if necessary. A photoactive Fe(III)-citrate complex is formed. Hydrogen peroxide is not added, however it is assumed to form in small amounts in the system from superoxide radicals, which are generated from a follow-up reaction to Fe(III)-citrate photolysis. The citrate becomes consumed in the photolysis reactions and once it is sufficiently depleted, Fe(III)-oxyhydroxide precipitates form, induced by the photoproduct of citrate (3-oxoglutaric acid). The As(V) adsorbs onto the precipitates, and can be removed from the system by settling and decantation. The process can be carried out using readily available non-toxic substances (such as lemon juice), sunlight, and simple reactors (transparent plastic bottles), hence it is suitable for use on a small scale in developing countries.

7.4.6 Commercial Applications

7.4.6.1 Industrial wastewater treatment plants

Ex-situ Fenton processes can often make use of general wastewater treatment equipment. The following companies supply wastewater treatment plants that are specifically designed for Fenton type processes:

- USP Technologies (formerly known as US Peroxide), USA (http://www.h2o2.com/)
- Lenntech, The Netherlands (http://www.lenntech.com/fenton-reaction.htm)
- Enviolet, Germany (Aqua Concept) (http://www.enviolet.com/en/uv-oxidation/uvoxidation/photo-fenton-reaction.html)

7.4.6.2 In-situ soil and groundwater remediation

In-situ chemical oxidation (ISCO) processes involve injection of solutions containing H_2O_2 into the soil or groundwater. These processes are also called catalyzed H_2O_2 propagations (CHP) and involve Fenton type chemistry. Several patented technologies involving the Fenton reaction for remediation of contaminated soil and groundwater by ISCO are available, including:

- The Geo-Cleanse Process® (in-situ treatment of subsurface environments using hydrogen peroxide and metallic salts) by Geo-Cleanse International, Inc. (http://www.geocleanse.com/chemical-remediation/reagents).
- The CleanOX® ISCO Process (in-situ treatment of soil and groundwater using hydrogen peroxide and naturally occurring iron, with supplemental iron if required) by MEC^X (http://www.mecx.net).
- ISOTECH^{SM} in-situ treatment using modified Fenton's reagent (iron chelators at neutral pH) (http://www.isotec-inc.com).

- The BIOX® Process (in-situ coupled chemical oxidation – enhanced biodegradation, using iron sulphate) by Innovative Remediation Technologies (http://irtechllc.com/in-situ-chemical-oxidation-technology/).
- Soil and groundwater remediation using ISCO by In-Situ Technieken b.v., The Netherlands (http://insitu.nl/en/).

ISCO processes involve the injection of hydrogen peroxide and a catalyst (a proprietary mixture of non-hazardous metallic salts) into the subsurface environment via various techniques. The remediation processes usually take between several hours and several months to complete.

One of the disadvantages of ISCO is that a violent exothermic reaction may occur on injection of the reagents. In addition, since the Fenton reaction performance is optimal at around pH 3, many applications are precluded due to the presence of alkaline rock, which would neutralise the added acid. This problem is overcome in many of the above treatment processes by using iron complexing agents that allow operation in the near to neutral pH range and with minimal temperature change.

The BIOX® Process takes place at neutral pH and ambient groundwater temperature. The process applies a proprietary reagent (containing hydrogen peroxide, iron sulphate, calcium or magnesium oxide, and other substances) to the soil. It does not involve the use of injection wells or acid addition. The organic degradation products from the process (which have been rendered less toxic) are subsequently consumed by the naturally occurring soil bacteria (Holish *et al.* 2000).

The ISOTECH[SM] Modified Fenton's Process uses a proprietary catalytic agent to delay the formation of hydroxyl radicals to allow adequate dispersion of the reagent through the contaminant plume. The catalyst contains an iron complexing agent that keeps the iron from precipitating and enables operation at circumneutral pH (Greenberg *et al.* 1998).

The Geo-Cleanse Process® operates under slightly acidic conditions (pH 4 to 6) and is not recommended for groundwater with a pH above 8 or for soil containing peat. When the chemical oxidation process is finished, naturally occurring microorganisms are used to complete the site remediation (Casey & Bergren, 1999).

7.4.7 Equipment design and economic considerations

The capital cost of a Fenton type process includes equipment such as reaction vessels, pumps and dosing equipment, mixers, and light sources (such as UVA lamps) or solar collectors, as well as site costs. The operating costs include chemicals, electricity, labour, equipment maintenance, and disposal costs. Most processes make use of an inexpensive iron salt (such as $FeSO_4 \cdot 7H_2O$) as the catalyst, as well as hydrogen peroxide, acids, and bases, however some processes require the use of more expensive catalysts.

The cost associated with increased energy consumption (such as by using a stronger light source) may be offset by the benefits of reducing the treatment time, which reduces the capital cost relative to the plant output.

Quotes for treatment costs are usually quoted as a price for unit volume of water treated. This does not take into account the additional cost associated with an increase in contaminant concentration. The Figures-of-Merit approach, proposed by Bolton *et al.* (2001), can be used as a standardized basis for comparison of the energy consumption of AOPs used for wastewater treatment by using an efficiency rating that also takes into account the concentration of the target contaminants. Formulas have been developed for two types of systems, electric energy-driven and solar energy-driven, each of which is independent of the type of AOP. For electric energy-driven systems, the figure-of-merit suitable for wastewater with a high concentration of contaminants is electric energy per mass removed (E_{EM}), while for wastewater with a low concentration of contaminants it is electric energy per order of magnitude of removal (E_{EO}). For solar energy-driven systems, the corresponding figures-of-merit are collector area per mass removed (A_{CM}) and collector area per order of magnitude of removal (A_{CO}).

Consideration of the capital and operating costs of a process does not take into account the environmental impact of using the process. Muñoz *et al.* (2005) have compared the use of solar energy and UVA lamps as sources of light for several AOPs used for treating kraft mill bleaching wastewaters. They took into account the production of the catalysts and other reagents as well as the electricity in terms of the following eight environmental impact categories: global warming, ozone depletion, aquatic eutrophication, acidification, human toxicity, freshwater aquatic toxicity, photochemical ozone formation, and abiotic resource depletion. They found that the electricity consumption has the highest effect on the environmental impact, with the impact of producing the catalysts and other reagents being comparatively low. The use of solar energy reduced the environmental impact of the process by over 90% when compared with using UVA lamps.

7.4.8 Process integration

Fenton type processes are often used in combination with other water treatment processes such as biodegradation, coagulation and settling, or pressure-driven membrane processes to provide an effective treatment system (Comninellis, 2008). Although the Fenton process could potentially be carried out either before or after the other process, the type of waste being treated and the treatment objectives usually determine which order would be most effective.

7.4.8.1 Integration with coagulation and settling

The iron used in the Fenton process may be used to replace some of the ferric iron that is dosed during coagulation (Dolejs *et al.* 1994). The Fenton process leads to lower sludge production than conventional water treatment processes; hence Fenton treatment may be feasible for treating water that is contaminated by organic compounds using existing coagulation-flocculation facilities. However, the low pH requirements of Fenton processes may preclude this due to possible corrosion of the plant. It should be noted that some of the organic matter that is not degraded in the Fenton process may be removed in the coagulation and settling process. Although this may be seen as beneficial, this may result in the formation of toxic sludge that is difficult to dispose.

If the water contains suspended matter then this would obscure the light, hence only the dark Fenton process may be feasible. Since Fenton type processes typically do not completely mineralize all the organic matter the product water may still need to be treated by a biological process prior to reuse or disposal to a natural waterway. Neutralization followed by a further settling process would be required to remove the iron containing sludge from the product water.

7.4.8.2 Integration with biodegradation

The Fenton process will usually partially mineralize both biodegradable and non-biodegradable organics. If the wastewater contains very high amounts of non-biodegradable organics compared with biodegradable organics, particularly those containing highly toxic organics that are inhibitory to the biological process, then there is an advantage in using a Fenton type process first. This would allow these compounds to be degraded into more biodegradable compounds. A disadvantage of this process order is that, since the photo-Fenton process is more efficient at low pH, the pH of the water entering the biodegradation stage would need to be adjusted. This is an additional expense and leads to an increase in the salinity of the treated water. If the water to be treated has a high concentration of biodegradable organics and a low concentration of non-biodegradable organics which are relatively non-toxic to the organisms in the bioreactor, then it would be possible to use the biodegradation process first.

If the biodegradation process is inhibited by the presence of toxic organic compounds then biological activity may be increased in the water or soil after the Fenton process. However conditions during the Fenton process are very aggressive due to the presence of excess hydrogen peroxide and other oxidants and a low pH (pH < 4) is usually encountered; hence the medium often becomes sterilized. Disinfection is an advantage where pathogens are present, however if subsequent biodegradation is required, the water must be neutralized and residual hydrogen peroxide must be allowed to react. Pulgarin *et al.* (1999) found that the problem of high residual hydrogen peroxide in the Fenton treated water could be overcome by operating in the semi-continuous mode.

De la Cruz *et al.* (2012) found that complete or significant removal of micropollutants by Fenton and photo-Fenton processes occurred in water that had already been treated by the activated sludge process (an aerobic biological process). This water contained dissolved organic matter (DOM) in the order of 2000 times as high as the micropollutants and the processes were carried out at close to neutral pH. The DOM may have acted as a photosensitizer or as a complexing agent, keeping the iron dissolved.

In many cases it is necessary to innoculate the soil or water with suitable organisms after treatment. This can be done by mixing in uncontaminated soil or water containing the desired organisms. If conditions are good the biological population will become as high, or more commonly higher, than before due to the removal of toxic materials.

Miller *et al.* (1996) have found that certain species of bacteria (particularly *Pseudomonas*) will proliferate in the treated soil or water. The presence of these species is desirable since they enhance the remediation of contaminated soils. The diversity of heterotrophic bacteria is decreased due to sterilization; however this enables species that are usually overshadowed to proliferate once the oxidant has decayed.

The right balance between treatment times in the Fenton process and the biological process (relatively lower cost) should be determined to optimise the process (Pulgarin *et al.* 1999). If the Fenton process does not proceed far enough the water will still be toxic to the biological process, however increasing the time in the Fenton stage increases the cost of treatment more than an equivalent time in the biological stage. If the photo-Fenton process is carried out under optimal conditions it may not be necessary to use a biodegradation process.

7.4.8.3 Integration with membrane processes

The use of the photo-Fenton process with a pressure-driven membrane process such as reverse-osmosis, nanofiltration, or ultrafiltration can be more effective than using either process on its own. The disposal problem posed by the retentate from a membrane process, which contains high concentrations of non-biodegradable organic matter, can be solved by treatment with a Fenton type process. This combination of processes is most suitable when there are trace levels of organic contaminants in the wastewater (i.e. micropollutants, in the $ng \cdot L^{-1}$ to $\mu g \cdot L^{-1}$ range) and the selected membrane exhibits a high retention for the organic contaminants. Use of a membrane process to treat water that has been treated by a Fenton type process is usually not feasible because the residual iron in the treated water would cause membrane fouling. Even if an intermediate coagulation/settling process was used the process may not be feasible since the organic matter would be reduced in size by the oxidative degradation and hence less likely to be rejected by a nanofiltration or ultrafiltration membrane.

A study by Miralles-Cuevas *et al.* (2014) showed that for a synthetic wastewater initially containing $15\ \mu g \cdot L^{-1}$ of each of five pharmaceuticals (carbamazepine, flumequine, ibuprofen, ofloxacin, and sulfamethoxazole) dissolved in natural water, treatment of the nanofiltration retentate by the solar photo-Fenton process reduced the treatment time by 88% and used 89% less H_2O_2 compared with treating the original wastewater. The nanofiltration process had a retention of greater than 95% and produced a permeate that contained less than 1.5% of the initial concentration of pharmaceuticals. The retentate, containing about 150 μg · L^{-1} of each pharmaceutical, was treated with 5 mg · L^{-1} of Fe^{2+} and 25 mg · L^{-1} of H_2O_2. The process could be carried out at around pH 5. A relatively low amount of H_2SO_4 was required for carbonate removal (using acidification followed by agitation) because there was a lower total amount of carbonate in the retentate than in the original wastewater.

7.4.8.4 Co-treatment of wastes

Some wastes are difficult to treat by Fenton type processes without the addition of other chemicals such as acids or complexing agents. Co-treatment of waste streams should be considered when one waste contains substances that may assist in the Fenton treatment of another waste.

For example, Rivas *et al.* (2002) have recommended that some types of foodstuff wastes containing phenolic substances be combined with wastewater containing pesticides. The phenolic substances form a complex with Fe(III), producing a more effective catalyst. Other foodstuff wastes that could be used include those containing aliphatic organic acids such as citric acid. The use of these compounds may make the final effluent after Fenton treatment more amenable to further degradation by microorganisms.

Chou *et al.* (2001) have carried out pilot-plant studies on the co-treatment of dying-finishing wastewater and benzoic acid using supported iron oxyhydroxide and Fenton's reagent in a fluidised-bed reactor (FBR) of height 7 m and diameter 1 m. The process involves the simultaneous mineralisation of organic compounds and precipitation of Fe(III). The FBR-Fenton method was shown to be better than the conventional Fenton treatment of benzoic and textile dying/finishing wastewater in terms of mineralisation and Fe(III) precipitation rates. This method can reduce the sludge production compared with the conventional Fenton's method by more than 60%.

7.5 FUTURE RESEARCH NEEDS

The number of studies on Fenton, extended Fenton, and Fenton-like processes that have now been reported in the literature is astounding, with this activity, in part, evidence of the potential that these processes have for contaminant degradation. Many of the studies, however, especially in recent years, provide limited advances in the understanding of the processes underpinning the reported degradation performance and provide limited insight into the viability of the proposed technology at full scale. In light of this, we emphasise the need for more thorough, systematic, and mechanistically-driven research on Fenton-related technologies. Progress in many promising areas is hindered by a lack of sufficient understanding of how some systems work (and, equally importantly, the underlying reasons for the limitations of others). Deeper, systematic, inquiring research would allow for a more coherent and holistic understanding of the many parallels between all these systems, providing a better foundation to push towards more cost-effective and practical solutions.

Every study is quick to claim that their particular invocation of a technique is "unprecedented" or "exceptionally reactive". Adoption of particular test-cases or representative scenarios to allow a genuine, honest, and fair comparison between methodologies would allow the leading processes to be more readily determined and accelerate the development of these truly promising technologies.

7.6 REFERENCES

Ammar S., Oturan M. A., Labiadh L., Guersalli A., Abdelhedi R., Oturan N. and Brillas E. (2015). Degradation of tyrosol by a novel electro-Fenton process using pyrite as heterogeneous source of iron catalyst. *Water Research*, **74**, 77–87.

Andreozzi R., Caprio V., Insola A. and Marotta R. (1999). Advanced oxidation processes (AOP) for water purification and recovery. *Catalysis Today*, **53**, 51–9.

Aplin R. (2001). Degradation of Reactive Dyes by a Modified Photo-Fenton Process. PhD Thesis, School of Civil and Environmental Engineering, The University of New South Wales, Australia.

Ardo S. G., Nélieu S., Ona-Nguema G., Delarue G., Brest J., Pironin E. and Morin G. (2015). Oxidative degradation of nalidixic acid by nano-magnetite via Fe^{2+}/O_2-mediated reactions. *Environmental Science & Technology*, **49**(7), 4506–14.

Armstrong W. A. (1969). Relative rate constants for reactions of hydroxyl radicals from the reaction of Fe(II) or Ti(III) with H_2O_2. *Canadian Journal of Chemistry*, **47**(20), 3737–44.

Arnold S. M., Hickey W. J. and Harris R. F. (1995). Degradation of atrazine by Fenton's reagent: Condition optimisation and product quantification. *Environmental Science & Technology*, **29**, 2083–9.

Aruoma O. I., Halliwell B., Gajewski E. and Dizdaroglu M. (1989). Damage to the bases in DNA induced by hydrogen peroxide an ferric ion chelates. *Journal of Biological Chemistry*, **264**(34), 20509–12.

Balanosky E., Fernandez J., Kiwi J. and Lopez A. (1999). Degradation of membrane concentrates of the textile industry by Fenton like reactions in iron-free solutions at biocompatible pH values (pH ≈ 7-8). *Water Science & Technology*, **40**(4-5), 417–24.

Ballesteros Martín M. M., Sanchez Perez J. A., Acien Fernandez F. G., Casas López J. L., García-Ripoll A. M., Arques A., Oller I. and Malato Rodriguez S. (2008). Combined photo-Fenton and biological oxidation for pesticide degradation: effect of photo-treated intermediates on biodegradation kinetics. *Chemosphere*, **70**(8), 1476–83.

Balmer M. E. and Sultzberger B. (1999). Atrazine degradation in irradiated iron/oxalate systems: Effects of pH and oxalate. *Environmental Science & Technology*, **33**, 2418–24.

Bandala E. R., Peláez M. A., Dionysiou D. D., Gelover S., Garcia J. and Macías D. (2007). Degradation of 2,4-dichlorophenoxyacetic acid (2,4-D) using cobalt-peroxymonosulfate in Fenton-like process. *Journal of Photochemistry and Photobiology A*, **186**(2–3), 357–63.

Bañuelos J. A., El-Ghenymy A., Rodríguez F. J., Manríquez J., Bustos E., Rodríguez A., Brillas E. and Godínez L. A. (2014). Study of an air diffusion activated carbon packed electrode for an electro-Fenton wastewater treatment. *Electrochimica Acta*, **140**, 412–8.

Barb W. G., Baxendale J. H., George P. and Hargrave K. R. (1951a). Reactions of ferrous and ferric ions with hydrogen peroxide. Part I.—The ferrous ion reaction. *Transactions of the Faraday Society*, **47**, 462–500.

Barb W. G., Baxendale J. H., George P. and Hargrave K. R. (1951b). Reactions of ferrous and ferric ions with hydrogen peroxide. Part II.—The ferric ion reaction. *Transactions of the Faraday Society*, **47**, 591–616.

Barbusiński K. (2009). Fenton reaction-controversy concerning the chemistry. *Ecological Chemistry and Engineering. S*, **16**, 347–58.

Barreiro J. C., Capelato M. D., Martin-Neto L. and Bruun Hansen H. C. (2007). Oxidative decomposition of atrazine by a Fenton-like reaction in a H_2O_2/ferrihydrite system. *Water Research*, **41**(1), 55–62.

Bataineh H., Pestovsky O. and Bakac A. (2012). pH-induced mechanistic changeover from hydroxyl radicals to iron(IV) in the Fenton reaction. *Chemical Science*, **3**(5), 1594–9.

Benitez F. J., Beltran-Heredia J., Acero J. L. and Rubio F. J. (2001). Oxidation of several chlorophenolic derivatives by UV irradiation and hydroxyl radicals. *Journal of Chemical Technology and Biotechnology*, **76**, 312–20.

Berdnikov V. M., Zhuravleva O. S. and Terent'eva L. A. (1977). Kinetics and mechanism of elementary electron transfer in the oxidation of alcohol radicals by hydrated Fe(III) ions. *Bulletin of the Academy of Sciences of the USSR, Division of Chemical Science*, **26**(10), 2050–6.

Biaglow J. E. and Kachur A. V. (1998). The generation of hydroxyl radicals in the reaction of molecular oxygen with polyphospate complexes of ferrous ion. *Radiation Research*, **148**(2), 181–7.

Boland D. D., Collins R. N., Miller C. J., Glover C. J. and Waite T. D. (2014). Effect of solution and solid-phase conditions on the Fe(II)-accelerated transformation of ferrihydrite to lepidocrocite and goethite. *Environmental Science & Technology*, **48**(10), 5477–85.

Bolton J. R., Bircher K. G., Tumas W. and Tolman C. A. (2001). Figures-of-merit for the technical development and Application of advanced oxidation technologies for both electric- and solar-driven systems (IUPAC Technical Report). *Pure and Applied Chemistry*, **73**(4), 627–37.

Bossmann S. H., Oliveros E., Göb S., Siegwart S., Dahlen E. P., Payawan (Jr) L., Straub M., Wörner M. and Braum A. M. (1998). New evidence against hydroxyl radicals as reactive intermediates in the thermal and photochemically enhanced Fenton reactions. *Journal of Physical Chemistry A*, **102**, 5542–50.

Boye B., Dieng M. M. and Brillas E. (2002). Degradation of herbicide 4-chlorophenoxyacetic acid by advanced electrochemical oxidation methods. *Environmental Science & Technology*, **36**, 3030–5.

Bozzi A., Yuranova T., Mielczarski E., Mielczarski J., Buffat P. A., Lais P. and Kiwi, J. (2003). Superior biodegradability mediated by immobilized Fe-fabrics of waste waters compared to Fenton homogeneous reactions. *Applied Catalysis B*, **42**(3), 289–303.

Bray W. C. and Gorin M. H. (1932). Ferryl ion, A compound of tetravalent iron. *Journal of the American Chemical Society*, **54**(5), 2124–5.

Brillas E. (2011). Fenton-Electrochemical Treatment of Wastewaters for the Oxidation of Organic Pollutants Using BDD. In: "Synthetic Diamond Films: Preparation, Electrochemistry, Characterization, and Applications", E. Brillas and C. A. Martinez-Huitle (eds.), Wiley, Hoboken, New Jersey, USA, pp. 405–35.

Brillas E. and Casado J. (2002). Aniline degradation by Electro-Fenton® and peroxi-coagulation processes using a flow reactor for wastewater treatment. *Chemosphere*, **47**(3), 241–6.

Brillas E. and Martinez-Huitle C. A. (2015). Decontamination of wastewaters containing synthetic organic dyes by electrochemical means. An updated review. *Applied Catalysis B*, **166-7**, 603–45.

Brillas E., Mur E. and Casado J. (1996). Iron(II) catalysis of the mineralization of aniline using a carbon-PTFE O_2-fed cathode. *Journal of The Electrochemical Society*, **143**(3), L49–L53.

Brillas E., Calpe J. C. and Casado J. (2000). Mineralisation of 2,4-D by advanced electrochemical oxidation processes. *Water Research*, **34**, 2253–62.

Brillas E., Sirés I. and Oturan M. A. (2009). Electro-Fenton process and related electrochemical technologies based on Fenton's reaction chemistry. *Chemical Reviews*, **109**(12), 6570–631.

Bucheli-Witschel M. and Egli T. (2001). Environmental fate and microbial degradation of aminopolycarboxylic acids. *FEMS Microbiology Reviews*, **25**(1), 69–106.

Buettner G. R., Doherty T. P. and Patterson L. K. (1983). The kinetics of the reaction of superoxide radical with Fe(III) complexes of EDTA, DETAPAC and HEDTA. *FEBS Letters*, **158**(1), 143–6.

Burns J. M., Craig P. S., Shaw T. J. and Ferry J. L. (2010). Multivariate examination of Fe(II)/Fe(III) cycling and consequent hydroxyl radical generation. *Environmental Science & Technology*, **44**(19), 7226–31.

Buxton G. V. and Green J. C. (1978). Reactions of some simple α- and β-hydroxyalkyl radicals with Cu^{2+} and Cu^{+} ions in aqueous solution. A radiation chemical study. *Journal of the Chemical Society, Faraday Transactions 1*, **74**, 697–714.

Buxton G. V., Greenstock C. L., Helman W. P. and Ross A. B. (1988). Critical review of rate constants for reactions of hydrated electrons, hydrogen atoms and hydroxyl radicals ($^{\bullet}OH/^{\bullet}O^{-}$) in aqueous solution. *Journal of Physical and Chemical Reference Data*, **17**(2), 513–886.

Cañizares P., Paz R., Sáez C. and Rodrigo M. A. (2007). Costs of the electrochemical oxidation of wastewaters: A comparison with ozonation and Fenton oxidation processes. *Journal of Environmental Management*, **90**(1), 410–20.

Casey C. and Bergren C. (1999). Chemical oxidation, natural attenuation drafted in Navy cleanup. *Pollution Engineering*, March 1999 Supplement.

Chahbane N., Popescu D. L., Mitchell D. A., Chanda A., Lenoir D., Ryabov A. D., Schramm, K.-W. and Collins, T. J. (2007). Fe III–TAML-catalyzed green oxidative degradation of the azo dye Orange II by H_2O_2 and organic peroxides: products, toxicity, kinetics, and mechanisms. *Green Chemistry*, **9**(1), 49–57.

Chan K. H. and Chu W. (2005). Model applications and mechanism study on the degradation of atrazine by Fenton's system. *Journal of Hazardous Materials*, **118**(1), 227–37.

Che H. and Lee W. (2011). Selective redox degradation of chlorinated aliphatic compounds by Fenton reaction in pyrite suspension. *Chemosphere*, **82**(8), 1103–8.

Chen L., Ma J., Li X., Zhang J., Fang J., Guan Y. and Xie P. (2011). Strong enhancement on Fenton oxidation by addition of hydroxylamine to accelerate the ferric and ferrous iron cycles. *Environmental Science & Technology*, **45**(9), 3925–30.

Chevallier E., Jolibois R. D., Meunier N., Carlier P. and Monod A. (2004). "Fenton-like" reactions of methylhydroperoxide and ethylhydroperoxide with Fe^{2+} in liquid aerosols under tropospheric conditions. *Atmospheric Environment*, **38**(6), 921–33.

Choi K., Bae S. and Lee W. (2014). Degradation of off-gas toluene in continuous pyrite Fenton system. *Journal of Hazardous Materials*, **280**, 31–7.

Chou S., Huang Y.-H., Lin H.-L. and Liao C.-C. (2001). Treatment of dying-finishing wastewater using iron oxyhydroxide and Fenton's reagent in a fluidised bed reactor. The Seventh International Conference on Advanced Oxidation Technologies for Water and Air Remediation, Niagara Falls, Ontario, Canada, 25–29 June, pp. 89–91.

Comninellis C., Kapalka A., Malato S., Parsons S. A., Poulios I. and Mantzavinos D. (2008). Advanced Oxidation Processes for Water Treatment: Advances and Trends for R&D. *Journal of Chemical Technology and Biotechnology*, **83**, 769–76.

Costa R. C. C., Lelis M. F. F., Oliveira L. C. A., Fabris J. D., Ardisson J. D., Rios R. R. V. A., Silva C. N. and Lago R. M. (2006). Novel active heterogeneous Fenton system based on $Fe_{3-x}M_xO_4$ (Fe, Co, Mn, Ni): The role of M^{2+} species on the reactivity towards H_2O_2 reactions. *Journal of Hazardous Materials*, **129**(1-3), 171–8.

Costa R. C. C., Moura F. C. C., Ardisson J. D., Fabris J. D. and Lago R. M. (2008). Highly active heterogeneous Fenton-like systems based on Fe^0/Fe_3O_4 composites prepared by controlled reduction of iron oxides. *Applied Catalysis B: Environmental*, **83**(1-2), 131–9.

Croft S., Gilbert B. C., Smith J. R. L. and Whitwood, A. C. (1992). An E.S.R. investigation of the reactive intermediate generated in the reaction between Fe^{II} and H_2O_2 in aqueous solution. Direct evidence for the formation of the hydroxyl radical. *Free Radical Research*, **17**(1), 21–39.

Da Silva S. S., Chiavone-Filho O., de Barros Neto E. L. and Nascimento C. A. (2012). Integration of processes induced air flotation and photo-Fenton for treatment of residual waters contaminated with xylene. *Journal of Hazardous Materials*, **199**, 151–7.

De Laat J. and Gallard H. (1999). Catalytic decomposition of hydrogen peroxide by Fe(III) in homogeneous aqueous solution: mechanism and kinetic modeling. *Environmental Science & Technology*, **33**(16), 2726–32.

De Laat J. and Le G. T. (2005). Kinetics and modeling of the Fe(III)/H_2O_2 system in the presence of sulfate in acidic aqueous solutions. *Environmental Science & Technology*, **39**(6), 1811–8.

De Laat J., Le G. T. and Legube B. (2004). A comparative study of the effects of chloride, sulfate and nitrate ions on the rates of decomposition of H_2O_2 and organic compounds by Fe(II)/H_2O_2 and Fe(III)/H_2O_2. *Chemosphere*, **55**(5), 715–23.

De la Cruz N., Giménez J., Esplugas S., Grandjean D., De Alencastro L. F. and Pulgarin C. (2012). Degradation of 32 emergent contaminants by UV and neutral photo-Fenton in domestic wastewater effluent previously treated by activated sludge. *Water Research*, **46**(6), 1947–57.

Deng Y. and Englehardt J. D. (2006). Treatment of landfill leachate by the Fenton process. *Water Research*, **40**(20), 3683–94.

Dhananjeyan M. R., Kiwi J., Albers P. and Enea O. (2001a). Photo-assisted immobilized Fenton Degradation up to pH 8 of Azo Dye Orange II mediated by Fe^{3+}/*Nafion*/Glass Fibers. *Helvetica Chimica Acta*, **84**, 3433–41.

Dhananjeyan M. R., Mielczarski E., Thampi K. R., Buffat Ph., Bensimon M., Kulik A., Mielczarski J. and Kiwi J. (2001b). Photodynamics and surface characterization of TiO_2 and Fe_2O_3 photocatalysts immobilized on modified polyethylene films. *Journal of Physical Chemistry B*, **105**, 12046–55.

Dolejs P., Kalousková N., Prados M. and Legube B. (1994). Coagulation of humic water with Fenton reagent. In: Chemical Water and Wastewater Treatment III, R. Klute and H. H. Hahn (eds), Springer-Verlag, Berlin, pp. 117–29.

Dorfman L. M., Taub I. A. and Buhler R. E. (1962). Pulse radiolysis studies. I. Transient spectra and reaction-rate constants in irradiated aqueous solutions of benzene. *Journal of Chemical Physics*, **36**(11), 3051–61.

Duesterberg, C. K. and Waite, T. D. (2006). Process optimization of Fenton oxidation using kinetic modeling. *Environmental Science & Technology*, **40**(13), 4189–95.

Duesterberg C. K. and Waite T. D. (2007). Kinetic modeling of the oxidation of *p*-hydroxybenzoic acid by Fenton's reagent: Implications of the role of quinones in the redox cycling of iron. *Environmental Science & Technology*, **41**(11), 4103–10.

Duesterberg C. K., Cooper W. J. and Waite T. D. (2005). Fenton-mediated oxidation in the presence and absence of oxygen. *Environmental Science & Technology*, **39**, 5052–8.

Dunford H. B. (2002). Oxidations of iron(II)/(III) by hydrogen peroxide: from aquo to enzyme. *Coordination Chemistry Reviews*, **233–234**, 311–8.

Eberhardt M. K. (1984). Formation of olefins from alkyl radicals with leaving groups in the β-position. *Journal of Organic Chemistry*, **49**(20), 3720–5.

EPA (2014a). Water: Contaminants of Emerging Concern. http://water.epa.gov/scitech/cec/ (accessed 17 December 2014).

EPA (2014b). Water: Industry Effluent Guidelines. http://water.epa.gov/scitech/wastetech/guide/laws.cfm (accessed 17 December 2014).

EPA (2015). Persistent Organic Pollutants: A Global Issue, a Global Response. http://www2.epa.gov/international-cooperation/persistent-organic-pollutants-global-issue-global-response (accessed 23 February 2015)

European Commission (2015). Priority substances under the Water Framework Directive, updated: 04/02/2015. http://ec.europa.eu/environment/water/water-dangersub/pri_substances.htm (accessed 23 February 2015).

Fang G.-D., Zhou D.-M. and Dionysiou D. D. (2013). Superoxide mediated production of hydroxyl radicals by magnetite nanoparticles: Demonstration in the degradation of 2-chlorobiphenyl. *Journal of Hazardous Materials*, **250-251**, 68–75.

Fang X., Pan X., Rahmann A., Schuchmann H.-P. and von Sonntag C. (1995). Reversibility in the reaction of cyclohexadienyl radicals with oxygen in aqueous solution. *Chemistry – A European Journal*, **1**(7), 423–9.

Fatta-Kassinos D., Vasquez M. I. and Kümmerer K. (2011). Transformation products of pharmaceuticals in surface waters and wastewater formed during photolysis and advanced oxidation processes–degradation, elucidation of byproducts and assessment of their biological potency. *Chemosphere*, **85**(5), 693–709.

Faust B. C. and Hoigné J. (1990). Photolysis of iron(III)-hydroxy complexes as sources of hydroxyl radicals in clouds, fog and rain. *Atmospheric Environment*, **24A**(1), 79–89.

Feng J., Hu X. and Yue P. L. (2006). Effect of initial solution pH on the degradation of Orange II using clay-based Fe nanocomposites as heterogeneous photo-Fenton catalyst. *Water Research*, **40**(4), 641–6.

Fenton H. J. H. (1876). On a new reaction of tartaric acid. *Chemistry News*, **33**, 190.

Fenton H. J. H. (1894). Oxidation of tartaric acid in the presence of iron. *Journal of the Chemical Society*, **65**, 899–910.

Fenton H. J. H. (1894). LXXIII. – Oxidation of tartaric acid in presence of iron. *Journal of the Chemical Society, Transactions*, **65**, 899–910.

Fernandez J., Bandara J., Lopez A., Buffat, P. and Kiwi J. (1999). Photoassisted Fenton degradation of nonbiodegradable azo dye (Orange II) in Fe-free solutions mediated by cation transfer membranes. *Langmuir*, **15**, 185–92.

Fernandez J., Dhananjeyan M. R., Kiwi J., Senuma Y. and Hilborn J. (2000). Evidence for Fenton photoassisted processes mediated by encapsulated Fe ions at biocompatible pH values. *Journal of Physical Chemistry B*, **104**, 5298–301.

Fernández-Alba A. R., Hernando D., Agüera A., Cáceres J. and Malato S. (2002). Toxicity assays: a way for evaluating AOPs efficiency. *Water Research*, **36**, 4255–62.

Garrido-Ramírez E. G., Theng B. K.G. and Mora, M. L. (2010). Clays and oxide minerals as catalysts and nanocatalysts in Fenton-like reactions – a review. *Applied Clay Science*, **47**(3), 182–92.

Gernjak W., Krutzler T., Glaser A., Malato S., Caceres J., Bauer R. and Fernández-Alba A. R. (2003). Photo-Fenton treatment of water containing natural phenolic pollutants. *Chemosphere*, **50**(1), 71–8.

Gonzales-Olmos R., Kopinke F-D., Mackenzie K. and Georgi A. (2013). Hydrophobic Fe-zeolites for removal of MTBE from water by combination of adsorption and oxidation. *Environmental Science & Technology*, **47**(5), 2353–60.

Gözmen B., Oturan M. A., Oturan N. and Erbatur O. (2003). Indirect electrochemical treatment of bisphenol A in water via electrochemically generated Fenton's reagent. *Environmental Science & Technology*, **37**(16), 3716–23.

Greenberg R. S., Andrews T., Kakarla P. K. C., Watts R. J. (1998). In-situ Fenton-like oxidation of volatile organics: Laboratory, pilot, and full-scale demonstrations. *Remediation*, **8**(2), 29–42.

Guimaraes J. R., Farah C. R.T., Maniero M. G. and Fadini P. S. (2012). Degradation of formaldehyde by advanced oxidation processes. *Journal of Environmental Management*, **107**, 96–101.

Gurol M. D. and Lin S. (1998) Continuous Catalytic Oxidation Process. US Patent No. 5,755,977.

Gutteridge J. M. C. (1990). Superoxide-Dependent Formation of Hydroxyl Radicals from Ferric-Complexes and Hydrogen Peroxide: An Evaluation of Fourteen Iron Chelators. *Free Radical Research*, **9**(2), 119–25.

Haber F. and Weiss J. (1932). Über die katalyse des hydroperoxydes (About the catalysis of hydroperoxides). *Naturwissenschaften*, **20**(51), 948–50.

Haber F. and Weiss J. (1934). The catalytic decomposition of hydrogen peroxide by iron salts. *Proceedings of the Royal Society, Series A*, **147**(861), 332–51.

Haber F. and Willstätter R. (1931). Unpaarigkeit und radikalketten im reaktionsmechanismus organischer und enzymatischer vorgänge (Unpaired and radical chain reaction mechanisms in organic and enzymatic processes). *Ber. Dtsch. Chem. Ges.*, **64**(11), 2844–56.

Halliwell B. and Gutteridge J. M. C. (1990). Role of free radicals and catalytic metal ions in human disease: An overview. *Methods in Enzymology*, **186**, 1–85.

Hammerschmidt C. R. and Fitzgerald W. F. (2010). Iron-mediated photochemical decomposition of methylmercury in an arctic Alaskan lake. *Environmental Science & Technology*, **44**(16), 6138–43.

Hanna K., Kone T. and Medjahdi G. (2008). Synthesis of the mixed oxides of iron and quartz and their catalytic activities for the Fenton-like oxidation. *Catalysis Communications*, **9**(5), 955–9.

Hatchard C. G. and Parker C. A. (1956). A new sensitive chemical actinometer. II. Potassium ferrioxalate as a standard chemical actinometer. *P. Roy. Soc. Lond. A Mat.*, **235**(1203), 518–36.

He D., Ma J., Collins R. N. and Waite T. D. (2016). Effect of structural transformation of nanoparticulate zero-valent iron on generation of reactive oxygen species. *Environmental Science & Technology*, **50**, 3820–3828.

He D., Miller C. J. and Waite T. D. (2014). Fenton-like zero-valent silver nanoparticle-mediated hydroxyl radical production. *Journal of Catalysis*, **317**, 198–205.

He J., Yang X., Men B., Bi Z., Pu Y. and Wang D. (2014). Heterogeneous Fenton oxidation of catechol and 4-chlorocatechol catalyzed by nano-Fe_3O_4: Role of the interface. *Chemical Engineering Journal*, **258**, 433–41.

Hislop, K. A. and Bolton, J. B. (1999). The photochemical generation of hydroxyl radicals in the UV-vis/ferrioxalate/H_2O_2 system. *Environmental Science & Technology*, **33**, 3119–26.

Holish L. L., Lundy W. L. and Nuttall H. E. (2000). Indiana "land banned" herbicide release. *Soil Sediment & Groundwater*, October/November, 14–8.

Huang W., Brigante M., Wu F., Hanna K. and Mailhot G. (2012). Development of a new homogenous photo-Fenton process using Fe (III)-EDDS complexes. *Journal of Photochemistry and Photobiology A*, **239**, 17–23.

Huang W., Brigante M., Wu F., Mousty C., Hanna K. and Mailhot G. (2013). Assessment of the Fe(III)–EDDS complex in Fenton-like processes: From the radical formation to the degradation of bisphenol A. *Environmental Science & Technology*, **47**(4), 1952–9.

Hug S. J., Canonica L., Wegelin W., Gechter G. and von Gunten, U. (2001). Solar oxidation and removal of arsenic at circumneutral pH in iron containing waters. *Environmental Science & Technology*, **35**, 2114–21.

Huston P. L. and Pignatello J. J. (1999). Degradation of selected pesticide active ingredients and commercial formulations in water by the photo-assisted Fenton reaction. *Water Research*, **33**(5), 1238–46.

Jiang Y. and Waite T. D. (2003). Degradation of trace contaminants using coupled sonochemistry and Fenton's reagent. *Water Science & Technology*, **47**(10), 85–92.

Jones A. M., Griffin P. J., Collins R. N. and Waite T. D. (2014). Ferrous iron oxidation under acidic conditions – The effect of ferric oxide surfaces. *Geochimica et Cosmochimica Acta*, **145**, 1–12.

Keenan C. R. and Sedlak D. L. (2008). Factors affecting the yield of oxidants from the reaction of nanoparticulate zero-valent iron and oxygen. *Environmental Science & Technology*, **42**(4), 1262–7.

Khaikin G. I., Alfassi Z. B., Huie R. E. and Neta P. (1996). Oxidation of ferrous and ferrocyanide ions by peroxyl radicals. *Journal of Physical Chemistry*, **100**(17), 7072–7.

Kinsela A. S., Jones A. M., Bligh M. W., Pham A. N., Collins R. N., Harrison J. J., Wilsher K. L., Payne T. E. and Waite T. D. (2016). Influence of dissolved silicate on rates of Fe(II) oxidation. *Environmental Science & Technology*, **50**, 11663–11671.

Kishore K., Moorthy P. N. and Rao K. N. (1987). Reactivity of H_2O_2 with radiation produced free radicals: Steady state radiolysis methods for estimating the rate constants. *International Journal of Radiation Applications and Instrumentation Part C*, **29**(4), 309–13.

Kiwi J., Lopez A. and Nadtochenko V. (2000). Mechanism and kinetics of the OH-radical intervention during Fenton oxidation in the presence of a significant amount of radical scavenger (Cl^-). *Environmental Science & Technology*, **34**(11), 2162–8.

Kocot P., Szaciłowski K. and Stasicka Z. (2007). Photochemistry of the $[Fe^{III}(edta)(H_2O)]^-$ and $[Fe^{III}(edta)(OH)]^{2-}$ complexes in presence of environmentally relevant species. *Journal of Photochemistry and Photobiology A*, **188**(1), 128–34.

Kremer M. L. (1999). Mechanism of the Fenton reaction. Evidence for a new intermediate. *Physical Chemistry Chemical Physics*, **1**, 3595–605.

Kremer M. L. (2003). The Fenton Reaction. Dependence of the Rate on pH. *Journal of Physical Chemistry*, **107**(11), 1734–41.

Kundu S., Chanda A., Khetan S. K., Ryabov A. D. and Collins T. J. (2013). TAML activator/peroxide-catalyzed facile oxidative degradation of the persistent explosives trinitrotoluene and trinitrobenzene in micellar solutions. *Environmental Science & Technology*, **47**(10), 5319–26.

Kurt U., Apaydin O. and Gonullu M. T. (2007). Reduction of COD in wastewater from an organized tannery industrial region by Electro-Fenton process. *Journal of Hazardous Materials*, **143**(1), 33–40.

Lapertot M., Pulgarín C., Fernández-Ibáñez P., Maldonado M. I., Pérez-Estrada L., Oller I., Gernjak W. and Malato S. (2006). Enhancing biodegradability of priority substances (pesticides) by solar photo-Fenton. *Water Research*, **40**(5), 1086–94.

Lee H., Lee H.-J., Sedlak D. L. and Lee C. (2013). pH-Dependent reactivity of oxidants formed by iron and copper-catalyzed decomposition of hydrogen peroxide. *Chemosphere*, **92**(6), 652–8.

Lehni M. and Fischer H. (1983). Effects of diffusion on the self-termination kinetics of isopropylol radicals in solution. *International Journal of Chemical Kinetics*, **15**(8), 733–57.

Liang X., Zhong Y., He H., Yuan P., Zhu J., Zhu S. and Jiang Z. (2012). The application of chromium substituted magnetite as heterogeneous Fenton catalyst for the degradation of aqueous cationic and anionic dyes. *Chemical Engineering Journal*, **191**, 177–84.

Lin S. H., Lin c.M. and Leu H. G. (1999). Operating characteristics and kinetic studies of surfactant wastewater treatment by Fenton oxidation. *Water Research*, **33**(7), 1735–41.

Lister M. W. (1956). Decomposition of sodium hypochlorite: The catalyzed reaction. *Canadian Journal of Chemistry*, **34**(4), 479–88.

Liu H., Li X. Z., Leng Y. J. and Wang, C. (2007a). Kinetic modeling of electro-Fenton reaction in aqueous solution. *Water Research*, **41**(5), 1161–7.

Liu H., Wang C., Li X., Xuan X., Jiang C. and Cui H. (2007b). A novel electro-Fenton process for water treatment: Reaction-controlled pH adjustment and performance assessment. *Environmental Science & Technology*, **41**(8), 2937–42.

Liu X., Wang F., Chen Z., Megharaj M. and Naidu R. (2014). Heterogeneous Fenton oxidation of Direct Black G in dye effluent using functional kaolin-supported nanoscale zero iron. *Environmental Science and Pollution Research*, **21**, 1936–43.

Ma J., He D., Collins R. N., He C. and Waite T. D. (2016). The tortoise versus the hare – Possible advantages of microparticulate zerovalent iron (mZVI) over nanoparticulate zerovalent iron (nZVI) in aerobic degradation of contaminants. *Water Research*, **105**, 331–340.

Ma Y.-S. (2012). Short review: Current trends and future challenges in the application of sono-Fenton oxidation for wastewater treatment. *Sustainable Environment Research*, **22**(5), 271–8.

Magalhães F., Pereira M. C., Botrel S. E. C., Fabris J. D., Macedo W. A., Mendonça R., Lago R. M. and Oliveira L. C. A. (2007). Cr-containing magnetites $Fe_{3-x}Cr_xO_4$: The role of Cr^{3+} and Fe^{2+} on the stability and reactivity towards H_2O_2 reactions. *Applied Catalysis A: General*, **332**(1), 115–23.

Maletzky P., Bauer R., Lahnsteiner J. and Pouresmael B. (1999). Immobilisation of iron ions on Nafion and its applicability to the photo-Fenton method. *Chemosphere*, **38**(10), 2315–25.

Matta R., Hanna K. and Chiron S. (2007). Fenton-like oxidation of 2,4,6-trinitrotoluene using different iron minerals. *Science of the Total Environment*, **385**, 242–51.

Merz, J. H. and Waters, W. A. (1947). A – Electron-transfer reactions. The mechanism of oxidation of alcohols with Fenton's reagent. *Discussions of the Faraday Society*, **2**, 179–88.

Merz J. H. and Waters W. A. (1949). S 3. Some oxidations involving the free hydroxyl radical. *Journal of the Chemical Society*, S15–25.

Miller C. M., Valentine R. L., Roehl M. E. and Alvarez P. J. J. (1996). Chemical and microbiological assessment of pendimethalin-contaminated soil after treatment with Fenton's reagent. *Water Research*, **30**(11), 2579–86.

Miller C. J., Rose A. L. and Waite T. D. (2013). Hydroxyl radical production by H_2O_2-mediated oxidation of Fe(II) complexed by Suwannee River fulvic acid under circumneutral freshwater conditions. *Environmental Science & Technology*, **47**(2), 829–35.

Miller C. J., Rose A. L. and Waite T. D. (2016). Importance of iron complexation for Fenton-mediated hydroxyl radical production at circumneutral pH. *Frontiers in Marine Science*, **3**, 134.

Miralles-Cuevas S., Oller I., Aguirre A. R., Pérez J. S. and Rodríguez S. M. (2014). Removal of pharmaceuticals at microg L^{-1} by combined nanofiltration and mild solar photo-Fenton. *Chemical Engineering Journal*, **239**, 68–74.

Molkenthin M., Olmez-Hanci T., Jekel M. R. and Arslan-Alaton, I. (2013). Photo-Fenton-like treatment of BPA: Effect of UV light source and water matrix on toxicity and transformation products. *Water Research*, **47**(14), 5052–64.

Muñoz I., Rieradevall J., Torrades F., Peral J. and Domènech X. (2005). Environmental assessment of different solar driven advanced oxidation processes. *Solar Energy*, **79**, 369–75.

Murov, S. L., Carmichael, I. and Hug, G. L. (1993). Handbook of photochemistry. 2nd edn, Marcel Dekker, New York, 299–305.

Neta P. and Dorfman L. M. (1968). Pulse radiolysis studies. XIII. Rate constants for the reaction of hydroxyl radicals with aromatic compkounds in aqueous solutions. Radiation Chemistry. American Chemical Society, Washington, D. C., 222–30.

Neta P., Huie R. E. and Ross A. B. (1990). Rate constants for reactions of peroxyl radicals in fluid solutions. *Journal of Physical and Chemical Reference Data*, **19**(2), 413–513.

Neta P., Grodkowski J. and Ross A. B. (1996). Rate constants for reactions of aliphatic carbon-centred radicals in aqueous solutions. *Journal of Physical and Chemical Reference Data*, **25**(3), 709–1050.

Nidheesh P. V., Gandhimathi R. and Ramesh S. T. (2013). Degradation of dyes from aqueous solution by Fenton processes: a review. *Environmental Science and Pollution Research*, **20**, 2099–132.

Oturan M. A. (2000). An ecologically effective water treatment technique using electrochemically generated hydroxyl radicals for in situ destruction of organic pollutants: Application to herbicide 2,4-D. *Journal of Applied Electrochemistry*, **30**, 475–82.

Olvera-Vargas H., Oturan N., Oturan M. A. and Brillas E. (2015). Electro-Fenton and solar photoelectro-Fenton treatments of the pharmaceutical ranitidine in pre-pilot flow plant scale. *Separation and Purification Technology*, **146**, 127–35.

Pestunova O. P., Elizarova G. L., Ismagilov Z. R., Kerzhentsev M. A. and Parmon V. N. (2002). Detoxication of water containing 1,1-dimethylhydrazine by catalytic oxidation with dioxygen and hydrogen peroxide over Cu- and Fe-containing catalysts. *Catalysis Today*, **75**, 219–25.

Pereira M. C., Oliveira L. C. A. and Murad E. (2012). Iron oxide catalysts: Fenton and Fenton-like reactions - a review. *Clay Minerals*, **47**(3), 285–302.

Pham A. L.-T., Doyle F. M. and Sedlak D. L. (2012). Kinetics and efficiency of H_2O_2 activation by iron-containing minerals and aquifer materials. *Water Research*, **46**(19), 6454–62.

Pham A. N., Xing G., Miller C. J. and Waite T. D. (2013). Fenton-like copper redox chemistry revisited: Hydrogen peroxide and superoxide mediation of copper-catalyzed oxidant production. *Journal of Catalysis*, **301**, 54–64.

Pignatello J. J., Oliveros E. and MacKay A. (2006). Advanced oxidation processes for organic contaminant destruction based on the Fenton reaction and related chemistry. *Critical Reviews in Environmental Science and Technology*, **36**(1), 1–84.

Pribush A. G., Brusentseva S. A., Shubin V. N. and Dolin P. I. (1975). Mechanism of the reaction of α-hydroxyisopropyl radicals with ferrous ions. *High Energy Chemistry*, **9**, 206–8.

Prucek R., Hermanek M. and Zbořil R. (2009). An effect of iron(III) oxides crystallinity on their catalytic efficiency and applicability in phenol degradation—A competition between homogeneous and heterogeneous catalysis. *Applied Catalysis A: General*, **366**(2), 325–32.

Pulgarin C., Invernizzi M., Parra S., Sarria V., Polania R. and Péringer P. (1999). Strategy for the coupling of photochemical and biological flow reactors useful in mineralization of biorecalcitrant industrial pollutants. *Catalysis Today*, **54**, 341–52.

Pulgarin C., Peringer P., Albers P. and Kiwi J. (1995). Effect of Fe-ZSM-5 zeolite on the photochemical and biochemical degradation of 4-nitrophenol. *Journal of Molecular Catalysis A: Chemical*, **95**(1), 61–74.

Qiu S., He D., Ma J., Liu T. and Waite T. D. (2015). Kinetic modeling of the electro-Fenton process: quantification of reactive oxygen species generation. *Electrochimica Acta*, **176**, 51–8.

Reisz E., Schmidt W., Schuchmann H. P. and von Sonntag C. (2003). Photolysis of ozone in aqueous solutions in the presence of tertiary butanol. *Environmental Science & Technology*, **37**(9), 1941–8.

Remucal C. K. and Sedlak D. L. (2011). The role of iron coordination in the production of reactive oxidants from ferrous iron oxidation by oxygen and hydrogen peroxide. In: Aquatic Redox Chemistry, P. G. Tratnyek, T. J. Grundl and S. B. Haderlein (eds.) ACS Symp. Ser., **1071**, 177–97.

Rivas F. J., Beltrán F. J., Garcia-araya J. F., Navarrete V and Gimeno, O. (2002). Co-oxidation of *p*-hydroxybenzoic acid and atrazine by the Fenton's like system Fe(III)/H_2O_2. *Journal of Hazardous Materials*, **91**(1–3), 143–57.

Ruan X., Gu X., Lu S., Qiu Z. and Sui Q. (2015). Trichloroethylene degradation by persulphate with magnetite as a heterogeneous activator in aqueous solution. *Environmental Technology*, **36**(11), 1389–97.

Rush J. D. and Bielski B. H. J. (1985). Pulse radiolytic studies of the reaction of perhydroxyl/superoxide O_2^- with iron(II)/iron(III) ions. The reactivity of HO_2/O_2^- with ferric ions and its implication on the occurrence of the Haber-Weiss reaction. *Journal of Physical Chemistry*, **89**(23), 5062–6.

Ryabov A. D. and Collins T. J. (2013). Green challenges of catalysis via iron(IV)oxo and iron(V)oxo species. In: Adv. Inorg. Chem., R. van Eldik and C. D. Hubbard (eds), Academic Press, USA, **61**, 471–521.

Sabhi S. and Kiwi J. (2001). Degradation of 2,4-dichlorophenol by immobilised iron catalysts. *Water Research*, **35**(8), 1994–2002.

Safarzadeh-Amiri, A., Bolton, J. R. and Cater, S. R. (1996). The use of iron in advanced oxidation processes. *Journal of Advanced Oxidation Technologies*, **1**(1), 18–26.

Sawyer D. T., Sobkowiak A. and Matsushita T. (1996). Metal [ML_x; M = Fe, Cu, Co, Mn]/hydroperoxide-induced activation of dioxygen for the oxygenation of hydrocarbons: Oxygenated Fenton chemistry. *Accounts of Chemical Research*, **29**(9), 409–16.

Schuchmann M. N. and von Sonntag C. (1979). Hydroxyl radical-induced oxidation of 2-methyl-2-propanol in oxygenated aqueous solution. A product and pulse radiolysis study. *Journal of Physical Chemistry*, **83**(7), 780–4.

Shappell N. W., Vrabel M. A., Madsen P. J., Harrington G., Billey L. O., Hakk H., Larsen G. L., Beach E. S., Horwitz C. P., Ro K., Hunt P. G. and Collins T. J. (2008). Destruction of estrogens using Fe-TAML/peroxide catalysis. *Environmental Science & Technology*, **42**(4), 1296–300.

Simic M., Neta P. and Hayon E. (1969). Pulse radiolysis study of alcohols in aqueous solution. *Journal of Physical Chemistry*, **73**(11), 3794–800.

Southworth B. A. and Voelker B. M. (2003). Hydroxyl radical production via the photo-Fenton reaction in the presence of fulvic acid. *Environmental Science & Technology*, **37**(6), 1130–6.

Stefánsson A. (2007). Iron(III) hydrolysis and solubility at 25 °C. *Environmental Science & Technology*, **41**(17), 6117–23.

Sun Y. and Pignatello, J. J. (1992). Chemical Treatment of pesticide wastes. Evaluation of Fe(III) chelates for catalytic hydrogen peroxide oxidation of 2,4-D at circumneutral pH. *Journal of Agricultural and Food Chemistry*, **40**(2), 322–7.

Sun Y. and Pignatello J. J. (1993). Activation of hydrogen peroxide by iron(III) chelates for abiotic degradation of herbicides and insecticides in water. *Journal of Agricultural and Food Chemistry*, **41**(2), 308–12.

Torres R. A., Pétrier C., Combet E., Moulet F. and Pulgarin C. (2007). Bisphenol A mineralization by integrated ultrasound-UV-iron (II) treatment. *Environmental Science & Technology*, **41**, 297–302.

Valentine R. L., Wang H. C. A. (1998). Iron oxide surface catalysed oxidation of quinoline by hydrogen peroxide. *Journal of Environmental Engineering*, **124**(1), 31–8.

Ventura A., Jacquet G., Bermond A. and Camel V. (2002). Electrochemical generation of the Fenton's reagent: application to atrazine degradation. *Water Research*, **36**, 3517–22.

von Sonntag C. (2008). Advanced oxidation processes: mechanistic aspects. *Water Science & Technology*, **58**(5), 1015–21.

von Sonntag C., Dowideit P., Xingwang F., Mertens R., Xianming P., Schuchmann M. N. and Schuchmann, H.-P. (1997). The fate of peroxyl radicals in aqueous solution. *Water Science & Technology*, **35**(4), 9–15.

Walling C. and Camaioni D. M. (1978). Role of silver(II) in silver-catalyzed oxidations by peroxydisulfate. *The Journal of Organic Chemistry*, **43**(17), 3266–71.

Walling C., Partch R. E. and Weil T. (1975). Kinetics of the decomposition of hydrogen peroxide catalyzed by ferric ethylenediaminetetraacetate complex. *Proceedings of the National Academy of Sciences*, **72**(1), 140–2.

Wang L., Wang F., Li P. and Zhang L. (2013). Ferrous-tetrapolyphosphate complex induced dioxygen activation for toxic organic pollutants degradation. *Separation and Purification Technology*, **120**, 148–55.

Wang L., Cao M., Ai Z. and Zhang L. (2014). Dramatically enhanced aerobic atrazine degradation with $Fe@Fe_2O_3$ core-shell nanowires by tetrapolyphosphate. *Environmental Science & Technology*, **48**(6), 3354–62.

Wang L., Cao M., Ai Z. and Zhang L. (2015). Design of a highly efficient and wide pH electro-Fenton oxidation system with molecular oxygen activated by ferrous-tetrapolyphosphate complex. *Environmental Science & Technology*, **49**(5), 3032–9.

Wang Q. and Lemley A. T. (2001). Kinetic model and optimization of 2,4-D degradation by anodic Fenton treatment. *Environmental Science & Technology*, **35**, 4509–14.

Watts R., Sarasa J., Loge F. and Teel A. (2005). Oxidative and reductive pathways in manganese-catalyzed Fenton's reactions. *Journal of Environmental Engineering*, **131**(1), 158–64.

Wu Y., Passananti M., Brigante M., Dong W. and Mailhot G. (2014). Fe(III)–EDDS complex in Fenton and photo-Fenton processes: from the radical formation to the degradation of a target compound. *Environmental Science and Pollution Research*, **21**(21), 12154–62.

Xu L. and Wang J. (2011). A heterogeneous Fenton-like system with nanoparticulate zero-valent iron for removal of 4-chloro-3-methyl phenol. *Journal of Hazardous Materials*, **186**, 256–64.

Yang Y., Wang P., Shi S. and Liu Y. (2009). Microwave enhanced Fenton-like process for the treatment of high concentration pharmaceutical wastewater. *Journal of Hazardous Materials*, **168**, 238–45.

Yang X.-j., Xu X.-m., Xu J. and Han Y.-f. (2013). Iron oxychloride (FeOCl): An efficient Fenton-like catalyst for producing hydroxyl radicals in degradation of organic contaminants. *Journal of the American Chemical Society*, **135**(43), 16058–61.

Yang X.-j., Tian P.-f., Zhang X.-m., Yu X., Wu T., Xu J. and Han Y.-f. (2015). The generation of hydroxyl radicals by hydrogen peroxide decomposition on FeOCl/SBA-15 catalysts for phenol degradation. *AIChE Journal*, **61**(1), 166–76.

Yip A. C.-K., Lam F. L.-Y. and Hu X. (2005). Chemical-vapor-deposited copper on acid-activated bentonite clay as an applicable heterogeneous catalyst for the photo-Fenton-like oxidation of textile organic pollutants. *Industrial & Engineering Chemistry Research*, **44**(21), 7983–90.

Yu X.-Y. (2004). Critical evaluation of rate constants and equilibrium constants of hydrogen peroxide photolysis in acidic aqueous solutions containing chloride ions. *Journal of Physical and Chemical Reference Data*, **33**(3), 747–63.

Zhang H. and Lemley A. T. (2006). Reaction mechanism and kinetic modeling of DEET degradation by flow-through anodic Fenton treatment. *Environmental Science & Technology*, **40**, 4488–94.

Zuo Y. and Hoigne J. (1992). Formation of hydrogen peroxide and depletion of oxalic acid in atmospheric water by photolysis of iron (III)-oxalato complexes. *Environmental Science & Technology*, **26**(5), 1014–22.

Chapter 8

Photocatalysis as an effective advanced oxidation process

Suresh C. Pillai, Niall B. McGuinness, Ciara Byrne, Changseok Han, Jacob Lalley, Mallikarjuna Nadagouda, Polycarpos Falaras, Athanassios G. Kontos, Miguel A. Gracia-Pinilla, Kevin O´Shea, Ramalinga V. Mangalaraja, Christophoros Christophoridis, Theodoros Triantis, Anastasia Hiskia, and Dionysios D. Dionysiou

8.1 INTRODUCTION

In 1972, a new type of electrochemical cell was designed, composed of titanium dioxide (TiO_2) and platinum black (Pt-black) electrodes, which could induce the photolysis of H_2O using visible light energy (Fujishima & Honda, 1972). This resulted in the vast development of this field, where research entails the decomposition of organic compounds *via* photocatalysis. This advanced oxidation process (AOP) employs materials, which can increase the rate of a chemical reaction *via* the absorption of photon energy of a specific wavelength (Banerjee *et al.* 2014; Pelaez *et al.* 2012; Schneider *et al.* 2014) and this is determined by the compounds band gap, along with the electronic structure (Hoffmann *et al.* 1995).

In recent days, human beings, animals, vegetation and the environment are affected by hazardous pollutants of emerging concern identified in air and water. These pollutants include nitrogen oxides (NO_x), volatile organic compounds (VOCs), industrial and pharmaceutical compounds, toxins, hormones, endocrine disruptors, and pathogenic microorganisms (Keane *et al.* 2014). There is a need for worldwide coordinated research aiming at pollutant detection, assessment, and abatement, especially concerning new extremely harmful emerging contaminants. In this direction, the development of novel nanomaterials and the implementation of innovative AOPs, technologies (AOTs) and nanotechnologies (AONs) for their elimination are one of the highest priorities for the protection and the remediation of the environment.

Photocatalytic materials can be synthesized using various methods such as sol-gel processes, microwave assisted synthesis, and electrochemical, hydrothermal and solvothermal synthesis. They have been applied to a vast array of surfaces as they possess many desirable properties such as self-cleaning and antimicrobial activity, resulting in many industrial and commercial applications. Photocatalytic related applications include electrochromics, self-cleaning materials, solar cells, water splitting, pollutant degradation processes and water purification technologies to name a few (Etacheri *et al.* 2012a; Etacheri *et al.* 2011; Kamat, 2007; Keane *et al.* 2014; Mills & Lee, 2002; Pelaez *et al.* 2012; Periyat *et al.* 2010; Pillai *et al.* 2007).

8.2 PROCESS PRINCIPLES INCLUDING THE MOST RECENT SCIENTIFIC FINDINGS AND INTERPRETATION

8.2.1 Nanotubular titania-based materials for photocatalytic water and air purification

Titania-based nanomaterials possess unique structural, morphological and optoelectronic properties that make them ideal candidates for significant light-driven applications in the fields of environmental protection and energy conversion (Hoffmann *et al.* 1995; O'Regan *et al.* 1991; Khan *et al.* 2002; Babu *et al.* 2015). Innovative synthetic strategies permit the efficient control of the preparation conditions and lead to the optimization of the materials functional parameters including size, shape, structural modification, surface area, light absorption (including band gap engineering) and electronic conduction. Thus, TiO_2 catalysts in the form of slurries, thin films and composites present excellent photocatalytic activity and have been used with success for the degradation of a great number of organic and inorganic pollutants in natural environments (Fujishima *et al.* 2008).

Different approaches including template-assisted, electrochemical anodic oxidation and alkaline hydrothermal treatment are used to prepare the TiO_2-based nanotubes (NTs) (Pang *et al.* 2014). Self-organized TiO_2 nanotubular films prepared by titanium foil anodization in corrosive media (usually F-containing aqueous or organic solutions) are endowed with a highly ordered and open porous arrangement with chemical stability (Figure 8.1).

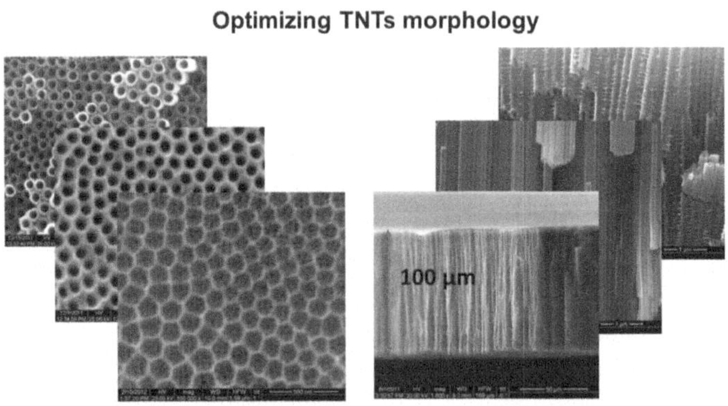

Figure 8.1 Effective control of the anodization conditions (potentiostatic or galvanostatic anodic oxidation) in organic electrolyte (ethylene glycol (EG)) with the presence of water and fluoride ions leads to a variety of morphologies (top surface, tubes aspect and length, wall thickness). Image courtesy from N. Vaenas' PhD Thesis, NCSR Demokritos, 2014.

Such one-dimensional high aspect (surface-volume) ratio nanostructures, having packed and vertically aligned morphologies outperform TiO_2 powders/slurries by avoiding the problems of particle agglomeration and additional separation steps. Furthermore, by offering the additional advantages of easy operation and external bias support they open new directions for the development of innovative photocatalytic reactors

(Zhang *et al.* 2011). The critical factors and the key parameters controlling the anodization process (applied potential, concentration, pH and type of corrosive solutions, anodization time, bath temperature and crystallization conditions), that lead to the spectacular self-ordering which occurs during the growth of the TiO_2 NT arrays, were explicitly discussed in the literature (Zhou *et al.* 2014; Ghicov *et al.* 2009). These parameters influence mechanical, structural, chemical and electrical features of porous TiO_2. It has also been confirmed that the photocatalytic activity of self-organized TiO_2 NTs possessing a packed, vertically aligned morphology is related to surface wetting properties expressed *via* the UV-induced superhydrophilic effect. In addition, it was further found that the photocatalytic activity quantitatively correlates with the variation of the geometric roughness factor, verifying the strong impact of morphology on the photoinduced properties of the vertically oriented TiO_2 tubular architectures (Kontos *et al.* 2009). The main applications of TiO_2 NT arrays, i.e. removal of pollutants, environmental analytical chemistry, water splitting, solar cells and CO_2 conversion, have been recently reviewed, confirming the high scientific and technological potential of these porous nanostructures (Zhou *et al.* 2015). Particular attention has been paid to self-organized TiO_2 NT arrays and their application for the photocatalytic decomposition of toxic and hazardous organic pollutants in contaminated water and air. In this case, the nanotubular materials, fabricated by the anodization of the titanium metal substrate, yield structures of one orientation with a tunable pore size, a large internal surface area, and demonstrate excellent recovery and reutilization. The structural parameters of TiO_2 nanotubular films (wall thickness, tube length and crystalline phase) can be finely tuned and have important effects on the photocatalytic activity. The photocatalytic properties of anodized anatase TiO_2 nanotubular films have been tested for the degradation of toluene and benzene volatile organic pollutants. Examining ppb concentrations, while under conditions corresponding to the real outdoor environment, it has been shown that the photocatalytic activity depends on the tubes length and morphology. The titania NTs presented high photocatalytic decomposition rates and most importantly outperformed commercially available nanoparticulate Evonic P-25 films of similar thickness (Kontos *et al.* 2010). In addition, the photocatalytic oxidation of NO pollutants at ppb concentrations was achieved in a continuous flow photocatalytic reactor under dynamic experimental conditions corresponding to the real environment, (Kontos *et al.* 2012). All the above results verify the advantages for the employment of self-ordered nanoporous TiO_2 NT arrays during the photocatalytic process along with their application potential for air purification.

Besides air pollutants, TiO_2 NTs can be used for water purification purposes. High photocatalytic activity has been observed for TiO_2 nanotubular films prepared in a solution of ethylene glycol (EG) containing ionic liquids and water, when utilized for the degradation of the azo-dye methyl orange (MO) under UV light (Wender *et al.* 2011). It has been confirmed that the optimization of thermal treatment conditions leads to large specific NT surface area (permitting greater pollutant adsorption) and high pore volume (accelerating species diffusion), which both enhance the photocatalytic reaction rate for the degradation of the pollutant 2,3-dichlorophenol (2,3-DCP) under UV illumination (Liang *et al.* 2009). Anatase TiO_2 NT powders, produced using anodic oxidation and ultrasonication, present high crystallinity and enhanced photocatalytic performance for methylene blue dye degradation (Lin *et al.* 2015). TiO_2 NTs and other titania anisotropic nanomaterials (nanorods, nanoplates, nanospheres, and nanoparticles) with controlled structural and textural properties have been evaluated during the photocatalytic degradation of phenol under UV illumination conditions (Turki *et al.* 2015). It has been demonstrated that a good compromise between anatase crystallinity, crystallites sizes, and specific surface area can, however improve the photocatalytic activity of the TiO_2 nanomaterials. Anodized TiO_2 nanotubular films and membranes can be detached from the metallic foil and transferred onto glass/conductive glass substrates. This enables the illumination of both sides (front and back) and amplifies the materials potential in novel applications. Thus, significantly enhanced photocatalytic activity for glucose photo-oxidation was observed (Zhang *et al.* 2011). In addition, micro-Raman spectroscopic analysis using polarized light confirmed the existence of an antenna effect, indirect proof of vectorial electron transfer along with the NT axis (Likodimos *et al.* 2008). Such an antenna Raman effect (Figure 8.2) justifies expectations for higher charge (electron) mobility, a fact that also warrants lower recombination rates and therefore could justify the increased photocatalytic and photoelectrocatalytic performance of anodized TiO_2 nanotubular films and membranes (Liu *et al.* 2008).

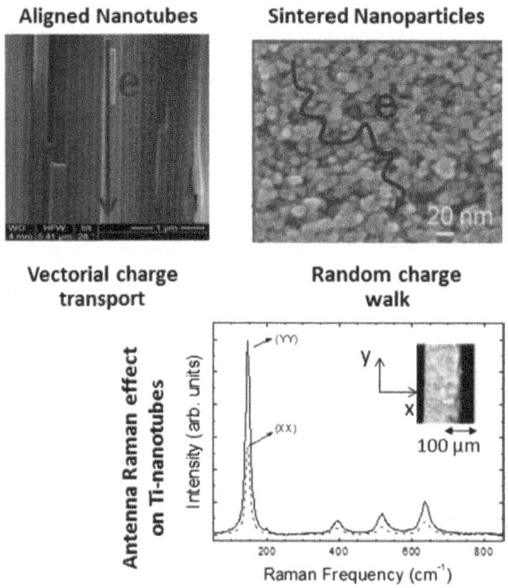

Figure 8.2 Comparing electron transport and recombination between nanotubular and nanoparticulate TiO_2 films.

The NTs modification (*via* doping, metal deposition, and sensitization) insures a visible light response and decreases electron/hole recombination. Thus, the removal of herbicide 2,4-dichlorophenoxyacetic acid from water using silver/reduced Graphene Oxide (Ag/RGO) co-decorated TiO_2 NT arrays has been reported (Tang *et al.* 2012). The same authors also demonstrated that RGO and PbS nanoparticles co-modified TiO_2 NTs are a recyclable and stable photocatalyst for the efficient degradation of pentachlorophenol (Zhang *et al.* 2013). The highly efficient photocatalytic degradation of *p*-nitrophenol was obtained under artificial solar light on a Cu_2O/TiO_2 p-n heterojunction network catalyst, fabricated by anodizing CuO particle-loaded TiO_2 NTs (Yang *et al.* 2010). In the same direction, the high photocatalytic removal of both dye acid orange 7 (AO7) and the industrial chemical *p*-nitrophenol while using $Ag–Ag_2O/TiO_2$ NTs under visible light irradiation has been recently reported (Liu *et al.* 2015). The increased visible-light-driven photocatalytic properties of Cr,N co-doped titania NTs were confirmed using MO dye as a model pollutant (Fan *et al.* 2015). Finally, the photocatalytic degradation and toxicity evaluation of the nonsteroidal anti-inflammatory drug diclofenac was investigated using nanotubular titanium dioxide-Polyethersulfone (PES) membranes in a static and continuous setup (Fischer *et al.* 2015), confirming that this industrially compatible approach can lead to complete degradation of pollutants and avoid the release of highly toxic intermediate products into surface waters.

Following a holistic approach, beginning from the level of the nanomaterials synthesis to device design and manufacture, and then passing through process engineering and optimization, an innovative and efficient environmental purification nanotechnology for the destruction of extremely hazardous toxins and emerging pollutants in polluted water and air environments can be developed. This can be successfully achieved (Figure 8.3) through the use of organized and self-assembled TiO_2 NTs and solar irradiation, which has a positive impact on the environment. The new technology, in contrast to conventional separation methods, focuses on AOPs for global environmental applications and brings about the photocatalytic degradation of contaminants. It thus permits air/water purification and environmental quality enhancement under normal solar light conditions, with high efficiency and low cost (Pelaez *et al.* 2012; Moustakas *et al.* 2013; Romanos *et al.* 2013; Moustakas *et al.* 2014; Banerjee *et al.* 2014).

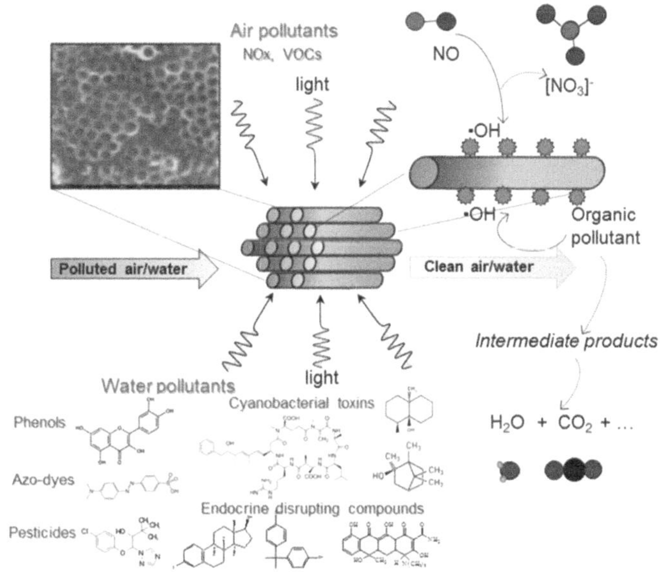

Figure 8.3 Photocatalytic purification of air and water pollutants flowing through tubular TiO_2 nanostructures.

8.2.2 Magnetically separable photocatalysts

Iron-based materials have been widely used for environmental remediation (Casbeer *et al.* 2012; Ren *et al.* 2013). In particular, spinel ferrites (MFe_2O_4, M = Ni, Co etc.) with a relatively narrow band gap (approx. 2.0 eV) have been extensively studied as magnetically separable photocatalysts since they have demonstrated photocatalytic activities for the decomposition of contaminants in water (Casbeer *et al.* 2012; He, 2015). Moreover, due to the magnetic properties of iron-based materials, such as ferrites, maghemite (γ-Fe_2O_3) and magnetite (Fe_3O_4), different materials, including TiO_2, graphene, and graphitic carbon nitride (g-C_3N_4), were deposited on or composited into the iron-based materials, with the purpose of developing magnetically separable photocatalysts exhibiting high photocatalytic activity (Larumbe *et al.* 2014; Xuan *et al.* 2009; Chen & Sun, 2012; Ye *et al.* 2013; Fu & Wang, 2011).

Nickel-cobalt ferrite ($Ni_{1-x}Co_xFe_2O_4$) was prepared by He (2015) *via* a hydrothermal process for the effective degradation of malachite green under solar light irradiation. Using this method, nanostructured $Ni_{1-x}Co_xFe_2O_4$ particles with average sizes ranging from 14–29 nm were synthesized. A decrease in band gap to within the range of 1.75–1.91 eV, depending on the cobalt content, was also observed. All synthesized samples were ferrimagnetic and their saturation magnetization ranged from 50.7–64.2 emu g^{-1}. Casbeer *et al.* (2012) reviewed literature on visible light-induced ferrite photocatalysis. Calcium, zinc, and manganese ferrites demonstrated the degradation of methylene blue (MB) dye under light irradiation with wavelengths longer than 400 nm. Methyl orange (MO) dye was also effectively decomposed using ferrite photocatalysis under UV, visible, and solar light illumination. Moreover, zinc ferrite ($ZnFe_2O_4$) NT arrays were synthesized and evaluated for the decomposition of 4-chlorophenol (4-CP) under visible-light irradiation (Li *et al.* 2011). The $ZnFe_2O_4$ NT arrays with a length of 60 μm and a diameter of 200 nm, synthesized using a sol-gel based method, effectively decomposed 4-CP after 6 hours of visible light illumination. The sample showed consistent efficiency for the photocatalytic degradation of 4-CP during five consecutive experiments, indicating that $ZnFe_2O_4$ NT arrays are reusable.

In addition to ferrite photocatalysis, as mentioned above, iron-based materials have been used as a core to prepare magnetically separable photocatalysts with high photocatalytic activity. TiO_2 is the main photocatalyst used during the preparation since its photocatalytic activity has already been proven to decompose many contaminants (Xuan *et al.* 2009; Larumbe *et al.* 2014; Han *et al.* 2011, 2014a). Details on TiO_2 photocatalysis are discussed in Section 8.2. Magnetic Fe_3O_4/TiO_2 hollow spheres, with a saturation magnetization value of 42.7 emu g^{-1}, were synthesized using poly(styrene-acrylic acid) (Xuan *et al.* 2009). The wall thickness of Fe_3O_4/TiO_2 was approximately 50 nm, while the anatase TiO_2, possessing a calculated crystal size of 10 nm (Scherrer equation), formed within the hollow spheres. The synthesized hollow spheres were employed to decompose Rhodamine B (RhB) dye under UV irradiation. After 80 minutes of UV illumination, 96% of the RhB dye was degraded and furthermore the hollow spheres demonstrated high photocatalytic activity for RhB dye decomposition during six cycle experiments. Chung *et al.* (2004) prepared spherical TiO_2-coated $NiFe_2O_4$ (TiO_2-$NiFe_2O_4$) using ultrasonic spray pyrolysis. Since the unfavourable heterojunction of the TiO_2 and iron-based materials decreased the efficiency of TiO_2 photocatalysis, a layer of silicon dioxide (SiO_2) was added between the TiO_2 coating and the iron core. The addition of a SiO_2 layer can prevent drawbacks, including the photo-dissolution of iron and an increase in the recombination of photo-generated electrons and holes which are directly associated with a decrease in TiO_2 photocatalysis. The content of $NiFe_2O_4$ in TiO_2-$NiFe_2O_4$ varied from 0–100%. All samples were ferromagnetic, had an average diameter of 1.3 μm and their magnetic strength was proportional to the content of ferrite in the samples. In order to investigate the photocatalytic activity of the samples, MB dye degradation was performed under UV irradiation. The samples were reused three times for MB dye degradation and their photocatalytic activity during each cycle of the test was found to be similar, indicating that the TiO_2 coating on the ferrite is very stable. This may result from the dilution effect of the Ti and Si precursors of TiO_2-$NiFe_2O_4$ and the inner layer of SiO_2, respectively. For visible light-induced TiO_2 photocatalysis, Pelaez *et al.* (2013) synthesized a composite, N-$TiO_2/NiFe_2O_4$, with N-TiO_2 being coated on the surface of the $NiFe_2O_4$ core. Under visible light illumination, N-$TiO_2/NiFe_2O_4$ effectively decomposed microcystin-LR (MC-LR). After 5 hours of visible light irradiation, MC-LR was completely degraded while N-TiO_2 alone decomposed 75% of the MC-LR. This may result from the role of $NiFe_2O_4$ as a recombination center for electrons and holes during the visible light-induced N-TiO_2 photocatalysis. However, its photocatalytic activity decreased to 70% of the initial activity after three cycle experiments, indicating that the N-TiO_2 coating is not very stable, although the samples showed appropriate magnetic properties for easy separation.

In addition to the employment of TiO_2 photocatalysts, different carbon-based materials such as graphene, CNTs, and carbon nitride were introduced to iron-based materials for the synthesis of magnetically separable photocatalysts (Fu & Wang, 2011; Fu *et al.* 2012; Ye *et al.* 2013; Xiong *et al.* 2012). CNTs were employed to develop magnetically separable photocatalysts with $NiFe_2O_4$ due to their high surface area, high chemical stability, and electrical conductivity (Xiong *et al.* 2012). In their study, non photocatalytically active $NiFe_2O_4$ nanoparticles (NPs) (<10 nm) were deposited on multi-walled carbon nanotubes (MWCNTs) using a one-step hydrothermal method. The light absorption edge of samples was redshifted and the samples demonstrated high photocatalytic activity for the degradation of phenol. During three cycle experiments, the samples decomposed >80% phenol in each experiment, indicating that the synthesized materials are mechanically stable. Moreover, graphene was newly introduced to ferrites for visible light-induced photocatalysis due to its properties such as high surface area, high chemical and thermal stability, and remarkable electrical conductivity (Fu & Wang, 2011; Fu *et al.* 2012). In the studies, ferrite (i.e., $ZnFe_2O_4$ and $NiFe_2O_4$) NPs (<10 nm) were deposited on graphene using a simple one-step hydrothermal method. In both samples, $ZnFe_2O_4$ and $NiFe_2O_4$, a redshift of the light absorption edge was observed, implying a band gap narrowing of the samples. MB dye was effectively decomposed under visible light illumination in the presence of ferrite-deposited graphene. Recently, g-C_3N_4 was used to develop magnetically separable photocatalysts since g-C_3N_4 has gained much attention for its use in the photocatalytic degradation of organic pollutants (Ye *et al.* 2013). They reported that mainly magnetic γ-Fe_2O_3 was formed during their synthesis method and that the content of Fe_2O_3 varied from 2.8–11.6% in the synthesized samples. The highest saturation magnetization value was 1.56 emu g^{-1} when the Fe_2O_3 content was 11.6 wt% and its calculated band gap was 1.66 eV due to a shift of light absorption towards longer wavelengths. In order to evaluate the photocatalytic activity of the samples, RhB dye degradation experiments were performed under visible light. The highest photocatalytic activity was observed for the sample containing 2.8% Fe_2O_3 and the activity was inversely proportional to the content of Fe_2O_3, which may result from increasing electron-hole recombination due to excess Fe_2O_3 in the samples. Therefore, the content of Fe_2O_3 has to be carefully controlled to develop highly active photocatalysts using Fe_2O_3.

8.2.3 Improving the photocatalytic activity

Improvements in the properties of photocatalytic materials, such as TiO_2, have been achieved *via* doping of the semiconductor with both metal and/or non-metal species such as iron, (Niu *et al.* 2013; Yang *et al.* 2010) neodymium, (Wu *et al.* 2013) vanadium, (Liu *et al.* 2011) molybdenum, (Li *et al.* 2012) copper, (Fisher *et al.* 2013) tungsten (Thind *et al.* 2012) nitrogen, (Fisher *et al.* 2013; Hamilton *et al.* 2014; Liu *et al.* 2011; Nolan *et al.* 2012; Thind *et al.* 2012; Yang *et al.* 2010) carbon, (Li *et al.* 2012; Wu *et al.* 2013) sulfur, (Niu *et al.* 2013) and fluorine (Hamilton *et al.* 2014). Other popular semiconductors, such as ZnO and ZnS, have also been subjected to doping using metals (Etacheri *et al.* 2012a; Kudo & Sekizawa, 2000).

Researchers have produced nanoscale heterojunctions *via* the deposition of various metal oxides as in the form of small particles in a thin film (Patrocinio *et al.* 2014). Also, metal-inorganic heterojunctions have been fabricated where metals, such as Ag, dope the metal oxide, trapping photoexcited electrons from the semiconductors conduction band (CB) (Georgekutty *et al.* 2008). The stable polymeric semiconductor, g-C_3N_4, exhibits a visible light photocatalytic response and has many applications (Gondal *et al.* 2015; C. Wang *et al.* 2014). The incorporation of graphene into semiconductor systems is becoming well documented due to the fabrication of graphene-TiO_2 (Cao *et al.* 2014; Sayed *et al.* 2014; Wang *et al.* 2014) and graphene-ZnO (Chang *et al.* 2015; Fu *et al.* 2012; L. Zhang *et al.* 2013) composites, which possess an enhanced charge separation and demonstrate a long recombination time for the photoinduced electron-hole pairs (Fan *et al.* 2012; Xu *et al.* 2011; H. Zhang *et al.* 2010; J. Zhang *et al.* 2013).

8.2.3.1 *Metal and non-metal doped TiO_2*

The doping of TiO_2 with various species has been studied for some time due to the attainment of a desirable narrowed band gap and the increase in recombination time of the photoinduced electron-hole pairs, therefore allowing for visible light photoactivation. Nitrogen and copper co-doped TiO_2 transparent thin coatings have been adhered to the inside of glass bottles using sol-gel chemistry and these modified coatings were capable of decomposing both bacteria (*Escherichia coli (E. coli)* and *Enterococcus faecalis*) and chemical (MB dye) contaminants *via* solar irradiation. These effects were found to be enhanced when compared to undoped TiO_2 and the effects were also observed to continue in the absence of UV light (Fisher *et al.* 2013). Also employing sol-gel chemistry, co-doping TiO_2 with iron and sulfur, while using a solvothermal method to narrow the band gap, extended absorption into the visible light region. The Fe^{3+} and SO_4^{2-} species on the surface enhanced the separation of photoinduced electron-hole pairs, smothering recombination by trapping the photoexcited electrons. The doping of TiO_2 with iron and sulfur produced particles of a small crystallite size and large surface area that absorbed water onto its surface. This further encouraged hydroxyl radicals (•OH) formation, enhancing the photocatalytic decomposition of phenol (99.4% degradation) in water, when exposed to visible light for 10 hours (Niu *et al.* 2013).

Hydrothermally synthesised anatase TiO_2 NPs containing nitrogen and iron co-dopants have been synthesised. Utilising visible light irradiation, doping with nitrogen was found to increase the materials photocatalytic efficiency two-fold, but doping with Fe or N-Fe was found to diminish activity when compared to pure TiO_2. Incorporation of Fe^{3+} narrows the band gap effectively, but produces little hydroxyl radicals and possesses a high photoluminescence efficiency, causing an increase in the electron-hole recombination (Yang *et al.* 2010). Using a two-step hydrothermal impregnation technique, TiO_2 nano-catalysts were synthesised. Firstly, they were doped with vanadium (Figure 8.4 (left)) and then secondly with nitrogen (Figure 8.4 (right)), which after treatment, exhibited the formation of smaller crystallites, accompanied by a greater surface area. The impregnation of vanadium produced a redshift in absorption to the visible light region, due to the formation of a V 2p state, which was induced by the presence of V^{4+} species inside the lattice. Furthermore, the formation of both substitutional N 2p and interstitial N–O states, along with the NO_x species and vacant oxygen atom sites, contribute to the strong absorption shift for V-N-TiO_2. The photocatalytic degradation rate of harmful sodium pentachlorophenate (PCP–Na) was examined in the presence of the co-doped V-N-TiO_2 material and this was found to be 3.1 times greater than that for the vanadium impregnated TiO_2 which only underwent one hydrothermal treatment (Liu *et al.* 2011).

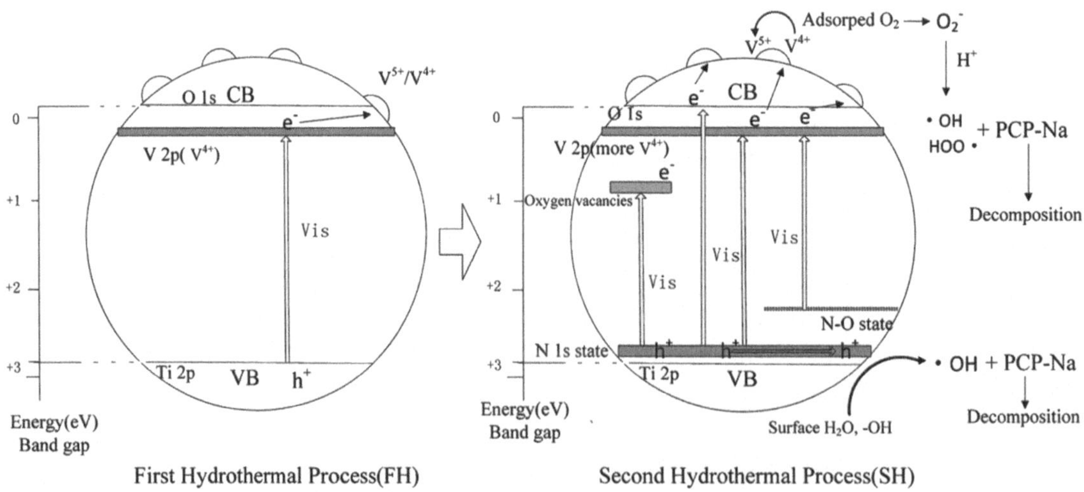

Figure 8.4 The two-step hydrothermal sequence (left: first process and right: second process) and the visible-light-induced photocatalytic reaction. (Reprinted with permission from Liu *et al.* 2011).

Carbon and molybdenum co-doped TiO_2 materials have been synthesised *via* the hydrothermal method. After calcination of this material, carbonaceous species were discovered on the surface, which permits the enhanced absorption of visible light, while within the lattice, Mo^{6+} species were substituted for titanium, yielding a doping energy level which increased photogenerated electron-hole separation. Using visible light irradiation, the photocatalytic decompositions of RhB dye and acetone in water were found to be most efficient in the co-doped material as opposed to both C-TiO_2, Mo-TiO_2 and pure TiO_2 (Li *et al.* 2012). TiO_2 was co-doped with neodymium and carbon using a similar solvothermal method and this yielded a material with absorption properties spanning the total UV and visible light range, that is, wavelengths from 200–900 nm. This was possibly due to the presence of the unoccupied Nd^{3+} 4f energy level below the TiO_2 CB, therefore permitting the facile transfer of electrons from the TiO_2 Valance Band (VB) to the unoccupied neodymium level, making visible light absorption possible. Also, neodymium doping within the TiO_2 matrix yields diminished particle size and therefore greater specific surface area which decreases the distance photoexcited electrons need to travel to reach the surface reaction sites. Furthermore, the material exhibited an extremely effective photocatalytic degradation of gaseous NO_x and MO in water (Wu *et al.* 2013). Employing a solution combustion process, a high specific surface area, mesoporous anatase TiO_2 was produced that was co-doped with nitrogen and tungsten and possessed a narrowed band gap (approx. 2.7 eV) due to the effective incorporation of these dopants into the TiO_2 lattice. While using the co-doped material under visible light irradiation ($\lambda > 420$ nm), the rate of photocatalytic decomposition of RhB dye in water was found to be fourteen-fold greater than that of Degussa P-25 (Thind *et al.* 2012).

8.2.3.2 Nano-heterojunctions

Heterojunctions are usually composed of two different semiconductor compounds, which supply the material with an unequal band gap, suitable for inducing a long recombination time for electron-hole pairs. TiO_2 anatase-rutile heterojunctions, doped with nitrogen, have been fabricated *via* an ethylene-diamine-tetra-acetic acid (EDTA) modified sol-gel method, which displayed a nine-fold increase in visible light photocatalytic activity, when compared to the commercial photocatalyst Degussa P-25. Analysis confirmed effective electron-hole separation promoted by the transfer of photoexcited electrons from the anatase CB to the rutile CB, instigating a greater photocatalytic activity under visible light irradiation (Etacheri *et al.* 2010). The authors also synthesised an anatase-rutile heterojunction co-doped with sulfur and nitrogen which exhibited a photocatalytic activity that was eight-fold greater than that of Degussa P-25. Isolated S 3p, N 2p and π antibonding N–O states formed between the VB and CB, narrowed the band gap and permitted the enhanced absorption of visible light (Etacheri *et al.* 2012b). Another method developed consisted of the nonhydrothermal low power (300 W) microwave assisted synthesis of an anatase-brookite TiO_2 nano-heterojunction at the low temperature of 100°C. Due to doping with carbon, inter-band C 2p states were formed between the VB and CB causing narrowing of the TiO_2 band gaps, resulting in a photocatalytic activity that was two-fold greater than Degussa P-25. Furthermore, this material possessed visible light induced photocatalytic activity and exhibited antibacterial activity for *S. aureus*. This was attributed to the transfer of photoexcited electrons from the CB of the brookite phase to that of the anatase, yielding the effective separation of photoinduced electron-hole pairs (Etacheri *et al.* 2013).

TiO_2/WO_3 films have been synthesised, which act as nanoscale heterojunctions, *via* a layer-by-layer deposition method (Figure 8.5). Sol-gel chemistry was utilised to form the metal oxide NPs on fluorine doped tin oxide (FTO) glass substrates. The nanosized WO_3 components within the film were found to decrease the optical band gap of the bilayers when compared to analogous pure TiO_2 films. The UV light enhanced superhydrophilic films were utilised to oxidise acetaldehyde in the gas phase at photonic efficiencies two-fold greater than analogous TiO_2 films, deeming them excellent antifogging and self-cleaning materials (Patrocinio *et al.* 2014).

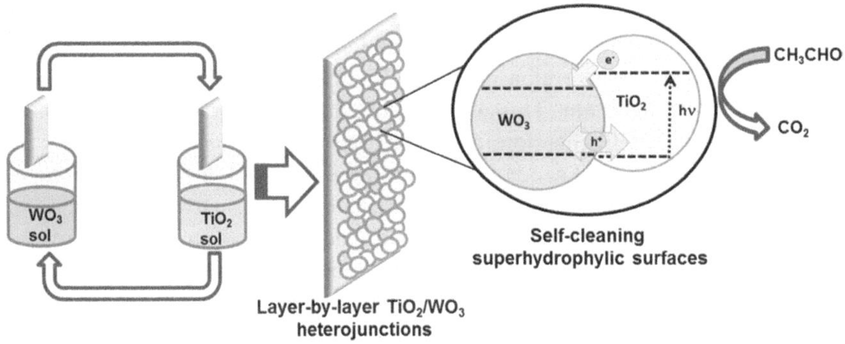

Figure 8.5 Demonstration of the formation of the TiO_2/WO_3 heterojunctions using sol-gel chemistry, succeeded by photoinduced electron-hole separation on a layer-by-layer TiO_2/WO_3 film, resulting in an efficient self-cleaning surface. (Reprinted with permission from Patrocinio *et al.* 2014).

Single crystal p-NiO NPs, adhered to n-ZnO nanowires, have also been fabricated to enhance UV photogain to approximately 2.8×10^8, a value that is one thousand-fold greater than that of pristine ZnO nanowires. This dramatic enhancement has been attributed to the formation of p-n nano-heterojunctions (approximately 70% surface coverage) on the ZnO nanowire surface. This property promotes the separation of photoinduced electron-hole pairs by upward surface band bending, which was confirmed by the shift in gate voltage, yielding an effective way to alter the performance of devices by using surface engineering (Retamal *et al.* 2014).

8.2.3.3 Metal-inorganic hetero-structures

The photocatalytic activity of Ag-ZnO has been studied *via* the analysis of organic dye degradation in water, specifically Rhodamine 6G dye. At certain mol% and temperature treatments, these materials exhibited significantly improved photocatalytic activity, three-fold and five-fold greater than that of pure ZnO and Degussa P-25, respectively. Furthermore, under visible light irradiation the Ag-ZnO sample possessed a five-fold increase in activity as opposed to the pure ZnO material. The introduction of the crystallite Ag permits the utilisation of the ZnO e^-_{CB} by trapping it, which in turn, enhances the photocatalytic activity (Georgekutty *et al.* 2008). Graphene-WO_3-Au hybrid membranes on Fluorinated doped Tin Oxide (FTO) have demonstrated superior photocatalytic activity for enzyme mediated oxidation of glucose to gluconic acid (see Figure 8.6 (top-left and top-right)). WO_3 NPs generate electron-hole pairs, which can oxidize biomolecules *via* flavin adenine dinucleotide/ reduced flavin adenine dinucleotide (FAD/$FADH_2$) while also reducing H_2O to H_2. In this system (Figure 8.6 (bottom)), the plasmonic Au NPs reinforce the visible light photocatalytic activity of the graphene-WO_3 membrane by promoting charge collection at its interface *via* Schottky junction formation (Electron Route I). In tandem, the catalytic activity of the immobilised glucose oxidase (GOD) enzyme is also enhanced (Electron Route II) (Devadoss *et al.* 2014).

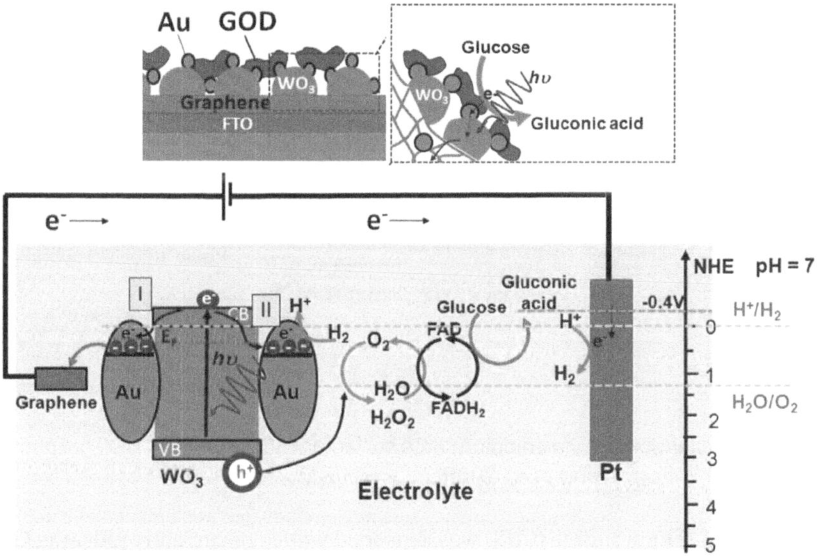

Figure 8.6 (top-left and top-right) The triplet junction of graphene-WO_3-Au for glucose sensing. (Bottom) Mechanism of glucose oxidation and the energy levels of the graphene-WO_3-Au photoelectrode. (Reprinted with permission from Devadoss *et al.* 2014).

A liquid metal/metal oxide framework has been synthesised by employing a solvothermal technique. This framework consisted of micro to nanosized plasmonic galistan material, while also incorporating γ-Ga_2O_3 NPs. Galistan is an alloy of gallium, indium and tin with a naturally forming oxide layer on the surface. In this study the optimised system demonstrated approximately 100% photocatalytic efficiency per hour. Repeated catalytic cycling was shown to yield high reusability, when examining the photocatalytic degradation of the organic dye Congo Red in water under simulated solar light. Examination of the band structures showed the potential formation of pseudo-ohmic contacts between the NPs and the liquid metal constituent. This reduces the hole injection barrier, causing an increase in the oxidation potential and the concentration of excess holes at the γ-Ga_2O_3 valance band, which can react with other species (W. Zhang *et al.* 2015). Other materials synthesised are Pt-CdSe-Pt nanodumbbells, which have been found to demonstrate a two to three-fold catalytic activity increase for the oxidation of CO in contrast to pure platinum NPs, when irradiated by light possessing energy greater than the CdSe band gap. Therefore, it was found in this study that the photoinduced hot electrons are injected into the Pt NPs from the CdSe semiconductor and are involved in the oxidation of CO (S. M. Kim *et al.* 2013).

8.2.3.4 Graphene composites

Research has demonstrated that the photocatalytic self-cleaning efficiency of graphene/TiO_2 composites increases two-fold by incorporating larger concentrations of graphene within the material. Furthermore, the hydrophilic conversion efficiency also increases in contrast to TiO_2 films (Anandan *et al.* 2013). The self-assembly of rutile and anatase nanocrystalline TiO_2 with graphene by using the anionic surfactant, sodium dodecyl sulfate has been performed (Figure 8.7). Under visible light irradiation, Graphene Oxide (GO) (5 wt% relative to the nanohybrid rod) has been shown to be necessary for maximum photocatalytic efficiency. The decrease at higher concentrations is caused by both graphene and the TiO_2 competing for the absorption of light (Lee *et al.* 2012). After simultaneous exposure to visible (absorption *via* graphene) and UV (absorption *via* graphene and TiO_2) light, layer-by-layer self-assembled graphene/TiO_2 composite films can exhibit superhydrophilic behaviour. This property is yielded by the higher concentration of photoinduced electron-hole pairs generated which underwent slow recombination (Zhu & He, 2012).

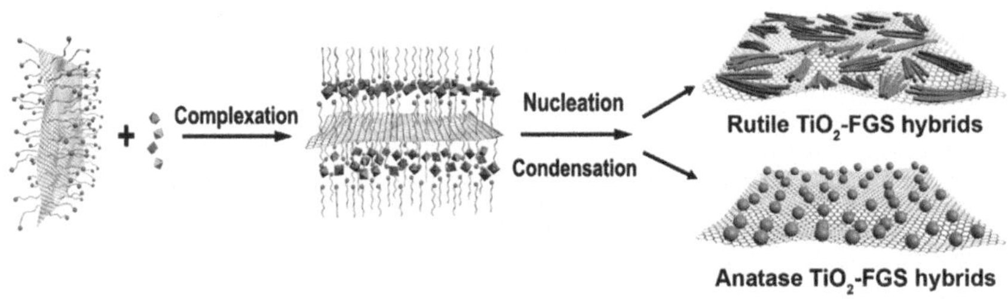

Figure 8.7 Stabilisation of graphene *via* anionic sulfate surfactant and growth of TiO_2-graphene nanohybrid. (Reprinted with permission from Wang *et al.* 2009).

The semiconductor cadmium sulfide (CdS) was deposited while concurrently reducing GO to graphene and this one step procedure yielded a single layer graphene-CdS nanocomposite, which exhibited very fast picosecond electron transfer between the CdS and the graphene sheet (Cao *et al.* 2010). MoS_2 quantum dots have been incorporated into graphene TiO_2 composites while utilising a one-pot solvothermal method, resulting in a greater surface area with an increased number of reaction sites, thus improving photocatalytic degradation. The enhanced degradation of species was attributed to the absorption of visible light and improved charge separation (Gao *et al.* 2015). The photoresponse of bismuth vanadate ($BiVO_4$) has been improved using a single step visible light photocatalytic reaction. This was performed effectively while simultaneously reducing GO to form the $BiVO_4$-RGO composite. Compared with pure $BiVO_4$, the composite exhibited a ten-fold improvement in the photoelectrochemical water splitting reaction when exposed to visible light irradiation. The enhancement was found to be caused by the lengthier lifetime of the excited $BiVO_4$ e^-_{CB}, since the majority of these are immediately transferred into the RGO upon formation, circumventing charge recombination (Ng *et al.* 2010). A one-step, low cost, low temperature method to produce a graphene/TiO_2 composite has been developed, which involves the reduction of GO to graphene and loading of TiO_2 NPs on the graphene surface, while the formation of O-Ti-C/Ti-O-C structures and self-doping of Ti^{3+} on TiO_2 occurs. This provides narrowing of the TiO_2 band gap permitting excellent efficiency for light-to-electricity conversion within the visible light region (Qiu *et al.* 2015).

8.2.3.5 Graphitic carbon nitride (g-C_3N_4)

The *g*-C_3N_4 composites are very versatile photocatalysts that have been applied to a varied array of applications. TiO_2/*g*-C_3N_4 has been employed for the effective photocatalytic oxidation (98.9% conversion) of dibenzothiophene and other sulfur compounds (fuel oils) to their corresponding sulfone, which are then ready for extraction (Wang *et al.* 2014). Certain composites have been used for the photocatalytic decomposition of organic dyes in water. Upon exposure to visible light, CeO_2/*g*-C_3N_4 (5 wt%) exhibited the photocatalytic degradation of MB dye that was eight-fold that of pure *g*-C_3N_4 (She *et al.* 2015) and a WO_3/*g*-C_3N_4 (10 wt%) composite demonstrated superb stability and efficiency, lasting for over five cycles when employed for the photocatalytic degradation of RhB dye (Gondal *et al.* 2015). Under solar irradiation, a Ag_3PO_4/*g*-C_3N_4 composite has been developed for converting CO_2 to fuels which possibly employ a Z-scheme mechanism (Figure 8.8). In this process, the Ag NP can act as a charge transfer bridge between the Ag_3PO_4 and g-C_3N_4 components. The photoexcited e^-_{CB} of Ag_3PO_4 and the photoinduced h^+_{VB} of g-C_3N_4 both migrate to the Ag NP, where they combine. This facilitates the greater separation of the g-C_3N_4 photoexcited e^-_{CB} and Ag_3PO_4 photoinduced h^+_{VB}. Furthermore, the g-C_3N_4 photoexcited e^-_{CB} possesses strong reducibility due to the negativity of the g-C_3N_4 CB. The material showed a six-fold conversion rate in contrast to pure g-C_3N_4, as the Ag_3PO_4 heterojunction aids light absorption and promotes CO_2 reduction (He *et al.* 2015). Composites composed of two phases have also been fabricated. In the correct ratio, metastable tetragonal phase lanthanum orthovanadate (t-$LaVO_4$) and monoclinic phase lanthanum orthovanadate (m-$LaVO_4$) have been shown to enhance the photocatalytic activity of g-C_3N_4 for the degradation of RB, *via* suppression of electron-hole pair recombination. t-$LaVO_4$ is known to undergo greater dispersion throughout the g-C_3N_4 composite as it forms a smaller particle than m-$LaVO_4$ which results in a larger number of heterojunctions within the composite (He *et al.* 2014).

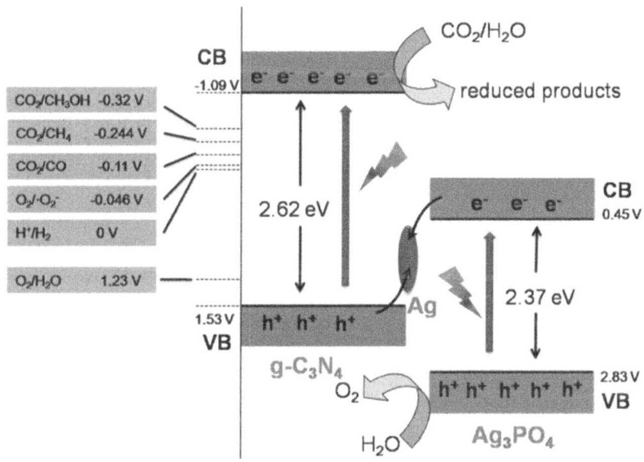

Figure 8.8 Photocatalytic mechanism of the Ag_3PO_4/g-C_3N_4 composite. (Reprinted with permission from He *et al.* 2015).

8.3 CLASSES OF COMPOUNDS SUITABLE TO TREATMENT AND EXAMPLES OF REACTION MECHANISMS

AOPs have been extensively studied to remove recalcitrant contaminants in water since they can produce reactive oxygen species (ROS), i.e., $^\bullet OH$, $^1O_2$, and $O_2^{-\bullet}$. The ROS are able to attack and decompose a multitude of organic contaminants *via* AOPs (Han *et al.* 2013; Han *et al.* 2014b). Semiconductors such as ZnO and TiO_2 have gained much attention in the field of environmental applications as photocatalysts in AOPs. Due to its unique physicochemical properties, TiO_2 has been widely used in AOPs to remove contaminants in water (Comninellis *et al.* 2008; Poyatos *et al.* 2010; Han *et al.* 2014a). Under UV irradiation, TiO_2 produces ROS and many studies have reported the efficient degradation of water contaminants using TiO_2 photocatalysis (Choi *et al.* 2006; Giraldo *et al.* 2010; Antoniou *et al.* 2009). Different contaminants including azo-dyes (methylene blue/MB), antibiotics (oxolinic acid/OA), and cyanotoxins (microcystin-LR/MC-LR) were effectively decomposed using UV-induced TiO_2 photocatalysis. Choi *et al.* (2006) reported that the morphological properties of TiO_2 were influenced by a ratio of surfactants used to provide high porosity and surface area for high photocatalytic activity of TiO_2. The ratio of 1.0 was an optimal condition to prepare the most effective, anatase TiO_2 film with a high Brunauer, Emmett and Teller (BET) surface area of 147 $m^2 g^{-1}$ compared to TiO_2 synthesized without a surfactant (22.7 $m^2 g^{-1}$). MB dye was effectively decomposed under UV illumination using this photocatalyst.

Giraldo *et al.* (2010) studied the degradation of OA using a commercial TiO_2 (Degussa P-25). In the study, experimental conditions were optimized (pH = 7.5 and TiO_2 loading = 1.0 g L^{-1}) to achieve the highest photocatalytic activity for the degradation of OA. After 30 minutes of UV illumination, 100% of OA was decomposed while 50% of both chemical oxygen demand and dissolved oxygen demand decreased after 60 minutes of UV irradiation, which may imply that the antibacterial activity of reaction intermediates remains. Therefore, the residual antibacterial activity was investigated using *E. coli*. After 30 minutes of UV illumination, antibacterial activity was completely removed, which indicates TiO_2 photocatalysis is very effective at decomposing the antibiotic OA. Antoniou *et al.* (2009) decomposed the cyanotoxin MC-LR using UV active TiO_2 photocatalytic films. The TiO_2 films, with a BET surface area of 147 $m^2 g^{-1}$, porosity of 46%, and anatase crystal size of 9.2 nm, were prepared using a sol-gel method. The degradation of MC-LR was investigated under UV illumination by examining the effect of initial MC-LR concentration, initial solution pH, film thickness, and area of TiO_2 coated on a substrate. The solution pH was a significant parameter since TiO_2 and MC-LR were negatively or positively charged, depending on the solution pH. At pH = 3.0, the highest MC-LR degradation was observed as TiO_2 is positively charged and MC-LR is negatively charged. Additionally, the initial degradation rate was investigated using different MC-LR concentrations ranging from 0.84 to 5.38 μM. The reaction followed *pseudo-first order* kinetics and the rate was proportional to the initial MC-LR concentration, which ranged from 1.48×10^{-2} μM min^{-1} to 3.82×10^{-2} μM min^{-1}. The effect of TiO_2 film area on a substrate with regard to MC-LR degradation was investigated by preparing films with different areas of TiO_2 coated on substrates, e.g., 22.5, 45.0, and 67.5 cm^2. The TiO_2 coated area significantly affected MC-LR degradation since larger areas may provide more active sites for TiO_2 photocatalysis. The results showed that the degradation of MC-LR increased when larger TiO_2 areas were coated on the substrates. Moreover, the effect of film thickness was investigated on MC-LR degradation as film thickness is directly associated with adsorption capacity. Also, the thickness is related to the mechanical stability of films for the evaluation of film reusability. The thickness of the films was controlled by repeating TiO_2 film coating procedures. Peeling of the films was observed when excessive TiO_2 coatings were applied (over 10 times); while a maximum adsorption capacity was observed for the film with three TiO_2 coatings. This indicates that the outer layers of TiO_2 were utilized for TiO_2 photocatalysis due to the limited penetration of MC-LR into the inner layers of TiO_2.

As discussed above, UV-induced TiO_2 photocatalysis is very effective for the decomposition of many contaminants in water but it also possesses limitations. The large band gap of conventional TiO_2 (3.2 eV for anatase) requires UV light for TiO_2 activation and the whole solar spectrum is comprised of only 4–5% UV light. In order to extend the region for light absorption towards visible light (approx. 45% of the whole solar spectrum), non-metals and noble metals are used to narrow the band gap of conventional TiO_2. In addition, localized midgap energy states were introduced into the band gap of conventional TiO_2, or metals that possess surface plasmon resonance such as silver (Ag) or gold (Au), were added onto the surface of conventional TiO_2 (Pelaez *et al.* 2009; Han *et al.* 2011, 2014a, 2014b; Khan *et al.* 2014; Sacco *et al.* 2015). Han *et al.* (2011) synthesized visible light active sulfur-doped TiO_2 (S-TiO_2) films using a simple sol-gel method to decompose MC-LR. The effect of calcination temperature was investigated to synthesize S-TiO_2 films possessing a high photocatalytic activity. Anatase S-TiO_2, which exhibited the highest photocatalytic activity for MC-LR degradation, was synthesized at a calcination temperature of 350 °C. The calculated band gap, BET surface area, porosity, and sulfur content were determined to be 2.94 eV, 179 $m^2 g^{-1}$, 33.3%, and 4.1%, respectively. The sulfur content in the samples decreased at higher calcination temperatures and the films demonstrated high mechanical stability during three consecutive experiments, indicating that the films are reusable. In addition to the investigation of calcination temperatures used for the synthesis of S-TiO_2 films, the effect of solvents on the sol-gel method producing different properties for the S-TiO_2 films was investigated (Han *et al.* 2014a). Using different solvents, the structural properties of S-TiO_2 films including surface roughness, BET surface area, pore size distribution, and crystal size were significantly affected. Surface roughness, crystal size, and pore size distribution were proportional to the dielectric constant (*D*-value) of each solvent, while BET surface area decreased with increasing *D*-values. All films prepared using different solvents effectively decomposed MC-LR under visible light, which indicated that the optical and electrical properties of S-TiO_2 were not influenced by the choice of solvents during S-TiO_2 film preparation.

Co-doping with two different dopants was introduced by Pelaez *et al.* (2009) and Khan *et al.* (2014) for the development of highly visible light-active, nitrogen-fluorine co-doped TiO_2 (NF-TiO_2) and phosphorus-fluorine co-doped TiO_2 (PF-TiO_2) nanoparticles (NPs), respectively. Anatase PF-TiO_2 NPs with a calculated band gap of 2.75 eV, BET surface area of 141 m^2 g^{-1}, porosity of 49%, and pore size of 2–10 nm demonstrated the highest photocatalytic activity for the decomposition of MC-LR under the experimental conditions. Pelaez *et al.* (2009) reported that nitrogen doping contributed to TiO_2 band gap narrowing and fluorine doping produced surface oxygen vacancies. Both nitrogen and fluorine co-doping provided synergistic effects on the visible light utilization of TiO_2 photocatalysis. Additionally, Khan *et al.* (2014) synthesized PF-TiO_2 NPs exhibiting a calculated band gap of 2.70 eV, BET surface area of 212 m^2 g^{-1}, porosity of 36.5%, and crystal size of 5.9 nm using a simple sol-gel method. PF-TiO_2 showed higher photocatalytic degradation of atrazine than single-doped TiO_2, i.e., P-TiO_2 and F-TiO_2, under UV-visible light illumination. This indicates that the co-doping provided synergistic effects, which resulted in the enhanced UV-visible light-induced TiO_2 photocatalysis.

In a study performed by Sacco *et al.* (2015), visible light-active nitrogen-doped TiO_2 (N-TiO_2) NPs, with a calculated effective band gap of 2.5 eV, were immobilized on micro sized ZnS-based phosphors (ZSP) that can emit a luminescence with a wavelength of 440 nm, when irradiated by a wavelength of 365 nm. The loading of N-TiO_2 on ZSP varied from 15 to 50 wt% to enhance TiO_2 photocatalysis for the degradation of the pesticide atrazine under UV irradiation. Due to the emission from ZSP, the photocatalytic activity was improved and the 30 wt% loaded sample demonstrated the highest photocatalytic activity for the decomposition of atrazine under the experimental conditions. The optimal concentrations of the photocatalyst and solution pH for achieving the highest efficiency of the TiO_2 photocatalysis were 0.5 g L^{-1} and 5.8, respectively. When the N-TiO_2 loading on ZSP was increased beyond the optimal concentration, this thick layer of N-TiO_2 on ZSP interfered with UV light penetration, which is required for the ZSP luminescence.

Moreover, Ag NPs, with an average diameter of 5.9 ± 1.2 nm, were decorated on monodisperse TiO_2 aggregates with an average diameter of 350 nm for their utilization of visible light from solar light (Han *et al.* 2014b). The free electrons produced by the localized surface plasmon resonance of noble metals in visible light can be used to produce ROS for the decomposition of pollutants in water (Wang & Lim, 2013). To obtain the highest photocatalytic activity of the samples, Ag loadings on the surface of monodisperse TiO_2 aggregates were varied from 0.8 to 5.5 wt%. It was found that the visible light absorption of the samples increased upon increasing the amount of Ag deposited on the surface of monodisperse TiO_2 aggregates. The sample containing 1.9 wt% Ag demonstrated the highest photocatalytic activity, under UV-visible light illumination, for the decomposition of the pharmaceutical oxytetracycline under their experimental conditions. When the Ag loading was increased beyond the optimal conditions it was shown that the photocatalytic activity decreased, which may result from an increased electron-hole recombination, along with the blocking of active sites of TiO_2 due to excessive Ag deposition. Moreover, the total amount of Ag leached from the sample, which demonstrated the highest photocatalytic activity, was much lower than the national secondary drinking water regulation of 1.0 mg L^{-1}. Recently, carbon-based materials, including carbon nanotubes (CNTs) and GO, were employed to enhance TiO_2 photocatalytic activity under simulated solar light (Sampaio *et al.* 2015). The sample containing 4 wt% GO showed the highest photocatalytic activity for the removal of the cyanotoxin microcystin-LA (MC-LA) under the experimental conditions applied in their study. After 5 minutes of simulated solar light, MC-LA was completely decomposed, while 88% of MC-LA was degraded after 2 hours of visible light illumination. Due to the optimal interfacial coupling of GO and TiO_2, the solar light-induced TiO_2 photocatalysis was significantly enhanced. Hydroxyl radicals effectively attacked and broke down the conjugated diene structure of the Adda moiety in MC-LA which is associated with its biological toxicity.

Sleiman *et al.* (2008) investigated the use of titanium dioxide as a photocatalyst for removing pesticides in the gas phase including dichlorvos. Static and dynamic reaction system experiments were carried out to examine the various aspects of the process, such as the relative humidity (RH) and the association of TiO_2 with activated carbon (AC). Direct charge transfer and chlorine radical (•Cl) attack were needed for the reaction at low RH, while •OH radicals were performed as the active species when the RH was high. There are two pathways that occur depending on the RH level, as illustrated in Figure 8.9 (Sleiman *et al.* 2008).

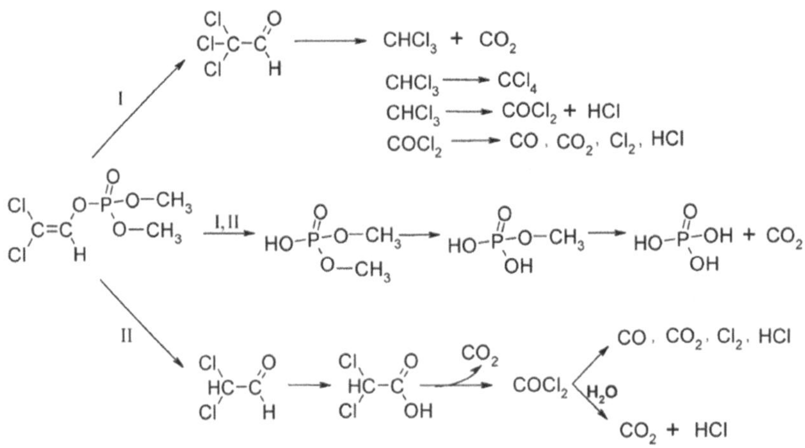

Figure 8.9 Reaction scheme for the degradation of dichlorvos in the gas phase by photocatalysis. Pathway I is favoured at 0% RH and pathway II is favoured at 40% RH. (Sleiman *et al.* 2008).

Pathway I: Without the presence of water, there is limited •OH radical production. The electron transfer on the C=C in dichlorvos forms radicals. These radicals react with •Cl and $^{\bullet}O_2$ resulting in trichloroacetaldehyde (TCA) and dimethyl phosphate (DMP), *via* a Russel mechanism as an intermediate step A similar mechanism has been suggested for the degradation of trichloroethylene and perchloroethene (Ou & Lo, 2007; Hegeds & Dombi, 2004; Sleiman *et al.* 2008). TCA is oxidized and produces CO_2 and $CHCl_3$. The $CHCl_3$ can react with •Cl which results in CCl_4 or it can react with $^{\bullet}O_2^-$ and form HCl and phosgene ($COCl_2$). CO, CO_2, Cl_2 and HCl are products of the oxidation of the phosgene.

Pathway II: When the RH values are high, water molecules are adsorbed to the hole (h^+), this forms •OH radicals. The •Cl radicals are terminated by reaction with water, producing HCl. The reaction pathways of dichlorvos with •OH was experimentally examined under atmospheric conditions (Feigenbrugel *et al.* 2006). Feigenbrugel *et al.* (2006) have proposed two rotues: (i) the removal of H^+ from the methoxy group and (ii) the addition of •OH radicals to the C=C bond, forming carbon monoxide and phosgene respectively.

Differences were noted in Sleiman *et al.* and Feigenbrugal *et al.* research when comparing the reaction products and their proportions. Sleiman *et al.* have two main reasons that account for these differences. Feigenbrugal *et al.* did not quantify or identify most of the intermediates during the degradation. The second reason stated is as surface chemistry and adsorption processes had a part in the reaction, the homogenous gas phase OH-oxidation and the OH-initiated oxidation by heterogeneous photocatalysis had different reaction mechanisms. Sleiman *et al.* concluded that the route that reaction followed at high RH was the addition •OH to C=C in dichlorvos, leading to DCA and DMP as products. It was found that in first route, the removal of H^+, was only a minor one as there was very low concentrations of CO found in the gas phase. The DCA reacts with •OH and is oxidized resulting in DCAA. This is broken down on the surface of TiO_2 producing CO_2 and phosgene. The final step for this reaction is both the oxidation and hydrolysis of phosgene with CO, CO_2, Cl_2 and HCl as resulting products. The degradation of the DMP is by the removal of H from CH_3O moieties, with the final product of phosphates. While the experimental data confirm the proposed reaction scheme, Sleiman *et al.* state that further research is required to verify the reaction intermediates.

Grandos-Oliveros *et al.* (2009) studied the degradation of the herbicide atrazine, which is used for the control of broadleaf and grassy weeds. TiO_2 was used as the photocatalyst, the surface was modified with tetra(4-carboxyphenyl)porphyrin on various centers (Zn(II), Cu(II), Fe(III) and metal-free). Initially there was no photocatalytic activity or degradation of the atrazine. It was not until hydrogen peroxide was added that activity and degradation began. From the four types of centers, Cu(II) porphyrin showed the highest level of photocatalytic activity. When H_2O_2 concentration increases from 0.015 to 0.05 mol L^{-1} there is an increased rate of degradation of atrazine. Above 0.05 mol L^{-1} H_2O_2 the rate of degradation plateaux. Grandos-Oliveros *et al.* also proposed a reaction mechanism and reaction intermediates.

Reaction pathway: In order to examine the role of •OH as the oxidant that was produced and to determine the degradation of atrazine pathway, the samples were oxidized under visible light irradiation, with H_2O_2, molecular oxygen and $TcPPCu/TiO_2$ (Granados-Oliveros *et al.* 2009). After one hour of irradiation with visible light, 82% atrazine was degraded. The •OH radicals attacked the aminoalkyl groups in atrazine to initiate the degradation process. This produced organic free radicals by the rapid removal of H^+ from organic substrates. Intermediates I, II and III were formed when alkylic oxidation occurred. This was followed up by the dealkylation which resulted in intermediates IV, V and VI (Table 8.1 and Figure 8.10). The presence of VI could mean the carbon atom next to the amino group is oxidised at the early stage of the degradation. The reaction mechanism suggested by Grandos-Oliveros *et al.* relates to other previous studies (Chan & Chu, 2005; Maurino *et al.* 1999).

Table 8.1 Intermediates of atrazine during degradation with visible light irradiation (Granados-Oliveros *et al.* 2009).

Compound	Name
I	2-Chloro-4-(ethylamino)-6-(isopropylamine)-triazine
II	2-Chloro-4-acetamido-6-isopropylamine-1,3,5-triazine
III	2-Chloro-4-amido-6-isopropylamine-1,3,5-triazine
IV	2-Chloro-4-ethylamino-6-(2-proplanol)amino-1,3,5-triazine
V	Desethylatrazine
VI	Edisopropylatrazine
VII	Desethyldesisopropyl atrazine

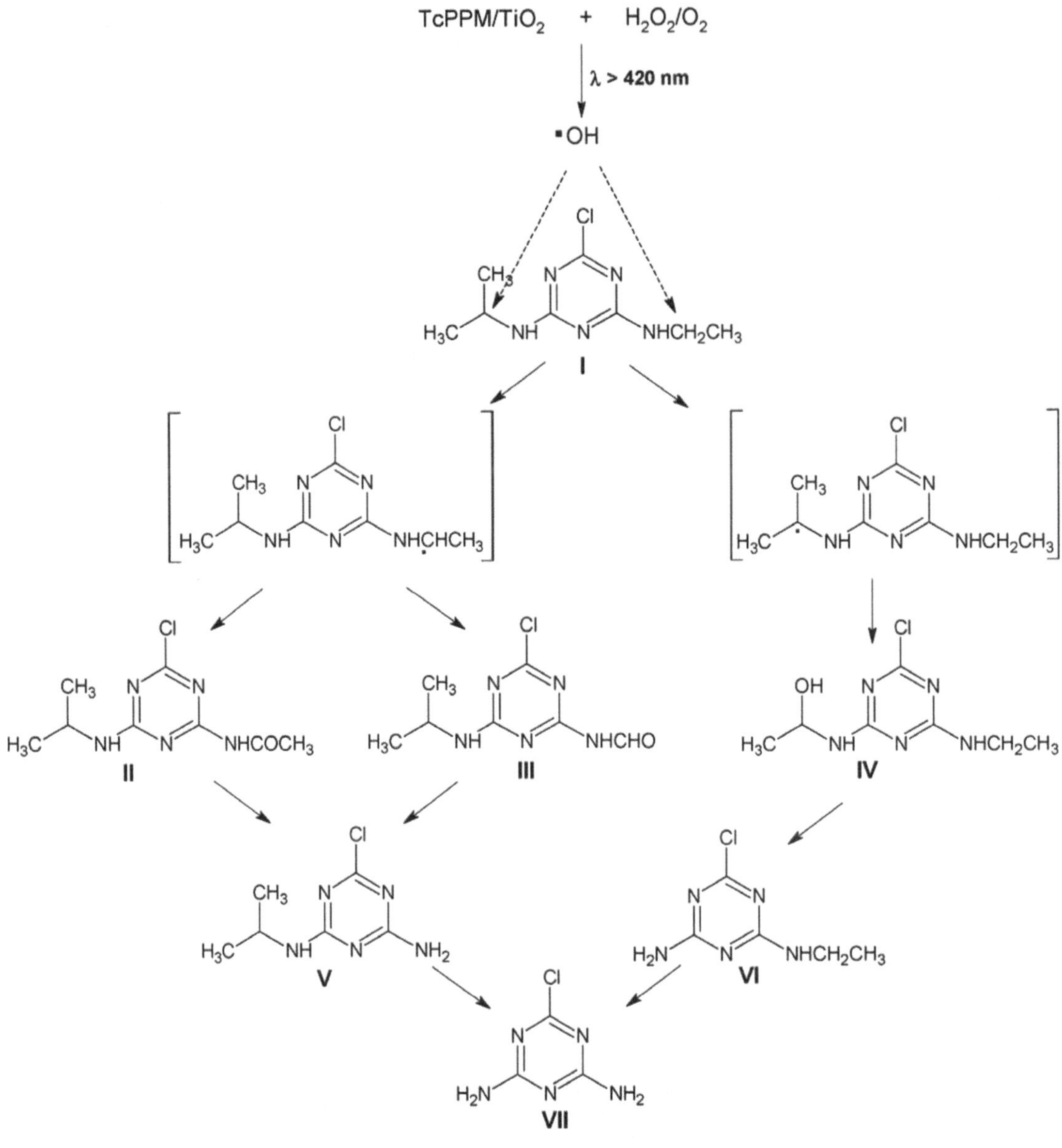

Figure 8.10 Pathways for the degradation of atrazine (Granados-Oliveros *et al.* 2009).

Li *et al.* (2011) investigated the degradation of pentachlorophenol (PCP) by photocatalysis with the use of bismuth silicate under xenon lap irradiation. It has been shown that $^{\bullet}O_2^-$ was responsible for degradation of the PCP. Thirteen intermediates were observed when examined with GC/MS, including tetrachlorophenols (TeCP), trichlorophenols (TCP), dichlorophenols (DCP), chlorophenols (CP) and phenol. Based on the intermediates and the changes in concentration, the degradation pathways for PCP were proposed, Figure 8.11 (Li *et al.* 2011).

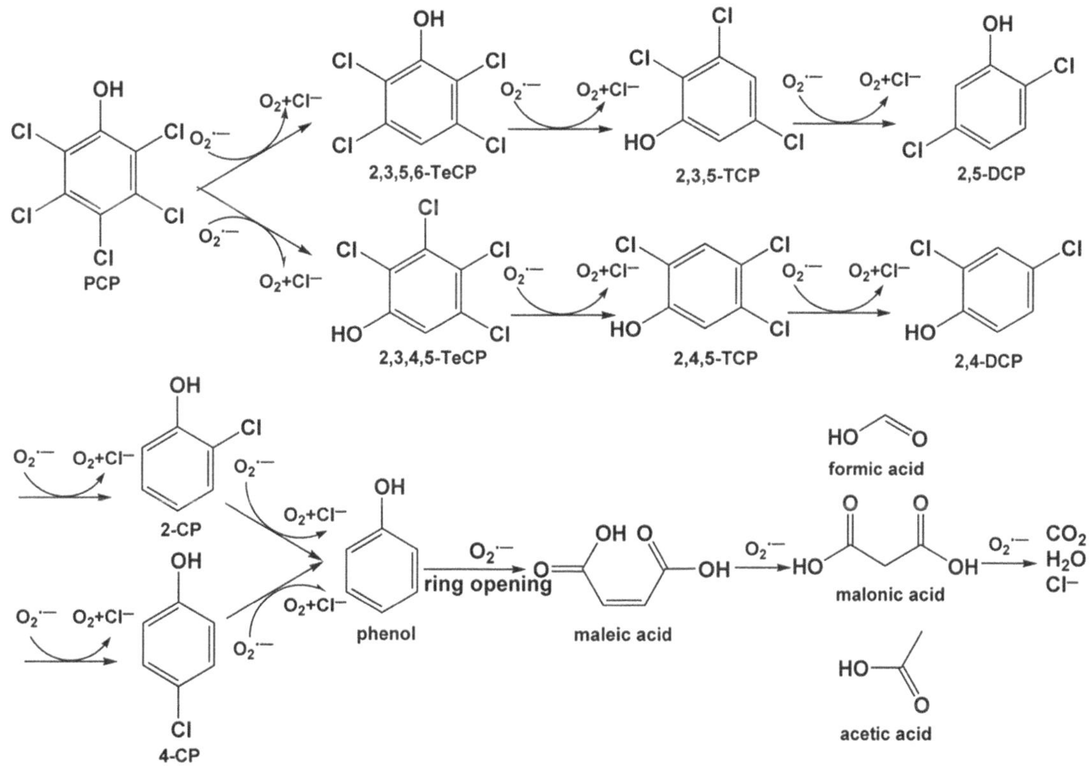

Figure 8.11 Reaction pathways for PCP degradation (Li *et al.* 2011).

8.4 KINETIC ASPECTS, REACTION MODELLING, QUANTITATIVE STRUCTURE-ACTIVITY RELATIONSHIP (QSAR)

In the photocatalytic process, a semiconductor metal oxide is exposed to light energy (Figure 8.12). If the energy supplied to the material is equal or greater than that of the material band gap energy, the photon absorption results in the excitation of an electron (e^-_{CB}) to CB from VB, where a hole (h^+_{VB}) is created (Equation 8.1). Recombination of the electron-hole pair can occur releasing energy (Equation 8.2) and the low efficiency of the semiconductor (e.g., TiO_2) light-to-energy conversion is attributed to this facile process (Etacheri *et al.* 2012a).

$$TiO_2 + h\nu \rightarrow h^+_{VB} + e^-_{CB} \tag{8.1}$$

$$h^+_{VB} + e^-_{CB} \rightarrow \text{energy} \tag{8.2}$$

If the recombination does not occur, the electron and hole charge carriers travel to the materials surface, where they can react with adsorbed compounds (Figure 8.13). The photoexcited e^-_{CB} can reduce atmospheric oxygen to produce reactive oxygen species (ROS), such as the superoxide radical anion ($O_2^{-\bullet}$) (Equation 8.3). Concurrently, the photoinduced h^+_{VB} can oxidize water to produce a hydroxyl radical ($^\bullet OH$) (Equation 8.4). Both of these species, along with others generated, can react with organic compounds yielding intermediate byproducts which under prolonged exposure to high energy UV radiation could be mineralized to carbon dioxide (CO_2), water (H_2O), and inorganic ions such as halides, nitrate, and sulfate (Equation 8.5 and 8.6).

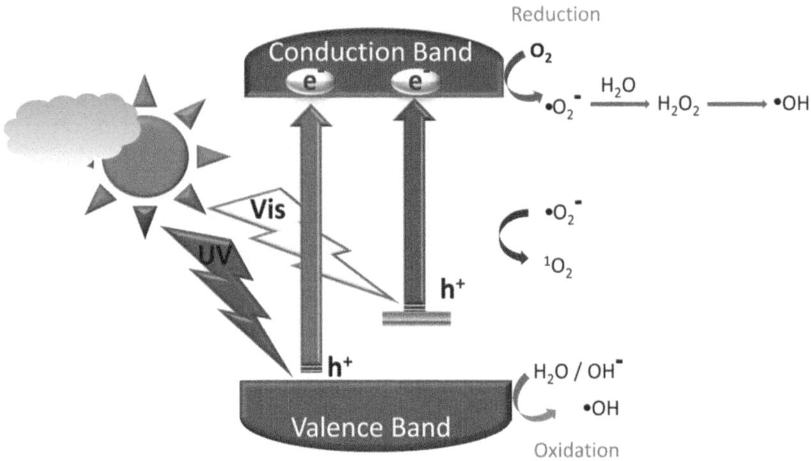

Figure 8.12 Mechanism of photocatalysis. (Reprinted with permission from Banerjee *et al.* 2014).

$$O_2 + e^-_{CB} \rightarrow O_2^{-\bullet} \tag{8.3}$$

$$H_2O + h^+_{VB} \rightarrow {}^\bullet OH + H^+ \tag{8.4}$$

$$O_2^{-\bullet} + \text{Pollutants} \rightarrow\rightarrow\rightarrow\rightarrow H_2O + CO_2 \tag{8.5}$$

$$^\bullet OH + \text{Pollutants} \rightarrow\rightarrow\rightarrow\rightarrow H_2O + CO_2 \tag{8.6}$$

The facile recombination of the photoinduced electron-hole pairs of the metal oxide yields a low photon efficient photocatalytic semiconductor. Furthermore, the wide band gap exhibited by semiconductors, for example TiO_2 (>3.0 eV), does not permit the absorption of visible light (>390 nm), thus, restricting the application of these potentially very useful materials. The doping of the semiconductor material with metals or non-metals, employing the use of heterojunctions along with incorporating advanced materials, such as graphene, or *g*-C_3N_4, have been demonstrated as excellent methods to circumvent these issues.

Recently, research regarding the photocatalytic mechanism of TiO_2 co-doped with nitrogen and fluorine has been investigated by examining the correlation between the optical absorption spectra and open-circuit photopotential measurements under solar visible light. It was shown that after analysing the low visible light photocurrent and open-circuit photopotential, electrons excited from the nitrogen midgap state to the TiO_2 CB are trapped by the defect levels below this. Furthermore, it is probable that the oxidation of the $O_2^{-\bullet}$ to singlet oxygen ($^1O_2$) can occur *via* holes in the nitrogen mid-gap state, while reduction of O_2 to ROS *via* interaction with e^-_{CB} leads to the oxidative decomposition of pollutants composed of an organic nature (Figure 8.14) (Hamilton *et al.* 2014).

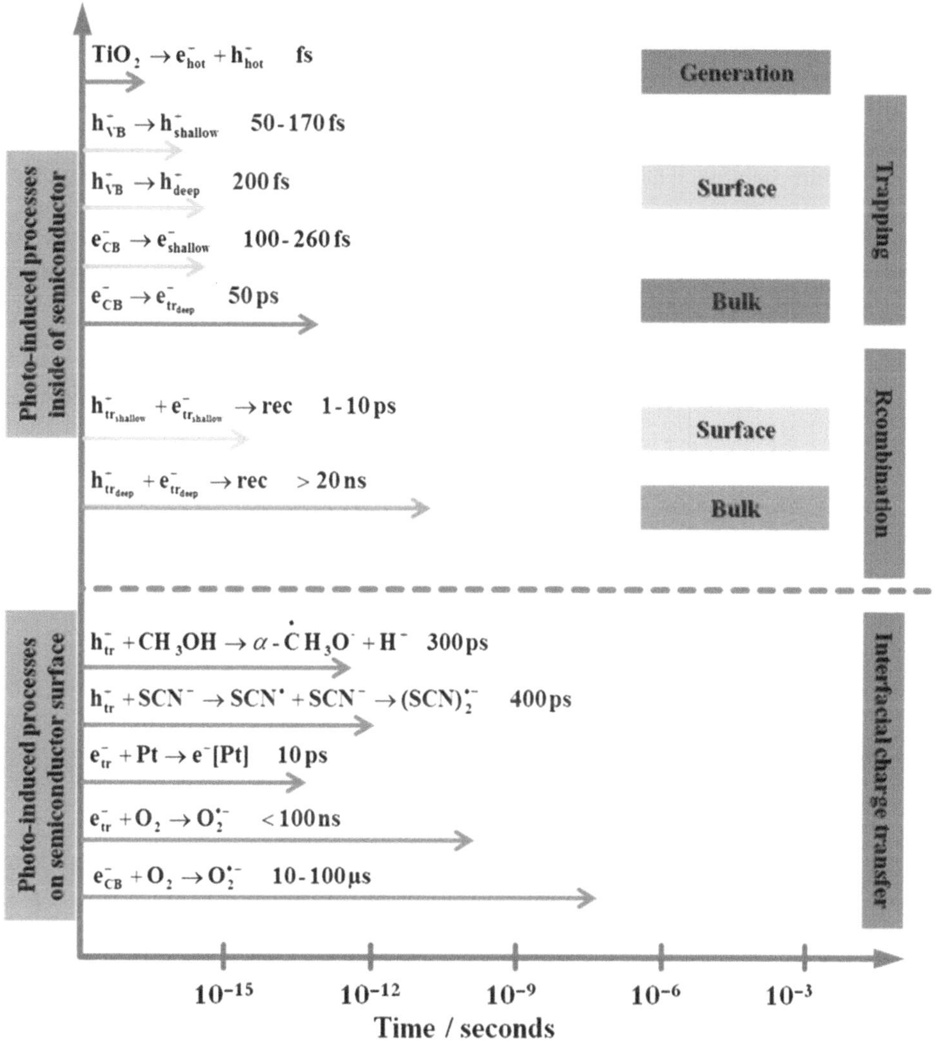

Figure 8.13 Photoinduced processes which occur both on the surface and inside of TiO_2 photocatalyst accompanied by the corresponding time scales. (Reprinted with permission from Schneider *et al.* 2014).

When zinc sulfide (ZnS) nanocrystalline photocatalysts (4–6 nm) were synthesised *via* a microwave assisted method, they exhibited photocatalytic degradation of *Staphylococcus aureus* and *Escherichia coli* when exposed to a 60 W light bulb. It was found that the material possessed both bacteriostatic and bactericidal (5 hours exposure produced an 88% decrease in bacteria concentration) activity. During this study, point defects were identified within the ZnS material (interstitial zinc along with sulfur and zinc vacancies), which act as inter band donor levels, decreasing the amount of energy necessary to promote electrons from the VB to the CB. These donor levels produce energy gaps (for example 2.5 eV) which permits the transfer of electrons upon irradiation with wavelengths shorter than 500 nm, which also include a small portion of the visible light range (Synnott *et al.* 2013). Using their results from laser flash photolysis, researchers in 1995 proposed a mechanism which involved two types of processes. Firstly, there was competition between the trapping and recombination of the electron-hole pairs and also between the interfacial charge transfer and recombination of these trapped species (Hoffmann *et al.* 1995). It has been demonstrated in the literature that narrowing of the band gap is the dominant factor enabling visible light photocatalytic activity and that the presence of oxygen defects within the TiO_2 yields a material with high temperature stability. This enables the crystalline structure of the material to remain in the photocatalytically active anatase form at temperatures of up to 900°C, whereas the phase transition from photocatalytically active anatase to less active rutile usually occurs at approximately 600°C, once the anatase particles grow to their less stable, critical particle-size of approximately 14 nm (Etacheri *et al.* 2011; Pillai *et al.* 2007).

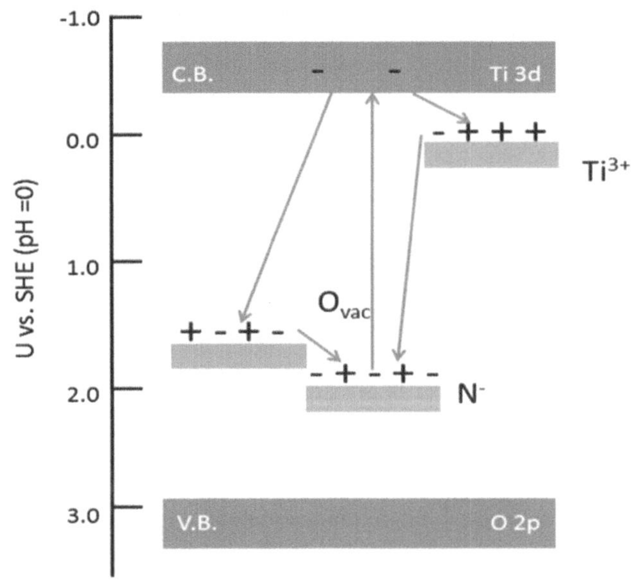

Figure 8.14 Visible light excitation of N-TiO_2 and refilling of empty N states by electron transfer from either Ti^{3+} or $O_{vac.}$ (Reprinted with permission from Hamilton *et al.* 2014).

Heterogeneous photocatalysis has been widely used for environmental applications to decontaminate water and air (Wang *et al.* 2003; Zhou *et al.* 2006; Valencia *et al.* 2011). Photocatalysis using TiO_2 has been extensively studied for the decomposition of contaminants and models have been used to understand kinetics of the photocatalysis (Upadhya & Ollis, 1997; Monllor-Satoca *et al.* 2007; Ollis, 2005). To address the photocatalytic degradation of environmental contaminants, the most common kinetic model is the *Langmuir-Hinshelwood* (*L-H*) model based on the assumptions: (1) limited surface adsorption sites, (2) only single layer adsorption, and (3) no interactions between molecules after adsorption (Allen *et al.* 1989; Hoffmann *et al.* 1995; Fox & Dulay, 1993). It is expressed as (Valencia *et al.* 2011):

$$\frac{dC}{dt} = \frac{k_{obs}K_L C}{1 + K_L C}$$

where $\frac{dC}{dT}$ is the rate of product formation, k_{obs} is the apparent rate constant (the maximum rate), K_L is the Langmuir adsorption/desorption constant, and C is the reactant concentration. Based on the *L-H* model, the reaction rate depends on the surface area and the concentration of reactants. The rate increases at higher reactant concentrations and surface areas. Many studies reported that various compounds were decomposed by TiO_2 photocatalysis following the *L-H* model (Daneshvar *et al.* 2004; Dung *et al.* 2005; Kim & Hong, 2002). However, the model does not have a parameter which can explain the degradation of unadsorbed contaminants on the surface of catalysts (Minero, 1999; Valencia *et al.* 2011). Also, the influence of the photon flux on the reaction rate cannot be explained with the model, even though K_L is dependent on the light intensity (Mora-Seró *et al.* 2005; Montoya *et al.* 2009; Davydov & Smirniotis, 2000).

A study by Ollis (2005) reported that the catalytic rate constant and apparent adsorption and desorption binding constants were consistent with those of the *L-H* model under an assumption. It was assumed that there is no adsorption/desorption equilibrium because adsorbed compounds on the photocatalytic surface were continuously displaced due to reactions with active centers such as electrons, holes, and hydroxyl radicals. Additionally, Monllor-Satoca *et al.* (2007) carried out an alternative kinetic approach with the "Direct-Indirect" model considering two mechanisms of direct/indirect interfacial charge transfer. For the direct transfer mechanism, organic compounds directly react with delocalized valance band holes. For the indirect transfer mechanism, contaminants are photooxidized by the surface trapped holes. It is not easy to distinguish a photooxidation mechanism because there are many parameters to be considered, including photon flux, contaminant concentration, and specific adsorption of contaminants. Montoya *et al.* (2009) reanalyzed literature data for the degradation of phenol and formic acid with the "Direct-Indirect" model. For phenol, the model fit well to the degradation results when the indirect transfer mechanism was dominant since phenol was not adsorbed on the surface of TiO_2, specifically. The degradation results of formic acid appropriately fit to the "Direct-Indirect" model under the condition that the direct charge transfer mechanism dominates. Recently, Montoya *et al.* (2014) reported a theoretical approach to analyze TiO_2 photocatalysis using the "Direct-Indirect" model. It is expressed as:

$$r_{ox}^{IT} = \left[\left(\alpha^{IT}[C_{liq}]\right)^2 + 2\alpha^{IT}k_0\rho[C_{liq}]\right]^{1/2} - \alpha^{IT}\left[C_{liq}\right]$$

For the indirect transfer mechanism

where $\alpha^{IT} = \frac{k_{ox}^{IT}k'_{red}[(O_2)_{liq}]}{4k'_{r,1}}$, r_{ox}^{IT} is initial, indirect transfer photo-oxidation rate of semiconductor physisorbed contaminants, α^{IT} is indirect transfer parameter, C_{liq} is concentration of dissolved contaminants in the liquid phase, k_0 is empirical photon absorption constant, ρ is photon flux intensity, k_{ox}^{IT} is indirect transfer photo-oxidation rate constant, k'_{red} is recombination rate constant of conduction band free electrons with electrolyte dissolved O_2, $(O_2)_{liq}$ is concentration of dissolved O_2 in the liquid phase, and $k'_{r,1}$ is rate constant for recombination of conduction band free electrons with surface trapped holes by terminal oxygen and bridging oxygen radicals.

For the indirect transfer mechanism,

$$r_{ox}^{DT} = \frac{k_o k_{ox}^{DT}[C_{ads}]}{k'_1\left[O_s^{2-}\right] + k_{ox}^{DT}[C_{ads}]}\rho$$

where $C_{ads} = \frac{k_{abs}bC_{liq}}{1 + k_{abs}C_{liq}}$, C_{ads} is semiconductor surface concentration of adsorbed contaminants, k_{abs} is the Langmuir adsorption constant, b is the maximum semiconductor surface area available for contaminant adsorption, r_{ox}^{DT} is initial, direct transfer photo-oxidation rate of semiconductor chemisorbed contaminants, k_{ox}^{DT} is direct transfer photo-oxidation rate constant, k_1' is diffusion constant of dissolved contaminant from the electrolyte toward the semiconductor surface, and O_s^{2-} is terminal oxygen ions.

For combination of the direct and indirect transfer mechanism under adsorption-desorption equilibrium of dissolved contaminants, the rate is provided by the expression below:

$$r_{ox} = r_{ox}^{IT} + r_{ox}^{DT} \approx \left[\left(\alpha^{IT}[C_{liq}]\right)^2 + 2\alpha^{IT}k_0\rho[C_{liq}]\right]^{\frac{1}{2}} - \alpha^{IT}\left[C_{liq}\right] + \frac{k_o k_{ox}^{DT}[C_{ads}]}{k_1'\left[O_s^{2-}\right] + k_{ox}^{DT}[C_{ads}]}\rho$$

A recent study reported the development of a new kinetic model for unadsorbed compounds on the surface of photocatalysts including a back reaction (Valencia *et al.* 2011). This model specifically addressed the photocatalytic degradation mechanism of non-specific adsorbed molecules on the surface of photocatalysts when a back reaction was involved. In particular, for the photocatalytic degradation of phenol, the "Direct-Indirect" model was not the best one to fit the degradation results since the model does not explain the back reaction.

8.5 WATER QUALITY IMPACT ON PROCESS PREFORMANCE, PRACTICAL CONSIDERATIONS ON PROCESS PARAMETER SELECTION CRITERIA

Water is adversely affected in quality by diverse anthropogenic, industrial, commercial, agricultural and domestic activities. Each target group of wastewater presents its own difficulties in their treatment by applying AOPs. For example, urban wastewater generated from washing activities in a household (greywater) is considerably less polluted than industrial wastewater but has a broad range of organic loads including preservatives, fragrances and UV filters and special requirements in respect to the type of water reuse, quantity and rate of detoxification etc. The effectiveness of the photocatalytic process is strongly related to the water quality, as the presence of organic matter, turbidity, microorganisms and pH may significantly affect the photocatalytic degradation performance, especially when the treatment targets emerging pollutants and persistent organic compounds. To optimize the photocatalytic efficiency, an additional step (pre-treatment or post-treatment) is necessary that consists of eliminating or/and normalizing the water quality features and fine-tuning the experimental parameters.

The removal of turbidity and natural organic matter (NOM) can be effectively performed by coagulation (Chekli *et al.* 2015). For this purpose, Ti-based materials (e.g. Titanium tetrachloride -$TiCl_4$) are the most promising, combining high reduction of chemical oxygen demand (COD) and dissolved organic carbon (DOC) with a significant advantage over conventional coagulants (e.g., polyaluminum chloride-PACl). In fact, the resulting non-toxic flocculated sludge can be recovered and recycled to produce functional TiO_2 photocatalysts, whose efficiency is comparable to that of Evonic P25 commercial reference (Zhao *et al.* 2011).

Slurry type photocatalytic reactors need subsequent particle separation to retain the catalyst in the system and allow the discharge of treated effluent. In-line coagulation offers a simple yet effective means to improve the performance of slurry type photocatalytic-MF/UF hybrid systems for advanced water and wastewater treatment applications (Erdei *et al.* 2008). Using a bench-scale hybrid treatment system comprised of a slurry (suspension) type continuous photocatalytic (CP) system and an immersed hollow fibre membrane micro-ultrafilter (MF/UF) unit, they achieved efficient purification of a synthetic wastewater (representing biologically treated sewage effluent), without any prior slurry settling step. Membrane feed pre-treatment with pH adjustment and particle charge neutralization with aluminum chloride coagulant led to improved critical membrane fluxes, producing near zero turbidity permeate, completely retaining the photocatalyst, improving the efficiency of DOC removal and controlling membrane fouling.

TiO_2-catalyst suspensions work efficiently in photocatalysis for wastewater treatment. Nevertheless, once photocatalysis is complete, separation of the catalyst from solution becomes the main problem. Enhanced catalyst recovery was achieved (Fernandez-Ibanez *et al.* 2003) through charge neutralization and coagulation with electrolytes at lab and pilot-plant scale and evaluated the potential for its separation after photocatalytic degradation of tetrachloroethylene (C_2Cl_4) pollutant. Zeta-potential analysis showed that the isoelectric point (IEP) of TiO_2 suspensions is near pH 7, where settling rates and hydrodynamic diameter of TiO_2 particles are maxima. They concluded that photocatalytic efficiency worsens with successive runs when catalyst and water are reused without separation, whereas, when TiO_2 is separated, the photocatalyst is not deactivated.

Photocatalytic performance efficiency and results reproducibility are related to the ability of the reaction medium to absorb (qualitatively and quantitatively) the incident light, parameters that depend on the solution composition. Investigation of TiO_2 aqueous suspensions has shown that their optical spectra can be unstable, with instability associated to both the ionic strength of the suspension and the charge of TiO_2 particles (Martyanov *et al.* 2003). It has been observed that the increase of the TiO_2 particle charge *via* pH adjustment (shifting pH in acidic or basic directions) stabilizes the suspension's optical spectra, and that ultrasonic treatment promotes optical density recovery. This behavior was attributed to alteration in the size of the suspension aggregates.

Reverse osmosis (RO) appears to be an attractive technology for the reclamation of the highly-polluted recalcitrant effluents by-produced in olive oil industries. Combining RO with feedstock pretreatment additionally with UV/TiO_2 photocatalysis increased threshold flux values were obtained and feed recovery rates were ensured, together with very significant reduction of the long-term fouling index (Ochando-Pulido *et al.* 2014). This could lead to lower energy costs and significantly minor required membrane area and, as a consequence, reduced capital costs in case of batch membrane processes. The produced water fulfills irrigation water quality standards and the obtained results enable the reuse of the regenerated effluent in the production process (in the olives washing machines).

Photocatalysis has sometimes limited potential for the degradation of biorecalcitrant and toxic organic compounds in water. The process efficiency can be increased by intimate coupling of photocatalysis and biodegradation (ICPB). In fact, Li *et al.* (2011) reported on a novel sponge-type, TiO_2 coated biofilm carrier that showed significant adherence of TiO_2 and ability to accumulate biomass in its interior. This carrier was tested for ICPB in a continuous-flow photocatalytic circulating-bed biofilm reactor (PCBBR) to mineralize 2,4,5-trichlorophenol (TCP), a biorecalcitrant compound. Photocatalysis produced TCP-degradation products that could be mineralized and the strong adsorption of TCP to the carrier enhanced biodegradation by relieving toxicity.

AOPs have been demonstrated to be very efficient in decreasing the pathogen load in contaminated water. Polo-Lopez *et al.* (2014) evaluated several solar-driven AOPs, i.e., photo-Fenton (Fe^{2+}, Fe^{3+}) at low reagent concentration, heterogeneous photocatalysis (TiO_2), and solar photoassisted H_2O_2 treatment for removal of the spores of Fusarium sp., a worldwide phytopathogen. Working in a pilot solar photoreactor with Compound Parabolic Collector (CPC), they confirmed that the inactivation rates varied and depended on the water matrix, and disinfection was fastest in distilled water followed by simulated municipal wastewater effluent (SMWWE) and real municipal wastewater effluents (RMVVWE). The best solar inactivation rate was with photo-Fenton treatment at pH 3, followed by H_2O_2/Solar and finally TiO_2/Solar was the slowest. These results underline the importance of solar AOPs and the CPC reactor technology as a good option for the killing/inactivation of waterborne pathogens.

Investigating the adsorption of viruses on iron-oxide coated sand (IOCS), Pecson *et al.* (2012) indicated that combining an IOCS matrix with sunlight exposure could improve wetland performance with respect to virus inactivation, taking advantage of the photoreactivity of the iron oxide surface to adsorption enhancement, maintaining the viruses closer to the inactivating surface. To take advantage of this mechanism, they proposed an innovative design of wetlands able to enhance sunlight exposure to the IOCS, e.g., constructing vertical flow wetlands with the IOCS in the top layer. Under such conditions, it may be possible to achieve the 4-log inactivation that is typically required for drinking water disinfection.

The wastewaters resulting from textile dyeing have a complex pollutant load characterized by quality indicators (pH, TDS, BOD_5, COD, TOC, color, and total chromium content) and their treatment can be done by adsorption and/or photocatalysis, on a well-chosen substrate and in optimized process conditions. Visa *et al.* (2011) confirmed that industrial wastewater treatment on fly-ash (FA) and on FA/TiO_2 mixtures can act as preliminary steps, with efficiencies up to 40%, and should be followed by biological or advanced chemical treatment. Thus, although photocatalysis represents an efficient treatment for single-dye solutions, industrial waste waters raise more complex problems, especially when mineralization is not fully possible, for all the organic components. Competitive reactions of dyes and by-products may result in much lower overall efficiencies compared to those obtained in single-dye solutions.

Taking as case study the photocatalytic degradation of tartrazine under UV light, Gupta *et al.* (2009) concluded that the successful removal of hazardous dyes from aqueous solutions needs the optimization of process conditions including water pH, catalyst (titanium dioxide) dose and dye concentration. Observing a significant decrease of the COD, they also confirmed a net improvement in the waste water quality by the addition of electron acceptors (e.g., H_2O_2) that simultaneously boosts hydroxyl radical formation and retard electron-hole recombination.

The photocatalytic decomposition of pollutants of emerging concern (e.g., cyanotoxins) is very sensitive to both photocatalyst nature and the kind of light source (UV, visible, solar). Zhao *et al.* (2014) demonstrated that anion doped NF-TiO_2 VLA photocatalysts are effective under UV and visible irradiation and over a range of water qualities. Probing the roles of different ROS, they found that the cyanotoxin contaminated water treatment is improved by increasing pH and that the presence of humic acid (HA), Fe^{3+} and Cu^{2+} can enhance the degradation rate.

8.6 PROCESS LIMITATIONS AND BYPRODUCT FORMATION; STRATEGIES TO MITIGATE THE ADVERSE EFFECTS ON THE TREATED WATER QUALITY

TiO_2 is considered an efficient, low cost and highly stable semiconductor photocatalyst, widely used in water treatment applications. The efficacy of heterogenous photocatalytic processes is generally dependent on various operational parameters, which can either introduce limitations or contribute to the overall performance (Chong *et al.* 2010).

The concentration of TiO_2 shows a linear proportional dependency to the reaction rate of the photocatalytic process, until the amount of TiO_2 surpasses a certain saturation level and photocatalytic efficiency is eventually reduced. This is attributed to the limited light photon absorption on the surface of the photocatalyst, due to the excess TiO_2 particles which hinder light penetration and therefore minimize the available TiO_2 surface area exposed to light irradiation (Chong *et al.* 2010; Gaya & Abdullah, 2008). Related studies have shown that the optimal TiO_2 concentration is related to the design and dimension of each photocatalytic reactor (Chong *et al.* 2009; Herrmann, 1999). In continuous flow photoreactors, TiO_2 loading is also related to hydrodynamics and flow considerations, since smaller reactor diameters are necessary for effective light penetration. On the other hand, for the minimum deposition of catalyst in parts of the reactor, a turbulent flow is necessary, which cannot be achieved in diameters smaller than 20 mm (Malato *et al.* 2009; Malato *et al.* 2003). Therefore, the photocatalytic reactor design should take under consideration the effect of catalyst loading and its relation to hydrodynamic parameters.

In heterogeneous photocatalysis, pH can significantly influence the efficiency of the process, since it can change the surface charge of the photocatalyst. The point of zero charge (PZC) is the condition where the surface charge of the catalyst is neutral. For TiO_2 catalysts, PZC values range between 4.0–7.0, depending on the specific catalyst (Ahmed, 2012; Chong *et al.* 2010). At pH < PZC (TiO_2) the catalytic surface is positively charged ($TiOH_2^+$) attracting negatively charged compounds, which can be sorbed on the surface and subsequently photodegraded. At pH > PZC (TiO_2) the surface is negatively charged (TiO^-), promoting the sorption of positively charged compounds (Ahmed, 2012; Rincón & Pulgarin, 2004). Changes in surface charge of TiO_2 also affect the interaction of catalyst particles themselves. At the point of zero charge, the catalyst surface is neutrally charged, leading to particle aggregation and sedimentation (Gaya & Abdullah, 2008). This property is utilized in Photocatalytic Membrane Reactors (PMRs) where initial water pH is adjusted to the range of TiO_2 particle PZC for maximum aggregation, sedimentation and collection of catalyst particles. Appropriate pH adjustment is necessary in various steps of the photocatalytic water treatment processes, for optimum reaction efficiency and adequate removal of the catalyst particles.

The presence of Dissolved Oxygen (DO) in the photocatalytic process ensures the presence of sufficient electron scavengers, which inhibit the recombination of excited conduction-band electrons and holes. DO also contributes in the production of various ROS and the stabilization of radical intermediates. Care should be given during the photocatalytic process, to ensure the presence of abundant DO, using air or oxygen sparging which could also promote TiO_2 suspension in the reactor (Pichat, 2013).

Another parameter limiting the process efficiency is the applied illumination wavelength and intensity. TiO_2 photocatalysts present variations in their band-gap energies, requiring different ranges of light wavelength for photon activation (Bahnemann, 2004; Herrmann, 1999). A careful selection of the applied photocatalyst for optimum utilization of irradiation wavelength is therefore necessary. For most TiO_2 photocatalysts the UV-A range (315–400 nm) provides sufficient irradiation energy for TiO_2 photon excitation. This also presents a limitation, since the intensity of natural UV irradiation reaching the surface of the earth is low, therefore the need for artificial light sources of adequate intensity is evident (Malato *et al.* 2009; Pichat, 2013; Spasiano *et al.* 2015). This raises the importance of light intensity in the overall efficiency of the photocatalytic process. It has been proven that at low light intensities there is a linear relation between reaction rates and radiant flux (Φ), whereas when intensity increases a square root dependency is observed above a specific value ($\Phi^{0.5}$) (Gaya & Abdullah, 2008; Herrmann, 1999; Malato *et al.* 2009; Pichat, 2013). At high intensities, the only limitation is mass transfer between the catalytic surface and bulk solution. For photocatalytic water treatment applications, a relatively high intensity is necessary, in order to sufficiently provide enough irradiation energy for the excitation of the catalyst (Chong *et al.* 2010).

Turbidity of the treated water, caused by suspended particulate matter, also inhibits the penetration of UV light, due to the light scattering effect and absorption of irradiation energy, thus introducing a limiting factor to the efficiency of the photocatalytic water treatment process (Ahmed, 2012; Haque & Muneer, 2007). Ensuring that turbidity remains low (below 30 NTU), enables better UV penetration and increases reaction rates (Gelover *et al.* 2006; Rincón & Pulgarin, 2003). A conventional physical treatment process (coagulation, filtration, etc.) should be applied before the photocatalytic treatment process, in order to reduce turbidity and provide water of increased clarity.

The presence of inorganic ions in water can affect the rate of photocatalytic reactions either through inhibition of ROS production or through deactivation of catalyst surface due to sedimentation and fouling of the surface (Ahmed, 2012; Guillard *et al.* 2003a, 2003b; Habibi *et al.* 2005; Mahvi *et al.* 2009; Muruganandham *et al.* 2006; Naeem & Feng, 2008; Özkan *et al.* 2004; Rincón & Pulgarin, 2004). As already mentioned, pH can greatly affect the surface charge of TiO_2 and selectively attract or repel inorganic ions from the surface (Guillard *et al.* 2003b). Although Fe^{2+} can catalyze photo-Fenton reactions, in increased concentrations it may have detrimental effect on the efficiency of the catalyst, since at neutral pH it can form $Fe(OH)_3$ on the surface of the catalyst (Choi *et al.* 1994). PO_4^{3-} and SO_4^{2-} have shown to be strongly adsorbed on the surface and further inhibit the activity of the catalyst (Abdullah, 1990). The fouling effect of inorganic ions could be explained by different mechanisms such as screening of the UV irradiation, surface deposition of precipitates, radical and hole scavenging, and competition for active surface sites (Abdullah, 1990; Burns, 1999; Chong *et al.* 2010; Guillard *et al.* 2003b). Cl^-, HCO_3^-, SO_4^{2-} and PO_4^{3-} have been known to act as radical and hole scavengers (Abdullah, 1990; Chong *et al.* 2010; Muruganandham *et al.* 2006; Pichat, 2013). Fouling of TiO_2 surface due to the presence of inorganic ions in water should be prevented using various water pretreatment techniques, including sedimentation, complexation or ion exchange in order to minimize the effect on the photochemical process (Burns, 1999). Also, regeneration of the fouled catalyst could be carried out, using various chemicals to rinse the catalyst, dissolute surface precipitates and reuse the catalyst (Abdullah, 1990).

The presence of NOM can limit the efficiency of the photocatalytic process, since it can occupy part of the active sites on the surface of the catalyst. Coupling photocatalytic oxidation with physical adsorption offers a solution to this problem. Appropriate adsorbents can be activated carbon, zeolites, clays, etc. The toxic organic contaminants are retained on the surface of the adsorbent, which is subsequently mixed with TiO_2. Desorbed compounds are photocatalytically degraded under irradiation and the adsorbent can be further used in multiple cycles through regeneration. A different strategy is the immobilization of catalyst on the surface of the adsorbent, producing a permeable mixed material. Applications of the process include phenol and chlorophenol degradation among others (Crittenden *et al.* 1997; Herrmann *et al.* 1999).

Two types of photocatalytic reactors have been widely used: reactors using suspended TiO_2 particles in slurry and reactors with photocatalyst immobilized on the surface of an inert surface.

TiO_2 is widely used in photocatalytic reactors as a suspension of powder in slurry, based on the observation that the rate of ROS generation is related to the amount of surface active sites of the suspended catalyst (Pozzo *et al.* 1997). This configuration provides improved photocatalytic activity compared to systems using immobilized photocatalysts, since it provides high surface area of catalyst per unit volume. The degradation efficiency of the fixed photocatalytic systems is defined by limitations of pollutant mass transfer from the bulk solution to the active sites on the surface of the catalyst and by decreased photocatalytic activity since photons do not reach the available catalytic sites (Augugliaro *et al.* 2006; Pozzo *et al.* 1997). Moreover the illumination path used in the photocatalytic reactor is important to the overall efficiency of the process. Direct photon activation or indirect illumination *via* parabolic light reflectors are used in combination to correct light source positioning in order to achieve uniform light distribution and symmetrical transmission (Bahnemann, 2004; Chong *et al.* 2010; Malato *et al.* 2009).

The disadvantage of the slurry reactor is the necessity for an additional post-treatment step, for the separation of the suspended photocatalyst from the treated water, in order to avoid catalyst loss and prevent pollution of treated water with TiO_2 particles. The separation and catalyst recovery can be achieved using conventional physical-chemical treatment processes such as sedimentation, coagulation, flotation and filtration (Chong *et al.* 2010; Fernández-Ibáñez *et al.* 2003), especially when the treatment parameters are optimized. Coagulation of TiO_2 particles can be achieved with pH adjustment close to the PZC or the addition of coagulants, but it is a long lasting process with additional necessary steps (filtration and removal of excess coagulant). Since TiO_2 particles usually have diameter of 20–40 nm, membrane filtration is necessary for efficient solid-liquid separation (Doll & Frimmel, 2005; Malato *et al.* 2009; Xi & Geissen, 2001; Zhang *et al.* 2008; Zhao *et al.* 2002). Several parameters should be taken under consideration for the application of the microfiltration and ultrafiltration processes including membrane fouling, pore blockage and regeneration steps (Lee *et al.* 2001; Molinari *et al.* 2002; Xi & Geissen, 2001).

Combinations of slurry with membrane filtration, i.e. membrane reactors (MRs), utilize the advantages of both processes. Nevertheless, the application of photocatalytic small-sized particles, require that pollutants must initially diffuse on the surface of the particles, which could prove to be a slow process. Moreover, immobilizing TiO_2 particles on a membrane could cause severe damage to the structure of the membrane due to close interaction with UV irradiation and hydroxyl radicals.

A hybrid photocatalytic membrane reactor (PMR) incorporates photocatalytic material on the surface and pores of the membrane, therefore the membrane performs simultaneous selective separation and photocatalytic degradation of pollutants, minimizing mass transfer limitations (Augugliaro *et al.* 2006; Herz, 2004; Molinari *et al.* 2004). Photocatalytic membranes have been prepared from semiconductor materials, TiO_2 supported on polymers or doted polymer membranes (Bellobono *et al.* 2005; Bosc *et al.* 2005; Molinari *et al.* 2004). They have been extensively used for the removal of trichloroethylene, 4-nitrophenol, MB, humic acids and bisphenol A in water treatment processes (Artale *et al.* 2001; Chin *et al.* 2006; Ryu *et al.* 2005; Tsuru *et al.* 2003; Tsuru *et al.* 2005).

A significant parameter, determining the treatment cost in such systems is the transmembrane pressure. Adjustment of water pH and the addition of electrolytes could also help in prolonging the membrane's life span by preventing pore blockage (Xi & Geissen, 2001). Recent trends point towards the use of selective organophilic membrane in combination with photocatalysts (pervaporation) (Augugliaro *et al.* 2006; Camera-Roda & Santarelli, 2007), which shows promise for applications focusing on degradation of trace organic pollutants, since it acts as a selective barrier for target molecules.

Photocatalytic processes generate highly reactive radical species, able to partially degrade and even mineralize water pollutants. Nevertheless, various transformation products (TPs) and intermediate degradation products are formed during the process up to the point of mineralization. Identification of these products is essential for the assessment of their environmental impact. Some of these TPs could be more persistent than the initial parent compound and could present higher toxicity. Although the identification of TPs and the assessment of their ecotoxicity has recently raised scientific interest, however, there is yet little available information on their environmental risk. In order to reduce the risk of their presence in treated water, numerous promising technologies can be applied for their abatement (membrane filtration, reverse osmosis, etc.) and their combination with various AOPs in order to promote full mineralization of recalcitrant TPs (Konstantinou *et al.* 2014).

There is scant information in past literature concerning the toxicity of TiO_2. Nevertheless, a recent study (Adams *et al.* 2006) has proven that nano-scale TiO_2 has caused bacterial growth inhibition to Gram-positive *Bacillus subtilis* and Gram-negative *Escherichia co*li used as test organisms. Bacterial growth inhibition was also observed under dark conditions. The size of material particles was higher than initially though, since aggregation took place. In a study conducted by Aruoja *et al.* (Aruoja *et al.* 2009), algal growth inhibition tests on *Pseudokirchneriella subcapitata* showed significant toxicity of nano-TiO_2 particles even under shaded light conditions. Aggregates of nano-particles entrapped algal cells contributing to overall toxicity. Heinlaan *et al.* (Heinlaan *et al.* 2008) has proven that nano- and bulk-TiO_2 particles were not found to be toxic to bacteria *Vibrio fischeri* and crustaceans *Daphnia magna* and *Thamnocephalus platyurus*, up to concentrations of 20 g L^{-1}.

8.7 REACTOR/EQUIPMENT DESIGN AND ECONOMIC CONSIDERATIONS, FIGURES-OF-MERIT

The design of photocatalytic reactors play a major role in reducing the cost required for the commercialization of the techniques. The reactors developed for the photocatalytic degradation of various environmental pollutants are shown in Figure 8.15 which is adapted from Chin *et al.* (2004). The tubular photocatalytic reactor with circulation mode, the semiconductor coated tubular photoreactor, and the batch photocatalytic reactor are few of the examples of photocatalytic reactors developed to study the photocatalytic degradation of pollutants. The design of the photocatalytic reactor with low cost or the replacement of expensive components of the photocatalytic reactor can reduce the cost required for commercialization. The utilization of solar energy as a source for the photocatalytic degradation process significantly enhances the efficiency and allows elimination of the need for the ultraviolet light source used in traditional photocatalytic reactors. Moreover, the modification of photocatalysts with suitable dopants allows activation by visible light and improves the photocatalytic efficiency. Therefore the development of sunlight assisted photocatalytic reactors received much attention and consideration for cost-effective construction and enchanted efficiency. In addition, the number of reactive oxygen species generated in the process depends on the geometry of the photocatalytic reactors. The detailed analysis about the energy consumption and operational cost and the advantages of the photocatalytic reactors needs to be thoroughly investigated before its commercialization.

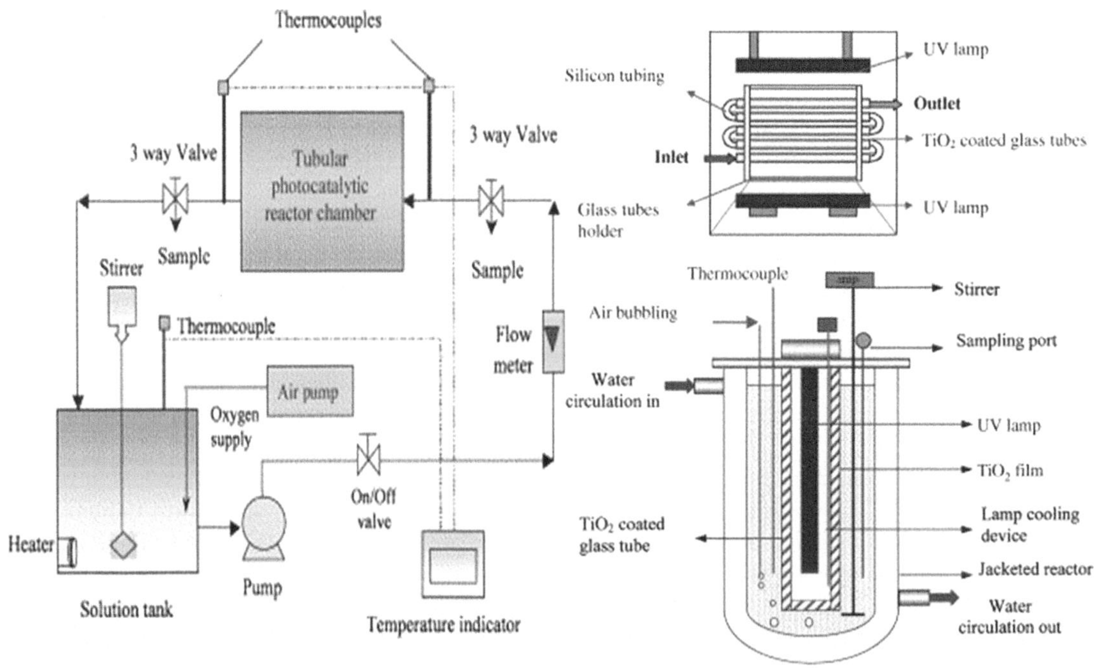

Figure 8.15 (a) Schematic diagram of the tubular photocatalytic reactor with circulation mode; (b) TiO_2 coated glass tubes in the tubular photoreactor chamber and (c) batch photocatalytic reactor (cross-section view). Reprinted with premission from Chin *et al.* (2004). AQ1

8.8 CASE STUDIES RELEVANT TO PROCESS IMPLEMENTATION TO WATER TREATMENT

Nowadays, widespread concerns continue to be raised about the increasing presence of emerging contaminants in the environment. AOPs encompass treatment technologies that rely on the production of hydroxyl radicals, which act as strong oxidants capable of degrading recalcitrant compounds. Among these AOPs, heterogeneous photocatalysis using TiO_2 is a promising technology and TiO_2 photocatalyst has been considered a key material for the complete destruction of recalcitrant organic pollutants (Fujishima *et al.* 2008). Commercialization level of semiconductor photocatalysis varies, depending on the application field and ranges from basic research to mature market. At this time, the total global market for photocatalytic products exhibited an annual growth rate of 14.3% during the last 5 years, including more than 1000 companies that sell products for a wide variety of applications (Gagliardi, 2010; Portela & Hernandez-Alonso, 2013). Major applications of photocatalysis include areas of self-cleaning surfaces, air purification/sterilization and water treatment. Although the growth of the photocatalytic products market was impressive during the last decade, a limited number of commercially available water purification systems have been developed. It seems that the major water treatment companies have not yet adopted photocatalysis as an economically viable alternative to current methods such as chlorination and ozonolysis. However, photocatalysis systems for water purification are the choice of companies with small and specific water purification needs such as electronics, textile and automotive industries, and also military and petrochemical industries (effluent or contaminated sites clean up).

The major companies promoting water purification systems have been presented by Mills and Lee in a web-based review (Mills & Lee, 2002) and more recently by J. Saari *et al.* in a briefing report (Saari *et al.* 2010). Many photocatalytic water treatment products are still in the development or demonstration phase like the Panasonic's demonstration machine of photocatalytic water treatment technology presented in Tokyo at Eco-Products 2014 (Panasonic, 2014) or the SOWARLA system (Saari *et al.* 2010; SOWARLA GmbH), a solar photocatalytic water purification system realized in a demonstration plant at the DRL location in Lampoldshusen, Germany. Also, other renowned solar photocatalytic plants located at the INETI (Instituto Nacional de Engenharia, Technologia Industrial eInovacao, Portugal) and PSA (Plataforma Solar de Almeria, Spain) have been established in a pilot scale. Considering the limited number of commercially available water treatment systems using semiconductor photocatalysis, the number of case studies is also limited and rarely company's web pages contain detailed information concerning the product and the operational and/or process conditions.

Despite this dearth of detailed documentation, the following case studies offer valuable information on the technological progress made so far in the field of semiconductor photocatalysis and how commercial water purification systems have met and overcome real-world challenges. Case studies presented previously by others have been omitted (Mills & Lee, 2004).

Among all companies mentioning in their web site that offer photocatalytic water purification systems, Purifics® ES Inc. is the largest supplier which reports a number of interesting case histories for its Photo-Cat® system. This system utilizes the TiO_2 in the form of a slurry which is activated by full spectrum artificial UV light. The Photo-Cat® system comprises pre-filters to remove fine solid materials, a coalescor to remove emulsified oils or greases, a Photo-Cat® reactor consisting of a set of stainless-steel tubes connected together to form a series of racks and a patented ceramic membrane separation process which allows the catalyst to be completely recaptured and reintroduced into the inlet stream (Purifics ES Inc. DOC3006R5, 2011). Three brief case histories for this system are presented below.

8.8.1 Contaminated groundwater with 1,4-dioxane and volatile organic solvents, Sarasota, Florida, USA (2013)

Lockheed Martin Corporation acquired the former American Beryllium Co facility located in Sarasota, Florida, USA in 1996 and discovered that the ground water was contaminated with 1,4-dioxane, volatile organic compounds (VOCs) (trichloroethene, TCE; tetrachloroethene, PCE; 1,1-Dichloroethene, DCE; 1,1-Dichloroethane, DCA & vinyl chloride, VC), aluminum and elevated levels of iron. The company has assumed responsibility for the clean-up operation and used an interim Photo-Cat system to purify contaminated groundwater from 10 source extraction wells at this site from 2006 to 2012. These remediation activities have been implemented in the frame of an Interim Remedial Action Plan approved by Florida Department of Environmental Protection (FDEP). In 2013, Purifics commissioned an integrated and fully automated ground water purification system (Photo-cat) capable to remove metals, such as iron and aluminum, and to destroy dissolved organic chemicals (i.e., 1,4-dioxane and VOCs). According to the firm, this chemical-free process is capable of treating contaminated groundwater pumped from 80 extraction wells with a capacity of 1900 m^3 per day (0.5 million gallons per day, MGD). This system is fully automated and can be operated unattended for long periods of time. In addition, it treats contaminants of concern to levels below the detection limit (3 ppb), without the generation of a waste stream (Purifics ES Inc. DOC4016R6, 2013). A Lockheed Martin information bulletin issued about the Photo-Cat process on this site states: "The extracted groundwater is … treated using advanced oxidation technology [Photo-Cat], which Lockheed Martin has found to be the most efficient, safe and cost effective way to treat site-specific chemicals of concern (COCs), including trichloroethylene (TCE) and 1,4-dioxane" (Purifics ES Inc., 2011).

8.8.2 1,4-Dioxane and VOCs destruction in drinking water, Southern US water district (2013)

Municipal groundwater wells in a Southern US water district have been contaminated with 1,4-dioxane and carcinogenic VOCs. The concentration of these compounds is required to be below 1 ppb for 1,4-dioxane and below their respective maximum contaminant level (MCL) for other VOCs with a goal of non-detectable concentrations. Before the final selection of an appropriate Photo-Cat system, in-house bench and pilot tests have been performed achieving water treatment efficiency optimization. In August 2013, two NSF-61 certified Photo-Cat systems installed in the site with a treatment capacity of one million gallons per day (1 MGD). The system started up in October 2013 and produced high quality water at design flow (Purifics ES Inc. DOC4020R8, 2013).

8.8.3 Removal of chromium (Cr^{6+}) in groundwater, Superfund site in Odessa, Texas, USA (2013)

Contaminated groundwater with hexavalent chromium (Cr^{6+}) at a US EPA Superfund Site in Odessa, Texas had been treated with conventional ion exchange technology to reduce Cr^{6+} from 400 ppb levels to 50 ppb discharge limit. This treatment process was not able to treat contaminated water in a sustainable way and achieve effluent concentration levels of Cr^{6+} below the pending standard of 1 ppb. Consequently, the EPA examined alternative options such as Purifics Photo-Cat platform. Initially, a pilot testing period of four months was conducted in order to optimize the photocatalytic process and the results showed that the pilot system achieved a cost effective removal of Cr^{6+} below 1 ppb. After this successful pilot test, a fully automated Photo-Cat system with a treatment capacity of 2180 m^3 per day (400 gallons per minute, gpm) procured and installed during the spring of 2013 (Purifics ES Inc. DOC4015R10, 2013). The water treatment system composed of the photocatalytic unit (Photo-cat) and a dewatering recovery system (DeWRS) for TiO_2 cleaning and recovery of chromium. During the photocatalytic process, hexavalent chromium reduces to less toxic Cr^{3+} which adsorbs onto TiO_2. The water passes through membrane filters which separate the flow stream from the TiO_2 catalyst. The decontaminated water exits the system and then is injected again to the aquifer. The TiO_2 slurry is reused to treat incoming water while a slipstream of the TiO_2 is continuously removed and entered in the dewatering recovery system. After a regeneration process of the spent TiO_2 through acid washing, cleaned TiO_2 is reused in the Photo-Cat system and Cr^{3+} is recovered as a highly concentrated wet solid ($Cr(OH)_3$) (Purifics ES Inc. DOC4015R10; Endlund, 2013). The effectiveness of this technology for removing all chromium species from water was also confirmed by an independent study using a commercially available integrated UV reactor system (Photo-Cat serial 0700 system; Purifics ES, Inc.) and commercial grade titanium dioxide (Stancl *et al.* 2015).

8.9 COMMERCIAL APPLICATIONS

8.9.1 Global market and standards

The global market for catalysts with energy and environmental applications was found to be 25.7 billion USD in 2014. This market is estimated to climb at a compound annual growth rate (CAGR) of 5.6% from 27.3 billion USD in 2015 to 35.8 billion USD by 2020. The fastest growing section of this market is environmental applications while the global market for environmental catalysts alone is calculated to increase by a CAGR of 4.4%, from 19.8 billion USD in 2015 to 24.7 billion USD in 2020. Also, the global market for energy catalysts is anticipated to increase by a CAGR of 8.6% from 7.4 billion USD in 2015 to 11.1 billion USD by 2020, (McWilliams, 2015).

8.9.2 Drinking water regulations driving the process implementation

Since there has been an increasing amount of developments regarding photocatalyst technologies, a number of standards have been put in place. These standards relate to photocatalysts in general, along with the utilisation of photocatalysts during water treatment and purification. The International Organization of Standardization (ISO) has published several standards in relation to photocatalysts and water. Specific standards include ISO 10676:2010 (2010) which describes the testing of photocatalytic surfaces for the decomposition and elimination of pollutants in water while under UV irradiation, ISO 10677:2011 (2011) and ISO 14605:2013 (2013) describe the appropriate method for measuring radiation intensity when performing the laboratory testing of photocatalytic materials while exposed to various light sources. ISO 10678:2010 (2010) specifies the method for determining the photocatalytic activity of surfaces. This is achieved by observing the degradation of a dye molecule, MB, in aqueous solution using UV light. ISO 13125:2013 (2013) specifies a test method for determining the antifungal activity of photocatalytic materials. This is performed by analysing the number of pre-incubated fungal spores that survive UV-A light exposure. ISO 27447:2009 (2009) provides a test method for determining the antibacterial activity of photocatalytic materials by measuring the enumeration of bacteria after UV light irradiation. Upon developing a new photocatalytic technology directed at water cleaning for commercialisation purposes the technical committees that are responsible for introducing new standards/amendments are water quality (ISO/TC 147), fine ceramics (ISO/TC 206), service activities relating to drinking water supply systems and wastewater systems (ISO/TC 224), nanotechnologies (ISO/TC 229) and water reuse (ISO/TC 282).

8.9.3 Commercialization technologies

Photocatalytic reaction at a commercial scale utilize a large photoreactor to conduct the reaction in the presence of light. The photoreactor must be designed in such a way that all of the chemical mixture interacts with light to produce the necessary product. The designs for photoreactors can be divided into three main categories, that is the parabolic trough collector (PTC), the non-concentrating collector (NCC) and the compound parabolic collector (CPC), observed in Figure 8.16 (a, b), (c, d) and (e, f), respectively.

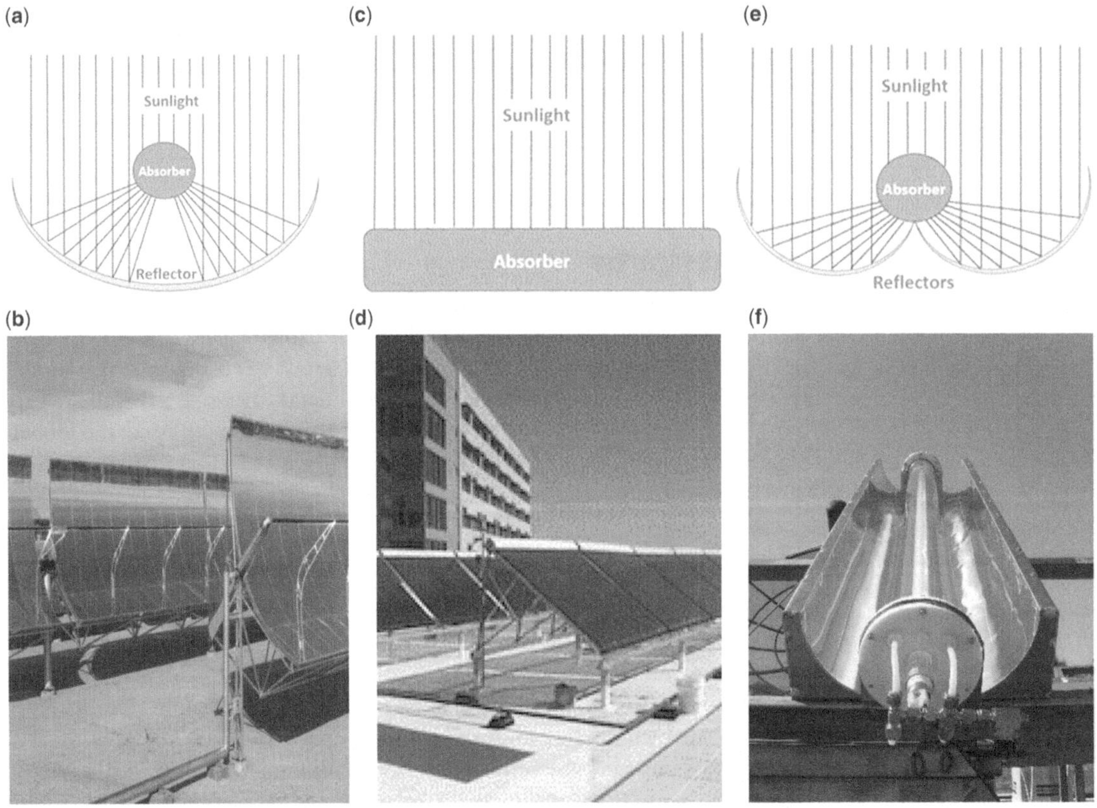

Figure 8.16 The method for light collection in the three main variations of photoreactor, (a) and (b) parabolic trough collectors (PTC), (c) and (d) non-concentrating collectors (NCC) and (e) and (f) compound parabolic collectors (CPC). SkyFuel (2015), Apricus (2015), PSA (2015).

In 1989, PTCs, capable of concentrating sunlight ×50, were installed at the Solar Thermal Test Facility in Sandia, New Mexico. The photocatalytic treatment of water polluted with trichloroethylene (TCE) consisted of six PTCs with single-axis solar tracking. The system had a total reactor length and trough aperture area of 218 m and 465 m^2, respectively (Pacheco *et al.* 1990). Again in 1993, through collaboration between the Sandia National Laboratory, the National Renewable Energy Laboratory, and the Lawrence Livermore National Laboratory, polluted water from the Superfund site at the Lawrence Livermore National Laboratory in California was treated within 158 m^2 of PTC for the decomposition of

TCE. Results from the study determined that the processing costs for the treatment of the polluted water would be in comparison to conventional methods (Mehos & Turchi, 1993).

NCCs have been employed by solar pilot scale plants for the photocatalytic treatment of the commercial azo dye, Yellow Cibacron FN-2R (Zayani *et al.* 2009), the organic pollutants, 4-chlorophenol, formetanate, indigo carmine and Congo red (Guillard *et al.* 2003), along with phenol and dichloroacetic acid (Feitz *et al.* 2000), while also being employed for the inactivation of *Aeromonas hydrophila* ATCC 35654, (Khan *et al.* 2012).

CPCs have been employed at the pilot scale for the degradation of the phenolic compounds, vanillin, protocatechuic acid, syringic acid, *p*-coumaric acid, gallic acid and L-tyrosine. This treatment was performed using the photo-Fenton reaction, under sunlight at the Plataforma Solar de Almeria in Spain and yielded completely mineralised compounds as degradable intermediates were achieved during processing (Gernjak *et al.* 2003). Pilot plants have also utilised the photo-Fenton process for the pre-treatment of the reactive dyes Procion Red H-E7B and Cibacron Red FN-R. The treated dyes were found to exist at sufficient concentrations (>82% Dissolved Organic Carbon (DOC) removal) for their introduction into an immobilised biomass reactor, ready for biological treatment (García-Montaño *et al.* 2008).

Pilot plants have utilised both 35 L and 75 L CPCs for the solar photo-Fenton and TiO_2 photocatalytic treatment of the water soluble pesticides, cymoxanil, methomyl, oxamyl, dimethoate, pyrimethanil and telone. This has permitted the complete degradation of these compounds, while also observing near complete mineralisation (Oller *et al.* 2006). Solar TiO_2 photocatalysis and solar photo-Fenton processes were compared at a plant using CPCs with a photoreactor volume of 35 L. The non-biodegradable chlorinated solvents (NBCSs) dichloromethane, dichloroethane and trichloromethane where dissolved in water at a concentration of 50 mg L^{-1} and it was assessed that the photo-Fenton process was a more appropriate treatment upon analysis of the chloride release and TOC mineralisation results (Rodríguez *et al.* 2005). Solar TiO_2 photocatalysis via CPC possessing a photoreactor volume of 5.4 L and total plant volume of 11.0 L was employed for the disinfection of water. The water present was contaminated with pure cultures of *Escherichia coli*. During this treatment the results showed that this method was not totally effective, as regrowth of the *E. coli* was observed (Fernández *et al.* 2005). An industrial plant possessing an initial 250 L h^{-1} treatment capacity was examined for the decomposition of the non-biodegradable pharmaceutical precursor, α-methyl phenyl glycine (MPG). The system consisted of photo-Fenton CPC collectors with an area of 100 m^2 which treated water containing MPG at a concentration of 500 mg L^{-1}. Results showed that the cost per m^3 of MPG effluent treated was between 7–10 USD (30 and 70% capital and operational costs, respectively) (Malato *et al.* 2007).

CPCs have been used for the treatment of leachates originating from sanitary landfills. After lagooning pre-treatment, a solar photo-Fenton oxidation system, composed of 39.5 m^2 CPCs, was utilised for the degradation of organic compounds (humic compounds) of the most recalcitrant nature. This step in the process led to a biodegradability of 70% for the organic compounds according to Zahn-Wellens test (Silva *et al.* 2013). A combination of solar TiO_2 photocatalysis and solar photo-Fenton systems have been employed to treat undiluted olive mill wastewater containing high levels of polyphenols in the g L^{-1} range. The results from the treatments suggested that the treated wastewater, after flocculation for degradation of large solids, could be used for irrigation and fertilisation due to its high potassium and phosphate levels (Gernjak *et al.* 2004). CPCs possessing a total reactor volume of 35 L and utilising both heterogeneous and homogeneous solar photocatalytic processes, at a pilot plant at the Plataforma Solar de Almeria, Spain, were employed for the degradation of a synthetic municipal wastewater containing peptone, meat extract, urea, K_2HPO_4, NaCl, $CaCl_2$, $2H_2O$ and $Mg_2SO_4{\cdot}7H_2O$ at a DOC of 200 mg L^{-1}. The photo-Fenton method was found to be more effective, exhibiting an 80% degradation of organic content after 4 hours of illumination, while the $TiO_2/Na_2S_2O_8$ system exhibited a 73% decrease in organic content after 5 hours of illumination (Kositzi *et al.* 2004). Real industrial wastewater containing the commercial pesticides Couraze®, Ditimur-40®, Metomur®, Scala® and Vydate®, at a DOC of 500 mg L^{-1}, has been treated using the solar photo-Fenton process. This was performed at a plant (150 m^2 of CPCs with a total reactor volume of 1060 L) where the treatment was found to enhance the biodegradability of the biorecalcitrant compounds resulting in a decrease of DOC from 500 to 325 mg L^{-1} and a decrease in active pesticide ingredient from 325 down to 0 mg L^{-1} after 9 hours of illumination time (Zapata *et al.* 2010). The European consortium, SOLARDETOX has utilised solar TiO_2 photocatalysis and CPCs for the degradation of dichloroacetic acid and cyanide at two plants. The plants possessed collector surfaces of 9.9 and 3.0 m^2 with photoreactor volumes of 247 and 22 L, respectively, yielding results which showed that cyanide, at a high concentration of 1 g L^{-1}, could be completely degraded successfully (Malato *et al.* 2002).

8.9.4 Companies and products

Degussa P-25 (now known as AEROXIDE® TiO_2 P25), produced by Evonik Industries, is by far the most commonly used commercially available photocatalytically active material. This product consists of a mixture of approx. 75% anatase and 25% rutile and possesses a specific surface area of 30–40 m^2 g^{-1}. Hombifine N, produced by Sachtleben Chemie GmbH, is another very common TiO_2 powder. Cristal Chemical Company is the second largest producer of TiO_2 in the world, while being the largest of titanium chemicals. The company produces CristalACTiV™ ultrafine TiO_2 which can be used for the purification of water. Green Millenium Incorporated have manufactured sols, which consist of a mixed solution of peroxotitanic acid (PTA) and TiO_2 (peroxo-modified anatase), termed TPX-85 and TPX-220 respectively, both containing particles of approx. 10 nm. These neutral solutions can be applied to a variety of surfaces, including both metals and resin, yielding a coating that is completely insoluble in water. Zixilai Environmental Protection Technology uses TiO_2 nanoparticle photocatalysts to decompose organic and inorganic toxic pollutants from a wide range of industries including alcohol and chemical plants, biological drug manufacturers, pesticide manufacturers, paper mills and print works. Their nanoparticle product has exhibited a COD decomposition rate of >90% within one hour. Purifics Environmental Technologies Incorporated produce the Photo-Cat(R) system, which is composed of a TiO_2 slurry along with low pressure Hg lamps. The company is the largest supplier of photocatalytic treatment systems designed for the processing of industrial water. Its markets in the USA, Canada and Korea predominately treat contaminated ground water as well as industrial wastewater and lagoons (Oller *et al.* 2007).

8.10 FUTURE RESEARCH NEEDS

The structural, morphological, polarity and optoelectronic properties of TiO_2 nanomaterials being used can affect its performance regarding its applications to environmental improvement and energy conversion. Parameters such as applied potential, precursor concentration, pH and type of corrosive solutions, and crystallization conditions, can be altered during the preparation of samples in order to optimise the properties of TiO_2 that are required, e.g. producing TiO_2 NTs. TiO_2 NTs have been shown to have high photocatalytic activity and decomposition rates, however the length of the NTs has an effect on the materials activity. A method for applying high efficiency and cost effective TiO_2 NTs can be developed for air and water purification, especially in relation to the photocatalytic degradation of hazardous toxins and pollutants.

Various methods (e.g., doping with metals or non-metals, formation of heterojunctions) improves the treatment of water by photocatalytic activity. It is, however, difficult to compare the resulting photocatalysts as they can have different morphologies and compositions and are often tested under different operating conditions such as irradiation wavelength and flux, solution pH, contaminant concentration, etc. (Banerjee *et al.* 2015). Another method for increasing photocatalytic activity in water treating materials is the modification, by chemical treatment, of materials in order to increase the surface roughness. Investigations into the understanding structure activity relationships and excited state behaviour of photocatalysts are required in the future (Banerjee *et al.* 2015). At present, researchers are mainly examining the development of photocatalytic wetting materials. However, the ability to produce underwater superoleophobic materials shows attention because of their applications, separation of oil-water emulsions, and wastewater treatment. Research is needed to comprehend the structure-wettability patterns of these materials.

Iron-based materials have also been examined for the use in environmental remediation. Spinel ferrites possessing a narrow band gap have been examined as magnetically separable photocatalysts since they have the photocatalytic properties needed for the degradation of contaminants in water. Various materials, such as TiO_2, graphene or nickel-cobalt, have been used in combination in order to improve photocatalytic performance. Some of these novel composite materials offer enhanced potential for various photocatalytic applications when compared with existing photocatalysts. However, the efficiency of all these novel materials under various environmental conditions have not been comprehensively tested and additional research will be required to address issues such as stability and robustness of these catalysts under various conditions (e.g., change of pH or weathering effects). Nevertheless, the rapid growth and continued progresses in the development of novel photocatalytic materials offers great promise for the future of advanced oxidation process.

Fagan *et al.* (2016) discussed in great detail the development of active solar and visible light photocatalysis for environmental remediation, including the treatment of contaminants of emerging concern (CECs), resistant pathogenic microorganisms, and endocrine disrupting compounds (EDCs). Standard water treatment processes are often ineffective to treat these contaminants and result in incomplete bio-degradation of the resulting products. It is therefore highly important to develop more efficient technologies to eliminate such recalcitrant pollutants. Such efforts could also include technologies that make use of both ultraviolet and visible spectra of the solar light, such as photocatalytic technologies discussed in this chapter. There is also the need for improving the effectiveness of these photocatalytic materials for bulk industrial use (Fagan *et al.* 2016).

To date the research completed shows the need for further developments in the synthesis of photocatalytic materials activated by solar light (i.e., utilizing effectively both the UV and visible spectra of solar light) and effective reduction in the electron-hole recombination pathway in order to enhance photocatalyst activity as well as to solve practical reactor design challenges associated with catalyst reuse/retaining, catalyst fouling/attrition, and process operation. All these will be necessary steps towards further industrialization and large scale implementation of photocatalytic technologies for water treatment applications.

8.11 DISCLAIMER

The U.S. Environmental Protection Agency, through its Office of Research and Development, funded and managed, or partially funded and collaborated in, the research described herein. It has been subjected to the Agency's administrative review and has been approved for external publication. Any opinions expressed in this paper are those of the author(s) and do not necessarily reflect the views of the Agency, therefore, no official endorsement should be inferred. Any mention of trade names or commercial products does not constitute endorsement or recommendation for use.

8.12 ACKNOWLEDGEMENTS

The authors wish to acknowledge financial support under the U.S.-Ireland R&D Partnership Initiative from the Science Foundation Ireland (SFI-grant number 10/US/I1822(T)) and U.S. National Science Foundation-CBET (Award, 1033317). One of the authors (Ciara Byrne) wishes to acknowledge The Institute of Technology Sligo President's Bursary for providing financial support. "This work was performed in the context of the European COST Action MP1302 Nanospectroscopy." The research has been co-financed by the European Union (European Social Fund – ESF) and Greek national funds through the Operational Program "Education and Lifelong Learning" of the National Strategic Reference Framework (NSRF) – Research Funding Program: Thales "AOP-NanoMat" (MIS, 379409). D. D. Dionysiou acknowledges support from the University of Cincinnati through an UNESCO Co-Chair position on "Water Access and Sustainability". P. Falaras acknowledges funding by Prince Sultan Bin Abdulaziz International Prize for Water (PSIPW)-Alternative Water Resources Prize 2014.

8.13 REFERENCES

Abdullah M. (1990). Effects of common inorganic anions on rates of photocatalytic oxidation of organic carbon over illuminated titanium dioxide. *Journal of Physical Chemistry*, **94**(17), 6820–6825.

Adams L. K., Lyon D. Y. and Alvarez P. J. J. (2006). Comparative eco-toxicity of nanoscale TiO_2, SiO_2, and ZnO water suspensions. *Water Research*, **40**(19), 3527–3532.

Ahmed S. (2012). Impact of operating conditions and recent developments in heterogeneous photocatalytic water purification process. *Critical Reviews in Environmental Science and Technology*, **42**(6), 601–675.

Allen S. J., McKay G. and Khader K. Y. H. (1989). Equilibrium adsorption isotherms for basic dyes onto lignite. *Journal of Chemical Technology and Biotechnology*, **45**, 291–302.

Anandan S., Narasinga Rao T., Sathish M., Rangappa D., Honma I. and Miyauchi M. (2013). Superhydrophilic graphene-loaded TiO_2 thin film for self-cleaning applications. *ACS Applied Materials & Interfaces*, **5**, 207–212.

Antoniou M. G., Nicolaou P. A., Shoemaker J. A., de la Cruz A. A. and Dionysiou D. D. (2009). Impact of the morphological properties of thin TiO_2 photocatalytic films on the detoxification of water contaminated with the cyanotoxin, microcystin-LR. *Applied Catalysis B: Environmental*, **91**, 165–173.

Apricus (2015). Evacuated Tube Solar Collectors. http://www.apricus.com/html/solar_collector.htm#.VfbW4NLBwXA (accessed 14 November 2015).

Artale M. A., Augugliaro V., Drioli E., Golemme G., Grande C., Loddo V., Molinari R., Palmisano L. and Schiavello M. (2001). Preparation and characterisation of membranes with entrapped TIO_2 and preliminary photocatalytic tests. *Annali di Chimica*, **91**(3–4), 127–136.

Aruoja V., Dubourguier H.-C., Kasemets K. and Kahru A. (2009). Toxicity of nanoparticles of CuO, ZnO and TiO_2 to microalgae Pseudokirchneriella subcapitata. *Science of The Total Environment*, **407**(4), 1461–1468.

Augugliaro V., Litter M., Palmisano L. and Soria J. (2006). The combination of heterogeneous photocatalysis with chemical and physical operations: A tool for improving the photoprocess performance. *Journal of Photochemistry and Photobiology C: Photochemistry Reviews*, **7**(4), 127–144.

Babu V. J., Vempati S., Uyar T. and Ramakrishna S. (2015). Review of one-dimensional and two-dimensional nanostructured materials for hydrogen generation. *Physical Chemistry Chemical Physics*, **17**, 2960–2986.

Bahnemann D. (2004). Photocatalytic water treatment: solar energy applications. *Solar Energy*, **77**(5), 445–459.

Banerjee S., Pillai S., Falaras P., O'Shea K., Byrne J. and Dionysiou D. (2014). New insights into the mechanism of visible light photocatalysis. *The Journal of Physical Chemistry Letters*, **5**, 2543–2554.

Banerjee S., Dionysiou D. D. and Pillai S. C. (2015). Self-cleaning applications of TiO_2 by photo-induced hydrophilicity and photocatalysis. *Applied Catalysis B: Environmental*, **176**, 396–428.

Bellobono I. R., Morazzoni F., Bianchi R., Mangone E. S., Stanescu R., Costache C. and Tozzi P. M. (2005). Solar energy driven photocatalytic membrane modules for water reuse in agricultural and food industries. Pre-industrial experience using s-triazines as model molecules. *International Journal of Photoenergy*, **7**(2), 87–94.

Bosc F., Ayral A. and Guizard C. (2005). Mesoporous anatase coatings for coupling membrane separation and photocatalyzed reactions. *Journal of Membrane Science*, **265**(1–2), 13–19.

Burns R. A. (1999). Effect of inorganic ions in heterogeneous photocatalysis of TCE. *Journal of Environmental Engineering*, **125**(1), 77–85.

Camera-Roda G. and Santarelli F. (2007). Intensification of water detoxification by integrating photocatalysis and pervaporation. *Journal of Solar Energy Engineering, Transactions of the ASME*, **129**(1), 68–73.

Cao A., Liu Z., Chu S., Wu M., Ye Z., Cai Z., Chang Y., Wang S., Gong Q. and Liu Y. (2010). A facile one-step method to produce graphene-CdS quantum dot nanocomposites as promising optoelectronic materials. *Advanced Materials (Weinheim, Germany)*, **22**, 103–106.

Cao S., Chen C., Liu T., Zeng B., Ning X., Chen X., Xie X. and Chen W. (2014). Superfine and closely-packed TiO_2/Bi_2O_3 lamination on graphene nanoplates with high photocatalytic activity. *Catalysis Communications*, **46**, 61–65.

Casbeer E., Sharma V. K. and Li X. Z. (2012). Synthesis and photocatalytic activity of ferrites under visible light: A review. *Separation and Purification Technology*, **87**, 1–14.

Chan K. and Chu W. (2005). Atrazine removal by catalytic oxidation processes with or without UV irradiation: Part I–quantification and rate enhancement via kinetic study. *Applied Catalysis B: Environmental*, **58**(3), 157–163.

Chang Q., Ma Z., Wang J., Yan Y., Shi W., Chen Q., Huang Y., Yu Q. and Huang L. (2015). Graphene nanosheets@ZnO nanorods as three-dimensional high efficient counter electrodes for dye sensitized solar cells. *Electrochimica Acta*, **151**, 459–466.

Chekli L., Galloux J., Zhao Y. X., Gao B. Y. and Shon H. K. (2015). Coagulation performance and floc characteristics of polytitanium tetrachloride (PTC) compared with titanium tetrachloride ($TiCl_4$) and iron salts in humic acid-kaolin synthetic water treatment. *Separation and Purification Technology*, **142**, 155–161.

Chin M. L., Mohamed A. R. and Bhatia S. (2004). Performance of photocatalytic reactors using immobilized TiO_2 film for the degradation of phenol and methylene blue dye present in water stream. *Chemosphere*, **57**(7), 547–554.

Chin S. S., Chiang K. and Fane A. G. (2006). The stability of polymeric membranes in a TiO_2 photocatalysis process. *Journal of Membrane Science*, **275**(1–2), 202–211.

Chio H., Stathatos E. and Dionysiou D. D. (2006). Synthesis of nanocrystalline photocatalytic TiO_2 thin films and particles using sol-gel modified with nonionic surfactants. *Thin Solid Films*, **510**, 107–114.

Choi W., Termin A. and Hoffmann M. R. (1994). The role of metal ion dopants in quantum-sized TiO_2: correlation between photoreactivity and charge carrier recombination dynamics. *Journal of Physical Chemistry*, **98**(51), 13669–13679.

Chong M. N., Lei S., Jin B., Saint C. and Chow C. W. K. (2009). Optimisation of an annular photoreactor process for degradation of Congo Red using a newly synthesized titania impregnated kaolinite nano-photocatalyst. *Separation and Purification Technology*, **67**(3), 355–363.

Chong M. N., Jin B., Chow C. W. K. and Saint C. (2010). Recent developments in photocatalytic water treatment technology: A review. *Water Research*, **44**(10), 2997–3027.

Chung Y. S., Park S. B. and Kang D.-W. (2004). Magnetically separable titania-coated nickel ferrite photocatalyst. *Materials Chemistry and Physics*, **86**, 375–381.

Comninellis C., Kapalka A., Malato S., Parsons S. A., Poulios I. and Mantzavinos D. (2008). Advanced oxidation processes for water treatment: advances and trends for R&D. *Journal of Chemical Technology and Biotechnology*, **83**, 769–776.

Crittenden J. C., Suri R. P. S., Perram D. L. and Hand D. W. (1997). Decontamination of water using adsorption and photocatalysis. *Water Research*, **31**(3), 411–418.

Daneshvar N., Rabbani M., Modirshahla N. and Behnajady M. (2004). Kinetic modeling of photocatalytic degradation of acid red 27 in UV/TiO_2 process. *Journal of Photochemistry and Photobiology A: Chemistry*, **168**, 39–45.

Davydov L. and Smirniotis P. G. (2000). Quantification of the primary processes in aqueous heterogeneous photocatalysis using single-stage oxidation reactions. *Journal of Catalysis*, **191**, 105–115.

Devadoss A., Sudhagar P., Das S., Lee S. Y., Terashima C., Nakata K., Fujishima A., Choi W., Kang Y. S. and Paik U. (2014). Synergistic metal-metal oxide nanoparticles supported electrocatalytic graphene for improved photoelectrochemical glucose oxidation. *ACS Applied Materials & Interfaces*, **6**, 4864–4871.

Doll T. E. and Frimmel F. H. (2005). Cross-flow microfiltration with periodical back-washing for photocatalytic degradation of pharmaceutical and diagnostic residues-evaluation of the long-term stability of the photocatalytic activity of TiO_2. *Water Research*, **39**(5), 847–854.

Dung N. T., Van Khoa N. and Herrmann J.-M. (2005). Photocatalytic degradation of reactive dye RED-3BA in aqueous TiO_2 suspension under UV-visible light. *International Journal of Photoenergy*, **7**, 11–15.

Endlund C. E. (2013). Second Five-Year Review the Sprague Road Groundwater Plume Superfund Site, EPA ID# TX0001407444, Odessa, Ector County, Texas, USA. http://www.epa.gov/superfund/sites/fiveyear/f2013060004657.pdf (accessed 25 August 2015).

Erdei L., Arecrachakul N. and Vigneswaran S. (2008). A combined photocatalytic slurry reactor-immersed membrane module system for advanced wastewater treatment. *Separation and Purification Technology*, **62**, 382–388.

Etacheri V., Seery M. K., Hinder S. J. and Pillai S. C. (2010). Highly visible light active $TiO_{2-x}N_x$ heterojunction photocatalysts. *Chemistry of Materials*, **22**, 3843–3853.

Etacheri V., Seery M. K., Hinder S. J. and Pillai S. C. (2011). Oxygen rich titania: a dopant free, high temperature stable, and visible-light active anatase photocatalyst. *Advanced Functional Materials*, **21**, 3744–3752.

Etacheri V., Roshan R. and Kumar V. (2012a). Mg-doped ZnO nanoparticles for efficient sunlight-driven photocatalysis. *ACS Applied Materials & Interfaces*, **4**, 2717–2725.

Etacheri V., Seery M. K., Hinder S. J. and Pillai S. C. (2012b). Nanostructured $Ti_{1-x}S_xO_{2-y}N_y$ heterojunctions for efficient visible-light-induced photocatalysis. *Inorganic Chemistry*, **51**, 7164–7173.

Etacheri V., Michlits G., Seery M. K., Hinder S. J. and Pillai S. C. (2013). A highly efficient $TiO_{2-x}C_x$ nano-heterojunction photocatalyst for visible light induced antibacterial applications. *ACS Applied Materials & Interfaces*, **5**, 1663–1672.

Fagan R., Mccormack D. E., Dionysiou D. D. and Pillai S. C. (2016). A review of solar and visible light active TiO_2 photocatalysis for treating bacteria, cyanotoxins and contaminants of emerging concern. *Materials Science in Semiconductor Processing*, **42**(1), 2–14.

Fan H., Zhao X., Yang J., Shan X., Yang L., Zhang Y., Li X. and Gao M. (2012). ZnO-graphene composite for photocatalytic degradation of methylene blue dye. *Catalysis Communications*, **29**, 29–34.

Fan J. M., Zhao Z. H., Wang J. Y. and Zhu L. X. (2015). Synthesis of Cr,N-codoped titania nanotubes and their visible-light-driven photocatalytic properties. *Applied Surface Science*, **324**, 691–697.

Feigenbrugel V., Person A. L., Calvé S. L., Mellouki A., Munoz A. and Wirtz K. (2006). Atmospheric fate of dichlorvos: photolysis and OH-initiated oxidation studies. *Environmental science & technology*, **40**(3), 850–857.

Feitz A. J., Boyden B. H. and Waite T. D. (2000). Evaluation of two solar pilot scale fixed-bed photocatalytic reactors. *Water Research*, **34**(16), 3927–3932.

Fernández P., Blanco J., Sichel C. and Malato S. (2005). Water disinfection by solar photocatalysis using compound parabolic collectors. *Catalysis Today*, **101**(3), 345–352.

Fernandez-Ibanez P., Blanco J., Malato S. and de las Nieves F. J. (2003). Application of the colloidal stability of TiO_2 particles for recovery and reuse in solar photocatalysis. *Water Research*, **37**(13), 3180–3188.

Fischer K., Kuhnert M., Glaser R. and Schulze A. (2015). Photocatalytic degradation and toxicity evaluation of diclofenac by nanotubular titanium dioxide-PES membrane in a static and continuous setup. *RSC Advances*, **5**, 16340–16348.

Fisher M. B., Keane D. A., Fernandez-Ibanez P., Colreavy J., Hinder S. J., McGuigan K. G. and Pillai S. C. (2013). Nitrogen and copper doped solar light active TiO_2 photocatalysts for water decontamination. *Applied Catalysis, B: Environmental*, **130–131**, 8–13.

Fox M. A. and Dulay M. T. (1993). Heterogeneous photocatalysis. *Chemical Reviews*, **93**, 341–357.

Fu D., Han G., Chang Y. and Dong J. (2012). The synthesis and properties of ZnO-graphene nano hybrid for photodegradation of organic pollutant in water. *Materials Chemistry and Physics*, **132**, 673–681.

Fu Y. and Wang X. (2011). Magnetically separable $ZnFe_2O_4$-Graphene catalyst and its high photocatalytic performance under visible light irradiation. *Industrial & Engineering Chemistry Research*, **50**, 7210–7218.

Fu Y., Chen H., Sun X. and Wang X. (2012). Graphene-supported nickel ferrite: a magnetically separable photocatalyst with high activity under visible light. *AIChE Journal*, **58**, 3298–3305.

Fujishima A. and Honda K. (1972). Electrochemical photolysis of water at a semiconductor electrode. *Nature (London, United Kingdom)*, **238**, 37–38.

Fujishima A., Zhang X. and Tryk D. A. (2008). TiO_2 photocatalysis and related surface phenomena. *Surface Science Reports*, **63**(12), 515–582.

Gagliardi M. M. (2010). Photocatalysts: Technologies and Global Market. BCC Research Report, Report Code: AVM069A.

Gao W., Wang M., Ran C. and Li L. (2015). Facile one-pot synthesis of MoS_2 quantum dots-graphene-TiO_2 composites for highly enhanced photocatalytic properties. *Chemical Communications (Cambridge, United Kingdom)*, **51**, 1709–1712.

García-Montaño J., Pérez-Estrada L., Oller I., Maldonado M. I., Torrades F. and Peral J. (2008). Pilot plant scale reactive dyes degradation by solar photo-Fenton and biological processes. *Journal of Photochemistry and Photobiology A: Chemistry*, **195**(2), 205–214.

Gaya U. I. and Abdullah A. H. (2008). Heterogeneous photocatalytic degradation of organic contaminants over titanium dioxide: a review of fundamentals, progress and problems. *Journal of Photochemistry and Photobiology C: Photochemistry Reviews*, **9**(1), 1–12.

Gelover S., Gómez L. A., Reyes K. and Teresa Leal M. (2006). A practical demonstration of water disinfection using TiO_2 films and sunlight. *Water Research*, **40**(17), 3274–3280.

Georgekutty R., Seery M. K. and Pillai S. C. (2008). A highly efficient Ag-ZnO photocatalyst: Synthesis, properties, and mechanism. *Journal of Physical Chemistry C*, **112**, 13563–13570.

Gernjak W., Krutzler T., Glaser A., Malato S., Caceres J., Bauer R. and Fernández-Alba A. R. (2003). Photo-Fenton treatment of water containing natural phenolic pollutants. *Chemosphere*, **50**(1), 71–78.

Gernjak W., Maldonado M. I., Malato S., Cáceres J., Krutzler T., Glaser A. and Bauer R. (2004). Pilot-plant treatment of olive mill wastewater (OMW) by solar TiO_2 photocatalysis and solar photo-Fenton. *Solar Energy*, **77**(5), 567–572.

Ghicov A., Albu S. P., Hahn R., Kim D., Stergiopoulos T., Kunze J., Schiller C.-A., Falaras P. and Schmuki P. (2009). TiO_2 nanotubes in dye-sensitized solar cells: critical factors for the conversion efficiency. *Chemistry, An Asian Journal*, **4**(4), 520–525.

Giraldo A. L., Peñuela G. A., Torres-Palma R. A., Pino N. J., Palominos R. A. and Mansilla H. D. (2010). Degradation of the antibiotic oxolinic acid by photocatalysis with TiO_2 in suspension. *Water Research*, **44**, 5158–5167.

Gondal M. A., Adesida A. A., Rashid S. G., Shi S., Khan R., Yamani Z. H., Shen K., Xu Q., Seddigi Z. S. and Chang X. (2015). Preparation of $WO_3/g\text{-}C_3N_4$ composites and their enhanced photodegradation of contaminants in aqueous solution under visible light irradiation. *Reaction Kinetics, Mechanisms and Catalysis*, **114**, 357–367.

Granados-Oliveros G., Páez-Mozo E. A., Ortega F. M., Ferronato C. and Chovelon J.-M. (2009). Degradation of atrazine using metalloporphyrins supported on TiO_2 under visible light irradiation. *Applied Catalysis B: Environmental*, **89**(3), 448–454.

Guillard C., Disdier J., Monnet C., Dussaud J., Malato S., Blanco J., Maldonado M. I. and Herrmann J. M. (2003a). Solar efficiency of a new deposited titania photocatalyst: Chlorophenol, pesticide and dye removal applications. *Applied Catalysis B: Environmental*, **46**(2), 319–332.

Guillard C., Lachheb H., Houas A., Ksibi M., Elaloui E. and Herrmann J. M. (2003b). Influence of chemical structure of dyes, of pH and of inorganic salts on their photocatalytic degradation by TiO_2 comparison of the efficiency of powder and supported TiO_2. *Journal of Photochemistry and Photobiology A: Chemistry*, **158**(1), 27–36.

Gupta V. K., Jain R., Nayak A., Agarwal S. and Shrivastava M. (2011). Removal of the hazardous dye-Tartrazine by photodegradation on titanium dioxide surface. *Materials Science & Engineering C-Materials for Biological Applications*, **31**, 1062–1067.

Habibi M. H., Hassanzadeh A. and Mahdavi S. (2005). The effect of operational parameters on the photocatalytic degradation of three textile azo dyes in aqueous TiO_2 suspensions. *Journal of Photochemistry and Photobiology A: Chemistry*, **172**(1), 89–96.

Hamilton J. W. J., Byrne J. A., Dunlop P. S. M., Dionysiou D. D., Pelaez M., O'Shea K., Synnott D. and Pillai S. C. (2014). Evaluating the mechanism of visible light activity for N,F-TiO_2 using photoelectrochemistry. *Journal of Physical Chemistry C*, **118**, 12206–12215.

Han C., Pelaez M., Likodimos V., Kontos A. G., Falaras P., O'Shea K. and Dionysiou D. D. (2011). Innovative visible light-activated sulfur doped TiO_2 films for water treatment. *Applied Catalysis B: Environmental*, **107**, 77–87.

Han C., Choi H. and Dionysiou D. D. (2013). Green chemistry for environmental remediation. In: An Introduction to Green Chemistry Methods, R. Luque and J. C. Colmenares (eds), 1st edn, Future Science, London, UK, pp. 148–166.

Han C., Andersen J., Likodimos V., Falaras P., Linkugel J. and Dionysiou D. D. (2014a). The effect of solvent in the sol-gel synthesis of visible light-activated, sulfur-doped TiO_2 nanostructured porous films for water treatment. *Catalysis Today*, **224**, 132–139.

Han C., Likodimos V., Khan J. A., Nadagouda M. N., Andersen J., Falaras P., Rosales-Lombardi P. and Dionysiou D. D. (2014b). UV-visible light-activated Ag-decorated, monodisperse TiO_2 aggregates for treatment of the pharmaceutical oxytetracycline. *Environmental Science and Pollution Research*, **21**, 11781–11793.

Haque M. M. and Muneer M. (2007). TiO_2-mediated photocatalytic degradation of a textile dye derivative, bromothymol blue, in aqueous suspensions. *Dyes and Pigments*, **75**(2), 443–448.

Hashimoto K., Irie H. and Fujishima A. (2005). TiO_2 photocatalysis: a historical overview and future prospects. *Japanese Journal of Applied Physics*, **44**(12R), 8269–8285.

He H.-Y. (2015). Photocatalytic degradation of malachite green on magnetically separable Ni1-xCoxFe$_2$O$_4$ nanoparticles synthesized by using a hydrothermal process. *American Chemical Science Journal*, **6**, 58–68.

He Y., Cai J., Zhang L., Wang X., Lin H., Teng B., Zhao L., Weng W., Wan H. and Fan M. (2014). Comparing two new composite photocatalysts, *t*-$LaVO_4$/*g*-C_3N_4 and *m*-$LaVO_4$/*g*-C_3N_4, for their structures and performances. *Industrial & Engineering Chemistry Research*, **53**, 5905–5915.

He Y., Zhang L., Teng B. and Fan M. (2015). New application of Z-scheme Ag_3PO_4/*g*-C_3N_4 composite in converting CO_2 to fuel. *Environmental Science & Technology*, **49**, 649–656.

Hegedűs M. and Dombi A. (2004). Gas-phase heterogeneous photocatalytic oxidation of chlorinated ethenes over titanium dioxide: Perchloroethene. *Applied Catalysis B: Environmental*, **53**(3), 141–151.

Heinlaan M., Ivask A., Blinova I., Dubourguier H.-C. and Kahru A. (2008). Toxicity of nanosized and bulk ZnO, CuO and TiO_2 to bacteria Vibrio fischeri and crustaceans Daphnia magna and Thamnocephalus platyurus. *Chemosphere*, **71**(7), 1308–1316.

Herrmann J. M. (1999). Heterogeneous photocatalysis: fundamentals and applications to the removal of various types of aqueous pollutants. *Catalysis Today*, **53**(1), 115–129.

Herrmann J. M., Matos J., Disdier J., Guillard C., Laine J., Malato S. and Blanco J. (1999). Solar photocatalytic degradation of 4-chlorophenol using the synergistic effect between titania and activated carbon in aqueous suspension. *Catalysis Today*, **54**(2–3), 255–265.

Herz R. K. (2004). Intrinsic kinetics of first-order reactions in photocatalytic membranes and layers. *Chemical Engineering Journal*, **99**(3), 237–245.

Hoffmann M. R., Martin S. T., Choi W. Y. and Bahnemann D. W. (1995). Environmental applications of semiconductor photocatalysis. *Chemical Reviews*, **95**(1), 69–96.

ISO 27447:2009 (2009). Fine ceramics (advanced ceramics, advanced technical ceramics) – Test method for antibacterial activity of semiconducting photocatalytic materials.

ISO 10676:2010 (2010a). Fine ceramics (advanced ceramics, advanced technical ceramics) – Test method for water purification performance of semiconducting photocatalytic materials by measurement of forming ability of active oxygen.

ISO 10678:2010 (2010b). Fine ceramics (advanced ceramics, advanced technical ceramics) – Determination of photocatalytic activity of surfaces in an aqueous medium by degradation of methylene blue.

ISO 10677:2011 (2011). Fine ceramics (advanced ceramics, advanced technical ceramics) – Ultraviolet light source for testing semiconducting photocatalytic materials.

ISO 13125:2013 (2013a). Fine ceramics (advanced ceramics, advanced technical ceramics) – Test method for antifungal activity of semiconducting photocatalytic materials.

ISO 14605:2013 (2013b). Fine ceramics (advanced ceramics, advanced technical ceramics) – Light source for testing semiconducting photocatalytic materials used under indoor lighting environment.

Kamat P. V. (2007). Meeting the clean energy demand: Nanostructure architectures for solar energy conversion. *Journal of Physical Chemistry C*, **111**, 2834–2860.

Keane D. A., McGuigan K. G., Ibanez P. F., Polo-Lopez M. I., Byrne J. A., Dunlop P. S. M., O'Shea K., Dionysiou D. D. and Pillai S. C. (2014). Solar photocatalysis for water disinfection: materials and reactor design. *Catalysis Science & Technology*, **4**, 1211–1226.

Khan J. A., Han C., Shah N. S., Khan H. M., Nadagouda M. N., Likodimos V., Falaras P., O'Shea K. and Dionysiou D. D. (2014). Ultraviolet-visible light-sensitive high surface area phosphorous-fluorine-co-doped TiO_2 nanoparticles for the degradation of atrazine in water. *Environmental Engineering Science*, **31**, 435–446.

Khan S. J., Reed R. H. and Rasul M. G. (2012). Thin-film fixed-bed reactor (TFFBR) for solar photocatalytic inactivation of aquaculture pathogen Aeromonas hydrophila. *BMC microbiology*, **12**(1), 5.

Khan S. U. M., Al-Shahry M. and Ingler W. B. (2002). Efficient photochemical water splitting by a chemically modified n-TiO_2. *Science*, **297**(5590), 2243–2245.

Kim S. B. and Hong S. C. (2002). Kinetic study for photocatalytic degradation of volatile organic compounds in air using thin film TiO_2 photocatalyst. *Applied Catalysis B: Environmental*, **35**, 305–315.

Kim S. M., Lee S. J., Kim S. H., Kwon S., Yee K. J., Song H., Somorjai G. A. and Park J. Y. (2013). Hot carrier-driven catalytic reactions on Pt-CdSe-Pt nanodumbbells and Pt/GaN under light irradiation. *Nano Letters*, **13**, 1352–1358.

Konstantinou I. K., Antonopoulou M. and Lambropoulou D. A. (2014). Transformation products of emerging contaminants formed during advanced oxidation processes. In: Transformation Products of Emerging Contaminants in the Environment, Lambropoulou D. A. and Nollet L. M. L. (eds), John Wiley and Sons Ltd, West Sussex, UK, pp. 179–228.

Kontos A. G., Kontos A. I., Tsoukleris D. S., Likodimos V., Kunze J., Schmuki P. and Falaras P. (2009). Photo-induced effects on self-organized TiO_2 nanotube arrays: the influence of surface morphology. *Nanotechnology*, **20**(4), 045603.

Kontos A. G., Katsanaki A., Maggos T., Likodimos V., Ghicov A., Kim D., Kunze J., Vasilakos C., Schmuki P. and Falaras P. (2010). Photocatalytic degradation of gas pollutants on self-assembled titania nanotubes. *Chemical Physics Letters*, **490**(1–3), 58–62.

Kontos A. G., Katsanaki A., Likodimos V., Maggos T., Kim D., Vasilakos C., Dionysiou D. D., Schmuki P. and Falaras P. (2012). Continuous flow photocatalytic oxidation of nitrogen oxides over anodized nanotubular titania films. *Chemical Engineering Journal*, **179**, 151–157.

Kositzi M., Poulios I., Malato S., Caceres J. and Campos A. (2004). Solar photocatalytic treatment of synthetic municipal wastewater. *Water Research*, **38**(5), 1147–1154.

Kudo A. and Sekizawa M. (2000). Photocatalytic H2 evolution under visible light irradiation on Ni-doped ZnS photocatalyst. *Chemical Communications (Cambridge, United Kingdom)*, **15**, 1371–1372, doi: 10.1039/b003297m.

Larumbe S., Monge M. and Gómez-Polo C. (2012). Magnetically separable photocatalyst $Fe_3O_4/SiO_2/N\text{-}TiO_2$ hybrid nanostructures. *IEEE Transactions on Magnetics*, **50**, 2302404.

Lee E., Hong J.-Y., Kang H. and Jang J. (2012). Synthesis of TiO_2 nanorod-decorated graphene sheets and their highly efficient photocatalytic activities under visible-light irradiation. *Journal of Hazardous Materials*, **219–220**, 13–18.

Lee S. A., Choo K. H., Lee C. H., Lee H. I., Hyeon T., Choi W. and Kwon H. H. (2001). Use of ultrafiltration membranes for the separation of TiO_2 photocatalysts in drinking water treatment. *Industrial and Engineering Chemistry Research*, **40**(7), 1712–1719.

Li G., Park S., Kang D. W., Krajmalnik-Brown R. and Rittmann B. E. (2011). 2,4,5-Trichlorophenol degradation using a novel TiO_2-coated biofilm carrier: Roles of adsorption, photocatalysis, and biodegradation. *Environmental Science & Technology*, **45**, 8359–67.

Li X., Hou Y., Zhao Q., Teng W., Hu X. and Chen G. (2011). Capability of novel $ZnFe_2O_4$ nanotube arrays for visible-light induced degradation of 4-chlorophenol. *Chemosphere*, **82**, 581–586.

Li Y., Niu J., Yin L., Wang W., Bao Y., Chen J. and Duan Y. (2011). Photocatalytic degradation kinetics and mechanism of pentachlorophenol based on superoxide radicals. *Journal of Environmental Sciences*, **23**(11), 1911–1918.

Li Y.-F., Xu D., Oh J. I., Shen W., Li X. and Yu Y. (2012). Mechanistic study of codoped titania with nonmetal and metal ions: a case of C + Mo codoped TiO_2. *ACS Catalysis*, **2**, 391–398.

Liang H. C. and Li X. Z. (2009). Effects of structure of anodic TiO_2 nanotube arrays on photocatalytic activity for the degradation of 2,3-dichlorophenol in aqueous solution. *Journal of Hazardous Materials*, **162**(2–3), 1415–1422.

Likodimos V., Stergiopoulos T., Falaras P., Kunze J. and Schmuki P. (2008). Phase composition, size, orientation, and antenna effects of self-assembled anodized titania nanotube arrays: A polarized micro-raman investigation. *The Journal of Physical Chemistry C*, **112**(33), 12687–12696.

Lin J., Liu X. L., Zhu S., Liu Y. S. and Chen X. F. (2015). Anatase TiO_2 nanotube powder film with high crystallinity for enhanced photocatalytic performance. *Nanoscale Research Letters*, **10**, 110.

Liu C., Cao C., Luo X. and Luo S. (2015). Ag-bridged Ag_2O nanowire network/TiO_2 nanotube array p–n heterojunction as a highly efficient and stable visible light photocatalyst. *Journal of Hazardous Materials*, **285**, 319–324.

Liu J., Han R., Zhao Y., Wang H., Lu W., Yu T. and Zhang Y. (2011). Enhanced photoactivity of V-N codoped TiO_2 derived from a two-step hydrothermal procedure for the degradation of PCP-Na under visible light irradiation. *Journal of Physical Chemistry C*, **115**, 4507–4515.

Liu Z., Zhang X., Nishimoto S., Jin M., Tryk D. A., Murakami T. and Fujishima A. (2008). Highly ordered TiO_2 nanotube arrays with controllable length for photoelectrocatalytic degradation of phenol. *The Journal of Physical Chemistry C*, **112**(1), 253–259.

Malato S., Blanco J., Maldonado M. I., Oller I., Gernjak W. and Pérez-Estrada L. (2007). Coupling solar photo-Fenton and biotreatment at industrial scale: Main results of a demonstration plant. *Journal of Hazardous Materials*, **146**(3), 440–446.

Maurino V., Minero C., Pelizzetti E. and Vincenti M. (1999). Photocatalytic transformation of sulfonylurea herbicides over irradiated titanium dioxide particles. *Colloids and Surfaces A: Physicochemical and Engineering Aspects*, **151**(1), 329–338.

McWilliams A. (2015). Catalysts for Environmental and Energy Applications, Report CHM020E, BCC Research, Wellesley, MA 02481, USA.

Mehos M. S. and Turchi C. S. (1993). Field testing solar photocatalytic detoxification on TCE-contaminated groundwater. *Environmental Progress*, **12**(3), 194–199.

Mills A. and Lee S.-K. (2002). A web-based overview of semiconductor photochemistry-based current commercial applications. *Journal of Photochemistry and Photobiology, A: Chemistry*, **152**, 233–247.

Mills A. and Lee S.-K. (2004). Semiconductor photocatalysis. In: Advanced Oxidation Processes for Water and Wastewater Treatment, S. Parsons (ed.), IWA Publishing, London, pp. 137–166.

Mahvi A. H., Ghanbarian M., Nasseri S. and Khairi A. (2009). Mineralization and discoloration of textile wastewater by TiO_2 nanoparticles. *Desalination*, **238**(1–3), 309–316.

Malato S., Blanco J., Vidal A., Fernández P., Cáceres J., Trincado P., Mezcua M. and Vincent M. (2002). New large solar photocatalytic plant: Set-up and preliminary results. *Chemosphere*, **47**(3), 235–240.

Malato S., Blanco J., Campos A., Cáceres J., Guillard C., Herrmann J. M. and Fernández-Alba A. R. (2003). Effect of operating parameters on the testing of new industrial titania catalysts at solar pilot plant scale. *Applied Catalysis B: Environmental*, **42**(4), 349–357.

Malato S., Fernández-Ibáñez P., Maldonado M. I., Blanco J. and Gernjak W. (2009). Decontamination and disinfection of water by solar photocatalysis: Recent overview and trends. *Catalysis Today*, **147**(1), 1–59.

Martyanov I. N., Savinov E. N. and Klabunde K. J. (2003). Influence of solution composition and ultrasonic treatment on optical spectra of TiO_2 aqueous suspensions. *Journal of Colloid and Interface Science*, **267**, 111–116.

Minero C. (1999). Kinetic analysis of photoinduced reactions at the water semiconductor interface. *Catalysis Today*, **54**, 205–216.

Mitsubishi Mat. Corp., EP 786,283.

Molinari R., Palmisano L., Drioli E. and Schiavello M. (2002). Studies on various reactor configurations for coupling photocatalysis and membrane processes in water purification. *Journal of Membrane Science*, **206**(1–2), 399–415.

Molinari R., Pirillo F., Falco M., Loddo V. and Palmisano L. (2004). Photocatalytic degradation of dyes by using a membrane reactor. *Chemical Engineering and Processing: Process Intensification*, **43**(9), 1103–1114.

Monllor-Satoca D., Gómez R., González-Hidalgo M. and Salvador P. (2007). The "Direct-Indirect" model: An alternative kinetic approach in heterogeneous photocatalysis based on the degradation of interaction of dissolved pollutant species with the semiconductor surface. *Catalysis Today*, **129**, 247–255.

Montoya J., Velásquez J. and Salvador P. (2009). The direct-indirect kinetic model in photocatalysis: A reanalysis of phenol and formic acid degradation rate dependence on photon flow and concentration in TiO_2 aqueous dispersions. *Applied Catalysis B: Environmental*, **88**, 50–58.

Montoya J. F., Peral J. and Salvador P. (2014). Comprehensive kinetic and mechanistic analysis of TiO_2 photocatalytic reactions according to the Direct-Indirect Model: (I) Theoretical approach. *The Journal of Physical Chemistry C*, **118**, 14266–14275.

Mora-Seró I., Villareal L., Bisquert J., Pitarch A., Gómez R. and Salvador P. (2005). Photoelectrochemical behavior of nanostructured TiO_2 thin-film electrodes in contact with aqueous electrolytes containing dissolved pollutants: a model for distinguishing between direct and indirect interfacial hole transfer from photocurrent measurements. *The Journal of Physical Chemistry B*, **108**, 3371–3380.

Moustakas N. G., Kontos A. G., Likodimos V., Katsaros F., Boukos N., Tsoutsou D., Dimoulas A., Romanos G. E., Dionysiou D. D. and Falaras P. (2013). Inorganic-organic core-shell titania nanoparticlesfor efficient visible light activated photocatalysis. *Applied Catalysis B: Environmental*, **130–131**, 14–24.

Moustakas N. G., Katsaros F. K., Kontos A. G., Romanos G. E., Dionysiou D. D. and Falaras P. (2014). Visible light active TiO_2 photocatalytic filtration membranes with improved permeability and low energy consumption. *Catalysis Today*, **224**, 56–69.

Muruganandham M., Shobana N. and Swaminathan M. (2006). Optimization of solar photocatalytic degradation conditions of Reactive Yellow 14 azo dye in aqueous TiO_2. *Journal of Molecular Catalysis A: Chemical*, **246**(1–2), 154–161.

Ng Y. H., Iwase A., Kudo A. and Amal R. (2010). Reducing graphene oxide on a visible-light $BiVO_4$ photocatalyst for an enhanced photoelectrochemical water splitting. *Journal of Physical Chemistry Letters*, **1**, 2607–2612.

Nolan N. T., Synnott D. W., Seery M. K., Hinder S. J., Van Wassenhoven A. and Pillai S. C. (2012). Effect of N-doping on the photocatalytic activity of sol-gel TiO_2. *Journal of Hazardous Materials*, **211–212**, 88–94.

Naeem K. and Feng O. (2008). Parameters effect on heterogeneous photocatalysed degradation of phenol in aqueous dispersion of TiO_2. *Journal of Environmental Sciences*, **21**, 527–533.

O'Regan B. and Gratzel M. (1991). A low-cost, high-efficiency solar cell based on dye-sensitized colloidal TiO_2 films. *Nature*, **353**, 737–740.

Ochando-Pulido J. M., Stoller M., Di Palma L. and Martinez-Ferea A. (2014). Threshold performance of a spiral-wound reverse osmosis membrane in the treatment of olive mill effluents from two-phase and three-phase extraction processes. *Chemical Engineering and Processing: Process Intensification*, **83**, 64–70.

Oller I., Gernjak W., Maldonado M. I., Pérez-Estrada L. A., Sánchez-Pérez J. A. and Malato S. (2006). Solar photocatalytic degradation of some hazardous water-soluble pesticides at pilot-plant scale. *Journal of Hazardous Materials*, **138**(3), 507–517.

Ollis D. F. (2005). Kinetics of liquid phase photocatalyzed reaction: an illuminating approach. *The Journal of Physical Chemistry B*, **109**, 2439–2444.

Ou H.-H. and Lo S.-L. (2007). Photocatalysis of gaseous trichloroethylene (TCE) over TiO_2: The effect of oxygen and relative humidity on the generation of dichloroacetyl chloride (DCAC) and phosgene. *Journal of Hazardous Materials*, **146**(1), 302–308.

Özkan A., Özkan M. H., Gürkan R., Akçay M. and Sökmen M. (2004). Photocatalytic degradation of a textile azo dye, Sirius Gelb GC on TiO_2 or Ag-TiO_2 particles in the absence and presence of UV irradiation: The effects of some inorganic anions on the photocatalysis. *Journal of Photochemistry and Photobiology A: Chemistry*, **163**(1–2), 29–35.

Pacheco J. E., Prairie M., Evans L. and Yellowhorse L. (1990). Engineering-scale experiements of solar photocatalytic oxidation of trichloroethylene. Proceedings 25th Intersociety Energy Conversion Engineering Conference, Reno, Nevada, 141–145.

Panasonic (2014). Panasonic Develops 'Photocatalytic Water Purification Technology' – Creating Drinkable Water with Sunlight and Photocatalysts. http://news.panasonic.com/global/stories/2014/30520.html (accessed 25 August 2015).

Pang Y. L., Lim S., Ong H. C. and Chong W. T. (2014). A critical review on the recent progress of synthesizing techniques and fabrication of TiO_2-based nanotubes photocatalysts. *Applied Catalysis A-General*, **481**, 127–142.

Patrocinio A. O. T., Paula L. F., Paniago R. M., Freitag J. and Bahnemann D. W. (2014). Layer-by-layer TiO_2/WO_3 thin films as efficient photocatalytic self-cleaning surfaces. *ACS Applied Materials & Interfaces*, **6**, 16859–16866.

Pecson B. M., Decrey L. and Kohn T. (2012). Photoinactivation of virus on iron-oxide coated sand: Enhancing inactivation in sunlit waters. *Water Research*, **46**, 1763–1770.

Pelaez M., de la Cruz A. A., Stathatos E., Falaras P. and Dionysiou D. D. (2009). Visible light-activated N-F-codoped TiO_2 nanoparticles for the photocatalytic degradation of microcystin-LR in water. *Catalysis Today*, **144**, 19–25.

Pelaez M., Nolan N., Pillai S. C., Seery M., Falaras P., Kontos A. G., Dunlop P. S. M., Byrne J. A., O'Shea K., Entezari M. H. and Dionysiou D. D. (2012). A review on the visible light active titanium dioxide photocatalysts for environmental applications. *Applied Catalysis B: Environmental*, **125**, 331–349.

Pelaez M., Baruwati B., Varma R. S., Luque R. and Dionysiou D. D. (2013). Microcystin-LR removal from aqueous solutions using a magnetically separable N-doped TiO_2 nanocomposite under visible light irradiation. *Chemical Communications*, **49**, 10118–10120.

Periyat P., Leyland N., McCormack D. E., Colreavy J., Corr D. and Pillai S. C. (2010). Rapid microwave synthesis of mesoporous TiO_2 for electrochromic displays. *Journal of Materials Chemistry*, **20**, 3650–3655.

Pichat P. (2013). Photocatalysis and Water Purification: From Fundamentals to Recent Applications. Wiley-VCH Verlag GmbH & Co. KGaA, Weinheim, Germany.

Pillai S. C., Periyat P., George R., McCormack D. E., Seery M. K., Hayden H., Colreavy J., Corr D. and Hinder S. J. (2007). Synthesis of high-temperature stable anatase TiO_2 photocatalyst. *Journal of Physical Chemistry C*, **111**, 1605–1611.

Portela R. and Hernandez-Alonso M. D. (2013). Environmental applications of photocatalysis. In: Design of Advanced Photocatalytic Materials for Energy and Environmental Applications, J. M. Coronado, F. Fresno, M. D. Hernandez-Alonso, R. Portela, (eds), Springer-Verlag, London, pp. 35–66.

Polo-Lopez M. I., Castro-Alferez M., Oller I. and Fernandez-Ibanez P. (2014). Assessment of solar photo-Fenton, photocatalysis, and H_2O_2 for removal of phytopathogen fungi spores in synthetic and real effluents of urban wastewater. *Chemical Engineering Journal*, **257**, 122–130.

Poyatos J. M., Muñio M. M., Almecija M. C. Torres J. C., Hontoria E. and Osorio F. (2010). Advanced oxidation processes for wastewater treatment: state of the art. *Water, Air & Soil Pollution*, **205**, 187–204.

Pozzo R. L., Baltanás M. A. and Cassano A. E. (1997). Supported titanium oxide as photocatalyst in water decontamination: State of the art. *Catalysis Today*, **39**(3), 219–231.

PSA (2015) Solar Chemical Facilities. http://www.sollab.eu/psa.html (accessed 15 April 2016).

Purifics ES Inc. (2011). Photo-Cat cleans up contaminated groundwater. *Membrane Technology*, **2011**(9), 6.

Purifics ES Inc. DOC3006R5. (2011). Photo-Cat Water Purification. http://www.purifics.com/lwdcms/doc-view.php?module=documents&module_id=376&doc_name=doc (accessed 25 August 2015).

Purifics ES Inc. DOC4015R10. (2013). Case History: Photo-Cat Chromium (Cr6) Removal to <1ppb. http://www.purifics.com/lwdcms/doc-view.php?module=documents&module_id=468&doc_name=doc (accessed 25 August 2015).

Purifics ES Inc. DOC4016R6. (2013). Chemical Free Case History: 1,4-Dioxane Groundwater Purification for Lockheed Martin. http://www.purifics.com/lwdcms/doc-view.php?module=documents&module_id=379&doc_name=doc (accessed 25 August 2015).

Purifics ES Inc. DOC4020R8. (2013). Case History: 1,4-Dioxane and cVOC Destruction in Drinking Water. http://www.purifics.com/lwdcms/doc-view.php?module=documents&module_id=461&doc_name=doc (accessed 25 August 2015).

Qiu B., Zhou Y., Ma Y., Yang X., Sheng W., Xing M. and Zhang J. (2015). Facile synthesis of the Ti^{3+} self-doped TiO_2-graphene nanosheet composites with enhanced photocatalysis. *Scientific Reports*, **5**, 8591.

Ren B., Han C., Al Anazi A. H., Nadagouda M. N. and Dionysiou D. D. (2013). Iron-based nanomaterials for the treatment of emerging environmental contaminants. In: Interactions of Nanomaterials with Emerging Environmental Contaminants, R. Doong, V. K. Sharma and H. Kim (ed.), ACS Symposium Series 1150, American Chemical Society, Washington, DC, pp. 135–146.

Retamal J. R. D., Chen C. Y., Lien D. H., Huang M. R., Lin C. A., Liu C. P. and He J. H. (2014). Concurrent improvement in photogain and speed of a metal oxide nanowire photodetector through enhancing surface band bending via incorporating a nanoscale heterojunction. *ACS Photonics*, **1**(4), 354–359.

Rincón A. G. and Pulgarin C. (2003). Photocatalytical inactivation of *E. coli*: effect of (continuous-intermittent) light intensity and of (suspended-fixed) TiO_2 concentration. *Applied Catalysis B: Environmental*, **44**(3), 263–284.

Rincón A. G. and Pulgarin C. (2004). Effect of pH, inorganic ions, organic matter and H_2O_2 on *E. coli* K12 photocatalytic inactivation by TiO_2: Implications in solar water disinfection. *Applied Catalysis B: Environmental*, **51**(4), 283–302.

Rodríguez S. M., Gálvez J. B., Rubio M. I. M., Ibáñez P. F., Gernjak W. and Alberola I. O. (2005). Treatment of chlorinated solvents by TiO_2 photocatalysis and photo-Fenton: influence of operating conditions in a solar pilot plant. *Chemosphere*, **58**(4), 391–398.

Romanos G., Athanasekou C., Likodimos V., Aloupogiannis P. and Falaras P. (2013). Hybrid ultrafiltration/photocatalytic membranes for efficient water treatment. *Industrial & Engineering Chemistry*, **52**, 13938–13947.

Ryu J., Choi W. and Choo K-H. (2005). A pilot-scale photocatalyst-membrane hybrid reactor: Performance and characterization. *Water Science and Technology*, **51**(6–7), 491–497.

Saari J., Muller N. and Nowack B. (2010). Photocatalysis for water treatment. *Observatory Nano Briefing*, **2**, 1–4.

Sacoo O., Vaiano V., Han C., Sannino D. and Dionysiou D. D. (2015). Photocatalytic removal of atrazine using N-doped TiO_2 supported on phosphors. *Applied Catalysis B: Environmental*, **164**, 462–474.

Sampaio M. J., Silva C. G., Silva A. M. T., Pastrana-Martínez L. M., Han C., Morales-Torres S., Figueiredo J. L., Dionysiou D. D. and Faria J. (2015). Carbon-based TiO_2 materials for the degradation of microcystin-LA. *Applied Catalysis B: Environmental*, **170**, 74–82.

Sayed F. N., Sasikala R., Jayakumar O. D., Rao R., Betty C. A., Chokkalingam A., Kadam R. M., Jagannath, Bharadwaj S. R., Vinu A. and Tyagi A. K. (2014). Photocatalytic hydrogen generation from water using a hybrid of graphene nanoplatelets and self doped TiO_2-Pd. *RSC Advances*, **4**, 13469–13476.

Schneider J., Matsuoka M., Takeuchi M., Zhang J., Horiuchi Y., Anpo M. and Bahnemann D. W. (2014). Understanding TiO_2 photocatalysis: Mechanisms and materials. *Chemical Reviews (Washington, DC, United States)*, **114**, 9919–9986.

She X., Xu H., Wang H., Xia J., Song Y., Yan J., Xu Y., Zhang q., Du D. and Li H. (2015). Controllable synthesis of CeO_2/*g*-C_3N_4 composites and their applications in environment. *Dalton Transactions*, **44**(15), 7021–7031.

Silva T. F., Silva M. E. F., Cunha-Queda A. C., Fonseca A., Saraiva I., Boaventura R. A. and Vilar V. J. (2013). Sanitary landfill leachate treatment using combined solar photo-Fenton and biological oxidation processes at pre-industrial scale. *Chemical Engineering Journal*, **228**, 850–866.

SkyFuel (2015). Skytrough, High Performance, Low Cost. http://www.skyfuel.com/skytrough.shtml (accessed 14 November 2015).

Sleiman M., Ferronato C. and Chovelon J.-M. (2008). Photocatalytic removal of pesticide dichlorvos from indoor air: A study of reaction parameters, intermediates and mineralization. *Environmental Science & Technology*, **42**(8), 3018–3024.

SOWARLA GmbH. The SOWARLA SUN system. http://www.sowarla.de/sowarla-sun-1.html (accessed 25 August 2015).

Spasiano D., Marotta R., Malato S., Fernandez-Ibañez P. and Di Somma I. (2015). Solar photocatalysis: Materials, reactors, some commercial, and pre-industrialized applications. A comprehensive approach. *Applied Catalysis B: Environmental*, **170–171**(0), 90–123.

Stancl H., Hristovski K. and Westerhoff P. (2015). Hexavalent Chromium Removal Using UV-TiO_2/Ceramic Membrane Reactor. *Environmental Engineering Science*, **32**(8), 676–683.

Synnott D. W., Seery M. K., Hinder S. J., Michlits G. and Pillai S. C. (2013). Anti-bacterial activity of indoor-light activated photocatalysts. *Applied Catalysis, B: Environmental*, **130–131**, 106–111.

Tang Y., Luo S., Teng Y., Liu C., Xu X., Zhang X. and Chen L. (2012). Efficient removal of herbicide 2,4-dichlorophenoxyacetic acid from water using Ag/reduced graphene oxide co-decorated TiO_2 nanotube arrays. *Journal of Hazardous Materials*, **241–242**, 323–330.

Thind S. S., Wu G. and Chen A. (2012). Synthesis of mesoporous nitrogen-tungsten co-doped TiO_2 photocatalysts with high visible light activity. *Applied Catalysis, B: Environmental*, **111–112**, 38–45.

Tsuru T., Toyosada T., Yoshioka T. and Asaeda M. (2003). Photocatalytic membrane reactor using porous titanium oxide membranes. *Journal of Chemical Engineering of Japan*, **36**(9), 1063–1069.

Tsuru T., Ohtani Y., Yoshioka T. and Asaeda M. (2005). Photocatalytic membrane reaction of methylene blue on nanoporous titania membranes. *Kagaku Kogaku Ronbunshu*, **31**(2), 108–114.

Turki A., Guillard C., Dappozze F., Ksibi Z., Berhault G. and Kochkar H. (2015). Phenol photocatalytic degradation over anisotropic TiO_2 nanomaterials: Kinetic study, adsorption isotherms and formal mechanisms. *Applied Catalysis B-Environmental* **163**, 404–414.

Upadhya S. and Ollis D. F. (1997). Simple photocatalysis model for photoefficiency enhancement via controlled, periodic illumination. *The Journal of Physical Chemistry B*, **101**, 2625–2631.

Valencia S., Cataño F., Rios L., Restrepo G. and Marín J. (2011). A new kinetic model for heterogeneous photocatalysis with titanium dioxide: Case of non-specific adsorption considering back reaction. *Applied Catalysis B: Environmental*, **104**, 300–304.

Visa M., Pricop F. and Duta A. (2011). Sustainable treatment of wastewaters resulted in the textile dyeing industry. *Clean Technologies and Environmental Policy*, **13**, 855–861.

Wang C., Böttcher C., Bahnemann D. and Dohrmann J. (2003). A comparative study of nanometer sized Fe(III)-doped TiO_2 photocatalysts: Synthesis, characterization and activity. *Journal of Materials Chemistry*, **13**, 2322–2329.

Wang C., Zhu W., Xu Y., Xu H., Zhang M., Chao Y., Yin S., Li H. and Wang J. (2014). Preparation of TiO_2/g-C_3N_4 composites and their application in photocatalytic oxidative desulfurization. *Ceramics International*, **40**, 11627–11635.

Wang D., Choi D., Li J., Yang Z., Nie Z., Kou R., Hu D., Wang C., Saraf L. V., Zhang J., Aksay I. A. and Liu J. (2009). Self-assembled TiO_2-graphene hybrid nanostructures for enhanced Li-ion insertion. *ACS Nano*, **3**, 907–914.

Wang R., Wu Q., Lu Y., Liu H., Xia Y., Liu J., Yang D., Huo Z. and Yao X. (2014). Preparation of nitrogen-doped TiO_2/graphene nanohybrids and application as counter electrode for dye-sensitized solar cells. *ACS Applied Materials & Interfaces*, **6**, 2118–2124.

Wang X. and Lim T.-T. (2013). Highly efficient and stable Ag-AgBr/TiO_2 composites for destruction of Escherichia coli under visible light irradiation. *Water Research*, **47**, 4148–4158.

Wender H., Feil A. F., Diaz L. B., Ribeiro C. S., Machado G. J., Migowski P., Weibel D. E., Dupont J. and Teixeira S. R. (2011). Self-organized TiO_2 nanotube nrrays: Synthesis by anodization in an ionic liquid and assessment of photocatalytic properties. *Applied Materials & Interfaces*, **3**(4), 1359–1365.

Wu X., Yin S., Dong Q., Guo C., Kimura T., Matsushita J.-i. and Sato T. (2013). Photocatalytic properties of Nd and C codoped TiO_2 with the whole range of visible light absorption. *Journal of Physical Chemistry C*, **117**, 8345–8352.

Xi W. and Geissen S. U. (2001). Separation of titanium dioxide from photocatalytically treated water by cross-flow microfiltration. *Water Research*, **35**(5), 1256–1262.

Xie K., Sun L., Wang C., Lai Y., Wang M., Chen H. and Lin C. (2010). Photoelectrocatalytic properties of Ag nanoparticles loaded TiO_2 nanotube arrays prepared by pulse current deposition. *Electrochimica Acta*, **55**, 7211–7218.

Xiong P., Fu Y., Wang L. and Wang X. (2012). Multi-walled carbon nanotubes supported nickel ferrite: A magnetically recyclable photocatalyst with high photocatalytic activity on degradation of phenols. *Chemical Engineering Journal*, **195**, 149–157.

Xu T., Zhang L., Cheng H. and Zhu Y. (2011). Significantly enhanced photocatalytic performance of ZnO via graphene hybridization and the mechanism study. *Applied Catalysis, B: Environmental*, **101**, 382–387.

Xuan S., Jiang W., Gong X., Hu Y. and Chen Z. (2009). Magnetically separable Fe_3O_4/TiO_2 hollow sphere: fabrication and photocatalytic activity. *The Journal of Physical Chemistry C*, **113**, 553–558.

Yang L., Luo S., Li Y., Xiao Y., Kang Q. and Cai Q. (2010). High efficient photocatalytic degradation of *p*-nitrophenol on a unique Cu_2O/TiO_2 *p–n* heterojunction network catalyst. *Environmental Science & Technology*, **44**, 7641–7646.

Yang M., Hume C., Lee S., Son Y.-H. and Lee J.-K. (2010). Correlation between photocatalytic efficacy and electronic band structure in hydrothermally grown TiO_2 nanoparticles. *Journal of Physical Chemistry C*, **114**, 15292–15297.

Ye S., Qiu L.-G., Yuan Y.-P., Zhu Y.-J., Xia J. and Zhu J.-F. (2013). Facile fabrication of magnetically separable graphitic carbon nitride photocatalysts with enhanced photocatalytic activity under visible light. *Journal of Materials Chemistry A*, **1**, 3008–3015.

Zapata A., Oller I., Sirtori C., Rodríguez A., Sánchez-Pérez J. A., López A., Mezcua M. and Malato S. (2010). Decontamination of industrial wastewater containing pesticides by combining large-scale homogeneous solar photocatalysis and biological treatment. *Chemical Engineering Journal*, **160**(2), 447–456.

Zayani G., Bousselmi L., Mhenni F. and Ghrabi A. (2009). Solar photocatalytic degradation of commercial textile azo dyes: Performance of pilot plant scale thin film fixed-bed reactor. *Desalination*, **246**(1), 344–352.

Zhang A., Zhou M., Han L. and Zhou Q. (2011). The combination of rotating disk photocatalytic reacotr and TiO_2 nanotube arrays for environmental polluntants removal. *Journal of Hazardous Materials*, **186**(2–3), 1374–1383.

Zhang H., Lv X.-J., Li Y.-M., Wang Y. and Li J.-H. (2010). P25-graphene composite as a high performance photocatalyst. *ACS Nano*, **4**, 380–386.

Zhang J., Zhu Z., Tang Y. and Feng X. (2013). Graphene encapsulated hollow TiO_2 nanospheres: Efficient synthesis and enhanced photocatalytic activity. *Journal of Materials Chemistry A: Materials for Energy and Sustainability*, **1**, 3752–3756.

Zhang L., Du L., Cai X., Yu X., Zhang D., Liang L., Yang P., Xing X., Mai W., Tan S., Gu Y. and Song J. (2013). Role of graphene in great enhancement of photocatalytic activity of ZnO nanoparticle-graphene hybrids. *Physica E: Low-Dimensional Systems & Nanostructures (Amsterdam, Netherlands)*, **47**, 279–284.

Zhang S., Qiu J., Han J., Zhang H., Liu P., Zhang S., Peng F. and Zhao H. (2011). A facile one-step preparation of hierarchically-structured TiO_2 nanotube array photoanodes with enhanced photocatalytic activity. *Electrochemistry Communications*, **13**(11), 1151–1154.

Zhang W., Naidu B. S., Ou J. Z., O'Mullane A. P., Chrimes A. F., Carey B. J., Wang Y., Tang S.-Y., Sivan V., Mitchell A., Bhargava S. K. and Kalantar-zadeh K. (2015). Liquid metal/metal oxide frameworks with incorporated Ga_2O_3 for photocatalysis. *ACS Applied Materials & Interfaces*, **7**, 1943–1948.

Zhang X., Du A. J., Lee P., Sun D. D. and Leckie J. O. (2008). TiO_2 nanowire membrane for concurrent filtration and photocatalytic oxidation of humic acid in water. *Journal of Membrane Science*, **313**(1–2), 44–51.

Zhang X., Tang Y., Li Y., Wang Y., Liu X., Liu C. and Luo S. (2013). Reduced graphene oxide and PbS nanoparticles co-modified TiO_2 nanotube arrays as a recyclable and stable photocatalyst for efficient degradation of pentachlorophenol. *Applied Catalysis A: General*, **457**, 78–84.

Zhao Y., Zhong J., Li H., Xu N. and Shi J. (2002). Fouling and regeneration of ceramic microfiltration membranes in processing acid wastewater containing fine TiO_2 particles. *Journal of Membrane Science*, **208**(1–2), 331–341.

Zhao Y. X., Gao B. Y., Cao B. C., Yang Z. L., Yue Q. V., Shon H. K. and Kim J. H. (2011). Floc characteristics of titanium tetrachloride ($TiCl_4$) compared with aluminum and iron salts in humic acid-kaolin synthetic water treatment. *Chemical Engineering Journal*, **166**, 544–550.

Zhao Y. X., Gao B. Y., Shon H. K., Wang Y., Kim J. H. and Yue Q. Y. (2011). The effect of second coagulant dose on the regrowth of flocs formed by charge neutralization and sweep coagulation using titanium tetrachloride ($TiCl_4$). *Journal of Hazardous Materials*, **198**, 70–77.

Zhou M., Yu J. and Cheng B. (2006). Effects of Fe-doping on the photocatalytic activity of mesoporous TiO_2 powders prepared by an ultrasonic method. *Journal of Hazardous Meterials*, **137**, 1838–1847.

Zhou Q. X., Fang Z., Li J. and Wang M. Y. (2015). Review: Applications of TiO_2 nanotube arrays in environmental and energy fields: A review. *Microporous and Mesoporous Materials*, **202**, 22–35.

Zhou X. M., Nhat N. T., Oezkan S. and Schmuki P. (2014). Anodic TiO_2 nanotubes layers: Why does self-organized growth occur – a mini review. *Electrochemistry Communications*, **46**, 157–162.

Zhu J. and He J. (2012). Self-assembly fabrication of graphene-based materials with optical-electronic, transient optical and electrochemical properties. *International Journal of Nanoscience*, **11**(6), 1240032.

Chapter 9

UV/Chlorine process

Joseph De Laat and Mihaela Stefan

9.1 INTRODUCTION

Since the early 1970s, an increasing number of research studies have been carried out to develop advanced oxidation processes (AOPs) for the degradation of organic pollutants refractory to the oxidants conventionally used in water treatment (chlorine, ozone, chlorine dioxide). AOPs are based on the *in situ* generation of the hydroxyl radical (•OH) which is a powerful oxidant and a highly reactive species with most organic pollutants. After the first studies on Fenton reaction (Fe^{2+}/H_2O_2) for treatment of industrial wastewaters, other advanced oxidation processes have been explored at laboratory- and/or pilot-scale, and full-scale AOPs were implemented at water treatment facilities over the past three decades. Hydroxyl radicals can be produced at room temperature and atmospheric pressure by various processes based on chemical, photochemical, electrochemical or sonochemical reactions (Legrini *et al.* 1993; Oturan & Aaron, 2014). Recently, the UV/Chlorine process has been investigated as an alternative to the UV/H_2O_2 process and was tested at a few water treatment utilities (pilot- or full-scale) for water reuse, drinking water production, or groundwater decontamination.

As mentioned by Allmand *et al.* (1925), the photodecomposition of chlorine in water has been investigated since the 1850s. Buxton and Subhani (1972b) showed that the photodecomposition mechanism of hypochlorite ion is very complex and depends on the irradiation wavelength. Photodecomposition of hypochlorite ion leads to chloride, chlorate, chlorite, and oxygen as end-products and to the *in situ* generation of •OH and Cl• as well as many other unstable intermediates. A few research works performed in the 1970s showed that UV/Cl_2 can oxidize or mineralize completely organic compounds which are resistant to chlorine in the dark and to direct photolysis. These reactions have been demonstrated in the case of ethanol, n-butanol and benzoic acid at pH < 10 and at ~350 nm (Oliver & Carey, 1977), ethylene glycol dimethyl ether and related substrates, acetic and propionic acids at pH > 12 with a high pressure Hg lamp (Ogata *et al.* 1978; Ogata *et al.* 1979), benzoic acid at pH ≥ 12 and $\lambda = 253.7$ nm and $\lambda \geq 350$ nm (Ogata & Tomizawa, 1984), benzoic acid and nitrobenzene at pH 6 and $\lambda \geq 350$ nm (Nowell & Crosby, 1985), 1-chlorobutane, n-octanol and nitrobenzene at sunlight wavelengths (Nowell & Hoigné, 1992b). Since 2005, a few research groups have performed studies to determine quantum yields of photolysis of free chlorine species (HOCl and ClO^-) at various wavelengths, to examine the effects of various experimental parameters on the rate of photodecomposition of free chlorine, and to evaluate the performance of the UV/Cl_2 process as an AOP.

This chapter will present first the fundamental aspects of the photochemical decomposition of aqueous solutions of free chlorine into $^•OH$ and chlorine atoms. The reactivity of chlorine atoms ($Cl^•$) and of dichlorine anion radicals ($Cl_2^{•-}$) towards organic and inorganic compounds will then be reviewed. In the last part of this chapter, the recent works on the degradation of organic pollutants by the UV/Cl_2 AOP will be discussed. In this context, pilot-scale and full-scale studies will be emphasized, and the performances of the UV/Cl_2 and UV/H_2O_2 oxidation processes will be compared.

9.2 PHOTODECOMPOSITION OF FREE CHLORINE BY UV LIGHT

9.2.1 Distribution of free chlorine species

The predominant free chlorine species in dilute aqueous solutions are aqueous dichlorine (Cl_2), hypochlorous acid (HOCl) and hypochlorite ion (ClO^-):

$$Cl_2 + H_2O \rightleftharpoons HOCl + H^+ + Cl^- \quad K_d = 3.94 \times 10^{-4}\ M^2 \text{ at } 25°C \text{ and } I = 0 \text{ (Connick \& Chia, 1959)} \tag{9.1}$$

$$HOCl \rightleftharpoons ClO^- + H^+ \quad pK_a = 7.537 \text{ at } 25°C \text{ and } I = 0 \text{ (Morris, 1966)} \tag{9.2}$$

The distribution of Cl_2, HOCl and ClO^- species can be calculated from the disproportionation constant of Cl_2 (K_d) and the pK_a value of HOCl. At a given temperature and zero ionic strength ($I = 0$), K_d and pK_a can be calculated from expressions 9.3 (Connick & Chia, 1959) and 9.4 (Morris, 1966), respectively:

$$\log K_d = -982798/T^2 + 5485.7/T - 10.7484 \text{ at temperatures from 0 to 45°C} \tag{9.3}$$

$$pK_a = 3000/T - 10.0686 + 0.0253T \text{ at temperatures from 0 to 35°C} \tag{9.4}$$

where T is the absolute temperature in Kelvin degrees (0°C = 273.15 K).

The K_d values determined from equation 9.3 vary from $1.45 \times 10^{-4}\ M^2$ at 0°C to $6.09 \times 10^{-4}\ M^2$ at 45°C, and the pK_a values of HOCl calculated from equation 9.4 decrease from 7.825 at 0°C to 7.463 at 35°C. At 25°C and $I = 0$, the values of K_d and pK_a are equal to $3.94 \times 10^{-4}\ M^2$ and 7.537, respectively.

As indicated by reactions 9.1 and 9.2, the distribution of free chlorine species depends on chloride concentration, temperature, and pH; of all, pH is the most impactful parameter. Figure 9.1 presents the distribution patterns of chlorine species calculated at 25°C and for a chloride concentration of 2 mM.

Cl_2 is only present at low pH values (pH < 3). At 25°C, HOCl is the predominant free chlorine species at pH < 7.5, and ClO^- at pH > 7.5. More than 99% of free chlorine is present as HOCl in the pH range of 3–5.5, and as ClO^- at pH > 9.5.

9.2.2 Absorption spectra of free chlorine species in water

HOCl and ClO^- absorb UV light at wavelengths ranging from 200 to 375 nm (Figure 9.2) and therefore, free chlorine can be photolyzed by UV-C light, as well as by the UV-B and UV-A radiation from the solar light (Buxton & Subhani, 1972b; Nowell & Hoigné, 1992a). Molar absorption coefficients (ε) range from 0 to 370 $M^{-1}\ cm^{-1}$. Absorption spectra show a maximum absorption band centered at 236 nm for HOCl

($\varepsilon = 101 \pm 2$ M^{-1} cm^{-1}) and at 292 nm for ClO^- ($\varepsilon = 365 \pm 8$ M^{-1} cm^{-1}). The molar absorption coefficients of HOCl and ClO^- at 253.7 nm are equal to 59 ± 1 and 66 ± 1 M^{-1} cm^{-1}, respectively (Feng *et al.* 2007; De Laat & Berne, 2009).

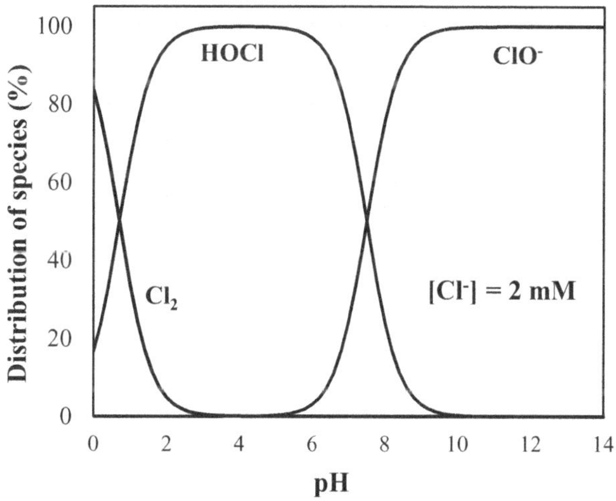

Figure 9.1 Distribution of Cl_2, HOCl and ClO^- in water as a function of pH at 25°C. (Distribution curves calculated for $[Cl^-] = 2$ mM, $K_d = 3.94 \times 10^{-4}$ M^2 and $pK_a = 7.537$; De Laat, 2016).

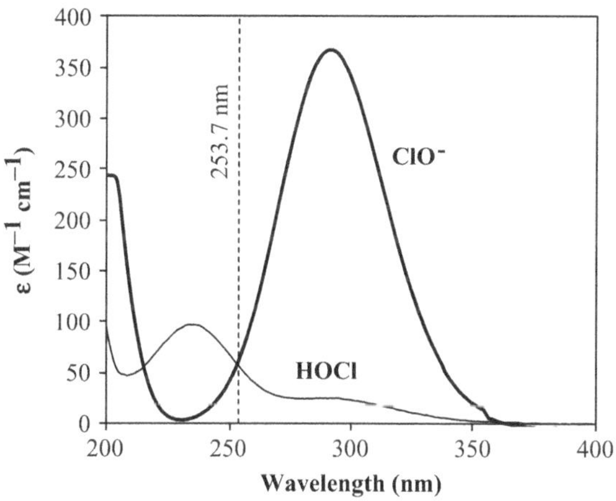

Figure 9.2 UV spectra of HOCl and ClO^- in water (De Laat, 2016).

9.2.3 Radical species, quantum yields and degradation mechanisms of free chlorine

While the photodecomposition of chlorine in water has been investigated since the 1850s (Allmand *et al.* 1925), the potential use of UV/Cl_2 as an AOP has been examined only in the recent years (Watts *et al.* 2007a, b; Jin *et al.* 2011; Sichel *et al.* 2011; Watts *et al.* 2012; Wang *et al.* 2012; Shu *et al.* 2014; Wang *et al.*

2015a). The photodecomposition of chlorine in aqueous solution leads to chloride, chlorate, chlorite, and oxygen as end-products. The yields of these products are both wavelength- and chlorine species-dependent (Buxton & Subhani, 1972b; Cooper *et al.* 2007). Molar yields for the production of chlorate vary from 0.02 to 0.3 mole of chlorate/mole of chlorine decomposed and depend on the irradiation wavelength, pH, and chlorine concentration (Buxton & Subhani, 1972b; Karpel Vel Leitner *et al.* 1992a; Feng *et al.* 2010; Rao *et al.* 2010; Wang *et al.* 2012). Recently, it was also shown that perchlorate can be produced at yields less than 0.01 mole of ClO_4^- per mole of chlorine photodecomposed (Kang *et al.* 2006; Rao *et al.* 2010). All these end-products are formed from complex reactions initiated by the primary photo-products of HOCl or ClO^-; numerous reactive intermediates are also produced.

9.2.3.1 Primary quantum yields of photolysis of hypochlorite ion and hypochlorous acid

From steady-state and flash photolysis experiments on sodium hypochlorite solutions at alkaline pH, Buxton and Subhani (1972b) showed that the primary photoproducts of hypochlorite ion as well as the primary quantum yields of photoproduct formation depend on the irradiation wavelength (Table 9.1). The primary products of photolysis of hypochlorite ion are chloride ion, hydroxyl radicals ($^{\bullet}OH/O^{\bullet-}$, $pK_a = 11.9$; Buxton *et al.* 1988), $Cl^{\bullet}$ and $O(^3P)$ at 253.7, 313 and 365 nm. $O(^1D)$ is also produced at 253.7 nm and 313 nm.

Table 9.1 Primary quantum yields of photodecomposition reactions of hypochlorite ion at pH > 11.5 (Buxton & Subhani, 1972b).

Initiation step	Reaction	253.7 nm	313 nm	365 nm
$ClO^- + h\nu \rightarrow Cl^- + O(^3P)$	9.5	0.074 ± 0.019	0.075 ± 0.015	0.28 ± 0.03
$ClO^- + h\nu \rightarrow Cl^{\bullet} + O^{\bullet-}$	9.6	0.278 ± 0.016	0.127 ± 0.014	0.08 ± 0.02
$ClO^- + h\nu \rightarrow Cl^- + O(^1D)$	9.7	0.133 ± 0.017	0.020 ± 0.015	0

As shown in Table 9.1, the quantum yields of formation of ($^{\bullet}OH/O^{\bullet-}$) and $Cl^{\bullet}$ from ClO^- photolysis markedly decrease when irradiation wavelength increases, e.g., from 0.28 at 253.7 nm to 0.08 at 365 nm.

Similarly, $^{\bullet}OH$, $Cl^{\bullet}$ and oxygen atoms are formed by UV photolysis of HOCl (Kläning *et al.* 1984; Thomsen *et al.* 2001; Herrmann, 2007). $Cl^{\bullet}$ exhibits a strong absorption band in the UV region with a maximum at around 310–320 nm (Figure 9.3) (Kläning & Wolff, 1985; Buxton *et al.* 2000; Thomsen *et al.* 2001). From femtosecond photolysis experiments of HOCl (~1.5 M) at 266 nm, Thomsen *et al.* (2001) showed that HOCl molecules photodissociate within 1 picosecond into $^{\bullet}OH$ and $Cl^{\bullet}$ with a near-unity quantum yield:

$$HOCl \xrightarrow{h\nu} {}^{\bullet}OH + Cl^{\bullet} \tag{9.8}$$

The authors also observed higher yield of $^{\bullet}OH$ than of $Cl^{\bullet}$ because a fraction of $Cl^{\bullet}$ (~10%) reacted with HOCl to form $^{\bullet}OH$ under the experimental conditions used (high HOCl concentration):

$$Cl^{\bullet} + HOCl \rightarrow Cl_2 + {}^{\bullet}OH \tag{9.9}$$

By monitoring the concentrations of photoproducts by ultrafast transient absorption spectroscopy, Thomsen *et al.* (2001) observed that the concentration of $Cl^{\bullet}$ decayed rapidly within 50 ps due to back geminate recombination with $^{\bullet}OH$ within the solvent "cage", to re-form HOCl (Figure 9.4).

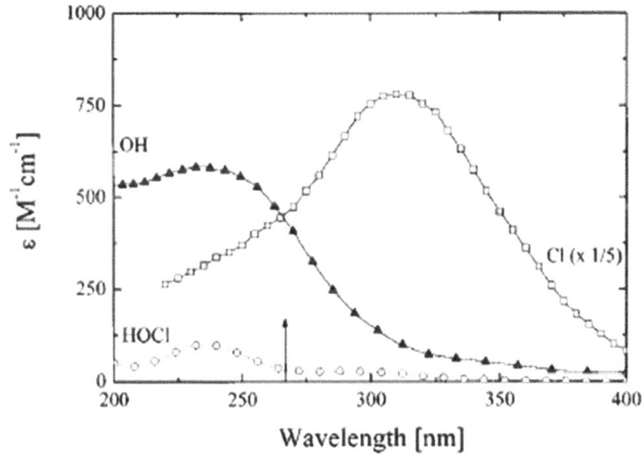

Figure 9.3 Absorption spectra of HOCl, •OH and Cl• (×1/5). Reprinted with permisson from Thomsen *et al.* (2001). Copyright 2016 AIP Publishing.

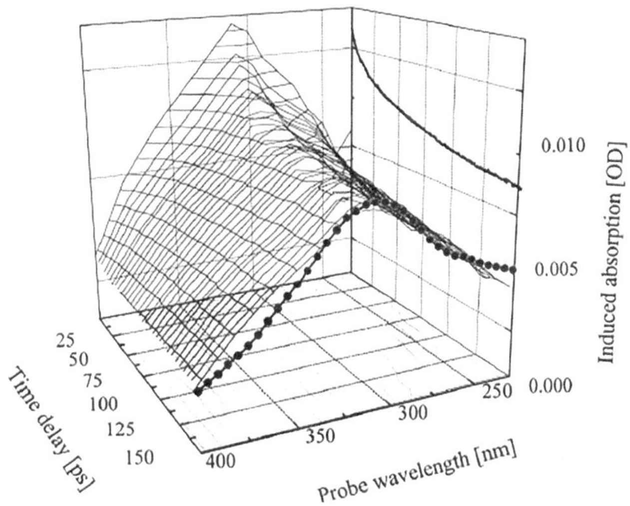

Figure 9.4 Full transient absorption spectrum following photolysis of aqueous HOCl at 266 nm. The steady state spectrum of Cl• is shown in the wavelength-transient absorption plane, and the spectrally integrated dynamics is shown in the time-transient absorption plane. Reprinted with permission from Thomsen *et al.* (2001). Copyright 2016 AIP Publishing.

Under the assumption that the decrease in the absorbance at 340 nm is proportional to the concentration of Cl•, the authors determined the quantum yield of Cl• that escaped recombination from the absorbance values measured at 1 ps and 150 ps, using equation 9.10:

$$\Phi(\mathrm{Cl}^\bullet) = \frac{\Delta\mathrm{OD}(t = 150\ \mathrm{ps})}{\Delta\mathrm{OD}(t = 1\ \mathrm{ps})} \tag{9.10}$$

The authors obtained a $\Phi(Cl^\bullet)$ of 0.55 ± 0.05 at 266 nm. Therefore, the primary quantum yields for HOCl photolysis (Φ(-HOCl) and for $^\bullet$OH- and $Cl^\bullet$-formation ($\Phi(HO^\bullet)$ and $\Phi(Cl^\bullet)$) are equal to 0.55 ± 0.05 at 266 nm.

A lower value has been reported in a more recent study (Herrmann, 2007) from excimer laser photolysis experiments at 248 nm. Experiments were carried at pH 1.5 in O_2-free solutions, with a concentration of chlorine of 4.4×10^{-4} M. Under the experimental conditions used by the authors, the concentrations of HOCl, Cl^-, Cl_2 and Cl_3^- were roughly 4.4×10^{-4}, 4.4×10^{-4}, 2×10^{-5} and 2×10^{-9} M, respectively. The absorption of UV (248 nm) light in the reaction cell by Cl_2 and Cl_3^- could be neglected. The $\Phi(Cl^\bullet)$ and $\Phi(^\bullet OH)$ values (0.25 ± 0.05; Herrmann, 2007) were determined from the initial concentration of $Cl^\bullet$ formed through HOCl photolysis at $t = 50$ ns and from the amount of $Cl^\bullet$ which has been consumed during the first 50 ns through the reactions with $Cl^\bullet$, HOCl, Cl^- and H_2O.

In order to assess the formation of $^\bullet$OH in the free chlorine photolysis process, Nowell and Hoigné (1992b) introduced the 'yield factor' which represents the released $^\bullet$OH per free chlorine photodecomposed, on a molar basis. The authors determined the pH-dependent $^\bullet$OH-yield factors in the photodecomposition of free chlorine under simulated solar radiation and at 253.7 nm. At pH 5 and in the absence of scavengers other than the probe compound, the $^\bullet$OH yield factors were determined as 0.85 and 0.70, at 253.7 nm and under simulated solar radiation, respectively; at pH 10, the yield factors were 0.12 and 0.10, respectively.

Watts and Linden (2007a) determined a quantum yield of 1.40 ± 0.18 (253.7 nm) for $^\bullet$OH formation from HOCl photolysis at pH 5, and compared it to the quantum yield of $^\bullet$OH formation ($\Phi_{\bullet OH} \sim 1.0$) in the H_2O_2 photolysis in aqueous solutions. Wang *et al.* (2012) reported $\Phi_{\bullet OH}$ (253.7 nm) = (0.79 ± 0.01) at pH 5, which is larger than that determined by Jin *et al.* (2011), i.e., $\Phi_{\bullet OH} = 0.46 \pm 0.09$. At pH 10, the $^\bullet$OH quantum yield reported by Wang *et al.* (2012) was 1.18 ± 0.12 (253.7 nm), almost twice as large as that determined by Chan *et al.* (2012), $\Phi_{\bullet OH} = 0.61$. Based on the $^\bullet$OH yield factor determined for ClO^- photolysis (0.70 ± 0.02), Chan *et al.* (2012) concluded that 30% of hypochlorite photolysis does not result in OH radicals.

9.2.3.2 Degradation pathways of hypochlorite and hypochlorous acid in organic-free water

Once formed, the primary photoproducts of ClO^- initiate decomposition reactions of ClO^- into unstable transient species ($ClO^\bullet$, Cl_2O_2, Cl_2O), secondary oxidants (ClO_2, O_3, H_2O_2) and stable by-products (Cl^-, ClO_2^-, ClO_3^-, ClO_4^- and O_2). The yields of by-products highly depend on the experimental conditions such as irradiation wavelength, pH and concentration of dissolved oxygen.

Table 9.2 lists the quantum yields of byproduct formation as determined by Buxton and Subhani (1972b) in deoxygenated solutions of ClO^-. The main by-products of photodecomposition of ClO^- at 253.7 nm are chloride (~0.82 mol/mol of ClO^-) and chlorate (~0.18 mol/mol of ClO^-). The yields of chloride and chlorate were roughly the same at 308 nm. At 365 nm, the molar yields of chloride, chlorite and chlorate were 0.60, 0.13 and 0.27 mol/mol of ClO^-, respectively.

Buxton and Subhani (1972b) proposed the sequence of reactions listed in table 9.3 to explain the formation of Cl^-, ClO_2^-, ClO_3^-, O_2 and O_3 by photolysis of hypochlorite ion. These reactions are initiated by the primary photo-products of ClO^- (reactions 9.5–9.7 in Table 9.1).

$^\bullet OH/O^{\bullet -}$ and $Cl^\bullet$ oxidize hypochlorite ion to $ClO^\bullet$ (Reactions 9.11–9.13). Dichlorine peroxide (Cl_2O_2 or ClOOCl) produced by the very fast $ClO^\bullet$ dimerization (reaction 9.14) decomposes *via* parallel pathways to produce chloride, chlorate and oxygen as stable end-products and chlorite (reactions 9.15–9.18). Chlorite has been detected only in hypochlorite solutions irradiated at 365 nm (Table 9.2). In a more recent study on the photolysis of hypochlorite ($[Cl_{2,T}]_0 \approx$ 16–18 mM, initial pH > 10.6) at 253.7, 300,

Table 9.2 Quantum yields of ClO^- photodecomposition and of by-product formation (experimental values) determined by Buxton and Subhani (1992b) at 253.7, 313 and 365 nm.

Conditions	253.7 nm pH 11.5 $[ClO^-] = 1$ mM	313 nm pH 12.0 $[ClO^-] = 1$ mM	365 nm pH 12.1 $[ClO^-] = 7–10$ mM
Φ-ClO^- photodecomposition	0.85 ± 0.02	0.39 ± 0.01	0.60 ± 0.05
Φ-ClO_2^- formation	0	0	0.160 ± 0.005
Φ-ClO_3^- formation	0.15 ± 0.02	0.08 ± 0.02	0.08 ± 0.02
Φ-Cl^- formation	0.70 ± 0.03	0.27 ± 0.02	0.36 ± 0.03
Φ-O_2 formation	0.200 ± 0.005	0.069 ± 0.005	0.04 ± 0.02

Table 9.3 Photodecomposition pathways of hypochlorite ion proposed by Buxton and Subhani (1972a and 1972b).

Reactions		Rate Constants and Comments
$^\bullet OH + ClO^- \rightarrow ClO^\bullet + OH^-$	9.11	$(9.0 \pm 0.5) \times 10^9$ M^{-1} s^{-1} (Buxton & Subhani, 1972a)
$O^{\bullet-} + ClO^- \rightarrow ClO^\bullet + O^{2-}$	9.12	$(2.4 \pm 0.1) \times 10^8$ M^{-1} s^{-1} (Buxton & Subhani, 1972a)
$Cl^\bullet + ClO^- \rightarrow ClO^\bullet + Cl^-$	9.13	8.2×10^9 M^{-1} s^{-1} (Kläning & Wolff, 1985)
$2\ ClO^\bullet \rightleftharpoons Cl_2O_2$	9.14	$2k = 1.5 \times 10^{10}$ M^{-1} s^{-1} (Buxton & Subhani, 1972a)
$Cl_2O_2 + H_2O \rightarrow ClO_2^- + ClO^- + 2\ H^+$	9.15	
$Cl_2O_2 + H_2O \rightarrow Cl^- + O_2 + ClO^- + 2\ H^+$	9.16	$k_{9.15}/k_{9.16} = 1.93 \pm 0.24$ (Buxton & Subhani, 1972b)
$Cl_2O_2 + ClO_2^- \rightarrow ClO_3^- + Cl_2O$	9.17	$k_{9.17}/k_{9.15} = (1.3 \pm 0.6) \times 10^5$ M^{-1} (Buxton & Subhani, 1972b)
$Cl_2O + H_2O \rightleftharpoons 2\ HOCl$	9.18	
$^\bullet OH + ClO_2^- \rightarrow ClO_2 + HO^-$	9.19	$(6.3 \pm 0.5) \times 10^9$ M^{-1} s^{-1} (Buxton & Subhani, 1972a)
$O^{\bullet-} + ClO_2^- \rightarrow ClO_2 + O^{2-}$	9.20	$(1.9 \pm 0.1) \times 10^8$ M^{-1} s^{-1} (Buxton & Subhani, 1972a)
$^\bullet OH + ClO_3^- \rightarrow$ Products	9.21	$<10^6$ M^{-1} s^{-1} (Buxton & Subhani, 1972a)
$O^{\bullet-} + ClO_3^- \rightarrow$ Products	9.22	$<10^6$ M^{-1} s^{-1} (Buxton & Subhani, 1972a)
$O(^1D) + H_2O \rightarrow H_2O_2$	9.23	$O(^1D)$ only produced at $\lambda = 313$ and 253.7 nm
$ClO^- + H_2O_2 \rightarrow O_2 + Cl^- + H_2O$	9.24	
$O(^3P) + ClO^- \rightarrow ClO_2^-$	9.25	
$O(^3P) + ClO^- \rightarrow Cl^- + O_2$	9.26	$k_{9.26}/k_{9.25} = 0.17 \pm 0.09$ (Buxton & Subhani, 1972b)
$O(^3P) + ClO_2^- \rightarrow ClO_3^-$	9.27	
$O(^3P) + ClO_2^- \rightarrow Cl^- + O_2 + O(^3P)$	9.28	$k_{9.28}/k_{9.27} = 1.50 \pm 0.27$ (Buxton & Subhani, 1972b)
$O(^3P) + O_2 \rightarrow O_3$	9.29	4×10^9 M^{-1} s^{-1} (Kläning *et al.* 1984)
$O^{\bullet-} + O_2 \rightleftharpoons O_3^{\bullet-}$	9.30	3.6×10^9 M^{-1} s^{-1} (Buxton *et al.* 1988)
$ClO^\bullet + O_3^{\bullet-} \rightarrow O_3 + ClO^-$	9.31	1×10^9 M^{-1} s^{-1} (Kläning *et al.* 1984)

and 350 nm, Rao *et al.* (2010) also confirmed that chlorite was only produced at 350 nm. Buxton and Subhani (1972b) explained the absence of chlorite in irradiated solutions of ClO^- at 253.7 and 313 nm by the low yield of $O(^3P)$ and by destruction of any chlorite ion produced from Cl_2O_2 (reaction 9.17). Chlorite can also be oxidized to chlorine dioxide by $^{\bullet}OH/O^{\bullet-}$ (reactions 9.19 and 9.20). Other studies showed that chlorine dioxide can be oxidized to chlorate by $^{\bullet}OH/O^{\bullet-}$ (Kläning *et al.* 1985):

$$^{\bullet}OH + ClO_2 \rightarrow ClO_3^- + H^+ \quad (k = 4.0 \pm 0.4) \times 10^9\ M^{-1}\ s^{-1};\ \text{Kläning } et\ al.\ 1985) \tag{9.32}$$

$$O^{\bullet-} + ClO_2 \rightarrow ClO_3^- \quad (k = 2.7 \pm 0.4) \times 10^9\ M^{-1}\ s^{-1};\ \text{Kläning } et\ al.\ 1985) \tag{9.33}$$

The oxygen atoms $O(^1D)$ and $O(^3P)$ produced through hypochlorite photolysis (reactions 9.5 and 9.7, Table 1) also contribute to the decomposition of hypochlorite. Most of $O(^1D)$ will react with water molecules to generate H_2O_2 (reaction 9.23) which further reacts with hypochlorite to form dioxygen ($^1O_2$) and chloride (reaction 9.24). H_2O_2 has also been detected during electrochemical oxidation of hypochlorite solutions at pH 9.5 by Macounová *et al.* (2015) and its formation has been attributed to the following set of reactions initiated by the hydrolysis of $ClO^{\bullet}$ to $HO_2^{\bullet}/O_2^{\bullet-}$ radicals:

$$ClO^{\bullet} + H_2O \rightarrow HO_2^{\bullet} + Cl^- + H^+ \tag{9.34}$$

$$ClO^{\bullet} + OH^- \rightarrow O_2^{\bullet-} + Cl^- + H^+ \tag{9.35}$$

$$O_2^{\bullet-} + HOCl \rightarrow O_2 + Cl^- + {^{\bullet}OH} \tag{9.36}$$

$$HO_2^{\bullet} \rightleftharpoons O_2^{\bullet-} + H^+ \tag{9.37}$$

$$HO_2^{\bullet} + O_2^{\bullet-} + H_2O \rightleftharpoons H_2O_2 + O_2 + OH^- \tag{9.38}$$

$O(^3P)$ atoms can both reduce hypochlorite and chlorite to chloride (reactions 9.26 and 9.28) and oxidize hypochlorite and chlorite to chlorite and chlorate (reactions 9.25 and 9.27), respectively. The distribution of stable photoproducts (Cl^-, ClO_3^- and O_2) will therefore depend on the quantum yields of formation of $O(^1D)$ and $O(^3P)$ which are wavelength dependent (Table 9.1), and on the absolute and relative rate constants of reactions 9.25–9.28.

Buxton and Subhani (1972b) observed the formation of ozone (O_3) and ozonide anion radicals ($O_3^{\bullet-}$) in flash photolysis experiments of O_2-saturated solutions of ClO^-, which was attributed to reactions 9.29–9.30. Ozone can be involved in reactions with ClO^-, ClO_2^- and with chlorine containing free radicals (e.g., $Cl^{\bullet}$, $ClO^{\bullet}$) (Siddiqui, 1996; von Sonntag & von Gunten, 2012).

In addition, chlorite and chlorine dioxide can also be photodecomposed to chloride and chlorate as major end-products (Buxton & Subhani 1972a; Karpel Vel Leitner *et al.* 1992a, 1992b).

The photodecomposition pathways of HOCl in water are not known. $^{\bullet}OH$ and $Cl^{\bullet}$ generated from the primary step of HOCl photolysis react with HOCl to form $ClO^{\bullet}$ (reactions 9.39 and 9.40 in Table 9.4). The self-reaction of $ClO^{\bullet}$ forms Cl_2O_2 which decomposes to Cl^-, ClO_2^-, ClO_3^- and O_2, whereas secondary reactions of $^{\bullet}OH$ with ClO_2^- result in ClO_2 and ClO_3^-.

As shown in Table 9.4, the rate constants for the reaction of $^{\bullet}OH$ and $Cl^{\bullet}$ with free chlorine (HOCl and ClO^-) are not well documented in the literature and the published values for the reaction of $HO^{\bullet}$ with HOCl differ by several orders of magnitude (from $8.46 \times 10^4\ M^{-1}\ s^{-1}$ to $2 \times 10^9\ M^{-1}\ s^{-1}$).

Table 9.4 Rate constants for the reaction of $^{\bullet}OH$ and $Cl^{\bullet}$ with HOCl and ClO^-.

Reactions		Rate Constant ($M^{-1}\ s^{-1}$)	Reference
$^{\bullet}OH + HOCl \rightarrow ClO^{\bullet} + H_2O$	9.39	1.4×10^8	Zuo *et al.* (1997); pH 0.
		2.0×10^9	Matthew & Anastasio (2006)
		8.46×10^4	Watts & Linden (2007a)
$^{\bullet}OH + ClO^- \rightarrow ClO^{\bullet} + HO^-$	9.11	$(9.0 \pm 0.5) \times 10^9$	Buxton & Subhani (1972a)
		2.7×10^9	Zuo *et al.* (1997)
$Cl^{\bullet} + HOCl \rightarrow ClO^{\bullet} + HCl$	9.40	3.0×10^9	Kläning & Wolff (1985)
$Cl^{\bullet} + ClO^- \rightarrow ClO^{\bullet} + Cl^-$	9.13	8.2×10^9	Kläning & Wolff (1985)

It has also been shown that trace levels of perchlorate (ClO_4^-) can be formed from the photodecomposition of ClO^- at pH > 10 (Kang *et al.* 2006) or of free chlorine ($HOCl + ClO^-$) at initial pH values ranging from 2.6 to 11.4 (Rao *et al.* 2010). As $^{\bullet}OH$ cannot oxidize ClO_3^-, perchlorate production has been attributed to the oxidation of ClO_3^- by $Cl^{\bullet}$ to $ClO_3^{\bullet}$ radicals, which dimerize to give dichlorine hexaoxide Cl_2O_6. The subsequent hydrolysis of Cl_2O_6 results in ClO_4^- and ClO_3^- (Kang *et al.* 2006; Rao *et al.* 2010):

$$ClO_3^- + Cl^{\bullet} \rightarrow ClO_3^{\bullet} + Cl^- \tag{9.41}$$

$$ClO_3^{\bullet} + ClO_3^{\bullet} \rightarrow Cl_2O_6 \tag{9.42}$$

$$Cl_2O_6 + H_2O \rightarrow ClO_4^- + ClO_3^- + 2H^+ \tag{9.43}$$

Rao *et al.* (2010) found that the amount of ClO_4^- formed depends on the initial concentrations of free chlorine species, the background concentrations of Cl^-, ClO_2^- and ClO_3^-, wavelength and pH, but it is independent of light intensity. However, the concentration of ClO_4^- would not exceed 0.1 µg/L in the UV/Cl_2 applications to drinking water treatment ($[Cl_2] < 0.1$ mM; pH 7) (Rao *et al.* 2010).

9.2.3.3 Reaction quantum yields of chlorine photodecomposition in the absence of organic compounds

Table 9.5 summarizes the overall quantum yield values of photodecomposition of HOCl (Φ-HOCl, acidic pH), ClO^- (Φ-ClO^-, alkaline pH), and of their combination at pH values where the two species co-exist (Φ-Chlorine) as reported in the literature.

Feng *et al.* (2010) determined the overall quantum yield of photodecomposition of free chlorine at 253.7 nm using the expression developed by Bolton and Stefan (2002):

$$\ln\frac{[C]_0}{[C]_F} = k_1'F = \frac{\Phi\varepsilon(\lambda)\ln(10)}{10U(\lambda)}F \tag{9.44}$$

$$\Phi_\lambda = \frac{k_1' \times 10 \times U(\lambda)}{\varepsilon(\lambda)\ln(10)} \tag{9.45}$$

Table 9.5 Overall quantum yields of photodecomposition of HOCl (Φ-HOCl), ClO^- (Φ-ClO^-) and their mixture (Φ-Chlorine) in organic-free water.

Chlorine Species	Experimental Conditions			Quantum Yield (Φ)	Quantum Yield	References
	pH	Wavelength (nm)	[Chlorine] (mM)			
ClO^-	11.5	253.7	1	Φ-ClO^-	0.86 ± 0.07	Buxton & Subhani (1972b)
ClO^-	12.0	313	1	Φ-ClO^-	0.43 ± 0.06	Buxton & Subhani (1972b)
ClO^-	12.1	365	7–10	Φ-ClO^-	0.64 ± 0.05	Buxton & Subhani (1972b)
HOCl + ClO^-	5, 7 and 9	240	5.71	Φ-Chlorine	1.21 (pH 5) 1.24 (pH 7) 1.50 (pH 9)	Cooper *et al.* (2007)
HOCl + ClO^-	5, 7 and 9	253.7	5.71	Φ-Chlorine	1.64 (pH 5) 1.51 (pH 7) 0.84 (pH 9)	Cooper *et al.* (2007)
HOCl + ClO^-	5, 7 and 9	265.2	5.71	Φ-Chlorine	1.39 (pH 5) 1.10 (pH 7) 0.92 (pH 9)	Cooper *et al.* (2007)
HOCl + ClO^-	5, 7 and 9	296.7	5.71	Φ-Chlorine	0.78 (pH 5) 0.76 (pH 7) 0.80 (pH 9)	Cooper *et al.* (2007)
HOCl + ClO^-	5, 7 and 9	313	5.71	Φ-Chlorine	0.80 (pH 5) 0.71 (pH 7) 0.80 (pH 9)	Cooper *et al.* (2007)
HOCl + ClO^-	5 and 7	334	5.71	Φ-Chlorine	0.69 (pH 5) 0.61 (pH 7)	Cooper *et al.* (2007)
HOCl + ClO^-	5 and 7	365	5.71	Φ-Chlorine	0.73 (pH 5) 0.55 (pH 7)	Cooper *et al.* (2007)
HOCl	5	253.7	< 1	Φ-HOCl	1.0 ± 0.1	Feng *et al.* (2007)
HOCl	5	253.7	19	Φ-HOCl	4.5 ± 0.2	Feng *et al.* (2007)
ClO^-	10	253.7	0.05–9	Φ-ClO^-	0.9 ± 0.1	Feng *et al.* (2007)
HOCl	4	253.7	0.014–0.056	Φ-HOCl	1.5	Watts & Linden (2007a)
HOCl	4	MP UV (200–350)	0.014–0.056	Φ-HOCl	3.7	Watts & Linden (2007a)
ClO^-	10	253.7	0.014–0.056	Φ-ClO^-	1.2	Watts & Linden (2007a)
ClO^-	10	MP lamp (200–350)	0.014–0.056	Φ-ClO^-	1.7	Watts & Linden (2007a)
HOCl + ClO^-	7.1–7.9	253.7	0.014–0.056	Φ-Chlorine	1.3–1.7	Watts & Linden (2007a)

(*Continued*)

Table 9.5 Overall quantum yields of photodecomposition of HOCl (Φ-HOCl), ClO^- (Φ-ClO^-) and their mixture (Φ-Chlorine) in organic-free water. (*Continued*)

Chlorine species	Experimental conditions			Quantum yield (Φ)	Quantum yield	References
	pH	Wavelength (nm)	[Chlorine] (mM)			
HOCl + ClO^-	7.1–7.9	MP UV (200–350)	0.014–0.056	Φ-Chlorine	1.2–1.7	Watts & Linden (2007a)
HOCl	5	253.7 nm	1.41	Φ-ClO^-	1.0 ± 0.1	Jin *et al.* (2011)
ClO^-	10	253.7 nm	1.41	Φ-ClO^-	1.15 ± 0.08	Jin *et al.* (2011)
ClO^-	10	303 ± 8 nm	0–4.23	Φ-ClO^-	0.87 ± 0.01	Chan *et al.* (2012)
HOCl	5	MP lamp	0.15	Φ-HOCl	1.06 ± 0.01	Wang *et al.* (2012)
ClO^-	10	MP lamp	0.15	Φ-ClO^-	0.89 ± 0.08	Wang *et al.* (2012)
HOCl	5	253.7 nm	0.01–0.1	Φ-HOCl	1.45 ± 0.06	Fang *et al.* (2014)
ClO^-	10	253.7 nm	0.01–0.1	Φ-ClO^-	0.97 ± 0.05	Fang *et al.* (2014)

where

$[C]_0$ and $[C]_t$ are the concentrations (M) of free chlorine before and after exposure to UV light, respectively,
F is the fluence at the wavelength λ (J m^{-2}),
k'_1 is the fluence-based first-order rate constant (m^2 J^{-1}),
Φ (λ) is the overall quantum yield of photodecomposition of free chlorine at wavelength λ,
ε (λ) is the molar absorption coefficient (M^{-1} cm^{-1}) of free chlorine at wavelength λ,
U (λ) is the molar photon energy (J $einstein^{-1}$) at wavelength λ.

When irradiated with the polychromatic radiation emitted by the medium pressure (MP) UV lamps, the free chlorine species can be photodegraded by all radiations absorbed in the 200–375 nm range (Figure 9.2). Under MP-UV irradiation, an apparent quantum yield of photodecomposition of free chlorine ($\Phi_{300-375\ nm}$) can be determined from the photolysis *pseudo*-first order rate constant (k_1, s^{-1}) and the specific rate of light absorption by free chlorine species summed over all wavelengths ($\Sigma\ k_s(\lambda)$, einstein mol^{-1} s^{-1}) using the expressions developed by Schwazenbach *et al.* (1993):

$$\Phi_{200-375nm} = \frac{k_1}{\sum k_S(\lambda)} \tag{9.46}$$

$$k_S(\lambda) = \frac{E_p^0(\lambda)\varepsilon(\lambda)[1-10^{-a(\lambda)z}]}{a(\lambda)z} \tag{9.47}$$

where

E_P^0 is the incident photon irradiance (10^{-3} einstein cm^{-2} s^{-1}),
a (λ) represents the solution absorption coefficient (cm^{-1}) at wavelength λ,
z is the depth of solution layer (cm).

Despite the disparity of the published values (Table 9.2), most quantum yields for the decomposition of HOCl (pH ≈ 5) in dilute aqueous solution ($[Cl_2]_{0,T}$ < 1 mM) are in the range 0.90–1.64 at radiation wavelengths in the range 240 to 265 nm. If each $^{\bullet}OH$ and $Cl^{\bullet}$ generated from the primary step of photolysis of HOCl (reaction 9.8) decomposes another molecule of HOCl, the overall quantum yield of photodecomposition of HOCl should be equal to $3\Phi_{primary}$ (if no free chlorine is generated by subsequent reactions), where $\Phi_{primary}$ represents the primary quantum yield of photolysis of HOCl. The highest overall quantum values reported in Table 9.5 for the photodecomposition of HOCl in the range 240–365 nm ($1.45 < \Phi < 1.64$) seem to be consistent with the primary quantum yield value of photolysis of HOCl at 266 nm ($\Phi_{primary} \approx 0.5 \pm 0.05$) determined by Thomsen *et al.* (2001). For radiation wavelengths longer than 290 nm, the published overall quantum yields of photolysis of HOCl at monochromatic radiation are in the range 0.69–0.80 (Table 9.5). Similar experiments conducted with free chlorine solution at pH ≈ 9 (ClO^- present as the predominant free chlorine species) or with aqueous solution of hypochlorite (pH > 11) show that the overall quantum yields of photodecomposition of ClO^- are roughly equal to 0.75 ± 0.10.

Under polychromatic irradiation (MP UV lamp; 200–350 nm), overall quantum yields close to unity were determined by Wang *et al.* (2012) at pH 5 and 10, whereas higher values were obtained by Watts and Linden (2007a) at pH 4 ($\Phi = 3.7$) and at pH 7.1–7.9 ($\Phi = 1.2$–1.7) (Table 9.5).

Cooper *et al.* (2007) examined the rates of photodecomposition of free chlorine ($[Cl_2]_{0,T} = 5.1$ mM) at various wavelengths (from 240 to 365 nm) and at various pH (pH 5, 7 and 9). The authors isolated *quasi*-monochromatic radiation from a Xe-Hg lamp using narrow band cut-off filters (10 nm band-width). The data obtained at pH 5 and 7 show a decrease of the quantum yields with increasing wavelength.

Feng *et al.* (2007) showed that the quantum yield of photodecomposition of HOCl (pH 5) at 253.7 nm increased linearly when the concentration of HOCl increased from 70 to 1350 mg chlorine/L ($\Phi = 1.0 \pm 0.1$ for [HOCl] < 70 mg/L and $\Phi = 4.5 \pm 0.2$ for [HOCl] = 1350 mg/L) whereas the quantum yield of ClO^- photolysis (pH 10) did not vary with the concentration of chlorine ($\Phi = 0.9 \pm 0.1$ for $3.5 < [ClO^-] < 640$ mg/L). The increase in the quantum yield of HOCl photodecomposition with the concentration of HOCl has been attributed to chain reactions initiated by $^{\bullet}OH$ and $Cl^{\bullet}$ radicals (Table 9.4). However, the concentration-independent quantum yield reported by Feng *et al.* (2007) for the photodecomposition of ClO^- remains unexplained providing that $^{\bullet}OH$ and $Cl^{\bullet}$ can also decompose ClO^- ($k > 1 \times 10^9\ M^{-1}\ s^{-1}$; Table 9.4).

9.2.3.4 Reaction quantum yields of chlorine photodecomposition in the presence of organic compounds

Addition of methanol was found to accelerate the photodecomposition of chlorine at sunlight wavelengths, especially at low pH (Nowell & Crosby, 1985). Feng *et al.* (2007) also demonstrated that addition of methanol increased the photodecomposition rate of HOCl at 253.7 nm. As shown in Figure 9.5, the quantum yield of photodecomposition of HOCl in the absence and in the presence of 120 mM of methanol were equal to 1.0 ± 0.1 and ~55, respectively. In another study, the quantum yield of HOCl photolysis at pH 5 and 253.7 nm increased from (1.0 ± 0.1) to (16.3 ± 0.3) when the concentration of methanol increased from 0 to 86.5 mM (Jin *et al.* 2011). By contrast, addition of methanol (0–50 mM) did not affect the rate of photodecomposition of ClO^- (pH = 10) as reported by Feng *et al.* (2007) and Jin *et al.* (2011), i.e. (1.20 ± 0.20) at $20 < [CH_3OH]_0 < 50$ mM) and (1.15 ± 0.08) at $0 < [CH_3OH]_0 < 61.8$ mM, respectively.

The increase of the quantum yield of HOCl in the presence of methanol has been explained by chain reactions involving organic radicals as also suggested by Oliver and Carey (1977):

$$RH + {}^{\bullet}OH \rightarrow R^{\bullet} + H_2O \tag{9.48}$$

$$RH + Cl^{\bullet} \rightarrow R^{\bullet} + H^+ + Cl^- \tag{9.49}$$

$$R^{\bullet} + HOCl \rightarrow ROH + Cl^{\bullet} \tag{9.50}$$

$$R^{\bullet} + HOCl \rightarrow RCl + {}^{\bullet}OH \tag{9.51}$$

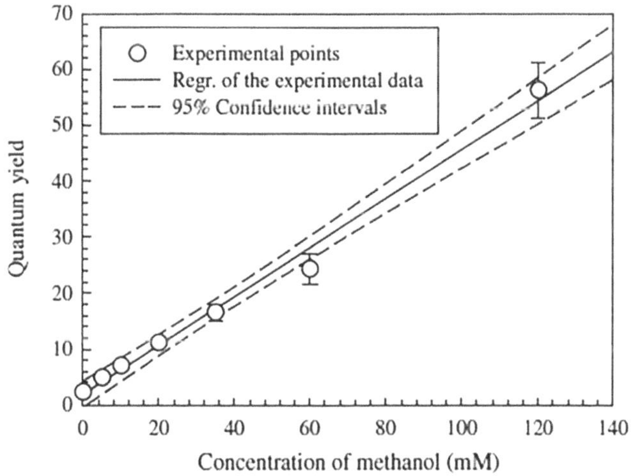

Figure 9.5 Quantum yield of HOCl photolysis (253.7 nm, pH 5) as a function of methanol concentration (Feng *et al.* 2007).

This accelerating effect on the rate of photodecomposition of free chlorine was also observed with more complex molecules like those constituting the natural organic matter (NOM) present in drinking water (Ormeci *et al.* 2005). By dissolving a mixture of humic acid and alginic acid in deionized water (pH ~ 7.2) followed by chlorination for 5 hours to minimize the background chlorine demand during the photolysis experiments, Feng *et al.* (2010) showed that the quantum yields of free chlorine photolysis at 253.7 nm increased linearly with TOC concentration ($\Phi = 1.0 \pm 0.1$ for TOC < 0.4 mg/L; $\Phi = 4.9 \pm 0.4$ for TOC ~ 6.8 mg/L).

Feng *et al.* (2007) determined similar fluence-based rate constants (~2.9×10^{-5} m^2 J^{-1}) for 71 mg/L (1 mM) free chlorine decomposition at pH 5 and pH 10 in deionized water, under exposure to the 253.7 nm radiation. That would correspond to UV doses of ~2.4×10^4 J m^{-2} and ~8×10^4 J m^{-2} required to decompose 50% and 90% of free chlorine, respectively. At a UV dose of 400 J m^{-2} typically used for UV disinfection of drinking water, the decomposition of free chlorine would be nearly 1% (pH independent), when the initial concentration of free chlorine is lower than 71 mg/L.

Lower UV doses (2×10^3–6×10^3 J m^{-2}) are needed to reduce the free chlorine concentration by 50% in surface water or in drinking water containing about 2–4 mg/L TOC (Ormeci *et al.* 2005).

Wang *et al.* (2012) determined the fluence-based rate constants (m^2 J^{-1}) for photodecomposition of free chlorine species under polychromatic radiation emitted from a MP lamp as $(15.0 \pm 0.1) \times 10^{-6}$, $(33.7 \pm 0.4) \times 10^{-6}$ and $(58.7 \pm 1.5) \times 10^{-6}$ at pH 5, 7.5, and 10, respectively. These rate constants are up to one order of magnitude larger than those determined for H_2O_2 photolysis under the same pH and UV radiation conditions. Fluence-based rate constants for free chlorine and monochloramine photodegradation at pH 6 and 9 under monochromatic (253.7 nm) and polychromatic (200–400 nm) radiation as function of chlorine concentration and water quality were reported by Watts and Linden (2007a).

9.3 REACTIVITY AND FATE OF CHLORINE RADICALS

9.3.1 Equilibria involving the $Cl^•$, $Cl_2^{•-}$ and $^•OH$ species

As chloride is always present in free chlorine solutions, $Cl^•$ and $^•OH$ generated from the photolysis of HOCl and ClO^- can react with chloride to give $Cl_2^{•-}$, $HClOH^•$ and $ClOH^{•-}$. The chlorine-containing free radicals and $^•OH$ are involved in very fast equilibria. The rate constants for various reactions have been reported in the literature and are summarized in Table 9.6. Of note, the reactions given in Table 9.6 do not lead to a decrease of the total concentration of radicals.

Table 9.6 Rate constants for various equilibrium reactions involving $Cl^•$ and $^•OH$ in the presence of chloride.

Reaction		Rate constants
$Cl^• + Cl^- \rightarrow Cl_2^{•-}$	9.52	4.1×10^9 M^{-1} s^{-1} (Jayson *et al.* 1973) 6.5×10^9 M^{-1} s^{-1} (Kläning & Wolff, 1985) $(8.5 \pm 0.7) \times 10^9$ M^{-1} s^{-1} (Buxton *et al.* 1998) $(7.8 \pm 0.8) \times 10^9$ M^{-1} s^{-1} (Yu & Barker, 2003) $(7.8 \pm 0.8) \times 10^9$ M^{-1} s^{-1} (Yu, 2004)
$Cl_2^{•-} \rightarrow Cl^• + Cl^-$	9.53	$(1.1 \pm 0.4) \times 10^5$ s^{-1} (Jayson *et al.* 1973) $(6.0 \pm 0.5) \times 10^4$ s^{-1} (Buxton *et al.* 1998) $(5.2 \pm 0.3) \times 10^4$ s^{-1} (Yu & Barker, 2003) $(5.7 \pm 0.4) \times 10^4$ s^{-1} (Yu, 2004)
$Cl^• + H_2O \rightarrow HClOH^•$	9.54	$(2.5 \pm 0.2) \times 10^5$ s^{-1} (McElroy, 1990)
$HClOH^• \rightarrow Cl^• + H_2O$	9.55	$<1 \times 10^3$ s^{-1} (McElroy, 1990)
$Cl^• + H_2O \rightarrow ClOH^{•-} + H^+$	9.56	7.2×10^4 s^{-1} (Jayson *et al.* 1973) 1.6×10^5 s^{-1} (Kläning & Wolff, 1985) $(2.5 \pm 0.2) \times 10^5$ s^{-1} (McElroy, 1990) $(1.8 \pm 0.6) \times 10^5$ s^{-1} (Yu, 2004)
$ClOH^{•-} + H^+ \rightarrow Cl^• + H_2O$	9.57	$(2.1 \pm 0.7) \times 10^{10}$ M^{-1} s^{-1} (Jayson *et al.* 1973) 2.1×10^{10} M^{-1} s^{-1} (Kläning & Wolff, 1985) $(2.6 \pm 0.6) \times 10^{10}$ M^{-1} s^{-1} (Yu & Barker, 2003) $(2.4 \pm 0.4) \times 10^{10}$ M^{-1} s^{-1} (Yu, 2004)
$^•OH + Cl^- \rightarrow ClOH^{•-}$	9.58	$(4.3 \pm 0.4) \times 10^9$ M^{-1} s^{-1} (Jayson *et al.* 1973) $(4.2 \pm 0.2) \times 10^9$ M^{-1} s^{-1} (Yu, 2004)
$ClOH^{•-} \rightarrow {}^•OH + Cl^-$	9.59	$(6.1 \pm 0.8) \times 10^9$ s^{-1} (Jayson *et al.* 1973) $(6.0 \pm 1.1) \times 10^9$ s^{-1} (Yu, 2004)
$Cl_2^{•-} + H_2O \rightarrow HClOH^• + Cl^-$	9.60	1.3×10^3 s^{-1} (McElroy, 1990)
$HClOH^• + Cl^- \rightarrow Cl_2^{•-} + H_2O$	9.61	5×10^9 M^{-1} s^{-1} (McElroy, 1990)
$Cl_2^{•-} + HO^- \rightarrow ClOH^{•-} + Cl^-$	9.62	4.5×10^7 M^{-1} s^{-1} (Grigorev *et al.* 1987)
$ClOH^{•-} + Cl^- \rightarrow Cl_2^{•-} + HO^-$	9.63	1.0×10^5 M^{-1} s^{-1} (Grigorev *et al.* 1987)
$Cl^• + OH^- \rightarrow ClOH^{•-}$	9.64	1.8×10^{10} M^{-1} s^{-1} (Kläning & Wolff, 1985)
$ClOH^{•-} \rightarrow {}^•OH + Cl^-$	9.65	23 s^{-1} (Kläning & Wolff, 1985)
$HClOH^• \rightarrow ClOH^{•-} + H^+$	9.66	1.0×10^8 s^{-1} (McElroy, 1990)
$Cl_2^{•-} + H_2O \rightarrow ClOH^{•-} + H^+ + Cl^-$	9.67	$<1 \times 10^2$ (Yu & Baker, 2003)

Not all reactions in Table 9.6 are well-known or well-understood, and discrepancies still exist between reported rate constants and measured equilibrium constants (Yu, 2004). Such uncertainties impact severely the accuracy of the kinetic models for the UV/Chlorine AOP reported in the literature, as well as the calculations of the concentrations of reactive radicals (•OH and Cl-based radicals) in this process. Armstrong *et al.* (2015) recommend the following values of equilibrium constants and uncertainties for the reactions involving Cl-based radicals and •OH:

$$Cl^{\bullet} + Cl^- \rightleftharpoons Cl_2^{\bullet -} \quad K = (1.4 \pm 0.2) \times 10^5\,M^{-1} \tag{9.52–9.53}$$

$$Cl^{\bullet} + H_2O \rightleftharpoons ClOH^{\bullet -} + H^+ \quad K = 5.0 \times 10^{-6}\,M \text{ (uncertainty within a factor of two)} \tag{9.56–9.57}$$

$$^{\bullet}OH + Cl^- \rightleftharpoons ClOH^{\bullet -} \quad K = (0.70 \pm 0.13)\,M^{-1} \tag{9.58–9.59}$$

$$^{\bullet}OH + Cl^- + H^+ \rightleftharpoons Cl^{\bullet} + H_2O \quad K = 9.1 \times 10^4\,M^{-2} \text{ (uncertainty within a factor of two)} \tag{9.68}$$

$$Cl_2^{\bullet -} + H_2O \rightleftharpoons {}^{\bullet}OH + H^+ + 2Cl^- \quad K = (6.1 \pm 0.7) \times 10^{-11}\,M^3 \tag{9.69}$$

Reactions 9.52–9.69 show that the types and concentrations of radicals in organic-free water will depend on pH and on the concentration of chloride. For example, based on the above reported bimolecular rate constants and equilibrium constants, one can conclude that $Cl^{\bullet}$ can be converted into $Cl_2^{\bullet -}$ in the presence of chloride at concentrations higher than ~1 mM under acidic pH or into •OH at neutral pH. However, more studies are required to understand and to validate the contribution of each reactive radical species to the degradation of micropollutants *via* UV/Chlorine process.

9.3.2 Termination reactions of •OH, Cl• and $Cl_2^{\bullet -}$ in water

Termination reactions involving •OH, $Cl^{\bullet}$ and $Cl_2^{\bullet -}$ decrease the concentrations of free radicals and lead to the formation of free chlorine (reactions 9.71–9.74 in Table 9.7) and hydrogen peroxide (reaction 9.70 in Table 9.7). H_2O_2 can then reduce free chlorine to chloride and O_2 (reactions 9.77 and 9.78; Held *et al.* 1978).

Table 9.7 Termination reactions involving •OH, Cl• and $Cl_2^{\bullet -}$ and subsequent reactions.

Reactions		Rate Constants ($M^{-1}\,s^{-1}$)	Reference
$^{\bullet}OH + {}^{\bullet}OH \rightarrow H_2O_2$	9.70	5.5×10^9	Buxton *et al.* (1988)
$Cl^{\bullet} + Cl^{\bullet} \rightarrow Cl_2$	9.71	8.8×10^7	Wu *et al.* (1980)
$Cl_2^{\bullet -} + Cl_2^{\bullet -} \rightarrow Cl_2 + 2\,Cl^-$	9.72	$\log(k) = 8.8 + 1.6\,I^{0.5}/(I^{0.5} + 1)$ $k = 6.3 \times 10^8\,M^{-1}\,s^{-1}$ for $I = 0$	Alegre *et al.* (2000)
$Cl_2^{\bullet -} + {}^{\bullet}OH \rightarrow HOCl + Cl^-$	9.73	1×10^9	Wagner *et al.* (1986)
$Cl_2^{\bullet -} + Cl^{\bullet} \rightarrow Cl_2 + Cl^-$	9.74	2.1×10^9	Yu *et al.* (2004)
$Cl_2 + H_2O \rightarrow HOCl + HCl$	9.1		
$HOCl \rightleftharpoons ClO^- + H^+$	9.2		
$^{\bullet}OH + HOCl/ClO^- \rightarrow ClO^{\bullet} + H_2O/HO^-$	9.39/9.11	Table 9.4	
$Cl^{\bullet} + HOCl/ClO^- \rightarrow ClO^{\bullet} + HCl/Cl^-$	9.40/9.13	Table 9.4	
$Cl_2^{\bullet -} + ClO^- \rightarrow ClO^{\bullet} + 2\,Cl^-$	9.75	5.4×10^8 ($I = 5$ M)	Zuo *et al.* (1997)
$H_2O_2 \rightleftharpoons HO_2^- + H^+$	9.76	$pK_a = 11.7$	
$H_2O_2 + ClO^- \rightarrow Cl^- + {}^1O_2 + H_2O$	9.77	3.4×10^3	Held *et al.* (1978)
$HO_2^- + HOCl \rightarrow Cl^- + {}^1O_2 + H_2O$	9.78	4.4×10^7	Held *et al.* (1978)

The role of radical – radical termination reactions on the decomposition pathways of free chlorine in the UV/Cl_2 process is relatively minor because the concentrations of free radicals are much lower than the concentrations of free chlorine species and therefore most of the $^\bullet OH$, $Cl^\bullet$ and $Cl_2^{\bullet-}$ will be consumed by the molecular chlorine species or by other radical scavengers present in water.

9.3.3 Reactivity of $Cl^\bullet$ and $Cl_2^{\bullet-}$ towards organic and inorganic compounds

9.3.3.1 Methods for determination of $Cl^\bullet$ and $Cl_2^{\bullet-}$ rate constants

$Cl^\bullet$ and $Cl_2^{\bullet-}$ are strong oxidants with $E°(Cl^\bullet/Cl^-) = 2.43$ V and $E°(Cl_2^{\bullet-}/2Cl^-) = 2.13$ V *vs* NHE (Armstrong *et al.* 2015). $Cl^\bullet$ and $Cl_2^{\bullet-}$ react with organic and inorganic compounds by H-atom abstraction, electron transfer or addition to unsaturated bonds. The rate constants for the reactions of $Cl^\bullet$ and $Cl_2^{\bullet-}$ with solutes in water are very scarce.

The bimolecular rate constants of $Cl^\bullet$ and $Cl_2^{\bullet-}$ with solutes are usually determined from pulse radiolysis or laser flash photolysis experiments by monitoring the decay of transient absorbance at 340 nm. Both $Cl^\bullet$ and $Cl_2^{\bullet-}$ display an absorption band with molar absorption coefficients at 340 nm of ~3700 and ~8800 M^{-1} cm^{-1}, respectively (Thomsen *et al.* 2001; Yu *et al.* 2004).

$Cl^\bullet$ can be generated directly by laser flash photolysis of chloroacetone (Buxton *et al.* 2000, reaction 9.79), and indirectly, by pulse radiolysis of aqueous solutions of sodium chloride (reactions 9.80, 9.56–9.59). Chlorine atoms can be also generated during flash photolysis of sodium persulfate solution in the presence of chloride ions (reactions 9.81–9.82):

$$CH_3C(O)CH_2Cl + h\nu \rightarrow CH_3C(O)CH_2^\bullet + Cl^\bullet \quad (9.79)$$

$$H_2O \rightsquigarrow H^\bullet, {}^\bullet OH, e_{aq}^-, H^+, H_2O_2, H_2 \quad (9.80)$$

$$^\bullet OH + Cl^- \rightleftharpoons ClOH^{\bullet-} \quad (9.58\text{–}9.59)$$

$$ClOH^{\bullet-} + H^+ \rightleftharpoons Cl^\bullet + H_2O \quad (9.56\text{–}9.57)$$

$$S_2O_8^{2-} + h\nu \rightarrow 2SO_4^{\bullet-} \quad (9.81)$$

$$SO_4^{\bullet-} + Cl^- \rightleftharpoons Cl^\bullet + SO_4^{2-} \quad (9.82)$$

Chlorine and dichlorine radicals oxidize the organic or inorganic solutes, S:

$$Cl^\bullet + S \rightarrow \text{Products} \quad (9.83)$$

$$Cl_2^{\bullet-} + S \rightarrow \text{Products} \quad (9.84)$$

The system is selected such that the substrate S decays exclusively through $Cl^\bullet$ and $Cl_2^{\bullet-}$ radicals:

$$-d[S]/dt = (k_1[Cl^\bullet] + k_2[Cl_2^{\bullet-}])[S] \quad (9.85)$$

where k_1 and k_2 are the second-order rate constants (M^{-1} s^{-1}) for the reaction of $Cl^\bullet$ and $Cl_2^{\bullet-}$ with S, respectively.

If $[S] \gg [Cl^\bullet]$ and $([S] \gg [Cl_2^{\bullet-}]$, the degradation of S follows *pseudo*-first order kinetics with an apparent rate constant k_{app} (s^{-1}):

$$k_{app} = \alpha_1 k_1 + \alpha_2 k_2,\ i.e.,\quad k_{app} = \frac{1}{1+K[\mathrm{Cl}^-]}k_1 + \frac{K[\mathrm{Cl}^-]}{1+K[\mathrm{Cl}^-]}k_2 \qquad (9.86)$$

where, α_1 and α_2 are the molar fractions of $Cl^\bullet$ and $Cl_2^{\bullet-}$ radicals, respectively, and $K = [Cl_2^{\bullet-}]/([Cl^-][Cl^\bullet]) = 1.4 \times 10^5\ M^{-1}$ (Buxton *et al.* 1998).

Laser flash photolysis of chloroacetone represents the best method to generate Cl atoms for the determination of $Cl^\bullet$ rate constants because $Cl_2^{\bullet-}$ is not present at the beginning of the reaction (Buxton *et al.* 2000). Determination of rate constants for $Cl^\bullet$ and $Cl_2^{\bullet-}$ reactions from laser photolysis of $Na_2S_2O_8$/NaCl solutions is more complicated because chloride concentration must be adjusted in order to minimize reaction of $SO_4^{\bullet-}$ with S, thus to have a selective degradation of S by $Cl^\bullet$ or by $Cl_2^{\bullet-}$. Literature data show that some kinetic constants for the reaction of $Cl_2^{\bullet-}$ with organic compounds were significantly overestimated due to an incomplete conversion of $Cl^\bullet$ to $Cl_2^{\bullet-}$ (Wicktor *et al.* 2003). Furthermore, the determination of rate constants for reactions of $Cl_2^{\bullet-}$ requires high chloride concentrations (0.5–1 M), and ionic strength corrections.

9.3.3.2 Reactions of Cl• with organic compounds

Table 9.8 provides rate constants for the reaction of $Cl^\bullet$ with selected aliphatic (Gilbert *et al.* 1988; Buxton *et al.* 2000; Wicktor *et al.* 2003; Mertens & von Sonntag, 1995; Zhu *et al.* 2005) and aromatic compounds (Alegre *et al.* 2000; Mártire *et al.* 2001). An excellent database on the rate constants for the reactions of selected inorganic radicals, including $Cl_2^{\bullet-}$, with inorganic and organic compounds was compiled by Neta *et al.* (1988).

Rate constants for the reactions of $Cl^\bullet$ with most aliphatic compounds (alcohols, aldehydes, ketones and acids) are in the range of 10^7 and $10^{10}\ M^{-1}\ s^{-1}$ and are of the same order of magnitude as those of hydroxyl radical. $Cl^\bullet$ atoms react with aromatic compounds at near diffusion-controlled rates ($k \sim (1\text{–}2) \times 10^{10}\ M^{-1}\ s^{-1}$).

$Cl^\bullet$ reacts with oxygenated aliphatic compounds more likely by H-abstraction from a C—H bond. This mechanism is supported by ESR spectroscopy (Gilbert *et al.* 1988), and by the correlation of the rate constant with the bond dissociation enthalpy for the weakest C—H bond (Wicktor *et al.* 2003). H-abstraction from O—H group *via* an electron transfer mechanism has also been proposed (Gilbert *et al.* 1988):

$$(CH_3)_3COH + Cl^\bullet \rightarrow [Cl^- + (CH_3)_3C(+)OH] \rightarrow H^+ + Cl^- + (CH_3)_3CO^\bullet \rightarrow (CH_3)_2CO + CH_3^\bullet \qquad (9.87)$$

$Cl^\bullet$ reacts by direct addition to the C=C bond of unsaturated aliphatic compounds. For example, $Cl^\bullet$ adds to the C=C double bond of tetrachloroethene to form a pentachloroethyl radical which is rapidly converted into a peroxyl radical in the presence of oxygen (Mertens & von Sonntag, 1995):

$$Cl^\bullet + CCl_2{=}CCl_2 \rightarrow CCl_3{-}CCl_2^\bullet \qquad (9.88)$$

$$CCl_3{-}CCl_2^\bullet + O_2 \rightarrow CCl_3{-}CCl_2OO^\bullet \qquad (9.89)$$

$$CCl_3{-}CCl_2OO^\bullet \rightarrow \text{Products (trichloroacetic acid, } CO_2 \text{ and HCl)} \qquad (9.90)$$

Table 9.9 gives the identified oxidation by-products of benzene, benzoic acid, toluene, and chlorobenzene from the reactions of these compounds with $Cl^\bullet$ and $Cl_2^{\bullet-}$ generated by laser and conventional flash photolysis of $Na_2S_2O_8$ aqueous solutions containing chloride ions (Alegre *et al.* 2000; Mártire *et al.* 2001). Absorption spectra of transient species showed that the addition of $Cl^{\bullet}$ to aromatic rings to form chlorocyclohexadienyl radicals (Cl-CHD) is the predominant mechanism. The disproportionation reactions of Cl-CHD radicals lead to chlorobenzoic acid. In air saturated solutions, the Cl-CHD radicals decompose to form both chlorinated and oxidation by-products (Table 9.9).

Table 9.8 Rate constants for reactions of $Cl^{\bullet}$ and $Cl_2^{\bullet-}$ with selected organic compounds in water.

Organic Compound	$Cl^{\bullet}$ (M^{-1} s^{-1})	$Cl_2^{\bullet-}$ (M^{-1} s^{-1})
Methanol	$(1.0 \pm 0.2) \times 10^{9}$[a]; $(1.0 \pm 0.1) \times 10^{9}$[b]	3.5×10^{3}[c]; $(0.3 \pm 1.3) \times 10^{4}$[b] $(5.1 \pm 0.8) \times 10^{4}$[d]
Ethanol	$(1.7 \pm 0.7) \times 10^{9}$[a]; $(2.2 \pm 0.7) \times 10^{9}$[b]	4.5×10^{4}[c]; $(1.2 \pm 0.2) \times 10^{5}$[d]
2-propanol	$(1.5 \pm 0.1) \times 10^{9}$[a]; $(3.2 \pm 0.7) \times 10^{9}$[b]	1.2×10^{5}[c]; $(1.9 \pm 0.3) \times 10^{5}$[d]
Tert-butanol	$(6.2 \pm 0.5) \times 10^{8}$[a]; $(1.5 \pm 0.1) \times 10^{9}$[b]	$(2.6 \pm 0.5) \times 10^{4}$[d]
Diethyl ether	$(1.3 \pm 0.1) \times 10^{9}$[b]	$(4.0 \pm 0.2) \times 10^{5}$[d]; $(3.6 \pm 0.2) \times 10^{5}$[b]
Methyl-*tert*-butyl ether (MTBE)	$(1.3 \pm 0.1) \times 10^{9}$[b]	$(7 \pm 1) \times 10^{4}$[d]; $(1.6 \pm 1.5) \times 10^{4}$[b]
Tetrahydrofuran	$(2.6 \pm 0.4) \times 10^{9}$	$(5.3 \pm 0.6) \times 10^{5}$[b]
Formic acid ($pK_a = 3.75$)	$(1.3 \pm 0.3) \times 10^{8}$ (pH 1)[a]; $(4.2 \pm 0.5) \times 10^{9}$[a]; $(2.8 \pm 0.3) \times 10^{9}$(pH 1)[b]	6.7×10^{3} (pH 1)[c]; $(8.0 \pm 1.4) \times 10^{4}$ (pH 4)[d]
Acetic acid ($pK_a = 4.75$)	$(3.2 \pm 0.2) \times 10^{7}$(pH 1)[a]; $(3.7 \pm 0.4) \times 10^{9}$[a]; $(1.0 \pm 0.2) \times 10^{8}$ (pH 1)[b]	$<10^{4}$ (pH 7)[c]; $(1.5 \pm 0.8) \times 10^{3}$ (pH 0.4)[d]
Hydrated formaldehyde	$(1.4 \pm 0.1) \times 10^{9}$(pH 1)[a]; $(3.2 \pm 0.2) \times 10^{7}$ (pH 1)[a]; $(1.4 \pm 0.3) \times 10^{9}$[b]	
Acetone	$<5 \times 10^{6}$[a]; $(7.8 \pm 0.7) \times 10^{7}$[b]	1.4×10^{3}[c,d]
Dichloromethane	$(9.3 \pm 0.3) \times 10^{6}$[b]	
Trichloromethane	$(2.3 \pm 0.5) \times 10^{8}$[b]	
Tetrachloroethene	2.8×10^{8}[e]	
Chlorpromazine		5×10^{9}[i]
Dimethylsulfoxide (DMSO)	$(6.3 \pm 0.6) \times 10^{9}$[f]	1.6×10^{7}[f]
Dimethylsulfone	$(8.2 \pm 1.6) \times 10^{5}$[f]	
Benzene	$(0.6–1.2) \times 10^{10}$[g]	$<1 \times 10^{5}$[h]
Toluene	$(1.8 \pm 0.3) \times 10^{10}$[h]	$<1 \times 10^{6}$[h]
Chlorobenzene	$(1.8 \pm 0.3) \times 10^{10}$[h]	$<1 \times 10^{6}$[h]
Benzoic acid ($pK_a = 4.2$)	$(1.8 \pm 0.3) \times 10^{10}$[h]	2×10^{6} (pH 7)[c]; $<10^{6}$ (pH 4)[h]
p-Chlorobenzoic acid ($pK_a = 4.0$)		3×10^{6} (pH 7)[c]
p-Hydroxybenzoic acid ($pK_{a1} = 4.6$; $pK_{a2} = 9.2$)		2.8×10^{8} (pH 7)[c]
Phenol		2.5×10^{8}[c]; 5×10^{8}[i]

[a]Buxton *et al.* (2000); [b]Wicktor *et al.* (2003); [c]Hasegawa & Neta (1978); [d]Jacobi *et al.* (1999); [e]Mertens & von Sonntag (1995); [f]Zhu *et al.* (2005); [g]Alegre *et al.* (2000); [h]Mártire *et al.* (2001); [i]Willson (1973).

Table 9.9 By-products from the degradation of aromatic compounds by chlorine atoms.

Parent Compound	Identified Oxidation By-products	Reference
Benzene	Chlorobenzene (10%) 2,4-Hexadiene-aldehyde	Alegre *et al.* (2000)
Chlorobenzene	Phenol 2-Chlorophenol Trichlorophenols	Mártire *et al.* (2001)
Toluene	1-Chloro-2-methylbenzene, 1-Chloro-4-methylbenzene, Benzaldehyde, benzyl alcohol Benzyl chloride, biphenyl	Mártlre *et al.* (2001)
Benzoic acid	3-Chlorobenzoic acid and 4-chlorobenzoic acid (estimated yield: 30%) Chlorobenzene (yield < 0.4%)	Mártire *et al.* (2001)

In the case of chlorobenzene, cyclohexadienyl radicals may also be formed by charge transfer from the aromatic ring to $Cl^{\bullet}$ to give an unstable cation radical which hydrolyses to give hydroxycyclohexadienyl radicals (HO-CHD) (Figure 9.6). HO-CHD radicals undergo disproportionation reactions to yield chlorophenol and unstable chloroxyclohexadienes which decompose to phenol or chlorobenzene. The mechanism of formation of tri- and tetrachlorophenols involve peroxyl radicals as intermediate species, which decompose to endoperoxides and further to chlorophenols.

9.3.3.3 Reactions of $Cl_2^{\bullet-}$ with organic compounds

Literature data show that $Cl_2^{\bullet-}$ is 2–5 orders of magnitude less reactive than $Cl^{\bullet}$ (Table 9.8). Rate constants for the reaction of $Cl_2^{\bullet-}$ with aliphatic compounds are in the range 10^3–10^5 M^{-1} s^{-1} for saturated alcohols, ketones and acids, 10^5–10^6 M^{-1} s^{-1} for amino-acids, 10^6–10^8 M^{-1} s^{-1} for unsaturated aliphatic compounds (e.g., fumaric acid, acrylic acid and allyl alcohol). Rate constants for $Cl_2^{\bullet-}$ reactions with aromatic compounds range from ~10^6 M^{-1} s^{-1} for benzoic acid derivatives to 10^7–10^9 for phenol, aniline and anisole derivatives. $Cl_2^{\bullet-}$ reacts with organic solutes by H-abstraction from aliphatic compounds and aromatic rings, with the NH_2 group of amino acids, with the unsaturated bonds in aliphatic compounds by $Cl^{\bullet}$ addition, and by one electron transfer from aromatic compounds containing electron-rich substituents like OH, NH_2 or OCH_3 groups (Hasegawa & Neta, 1978). For example, $Cl_2^{\bullet-}$ can abstract one electron from phenol to produce a phenol cation radical which evolves to a phenoxyl radical. The subsequent reaction of $Cl_2^{\bullet-}$ with phenoxyl radical has been hypothesized to explain the formation of chlorophenols (Vione *et al.* 2015):

$$H - Ph - OH + Cl_2^{\bullet-} \rightarrow H - Ph^{\bullet+} - OH + 2\,Cl^- \tag{9.91}$$

$$H - Ph^{\bullet+} - OH \rightleftharpoons H - Ph - O^{\bullet} + H^+ \tag{9.92}$$

$$H - Ph - O^{\bullet} + Cl_2^{\bullet-} \rightarrow Cl - Ph - OH + Cl^- \tag{9.93}$$

The lifetime of $Cl_2^{\bullet-}$ is much longer than that of $Cl^{\bullet}$, thus, it accumulates in solution at higher concentrations than those of $Cl^{\bullet}$. Therefore, the bimolecular reactions of $Cl_2^{\bullet-}$ may not be negligible in the UV/Chlorine process applications.

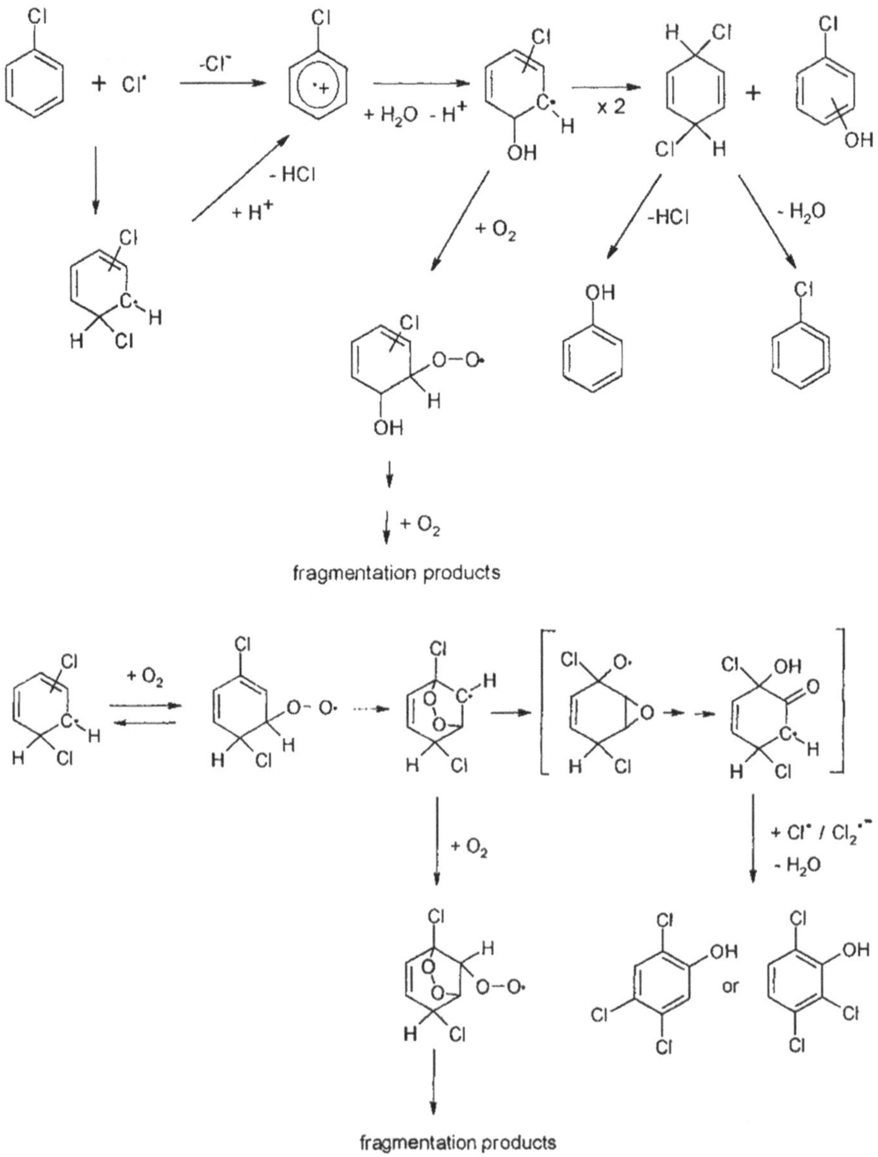

Figure 9.6 Degradation pathways of chlorobenzene by Cl•. Adapted with permission from Mártire *et al.* (2001). Copyright 2016 American Chemical Society.

9.3.3.4 Reactions of $Cl^{\bullet}$ and $Cl_2^{\bullet-}$ with inorganic compounds present in natural waters

Reduced forms of inorganic compounds such as Fe^{2+}, Mn^{2+}, NO_2^- and H_2S which can be present in groundwater under anoxic and anaerobic conditions can be oxidized by $Cl^{\bullet}$ and $Cl_2^{\bullet-}$ as well as by molecular chlorine. Bicarbonate and carbonate ions which are present in groundwater and surface waters at concentrations ranging from 0.5 to 5 mM are well-known •OH scavengers. The few rate constants reported in literature indicate that these ions can also scavenge $Cl^•$ and $Cl_2^{•-}$ and that $Cl^•$ and $Cl_2^{•-}$ can react faster than $^•OH$ with the bicarbonate ion which is the predominant species at neutral pH (Table 9.10).

Table 9.10 Rate constants for the reactions of $Cl^•$, $Cl_2^{•-}$ and $^•OH$ with bicarbonate and carbonate ions.

Reactions		Rate Constant (M^{-1} s^{-1})	References
$HCO_3^- + Cl^• \rightarrow CO_3^{•-} + HCl$	9.94	2.2×10^8	Mertens & von Sonntag (1995)
$CO_3^{2-} + Cl^• \rightarrow CO_3^{•-} + Cl^-$	9.95	5.0×10^8	Mertens & von Sonntag (1995)
$HCO_3^- + Cl_2^{•-} \rightarrow CO_3^{•-} + H^+ + 2\ Cl^-$	9.96	8.0×10^7	Matthew & Anastasio (2006)
$CO_3^{2-} + Cl_2^{•-} \rightarrow CO_3^{•-} + 2\ Cl^-$	9.97	1.6×10^8	Matthew & Anastasio (2006)
$HCO_3^- + {}^•OH \rightarrow CO_3^{•-} + H_2O$	9.98	8.5×10^6	Buxton *et al.* (1988)
$CO_3^{2-} + {}^•OH \rightarrow CO_3^{•-} + HO^-$	9.99	3.9×10^8	Buxton *et al.* (1988)

During the application of the UV/Cl_2 AOP, the highly reactive hydroxyl radicals and chlorine atoms generated from HOCl/ClO photolysis are involved in pH-dependent equilibria and reactions with water constituents (e.g. chloride ion, bicarbonate ion, dissolved organic matter, chemical pollutants, etc.) as well as with the free chlorine species ($HOCl/ClO^-$). Some of these reactions are shown in Tables 9.4, 9.6 and 9.7, and a simplified reaction scheme is given in Figure 9.7, where S_i represents any radical scavenger in the water matrix other than the target pollutant and free chlorine species.

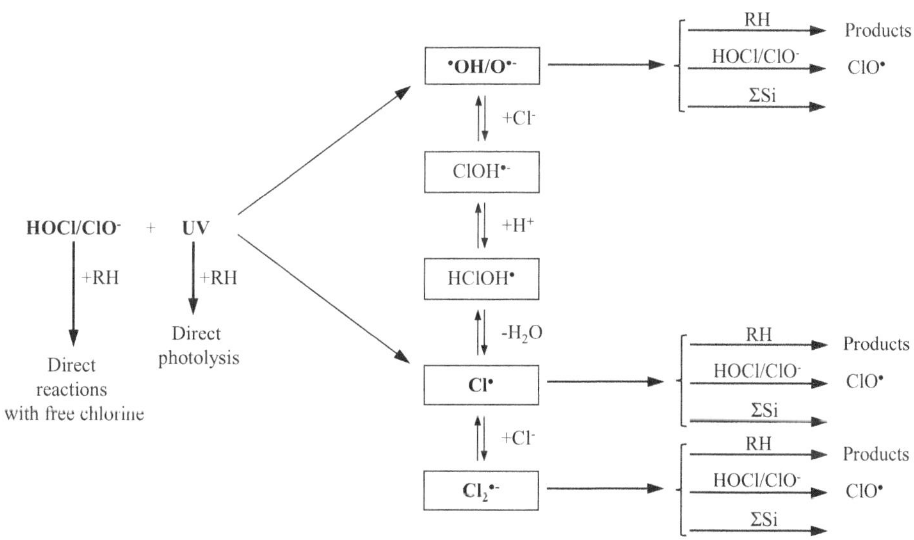

Figure 9.7 Simplified reaction scheme showing the different degradation pathways of an organic solute (RH) by the UV/Cl_2 process (ΣSi represents free radical scavengers) (De Laat, 2016).

The individual contribution of $^•OH$, $Cl^•$ and $Cl_2^{•-}$ reactions to the overall oxidation of a given pollutant RH will depend on parameters that affect the equilibrium reactions involving these free radicals such as pH and chloride ion concentration (reactions in Table 9.6), on the steady-state concentrations of free radicals and on the rate constant values for the reactions of $^•OH$, $Cl^•$ and $Cl_2^{•-}$ with RH.

9.4 UV/Cl_2 PROCESS FOR CONTAMINANT REMOVAL FROM WATER

9.4.1 Degradation pathways of organic compounds

The degradation pathways of organic pollutants during the UV/Cl_2 AOP can follow parallel reactions e.g. direct photolysis if the organic compounds absorb the radiation, oxidation *via* free radicals (•OH and Cl• and to a lesser extent $Cl_2^{\bullet-}$ and other radical species formed in the system), and dark ("thermal") reactions of free chlorine with the organic pollutants. Chlorination reactions can occur before, inside, and after the UV reactor. The relative contribution of direct photolysis, dark chlorination, and free radical reactions to the degradation of organic pollutants depends on many parameters among which the chemical and photochemical properties of organic compounds (e.g., quantum yields and molar absorption coefficients at the irradiation wavelengths, rate constants for the reactions of organic pollutants with free chlorine and with oxidizing radicals), UV system (lamp type and spectral power distribution, fluence rate, mixing efficiency), water composition (dissolved organic carbon, alkalinity, other water matrix constituents which can compete with the target pollutant for photons or/and for the reactive radical species), UV/Cl_2 process operating conditions (pH, concentration of free chlorine).

Some studies on the UV/Cl_2 process showed that organic compounds resistant to both molecular chlorine oxidation and UV photolysis (LP and MP lamps, solar radiation-simulating lamps) could be degraded by the combination of chlorine and UV radiation, thus demonstrating indirectly the generation of highly reactive radicals (•OH, Cl•) during photolysis of HOCl or ClO^-. These degradation reactions have been observed for ethanol and *n*-butanol (Oliver & Carey, 1977), ethylene glycol dimethyl ether and related substrates, acetic and propionic acids (Ogata *et al.* 1978; Ogata *et al.* 1979), 1-chlorobutane and n-octanol (Nowell & Hoigné, 1992b) and aromatic compounds containing electron-withdrawing groups such as nitrobenzene and benzoic acid (Oliver & Carey, 1977; Ogata & Tomizawa, 1984; Nowell & Crosby, 1985). More recently, *p*-chlorobenzoic acid (pCBA) has been used in UV/Cl_2 experiments as a probe compound to quantify the steady-state concentrations of •OH (Watts & Linden, 2007a; Jin *et al.* 2011). Methanol has also been used in UV/Cl_2 experiments to estimate hydroxyl radical yield factor, $\eta = \Delta[^{\bullet}OH]/\Delta[\text{Free chlorine}]$ (Jin *et al.* 2011).

The observed by-products indicate that the degradation mechanisms of organic compounds by UV/Cl_2 AOP involve hydroxylation, oxidation, and halogenation reactions. For example, the oxidation of nitrobenzene by UV/Cl_2 and by UV/H_2O_2 yields 2-, 3- and 4-nitrophenol as initial by-products (Nowell & Crosby, 1985; Watts *et al.* 2007b) but in the case of UV/Cl_2 other by-products (e.g. chlorophenols) are rapidly formed (Watts *et al.* 2007b).

In a study on the herbicide chlortoluron degradation by UV(253.7 nm)/Cl_2 AOP ($[\text{Chlortoluron}]_0 = 5\ \mu M$; $[\text{Chlorine}]_0 = 25–100\ \mu M$), Guo *et al.* (2016) showed that the attack of •OH on chlortoluron ($k = (2.18 \pm 0.83) \times 10^9\ M^{-1}s^{-1}$) was the predominant degradation pathway. The authors identified ten aromatic degradation by-products. Among the identified intermediates, two were also formed through direct photolysis (C—Cl bond scission) followed by hydroxylation of the aromatic ring. The other products indicated that the UV/Cl_2 AOP leads to hydroxylation and chlorination of the aromatic ring and of methyl groups at the amide moiety. Post-chlorination experiments ($[\text{Chlortoluron}]_0 = 100\ \mu M$; ($[\text{Chlorine}]_0 = 500\ \mu M$) showed that UV/$Cl_2$ enhanced the formation of C-DBPs (1,1-dichloroacetone, 1,1,1-trichloroacetone, chloroform, chloroacetaldehyde) and of N-DBPs (dichloroacetonitrile and trichloronitromethane) compared to UV radiation alone.

Qin *et al.* (2014) studied the oxidation of antiprotozoal agent ronidazole (RNZ) by UV/Cl_2 ($[\text{RNZ}]_0 = 0.1$ mM; $[Cl_2]_{T,0} = 1$ mM; pH 7; 253.7 nm) and identified two chlorinated by-products which were also formed by chlorination alone and five hydroxylated by-products, four of which were also formed during UV photolysis. Post-chlorination ($[\text{post-}Cl_2]_0 = 1$ mM; pH 7; 7 days) showed that UV/Cl_2 generated chloroform and trichloronitromethane (TCNM) precursors (such as amines and imines) through the •OH-induced degradation of heterocyclic ring of RNZ.

Deng *et al.* (2014) showed that the UV(253.7 nm)/Cl_2 process promoted the formation of TCNM precursors (methylamine and dimethylamine) during the treatment of aqueous solutions of polyamine. TCNM concentrations decreased under extended UV exposure through direct photolysis. The UV/Cl_2-enhanced degradation of polyamine to methylamine and dimethylamine and probably to other TCNM precursors has been attributed to the contribution of free radicals (•OH and Cl•) and of reactive species generated from photodegradation of chloramine intermediates. The addition of *tert*-butanol (free radical scavenger) completely inhibited the formation of TCNM from methylamine and partially decreased the formation of TCNM from dimethylamine and polyamine. Based on experimental data, the authors proposed two different reaction schemes for the formation of TCNM from methylamine and dimethylamine. The higher levels of TCNM formed from methylamine by UV/Cl_2 as compared to the dark chlorination, were explained by the OH and Cl radical-enhanced rate of formation of nitromethane (CH_3NO_2) from chloramines. Other mechanisms leading to TCNM from dimethylamine involve methyl-imine as an intermediate. TCNM formation *via* the UV/Cl_2 process likely involves photo-enhanced chlorination reactions (Deng *et al.* 2014).

Xiang *et al.* (2016) showed that the degradation of the anti-inflammatory drug ibuprofen (IBP) by UV(253.7 nm)/Cl_2 AOP generated significantly more total organic chlorine ([TOCl] = 31.6 μM) than dark chlorination ([TOCl] = 0.2 μM) after 30 min reaction time. Aliphatic chlorinated by-products (chloroform, chloral hydrate, 1,1-dichloro-2-propanone, 1,1,1-trichloropropanone, dichloroacetic and trichloroacetic acids) accounted for 17.4% to the TOCl. Based on the identified hydroxylated and chlorinated by-products, the authors proposed the degradation pathways for ibuprofen by UV/Cl_2 AOP presented in Figure 9.8.

The chlorinated products shown in brackets were identified for the first time. The analyses showed that the primary steps of the degradation of IBP lead to the formation of six hydroxylated derivatives and one monochlorinated by-product at pH 6. Similar experiments conducted at pH 0 where only Cl• is formed from photolysis of Cl_2, demonstrated that degradation of IBP yielded monochlorinated byproducts without formation of hydroxylated intermediates. From these data, the authors concluded that the formation of hydroxylated compounds can be attributed to •OH attack only. Further degradation reactions lead to the formation of several aromatic intermediates *via* hydroxylation, decarboxylation, demethylation and chlorination reactions and to the formation aliphatic by-products *via* ring opening reactions (Figure 9.8).

In a comparative study of the degradation of herbicide atrazine (ATZ) by UV alone, UV/Cl_2 and UV/H_2O_2 AOPs ($[ATZ]_0 = 5\ \mu M$; $[Chlorine]_0$ or $[H_2O_2] = 0$ or 70 μM; pH 7; 253.7 nm), Kong *et al.* (2016) identified five, eleven, and ten atrazine by-products in UV alone, UV/Cl_2 and UV/H_2O_2, respectively. As the identified by-products of atrazine in the UV/Cl_2 process did not reveal the addition of Cl atom to atrazine structure, the authors proposed that oxidation by •OH and Cl• rather than chlorine substitution was the degradation pathway. Quantification of the three primary by-products of atrazine (desethyl-atrazine (DEA), desisopropyl-atrazine (DIA) and desethyl-desisopropyl-atrazine (DEIA)) at the beginning of the reaction showed that the DEA:DIA yield ratios were ~4 and ~1 for UV/Cl_2 and UV/H_2O_2, respectively. This was attributed to different reaction mechanisms of •OH and Cl• with atrazine.

Wang *et al.* (2016a) conducted an extensive study on UV(253.7 nm)/Cl_2–based degradation of the antiepileptic drug carbamazepine (CBZ), which included by-product identification and mechanistic pathways, kinetic modeling, and process economics (electrical energy per order, E_{EO}). No significant direct photolysis was observed, and direct (thermal) chlorination accounted for only 5% of the overall degradation yield of CBZ by the UV/Cl_2 process. The degradation rate constants were determined for various experimental conditions, including pH (4.5–10.5), chloride and bicarbonate concentrations (0–50 mM), free chlorine dose (0.03–0.63 mM), and UV irradiance. The authors concluded that chloride ion effect was negligible; instead, bicarbonate ion concentration reduced the observed rate constant. The kinetic model is similar to that developed by Fang *et al.* (2014), but some key radical rate constants used in the two works are largely different; e.g., $k_{Cl^\bullet,HCO_3^-} = 2.2 \times 10^8\ M^{-1}\ s^{-1}$ (Fang *et al.* 2014), $k_{Cl^\bullet,HCO_3^-} = 2.9 \times 10^9\ M^{-1}\ s^{-1}$ (Wang *et al.* 2016a); $k_{Cl^\bullet,TBA} = 3.0 \times 10^8\ M^{-1}\ s^{-1}$ (Fang *et al.* 2014), $k_{Cl^\bullet,TBA} = 1.9 \times 10^9\ M^{-1}\ s^{-1}$ (Wang *et al.* 2016a). Although no chlorinated byproducts were identified in the product study, the authors did not exclude the Cl atom reactions and determined the relative contribution of $^\bullet$OH and Cl$^\bullet$ to the overall rate constant (k_{obs}), as a function of pH.

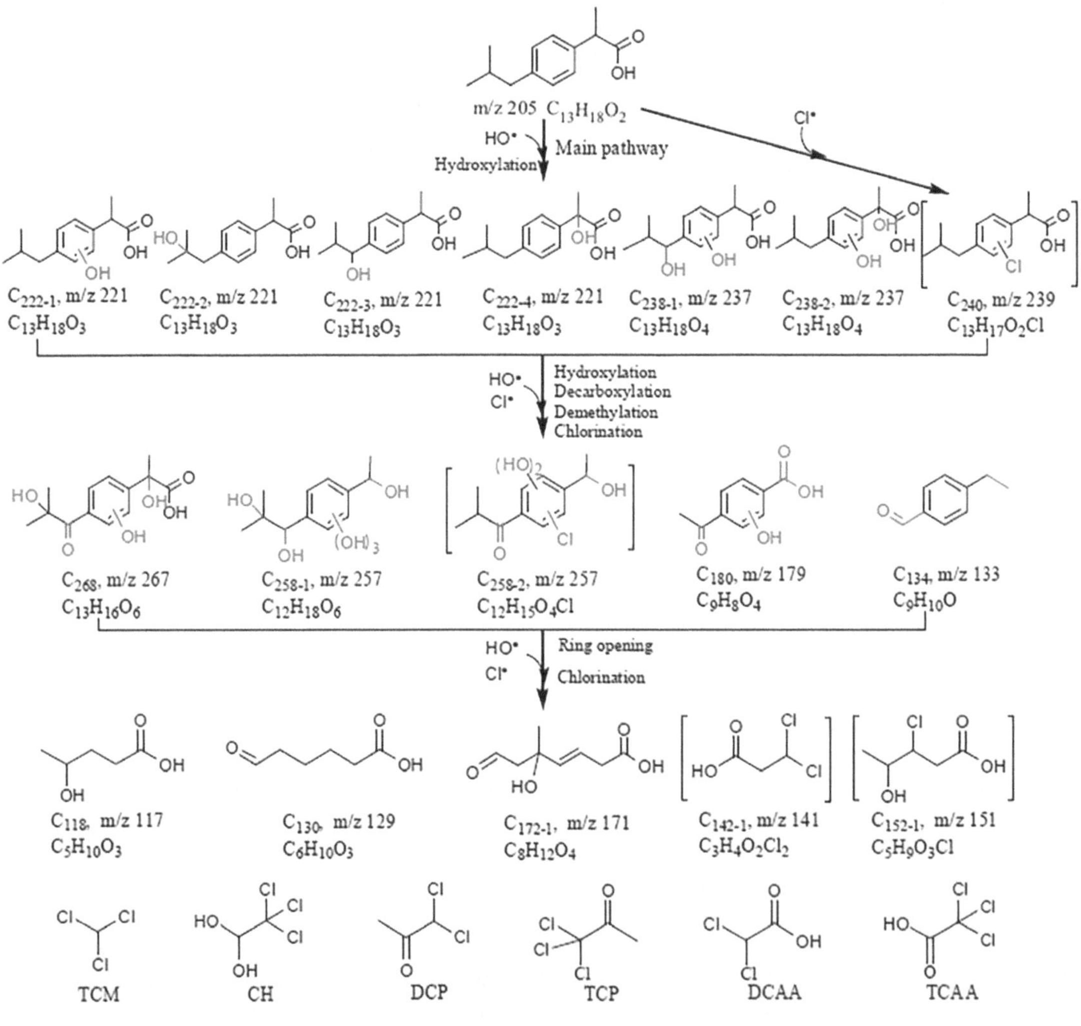

Figure 9.8 Proposed pathways for the degradation of ibuprofen by UV/Cl_2 AOP. Reprinted with permission from Xiang *et al.* (2016). Copyright 2016 Elsevier.

Competition kinetics experiments with nitrobenzene and benzoic acid as chemical probes allowed the determination of $k_{Cl^\bullet,CBZ} = (5.6 \pm 1.6) \times 10^{10}$ M^{-1} s^{-1}, which is larger than typical diffusion rate coefficients. The authors recognized that more investigations are required in order to properly understand the $Cl^\bullet$ reactivity, i.e., transient kinetics in complex systems. The E_{EO} data were calculated from collimated beam experiments in deionized water and in wastewater effluent samples for a chlorine dose of 0.28 mM (~20 mg/L). The authors included the energy requirements associated with both chlorine and UV power consumption, and were found as 0.32 and 0.44 kWh/m^3/order for CBZ removal in deionized water (pH 7) and wastewater effluent (3.1 mg/L TOC; 62.2 mg/L IC; 7.9 mg-N/L; 24 mg/L NO_3^-; pH 7), respectively.

9.4.2 Kinetic modeling of UV/Cl_2 AOP

As compared to the UV/H_2O_2 AOP, the photochemical and chemical reactions involved in the UV/Cl_2 process are much less elucidated, and many rate constants are either unknown or not well documented. Some studies on micropollutant degradation with the UV/Cl_2 AOP have attempted kinetic modeling of the experimental data. However, the key input parameters, including the apparent photolysis quantum yield of free chlorine species and radical yields, rate constants for the reactions of radicals with the free chlorine species, as well as the radical species considered in the oxidative degradation mechanisms vary from one model to another.

Fang *et al.* (2014) investigated the degradation of benzoic acid by UV(253.7 nm)/Cl_2 process under various experimental conditions, including pH and initial concentrations of benzoic acid, free chlorine, chloride ion, bicarbonate ion, *tert*-butanol and Suwannee River Natural Organic Matter. Benzoic acid was not degraded by either UV alone or 'dark' chlorination. The degradation with the UV/Cl_2 process followed first-order kinetics with respect to the concentration of benzoic acid. The authors developed a complex kinetic model which accounted for HOCl and ClO^- photolysis to $^\bullet OH$ /$O^{\bullet -}$ and $Cl^\bullet$ radicals, pH and chloride-dependent equilibria between the highly reactive species ($^\bullet OH$, $O^{\bullet -}$, $Cl^\bullet$, $Cl_2^{\bullet -}$, HClOH and $ClOH^{\bullet -}$, Table 9.6), radical ($^\bullet OH$, $Cl^\bullet$) reactions with free chlorine species (HOCl, ClO^-) yielding the oxychlorine radical $ClO^\bullet$, and reactions of radical species with benzoic acid and water matrix constituents, such as bicarbonate and chloride ions, and Suwannee River NOM. In order to elucidate the role of each radical species in the degradation of benzoic acid, Fang *et al.* (2014) used chemical probes, such as nitrobenzene (apparently non-reactive towards $Cl^\bullet$) and *tert*-butanol (reactive with both to $^\bullet OH$ and $Cl^\bullet$).

The kinetic model simulated reasonably well the degradation of benzoic acid under a variety of experimental conditions, some of which are exemplified in the next section. Computer simulations showed that the degradation of benzoic acid could mainly be attributed to $^\bullet OH$ and $Cl^\bullet$ given that the rate constants of $Cl_2^{\bullet -}$ and $O^{\bullet -}$ with benzoic acid are several orders of magnitude smaller than those of $^\bullet OH$ and $Cl^\bullet$ and the fraction of hydroxyl radicals present as $O^{\bullet -}$ can be neglected at pH < 9. The model was also used to determine the relative contributions of $^\bullet OH$, $O^{\bullet -}$, $Cl^\bullet$ and $Cl_2^{\bullet -}$ radicals to the degradation of benzoic acid. From the theoretical expression and experimental value of the *pseudo*-first order rate constant of benzoic acid decay in the presence of NOM, a rate constant of $k_{Cl\bullet,NOM} = 1.3 \times 10^4$ (mg C/L)$^{-1}$ s^{-1} was computed for the reaction of $Cl^\bullet$ with NOM. This value is smaller than the frequently reported rate constant of $^\bullet OH$ with NOM ($k_{\bullet OH,NOM} = 2.5 \times 10^4$ (mg C/L)$^{-1}$ s^{-1} which indicates that NOM is a stronger scavenger of $^\bullet OH$ than of $Cl^\bullet$. However, the computed $k_{Cl\bullet,NOM}$ rate constant must be validated experimentally.

Fang *et al.*'s kinetic model appears to be the most comprehensive model to-date for the UV/Cl_2 AOP, although it does not consider the fate of oxychlorine radical $ClO^\bullet$ which is formed with high yields in this process and could impact the real-time oxidant and radical concentrations in the UV/Cl_2 process. This kinetic model was recently used by Aghdam *et al.* (2016) for the UV/Cl_2-based degradation of N,N′-diethyl-*meta*-toluamide (DEET), an insect repellent. In this study, the authors estimated the rate constant for the chlorine atom reaction with DEET as 8×10^9 M^{-1} s^{-1}, which is almost twice the •OH rate constant (4.95×10^9 M^{-1} s^{-1}, Song *et al.* 2009). It was reported that the contribution of Cl• to the overall degradation of DEET at pH 6 and 7 was approx. 30 and 45%, respectively. The basics of Fang *et al.*'s kinetic model were employed in other recent UV/Cl_2 studies to estimate the contribution of direct UV photolysis and •OH and Cl• species to the ibuprofen degradation (Xiang *et al.* 2016) and to model carbamazepine degradation (Wang *et al.* 2016a).

Qin *et al.* (2014) proposed a kinetic model for the UV(253.7 nm)/Cl_2 degradation of ronidazole (RNZ) in deionized water in the absence of radical scavengers other than the oxidant species and target compound. The authors state that in the UV/Cl_2 process the dominant oxidant is the •OH, and used a quantum yield of ~1.0 for both HOCl and ClO^- photolysis. The model accounts for 'dark' chlorination of RNZ, but disregards the Cl• reactions with the oxidant species and RNZ. Both Fang *et al.*'s (2014) and Qin *et al.*'s (2014) UV/Cl_2 kinetic models used the $k_{HO\bullet,HOCl} = 2 \times 10^9$ M^{-1} s^{-1} (Matthew & Anastasio, 2006).

Wang *et al.* (2012) built a numerical model in MatLab® to simulate the TCE photodegradation with the UV/Cl_2 process in deionized water exposed to polychromatic radiation from a MP lamp. The authors determined the experimental fluence rate-based rate constants and compared those to the model-computed data. The model accounted for TCE direct photolysis and TCE reaction with Cl• formed from TCE photolysis, but did not include the reactions of Cl• formed from HOCl and ClO^- photolysis with either the oxidant species or TCE. The authors used the observed •OH yields in their study (0.79 and 1.18, for HOCl and ClO^-, respectively), which were different from those reported in other studies (Jin *et al.* 2011; Chan *et al.* 2012; Watts & Linden, 2007a). The ClO• reactions were not considered in the model. Wang *et al.*'s kinetic model used $k_{HO\bullet,HOCl} = 8.46 \times 10^4$ M^{-1} s^{-1} (Watts & Linden, 2007a), which is almost five orders of magnitude smaller than that used with the models discussed above. For the input kinetic parameters used in their model, Wang *et al.* found a good agreement between the experimental and computed fluence based-rate constants of TCE degradation by UV (polychromatic)/Chlorine process.

The UV/Cl_2 AOP kinetic models discussed above use different values for the same key input parameters (e.g., quantum yields, •OH and Cl• rate constants) – all of which are in the published literature. Large discrepancies are between such key parameters reported in the literature, which reflects rather limited current understanding of the fundamentals of the UV/Cl_2 process and of the fate and contribution of Cl• and ClO• species to the reaction mechanisms, as well as the need for more research on the UV/Cl_2 process.

9.4.3 The impact of selected parameters on UV/Cl_2 process performance

The oxidation rates of organic micropollutants in aqueous solutions follow *pseudo*-first-order decay with respect to the concentration of the organic solute (Nowell & Hoigné 1992b; Wang *et al.* 2012; Fang *et al.* 2014; Qin *et al.* 2014; Guo *et al.* 2016; Xiang *et al.* 2016) and are proportional to the fluence rate (Qin *et al.* 2014; Guo *et al.* 2016). The degradation rate of a target micropollutant by UV/Cl_2 AOP depends on various parameters including radiation wavelength, pH, chlorine dose, concentration of radical scavengers, and UV water absorption coefficient. The degradation rates also depend on pollutant chemical properties such as reactivity towards free radicals (in particular •OH and Cl•) and free chlorine, and on micropollutant photostability (molar absorption coefficient and photolysis quantum yield at the radiation wavelength).

9.4.3.1 Effect of pH

The UV/Cl_2 process is more efficient in mild acidic conditions (pH ~ 5) than at neutral pH and in alkaline media (pH > 7). The decrease of the efficiency of the UV/Cl_2 AOP with increasing pH was reported in buffered solutions prepared in deionized water. Some examples include the oxidation of *p*-chlorobenzoic acid (LP and MP lamps; Watts & Linden, 2007a), nitrobenzene (Watts *et al.* 2007b), benzoic acid (Fang *et al.* 2014; Figure 9.9), trichloroethene (MP lamp; Wang *et al.* 2012), 1,4-dioxane (λ = 253.7 nm; Kishimoto & Nishimura, 2015), ronidazole (λ = 253.7 nm; Qin *et al.* 2015), chlortoluron (λ = 253.7 nm; Guo *et al.* 2016), ibuprofen (λ = 253.7 nm; Xiang *et al.* 2016), atrazine (λ = 253.7 nm; Kong *et al.* 2016) and carbamazepine (λ = 253.7 nm; Wang *et al.* 2016a). Similar trends were observed for the oxidation of micropollutants in natural waters as reported for 2-methylisoborneol (Rosenfeldt *et al.* 2013; Wang *et al.* 2015a), trichloroethene (Wang *et al.* 2011), geosmin and caffeine (Wang *et al.* 2015a), and chlorinated volatile organic compounds (Boal *et al.* 2015).

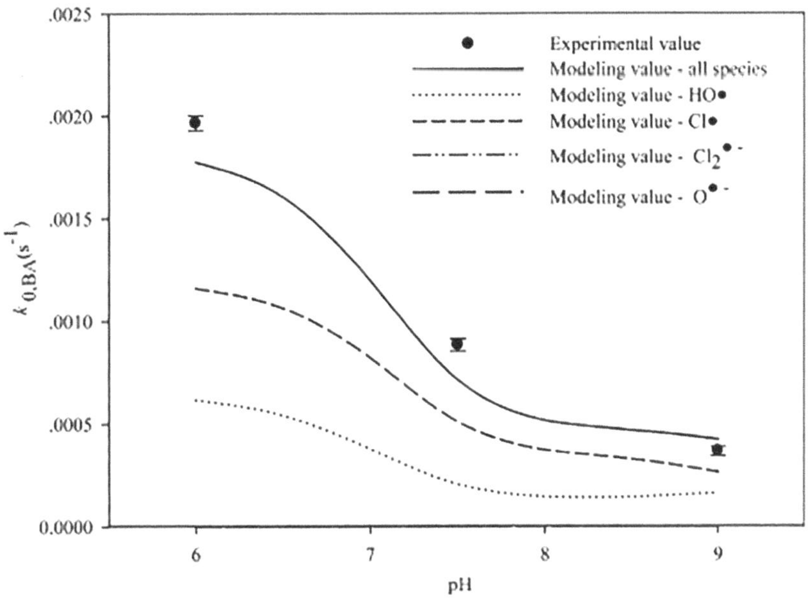

Figure 9.9 The pH effect on the *pseudo*-first-order rate constant of benzoic acid degradation by UV/Cl_2 AOP ([Benzoic acid]$_0$ = 5 μM, [Free chlorine]$_0$ = 70 μM). Reprinted with permission from Fang *et al.* (2014). Copyright 2016 American Chemical Society.

The efficiency of the UV/Cl_2 process is optimal at slightly acidic pH where HOCl is the predominant free chorine species. Watts and Linden (2007b) and Watts *et al.* (2010) attributed the higher efficiency of the UV/Cl_2 process at pH ≈ 5 to higher photo-generation rates of free radicals from HOCl than from ClO^- and to a much lower scavenging rate of $^•OH$ by HOCl ($k = 8.5 \times 10^4$ M^{-1} s^{-1}) than by ClO^- ($k = 8 \times 10^9$ M^{-1} s^{-1}). As illustrated in Figure 9.10, Fang *et al.* (2014) could simulate reasonably well the effect of pH (6 < pH < 9) on the apparent first-order rate constant of degradation of benzoic acid by using rate constants of 2×10^9 and 8×10^9 M^{-1} s^{-1} for the reaction of $^•OH$ with HOCl and ClO^-, respectively. Simulation curves for the degradation of benzoic acid by $^•OH$ and $Cl^•$ also confirmed that the decrease of the rate of degradation of benzoic acid with increasing pH can be attributed to a decrease of the quantum yields of radical formation and to an increase in the scavenging capacity of free chlorine for both •OH and Cl•. In the pH range of 6 to 9, the contribution of $Cl_2^{\bullet-}$ and $O^{\bullet-}$ to the oxidation of benzoic acid can be neglected. The need for chemicals (acids and bases) for pH adjustment in the full-scale applications of the UV/Cl_2 AOP to decontamination of natural waters would increase considerably the operating costs and complexity.

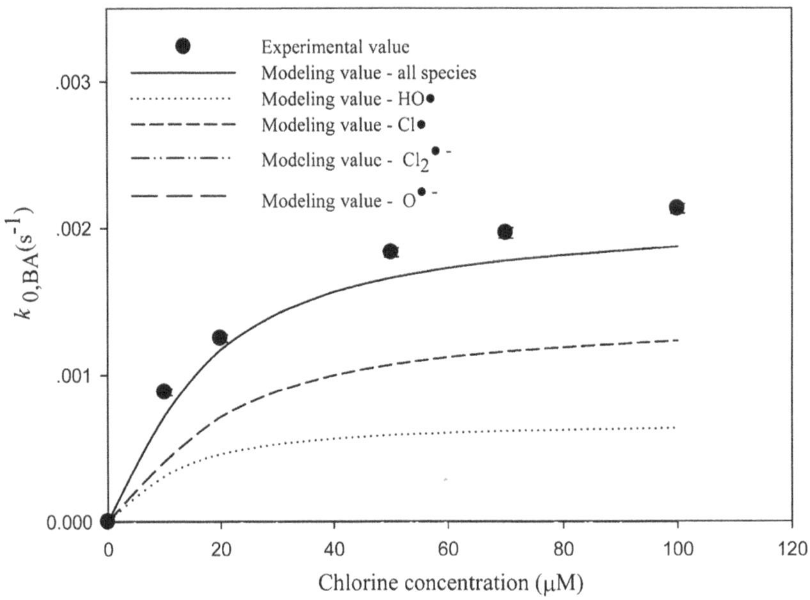

Figure 9.10 *Pseudo*-first-order rate constants for the degradation of benzoic acid by UV/Cl_2 as a function of chlorine dose; [Benzoic acid]$_0$ = 5 μM, pH 6.0. Reprinted with permission from Fang *et al.* (2014). Copyright 2016 American Chemical Society.

9.4.3.2 Effect of free chlorine dose

In the UV/Cl_2 AOP, increasing free chlorine dosage will increase the photo-generation rates of oxidizing radicals and the consumption rates of free radicals by residual free chlorine. Therefore, there exists an optimum dose of free chlorine for a given UV/Cl_2 application, which is related to water quality and process conditions.

Laboratory scale UV (253.7 nm)/Cl_2 experiments showed that the *pseudo*-first order rate constants for the degradation of nitrobenzene ([Nitrobenzene]$_0$ = 5 μM; [Free chlorine]$_0$ = 30–150 μM, pH 5; Watts *et al.* 2007b), chlortoluron ([Chlortoluron]$_0$ = 5 μM; [Free chlorine]$_0$ = 25–100 μM, pH 7; Guo *et al.* 2016) and carbamazepine ([Carbamazepine]$_0$ = 8.5 μM; [Free chlorine]$_0$ = 30–630 μM, pH 7; Wang *et al.* 2016a) increased linearly with increasing concentration of free chlorine. In these cases, the optimum chlorine dose was higher than the chlorine doses tested by the authors. On the other hand, the experimental and modeling results obtained by Fang *et al.* (2014) for the degradation rates of benzoic acid show two kinetic regimes (Figure 9.10), with faster rate increase at low chlorine doses than at higher chlorine doses. This effect can be explained by the higher •OH and Cl• scavenging capacity of the oxidant (free chlorine) at large doses than and low doses.

Xiang *et al.* (2016) also showed that the degradation rate of ibuprofen ([Ibuprofen]$_0$ = 10 µM; [Free chlorine]$_0$= 0–100 µM; pH 6; 253.7 nm) increased more gradually with increasing chlorine dose at chlorine dose higher than 70 µM.

Depending on the chemical and photochemical properties of the organic micropolluant, the chlorine dose would impact the contribution of direct photolysis reactions (for compounds which can be photolyzed at the irradiation wavelength) or of direct oxidation reactions with chlorine (for compounds very reactive towards molecular chorine) to the overall micropollutant degradation rates.

9.4.3.3 Effect of chloride ion concentration

Chloride is present in free chlorine solutions and in natural waters. Chloride reacts reversibly with $^{\bullet}OH$ and $Cl^{\bullet}$ to produce $HClOH^{\bullet}/ClOH^{\bullet-}$ and $Cl_2^{\bullet-}$ radicals (reactions 9.52–9.68), thus the presence of chloride may affect the degradation rates of organic micropollutants by the UV/Cl_2 AOP. However, the experimental data obtained at pH 6 for benzoic acid ([Cl^-]: 0–20 mM; Fang *et al.* 2014) and for ibuprofen ([Cl^-]: 0–10 mM; Xiang *et al.* 2016), and at pH 7 for atrazine ([Cl^-]: 0–10 mM; Kong *et al.* 2016) and carbamazepine ([Cl^-]: 0–50 mM; Wang *et al.* 2016a) show that the degradation rates of these micropollutants were not significantly affected by the chloride ion concentration under the tested conditions. The modeling data obtained by Fang *et al.* (2014) for the oxidation of benzoic acid at pH 6 showed that an increase of the chloride concentration from 0 to 20 mM did not change the contribution of $^{\bullet}OH$ and $Cl^{\bullet}$ to the degradation of benzoic acid but slightly increased the contribution of $Cl_2^{\bullet-}$.

9.4.3.4 Effect of alkalinity

As for all the other AOPs which are operated at neutral pH, the degradation rates of micropollutants by the UV/Cl_2 process also decrease in the presence of increasing alkalinity. Fang *et al.* (2014) showed that the *pseudo*-first order rate constants of degradation of benzoic acid decreased by a factor of 2.6 when the concentration of bicarbonate increased from 0 to 4 mM. Similar experiments conducted by Xiang *et al.* (2016) for the degradation of ibuprofen showed that the rate constant decreased only by a factor of 1.25 when the concentration of bicarbonate increased from 0.2 to 4 mM. As shown in Table 9.10, bicarbonate reacts much faster with $Cl^{\bullet}$ than with $^{\bullet}OH$. Therefore, the inhibiting effect of alkalinity on the efficiency of the UV/Cl_2 AOP would be more pronounced for micropollutants which have large second-order reaction rate constants with $Cl^{\bullet}$, and whose radical degradation mechanism is dominated by $Cl^{\bullet}$ rather than by $^{\bullet}OH$. In natural waters, the alkalinity exerts a buffering capacity and the water pH could be within the 6.8–8.2 range. In that range the free chlorine speciation varies significantly, and that impacts the UV/Cl_2 process performance.

9.4.3.5 Effect of natural organic matter (NOM)

Natural organic matter (NOM) can affect the efficiency of UV/Cl_2 AOP in various ways. NOM absorbs UV light (absorption coefficient at 253.7 nm ≈ 3 L m^{-1} (mg DOC)$^{-1}$), thus, acts as an inner filter reducing the photolysis rate of free chlorine to oxidizing radicals ($^{\bullet}OH$ and $Cl^{\bullet}$). It has also been shown that NOM can significantly increase the overall quantum yields of photodecomposition of free chlorine through radical chain reactions which are undesired in AOP applications (Ormeci *et al.* 2005; Feng *et al.* 2010). NOM competes with the target organic micropollutants for the $^{\bullet}OH$ and $Cl^{\bullet}$, thus lowering the steady-state concentrations of oxidizing radicals and the degradation rates of organic micropollutants. The rate constants for the reaction of $^{\bullet}OH$ with NOM extracts are in the range 1.2×10^4–3.8×10^4 L (mg DOC)$^{-1}$ s^{-1}

(Westerhoff *et al.* 2007). The direct reaction of free chlorine with NOM (water chlorine demand) may also lead to an instant consumption of free chlorine and to the formation of chlorination by-products (Westerhoff *et al.* 2004).

In their detailed study on the degradation of benzoic acid by the UV/Cl_2 AOP, Fang *et al.* (2014) showed that the *pseudo*-first order rate constant decreased from 2×10^{-3} s^{-1} to 1.2×10^{-3} s^{-1} when the concentration of NOM (Suwannee River NOM isolate, SW NOM) increased from 0 to 10 mg C/L. The authors estimated the $Cl^{\bullet}$ rate constant for the reaction with SW NOM as $k_{Cl\bullet,NOM} = 1.3 \times 10^4$ (mg C/L)$^{-1}$ s^{-1}, which is roughly half of that for the $^{\bullet}OH$ reaction.

9.4.4 UV/Cl_2 versus UV/H_2O_2

Similarly to the UV/H_2O_2 process, the UV/Cl_2 AOP may only be used to treat water that absorbs weakly the UV radiation wavelengths emitted by LP lamps (253.7 nm) or MP lamps (200–400 nm). The potential applications of these two AOPs therefore concern the decontamination of groundwater, drinking water sources, and tertiary wastewater effluents for direct or indirect potable reuse. Literature is available on bench-scale studies comparing the UV/Cl_2 and UV/H_2O_2 AOP efficiencies, most of which conducted in deionized water. Comparative studies on pilot- and full-scale UV/Cl_2 and UV/H_2O_2 at water treatment facilities are very limited, and were published almost exclusively in conference proceedings rather than in peer-reviewed journals. Selected examples from the published literature are discussed in this section.

9.4.4.1 Oxidation of nitrobenzene

A lab-scale comparative study on the degradation of a probe compound (nitrobenzene, NB, $k_{\bullet OH} = 3.9 \times 10^9$ M^{-1} s^{-1}) by the UV/Cl_2 and UV/H_2O_2 processes has been performed by Watts *et al.* (2007b) under identical experimental conditions ($[NB]_0 = 5$ μM in deionized water; [Free chlorine]$_0$ or $[H_2O_2]_0 = 0.03$–0.35 mM), using a *quasi*-collimated beam apparatus with LP lamps (253.7 nm). The data showed that the degradation rates of NB ([Oxidant]$_0 = 0.11$ mM) increased in the following order: UV/Cl_2 (pH 7) < UV/H_2O_2 (pH 5 and 7) < UV/Cl_2 (pH 6) < UV/Cl_2 (pH 5). The degradation rate of NB by UV/H_2O_2 at pH 6 was faster than that by UV/Cl_2 at pH 7 at oxidant dose > 0.22 mM. Similar experiments performed with a reservoir water (TOC = 4.32 mg/L; Alkalinity = 80 mg $CaCO_3$/L; $UV_{253.7\ nm} = 0.21$ cm^{-1}) and a post-sand filtered water (TOC = 2.2 mg/L; Alkalinity = 77.5 mg $CaCO_3$/L; $UV_{253.7\ nm} = 0.05$ cm^{-1}) showed that the degradation rates of NB ($[NB]_0 = 5$ μM) by the two AOPs (pH 5, 7 and 9.5) were much slower than in deionized water because of the presence of photon absorbers and of radical scavengers in the natural waters. Data obtained with sand-filtered water showed that the UV/Cl_2 AOP at pH 5 was the most efficient process and that a 90% removal of NB was achieved with a chlorine dose of 7.7 mg/L and a UV dose of about 1200 mJ cm^{-2}.

9.4.4.2 Removal of volatile organic compounds (VOCs)

Middleton Water Supply Region of Waterloo (ON, Canada) uses the UV/Cl_2 process with TrojanUVSwift™ECT 16L30 reactors for groundwater remediation. The primary contaminant is trichloroethene (TCE, ~5 μg/L), but very low levels of 1,4-dioxane are occasionally detected in the water. Wang *et al.* (2011) performed a series of tests at this water facility to examine the UV/Cl_2 treatment performance at various flowrates (50, 55, 80, and 95 L/s) and to compare the UV/Cl_2 and UV/H_2O_2 processes at one flowrate (55 L/s). The oxidant concentrations were fixed for all tests, i.e., 9 mg/L (127 μM) free chlorine and 8 mg/L (235 μM) H_2O_2; TCE concentration in the groundwater was within the 4–6 μg/L range. Groundwater contained very low TOC (~0.65 mg/L), relatively high alkalinity

(~288 mg/L as $CaCO_3$), and pH was 7.55. The authors reported TCE data for both direct photolysis (no oxidant) and UV-AOPs. Free chlorine was almost completely removed in the UV reactor (0.06 mg Cl_2/L residual) in all tests. Depending on the flowrate, 50 to 70% TCE was destroyed by UV alone. At 55 L/s, ~64% TCE was removed by direct photolysis, whereas ~90% and 75% removal was achieved with the UV/Cl_2 and the UV/H_2O_2 processes, respectively. A 90% removal yield was determined for the UV/H_2O_2 process at a flowrate of 30 L/s. The cost estimates for the same TCE treatment level were 13 and 18 cents per m^3 water treated by UV/Cl_2 and UV/H_2O_2 process, respectively. No process optimization (e.g., oxidant dose) was attempted in this study, such that the outcomes may not reflect correctly the AOP performance.

Bench-scale experiments conducted by Wang *et al.* (2012) using a MP collimated beam apparatus (UV doses up to 2000 mJ cm^{-2}), TCE (1.1 μM) in ultra-pure water, and identical initial concentrations of free chlorine and H_2O_2 (0.15 mM) showed that UV/Cl_2 was 2.3-fold more efficient for TCE removal than the UV/H_2O_2 process at pH 5.5, whereas the UV/H_2O_2 process became more efficient than UV/Cl_2 at pH 7.5 (4-fold) and pH 10. In the UV/Cl_2 process at pH 7.5 and pH 10, direct photolysis contributed approx. 71–74% to the overall rate constant, with less than ~30% contribution of the radical-induced degradation. The comparison of the two AOPs is based on the fluence-based rate constants calculated for each set of conditions. Identification of TCE degradation by-products was not undertaken in these two studies but it is conceivable to assume that the routes of degradation of TCE by UV/Cl_2 are as complex as those observed with the UV/H_2O_2 process (Li *et al.* 2007).

Boal *et al.* (2015) reported comparative studies on UV/Cl_2 and UV/H_2O_2 AOPs at two Aerojet Rocketdyne (Sacramento County, CA, USA) groundwater extraction and treatment (GET) plants. The two GET facilities use different treatment technologies to meet the water remediation goals. Both water facilities implemented the UV/H_2O_2 AOP with Calgon Carbon RAYOX™ Medium Pressure Ultraviolet (MP-UV) reactors (Calgon Carbon Corporation, Pittsburgh, Pennsylvania) to remove NDMA and VOCs (TCE, 1,2-DCE, 1,1-DCE, and vinyl chloride) from contaminated groundwater. One facility operates at low flowrate and high chemical dose (LFHC), whereas the other plant operates at high flowrate and low chemical dose (HFLC). The UV-AOP effluent from the HFLC facility is passed through GAC filters to enhance the UV/H_2O_2 treatment prior to groundwater recharge. The Electrical Energy Dose was 5.89 kWh/kgal (1.56 kWh/m^3) and 0.97 kWh/kgal (0.26 kWh/m^3) at the LFHC and HFLC facility, respectively. Hypochlorite for the UV/Cl_2 process was produced with a MIOX VAULT H25 on-site generation system. The UV/H_2O_2 process was tested without any pH adjustment and concentration ranges of 6.5–8.2 mg/L (LFHC plant) and 6.6–6.9 mg/L (HFLC plant). The H_2O_2 residual ranged from 3.3 to 3.5 mg/L (LFHC), and from 5.3 to 6.0 mg/L (HFLC). UV/Cl_2 tests were conducted at chlorine doses ranging from 0.8 to 7.7 mg/L at LFHC facility, and 0.9 to 5.7 mg/L at HFLC plant. NDMA was removed below method detection limit (MDL) of 2 ng/L in all tests at both water treatment plants, except in the 7.8 mg/L H_2O_2 run (23 ng/L). VOCs were all below MDL in the UV/H_2O_2 tests. In the UV/Cl_2 process, at LFHC facility, the VOCs were below MDL in all tests except for the 0.8 mg/L chlorine test where TCE (0.53 μg/L) approached the MDL (0.5 μg/L), and 1,1-DCE residual was 0.66 μg/L. At the HFLC plant, where the water pH was 7.7, all VOCs except TCE were removed under all free chlorine dose conditions.

The data showed that the optimum chlorine dose for TCE removal was about 2.5 mg/L at the LFHC plant (pH ~ 7.0), and no benefit was observed from increasing the chlorine dose from 3.6 mg/L to 5.7 mg/L at the HFLC plant (pH 7.69). The process economics indicated that the chemical costs for the UV/Cl_2 AOP were approximately one-fourth to one-half of the costs of the UV/H_2O_2 AOP which is the current process used at the two facilities. Table 9.11 summarizes the key outcomes from this study.

The UV/Cl_2 and UV/H_2O_2-effluents were not acutely toxic to the aquatic organism *Ceriodaphnia dubia*, except one UV/Cl_2 effluent from LFHC facility which affected 10% of the organisms.

Table 9.11 NDMA and VOC treatment data for the UV/H_2O_2 and UV/Cl_2 processes and associated treatment costs at the two facilities (Boal *et al.* 2015).

	LFHC Facility			HFLC Facility		
	Raw Water	UV/H_2O_2	UV/Cl_2	Raw Water	UV/H_2O_2	UV/Cl_2
Oxidant dose (mg/L)		6.5–8.2	2.8		6.6–6.9	3.6
Effluent oxidant (mg/L)		3.3–3.5	<0.02		5.3–6.0	<0.02
Electrical Energy Dose (kWh/1000 gallons)		5.89	5.89		0.97	0.97
pH	7.06			7.69		
Alkalinity (mg/L $CaCO_3$)	86			130		
NDMA (ng/L)	930–1300	<2; 23	<2	27–37	<2	<2
TCE (μg/L)	12–14	<0.5	<0.5	8–9.7	<0.5	1.8
1,1-DCE (μg/L)	16–19	<0.5	<0.5	<0.5	<0.5	<0.5
1,2 DCE (μg/L)	0.8–0.9	<0.5	<0.5	<0.5–0.55	<0.5	<0.5
Treatment cost ($/kgal)		0.067*	0.017*		0.073**	0.038**

*Treatment including only oxidant costs (7.4 mg/L H_2O_2; 2.5 mg/L Cl_2)
**Treatment including oxidant costs (7.4 mg/L H_2O_2; 3.0 mg/L Cl_2) and GAC costs.

9.4.4.3 Removal of emerging contaminants

Xiang *et al.* (2016) compared the degradation rates of ibuprofen (IBP) in deionized water by UV/Cl_2 and UV/H_2O_2 AOPs under identical experimental conditions ($[IBP]_0 = 10$ μM; $[\text{Free chlorine}]_0$ or $[H_2O_2] = 100$ μM; pH 6.0; UV (253.7 nm) fluence rate = 1.05 mW/cm²). The data showed that the degradation rate of IBP at pH 6 was 3.3 times faster for the UV/Cl_2 than for the UV/H_2O_2 process, which was attributed to the difference in photochemical characteristics of the two oxidants: $\varepsilon_{HOCl} \approx 60\ M^{-1}\ cm^{-1}$; $\varepsilon_{H2O2} = 19.6\ M^{-1}\ cm^{-1}$ and $\Phi_{HOCl} \approx 1$–1.5; $\Phi_{H2O2} \approx 1.0$.

Watts and Linden (2009) observed similar removal yields of tris(2-butylethyl) phosphate (TBEP, $[TBEP]_0 = 50$ μg/L) with the UV/Cl_2 and UV/H_2O_2 processes. The experiments were performed in simulated surface water (2 mg/L DOC; 30 mg/L bicarbonate; 0.3 mg/L nitrate) at pH 6.8, with oxidant doses of 6.1 mg/L and 3.4 mg/L for H_2O_2 and free chlorine, respectively, and UV fluence up to 1000 mJ cm^{-2}.

Sichel *et al.* (2011) compared the degradation yields of eight emerging contaminants (ECs) treated with the UV/Cl_2 and UV/H_2O_2 processes (Table 9.12). The treatment was performed in a flow-through mode using a low-pressure lamp UV reactor powered at 40, 80 or 200 W. The matrix was tap water as-is or enriched in dissolved organic carbon, which was spiked with a mixture of ECs at environmental levels (μg/L). The initial concentrations of free chlorine were 1 or 6 mg/L, whereas the H_2O_2 concentration was 5 mg/L. This concentration of H_2O_2 was selected to be similar to that used in the first full-scale application of the UV/H_2O_2 process at a drinking water treatment plant in Europe (Andijk, The Netherlands; Kruithof *et al.* 2007). The MP-UV (Trojan UVSwiftECT™ 16L30) reactors in Andijk are operated with 6 mg/L H_2O_2 and an electrical energy dose (E_{ED}) of approximately 0.5 kWh/m³ to remove a wide range of organic micropollutants from surface water.

For an electrical energy consumption of 0.32 kWh/m³, the addition of 1 mg Cl_2/L significantly increased the removals of all ECs as compared to UV alone. At a lower energy consumption (0.16 kWh/m³), the UV/Cl_2 process ($[Cl_2]_{inlet} = 6.1 \pm 0.1$ mg/L; $[Cl_2]_{outlet} = 5.4 \pm 0.2$ mg/L) was found to be more efficient than the UV/H_2O_2 process ($[H_2O_2] = 5.2$ mg/L). The electrical energy required to achieve 90%

reduction of ECs with the UV/H_2O_2 process under the conditions used in Sichel *et al.*'s work varied from 0.17 to 1.00 kWh/m^3, depending on the treated contaminant. The authors estimated 30–75% energy reduction and up to 30–50% operational process cost savings for the UV/Cl_2 process as compared to the UV/H_2O_2 process.

Table 9.12 Degradation yields of emerging contaminants (tap water, pH 7) under various treatment conditions. Data estimated from the figures given in Sichel *et al.* (2011).

	Cl_2 Alone 15 min Contact Time	UV Alone	UV/Cl_2	UV/H_2O_2	UV/Cl_2	UV/H_2O_2	E_{EO}(kWh/m^3/order) UV/H_2O_2
[Cl_2] (mg/L)	6	0	1	0	6	0	
[H_2O_2] (mg/L)	0	0	0	5	0	5	
E_{ED} (kWh/m^3)	0	0.32	0.32	0.32	0.16	0.16	
17α-ethinylestradiol		8	100	92			
Benzotriazole	1	11	94	71	71	25	0.52
Tolyltriazole	4	18	94	63	70	29	0.59
Desethylatrazine	17	13	21	48	23	19	1.00
Carbamazepine	4	3	48	58	90	32	0.62
Sulfamethoxazole	100	55	100	92	100	64	0.29
Diclofenac	32	61	100	100	100	88	0.17
Iopamidole	4	31	92	81	96	58	0.42

E_{ED} = Electrical Energy Dose
E_{EO} = Electrical energy per order

Yang *et al.* (2016) reported the removal yields for ten pharmaceuticals and personal care products (PPCP) treated with UV/Cl_2 and UV/H_2O_2 processes (253.7 nm radiation) in sand-filtered water samples collected from three water treatment facilities. The water quality parameters varied greatly; e.g., 3.5, 1.1, and 1.9 mg/L TOC; 0.013, 0.027, and 3.14 mg/L NH_3-N (important parameter in UV/Cl_2 process); 3.5, 47.4, 14.1 mg/L HCO_3^-. The transmittance (94–95% $T_{1\ cm,253.7\ nm}$) and pH (7.6–7.9) were *quasi*-similar in the three waters. The pH was adjusted to 7 in all tests. The authors observed ~100% degradation yields for triclosan, 2-ethylhexyl-4-methoxycinnamate (EHMC), ciprofloxacin, tetracycline, benzophenone-3 (BP3), and trimethoprim with chlorine alone (3 mg/L) and UV/chlorine (3 mg/L) at the same reaction time, in the high-TOC, low ammonia, and low alkalinity water. 90–98% degradation was determined in the same water with the UV/H_2O_2 (5 mg/L) process for triclosan, sulfamethoxazole, EHMC, and ciprofloxacin. Less than 20% removal was observed for caffeine for all processes used with this water. Among the three waters tested, the lowest removal yields with the UV/Cl_2 (5 mg/L) process were observed in the water with the highest ammonia content. The UV/Cl_2 process outperformed the UV/H_2O_2 process in two of the three waters, and showed similar performance in ammonia-rich water. The authors indicated that both $^\bullet$OH and Cl$^\bullet$ were the oxidizing radicals in the UV/Cl_2 process. Approx. 20% increase in total THM level was observed UV/Cl_2- and UV/H_2O_2-treated water (high ammonia, moderate alkalinity sample) upon 24h-incubation time and a free chlorine residual of ~1 mg/L, but the t-THMs did not exceed 40 µg/L. Chloral hydrate increased by ~30% post-UV/Cl_2 process applied to high-TOC water, relative to untreated and UV/H_2O_2-treated water samples.

9.4.4.4 *Removal of taste and odor-causing compounds (T&O) from drinking water sources*

A pilot-scale study was performed at Richard Miller Water Treatment Plant of Greater Cincinnati Water Works to compare the efficiency of the UV/Cl_2 and UV/H_2O_2 AOPs for the removal of 2-methylisoborneol (MIB) (Rosenfeldt *et al.* 2013). MIB was spiked at 30 ng/L to the pretreated water (0.78 mg/L TOC; 68 mg/L alkalinity as $CaCO_3$). The data showed that the MP UV/Cl_2 (1 or 5 mg/L) AOP was able to remove 80–90% MIB at pH 6, and 45–70% MIB at pH 7.5 at a UV dose of 250 mJ cm^{-2}. The authors observed that the UV/Cl_2 process outperformed the UV/H_2O_2 AOP at both low and high oxidant concentrations when pH was in the acidic range. Operating costs (UV + chemicals, \$/1000 gal of treated water) for 1-log removal of MIB were lower for the UV/Cl_2 AOP at pH 6 (0.07–0.14 and 0.11–0.15 for chlorine doses of 2 and 5 mg/L, respectively) than for the UV/H_2O_2 AOP at pH 6 or 7.5 (0.26 and 0.18–0.19 for H_2O_2 doses of 2 and 5 mg/L, respectively). The costs for the UV/H_2O_2 process would increase to \$0.30/kgal and \$0.27-\$0.28 per kgal of treated water for 2 and 5 mg/L H_2O_2, respectively, if the costs associated with the removal of H_2O_2 residual are considered.

A recent study compared the UV/Cl_2 and the UV/H_2O_2 AOPs with respect to the removal of geosmin, MIB and caffeine from drinking water at full-scale UV installation (Wang *et al.* 2015a). The tests were conducted at Cornwall Water Purification Plant (Ontario, Canada) which uses the UV/H_2O_2 AOP with MP-UV reactors (Trojan UVSwiftECT™ 8L24) to control T&O-causing compounds occurring in the surface water during the seasonal algal blooming events. Pre-treated (pre-chlorination, coagulation, flocculation, sand/anthracite filtration) St. Lawrence River water (pH 7.9; 1.5 mg/L TOC; 92 mg/L alkalinity as $CaCO_3$; 0.3 mg/L chlorine residual) was passed through one MP-UV reactor at a flowrate of 100 L/s (estimated UV dose of 2000 ± 150 mJ cm^{-2}; 83.5 kW reactor power; $E_{ED} = 0.23$ kWh/m^3). In the first set of tests, geosmin and MIB were spiked to the water upstream to the UV reactor at ≈ 400 ng/L), and the oxidant doses were either 2, 6 or 10 mg/L free chlorine (UV/Cl_2) or 1.0, 2.9 or 4.8 mg/L H_2O_2 (UV/H_2O_2). The second set was scheduled during a T&O event (~18 ng/L geosmin in water) and only the UV/Cl_2 AOP was tested. Water pH was adjusted to 6.5, 7.5 and 8.5 in both sets. The MIB and geosmin treatment data from the first set are shown in Figure 9.11.

Wang *et al.* (2015a) calculated the electrical energy per order (E_{EO}) for all treatment conditions illustrated in Figure 9.11. The lowest E_{EO} were observed at pH 6.5 and for the highest oxidant dose tested in both AOPs. The E_{EO} values at pH 6.5 for geosmin were 0.16 and 0.23 kWh/m^3/order for the UV/Cl_2 (10 mg/L) AOP and UV/H_2O_2 (4.8 mg/L), respectively. At pH 7.5 and 8.5, the geosmin E_{EO} values increased for both AOPs at all oxidant doses. The reported E_{EO} values at pH 6.5 for MIB were 0.22 and 0.31 kWh/m^3/order for the UV/Cl_2 (10 mg/L) and UV/H_2O_2 (4.8 mg/L), respectively; similar E_{EO} – pH patterns as for geosmin were observed.

In addition to the full-scale tests described above, the authors performed pilot-scale tests on caffeine treatment under oxidant dose conditions similar to full-scale. Post-filtered Lake Simcoe water from Keswick WTP (ON, Canada) – 3.5 mg/L TOC, 123 mg/L alkalinity as $CaCO_3$, 0.04 cm^{-1} (254 nm) absorption coefficient – was treated in a 40L completely mixed batch reactor (Rayox®, Calgon Carbon Corporation, Pittsburgh, PA) equipped with 1 kW MP lamp. Larger caffeine E_{EO} data were reported for pilot- than full-scale tests, which is explained by the difference in the water quality used in the two treatment settings. The authors concluded that caffeine could be a good surrogate to estimate the treatment of T&O compounds with UV/Cl_2 and UV/H_2O_2 processes, and recommended pilot- or full-scale tests be conducted at water utilities to confirm site-specific performance of UV/Cl_2 process prior to implementation.

9.4.4.5 *UV/Chlorine AOP for Water Reuse: Terminal Island Water Reclamation Plant (TIWRP) Case Study*

In recent years, the UV/Cl_2 AOP has been evaluated at pilot- and full-scale as a potential cost-effective advanced treatment of tertiary wastewater effluents in water reuse applications. The City of Los Angeles

Department of Public Works (DPW) Bureau of Sanitation (LASAN) is expanding the TIWRP Advanced Water Purification Facility (AWPF) capacity for advanced treatment of tertiary effluent from 15.77 m^3/min (6 MGD) to 31.55 m^3/min (12 MGD) (Aflaki *et al.* 2015). The advanced treated water will be used for groundwater recharge and to supply the local industries and irrigation facilities, thus, replacing and reducing the potable water consumption for industrial purposes. The AWPF Expansion will include an additional microfiltration (MF) system, a reverse osmosis (RO) system followed by an AOP for disinfection and removal of the organic micropollutants passing through RO membranes, a tertiary effluent equalization tank (7570 m^3 (2MG)) upstream of AWTP, and a series of upgrades to the existing pumping and chemical dosing facilities.

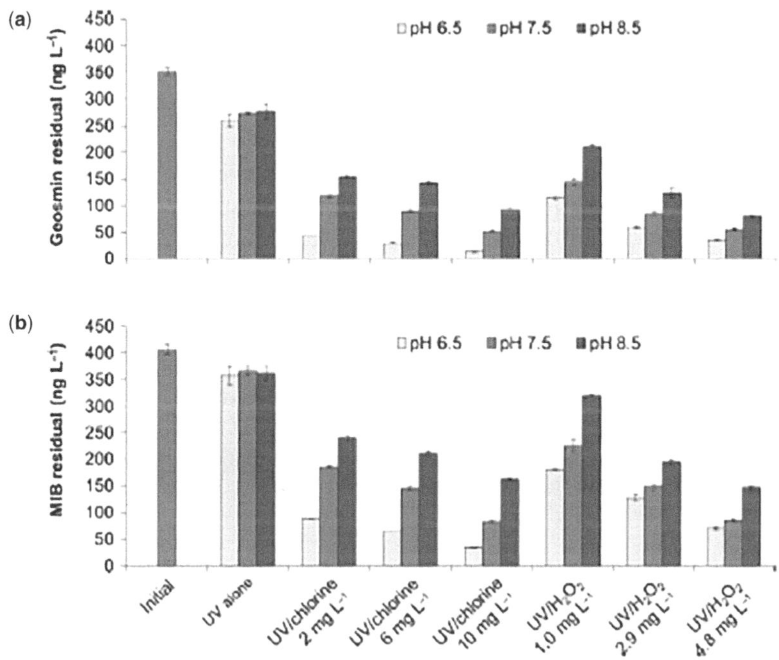

Figure 9.11 Geosmin (a) and MIB (b) removal in the 1st full-scale test at the Cornwall Water Purification Plant (ON, Canada). Reprinted with permission from Wang *et al.* (2015a). Copyright 2016 Elsevier.

Advanced treatment of tertiary effluents is currently practiced in (indirect or direct) potable water reuse applications at water utilities around the world, with UV/H_2O_2 as an AOP for microorganism (viruses, bacteria and protozoa) control and NDMA, 1,4-dioxane, and other micropollutant treatment (see Chapter 14 for extended information). NDMA levels in tertiary effluents at TIWRP are below CA notification limit of 10 ng/L, such that this compound was not the primary consideration for AOP implementation at AWPF. LASAN's AOP investment decision for the TIWRP AWPF Ultimate Expansion project was based on 18-month extensive bench- and pilot-scale studies, which allowed LASAN to select the best AOP technology for this specific project and the AOP system scale-up approach for full compliance with California Department of Drinking Water (DDW) and Groundwater Replenishment Reuse Regulations (GRRR). CA DWW, GRRRs of 2014 require 12-log virus-, 10-log *Giardia lamblia*, and 10-log *Cryptosporidium parvum* inactivation, <10 ng/L NDMA, and ≥0.5-log 1,4-dioxane removal (or equivalent treatment) in RO-AOP treated water, and were used by LASAN to set design criteria for the bench- and pilot-scale tests.

Four AOPs were investigated at bench-scale on RO product water: UV/H_2O_2, UV/Cl_2, O_3/H_2O_2, and $H_2O_2/O_3/UV$. The UV/H_2O_2 and UV/Cl_2 AOPs were tested with both low- and medium-pressure mercury vapor lamps, whereas the O_3/H_2O_2 process was tested for both H_2O_2 injection before O_3 and H_2O_2 injection after ozone. Specifics on these preliminary tests are available in Aflaki *et al.* (2015). The bench-scale data showed that all tested AOPs met the microorganism removal criteria set by LASAN for each AOP, THMs and HAAs were not formed as a result of UV/Cl_2 AOP, chlorate was not detected in the AOP effluents, and bromate was formed in O_3/H_2O_2 AOP only under specific pH and $[O_3]:[H_2O_2]$ molar ratio conditions, which were further adjusted in the pilot trials.

Pilot testing was considered absolutely required in order to (*a*) confirm the bench-scale results, (*b*) establish a correct relationship between bench-, pilot- and, thereafter, full-scale data, (*c*) establish relationship between chemical dose and speciation and hydraulic conditions, (*d*) refine design criteria and costs for each AOP, (*e*) assess byproduct formation under refined chemical and hydraulic conditions, (*f*) determine scalable conditions/criteria and select the AOP for full-scale system.

Pilot-scale testing was conducted for 12 months using WEDECO MiPRO AOP pilot system designed to operate all abovementioned AOPs, with fully automated PLC and operator interface, remote monitoring and data-logging and online water quality monitoring instrumentation. Neither bench- nor pilot-scale data are available in Aflaki *et al.* (2015) except some UV (253.7 nm)/Cl_2 process data which are shown in Figure 9.12. The O_3/H_2O_2 AOP removed effectively 1,4-dioxane, but failed to remove NDMA from 30 ng/L (spiked to RO product water for the purpose of these studies) to below 10 ng/L. MP UV-based AOPs met all design criteria, but was not considered due to high life cycle costs as compared to LP UV (see Aflaki *et al.* 2015 for life cycle cost comparison). Both LP – UV/Cl_2 and -UV/H_2O_2 were effective for 1,4-dioxane and NDMA treatment, but higher H_2O_2 concentrations were required.

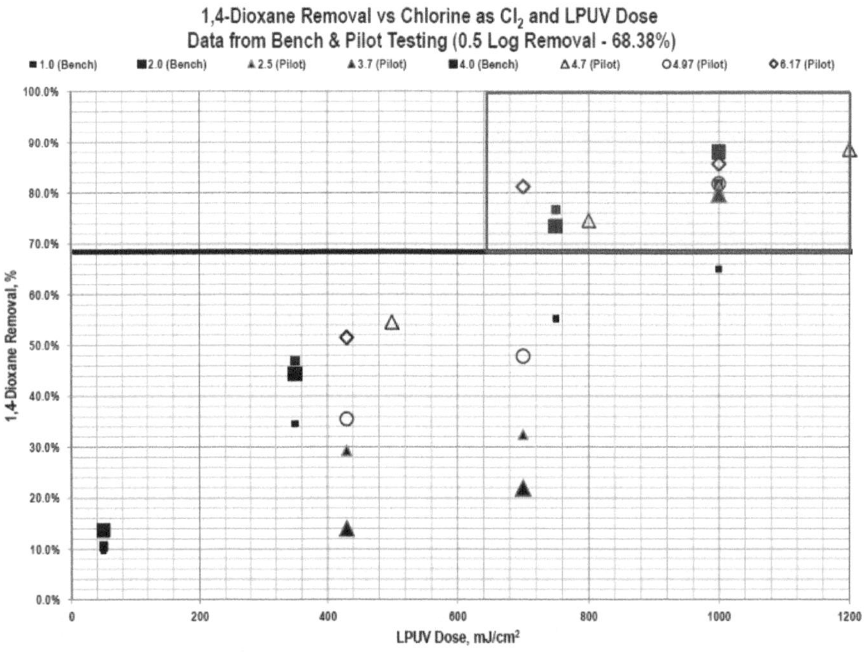

Figure 9.12 Selected UV(253.7 nm)/Chlorine pilot test data (courtesy of Dr. R. Aflaki).

Under all chemical dose and UV dose conditions shown in the squared area in Figure 9.12, LP-UV/chlorine achieved more than 0.5-log 1,4-dioxane (>68.38%) removal in RO product water.

The LP-UV/Cl_2 process was the most cost-effective among tested AOPs. In addition to meeting the performance criteria and low life cycle cost, the selection of the UV/Cl_2 AOP over the UV/H_2O_2 AOP presents the following advantages: NaOCl is already used in the existing plant as secondary disinfectant; lower dosage of oxidant; the chlorine residual is used for water disinfection in the distribution network; insignificant DBP formation; easy-to-understand, -to operate, and -to monitor process.

The LP-UV/Chlorine AOP was selected for full-scale implementation at TIWRP AWPF. Chemical cost savings of \$3.3 M over 20 year-operation is expected based on H_2O_2 and free chlorine cost calculations.

The AOP design basis are: flowrate of 3 to 12 MGD, TOC < 0.25 mg/L, %*T*(1 cm, 254 nm) > 96%, California groundwater recharge regulations.

The AOP design criteria are: 6-log virus inactivation, 0.5-log 1,4-dioxane removal, <10 ng/L NDMA, UV dose of 920 mJ cm^{-2}, free chlorine dose of 2–4 mg/L. The project was awarded to Xylem/Wedeco, with the startup in late 2016. Two Wedeco K143 Series LP UV reactors with 17 rows of lamps (one UV intensity sensor per row) and 12 lamps (600 W/ lamp) per row in each reactor will be installed and operated at TIWRP AWPF. The K143 reactor is modular allowing for more rows for linear expansion. The system is equipped with real-time monitoring of transmittance, flowrate, and intensity, which allows real-time UV dose calculations and operational adjustments to the UV dose set point. Of note, the UV dose is the design criterion and the process performance monitoring parameter. Online monitoring instrumentation of pH, free and total chlorine will ensure optimal oxidant dosing and treatment performance. The LASAN's TIWRP AWPF will be the first advanced treatment facility in the world to install the LP-UV/chlorine AOP for water reuse.

9.4.4.6 Solar radiation-based UV/Cl_2 for remediation of oil sands process-affected water

The possibility of using the solar-driven UV/Cl_2 AOP in water and wastewater treatment has been investigated by Chan *et al.* (2012) for the degradation of methylene blue (MB) and cyclohexanoic acid (CHA), and by Shu *et al.* (2014) for detoxification of oil sands process-affected water remediation (OSPW). These two studies were carried out under actual sunlight on the campus of the University of Alberta (Edmonton, AB, Canada). Chan *et al.* (2012) demonstrated that chlorination under sunlight UV led to MB photobleaching (0.039 mM MB; 1.37 mM ClO^-) and to CHA degradation (0.23 mM CHA; 1.55 mM ClO^-). Shu *et al.* (2014) studied the solar UV radiation-induced photolysis of free chlorine as a means to degrade naphthenic acids (NAs) and fluorophore organic compounds, and to decrease the acute toxicity of OSPW. The main characteristics of the OSPW were as follows: pH 8.3–8.6; 277 NTU; 47 mg/L TOC; 683 ± 32 mg/L bicarbonate; 451 mg/L Cl^-; approx. 33 mg/L total NAs. Four experiments were conducted. Upon exposure for up to 7 hours to solar radiation in the presence of either 200 or 300 mg/L NaOCl, up to 84% NAs and fluorophore organic compounds (e.g., petroleum polyaromatic hydrocarbons) were removed from OSPW. The process reduced the acute toxicity of OSPW toward *Vibrio fischeri*, but no change in the toxicity toward goldfish primary kidney macrophages was observed relative to untreated OSPW. The study concluded that the solar-driven UV/Cl_2 AOP is a promising approach to decontamination of OSPW but further studies are needed to optimize the process for toxicity removal.

9.4.5 Byproduct formation in the UV/Cl_2 AOP

Chlorine is widely used for drinking water disinfection. The applied chlorine dose satisfies the water background chlorine demand and ensures 0.3–0.5 mg/L residual in the water distribution network. The reactions of free chlorine with the organic matter present in the water lead to a series of disinfection by-products (DBPs) such as trihalomethanes (THMs), haloacetic acids (HAAs), haloacetonitriles (HANs), haloketones (HKs) and halopicrins (Richardson *et al.* 2007).

Application of UV/Cl_2 AOP to drinking water treatment could generate chlorinated organic (e.g., halogenated NOM, DBPs) and inorganic (chlorite, chlorate, bromate) byproducts. Unlike the drinking water chlorination conditions (0.5–2 mg/L Cl_2; 0.5–4 h contact time in the reservoirs, up to 72 h in the distribution network), the UV/Cl_2 AOP would require high chlorine doses (3–6 mg/L as a practical range) and very short contact time ($\ll$1 min) under high UV energy exposures in the UV reactors. Chlorinated compounds can be formed inside the UV reactors through the reactions of chlorine atoms with NOM. Additionally, the UV radiation and the hydroxyl and chlorine radicals generated from the photolysis of free chlorine change the structure of natural organic matter to DBP precursors prone to chlorination in the reservoirs and water distribution system. Selected publications on byproduct formation as a result of UV/Cl_2 process are discussed in the next few sections. Additional information on the byproduct formation topic is provided in the studies on micropollutant removal with the UV/Cl_2 AOP exemplified in Section *9.4.4*.

9.4.5.1 NOM oxidation and Disinfection byproducts

Zhao *et al.* (2011) examined the formation of halogenated organics (TOX) upon exposure to MP-UV in combination with free chlorine, and compared the data to 'dark' chlorination. The authors used probe compounds and showed that nitrobenzene (NB) and benzoic acid (BA), which are compounds bearing electron-withdrawing groups, were largely converted to chlorinated products upon exposure to MP-UV/Cl_2 (0.28 mM NaOCl, pH 6.5, ~220 mJ cm^{-2}). Increased TOX relative to dark chlorination were observed also in the presence of bromide, with bromine being incorporated into the products. At pH 8.5, no TOX were observed from NB, and lower TOX yields from BA were measured in the absence of bromide. Similar experiments conducted with Suwannee River NOM (5 mg C/L) showed significant conversions of NOM to TOX both under dark chlorination and MP-UV/Cl_2 exposure; however, no significant difference was observed between the levels measured for the two processes. In the presence of bromide, MP-UV/Cl_2 led to slightly lower TOX levels than chlorination alone, at both pHs. Structural changes in NOM were observed in the ESI-tqMS spectra recorded for UV/Cl_2-treated NOM, particularly in polar and ionizable fractions of NOM. Other changes in the solution properties – e.g., UV absorbance – were noted. The study reflects the role of halogen atoms in the transformation of UV exposed organic matter.

Pisarenko *et al.* (2013) investigated the impact of UV/Cl_2 AOP on NOM oxidation and DBP formation on pre-chlorinated Colorado River water collected from a water utility. The water samples (2.6 mg/L TOC; pH 8.1, adjusted to 6.0, 7.5, and 9.0; $UV_{253.7\,nm}$ = 0.045 cm^{-1}) were spiked at various free chlorine levels (0–50 mg/L NaOCl), and further exposed to either monochromatic UV-C (253.7 nm, ~3900 mJ/cm^2) or polychromatic UV-A (310–410 nm; max. at 365 nm, ~7000 mJ cm^{-2} radiation. Chlorine doses and UV fluences used in this study are far larger than those practiced in real water treatment applications. Hydrochloric acid used to adjust pH to 6 and 7.5 raised the chloride concentration and increased the $Cl^{\bullet}$ water demand *via* Cl^- reactions. Insignificant NOM mineralization (TOC loss < 0.3 mg/L) was observed, while ~50% and 40% reduction of UV_{254} absorbance at pH 6 and 7.5, respectively, was achieved with the UV-C/Cl_2 process. Fluorescence excitation-emission matrix and SEC-UV fluorescence data indicated major structural changes in NOM and reduction of aromaticity (chromophoric NOM). At 10 mg/L free chlorine and pH 7.5, the UV-C/Cl_2 process did not affect the total (four) THM level, increased the total (nine) HAA level and decreased the TOX, relative to the control (dark chlorination). Chloroform was the overwhelming THM with UV-C/Cl_2, whereas chloro-bromo-THMs prevailed in both dark chlorination and UV-A/Cl_2 processes. Dichloro- and trichloroacetic acids were the major HAAs. As free chlorine was almost completely decomposed over the 2-h exposure to either LP or MP radiation, the 2-h 'dark chlorination' where large Cl_2 residual is expected, may have not been a true 'control' for these tests. The high UV doses may have degraded the DBP precursors or/and the DBPs generated in the process.

Wang *et al.* (2015b) undertook full-scale and pilot-scale byproduct formation studies during the application of UV/Cl_2 and UV/H_2O_2 processes for T&O and caffeine treatment. Process details are given in Section *9.4.4.4*. The chlorinated and brominated organic byproducts monitored in these studies were the four THMs, nine HAAs, 1,1-dichloro- and 1,1,1-trichloro-2-propanone (HKs), four haloacetonitriles (HANs), chloropicrin, and adsorbable organic halides (AOX), whereas the inorganic byproducts monitored included chlorite, chlorate, perchlorate, and bromate. The inorganic byproducts from this work are discussed in Sections *9.4.5.2* and *9.4.5.3*. No change in t-THMs relative to untreated water (18 μg/L) was observed in any UV/Cl_2- and UV/H_2O_2-effluents from full-scale tests (Cornwall WTP). That is, THMs were not formed during the short residence time (30 s) in the UV reactor (~1800 mJ cm^{-2}). In Lake Simcoe water (pilot), which was not pre-chlorinated at the intake, UV/Cl_2-treated samples (60 s) led to more than 100% increase of t-THMs relative to untreated or UV alone-treated waters. Upon 24-h incubation time, THMFP increased by 30–100% in the UV/Cl_2 (10 mg/L)- and UV/H_2O_2(4.8 mg/L)-treated full-scale effluents, with more t-THMs being formed in UV/Cl_2 process at pH 6.5 than in all other samples. Much larger THMFP increase was observed in pilot- than at full-scale tests. HAA9 levels increased in full-scale UV/Cl_2 (pH 6.5 and 7.5) effluents (10–40%) relative to untreated water; no increase was observed for the UV/H_2O_2-treated effluents. The 24-h HAA9FP increased by 40–110%, and 20–90%, in Cornwall and Lake Simcoe water, respectively, for both AOPs. That shows that HAA precursors are formed from radical-induced degradation of dissolved NOM. Dichloro- and trichloroacetic acids accounted for approx. 60% of the t-HAAs.

HKs and chloropicrin were not formed in either UV-AOP at full-scale treatment; up-to 4-fold increase in HANs was observed only in the UV/Cl_2 process at pH 6.5 and 7.5, but the levels were very low (up to 4 μg/L). Similar fold-increase yields were observed for HANFP in both UV/Cl_2 and UV/H_2O_2-treated samples, with levels up to 15 μg/L. Much larger HAN levels were formed in UV/Cl_2-treated Lake Simcoe water at pilot-scale than in UV/Cl_2-treated water at full-scale (Cornwall WTP), and the HANFP at pH 6.5 was determined as ~30 μg/L. Dichloro- and bromodichloro-acetonitriles were the major HAN species. AOX and AOX-FP increased marginally if at all in the UV-AOP effluents at full-scale relative to the original water. Up to ~5-fold increased AOX levels (~100 μg C/L) were quantified in UV/Cl_2-treated Lake Simcoe water relative to untreated water (~20 μg C/L). Since no such patterns were observed in the UV/H_2O_2 tests, the authors suggested that the adsorbable halogenated organics formed in the UV/Cl_2 process resulted from $Cl^\bullet$ reactions with NOM and that $Cl^\bullet$ species originated from chlorine photolysis.

Iodinated DBPs (I-DBPs) can be formed in iodide-containing waters during the chlorination process. Iodinated X-ray contrast agents (e.g., iopromide, iohexol, iopamidol) are emerging contaminants in waste- and surface waters. Both UV alone and AOP-driven degradation of iodinated compounds results in iodide which is further oxidized to hypoiodous acid (HOI) during the chlorination practices. HOI could generate I-DBPs in reactions with dissolved organic matter. Wang *et al.* (2016b) examined the degradation of iohexol by UV alone and UV/Cl_2 AOP, in both laboratory water and water collected from two WTPs. For a given water matrix, the yield and speciation of the observed I-THMs depended on the process, chlorine concentration, and pH. Post-UV/Cl_2 treatment chlorination resulted in less I-THMs than post-UV (alone) chlorination.

Zhang *et al.* (2015) investigated the role of UV/chlorine for ammonia and DBP reduction. In their study, chlorine was added to ammonia-rich water to a 0.8 Cl_2/N molar ratio, which in the presence of UV (253.7 nm) translated into a UV/chloramine process rather than UV/(free) chlorine AOP. The water samples were collected from a municipal drinking water treatment plant from Harbin, China. Monochloramine absorbs the 253.7 nm radiation much stronger than free chlorine and photolyzes to $Cl^•$; nitrite and nitrate are among final degradation products. Free ammonia (0.07 mM) was removed by ~50% upon exposure to UV/chlorine (as NH_2Cl) at 800 mJ cm^{-2}. The authors postulated H-atom abstraction from NH_3 by $Cl^•$ followed by a sequence of reactions starting from the aminyl ($NH_2^•$) radical. Post-UV/chlorine water chlorination showed a lower Cl_2 demand and generated less THMs and HAAs than untreated water. The process formed more HANs than 'dark' chlorination of untreated water. This study is relevant to water reuse applications where free ammonia can be present in RO permeates, thus, an increased Cl_2 demand would be observed. Chloramine photolysis will occur in UV/Cl_2 water reuse applications, such that more complex chemistry and photochemistry than in the absence of chloramines is expected, and undesirable process byproducts could be formed.

9.4.5.2 Chlorite, chlorate, and perchlorate

Chloride, chlorite, and chlorate are the only reported photodegradation products of free chlorine. Laboratory studies performed in deionized water showed that 9% to 30% (by mass) of chlorine can convert to chlorate at 253.7 nm (Buxton & Subhani, 1972b; Feng *et al.* 2010). In a full-scale study on the degradation of TCE in groundwater by MP–UV/Cl_2 (9 mg/L; pH 7.55; 90% TCE reduction) at Middleton WTP, ON, Canada, Wang *et al.* (2011) showed that the concentration of chlorate in treated water was approximately 0.95 mg/L. This concentration exceeded the WHO drinking water guideline for chlorate of 0.7 mg/L, and was marginally lower than Health Canada drinking water guideline of 1 mg/L. In their full-scale (Trojan UVSwiftECT™ 8L24) studies conducted at the drinking water treatment plant in Cornwall (see test conditions in Section *9.4.4.4*), Wang *et al.* (2015b) showed that 2 to 17% (by mass) of free chlorine photolyzed in the reactor was converted to chlorate. The highest values were measured at pH 8.5 and 10 mg/L Cl_2. Chlorite and perchlorate were not detected. However, chlorite is a product of chlorine photolysis at 365 nm, which is one of the strong lines of MP lamp emission spectrum (Buxton & Subhani, 1972b). As the maximum concentration of chlorate in drinking water has been regulated at 1 mg/L by Health Canada (2012), the formation of chlorate would be the predominant limiting factor for the use of the UV/Cl_2 AOP for water treatment with the purpose of drinking water production. Additionally, chlorate may also be introduced in water with commercial NaOCl solutions (Stanford *et al.* 2011).

9.4.5.3 Bromate

Bromate ion is a potential human carcinogen and its concentration in drinking water is regulated at 10 μg/L. Bromate is typically produced at drinking water treatment plants which use ozonation of natural waters containing bromide ions (von Gunten & Oliveras, 1998). Formation of hydroxyl radicals in the free chlorine photolysis may promote the formation of bromate in UV reactors during the treatment of bromide-rich waters.

Laboratory scale experiments carried out in a batch LP-UV reactor with bromide solutions prepared in deionized water (80 μg/L Br^-; pH ≈ 7.8) and filtered Yangtze River water (56.3 μg/L Br^-; 1.7 mg/L DOC; pH ≈ 7.5) showed that the UV/Cl_2 process (5 mg/L Cl_2) led to bromate formation (Huang *et al.* 2008). The bromate yields increased with the applied UV dose. Bromate levels of ~10 μgL (maximum permissible level in drinking water) were measured at ~200 mJ cm^{-2} in deionized water, and at ~700 mJ cm^{-2} in the river water. For the highest UV dose tested (~1450 mJ cm^{-2}), residual free chlorine was <0.5 mg/L, and bromate was quantified as ~18 and ~12 μg/L in deionized water and in river water, respectively (0.01–0.14 mole BrO_3^-/mole Br^-). Huang *et al.*'s study showed that bromate formation was much higher in the UV/Cl_2 process than in dark chlorination, and depended also on pH and chlorine dose. Bromate formation in the order of 0.1 to 2 μg/L, with higher levels at lower pH (consistent with Huang *et al.*'s findings), were measured by Wang *et al.* (2015b) in the full-scale study on the UV/Cl_2 AOP at Cornwall, ON, WTP, even at the low concentrations of bromide (2–3 μg/L) present in the surface water. More full-scale studies with water containing various bromide levels (levels as large as 700 μg/L are reported in surface waters treated for drinking water production) are needed in order to assess whether bromate formation in the UV/Cl_2 AOP should be a concern.

9.5 RESEARCH NEEDS

This literature review has shown that the mechanisms for the overall photodecomposition of free chlorine in pure water are very complex and depend on many parameters, among which, pH and radiation wavelength. The kinetic studies available in the public domain provide quantum yield values for the overall photodecomposition of free chlorine species, but there is no comprehensive kinetic model for the UV/Cl_2 AOP which includes all initiation, propagation, and termination reactions occurring in this process. These reactions must be captured in a kinetic model in order to predict correctly the steady state concentrations of the oxidizing species and the micropollutant degradation rates. Fundamental research is needed to elucidate reaction mechanisms, to understand the role of reactive species characteristic to the UV/Cl_2 process such as $Cl^\bullet$, $ClO^\bullet$, $Cl_2^{\bullet-}$, to determine accurate rate constants for the reactions of $^\bullet OH$ and $Cl^\bullet$ with HOCl, water constituents and micropollutants targeted in the water sources.

In water reuse, the RO permeates contain chloramines and could contain free ammonia. Chloramines absorb the UV radiation and undergo photolysis with large and pH- and oxygen-dependent quantum yields, with formation of radical species and stable products which interfere with free chlorine chemistry and photochemistry. Should ammonia be present in the water, free chlorine consumption and additional chloramine formation should be expected. The role of chloramines in the UV/Cl_2 AOP is not studied and its implications on process performance are unknown.

Research is needed to clarify the contribution of the various radicals generated during the UV/chlorine process to the degradation of micropolluants, to determine the rate constants for the reactions of $Cl^\bullet$ and $Cl_2^{\bullet-}$ with organic and inorganic compounds, to better understand the fate of organic radicals ($R^\bullet$, $ROO^\bullet$) and the secondary reactions involving HOCl and ClO^-, and to know the impact of water matrix constituents on the efficiency of the UV/Cl_2 process.

The UV/Cl_2 AOP is insufficiently studied with respect to its implications on the treated water quality, including toxicity, mutagenicity, and genotoxicity, and the emerging disinfection byproduct formation.

9.6 CONCLUSIONS

As compared to other AOPs, the studies on the oxidation of organic compounds by UV/Cl_2 process are very recent and scarce. However, pilot- and full-scale studies performed at drinking water and water reuse treatment plants over the past few years indicate that the UV/Chlorine process may represent a promising alternative to the UV/H_2O_2 AOP, particularly from economic and process operating perspectives. For the purpose of potable water production and water reuse, the formation of undesirable by-products (chlorate, bromate, TOX, THMs, HAAs, HANs) must be predicted and controlled to ensure that the treated water quality meets the regulatory standards.

The use of the UV/Cl_2 AOP to treat groundwater or surface water contaminated with organic micropollutants requires chlorine doses of 5 to 10 mg /L and UV doses of 500–2000 mJ cm^{-2}. Given the complexity of reaction mechanisms and the diversity of micropollutant degradation pathways (direct photolysis, reaction with free chlorine, oxidation by radicals), bench-scale research and preliminary pilot trials are needed in order to determine the optimal process conditions (UV dose, chlorine dose, pH), to select and to size the UV equipment for specific applications, and to estimate the treatment costs. The chlorine dosage in the UV/Cl_2 applications must be optimized in order to obtain an acceptable residual concentration of free chlorine for secondary disinfection of treated water in the reservoirs and distribution network, thus, avoiding the use of a quenching reagent for free chlorine.

9.7 ACKNOWLEDGEMENT

Special thanks to Dr. Roshanak Aflaki from City of Los Angeles Department of Public Works, CA, for providing the figure with the TIWRP pilot data adapted from Aflaki *et al.* (2015).

9.8 REFERENCES

Aflaki R., Hammond S., Tag Oh S., Hokanson D., Trussell S. and Bazzi A. (2015). Scaling-up step-by-step and AOP investment decision. Proceedings of the Water Environment Federation's Technical Exhibition and Conference (WEFTEC), September 26–30, 2015, Chicago, IL, USA, pp. 9–21.

Aghdam E., Sun J. and Shang C. (2016). DEET degradation by the UV/chlorine process: kinetics, contributions of radicals and byproduct formation. *IUVA World Congress*, January 31–February 3, 2016, Vancouver, BC, Canada. https://iuva.wildapricot.org/resources/Documents/world%20conference%202016/tuesday/5/Ehsan%20Aghdam%20IUVA.pdf

Alegre M. L., Gerones M., Rosso J. A., Bertolitti S. G., Braun A. M., Martire D. O. and Gonzalez M. C. (2000). Kinetic study of the reactions of chlorine atoms and $Cl_2^{\bullet-}$ radical anions in aqueous solutions. 1. Reaction with Benzene. *Journal of Physical Chemistry A*, **104**(14), 3117–3125.

Allmand A. J., Cunliffe P. W. and Maddison R. E. W. (1925). The photodecomposition of chlorine water and of aqueous hypochlorous acid solutions. Part I. *Journal of the Chemical Society, Transactions*, **127**, 822–840.

Armstrong D. A., Huie R. E., Koppenol W. H., Lymar S. V., Merényi G., Neta P., Ruscic B., Stanbury D. M., Steenken S. and Wardman P. (2015). Standard electrode potentials involving radicals in aqueous solution: inorganic radicals (IUPAC Technical Report). *Pure and Applied Chemistry*, **87**(11–12), 1139–1150.

Boal A. K., Rhodes C. and Garcia S. (2015). Pump-and-treat groundwater remediation using chlorine/ultraviolet advanced oxidation processes. *Groundwater Monitoring & Remediation*, **35**(2), 93–100.

Bolton J. R. and Stefan M. I. (2002). Fundamental photochemical approach to the concepts of fluence (UV dose) and electrical energy efficiency in photochemical degradation reactions. *Research on Chemical Intermediates*, **28**(7), 857–870.

Buxton G. V. and Subhani M. S. (1972a). Radiation chemistry and photochemistry of oxychlorine ions. Part 1. -Radiolysis of aqueous solutions of hypochlorite and chlorite ions. *Journal of the Chemical Society, Faraday Transactions 1: Physical Chemistry in Condensed Phases*, **68**, 947–957.

Buxton G. V. and Subhani M. S. (1972b). Radiation chemistry and photochemistry of oxychlorine ions. Part 2. -Photodecomposition of aqueous solutions of hypochlorite ions. *Journal of the Chemical Society, Faraday Transactions 1: Physical Chemistry in Condensed Phases*, **68**, 958–969.

Buxton G. V., Bydder M. and Salmon G. A. (1998). Reactivity of chlorine atoms in aqueous solution. Part I: the equilibrium $Cl^{\bullet} + Cl^- \rightleftharpoons Cl_2^{\bullet-}$. *Journal of the Chemical Society, Faraday Transactions*, **94**(5), 653–657.

Buxton G. V., Greenstock C. L., Helman W. P. and Ross A. B. (1988). Critical review of rate constants for reactions of hydrated electrons, hydrogen atoms and hydroxyl radicals ($OH^{\bullet}/O^{\bullet-}$) in aqueous solution. *Journal of Physical and Chemical Reference Data*, **17**(2), 513–886.

Buxton G. V., Bydder M., Salmon G.A. and Williams J. E. (2000). The reactivity of chlorine atoms in aqueous solution. Part III: the reaction of Cl$^•$ with solutes. *Physical Chemistry Chemical Physics*, **2**, 237–245.

Chan P. Y., El-Din M. G. and Bolton J. R. (2012). A solar-driven UV/chlorine advanced oxidation process. *Water Research*, **46**(17), 5672–5682.

Connick R. E. and Chia, Y. (1959). The hydrolysis of chlorine and its variation with temperature. *Journal of the American Chemical Society*, **81**(6), 1280–1284.

Cooper W. J., Jones A. C., Whitehead R. F. and Zika R. G. (2007). Sunlight-induced photochemical decay of oxidants in natural waters: implications in ballast water treatment. *Environmental Science & Technology*, **41**(10), 3728–3733.

De Laat J. (2016). Personal communication.

De Laat J. and Berne F. (2009). La déchloramination des eaux de piscines par irradiation UV. Etude bibliographique. Theoretical and practical aspects of the dechloramination of swimming pool water by UV irradiation. *European Journal of Water Quality*, **40**(2), 129–149.

Deborde M. and von Gunten U. (2008). Reactions of chlorine with inorganic and organic compounds during water treatment – Kinetics and mechanisms: a critical review. *Water Research*, **42**(1–2), 13–51.

Deng L., Huang C. H. and Wang Y. L. (2014). Effects of combined UV and chlorine treatment on the formation of trichloronitromethane from amine precursors. *Environmental Science & Technology*, **48**(5), 2697–2705.

Fang J., Fu Y. and Shang C. (2014). The roles of reactive species in micropollutant degradation in the UV/Free chlorine system. *Environmental Science & Technology*, **48**(3), 1859–1868.

Feng Y., Smith, D. W. and Bolton J. R. (2007). Photolysis of aqueous free chlorine species (HOCl and OCl) with 254 nm ultraviolet light. *Journal of Environmental Engineering and Science*, **6**(3), 277–284.

Feng Y., Smith, D. W. and Bolton J. R. (2010). A potential new method for determination of the fluence (UV dose) delivered in UV reactors involving the photodegradation of free chlorine. *Water Environment Research*, **82**(4), 328–334.

Gilbert B. C., Stell J. K., Peet W. J. and Radford K. J. (1988). Generation and reactions of the chlorine atom in aqueous solution. *Journal of the Chemical Society, Faraday Transactions 1: Physical Chemistry in Condensed Phases*, **84**(10), 3319–3330.

Grigorev A. E., Makarov I. E. and Pikaev A. K. (1987). Formation of Cl_2 in the bulk solution during the radiolysis of concentrated aqueous solutions of chloride. *High Energy Chemistry*, **21**, 99–102.

Guo Z. B., Lin Y. L., Xu B., Huang H., Zhang T. Y., Tian F. X. and Gao Y. (2016). Degradation of chlortoluron during UV irradiation and UV/chlorine processes and formation of disinfection by-products in sequential chlorination. *Chemical Engineering Journal*, **283**, 412–419.

Hasegawa K. and Neta P. (1978). Rate constants and mechanisms of reaction of Cl_2^- radicals. *Journal of Physical Chemistry*, **82**(8), 854–857.

Health Canada (2012). Guidelines for Canadian Drinking Water Quality – Summary Table. Water, Air and Climate Change Bureau. Healthy Environments and Consumer Safety Branch, Health Canada, Ottawa, Ontario, Canada.

Held A. M., Halko D. J. and Hurst J. K. (1978). Mechanisms of chlorine oxidation of hydrogen peroxide. *Journal of American Chemical. Society*, **100**(18), 5732–5740.

Herrmann H. (2007). On the photolysis of simple anions and neutral molecules as sources of O^-/OH, SO_x- and Cl in aqueous solution. *Physical Chemistry Chemical Physics*, **9**(30), 3925–3964.

Huang X., Gao N. and Deng Y. (2008). Bromate ion formation in dark chlorination and ultraviolet/chlorination processes for bromide-containing water. *Journal of Environmental Sciences*, **20**(2), 246–251.

Jacobi H. W., Wicktor F., Herrmann H. and Zellner R. (1999). A laser flash photolysis kinetic study of reactions of the Cl_2^- radical anion with oxygenated hydrocarbons in aqueous solution. *International Journal of Chemical Kinetics*, **31**(3), 169–181.

Jayson G. G., Parsons B. J. and Swallow A. J. (1973). Some simple, highly reactive, inorganic chlorine derivatives in aqueous solution. Their formation using pulses of radiation and their role in the mechanism of the Fricke dosimeter. *Journal of the Chemical Society, Faraday Transactions 1: Physical Chemistry in Condensed Phases*, **69**(0), 1597–1607.

Jin J., El-Din M. G. and Bolton J. R. (2011). Assessment of the UV/chlorine process as an advanced oxidation process. *Water Research*, **45**(4), 1890–1896.

Kang N., Anderson T. A. and Jackson W. A. (2006). Photochemical formation of perchlorate from aqueous oxychlorine anions. *Analytica Chimica Acta*, **567**(1), 48–56.

Karpel Vel Leitner N., De Laat J. and Doré M. (1992a). Photodécomposition du bioxyde de chlore et des ions chlorite par irradiation U.V. en milieu aqueux – Partie I. Sous-produits de réaction. Photodecomposition of chlorine dioxide and chlorite by U.V.-Irradiation—Part I. Photo-products. *Water Research*, **26**(12), 1655–1664.

Karpel Vel Leitner N., De Laat J. and Doré M. (1992b). Photodécomposition du bioxyde de chlore et des ions chlorite par irradiation U.V. en milieu aqueux – Partie II. Etude cinétique. Photodecomposition of chlorine dioxide and chlorite by U.V.-Irradiation—Part II. Kinetic study. *Water Research*, **26**(12), 1665–1672.

Kishimoto N. and Nishimura H. (2015). Effect of pH and molar ratio of pollutant to oxidant on a photochemical advanced oxidation process using hypochlorite. *Environmental Technology*, **36**(19), 2436–2442.

Kläning U. K. and Wolff T. (1985). Laser flash photolysis of HClO, ClO^-, HBrO, and BrO^- in aqueous solution. Reactions of Cl^- and Br^- atoms. *Berichte der Bunsengesellschaft für Physikalische Chemie*, **89**, 243–245.

Kläning U. K., Sehested K. and Wolff T. (1984). Ozone formation in laser flash photolysis of oxoacids and oxoanions of chlorine and bromine. *Journal of the Chemical Society, Faraday Transactions 1*, 80, 2969–2979.

Kläning U. K., Sehested K. and Holcman J. (1985). Standard Gibbs energy of formation of the hydroxyl radical in aqueous solution. Rate constants for the reaction $ClO_2^- + O_3 \rightleftarrows O_3^- + ClO_2$. *Journal of Physical Chemistry*, **89**(5), 760–763.

Kong X., Jiang J., Ma J., Yang Y., Liu W. and Liu Y. (2016). Degradation of atrazine by UV/chlorine: efficiency, influencing factors, and products. *Water Research*, **90**, 15–23.

Kruithof J. C., Kamp P. C. and Martijn B. J. (2007). UV/H_2O_2 Treatment: a practical solution for organic contaminant control and primary disinfection. *Ozone: Science & Engineering: The Journal of the International Ozone Association*, **29**(4), 273–280.

Legrini O., Oliveros, E. and Braun, M. (1993). Photochemical processes for water treatment. *Chemical Reviews*, **93**(2), 671–689.

Li K., Stefan M. I. and Crittenden J. C. (2007). Trichloroethene degradation by UV/H_2O_2 advanced oxidation process: product study and kinetic modelling. *Environmental Science & Technology*, **41**(5), 1696–1703.

Macounová K. M., Simic N., Ahlberg E. and Krtil P. (2015). Electrochemical water-splitting based on hypochlorite oxidation. *Journal of the American Chemical Society*, **137**(23), 7262–7265.

Mártire D. O., Rosso J. A., Bertolotti S., Le Roux G. C., Braun A. M. and Gonzalez M. C. (2001). Kinetic Study of the reactions of chlorine atoms and $Cl_2^{\bullet-}$ radical anions in aqueous solutions. II. Toluene, benzoic acid and chlorobenzene. *Journal of Physical Chemistry A*, **105**(22), 5385–5392.

Matthew B. M. and Anastasio C. (2006). A chemical probe technique for the determination of reactive halogen species in aqueous solution: part 1 – Bromide solutions. *Atmospheric Chemistry and Physics*, **6**(9), 2423–2437.

McElroy W. J. (1990). A laser photolysis study of the reaction of SO_4^- with Cl^- and the subsequent decay of Cl_2^- in aqueous solution. *Journal of Physical Chemistry*, **94**, 2435–2441.

Mertens M. and von Sonntag C. (1995). Photolysis ($\lambda = 254$ nm) of tetrachloroethene in aqueous solutions. *Journal of Photochemistry and Photobiology A: Chemistry*, **85**(1), 1–9.

Morris J. C. (1966). The acid ionization constant of HOCl from 5 to 35°. *Journal of Physical Chemistry*, **70**(12), 3798–3805.

Neta P., Huie R. E. and Ross A. B. (1988). Rate constants for reactions of inorganic radicals in aqueous solution. *Journal of Physical Chemistry and Reference Data*, **17**(3), 1027–1284.

Nowell L. H. and Crosby D. G. (1985). Photodegradation of water pollutants in chlorinated water. In: Water Chlorination: Chemistry, Environmental Impact and Health Effects, R. Jolley W. Davis S. Katz M. Jr, Roberts and V. Jacobs (eds), Lewis Publishers Inc., Chelsea, Michigan, pp. 1055–1062.

Nowell L. H. and Hoigné, J. (1992a). Photolysis of aqueous chlorine at sunlight and ultraviolet wavelengths I. Degradation rates. *Water Research*, **26**(5), 593–598.

Nowell L.H. and Hoigné, J. (1992b). Photolysis of aqueous chlorine at sunlight and ultraviolet wavelengths II. Hydroxyl radical production. *Water Research*, **26**(5), 599–605.

Ogata Y. and Tomizawa K. (1984). Photoreaction of benzoic acid with sodium hypochlorite in aqueous alkali. *Journal of the Chemical Society, Perkin Transactions 2*, 6, 985–988.

Ogata Y., Takagi K. and Susuki T. (1978). Photolytic oxidation of ethylene glycol, dimethyl ether and related compounds by aqueous hypochlorite. *Journal of the Chemical Society, Perkin Transactions 2*, **6**, 562–567.

Ogata Y., Suzuki T. and Takagi K. (1979). Photolytic oxidation of aliphatic acids by aqueous sodium hypochlorite. *Journal Chemical Society, Perkin Transactions 2*, **12**, 1715–1719.

Oliver B. G. and Carey J. H. (1977). Photochemical production of chlorinated organic in aqueous solutions containing chlorine. *Environmental Science & Technology*, **11**(9), 893–895.

Ormeci B., Ducoste J. J. and Linden K. G. (2005). UV disinfection of chlorinated water: impact on chlorine concentration and UV dose delivery. *Journal of Water Supply: Research and Technology-AQUA*, **54**(3), 189–199.

Oturan M. A. and Aaron J.-J. (2014). Advanced oxidation processes in water/wastewater treatment: principles and applications. A Review. *Critical Reviews in Environmental Science and Technology*, **44**(23), 2577–2641.

Pisarenko A. N., Stanford B. D., Snyder S. A., Rivera S. B. and Boal A. K. (2013). Investigation of the use of chlorine based advanced oxidation in surface water: oxidation of natural organic matter and formation of disinfection byproducts. *Journal of Advanced Oxidation Technologies*, **16**(1), 137–150.

Qin L., Lin Y. L., Xu B., Hu C. Y., Tian F. X., Zhang T. Y., Zhu W. Q., Huang H. and Gao N. Y. (2014). Kinetic models and pathways of ronidazole degradation by chlorination, UV irradiation and UV/chlorine processes. *Water Research*, **65**, 271–281.

Rao B., Estrada N., McGee S., Mangold J., Gu B. and Jackson W. A. (2010). Perchlorate production by photodecomposition of aqueous chlorine solutions. *Environmental Science & Technology*, **46**(21), 11635–11643.

Richardson S. D., Plewa M. J., Wagner E. D., Schoeny R. and Demarini D. M. (2007). Occurrence, genotoxicity, and carcinogenicity of regulated and emerging disinfection by-products in drinking water: a review and roadmap for research. *Mutation Research*, **636**(1–3), 178–242.

Rosenfeldt E., Boal A. K., Springer J., Stanford B., Rivera S., Mashinkunti R. D. and Metz D. H. (2013). Comparison of UV-mediated advanced oxidation. *Journal of the American Water Works Association*, **105**(7), 29–33.

Schwarzenbach R.P., Gschwend P.M. and Imboden D.M. (1993). Chapter 13 in Environmental Organic Chemistry. Wiley-Interscience: John Wiley and Sons, New York.

Shu Z., Li C., Belosevic M., Bolton J. R. and El-Din G. M. (2014). Application of a solar UV/solar advanced oxidation process to oil sands process-affected water remediation. *Environmental Science & Technology*, **48**(16), 9692–9701.

Sichel C., Garcia C. and Andre K. (2011). Feasibility studies: UV/chlorine advanced oxidation treatment for the removal of emerging contaminants. *Water Research*, **45**(19), 6371–6380.

Siddiqui M. S. (1996). Chlorine-ozone interactions: formation of chlorate. *Water Research*, **30**(9), 2160–2170.

Song W., Cooper W. J., Peake B. M., Mezyk S. P., Nickelsen M. G. and O'Shea K. E. (2009). Free-radical-induced oxidative and reductive degradation of N,N'-diethyl-*m*-toluamide (DEET): kinetic studies and degradation pathway. *Water Research*, **43**(3), 635–642.

Stanford B. D., Pisarenko, A. N., Snyder, S. A. and Gordon, G. (2011). Perchlorate, bromate, and chlorate in hypochlorite solutions: guidelines for utilities. *Journal of the American Water Works Association*, **103**(6), 71–83.

Thomsen C. L., Madsen D., Poulsen J. Aa., Thøgersen J., Knak Jensen S. J. and Keiding S. R. (2001). Femtosecond photolysis of aqueous HOCl. *The Journal of Chemical Physics*, **115**(20), 9361–9369.

Vione D., Maurino V., Minero C., Calza P. and Pelizzetti E. (2005). Phenol chlorination and photochlorination in the presence of chloride ions in homogeneous aqueous solution. *Environmental Science & Technology*, **39**(13), 5066–5075.

Von Gunten U. and Oliveras Y. (1998). Advanced oxidation of bromide-containing waters: bromate formation mechanisms. *Environmental Science & Technology*, **32**(1), 63–70.

von Sonntag C. and von Gunten, U. (2012). Chemistry of Ozone in Water and Wastewater Treatment. IWA Publishing, London, UK.

Wagner I., Karthäuser J. and Strehlow H. (1986). On the decay of the dichloride anion Cl_2^- in aqueous solution. *Berichte der Bunsengesellschaft für Physikalische Chemie*, **90**(10), 861–867.

Wang D., Walton T., McDermott L. and Hofmann R. (2011). Control of TCE using UV combined with hydrogen peroxide or chlorine. Proceedings of IOA-IUVA World Congress & Exhibition, 23–27 May, 2011, Paris, France, pp. 152–158.

Wang D., Bolton J. R. and Hofmann R. (2012). Medium pressure UV combined with chlorine advanced oxidation for trichloroethylene destruction in a model water. *Water Research*, **46**(15), 4677–4686.

Wang D., Bolton J. R., Andrews S. A. and Hofmann R. (2015a). UV/chlorine control of drinking water taste and odour at pilot and full-scale. *Chemosphere*, **136**, 239–244.

Wang D., Bolton J. R., Andrews S. A. and Hofmann R. (2015b). Formation of disinfection by-products in the ultraviolet/chlorine advanced oxidation process. *Science of the Total Environment*, **518**–519, 49–57.

Wang W. L., Wu Q. Y., Huang N., Wang T. and Hu H. Y. (2016a). Synergistic effect between UV and chlorine (UV/chlorine) on the degradation of carbamazepine: influence factors and radical species. *Water Research*, **98**, 190–198.

Wang Z., Lin, Y. L., Xu, B., Xia, S. J., Zhang, T. Y. and Gao N. Y. (2016b). Degradation of iohexol by UV/chlorine process and formation of iodinated trihalomethanes during post-chlorination. *Chemical Engineering Journal*, **283**, 1090–1096.

Watts, M. J. and Linden K. G. (2007). Chlorine photolysis and subsequent OH radical production during UV treatment of chlorinated water. *Water Research*, **41**(13), 2871–2878.

Watts M. J. and Linden K. G. (2009). Advanced oxidation kinetics of aqueous trialkyl phosphate flame retardants and plasticizers. *Environmental Science & Technology*, **43**(8), 2937–2942.

Watts M. J. Rosenfeldt E. J. and Linden K. G. (2007). Comparative OH radical oxidation using UV-Cl_2 and UV-H_2O_2 processes. *Journal of Water Supply: Research and Technology-AQUA*, **56**(8), 469–477.

Watts M. J., Hofmann R. and Rosenfeldt E. J. (2012). Low-pressure UV/C12 for advanced oxidation oxidation of taste and odor. *Journal of American Water Works Association*, **104**(1), E58–E65.

Westerhoff P., Chao P. and Mash H. (2004). Reactivity of natural organic matter with aqueous chlorine and bromine. *Water Research*, **38**(6), 1502–1513.

Westerhoff P., Mezyk S. P., Cooper W. J. and Minakata D. (2007). Electron pulse radiolysis determination of hydroxyl radical rate constants with Suwannee River fulvic acid and other dissolved organic matter isolates. *Environmental Science & Technology*, **41**(13), 4640–4646.

Wicktor F., Donati A., Herrmann H. and Zellner R. (2003). Laser based spectroscopic and kinetic investigations of reactions of the Cl atom with oxygenated hydrocarbons in aqueous solution. *Physical Chemistry Chemical Physics*, **5**(12), 2562–2572.

Willson R. L. (1973). Free-radical chain oxidation by carbon tetrachloride and related compounds: model pulse-radiolysis studies. *Biochemical Society Transactions*, **1**, 929–931.

Wu D., Wong D. and Di Bartolo B. (1980). Evolution of Cl_2^- in aqueous NaCl solutions. *Journal of Photochemistry*, **14**(4), 303–310.

Xiang Y., Fang J. and Shang C. (2016). Kinetics and pathways of ibuprofen degradation by the UV/chlorine advanced oxidation process. *Water Research*, **90**, 301–308.

Yang X., Sun J., Fu W., Shang C., Li Y., Chen Y., Gan W., and Fang J. (2016). PPCP degradation by UV/chlorine treatment and its impact on DBP formation potential in real waters. *Water Research*, **98**, 309–318.

Yu X. Y. and Barker J. R. (2003). Hydrogen peroxide photolysis in acidic solutions containing chloride ions. I. Chemical mechanism. *Journal of Physical Chemistry A*, **107**(9), 1313–1324.

Yu X. Y., Bao Z. C. and Barker J. R. (2004). Free radical reactions involving $Cl^{\bullet}$, $Cl_2^{-\bullet}$, and $SO_4^{-\bullet}$ in the 248 nm photolysis of aqueous solutions containing $S_2O_8^{2-}$ and Cl^-. *Journal of Physical Chemistry A*, **108**(2), 295–308.

Zhang X., Li W., Blatchley III E. R., Wang X. and Ren P. (2015). UV/chlorine process for ammonia removal and disinfection by-product reduction: comparison with chlorination. *Water Research*, **68**, 804–811.

Zhao Q., Shang C., Zhang X., Ding G. and Yang X. (2011). Formation of halogenated organic byproducts during medium-pressure UV and chlorine coexposure of model compounds, NOM and bromide. *Water Research*, **45**(19), 6545–6554.

Zhu L., Nicovich J. M. and Wine P. H. (2005). Kinetics studies of aqueous reactions of Cl atoms and Cl_2^- radicals with organic sulfur compounds of atmospheric interest. *Journal of Physical Chemistry A*, **109**(17), 3903–3911.

Zuo Z., Katsumura Y., Ueda K. and Ishigure K. (1997). Reactions between some inorganic radicals and oxychlorides studied by pulse radiolysis and laser photolysis. *Journal of the Chemical Society, Faraday Transactions*, **93**(10), 1885–1891.

Chapter 10

Sulfate radical ion – based AOPs

Nathalie Karpel Vel Leitner

10.1 INTRODUCTION

As extensively discussed in the other chapters, the information on HO•-based processes from fundamental and applied research was acquired over more than a century. It started in 1894 with Fenton's discovery on metal-activated hydrogen peroxide decomposition resulting in tartaric acid oxidation. The process was further investigated in the 1930's by Haber and Weiss who proved the involvement of hydroxyl radical as the reactive species. But it was in 1977 when the UV/H_2O_2 AOP emerged based on Koubek's publication on the destruction of a number of organic acids through UV photolysis of H_2O_2. Ever since, the use of this process for water decontamination was extensively investigated such that today it is implemented at several water treatment plants around the world.

The 'history' of sulfate radical anion ($SO_4^{\bullet-}$) is more recent and covers about half a century of research, with an increasing interest into sulfate radical-driven oxidation particularly during the last decade when a significant number of laboratory scale studies have been reported in the scientific literature.

The sulfate radical-based processes have been studied extensively in the academic settings, but their industrial implementations are rare. Only a few niche applications were implemented to-date for soil and groundwater remediation with persulfate or sulfate radicals generated through persulfate activation on site; this technology is known as *In Situ* Chemical Oxidation (ISCO).

This chapter gives an overview on the state-of-the-art in research and applications of sulfate radical-based AOPs. The mechanisms of sulfate radical generation and the reactions with organic and inorganic compounds are also described.

10.2 METHODS FOR SULFATE RADICAL GENERATION

Several processes are described in the published literature on the generation of sulfate radical. Those could be homogeneous or heterogeneous, chemical and/or thermal, photochemical, or physical processes. The most common sulfate radical anion generation means are based on persulfate activation. Persulfate ($S_2O_8^{2-}$) is a strong oxidant ($E^0 = 2.1$ eV) which is relatively stable during storage and handling, but can be activated by various agents in order to generate highly reactive sulfate radicals which are characterized by a higher oxidative potential (2.6 eV) than that of persulfate ion. In practice, persulfate (peroxodisulfate, PDS) can be converted to sulfate radicals through three main processes: direct UV photolysis, thermal-, and metal-activation. The activation of peroxymonosulfate (Oxone, PMS) is also an efficient process for $SO_4^{\bullet -}$ generation.

10.2.1 Mild-thermal and base activation of persulfate

Persulfate heated in the range of 30–90°C was found to be effective for the destruction of various contaminants in water, among which, volatile organic compounds (VOCs) (Huang *et al.* 2005), hydroperfluorocarboxylic acids (Hori *et al.* 2010), and compounds inert toward HO•. Under thermal activation, PDS converts into the powerful oxidizing sulfate radicals.

The scission of the O—O bond in $S_2O_8^{2-}$ molecule (reaction 10.1) can be slowly initiated at 30°C and can be extremely rapid at 90°C with an activation energy of about 29 kcal mol^{-1} in neutral solutions (Kolthoff & Miller, 1951).

$$S_2O_8^{2-} \xrightarrow{\text{heat}} 2\,SO_4^{\bullet -} \tag{10.1}$$

It has been reported also (Furman *et al.* 2010) that alkaline pH conditions catalyze hydrolysis of persulfate to hydroperoxide anion and sulfate ion. The reduction of another persulfate molecule by the hydroperoxide ion leads to sulfate radical and sulfate anion. The net reaction (10.2) for base-activation of persulfate generates also superoxide radical anion ($O_2^{\bullet -}$).

$$2S_2O_8^{2-} + 2H_2O \xrightarrow{OH^-} 3SO_4^{2-} + SO_4^{\bullet -} + O_2^{\bullet -} + 4H^+ \tag{10.2}$$

Until now, alkali would be the most commonly used activators of persulfate for treatment of contaminated groundwater by *in situ* chemical oxidation (ISCO; Siegrist *et al.* 2011).

10.2.2 Photochemical processes

10.2.2.1 Persulfate photolysis

Under UV irradiation, free $SO_4^{\bullet -}$ radicals are formed through homolytic cleavage of peroxide bond in the peroxydisulfate dianion (reaction 10.3).

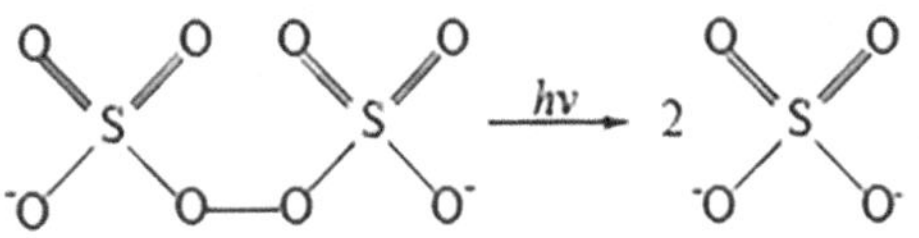

(10.3)

This reaction is a clean source of sulfate radicals with high and pH-independent quantum yields. With similar absorption spectra, the molar absorption coefficients of persulfate ion (20–22 M^{-1} cm^{-1}; Heidt, 1942; Mark *et al.* 1990) and hydrogen peroxide (18.6 M^{-1} cm^{-1} at pH 7.5; Hochanadel, 1962) at 254 nm are comparable.

The quantum yield of sulfate radical formation from persulfate photolysis within the 248–254 nm wavelength range is 1.4 ± 0.3 (Herrmann, 2007; Mark *et al.* 1990; Figure 10.1), much larger than that of HO• formation from hydrogen peroxide photolysis (1.0; Hochanadel, 1962).

The rate constant for the reaction of the generated radical with the oxidant (i.e., $SO_4^{\bullet -} + S_2O_8^{2-}$ *vs* $HO^\bullet + H_2O_2$) is smaller in the $S_2O_8^{2-}$/UV system ($k_{SO_4^{\bullet -}/S_2O_8^{2-}} = (6.1 \pm 0.6) \times 10^5\ \text{L mol}^{-1}\,\text{s}^{-1}$; McElroy & Waygood, 1990) than in the UV/H_2O_2 process ($k_{HO^\bullet/H_2O_2} = 2.7 \times 10^7\ \text{L mol}^{-1}\,\text{s}^{-1}$; Buxton *et al.* 1988).

The low competition for radicals from the oxidant would favor the efficiency of the $S_2O_8^{2-}$/UV process over the UV/H_2O_2 process, under conditions where the water matrix and target contaminants display the same 'radical scavenging capacity' in the two processes.

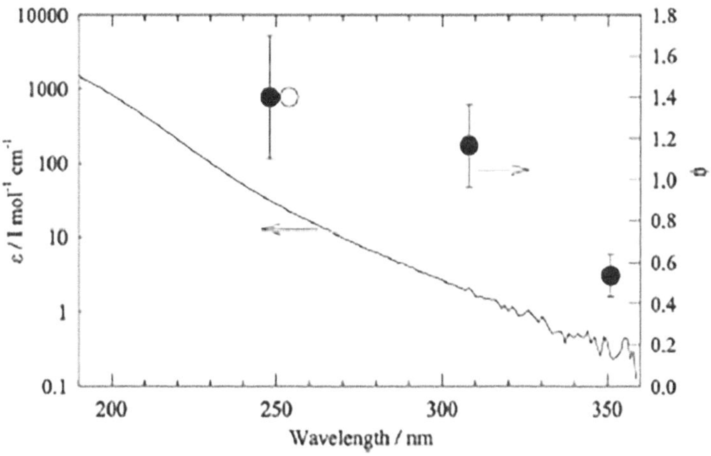

Figure 10.1 UV-absorption spectrum of $S_2O_8^{2-}$ and $SO_4^{\cdot-}$ quantum yields [(o) Mark *et al.*, 1990; (•) Herrmann, 2007] for the photolysis of $S_2O_8^{2-}$ in aqueous solution. Reproduced with permission from Herrmann (2007). Copyright The Royal Society of Chemistry, 2016.

10.2.2.2 Peroxomonosulfate photolysis

The molar absorption coefficient of peroxomonosulfate (PMS) in monoprotonated form (HSO_5^-) at $\lambda = 254$ nm was determined as 13.8–14 M^{-1} cm^{-1} at pH 6–7 (Herrmann, 2007; Guan *et al.* 2011) with a higher value for the dissociated form SO_5^{2-} ($\varepsilon = 149.5$ M^{-1} cm^{-1}). PMS photolyzes at 253.7 nm yielding $HO^\bullet$ and $SO_4^{\bullet-}$ radicals through the cleavage of peroxo bond. The apparent quantum yield for $SO_4^{\bullet-}$ in UV/PMS system at 254 nm was calculated as 0.52 ± 0.01 at pH 7, and the apparent quantum yield for total radical formation was estimated to be 1.04 based on the assumption that $HO^\bullet$ and $SO_4^{\bullet-}$ were produced equally (Guan *et al.* 2011); therefore, the total radical yield in UV/PMS process is similar to $HO^\bullet$ yield from the UV/H_2O_2 process.

$$HSO_5^- /SO_5^{2-} + h\nu \rightarrow SO_4^{\bullet-} + HO^\bullet \tag{10.4}$$

The rate of PMS photolysis with the $HO^\bullet$ and $SO_4^{\bullet-}$ formation increases with pH over the pH range of 8–10 (Guan *et al.* 2011).

10.2.3 Transition metal-activated decomposition of persulfate salts

Many transition metals, especially divalent metals which are commonly encountered in soil and groundwater, may act as electron donors to catalyze the decomposition of persulfate through one-electron transfer reaction with generation of sulfate radicals:

$$S_2O_8^{2-} + M^{n+} \rightarrow M^{(n+1)+} + SO_4^{\bullet-} + SO_4^{2-} \tag{10.5}$$

Similarly to Fenton's reagent (ferrous ion-activated hydrogen peroxide decomposition), persulfate is activated by Fe^{2+}:

$$Fe^{2+} + S_2O_8^{2-} \rightarrow SO_4^{\bullet-} + SO_4^{2-} + Fe^{3+} \quad (10.6)$$

However Fe(II) may not be an ideal activating agent since it can also act as a radical and PDS scavenger (reactions 10.7 and 10.8) resulting in a decrease in contaminant destruction efficiency (Liang *et al.* 2004a).

$$SO_4^{\bullet-} + Fe^{2+} \rightarrow Fe^{3+} + SO_4^{2-} \quad k = 4.6 \times 10^9 \text{ L mol}^{-1} \text{ s}^{-1} \text{ (22°C)} \quad (10.7)$$

$$S_2O_8^{2-} + 2Fe^{2+} \rightarrow 2Fe^{3+} + 2SO_4^{2-} \quad (10.8)$$

By minimizing the undesired effects of competing reactions, sequential addition of Fe^{2+} in small increments resulted in an increased pollutant removal efficiency (Liang *et al.* 2004a; Killian *et al.* 2007).

Also, the use of chelated ferrous ion is far superior to the use of free ferrous ion as an activator. Persulfate activation by Fe(II) in the presence of added chelating agents such as EDTA and citric acid, prevents the $SO_4^{\bullet-}$ scavenging by Fe^{2+} (Liang *et al.* 2004b; Rastogi *et al.* 2009; Liang *et al.* 2009). The availability of ferrous ion was found to be controlled by the molar ratio of chelate/Fe^{2+}.

Oh *et al.* (2010) showed that zero-valent iron (Fe^0) can activate persulfate through a mechanism which does not involve aqueous Fe^{2+}. The suggested mechanism involved the heterogeneous activation of persulfate ion *via* electron transfer from Fe^0.

$$Fe^0_{(s)} + 2S_2O_8^{2-}{}_{(aq)} \rightarrow Fe^{2+}{}_{(aq)} + 2SO_4^{\bullet-}{}_{(aq)} + 2SO_4^{2-}{}_{(aq)} \quad (10.9)$$

However, Fe^0 could also serve as a slow-releasing source of dissolved ferrous ions (reaction 10.10) which would subsequently activate PDS to produce $SO_4^{\bullet-}$.

$$Fe^0 + S_2O_8^{2-} \rightarrow Fe^{2+} + 2SO_4^{2-} \quad (10.10)$$

In this system, the recycling of ferric ion at the zero valent iron surface could avoid the accumulation of ferrous ion and reduce the precipitation of iron hydroxides.

Similarly to PDS activation, variant valence transition metal ions can activate PMS to form sulfate (majority) but also hydroxyl radicals:

$$M^{n+} + HSO_5^- \rightarrow M^{(n+1)+} + OH^- + SO_4^{\bullet-} \quad (10.11)$$

$$HSO_5^- + M^{n+} \rightarrow M^{(n+1)+} + SO_4^{2-} + HO^\bullet \quad (10.12)$$

The transition metal efficiency for the activation of PMS and the distribution of the two primary radical species formed ($SO_4^{\bullet-}$ *vs* $HO^\bullet$) depends on the nature and speciation of the metal, but $SO_4^{\bullet-}$ formation would be favored in most cases (Anipsitakis & Dionysiou, 2004).

10.2.4 Miscellaneous processes

10.2.4.1 Peroxymonosulfatelozone combination

Recently, the simultaneous generation of hydroxyl radical and sulfate radical in the reaction of ozone with deprotonated peroxymonosulfate (SO_5^{2-}) has been demonstrated by Yang *et al.* (2015). In the proposed reaction mechanism, the formation of an adduct ($^-O_3SO_5^-$) between the reacting species was suggested by the authors (reactions 10.13–10.18).

$$SO_5^{2-} + O_3 \rightarrow {}^-O_3SO_5^- \quad (10.13)$$

$$^-O_3SO_5^- \rightarrow SO_5^{\bullet-} + O_3^{\bullet-} \quad (10.14)$$

$$^-O_3SO_5^- \rightarrow SO_4^{2-} + 2O_2 \quad (10.15)$$

$$O_3^{\bullet-} \rightleftharpoons O^{\bullet-} + O_2 \quad (10.16)$$

$$O^{\bullet-} + H_2O \rightleftharpoons HO^{\bullet} + OH^- \quad (10.17)$$

$$SO_5^{\bullet-} + O_3 \rightarrow SO_4^{\bullet-} + 2O_2 \quad (10.18)$$

10.2.4.2 Photocatalysis in presence of persulfate

It was observed that like in the case of H_2O_2, the $S_2O_8^{2-}$ or HSO_5^- addition was beneficial in the photocatalysis oxidation and mineralisation of dyes. The sulfate radical formed and reactions of persulfate with the photogenerated conduction band electrons (reaction 10.19) can exert a dual function: as strong oxidant with production of additional oxidizing species and as electron scavenger, thus inhibiting the electron-hole recombination at the semiconductor surface (Konstantinou & Albanis, 2004; Madhavan *et al.* 2006).

$$S_2O_8^{2-} + e^-_{CB} \rightarrow SO_4^{\bullet-} + SO_4^{2-} \quad (10.19)$$

Most results demonstrate that peroxydisulfate significantly enhances the photocatalytic activity. However, an excessive dosage of persulfate may result in the saturation of reaction rate due to adsorption of the SO_4^{2-} end-product onto the TiO_2 surface (Syoufian & Nakashima, 2007).

10.2.4.3 Persulfate activation by reaction with species from water molecules

Pulse radiolysis of water was used for the determination of many rate constants (Neta *et al.* 1988). The radiolysis of water forms short-lived intermediates i.e. hydrated electrons, hydrogen atoms and hydroxyl radicals. Numerous kinetic studies involving the sulfate radical have used the reduction of peroxydisulfate ions with the hydrated electron (reaction 10.20) as a means of radical generation.

$$S_2O_8^{2-} + e^-_{aq} \rightarrow SO_4^{\bullet-} + SO_4^{2-} \quad k = 1.2 \times 10^{10}\ L\,mol^{-1}\,s^{-1} \quad \text{(Buxton et al. 1988)} \quad (10.20)$$

Recently, Criquet and Leitner (2011, 2012) showed that the oxidizing properties of the sulfate radicals formed in the reaction of peroxydisulfate ions with the hydrated electrons during electron beam irradiation of aqueous solutions containing carboxylic acids are responsible for enhancing the degradation and mineralization of these compounds.

In sonochemical systems, acoustic cavitation is also known to generate OH and H radicals due to thermal decomposition of water. Thus, these species, especially the reductive $H^{\bullet}$ can react with peroxydisulfate ions to produce the sulfate radicals in ultrasound/$S_2O_8^{2-}$ processes. In addition, peroxydisulfate ions can also be decomposed into sulfate radicals by the heat generated within the bubbles formed in such sonochemical systems (Neppolian *et al.* 2010).

Additional persulfate activation processes and techniques for organic contaminant treatment (e.g. activation by electrochemistry or by activated carbon) are covered in a recent review by Matzek and Carter (2016).

10.3 PROPERTIES AND STABILITY OF SULFATE RADICAL IN PURE WATER

10.3.1 Oxidation-reduction potential

In aqueous solutions, the standard oxidation-reduction potentials (E^0) of ozone ($E^0 = 2.07$ V), hydrogen peroxide ($E^0 = 1.78$ V), permanganate ($E^0 = 1.70$ V) and persulfate (~2.01 V), are close. The sulfate radical has an even higher redox potential ($E^0 = 2.6$ V). Once formed, the sulfate radical is a short-lived but very strong oxidizing agent. $SO_4^{\bullet-}$ is the strongest oxidant among the sulfur-oxygen radicals $SO_n^{\bullet-}$ (with n = 3, 4, and 5 and redox potentials of 0.63, 2.5–3.1, and 1.1 V respectively at pH 7; Neta *et al.* 1988; Antoniou *et al.* 2010a).

Table 10.1 summarizes the radical rate constants of a few classes of organic compounds. It is clear that the reactivity of $SO_4^{\bullet-}$ is generally lower than that of the $HO^\bullet$ except in the case of aromatics.

Table 10.1 Rate constants (k/L mol^{-1} s^{-1}) for the reaction of $HO^\bullet$ and $SO_4^{\bullet-}$ radicals with selected carboxylic acids, methyl ether, alcohols, aromatics and amines in aqueous solution at 292 K.

Compound	$HO^\bullet$	$SO_4^{\bullet-}$
(a) HCO_2^-	4.3×10^9	1.1×10^8
(a) HCOOH	1.3×10^8	4.6×10^5
(a) CH_3COO^-	7.5×10^7	3.7×10^6
(a) CH_3COOH	1.8×10^7	1.4×10^4
(a) $CH_3CO_2CH_3$	1.2×10^8	5.0×10^4
(b) methanol	9.7×10^8	3.2×10^6
(b) ethanol	$1.2–2.8 \times 10^9$	$1.6–7.7 \times 10^7$
(b) propanol	2.8×10^9	6×10^7
(b) *tert*-butyl alcohol	$3.8–7.6 \times 10^8$	$4–9.1 \times 10^5$
Aromatics		
(b) Benzene	7.8×10^9	$2.4–3 \times 10^9$
(b) Nitrobenzene	$3.0–3.9 \times 10^9$	$<10^6$
(b) Phenol	6.6×10^9	8.8×10^9
(b) Benzoic acid	4.2×10^9	1.2×10^9
(b) Anisole	7.8×10^9	4.9×10^9
(c) 4-Chlorotoluene	5.5×10^9	1.1×10^9
(c) 4-Bromotoluene	2.9×10^9	1.0×10^9
Amines and Amino acids		
(d) Alanine	7.7×10^7	$(4.9 \pm 0.1) \times 10^6$
(d) Glycine	1.7×10^7	$(3.7 \pm 0.1) \times 10^6$
(d) Tryptophan		2.0×10^9
(d) Tyrosine		5.8×10^8

(a) Buxton *et al.* (2001); (b) cited in Liang and Su (2009); (c) Choure *et al.* (1997); (d) Bosio *et al.* (2005).

10.3.2 pH dependence

In pure water, the sulfate radical concentration can decrease due to the recombination and the reaction with the oxidant:

$$SO_4^{\bullet -} + SO_4^{\bullet -} \rightarrow S_2O_8^{2-} \quad 2k = (8.9 \pm 0.3) \times 10^8\,M^{-1}\,s^{-1} \quad \text{(McElroy \& Waygood, 1990)} \tag{10.21}$$

$$S_2O_8^{2-} + SO_4^{\bullet -} \rightarrow S_2O_8^{\bullet -} + SO_4^{2-} \quad k < 1 \times 10^4 - 7 \times 10^5\,M^{-1}\,s^{-1} \quad \text{(Chitose } et\ al. \text{ 1999)} \tag{10.22}$$

In aqueous solution, the sulfate radicals can also participate in pH-dependent reactions to produce hydroxyl radicals (Liang & Su, 2009):

$$SO_4^{-\bullet} + H_2O \rightarrow H^+ + SO_4^{2-} + OH^\bullet \quad \text{(over the entire pH range)} \quad k[H_2O] < 2 \times 10^{-3}\,s^{-1} \tag{10.23}$$

$$SO_4^{\bullet -} + OH^- \rightarrow SO_4^{2-} + OH^\bullet \quad \text{(predominant in alkaline media)} \quad k = (6.5 \pm 1.0) \times 10^7\,M^{-1}\,s^{-1} \tag{10.24}$$

While the contribution of reaction 10.23 is small in most reaction systems (Norman *et al.* 1970), in alkaline solutions, the reaction of sulfate radical with hydroxide ion occurs very rapidly (reaction 10.24).

Therefore, the dominant radical species ($SO_4^{\bullet -}$ *vs* $OH^\bullet$) changes depending on the solution pH. Figure 10.2 exemplifies the relative distribution of sulfate and hydroxyl radicals in water as a function of pH.

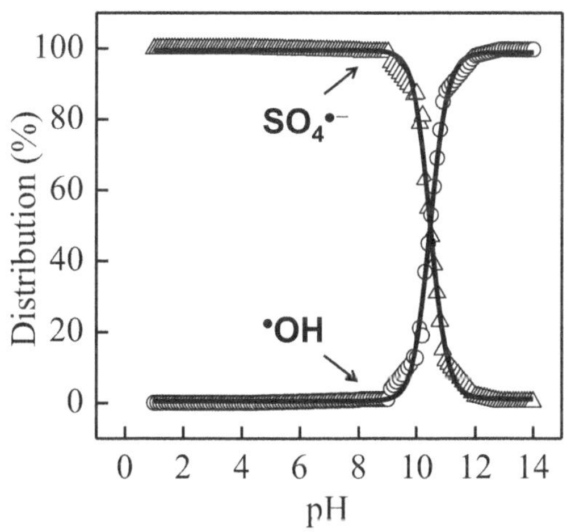

Figure 10.2 Effect of pH on the distribution of radical species in pure water. Adapted from Fang *et al.* (2012); courtesy of Dr. Guodong Fang, 2016.

Dogliotti and Hayon (1967) monitored the absorbance of sulfate radical ion at 455 nm during the flash photolysis of persulfate solutions of varying pH; virtually no absorption was measured at pH > 10.8, which indicates the total conversion of $SO_4^{\bullet -}$ into $OH^\bullet$ in strong alkaline media (Figure 10.3).

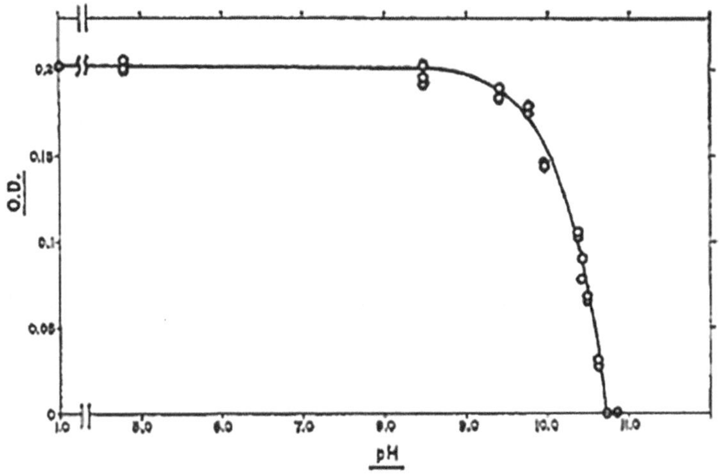

Figure 10.3 Variation of optical density of $SO_4^{\bullet-}$ transient at 455 nm with pH in the flash photolysis of 10^{-2} M persulfate solutions. Reprinted with permission from Dogliotti and Hayon (1967). Copyright 2016 American Chemical Society.

The water pH impacts the effectiveness of contaminant removal with $SO_4^{\bullet-}$ based AOTs through the following parameters:

- rate of generation of radical species, which can depend on the oxidant speciation for certain activating systems; thus when considering the elimination of benzoic acid in the UV/PMS system, Guan *et al.* (2011) concluded that the production of $SO_4^{\bullet-}$ and $OH^{\bullet}$ increases with pH in the range 8–11 due to the increase of the form SO_5^{2-} which has a higher molar absorption coefficient than the form HSO_5^-. This observation has similarity with the UV/H_2O_2 system whereas it does not apply in the case of the $S_2O_8^{2-}/UV$ process (see § 10.2.2).
- relative distribution of radical species in water (e.g. $SO_4^{\bullet-}$ *vs* $OH^{\bullet}$; Figure 10.2; or other like $Cl^{\bullet}/Cl_2^{\bullet-}$ *vs* $OH^{\bullet}$; see § 10.6 and Chapter 9);
- ionization state of some organic compounds, i.e. protonated or deprotonated, given that these forms display different reactivity towards the oxidizing radical species;
- radical scavenging capacity of the water matrix and its pH-dependence (e.g. reactions with bicarbonates/carbonates, dissolved organic matter).

The overall pH impact on the contaminant degradation yield would be dictated by the combination of all the above conditions.

It should also be noted that in unbuffered or low alkalinity waters, an acidification is usually observed in the PDS-activated AOPs due to HSO_4^- formed as a primary stable reaction product and to the presence of acid transformation products and protons.

10.4 REACTION MECHANISMS WITH ORGANIC MOLECULES IN PURE WATER

The sulfate radical anion reacts with organic compounds of various classes following three basic processes: hydrogen atom abstraction, one-electron transfer, or addition to unsaturated bonds.

Hydrogen atom abstraction:

$$RH + SO_4^{\bullet -} \rightarrow R^{\bullet} + HSO_4^{-} \quad (10.25)$$

One electron oxidation:

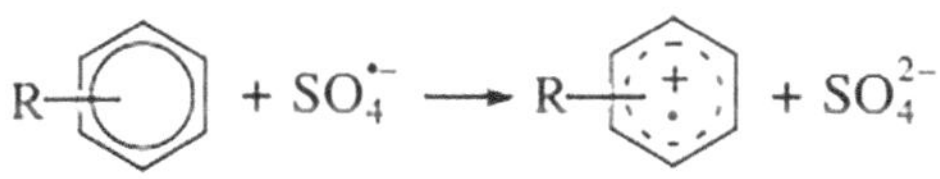

(10.26)

Addition to an unsaturated bond:

$$H_2C{=}CHR + SO_4^{\bullet -} \rightarrow {}^{-}OSO_2OCH_2{-}C^{\bullet}HR \quad (10.27)$$

Sulfate radical is more selective than the HO radical in the reactions with organic and inorganic compounds. Mechanisms of $SO_4^{\bullet -}$ reactions differ somewhat from those of hydroxyl radical reactions; for example, sulfate radical reacts more readily by electron transfer than $HO^{\bullet}$ but slower by H-abstraction and addition. Thus, the products of sulfate radical-based oxidation may be different than those of OH radical-induced oxidations. The following sub-sections provide examples of reactions of sulfate radical anion with various classes of organic compounds.

10.4.1 Hydrogen-abstraction reactions

10.4.1.1 Alkanes and alcohols

The reactions of alcohols, alkanes and ethers with $SO_4^{\bullet -}$ proceed *via* H-abstraction. The rate constants for H-abstraction reactions of $SO_4^{\bullet -}$ are about one to two orders of magnitude smaller than those of $HO^{\bullet}$ reactions.

In the absence of oxygen, the radicals formed in the H-abstraction reactions can be involved in chain reactions as described by Mendez-Diaz *et al.* (2010):

$$SO_4^{\bullet -} + RH_2 \rightarrow SO_4^{2-} + H^+ + RH^{\bullet} \quad (10.28)$$

$$RH^{\bullet} + S_2O_8^{2-} \rightarrow R + SO_4^{2-} + H^+ + SO_4^{\bullet -} \quad (10.29)$$

$$SO_4^{\bullet -} + RH \rightarrow R^{\bullet} + SO_4^{2-} + H^+ \quad (10.30)$$

$$2R^{\bullet} \rightarrow RR \text{ (dimer)} \quad (10.31)$$

$$SO_4^{\bullet -} + H_2O \rightarrow HSO_4^{-} + HO^{\bullet} \quad (10.32)$$

$$HO^{\bullet} + S_2O_8^{2-} \rightarrow HSO_4^{-} + SO_4^{\bullet -} + \tfrac{1}{2}O_2 \quad (10.33)$$

Levey and Hart (1975) suggested the following chain mechanism for the oxidation of alcohols by sulfate radical:

$$SO_4^{-\bullet} + R_2CHOH \rightarrow HSO_4^{-} + R_2C^{\bullet}OH \quad (10.34)$$

$$R_2C^{\bullet}OH + S_2O_8^{2-} \rightarrow HSO_4^{-} + SO_4^{\bullet -} + R_2C = O \quad (10.35)$$

In the presence of oxygen, organic peroxyl radicals ($ROO^\bullet$) are formed. Mark *et al.* (1990) provided a detailed mechanism of *tert*-butanol (*tert*-butyl alcohol, TBA) oxidation in the presence of oxygen *via* a bimolecular decay of *tert*-butanol-derived peroxyl radical with formation of carbonyl compounds:

$$SO_4^{\bullet-} + (CH_3)_3COH \rightarrow SO_4^{2-} + H^+ + {}^\bullet CH_2C(CH_3)_2OH \tag{10.36}$$

$${}^\bullet CH_2C(CH_3)_2OH + O_2 \rightarrow {}^\bullet O_2CH_2C(CH_3)_2OH \tag{10.37}$$

$$2{}^\bullet O_2CH_2C(CH_3)_2OH \rightarrow HOC(CH_3)_2CH_2O_4CH_2C(CH_3)_2OH \quad (RO_4R) \tag{10.38}$$

$$RO_4R \rightarrow HOC(CH_3)_2CH_2OH + O_2 + HOC(CH_3)_2CHO \tag{10.39}$$

$$RO_4R \rightarrow 2\,HOC(CH_3)_2CHO + H_2O_2 \tag{10.40}$$

$$RO_4R \rightarrow 2\,CH_2O + 2{}^\bullet C(CH_3)_2OH + O_2 \tag{10.41}$$

$${}^\bullet C(CH_3)_2OH + O_2 \rightarrow HO_2^\bullet + (CH_3)_2CO \tag{10.42}$$

This reaction mechanism is similar to that known from $HO^\bullet$ attack on TBA.

10.4.2 Electron transfer reactions

10.4.2.1 Carboxylic acids

Hydroxyl radicals have been reported to react with substituted acetates by hydrogen abstraction (electrophilic attack) from the substituted methyl group ($—CR_1R_2H$):

$$HO^\bullet + R_1—CHR_2COO^- \rightarrow H_2O + R_1—C^\bullet R_2COO^- \tag{10.43}$$

Oxidation of carboxylic acids by $HO^\bullet$ results in minor decarboxylation.

Unlike the $OH^\bullet$ reaction, the H-abstraction reaction of $SO_4^{\bullet-}$ with aliphatic carboxylic acids is not the predominant degradation pathway of these compounds. Instead, electron-transfer reaction with the carboxylate ion appears to be more favored, and leads to decarboxylation (Madhavan *et al.* 1978). For instance, the sulfate radicals are known to react by an electron transfer mechanism with the carboxylate group of acetate and halogenated acetates, yielding carboxy radicals. The carboxy radicals undergo further decarboxylation to the $C^\bullet R_1R_2H$ radicals (reactions 10.44, 10.45 where R_1, R_2: CH_3, H, Br, Cl, F, NH_3^+; Chawla & Fessenden, 1975; Madhavan *et al.* 1978; Bosio *et al.* 2005):

$$SO_4^{\bullet-} + R_1CHR_2COO^- \rightarrow SO_4^{2-} + R_1CHR_2COO^\bullet \tag{10.44}$$

$$R_1CHR_2COO^\bullet \rightarrow {}^\bullet CR_1R_2H + CO_2 \tag{10.45}$$

Criquet and Leitner (2009) compared acetate and TOC removal in the H_2O_2/UV and $S_2O_8^{2-}$/UV processes and observed similar acetate decay patterns, but greater TOC abatement during the UV/persulfate process. The authors explained the better performance of mineralization through the decarboxylation reaction *via* sulfate radicals (reaction 10.46) and the oxidation of intermediates (methanol, formaldehyde and formate ions). Unlike the UV/persulfate process, the H_2O_2/UV-induced degradation of acetate results in the formation of glyoxylate, glycolate, and oxalate ions.

$$SO_4^{\bullet-} + CH_3COO^- \rightarrow SO_4^{2-} + {}^{\bullet}CH_3 + CO_2 \qquad (10.46)$$

Acetate, malonate, propionate, and succinate ions undergo decarboxylation with a minimum 80% efficiency (Madhavan *et al.* 1978). Similarly to the OH radical, the sulfate radical reacts slower with protonated species than with deprotonated species.

10.4.2.2 Aminoacids and amines

Studies by Madhavan *et al.* (1978) and Bosio *et al.* (2005) on amino acids (and peptides) degradation by sulfate radicals showed high decarboxylation yields, which points to the carboxyl group as the favored attack site for $SO_4^{\bullet-}$. Peptides are much more reactive than their parent amino acids, and the aromatic moieties confer a higher reactivity. In oxygen-depleted conditions, the carbon-centered radical is involved in the reaction with the oxidant (PDS) as exemplified below for glycine:

$$H_2N{-}CH_2CO_2^- + SO_4^{\bullet-} \rightarrow H_2N{-}C^{\bullet}H_2 + CO_2 + SO_4^{2-} \qquad (10.47)$$

$$H_2NC^{\bullet}H_2 + S_2O_8^{2-} \rightarrow HN{=}CH_2 + H^+ + SO_4^{2-} + SO_4^{\bullet-} \qquad (10.48)$$

Oxidation of amine-based molecules by sulfate radicals would also proceed through electrophilic attack at the nitrogen atoms with formation of N-centered radicals *via* one electron transfer, as in the case of cytidine (Aravindakumar *et al.* 2003). Thus, the aniline-derivative oxidation with $SO_4^{\bullet-}$ starts with one-electron transfer to the sulfate radical with formation of a N-centered radical cation (R-$NH_2^{\bullet+}$) which hydrolyzes rapidly to hydroxylamine derivatives (R-NHOH), and not through radical addition to the aromatic ring as commonly observed in $OH^{\bullet}$ oxidation (Mahdi-Ahmed *et al.* 2012).

10.4.2.3 Aromatics

In the reaction of sulfate radical with electron-rich aromatic compounds, the electron transfer from the aromatic ring to $SO_4^{\bullet-}$ is the preferred reaction pathway. Zemel and Fessenden (1978) studied the products of the reactions of $SO_4^{\bullet-}$ with some benzene derivatives using ESR and optical absorption spectrometry. All reactions are consistent with the oxidation of the benzene ring to an intermediate cationic species:

$$C_6H_4(R_1R_2) + SO_4^{\bullet-} \rightarrow {}^{\bullet+}C_6H_4(R_1R_2) + SO_4^{2-} \qquad (10.49)$$

This high selectivity of the electron transfer reaction from the aromatic ring to sulfate radical was confirmed by Neta *et al.* (1977), who observed a linear correlation between the rate constants for the reaction of $SO_4^{\bullet-}$ with a number of substituted benzenes and the Hammett substituent constant σ which reflects the electron charge distribution on the ring. The authors determined the bimolecular rate constants using pulse radiolysis.

The radical cation formed reacts rapidly with water molecule to form the hydroxycyclohexadienyl radical, which is the same intermediate formed upon reaction of $HO^{\bullet}$ with aromatic rings. Parallel reactions also occur depending on the substituents on the aromatic ring, i.e. either loss of a proton (from hydroxyl or methyl group) or loss of CO_2 from the carboxyl functional groups (Zemel & Fessenden, 1978).

Thus, with a very short lifetime, the cation radical intermediate species resulting from sulfate radical reaction with benzoate ion can follow two pathways, i.e., decarboxylation to a phenyl radical, and hydrolysis and protonation to hydroxycyclohexadienyl radical (Zemel & Fessenden, 1978). The branching ratio of the two pathways is 70:30 (Figure 10.4), indicating that the loss of CO_2 from the oxidized form is the most important pathway (70%) in benzoic acid degradation *via* $SO_4^{\bullet-}$, whereas the decarboxylation yield resulting from the hydroxyl radical reaction with benzoate ion represents less than 10% of the overall degradation mechanism. Similar results were reported by Criquet and Leitner (2015) in a recent study on the degradation of *p*-hydroxybenzoic acid by sulfate radicals generated by electron beam irradiation. Benzoquinone was the main transformation product detected in the presence of persulfate with no polyhydroxybenzoic acids being formed, whereas polyhydroxylated derivatives of *p*-hydroxybenzoic acid were found in the OH• radical-based ionizing radiation system in absence of persulfate.

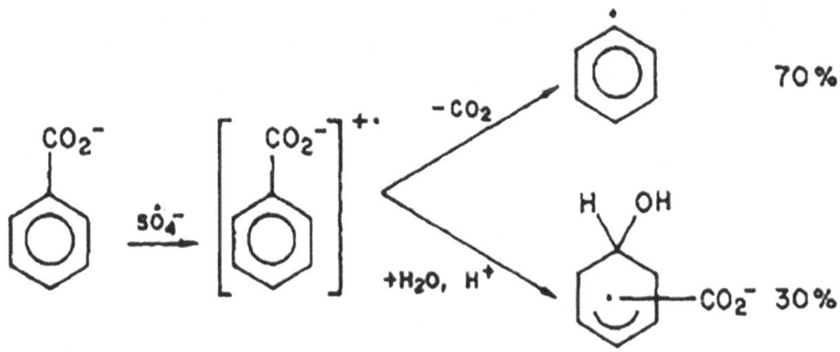

Figure 10.4 Mechanism of reaction of $SO_4^{\bullet-}$ with benzoate proposed by Zemel and Fessenden (1978). Reprinted with permission. Copyright 2016 American Chemical Society.

Indeed, the primary radical cation produced in the reaction of the sulfate radical with aromatics follows various paths to more stable intermediate radicals. Zemel and Fessenden (1978) studied the products of the reaction of $SO_4^{\bullet-}$ with three toluate isomers and concluded on the following pathways:

- *o*-toluate gives mainly the *o*-carboxybenzyl radical by loss of H^+ from the methyl group;
- *m*-toluate would lead to the formation of a mixture of OH adducts (hydroxycyclohexadienyl radicals) through hydroxylation *via* reaction of primary radical cation with water molecules with no trace of *m*-carboxybenzyl radicals;
- *p*-toluate: 25–35% and 25–45% of primary radical cation evolve to *p*-carboxybenzyl and two methyl-substituted hydroxycyclohexadienyl radicals, respectively, the other 20–50% would decarboxylate to *p*-methylphenyl radical.

Figure 10.5 presents the evolution of the primary radical cation to the intermediate radicals.

Even though the radical intermediates generated through sulfate radical reactions with aromatics are often different than those formed in the hydroxyl-radical mediated degradation reactions of organic contaminants, many transformation products are identical. For example, Anipsitakis *et al.* (2006) identified catechol, hydroquinone, and quinone from the sulfate radical attack of phenol which are also common transformation products found in OH•-based AOTs. The authors explained these results by the formation of the dihydroxycyclohexadienyl radical (the same as that involved in the attack of aromatic rings by the OH radical) from the primary radical cation produced in the initial reaction of the sulfate radical on phenol.

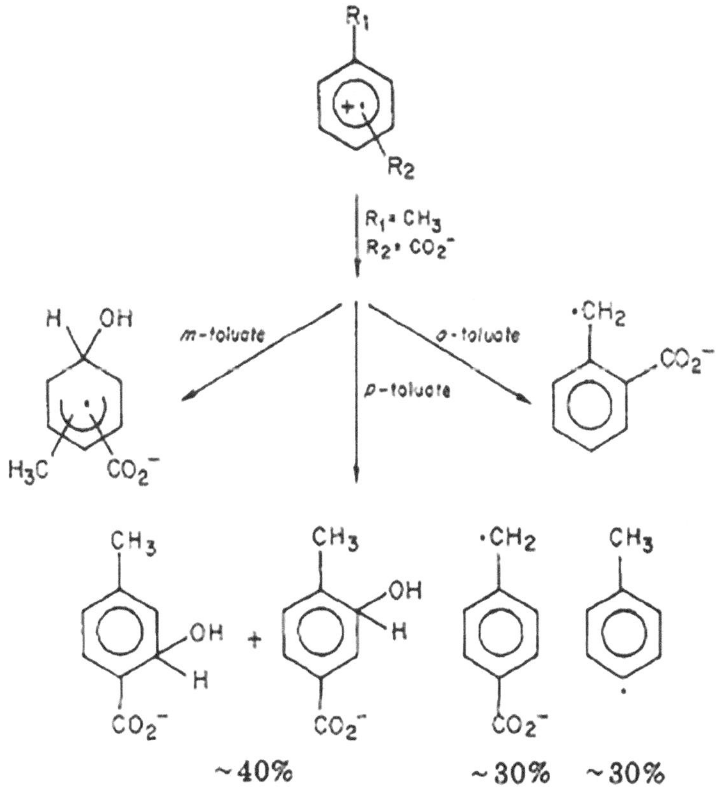

Figure 10.5 Product yields in the reaction of $SO_4^{\bullet-}$ with three isomeric toluate ions. Adapted with permission from Zemel and Fessenden (1978). Copyright 2016 American Chemical Society.

10.4.3 Addition to unsaturated bonds

Norman *et al.* (1970) suggested that the sulfate radical anion reacts with the unsaturated structure of 3,3-dimethylacrylic acid or its anion either by addition to the double bond followed by heterolysis (reactions 10.50 and 10.51) or by electron transfer (reaction 10.52):

$$(CH_3)_2C{=}CH{-}COOH + SO_4^{\bullet-} \rightarrow (CH_3)_2C(OSO_2O^-){-}C^{\bullet}H{-}COOH \rightarrow (CH_3)_2C^+{-}C^{\bullet}H{-}COOH + SO_4^{2-} \quad (10.50)$$

$$(CH_3)_2C{-}CH \quad COOH + SO_4^{\bullet-} \rightarrow (CH_3)_2C^{\bullet}{-}C(OSO_2O^-)H{-}COOH \rightarrow (CH_3)_2C^{\bullet}{-}C^+H{-}COOH + SO_4^{2-} \quad (10.51)$$

$$(CH_3)_2C{=}CH{-}COOH + SO_4^{\bullet-} \rightarrow (CH_3)_2C^+{-}C^{\bullet}H{-}COOH + SO_4^{2-} \quad (10.52)$$

The radical cations resulted from the two pathways hydrolyze to form the hydroxylated radicals ($(CH_3)_2C(OH){-}C^{\bullet}H{-}COOH$ and $(CH_3)_2C^{\bullet}{-}C(OH)H{-}COOH$), which are similar to those arising from the initial $HO^{\bullet}$ attack.

Similarly, when sulfate radicals are involved in the oxidation of trichloroethylene (TCE), an electron-transfer mechanism from olefin to $SO_4^{\bullet-}$ was proposed (Liang *et al.* 2009), followed by decomposition of the radical cation in the aqueous medium to form the hydroxyalkyl radical (analogous to the one formed directly by OH radical addition):

$$Cl_2C{=}CHCl + SO_4^{\bullet-} \rightarrow Cl_2C^{\bullet}{-}C^{+}HCl + SO_4^{2-} \quad (10.53)$$

$$Cl_2C^{\bullet}{-}C^{+}HCl + H_2O \rightarrow (OH)CHCl{-}C^{\bullet}Cl_2 + H^{+} \quad (10.54)$$

10.5 SULFATE RADICAL-BASED TREATMENT OF WATER MICROPOLLUTANTS

There is an abundant literature on the potential use of sulfate radical-based advanced oxidation systems for the degradation of many organic contaminants of concern over their impact on environment, aquatic wildlife, and human health, given their occurrence and persistence. A compilation of these studies has been published recently (Zhang *et al.* 2015). Many studies at laboratory scale were focused on the elimination of contaminants frequently detected in aquatic environment and drinking water sources and transformation product formation. Only selected classes of micropollutants will be exemplified in this section. Table 10.2 presents the $SO_4^{\bullet-}$ second order rate constants for the reactions with a series of environmental contaminants, as collected from the recently published literature. The kinetic data as well as the full-studies indicate that sulfate radical-based AOTs are effective at degrading the studied contaminants at comparable or higher rates than those determined in hydroxyl radical-based AOTs.

Table 10.2 Bimolecular rate constants for $SO_4^{\bullet-}$ reactions with relevant environmental micropollutants.

Micropollutant (P)	$k_{SO_4^{\bullet-},P}$ /10^9 M^{-1} s^{-1}	Reference
	Pesticides	
Atrazine	3 (pH5)	Manoj *et al.* (2007)
Atrazine	3.5 ± 0.08 (pH7-8)	Lutze *et al.* (2015a)
Atrazine	2.59	Khan *et al.* (2014)
Deisopropyl-atrazine	2.0 ± 0.57 (pH7-8)	Lutze *et al.* (2015a)
Desethyl-atrazine	0.96 ± 0.017 (pH7-8)	Lutze *et al.* (2015a)
Desethyl-desisopropyl-atrazine	0.15 ± 0.008 (pH7-8)	Lutze *et al.* (2015a)
Terbuthylazine	3.0 ± 0.05 (pH7-8)	Lutze *et al.* (2015a)
Propazine	2.2 ± 0.05 (pH7-8)	Lutze *et al.* (2015a)
Desethyl-terbuthylazine	0.36 ± 0.023 (pH7-8)	Lutze *et al.* (2015a)
1,3,5-triazine	0.081 (pH5)	Manoj *et al.* (2007)
2,4,6-trimethoxy-1,3,5-triazine	0.051 (pH5)	Manoj *et al.* (2007)
2,4-dioxohexahydro-1,3,5-triazine	0.046	Manoj *et al.* (2007)
Amitrol	0.377 ± 0.017 (pH7)	Orellana-Garcia *et al.* (2015)
	0.258 ± 0.028 (pH3)	
	0.292 ± 0.013 (pH12)	

(Continued)

Table 10.2 Bimolecular rate constants for $SO_4^{\bullet-}$ reactions with relevant environmental micropollutants. (*Continued*)

Micropollutant (P)	$k_{SO_4^{\bullet-},P}$ **/10^9 M^{-1} s^{-1}**	**Reference**
Glyphosate	0.16 ± 0.01 (pH4-5)	Diaz Kirmser *et al.* (2010)
Paraquat	1.2 ± 0.3 (pH4-5)	Diaz Kirmser *et al.* (2010)
Clomazone	0.94 ± 0.04 (pH4.4)	Gara *et al.* (2009)
	Algal toxins and T&O causing compounds	
Cylindrospermopsin	4.5 (pH 7.4)	He *et al.* (2013)
2-methylisoborneol (2-MIB)	0.42 ± 0.06 (pH 4-7)	Xie *et al.* (2015)
Geosmin	0.76 ± 0.06 (pH 7.0)	Xie *et al.* (2015)
	Polychlorinated biphenyls (pH 5)	
2-Monochlorobiphenyl	0.21	Fang *et al.* (2012)
4-Monochlorobiphenyl	0.199	Fang *et al.* (2012)
2,4-Dichlorobiphenyl	0.089	Fang *et al.* (2012)
2,4′-Dichlorobiphenyl	0.0688	Fang *et al.* (2012)
2,2′-Dichlorobiphenyl	0.0265	Fang *et al.* (2012)
4,4′-Dichlorobiphenyl	0.131	Fang *et al.* (2012)
2,4,4′-Trichlorobiphenyl	0.025	Fang *et al.* (2012)
	Pharmaceuticals	
Amoxicillin*	3.48 ± 0.05 (pH 7.4)	Rickman and Mezyk (2010)
Penicillin G*	2.08 ± 0.04 (pH 7.4)	Rickman and Mezyk (2010)
Penicillin V*	2.89 ± 0.05 (pH 7.4)	Rickman and Mezyk (2010)
Piperacillin*	1.74 ± 0.11 (pH 7.4)	Rickman and Mezyk (2010)
Cefaclor*	2.44 ± 0.04 (pH 7.4)	Rickman and Mezyk (2010)
Sulfamethoxazole	12.5 ± 3.1 (pH7)	Mahdi-Ahmed *et al.* (2012)
Diclofenac	9.2 ± 2.6 (pH7)	Mahdi-Ahmed *et al.* (2012)
Carbamazepine	1.92 ± 0.01 (pH3)	Matta *et al.* (2011)
	2.63	Rao *et al.* (2013)
Ciprofloxacin	1.2 ± 0.1 (pH7)	Mahdi-Ahmed and Chiron (2014)
Atenolol	13 (pH7)	Liu *et al.* (2013a)
	22	Liu *et al.* (2013b)
Iopromide	(1.5 ± 0.5) × 10^4 M^{-1} s^{-1}	Chan *et al.* (2010)

**Beta-lactam antibiotics*

Usually, $SO_4^{\bullet-}$ rate constants have been determined by competition kinetics, laser flash photolysis, pulse radiolysis or by solving equations of the reaction system based on the radical steady-state assumption. In a recent study, Xiao *et al.* (2015) developed a Quantitative Structure Activity Relationship (QSAR) model for $SO_4^{\bullet-}$ reactivity prediction. An empirical relationship with two descriptors, i.e. E_{LUMO} (Energy of the

Lowest Unoccupied Molecular Orbital) and E_{HOMO} (Energy of the Highest Occupied Molecular Orbital) energy gap and the ratio of oxygen atoms to carbon atoms, was proposed to provide a reliable mean to predict $k_{SO_4^{-\bullet}}$ of trace organic compounds.

10.5.1 Pesticides

The rate constants of $SO_4^{\bullet-}$ with many substituted triazine pesticides and their metabolites were determined by Manoj *et al.* (2007) and Lutze *et al.* (2015a). The rate constants for $SO_4^{\bullet-}$ reactions (Table 10.2) are smaller than those for the $OH^\bullet$ reactions with similar triazines except that for atrazine. The degradation products of $SO_4^{\bullet-}$ reaction with atrazine were identified, and potential degradation pathways were proposed (Lutze *et al.* 2015a; Khan *et al.* 2014). From transient absorption spectra, intermediate species were assigned to carbon centered radicals, OH-adducts and N-centered radicals. Among various possible mechanisms, dealkylation appears to predominate, with the removal of ethyl group being favored over the removal of the isopropyl substituent (Lutze *et al.* 2015a).

Degradation studies on other pesticides, such as diuron, amitrol, endosulfan, clomazone, paraquat, and glyphosate focused on kinetic aspects and on the role of experimental parameters (temperature, pH, oxidant dose, initial pollutant concentration, alkalinity and inorganic anions, organic matter) (Tan *et al.* 2012; Orellana-Garcia *et al.* 2015; Shah *et al.* 2013; Diaz Kirmser, 2010; Gara *et al.* 2009). Along with other environmental chemical pollutants, pesticides were proven to interfere with the endocrine system, and can produce adverse developmental, reproductive, neurological, cardiovascular, metabolic, and immune effects in humans (Schug *et al.* 2011).

10.5.2 Pharmaceuticals

The reactions of sulfate radical with pharmaceuticals of various structures and biochemical activities are also well represented in the literature. Such studies include antimicrobial agents (sulfamethoxazole, sulfamethazine, ciprofloxacin, *β*-lactam antibiotics), anti-inflammatory drugs and analgesics (ibuprofen, diclofenac), X-ray contrast media (iopromide), antiepileptic drugs (carbamazepine), and *β*-blockers (atenolol) (Ghauch *et al.* 2013; Mahdi-Ahmed *et al.* 2012; Gao *et al.* 2012; Mahdi-Ahmed & Chiron, 2014; Rickmann & Mezyk, 2010; Roshani & Leitner, 2011; Chan *et al.* 2010; Deng *et al.* 2013; Matta *et al.* 2011; Rao *et al.* 2013; Liu *et al.* 2013a, 2013b). The successful reduction of most of these pollutants in water (initial concentration range: 20–500 μM) was achieved and the impact of various water quality and process parameters on the performance of sulfate radical-induced oxidation as well as on the yield of $SO_4^{\bullet-}$ generation were addressed in these and other literature studies.

The reactivity of aniline-based pharmaceuticals toward sulfate radical anion was studied using diclofenac and sulfamethoxazole as probe compounds (Mahdi-Ahmed *et al.* 2012). It was reported that these compounds react with sulfate radical anion at diffusion-controlled kinetics, with second order rate constants in the $9 \times 10^9 - 13 \times 10^9\ M^{-1}\ s^{-1}$ range. It was suggested that selective oxidative reaction started with the formation of a N-centered radical through one-electron transfer from the amino group of pharmaceutical to sulfate radical anion. This was followed by a sequence of degradation reactions involving decarboxylation, hydroxylation, and bond cleavage.

10.5.3 Algal toxins and taste-and-odor (T&O) causing compounds

Algal toxins (powerful neurotoxins and hepatotoxins) and T&O compounds are currently treated in surface water for drinking water production using the UV/H_2O_2 AOP. To the best of our knowledge only few studies explored the potential use of sulfate radical-based AOPs for the degradation of these compounds. He *et al.* (2013) found comparable second-order rate constants of the cyanobacterial toxin cylindrospermopsin with hydroxyl and sulfate radicals (5.1×10^9 M^{-1} s^{-1} and 4.5×10^9 M^{-1} s^{-1}, respectively). The other few studies (Antoniou *et al.* 2010a, 2010b) have shown that sulfate radical-based AOPs are capable at degrading the cyanotoxin microcystin-LR (1 μM) at comparable and/or higher rates than selected hydroxyl radical-based AOPs, and that a complex mixture of intermediate products are formed from sulfate radical attack on different sites of the MC-LR structure (benzene ring, diene bonds, unsaturated C-bond, oxidative bond cleavage).

2-Methylisoborneol (2-MIB) and geosmin are also produced from *Cyanobacteria* during the algae bloom seasons, and, although non-toxic, they affect the water aesthetics (odor). The degradation of 2-MIB and geosmin using the UV/persulfate process has been satisfactorily modeled and accounted for the $SO_4^{\bullet-}$ reactions with carbonate species and the organic matter present in the solution (Xie *et al.* 2015).

10.5.4 Volatile organic compounds (VOCs)

Many studies concerned with the role of various factors such as oxidant/compound molar ratio and water matrix characteristics (e.g., pH, Cl^- and HCO_3^- concentrations, presence of humic acid) on the removal of VOCs, among which methyl *tert*-butyl ether (MTBE), 1,1,1-trichloroethane (TCA), 1,4-dioxane, trichloroethylene, BTEX, whereas other studies focused on degradation intermediates and possible oxidation pathways. One of the earliest studies investigated the UV-light assisted degradation of the industrial solvent 1,4-dioxane (400 μM) in the presence of peroxydisulfate, and showed the formation of ethylene glycol diformate, formic acid, and formaldehyde as intermediates indicating an oxidative ring opening mechanism (Maurino *et al.* 1997). These compounds were also postulated mechanistically and observed experimentally in the 1,4-dioxane degradation *via* OH radical oxidation during the UV/H_2O_2 AOP (Stefan & Bolton, 1998).

In a study comparing H_2O_2/UV and $S_2O_8^{2-}$/UV performance for the degradation of 1,1,1-trichloroethane (TCA; 0.15–0.75 mM), Gu *et al.* (2012) identified 1,1,1,2-tetrachloroethane, carbon tetrachloride, chloroform, tetrachloroethylene, 1,1-dichloroethylene, and tri- and dichloroacetic acids during TCA oxidation in the $S_2O_8^{2-}$/UV process while only monochloroacetic acid and no volatile intermediates were detected in the H_2O_2/UV process. The numerous toxic intermediate produced in the $S_2O_8^{2-}$/UV process determined the authors to conclude that despite the higher efficiency of the $S_2O_8^{2-}$/UV process, the H_2O_2/UV process was more environmentally friendly. However, in the course of TCA removal by the sonolysis-activated persulfate process, Li *et al.* (2013) observed that the concentrations of most intermediates reached a maximum and then gradually decreased to near zero with the released chloride ion accounting for 91% of the total chlorine initially present in the TCA $(TCA)_0 = 85$ mg/L). Almost all initial TCA-organic chlorine was mineralized as chloride ion except for that bound in carbon tetrachloride and 1,1,1,2-tetrachloroethane indicating that sufficient oxidation time (energy) can overcome the release of toxic intermediates from the process. Moreover, the chloride ions can be oxidized by sulfate radicals to chlorine radicals such that chlorine radical reactions would play an important role in the mechanism of TCA degradation (Figure 10.6).

The ability of thermally activated persulfate oxidation to degrade 59 VOCs in their mixture (0.6–2 μM for each VOC) including chlorinated ethenes, BTEX and trichloroethanes, was investigated by Huang *et al.* (2005). The authors found that compounds with "C=C" bonds or with aromatic rings substituted with reactive functional groups were readily degraded, whereas the saturated hydrocarbons and halogenated alkanes were much more resistant to oxidation.

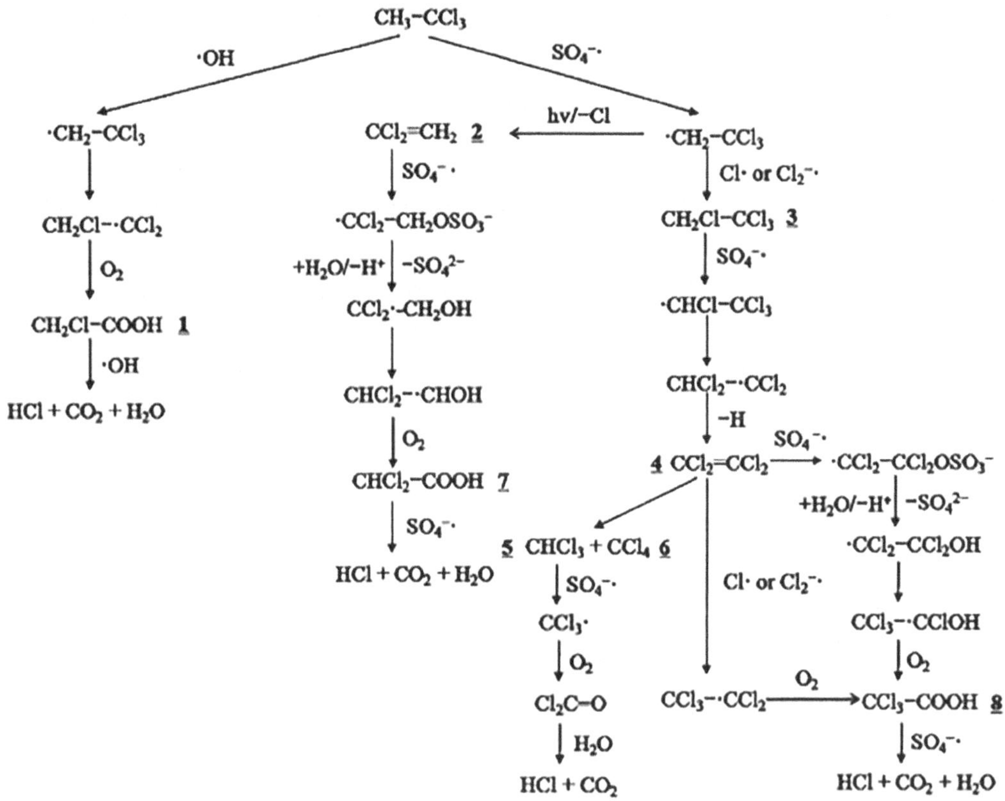

Figure 10.6 Proposed TCA oxidation pathway in UV/H_2O_2 and UV/$S_2O_8^{2-}$ processes. Reprinted with permission from Gu *et al.* (2012). Copyright 2016 American Chemical Society.

10.5.5 Perfluorinated compounds

Perfluorinated acids, especially perfluorocarboxylic acids (PFCAs), hydroperfluorocarboxylic acids (H-PFCAs), perfluoroether carboxylic acids and perfluorooctane sulfonate (PFOS) raised a particular interest, given the widespread contamination of water and soil with these toxic pollutants as well as their resistance to treatment *via* $OH^{\bullet}$-based AOPs. On May 19, 2016, the US EPA has issued the health advisory levels of 70 ng/L (ppt) for PFOS and perfluorooctanoic acid (PFOA) in order to provide Americans, including the most sensitive populations, with a margin of protection from a lifetime exposure to these contaminants from drinking water (https://www.epa.gov/ground-water-and-drinking-water/drinking-water-health-advisories-pfoa-and-pfos). These advisory levels were determined based on the agency's assessment of the latest peer-reviewed science on the health risks from exposure to these chemicals.

The effective decomposition of perfluorocarboxylic acids (PFCAs, $C_nF_{2n+1}COOH$; 0.5 μM-1.35 mM) bearing C_4-C_8 perfluoroalkyl groups with $SO_4^{\bullet-}$ radical anions formed from activated persulfate processes was demonstrated (Hori *et al.* 2005, 2010; Lee *et al.* 2009; Liu *et al.* 2012). Fluoride and CO_2 were reported as the major products during the breakdown of fluorinated chains, along with shorter-chain PFCAs through the loss of CF_2 units, with no formation of CF_4 and CF_3H fluorocarbons responsible for ozone layer depletion.

The mechanism of PFCA decomposition induced by $SO_4^{\bullet-}$ was studied and reported in the literature. Reactions 10.55–10.61 exemplify succinctly the degradation of PFOA (Lee *et al.* 2009; Hori *et al.* 2010). Decarboxylation with formation of unstable perfluoroalkyl radicals ($C_nF_{2n+1}^{\bullet}$; reactions 10.55, 10.56) which hydrolyze to perfluorinated alcohols ($C_nF_{2n+1}OH$; reaction 10.57) are the first steps of $SO_4^{\bullet-}$-induced degradation of PFCAs. The perfluorinated alcohol eliminates HF to form $C_{n-1}F_{2n-1}COF$ (reaction 10.60). Liu *et al.* (2012) suggested that the perfluoroalkyl radical could react with $SO_4^{\bullet-}$ to form the unstable intermediate adduct ($C_nF_{2n+1}OSO_3^-$), which hydrolyzes to the corresponding alcohol ($C_nF_{2n+1}OH$; reactions 10.58, 10.59). Hydrolysis of the acid fluoride ($C_{n-1}F_{2n-1}COF$) would generate a PFCA shortened by one CF_2 unit (i.e. $C_{n-1}F_{2n-1}COOH$; reaction 10.61). The shortened PFCAs would undergo a similar mechanism, essentially leading to shorter and shorter PFCAs through the loss of CF_2 units and mineralization to CO_2 and F^-. Short-chain PFCA species were found to degrade and mineralize more easily than long chains (Lee *et al.* 2009). The $SO_4^{\bullet-}$ generated from persulfate ($S_2O_8^{2-}$) was quantitatively converted into sulfate ion (SO_4^{2-}).

$$SO_4^{\bullet-} + C_7F_{15}COOH \rightarrow SO_4^{2-} + C_7F_{15}COOH^{\bullet+} \qquad (10.55)$$

$$C_7F_{15}COOH^{\bullet+} \rightarrow C_7F_{15}^{\bullet} + CO_2 + H^+ \qquad (10.56)$$

$$C_7F_{15}^{\bullet} + H_2O \rightarrow C_7F_{15}OH + H^{\bullet} \qquad (10.57)$$

$$C_7F_{15}^{\bullet} + SO_4^{\bullet-} \rightarrow C_7F_{15}OSO_3^- \qquad (10.58)$$

$$C_7F_{15}OSO_3^- + H_2O \rightarrow C_7F_{15}OH + HSO_4^- \qquad (10.59)$$

$$C_7F_{15}OH \rightarrow C_6F_{13}COF + F^- + H^+ \qquad (10.60)$$

$$C_6F_{13}COF + H_2O \rightarrow C_6F_{13}COOH + F^- + H^+ \qquad (10.61)$$

The $S_2O_8^{2-}$/UV process was successfully applied to the decomposition of perfluorononanoic acid (3.25 μM) contained in a floor wax wastewater (Hori *et al.* 2005).

The decomposition mechanism of hydroperfluorocarboxylic acids (H-PFCAs; $HC_nF_{2n}COOH$) differs from that of PFCAs. H-PFCAs did not form shorter-chain H-PFCAs (Hori *et al.* 2010). Perfluorodicarboxylic acids with various chain lengths were found to be the reaction intermediates. The reaction mechanism for H-PFCAs degradation can be explained by nucleophilic substitution by $SO_4^{\bullet-}$ at the carbon atom attached to the ω-H atom of the H-PFCAs which releases F^-, followed by formation of perfluorodicarboxylic acids ($HOOCC_{n-1}F_{2n-2}COOH$) which react with $SO_4^{\bullet-}$ to give shorter-chain perfluorodicarboxylic acids (Hori *et al.* 2010). The initial decomposition rates of H-PFCAs were higher than those for the corresponding PFCAs (Hori *et al.* 2010).

The literature concludes that sulfate radical-based processes are promising options for the removal of these persistent micropollutants from wastewater effluents and drinking water sources.

10.6 REACTIONS WITH WATER MATRIX CONSTITUENTS IN SULFATE RADICAL-DRIVEN OXIDATIONS

10.6.1 Reactions with inorganic compounds

10.6.1.1 Carbonate / bicarbonate ions

Bicarbonate ion (HCO_3^-) is an important water matrix constituent over the pH range characteristic to natural waters (6.8–8.2), and is a known 'scavenger' of OH radical, thus, impacting the micropollutant treatment performance. The rate constants for the reactions of $HO^{\bullet}$ and $SO_4^{\bullet-}$ with bicarbonate ion reported in the literature are 8.5×10^6 M^{-1} s^{-1} and 1.6×10^6 M^{-1} s^{-1}, respectively, whereas the rate constants for the reactions with carbonate ion are 3.9×10^8 M^{-1} s^{-1} and 6.1×10^6 M^{-1} s^{-1}, for $HO^{\bullet}$ and $SO_4^{\bullet-}$, respectively (Buxton *et al.* 1988; Zuo *et al.* 1999). The sulfate radical reactions with these species are electron transfer reactions.

$$HCO_3^- + SO_4^{\bullet-} \rightarrow CO_3^{\bullet-} + SO_4^{2-} + H^+ \tag{10.62}$$

$$CO_3^{2-} + SO_4^{\bullet-} \rightarrow CO_3^{\bullet-} + SO_4^{2-} \tag{10.63}$$

Yang *et al.* (2014) showed that the presence of 2.3 mM bicarbonate did not have a significant impact on the degradation kinetics of benzoic acid, 3-cyclohexene-1-carboxylic acid, and cyclohexanecarboxylic acid through the $S_2O_8^{2-}$/UV process. The three compounds were selected as model compounds for the aromatic, alkene, and alkane constituents of naphthenic acids, and were dosed at 10–100 μM. Lutze *et al.* (2015b) observed a significant decrease in the atrazine and *p*-nitrobenzoic acid removal rates with the $S_2O_8^{2-}$/UV process in the presence of HCO_3^- in river Ruhr water samples. However, the authors ascribed the main loss in oxidation capacity to the presence of chloride ions rather than to the scavenging of $SO_4^{\bullet-}$ by HCO_3^-. Moreover, Bennedsen *et al.* (2012) observed that the effectiveness of various persulfate activation processes for the dyestuff *p*-nitrosodimethylaniline removal was impacted differently by the presence of carbonate and hydrogen carbonate anions. Results showed that carbonates (10–100 mM) could enhance the bleaching rate in the alkaline persulfate activation system and should not be strictly considered as radical scavengers. Reactive carbonate radicals produced by oxidation with sulfate radicals may play an important role in the system.

10.6.1.2 Reaction with chloride ion

Sulfate radical-based AOPs are more sensitive to the presence of chloride ions in the water matrix than the $HO^{\bullet}$-based AOPs since the reaction of $SO_4^{\bullet-}$ with Cl^- is pH-independent, whereas the OH radical reaction with chloride ion is acid-catalyzed. Sulfate radicals react moderately fast with Cl^- ($k = (3.6 \pm 0.1) \times 10^8$ $M^{-1}s^{-1}$; Zuo *et al.* 1997), and the reaction kinetics depend on the ionic strength of the medium (Yu *et al.* 2004). The reaction yields chlorine atoms ($Cl^{\bullet}$), which further react with Cl^- ($k = 7.8 \pm 0.8) \times 10^9$ M^{-1} s^{-1}; Yu & Barker, 2003) to form the dichlorine radical anion ($Cl_2^{\bullet-}$), and an equilibrium is established between all species (reactions 10.64, 10.65). The equilibrium constant ($K = (1.4 \pm 0.2) \times 10^5$ M^{-1}; Yu & Barker, 2003) implies that $Cl_2^{\bullet-}$, not $Cl^{\bullet}$ will be the predominant chlorine radical species in water matrices containing chloride ions. The fast reacting species are involved in reactions with pH and chloride concentration-dependent product patterns (Yu *et al.* 2004; selected reactions 10.66–10.68; for details see also Chapter 9). The $HO^{\bullet}$ is formed at neutral pH from $Cl^{\bullet}/Cl_2^{\bullet-}$ reaction with water. Thus, at moderate to high Cl^- concentrations (mM range) the sulfate radical-based process may be converted into a conventional hydroxyl radical-based AOP (Lutze *et al.* 2015b).

$$Cl^- + SO_4^{\bullet-} \rightleftharpoons Cl^{\bullet} + SO_4^{2-} \tag{10.64}$$

$$Cl^{\bullet} + Cl^- \rightleftharpoons Cl_2^{\bullet-} \tag{10.65}$$

$$Cl^{\bullet}/Cl_2^{\bullet-} + H_2O \rightarrow ClOH^{\bullet-} + H^+/ + Cl^- \tag{10.66}$$

$$ClOH^{\bullet-} \rightleftharpoons OH^{\bullet} + Cl^- \tag{10.67}$$

$$Cl_2^{\bullet-} + Cl_2^{\bullet-} \rightarrow Cl_2 + 2Cl^- \tag{10.68}$$

At pH 3 a significant formation of chlorate was also observed in $S_2O_8^{2-}$/UV experiments on chloride oxidation in pure water (Lutze *et al.* 2015b).

In natural waters, considering the rate constants for the reactions of chloride and bicarbonate ions with sulfate radical, the scavenging of $SO_4^{\bullet-}$ by chloride will predominate.

Following a complex reaction system, bicarbonate ion scavenges $Cl^{\bullet}$ and $Cl_2^{\bullet-}$ (reactions 10.69 and 10.70; $k = 2.2 \times 10^8$ M^{-1} s^{-1} and $k = 8.0 \times 10^7$ M^{-1} s^{-1} respectively; Mertens & von Sonntag, 1995; Matthew & Anastasio, 2006) which are the key radical intermediates in the formation of OH radicals, thus impacting the conversion of $SO_4^{\bullet-}$ into $HO^{\bullet}$. The resulting carbonate radical anion is much less reactive toward organic compounds than the OH radical.

$$Cl^{\bullet} + HCO_3^- \rightarrow CO_3^{\bullet-} + Cl^- + H^+ \tag{10.69}$$

$$Cl_2^{\bullet-} + HCO_3^- \rightarrow CO_3^{\bullet-} + 2Cl^- + H^+ \tag{10.70}$$

The overall oxidation efficiency and reaction pathways are affected differently by Cl^- and HCO_3^- depending on the $SO_4^{\bullet-}$-based AOP, the target compounds and the experimental conditions (Bennedsen *et al.* 2012).

10.6.1.3 Reaction with bromide ion

The reaction between Br^- and $SO_4^{\bullet-}$ ($k_{Br^-,SO_4^{\bullet-}} = 3.5 \times 10^9$ M^{-1} s^{-1}, Neta *et al.* 1988) forms bromine atoms ($Br^{\bullet}$) (reaction 10.71). The rate constant $k_{Br^-,SO_4^{\bullet-}}$ is one order of magnitude larger than $k_{Cl^-,SO_4^{\bullet-}}$ ((3.6 ± 0.1) × 10^8 M^{-1}s^{-1}; Zuo *et al.* 1997).

$$Br^- + SO_4^{\bullet-} \rightarrow Br^{\bullet} + SO_4^{2-} \tag{10.71}$$

Once $Br^{\bullet}$ is formed, it quickly reacts with bromide ion to produce dibromine anion radical ($Br_2^{\bullet-}$). $Br^{\bullet}$ and $Br_2^{\bullet-}$ can go through a series of reactions with Br^- and H_2O to form $BrOH^{\bullet-}$ and HOBr/BrO^- (Wang *et al.* 2014; Lutze *et al.* 2014). In ultrapure water at pH 7, 5.8 to 100% conversion of an initial concentration of 20 µM bromide into bromate by UV/persulfate was observed for persulfate dosages of 20 to >300 µM (Fang & Shang, 2012). The formation of bromate via sulfate radical-initiated reactions resembles the well described mechanism of the hydroxyl radical-based bromate formation. Hypobromous acid is a requisite intermediate in the formation of bromate (Figure 10.7; Fang & Shang, 2012).

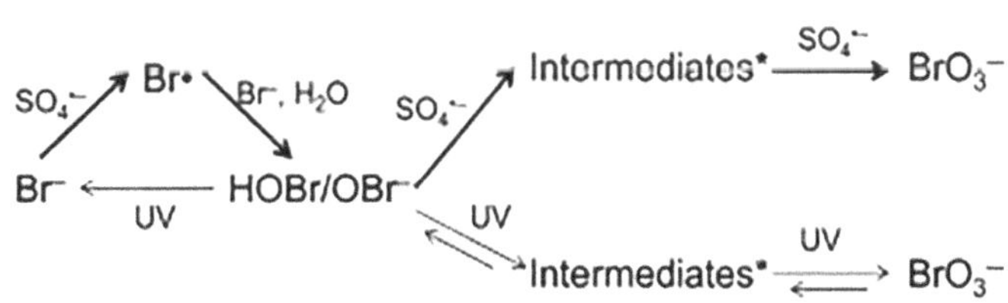

Figure 10.7 Proposed pathways of bromate formation from bromide oxidation by UV/Persulfate. Reprinted with permission from Fang and Shang (2012). Copyright 2016 American Chemical Society.

Bromide is present in water sources worldwide at concentrations in the 0–2.3 mg L^{-1} range, such that the conversion to bromate could constitute a disadvantage of sulfate radical-based AOPs if considered to treat bromide-containing drinking waters. It has been shown (Fang & Shang, 2012) that for controlling the bromate formation, elevated pH (>8) may be an option for degradation of micropollutants *via* AOPs which are not negatively impacted by alkaline conditions.

10.6.2 Reactions in natural waters

The $SO_4^{\bullet-}$-based treatment efficiency depends upon the competing reactions that scavenge the free radicals. In natural waters, the reactions of sulfate radical with bicarbonate and chloride ions compete with those of target micropollutants, and could limit the sulfate radical process efficiency. The dissolved organic matter present in all natural waters will also have a significant impact on the performance of sulfate radical-based AOPs.

10.6.2.1 Reactions with dissolved organic matter

The rate constants for the reactions of sulfate and hydroxyl radicals with humic acids differ significantly. A value of 6.8×10^3 L mg-C^{-1} s^{-1} for the rate constant of humic acids with sulfate radicals was reported recently by Lutze *et al.* (2015a). The rate constants cited in the literature for the reaction of $HO^\bullet$ with NOM are in the range $(2.2–6.7) \times 10^4$ L mg-C^{-1} s^{-1} (Westerhoff *et al.* 1999) thus, higher than that for $SO_4^{\bullet-}$. The lower reactivity of $SO_4^{\bullet-}$ toward organic matter compared with $HO^\bullet$ was explained by the fact that H-abstraction reactions involving $SO_4^{\bullet-}$ are slower than those exerted by $^\bullet OH$ (Lutze *et al.* 2015a). Thus, saturated moieties of the organic matter react faster with $HO^\bullet$. For water treatment purposes, this will lead to a weaker $SO_4^{\bullet-}$ scavenging capacity of the dissolved organic matter, which would contribute to an increased degradation efficiency of pollutants by the sulfate radical-based processes compared to $^\bullet OH$–based AOPs, and could counterbalance the lower reactivity of some compounds toward sulfate radical anion. Moreover, since the dissociated forms of acids usually react faster than their protonated forms, the scavenging of sulfate radicals is expected to decrease with decreasing pH. Thus, in the presence of humic acids, many compounds are degraded more efficiently by sulfate radicals than by hydroxyl radicals. As a consequence, sulfate radical based technologies would be promising for the removal of contaminants from wastewater treatment plant effluents because of the higher selectivity of sulfate radical anion over hydroxyl radical, limiting radical scavenging by natural organic matter and allowing for higher abatement rates.

10.6.2.2 Chlorinated transformation byproducts

Both enhancing and inhibitory effects of chloride ion on the degradation of pollutants in sulfate radical-driven processes are reported in the published literature. Figure 10.8 summarizes the key reactions in the UV/persulfate process in the presence of chloride ion and the pH-dependent species involved in those reactions. The distribution of radicals in the system depicted in Figure 10.8 depends on:

(i) scavenging of the reactive sulfate radicals;
(ii) formation of reactive halogen radicals;
(iii) sulfate radical reactions leading to the less selective HO radicals.

Using a model freshwater matrix with added salts, Yang *et al.* (2014) reported that the degradation yields of benzoic acid and cyclohexanecarboxylic acid (10–100 μM) were higher with the $UV/S_2O_8^{2-}$ AOP than with the UV/H_2O_2 process because of the higher quantum efficiency of $S_2O_8^{2-}$ than H_2O_2 photolysis at 253.7 nm, i.e. 0.7 and 0.5, respectively. However in simulated saline water, the authors observed that the UV/$S_2O_8^{2-}$ AOP was more inhibited by high concentrations of Cl^- (540 mM) than the UV/H_2O_2 AOP because of the more favorable oxidation of Cl^- by $SO_4^{\bullet-}$ than $HO^\bullet$ at pH 7. However, the reactive halogen species formed in the UV/$S_2O_8^{2-}$/Cl^- system are also capable of oxidizing organics. The contribution of chlorine reactive species to the contaminant degradation will depend on the structural characteristics of those compounds, as Cl species are more selective in their reactions than the sulfate and hydroxyl radicals (Yang *et al.* 2014), on the chloride concentration, and on the sulfate radical generation system (Yuan *et al.* 2011; Wang *et al.* 2011). Reactive halogen species would react efficiently with the electron-rich moieties of the organic compounds (Yang *et al.* 2014).

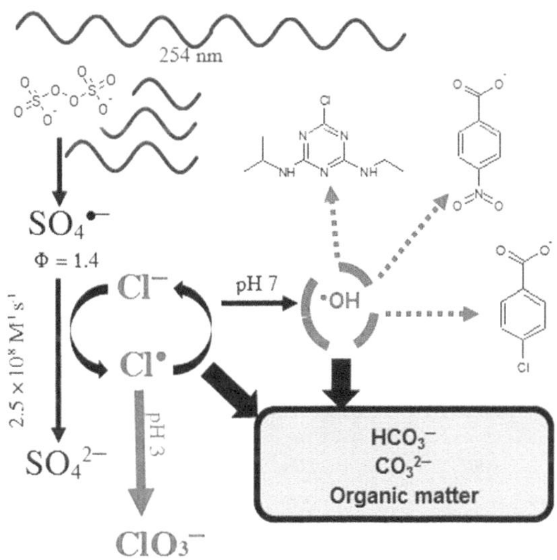

Figure 10.8 Simplified reaction scheme of sulfate radical interactions with chloride ion in the UV/persulfate process. Reprinted from Lutze *et al.* (2015b). Copyright 2016 with permission from Elsevier.

In the presence of high concentrations of chloride, some undesirable chlorinated compounds can be formed. Anipsitakis *et al.* (2006) reported the formation of chlorinated intermediates (maximum concentration 46 μM) in cobalt-activated peroxomonosulfate oxidation of phenol and 2,4-dichlorophenol (3–30 mM). The authors observed that chloride from either the cobalt counterion (2–15 mM) or the substrate oxidation itself is involved in the production of chlorine radicals and free chlorine species from reaction with sulfate radicals, which are responsible for the formation of the detected chlorinated intermediates. In a study on the degradation of an azo dye (0.1–0.2 mM) by the sulfate radical-based cobalt/peroxymonosulfate process, Yuan *et al.* (2011) also identified some chlorinated aromatic compounds (Figure 10.9). The authors reported a dual effect of chloride (0.2–200 mM) (i) inhibition of sulfate radical-induced oxidation due to scavenging by chloride, and (ii) accelerated degradation of the azo dye *via* chlorine radical/active chlorine species (HOCl) formed in reaction of Cl^- with sulfate radical and peroxomonosulfate, respectively. From a kinetic model of the sulfate radical generation system based on persulfate decomposition, Fang *et al.* (2012) calculated that the total radical concentration was significantly higher (fifty times) in the presence of 100 mM Cl^- (pH < 6) than that without Cl^-. These aspects could have significant impact on the selection of sulfate radical-based advanced oxidation process for detoxification of chloride-rich wastewaters (like in the textile industry).

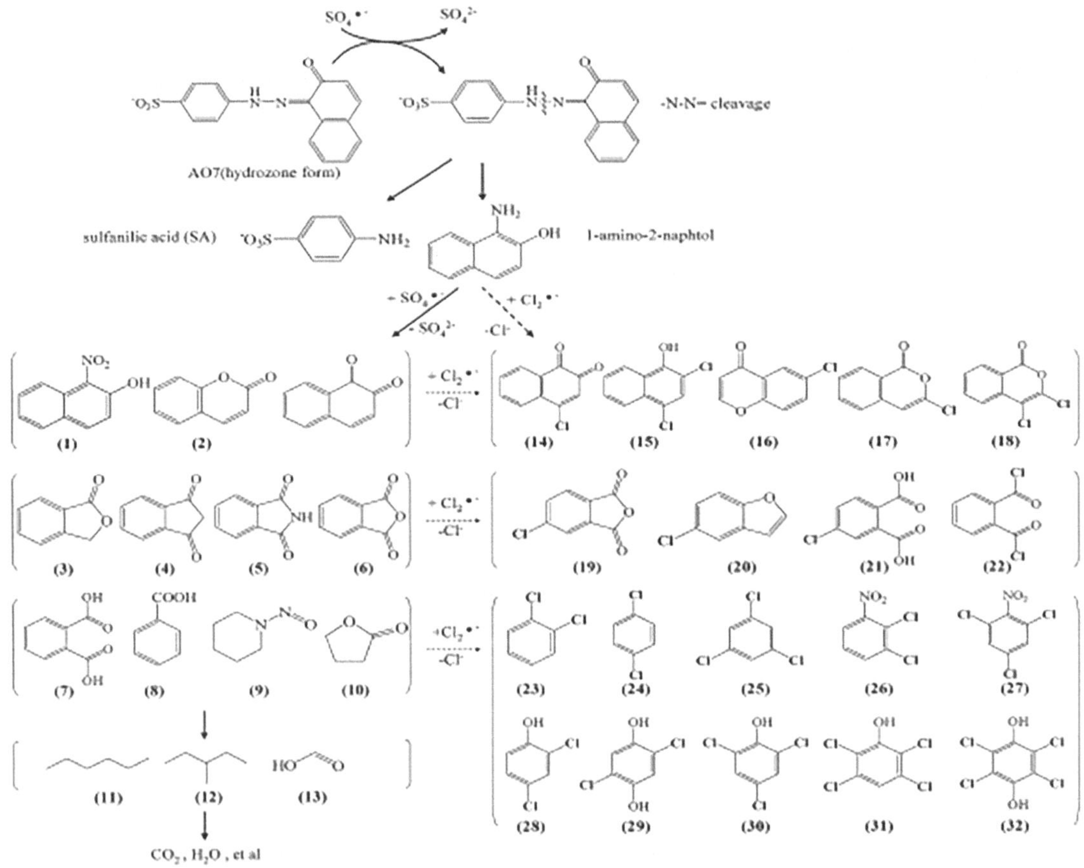

Figure 10.9 Byproducts identified (numbered in parentheses) and proposed reaction pathways for $SO_4^{\bullet-}$ radical-based oxidation of Acid Orange 7 in the presence of chloride ions. Reprinted with permission from Yuan *et al.* (2011). Copyright Elsevier 2016.

10.6.2.3 Brominated by-products

Water matrix constituents such as natural organic matter (NOM) are not only diminishing the oxidation efficiency by consumption of oxidants but can also affect the reaction mechanisms. Significantly reduced bromate formation was observed in real waters as compared to laboratory water. This inhibition was attributed to the role of NOM and inorganic compounds as competitors by reacting with the key intermediates in bromate formation (i.e., $HOBr/BrO^-$ and $Br^\bullet$; Fang & Shang, 2012). The effect of NOM in the control of bromate formation was also explained by formation of superoxide in the reaction of sulfate radicals with aromatic moieties of organic matter (Lutze *et al.* 2014). In general, the superoxide radical anion is a mild reducing agent, but it reduces hypobromous acid/hypobromite efficiently (reactions

10.72 and 10.73; $k_{O_2^{\bullet-}, HOBr} = 3.5 \times 10^9\ M^{-1}\ s^{-1}$ and $k_{O_2^{\bullet-}, BrO^-} < 2 \times 10^8\ M^{-1}\ s^{-1}$, Schwarz & Bielski, 1986) yielding bromine atoms and bromide ions.

$$HOBr/Br^- + O_2^{\bullet-} \rightarrow O_2 + OH^- + Br_2^{\bullet-} \quad (10.72)$$

$$BrO^-/Br^- + O_2^{\bullet-} + H_2O \rightarrow O_2 + 2OH^- + Br_2^{\bullet-} \quad (10.73)$$

The chain reactions involving reactive bromine species in the presence of bromide can be terminated by reactions with NOM. These reactions may lead to the formation of bromine-containing by-products of potential health risk concern (Wang *et al.* 2014). Wang *et al.* (2014) studied the formation of total organic bromine (TOBr) and brominated disinfection byproducts (Br-DBPs) from isolated natural organic matter oxidation by sulfate radicals generated from PMS activation in the presence of bromide ion (2.5–6.5 μM). Bromoform and dibromoacetic acid were identified as major Br-DBPs (3.5–6 and 2–7 μg mg^{-1} C, respectively) at pH 7.5. However, it was found that the sulfate radical AOP produced much less TOBr than the bromination process.

10.6.2.4 Sulfate ion as a byproduct

In sulfate radical-based AOPs, sulfate ions will be formed as the end-product, which leads to an increase in salt content in the media. The release of sulfate ion in waters from using oxidants such as persulfate or peroxymonosulfate might affect the quality of the water. However, the SO_4^{2-} is practically inert and is not considered to be a pollutant. The US EPA has listed sulfate ion under the National Secondary Drinking Water Standards with a maximum concentration of 250 mg/L (1.43 mM), based on aesthetic reasons. The literature studies on UV/persulfate process for contaminant treatment indicate that the concentration of the sulfate ion released from the process is very low and below the secondary drinking water standard. Furthermore, when the degradation of concentrated pollutants is pursued, the high concentration of sulfate ions from the oxidant can be reduced or eliminated by implementing technologies available for sulfate ion removal (e.g. ion exchange resins).

10.7 COMMERCIAL APPLICATIONS

The sulfate radical-based processes have received an increasing interest from the academic communities. Most of the published literature on sulfate radical-based AOPs reports laboratory studies. Optimization of process efficiency is rarely within the scope of those studies. Industrial implementations are limited.

10.7.1 Total organic carbon (TOC) analyzers

The oldest application concerns TOC analyzers, and dates back to late 1960s when Ehrhardt (1969) first described a new method combining the radiation from a mercury vapor high pressure arc lamp and persulfate solutions for measurement of dissolved organic carbon in sea water by oxidation into carbon dioxide. In that publication, PDS was only considered as an oxygen donor to enable the complete oxidation of dissolved organic carbon. TOC analyzers were developed later on the principle of persulfate activated by UV (253.7 nm) and more recently a heated-UV-persulfate TOC system was commercialized.

10.7.2 *In Situ* chemical oxidation (ISCO)

Over the past few decades, there has been an increasing interest in sodium persulfate as an oxidant for the destruction of a broad range of soil and groundwater contaminants, conceptually developed as an *In Situ*

Chemical Oxidation (ISCO) process. Persulfate is widely used in ISCO remediation of contaminated environment with organic pollutants, such as chlorinated solvents, BTEXs, PAHs (Siegrist *et al.* 2011). The remediation of groundwater contamination using ISCO involves injecting oxidants (with potentially co-amendments) directly into the contaminated source zone.

The common methods for persulfate activation into sulfate radicals in ISCO technology are highly alkaline media, heat, and the environmental friendly transition metal iron.

10.7.3 Other applications

Some recent presentations report the UV/persulfate process implementation for urea, isopropyl alcohol, acetone, atrazine, and chloroform destruction in high purity water production for microelectronics industry (Knapp, 2009; Feng, 2011; Coulter *et al.* 2014). The data emphasize the process efficiency relative to UV (and $VUV_{185\ nm}$) oxidation process commonly used in Ultrapure Water (UPW) production. In particular, the combination with persulfate (Siemens Vanox^{TM} AOP) has been proven to be effective for urea oxidation, one of contaminants poorly treated by $HO^{\bullet}$ and not effectively removed with conventional UPW treatment processes. With such treatment options, the TOC is removed to very low levels (usually to less than 1 microgram/L) and the water can be recycled (as ultrapure water) in the microelectronics industry. The UV/persulfate process was also found superior to UV(253.7 nm)/VUV(184.9 nm) AOP for trace TOC removal in water contaminated with isopropyl alcohol (IPA) a common cleaning agent and its major degradation byproduct, acetone. The hydroxyl radicals formed through water photolysis at 184.9 nm are effective at the removal of IPA ($k_{OH,IPA} = 1.9 \times 10^9\ M^{-1}\ s^{-1}$, Buxton *et al.* 1988), but are not very reactive towards acetone ($k_{OH,acetone} = 1.1 \times 10^8\ M^{-1}\ s^{-1}$, Buxton *et al.* 1988).

10.8 FUTURE RESEARCH NEEDS

The sulfate radical-based AOPs are still at the research stage, with only a few, limited niche applications (e.g., soil decontamination), and virtually no full-scale implementation.

The following could be identified as future research needs at laboratory scale:

- In view of the complexity of the reaction systems involving sulfate radicals in natural waters and wastewaters ($SO_4^{\bullet -}$ interaction with chlorides and bicarbonates, contribution of the $^{\bullet}OH$ and $Cl_2^{\bullet -}$ in the oxidation process, impact of organic matter, etc.), the simulation of the impact of various parameters by kinetic modeling is necessary for a better understanding of the system and process optimization (pollutant removal, oxidant dosage and consumption, byproduct mitigation strategies). Depending on the application areas, this approach would help with the quantification of by-products and would allow assessments on the potential risk of formation of undesirable compounds (halogenated organics, bromate, chlorate, etc.). This would also allow a better knowledge of the classes of compounds which may be degraded by the use of the sulfate radical-based AOPs, as well as the impact of water matrix constituents on the process efficiency with respect to the reactive species (e.g., $SO_4^{\bullet -}$, $^{\bullet}OH$, $Cl^{\bullet}$, $Cl_2^{\bullet -}$) and their reactivity toward the target micropollutants.
- Systematic approach to the role of various physico-chemical characteristics of water (pH, organic and mineral composition, and UV light absorption properties) on the sulfate radical-based AOP treatment performance for various classes of micropollutants, as well as to the process optimization and comparison with the OH radical-based AOPs are needed.
- Research on oxidant dose optimization is rare and removal of oxidant residual from the treated water is not represented in the published literature.

Practical implementation of sulfate radical-based AOPs would require extensive pilot work, reactor design and scale-up kinetic and computational fluid dynamics models. Life cycle assessment and cost studies of water treatment with sulfate-based AOPs would be also required, should the sulfate radical-based AOP become a competitive process for water remediation.

10.9 CONCLUSIONS

Although the first works date back more than half a century, the sulfate radical-based processes have attracted increasing interest over the past decade as evidenced by the increasing number of publications. At laboratory scale, persulfate and peroxomonosulfate are commonly used with UV light or transition metals in order to initiate the sulfate radical oxidation mechanisms. The reactive sulfate radicals, often associated with formed hydroxyl radicals, are of interest due to their high selectivity and reactivity toward numerous micropollutants and their intermediate oxidation products. $SO_4^{\bullet-}$-based oxidation may complement more common $HO^\bullet$-based AOPs. However, the technology should be carefully optimized in practice considering the organic and inorganic water constituents to maximize the contaminant removal efficiency and to minimize the risk of formation of potential harmful transformation products.

10.10 ACKNOWLEDGEMENTS

Special thanks to Dr Mihaela Stefan (Trojan Technologies, London, ON, Canada) for her valuable comments on the manuscript and warm support throughout this work.

10.11 REFERENCES

Anipsitakis G. P. and Dionysiou D. D. (2004). Radical generation by the interaction of transition metals with common oxidants. *Environmental Science & Technology,* **38**, 3705–3712.

Anipsitakis G. P., Dionysiou D. D. and Gonzalez M. A. (2006). Cobalt-mediated activation of peroxymonosulfate and sulfate radical attack on phenolic compounds. Implications of chloride ions. *Environmental Science & Technology*, **40**, 1000–1007.

Antoniou M. G., de la Cruz A. A. and Dionysiou D. D. (2010a). Degradation of microcystin-LR using sulfate radicals generated through photolysis, thermolysis and e- transfer mechanisms. *Applied Catalysis B: Environmental*, **96**, 290–298.

Antoniou M. G., de la Cruz A. A. and Dionysiou D. D. (2010b). Intermediates and reaction pathways from the degradation of Microcystin-LR with sulfate radicals. *Environmental Science & Technology*, **44**, 7238–7244.

Aravindakumar, C. T., Schuchmann, M. N., Rao, B. S. M., von Sonntag, J. and von Sonntag, C. (2003). The reactions of cytidine and 2′-deoxycytidine with $SO_4^{\bullet-}$ revisited. Pulse radiolysis and product studies. *Organic & Biomolecular Chemistry*, **1**, 401–408.

Bennedsen L. R., Muff J. and Sogaard E. G. (2012). Influence of chloride and carbonates on the reactivity of activated persulfate. *Chemosphere*, **86**, 1092–1097.

Bosio G. Criado S., Massad W., Nieto F. J. R., Gonzalez M. C., Garcia N. A. and Martire D. O. (2005). Kinetics of the interaction of sulfate and hydrogen phosphate radicals with small peptides of glycine, alanine, tyrosine and tryptophan, *Photochem. Photobiol. Sci.*, **4**, 840–846.

Buxton G. V., Greenstock C. L., Helman W. P. and Ross A. B. (1988). Critical review of rate constants for reactions of hydrated electrons, hydrogen atoms and hydroxyl radicals ($^\bullet OH/^\bullet O^-$) in aqueous solution. *Journal of Physical and Chemical Reference Data*, **17**, 513–886.

Buxton G. V., Wang J. and Salmon G. A. (2001). Rate constants for the reactions of $NO_3^\bullet$, $SO_4^{\bullet-}$ and $Cl^\bullet$ radicals with formate and acetate esters in aqueous solution. *Physical Chemistry Chemical Physics*, **3**, 2618–2621.

Chan T. W., Graham N. J. D. and Chu W. (2010). Degradation of iopromide by combined UV irradiation and peroxydisulfate. *Journal of Hazardous Materials*, **181**, 508–513.

Chawla O. P. and Fessenden R. W. (1975). Electron Spin resonance and pulse radiolysis studies of some reactions of $SO_4^{\bullet-}$. *Journal of Physical Chemistry*, **79**, 2693–2700.

Chitose N., Katsumura Y., Domae M., Zuo Z. and Murakami T. (1999). Radiolysis of aqueous solutions with pulsed helium ion beams – 2. Yield of $SO_4^{\bullet-}$ formed by scavenging hydrated electron as a function of $S_2O_8^{2-}$ concentration. *Radiation Physics and Chemistry*, **54**, 385–391.

Choure S. C., Bamatraf M. M. M., Rao B. S. M., Das R., Mohan H. and Mittal J. P. (1997). Hydroxylation of chlorotoluenes and cresols: a pulse radiolysis, laser flash photolysis, and product analysis study. *Journal of Physical Chemistry*, **101**, 9837–9845.

Coulter B., Sundstrom G., Hall C. and Doung S. (2014). An advanced oxidation process update for removal of low organic levels. http://www.evoqua.com/en/brands/IPS/productinformationlibrary/aop-removal-low-organics-presentation.pdf

Criquet J. and Leitner N. K. V. (2009). Degradation of acetic acid with sulfate radical generated by persulfate ion photolysis. *Chemosphere*, **77**, 194–200.

Criquet J. and Leitner N. K. V. (2011). Radiolysis of acetic acid aqueous solutions-Effect of pH and persulfate addition. *Chemical Engineering Journal*, **174**, 504–509.

Criquet J. and Leitner N. K. V. (2012). Electron beam irradiation of citric acid aqueous solutions containing persulfate. *Separation and Purification Technology*, **88**, 168–173.

Criquet J. and Leitner N. K. V. (2015). Reaction pathway of the degradation of the p-hydroxybenzoic acid by sulfate radical generated by ionizing radiations. *Radiation Physics and Chemistry*, **106**, 307–314.

Deng J., Shao Y., Gao N., Xia S., Tan C., Zhou S. and Hu X. (2013). Degradation of the antiepileptic drug carbamazepine upon different UV-based advanced oxidation processes in water. *Chemical Engineering Journal*, **222**, 150–158.

Diaz Kirmser E. M., Martire D. O., Gonzalez M. C. and Rosso J. A. (2010). Degradation of the herbicides Clomazone, Paraquat, and Glyphosate by thermally activated peroxydisulfate. *Journal of Agricultural and Food Chemistry*, **58**, 12858–12862.

Dogliotti L. and Hayon E. (1967). Flash photolysis of persulfate in aqueous solutions. Study of the sulfate and ozonide radical anions. *Journal of Physical Chemistry*, **71**, 2511–2516.

Ehrhardt M. (1969). A new method for the automatic measurement of dissolved organic carbon in sea water. *Deep-Sea Research*, **16**, 393–397.

Fang G.-D., Dionysiou D. D., Wang Y., Al-Abed S. R. and Zhou D.-M. (2012). Sulfate radical-based degradation of polychlorinated biphenyls: effects of chloride ion and reaction kinetics. *Journal of Hazardous Materials*, **227–228**, 394–401.

Fang J. Y. and Shang C. (2012). Bromate formation from bromide oxidation by the UV/persulfate process. *Environmental Science & Technology*, **46**, 8976–8983.

Feng J. (2011). UV oxidation with persulfate for High-Purity Water in microelectronics and pharmaceutical applications. CD Conference Proceedings: Ultrapure Water Asia Conference. Singapore Water Week, Singapore, July 6–7, 2011.

Furman O. S., Teel A. L. and Watts R. J. (2010). Mechanism of base activation of persulfate. *Environmental Science & Technology*, **44**, 6423–6428.

Gao Y.-Q., Gao N.-Y., Deng Y., Yang Y.-Q. and Ma Y. (2012). Ultraviolet (UV) light-activated persulfate oxidation of sulfamethazine in water. *Chemical Engineering Journal*, **195–196**, 248–253.

Gara P. M. D., Bosio G. N., Arce V. B., Poulsen L., Ogilby P. R., Giudici R., Gonzalez M. C. and Martire D. O. (2009). Photoinduced degradation of the herbicide Clomazone model reactions for natural and technical systems. *Photochemistry and Photobiology*, **85**, 686–692.

Ghauch A., Ayoub G. and Naim S. (2013). Degradation of sulfamethoxazole by persulfate assisted micrometric Fe^0 in aqueous solution. *Chemical Engineering Journal*, **228**, 1168–1181.

Gu X., Lu S., Qiu Z., Sui Q., Miao Z., Lin K., Liu Y. and Luo Q. (2012). Comparison of photodegradation performance of 1,1,1-Trichloroethane in aqueous solution with the addition of H_2O_2 or $S_2O_8^{2-}$ oxidants. *Industrial & Engineering Chemistry Research*, **51**, 7196–7204.

Guan Y.-H., Ma J., Li X.-C., Fang J.-Y. and Chen L.-W. (2011). Influence of pH on the formation of sulfate and hydroxyl radicals in the UV/peroxymonosulfate system. *Environmental Science & Technology*, **45**, 9308–9314.

He X., de la Cruz A. A. and Dionysiou D. D. (2013). Destruction of cyanobacterial toxin cylindrospermopsin by hydroxyl radicals and sulfate radicals using UV-254 nm activation of hydrogen peroxide, persulfate and peroxymonosulfate. *Journal of Photochemistry and Photobiology A-Chemistry*, **251**, 160–166.

Heidt L. J. (1942). The photolysis of persulfate. *Journal of Chemical Physics*, **10**(5), 297–302.

Herrmann H. (2007). On the photolysis of simple anions and neutral molecules as sources of O^-/OH, SO_x^- and Cl in aqueous solution. *Physical Chemistry Chemical Physics.*, **9**, 3935–3964.

Hochanadel C. J. (1962). Photolysis of dilute hydrogen peroxide solution in the presence of dissolved hydrogen and oxygen. Evidence relating to the nature of the hydroxyl radical and the hydrogen atom produced in the radiolysis of water. *Radiation Research*, **17**, 286–301.

Hori H., Yamamoto A., Hayakawa E., Taniyasu S., Yamashita N. and Kutsuna S. (2005). Efficient decomposition of environmentally persistent perfluorocarboxylic acids by use of persulfate as a photochemical oxidant. *Environmental Science & Technology*, **39**, 2383–2388.

Hori H., Murayama M., Inoue N., Ishida K. and Kutsuna S. (2010). Efficient mineralization of hydroperfluorocarboxylic acids with persulfate in hot water. *Catalysis Today*, **151**, 131–136.

Huang K. C., Zhao Z., Hoag G. E., Dahmani A. and Block P. A. (2005). Degradation of volatile organic compounds with thermally activated persulfate oxidation. *Chemosphere*, **61**, 551–560.

Khan J. A., He X., Shah N. S., Khan H. M., Hapeshi E., Fatta-Kassinos D. and Dionysiou D. D. (2014). Kinetic and mechanism investigation on the photochemical degradation of atrazine with activated H_2O_2, $S_2O_8^{2-}$ and HSO_5^-. *Chemical Engineering Journal*, **252**, 393–403.

Killian P. F., Bruell C. J., Liang C. and Marley M. C. (2007). Iron(II) activated persulfate oxidation of MGP contaminated soil. *Soil & Sediment Contamination*, **16**, 523–537.

Knapp A. G. (2009). Advanced oxidation for the removal of organic contaminants in industrial waters. CD Conference Proceedings: Ultrapure Water Conference. Portland, OR, November 4–5, 2009.

Kolthoff I. M. and Miller I. K. (1951). The chemistry of persulfate. I. The kinetics and mechanism of the decomposition of the persulfate ion in aqueous medium. *Journal of the American Chemical Society*, **73**, 3055–3059.

Konstantinou I. K. and Albanis T. A. (2004). TiO_2-assisted photocatalytic degradation of azo dyes in aqueous solution: kinetic and mechanistic investigations A review. *Applied Catalysis B.*, **49**, 1–14.

Lee Y. C., Lo S. L., Chiueh P. T. and Chang D. G. (2009). Efficient decomposition of perfluorocarboxylic acids in aqueous solution using microwave-induced persulfate. *Water Research*, **43**, 2811–2816.

Levey G. and Hart E. J. (1975). γ-Ray and electron pulse radiolysis studies of aqueous peroxodisulfate and peroxodiphosphate ions. *Journal of Physical Chemistry*, **79**, 1642–1646.

Li B., Li L., Lin K., Zhang W., Lu S. and Luo Q. (2013). Removal of 1,1,1-Trichloroethane from aqueous solution by a sono-activated persulfate process. *Ultrasonics Sonochemistry*, **20**, 855–863.

Liang C. and Su H. W. (2009). Identification of sulfate and hydroxyl radicals in thermally activated persulfate. *Industrial & Engineering Chemistry Research*, **48**, 5558–5562.

Liang C., Bruell C. J., Marley M. C. and Sperry K. L. (2004a). Persulfate oxidation for *in situ* remediation of TCE. I. Activated by ferrous ion with and without a persulfate-thiosulfate redox couple. *Chemosphere*, **55**, 1213–1223.

Liang C., Bruell C. J., Marley M. C. and Sperry K. L. (2004b). Persulfate oxidation for *in situ* remediation of TCE. II. Activated chelated by ferrous ion. *Chemosphere*, **55**, 1225–1233.

Liang C., Liang C.-P. and Chen C.-C. (2009). pH dependence of persulfate activation by EDTA/Fe(III) for degradation of trichloroethylene. *Journal of Contaminant Hydrology*, **106**, 178–182.

Liu C. S., Higgins C. P., Wang F. and Shih K. (2012). Effect of temperature on oxidative transformation of perfluorooctanoic acid (PFOA) by persulfate activation in water. *Separation and Purification Technology*, **91**, 46–51.

Liu X., Zhang T., Zhou Y., Fang L. and Shao Y. (2013a). Degradation of atenolol by UV/peroxymonosulfate: kinetics, effect of operational parameters and mechanism. *Chemosphere*, **93**, 2717–2724.

Liu X., Fang L., Zhou Y., Zhang T. and Shao Y. (2013b). Comparison of UV/PDS and UV/H_2O_2 processes for the degradation of atenolol in water. *Journal of Environmental Sciences*, **25**, 1519–1528.

Lutze H. V., Bakkour R., Kerlin N., von Sonntag C. and Schmidt T. C. (2014). Formation of bromate in sulfate radical based oxidation: mechanistic aspects and suppression by dissolved organic matter. *Water Research*, **53**, 370–377.

Lutze H. V., Bircher S., Rapp I., Kerlin N., Bakkour R., Geisler M., von Sonntag C. and Schmidt T. C. (2015a). Degradation of chlorotriazine pesticides by sulfate radicals and the influence of organic matter. *Environmental Science & Technology*, **49**, 1673–1680.

Lutze H. V., Kerlin N. and Schmidt T. C. (2015b). Sulfate radical-based water treatment in presence of chloride: formation of chlorate, inter-conversion of sulfate radicals into hydroxyl radicals and influence of bicarbonate. *Water Research*, **72**, 349–360.

Madhavan J., Muthuraaman B., Murugesan S., Anandan S. and Maruthamuthu P. (2006). Peroxomonosulphate, an efficient oxidant for the photocatalysed degradation of a textile dye, acid red 88. *Solar Energy Materials & Solar Cells*, **90**, 1875–1887.

Madhavan V., Levanon H. and Neta P. (1978). Decarboxylation by $SO_4^{\bullet-}$ radicals. *Radiation Research*, **76**, 15–22.

Mahdi-Ahmed M. and Chiron S. (2014). Ciprofloxacin oxidation by UV-C activated peroxymonosulfate in wastewater. *Journal of Hazardous Materials*, **265**, 41–46.

Mahdi-Ahmed M., Barbati S., Doumenq P. and Chiron S. (2012). Sulfate radical anion oxidation of diclofenac and sulfamethoxazole for water decontamination. *Chemical Engineering Journal*, **197**, 440–447.

Manoj P., Prasanthkumar K. P., Manoj V. M., Aravind U. K., Manojkumar T. K. and Aravindakumar C. T. (2007). Oxidation of substituted triazines by sulfate radical anion ($SO_4^{\bullet-}$) in aqueous medium: a laser flash photolysis and steady state radiolysis study. *Journal of Physical Organic Chemistry*, **20**, 122–129.

Mark G., Schuchmann M. N., Schuchmann H. P. and von Sonntag C. (1990). The photolysis of potassium peroxodisulphate in aqueous solution in the presence of *tert*-butanol: a simple actinometer for 254 nm radiation. *Journal of Photochemistry and Photobiology A-Chemistry*, **55**, 157–168.

Matta R., Tlili S., Chiron S. and Barbati S. (2011). Removal of carbamazepine from urban wastewater by sulfate radical oxidation. *Environmental Chemistry Letters*, **9**, 347–353.

Matthew B. M. and Anastasio C. (2006). A chemical probe technique for the determination of reactive halogen species in aqueous solution: Part 1- Bromide solutions. *Atmospheric Chemistry and Physics*, **6**, 2423–2437.

Matzek L. W. and Carter K. E. (2016). Activated persulfate for organic chemical degradation: a review. *Chemosphere*, **151**, 178–188.

Maurino V., Calza P., Minero C., Pelizzetti E. and Vincenti M. (1997). Light-assisted 1,4-Dioxane degradation. *Chemosphere*, **35**, 2675–2688.

McElroy W. J. and Waygood S. J. (1990). Kinetics of the reactions of the $SO_4^{\bullet-}$ radical with $SO_4^{\bullet-}$, $S_2O_8^{2-}$, H_2O and Fe^{2+}. *Journal of the Chemical Society Faraday Transactions*, **86**, 2557–2564.

Mendez-Diaz J., Sanchez-Polo M., Rivera-Utrilla J., Canonica S. and von Gunten, U. (2010). Advanced oxidation of the surfactant SDBS by means of hydroxyl and sulphate radicals. *Chemical Engineering Journal*, **163**, 300–306.

Mertens M. and von Sonntag C. (1995). Photolysis ($\lambda = 254$ nm) of tetrachloroethene in aqueous solutions. *Journal of Photochemistry and Photobiology A-Chemistry*, **85**, 1–9.

Neppolian B., Doronila A. and Ashokkumar M. (2010). Sonochemical oxidation of arsenic(III) to arsenic(V) using potassium peroxydisulfate as an oxidizing agent. *Water Research*, **44**, 3687–3695.

Neta P., Madhavan V., Zemel H. and Fessenden R. W. (1977). Rate constants and mechanism of reaction of $SO_4^{\bullet-}$ with aromatic compounds. *Journal of the American Chemical Society*, **99**, 163–164.

Neta P., Huie R. E. and Ross A. B. (1988). Rate constants for reactions of inorganic radicals in aqueous solution. *Journal of Physical and Chemical Reference Data*, **17**, 1027–1247.

Norman R. O. C., Storey P. M. and West P. R. (1970). Electron spin resonance studies. Part XXV. Reactions of the sulphate radical anion with organic compounds. *Journal of the Chemical Society (B)*, 1087–1095.

Oh S.-Y., Kang S.-G. and Chiu P. C. (2010). Degradation of 2,4-dinitrotoluene by persulfate activated with zero-valent iron. *Science of the Total Environment*, **408**, 3464–3468.

Orellana-Garcia F., Alvarez M. A., Lopez-Ramon M. V., Rivera-Utrilla J. and Sanchez-Polo M. (2015). Effect of $HO^{\bullet}$, $SO_4^{\bullet-}$ and $CO_3^{\bullet-}$ /$HCO_3^{\bullet}$ radicals on the photodegradation of the herbicide amitrole by UV radiation in aqueous solution. *Chemical Engineering Journal*, **267**, 182–190.

Rao Y. F., Chu W. and Wang Y. R. (2013). Photocatalytic oxidation of carbamazepine in triclinic-WO_3 suspension: role of alcohol and sulfate radicals in the degradation pathway. *Applied Catalysis A*, **468**, 240–249.

Rastogi A., Al-Abed S. R. and Dionysiou D. D. (2009). Effect of inorganic, synthetic and naturally occurring chelating agents on Fe(II) mediated advanced oxidation of chlorophenols. *Water Research*, **43**, 684–694.

Rickman K. A. and Mezyk S. P. (2010). Kinetics and mechanisms of sulfate radical oxidation of β-lactam antibiotics in water. *Chemosphere*, **81**, 359–365.

Roshani B. and Leitner N. K. V. (2011). The influence of persulfate addition for the degradation of micropollutants by ionizing radiation. *Chemical Engineering Journal*, **168**, 784–789.

Schug T. T., Janesick A., Blumberg B. and Heindel J. J. (2011). Endocrine disrupting chemicals and disease susceptibility. *Journal of Steroid Biochemistry and Molecular Biology*, **127**(3–5), 204–215.

Schwarz H. A. and Bielski B. H. J. (1986). Reactions of hydroperoxo and superoxide with iodine and bromine and the iodide (I_2^-) and iodine atom reduction potentials. *Journal of Physical Chemistry*, **90**, 1445–1448.

Shah N. S., He X., Khan H. M., Khan J. A., O'Shea K. E., Boccelli D. L. and Dionysiou D. D. (2013). Efficient removal of endosulfan from aqueous solution by UV-C/peroxides: a comparative study. *Journal of Hazardous Materials*, **263**, 584–592.

Siegrist R. L., Crimi M. and Simpkin T. J. (2011). *In Situ* Chemical Oxidation for Groundwater Remediation. Springer, New York.

Stefan M. I. and Bolton J. R. (1998). Mechanism of the degradation of 1,4-Dioxane in dilute aqueous solution using the UV/Hydrogen peroxide process. *Environmental Science & Technology*, **32**(11), 1588–1595.

Syoufian A. and Nakashima K. (2007). Degradation of methylene blue in aqueous dispersion of hollow titania photocatalyst: optimization of reaction by peroxydisulfate electron scavenger. *Journal of Colloid and Interface Science*, **313**, 213–218.

Tan C., Gao N., Deng Y., An N. and Deng J. (2012). Heat-activated persulfate oxidation of diuron in water. *Chemical Engineering Journal*, **203**, 294–300.

Wang P., Yang S., Shan L., Niu R. and Shao X. (2011). Involvements of chloride ion in decolorization of Acid Orange 7 by activated peroxydisulfate or peroxymonosulfate oxidation. *Journal of Environmental Sciences*, **23**, 1799–1807.

Wang Y., Le Roux J., Zhang T. and Croue J. P. (2014). Formation of brominated disinfection byproducts from natural organic matter isolates and model compounds in a sulfate radical-based oxidation process. *Environmental Science & Technology*, **48**, 14534–14542.

Westerhoff P., Aiken G., Amy G. and Debroux J. (1999). Relationships between the structure of natural organic matter and its reactivity towards molecular ozone and hydroxyl radicals. *Water Research*, **33**, 2265–2276.

Xiao R., Ye T., Wei Z., Luo S., Yang Z. and Spinney R. (2015). Quantitative Structure-Activity Relationship (QSAR) for the oxidation of trace organic contaminants by sulfate radical. *Environmental Science & Technology*, **49**, 13394–13402.

Xie P., Ma J., Liu W., Zou J., Yue S., Li X., Wiesner M. R. and Fang J. (2015). Removal of 2-MIB and geosmin using UV/persulfate: contributions of hydroxyl and sulfate radicals. *Water Research*, **69**, 223–233.

Yang Y., Pignatello J. J., Ma J. and Mitch W. A. (2014). Comparison of halide impacts on the efficiency of contaminant degradation by sulfate and hydroxyl radical-based advanced oxidation processes (AOPs). *Environmental Science & Technology*, **48**, 2344–2351.

Yang Y., Jiang J., Lu X., Ma J. and Liu Y. (2015). Production of sulfate radical and hydroxyl radical by reaction of ozone with peroxymonosulfate: a novel Advanced Oxidation Process. *Environmental Science & Technology*, **49**, 7330–7339.

Yu X.-Y. and Barker J. R. (2003). Hydrogen peroxide photolysis in acidic aqueous solutions containing chloride ions. I. Chemical mechanism. *Journal of Physical Chemistry A*, **107**, 1313–1324.

Yu X.-Y., Bao Z.-C. and Barker J. R. (2004). Free radical reactions involving $Cl^{\bullet}$, $Cl_2^{\bullet-}$, and $SO_4^{\bullet-}$ in the 248 nm photolysis of aqueous solutions containing $S_2O_8^{2-}$ and Cl^-. *Journal of Physical Chemistry A*, **108**, 295–308.

Yuan R., Ramjaun S. N., Wang Z. and Liu J. (2011). Effects of chloride ion on degradation of Acid Orange 7 by sulfate radical-based advanced oxidation process: implications for formation of chlorinated aromatic compounds. *Journal of Hazardous Materials*, **196**, 173–179.

Zhang B.-T, Zhang Y., Teng Y. and Fan M. (2015). Sulfate radical and its application in decontamination technologies. *Critical Reviews in Environmental Science and Technology*, **45**, 1756–1800.

Zemel H. and Fessenden R. W. (1978). The mechanism of reaction of $SO_4^{\bullet-}$ with some derivatives of benzoic acid. *Journal of Physical Chemistry*, **82**, 2670–2676.

Zuo Z., Katsumura Y., Ueda K. and Ishigure K. (1997). Reactions between some inorganic radicals and oxychlorides studied by pulse radiolysis and laser photolysis. *Journal of the Chemical Society Faraday Transactions*, **93**, 1885–1891.

Zuo Z., Cai Z., Katsumura Y., Chitose N. and Muroya Y. (1999). Reinvestigation of the acid-base equilibrium of the (bi)carbonate radical and pH dependence of its reactivity with inorganic reactants. *Radiation Physics and Chemistry*, **55**, 15–23.

Chapter 11

Ultrasound wave-based AOPs

O. A. Larpparisudthi, T. J. Mason and L. Paniwnyk

11.1 INTRODUCTION

Research into the use of ultrasound in environmental protection has received a considerable amount of attention with the majority of investigations focusing on the harnessing of cavitational effects, both mechanical and chemical, for the destruction of pollutants in water and for the processing of sewage (Mason & Tiehm, 2001). The field of applications is much broader than this however and a summary of current topics of interest is given in Table 11.1. This chapter summarizes the various ways in which ultrasound can be used alone or in combination with other advanced oxidation processes for the treatment of water.

Table 11.1 General uses of ultrasound in environmental protection.

Removal of chemical and biological contamination from water	Oxidation of chemical and biological decontamination using ultrasound alone or in combination with other AOP techniques e.g. ozonation, uv light, electrochemistry Direct mechanical action on biological cells inducing cell rupture, the break-up of bacterial clumps and increased cell permeability to bactericide
Surface cleaning	Removal of surface contamination and biofilms
Soil washing	Removal of organic and inorganic contamination
Control of air-borne contamination	Agglomeration of smokes and aerosols Defoaming of liquids
Sewage treatment	Accelerated anaerobic digestion Dewatering of sludge

11.2 PRINCIPLES OF SONOCHEMISTRY

The use of ultrasound in water treatment is an active subject for research and development. The effects of sonochemistry derive from the result of ultrasonically induced acoustic cavitation (see later) and so the sound ranges employed are those which can generate cavitation i.e. generally in the lower frequency range of ultrasound below 1 MHz (Figure 11.1). The use of frequencies above 1 MHz is less common because there are very few commercial manufacturers that produce equipment in this range (Mason & Lorimer, 2002). Nevertheless interest is increasing in using higher frequencies because there is a greater chance of producing free radicals i.e. chemical reactions at higher frequencies (Mason *et al.* 2011).

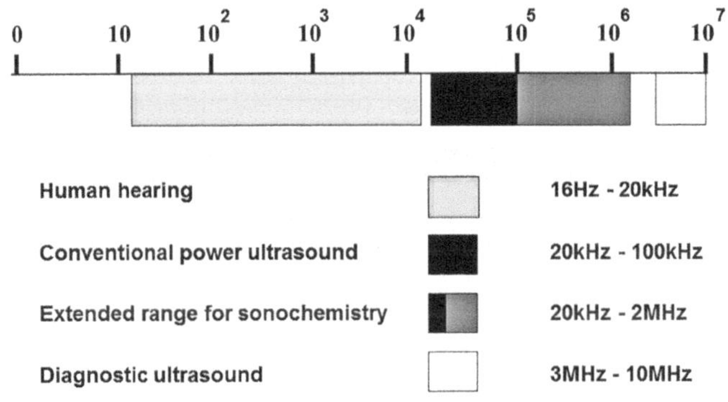

Figure 11.1 Frequency ranges of sound.

The origins of the use of power ultrasound in water treatment and processing can be traced back to 1927 when both the biological and chemical effects of sonication were reported.

- Wood and Loomis identified the biological effects of "supersound" and identified two contrasting effects, e.g. either stimulation or lethal effects on unicellular organisms, tissues, small fish, and animals (Wood & Loomis, 1927). Two years later, Harvey and Loomis examined the reduction in light emission (a factor related to bacterial kill) from a seawater suspension of rod shaped Bacillus fisheri caused by sonication at 375 kHz and 19°C (Harvey & Loomis, 1929). The final sentence of the paper predicted a poor future for the commercial exploitation of sonication (which they referred to as "raying") and this read: "In conclusion we can state that, under proper conditions of raying, luminous bacteria can be broken up and killed by sound waves of approximately 400,000 cycles per second and the solutions sterilized, but the method is not one of any practical or commercial importance because of the expense of the process". At the time this was written the conclusion was probably valid in that ultrasonic equipment was specialist, large and expensive. Today that situation has changed, ultrasonic technology is more commonplace, capital costs have been reduced and applications are more economic.
- Richards and Loomis published a paper (also in 1927) on the chemical effects of high frequency sound waves which described the development of power ultrasound for use in a range of applications including emulsification and surface cleaning (Richards & Loomis, 1927). In 1929 Schmitt *et al.* discovered that cavitation was found to induce oxidation when iodine was liberated during sonication of aqueous potassium iodide. The oxidation was thought to arise from the formation of hydrogen peroxide (Schmitt *et al.* 1929).
- In the 1930's the effects of ultrasound were becoming more familiar and the subject of many reports. These were reviewed by Richards in an excellent paper entitled "Supersonic Phenomena" that contained 348 references (Richards, 1939). Ultrasound was also recognised as an important factor for the improvement of electrochemical processes and Yeager reviewed these effects in terms of (1) the effects of ultrasonic waves on electrode processes (2) electrokinetic phenomena involving

ultrasonic waves and (3) ultrasonic waves as a tool in the study of the structure of electrolytic solutions (Yeager & Hovorka, 1953). Shortly afterwards Rich published a paper which has proved to be quite fundamental in subsequent studies leading to modern ultrasonically assisted plating (Rich, 1955).

- From the 1950's to the present day ultrasonically generated cavitation and its effects have continued to be a research topic of great interest from the point of view of its fundamental (Noltingk & Neppiras, 1950), applied (Knapp & Hollander, 1948) and chemical aspects (Fitzgerald *et al.* 1956). By the 1960's the industrial uses of power ultrasound were well accepted (Brown & Goodman, 1965; Frederick, 1965; Neppiras, 1972) and this interest has since continued to expand (Abramov, 1998; Mason & Lorimer, 2002; Feng *et al.* 2011; Gallego-Juárez & Graff, 2015).

11.3 ACOUSTIC CAVITATION, THE DRIVING FORCE FOR SONOCHEMISTRY

The origin of the effects of sonochemistry is the collapse of cavitation bubbles generated in an ultrasonic field. Like any sound wave, ultrasound is propagated *via* a series of compression and rarefaction waves induced in the molecules of the medium through which it passes. At sufficiently high power the rarefaction cycle may exceed the attractive forces of the molecules of the liquid and cavitation bubbles will form. Such bubbles grow by a process known as rectified diffusion i.e. small amounts of vapour (or gas) from the medium enters the bubble during its expansion phase and is not fully expelled during compression. The bubbles grow over the period of a few cycles to an equilibrium size for the particular frequency applied. It is the fate of these bubbles when they collapse in succeeding compression cycles that generates the energy for chemical and mechanical effects (Figure 11.2). Cavitation bubble collapse is a remarkable phenomenon induced throughout the liquid by the power of sound. In aqueous systems at an ultrasonic frequency of 20 kHz each cavitation bubble collapse acts as a localised "hotspot" generating temperatures of about 4000 K and pressures in excess of 1000 atmospheres (Fitzgerald *et al.* 1956; Neppiras, 1984; Henglein, 1987; Suslick, 1990; Ashokkumar & Mason, 2007; Nikitenko, 2014).

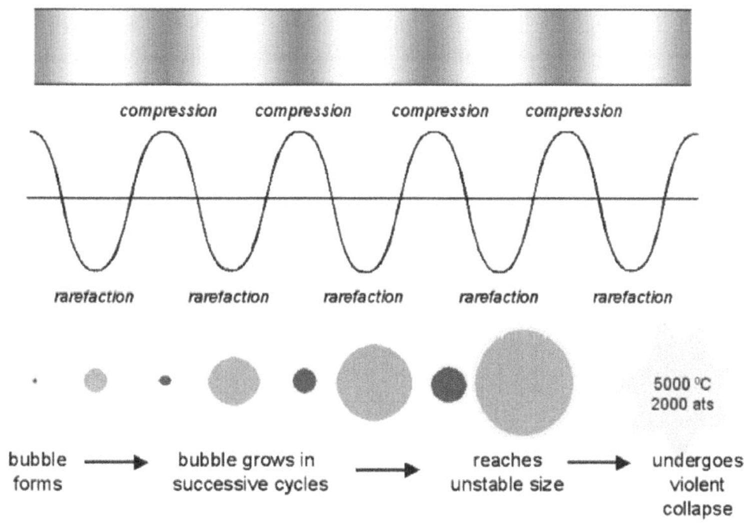

Figure 11.2 The acoustic generation of a cavitation bubble.

Whatever application of sonochemistry is to be studied or developed there are two essential components, a liquid medium and a source of high-energy vibrations. The liquid medium is necessary because sonochemistry is driven by acoustic cavitation and this can only occur in liquids. The source of the vibrational energy is the transducer of which there are three main types: fluid driven (liquid whistle), or electromechanical (magnetostrictive and piezoelectric transducers). The two most common pieces of equipment that are used for the generation of acoustic cavitation in the laboratory are the ultrasonic bath often a simple cleaning bath and the more powerful probe system (Figure 11.3) (Mason & Peters, 2002).

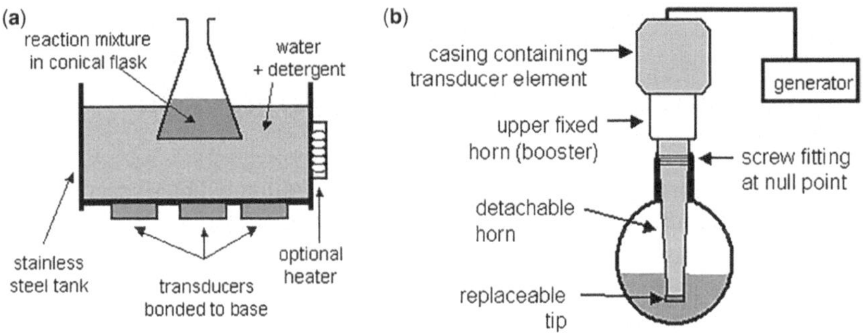

Figure 11.3 Common laboratory ultrasonic equipment (a) bath (b) probe.

The cavitation bubble has a variety of effects within the liquid medium depending upon the type of system in which it is generated. These systems can be broadly divided into homogeneous liquid, heterogeneous solid/liquid and heterogeneous liquid/liquid. Each of these classifications is of interest in water treatment.

11.3.1 Homogeneous liquid-phase systems

It is not strictly correct to describe any system within which cavitation occurs as homogeneous since bubbles will be present. However, in this case "homogeneous" refers to the state of the system before cavitation, i.e. a homogeneous liquid or solution. Cavitation collapse in such systems can affect the bulk liquid immediately surrounding the bubble where the rapid collapse of the bubble generates powerful shear forces which can produce mechanical effects such as the breaking of chemical bonds in polymeric materials dissolved in the fluid (Price, 1990). Such forces can also provide extremely efficient mixing which is of use for rapid emulsification and dissolution of chemicals or gases. Inside the collapsing bubble the extreme conditions of temperature and pressure can induce chemical effects resembling pyrolytic decomposition of any pollutant material contained within it. In the case of water itself bubble collapse causes rupture of the O−H bond. This results in the formation of radical species and the production of oxygen gas and hydrogen peroxide (Scheme 11.1).

$$
\begin{gathered}
H_2O \rightarrow H\bullet + OH\bullet \\
OH\bullet + OH\bullet \rightarrow H_2O_2 \\
OH\bullet + OH\bullet \rightarrow H_2 + O_2 \\
H\bullet + O_2 \rightarrow HO_2\bullet \\
HO_2\bullet + H\bullet \rightarrow H_2O_2 \\
HO_2\bullet + HO_2\bullet \rightarrow H_2O_2 + O_2
\end{gathered}
$$

Scheme 11.1 Ultrasonically induced decomposition of water.

In the reaction scheme above, radical species such as H•, O•, OH• and HO_2• are produced through bond dissociation. The initial dissociation of water *via* the cleavage of the H−O bond to form H and OH radicals has a bond dissociation energy requirement of 460 kJ mol^{-1}. In the scheme above the lowest bond energy requirement for the formation of OH radicals is from the dissociation of HO−OH from hydrogen peroxide which has a far lower energy requirement of 180 kJ mol^{-1}. The bond dissociation energy for H_2 to form H radicals is 436 kJ mol^{-1} and the bond dissociation energy for O_2 to form O radicals is 497 kJ mol^{-1}. The existence of OH radicals from the dissociation of water, and also subsequent dissociation of hydrogen peroxide would be the most likely source of oxidative species. The formation of O radicals from either the dissociation of water or from O_2 has the highest energy requirement for formation and so would be a less likely route. H radicals although present to some extent are reductive species and so not involved in oxidation pathways.

In a key study using electron spin resonance (ESR), Makino *et al.* reported the first conclusive evidence for the formation of OH• and H• in the cavitation bubbles produced by ultrasound in aqueous solutions. The formation of these species was confirmed by investigating reactions in which OH• and H• scavengers [formate, thiocyanate, benzoate, methanol, ethanol, 1-propanol, **2**-methyl-2-propanol, acetone, 2-methyl-2-nitrosopropane (MNP)] compete with the spin traps for hydroxyl radicals and hydrogen atoms (Makino *et al.* 1983).

For all practical purposes the two main oxidising species of interest in water remediation are the highly active and short lived hydroxyl radical and hydrogen peroxide. The former has been implicated in a range of oxidative degradations of model pollutants such as *p*-nitrophenol, carbon tetrachloride, parathion, 4-nitrophenyl acetate and trinitrotoluene (Hoffmann *et al.* 1996). Hydroxyl radicals can react with material within the cavitation bubble and also have a transitory existence in the bulk liquid immediately surrounding it. It has also been found to be effective in improving the results of microbiological contamination using chemical biocides, ozone or UV light (Joyce & Mason, 2008).

The production of hydrogen peroxide via cavitation has been of interest for many years (Anbar & Pecht, 1964). The yield depends on frequency and, in 1997, Petrier reported that 200 kHz gave a yield of 0.6 mM after 2 hours sonication that was better than that obtained using 20, 500 or 850 kHz (Pétrier & Francony, 1997). The production of hydrogen peroxide has been used as a monitor of sonochemical yield (Gong & Hart, 1998). It is one of the oxidants, together with radical species, involved in sonochemical oxidation.

11.3.2 Heterogeneous solid surface-liquid systems

Unlike cavitation bubble collapse in the bulk liquid, collapse on or near to a surface is unsymmetrical because the surface provides resistance to liquid flow from that side. The result is an inrush of liquid predominantly from the side of the bubble remote from the surface resulting in a powerful liquid jet being formed, targeted at the surface. The effect is equivalent to high pressure jetting and this is the reason why ultrasound is used extensively for cleaning and the removal of biofilms (Bulat, 1974). This has been implemented in the food industry for example as a means of improving the efficacy of fresh vegetable washing chemical treatments are combined with other methods of removal such as ultrasound (Warning & Datta, 2013). More generally, ultrasound is finding many applications in decontamination procedures in food processing (Chandrapala *et al.* 2012).

Ultrasonic cleaning has developed and is used not only for general cleaning at frequencies around 40 kHz but also for fragile materials such as silicon wafers where frequencies of around 1 MHz are used in what is called megasonic cleaning (Mason, 2016). This effect can also increase mass and heat transfer to the surface by disruption of the interfacial boundary layers. It has found applications in electrochemistry where mass transport effects and surface cleaning are important (Mason & Bernal, 2012). This means that the use electrochemistry in environmental remediation can be significantly improved with the use of ultrasound because it provides a means of continuously removing contamination from the electrode surface.

11.3.3 Heterogeneous particle-liquid systems

Acoustic cavitation can produce dramatic effects on particulate or agglomerated material suspended in a liquid. Cavitation can achieve particle size reduction, deagglomeration and efficient dispersion. Cavitation bubble collapse in the liquid phase near to a particle can force it into rapid motion and the general dispersive effect is accompanied by inter-particle collisions that can lead to erosion, surface cleaning, and wetting. Such properties are very valuable for the enhancement of remediation using heterogeneous photocatalysis (Augugliaro *et al.* 2006).

11.3.4 Heterogeneous liquid-liquid systems

In heterogeneous liquid/liquid reactions, cavitational collapse at or near the liquid interface will cause disruption and mixing, resulting in the formation of very fine emulsions. In situations where remediation involves insoluble liquid pollutants the process of emulsification can facilitate any interactions with short-lived sonochemically produced radicals. Applications for this type of technology are to be found in the oil industry (Hu *et al.* 2013).

11.4 HISTORICAL INTRODUCTION ON THE OXIDATIVE PROPERTIES OF ULTRASOUND IN WATER

Ever since the earliest observation that ultrasound could promote chemical reactions it was recognised that the propagation of a sound wave of sufficient intensity in water can induce oxidation reactions. As far back as 1929; Schmitt *et al.* reported the following observations (Schmitt *et al.* 1929):

- Upon ultrasonic irradiation at 750 kHz iodine can be liberated from a potassium iodide solution saturated with air indicating the generation of an oxidising species.
- Under the same conditions both potassium chloride and potassium bromide solutions can oxidise a potassium iodide starch reagent.
- Ultrasonic irradiation of titanium sulphate solution produces a yellow colour indicating the formation of hydrogen peroxide in the process.
- The effects caused by ultrasound only occurred when the energy was sufficient to cause the formation of cavitation bubbles in the liquid.
- When carbon tetrachloride is added to the potassium iodide/starch reaction mixture the rate of reaction was increased, the authors suspected that this was due to oxygen cleavage to form "active oxygen" in the oxidation process. Some 20 years later the addition of CCl_4 to enhance the oxidation of potassium iodide in solution was characterised by Weissler *et al.* and this reagent combination has since become known as "Weissler's solution" and used extensively to assess the level of acoustic cavitation (Weissler *et al.* 1950).

Extending the original studies of Schmitt, several groups confirmed experimentally the formation of hydrogen peroxide and direct activation of oxidation through ultrasonically induced cavitation (Liu & Wu, 1934; Flosdorf *et al.* 1936; Schulter & Gohr, 1936; Loiseleur, 1944). These authors also confirmed that carbon tetrachloride or chloroform react with the so-called "active oxygen" formed during cavitation.

Subsequently, in 1949 the very important observation was made that oxygen was not necessary for the oxidative reaction because oxidation also occurred in deoxygenated water saturated with argon (Prudhomme & Grabar, 1949). Water itself was however essential for the process. Under cavitating conditions, the ultrasonic treatment of water in the absence of oxygen was found to produce molecular oxygen and hydrogen in addition to hydrogen peroxide.

The formation of hydrogen peroxide in cavitating water was reinvestigated by Anbar and Pecht who used labelled water H_2O^{18} in presence of hydrogen peroxide $H_2O_2^{16,16}$ in a medium which was saturated either with argon or with labelled molecular oxygen $O_2^{18,18}$(Anbar & Pecht, 1964). Under an argon atmosphere the formation of pure $H_2O_2^{18,18}$ supported the splitting of water. However experiments conducted with water H_2O^{18} saturated with $O_2^{16,16}$, or water H_2O^{16} saturated with $O_2^{18,18}$ led to the formation of both to $H_2O_2^{18,18}$ and $H_2O_2^{16,18}$. This suggested that in the case of a medium saturated with oxygen that there were three possible sources of H_2O_2: water, molecular oxygen and oxygen atoms. The question of the source of hydrogen peroxide was revisited in 1986 using the same technique i.e. using water saturated with various proportions of $O_2^{18,18}$ and argon (Fischer *et al.* 1986). In this study they took into account the stability of the intermediates in the high temperature zone of the bubble. Their discussion emphasized that the formation of H_2O_2 occurred in the low temperature interfacial zone between bubble and bulk solution mainly through the recombination of hydroxyl radicals. However the reaction between atomic oxygen and H_2O occurs mainly in the high temperature zone inside the bubble and should generate only OH radical.

Direct confirmation of the production of radicals in a medium subjected to acoustic cavitation was reported by Makino, Mossoba and Riesz using electron spin resonance (Makino *et al.* 1982; Makino *et al.* 1983). Using the spin trap DMPO (5,5-dimethyl-1-pyrroline N-oxide) they observed the spectrum of the stable spin adducts HO-DMPO and H-DMPO from sonicated water saturated with argon. This was the first time that direct proof was presented for the identities of the radicals produced from the sonochemical splitting of water.

The generation of OH radicals by acoustic cavitation in water can be monitored by reaction with terephthalic acid (TA) anion to produce fluorescent hydroxyterephthalate ions. This method has been used to compare radical production in water using a probe system (20, 40 and 60 kHz) (Mason *et al.* 1994). The fluorescence yield (fluorescence intensity/ultrasound dosage) was highest at 60 kHz and in all cases was directly proportional to power input and the concentration of TA. The same method was used to compare the formation of hydroxyl radicals in cavitating water under batch and continuous flow conditions (Juretic *et al.* 2015). At the same acoustic energy input for batch reactions the efficiency of radical production depended upon reactor volume and for the flow system on the flow rate.

For more than 70 years there have been attempts to develop ideas of what is actually happening to the material inside of the collapsing bubble that would help to explain the chemical effects of cavitation. Over the years there have been a range of suggestions involving electrical or corona discharges, plasma formation, supercritical fluid formation, hotspots and chemical reactions resulting from liquid droplets injected inside the bubble.

Various electrical theories were developed the earliest of which dates back to the work of Frenkel who suggested that electrical charges could be built up on opposite faces of a cavitation bubble as it was formed (Frenkel, 1940). The discharge of such charges during bubble collapse would then lead to ionisation and bond cleavage of any vapour in the cavitation bubble. This was in agreement with the hypothesis of Weiss in relation to ionising radiation (Weiss, 1944). An alternative electrical theory was suggested decades later based upon the production of an electric discharge during bubble fragmentation during collapse (Margulis, 1985). The electrical theory was the main physico-chemical support used to explain sonochemical effects up until around the middle of the 20th century.

In 1950 Griffing suggested that water dissociation could occur upon thermal reaction which occurs during the final step of the adiabatic collapse of the cavitation bubble (Griffing, 1950). This explanation has become known as the "Hot spot theory" can account for radical production from ultrasonic action and hence to the promotion of oxidation processes. This theory has received experimental support and has superceded the electrical theory in terms of scientific acceptance (Noltingk & Neppiras, 1950; Fitzgerald *et al.* 1956; Henglein, 1987).

Alternatives to the hot spot theory have been postulated since then including the injection of sprays from the surrounding liquid into the cavity (Lepoint & Mullie, 1994; Lepoint-Mullie *et al.* 1996). As a result partial ionisation of the bubble content could be produced and thus suggest the possibility of plasma chemistry. Indeed plasma formation during bubble collapse has received some more recent support (Nikitenko, 2014). Plasma technology is an active area of research in wastewater remediation (Jiang *et al.* 2014).

11.5 SONOCHEMICAL DECONTAMINATION OF AQUEOUS SYSTEMS

Advanced oxidation processes (AOPs) generally rely on the production of hydroxyl radicals (HO•) which are able to oxidize organic compounds of environmental concern into simpler and harmless end products (Andreozzi *et al.* 1999; Pera-Titus *et al.* 2004). Generally there are three steps involved:

(a) Primary oxidation of the pollutant molecule through the attack of $HO^•$. The major reactions of the OH radical with chemical pollutants consist in H-abstraction, addition to unsaturated bonds, and electron transfer.
(b) The radicals formed from pollutant molecule in step one, evolve toward formation of stable molecules, which, depending on the original structure of pollutant, could have lower molecular weights than the parent compound.
(c) Mineralization of the original organic carbon and heteroatoms to CO_2, H_2O and inorganic ions (e.g., halide, nitrate, sulfate) could occur through further oxidation of the intermediate by-products.

In some cases the degradation pathways may generate products of the same or greater toxicity than the original pollutants. Although the break-down of the parent molecule may only involve partial oxidation, this is sometimes sufficient to turn specific compounds into those more amenable to subsequent biological treatment or to reduce their toxicity (Metcalf & Eddy, 2004).

A variety of technologies are available to produce $HO^•$ in the aqueous phase including Fenton's reagent, ozone or photocatalysis using TiO_2. The efficiency of these processes can be enhanced with the use of ultrasound and in some cases ultrasound alone may be sufficient to induce pollutant break-down.

11.5.1 AOP involving ultrasound alone

The majority of systems studied have been aqueous solutions of either aromatic compounds or halogenated hydrocarbons. Such materials represent models for the major classes of organic pollutants in waste and ground water. In a classic study Berlan *et al.* reported that the primary products from the sonochemical treatment of phenol at 541 kHz (27°C with bubbled air) are hydroquinone and catechol (Berlan *et al.* 1994). These compounds are easy to monitor and are clearly seen to be intermediates that disappear as the reaction progresses (Figure 11.4). Similarly, the sonolysis of aqueous 4-chlorophenol leads to products mainly characteristic of oxidation by OH radical e.g. 4-chlorocatechol but in both cases the final organic products are CO, CO_2 and HCOOH (Figure 11.5) (Petrier *et al.* 1994; Serpone *et al.* 1994).

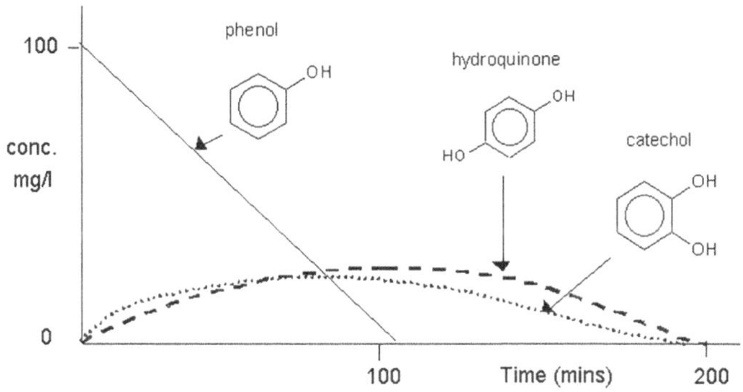

Figure 11.4 Sonochemical decomposition phenol in aerated water.

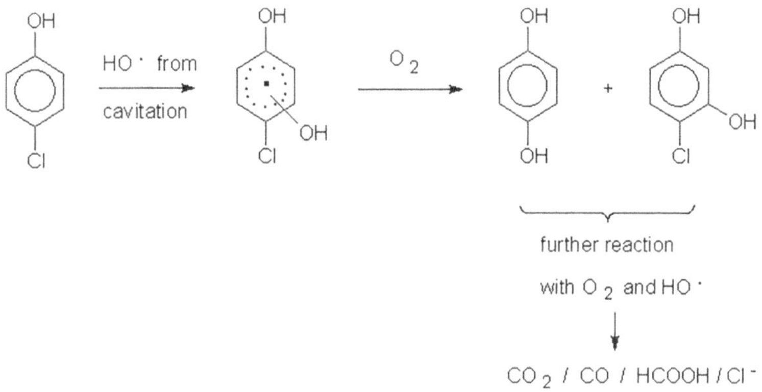

Figure 11.5 Sonochemical decomposition of 4-chlorophenol in aerated water.

After an irradiation time of 300 min, (500 kHz, 30 W, 20°C), the hydroxylated intermediates disappear and the chlorine atoms are completely mineralized to chloride ions. The concentrations of the final products (CO and CO_2) rise slowly and after 400 min they represent 21% of the starting chlorophenol and 18% of the starting phenol. These studies have been extended to the destruction of a number of chloroaromatics (Table 11.2) (Petrier & Casadonte, 2001).

Table 11.2 Initial concentration and chloride yield after 150 min irradiation for the degradation of chloroaromatics.

Chloroaromatic Compound	Concentration (mM)	Yield of Cl^- (%)
1,2-dichlorobenzene	0.4	90
1,3-dichlorobenzene	0.05	89
1,4-dichlorobenzene	0.2	95
1,3,5-trichlorobenzene	0.02	95
1-chloronaphthalene	0.04	98

Volume treated: 250 cm^3, ultrasonic power: 30 W, electric power: 50 W, frequency 500 kHz.

In general chloroaromatic hydrocarbons decompose faster than the more hydrophilic compounds such as phenols. The destruction of these two kinds of pollutants occurs at different sites in or around the cavitation bubble and follows different pathways. The difference becomes even more pronounced when more volatile contaminants are used. This is neatly illustrated in a comparative study of the decomposition of phenol (to carboxylic acids) and carbon tetrachloride (to CO_2 and Cl^-) in water saturated with oxygen at 20, 200, 500 and 800 kHz (Table 11.3) (Pétrier & Francony, 1997). The results clearly show a difference in the behaviour of the chemical contaminants (original concentration 10^{-3}M) with the phenol degradation mirroring the peroxide formation indicating this reaction proceeding at the bubble interface or outside of the bubble. The volatile CCl_4 however is decomposed within the bubble and this process is accelerated with increasing frequency.

Table 11.3 Comparison between H_2O_2 production and the rates of phenol and carbon tetrachloride disappearance at different frequencies (μMmin^{-1}).

Frequency (kHz)	20	200	500	800
H_2O_2 formation	0.7	5.0	2.1	1.4
Phenol degradation	0.5	4.9	1.9	1.0
CCl_4 degradation	19	33	37	50

Table 11.3 shows that 200 kHz is the optimum frequency for peroxide formation and phenol removal. This was explained through a two-step reaction pathway for the oxidation process. In the first step the sonolysis of water produces radicals within the bubble. In step two the radicals migrate to the bubble interface or into the bulk aqueous medium to form peroxide or react with the phenolic substrate. The authors suggest that lower frequencies are more efficient for the decomposition of molecules inside the bubble. As the frequency increases the pulsation and collapse of the bubble occurs more rapidly and more radicals escape from the bubble. This effect is counteracted by a drop in the cavitation intensity with increase in frequency resulting in a reduction in the yield of radicals and consequently the number which reach the interface and bulk solution. These two factors give rise to the optimum frequency for the reaction.

The same optimum frequency was observed in the sonication of 4-chlorophenol in O_2 saturated solutions at different chlorophenol concentration, temperature and frequency (Jiang *et al.* 2006). The initial degradation rate increased with the increase of the concentration up to about 1000 μM. The authors suggested that at high concentration the degradation occurs predominantly at the liquid-gas bubble interface whereas at low concentrations the degradation occurs in the bulk of the solution.

A subsequent study of the oxygen-induced concurrent ultrasonic degradation of the volatile and non-volatile aromatic compounds chlorobenzene and 4-chlorophenol respectively revealed more information about reactions inside and outside of the bubble (Pétrier *et al.* 2007). These experiments show that within the collapsing cavitation bubble the degradation of volatile organic contaminants can produce hydroxyl radicals and that these additional radicals can then react with less volatile organic compounds in the water shell around the bubble. Thus the cavitation bubble and the shell of aqueous medium surrounding it can be regarded as a multifunctional microreactor especially in an oxygen-enriched medium. The decomposition of volatile compounds within the bubble will provide extra OH radicals at its surface available for the oxidation of non-volatile contaminants. This mechanism is of particular significance in the treatment of aqueous solutions containing a mixture of contaminants and suggests that the addition of oxygen could provide a new source of OH radicals from the degradation of volatiles within the cavitation bubbles to assist in the oxidation of less volatile compounds in the bulk solution.

The sonochemical degradation of 2-, 3-, 4-chlorophenol and pentachlorophenol at 200 kHz under air and argon atmosphere has been reported (Nagata *et al.* 2000). The degradation process here as in many other sonochemical AOPs followed first-order kinetics. Due to the electrophilic character of hydroxyl radicals, the attack site of hydroxyl radicals is expected to be the ortho and para orientations to Cl and OH group as generally observed in hydroxyl radical addition to aromatic compounds. As a result the decomposition rates of 3-chlorophenol were faster than those of 2-chlorophenol and 4-chlorophenol as 3-chlorophenol has three points of simultaneous ortho- and para-orientation. Addition of t-butanol as a radical scavenger inhibited the degradation although not completely. This suggested that although the main degradation pathways involve reaction with sonochemically generated hydroxyl radicals there is a possibility that some pyrolysis can take place within the bubble. Supporting evidence for OH radical providing the major source of oxidation of 4-chlorophenol was provided by the addition of a radical scavenger (t-butanol) to the reaction (Hamdaoui & Naffrechoux, 2008). In the presence of t-butanol the degradation was almost completely quenched.

Another class of compound that has been extensively investigated in terms of sonochemical treatment is chemical dyes in particular the azo dyes (Vinodgopal *et al.* 1998; Ince and Tezcanlı́, 2001; Tezcanli-Guyer & Ince, 2003; Rehorek *et al.* 2004; Tezcanli-Güyer & Ince, 2004; Okitsu *et al.* 2005; Vajnhandl & Le Marechal, 2007). Ultrasound is able to decolourise & mineralize chemical dyes through similar processes that occur in the removal of chlorophenols i.e. through the production of oxidising radicals. This has been supported by the addition of radical scavengers such as t-butanol which cause a reduction of the reaction rate (Rehorek *et al.* 2004; Okitsu *et al.* 2005). In the case of azo dyes the radicals initially attack the N=N bond to cause decolourisation (Comeskey *et al.* 2012). In this study a range of frequencies (20, 40, 380, 512, 850, 1000 and 1,176 kHz) were investigated and it was found that the highest levels of hydrogen peroxide and also the greatest amount of decolourisation occurred at 850 kHz. Although decolourisation may be relatively rapid complete mineralization requires long treatment times. Thus 33 µmol/l of Reactive Black 5 at 640 kHz was decolourised completely after 90 minutes sonication but only about 60% mineralization was observed after 6 hours of treatment (Vinodgopal *et al.* 1998). They further reported that the remaining components after ultrasonic treatment were oxalate, sulphate and nitrate ions. Rehorek *et al.* reported the same products but they also detected formate and acetate ions (Rehorek *et al.* 2004). The degradation rate of azo dyes is generally first order and is measured through changes to its maximum absorption in the visible range, the rate also increases with acidity (Tezcanli-Guyer & Ince, 2003; Dükkancı *et al.* 2014).

One of the questions regarding the use of cavitation in wastewater treatment is whether the sonochemically induced partial break-down of pollutants might generate species that could be more toxic or that can be digested more readily than the original pollutant. The degradation of 2,3,5-trichlorophenol was studied at 41, 206, 360, 618, 1068 and 3217 kHz (Tiehm & Neis, 2005). In their study, the greatest decomposition was found to occur at 360 kHz. The bioluminescence test was performed with *Vibrio fisheri* according to the European standard procedure EN ISO 11348. The marine bacterium *V. fisheri* emits light during physiological activity. A decreasing light intensity is correlated with an increasing toxicity of the sample tested. Toxicity in the bioluminescence test on samples treated with ultrasonication increased during the initial period and then decreased significantly. This pattern indicates that toxic by-products were formed during the first stage of treatment but were destroyed upon prolonged treatment by a break-down to non-toxic compounds. In order to further study the biodegradability of the sonicated samples, mineral salts suitable for microbial growth were added and the pH was adjusted to 7.0. Inoculation was done with activated sludge obtained from a municipal wastewater treatment plant, and biological degradation was monitored in a respirometer at 20°C. After a short ultrasonic pre-treatment, a faster biodegradation of the remaining organic pollutants was observed. In conclusion, the results demonstrate the potential of integrated ultrasonic/biological approaches for the treatment of wastewaters containing toxic pollutants.

In recent years a number of new and complex chemical compounds have been detected and identified in the effluent streams of wastewater processing plants as direct result of human activities. Generally referred to as "Pharmaceutically Active Compounds" these are an emerging problem in the aquatic environment for both wildlife and humans. Among the compounds causing a growing concern are the synthetic estrogens which are known as endocrine disruptor chemicals (EDCs). The discharge of EDCs into the environment impacts on aquatic wildlife and poses serious health threats to humans when surface water is used as a source of drinking water (Diamanti-Kandarakis *et al.* 2009). One of the first of these was bisphenol A (BPA) an additive in many plastics which leaches into the environment. The removal of this material by sonication was investigated at ultrasonic frequencies between (300–800 kHz) and powers of (20–80 W) under laboratory conditions (Torres *et al.* 2008). The best results were obtained at 300 kHz, 80 W, with oxygen as saturating gas where BPA was eliminated after about 90 min. Analyses of intermediates using HPLC-MS revealed several hydroxylated aromatic compounds showing that the main ultrasonic BPA degradation pathway is related to the reaction of BPA with OH radical. After 2 h, these early products were converted into biodegradable aliphatic acids.

The efficiency of the sonochemical removal of two other endocrine disrupting chemicals [17β-estradiol (E2) and 17α-ethinylestradiol (EE2)] has been investigated using laboratory scale ultrasonic baths at low power intensities. Higher ultrasonic frequencies were found to be more effective for pollutant degradation with 850 kHz the best: 9.0×10^{-1} mg/kW h for E2 and 6.8×10^{-1} mg/kW h for EE2 at initial concentrations of 1 ppm (Capocelli *et al.* 2012). The rate of ultrasonic degradation of E2 and EE2 in water and wastewater is influenced by the pH, power, air sparging and the dissolved organic content of the aqueous solutions (Ifelebuegu *et al.* 2014). The destruction of another EDC, 1-H-benzotriazole has been examined under varying conditions such as ultrasonic power, initial pollutant concentration, pH and the presence of co-existing chemical species (oxygen, nitrogen, ozone, and radical scavengers). It was determined that a high level of applied ultrasonic power enhanced the extent of 1-H-benzotriazole removal which was accelerated in the presence of ozone and oxygen, but inhibited by nitrogen. The most favourable pH for the ultrasonic degradation was acidic media, reaching a 90% pollutant removal in 2 h. It was concluded that degradation takes place at both the bubble-liquid interfacial region and in the bulk solution, and OH radicals were the main species responsible for the reaction (Zuniga-Benitez *et al.* 2014).

Another type of test for examining the relative exposure of cells to natural and xenoestrogens involves the use of Yeast Estrogen Screen (YES) assay (Arnold *et al.* 1996). This has been applied in a comparative study of ultrasonication, Fenton's oxidation and ferro-sonication for the degradation of carbamazepine (CBZ) in water. This chemical is an anticonvulsant and analgesic drug and was the most frequently

detected pharmaceutical residue in water bodies (Zhang *et al.* 2008). The majority of research on ultrasonication and Fenton's oxidation or combination of both (ferro-sonication) for degradation of CBZ has been carried out in water but a more recent study has applied these techniques using 20 kHz ultrasound to real wastewater (Mohapatra *et al.* 2013). The results showed that while Fenton's oxidation was the most efficient for the removal of CBZ the ultrasonic processes similar oxidative by-products of carbamazepine degradation (epoxycarbamazepine and hydroxycarbamazepine) were observed in all three and the YES assay results showed that the treated wastewater not toxic to the yeast i.e. not estrogenic.

Another common pharmaceutical waste that finds its way into wastewater is ibuprofen and the removal of this chemical has been studied under sonication at 300 kHz (Méndez-Arriaga *et al.* 2008). At 80 W of applied power 98% destruction was achieved in 30 min. Chemical and biological oxygen demands indicated that the process oxidizes the IBF to biodegradable substances that can be removed in a subsequent biological step.

Gasoline oxygenates are volatile branched ethers added to gasoline to improve its combustion efficiency. They find their way into groundwater as a result of gasoline spills or gasoline leakage from underground storage tanks. An ultrasonic treatment at 665 kHz has been investigated for the removal of gasoline oxygenates such as di-isopropyl ether (DIPE), ethyl tert-butyl ether (ETBE), and tert-amyl ether (TAME) which are as alternatives to the widely used oxygenate methyl tert-butyl ether (MTBE). In water saturated with oxygen pulsed sonication with a duration of 0.62 s and pulse repetition of 2.5 s was applied at a temperature regulated at $12 \pm 3°C$ (Kim *et al.* 2012). Over 60% of degradation of these gasoline oxygenates was observed after30 min treatment and almost complete degradation was observed after 6 h. The mechanisms proposed for the decomposition involved a combination of the abstraction of α-hydrogen atoms by hydroxyl radicals (formed during cavitation) and pyrolytic degradation (of some of the volatile material which entered the cavitation bubble).

Blue green algae (cyanobacteria) are a significant pollutant in surface waters worldwide especially in calm, nutrient-rich waters. Some species of cyanobacteria produce toxins that affect animals and humans through drinking contaminated water or by ingestion during recreational water contact. Algal cells also release compounds which affect the water aesthetics. Two of the chemicals produced by such algae are 2-Methylisoborneol (MIB) and geosmin (GSM) which are taste-and-odor causing compounds. These are semi-volatile compounds and their removal presents a significant challenge to drinking water providers. While conventional water treatments have proved ineffective for removing low concentrations of these chemicals ultrasonic irradiation at 640 kHz leads to rapid degradation of MIB and GSM (Song & O'Shea, 2007). The degradations followed first-order kinetics and the main mechanism involves hydroxyl radicals; GSM degraded faster than MIB due to its larger rate constant for the OH radical reaction than that of MIB, and also due to a higher vapour pressure. There have been studies which attempted to remove the cyanobacteria by direct exposure to ultrasound but this can be difficult when treating very large areas of water during periods of algal blooms (Wu *et al.* 2011).

11.5.2 AOP involving ultrasound combined with ozone

Ozone (O_3) is a very strong oxidant which has the potential to oxidize a wide range of organic compounds to carbon dioxide and water but the reaction rate depends on the structure of the pollutant. Ozonolysis can therefore be considered to be more selective than the hydroxyl radical. The combination of sonication and ozonation provides three sources of the very highly active OH radical (Weavers & Hoffmann, 1998):

(a) from the sonochemical decomposition of water
(b) from the normal chemical break-down of ozone leading to the formation of atomic oxygen and hydroxyl radical
(c) from the thermolytic decomposition of ozone in the acoustic cavitation bubble.

The thermolytic decomposition of ozone in the vapour phase of a cavitation bubble proceeds as follows (Weavers, 2001):

$$O_3 \xrightarrow{\Delta} O_2 + O(^3P) \tag{11.1}$$

This reaction yields atomic oxygen that can react with water to form hydroxyl radical:

$$O(^3P) + H_2O \rightarrow 2OH^\bullet \tag{11.2}$$

The reaction sequence (Eqns. 1 and 2) occurs in the vapour phase of cavitation bubbles and the short-lived products then migrate to the interface of the bubble and water and from there into the aqueous phase.

The combination of O_3 and ultrasound should thus be an effective oxidation system since 2 OH• molecules are formed per O_3 molecule consumed.

Weavers proposed a conceptual diagram of possible pathways during combined sonolysis and ozonolysis (Scheme 11.2) (Weavers, 2001). Pathways 1 and 2 indicate ozonolysis processes occurring in the absence of sonication including mass transfer of ozone from the gas phase to solution (pathway 1) and solution reactions of the compound with ozone or radicals from O_3 autodecomposition occurring in the absence of sonolysis (pathway 2). Pathways 6 and 7 are sonolysis reactions that occur in the absence of ozonolysis. Upon combining the processes, additional processes occur by pathways 3, 4, 5, and 8. In addition, pathways 2, 6, and 7 may be altered by the direct interaction of sonolysis and ozonolysis. For example, ozone decomposes in a cavitation bubble to form $OH^•$ that is available to react with C (pathway 3). Additionally, O_3 may react with $O(^3P)$ generated from the thermal decomposition of O_3 in the cavitation bubble resulting in O_2 (pathway 4). Both of these pathways reduce the O_3 available to react via pathway 2.

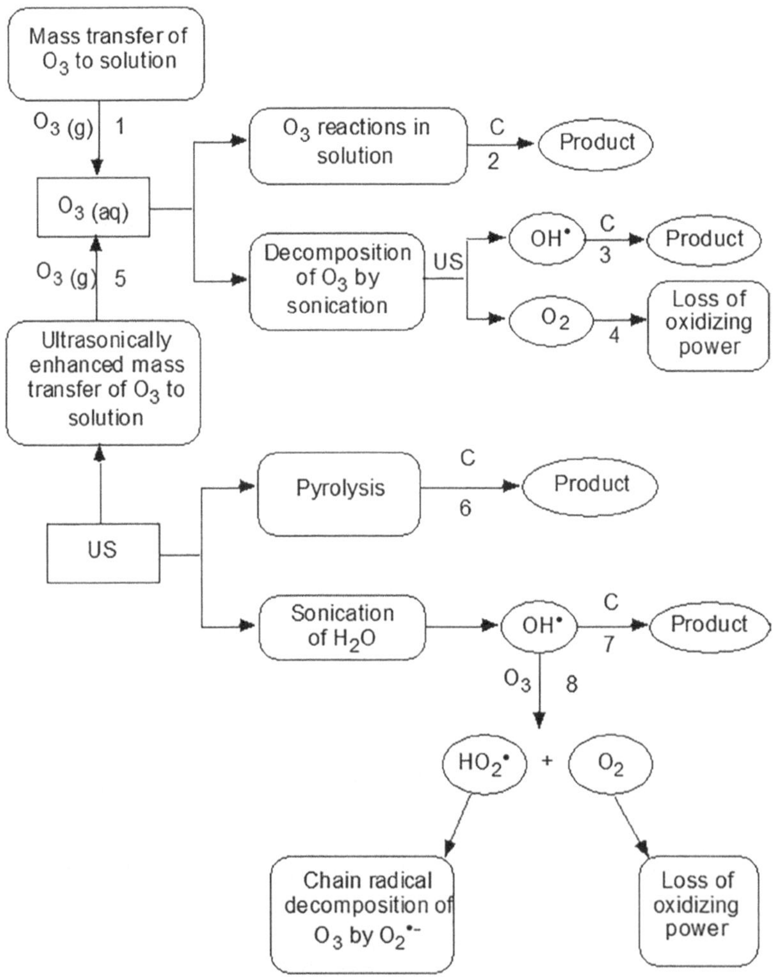

Scheme 11.2 Possible pathways during combined sonolysis and ozonolysis.

An environmental problem that has existed for a long time is the removal of colour from the effluent streams of textile factories since the presence of residual dyes is rather obvious even at low dilutions. There are several conventional approaches to the solution of this problem including absorption onto activated charcoal, flocculation, chemical oxidation, ozonolysis and irradiation with UV light. Sonication can be added to any of these and the combination of ultrasound with ozonolysis seems particularly efficient (Tezcanli-Güyer & Ince, 2004; Gültekin & Ince, 2006). Application of ultrasound with ozone was also reported to be an effective treatment to aid the decolourisation of C.I. Reactive Yellow 84 (He *et al.* 2007), CI Basic Red 9 (Martins *et al.* 2006) methyl orange (Zhang *et al.* 2006) and C.I. Acid Orange 7 (Zhang *et al.* 2008).

Removal of organic waste from industries which includes that from textile dying continues to be of concern. Aniline is one of the main pollutants from the textile dyeing industry and also from pharmaceutical and petroleum industries. This can be reduced in concentration and mineralized through the combined use of ultrasound and ozone. Thus aniline at a concentration of 100 mgl^{-1} can be reduced by only 6% (ultrasound alone) and 44% (ozone alone) but by 82% using a combination of the two (Song *et al.* 2007). The release of pharmaceutical waste into the environment comes not only from industry but also from the population that takes medicines or contraceptive pills and pollution from either source can alter the natural ecosystem. An ozone-ultrasound combination was used in an air-lift reactor to enhance the degradation of tetracycline (Wang *et al.* 2012). The authors reported that the drug removal followed pseudo-first-order kinetics and the removal rate increased with increasing ozone gas concentration (35.8–47 mg L^{-1}), gas flow rate (30–50 L h^{-1}) and power density (0–218.6 W L^{-1}). The use of 20 kHz ultrasound (600 W L^{-1}) significantly enhanced sulfamethooxazole degradation in the presence of ozone (3 g h^{-1}) through the increased production of hydroxyl radicals (Guo *et al.* 2015). Under different pH conditions the reaction rate increased by 6–26% as a result of ultrasonically enhanced cleavage of the S−N bond thus exposing it to attack by oxidising radicals which facilitated the overall degradation.

A combination of ultrasound (850 kHz), ozone and catalysts have been used to remove ibuprofen from aqueous solution (Ziylan & Ince, 2015). From a range of combinations the maximum removal of the drug coupled with TOC decay was observed after 40 min ozonation using ultrasound and zero-valent iron (ZVI) on graphite (100%, 58%) at pH 3.0. The synergy of US + ZVI at acidic pH was attributed to the contribution of hydrodynamic shear forces in the continuous enhancement and cleaning of the catalytic surfaces.

The effects of ultrasound on ozone treatment processes for bleaching cotton fabrics were also investigated and compared with the conventional hydrogen peroxide bleaching process (at 60°C over a period of 90 min). Two different ultrasonic and ozone treatments of cotton fabric samples were carried out. One employed ozone in an ultrasonic homogenizer and the second with ozone and an ultrasonic bath. The ozone and homogeniser process (30°C for 30 min) produced similar cotton fabric whiteness and yellowness to the classic peroxide bleaching process, with slightly less weight loss and dramatically reduced levels of COD in the process effluent leading to a much more economic and environmentally friendly process (Eren *et al.* 2014).

Sonolytic ozonation is an effective general treatment for the degradation of many pollutants in drinking and waste water as a result of the enhanced mass transfer of the ozone gas and the increase in hydroxyl radical formation. A group of pollutants arising from oil industry waste and that naturally occur in hydrocarbon deposits like oil sand, petroleum, bitumen and crude oil are naphthenic acids (Headley & McMartin, 2004). They are a complex mixture of organic components, some of which include saturated alkyl-substituted cycloaliphatic carboxylic acids and acyclic aliphatic acids. Synergistic effects were observed for the removal of dicyclohexyl acetic acid (DAA) by combining sonication with ozonation and resulted in 100% DAA removal with 98 ± 0.8% reduction in COD within 15 min at 3.3 mg L(−1) ozone concentration and 130 Watts ultrasonication power (Kumar *et al.* 2014). Dual frequency ultrasound of 28 kHz combined with 208 kHz, 495 kHz, 679 kHz or 890 kHz was used to degrade pesticide contaminated wastewater (Wang *et al.* 2014). The results suggested that the 28/208 kHz combination system was suitable for dissolving ozone and the 28/495 kHz system was suitable for the degradation of the pesticide acephate. Sonication at a frequency of 35 kHz (1.6 W cm^{-2}) in the presence of ozone was found to be able to reduce the visible colour of humic acid and tannic acid by 99% and 98%, respectively, in addition to reducing the molecular size of the humic and tannic acids and enhancing their mineralisation (Cui *et al.* 2011).

Gogate *et al.* have studied the development of a hybrid advanced oxidation reactor which can be effectively used for the treatment of various types of waste water. The reactor is based on the principle of intensifying degradation/disinfection using a combination of hydrodynamic cavitation, acoustic cavitation, ozone injection and electrochemical oxidation/precipitation. The authors suggest that the hybrid reactor is expected to intensify the normal conventional treatment process by 5–20 times, depending on the type of application, and suggest that it is due to the enhanced generation of hydroxyl radicals and enhanced contact of ozone and contaminants coupled with the elimination of mass transfer resistances during electrochemical oxidation/precipitation processes (Gogate *et al.* 2014).

Phenol degradation has been reported at three ultrasonic frequencies (20, 300 and 520 kHz) in combination with hydrogen peroxide, ozone, zero-valent iron and zero-valent copper (Chand *et al.* 2009). Ultrasound at 300 kHz resulted in maximum hydroxyl radical production and phenol reduction. A combined system of ultrasound at 300 kHz with zero-valent iron, H_2O_2 and air (as a sparge gas) was the most effective approach amongst the different combinations. Under these conditions phenol was completely removed and 37% of TOC mineralization was achieved within 25 minutes.

Ozone is an important disinfectant for drinking water and has proved to be a suitable replacement for chlorination in that it reduces disinfection by-products (von Gunten, 2003). The effectiveness of ozonation can be increased by sonication and this synergistic effect of ultrasonic cavitation and ozone significantly increases bacterial inactivation rates for through several processes:

- De-clumping of bacterial clusters to disperse bacteria as single cells which are more susceptible to oxidation by ozone
- Weakening of the bacterial cells due to breaking chemical bonds in the cell membrane
- Enhanced ozone decomposition rates in water during treatment

It can not only reduce the bacteria count but it can also reduce the total coliforms, faecal coliforms and faecal streptococci, which are considered as the indicators of pollution in drinking water (Jyoti & Pandit, 2004).

The effect of combined ozone and ultrasound treatment on *Escherichia coli* in saline suspension has been assessed using a commercially available USO3 system from Ultrasonic Systems Gmbh (Al-hashimi *et al.* 2015). The results show the advantage in using a combination treatment (Table 11.4)

Table 11.4 Percentage inactivation in E. coli bacteria 10^{11} cell/ml treated using the USO_3 system for 16 minutes analysed by TEM.

Bacterial Strain	Ultrasound (612 kHz)	Ozone (0.5 mg/l)	Treatment Time (min)	% inactivation
E. coli	—	YES	16	86
E. coli	YES	—	16	24
E. coli	YES	YES	16	95

11.5.3 AOP involving ultrasound combined with ultraviolet light

Studies of the combined application of ultrasound and ultraviolet light for the destruction of chemical pollutants in water have suggested that there is a synergistic improvement in the combined technology e.g. the removal of 1,1,1-trichloroethane from aqueous solutions using the combined application of ultrasound and light is more efficient than the application of either technique individually (Toy *et al.* 1990). From a study of the sonophotochemical degradation of phenol in water Wu *et al.* concluded that the synergistic effect was the result of enhanced generation of the OH radicals (Wu *et al.* 2001). They reported that the first intermediates of this reaction were hydroquinone, catechol, benzoquinone and resorcinol which clearly implicated that OH radicals were present in the sonophotochemical degradation mechanisms.

The majority of research reported on sonophotochemical remediation involves the presence of a catalyst (Gogate & Pandit, 2004; Adewuyi, 2005). Sonophotocatalysis involves chemical pollutant decomposition by UV light in the presence of a suspension of photoactive material such as TiO_2 where substrate absorption onto the UV activated surface can initiate chemical reactions e.g. the oxidation of sulphides to sulphones and sulphoxides (Sierka & Amy, 1985). This technology has been adapted for the destruction of polychlorobiphenyls (PCB's) in wastewater. Using pentachlorophenol as a model substrate in the presence of 0.2% TiO_2 UV irradiation is relatively efficient in dechlorination (Table 11.5) (Johnston & Hocking, 1993; Theron *et al.* 2001). When ultrasound is used in conjunction with photolysis, dechlorination is dramatically improved. This improvement is the result of three mechanical effects of sonochemistry namely surface cleaning, particle size reduction and increased mass transport to the powder surface.

Table 11.5 Photolysis of aqueous pentachlorophenol (2.4×10^{-4} M) containing 0.2% TiO_2).

Conditions	Cl^- yield (50 mins)	Cl^- yield (120 mins)
UV light	40%	no change
UV light/ultrasound	60%	100%

A similar combination of ultrasound and photocatalysis has also been reported for the destruction of 2,4,6-trichlorophenol in aqueous solution (Shirgaonkar & Pandit, 1998). An ultrasonic probe (22 kHz) with a UV light source (15 W) was used to examine the effect of changing such operating conditions as ultrasonic intensity, reaction temperature and UV transmittance. The experiments involved using 2,4,6-trichlorophenol (100 mg L^{-1}) and TiO_2 (0.1 gL^{-1}) and showed that the degradation rates increased with the temperature of the solution. The cumulative effect was more pronounced at lower ultrasonic intensities with little additional benefit derived at increased ultrasonic powers.

The individual ultrasonic degradation of trichloroacetic acid in water, ultraviolet light and catalyst TiO_2, and a combination of UV/TiO_2 and sonication was investigated. The extent of sonophotocatalytic degradation was higher than with sonolysis or photolysis alone and the degradation rate was increased at a higher catalyst loading, higher temperature and under acidic conditions with dissolved oxygen (Hu *et al.* 2014). The oxidative degradation of dinitrotoluenes (DNTs) and 2,4,6-trinitrotoluene (TNT) in wastewater under ultrasonic irradiation combined with UV/TiO_2 showed a synergistic effect (Chen & Huang, 2011). Nitrotoluene contaminants could be almost completely eliminated by sonophotocatalysis.

The degradation of aqueous solution of Rhodamine 6G has been reported using sonocatalytic and sonophotocatalytic treatments based on the use of CuO and TiO_2 as the solid catalysts (Bokhale *et al.* 2014). The maximum degradation achieved with CuO was 52.2% at a concentration of 1.5 g/L a similar maximum value of 51.2% was achieved with a TiO_2 loading of 4 g/L over similar treatment period. A combination of ultrasound irradiation with heterogeneous (TiO_2) and homogeneous photocatalysis (photo-Fenton) has been used to degrade malachite green in aqueous system (Berberidou *et al.* 2007). An ultrasonic horn irradiated at the frequency of 80 kHz was used with a 9 W UV-A lamp. Degradation of malachite green increased with the increase of ultrasound power intensity and decreased with the increase of initial concentration. Combination process of ultrasound with photocatalysis showed synergy on degradation and resulted in an even faster degradation. During the early stage of the reaction, degradation occurred due to radical attack on the central carbon atom and the dimethylamino group of malachite green. This initial attack is then followed by a series of demethylation/oxidation reactions to form smaller molecules and nitrates. compared The efficiency of degradation of Reactive Black 5 employing an ultrasound, UV and ferric system has been reported (Zhou *et al.* 2011). The presence of organic ligands such as oxalate, citrate, tartrate and succinate enhanced the degradation but nitrilotriacetic amine (NTA) and ethylenediamine tetraacetic acid (EDTA) strongly inhibited it. The process was also pH dependent and low pH (pH 3–4) showed the most rapid degradation. A lower pH was also reported to be advantageous for the sonocatalytic degradation of direct blue 71 in the presence of ZnO nanocatalyst (Ertugay & Acar, 2014). Addition of H_2O_2 further enhanced the degradation process.

In recent years there has been an increasing emphasis on the removal of drug and pharmaceutical waste from water. The degradation of paracetamol was studied by sonolysis, photocatalysis and sonophotocatalysis in the presence of homogeneous (Fe^{3+}) and heterogeneous (TiO_2) photocatalysts using 213 kHz ultrasound (US) (Jagannathan *et al.* 2013). The results indicated that the combination of sonolysis with photocatalysis (TiO_2 or Fe^{3+}) resulted in an additive effect from combining the two processes, with a synergy index value of ~1.0. There was no synergistic effect in total organic carbon (TOC) removal in sonophotocatalysis with TiO_2, whereas the mineralization process was synergistic when sonophotolysis was carried out in the presence of Fe^{3+}. Electrospray mass spectrometry analysis revealed the formation of a hydroxylated derivative of paracetamol during the sonolysis process.

Antipyrine, a pain releaving drug which is an emerging contaminant can be mineralized in aqueous solutions using a homogeneous sono-photocatalytic oxidation process (H_2O_2/UV/Fe II/Ultrasound) (Durán *et al.* 2013). Over 90% of TOC was removed after 50 min treatment of an aqueous solution containing 50 ppm of antipyrine employing 1500 ppm of H_2O_2, pH = 2.7, amplitude = 100%, pulse length (cycles) = 0.3 over 15 min. Experiments under OH radical scavenger-loaded conditions demonstrated that the antipyirine degradation proceeds mainly through a radical mechanism probably through the cleavage of the N–N bond of the penta heterocycle leading to the formation of aromatic acids (anthranilic and 1,4-benzenedicarboxylic acids) which is followed by the opening of phenyl ring to form small molecular organic acids (mainly 2-butenedioic, 4-oxo-pentanoic and butanedioic acids), which may be further mineralised.

A study of the degradation of the hormone estriol by sonolysis, photolysis and sonophotolysis at various UV wavelengths has been reported (Na *et al.* 2012). Degradation was monitored for UVA (365 nm), UVC (254 nm), or VUV (185 nm) irradiation and/or ultrasound exposure (283 kHz). The dominant reaction mechanism of estriol degradation in sonolysis was considered to be *via* a hydroxyl radical reaction. Photolytic and sonophotolytic estriol degradation rates were also higher at the shorter UV wavelengths (VUV) due to the higher energy of photons, higher molar absorption coefficient of estriol and increased hydroxyl radical generation from the homolysis of water. Smaller synergistic effects were observed for the sonophotolytic degradation with UVA and UVC irradiation.

A combination of sonolysis (US) and photo-oxidation (UV/H_2O_2), was used in a pilot-scale external loop airlift sonophotoreactor for the treatment of a synthetic pharmaceutical wastewater (Ghafoori *et al.* 2015). The effluent was characterized by the total organic carbon (TOC) percent removal which showed significant effects of H_2O_2 dosage and ultrasound power.

Ultrasound has also been used to enhance Fenton type oxidations of other chemical effluents an example of which is the destruction of dyes. Orange II dye removal from aqueous solution under irradiation at 850 kHz has been studied using both homogeneous Fenton type reagents ($FeSO_4/H_2O_2$), and heterogeneous reagents (Fe containing ZSM-5 zeolite/H_2O_2) with UV and ultrasound (Dükkancı *et al.* 2014). Degrees of decolourisation of 6.5% and 28.9% were observed using UV irradiation and ultrasound, respectively, whereas under the simultaneous irradiation of ultrasound and UV light, the decolourisation degree reached 47.8%, indicating a synergetic effect of ultrasound and UV light. The decolourisation was increased with the addition of Fenton's reagent with an optimal reagent ratio, Fe^{2+} :H_2O_2 of 1:50. In the combined process of ultrasound and UV light with the homogeneous Fenton system, 80.8% decolourisation could be achieved after 2 h indicating that UV light improves the Orange II dye degradation. The degree of decolourisation obtained using the heterogeneous sono-Fenton system (Fe containing ZSM-5 zeolite catalysts + H_2O_2 + ultrasound) were consistently lower than the traditional homogeneous ultrasound Fenton system. This can be attributed to the greater difficulty of the reaction between Fe ions and hydrogen peroxide. In all cases the Orange II ultrasonic decolourisation was found to follow first order kinetics. A combination of ultrasound, UV and Fe^{3+} proved to be better than Fe^{3+} with either UV or ultrasound alone by a factor of 2.5 for the degradation of Reactive Black 5 (Zhou *et al.* 2011). The presence of organic ligands such as oxalate, citrate, tartrate and succinate enhanced the degradation but NTA and EDTA strongly inhibited it. The process was also pH dependent and low pH (pH 3–4) showed the most rapid degradation.

11.5.4 AOP involving ultrasound combined with electrochemistry

The advantages of using ultrasound for the electrochemical transformation of organic compounds is long established (Walton & Mason, 1998) but the sonoelectrochemical decomposition of organic compounds is a developing technique among advanced oxidation processes (AOPs). It has the advantage over sonication alone that it increases the efficiency of the process in terms of a more rapid decrease in chemical oxygen demand (COD) and in total organic carbon (TOC) and accelerates electrochemical oxidation which normally requires a lengthy period of time to achieve significant mineralisation.

Saez *et al.* have reported a successful sonoelectrochemical removal of perchloroethylene in batch mode at 20 kHz in aqueous sodium sulfate solution (Sáez *et al.* 2010). The solvent is used commercially in dry cleaning and as a degreasing agent and it has been linked to liver problems and cancer. It was shown that not only was the decomposition improved over simple electrolysis but also the current efficiency was enhanced. A significant enhancement was also achieved when the process was run in the absence of the background electrolyte (Sáez *et al.* 2011).

Oxidative degradation of dinitrotoluene (DNT) and 2,4,6-trinitrotoluene (TNT) in wastewater using electrochemical and electro-Fenton processes, combined with ultrasound resulted in a synergistic effect (Chen & Huang, 2014). Various operating variables such as electrode potential, sonoelectrolytic temperature, acidity of wastewater, oxygen dosage, and dosage of ferrous ions were examined and it was determined that nitrotoluene contaminants could be completely decomposed when employing the sonoelectro-Fenton method. Hydrogen peroxide was generated in situ *via* cathodic reduction of oxygen, supplied partially by the anodic oxidation of water. During the sonoelectrolytic process, in spite of existence of degassing phenomenon, a high yield of hydrogen peroxide was produced due to the significantly enhanced mass transfer rate of oxygen toward the cathode, caused by the ultrasonic irradiation. Sonoelectrochemical degradation of phenol was studied using high frequency ultrasound (850 kHz) and stainless steel electrodes (Ren *et al.* 2013). The synergistic effect was reported to be 60%. The use of sodium sulphate as the electrolyte gave higher degradation compared with the use of sodium hydroxide or sulphuric acid. Increase of sodium sulphate concentration and electric voltage led to the increase of degradation. At the electrolyte concentration of 4.26 g L^{-1} Na_2SO_4 and the electric voltages of 30 V, a nearly complete degradation of phenol at an initial concentration of 0.5 and 1 mmol L^{-1} was obtained in 60 minutes.

The red food additive Amaranth dye is an endocrine disruptor in wastewater. Its sonoelectrochemical degradation has been investigated using a boron doped diamond anode (Steter *et al.* 2014). TOC removal values of between 92.1% and 95.1% were obtained after 90 min electrolyses using current densities in the 10–50 mAcm^{-2} range. The efficiency of decomposition of methylene blue by sonoelectrochemical decomposition using $Ti/Ta_2O_5–SnO_2$ electrodes has been reported (Shestakova *et al.* 2015). Decolourisation was used to assess the initial stages of decomposition and COD together with TOC was used as a measure of total degradation. Frequencies of 850 and 380 kHz and the use of higher powers were found more effective towards dye decolourisation. The time for complete MB degradation was reduced from 180 min using electrolysis and from 90 min while carrying out sonolysis to 45 min when conducting a combined sonoelectrocatalytic experiments.

A novel sonoelectrochemical catalytic oxidation-driven process using a nanocoated electrode has been used to treat aqueous methylene blue (Yang *et al.* 2014). The nanocoated electrode generated more hydroxyl radicals than the non-nano-scale electrode and the effect of the nanocoated electrode on the dye solution was enhanced by ultrasound. The optimum operating conditions of a current of 600 mA and a frequency of 45 kHz resulted in a removal efficiency of 92% for TOC.

A report on the treatment of highly polluted wastewater (COD = 10240 mg/L; total suspended solids (TSS) = 2860 mg/L originating from the wine-making industry by sonoelectrochemical methods has shown promise (Orescanin *et al.* 2013). The effluent was treated using electro oxidation and electrocoagulation with stainless steel, iron and aluminium electrode sets with simultaneous sonication and recirculation in a strong electromagnetic field. The effluent was post-treated with ozone combined with UV irradiation in the presence of added hydrogen peroxide. Following the combined treatment, the final removal efficiencies of the suspended solids and phosphates were over 99%, Fe, Cu and ammonia approximately 98%, while the removal of COD and sulphates was 77% and 62%, respectively.

Ultrasound, electrochemistry, ozone and combination thereof were employed for the destruction of 1,3-dinitrobenzene and 2,4-dinitrotoluene (Abramov *et al.* 2006). 1,3-dinitrobenzene and 2,4-dinitrotoluene were stable to ozonation even with ultrasound but slowly degraded under an electrochemical reduction that was enhanced with ultrasound. However, a concurrent treatment with ozone, ultrasound and electrochemistry led to the complete destruction of the pollutants within a short time.

The sonoelectrochemical degradation of a number of pharmaceutical products using ultrasound at frequencies approaching 1 MHz has been reported. Rapid removal of the antimicrobial agent triclosan in aqueous solutions has been achieved using electrolysis with 850 kHz ultrasound (Ren *et al.* 2014). Diamond coated niobium electrode showed the best results amongst a number of different electrode systems used and achieved a reduction of 92% in triclosan from 1 mg L^{-1} aqueous solution within 15 min. Batch scale degradation of ibuprofen was developed using a platinum anode (6×8 cm^2) and stainless-steel cathode (7×8 cm^2) and ultrasound (1000 kHz) (Thokchom *et al.* 2015). The system achieved 89.32%, 81.85% and 88.7% degradation using NaOH, H_2SO_4 and deionized water respectively over 1 hour.

11.6 ULTRASONIC EQUIPMENT AND PROSPECTS FOR SCALE UP

With the potential and clear benefits of using ultrasound in water processing, the availability of large scale ultrasonic equipment is on the increase. Some manufacturers already produce batch or flow systems and as the need more pilot scale or industrial scale, ultrasonic equipment is being developed, manufactured and sold. The majority of these systems can can be easily adapted for environmental remediation. In a comparative study of sonochemical reactors with different geometries Nikitenko *et al.* concluded that ultrasonic efficiency is dependent upon the total absorbed energy and is roughly independent of reactor geometry (Nikitenko *et al.* 2007). In their study of various 20 kHz laboratory scale reactors they determine that a decrease in sonochemical horn surface area results in an increase in efficiency, if the absorbed power remains the same. This impacts the development of large scale reactors as it suggests that the absorbed ultrasonic power is the main factor to be considered in any future designs.

Several ultrasound reactor configurations have been described in the literature and although they might originally have been intended for other purposes these basic designs can be adapted for water remediation. One example of a large scale batch reactor that was originally designed for the extraction of medicinal compounds from plants is a case in point (Vinatoru, 2001). It comprises of a stainless steel tank with a capacity of 1 cubic meter and a working volume of 700 to 850 litres (Figure 11.6).

Figure 11.6 Batch reactor for extraction.

Possible configurations and options for the scale up of other batch systems of various sizes from bench scale to 1000 litres have been developed for applications of processing with ultrasound in the food industry (Chemat *et al.* 2011). Batch systems can provide both intermittent and continuous ultrasound exposure of low intensity in a large volume reactor (0.01–0.1 W/cm^3). It is possible to generate much higher local powers (0.5–10 W/cm^3) in a recirculating flow cell coupled externally to a batch reactor (Figure 11.7).

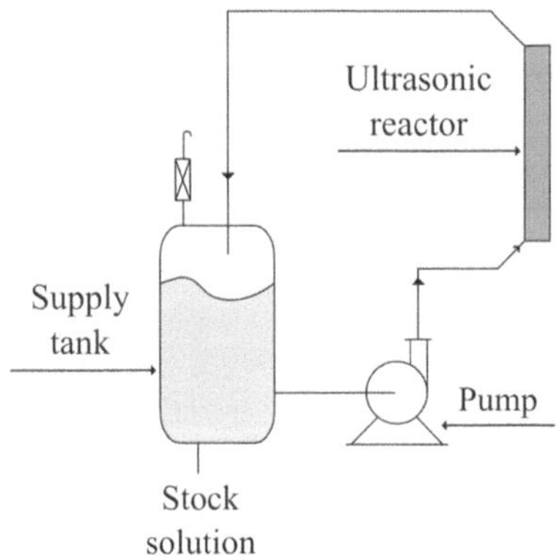

Figure 11.7 Schematic external flow loop system.

Several types of equipment are now available for the scale up of ultrasonic processes with different designs, all showing an increased benefit from the use of sonication (Leonelli & Mason, 2010).

The Prosonitron P500 system developed by Prosonix (Prosonix, 2015) (Figure 11.8) uses piezoelectric transducers attached to the outside surface of a metal tube through which the process liquid flows. The system has the advantage that the tube itself resonates thus delivering sound energy towards a focus in the centre of the flow.

Figure 11.8 Prosonitron system.

The Dual Frequency Reactor system (DFR) made by Advanced Sonic Processing Systems (Advanced Sonic Processing Systems 2015) (Figure 11.9). It is produced in variety of sizes but the system as shown is a typical configuration and consists of two parallel vertical plates through which the fluid to be processed is pumped (from the bottom). It uses magnetostrictive transducers with each plate operating at different frequencies (20 and 16 kHz) to produce high cavitational activity in the fluid flowing through. The gap between the plates can be adjusted.

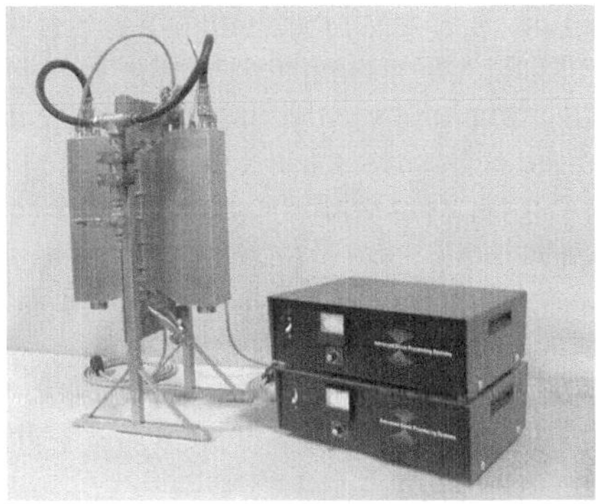

Figure 11.9 Dual frequency reactor.

Hielscher produce a flow through system UIP16000 incorporating a 20 kHz horn (Hielscher Ultrasonics gmbh 2015) (Figure 11.10) have produced which they claim to be currently the most powerful ultrasonic processor in the world operating at 16,000 W It is designed to work in clusters of three or more units, for large volume processing at up to 50 m^3/h.

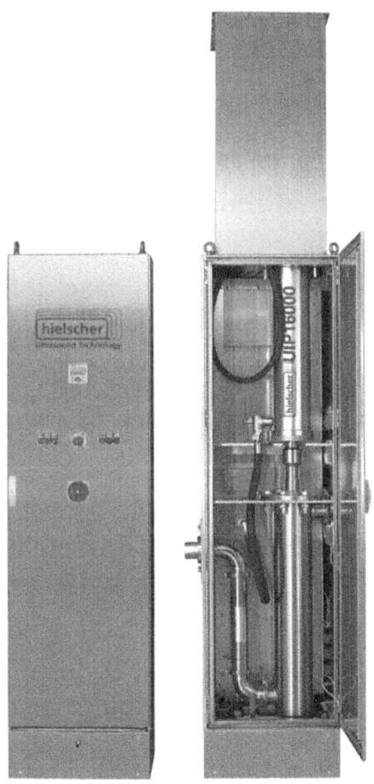

Figure 11.10 Heilscher UIP16000.

The scale-up of oxidation processes for the degradation of phenol under pilot scale operation (7 litres) has been described (Khokhawala & Gogate, 2010). The system could be used for treatments employing different modes of operation involving UV alone, ultrasound alone, UV/ultrasound, UV/TiO_2 (photocatalysis), UV/H_2O_2, UV/NaCl (NaCl was employed as an additive to determine if the use of additional salts would be beneficial to the degradation process), UV/ultrasound/TiO_2 and H_2O_2 assisted sonophotocatalysis (degradation via the formation of various oxidizing radicals) in order to achieve maximum phenol degradation.

The combination of UV irradiation (8 W UV tube) with sonication (rated power 1 kW and frequency of 25 kHz) proved to be more efficient when compared to the individual operations. The addition of hydrogen peroxide improved the degradation by providing an additional source of free radicals. The use of NaCl altered the distribution of the pollutants selectively at the site of radical generation and addition of TiO_2 resulted in an increase of degradation with the optimum concentration of TiO_2 being 2.0 g L^{-1}. The maximum phenol degradation achieved was 37.75% for H_2O_2 assisted photosonocatalysis at a pH of 2. The same laboratory has reported the degradation of *p*-nitrophenol involving ultrasound (25 kHz, acoustic power of 1 kW) at a pilot scale with an operating volume of 7 L (Mishra & Gogate, 2011). The maximum extent of degradation was achieved employing 10 ppm of *p*-nitrophenol using a combination of sonophotocatalysis and optimum quantity of H_2O_2. Removal of 2,4-dinitrophenol was also investigated using a pilot scale reactor (7 L) incorporating a radial sonicator (Bagal *et al.* 2013). It was found that removal was greater at lower pH (pH 2.5 and 4) and significantly increased to 98.7% when Fenton's reagent was added.

Two general methodologies adopted for the large scale decontamination of industrial wastewater containing oil and metal ions are flocculation and coagulation. A methodology has been reported for the production of ferrite (Fe_2O_3) particles to be used as coagulants in water purification by a galvanochemical reaction between iron and coke (Abramov *et al.* 2014). Ultrasonic processing of these particles reduces the particle size and provides surface cleaning making them more effective. Trials have proved their efficiency for the decontamination of wastewater prepared in a laboratory and also on real wastewater from a carriage cleaning station on the St. Petersburg Metro.

Gogate *et al.* have developed a commercial advanced oxidation reactor which utilizes a combination of hydrodynamic cavitation, acoustic cavitation, ozone injection and electrochemical oxidation/precipitation (Gogate *et al.* 2014). The reactor has been given the name Ozonix® and it has been tested, deployed and commercialized for the processing of recycled fluids at commercial sites on over 750 oil and natural gas wells around the United States. Equipment for the continuous treatment of a variety of wastewaters with concurrent ozone and ultrasound has been described (Poeschl & Oliveri, 2014). It has been used in a range of applications including disinfection, the removal of trace chemicals and the oxidation of alkaline sulphide residues.

A question that is commonly asked about sonochemical remediation is whether the process is economic. This is a difficult question to answer because the majority of the earlier discussions were based upon results obtained in laboratory size reactors (Löning *et al.* 2002). Often these reactors involve ultrasonic horns (probes) as the source of ultrasound and such results definitely cannot be applied to larger scale reactors which have very different modes of acoustic energy transfer (Leonelli & Mason, 2010). Nevertheless progress was made in determining energy transfer in reactors with different configurations (Nikitenko *et al.* 2007).

The economics associated with the running costs of a 1000 L/min capacity water treatment plant have been calculated (Mahamuni & Adewuyi, 2010). The calculations involved the rate constants of degradation of a range of pollutants, such as phenol, azo dyes, and trichloromethylene. Capital and operating costs were determined using various parameters and for the degradation of phenol indicative costs were \$15,576 when using ultrasound alone to treat 1000 gallons of water with costs dropping significantly to only \$89 when employing ultrasound with UV and Ozone indicating a hugely beneficial synergistic effect. Similar observations were made for the degradation of azo dyes whereas while trichloromethylene degradation did show some synergistic benefit of the hybrid techniques this was by no means as large as in the other cases.

11.7 CONCLUSIONS

Laboratory research into the uses of ultrasound for environmental protection is very broad and is showing enough promise to be attracting the interest of industry. It seems clear however that the use of ultrasound alone as an AOP is unlikely to be economic in comparison with existing processes however it is an attractive option when coupled with other oxidising techniques such as the use of UV, Ozone, Fenton type reagents or electrochemistry. Further research into ultrasound as an AOP must take into account these synergistic effects.

There is sufficient evidence that equipment manufacturers have moved into the field of the scale-up of ultrasonic systems and that many of these are applicable to AOP processes. Consideration must be therefore be given to the inclusion of other oxidisation techniques with some importance being allocated to the degree of absorbed ultrasonic power as a factor of making such advanced oxidation treatments both economic, beneficial and feasible in the future.

When the scale-up systems have been optimised for power delivery into water treatment there seems little doubt that the overall economics of power ultrasound usage will be justifiable in terms of environmental outcomes.

11.8 REFERENCES

Abramov O. V. (1998). High-Intensity Ultrasound: Theory and Industrial Applications. Gordon and Breach, London, U.K.

Abramov V. O., Abramov O. V., Gekhman A. E., Kuznetsov V. M. and Price G. J. (2006). Ultrasonic intensification of ozone and electrochemical destruction of 1,3-dinitrobenzene and 2,4-dinitrotoluene. *Ultrasonics Sonochemistry*, **13**(4), 303–307.

Abramov V. O., Abramova A. V., Keremetin P. P., Mullakaev M. S., Vexler G. B. and Mason T. J. (2014). Ultrasonically improved galvanochemical technology for the remediation of industrial wastewater. *Ultrasonics Sonochemistry*, **21**(2), 812–818.

Adewuyi Y. G. (2005). Sonochemistry in environmental remediation. 2. Heterogeneous sonophotocatalytic oxidation processes for the treatment of pollutants in water. *Environmental Science & Technology*, **39**(22), 8557–8570.

Advanced Sonic Processing Systems (2015). http://www.advancedsonicprocessingsystems.com

Al-hashimi A. M., Mason T. J. and Joyce E. M. (2015). The combined effect of ultrasound and ozone on bacteria in water. *Environmental Science & Technology*, **49**(19), 11697–11702.

Anbar M. and Pecht I. (1964). On the sonochemical formation of hydrogen peroxide in water. *Journal of Physical Chemistry*, **68**(2), 352–355.

Andreozzi R., Caprio V., Insola A. and Marotta R. (1999). Advanced oxidation processes (AOP) for water purification and recovery. *Catalysis Today*, **53**(1), 51 59.

Arnold S. F., Robinson M. K., Notides A. C., Guillette L. J. and McLachlan J. A. (1996). A yeast estrogen screen for examining the relative exposure of cells to natural and xenoestrogens. *Environmental Health Perspectives*, **104**(5), 544–548.

Ashokkumar M. and Mason T. J. (2007). Sonochemistry. In: Kirk-Othmer Encyclopedia of Chemical Technology, John Wiley & Sons, Inc. Doi: 10.1002/0471238961.1915141519211912.a01.pub2

Augugliaro V., Litter M., Palmisano L. and Soria J. (2006). The combination of heterogeneous photocatalysis with chemical and physical operations: a tool for improving the photoprocess performance. *Journal of Photochemistry and Photobiology C: Photochemistry Reviews*, **7**(4), 127–144.

Bagal M. V., Lele B. J. and Gogate P. R. (2013). Removal of 2,4-dinitrophenol using hybrid methods based on ultrasound at an operating capacity of 7 L. *Ultrasonics Sonochemistry*, **20**(5), 1217–1225.

Berberidou C., Poulios I., Xekoukoulotakis N. P. and Mantzavinos D. (2007). Sonolytic, photocatalytic and sonophotocatalytic degradation of malachite green in aqueous solutions. *Applied Catalysis B: Environmental*, **74**(1–2), 63–72.

Berlan J., Trabelsi F., Delmas H., Wilhelm A. M. and Petrignani J. F. (1994). Oxidative degradation of phenol in aqueous media using ultrasound. *Ultrasonics Sonochemistry*, **1**(2), S97–S102.

Bokhale N. B., Bomble S. D., Dalbhanjan R. R., Mahale D. D., Hinge S. P., Banerjee B. S., Mohod A. V. and Gogate P. R. (2014). Sonocatalytic and sonophotocatalytic degradation of rhodamine 6G containing wastewaters. *Ultrasonics Sonochemistry*, **21**(5), 1797–1804.

Brown B. and Goodman J. E. (1965). High Intensity Ultrasonics. Iliffe Books Ltd, London, U.K.

Bulat T. J. (1974). Macrosonics in industry: 3. Ultrasonic cleaning. *Ultrasonics*, **12**(2), 59–68.

Capocelli M., Joyce E., Lancia A., Mason T. J., Musmarra D. and Prisciandaro M. (2012). Sonochemical degradation of estradiols: incidence of ultrasonic frequency. *Chemical Engineering Journal*, **210**, 9–17.

Chand R., Ince N. H., Gogate P. R. and Bremner D. H. (2009). Phenol degradation using 20, 300 and 520 kHz ultrasonic reactors with hydrogen peroxide, ozone and zero valent metals. *Separation and Purification Technology*, **67**(1), 103–109.

Chandrapala J., Oliver C., Kentish S. and Ashokkumar M. (2012). Ultrasonics in food processing – food quality assurance and food safety. *Trends in Food Science & Technology*, **26**(2), 88–98.

Chemat F., Zilll H. and Khan M. K. (2011). Applications of ultrasound in food technology: processing, preservation and extraction. *Ultrasonics Sonochemistry*, **18**(4), 813–835.

Chen W.-S. and Huang S.-C. (2011). Sonophotocatalytic degradation of dinitrotoluenes and trinitrotoluene in industrial wastewater. *Chemical Engineering Journal*, **172**(2–3), 944–951.

Chen W.-S. and Huang C.-P. (2014). Decomposition of nitrotoluenes in wastewater by sonoelectrochemical and sonoelectro-Fenton oxidation. *Ultrasonics Sonochemistry*, **21**(2), 840–845.

Comeskey D., Larparadsudthi O. A., Mason T. J. and Paniwnyk L. (2012). The use of a range of ultrasound frequencies to reduce colouration caused by dyes. *Water Science and Technology*, **66**(10), 2251–2257.

Cui M., Jang M., Cho S.-H., Elena D. and Khim J. (2011). Enhancement in mineralization of a number of natural refractory organic compounds by the combined process of sonolysis and ozonolysis (US/O3). *Ultrasonics Sonochemistry*, **18**(3), 773–780.

Diamanti-Kandarakis E., Bourguignon J.-P., Giudice L. C., Hauser R., Prins G. S., Soto A. M., Zoeller R. T. and Gore A. C. (2009). Endocrine-disrupting chemicals: an endocrine society scientific statement. *Endocrine Reviews*, **30**(4), 293–342.

Dükkancı M., Vinatoru M. and Mason T. J. (2014). The sonochemical decolourisation of textile azo dye Orange II: effects of Fenton type reagents and UV light. *Ultrasonics Sonochemistry*, **21**(2), 846–853.

Durán A., Monteagudo J. M., Sanmartín I. and García-Díaz A. (2013). Sonophotocatalytic mineralization of antipyrine in aqueous solution. *Applied Catalysis B: Environmental*, 138–139(0), 318–325.

Eren H., Avinc O., Erişmiş B. and Eren S. (2014). Ultrasound-assisted ozone bleaching of cotton. *Cellulose*, **21**(6), 4643–4658.

Ertugay N. and Acar F. N. (2014). The degradation of Direct Blue 71 by sono, photo and sonophotocatalytic oxidation in the presence of ZnO nanocatalyst. *Applied Surface Science*, **318**(0), 121–126.

Feng H., Barbosa-Canovas G. and Weiss J. (2011). Ultrasound Technologies for Food and Bioprocessing. Springer, New York, NY.

Fischer C. H., Hart E. J. and Henglein A. (1986). Ultrasonic irradiation of water in the presence of oxygen 18, $^{18}O_2$: isotope exchange and isotopic distribution of hydrogen peroxide. *Journal of Physical Chemistry*, **90**(9), 1954–1956.

Fitzgerald M. E., Griffing V. and Sullivan J. (1956). Chemical effects of ultrasonics-'Hot Spot' chemistry. *Journal of Physical Chemistry*, **25**, 926–933.

Flosdorf E. W., Chambers L. A. and Malisoff W. M. (1936). Sonic activation in chemical systems: oxidations at audible frequencies. *Journal of the American Chemical Society*, **58**(7), 1069–1076.

Frederick J. R. (1965). Ultrasonic Engineering. John Wiley, London.

Frenkel Y. I. (1940). Electrical phenomena connected with cavitation caused by ultrasonic oscillation in a liquid. *Russian Journal of Physical Chemsitry*, **14**, 305–308.

Gallego-Juárez J. A. and Graff K. F. (2015). Power Ultrasonics. Woodhead Publishing, Oxford.

Ghafoori S., Mowla A., Jahani R., Mehrvar M. and Chan P. K. (2015). Sonophotolytic degradation of synthetic pharmaceutical wastewater: statistical experimental design and modeling. *Journal of Environmental Management*, **150**, 128–137.

Gogate P. R. and Pandit A. B. (2004). Sonophotocatalytic reactors for wastewater treatment: a critical review. *AIChE Journal*, **50**(5), 1051–1079.

Gogate P. R., Mededovic-Thagard S., McGuire D., Chapas G., Blackmon J. and Cathey R. (2014). Hybrid reactor based on combined cavitation and ozonation: from concept to practical reality. *Ultrasonics Sonochemistry*, **21**(2), 590–598.

Gong C. and Hart D. P. (1998). Ultrasound induced cavitation and sonochemical yields. *The Journal of the Acoustical Society of America*, **104**(5), 2675 2682.

Griffing V. (1950). Theoretical explanation of the chemical effects of ultrasonics. *Journal of Chemical Physics*, **18**(7), 997–998.

Gültekin I. and Ince N. H. (2006). Degradation of aryl-azo-naphthol dyes by ultrasound, ozone and their combination: effect of α-substituents. *Ultrasonics Sonochemistry*, **13**(3), 208–214.

Guo W.-Q., Yin R.-L., Zhou X.-J., Du J.-S., Cao H.-O., Yang S.-S. and Ren N.-Q. (2015). Sulfamethoxazole degradation by ultrasound/ozone oxidation process in water: kinetics, mechanisms, and pathways. *Ultrasonics Sonochemistry*, **22**(0), 182–187.

Hamdaoui O. and Naffrechoux E. (2008). Sonochemical and photosonochemical degradation of 4-chlorophenol in aqueous media. *Ultrasonics Sonochemistry*, **15**(6), 981–987.

Harvey E. N. and Loomis A. L. (1929). The destruction of luminous bacteria by high frequency sound waves. *Journal of Bacteriology*, **17**, 373–379.

He Z., Song S., Xia M., Qiu J., Ying H., Lü B., Jiang Y. and Chen J. (2007). Mineralization of C.I. Reactive Yellow 84 in aqueous solution by sonolytic ozonation. *Chemosphere*, **69**(2), 191–199.

Headley J. V. and McMartin D. W. (2004). A review of the occurrence and fate of naphthenic acids in aquatic environments. *Journal of Environmental Science and Health, Part A*, **39**(8), 1989–2010.

Henglein A. (1987). Sonochemistry: historical developments and modern aspects. *Ultrasonics*, **25**(1), 6–16.

Hielscher Ultrasonics gmbh. (2015). http://www.hielscher.com

Hoffmann M. R., Hua I. and Höchemer R. (1996). Application of ultrasonic irradiation for the degradation of chemical contaminants in water. *Ultrasonics Sonochemistry*, **3**(3), S163–S172.

Hu B., Wu C., Zhang Z. and Wang L. (2014). Sonophotocatalytic degradation of trichloroacetic acid in aqueous solution. *Ceramics International*, **40**(5), 7015–7021.

Hu G., Li J. and Zeng G. (2013). Recent development in the treatment of oily sludge from petroleum industry: a review. *Journal of Hazardous Materials*, **261**(0), 470–490.

Ifelebuegu A. O., Onubogu J., Joyce E. and Mason T. (2014). Sonochemical degradation of endocrine disrupting chemicals 17β-estradiol and 17α-ethinylestradiol in water and wastewater. *International Journal of Environmental Science and Technology*, **11**(1), 1–8.

Ince N. H. and Tezcanlí G. (2001). Reactive dyestuff degradation by combined sonolysis and ozonation. *Dyes and Pigments*, **49**(3), 145–153.

Jagannathan M., Grieser F. and Ashokkumar M. (2013). Sonophotocatalytic degradation of paracetamol using TiO_2 and Fe^{3+}. *Separation and Purification Technology*, **103**, 114–118.

Jiang B., Zheng J., Qiu S., Wu M., Zhang Q., Yan Z. and Xue Q. (2014). Review on electrical discharge plasma technology for wastewater remediation. *Chemical Engineering Journal*, **236**, 348–368.

Jiang Y., Petrier C. and Waite T. D. (2006). Sonolysis of 4-chlorophenol in aqueous solution: effects of substrate concentration, aqueous temperature and ultrasonic frequency. *Ultrasonics Sonochemistry*, **13**(5), 415–422.

Johnston A. J. and Hocking P. (1993). Ultrasonically accelerated photocatalytic waste treatment. *Emerging Technologies in Hazardous Waste Management III (American Chemical Society)*, **518**, 106–118.

Joyce E. M. and Mason T. J. (2008). Sonication used as a biocide a review: ultrasound a greener alternative to chemical biocides? Chimica Oggi – Chemistry Today, **26**(6), 12–15.

Juretic H., Montalbo-Lomboy M., van Leeuwen J., Cooper W. J. and Grewell D. (2015). Hydroxyl radical formation in batch and continuous flow ultrasonic systems. *Ultrasonics Sonochemistry*, **22**, 600–606.

Jyoti K. K. and Pandit A. B. (2004). Ozone and cavitation for water disinfection. *Biochemical Engineering Journal*, **18**(1), 9–19.

Khokhawala I. M. and Gogate P. R. (2010). Degradation of phenol using a combination of ultrasonic and UV irradiations at pilot scale operation. *Ultrasonics Sonochemistry*, **17**(5), 833–838.

Kim D. K., O'Shea K. E. and Cooper W. J. (2012). Oxidative degradation of alternative gasoline oxygenates in aqueous solution by ultrasonic irradiation: mechanistic study. *Science of The Total Environment*, **430**, 246–259.

Knapp R. T. and Hollander A. (1948). Laboratory investigations of the mechanism of cavitation. *Transactions of the American Society Mechanical Engineering*, **70**, 419–435.

Kumar P., Headley J., Peru K., Bailey J. and Dalai A. (2014). Removal of dicyclohexyl acetic acid from aqueous solution using ultrasound, ozone and their combination. *Journal Environmental Science Health A Toxic Hazardous Substances & Environmental Engineering*, **49**(13), 1512–1519.

Leonelli C. and Mason T. J. (2010). Microwave and ultrasonic processing: now a realistic option for industry. *Chemical Engineering and Processing: Process Intensification*, **49**(9), 885–900.

Lepoint-Mullie F., De Pauw D., Lepoint T., Supiot P. and Avni R. (1996). Nature of the 'Extreme Conditions' in single sonoluminescing bubbles. *Journal of Physical Chemistry*, **100**(30), 12138–12141.

Lepoint T. and Mullie F. (1994). What exactly is cavitation chemistry? *Ultrasonics Sonochemistry*, **1**(1), S13–S22.

Liu S.-C. and Wu H. (1934). Mechanism of oxidation promoted by ultrasonic radiation. *Journal of the American Chemical Society*, **56**(5), 1005–1007.

Loiseleur J. (1944). Sur l'activation de l'oxygène par les ultrasons. *C.R. Academy Science France*, **218**, 876–878.

Löning J.-M., Horst C. and Hoffmann U. (2002). Investigations on the energy conversion in sonochemical processes. *Ultrasonics Sonochemistry*, **9**(3), 169–179.

Mahamuni N. N. and Adewuyi Y. G. (2010). Advanced oxidation processes (AOPs) involving ultrasound for waste water treatment: a review with emphasis on cost estimation. *Ultrasonics Sonochemistry*, **17**(6), 990–1003.

Makino K., Mossoba M. M. and Riesz P. (1982). Chemical effects of ultrasound on aqueous solutions. Evidence for hydroxyl and hydrogen free radicals by spin trapping. *Journal of the American Chemical Society*, **104**(12), 3537–3539.

Makino K., Mossoba M. M. and Riesz P. (1983). Chemical effects of ultrasound on aqueous solutions. Formation of hydroxyl radicals and hydrogen atoms. *Journal of Physical Chemistry*, **87**(8), 1369–1377.

Margulis M. A. (1985). Sonoluminescence and sonochemical reactions in cavitation fields. A review. *Ultrasonics*, **23**(4), 157–169.

Martins A., de O., Canalli V. M., Azevedo C. M. N. and Pires M. (2006). Degradation of pararosaniline (C.I. Basic Red 9 monohydrochloride) dye by ozonation and sonolysis. *Dyes and Pigments*, **68**(2–3), 227–234.

Mason T. J. (2016). Ultrasonic cleaning: an historical perspective. *Ultrasonics Sonochemistry*, **29**, 519–523. Special Issue: Cleaning with bubbles; R. David Fernandez and V. Bram (guest eds).

Mason T. J. and Bernal V. S. (2012). An Introduction to Sonoelectrochemistry. In: Power Ultrasound in Electrochemistry, B. G. Pollet (ed.), John Wiley & Sons, Ltd, Chichester, UK, 21–44.

Mason T. J. and Lorimer J. P. (2002). Applied Sonochemistry. Wiley VCH, Weinheim, Germany.

Mason T. J. and Peters D. (2002). Practical Sonochemistry, Power Ultrasound uses and Applications, 2nd edn, Ellis Horwood Publishers, Chichester, UK.

Mason T. J. and Tiehm A. (2001). Ultrasound in Environmental Protection. Elsevier, Amsterdam, Netherlands.

Mason T. J., Lorimer J. P., Bates D. M. and Zhao Y. (1994). Dosimetry in sonochemistry: the use of aqueous terephthalate ion as a fluorescence monitor. *Ultrasonics Sonochemistry*, **1**(2), S91–S95.

Mason T. J., Cobley A. J., Graves J. E. and Morgan D. (2011). New evidence for the inverse dependence of mechanical and chemical effects on the frequency of ultrasound. *Ultrasonics Sonochemistry*, **18**(1), 226–230.

Méndez-Arriaga F., Torres-Palma R. A., Pétrier C., Esplugas S., Gimenez J. and Pulgarin C. (2008). Ultrasonic treatment of water contaminated with ibuprofen. *Water Research*, **42**(16), 4243–4248.

Metcalf S. and Eddy S. (2004). Wastewater Engineering – Treatment and Reuse. McGraw Hill Book Co, New York, NY.

Mishra K. P. and Gogate P. R. (2011). Intensification of sonophotocatalytic degradation of p-nitrophenol at pilot scale capacity. *Ultrasonics Sonochemistry*, **18**(3), 739–744.

Mohapatra D. P., Brar S. K., Tyagi R. D., Picard P. and Surampalli R. Y. (2013). A comparative study of ultrasonication, Fenton's oxidation and ferro-sonication treatment for degradation of carbamazepine from wastewater and toxicity test by Yeast Estrogen Screen (YES) assay. *Science of The Total Environment*, **447**, 280–285.

Na S., Park B., Cho E., Koda S. and Khim J. (2012). Sonophotolytic degradation of estriol at various ultraviolet wavelength in aqueous solution. Japanese Journal of Applied Physics, **51**(7S), 07GD11.

Nagata Y., Nakagawa M., Okuno H., Mizukoshi Y., Yim B. and Maeda Y. (2000). Sonochemical degradation of chlorophenols in water. *Ultrasonics Sonochemistry*, **7**(3), 115–120.

Neppiras E. A. (1972). Macrosonics in industry 1. Introduction. *Ultrasonics*, **10**(1), 9–13.

Neppiras E. A. (1984). Acoustic cavitation series: part one: acoustic cavitation: an introduction. *Ultrasonics*, **22**(1), 25–28.

Nikitenko S. I. (2014). Plasma formation during acoustic cavitation: toward a new paradigm for sonochemistry. *Advances in Physical Chemistry*, **2014**, 1–8.

Nikitenko S. I., Le Naour C. and Moisy P. (2007). Comparative study of sonochemical reactors with different geometry using thermal and chemical probes. *Ultrasonics Sonochemistry*, **14**(3), 330–336.

Noltingk B. E. and Neppiras E. A. (1950). Cavitation produced by Ultrasonics. *Proceedings of the Physical Society. Section B*, **63**(9), 674–685.

Okitsu K., Iwasaki K., Yobiko Y., Bandow H., Nishimura R. and Maeda Y. (2005). Sonochemical degradation of azo dyes in aqueous solution: a new heterogeneous kinetics model taking into account the local concentration of OH radicals and azo dyes. *Ultrasonics Sonochemistry*, **12**(4), 255–262.

Orescanin V., Kollar R., Nad K., Mikelic I. L. and Gustek S. F. (2013). Treatment of winery wastewater by electrochemical methods and advanced oxidation processes. *Journal of Environmental Science and Health, Part A*, **48**(12), 1543–1547.

Pera-Titus M., García-Molina V., Baños M. A., Giménez J. and Esplugas S. (2004). Degradation of chlorophenols by means of advanced oxidation processes: a general review. *Applied Catalysis B: Environmental*, **47**(4), 219–256.

Petrier C. and Casadonte D. (2001). The sonochemical degradation of aromatic and chloroaromatic contaminants. In: Advances in Sonochemistry: Theme issue – Ultrasound in Environmental Protection, T. J. Mason and A. Tiehm (eds), Elsevier, **6**.

Pétrier C. and Francony A. (1997). Ultrasonic waste-water treatment: incidence of ultrasonic frequency on the rate of phenol and carbon tetrachloride degradation. *Ultrasonics Sonochemistry*, **4**(4), 295–300.

Petrier C., Lamy M.-F., Francony A., Benahcene A., David B., Renaudin V. and Gondrexon N. (1994). Sonochemical degradation of phenol in dilute aqueous solutions: comparison of the reaction rates at 20 and 487 kHz. *Journal of Physical Chemistry*, **98**(41), 10514–10520.

Pétrier C., Combet E. and Mason T. (2007). Oxygen-induced concurrent ultrasonic degradation of volatile and non-volatile aromatic compounds. *Ultrasonics Sonochemistry*, **14**(2), 117–121.

Poeschl U. and Oliveri C. (2014). Method for Treatment of Sulphide-Containing Spent Caustic. US patent 20140346121 A1.

Price G. J. (1990). The use of ultrasound for the controlled degradation of polymer solutions. In: Advances in Sonochemistry, T. J. Mason (ed.), JAI Press, London U.K., **1**, 231–287.

Prosonix. (2015). http://www.prosonix.co.uk

Prudhomme R. O. and Grabar P. (1949). De l'action des ultrasons sur certaines solutions aqueuses. *Journal de Chimie Physique et de Physico-chimie Biologique*, **46**(7–8), 323–331.

Rehorek A., Tauber M. and Gübitz G. (2004). Application of power ultrasound for azo dye degradation. *Ultrasonics Sonochemistry*, **11**(3–4), 177–182.

Ren Y.-Z., Franke M., Anschuetz F., Ondruschka B., Ignaszak A. and Braeutigam P. (2014). Sonoelectrochemical degradation of triclosan in water. *Ultrasonics Sonochemistry*, **21**(6), 2020–2025.

Ren Y.-Z., Wu Z.-L., Franke M., Braeutigam P., Ondruschka B., Comeskey D. J. and King P. M. (2013). Sonoelectrochemical degradation of phenol in aqueous solutions. *Ultrasonics Sonochemistry*, **20**(2), 715–721.

Rich R. (1955). Improvement in electroplating due to ultrasonics. *Plating*, **42**, 1407–1411.

Richards W. T. (1939). Supersonic Phenomena. *Reviews of Modern Physics*, **11**(1) 36–64.

Richards W. T. and Loomis A. L. (1927). The Chemical effects of high frequency sound. *Journal of the American Chemical Society*, **49**(12), 3086–3100.

Sáez V., Esclapez M. D., Tudela I., Bonete P., Louisnard O. and González-García J. (2010). 20kHz sonoelectrochemical degradation of perchloroethylene in sodium sulfate aqueous media: influence of the operational variables in batch mode. *Journal of Hazardous Materials*, **183**(1–3), 648–654.

Sáez V., Tudela I., Esclapez M. D., Bonete P., Louisnard O. and González-García J. (2011). Sonoelectrochemical degradation of perchloroethylene in water: enhancement of the process by the absence of background electrolyte. *Chemical Engineering Journal*, **168**(2), 649–655.

Schmitt F. O., Johnson C. H. and Olson A. R. (1929). Oxidations promoted by ultrasonic radiation. *Journal of the American Chemical Society*, **51**(2), 370–375.

Schulter H. and Gohr H. (1936). Über chemische wirkungen des ultraschallwellen. *Zeits. Angew. Chem*, **49**(27), 420–423.

Serpone N., Terzian R., Hidaka H. and Pelizzetti E. (1994). Ultrasonic induced dehalogenation and oxidation of 2-, 3-, and 4-chlorophenol in air-equilibrated aqueous media. Similarities with irradiated semiconductor particulates. *Journal of Physical Chemistry*, **98**(10), 2634–2640.

Shestakova M., Vinatoru M., Mason T. J. and Sillanpää M. (2015). Sonoelectrocatalytic decomposition of methylene blue using Ti/Ta2O5–SnO2 electrodes. *Ultrasonics Sonochemistry*, **23**(0), 135–141.

Shirgaonkar I. Z. and Pandit A. B. (1998). Sonophotochemical destruction of aqueous solution of 2,4,6-trichlorophenol. *Ultrasonics Sonochemistry*, **5**(2), 53–61.

Sierka R. A. and Amy G. L. (1985). Catalytic effects of ultraviolet light and/or ultrasound on the ozone oxidation of humic acid and trihalomethane precursors. *Ozone: Science & Engineering*, **7**(1), 47–62.

Song S., He Z. and Chen J. (2007). US/O3 combination degradation of aniline in aqueous solution. *Ultrasonics Sonochemistry*, **14**(1), 84–88.

Song W. and O'Shea K. E. (2007). Ultrasonically induced degradation of 2-methylisoborneol and geosmin. *Water Research*, **41**(12), 2672–2678.

Steter J. R., Barros W. R. P., Lanza M. R. V. and Motheo A. J. (2014). Electrochemical and sonoelectrochemical processes applied to amaranth dye degradation. *Chemosphere*, **117**(0), 200–207.

Suslick K. S. (1990). Sonochemistry. *Science*, **247**(4949), 1439–1445.

Tezcanli-Guyer G. and Ince N. H. (2003). Degradation and toxicity reduction of textile dyestuff by ultrasound. *Ultrasonics Sonochemistry*, **10**(4–5), 235–240.

Tezcanli-Güyer G. and Ince N. H. (2004). Individual and combined effects of ultrasound, ozone and UV irradiation: a case study with textile dyes. *Ultrasonics*, **42**(1–9), 603–609.

Theron P., Pichat P., Petrier C. and Guillard C. (2001). Water treatment by TiO_2 photocatalysis and/or ultrasound: degradations of phenyltrifluoromethylketone, a trifluoroacetic-acid-forming pollutant, and octan-1-ol, a very hydrophobic pollutant. *Water Science and Technology*, **44**(5), 263–270.

Thokchom B., Kim K., Park J. and Khim J. (2015). Ultrasonically enhanced electrochemical oxidation of ibuprofen. *Ultrasonics Sonochemistry*, **22**(0), 429–436.

Tiehm A. and Neis U. (2005). Ultrasonic dehalogenation and toxicity reduction of trichlorophenol. *Ultrasonics Sonochemistry*, **12**(1–2), 121–125.

Torres R. A., Pétrier C., Combet E., Carrier M. and Pulgarin C. (2008). Ultrasonic cavitation applied to the treatment of bisphenol A. Effect of sonochemical parameters and analysis of BPA by-products. *Ultrasonics Sonochemistry*, **15**(4), 605–611.

Toy M. S., Carter M. K. and Passell T. O. (1990). Photosonochemical decomposition of aqueous 1,1,1 - trichloroethane. *Environmental Technology*, **11**(9), 837–842.

Vajnhandl S. and Le Marechal A. M. (2007). Case study of the sonochemical decolouration of textile azo dye Reactive Black 5. *Journal of Hazardous Materials*, **141**(1), 329–335.

Vinatoru M. (2001). An overview of the ultrasonically assisted extraction of bioactive principles from herbs. *Ultrasonics Sonochemistry*, **8**(3), 303–313.

Vinodgopal K., Peller J., Makogon O. and Kamat P. V. (1998). Ultrasonic mineralization of a reactive textile azo dye, remazol black B. *Water Research*, **32**(12), 3646–3650.

von Gunten U. (2003). Ozonation of drinking water: part II. Disinfection and by-product formation in presence of bromide, iodide or chlorine. *Water Research*, **37**(7), 1469–1487.

Walton D. J. and Mason T. J. (1998). Organic sonoelectrochemistry. In: Synthetic Organic Sonochemistry. J.-L. Luche, Plenum Press, New York, NY, 263–297.

Wang B., Zhu C. P., Yin Y. Z., Yao C., Chen B. Y., Yin C., Shan M. L., Ren Q. G., Gao Y. and Han Q. B. (2014). Dual-frequency ultrasonic assisted ozonation for degradation of pesticide wastewater. *Applied Mechanics & Materials*, **(587–589)**, 598–601.

Wang Y., Zhang H., Chen L., Wang S. and Zhang D. (2012). Ozonation combined with ultrasound for the degradation of tetracycline in a rectangular air-lift reactor. *Separation and Purification Technology*, **84**, 138–146.

Warning A. and Datta A. K. (2013). Interdisciplinary engineering approaches to study how pathogenic bacteria interact with fresh produce. *Journal of Food Engineering*, **114**(4), 426–448.

Weavers L. K. (2001). Sonolytic ozonation for the remediation of hazardous pollutants. In: Advances in Sonochemistry Theme issue – Ultrasound in Environmental Protection. T. J. Mason and A. Tiehm (eds), Elsevier, **6**, 111–140.

Weavers L. K. and Hoffmann M. R. (1998). Sonolytic decomposition of ozone in aqueous solution: mass transfer effects. *Environmental Science & Technology*, **32**(24), 3941–3947.

Weiss J. (1944). Radiochemistry of aqueous solutions. Nature, **153**, 748–750.

Weissler A., Cooper H. W. and Snyder S. (1950). Chemical effect of ultrasonic waves: oxidation of potassium iodide solution by carbon tetrachloride. *Journal of the American Chemical Society*, **72**(4), 1769–1775.

Wood R. W. and Loomis A. (1927). The physical and biological effects of high frequency sound waves of great intensity. *Philosophy Magazine*, **4**, 417.

Wu C., Liu X., Wei D., Fan J. and Wang L. (2001). Photosonochemical degradation of Phenol in water. *Water Research*, **35**(16), 3927–3933.

Wu X., Joyce E. M. and Mason T. J. (2011). The effects of ultrasound on cyanobacteria. *Harmful Algae*, **10**(6), 738–743.

Yang B., Zuo J., Tang X., Liu F., Yu X., Tang X., Jiang H. and Gan L. (2014). Effective ultrasound electrochemical degradation of methylene blue wastewater using a nanocoated electrode. *Ultrasonics Sonochemistry*, **21**(4), 1310–1317.

Yeager E. and Hovorka F. (1953). Ultrasonic waves and electrochemistry. I. a survey of the electrochemical applications of ultrasonic waves. *The Journal of the Acoustical Society of America*, **25**(3), 443–455.

Zhang H., Duan L. and Zhang D. (2006). Decolorization of methyl orange by ozonation in combination with ultrasonic irradiation. *Journal of Hazardous Materials*, **138**(1), 53–59.

Zhang H., Lv Y., Liu F. and Zhang D. (2008). Degradation of C.I. Acid Orange 7 by ultrasound enhanced ozonation in a rectangular air-lift reactor. *Chemical Engineering Journal*, **138**(1–3), 231–238.

Zhang Y., Geißen S.-U. and Gal C. (2008). Carbamazepine and diclofenac: removal in wastewater treatment plants and occurrence in water bodies. *Chemosphere*, **73**(8), 1151–1161.

Zhou T., Lim T.-T. and Wu X. (2011). Sonophotolytic degradation of azo dye reactive black 5 in an ultrasound/UV/ferric system and the roles of different organic ligands. *Water Research*, **45**(9), 2915–2924.

Ziylan A. and Ince N. H. (2015). Catalytic ozonation of ibuprofen with ultrasound and Fe-based catalysts. *Catalysis Today*, **240**, Part A: 2–8.

Zuniga-Benitez H., Soltan J. and Penuela G. (2014). Ultrasonic degradation of 1-H-benzotriazole in water. *Water Science and Technology*, **70**(1), 152–159.

Chapter 12

Electrical discharge plasma for water treatment

Selma Mededovic Thagard and Bruce R. Locke

12.1 INTRODUCTION – PLASMA PROCESSES FOR WATER TREATMENT

Plasma is considered the 4th state of matter and consists of ionized molecules, atoms and atomic fragments, as well as free electrons (Lieberman & Lichtenberg, 2005). The term *plasma* was coined by Irving Langmuir while studying electronic devices based on ionized gases (Tonks & Langmuir, 1929). There are many kinds of plasmas including both natural and man-made. They can be classified in various ways by the key properties of electron energy and electron density (Lieberman & Lichtenberg, 2005). Electron energy is usually reported in eV. Typical ranges for commercial plasma processes are 1 to 10 eV (Fridman & Kennedy, 2004). One important classification of plasma is based upon the relative energies or temperatures of the electrons in the plasma compared to the temperature of the background gas. In thermal plasma the energies of all species are the same and generally such equilibrium plasmas are of very high temperatures (greater than several thousand Kelvin). Arcs are examples of commercially useful thermal plasmas. A non-thermal plasma results when the energy of the electrons is much higher than the energy of the background gaseous species. The background gas temperature in such plasmas can vary from ambient to approximately one thousand Kelvin, while the energy of the electrons can vary from 1 to 10 eV (1 eV = 11,600 K). The dielectric barrier discharge (DBD) used for commercial ozone generation is a non-thermal plasma. Non-thermal plasmas of interest for water treatment can be generated by direct current (DC), alternating current (AC), pulsed discharges, and radiofrequency (RF) and microwave (MW) power supply sources (Fridman & Kennedy, 2004).

Electrical discharges can be credited with the development of the first Advanced Oxidation Process (AOP) for water treatment: ozonization. The method for producing ozone was invented in 1857 by Werner von Siemens while studying dielectric barrier discharges in oxygen and air. Its properties as a strong oxidant coupled with the efficient von Siemens process led to early 20th century adoption of ozone in municipal drinking water treatment processes in France, Russia, and Spain (Fridman, 2008; Kogelschatz & Eliasson, 1995). Ozone has been used ever since for water treatment in many places in the world, and while its primary use for water disinfection was displaced by the relatively cheaper chlorine-based methods, which have the advantage of longer residual lifetime in water delivery systems, many countries later abandoned chlorine-based methods due to issues of disinfection byproducts and taste. Because of concerns with the formation of disinfection byproducts from chlorine and the ineffectiveness of the treatment for a broad range of contaminants at common disinfection doses, significant research and development efforts have been put toward non-halogen chemical oxidation and AOPs such as ultraviolet light-hydrogen peroxide (UV/H_2O_2) (Ahmed *et al.* 2009) and ozone-hydrogen peroxide (O_3/H_2O_2) (Beltrán *et al.* 1996). These processes, however, require significant chemical additions and quenching of residual H_2O_2 which, if not removed, could promote biological re-growth in the distribution system due to released O_2 (Kavanaugh & Chowdhury, 2004). Additionally, AOPs involving UV are ineffective for waters with high turbidity due to reduced UV transmission (Ein-Mozaffari *et al.* 2009) whereas processes using O_3 require handling and storage of compressed oxygen (Li *et al.* 2010).

To overcome the problems associated with established and commercialized AOPs, new emerging oxidation technologies that involve TiO_2 catalysis (photocatalysis) (Pichat *et al.* 2000), cavitation (Kim *et al.* 2001), e-beam irradiation (Duarte *et al.* 2002), and Fenton's reaction (Ioan *et al.* 2007) have been considered for contaminant destruction. These AOPs yield hydroxyl radicals ($^{\bullet}OH$); however, their effectiveness can be limited depending on the type of hydroxyl radical precursor, catalyst, and the turbidity of the solution. Moreover, system complexity and mass transfer limitations of the reactants are additional reasons why a highly-efficient oxidation technology is still elusive. Although the chemistry and scale-up of these new technologies have received considerable interest, less attention has been given to the development of electrical discharge plasma reactors. Compared to other AOPs, electrical discharge has several advantages: it requires no chemical additions, can degrade a broader range of contaminants, and has the potential to be optimized for small treatment systems. Plasmas are capable of generating all the chemical species and effects found in the other AOPs as well as additional factors not typically found in those processes. These include the reactive oxygen species (ROS), e.g. $^{\bullet}OH$, H_2O_2, O_3, UV emissions, and shockwave formation (when the discharge occurs directly in the liquid). The abundance of these factors and the efficiencies of their generation vary with the plasma reactor design and the type of plasma. In addition to the generation of reactive oxygen-based species found in most AOPs, plasma processes can generate reactive nitrogen species, (RNS), when nitrogen and oxygen are present in the discharge, and various reducing species (e.g., H_2, $H^{\bullet}$, free and aqueous electrons) can also be formed (Locke *et al.* 2012; Lukes *et al.* 2012).

Many reviews of plasma processes for pollution control, including air contaminant treatment (Kim *et al.* 2006; Mizuno, 2007) and water treatment (Akiyama, 2000; Bruggeman & Locke, 2013; Hackam & Akiyama, 2000; Hijosa-Valsero *et al.* 2014; Jiang *et al.* 2014; Joshi & Thagard, 2013b; Locke *et al.* 2006; Malik *et al.* 2001; Sato, 2008; Sunka, 2001; Sunka *et al.* 1999), have been recently published. Detailed recent studies have also reviewed the fundamental physical and chemical principles of plasma interaction with liquid water (Joshi & Thagard, 2013a; Locke *et al.* 2012; Locke & Thagard, 2012; Lukes *et al.* 2012; Lukes *et al.* 2012). In this chapter, we seek to highlight the important key aspects and summarize the status of plasma applications for water and wastewater treatment as an Advanced Oxidation Process. The focus is on two types of treatment processes: indirect plasma and direct plasma. The *indirect plasma* is most typically represented by ozone generators whereby the plasma is used to generate the oxidant, and then the oxidant is delivered to a separate reactor for the water treatment. In the *direct plasma* treatment, the plasma contacts the liquid phase containing the contaminants directly. This chapter will also focus on the generation of ozone and other strong oxidants and examine reactive conditions produced by electrical discharge plasma processes with applications to water treatment.

12.2 INDIRECT PLASMA – OZONE GENERATION

Ozone generation from oxygen and air by electrical discharges has been extensively studied and reported in the literature (Eliasson *et al.* 1987; Eliasson & Kogelschatz, 1991; Lieberman & Lichtenberg, 2005). The key reactions involve electron impact dissociation of molecular oxygen (12.1 and 12.2) to form atomic oxygen species, which subsequently react with molecular oxygen (12.3) to form ozone by a three body collision.

$$O_2 + e^- \rightarrow e^- + O(^3P) + O(^3P) \tag{12.1}$$

$$O_2 + e^- \rightarrow e^- + O(^3P) + O(^1D) \tag{12.2}$$

$$O + O_2 + M \rightarrow O_3^* + M \rightarrow O_3 + M \tag{12.3}$$

M is a third collision body (e.g., O_2, O_3, or O in pure oxygen feed).

In the indirect plasma processes, a carrier gas such as oxygen or air is used to generate the reactive species, here ozone, that subsequently reacts with the target pollutants, typically at low concentrations, in a separate reaction chamber downstream of the plasma reactor. The reaction rate constant for species i for collisions with electrons, k_{ei}, for the case of oxygen in 12.1 and 12.2 can be determined from Equation 12.4 where the index i refers to O_2 (Lieberman & Lichtenberg, 2005).

$$k_{ei} = \int_0^\infty \left(\frac{\varepsilon}{2m}\right)^{1/2} \sigma_{ei}(\varepsilon) f(\varepsilon) \mathrm{d}\varepsilon \tag{12.4}$$

In Equation 12.4 ε is the electron energy, $\sigma(\varepsilon)$ is the collision cross-section, and $f(\varepsilon)$ is the electron energy distribution function. The electron energy distribution was historically assumed to follow a Maxwell distribution (Bell, 1974); however, more recent approaches that use accurate computational codes are capable of determining the exact distribution from the solution of the Boltzmann equation for a specific electrical discharge (e.g., BOLSIG (Hagelaar & Pitchford, 2005)). In ozone generation, the efficiency of ozone production from an oxygen carrier gas is only dependent upon the reduced electric field (E/N), where E is the electric field and N is the gas number density. This is due to the fact that the reactions 12.1 and 12.2 include only the direct electron collision with oxygen molecules followed by 12.3 for the three body collisions of O atoms with O_2 (Eliasson & Kogelschatz, 1986; Plank *et al.* 2014). In air, this dependence on E/N does not hold since other (non-electron collision) reactions involving nitrogen affect the ozone yield (Plank *et al.* 2014). While the basic reactions involved in the formation of ozone with plasma are very well known (Egli & Eliasson, 1989; Eliasson *et al.* 1987; Eliasson & Kogelschatz, 1991), there continues to be extensive research on various types of reactor configurations and plasma generation methods in order to improve the energy efficiency of ozone generation (Ono & Oda, 2007; Plank *et al.* 2014).

Of particular importance to ozone generation for environmental applications are the effects of molecular nitrogen on ozone generation since air is typically a cheaper reagent to use than pure oxygen. While electron collisions with molecular nitrogen can lead to additional atomic oxygen (12.5 to 12.7 (Kogelschatz *et al.* 1988)), subsequent reactions of ozone with such formed nitrogen oxides decrease the ozone yield (12.8 and 12.9). It must be noted that additional nitrogen oxides (N_2O, N_2O_5) may also be formed.

$$N_2 + e^- \rightarrow e^- + N + N \tag{12.5}$$

$$O_2 + N \rightarrow NO + O \tag{12.6}$$

$$N + NO \rightarrow N_2 + O \tag{12.7}$$

$$O_3 + NO \rightarrow NO_2 + O_2 \tag{12.8}$$

$$O_3 + NO_2 \rightarrow NO_3 + O_2 \tag{12.9}$$

The chemistry of nitrogen oxide formation as well as nitrogen oxide mitigation by plasma discharge has been extensively documented in the literature for air pollution control, and the injection of externally generated ozone into exhaust gas for nitrogen oxide removal has been quite successful (Mok, 2006).

The effects of humidity and water are important not only to the ozone generation processes, but also to the direct application of plasma to systems that contain vapor and liquid water molecules. The effect of water vapor to decrease the production of ozone has been shown by modeling (Chen & Wang, 2005; Peyrous, 1990a; Peyrous, 1990b) and experiments (Ono & Oda, 2003). Modeling suggests that atomic oxygen is removed by water molecules (Chen & Wang, 2005) and experiments show that ozone yield in air can decrease by 6 times with water vapor (Ono & Oda, 2003). While the combined effects of humidity and nitrogen species formation leads to acid formation and are detrimental to pure ozone generation, the formation of plasma activated water (PAW), a combination of hydrogen peroxide, nitrates, nitrites, and nitrogen oxide based acids, has applications in agriculture and medicine (Lindsay *et al.* 2014; Lukes *et al.* 2014; Naïtali *et al.* 2012).

Table 12.1 summarizes selected work showing the maximum achieved ozone energy yields by various plasma processes. The thermodynamic limit based upon the enthalpy of reaction for ozone generation in dry oxygen is 1220 g/kWh (Kogelschatz & Eliasson, 1995). As shown in Table 12.1, laboratory studies with nanosecond pulse discharges have achieved the highest energy yields of approximately 60% of the thermodynamic limit (725 g/kWh or 4×10^{-6} mol/J). On the other hand, typical commercial energy yields are 25 to 50 g/kWh and 50 to 70 g/kWh for air and oxygen fed ozonizers, respectively (Fridman, 2008; Kogelschatz & Eliasson, 1995). Kogelschatz and Eliasson note that approximately ½ of the operating costs for ozone synthesis commercially is from the supply of oxygen, and it can also be noted that most sources do not account for the cost of oxygen in reporting ozone energy yields (Kogelschatz & Eliasson, 1995). Fridman discusses earlier literature including studies where the highest yields obtained were for cases of cryogenic cooling of the reactor walls giving energy yields as high as 400 g/kWh; however, the cost of the cooling was not included (Fridman, 2008). It is also useful to note that ozone generated by electrochemical cells in various water electrolytes generally have much lower efficiencies than the plasma processes.

As previously stated, ozone generated by plasma reactors must be injected into the polluted water in order to react with target contaminants (von Sonntag & von Gunten, 2012; see also Chapters 3 and 4 of present book). Therefore, limitations of mass transfer as well as ozone solubility affect the amount of ozone that can react with a given target, and the properties of the liquid, including composition, pH, and temperature will affect degradation rates and extents. Many potential target compounds do not directly react with ozone but will react with the hydroxyl radical, $^{\bullet}OH$, formed by ozone decomposition in the liquid. While ozone may have some selectivity with certain organic compounds, the hydroxyl radical is not highly specific and has very high (near diffusion limits) reaction rates.

Like ozone, $^{\bullet}OH$ may also be produced in a downstream plasma reactor and injected in water to react with organic contaminants and microorganisms. In one such system, a strong electric field discharge (SED) uses a mixture of water vapor and oxygen to produce activated species such as O_2^+, H_2O^+, $O(^1D)$, $O(^3P)$, and H_2O_2 (Bai *et al.* 2010). Subsequent reactions of water molecules with O_2^+ and H_2O^+ and the radical propagation reactions of species produced by equilibrium shift reactions of H_2O_2 and $HO_2^{\bullet}$ are mainly responsible for the formation of $^{\bullet}OH$. The measured rate of $^{\bullet}OH$ production is ~5.9×10^3 µM/min, which is one of the highest values reported for non-thermal plasmas (Bai & Zhang, 2015). Once formed in the gas phase, $^{\bullet}OH$ is quickly injected into liquid water where it dissolves and undergoes reactions with target contaminants and microorganisms. Ozone concentration was unfortunately not measured so the relative contribution of ozonization to the overall treatment remains unknown.

Table 12.1 Summary of selected ozone generation by various plasma processes.

Gas	Ozone Concentration (gO_3/m^3)	Ozone Production Rate (gO_3/h)	Ozone Yield (g/kWh)	Discharge Type	Reference
Oxygen	15	0.9	725	Neg. ns pulse 60 kV, 5 pps	(Takamura *et al.* 2011)
Oxygen	10, 30	0.6, 1.8	570, 400	2 ns pulse, 40 and 50 kV	(Matsumoto *et al.* 2011)
Oxygen	5.8	0.7	274	Hybrid silent surface discharge	(Nomoto *et al.* 1995)
Oxygen	16.6	3.5	202	Pulsed DBD	(Samaranayake *et al.* 2000)
Oxygen	48	0.9	185	Silent discharge	(Garamoon *et al.* 2002)
Oxygen	61	3.7	173	Packed bed dielectric barrier	(Chen *et al.* 2006)
Oxygen	17	0.2	85	Micro-channel arrays of DBD	(Kim *et al.* 2013)
Oxygen	10–50	0.2–0.9	120–250	Shielded sliding discharge (+/–)	(Malik, 2014)
Air	4–13	0.07–0.2	75–110	Shielded sliding discharge (–)	(Malik, 2014)
Air	N/A	N/A	150–300	DC discharge, 50–122 Torr, 700–4600 V	(Plank *et al.* 2014)
Air	Up to 4	N/A	Up to 350	50 Hz, AC	(Buntat *et al.* 2009)
Air	Up to 18	Up to 1000	Up to 239	ns pulse, 50 kV, 20 pps	(Wang *et al.* 2010)
None	N/A	N/A	6 to 21	Various electrochemical cells containing different electrolytes	(Christensen *et al.* 2013)

While indirect $^{\bullet}OH$ generation may be an attractive alternative water treatment processing technique, many research and development efforts have sought to consider contacting the plasma with the contaminated liquid phase. These direct plasma treatment processes in which hydroxyl radicals are generated in or near the contaminated liquid are discussed in the next section.

12.3 DIRECT PLASMA – PLASMA DIRECTLY CONTACTS LIQUID SOLUTION

In contrast to the indirect plasma ozone generator discussed above, the direct plasma cases involve contacting the plasma directly with solutions (e.g., drinking or wastewater) containing the contaminants to be degraded. Perhaps the first study to contact a liquid phase with an electrical discharge was that of Gubkin in 1887 (Gubkin, 1887). Research in the field of glow discharge electrolysis and contact glow discharge electrolysis has its origins in the work of Gubkin and was followed by the extensive work of Hickling in the middle 20th century (Hickling, 1971; Hickling & Ingram, 1964). Work on corona-like discharges in liquids formed by pulsed power supplies has long been of interest for electrical insulation, but much of the pollution treatment work began in the late 1980s and early 1990s with pulsed discharges in water using point-plane electrode configurations (Clements *et al.* 1987; Sharma *et al.* 1993).

Figure 12.1 shows a selected variety of schemes by which a plasma discharge formed between two electrodes can be generated in or in contact with a liquid phase. Plasma can be generated directly in a liquid phase, completely in a gas phase, or in a gas phase such that the plasma propagates along the interface between the gas and the liquid phases. Reviews of these configurations and their performances, some shown in Figure 12.1, have been reported in the literature (Bruggeman & Locke, 2013; Bruggeman *et al.* 2016; Jiang *et al.* 2014; Joshi & Thagard, 2013b; Locke *et al.* 2006; Locke *et al.* 2012; Lukes *et al.* 2012).

Formation of an electrical discharge directly in the liquid phase, as in reactor types A and F of Figure 12.1, requires a very large electric field (on the order of 1 MV/cm) (Locke *et al.* 2006; Locke & Thagard, 2012; Sunka *et al.* 1999). A small radius needle or a very thin wire can be used with a pulsed discharge to generate such discharges. Diaphragm discharges (E) can also be used to generate discharges directly in the liquid through concentration of the electric field in a small opening separating the two electrodes. In contrast, when an electrical discharge is initiated in the gas phase, the breakdown fields are much lower, and, for example, in dry air the field at ambient pressure is on the order of 25 to 35 kV/cm. Therefore, the use of bubbles (B), gas gaps (C and D), and discharges completely in the gas phase (G, H, and I) simplifies the power supply requirements for discharges with liquid water. The breakdown field in a gas will decrease with humidity, and it can also be affected by the presence of the liquid phase as in reactor types C, D and I of Figure 12.1 where the discharge propagates along the interface between the gas and liquid phases. The energy requirements for the operation of electrical discharge plasma reactors range from ~1 J/pulse (Clements *et al.* 1987) to kJ/pulse (Hoffmann *et al.* 1997; Willberg *et al.* 1996a). It is likely that the temperatures of underwater plasmas generated by kJ pulses are quite high and that the plasma is thermal (Locke & Thagard, 2012).

Many reactor types shown in Figure 12.1 have been tested. It has been found that reactors with higher efficiency for chemical oxidation as well as hydrogen peroxide formation are those with water spray and/or water films where the discharge is formed in the gas phase and which may propagate along the gas-liquid interface (Bruggeman & Locke, 2013; Malik, 2010). Recent research with nanosecond pulses using submerged wire-to-cylinder electrodes (F) may lead to an exception to this finding (Banaschik *et al.* 2014).

The high efficiency of these "contact-oriented" reactors involving sprays and water films has been explained by their high interfacial area, that is, the area of the plasma in contact with water. In fact, for gas discharges contacting water, it has been shown that the contaminant removal rate is directly proportional to the interfacial area, which likely explains why gas discharges are more effective than liquid discharges for contaminant degradation (Stratton *et al.* 2015). Since (plasma) gas discharges contacting water represent a non-homogeneous medium, the chemical reactions responsible for the degradation of organic solutes must be treated as heterogeneous and not homogeneous reactions. Thus, for a heterogeneous irreversible pseudo-first-order reaction between a solute S and $^{\bullet}$OH, the rate expression is given by Equation 12.10 (Stratton *et al.* 2015).

$$r_S = \frac{A}{V \cdot (1/k_m + 1/k')} \cdot C_S = k_{\text{obs}} \cdot C_S \tag{12.10}$$

A (m^2) is the interfacial area, V (m^3) is the solution volume, k' (m · s^{-1}) is the pseudo-first-order reaction rate constant ($k' = k \cdot C_{\text{OH}}$), and k_m (m · s^{-1}) is the mass transfer coefficient of the solute ($k_m = D/L$, where D (m^2 · s^{-1}) is the solute diffusivity and L (m) is the thickness of the interfacial zone). k_{obs} is the measured pseudo-first-order reaction rate constant. As a result, the majority of scaled-up versions of electrical discharge plasma reactors utilize discharges in gases or gas bubbles to maximize the contact area between the plasma and the contaminated liquid. The effectiveness of these and bench scale plasma reactors for drinking water and wastewater treatment is discussed in Section 12.3.4.

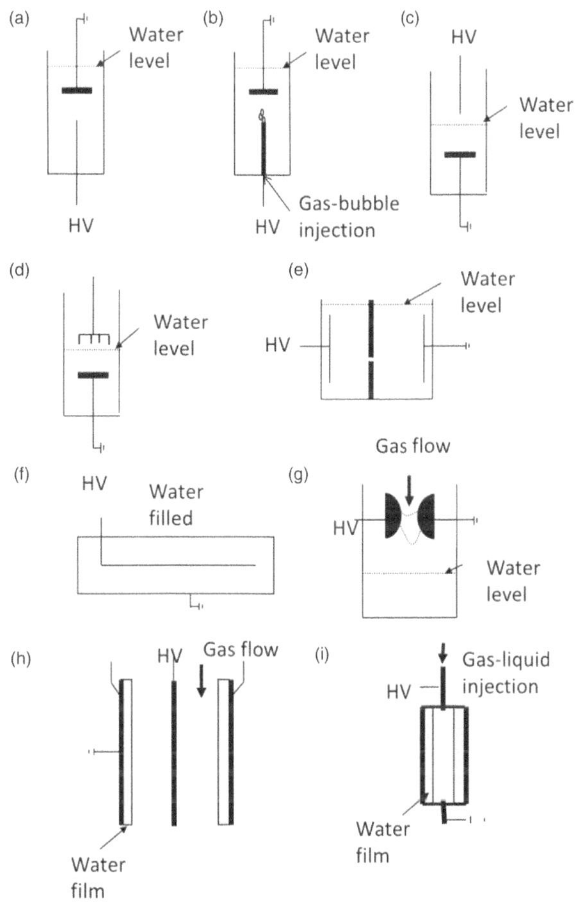

Figure 12.1 Schematics of various electrode configurations for liquid phase and liquid/gas phase electrical discharge plasma reactors. Modified and reprinted with permission from "Electrohydraulic Discharge and Nonthermal Plasma for Water Treatment", Industrial & Engineering Chemistry Research 2006, 45(3) 882–905. Copyright 2006 American Chemical Society.

12.3.1 Chemical species formed

Electrical discharges formed directly in the liquid phase, along a gas-liquid interface or in a humidified gas phase, dissociate water and, if present, dissociate or excite the processing gas. Figure 12.2 shows the optical emission spectrum for a plasma discharge formed in an argon-water film reactor schematically represented in Figure 12.1I. In this reactor water is flowing down the edges of the reactor wall, and the plasma discharge propagates along the gas-liquid interface as described in the literature (Wandell & Locke, 2014). The observed lines are originating from the emissions of oxygen, hydrogen, hydroxyl, and argon excited states. The OH peak can be used to estimate the temperature of the plasma and the H_β emissions to estimate electron density (Gigosos *et al.* 2003). These two parameters, along with the electron energy are critical to characterize the plasma and facilitate comparison of the many different means of generating the plasma. Subsequent recombination events involving the primary water dissociation products lead to the formation of more stable molecular species including O_2, H_2, and H_2O_2 (Kirkpatrick & Locke, 2005).

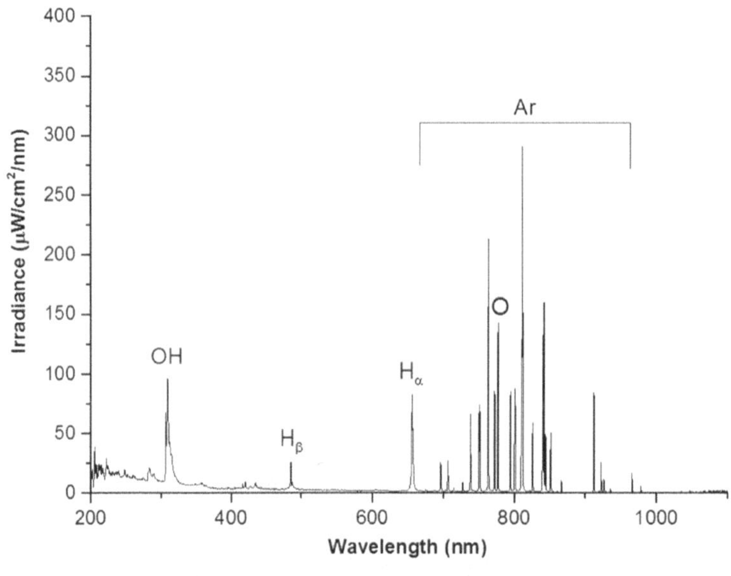

Figure 12.2 Optical emission spectrum of a plasma reactor with liquid water flowing at 0.25 mL/min in an argon carrier gas flowing at 0.4 L/min. Unpublished and provided by courtesy of Dr. Kevin Hsieh, Florida State University.

Molecular oxygen or other gases can be added by injecting a gas stream through the submerged electrode or by forming the discharge above the liquid in a gas. When pure oxygen or air is added to the system, ozone can be formed. The subsequent mass transfer and reactions of ozone in the liquid are based upon ozone chemistry in water. The simultaneous production of ozone in the gas phase and hydrogen peroxide in the liquid phase can occur when both gas and liquid discharges are produced in such reactors as in Figure 12.1B, C, D, G, H, and I (Grymonpre *et al.* 2004). If the gas phase contains nitrogen, then nitrogen oxides can be formed in the gas phase (12.5–12.9). Further reactions of these nitrogen oxides with $^{\bullet}OH$ or direct dissolution in the liquid water phase can lead to nitrite, nitrate, and the corresponding acids formation by 12.11 to 12.14.

$$NO + {}^{\bullet}OH \rightarrow HNO_2 \tag{12.11}$$

$$NO_2 + {}^{\bullet}OH \rightarrow HNO_3 \tag{12.12}$$

$$2NO_{2(aq)} + H_2O \rightarrow NO_2^- + NO_3^- + 2H^+ \tag{12.13}$$

$$NO_{(aq)} + NO_{2(aq)} + H_2O \rightarrow 2NO_2^- + 2H^+ \tag{12.14}$$

Other reactions are also involved. Of particular note in disinfection is the formation of peroxynitrite ($ONOO^-$) through $^{\bullet}OH$ reaction with nitrite ions and subsequent reactions to form reactive oxygen and reactive nitrogen species (Brisset & Hnatiuc, 2012; Lukes *et al.* 2014; Lukes *et al.* 2012).

12.3.2 H_2O_2 generation

The formation of H_2O_2 from water by various processes including atmospheric chemistry, radiation chemistry, and plasma chemistry has been reviewed extensively (Locke & Shih, 2011). The thermodynamic limit based upon the enthalpy of reaction at standard conditions is 400 g/kWh (Locke & Shih, 2011). The energy yields for H_2O_2 by various plasma processes are summarized in Table 12.2. This table is a condensed version of the complete data set available in (Locke & Shih, 2011).

Table 12.2 Summary of hydrogen peroxide generation by various plasma processes. Adapted from Bruggeman and Locke, 2013 and based upon Locke and Shih, 2011 (RF = Radio Frequency, MW = Microwave, and DBD = Dielectric Barrier Discharge).

Process	Medium	Generation Rate (g/h)	Energy Efficiency (g/kWh)
Spark discharge	Directly in liquid water	~8.6×10^{-2}	0.43–0.55
Pulsed corona	Directly in liquid water	0.01–0.21	0.96–3.64
Capillary/diaphragm discharge	Directly in liquid water	0.02–0.36	0.1–1
Contact glow discharge electrolysis	Directly in liquid water	0.03–0.64	0.8–1.6
RF/MW discharge	Directly in liquid water	~0.1	0.46–0.64
Discharges in bubbles	Air/Ar/O_2 in liquid water	2.3×10^{-3}–26	0.4–8.4
Hybrid discharges	Combined liquid water + air	2.7×10^{-3}–6.6×10^{-2}	0.37–1
Gas phase corona discharges	Air/Ar + water surface	5.7×10^{-5}–2.5×10^{-2}	0.13–5
DBD above liquid	Air/water surface	2.5×10^{-4}–0.12	0.04–2.7
MW	Steam	48	24
DBD	Humid gas	1.8×10^{-3}–1.6×10^{-2}	1.14–1.7
Gliding arc	Water droplets (in Ar)	0.02–0.14	0.57–80
Water film	Tubular reactors with water films	0.01	1 to 40
Electron beam	Liquid water	N/A	8.9
Ultrasound	Dissolved gas + liquid water	1.2–9.8×10^{-3}	0.01–0.12
Vacuum UV	Vapor or liquid water	N/A	13–33
Electrolysis	Liquid electrolyte solutions	N/A	112.4–227.3

As already mentioned, electrical discharges directly in the liquid phase (spark, pulsed corona) have relatively low efficiency in hydrogen peroxide generation, and water film and aerosol reactors tend to have higher efficiencies for the H_2O_2 production. Plasma generated in the gas phase has lower power requirements causing less heating and the contact of the liquid with the plasma allows the liquid to sequester the formed H_2O_2. Since H_2O_2 is highly soluble in liquid water and the plasma exists only in the surrounding phase, once H_2O_2 is in the liquid phase it is partially protected from degradation by the plasma. Large spatial and temporal gradients can occur at the plasma-liquid boundary, and these gradients can cause rapid quenching to reduce back reactions that destroy the desired product.

As in Equation 12.10, the overall rate of H_2O_2 formation is proportional to the area of the plasma-liquid interface. However, unlike Equation 12.10, the reaction is pseudo-zeroth-order and all reactants originate in the plasma interior; therefore, diffusion is negligible. These features are captured in Equation 12.15 (Stratton *et al.* 2015).

$$r_{H_2O_2} = \frac{A \cdot L}{V} \cdot k'_{H_2O_2} = k_{H_2O_2,obs} \tag{12.15}$$

A (m^2) is the interfacial area, V (m^3) is the solution volume, L (m) is the thickness of the interfacial zone, $k_{H_2O_2,obs}$ (mM · min^{-1}) is the observed rate constant calculated from $C_{H_2O_2}$ measurements, and $k'_{H_2O_2}$ is the pseudo-zeroth-order rate constant ($k'_{H_2O_2} = k_{H_2O_2} \cdot C^2_{OH}$, where $k_{H_2O_2}$ is the second-order reaction rate constant). It must be noted that Equation 12.15 is not technically a heterogeneous rate equation, but rather a homogeneous rate equation adapted to account for the fraction of the volume in which the reactions actually occur.

12.3.3 OH radical generation

The dissociation energy for water to form hydroxyl radicals by Reaction 12.16 is 117 kcal/mol. The energy yield can be given by the reciprocal of the dissociation energy as 2×10^{-6} mol/J (G value = 20 molecules/100 eV) (Ruscic *et al.* 2002).

$$H_2O \rightarrow H^{\bullet} + {}^{\bullet}OH \tag{12.16}$$

Table 12.3 gives the energy yields for the formation of $^{\bullet}OH$ by a range of processes including plasma in water, plasma above water, plasma in humid gas, electron beam in water, and sonolysis (Hsieh, 2015; Hsieh *et al.* 2017). The G value, defined as the number of product molecules formed per 100 eV of energy absorbed, may be somewhat dependent upon the method used to measure $^{\bullet}OH$ and the gas or liquid composition due to reactions with $^{\bullet}OH$. Nevertheless, typical energy yields of $^{\bullet}OH$ formation by plasma directly in water and by some of the humid plasma cases are quite low. The energy yields for pulsed discharges along a water film are comparable to electron beams directly in water and sonolysis with added gases. The highest energy yield for $^{\bullet}OH$ by plasma is 5 molecules/100 eV, which is 25% of the thermodynamic limit based upon the water dissociation energy. However, in order to be effective at treating liquid phase pollutants, the $^{\bullet}OH$ formed in such gas-liquid plasmas must contact and transport into the liquid, thereby requiring high rates of mass transfer and large surface areas. Reactions of $^{\bullet}OH$ with organic compounds typically are close to diffusion limits, and thus the reactions are likely to be close to or at the gas-liquid interface. It should be noted that in the absence of a chemical probe to react with the $^{\bullet}OH$, some of the $^{\bullet}OH$ recombines to form H_2O_2 which subsequently dissolves in the liquid, and additional $^{\bullet}OH$ are lost by other reactions. In the presence of a metal catalyst or a base, the dissolved H_2O_2 decomposes to form $^{\bullet}OH$ in the bulk liquid.

Table 12.3 Summary of $^{\bullet}OH$ production by various plasma processes. Adapted from (Hsieh, 2015).

Plasma Type	Water Source	Gas	$^{\bullet}OH$ Production Rate (mol/s)	Power (W)	G ($^{\bullet}OH$) (molecules/100 eV)	Reference
Gliding arc	Humid	Air	4.6×10^{-8}	1.0×10^{2}	4.4×10^{-3}	(Benstaali *et al.* 2002)
Pulsed arc	Humid	Air	2.3×10^{-6}	3.0×10^{5}	6.7×10^{-5}	(Ono & Oda, 2001)
AC arc	Humid	Air	5.8×10^{-6}	9.0×10^{2}	6.2×10^{-2}	(Srivastava *et al.* 2009)
Pulsed DBD	Vapor	Ar, H_2O	2.8×10^{-8}	0.4×10^{-1}	7.4×10^{-1}	(Hibert *et al.* 1999)
Atmospheric pressure glow discharge	Liquid tap water	Air	4.4×10^{-8}	1.0×10^{1}	4.3×10^{-2}	(Bruggeman *et al.* 2008)
Pulsed streamer	Liquid surface	Air	2.2×10^{-11}	2.8	7.6×10^{-5}	(Kanazawa *et al.* 2011)
Pulsed corona	Liquid film	Air	5.7×10^{-8}	3.4	1.6×10^{-1}	(Sano *et al.* 2003)
Pulsed corona	Liquid (150 μS/cm)	None	6.7×10^{-8}	64	1.0×10^{-2}	(Sahni & Locke, 2006b)
Pulsed corona	Liquid surface (150 μS/cm)	O_2	2.5×10^{-8}	64	3.7×10^{-3}	(Sahni & Locke, 2006b)
Pulsed corona	Liquid surface	Air	3.7×10^{-9}	0.6	6.0×10^{-2}	(Hoeben *et al.* 2000)
Electron beam irradiation	Liquid	None	1.7×10^{-2}	N/A	2.7	(Cooper *et al.* 1992)
Sonolysis	Liquid	Air	N/A	N/A	4.7	(Mark *et al.* 1998)
Sonolysis	Liquid	Ar	N/A	N/A	7.4	(Mark *et al.* 1998)
Pulsed corona	Vapor	Ar	3.1×10^{-7}	1.8	1.9	(Su *et al.* 2002)
Pulsed tubular discharge	Liquid film	Ar	2.0×10^{-7}–5.4×10^{-7}	0.6–2.8	0.9–5.0	(Hsieh, 2015)

12.3.4 Data on model compounds

12.3.4.1 Oxidation reactions

Of all the oxidative radical species that can be formed by the electron-induced and thermal decomposition of water molecules and which include $^{\bullet}OH$, O, $HO_2^{\bullet}$, and H_2O_2, $^{\bullet}OH$ is perhaps the most important one. It is one of the strongest oxidizers known ($E^{\circ}(^{\bullet}OH/H_2O) = 2.85$ V with respect to the standard hydrogen electrode (Tarr, 2003)) and its reactions with the majority of organic compounds are nearly diffusion-controlled and non-selective. The types of reactions $^{\bullet}OH$ undergoes with organic species include direct electron transfer, electrophilic addition to unsaturated systems, and hydrogen abstraction from typically alkyl or hydroxyl groups (Parvulescu *et al.* 2012). For a detailed mechanistic overview of these reactions, the reader is referred to (Parvulescu *et al.* 2012).

The oxygen atom O has been observed directly via optical emission spectroscopy (see Figure 12.2). Additionally, mathematical model results of primary reactions of underwater electrical discharge plasmas predict this species is formed by the recombination of hydroxyl radicals (Mededovic & Locke, 2007a):

$$^{\bullet}OH + {}^{\bullet}OH \rightarrow H_2O + O \tag{12.17}$$

Although O is a relatively strong oxidant ($E°(O/H_2O) = 2.42$ V with respect to the standard hydrogen electrode (Tarr, 2003)), the experimental evidence for the involvement of this species in the oxidation of organic compounds does not exist. This is because the chemical reactions of O with different organic compounds are very similar to those of $^{\bullet}OH$.

Hydroperoxyl radical ($HO_2^{\bullet}$) is an oxidative species produced in a reaction between molecular oxygen and hydrogen radicals. Its oxidation potential ($E°(HO_2^{\bullet}/H_2O)$) is 1.44 V with respect to the standard hydrogen electrode (Tarr, 2003) which makes it the weakest of all oxidative species produced in plasma. Similarly, as for O, chemical transformations of organic compounds involving $HO_2^{\bullet}$ have not been confirmed experimentally. Finally, while H_2O_2 has a relatively high oxidation potential ($E°(H_2O_2/H_2O) = 1.77$ V with respect to the standard hydrogen electrode (Tarr, 2003)) and could directly oxidize many organic compounds, in practice, these reactions are generally too slow to be of significance (Petri *et al.* 2011). However, indirect decomposition of H_2O_2 via photolysis or metal-based catalytic reactions has been shown to improve removal rates of many organic compounds by yielding additional $^{\bullet}OH$ in the bulk liquid.

12.3.4.1.1 Phenol and derivatives of phenol

Phenol is an important raw material in the production of phenolic resins, which are used in the plywood, construction, automotive, and appliance industries. Phenol is also used in the production of caprolactam and bisphenol A, which are intermediates in the manufacture of nylon and epoxy resins, respectively. The Environmental Protection Agency (EPA) has classified phenol as an extremely hazardous substance and regulates it under the Clean Water Act.

Phenol has been used as a model compound for the evaluation of many AOPs, and work with phenol in plasma processes dates to the early 1990's with many following studies on this compound (Hoeben *et al.* 2000; Joshi *et al.* 1995; Sharma *et al.* 1993). Phenol is a good representative compound for water pollutants because of its high solubility in the liquid, the presence of the aromatic ring, and the relatively well known degradation reaction pathways. However, phenol has some disadvantages for studying specific pathways because it can react with both ozone and hydroxyl radicals, and these reactions depend upon pH and the presence of other compounds in the liquid phase.

Table 12.4, adapted from (Bruggeman & Locke, 2013) summarizes the energy yields (amount reacted per energy input) for phenol degradation with various plasma processes. In general, energy yields are quite low for direct discharges in water. The addition of iron salts, which initiates the plasma-assisted Fenton's reaction or suspension of activated carbon in the solution, as well as the addition of oxygen, can significantly increase the energy yields for phenol oxidation and/or removal from solution. The highest energy yields of $1.6–2.6 \times 10^{-7}$ mol/J (Preis *et al.* 2013) have been reported for a water spray in a pulsed discharge with oxygen as a carrier gas. Various other plasma reactor configurations, discharge phases and working gases (Lukes & Locke, 2005a; Shin *et al.* 2000; Yan *et al.* 2006), operating parameters (Yuan

& Watanabe, 2011), and additives (Grymonpre *et al.* 1999) have been investigated for the degradation of phenol. Summaries of these studies can also be found in several recent reviews and books (Hijosa-Valsero *et al.* 2014; Jiang *et al.* 2014; Joshi & Thagard, 2013b; Parvulescu *et al.* 2012).

Table 12.4 Summary of energy yields (EY) for phenol degradation using different plasma processes with liquid water. Adapted from (Bruggeman & Locke, 2013).

Discharge Type	Discharge Phase	Additives	Gas	EY (mol/J)
Streamer-like	Directly in water	None	None	1×10^{-10}–2×10^{-9}
Streamer-like	Directly in water	None	O_2 bubbles	4×10^{-9}–1×10^{-8}
Streamer-like	Directly in water	Activated carbon	None	8×10^{-9}
Streamer-like	Directly in water	Fenton(iron)	None	8×10^{-10}
Streamer-like	Directly in water	None	O_3	4×10^{-9}
Dielectric barrier discharge	Falling film	None	O_2	3×10^{-9}
Corona	Over water	None	O_2	1×10^{-8}
Corona and streamer-like	Hybrid gas and liquid discharge	Activated carbon	O_2	1×10^{-8}
100 ns pulsed 20 kV	Water spray	None	O_2	3×10^{-7}
Indirect plasma	Bubbled	None	O_3/O_2	5.2×10^{-8} (Hoeben *et al.* 1999)
Dielectric barrier discharge	Falling film	None	Various	$(1–8) \times 10^{-8}$

Overall, the comparisons between plasma processes and other AOPs are fraught with difficulties due to differing solution conditions (pH, conductivity, temperature, presence of impurities and competing reactants) as well as extents of oxidation, levels of mineralization, and lack of knowledge of all plasma properties. Nevertheless, $2–3 \times 10^{-7}$ mol/J appears to be the best achieved energy yield for phenol decomposition in water. Generally, electrical discharges in gases over liquid films or with sprays or droplets are more efficient than direct discharges in the liquid phase. However, discharges directly in water may still have some advantages due to reduction of mass transfer limits, and recent work on nanosecond discharges in water may lead to improvements in energy yields (Banaschik *et al.* 2014).

Various studies identified degradation products and proposed chemical pathways for the degradation of phenol. For plasmas formed directly in water and those contacting water in argon gas, the primary mechanism by which phenol is degraded is •OH attack on the molecule (Bian *et al.* 2011; Lukes, 2001). In the presence of sufficient concentrations of oxygen, phenol may also directly react with ozone, a direct product of O_2 dissociation in plasma (Grabowski *et al.* 2006; Hayashi *et al.* 2000; Sugiarto & Sato, 2001). Phenol oxidation with •OH results in the formation of catechol, hydroquinone and 1,4 benzoquinone, which are the primary hydroxylated byproducts. All three compounds can be further oxidized to trihydroxybenzenes such as pyrogallol, hydroxyhydroquinone, and phloroglucinol (Parvulescu *et al.* 2012; Patai, 1971). Concurrently, intermediate phenyl radicals which are precursors to the formation of catechol, hydroquinone and 1,4 benzoquinone may undergo additional reactions with molecular oxygen to generate ring-opened products such as organic acids (maleic, formic, and oxalic acids are typically identified experimentally) and inorganic acids (Parvulescu *et al.* 2012).

Although the reaction rates of ozone and phenol are comparatively slower to those with •OH, the ozonization of phenol also yields catechol, hydroquinone, and 1,4-benzoquinone as primary hydroxylated aromatic products and additionally *cis, cis*-muconic acid and/or muconaldehyde as secondary products, all of which may undergo further ozonization (Lukes & Locke, 2005b). Lukes and Locke demonstrated the involvement of ozone in plasmochemical reactions by identifying *cis, cis*-muconic acid, the product specific for the ozone attack, for oxygen but not argon plasmas (Lukes & Locke, 2005c). Detailed chemical reaction mechanisms of the •OH and ozone attack on phenol molecule can be found in (Parvulescu *et al.* 2012). In general, the rate of phenol degradation and the types of byproducts formed will depend on the chemical composition of the headspace gas (for gas discharges in contact with liquid), solution pH, and the choice of operating parameters.

The degradation of derivatives of phenol including chlorophenols (e.g., 2-chlorophenol (Dojcinovic *et al.* 2008; Lukes *et al.* 2003; Lukes & Locke, 2005b), 3-chlorophenol (Lukes & Locke, 2005b; Tezuka & Iwasaki, 1998), 4-chlorophenol (Lukes & Locke, 2005b; Tezuka & Iwasaki, 1998), and pentachlorophenol (Brisset, 1997; Sharma *et al.* 2000)), hydroxylated phenols (resorcinol, catechol and hydroquinone (Lukes & Locke, 2005b)), nitrophenols (e.g., 2-nitrophenol (Lukes & Locke, 2005b)), 3-nitrophenol (Lukes & Locke, 2005b), 4-nitrophenol (Gao *et al.* 2004; Lukes & Locke, 2005b), and 2,4-dinitrophenol and 2,4,6-trinitrophenol (Gao *et al.* 2004; Willberg *et al.* 1996a)), and chloronitrobenzene (Liu, 2009), has also been investigated in different plasma systems. Identified products and degradation pathways that have been proposed to explain their formation indicate that the general mechanism for the degradation of substituted phenols is very similar to that of phenol and includes •OH and/or ozone attack, depending on the system. Primary transformation products include hydroxylated aromatic compounds although dehalogenation may also take place (Tezuka & Iwasaki, 1998). Organic acids are usually final degradation products. In the presence of air or nitrogen, nitrated compounds have also been observed (Parvulescu *et al.* 2012; Zhang *et al.* 2007).

When added to the aqueous solutions, homogeneous and heterogeneous catalysts usually enhance compounds' degradation rates. Different catalysts that have been tested for the degradation of phenol and its derivatives and have shown a positive effect on their removal include TiO_2 (Lukes *et al.* 2005; Wang & Chen, 2011), iron (Grymonpre *et al.* 2001; Liu & Jiang, 2005; Yan *et al.* 2006), zeolites (Kusic *et al.* 2005), carbon gels (Sano *et al.* 2006), activated carbon (Grymonpre *et al.* 1999), and carbonate ions (Li *et al.* 2012). In the case of iron, the effect has been attributed to the additional production of •OH in the bulk liquid as well as to direct reactions of phenol and its degradation products with ferrous and ferric ions. Similarly, for TiO_2, the photocatalytic interaction between the plasma-generated UV light and TiO_2 yields additional •OH and O. The mechanisms of action of zeolites, carbon, and inorganic ions are not well understood.

12.3.4.1.2 Organic dyes

The textile industry and its dye-containing wastewater is consider to be one of the main sources of colored aqueous effluents. In many instances, these wastewaters are directly discharged into natural water resources or wastewater treatment systems where, without adequate treatment they can remain for a long period of time. The discharge of colored effluents into the environment is undesirable because of their color as well as the toxic, carcinogenic and mutagenic properties of many dye constituents (Zaharia & Suteu, 2012).

Different dyes have been used as model compounds for the evaluation of numerous plasma systems since the 1980s. The effectiveness of different plasma reactors including electrode configuration, operating parameters, discharge polarity, solution pH and conductivity, and presence of additives has been evaluated for the degradation of numerous dyes (Baroch *et al.* 2008; Benetoli *et al.* 2012; Bozic *et al.* 2004; Burlica *et al.* 2004; Gao *et al.* 2001; Gao *et al.* 2008; Ghezzar *et al.* 2007; Hoeben, 2000; Ikoma *et al.* 2012; Krcma *et al.* 2010; Lu *et al.* 2010; Magureanu *et al.* 2008a; Magureanu *et al.* 2008b; Malik *et al.* 2002; Mededovic & Takashima, 2008; Minamitani *et al.* 2008; Nikiforov *et al.* 2011; Pawlat & Ihara, 2007; Reddy *et al.* 2013; Stara *et al.* 2008; Stratton *et al.* 2015; Sun *et al.* 2012; Xue *et al.* 2008).

Table 12.5 compares the energy yields of 27 different plasma reactor configurations for the treatment of various dyes. The most efficient are pulsed power reactors in which the plasma is formed in a gas and the treated liquid is sprayed into the plasma zone. The highest efficiency (3×10^{-7} mol/J) for indigo carmine dye decolorization occurs in a pulsed corona discharge over a thin water film, and this value is quite close to the highest efficiency discussed above for phenol. High efficiencies of these reactors were explained by the homogeneous distribution of the sprayed liquid droplets along the plasma zone and shorter distances active radicals must travel before they encounter the droplets' liquid surface to diffuse into and react with organic compounds.

As will be demonstrated in Section 12.3.6, the most successful examples of scaled up plasma reactors for water treatment are those in which the contact between the plasma and the treated solution is maximized. Overall, the rate of a dye degradation depends on the chemical composition of the processing gas (oxygen is often, but not always, the best), solution volume where smaller volumes are generally degraded more rapidly, solution pH (depends on the dye type), and initial dye concentration. In almost all cases, the addition of a catalyst such as TiO_2 (Zhang *et al.* 2010b), γ-Al_2O_3-TiO_2 (Liu *et al.* 2012a), zeolites (Vujevic *et al.* 2004), ferrous ions (Magureanu *et al.* 2013a; Reddy *et al.* 2013; Zhou *et al.* 2015), iron minerals (Benetoli *et al.* 2012), and activated carbon (Zhang *et al.* 2010a) enhanced the removal rates of various dyes by producing additional $^{\bullet}OH$. When present, sulfate ions are activated by the plasma and transformed into highly oxidative sulfate radical anions which alongside $^{\bullet}OH$ oxidize the dye (Reddy *et al.* 2013).

In general, the degradation of a dye is monitored by UV-Vis spectroscopy at the wavelength of maximum absorption of the particular dye, which is determined by the chemical composition of the chromophore. The intensity of this absorption decreases when the chromophore is chemically transformed by the oxidative radical and/or molecular attack. A very few studies identified dye degradation products and proposed chemical degradation pathways (Kozakova *et al.* 2010; Magureanu *et al.* 2007; Magureanu *et al.* 2008a; Minamitani *et al.* 2008; Mok *et al.* 2008; Moussa *et al.* 2007; Yan *et al.* 2008). For azo dyes, a class of dyes characterized by the presence of the azo (−N=N−) group, the dye degradation was mainly attributed to the reaction of $^{\bullet}OH$ with the carbon atom attached to the azo group (−C−N=N−) (Parvulescu *et al.* 2012). When oxygen was present and ozone was formed, ozone was assumed to react primarily with the aromatic rings of the dye and not the chromophore group (Parvulescu *et al.* 2012). In some instances, nitrogen-derived compounds such as peroxynitrite and nitrous acid also oxidized the dye (Moussa *et al.* 2007).

More than 150 studies that have investigated the degradation of other classes of dyes such as carbonyl dyes (containing −C=O chromophore), methane dyes (−CH= chromophore) and nitro dyes ($-NO_2$ chromophore) report very different results in terms of degradation mechanism and types of products formed (Parvulescu *et al.* 2012). For some dyes, ozone has been suggested as the main oxidant responsible for destroying the chromophore group with very little contributions of $^{\bullet}OH$ to the overall dye degradation (Grabowski *et al.* 2007). For other types of dyes, the situation is reversed (Velikonja *et al.* 2002). Primary degradation products of a dye degradation are generally derived from aromatic substituents attached to the chromophore group and may include aromatic amines, phenolic compounds, and naphthoquinones as well as various aliphatic and carboxylic acids (Parvulescu *et al.* 2012). From the available literature, it appears that general rules for predicting the dye reactivity with a particular oxidant are lacking. These rules will depend on the plasma system including operating conditions and physicochemical characteristics of the aqueous phase.

Table 12.5 Calculated energy yields and relative energy efficiencies of plasma reactors developed for water purification. Original source data given in (Malik, 2010). Table adapted to include energy yield based on molar amounts. (PCD – pulsed corona discharge, GDE – glow discharge electrolysis, CGDE – contact glow discharge electrolysis, RFD – radio frequency discharge, DD – diaphragm discharge, MWD – microwave discharge, DBD – dielectric barrier discharge, HS – hybrid series, SSD – streamer and spark discharges, SD – spark discharges).

Dye	Molecular Wt.	C_o (mg/L)	Conditions	g/kWh	mol/J
Reactive blue	901	20	PCD in water	0.034	1.0×10^{-11}
Eosin	692	12	GDE in water	0.029	1.2×10^{-11}
Reactive blue 137	901	20	AC-GlidArc in air	0.038	1.2×10^{-11}
Methyl orange	327	13	CGDE in water	0.024	2.0×10^{-11}
Chicago sky blue	993	10	PCD in water	0.081	2.3×10^{-11}
Reactive blue 137	901	20	HS-PCD in water and air	0.091	2.8×10^{-11}
Methylene blue	320	5	RFD in water	0.037	3.2×10^{-11}
Methylene blue	320	12	DD in water	0.042	3.6×10^{-11}
Reactive blue 137	901	20	Water spray in AC-GlideArc in O_2	0.130	4.0×10^{-11}
Rhodamine B	479	10	Pulsed-SSD in water	0.081	4.7×10^{-11}
Methylene blue	320	13	PCD in water	0.064	5.6×10^{-11}
Methyl orange	327	10	Pulsed-SSD in water	0.090	7.6×10^{-11}
Chicago sky blue	993	10	Pulsed-SSD in water	0.328	9.2×10^{-11}
Rhodamine B	479	50	Pulsed-SSD in water	0.202	1.2×10^{-10}
Eosin	691	12	DD in water	0.307	1.2×10^{-10}
Methylene blue	320	10	MWD in water with air bubbling	0.155	1.3×10^{-10}
Indigo carmine	262	20	PCD in air	0.149	1.6×10^{-10}
Methyl orange	327	10	Pulsed-DD with air bubbling	0.262	2.2×10^{-10}
Acid orange 7	350	20	PCD in water with air bubbling	0.370	2.9×10^{-10}
Methylene blue	320	13	PCD in water with O_2 bubbling	0.341	3.0×10^{-10}
Chicago sky blue	993	10	PCD and O_2 bubbling through pinhole	1.20	3.4×10^{-10}
Methyl orange	327	40	PCD in water with O_2 bubbling	0.585	5.0×10^{-10}
Indigo carmine	262	20	PCD in water with air bubbling	0.486	5.2×10^{-10}
Methylene blue	320	15	PCD on water surface in O_2	1.50	1.3×10^{-9}
Indigo carmine	262	20	PCD in water with O_2 bubbling	1.38	1.5×10^{-9}
Methyl orange	327	50	Pulsed-SD in water with air bubbling	1.79	1.5×10^{-9}
Rhodamine B	479	57	DC discharges in air over a 3 mm layer of water	4.86	2.8×10^{-9}
Acid orange 7	350	25	UV and O_3 from DBD in air	3.86	3.1×10^{-9}
Reactive blue 137	901	20	Water spray in pulsed-gliding arc in oxygen	12.7	3.9×10^{-9}
Indigo carmine	262	20	PCD on water surface in Ar	4.84	5.1×10^{-9}
Methyl orange	327	40	HS-PCD multiple pre-treatments in water and air	6.10	5.2×10^{-9}
Methyl orange	327	10	Pulsed-DBD in air over a thin film of water	9.38	8.0×10^{-9}
Rhodamine B	479	22	Water spray in pulsed-DBD in O_2	44.9	2.6×10^{-8}
Indigo carmine	262	20	PCD in air over a thin film of water	294	3.1×10^{-7}
Indigo carmine	262	20	PCD in oxygen over a thin film of water	566	6.0×10^{-7}
Indigo carmine	262	20	Water spray in PCD in air	622	6.6×10^{-7}

12.3.4.1.3 Pharmaceuticals and personal care products

Pharmaceuticals and personal care products (PPCPs), endocrine disrupting compounds (ECD), hormones, phytoestrogens, and other trace organic compounds have been detected in waters receiving treated wastewater (Snyder, 2008). Although wastewater is treated before being discharged into receiving waters, unit processes utilized in water treatment are designed for removal of known pathogens and priority contaminants and not drug residues. As a result, trace organic compounds tend to pass through wastewater treatment systems and can reach concentrations in receiving waters sufficient to produce deleterious biological effects on organisms living in the waters (Tyler *et al.* 1998). An array of these compounds has also been found in the drinking water supplies of at least 41 million people in the (U.S. Water and Wastes Digest, 2008).

Various types of electrical discharge plasma systems have been used to degrade PPCPs, ECDs, and other trace organic contaminants. The results of these investigations are summarized in Table 12.6, which reports the type and the initial concentration of the compound treated, treated volume, treatment time, % compound degraded, and energy costs. While the majority of the systems effectively degrade a broad spectrum of contaminants, the treatment times and energy costs vary significantly. It appears that the best performing plasma systems are again those in which gas plasma is contacting the liquid. For that purpose, mainly three gases were used: air, oxygen and argon. Both air and oxygen produce ozone, which enhances removal rates of many organics. Discharges in air additionally produce nitrogen and nitrogen oxide species and radicals, which may also participate in chemical reactions. Contrary to these general findings, in the case of pentoxifylline, Magureanu *et al* (Magureanu *et al.* 2010) found that only a small amount of ozone generated in the plasma is consumed in chemical reactions and that this compound is primarily oxidized with $^{\bullet}OH$. In the case of carbamazepine, Liu *et al.* (Liu *et al.* 2012b) found that for the same concentration of ozone, degradation of this compound is significantly faster for the ozonization process than the plasma process. Degradation products for the two processes were also different. This inhibitory effect of the plasma was attributed to the oxidation of carbamazepine not by ozone but by nitrogen radicals. Figure 12.3 compares the carbamazepine ozone and nitrogen oxidation pathways.

12.3.4.1.4 Bisphenol A

Bisphenol A (BPA) is an emerging contaminant widely used in manufacturing of epoxy, unsaturated polyester-styrene, polycarbonate resins, and flame retardants. BPA is released into the environment through industrial and municipal wastewater discharges and through landfill leachates (Chakma & Moholkar, 2014). BPA has been detected in baby feeding bottles, plastic waste, polycarbonate tubing, as well as surface water, and may adversely affect humans and aquatic wildlife via interactions with the endocrine system (Flint *et al.* 2012). BPA is active at very low concentrations and not easily degraded by conventional drinking water treatment processes. Additionally, many of BPA's degradation byproducts exhibit greater estrogenic activity than BPA itself.

A limited number of studies have assessed the effectiveness of plasma treatment in degrading BPA (Abdelmalek *et al.* 2008; Dai *et al.* 2015). Interestingly, they all compared the plasma process with the plasma-assisted Fenton process in which ferrous salts were added to the aqueous solution. As expected, the addition of iron salts improved the removal rate of BPA. This effect has been attributed to Fenton's reaction between Fe^{2+} and H_2O_2 produced by the electrical discharge. Different studies investigated the effect of varying ferrous (Fe^{2+}) and ferric ions (Fe^{3+}) and H_2O_2 concentration on BPA removal using different electrolytic solutions and working gases (Abdelmalek *et al.* 2008; Dai *et al.* 2015; Wang *et al.* 2008). Iron grounded electrode instead of iron salts has also been used to initiate the Fenton's reaction (Dai *et al.* 2015).

Table 12.6 Comparison of treatment efficacies and treatment parameters for the degradation of organic compounds. Adapted from (Banaschik *et al.* 2015).

Substances Investigated	Concentration	Treated Volume	Plasma System	Degradation and Treatment Time	Energy Cost	Reference
Clofibric acid Carbamazepine Iopromide	0.1 mM ((22–27) mg/L)	200 mL	Two barrier electrodes	>98%, 30 min	500 W	(Krause *et al.* 2009)
Meprobamate Dilantin Primidone Carbamazepine Atenolol Trimethoprim Atrazine	36–378 ng/L	150 L	Needle-plate in contact with water	>90%, 19 min	2.6–6.4 kWh/m^3 (EEO[a])	(Gerrity *et al.* 2010)
Pentoxifylline	100 mg/L	200 mL	Coaxial DBD[b]	>92.5%, 60 min	16 g/kWh	(Magureanu *et al.* 2010)
Amoxicillin Oxacillin Ampicillin	100 mg/L	200 mL	Coaxial DBD[b]	>90%, 30, 10, 20 min	27,105, 29 g/kWh	(Magureanu *et al.* 2011)
Carbamazepine	20 mg/L	100 mL	Coaxial DBD[b]	90.7%, 60 min	12 W	(Liu *et al.* 2012b)
Paracetamol β-oestradiol Salicylic acid Indomethacin Ibuprofen	3–100 mg/L	40–50 L	PCD[c]	70–99%, 30 min	1.5–150 g/kWh	(Panorel *et al.* 2013a; Panorel *et al.* 2013b)
Enalapril	50 mg/L	300 mL	Coaxial DBD[b]	>90%, 20 min	20.66 g/kWh	(Magureanu *et al.* 2013b)
Atrazine Chlorfenvinphos 2,4-dibromophenol Lindane	1–5 mg/L	175 mL	Coaxial DBD[b]	87–89%, 5 min	47–447 mg/kWh	(Hijosa-Valsero *et al.* 2013)
Diclofenac	50 mg/L	55 mL	PCD[c]	>99%, 15 min	0.76 g/kWh (90% conversion)	(Dobrin *et al.* 2013)
Carbamazepine Diatrizoate Diazepam Diclofenac Ibuprofen Trimethoprim 17α-ethinylestradiol	500 µg/L	300 mL	Coaxial directly in water	55–99%, 66 min	3–45 mg/kWh	(Banaschik *et al.* 2015)

[a]The EEO-value is defined as the amount of electrical energy (kWh) required to reduce the concentration of a pollutant by one order of magnitude (90%).
[b]Dielectric Barrier Discharge.
[c]Pulsed Corona Discharge.

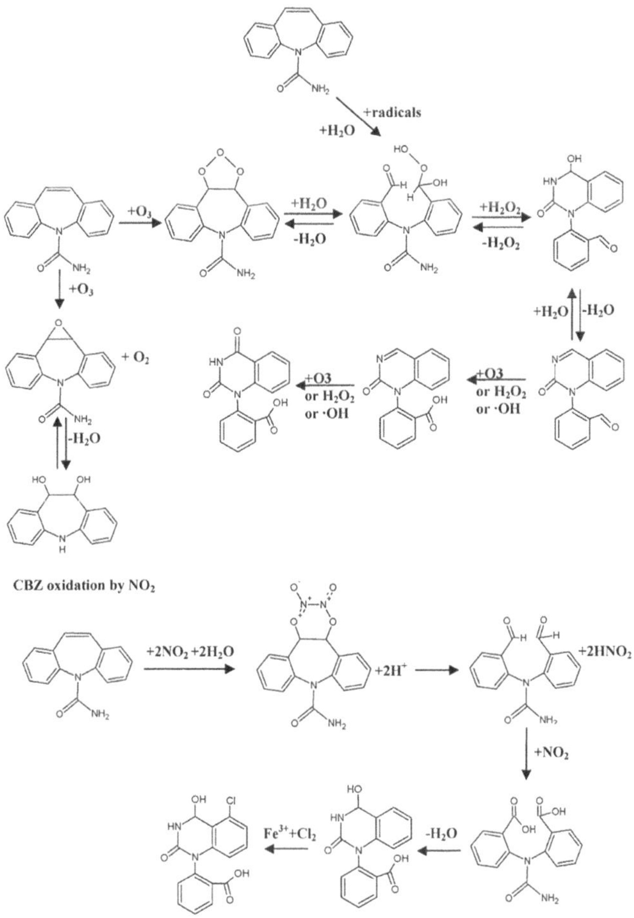

Figure 12.3 Carbamazepine degradation mechanism. Reprinted with permission from (Liu *et al.* 2012b). Copyright 2012 Elsevier.

Various BPA degradation products that have been identified indicate that peroxide-based AOPs form different, partially hydroxylated BPA analogues. Using Density Functional Theory (DFT) simulations and experimentally identified degradation byproducts, Dai *et al.* (Dai *et al.* 2015) proposed a general pathway by which BPA is degraded in the presence of trace amounts of air in an electrical discharge plasma reactor. The overall mechanism is shown in Figure 12.4 and the most likely steps (based on calculated activation energies and reaction rate constants) can be summarized as follows: (1) $^{\bullet}OH$ abstracts the H atom from the BPA's phenolic OH group and generates a BPAPXR radical, (2) BPAPXR radical recombines with OH and NO_2 radicals to form BPA2OH, BPA2NO2 and single-ring aromatic compounds, and (3) BPA2OH and BPA2NO2 undergo reactions similar to BPA to form a myriad of byproducts. In that study, all the identified byproducts have been shown to exhibit lower estrogenic activity than BPA. Dai and coauthors (Dai *et al.* 2015) also concluded that the Fenton-like plasma reaction system which utilizes the grounded electrode as a source of iron may be a viable treatment option for small water treatment systems looking to reduce concentrations of organic compounds in the absence of externally supplied chemicals.

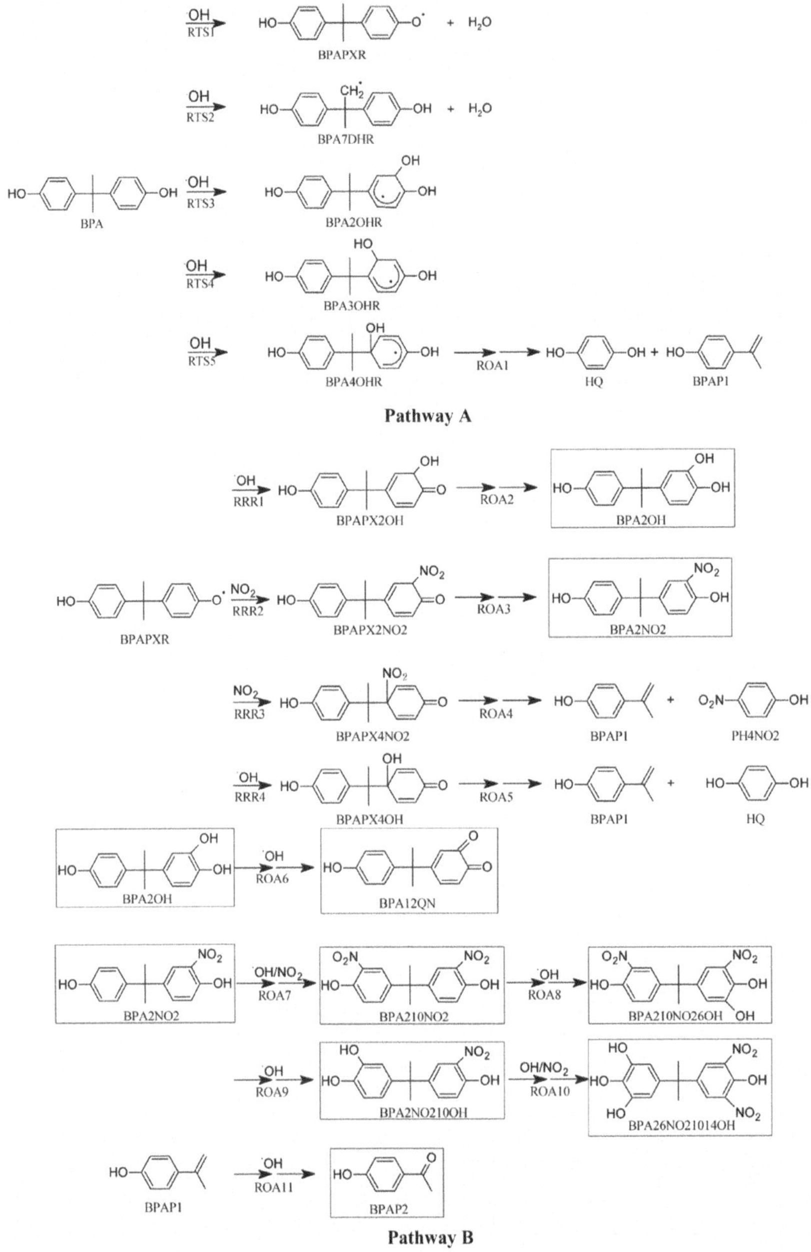

Figure 12.4 Suggested reaction pathways of BPA degradation by the Fenton-like plasma treatment. **Pathway A**: Primary degradation steps. **Pathway B**: Secondary and further degradation steps. Label RTS indicates that the reaction has a transition state which is a prerequisite for calculating activation energy and reaction rate constant. Label RRR indicates radical recombination reactions, which do not have transition states. ROA denotes an overall reaction. Squared structures have been identified experimentally. From Dai *et al.* (2015).

12.3.4.2 Reduction reactions

Direct collision of water molecules with electrons yields $H^•$, powerful reducing agents ($E°(H_2O/H^•) = -2.30$ V (Tarr, 2003)) which have been observed experimentally using optical emission spectroscopy. The chemical reactions of $H^•$ with organic compounds include hydrogen addition and hydrogen abstraction. For the hydrogen abstraction from an aliphatic compound or an addition to an aromatic ring, the reactions of $H^•$ usually follow the same pattern as those of $^•OH$. However, the reaction rate constants for the abstraction are usually higher for $^•OH$ than for $H^•$, and the selectivity of $^•OH$ is lower. Superoxide radical anion ($O_2^{•-}$) is perhaps the weakest reductive species constituting the plasma with $E°(H_2O/O_2^{•-})$ equal to −0.33 V with respect to the standard hydrogen electrode (Tarr, 2003). This species is in an acid-base equilibrium with hydroperoxyl radicals ($HO_2^•$) and will only be available in appreciable concentrations when pH is above 4.8 ($pK_A = 4.88$). Although $O_2^{•-}$ has been identified and quantified (Sahni & Locke, 2006c), there is limited direct experimental evidence that this species contributes to the degradation of organic molecules other than the reductive radical probe species. The superoxide radical anion has traditionally been viewed as a weak or slow reacting radical that is capable of degrading highly oxidized organic compounds including chlorometanes and nitromethanes (Siegrist *et al.* 2011).

Electrical discharges in and contacting water yield aqueous electrons (e_{aq}^-), the strongest reductive species in the plasma ($E°(H_2O/e_{aq}^-) = -2.77$ V with respect to the standard hydrogen electrode (Tarr, 2003)). For direct-in-liquid discharges this species has been identified in a reaction with silver ions (Mededovic Thagard *et al.* 2009), and for gas discharges contacting water detected directly using optical absorption spectroscopy (Rumbach *et al.* 2015). Aqueous electrons react with numerous organic compounds, and the extent of reactivity is usually determined by the electron density of a specific functional group. Research has implicated chemical reduction with aqueous electrons playing an important role in the degradation of acetophenone (Wen & Jiang, 2001). PCBs (Sahni *et al.* 2005) and trichloroethylene may degrade through reductive reactions but the pathways are not fully established (Sahni *et al.* 2002). Perhaps the most convincing case demonstrating the involvement of aqueous electrons in the degradation of pollutants is demonstrated on the plasma-assisted degradation of perfluoroalkyl substances (PFASs) described next.

PFASs have recently received considerable attention due to their ubiquitous presence and recalcitrance in the environment, and toxic properties (Houtz *et al.* 2013). Manufacture and disposal of PFAS-containing formulations and products, and use of aqueous firefighting foams at various sites has resulted in PFAS contamination of groundwater and drinking water supplies (Guelfo & Higgins, 2013). In particular, the presence of perfluoroalkyl acids (PFAAs, e.g., perfluorooctanoic acid (PFOA), $C_7F_{15}COOH$, and perfluorooctane sulfonate (PFOS), $C_8F_{17}SO_3H$) is problematic due to the lack of highly effective treatment technologies.

Due to the presence of the strong carbon-fluorine bond, only sulfate radical anions can oxidize PFOA and PFOS (Appleman *et al.* 2014). The oxidation with $^•OH$ is kinetically hindered whereas the chemical reduction of PFASs by either aqueous electrons (Zhang *et al.* 2014) or superoxide radicals (Schaefer *et al.* 2015) proceeds readily. However, the best treatment times for the majority of the oxidative and reductive processes in which these compounds are degraded below their detection limits are hours to tens of hours per liter of volume treated (Yasuoka *et al.* 2011).

Stratton *et al.* developed a plasma reactor capable of completely degrading PFOA in less than 30 minutes (Stratton *et al.* 2015). The reactor, laminar jet with bubbling, features a gas discharge in contact with foam created by bubbling argon gas through a submerged diffuser. For the PFOA removal, the plasma-based process was adapted for two cases, high removal rate and high removal efficiency. With a 30 minute treatment, the high rate process reduced the PFOA concentration in a 1.4 L solution by 90% with an input power of 76.5 W while the high efficiency process reduced the concentration by 25% using 4.1 W (unpublished).

As presented in Table 12.7, both plasma-based water treatment (PWT) processes have been shown to be superior in terms of removal and mineralization efficiencies compared to leading alternative technologies:

sonolysis, activated persulfate, and electrochemical treatment including the existing DC-driven plasma treatment in O_2 bubbles. For the high efficiency PWT, the PFOA removal efficiency, expressed as the ratio of the observed first-order removal rate constant (k_{obs}) and the power density (PD = input power/treated volume), was about eight times more efficient than activated persulfate, more than 57 times more efficient than sonolysis, 3.6 times more efficient than electrochemical treatment and more than 200 times more efficient than DC plasma treatment. Other performance indicators and corresponding experimental parameters for these five processes are also shown in Table 12.7. In addition to the PFOA removal efficiency, the efficiency of mineralization is important because degradation byproducts may still be harmful to the environment and be more mobile than the parent compound. The extent of mineralization can be represented by the degree of defluorination (Vecitis *et al.* 2009), thus the rate of mineralization is represented by F_{50}/t_{50}, where t_{50} is the time it takes to reduce the PFOA concentration by 50% ($t_{50} = -\ln(1/2)/k_{obs}$) and F_{50} is the percentage of fluorine removed from the degraded PFOA and detected as F^- ions. The mineralization efficiency is defined as (F_{50}/t_{50})/PD. Both PWT processes outperformed the alternative treatment methods with high efficiency PWT attaining mineralization efficiency about 29 times greater than that of activated persulfate, 11 times greater than that of sonolysis, 14% greater than electrochemical treatment and about 40 times greater than DC plasma treatment.

Table 12.7 Performance indicators and corresponding experimental parameters for high rate PWT (unpublished), high efficiency PWT (unpublished), UV-activated persulfate, sonolysis, electrochemical treatment, and DC plasma treatment. Data for UV-activated persulfate and sonolysis are from (Vecitis *et al.* 2009), data for electrochemical treatment is for a current density of 10 mA/cm^2 from (Schaefer *et al.* 2015) and data for DC plasma in bubbles from (Yasuoka *et al.* 2011). Unpublished and provided by courtesy of Gunnar Stratton, Plasma Research Laboratory, Clarkson University.

Treatment	$[PFOA]_0$ (μM)	PD (W/L)	k_{obs} (min^{-1})	$\frac{k_{obs}}{PD}\left(10^{-4}\cdot\frac{min^{-1}}{W/L}\right)$	$\frac{F_{50}}{t_{50}}\left(\frac{\%}{min}\right)$	$\frac{F_{50}/t_{50}}{PD}\left(10^{-2}\cdot\frac{\%/min}{W/L}\right)$	EY ($10^{-11}\cdot$ mol/J)
High rate PWT	20	54.6	0.074	14	2.08	3.8	140
High efficiency PWT	20	2.90	0.012	41	0.31	11	45
Sonolysis	20	250	0.018	0.72	2.47	0.99	2.4
UV-activated persulfate	50*	23	0.012	5.2	0.09	0.38	43
Electrochemical treatment	0.031*	5.0	0.0057	11	0.47	9.5	0.059
DC plasma in O_2 bubbles	100	1550	0.030	0.20	4.4	0.28	3.3

*Performance may be sensitive to initial PFOA concentration, thus comparisons are approximate.

Further investigation into the types of reactive species responsible for degrading PFOA confirmed that the traditional primary oxidants, hydroxyl and superoxide radicals, play no significant role. Instead it was found that aqueous electrons account for a sizable fraction of the degradation, with the rest attributed to free electrons and/or argon ions within the plasma interior. The involvement of aqueous electrons was confirmed in scavenging experiments with sodium nitrate, a well-known e_{aq}^- scavenger. The chemical reaction mechanisms that have been suggested to explain the transformation of PFAAs are generally hypothetical and, in some cases, contradictory. Nevertheless, those involving aqueous electrons agree on stepwise defluorination. A synergistic action between $^•OH$ and aqueous electrons has also been proposed. Different treatment processes, including PWT, identified F^- and shorter chain PFAAs such as $C_3F_7COO^-$ (PFBA), $C_4F_9COO^-$ (PFPnA), $C_5F_{11}COO^-$ (PFHxA), and $C_6F_{13}COO^-$ (PFHpA) as the transformation products of PFOA (Hori *et al.* 2004; Moriwaki *et al.* 2005; Yasuoka *et al.* 2011).

12.3.4.3 Other compounds

In the last 30 years, a wide variety of other EPA regulated and unregulated compounds have been degraded in electrical discharge plasma reactors (Hijosa-Valsero *et al.* 2014). These include formic acid (Grinevich *et al.* 2009), acetic acid (Bobkova *et al.* 2012; Sano *et al.* 2004), aniline (Tezuka & Iwasaki, 2001), gasoline components and its additives (Even-Ezraa *et al.* 2009; Johnson *et al.* 2003), methanol (Burlica *et al.* 2011; Thagard Mededovic *et al.* 2009; Yan *et al.* 2009), ethanol (Chernyak *et al.* 2008; Levko *et al.* 2011), cresols (Tomizawa & Tezuka, 2006), terephtalic acid (Sahni & Locke, 2006a), benzoic acid (Tezuka & Iwasaki, 1999), xylene (Gai, 2009; Manolache *et al.* 2004), dimethylsulfoxide (DMSO) (Lee *et al.* 2006), tetranitromethane (TNM) (Sahni & Locke, 2006c), n-hexane (Bresch *et al.* 2016), trichloroethylene (Even-Ezraa *et al.* 2009), chloroform and tributyl phosphate (Moussa & Brisset, 2003), trinitrotoluene (Lang *et al.* 1998), 4-aminodiethylaniline sulphate (Kravchenko *et al.* 2002), pentadecafluorooctanoic acid (Horikoshi *et al.* 2011), Forafac 1110, a perfluorinated non-ionic surfactant (Maroulf-Khelifa *et al.* 2008), sulfonol (Bobkova *et al.* 2015), pesticide nitenpyram (Li *et al.* 2013), s-triazine (Mededovic *et al.* 2007), and herbicide atrazine (Karpel Vel Leitner, 2005; Mededovic & Locke, 2007b).

12.3.5 Thermal plasma chemistry in direct water discharges

In addition to the electron-induced dissociation of water and organic contaminants, in plasmas with very high temperatures (>2000–4000 K) thermal dissociation can be important or even the predominant pathway for the formation of products. This is, for example, shown in Figure 12.5 for an aqueous solution of acetonitrile which during a pulsed direct-in-liquid electrical discharge thermally decomposes to form carbon-based atomic and diatomic excited radical products (Franclemont & Thagard, 2014, supplementary material). Similar spectra have been recorded for aqueous solutions of high concentrations (>0.1 M) of numerous other compounds (Franclemont & Thagard, 2014).

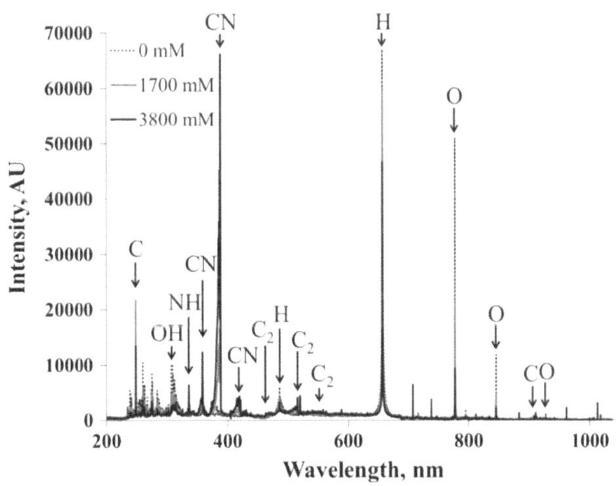

Figure 12.5 Optical emission spectra of plasma in an aqueous solution containing different concentrations of acetonitrile. From (Franclemont & Thagard, 2014, supplementary material).

While oxidation of organic compounds usually yields hydroxylated products (and eventually CO_2 and water), thermal degradation results in the formation of short-chained hydrocarbons. Indeed, in the case of underwater plasma treatment of an aqueous phenol solution (~19 g/L), the gaseous products that are formed include CO, CO_2, CH_4, C_2H_2, C_2H_4, and C_2H_6 (Fan & Mededovic Thagard, 2014). Furthermore, for very low phenol concentrations (< 1 mM), it has been shown that 75%–80% of the degradation occurs in the bulk liquid by radical reactions while at 10 mM, 50% of the degradation occurs by thermal pyrolysis (Parvulescu *et al.* 2012). For underwater electrical discharges it has been found that all organic compounds, regardless of their vapor pressure, readily diffuse into the plasma where they are oxidized and thermally degraded. For identical bulk liquid molar concentrations, it has been shown that higher concentrations of hydrophobic (those with higher K_{ow}, octanol-water partition coefficient) than hydrophilic compounds are found in the plasma interior (Franclemont & Thagard, 2014).

Finally, in some thermal plasma reactors the generation of large spatial and temporal gradients during the plasma propagation can lead to enhancements in the formation of stable molecular species such as H_2, O_2, and H_2O_2, for example (Fridman, 2008; Locke & Mededovic-Thagard, 2009). However, in very high temperature conditions species such as H_2O_2 are not stable, and hydroxyl radical reactions as well as pyrolysis may be more significant (Hoffmann *et al.* 1997; Willberg *et al.* 1996b).

12.3.6 Plasma process scale-up

Despite promising results of bench-scale studies and evident potential advantages of plasma water treatment, the technology has not yet reached a high enough level of development to be used in practice. A major obstacle hindering the progress in developing plasma reactors is associated with the absence of general principles to guide the design of new reactors with higher efficiency. While the most effective plasma reactors have been identified (Malik, 2010), the fundamental reasons behind their superior performances are not clear.

In an attempt to improve the feasibility of plasma-based water treatment technology and develop basic guidelines for reactor design and optimization, Stratton *et al.* (Stratton *et al.* 2015) conducted a study to identify and characterize design parameters and physical phenomena that influence treatment efficiency. The results of their study revealed that the single most important factor affecting the performance of a plasma reactor is the size of the contact area between the plasma and the treated solution. Building on this knowledge, they developed several "contact-oriented" reactors in which the goal was to maximize the plasma-liquid contact area. The best performing reactor, laminar jet with bubbling, was tested for the degradation of Rhodamine B dye. The laminar jet achieved a G_{50} of 11.4 g/kWh, which is nearly 150 times that of the liquid discharge and significantly greater than the Rhodamine B removal efficiency of other AOPs (Table 12.8).

Table 12.8 G_{50} and energy yield (EY) values for Rhodamine B dye degradation with different AOPs. Adapted from (Stratton *et al.* 2015).

AOP	G_{50} (g/kWh)	EY (10^{-11} · mol/J)
UV/H_2O_2	0.1	5.8
Ultrasound	0.2	12
Photocatalysis	0.2	12
Hydrodynamic cavitation	0.01	0.58
Ozonation	0.3	17
Electrical discharge plasma: liquid discharge (worst performing reactor)	0.078	4.5
Electrical discharge plasma: laminar jet with bubbling (best performing reactor)	11.4	660

Stratton and Mededovic Thagard have also scaled up the laminar jet reactor to yield the foaming plasma reactor shown in Figure 12.6 (Mededovic & Stratton, 2015). This reactor was tested for the treatment of groundwater containing perfluorinated compounds (PFCs) acquired from a contaminated naval research site in Warminster, PA. The results are shown in Table 12.9. The PFOA concentration was reduced to below its detection limit during one minute of treatment. With the exception of PFPnA, a 60-minute treatment resulted in a complete transformation of all other PFCs.

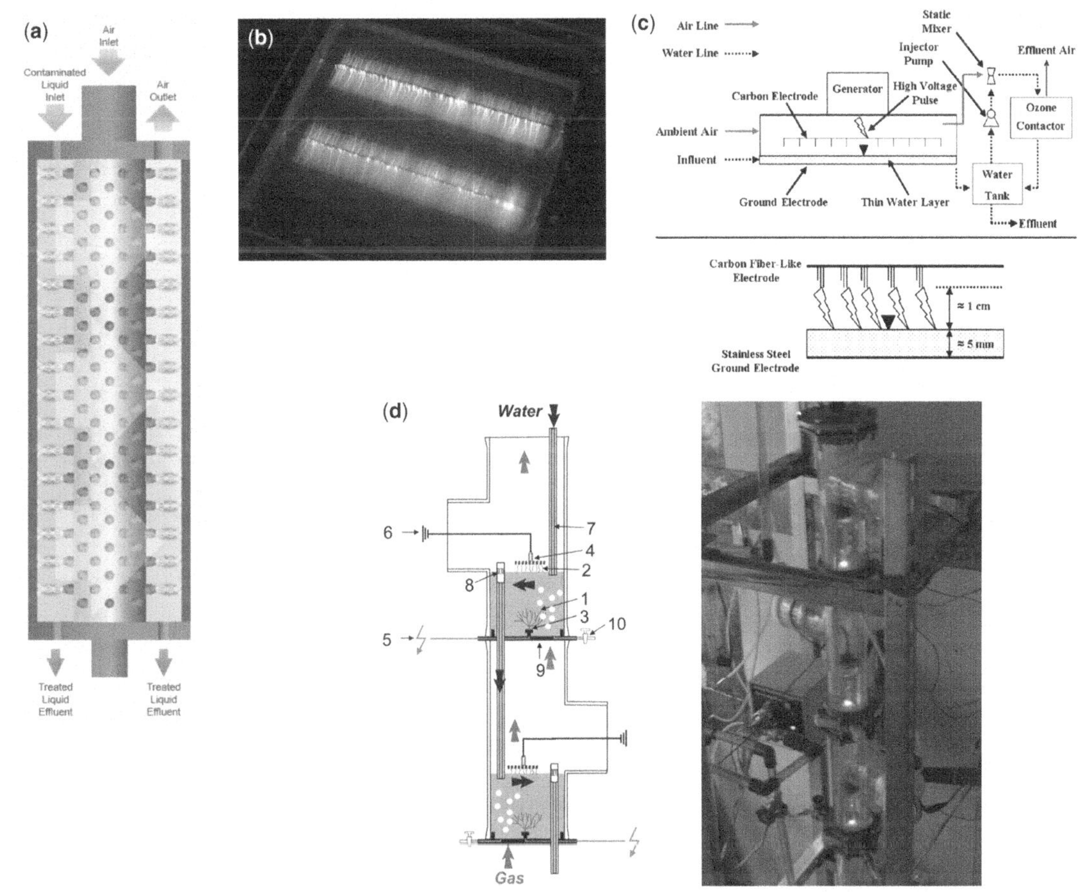

Figure 12.6 Scaled-up plasma reactors. (a) Commercial Symbios plasma reactor (www.symbiosplasma.com), (b) Foaming plasma reactor, Thagard Plasma Research Group, Clarkson University, (c) Pilot-scale gas discharge plasma reactor. Reprinted with permission from (Gerrity *et al.* 2010). Copyright 2010 Elsevier, and (d) Multistage gas-liquid electrical discharge column reactor. Reprinted with permission from (Holzer & Locke, 2008). Copyright 2008 American Chemical Society.

In the realm of wastewater treatment, particularly PFC-containing wastewater, adsorption processes based on activated carbon (AC) is the primary method for the removal of these compounds. AC is effective for the removal of many hazardous chemicals and surfactants, including PFCs; however, as shown in Table

12.10, treatment costs are high as a result of the need for post-treatment of saturated AC (Menger-Krug, 2011). There are as many as 10 other alternative treatment strategies proposed by various research groups for the degradation of PFOA and PFOS (Hori *et al.* 2015; Mitchell *et al.* 2014; Park *et al.* 2009; Zhang *et al.* 2014), but sonolysis (Cheng *et al.* 2008), the most effective/efficient of these processes (acknowledged as a potential treatment technology by the USEPA), is 35–73 times more costly than the foaming reactor plasma-based process (Table 12.10).

Table 12.9 Treatment of PFC-contaminated groundwater (naval research site, Warminster, PA) by the foaming plasma reactor (treatment volume = 4 L, input power = 100 W, gas flux = 35 mL/cm²min). Unpublished and provided by courtesy of Gunnar Stratton, Plasma Research Laboratory, Clarkson University.

Compound	$C_{0\,min}$ (μg/L) Direct Injection	$C_{1\,min}$ (μg/L) Direct Injection	$C_{60\,min}$ (μg/L) Solid-Phase Extraction	Removal (%)
Perfluorooctanoic acid (PFOA)	0.89	Below detect. limit	0.0035	99.6
Perfluorooctane sulfonate (PFOS)	0.18	N/A	0.0026	98.5
Perfluoroheptanoic acid (PFHpA)	0.11	N/A	0.0002	99.8
Perfluorohexanoic acid (PFHxA)	0.27	N/A	0.024	91.1
Perfluoropentanoic acid (PFPnA)	0.22	N/A	0.16	26.4
Perfluorohexane sulfonate (PFHxS)	0.32	N/A	0.0041	98.7

Table 12.10 Cost estimates for 75% removal of PFOA (initial concentration of 95 ng/L) and PFOS (initial concentration of 135 ng/L) using AC (Menger-Krug, 2011), the plasma-based process (Stratton *et al.* 2015) and the current top-performing alternative (Vecitis *et al.* 2009). Unpublished and provided by courtesy of Gunnar Stratton, Plasma Research Laboratory, Clarkson University.

Process	Contaminant	Treatment Costs (million $/kg)
AC	PFOA	5.5
	PFOS	4.4
Plasma	PFOA	2.6
	PFOS	1.5
Sonolysis	PFOA	190
	PFOS	320

Other work on plasma reactor scale up includes the development of the multistage gas-liquid column reactor (Holzer & Locke, 2008), tubular high density plasma reactor (Johnson *et al.* 2006), and multiple point-plane electrode reactor system (Gerrity *et al.* 2010), all shown in Figure 12.6. Holzer and Locke (Holzer & Locke, 2008) tested the performance of the multistage column reactor for the degradation of Reactive Blue 137 dye, phenol, 2 chloroethyl phenyl sulfide (2-CEPS), and diphenyl chlorophosphate (DPCP). The best removal energy yields were achieved while operating the reactor in the recirculation mode (as opposed to operating it in continuous flow mode). The energy yields were 2.08, 0.88, 2.42, and 0.63 g/kWh for the dye, phenol, 2-CEPS, and DPCP. After adding ferrous ions to the solution, the values for 2-CEPS and DPCP increased to 3.02 and 2.15 g/kWh, respectively.

The multiple point-plane electrode reactor was tested for the destruction of trace organic compounds such as pharmaceuticals and endocrine disrupting compounds (Gerrity *et al.* 2010). Based on the calculated energy efficiency index EEO (energy required for one-log contaminant removal in kWh/m^3-log), the plasma system was found to be similar to or more efficient than a pilot-scale UV/H_2O_2 process (Table 12.11). Because plasma reactors consume similar or less energy than other AOPs and do not require the addition of chemicals, the authors concluded that plasma-based technology is a viable alternative to the conventional AOPs.

Table 12.11 AOP-specific EEO values (kWh/m^3-log). EEO = electrical energy per order of magnitude degradation. Adapted from (Gerrity *et al.* 2010).

Contaminant	Batch Wastewater Process	Single-Pass Wastewater Process at Medium Energy Consumption (WW)	Single-Pass Spiked Surface Water Process at Medium Energy Consumption (SSW)	UV SSW	UV/H_2O_2 SSW	UV/TiO_2 SSW
Meprobamate	6.4	10	3.5	6.6	1.0	6.8
Dilantin	4.4	3.1	2.0	2.1	1.0	2.2
Primidone	5.8	4.3	2.2	3.7	0.6	3.9
Carbamazepine	2.2	1.8	< 0.7	2.3	0.4	2.1
Atenolol	3.8	2.9	1.0	1.4	0.5	2.0
Trimethoprim	3.2	2.2	< 0.7	0.8	0.4	1.5
Atrazine	N/A	N/A	3.7	3.3	1.2	4.7

12.3.7 Inactivation of biological species

Numerous methods for inactivating and killing various disease-causing microbes including bacteria, yeast, and viruses have been developed. The conventional processes rely on UV whereas other, emerging technologies cause inactivation by high energy radiation, high temperature and pressure, and various toxic chemicals. Plasma processes have been extensively studied for decontamination of water and solid surfaces for environmental and medical applications (Laroussi, 1996; Laroussi, 2002; Lukes *et al.* 2012). The reactive oxygen species formed in plasma processes as well as in other advanced oxidation methods are effective in inactivating many microbes through damage to DNA, and plasma processes may also have simultaneous and synergistic effects of high electric fields, UV light emissions, and in some cases shock wave formation in a liquid. The reaction pathways for both reactive oxygen species and reactive nitrogen species with various biomolecules that make up living cells have been recently reviewed (Lukes *et al.* 2012). Table 12.12 compares different plasma processes for the inactivation of *E. coli* where the D value (in J/ml per log order inactivation) indicates the relative efficiency. DBD and spark discharges in water appear to be the best and pulsed corona in water the worst plasma inactivation processes. Data on a wide range of other microorganisms have been reported and some are summarized in different reviews (Bruggeman & Locke, 2013; Lukes *et al.* 2012).

Table 12.12 Summary of *E. coli* inactivation by various plasma processes (*D value represents energy necessary (J) for 1 log reduction of cells per ml liquid). Adapted from (Bruggeman & Locke, 2013).

Plasma	D-Value* (J/ml)	Initial Bacterial Density (CFU/ml)	Reference
Surface discharge	0.3	10^6	(Anpilov *et al.* 2004)
Pulsed corona in water	3	N/A	(Abou-Ghazala *et al.* 2002)
Low frequency AC in air	23	10^5–10^6	(Chen *et al.* 2008)
Pulsed arc in water	18.7	10^7	(Ching *et al.* 2001)
DBD in air (bubbling)	0.29	N/A	(Gemo *et al.* 2012)
Pulsed corona in water	33.3	10^4–10^5	(Dors *et al.* 2008)
Pulsed arc in water	2.1	10^5–10^6	(Efremov *et al.* 2000)
Pulsed corona in water	45	10^6–10^7	(Fudamoto *et al.* 2008)
Capillary discharge in water	5.4	10^7	(Hong *et al.* 2010)
Corona in water	18	10^5	(Lukes *et al.* 2007)
PEF	<5	10^5	(Mazurek *et al.* 1995)
Streamers in air bubbles	13	10^5–10^6	(Marsili *et al.* 2002)
PEF	40	10^8–10^9	(Ohshima *et al.* 1997)
Spark-arc	0.6	10^6	(Rutberg, 2002)
Spark arc	1	4×10^4	(Rutberg *et al.* 2007)
Pulsed corona in air	0.1	10^7–10^8	(Satoh *et al.* 2007)
Low voltage 10 µs pulsed discharge in liquid	158	2.5×10^5	(Sakiyama *et al.* 2009)
Surface streamers	8.6	10^7	(Shmelev *et al.* 1996)
Spark discharges in water	0.1–0.4	10^4–10^6	(Yang *et al.* 2011)
Packed-bed air bubble discharge	9	10^6	(Zhang *et al.* 2006)

12.4 CONCLUSIONS

Plasma processes can be used both indirectly (to generate ozone) and directly to treat contaminated water. Applications include chemical destruction and biological inactivation. Indirect non-thermal plasma processes are the most efficient means to generate ozone from either pure oxygen or air. The first Advanced Oxidation Process was the treatment of water by ozone generated in a dielectric barrier plasma reactor, and in the last century there have been continuing further advances to increase the energy yields of ozone. Of significant note is recent work on nanosecond pulse discharges in the laboratory which have raised yields to close to 60% of the thermodynamic limit. Further development of these highly efficient ozone generators promises to provide significant improvements in Advanced Oxidation Process applications in general since ozone is one of the most effective means to oxidize organic compounds.

Many types of plasma processes have been developed to directly contact the plasma with the liquid phase containing a contaminant. The most efficient of these processes utilize gas phase plasma that contacts the liquid in the form of droplets or thin films. These plasma processes can generate $^{\bullet}OH$ and degrade model compounds (e.g., phenol, dyes, pharmaceuticals and personal care products, etc.) at energy yields of similar magnitude to or higher than conventional AOPs. Of particular note is that such plasmas may provide efficient means to generate H_2O_2 from water at competitive energy yields and that utilization of plasma with air leads to significant production of reactive nitrogen species which have antimicrobial effects. The continuing development of advanced power supplies for nanosecond pulse discharges in both the gas and the liquid phase promises to improve energy yields for not only ozone generation, but chemical degradation and biological inactivation.

In general, the performance of a plasma reactor will depend on the electrode geometry, discharge phase (gas vs. liquid), input power, means of power delivery, volume of the treated liquid, and the type and initial compound concentration. Research indicates that the best performing plasma reactors feature the following characteristics: (1) gas discharge is contacting the liquid as a spray and/or jet. Alternatively, gas bubbles are present in the liquid, (2) argon and/or oxygen are used as processing gases, and (3) thin layers of water are being treated. The plasma technology appears to be superior for the treatment of molecules that exhibit surfactant-like behavior and competitive for the removal of trace organic contaminants.

12.5 ACKNOWLEDGEMENTS

Dr. Bruce R. Locke would like to thank Dr. Kevin Hsieh for supplying the emission spectra and assistance with some data tables and both Dr. Hsieh and Mr. Robert Wandell for helpful discussions. Dr. Selma Mededovic Thagard would like to thank Mr. Gunnar Stratton for assistance with selected data tables.

12.6 REFERENCES

http://www.Symbiosplasma.Com/html/people.Php

www.Epa.Gov

(2008). U.S. Water and wastes digest, http://www.Wwdmag.Com/

Abdelmalek F., Torres R., Combet E., Petrier C., Pulgarin C. and Addou A. (2008). Gliding arc discharge (GAD) assisted catalytic degradation of bisphenol a in solution with ferrous ions. *Separation and Purification Technology*, **63**(1), 30–37.

Abou-Ghazala A., Katsuki S., Schoenbach K. H., Dobbs F. C. and Moreira K. R. (2002). Bacterial decontamination of water by means of pulsed-corona discharges. *IEEE Transactions on Plasma Science*, **30**, 1449–1453.

Ahmed B., Mohamed H., Limem E. and Nasr B. (2009). Degradation and mineralization of organic pollutants contained in actual pulp and paper mill wastewaters by a UV/H_2O_2 process. *Industrial and Engineering Chemistry Research*, **48**(7), 3370–3379.

Akiyama H. (2000). Streamer discharges in liquids and their applications. *IEEE Transactions on Dielectrics and Electrical Insulation*, **7**(5), 646–653.

Anpilov A., Barkhudarov E., Christofi N., Kopev V., Kossyi I. and Taktakishvili M. (2004). The effectiveness of a multi-spark electric discharge system in the destruction of microorganisms in domestic and industrial wastewaters. *Journal of Water and Health*, **2**(4), 267–277.

Appleman T. D., Higgins C. P., Quiñones O., Vanderford B. J., Kolstad C., Zeigler-Holady J. C. and Dickenson E. R. V. (2014). Treatment of poly- and perfluoroalkyl substances in U.S. Full-scale water treatment systems. *Water Research*, **51**, 246–255.

Bai M. and Zhang Z. (2015). Method for hydroxyl radical rapid production using a strong ionization discharge combined with effect of water jet cavitation. In: 250th American Chemical Society National Meeting & Exposition, August 16–20, 2015, Boston, MA.

Bai M., Zhang Z., Xue X., Yang X., Hua L. and Fan D. (2010). Killing effects of hydroxyl radical on algae and bacteria in ship's ballast water and on their cell morphology. *Plasma Chemistry and Plasma Processing*, **30**(6), 831–840.

Banaschik R., Koch F., Kolb J. F. and Weltmann K.-D. (2014). Decomposition of pharmaceuticals by pulsed corona discharges in water depending on streamer length. *IEEE Transactions on Plasma Science*, **42**(10), 2736–2737.

Banaschik R., Lukes P., Jablonowski H., Hammer M. U., Weltmann K. D. and Kolb J. F. (2015). Potential of pulsed corona discharges generated in water for the degradation of persistent pharmaceutical residues. *Water Research*, **84**, 127–135.

Baroch P., Saito N. and Takai O. (2008). Special type of plasma dielectric barrier discharge reactor for direct ozonization of water and degradation of organic pollution. *Journal of Physics D-Applied Physics*, **41**(8), 085207.

Bell A. T. (1974). Fundamentals of plasma chemistry. In: Techniques and Applications of Plasma Chemistry, J. Hollahan and A. T. Bell (eds), John Wiley & Sons, New York, pp. 1–56.

Beltrán F. J., Ovejero G. and Rivas J. (1996). Oxidation of polynuclear aromatic hydrocarbons in water. 4. Ozone combined with hydrogen peroxide. *Industrial and Engineering Chemistry Research*, **35**(3), 891–898.

Benetoli L. O. D., Cadorin B. M., Baldissarelli V. Z., Geremias R., de Souza I. G. and Debacher N. A. (2012). Pyrite-enhanced methylene blue degradation in non-thermal plasma water treatment reactor. *Journal of Hazardous Materials*, **237**, 55–62.

Benstaali B., Boubert P., Cheron B. G., Addou A. and Brisset J. L. (2002). Density and rotational temperature measurements of the OH and NO radicals produced by a gliding arc in humid air. *Plasma Chemistry and Plasma Processing*, **22**(4), 553–571.

Bian W. J., Song X. H., Liu D. Q., Zhang J. and Chen X. H. (2011). The intermediate products in the degradation of 4-chlorophenol by pulsed high voltage discharge in water. *Journal of Hazardous Materials*, **192**(3), 1330–1339.

Bobkova E. S., Isakina A. A., Grinevich V. I. and Rybkin V. V. (2012). Decomposition of aqueous solution of acetic acid under the action of atmospheric-pressure dielectric barrier discharge in oxygen. *Russian Journal of Applied Chemistry*, **85**(1), 71–75.

Bobkova E. S., Shishkina A. I., Borzova A. A. and Rybkin V. V. (2015). Influence of parameters of oxygen atmospheric-pressure dielectric barrier discharge on the degradation kinetics of sulfonol in water. *High Energy Chemistry*, **49**(5), 375–378.

Bozic A. L., Koprivanac N., Sunka P., Clupek M. and Babicky V. (2004). Organic synthetic dye degradation by modified pinhole discharge. *Czechoslovak Journal of Physics*, **54**, C958–C963.

Bresch S., Wandell R., Wang H., Alabugin I. and Locke B. (2016). Oxidized derivatives of n-hexane from a water/argon continuous flow electrical discharge plasma reactor. *Plasma Chemistry and Plasma Processing*, 36(**2**), 553–584.

Brisset J. L. (1997). Air corona removal of phenols. *Journal of Applied Electrochemistry*, **27**(2), 179–183.

Brisset J.-L. and Hnatiuc E. (2012). Peroxynitrite: a re-examination of the chemical properties of non-thermal discharges burning in air over aqueous solutions. *Plasma Chemistry and Plasma Processing*, **32**(4), 655–674.

Bruggeman P. and Locke B. R. (2013). Assessment of potential applications of plasma with liquid water. In: Low Temperature Plasma Technology: Methods and Applications, X. L. P. Chu (ed.), Taylor and Francis Group, Boca raton, FL, pp. 367–390.

Bruggeman P., Ribezl E., Maslani A., Degroote J., Malesevic A., Rego R., Vierendeels J. and Leys C. (2008). Characteristics of atmospheric pressure air discharges with a liquid cathode and a metal anode. *Plasma Sources Science and Technology*, **17**(2), 025012.

Bruggeman P., Kushner M. J., Locke B. R., *et al.* (2016). Plasma–liquid interactions: a review and roadmap. *Plasma Sources Science and Technology*, **25**(5), http://dx.doi.org/10.1088/0963-0252/25/5/053002

Buntat Z., Smith I. R. and Razali N. A. M. (2009). Ozone generation using atmospheric pressure glow discharge in air. *Journal of Physics D: Applied Physics*, **42**(23), 235202.

Burlica R., Kirkpatrick M., Finney W., Clark R. and Locke B. (2004). Organic dye removal from aqueous solution by glidarc discharges. *Journal of Electrostatics*, **62**(4), 309–321.

Burlica R., Shih K. Y., Hnatiuc B. and Locke B. R. (2011). Hydrogen generation by pulsed gliding arc discharge plasma with sprays of alcohol solutions. *Industrial and Engineering Chemistry Research*, **50**(15), 9466–9470.

Chakma S. and Moholkar V. S. (2014). Investigations in synergism of hybrid advanced oxidation processes with combinations of sonolysis plus Fenton process plus UV for degradation of bisphenol A. *Industrial and Engineering Chemistry Research*, **53**(16), 6855–6865.

Chen H. L., Lee H. M. and Chang M. B. (2006). Enhancement of energy yield for ozone production via packed-bed reactors. *Ozone Science and Engineering*, **28**(2), 111–118.

Chen H. L., Lee H. M., Chen S. H. and Chang M. B. (2008). Review of packed-bed plasma reactor for ozone generation and air pollution control. *Industrial and Engineering Chemistry Research*, **47**(7), 2122–2130.

Chen J. H. and Wang P. X. (2005). Effect of relative humidity on electron distribution and ozone production by dc coronas in air. *IEEE Transactions on Plasma Science*, **33**(2), 808–812.

Cheng J., Vecitis C. D., Park H., Mader B. T. and Hoffmann M. R. (2008). Sonochemical degradation of perfluorooctane sulfonate (PFOS) and perfluorooctanoate (PFOA) in landfill groundwater: environmental matrix effects. *Environmental Science and Technology*, **42**(21), 8057–8063.

Chernyak V. Y., Olszewski S. V., Yukhymenko V., Solomenko E. V., Prysiazhnevych I. V., Naumov V. V., Levko D. S., Shchedrin A. I., Ryabtsev A. V., Demchina V. P., Kudryavtsev V. S., Martysh E. V. and Verovchuck M. A. (2008). Plasma-assisted reforming of ethanol in dynamic plasma-liquid system: experiments and modeling. *IEEE Transactions on Plasma Science*, **36**(6), 2933–2939.

Ching W. K., Colussi A. J., Sun H. J., Nealson K. H. and Hoffmann M. R. (2001). *Escherichia coli* disinfection by electrohydraulic discharges. *Environmental Science and Technology*, **35**(20), 4139–4144.

Christensen P. A., Yonar T. and Zakaria K. (2013). The electrochemical generation of ozone: a review. *Ozone Science and Engineering*, **35**(3), 149–167.

Clements J. S., Sato M. and Davis R. H. (1987). Preliminary investigation of prebreakdown phenomena and chemical reactions using a pulsed high-voltage discharge in water. *IEEE Transactions on Industry Applications*, **23**, 224–235.

Cooper W. J., Nickelsen M. G., Meacham D. E., Cadavid E. M., Waite T. D. and Kurucz C. N. (1992). High energy electron beam irradiation: an innovation process for the treatment of aqueous bases organic hazardous wastes. *Journal of Environmental Science and Health*, **A27**, 219–244.

Dai F., Fan X., Stratton G. R., Bellona C. L., Holsen T. M., Crimmins B. S., Xia X. and Mededovic Thagard S. (2015). Experimental and density functional theoretical study of the effects of Fenton's reaction on the degradation of bisphenol A in a high voltage plasma reactor. *Journal of Hazardous Materials*, **308**, 419–429.

Dobrin D., Bradu C., Magureanu M., Mandache N. B. and Parvulescu V. I. (2013). Degradation of diclofenac in water using a pulsed corona discharge. *Chemical Engineering Journal*, **234**, 389–396.

Dojcinovic B. P., Manojlovic D., Roglic G. M., Obradovic B. M., Kuraica M. M. and Puric J. (2008). Plasma assisted degradation of phenol solutions. *Vacuum*, **83**(1), 234–237.

Dors M., Metel E., Mizeraczyk J. and Marotta E. (2008). *Coli* bacteria inactivation by pulsed corona discharge in water. *International Journal of Plasma Environmental Science and Technology*, **2**(1), 34–37.

Duarte C. L., Sampa M. H. O., Rela P. R., Oikawa H., Silveira C. G. and Azevedo A. L. (2002). Advanced oxidation process by electron-beam-irradiation-induced decomposition of pollutants in industrial effluents. *Radiation Physics and Chemistry*, **63**(3–6), 647–651.

Efremov N. M., Adamiak B. Y., Blochin V. I., Dadshev S. J., Dmitriev K. I., Semjonov V. N., Levashov V. F. and Jusbashev V. F. (2000). Experimental investigation of the action of pulsed eelctrical discharges in liquids on biological objects. *IEEE Transactions on Plasma Science*, **28**, 224–228.

Egli W. and Eliasson B. (1989). Numerical calculation of electrical breakdown in oxygen in a dielectric-barrier discharge. *Helvetica Physica Acta*, **62**, 302–305.

Ein-Mozaffari F., Mohajerani M. and Mehrvar M. (2009). An overview of the integration of advanced oxidation technologies and other processes for water and wastewater treatment. *International Journal of Engineering*, **3**, 120–146.

Eliasson B. and Kogelschatz U. (1986). Electron impact dissociation in oxygen. *Journal of Physics B: Atmospheric Molecular Physics*, **19**, 1241–1247.

Eliasson B. and Kogelschatz U. (1991). Modeling and applications of silent discharge plasmas. *IEEE Transactions on Plasma Science*, **19**, 309–323.

Eliasson B., Hirth M. and Kogelschatz U. (1987). Ozone synthesis from oxygen in dielectric barrier discharges. *Journal of Physics D: Applied Physics*, **20**, 1421–1437.

Even-Ezra I., Mizrahi A., Gerrity D., Snyder S., Salveson A. and Lahav O. (2009). Application of a novel plasma-based advanced oxidation process for efficient and cost-effective destruction of refractory organics in tertiary effluents and contaminated groundwater. *Desalination and Water Treatment*, **11**, 236–244.

Fan X. and Mededovic Thagard S. (2014). Quantum mechanical approach to studying degradation of phenol by electrical discharges in water. In: The 9th International Symposium on Non-Thermal/Thermal Plasma pollution Control Technology and Sustainable Energy, June 16–20, 2014, Dalian, China.

Flint S., Markle T., Thompson S. and Wallace E. (2012). Bisphenol A exposure, effects, and policy: a wildlife perspective. *Journal of Environmental Management*, **104**, 19–34.

Franclemont J. and Thagard S. M. (2014). Pulsed electrical discharges in water: can non-volatile compounds diffuse into the plasma channel? *Plasma Chemistry and Plasma Processing*, **34**(4), 705–719.

Fridman A. (2008). Plasma Chemistry. Cambridge University Press, Cambridge.

Fridman A. and Kennedy L. A. (2004). Plasma Physics and Engineering. Taylor and Francis, New York.

Fudamoto T., Namihira T., Katsuki S., Akiyama H., Imakubo T. and Majima T. (2008). Sterilization of *E. coli* by underwater pulsed streamer discharges in a continuous flow system. *Electrical Engineering in Japan*, **164**(1), 1–7.

Gai K. (2009). Anodic oxidation with platinum electrodes for degradation of p-xylene in aqueous solution. *Journal of Electrostatics*, **67**(4), 554–557.

Gao J. Z., Hu Z. G., Wang X. Y., Hou J. G., Lu X. Q. and Kang J. W. (2001). Oxidative degradation of acridine orange induced by plasma with contact glow discharge electrolysis. *Thin Solid Films*, **390**(1–2), 154–158.

Gao J. Z., Pu L. M., Yang W., Yu J. and Li Y. (2004). Oxidative degradation of nitrophenols in aqueous solution induced by plasma with submersed glow discharge electrolysis. *Plasma Processes and Polymers*, **1**(2), 171–176.

Gao J. Z., Yu J., Lu Q. F., He X. Y., Yang W., Li Y., Pu L. M. and Yang Z. M. (2008). Decoloration of alizarin red S in aqueous solution by glow discharge electrolysis. *Dyes and Pigments*, **76**(1), 47–52.

Garamoon A. A., Elakshar F. F., Nossair A. M. and Kotp E. F. (2002). Experimental study of ozone synthesis. *Plasma Sources Science and Technology*, **11**(3), 254–259.

Gemo N., Biasi P., Canu P. and Salmi T. O. (2012). Mass transfer and kinetics of H_2O_2 direct synthesis in a batch slurry reactor. *Chemical Engineering Journal*, **207–208**(0), 539–551.

Gerrity D., Stanford B. D., Trenholm R. A. and Snyder S. A. (2010). An evaluation of a pilot-scale nonthermal plasma advanced oxidation process for trace organic compound degradation. *Water Research*, **44**(2), 493–504.

Ghezzar M. R., Abdelmalek F., Belhadj M., Benderdouche N. and Addou A. (2007). Gliding arc plasma assisted photocatalytic degradation of anthraquinonic acid green 25 in solution with TiO_2. *Applied Catalysis B-Environmental*, **72**(3–4), 304–313.

Gigosos M. A., Gonzalez M. A. and Cardenoso V. (2003). Computer simulated Balmer-alpha, -beta and -gamma Stark line profiles for non-equilibrium plasmas diagnostics. *Spectrochimica Acta Part B-Atomic Spectroscopy*, **58**(8), 1489–1504.

Grabowski L., Van Veldhuizen E., Pemen A. and Rutgers W. (2006). Corona above water reactor for systematic study of aqueous phenol degradation. *Plasma Chemistry and Plasma Processing*, **26**(1), 3–17.

Grabowski L. R., van Veldhuizen E. M., Pemen A. J. M. and Rutgers W. R. (2007). Breakdown of methylene blue and methyl orange by pulsed corona discharge. *Plasma Sources Science and Technology*, **16**(2), 226–232.

Grinevich V. I., Plastinina N. A., Rybkin V. V. and Bubnov A. G. (2009). Oxidative degradation of formic acid in aqueous solution upon dielectric-barrier discharge treatment. *High Energy Chemistry*, **43**(2), 138–142.

Grymonpre D. R., Finney W. C. and Locke B. R. (1999). Aqueous-phase pulsed streamer corona reactor using suspended activated carbon particles for phenol oxidation: model-data comparison. *Chemical Engineering Science*, **54**, 3095–3105.

Grymonpre D. R., Sharma A., Finney W. and Locke B. (2001). The role of Fenton's reaction in aqueous phase pulsed streamer corona reactors. *Chemical Engineering Journal*, **82**(1–3), 189–207.

Grymonpre D., Finney W., Clark R. and Locke B. (2004). Hybrid gas-liquid electrical discharge reactors for organic compound degradation. *Industrial and Engineering Chemistry Research*, **43**(9), 1975–1989.

Gubkin J. (1887). Electrolytiche metallabscheidung an der freien oberfläche einer salzlösung. *Ann. Physik*, **268**(9), 114–115.

Guelfo J. L. and Higgins C. P. (2013). Subsurface transport potential of perfluoroalkyl acids at aqueous film-forming foam (AFFF)-impacted sites. *Environmental Science and Technology*, **47**(9), 4164–4171.

Hackam R. and Akiyama H. (2000). Air pollution control by electrical discharges. *IEEE Transactions on Dielectrics and Electrical Insulation*, **7**, 654–683.

Hagelaar G. J. M. and Pitchford L. C. (2005). Solving the Boltzmann equation to obtain electron transport coefficients and rate coefficients for fluid models. *Plasma Sources Science and Technology*, **14**(4), 722.

Hayashi D., Hoeben W. F. L. M., Dooms G., Veldhuizen E. M. v., Rutgers W. R. and Kroesen G. M. W. (2000). Influence of gaseous atmosphere on corona-induced degradation of aqueous phenol. *Journal of Physics D: Applied Physics*, **33**(21), 2769.

Hibert C., Gaurand I., Motret O. and Pouvesle J. M. (1999). [OH(X)] measurements by resonant absorption spectroscopy in a pulsed dielectric barrier discharge. *Journal of Applied Physics*, **85**(10), 7070–7075.

Hickling A. (1971). Electrochemical processes in glow discharge at the gas-solution interface. In: Modern Aspects of Electrochemistry, J. O. M. Bockris and B. E. Conway (eds), Plenum Press, New York, Vol. **6**, pp. 329–373.

Hickling A. and Ingram M. D. (1964). Contact glow-discharge electrolysis. *Transactions of the Faraday Society*, **60**, 783–793.

Hijosa-Valsero M., Molina R., Schikora H., Muller M. and Bayona J. M. (2013). Removal of priority pollutants from water by means of dielectric barrier discharge atmospheric plasma. *Journal of Hazardous Materials*, **262**, 664–673.

Hijosa-Valsero M., Molina R., Montràs A., Müller M. and Bayona J. M. (2014). Decontamination of waterborne chemical pollutants by using atmospheric pressure nonthermal plasma: A review. *Environmental Technology Reviews*, **3**(1), 71–91.

Hoeben W. F. L. M. (2000). Pulsed Corona-induced Degradation of Organic Materials in Water. PhD thesis, Eindhoven, Netherlands.

Hoeben W. F. L. M., van Veldhuizen E. M., Rutgers W. R. and Kroesen G. M. W. (1999). Gas phase corona discharges for oxidation of phenol in an aqueous solution. *Journal of Physics D: Applied Physics*, **32**, L133–L137.

Hoeben W. F. L. M., van Veldhuizen E. M., Rutgers W. R., Cramers C. A. M. G. and Kroesen G. M. W. (2000). The degradation of aqueous phenol solutions by pulsed positive corona discharges. *Plasma Sources Science and Technology*, **9**, 361–369.

Hoffmann M. R., Hua I., Hochemer R., Willberg D., Lang P. and Kratel A. (1997). Chemistry under extreme conditions in water induced electrohydraulic cavitation and pulsed-plasma discharges. In: Chemistry under Extreme or Non-Classical Conditions, R. van Eldik and C. D. Hubbard (eds), John Wiley and Sons, Inc., New York, pp. 429–477.

Holzer F. and Locke B. R. (2008). Multistage gas-liquid electrical discharge column reactor for advanced oxidation processes. *Industrial and Engineering Chemistry Research*, **47**(7), 2203–2212.

Hong Y. C., Park H. J., Lee B. J., Kang W.-S. and Uhm H. S. (2010). Plasma formation using a capillary discharge in water and its application to the sterilization of *E. coli*. *Physics of Plasmas*, **17**(5), 1–2.

Hori H., Hayakawa E., Einaga H., Kutsuna S., Koike K., Ibusuki T., Kiatagawa H. and Arakawa R. (2004). Decomposition of environmentally persistent perfluorooctanoic acid in water by photochemical approaches. *Environmental Science and Technology*, **38**(22), 6118–6124.

Hori H., Saito H., Sakai H., Kitahara T. and Sakamoto T. (2015). Efficient decomposition of a new fluorochemical surfactant: Perfluoroalkane disulfonate to fluoride ions in subcritical and supercritical water. *Chemosphere*, **129**, 27–32.

Horikoshi S., Sato S., Abe M. and Serpone N. (2011). A novel liquid plasma AOP device integrating microwaves and ultrasounds and its evaluation in defluorinating perfluorooctanoic acid in aqueous media. *Ultrasonics Sonochemistry*, **18**(5), 938–942.

Hsieh K. C., Wandell R. J., Bresch S. and Locke B. R. (2017). Analysis of hydroxyl radical formation in a gas-liquid electrical discharge plasma reactor utilizing liquid and gaseous radical scavengers. *Plasma Processes and Polymers*. DOI: 10.1002/ppap.201600171.

Houtz E. F., Higgins C. P., Field J. A. and Sedlak D. L. (2013). Persistence of perfluoroalkyl acid precursors in AFFF-impacted groundwater and soil. *Environmental Science and Technology*, **47**(15), 8187–8195.

Hsieh K. (2015). Analysis of Radicals in Gas–Liquid Electrical Discharges. PhD Dissertation, Tallahassee, Florida.

Ikoma S., Satoh K. and Itoh H. (2012). Decomposition of methylene blue in an aqueous solution using a pulsed-discharge plasma at atmospheric pressure. *Electrical Engineering in Japan*, **179**(3), 1–9.

Ioan I., Wilson S., Lundanes E. and Neculai A. (2007). Comparison of Fenton and sono-fenton bisphenol A degradation. *Journal of Hazardous Materials*, **142**(1–2), 559–563.

Jiang B., Zheng J., Qiu S., Wu M., Zhang Q., Yan Z. and Xue Q. (2014). Review on electrical discharge plasma technology for wastewater remediation. *Chemical Engineering Journal*, **236**, 348–368.

Johnson D. C., Shamamian V. A., Callahan J. H., Denes F. S., Manolache S. O. and Dandy D. S. (2003). Treatment of methyl tert-butyl ether contaminated water using a dense medium plasma reactor: a mechanistic and kinetic investigation. *Environmental Science and Technology*, **37**(20), 4804–4810.

Johnson D. C., Dandy D. S. and Shamamian V. A. (2006). Development of a tubular high-density plasma reactor for water treatment. *Water Research*, **40**(2), 311–322.

Joshi R. P. and Mededovic Thagard S. (2013a). Streamer-like electrical discharges in water: Part I. Fundamental mechanisms. *Plasma Chemistry and Plasma Processing*, **33**(1), 1–15.

Joshi R. P. and Mededovic Thagard S. (2013b). Streamer-like electrical discharges in water: Part II. Environmental applications. *Plasma Chemistry and Plasma Processing*, **33**(1), 17–49.

Joshi A., Locke B., Arce P. and Finney W. (1995). Formation of hydroxyl radicals, hydrogen peroxide and aqueous electrons by pulsed streamer corona discharge in aqueous solution. *Journal of Hazardous Materials*, **41**(1), 3–30.

Kanazawa S., Kawano H., Watanabe S., Furuki T., Akamine S., Ichiki R., Ohkubo T., Kocik M. and Mizeraczyk J. (2011). Observation of OH radicals produced by pulsed discharges on the surface of a liquid. *Plasma Sources Science and Technology*, **20**(3), 034010.

Karpel Vel Leitner N., Syoena G., Romat H., Urashima K. and Chang J.-S. (2005). Generation of active entities by the pulsed arc electrohydraulic discharge system and application to removal of atrazine. *Water Research*, **39**, 4705–4714.

Kavanaugh M., Chowdhury Z., Kommineni S., *et al.* (2004). Removal of MTBE with Advanced Oxidation Processes. IWA Publishing (International Water Assoc), Denver, CO.

Kim H.-H., Ogata A. and Futamura S. (2006). Application of plasma-catalyst hybrid processes for the control of NO_x and volatile organic compounds. In: Trends in Catalysis Research, L. P. Bevy (ed.), Nova Science Publishers, Inc., Hauggpauge, NY, pp. 1–50.

Kim I.-K., Huang C.-P. and Chiu P. C. (2001). Sonochemical decomposition of dibenzothiophene in aqueous solution. *Water Research*, **35**(18), 4370–4378.

Kim M. H., Cho J. H., Ban S. B., Choi R. Y., Kwon E. J., Park S. J. and Eden J. G. (2013). Efficient generation of ozone in arrays of microchannel plasmas. *Journal of Physics D-Applied Physics*, **46**(30), 305201.

Kirkpatrick M. and Locke B. (2005). Hydrogen, oxygen, and hydrogen peroxide formation in aqueous phase pulsed corona electrical discharge. *Industrial and Engineering Chemistry Research*, **44**(12), 4243–4248.

Kogelschatz U., Eliasson B. and Hirth M. (1988). Ozone generation from oxygen and air – discharge physics and reaction-mechanisms. *Ozone Science and Engineering*, **10**(4), 367–377.

Kogelschatz Y. and Eliasson B. (1995). Chapter 26, Ozone generation and applications. In: Handbook of Electrostatic Processes, J. S. Chang, A. J. Kelly and J. M. Crowley (eds), Marcel Dekker, Inc., New York, pp. 581–605.

Kozakova Z., Nejezchleb M., Krcma F., Halamova I., Caslavsky J. and Dolinova J. (2010). Removal of organic dye direct red 79 from water solutions by dc diaphragm discharge: analysis of decomposition products. *Desalination*, **258**(1–3), 93–99.

Krause H., Schweiger B., Schuhmacher J., Scholl S. and Steinfeld U. (2009). Degradation of the endocrine disrupting chemicals (EDCs) carbamazepine, clofibric acid, and iopromide by corona discharge over water. *Chemosphere*, **75**(2), 163–168.

Kravchenko A. V., Rudnitskii A. G., Nesterenko A. F. and Kublanovskii V. S. (2002). Purification of wastewater containing color developers and silver ions under conditions of low-temperature plasma electrolysis. *Russian Journal of Applied Chemistry*, **75**(8), 1265–1268.

Krcma F., Stara Z. and Prochazkova J. (2010). Diaphragm discharge in liquids: Fundamentals and applications. Third International Workshop and Summer School on Plasma Physics, June 30–July 5, 2008, Kiten, Bulgaria 207.

Kusic H., Koprivanac N. and Locke B. (2005). Decomposition of phenol by hybrid gas/liquid electrical discharge reactors with zeolite catalysts. *Journal of Hazardous Materials*, **125**(1–3), 190–200.

Lang P., Ching W., Willberg D. and Hoffmann M. (1998). Oxidative degradation of 2, 4, 6-trinitrotoluene by ozone in an electrohydraulic discharge reactor. *Environmental Science and Technology*, **32**(20), 3142–3148.

Laroussi M. (1996). Sterilization of contaminated matter with an atmospheric pressure plasma. *IEEE Transactions on Plasma Science*, **24**, 1188–1191.

Laroussi M. (2002). Nonthermal decontamination of biological media by atmospheric-pressure plasma: review, analysis and prospects. *IEEE Transactions on Plasma Science*, **30**(4), 1409–1415.

Lee C., Lee Y. and Yoon J. (2006). Oxidative degradation of dimethylsulfoxide by locally concentrated hydroxyl radicals in streamer corona discharge process. *Chemosphere*, **65**(7), 1163–1170.

Levko D., Shchedrin A., Chernyak V., Olszewski S. and Nedybaliuk O. (2011). Plasma kinetics in ethanol/water/air mixture in a 'tornado'-type electrical discharge. *Journal of Physics D-Applied Physics*, **44**(14), 145206.

Li S. P., Jiang Y. Y., Cao X. H., Dong Y. W., Dong M. and Xu J. (2013). Degradation of nitenpyram pesticide in aqueous solution by low-temperature plasma. *Environmental Technology*, **34**(12), 1609–1616.

Li W., Zhou Q. and Hua T. (2010). Removal of organic matter from landfill leachate by advanced oxidation processes: a review. *International Journal of Chemical Engineering*, **2010**, Article ID 270532, 10 pages.

Li X., Chen J. and Gao L. (2012). Enhanced degradation of phenol by carbonate ions with dielectric barrier discharge. *IEEE Transactions on Plasma Science*, **40**(1), 112–117.

Lieberman M. and Lichtenberg A. (2005). Principles of Plasma Discharge and Materials Processing. 2nd edn, Wiley, Hoboken, New Jersey.

Lindsay A., Byrns B., King W., Andhvarapou A., Fields J., Knappe D., Fonteno W. and Shannon S. (2014). Fertilization of radishes, tomatoes, and marigolds using a large-volume atmospheric glow discharge. *Plasma Chemistry and Plasma Processing*, **34**(6), 1271–1290.

Liu H., Du C. M., Wang J., Li H. X., Zhang L. and Zhang L. L. (2012a). Comparison of acid orange 7 degradation in solution by gliding arc discharge with different forms of TiO_2. *Plasma Processes and Polymers*, **9**(3), 285–297.

Liu Y. J. (2009). Aqueous p-chloronitrobenzene decomposition induced by contact glow discharge electrolysis. *Journal of Hazardous Materials*, **166**(2–3), 1495–1499.

Liu Y. J. and Jiang X. Z. (2005). Phenol degradation by a nonpulsed diaphragm glow discharge in an aqueous solution. *Environmental Science and Technology*, **39**(21), 8512–8517.

Liu Y. N., Mei S. F., Iya-Sou D., Cavadias S. and Ognier S. (2012b). Carbamazepine removal from water by dielectric barrier discharge: Comparison of ex situ and in situ discharge on water. *Chemical Engineering and Processing: Process Intensification*, **56**, 10–18.

Locke B. R. and Mededovic-Thagard S. (2009). Analysis of chemical reactions in gliding arc reactors with water spray. *IEEE Transactions on Plasma Science*, **37**(4), 494–501.

Locke B. R. and Shih K.-Y. (2011). Review of the methods to form hydrogen peroxide in electrical discharge plasma with liquid water. *Plasma Sources Science and Technology*, **20**(3), 034006.

Locke B. R. and Mededovic Thagard S. (2012). Analysis and review of chemical reactions and transport processes in pulsed electrical discharge plasma formed directly in liquid water. *Plasma Chemistry and Plasma Processing*, **32**(5), 875–917.

Locke B. R., Sato M., Sunka P., Hoffmann M. and Chang J. (2006). Electrohydraulic discharge and nonthermal plasma for water treatment. *Industrial and Engineering Chemistry Research*, **45**(3), 882–905.

Locke B. R., Lukes P. and Brisset J. L. (2012). Elementary chemical and physical phenomena in electrical discharge plasma in gas-liquid environments and in liquids. In: Plasma Chemistry and Catalysis in Gases and Liquids, M. M. V. I. Parvulescu and P. Lukes (ed.), Wiley-VCH Verlag GmbH & Co. KGaA, Weinheim, pp. 185–225.

Lu D. L., Chen J. R., Gao A. H., Zissis G., Hu S. B. and Lu Z. G. (2010). Decolorization of aqueous acid red B solution during the cathode process in abnormal glow discharge. *IEEE Transactions on Plasma Science*, **38**(10), 2854–2859.

Lukes P. (2001). PhD Dissertation, Institute of Chemical Technology, Prague, Czech Republic.

Lukes P. and Locke B. R. (2005a). Degradation of phenol in a hybrid series gas-liquid electrical discharge reactor. *Journal of Physics D: Applied Physics*, **38**, 4074–4081.

Lukes P. and Locke B. R. (2005b). Degradation of substituted phenols in a hybrid gas-liquid electrical discharge reactor. *Industrial and Engineering Chemistry Research*, **44**(9), 2921–2930.

Lukes P. and Locke B. R. (2005c). Plasmachemical oxidation processes in a hybrid gas-liquid electrical discharge reactor. *Journal of Physics D: Applied Physics*, **38**(22), 4074–4081.

Lukes P., Clupek M., Babicky V., Sunka P., Winterova G. and Janda V. (2003). Non-thermal plasma induced decomposition of 2-chlorophenol in water. *Acta Physica Slovaca*, **53**(6), 423–428.

Lukes P., Clupek M., Sunka P., Peterka F., Sano T., Negishi N., Matsuzawa S. and Takeuchi K. (2005). Degradation of phenol by underwater pulsed corona discharge in combination with TiO_2 photocatalysis. *Research on Chemical Intermediates*, **31**(4), 285–294.

Lukes P., Clupek M., Babicky V. and Vykouk T. (2007). 16th IEEE International Pulsed Power Conference (Vol. 1), June 17–22, 2007, Albuquerque, NM.

Lukes P., Brisset J. L. and Locke B. R. (2012a). Biological effects of electrical discharge plasma in water and in gas-liquid environments. In: Plasma Chemistry and Catalysis in Gases and Liquids, M. M. V. I. Parvulescu and P. Lukes (ed.), Wiley-VCH Verlag GmbH & Co. KGaA, Weinheim, pp. 309–337.

Lukes P., Locke B. R. and Brisset J. L. (2012b). Aqueous-phase chemistry of electrical discharge plasma in water and in gas-liquid environments. In: Plasma Chemistry and Catalysis in Gases and Liquids, M. M. V. I. Parvulescu and P. Lukes (ed.), Wiley-VCH Verlag GmbH & Co. KGaA, Weinheim, pp. 243–293.

Lukes P., Dolezalova E., Sisrova I. and Clupek M. (2014). Aqueous-phase chemistry and bactericidal effects from an air discharge plasma in contact with water: Evidence for the formation of peroxynitrite through a pseudo-second-order post-discharge reaction of H_2O_2 and HNO_2. *Plasma Sources Science and Technology*, **23**(1), 1–2.

Magureanu M., Mandache N. B. and Parvulescu V. I. (2007). Degradation of organic dyes in water by electrical discharges. *Plasma Chemistry and Plasma Processing*, **27**(5), 589–598.

Magureanu M., Piroi D., Gherendi F., Mandache N. B. and Parvulescu V. (2008a). Decomposition of methylene blue in water by corona discharges. *Plasma Chemistry and Plasma Processing*, **28**(6), 677–688.

Magureanu M., Piroi D., Mandache N. B. and Parvulescu V. (2008b). Decomposition of methylene blue in water using a dielectric barrier discharge: optimization of the operating parameters. *Journal of Applied Physics*, **104**(10), 1–2.

Magureanu M., Piroi D., Mandache N. B., David V., Medvedovici A. and Parvulescu V. I. (2010). Degradation of pharmaceutical compound pentoxifylline in water by non-thermal plasma treatment. *Water Research*, **44**(11), 3445–3453.

Magureanu M., Piroi D., Mandache N. B., David V., Medvedovici A., Bradu C. and Parvulescu V. I. (2011). Degradation of antibiotics in water by non-thermal plasma treatment. *Water Research*, **45**(11), 3407–3416.

Magureanu M., Bradu C., Piroi D., Mandache N. and Parvulescu V. (2013a). Pulsed corona discharge for degradation of methylene blue in water. *Plasma Chemistry and Plasma Processing*, **33**(1), 51–64.

Magureanu M., Dobrin D., Mandache N. B., Bradu C., Medvedovici A. and Parvulescu V. I. (2013b). The mechanism of plasma destruction of enalapril and related metabolites in water. *Plasma Processes and Polymers*, **10**(5), 459–468.

Malik M. A. (2010). Water purification by plasmas: which reactors are most energy efficient? *Plasma Chemistry and Plasma Processing*, **30**(1), 21–31.

Malik M. A. (2014). Ozone synthesis using shielded sliding discharge: effect of oxygen content and positive versus negative streamer mode. *Industrial and Engineering Chemistry Research*, **53**(31), 12305–12311.

Malik M. A., Ghaffar A. and Malik S. A. (2001). Water purification by electrical discharges. *Plasma Sources Science and Technology*, **10**(1), 82–91.

Malik M. A., Ubaid-ur-Rehman Ghaffar A. and Ahmed K. (2002). Synergistic effect of pulsed corona discharges and ozonation on decolourization of methylene blue in water. *Plasma Sources Science and Technology*, **11**(3), 236–240.

Manolache S., Shamamian V. and Denes F. (2004). Dense medium plasma-plasma-enhanced decontamination of water of aromatic compounds. *Journal of Environmental Engineering-ASCE*, **130**(1), 17–25.

Mark G. A., Tauber A., Laupert R., Schuchmann H. P., Schulz D., Mues A. and von Sonntag C. (1998). OH-radical formation by ultrasound in aqueous solution – part II: Terephthalate and ficke dosimetry and the influence of various conditions on the sonolytic yield. *Ultrasonics Sonochemistry*, **5**(2), 41–52.

Maroulf-Khelifa K., Abdelmalek F., Khelifa A. and Addou A. (2008). TiO_2-assisted degradation of a perfluorinated surfactant in aqueous solutions treated by gliding arc discharge. *Chemosphere*, **70**(11), 1995–2001.

Marsili L., Espie S., Anderson J. G. and MacGregor S. J. (2002). Plasma inactivation of food-related microorganisms in liquids. *Radiation Physics and Chemistry*, **65**, 507–513.

Matsumoto T., Wang D., Namihira T. and Akiyama H. (2011). Process performances of 2 ns pulsed discharge plasma. *Japanese Journal of Applied Physics*, **50**(8), 1–2.

Mazurek B., Lubicki P. and Staroniewicz Z. (1995). Effect of short HV pulses on bacteria and fungi. *IEEE Transactions on Dielectrics and Electrical Insulation*, **2**, 418–425.

Mededovic S. and Locke B. (2007a). Primary chemical reactions in pulsed electrical discharge channels in water. *Journal of Physics D: Applied Physics*, **40**(24), 7734–7746.

Mededovic S. and Locke B. (2007b). Side-chain degradation of atrazine by pulsed electrical discharge in water. *Industrial and Engineering Chemistry Research*, **46**(9), 2702–2709.

Mededovic S. and Stratton G. (2015). Enhanced Contact Electrical Discharge Plasma Reactor for Liquid and Gas Processing, 15/018780, USA.

Mededovic S. and Takashima K. (2008). Decolorization of indigo carmine dye by spark discharge in water. *International Journal of Plasma Environmental Science and Technology*, **2**(1), 56–66.

Mededovic S., Finney W. C. and Locke B. R. (2007). Aqueous-phase mineralization of s-triazine using pulsed electrical discharge. *International Journal of Plasma Environmental Science and Technology*, **1**(1), 82–90.

Mededovic Thagard S., Takashima K. and Mizuno A. (2009). Chemistry of the positive and negative electrical discharges formed in liquid water and above a gas-liquid surface. *Plasma Chemistry and Plasma Processing*, **29**(6), 455–473.

Menger-Krug E. (2011). Cohiba guidance document no. 4 – PFOS/PFOA. *COHIBA Project Consortium*.

Minamitani Y., Shoji S., Ohba Y. and Higashiyama Y. (2008). Decomposition of dye in water solution by pulsed power discharge in a water droplet spray. *IEEE Transactions on Plasma Science*, **36**(5), 2586–2591.

Mitchell S. M., Ahmad M., Teel A. L. and Watts R. J. (2014). Degradation of perfluorooctanoic acid by reactive species generated through catalyzed H_2O_2 propagation reactions. *Environmental Science and Technology Letters*, **1**(1), 117–121.

Mizuno A. (2007). Industrial applications of atmospheric non-thermal plasma in environmental remediation. *Plasma Physics and Controlled Fusion*, **49**(5A), A1–A15.

Mok Y. S. (2006). Absorption-reduction technique assisted by ozone injection and sodium sulfide for NO_x removal from exhaust gas. *Chemical Engineering Journal*, **118**(1–2), 63–67.

Mok Y. S., Jo J.-O. and Whitehead J. C. (2008). Degradation of an azo dye Orange II using a gas phase dielectric barrier discharge reactor submerged in water. *Chemical Engineering Journal*, **142**(1), 56–64.

Moriwaki H., Takagi Y., Tanaka M., Tsuruho K., Okitsu K. and Maeda Y. (2005). Sonochemical decomposition of perfluorooctane sulfonate and perfluorooctanoic acid. *Environmental Science and Technology*, **39**(9), 3388 3392.

Moussa D. and Brisset J.-L. (2003). Disposal of spent tributylphosphate by gliding arc plasma. *Journal of Hazardous Materials*, **102**(2), 189–200.

Moussa D., Doubla A., Karagang-Youbi G. and Brisset J. L. (2007). Postdischarge long life reactive intermediates involved in the plasma chemical degradation of an azoic dye. *IEEE Transactions on Plasma Science*, **35**(2), 444–453.

Naïtali M., Herry J.-M., Hnatiuc E., Kamgang G. and Brisset J.-L. (2012). Kinetics and bacterial inactivation induced by peroxynitrite in electric discharges in air. *Plasma Chemistry and Plasma Processing*, **32**(4), 675–692.

Nikiforov A. Y., Leys C., Li L., Nemcova L. and Krcma F. (2011). Physical properties and chemical efficiency of an underwater dc discharge generated in He, Ar, N_2 and air bubbles. *Plasma Sources Science and Technology*, **20**(3), 1–2.

Nomoto Y., Ohkubo T., Kanazawa S. and Adachi T. (1995). Improvement of ozone yield by a silent-surface hybrid discharge ozonizer. *IEEE Transactions on Industry Applications*, **31**(6), 1458–1462.

Ohshima T., Sato K., Terauchi H. and Sato M. (1997). Physical and chemical modifications of high-voltage pulse sterilization. *Journal of Electrostatics*, **42**, 159–166.

Ono R. and Oda T. (2001). OH radical measurement in a pulsed are discharge plasma observed by a LIF method. *IEEE Transactions on Industry Applications*, **37**(3), 709–714.

Ono R. and Oda T. (2003). Dynamics of ozone and OH radicals generated by pulsed corona discharge in humid-air flow reactor measured by laser spectroscopy. *Journal of Applied Physics*, **93**, 5876–5882.

Ono R. and Oda T. (2007). Ozone production process in pulsed positive dielectric barrier discharge. *Journal of Physics D-Applied Physics*, **40**(1), 176–182.

Panorel I., Preis S., Kornev I., Hatakka H. and Louhi-Kultanen M. (2013a). Oxidation of aqueous paracetamol by pulsed corona discharge. *Ozone Science and Engineering*, **35**(2), 116–124.

Panorel I., Preis S., Kornev I., Hatakka H. and Louhi-Kultanen M. (2013b). Oxidation of aqueous pharmaceuticals by pulsed corona discharge. *Environmental Technology*, **34**(7), 923–930.

Park H., Vecitis C. D., Cheng J., Choi W., Mader B. T. and Hoffmann M. R. (2009). Reductive defluorination of aqueous perfluorinated alkyl surfactants: effects of ionic headgroup and chain length. *The Journal of Physical Chemistry A*, **113**(4), 690–696.

Parvulescu V. I., Magureanu M. and Lukes P. (2012). Plasma Chemistry and Catalysis in Gases and Liquids. Wiley, Weinheim, Germany.

Patai S. (1971). The chemistry of the hydroxyl group. In: The Chemistry of Functional Groups, S. Patai (ed.), vol. **2**. Interscience, London, New York, pp. 1–51.

Pawlat J. and Ihara S. (2007). Removal of color caused by various chemical compounds using electrical discharges in a foaming column. *Plasma Processes and Polymers*, **4**(7–8), 753–759.

Petri B. G., Watts R. J., Teel A. L., Huling S. G. and Brown R. A. (2011). Fundamentals of ISCO using hydrogen peroxide. In: In Situ Chemical Oxidation for Groundwater Remediation, R. L. Siegrist, M. Crimi, and T. J. Simpkin (ed.), Springer, New York, pp. 33–88.

Peyrous R. (1990a). The effect of relative humidity on ozone production by corona discharge in oxygen or air – a numerical simulation, Part II: air. *Ozone Science and Engineering*, **12**, 40–64.

Peyrous R. (1990b). The effect of relative humidity on ozone production by corona discharge in oxygen or air – a numerical simulation. Part I: oxygen. *Ozone Science and Engineering*, **12**, 19–40.

Pichat P., Cermenati L., Albini A., Mas D., Delprat H. and Guillard C. (2000). Degradation processes of organic compounds over UV-irradiated TiO_2. Effect of ozone. *Research on Chemical Intermediates*, **26**(2), 161–170.

Plank T., Jalakas A., Aints M., Paris P., Valk F., Viidebaum M. and Jogi I. (2014). Ozone generation efficiency as a function of electric field strength in air. *Journal of Physics D: Applied Physics*, **47**(33), 1–2.

Preis S., Panorel I. C., Kornev I., Hatakka H. and Kallas J. (2013). Pulsed corona discharge: the role of ozone and hydroxyl radical in aqueous pollutants oxidation. *Water Science and Technology*, **68**(7), 1536–1542.

Reddy P. M. K., Raju B. R., Karuppiah J., Reddy E. L. and Subrahmanyam C. (2013a). Degradation and mineralization of methylene blue by dielectric barrier discharge non-thermal plasma reactor. *Chemical Engineering Journal*, **217**, 41–47.

Reddy P. M. K., Ramaraju B. and Subrahmanyam C. (2013b). Degradation of malachite green by dielectric barrier discharge plasma. *Water Science and Technology*, **67**(5), 1097–1104.

Rumbach P., Bartels D. M., Sankaran R. M. and Go D. B. (2015). The solvation of electrons by an atmospheric-pressure plasma. *Nature Communications*, **6**, 1–6.

Ruscic B., Wagner A. F., Harding L. B., Asher R. L., Feller D., Dixon D. A., Peterson K. A., Song Y., Qian X. M., Ng C. Y., Liu J. B. and Chen W. W. (2002). On the enthalpy of formation of hydroxyl radical and gas-phase bond dissociation energies of water and hydroxyl. *Journal of Physical Chemistry A*, **106**(11), 2727–2747.

Rutberg P. G. (2002). Some plasma environmental technologies developed in Russia. *Plasma Sources Science and Technology*, **11**(3A), A159–A165.

Rutberg P. G., Kolikov V. A., Kurochkin V. E., Panina L. K. and Rutberg A. P. (2007). Electric discharges and the prolonged microbial resistance of water. *IEEE Transactions on Plasma Science*, **35**(4), 1111–1118.

Sahni M. and Locke B. (2006a). Quantification of hydroxyl radicals produced in aqueous phase pulsed electrical discharge reactors. *Industrial and Engineering Chemistry Research*, **45**(17), 5819–5825.

Sahni M. and Locke B. R. (2006b). The effect of reaction conditions on hydroxyl radical production in gas-liquid pulsed electrical discharge reactors. *Plasma Processes and Polymers*, **3**, 668–681.

Sahni M. and Locke B. R. (2006c). Quantification of reductive species produced by high voltage electrical discharges in water. *Plasma Processes and Polymers*, **3**(4–5), 342–354.

Sahni M., Finney W. C., Clark R. J., Landing W. and Locke B. R. (2002). Degradation of Aqueous Phase Trichloroethylene Using Pulsed Corona Discharges. HAKONE VIII, International Symposium on High Pressure, Low Temperature Plasma Chemistry, July 21–25, 2002, Puhajarve, Estonia.

Sahni M., Finney W. and Locke B. (2005). Degradation of aqueous phase polychlorinated biphenyls (PCB) using pulsed corona discharges. *Journal of Advanced Oxidation Technologies*, **8**(1), 105–111.

Sakiyama Y., Tomai T., Miyano M. and Graves D. B. (2009). Disinfection of *E. coli* by nonthermal microplasma electrolysis in normal saline solution. *Applied Physics Letters*, **94**(16), 161501.

Samaranayake W. J. M., Miyahara Y., Namihira T., Katsuki S., Hackam R. and Akiyama H. (2000). Ozone production using pulsed dielectric barrier discharge in oxygen. *IEEE Transactions on Dielectrics and Electrical Insulation*, **7**(6), 849–854.

Sano N., Yamamoto D., Kanki T. and Toyoda A. (2003). Decomposition of phenol in water by a cylindrical wetted-wall reactor using direct contact of gas corona discharge. *Industrial and Engineering Chemistry Research*, **42**(22), 5423–5428.

Sano N., Fujimoto T., Kawashima T., Yamamoto D., Kanki T. and Toyoda A. (2004). Influence of dissolved inorganic additives on decomposition of phenol and acetic acid in water by direct contact of gas corona discharge. *Separation and Purification Technology*, **37**(2), 169–175.

Sano N., Yamamoto T., Takemori I., Kim S. I., Eiad-ua A., Yamamoto D. and Nakaiwa M. (2006). Degradation of phenol by simultaneous use of gas-phase corona discharge and catalyst-supported mesoporous carbon gels. *Industrial and Engineering Chemistry Research*, **45**(8), 2897–2900.

Sato M. (2008). Environmental and biotechnological applications of high-voltage pulsed discharges in water. *Plasma Sources Science and Technology*, **17**(2), 024021.

Satoh K., MacGregor S. J., Anderson J. G., Woolsey G. A. and Fouracre R. A. (2007). Pulsed-plasma disinfection of water containing *Escherichia coli*. *Japanese Journal of Applied Physics Part 1-Regular Papers Brief Communications and Review Papers*, **46**(3A), 1137–1141.

Schaefer C. E., Andaya C., Urtiaga A., McKenzie E. R. and Higgins C. P. (2015). Electrochemical treatment of perfluorooctanoic acid (PFOA) and perfluorooctane sulfonic acid (PFOS) in groundwater impacted by aqueous film forming foams (AFFFs). *Journal of Hazardous Materials*, **295**, 170–175.

Sharma A., Locke B., Arce P. and Finney W. (1993). A preliminary study of pulsed streamer corona discharge for the degradation of phenol in aqueous solutions. *Hazardous Waste and Hazardous Materials*, **10**(2), 209–219.

Sharma A. K., Josephson G. B., Camaioni D. M. and Goheen S. C. (2000). Destruction of pentachlorophenol using glow discharge plasma process. *Environmental Science and Technology*, **34**(11), 2267–2272.

Shin W.-T., Yiacoumi S., Tsouris C. and Dai S. (2000). A pulseless corona-discharge process for the oxidation of organic compounds in water. *Industrial and Engineering Chemistry Research*, **39**(11), 4408–4414.

Shmelev V. M., Evtyukhin N. V. and Che D. O. (1996). Water sterilization by pulse surface discharge. *Chemical Physics Reports*, **15**(3), 463–468.

Siegrist R. L., Crimi M., Simpkin T. J. (2011). Strategic Environmental Research and Development Program (U.S.), and Environmental Security Technology Certification Program (U.S.) "In situ chemical oxidation for groundwater remediation." SERDP and ESTCP remediation technology monograph series.

Snyder S. A. (2008). Occurrence, treatment, and toxicological relevance of EDCs and pharmaceuticals in water. *Ozone: Science and Engineering*, **30**(1), 65–69.

Srivastava N., Wang C. and Dibble T. S. (2009). A study of OH radicals in an atmospheric ac discharge plasma using near infrared diode laser cavity ringdown spectroscopy combined with optical emission spectroscopy. *European Physical Journal D*, **54**(1), 77–86.

Stara Z., Krcma F., Nejezchleb M. and Skalny J. D. (2008). Influence of solution composition and chemical structure of dye on removal of organic dye by DC diaphragm discharge in water solutions. *Journal of Advanced Oxidation Technologies*, **11**(1), 155–162.

Stratton G. R., Bellona C. L., Dai F., Holsen T. M. and Mededovic Thagard S. (2015). Plasma-based water treatment: conception and application of a new general principle for reactor design. *Chemical Engineering Journal*, **273**, 543–550.

Su Z. Z., Ito K., Takashima K., Katsura S., Onda K. and Mizuno A. (2002). OH radical generation by atmospheric pressure pulsed discharge plasma and its quantitative analysis by monitoring CO oxidation. *Journal of Physics D: Applied Physics*, **35**, 3192–3198.

Sugiarto A. T. and Sato M. (2001). Pulsed plasma processing of organic compounds in aqueous solution. *Thin Solid Films*, **386**(2), 295–299.

Sun B., Aye N. N., Gao Z. Y., Lv D., Zhu X. M. and Sato M. (2012). Characteristics of gas-liquid pulsed discharge plasma reactor and dye decoloration efficiency. *Journal of Environmental Sciences-China*, **24**(5), 840–845.

Sunka P. (2001). Pulse electrical discharges in water and their applications. *Physics of Plasmas*, **8**, 2587–2594.

Sunka P., Babicky V., Clupek M., Lukes P., Simek M., Schmidt J. and Cernak M. (1999). Generation of chemically active species by electrical discharges in water. *Plasma Sources Science and Technology*, **8**, 258–265.

Takamura N., Matsumoto T., Wang D., Namihira T. and Akiyama H. (2011). Ozone generation using positive- and negative-nano-seconds pulsed discharges. In: Pulsed Power Conference (PPC), 2011 IEEE, IEEE Conference Publications, pp. 1300–1303.

Tarr M. A. (2003). Chemical Degradation Methods for Wastes and Pollutants: Environmental and Industrial Applications. M. Dekker, New York.

Tezuka M. and Iwasaki M. (1998). Plasma induced degradation of chlorophenols in an aqueous solution. *Thin Solid Films*, **316**, 123–127.

Tezuka M. and Iwasaki M. (1999). Liquid-phase reactions induced by gaseous plasma. Decomposition of benzoic acids in aqueous solution. *Plasmas and Ions*, **1**, 23–26.

Tezuka M. and Iwasaki M. (2001). Plasma-induced degradation of aniline in aqueous solution. *Thin Solid Films*, **386**(2), 204–207.

Thagard Mededovic S., Takashima K. and Mizuno A. (2009). Electrical discharges in polar organic liquids. *Plasma Processes and Polymers*, **6**(11), 741–750.

Tomizawa S. and Tezuka M. (2006). Oxidative degradation of aqueous cresols induced by gaseous plasma with contact glow discharge electrolysis. *Plasma Chemistry and Plasma Processing*, **26**(1), 43–52.

Tonks L. and Langmuir I. (1929). A general theory of the plasma of an arc. *Physical Review*, **34**(6), 0876–0922.

Tyler C., Jobling S. and Sumpter J. (1998). Endocrine disruption in wildlife: a critical review of the evidence. *CRC Critical Reviews in Toxicology*, **28**(4), 319–361.

Vecitis C., Park H., Cheng J., Mader B. and Hoffmann M. (2009). Treatment technologies for aqueous perfluorooctanesulfonate (PFOS) and perfluorooctanoate (PFOA). *Frontiers of Environmental Science and Engineering in China*, **3**(2), 129–151.

Velikonja J., Bergougnou M. A., Castle G. S. P., Cairns W. L. and Inculet I. I. (2002). Ozone dissolution vs. aqueous methylene blue degradation in semi-batch reactors with dielectric barrier discharge over the water surface. *Ozone Science and Engineering*, **24**(3), 159–170.

von Sonntag C. and von Gunten U. (2012). Chemistry of Ozone in Water and Wastewater Treatment. IWA Publishing, London.

Vujevic D., Koprivanac N., Bozic A. L. and Locke B. R. (2004). The removal of direct orange 39 by pulsed corona discharge from model wastewater. *Environmental Technology*, **25**(7), 791–800.

Wandell R. J. and Locke B. R. (2014). Hydrogen peroxide generation in low power pulsed water spray plasma reactors. *Industrial and Engineering Chemistry Research*, **53**(2), 609–618.

Wang D. Y., Matsumoto T., Namihira T. and Akiyama H. (2010). Development of higher yield ozonizer based on nano-seconds pulsed discharge. *Journal of Advanced Oxidation Technologies*, **13**(1), 71–78.

Wang H. J. and Chen X. Y. (2011). Kinetic analysis and energy efficiency of phenol degradation in a plasma-photocatalysis system. *Journal of Hazardous Materials*, **186**(2–3), 1888–1892.

Wang L., Jiang X. Z. and Liu Y. J. (2008). Degradation of bisphenol A and formation of hydrogen peroxide induced by glow discharge plasma in aqueous solutions. *Journal of Hazardous Materials*, **154**(1–3), 1106–1114.

Wen Y.-Z. and Jiang X.-Z. (2001). Pulsed corona discharge-induced reactions of acetophenone in water. *Plasma Chemistry and Plasma Processing*, **21**(3), 345–354.

Willberg D. M., Lang P. S., Hochemer R. H., Kratel A. and Hoffmann M. R. (1996a). Degradation of 4-chlorophenol, 3,4-dichloroanilin, and 2,4,6-trinitrotoluene in an electrohydraulic discharge. *Environmental Science and Technology*, **30**, 2526–2534.

Willberg D. M., Lang P. S., Hochemer R. H., Kratel A. and Hoffmann M. R. (1996b). Electrohydraulic destruction of hazardous wastes. *Chemtech*, **26**, 52–57.

Xue J., Chen L. and Wang H. L. (2008). Degradation mechanism of alizarin red in hybrid gas-liquid phase dielectric barrier discharge plasmas: experimental and theoretical examination. *Chemical Engineering Journal*, **138**(1–3), 120–127.

Yan J., Du C. M., Li X., Cheron B., Ni M. and Cen K. (2006). Degradation of phenol in aqueous solutions by gas–liquid gliding arc discharges. *Plasma Chemistry and Plasma Processing*, **26**(1), 31–41.

Yan J. H., Liu Y. N., Bo Z., Li X. D. and Cen K. F. (2008). Degradation of gas-liquid gliding arc discharge on acid Orange II. *Journal of Hazardous Materials*, **157**(2–3), 441–447.

Yan Z. C., Li C. and Lin W. H. (2009). Hydrogen generation by glow discharge plasma electrolysis of methanol solutions. *International Journal of Hydrogen Energy*, **34**(1), 48–55.

Yang Y., Kim H., Starikovskiy A., Cho Y. I. and Fridman A. (2011). Note: an underwater multi-channel plasma array for water sterilization. *Review of Scientific Instruments*, **82**(9), 1–2.

Yasuoka K., Sasaki K. and Hayashi R. (2011). An energy-efficient process for decomposing perfluorooctanoic and perfluorooctane sulfonic acids using dc plasmas generated within gas bubbles. *Plasma Sources Science and Technology*, **20**(3), 034009.

Yuan M.-H. and Watanabe T. (2011). Decomposition mechanism of phenol in water plasmas by DC discharge at atmospheric pressure. *Chemical Engineering Journal*, **168**(3), 985–993.

Zaharia C. and Suteu D. (2012). Textile Organic Dyes – Characteristics, Polluting Effects and Separation/Elimination Procedures from Industrial Effluents – A Critical Overview, Organic Pollutants Ten Years After the Stockholm Convention – Environmental and Analytical Update, Dr. Tomasz Puzyn (ed.), InTech, DOI: 10.5772/32373.

Zhang R. B., Wang L. M., Wu Y., Guan Z. C. and Jia Z. D. (2006). Bacterial decontamination of water by bipolar pulsed discharge in a gas-liquid-solid three-phase discharge reactor. *IEEE Transactions on Plasma Science*, **34**(4), 1370–1374.

Zhang Y., Zhou M. H. and Lei L. C. (2007). Degradation of 4-chlorophenol in different gas-liquid electrical discharge reactors. *Chemical Engineering Journal*, **132**(1–3), 325–333.

Zhang Y. Z., Sun B. Y., Deng S. H., Wang Y. J., Peng H., Li Y. W. and Zhang X. H. (2010a). Methyl orange degradation by pulsed discharge in the presence of activated carbon fibers. *Chemical Engineering Journal*, **159**(1–3), 47–52.

Zhang Y. Z., Xiong X. Y., Han Y., Yuan H., Deng S. H., Xiao H., Shen F. and Wu X. B. (2010b). Application of titanium dioxide-loaded activated carbon fiber in a pulsed discharge reactor for degradation of methyl orange. *Chemical Engineering Journal*, **162**(3), 1045–1049.

Zhang Z., Chen J.-J., Lyu X.-J., Yin H. and Sheng G.-P. (2014). Complete mineralization of perfluorooctanoic acid (PFOA) by gamma irradiation in aqueous solution. *Scientific Reports*, **4**, 1–6.

Zhou Z. Y., Zhang X. Y., Liu Y., Ma Y. P., Lu S. J., Zhang W. and Ren Z. Q. (2015). Treatment of azo dye (acid Orange II) wastewater by pulsed high-voltage hybrid gas-liquid discharge. *RSC Advances*, **5**(88), 71973–71979.

Chapter 13

The role of photochemistry in the transformation of pollutants in surface waters

Douglas E. Latch

13.1 INTRODUCTION

Photochemical reactions play an important role in controlling the fates of many pollutants found in surface waters. These photochemical transformations proceed through one of two processes, namely *direct* photolysis and *indirect* photolysis. Several reviews have thoroughly covered the presence and photochemical fates of particular classes of pollutants in surface waters and the production and reactivity of transient species generated in photolysis reactions, and as such this chapter is designed to serve as a primer on direct and indirect photolysis pathways relevant to the fates of aquatic pollutants while also directing the reader to relevant prior work, including early seminal studies, manuscripts that report interesting photochemical results or analysis techniques, and pollutants of current concern. Topics include: pollutants commonly found in surface waters, how dissolved constituents such as natural organic matter and nitrate/nitrite contribute to photochemistry in surface waters, laboratory methods used to differentiate between various photochemical reaction pathways, and how to extrapolate results across seasons, latitude, and water conditions. The photochemical behavior of several important pollutants are presented to illustrate key concepts and to highlight photoreactions involved in the degradation of several particularly problematic aquatic contaminants.

13.2 SOLAR RADIATION AT THE EARTH'S SURFACE

13.2.1 The solar spectrum

Sunlight is the driver of photochemical reactions occurring in surface waters and it will be briefly described here. The sun emits photons from the vacuum UV (starting at $\lambda \approx 120$ nm) to the infrared region (to $\lambda \approx 3000$ nm) of the electromagnetic spectrum. The constituents of the atmosphere, however, reflect nearly half of the incident radiation and absorb significant amounts while constituents of the upper atmosphere also absorb certain wavelength regions (Larson & Weber, 1994). The primary components of the atmosphere that absorb sunlight are ozone, which absorbs a significant amount of the high energy UV radiation and thus filtering it from reaching the surface of the earth; water; and carbon dioxide. As a result of the reflection and absorption of light by atmospheric constituents, the solar irradiation reaching the earth is significantly different than that in extraterrestrial space: the radiation reaching the earth's surface generally contains very little light with $\lambda < 290$ nm. Figure 13.1 shows an example solar spectrum for a summer day at 45° latitude over the ultraviolet (UV) and visible region. For most organic compounds, including the natural organic matter found in the environment, only light in the UV ($\lambda < 400$ nm) and visible ($\lambda \approx 400$–800 nm) regions of the solar spectrum are energetic enough to cause transformations.

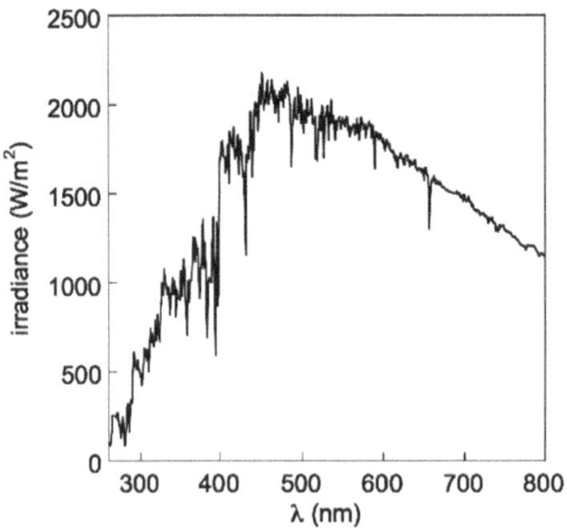

Figure 13.1. Surface level solar irradiance in the photochemically active UV and visible regions of the solar spectrum for a typical summer day at 45° latitude. (Adapted from Latch, 2005).

The energy of a photon E (λ) (kJ/mol) varies inversely with its wavelength, λ (nm), as shown in the following equation (Larson & Weber, 1994; Schwarzenbach *et al.* 2002):

$$E(\lambda) = \frac{119{,}600}{\lambda} \tag{13.1}$$

Light in the UV region of the solar spectrum is of higher energy than that in the visible range and is critically important in surface water photochemical processes because most organic molecules are weak absorbers at longer wavelengths. Another reason why UV light is particularly important in surface water photochemistry is that many constituents of natural waters (e.g. pollutants, nitrate ion, and natural organic matter) absorb UV light much more readily than photons from the visible region of the electromagnetic spectrum. These details will be described further in subsequent sections.

13.2.2 Diurnal, seasonal, and latitudinal variations

The solar spectrum varies with time of day, across seasons, and with latitude. These variations have been well documented over time and can be used to predict how a photochemical process occurring at one location or time of year would proceed at a different place or time (Zepp & Cline, 1977; Leifer, 1988; Schwarzenbach *et al.* 2002). The United States Environmental Protection Agency (EPA) offers the program GCSOLAR for predicting direct photolysis reaction rates across time and location, among other variables. This free program is available from their website and is designed to be used with personal computers (United States Environmental Protection Agency Exposure Assessment Models, GC Solar; http://www2.epa.gov/exposure-assessment-models/gcsolar). An additional photochemistry prediction model (Aqueous Photochemistry of Environmentally occurring Xenobiotics; APEX) has recently been developed and accounts for these seasonal and latitudinal variations in addition to several other surface water conditions pertinent to the photolysis of aquatic pollutants (Bodrato & Vione, 2014).

13.2.3 Light attenuation and depth dependence of photochemical reactions

As sunlight travels through surface water, it is subject to absorption and scattering due to the particulate and dissolved components in the system. In practice, it is difficult to fully account for the light attenuation that occurs in a natural water body, in part because the scattering of light by particles occurs in both the forward and reverse directions and these processes can be complicated to monitor simultaneously (Brezonik & Arnold, 2011; Schwarzenbach *et al.* 2002). A relatively simple method for estimating the effect of light attenuation in a given well-mixed non-turbid water body is to determine a light-screening factor at a given wavelength ($S(\lambda)$) as shown below (Schwarzenbach *et al.* 2002):

$$S(\lambda) = \frac{1 - 10^{-1.2\alpha(\lambda) z_{depth}}}{(2.303)(1.2)\alpha(\lambda) z_{depth}} \tag{13.2}$$

This light screening term depends on two variables: $\alpha(\lambda)$ and z_{depth}. The first term, $\alpha(\lambda)$, is a beam attenuation coefficient (accounting for absorption and scattering of light by all species present in the water, including the substrate) for that particular water body as measured by a spectrophotometer (i.e. it is a measure of how much the incident beam is reduced over the depth). The z_{depth} term is the depth (cm) of the photic zone (or depth of the water being studied) in a well-mixed water body. According to equation (13.2), $S(\lambda)$ has a maximum approaching unity for very shallow, clear waters and $S(\lambda)$ decreases as z_{depth} and/or $\alpha(\lambda)$ increase. The impact of light screening on a pollutant's expected direct photochemical half-life ($t_{1/2}$) over the z_{depth} region relative to the surface can be estimated as:

$$\text{relative } t_{1/2} \text{ for the mixed layer} = \frac{\text{surface } t_{1/2}}{S(\lambda)} \tag{13.3}$$

Zepp and Cline's (1977), Schwarzenbach *et al.* (2002), and Brezonik and Arnold (2011) provide more detailed accounts of light absorption and scattering processes and the impacts these events have on photochemical reaction rates at depth.

13.3 TYPES OF PHOTOCHEMICAL REACTIONS IN SURFACE WATERS

13.3.1 Direct photochemistry

In natural waters, one mechanism by which a compound is photochemically degraded is *direct photolysis*. In this process, the substrate absorbs light and undergoes a transformation of its chemical structure due to the energy gained in the absorption process. The direct photolysis occurs when the energy driving the transformation is *directly* absorbed by the substrate. Thus, for a pollutant molecule to be susceptible to a direct photolysis it must be able to absorb the incident radiation. The rate at which a pollutant degrades via direct photolysis depends on several factors (most of which are radiation wavelength-dependent), including the absorption properties of that specific molecule (i.e., molar absorption coefficients), the probability of the light absorption event to result in a photochemical transformation, (i.e., quantum yield), the fraction of the absorbed photons that lead to a chemical reaction, and the energy of the incident light.

In a sunlit surface water, a pollutant must absorb light of $\lambda > 290$ nm in order to degrade by direct photolysis (i.e. its absorption spectrum must overlap to some extent the solar spectrum at the Earth's surface). The light absorbing functional groups within a molecule are referred to as chromophores; important chromophores in environmental photochemistry are conjugated double bonds, substituted aromatic rings, and carbonyl, carboxylate, nitro, and phenol (particularly deprotonated phenoxide anion) groups. Many compounds found in the environment contain these functional groups, including natural organic matter (NOM) and anthropogenic pollutants such as pesticides, herbicides, natural and synthetic steroid hormones, pharmaceuticals, and chemicals arising from industrial processes, such as polycyclic aromatic hydrocarbons (Larson & Weber, 1994; Schwarzenbach *et al.* 2002).

The energy gained in the absorption process can be dissipated by a number of different processes, or it can lead to degradation and chemical reactions. The various reaction and relaxation pathways are summarized in the Jablonski diagram shown in Figure 13.2. This diagrams illustrates the absorption of radiation and the chemical, thermal and radiative relaxation pathways available to disperse the added energy, where S_0 is the singlet ground state, S_1 is the first electronic singlet excited state, T_1 is the first electronic excited triplet state, abs. is light absorption, fluor. is fluorescence, IC is internal conversion, ISC is intersystem crossing, phosph. is phosphorescence, $k_{rxn}[R]$ is chemical reaction with a reactant (R), $k_q[Q]$ is physical quenching with a quencher (Q), and $k_{degrade}$ is a substrate degradation process occurring from an excited state. The S (singlet) nomenclature indicates that all of the substrate's electrons are spin paired while the T (triplet) nomenclature is indicative of two unpaired (spin parallel) electrons. In general, under normal circumstances conversion between S and T states tends to occur slowly, as these transitions are spin-forbidden. According to Figure 13.2, an excited state may return to the singlet ground state by releasing the excess energy radiatively (by fluorescence or phosphorescence) and thermally (through internal conversion). These processes are not involved in degrading the substrate or in transferring energy to other species. Reactions, bond cleavage, and physical quenching of excited state substrates, however, lead to the degradation of the substrate or to the production of new excited state species that may be capable of transforming other chemicals. Direct photolysis reactions are shown as proceeding through S_1 and T_1 excited states in Figure 13.2.

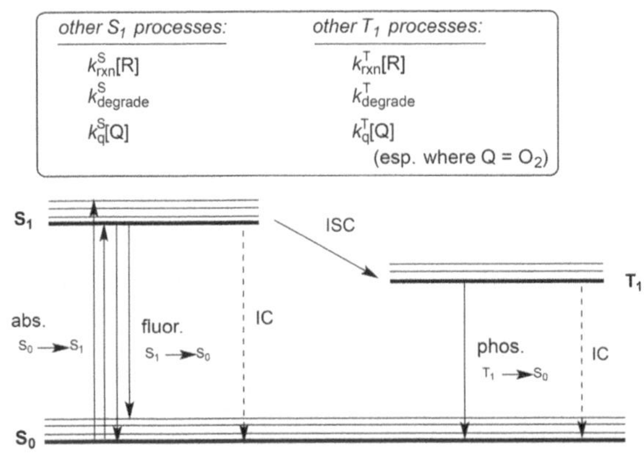

Figure 13.2. Simplified Jablonski diagram depicting light absorption and excited state relaxation, reaction, and quenching processes. Vibrational relaxations within S_0, S_1, and T_1 electronic states have been omitted for clarity. The box above the Jablonski diagram includes other common excited state processes including chemical reaction with another reactant (R), physical quenching by a quenching species (Q), and unimolecular degradation of the substrate (e.g. bond cleavage, intramolecular cyclization, rearrangement). (Adapted from Turro, 1991; Schwarzenbach *et al.* 2002; Klan & Wirz, 2010; Brezonik & Arnold, 2011).

Quantum yield is an important term in photochemistry, giving the fraction of time that a particular event (e.g. transformation or fluorescence) occurs per amount of light absorbed:

$$\Phi_x = \frac{\text{number of moles undergoing process x}}{\text{number of moles of photons absorbed}} \tag{13.4}$$

The fraction of photons that a substrate absorbs that lead to photodegradation is quantified as the substrate's direct photolysis quantum yield (Φ_{dir}):

$$\Phi_{dir} = \frac{\text{number of moles substrate transformed}}{\text{number of moles of photons absorbed}} \tag{13.5}$$

Substrates with high Φ_{dir} efficiently use the energy gained in the absorption process to cleave bonds or initiate some other reaction resulting in a change in molecular structure. Substances with high Φ_{dir} may not necessarily degrade rapidly if their molar absorption coefficients at the incident radiation wavelengths are small.

An essential component in measuring quantum yields is determining the amount of light that a substrate absorbs. The most common method for determination of the amount of light incident to a sample is chemical actinometry. In this method, a substrate with known direct photolysis quantum yield and absorption spectrum (the actinometer or reference compound) is irradiated alongside the substrate of interest. Both the reference compound and the substrate are used at relatively low concentrations (to ensure that the solutions are optically dilute). One can then determine the substrate's quantum yield ($\Phi_{dir,S}$) using the following equation:

$$\Phi_{dir,S} = \frac{k_{dir}\,\varepsilon_{act}}{k_{act}\,\varepsilon_{S}}\Phi_{dir,act} \tag{13.6}$$

Here, rate constants for the loss of the substrate (k_{dir}) and actinometer (k_{act}) are determined experimentally and inputted into the equation along with the molar absorptivities of the actinometer (ε_{act}) and substrate (ε_S) and the known $\Phi_{dir,act}$. Rates are usually measured in units of M s^{-1} and rate constants are typically given in units of s^{-1} and concentrations in molarity (M). This equation applies to photoreactions irradiated at a single wavelength. Quantum yields of fluorescence and intersystem crossing can be measured in an analogous fashion.

For experiments performed using broadband irradiation (such as sunlight) or multiple excitation lines (as is the case for Hg-vapor lamps that are commonly used in photochemistry experiments), the relevant equation becomes:

$$\Phi_{dir,S} = \frac{k_{dir}\,k_{a,act}}{k_{act}\,k_{a,dir}}\Phi_{dir,act} \tag{13.7}$$

In this case, $k_{a,act}$ and $k_{a,S}$ are integrated light absorption rates calculated over the wavelength range where the absorption spectra of the substrate and the actinometer overlap with the spectral output of the light source. The spectral output at each wavelength of the light source is also taken into account in determining $k_{a,act}$ and $k_{a,S}$, as is the light screening factor ($S(\lambda)$) over this range. These values can be easily calculated from absorption spectra and either measured or tabulated light source intensities (Leifer, 1988; Schwarzenbach *et al.* 2002). Two useful and frequently used actinometers (*para*-nitroanisole and *para*-nitroacetophenone) have been developed that allow for tuning the loss rate of the actinometer to match that of the substrate. This ensures that the substrate and actinometer receive the same amount of light over the same time interval (Dulin & Mill, 1982). This is particularly important for relatively slow photoreactions or when the light intensity of the source changes over time (e.g. as sunlight intensity varies over the course of a day). Zepp (1978) describes how to perform actinometry experiments and determine direct photolysis loss rates for pollutants in surface waters. The pesticide carbaryl is used as an example in this study to show how predictions of environmental half-lives can be made from direct photolysis quantum yields and molar absorptivities and sunlight intensities at different wavelengths. Additional details about chemical actinometry, including information about how to set up and run an actinometry experiment can be found in the following references: (Dulin & Mill, 1982; Leifer, 1988; Schwarzenbach *et al.* 2002). Arnold and McNeill (2007) reviewed the abiotic transformation reactions of many pharmaceuticals in the environment and tabulated quantum yields of direct photolysis for many of these pollutants.

The rate at which a substrate photodegrades by direct photolysis, $\mathrm{rate}_{\mathrm{dir}}$, is given below.

$$\mathrm{rate}_{\mathrm{dir}} = k_{\mathrm{dir}}[\mathrm{S}] = k_{\mathrm{abs}}\Phi_{\mathrm{dir}}[\mathrm{S}] \tag{13.8}$$

Here k_{dir} is the direct photolysis rate constant (s^{-1}), [S] is substrate concentration (M), Φ_{dir} is the quantum yield for the direct photolysis process, and k_{abs} is the rate of photon absorption (s^{-1}), which is dependent upon the spectrum of the incident radiation and the absorption spectrum of the substrate. Thus, the rate at which a substrate photodegrades in the environment can be calculated based on its absorption spectrum, the spectral distribution of solar radiation reaching the water, the substrate concentration, and Φ_{dir}. It should be noted that molar absorptivity values and k_{abs} are wavelength dependent while Φ_{dir} is usually consistent across wavelengths *within* an absorption band. Quantum yields, however, tend to vary across *different* absorption bands because each absorption band corresponds to a unique electronic transition. Rates of direct photolysis in surface waters account for all wavelengths of the surface solar spectrum that a substrate absorbs, the quantum yield of each relevant absorption band, and light attenuation over the depth of the water column. Given the very low concentrations of micropollutants in the aquatic environment, their degradation follows the first-order kinetics law. Because the absorption spectrum and Φ_{dir} of any substrate can be measured experimentally, and the seasonal and latitudinal variations in the solar spectrum are tabulated, measured direct photolysis half-lives can easily be extrapolated to different geographic regions and seasons.

The rate equation for direct photolysis reactions given above can be rearranged, integrated over time, and linearized by taking the natural logarithm to give

$$\ln\frac{[S]_t}{[S]_0} = -k_{\mathrm{dir}}t \tag{13.9}$$

If one then plots $\ln([\mathrm{S}]_t/[\mathrm{S}]_0)$ versus time, k_{dir} is given as the negative of the slope. The half-life of compound S which decays by direct photolysis is then easily calculated from the equation given below.

$$t_{1/2,\mathrm{dir}} = \frac{\ln 2}{k_{\mathrm{dir}}} \tag{13.10}$$

13.3.2 Indirect photochemistry

In indirect photochemistry, transformation of a substrate occurs through reaction with photochemically produced reactive intermediates (PPRIs) generated from photoexcited constituents of natural waters rather than through the direct photolysis process described in the preceding section. The substrate is thus *indirectly* transformed through reaction with PPRIs generated *photochemically* rather than from the energy gained through the direct absorption of light. The PPRIs are produced in natural waters through different pathways: (i) photolysis of chromophoric compounds contained in the water column; (ii) photosensitization in which chromophoric compounds (sensitizers) contained in the water absorb light and transfer energy to other compounds to form PPRIs. These PPRIs can then react with or transfer energy to pollutants in sunlit waters, leading to the transformation of the pollutants. The indirect photolysis nomenclature arises because the transformation does not require direct absorption of light by the substrate, but rather it depends on sunlight indirectly through the photochemical behavior of a sensitizing species. In this way, the term indirect photolysis is a bit of a misnomer when describing the loss of a pollutant due to reaction with PPRI, since this transformation reaction does not occur until *after* the light absorption process. Nevertheless, the indirect photolysis terminology is frequently used as a convenient way to describe the process of PPRI production and subsequent reaction with pollutants. Oftentimes the sensitization process depends on the excited state sensitizer undergoing an intersystem crossing to its excited triplet state (see Figure 13.2). The excited triplet state (T_1) is important in the sensitization process because it is longer lived than the excited singlet state due to the spin forbidden relaxation pathway to S_0. As the T_1 state of a photosensitizer is quenched by other compounds (e.g., micropollutants), energy or electrons are transferred to these "quenchers", often leading to subsequent degradation of the quenching species (such as pollutants) or to the production of other reactive intermediates (such as singlet oxygen).

The overall rate of indirect photolysis reactions ($\text{rate}_{\text{indir}}$) depends on the concentrations of individual PPRIs in natural waters and bimolecular rate constants for the reaction of the PPRIs with the pollutant of interest, as shown below.

$$\text{rate}_{\text{indir}} = \sum k_{\text{PPRI,S}}[\text{PPRI}]_{\text{SS}}[\text{S}] \tag{13.11}$$

Here, $k_{\text{PPRI,S}}$ is the bimolecular rate constant for a given PPRI with the substrate, $[\text{PPRI}]_{\text{SS}}$ is the steady-state concentration of the given PPRI, and [S] is the substrate concentration. According to the above equation, the overall indirect photolysis degradation rate is a summation of the individual transformation rates caused by each PPRI. Figure 13.3 shows a cartoon depiction of direct and some indirect photolysis pathways by which a pollutant may degrade in natural waters.

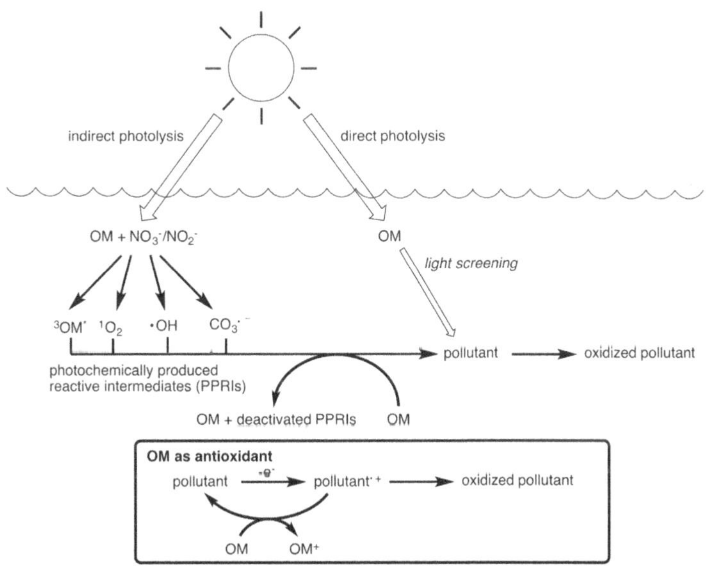

Figure 13.3. Schematic representation of the role of organic matter (OM) in direct and indirect photochemical processes. The scheme also illustrates that organic matter may also act as a quencher of some PPRIs as well as an antioxidant. (Adapted from Cooper *et al.* 1989; Blough & Zepp, 1995; Schwarzenbach *et al.* 2002; Canonica & Laubscher, 2008; Brezonik & Arnold, 2011; Wenk & Canonica, 2012; Wenk *et al.* 2013.)

13.4 LABORATORY METHODS AND TECHNIQUES FOR STUDYING POLLUTANT PHOTOCHEMISTRY

Light sources used in environmental photochemical studies include using natural sunlight, simulated broadband solar light (e.g. from filtered Xe lamps), and filtered line sources such as Hg-vapor lamps. These sources are typically used as steady state sources, where target pollutants are irradiated under a relatively constant, steady light intensity. More sophisticated experiments designed to determine reaction mechanisms are to probe a pollutant's reactivity with individual PPRI and may rely on time-resolved instrumentation where short bursts of intense laser light are used to excite the pollutant or generate PPRI. An advanced textbook on the photochemistry of organic chemicals from Klan and Wirz (2009) offers a good primer on commonly used photochemical equipment.

Photochemical degradation of target pollutants can be due to direct photolysis and/or reaction with one or more PPRIs. Delineating the extent to which a pollutant is susceptible to direct or indirect photodegradation pathways is critical in extrapolating the photochemical kinetic results obtained for a specific water body to other sites. To do so, a number of techniques have been employed. Estimating the degree to which a compound is degraded by direct photolysis in a natural water sample is relatively straightforward. Side-by-side photolysis of a filter-sterilized natural water sample with a DI water sample allows one to determine the photochemical reactivity of a substrate both in the presence and in the absence of potential sensitizing species contained in natural waters. Degradation in the DI water sample (as normalized to that measured in a dark control sample) is indicative of a direct photolysis mechanism. If the substrate degrades faster in the natural water sample, it indicates that an indirect photolysis mechanism is also involved. The extent to which the degradation rate is enhanced by indirect photolysis can be calculated based on the magnitude of the differences in the degradation rate constants in the two samples after accounting for light attenuation caused by constituents in the natural water (Schwarzenbach *et al.* 2002). Extrapolation of direct photodegradation kinetics in surface waters at a given pH is straightforward if the analyte's absorption spectrum and quantum yield are known; half-lives can be calculated rather simply in this scenario by accounting for variations in light spectrum, intensity, and filtering at different sites. Extrapolating indirect photochemical kinetic data is more complicated, as concentrations and distributions of PPRIs may be highly variable across different natural waters. This is because the chemical components dissolved in a given natural water body mediate the indirect photoprocesses, and the nature and concentrations of these constituents can widely vary from site to site.

Determining the specific PPRIs involved in indirect processes and extrapolating the kinetic results from one system to another is more complex than with direct photodegradation. The first step involves identification of the particular PPRI (or PPRIs) causing the degradation. Typically, various quenchers selective for particular PPRIs are added to the solutions prior to photolysis. If a quencher of a given PPRI does not reduce the loss rate of the substrate, a different transient species is likely to be causing the degradation. If a slower loss rate is observed, the PPRI being quenched is likely involved in the degradation of the pollutant being studied. For example, isopropanol is a good quencher of the PPRI hydroxyl radical ($^{\bullet}OH$). If addition of ca. 1% isopropanol does not alter the degradation rate of a given pollutant, $^{\bullet}OH$ is not likely to be involved in the pollutant loss in that particular water sample. If, however, a decrease in loss rate is observed upon addition of isopropanol, $^{\bullet}OH$ (or other PPRIs that may be quenched by isopropanol) is implicated in the indirect photolysis for the pollutant in that particular water sample.

To better measure the reactivity of a substrate with individual PPRIs, photochemical reactions are often performed in the laboratory with well-defined small molecule organic sensitizers, such as phenalenone for the production of singlet oxygen, $^1O_2$ (Schmidt *et al.* 1994). By choosing different sensitizers that are efficient at producing a particular PPRI, the reactivity of a substrate can be measured for a series of reactive transient species. The rate constants determined from this method, along with the concentrations measured for the individual PPRIs in natural waters, can then be used to extrapolate the rate at which a substrate will degrade by indirect photolysis processes ($rate_{indir}$) in various natural waters. An alternative approach is to use non-photochemical reactions to produce individual PPRI and measure their reactivity with target pollutants. Huang and Mabury (2000) describe such an approach for producing carbonate radical in alkaline solutions containing peroxynitrite. Other tests exist to implicate certain PPRIs in indirect photodegradation experiments. For example, $^1O_2$ involvement can be verified by diluting the solutions with either H_2O or D_2O; the lifetime of $^1O_2$ is lengthened in D_2O, leading to faster degradation rates for $^1O_2$-reactive substrates in the D_2O samples than in the H_2O diluted samples (Merkel *et al.* 1972; Wilkinson *et al.* 1995). Singlet oxygen can also be tracked by its faint phosphorescent signal at 1270 nm (Krasnovsky, 1976; Krasnovsky, 1977; Wilkinson *et al.* 1995; Nonell & Braslavsky, 2000).

The steady-state concentrations of individual PPRIs ($[PPRI]_{SS}$) are typically measured using molecular probes that have known bimolecular rate constants of reaction with the PPRI being studied ($k_{PPRI,probe}$) (Blough & Zepp, 1995; Zafiriou *et al.* 1990; Vione *et al.* 2010; Burns *et al.* 2012). Figure 13.4 depicts how probe molecules are used to determine $[PPRI]_{SS}$. Zepp (1977) provides an early example of the use of a probe molecule (dimethylfuran) used to trap and quantify singlet oxygen in sunlit surface waters.

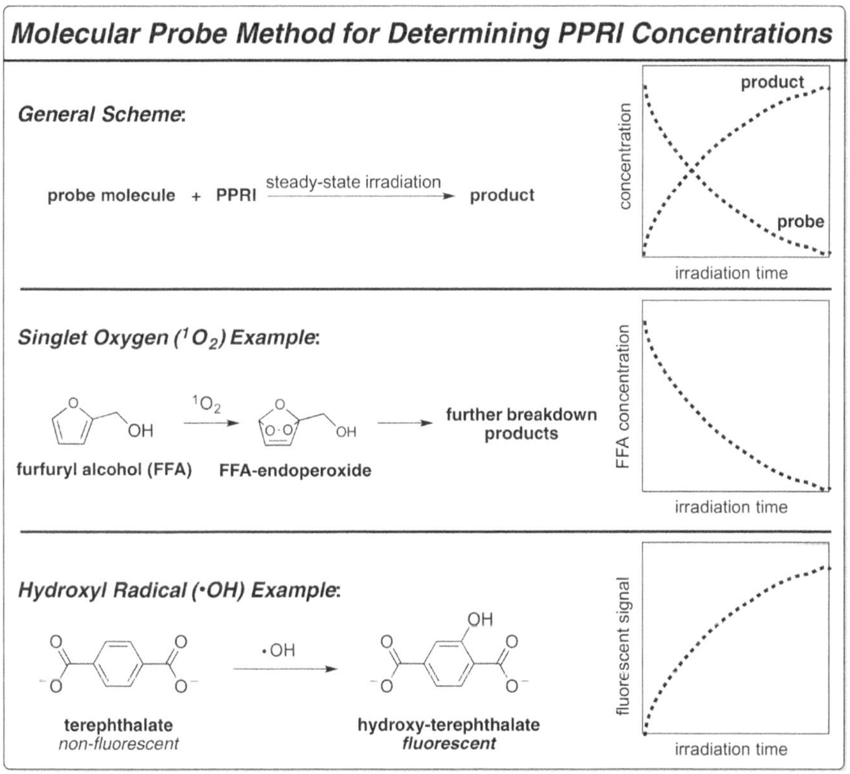

Figure 13.4. Schematic illustrating the use of molecular probes to determine PPRI concentrations under steady-state irradiation conditions. The loss of the probe molecules or the evolution of PPRI-initiated photoproducts are tracked to determine $[PPRIs]_{SS}$. (Adapted from Haag *et al.* 1984a; Haag & Hoigne, 1986; Zafiriou *et al.* 1990; Page *et al.* 2010; Burns *et al.* 2012.)

The loss of a probe molecule that is selective for a given PPRI is shown below.

$$\frac{-d[\text{probe}]}{dt} = k_{\text{PPRI,probe}}[\text{PPRI}]_{\text{SS}}[\text{probe}] = k_{\text{obs,probe}}[\text{probe}] \tag{13.12}$$

Equation (13.12) can be rearranged, integrated, and linearized to determine the observed reaction rate constant for the loss of the probe molecule ($k_{\text{obs,probe}}$).

$$\ln\left(\frac{[\text{probe}]_t}{[\text{probe}]_0}\right) = -k_{\text{obs,probe}}t \tag{13.13}$$

The steady-state PPRI concentration can then easily be computed by dividing $k_{\text{obs,probe}}$ (s^{-1}) by the known k_{PPRI} ($M^{-1}s^{-1}$) for the given probe: PPRI pair, as in the following equation.

$$[\text{PPRI}]_{\text{SS}} = \frac{k_{\text{obs,probe}}}{k_{\text{PPRI,probe}}} \tag{13.14}$$

Bimolecular rate constants for reaction of given PPRIs with pollutants can be determined in a similar manner by comparison to loss rates of molecular probes irradiated under identical conditions. If the probe and pollutant are present in solution at low enough concentrations, their effect on $[\text{PPRI}]_{\text{SS}}$ will be negligible. In this case, the probe molecule can be used to measure $[\text{PPRI}]_{\text{SS}}$ according to the equation given above. The reaction rate constant for the PPRI and the pollutant of interest can then be calculated according to equation 13.15.

$$k_{\text{PPRI,pollutant}} = \frac{k_{\text{obs,pollutant}}}{[\text{PPRI}]_{\text{SS}}} \tag{13.15}$$

This is useful, because if one knows k_{PPRI} ($M^{-1}s^{-1}$) for a variety of PPRIs, the extent of indirect photolysis can be predicted under widely varying environmental conditions (Lam *et al.* 2003; Al Housari *et al.* 2010; de Laurentiis *et al.* 2012; Vione *et al.* 2013; Bodrato & Vione, 2014). The half-life (s) for pollutant loss due to reaction with a particular PPRI can be determined with equation (13.17).

$$t_{1/2,\text{PPRI}} = \frac{\ln 2}{k_{\text{PPRI,pollutant}}[\text{PPRI}]_{\text{SS}}} \tag{13.16}$$

Table 13.1 shows estimated half-lives for several organic pollutants based on their reactivity toward the PPRI hydroxyl radical (•OH) at two different plausible environmental •OH concentrations.

Equation 13.17 can be used to estimate a pollutant's photochemical half-life when accounting for both direct and indirect photolysis reactions. This relatively simple equation includes potential transformation reactions from multiple PPRIs at a given depth and NOM concentration. Rate constants for direct and indirect photolysis, and thus substrates' half-lives, are expected to vary with depth and NOM concentration.

$$t_{1/2} = \frac{\ln 2}{k_{\text{obs,pollutant}}} \quad \text{where} \quad k_{\text{obs}} \approx k_{\text{dir}} + \sum\left(k_{\text{PPRI,pollutant}}[\text{PPRI}]_{\text{SS}}\right) \tag{13.17}$$

Table 13.1. Bimolecular Rate Constants for Reaction of Hydroxyl Radical (•OH) with Several Organic Pollutants and Estimated Pollutant Half-Lives at Two Environmental •OH Concentrations.

			Estimated $t_{1/2}$ (h)	
			$[^{\bullet}OH]_{SS}$	
Compound	**Structure**	**$k_{^{\bullet}OH}/10^9$ ($M^{-1}s^{-1}$)[a]**	**1 fM**	**0.01 fM**
diethylphthalate	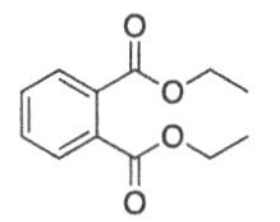	4	48	4800
2,3,3′,5,6-PCB	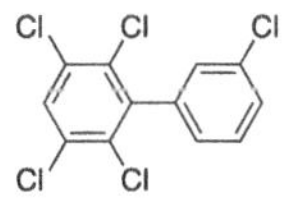	5	39	3900
benzo[a]pyrene	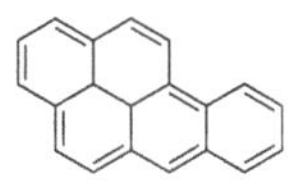	10	19	1900
2,3,7,8-tetrachlorodibenzo-*p*-dioxin	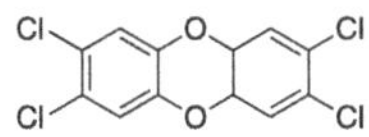	4	48	4813
lindane	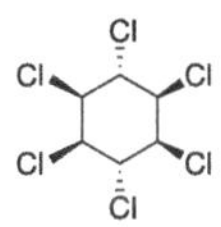	5.8	33	3300
toxaphene	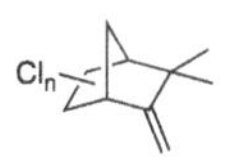	~4.6	42	4200
atrazine	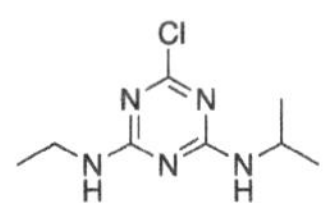	2.6	74	7400
alachlor	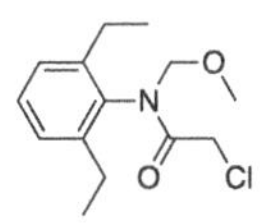	7	28	2800
2,4-dichlorophenoxyacetic acid	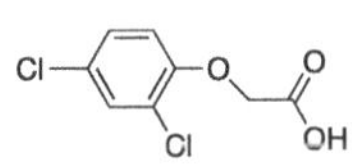	5	39	3900
aldicarb	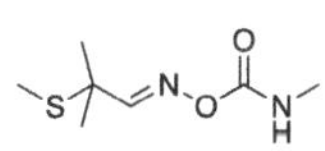	8.1	24	2400
glyphosate	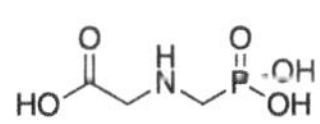	0.18	1100	11000
diquat	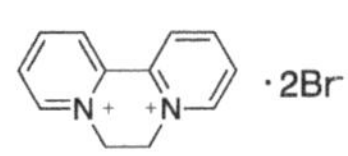	0.8	240	24000

[a]bimolecular reaction rate constants as reported in Haag and Yao (1992).

13.5 PHOTOCHEMICALLY PRODUCED REACTIVE INTERMEDIATES (PPRIs) AND THE ROLE OF ORGANIC MATTER IN INDIRECT PHOTOCHEMISTRY

A wide variety of PPRIs are produced in natural waters, including hydroxyl radical ($^{\bullet}OH$), excited state triplet organic matter (^3OM), singlet oxygen ($^1O_2$), hydrated electron (e_{aq}^-), superoxide radical anion ($O_2^{\bullet-}$), carbonate radical ($CO_3^{\bullet-}$), and organoperoxy radicals (Cooper *et al.* 1989; Blough & Zepp, 1995; Aguer *et al.* 1999). Organic matter (OM) is a critically important sensitizer in natural waters. Aquatic natural organic matter (NOM) is derived from the decomposition of terrestrial plant material ("terrestrial" or allochthonous OM) or produced within waters by algae and other microorganisms ("microbial" or autochthonous OM). Many surface waters in the United State and around the world are also impacted by effluent organic matter (EfOM) derived from wastewater treatment plants (Brooks *et al.* 2006). Effluent OM is similar in structure and reactivity to microbial OM (Nam & Amy, 2002; Drewes & Croue, 2006; Shon *et al.* 2006; Jarusutthiraka & Amy, 2007). The OM pool in most natural waters is dominated by terrestrial OM, but microbial OM may be important in settings where plant life is not sustained, in aquaculture settings, and in effluent-impacted surface waters (Brooks *et al.* 2006; Guerard *et al.* 2009a). Many studies have concluded that PPRIs originating from OM play a major role in the chemistry occurring in surface waters (Zepp *et al.* 1981; Cooper *et al.* 1989; Goldberg *et al.* 1992; Blough & Zepp, 1995; Richard & Canonica, 2005; Aguer & Richard, 1996; Scully *et al.* 1997; Aguer *et al.* 1999; Burns *et al.* 2012). PPRIs have been involved as reactive species in the degradation of many classes of pollutants found in sunlit natural waters. Figure 13.5 shows direct and indirect photolysis pathways and the production of various PPRIs from OM, nitrate, and nitrite. The following sections describe some of the important PPRIs found in sunlit surface waters.

13.5.1 Hydroxyl radical ($^{\bullet}OH$)

One of the most reactive PPRIs in natural waters is the hydroxyl radical ($^{\bullet}OH$). Although the photolysis of hydrogen peroxide (H_2O_2) under UV irradiation is a simple means of $^{\bullet}OH$ generation in the laboratory, this process is inefficient in the wavelength region of sunlight reaching the natural waters (Cooper *et al.* 1989). The primary $^{\bullet}OH$ sources in most natural waters are NO_3^- and NO_2^- (Brezonik & Fulkerson-Brekken, 1998; Larson & Weber, 1994; Schwarzenbach *et al.* 2002), although OM also has been shown to sensitize its production, as shown in Figure 13.5 (Hoigne *et al.* 1989). Estimates of the steady-state surface concentration of $^{\bullet}OH$ ($[^{\bullet}OH]_{ss}$) in natural waters at average noontime irradiation levels are about $0.15 - 50 \times 10^{-17}$ M (Cooper *et al.* 1989). Hydroxyl radical is a very non-selective reactant; it reacts at nearly diffusion controlled rates with most organic compounds (Buxton *et al.* 1988). The primary reaction pathways of $^{\bullet}OH$ are hydrogen atom abstraction or $^{\bullet}OH$ addition. Haag and Yao (1992) measured rate constants for the reaction of $^{\bullet}OH$ with several pollutants. Due to its high reactivity with a multitude of substrates, many different types of probe molecules have been employed in its detection (Mill *et al.* 1980; Cooper *et al.* 1989; Hoigne *et al.* 1989; Blough & Zepp, 1995; Zafiriou *et al.* 1990; Vione *et al.* 2010; Burns *et al.* 2012). Analysis of the reaction products arising from these molecular probes has become the general method for the analysis of $^{\bullet}OH$ in aquatic systems, because the products indicate whether $^{\bullet}OH$ or other radical species were involved in the degradation of the probes (Cooper *et al.* 1989; Hoigne *et al.* 1989; Blough & Zepp, 1995; Zafiriou *et al.* 1990). Vione *et al.* (2010) determined that benzene specificity for $^{\bullet}OH$ made it the superior probe out of several aromatic molecules tested. A terephthalate-based probe has been developed and used in recent studies assessing the formation and reactivity of $^{\bullet}OH$ in natural waters under both dark and irradiated conditions (Saran & Summer, 1999; Page *et al.* 2010; Page *et al.* 2011; Page *et al.* 2012). Advantages of the terephthalate probe are its excellent sensitivity and high reactivity with •OH. Two recent studies demonstrate that OM contained in WWTP effluent (EfOM) is also capable of sensitizing the production of •OH, which may be important in the indirect photodegradation of micropollutants downstream of WWTPs (Dong & Rosario-Ortiz, 2012; Lee *et al.* 2013).

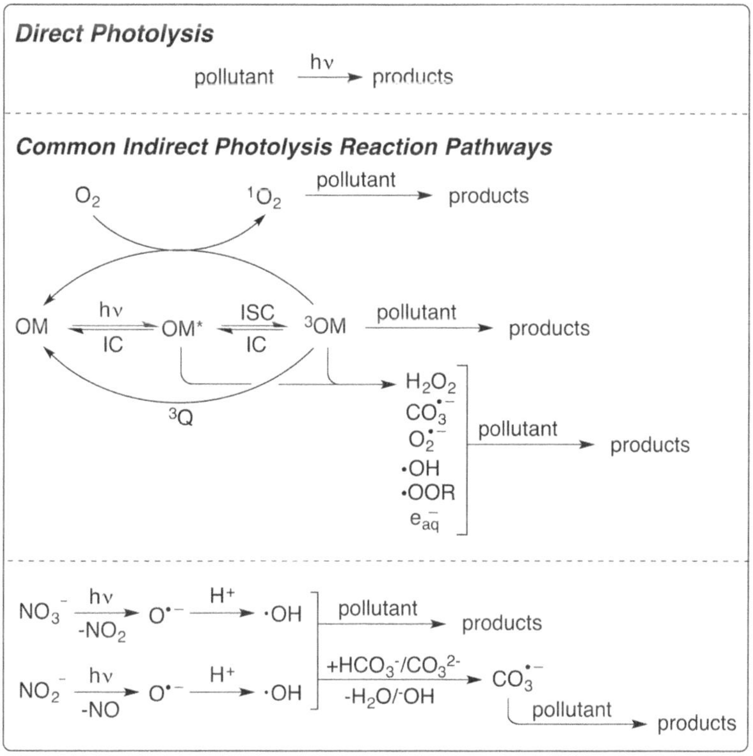

Figure 13.5. Scheme showing possible photochemical transformation reaction pathways for a given pollutant. The indirect photolysis portion shows a simplified scheme depicting the production of common PPRIs from organic matter (OM) and nitrate/nitrite. The bottom portion shows that hydroxyl radical (•OH) produces carbonate radical ($CO_3^{\bullet-}$) in alkaline waters in addition to rapidly reacting with aquatic pollutants. (Adapted from Cooper *et al.* 1989; Blough & Zepp, 1995; Schwarzenbach *et al.* 2002; Brezonik & Arnold, 2011; Huang & Mabury, 2000a, b; Canonica *et al.* 2005.)

13.5.2 Excited state triplet organic matter (^{3}OM)

Excited state triplet organic matter (^{3}OM) is a particularly important PPRI in natural waters, being an immediate precursor to other PPRIs (e.g. $^1O_2$ and $O_2^{\bullet-}$) (Cooper *et al.* 1989; Blough & Zepp, 1995). In addition to being the source of secondary PPRIs, ^{3}OM can also cause the transformation of organic pollutants through direct energy or electron transfer (Canonica & Hoigne, 1995; Canonica *et al.* 1995; Canonica *et al.* 2000; Canonica & Freiburghaus, 2001; Canonica, 2007; Felcyn *et al.* 2012; Grebel *et al.* 2012). Reactions with ^{3}OM (E = 1.36–1.95 V) may proceed via electron transfer, hydrogen abstraction, or energy transfer pathways (Canonica, 2007). Oxidation of pollutants by ^{3}OM, however, tends to occur via electron transfer

(Canonica *et al.* 2006; Canonica, 2007; Canonica & Laubscher, 2008). The formation of ^{3}OM occurs following intersystem crossing from singlet OM excited states (see the scheme shown in Figure 13.5). Zepp *et al.* have determined that half of all ^{3}OM formed in sunlit natural waters have excited state energies of >60 kcal/mol higher than ground state OM (Zepp *et al.* 1985). They also estimate ^{3}OM concentrations range from 10^{-15} to 10^{-13} M in surface waters. Reaction with ^{3}OM appears to be a particularly important reaction pathway in the fates of various phenolic compounds. This reactivity is robust enough that phenols have been used as probes for and quenchers of ^{3}OM in several laboratory studies, with trimethylphenol (TMP) being a frequently used ^{3}OM probe (Canonica *et al.* 1995; Canonica & Freiburghaus, 2001; Halladja *et al.* 2007). Evidence for ^{3}OM in photosensitization processes has also been determined by following the *cis-trans* isomerization of pentadiene, a ^{3}OM probe (Zepp *et al.* 1981), or by monitoring the effect of varying the concentration of dissolved oxygen, a potent triplet quencher, on the degradation kinetics of ^{3}OM-sensitized substrates (Canonica *et al.* 1995; Zepp *et al.* 1985; Felcyn *et al.* 2012). Grebel *et al.* (2011) demonstrated the usefulness of using sorbic acid (*trans-trans*-hexadienoic acid) as a ^{3}OM probe molecule. An advantage of using sorbic acid (SA) as a probe is that it is readily water soluble and can monitor energy transfer from ^{3}OM, whereas most other probes (such as TMP) monitor electron transfer. Figure 13.6 shows SA isomers that are monitored during the ^{3}OM-mediated transformation of SA.

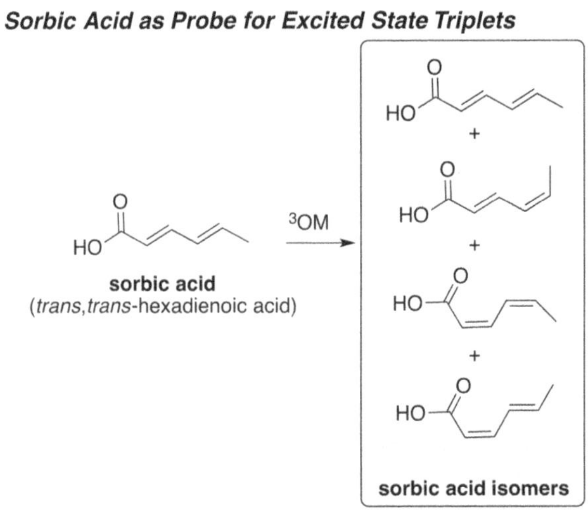

Figure 13.6. Sorbic acid as a probe for ^{3}OM. The distribution of sorbic acid isomers after irradiation is used to determine $[^3OM]_{SS}$ (adapted from Grebel *et al.* 2011).

It is worth noting here that in addition to being a sensitizer of PPRIs, OM also acts as a PPRI quencher and as an antioxidant toward partially oxidized pollutants produced via reaction with ^{3}OM. These processes are depicted near the bottom of Figure 13.5. Canonica and Laubscher (2008) demonstrated the ability of OM to quench partially oxidized intermediates before they could be completely oxidized to relatively stable products. In this process, OM donates an electron to partially oxidized pollutants thereby regenerating the pollutant, as depicted at the bottom of Figure 13.3. Wenk *et al.* (2011) further demonstrated this effect with a series of partially oxidized pollutants. In an effort to determine the moieties responsible for the antioxidant properties of OM, further studies demonstrated that various phenolic molecules inhibited the excited state triplet mediated oxidation of anilines and sulfonamide antibacterials (Wenk & Canonica,

2012). Another study found that OM from aquatic and terrestrial sources has much greater antioxidant capacity than OM from microbial sources, which means that ^3OM sensitized oxidation of pollutants may proceed more quickly when microbial OM dominates the OM pool due to the (Aeschbacher *et al.* 2012). Using time-resolved laser spectroscopy, Wenk *et al.* (2013) provide direct evidence of OM's ability to quench triplet excited states. Organic matter is also a sink for $^{\bullet}OH$ radicals.

13.5.3 Singlet oxygen ($^1O_2$)

Singlet oxygen ($^1O_2$) is formed in sunlit natural waters through the quenching of ^3OM by dissolved oxygen, which is a triplet in its ground state. This process is depicted in Figure 13.5. Its presence in natural waters was first reported by Zepp *et al.* (1977). In surface waters, $^1O_2$ concentrations tend to be in the range of 10^{-14}–10^{-12} M (Cooper *et al.* 1989; Blough & Zepp, 1995; Burns *et al.* 2012). Singlet oxygen is a much more selective oxidant than $^{\bullet}OH$, reacting primarily with alkenes, phenols, unsaturated heterocyclic ring systems, and sulfur compounds (Ackerman *et al.* 1971; Wilkinson *et al.* 1995). As an electrophile $^1O_2$ reacts more rapidly with electron-rich versions of the previously mentioned functional groups. Tratnyek and Hoigne (1991) demonstrate this behavior for a series of substituted phenols and deprotonated phenols. Furfuryl alcohol is a probe molecule that is frequently used in the quantification of $^1O_2$ (Haag *et al.* 1984a, b, 1986; Schwarzenbach *et al.* 2002). Sodium azide is commonly employed as a $^1O_2$ quencher in photochemistry studies (Haag & Mill, 1987; Burns *et al.* 2012). Paul *et al.* (2004) used a laser flash photolysis system to measure $^1O_2$ production quantum yields for several OM samples. They found that OM from terrestrial sources was slightly more efficient at producing $^1O_2$ than OM from microbial sources. Mostafa and Rosario-Ortiz demonstrated that effluent organic matter (EfOM) is also capable of generating $^1O_2$ upon irradiation, where it produces $[^1O_2]_{SS}$ of ca. 10^{-13} M at the surface under environmental conditions. It was also found to be slightly more efficient than NOM at producing $^1O_2$, giving quantum yields of $^1O_2$ production of 2.8–4.7% compared to 1.6–2.1% for NOM. A pair of studies shows that because $^1O_2$ is generated within OM molecules and has a limited diffusion distance in solution (due to quenching by water), a $^1O_2$ concentration gradient exists between OM molecules and the bulk aqueous phase (Latch & McNeill, 2006; Grandbois *et al.* 2008). Most pollutants and $^1O_2$ probe molecules are freely dissolved in the aqueous phase and "witness" the average aqueous phase $[^1O_2]_{SS}$. Hydrophobic pollutants that partition into the OM microenvironment, however, may observe apparent concentrations of $^1O_2$ that are two order of magnitude higher than that seen in the bulk aqueous phase, which could lead to much faster degradation rates for these types of pollutants.

13.5.4 Hydrated electron (e_{aq}^-), superoxide radical anion ($O_2^{\bullet-}$), and hydrogen peroxide

Hydrated electrons (e_{aq}^-) have also been detected in irradiated OM solutions. An elusive PPRI to monitor, e_{aq}^- has been reputed to arise from phenol moieties of OM (Zepp *et al.* 1987; Cooper *et al.* 1989; Blough & Zepp, 1995), though others have demonstrated that the quantum yield for e_{aq}^- production from OM is inversely proportional to phenolic content (Thomas-Smith & Blough, 2001). The authors of the latter study suggest that carbonyl groups are the dominant e_{aq}^- sensitizers in natural waters. Hydrated electrons are reactive with both organic and inorganic compounds containing electronegative atoms, though OM may be the major e_{aq}^- sink in natural waters (Cooper *et al.* 1989). Like ^3OM, e_{aq}^- is the source of secondary PPRIs, as it reacts with ground state oxygen to form superoxide anion. The reaction of O_2 and e_{aq}^- occurs quickly, $k \sim 2 \times 10^{10}$ M^{-1} s^{-1}, indicating that this is likely the primary mode of $O_2^{\bullet-}$ production in natural systems (Cooper *et al.* 1989). An early report of $O_2^{\bullet-}$ in natural water was provided by Baxter and Carey (1983). Estimates of the production of $O_2^{\bullet-}$ in natural water fall in the range of 10^{-11}–10^{-9} mol L^{-1} s^{-1} (Cooper *et al.*

1989). In seawater, $O_2^{\bullet-}$ is ubiquitous and has been postulated to be the dominant radical species present, accounting for up to one third of all radical species (Micinski *et al.* 1993). Compared to other PPRIs, $O_2^{\bullet-}$ is long-lived in natural waters due to its relatively low reactivity (with $t_{1/2} \sim 20$ min) (Petasne & Zika, 1987). It acts as a one-electron reductant, but primarily undergoes dismutation to form H_2O_2 (Petasne & Zika, 1987; Cooper *et al.* 1989; Blough & Zepp, 1995). Up to 60% of the $O_2^{\bullet-}$ formed in natural waters generate H_2O_2; because of its relative stability, H_2O_2 is typically detected at higher concentrations (up to 500 nM) in the environment than are other PPRIs (Petasne & Zika, 1987).

13.5.5 Carbonate radical ($CO_3^{\bullet-}$)

Carbonate radical is a relatively mild oxidant typically formed in high alkalinity waters via reactions of •OH with carbonate and bicarbonate ions as shown in Figure 13.5. It is capable of oxidizing pollutants containing N-heterocycles, sulfur species, phenols, and anilines, particularly if these species are electron-rich, as these species can effectively donate electrons and reduce $CO_3^{\bullet-}$ to the relatively stable CO_3^{2-} (Huang & Mabury, 2000a, c; Canonica, 2005). Field measurements using *N,N*-dimethylaniline as a probe molecule gave $[CO_3^{\bullet-}]_{SS}$ values of ca. 10^{-15}–10^{-13} M (Huang & Mabury, 2000c). Canonica *et al.* (2005) measured the reactivity of $CO_3^{\bullet-}$ with several classes of compounds, including anilines, deprotonated phenols, and phenylurea herbicides. Two studies by Huang and Mabury (2000a,c) demonstrated rapid reactions between $CO_3^{\bullet-}$ and electron-rich sulfur compounds and various pesticides. One of the studies presented a novel non-photochemical method for producing $CO_3^{\bullet-}$ via the thermal breakdown of peroxynitrite in the presence of carbonate or bicarbonate. This method is useful for independently assessing the reactivity of a given pollutant with $CO_3^{\bullet-}$ while avoiding direct photolysis or reaction with other PPRIs. Bimolecular reaction rate constants for the reaction of $CO_3^{\bullet-}$ with pollutant molecules can be determined via competition kinetics with various anilines (Huang & Mabury, 2000) or by directly tracking the change in $CO_3^{\bullet-}$ absorbance as pollutants are added in time-resolved experiments (Canonica *et al.* 2005; Burns *et al.* 2012).

13.5.6 Organoperoxyl radicals (•OOR)

Organoperoxy radicals (•OOR) are a class of PPRIs that are formed when extremely short-lived carbon-centered radicals react with O_2 (Cooper *et al.* 1989; Hoigne *et al.* 1989; Blough & Zepp, 1995). Due to the difficulty in differentiating their chemical behavior from •OH and in analytically detecting the individual species that make up $RO_2^{\bullet}$, they have not been as well studied as other PPRIs, especially regarding their role in the degradation of aquatic pollutants. The collective $RO_2^{\bullet}$ concentration has been estimated to be between 100 pM to and 10 nM (Cooper *et al.* 1989). Some electron-rich phenols exhibit rapid reaction with $RO_2^{\bullet}$ (Cooper *et al.* 1989; Hoigne *et al.* 1989; Blough & Zepp, 1995).

13.6 SALINITY EFFECTS ON PHOTOCHEMICAL REACTIONS IN NATURAL WATERS

The focus of this chapter is on the photochemical reactions in surface freshwaters because pollutants in these areas tend to be more problematic and better studied. A few studies will be highlighted here to show how the high salinity associated with estuaries and marine systems impact photolysis rates and pathways. The rate of direct photolysis of the fungicide fenarimol was found to decrease as salinity increases (Conceicao *et al.* 2000). Experiments designed to probe the reaction mechanism indicate that fenarimol degrades from its first singlet (S_1) excited state, and that halide ions may be quenching this S_1 excited state leading to the observed effect. In studying how halide ions impact advanced oxidation processes involving •OH, Grebel *et al.* (2010) found that bromide and chloride ions scavenged the •OH leading to reduced loss rates for aquatic pollutants and the production of reactive halogen species (RHS). The RHS produced are quite reactive with electron rich organic contaminants while reacting very slowly with electron poor molecules. Overall, the quenching of •OH by halides will thus be more impactful for electron poor pollutants. These results have implications for •OH reactive pollutants in sunlit saline waters.

Glover and Rosario-Ortiz (2013) measured quantum yields of production of various PPRIs from OM with various concentrations of added chloride and bromide. They found that the production of ^{3}OM increased as halide concentrations increased, likely due to enhanced intersystem crossing due to the halide ions. In support of this conclusion, a reduction in OM fluorescence was also observed, which is consistent with a reduction in excited singlet state OM (^{1}OM*) concentration that would be expected with increasing halide concentrations (since halide ions promote the conversion of ^{1}OM* to ^{3}OM). In surface waters, this may mean that pollutants that react with ^{3}OM or its downstream PPRIs may see rate enhancements in saline waters. Parker *et al.* (2013) also report that ^{3}OM concentrations increase with ionic strength, but that the consequences of such ^{3}OM concentration increases depend on the particular type of reaction being undertaken. Electron transfer reactions initiated by ^{3}OM actually decreased despite the elevated levels of ^{3}OM, due to putative slower electron transfer reactions within the OM itself. Energy transfer reactions initiated by ^{3}OM, on the other hand, suffered no such inhibition, and pollutants reacting through this pathway should see faster degradation rates due to the increases in concentration of ^{3}OM expected in estuarine and marine systems. A pair of studies, however, show that photobleaching of OM is expected to occur more rapidly in saline waters than in freshwater (Grebel *et al.* 2009; Grebel *et al.* 2012). This photobleaching effect appears to be due to some combination of generalized ionic strength effects and to reaction with reactive halogen species produced during the photolysis process. On a practical level, the enhanced photobleaching of OM in saline water will lead to a reduction in OM sensitized photoreactions due to the loss of OM chromophores. Grebel *et al.* (2012) show how this photobleaching effect impacts the photolysis of the estrogenic hormone 17β-estradiol (E2) and describe how E2's loss rate and reaction pathways are expected to change in moving from freshwater to more saline waters.

13.7 RANITIDINE AND CIMETIDINE: AN ILLUSTRATIVE SURFACE WATER PHOTOCHEMISTRY EXAMPLE

This section is devoted to the photochemical fates of two high-use pharmaceutical compounds (the antacids cimetidine and ranitidine) to illustrate some of the photochemical concepts and analysis techniques described in the preceding sections. The photochemical results reported in this section all come from Latch *et al.* (2003a). Cimetidine and ranitidine are high use over the counter medications that have been detected in several surface waters (Zuccato *et al.* 2000; Kopin *et al.* 2002; Calamari *et al.* 2003). Despite having similar therapeutic functions, cimetidine and ranitidine display very different photochemical behavior under solar irradiation conditions. Due to its nitroacetamidine chromophore, ranitidine absorbs a significant amount of radiation at $\lambda > 290$ nm, and a percentage of the light energy gained in the absorption process leads to chemical transformation. It was measured to have a Φ_{dir} of 0.005 and an environmental half-life of about 35 minutes at the surface of sunlit waters in midsummer sunlight at 45° latitude. Cimetidine, on the other hand, does not appreciably absorb solar light and is not susceptible to direct photolysis in deionized (DI) water. The top panel of Figure 13.7 shows the absorption spectra of ranitidine and cimetidine overlaid with the solar spectrum. The photochemical kinetic behavior of ranitidine and cimetidine under direct photolysis conditions is shown in the bottom panel. Cimetidine is readily transformed in natural waters exposed to sunlight, however, despite its inertness to direct photolysis and stability in samples stored in the dark, as seen in Figure 13.8. The photochemical behavior of cimetidine exemplifies indirect photolysis processes. Although cimetidine is not transformed by direct photolysis, it readily reacts with at least one PPRI produced in sunlit natural waters.

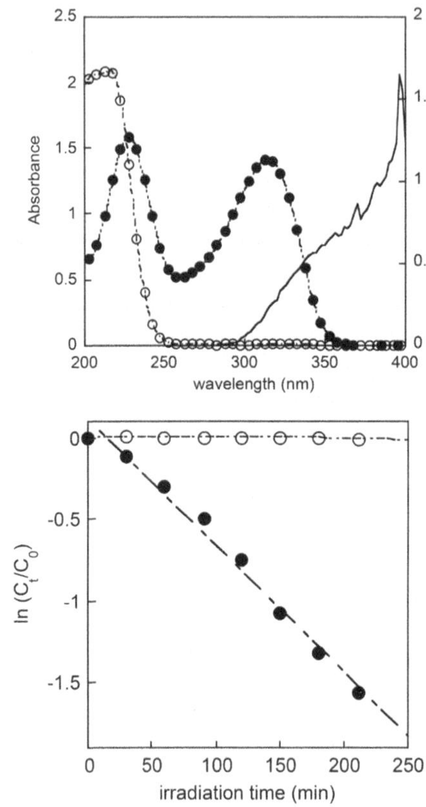

Figure 13.7. Top panel: Absorption spectra for ranitidine (filled circles) and cimetidine (open circles) plotted alongside the solar spectrum (solid line) to show spectral overlap. Bottom panel: Direct photolysis results for ranitidine (filled circles) and cimetidine (open circles). (Adapted from Latch *et al.* 2003a.)

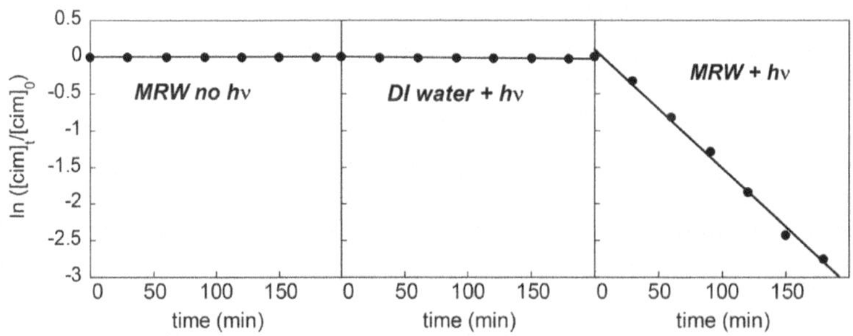

Figure 13.8. Plots showing the importance of indirect photolysis in the degradation of cimetidine (cim). The left panel is a dark control experiment in which cimetidine was dissolved in Mississippi River water (MRW) with no irradiation ($h\nu$), the center panel is the photochemical behavior of cimetidine in irradiated deionized (DI) water, and the right panel shows the degradation of cimetidine in irradiated MRW (adapted from Latch *et al.* 2003a).

Further experiments with ranitidine and cimetidine revealed additional details about their photochemical behavior. Experiments with various added quenchers revealed that $^1O_2$ was involved in the photodegradation of both ranitidine and cimetidine. For ranitidine, $^1O_2$ accelerated the loss rate by about 10% relative to its direct photolysis rate. The loss of cimetidine was dominated by reaction with $^1O_2$. Although only one photoproduct (cimetidine sulfoxide) could be conclusively identified, reaction sites could be determined by studying the photoreactivity of model compounds. Sulfide model compounds were somewhat reactive toward $^1O_2$, but models of the ranitidine and cimetidine heterocyclic rings (furan for ranitidine and imidazole for cimetidine) proved to be more reactive with $^1O_2$ and are likely the reaction sites for the parent compounds. The electron density of the cimetidine heterocycle was strongly influenced by pH, however. As the imidazole ring is protonated ($pK_a = 7.1$), its reactivity with $^1O_2$ plummets. This leads to a strong pH effect over the natural water range. Ranitidine also has an environmentally relevant pK_a value (8.2), but its photolysis rate in natural waters is not expected to vary much with pH. This is because the amine functionality that is the site of protonation at low pH is far removed from the nitroacetamidine chromophore driving the direct photolysis reaction. This was revealed through the absorption spectra and direct photolysis reaction rates of model compounds relative to that of the ranitidine parent molecule. A summary of the photochemical behavior of ranitidine and cimetidine is given in Figure 13.9.

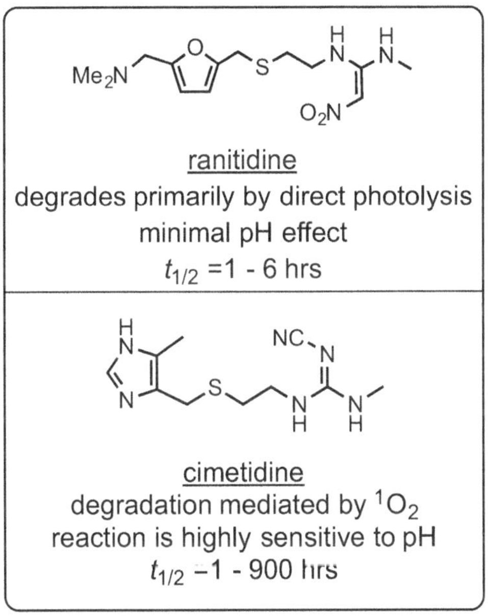

Figure 13.9. Summary of the photochemical kinetics behavior of ranitidine and cimetidine (adapted from Latch *et al.* 2003a).

13.8 SELECT PHOTOCHEMICALLY ACTIVE AQUATIC POLLUTANTS

The presence of pollutants in surface waters has attracted attention from the environmental science community and the general public (Richardson & Bowron, 1985; Buser *et al.* 1998; Christensen, 1998; Halling-Sorenson *et al.* 1998; Jobling *et al.* 1998; Belfroid *et al.* 1999; Daughton & Ternes, 1999; Jones *et al.* 2002; Kolpin *et al.* 2002; Singer *et al.* 2002; Kolodziej *et al.* 2003; Snyder *et al.* 20003; Holbrook *et al.*

2004; Lin & Reinhardt, 2005; Zuo *et al.* 2006). In a 1999–2000 national reconnaissance, the US Geological Survey (USGS) reported that 80% of the 139 studied waterways contained at least one pharmaceutical, hormone, or other wastewater contaminant compound in the receiving waters downstream from wastewater treatment plants (WWTPs; Kolpin *et al.* 2002). The results in that seminal USGS report and a subsequent AP news article (An AP Investigation: Pharmaceuticals found in drinking water) drew significant interest, because the chemicals that were targeted have been shown (indeed were *designed*) to elicit biological responses at low concentrations (Woodling *et al.* 2006; Martinovic *et al.* 2008; Thorpe *et al.* 2009).

A review in the journal *Science* outlined some of the challenges posed by micropollutants in natural waters (Schwarzenbach *et al.* 2006). The authors report that chemicals such as pesticides, fertilizers, and synthetic organic chemicals are produced in extremely high volumes and many of these chemicals have significant biological activities, may bioaccumulate, and could lead to bacterial resistance. Indeed, reports have shown that aquatic organisms living in contaminated systems have been negatively impacted, with endocrine disrupting chemicals such as hormones among the most important contaminants, as they have been shown to elicit profound negative physiological responses, including intersex development (Woodling *et al.* 2006; Kidd *et al.* 2007; Martinovic *et al.* 2008; Thorpe *et al.* 2009). Fortunately, natural processes, including photochemical reactions, in surface waters may reduce the concentrations of important pollutants and waterborne pathogens (Wegelin *et al.* 1994; Gurr & Reinhard, 2006). The following sections highlight some early studies on the photochemistry of aquatic pollutants and those that provide interesting details on the fates of important photoactive surface water pollutants that have been detected in natural waters. What follows is far from an exhaustive list, and references to several review articles on the fates of aquatic pollutants are included so that the reader can find information about other photolabile pollutants.

13.8.1 Pharmaceuticals

The photochemical fates of many pharmaceuticals and personal care products (PPCPs) have been examined. The impetus for this research has been the widespread detection of various PPCPs in natural waters throughout North and South America and Europe (Buser *et al.* 1998; Belfroid *et al.* 1999; Daughton & Ternes 1999; Stumpf *et al.* 1999; Kolpin *et al.* 2002; Boyd *et al.* 2003; Kolpin *et al.* 2003; Anderson *et al.* 2004; Boyd *et al.* 2004). Pharmaceuticals and personal care products enter the aquatic environment via numerous pathways (an overview of which is shown in Figure 13.10). The primary means for introducing PPCPs to the environment is through effluent leaving WWTPs (Daughton & Ternes, 1999). Other sources of PPCPs in the environment are from the dumping of unused medicinal products into solid-waste streams, disposal of unneeded or expired drugs, industrial waste, combined sewer overflows, from aquaculture, and by water-induced runoff from livestock fields (Halling-Sorenson *et al.* 1998; Daughton & Ternes, 1999; Kolodziej *et al.* 2004; Kolodziej & Sedlak, 2007; Mansell *et al.* 2011; Foreman *et al.* 2012; Phillips *et al.* 2012; Webster *et al.* 2012).

Khetan and Collins (2007) extensively reviewed the occurrence and fate of human pharmaceuticals in the environment. This review includes a description of how pharmaceuticals enter aquatic environments, reports on the environmental implications of the high use drugs, explores aquatic ecotoxicological effects of some particularly important pharmaceuticals, and highlights the photochemical behavior of several classes of drugs. Barcelo and Petrovic (2007) identify some of the knowledge gaps in regard to the fates of pharmaceuticals in the environment including a lack of information on pollutants contained in solid matrices or within biota, the need to identify transformation products, and the need to use bioassays or some other method to determine the activity of product mixtures. Addressing one of the points raised by Barcelo and Petrovic, the presence of PPCPs in fish tissue at low ng/g levels were reported at several sites throughout the United States (Ramirez *et al.* 2009). Oulton *et al.* (2010) reviewed the removal processes of PPCPs during wastewater treatment. In addition to describing the incomplete removal of these pollutants under most treatment conditions, they also highlight the potential problems associated with biologically active byproducts produced during the treatment processes, identifying the need for future research of these byproducts.

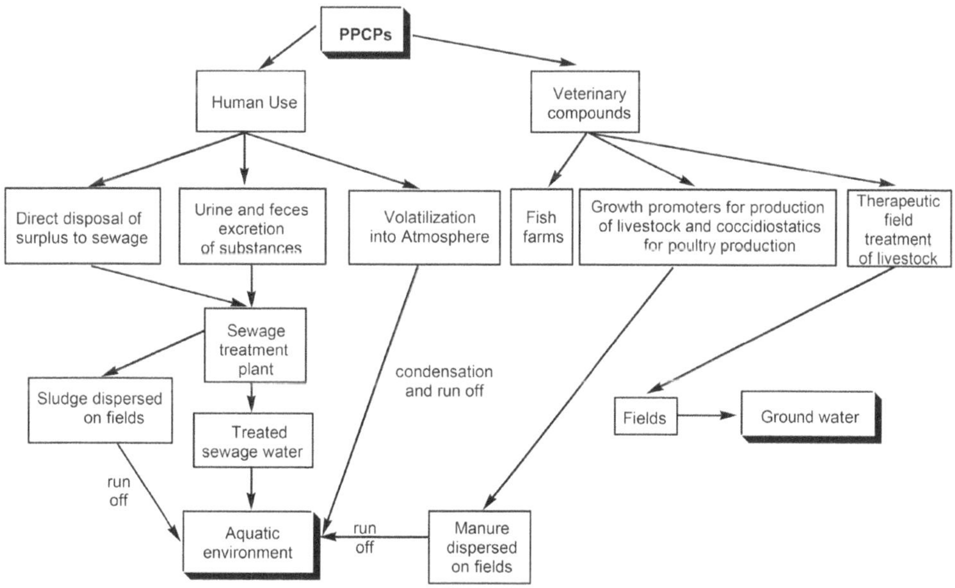

Figure 13.10. Processes by which PPCPs (and related pollutants) enter the aquatic environment (adapted from Halling-Sorensen *et al.* 1998).

A 2003 review by Boreen *et al.* has shown that many pharmaceuticals are photolabile, although many have not been studied under environmentally relevant conditions (i.e. in water with irradiation wavelengths >300 nm). In 2007, Arnold and McNeill reviewed the photochemistry of pharmaceuticals, reporting direct photolysis quantum yields and illustrating a direct/indirect photodegradation continuum for several common pollutants. Challis *et al.* (2014) recently wrote a review on the photolysis of pharmaceuticals in natural waters. Part of their focus was analyzing some of the most studied pollutants and assessing how well the experiments were designed and conducted; their review is worth consulting prior to initiating a pollutant photochemistry research program.

13.8.1.1 Antibiotics

Kummerer (2001) reviewed the occurrence and photochemical behavior of antibiotics in the environment. Typical concentrations in surface waters were in the low ng/L range, with some found at orders of magnitude higher levels (e.g. erythromycin was found as high as 1.7 μg/L). In a pair of extensive studies, Boreen *et al.* (2004, 2005) reported the direct and indirect photochemistry of ten sulfa drugs containing either five- or six-membered heterocyclic rings. They also performed experiments to quantify the importance of indirect photolysis pathways and identified major photoproducts. They determined that the sulfa drugs containing five-membered heterocycles degraded via direct photolysis (see Table 13.2) while the drugs containing six-membered rings underwent significantly slower direct photodegradation and were transformed primarily via reaction with ^3OM. The table also highlights the role of pH on the speciation and photochemistry of pollutants containing environmentally relevant pK_a values. The speciation of such pollutants can impact absorption spectra (and, therefore, k_{abs}), quantum yields, and rates of reaction with PPRIs. Table 13.3 shows half-life estimates for the sulfa drugs containing six-membered heterocycles across different seasons, latitudes, and $^{\bullet}OH$ concentrations. Guerard *et al.* (2009a) determined that the composition of the organic matter in aquaculture waters made a significant difference in the photochemical fate of the antibiotic sulfadimethoxine, with organic matter derived from microbial sources being a much better sensitizer than organic matter of terrestrial origin. Another study by Guerard and coworkers (2009b) confirmed that the indirect photolysis of sulfadimethoxine and the antibacterial agent triclocarban were accelerated when microbially derived organic matter was present relative to when terrestrial organic matter was used as photosensitizer. A similar result was reported for the photolysis of the antibacterial compounds sulfamethoxazole and trimethoprim (Ryan & Arnold, 2011). Their study focused on the use of lake water versus effluent organic matter, which contains a significant amount of microbial organic matter. Photolysis rate enhancements were evident to a much greater extent when effluent was sensitizer than in lake water. More recently, Bahnmuller *et al.* (2014) reported on the behavior of sulfadiazine and sulfamethoxazole in solutions containing natural organic matter, river water, and wastewater effluent. They found that direct and indirect photolysis pathways could be important depending on the specific pollutant and the identity of the OM. Reactions with $^{\bullet}OH$ and ^{3}EfOM were potentially important in the indirect photoprocess. In mesocosm experiments, sulfapyridine was rapidly photodegraded, with the loss attributed to some direct photolysis and through reaction with ^{3}OM and $^1O_2$ (Challis *et al.* 2013).

Amoxicillin is a potent antibiotic similar to penicillin that is found at low levels (ca. ng/L) in surface waters. It was found to degrade rapidly under simulated sunlight and natural water conditions, with the loss being attributed to a mix of some direct photolysis and through reaction with excited state OM (Xu *et al.* 2011).

Werner *et al.* (2006) studied the photochemical fate of tetracycline and found that it was sensitive to pH and concentrations of the divalent cations Mg^{2+} and Ca^{2+}, thus highlighting that water hardness can be an important variable in its photolysis (and similar pollutants capable of binding divalent cations). In the photochemistry of tylosin, a macrolide antibiotic used for veterinary purposes, photoisomerization to a less potent form preceded a direct photodegradation reaction (Werner *et al.* 2007).

13.8.1.2 Non-steroidal anti-inflammatory drugs (NSAIDs) and other analgesics

Packer *et al.* (2003) examined the photochemical behavior of four NSAIDs: ibuprofen, diclofenac, naproxen, and clofibric acid. All four were photolabile, with naproxen and diclofenac degrading primarily by direct photolysis. Clofibric acid and ibuprofen degraded by indirect photolysis via reaction with $^{\bullet}OH$ and other radicals.

A field study found that diclofenac was rapidly eliminated from a Swiss lake through direct photolysis, with a half-life of less than one hour (Buser *et al.* 1998). Tixier *et al.* (2003) monitored several pharmaceuticals in Lake Greifensee, Switzerland, and the depth profile obtained for diclofenac indicated that it was rapidly dissipated by photolysis at the surface of the lake, consistent with the finding of Buser *et al.* (1998). Their data suggest that in-lake photodegradation may also be important loss processes for naproxen and ketoprofen. They also identified several photoproducts. Ruggeri *et al.* (2013) also studied the photochemistry of ibuprofen, deducing that a mix of processes including direct photolysis and reactions with $^{\bullet}OH$ and ^{3}OM were involved in its transformation. They also determined that a major photoproduct, the toxic isobutylacetophenone (also detected by Jacobs *et al.* 2011), was formed in high yields. In addition, they modeled how the photochemical behavior of ibuprofen was expected to change based on water depth, OM concentration, carbonate/bicarbonate concentration, and nitrite content. The aquatic photochemical reactivity of acetaminophen was recently studied by De Laurentiis *et al.* (2014). Their work shows that acetaminophen can be expected to degrade via direct photolysis ($\Phi_{dir} = 0.046$), through reaction with $CO_3^{\bullet-}$ (especially under basic conditions), and reaction with ^3OM (only above about 5 mg DOC/L). They inputted various photochemical parameters to model acetaminophen fate under different environmental conditions (e.g. varying OM concentrations, water depths, and alkalinity) and found that summer photochemical half-lives are expected to range between days to several weeks. In addition, several photooxidation products were identified.

Table 13.2. Structures, Acid Dissociation Constants, Protonation State and Direct Photolysis Reaction Rate Constants and Quantum Yields of Several Sulfonamide Antibiotics (adapted from Boreen *et al.* 2004 and Boreen *et al.* 2005).

Compound	Structure	pK_{a1}	pK_{a2}	Protonation State[a]	$k_{dir}/10^{-6}(s^{-1})$	Φ_{dir}
sulfamethoxazole	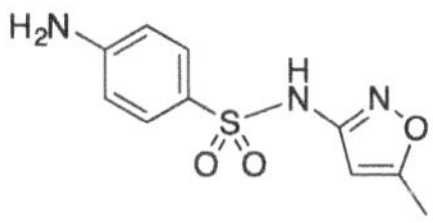	1.6	5.7	SH^+	<3	0
				S	60	0.5
				S^-	8	0.09
sulfisoxazole	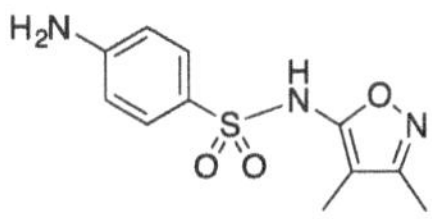	1.5	5.0	SH^+	110	0.7
				S	70	0.17
				S^-	21	0.07
sulfamethizole	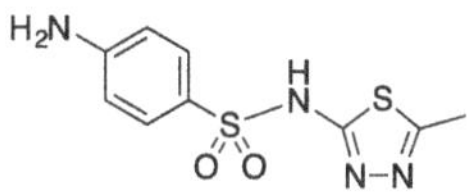	2.1	5.3	SH^+	<3	<0.01
				S	<3	<0.005
				S^-	13	0.5
sulfathiazole	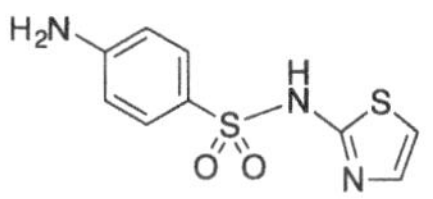	2.2	7.2	SH^+	6	0.02
				S	31	0.07
				S^-	140	0.40
sulfamethazine	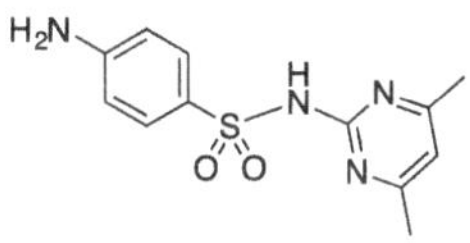	2.6	8	SH	10	0.0003
				S^-	7	0.005
sulfamerazine	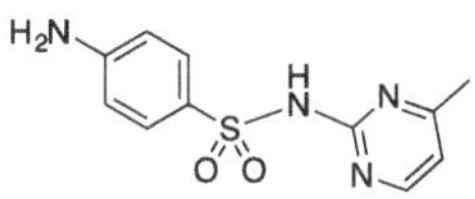	2.5	7	SH	5.4	0.00023
				S^-	6	0.003
sulfadiazine	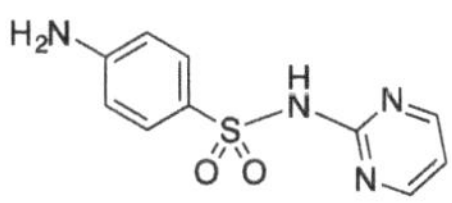	2	6.4	SH	8	0.0004
				S^-	10	0.0012
sulfachloropyridazine	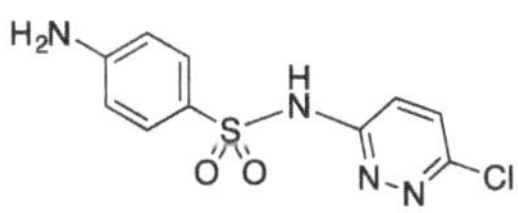	2	5.9	SH	10	0.0003
				S^-	34	0.0023
sulfadimethoxole	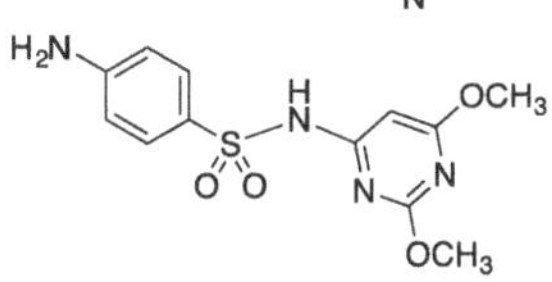	2.9	6.1	SH	20	0.000010
				S^-	4	0.00004

[a]The pH values used in determining the protonation states and measuring k_{dir} and Φ_{dir} are included in Boreen *et al.* (2004) and Boreen *et al.* (2005).

Table 13.3. Expected Seasonal and Latitudinal Variations in Photochemical Half-Lives and Indirect Photolysis Contribution for Several Sulfonamide Antibiotics in a pH 8 Surface Water at Two Different Environmental •OH Concentrations (adapted from Boreen *et al.* 2005).

Compound	$t_{1/2}$ (h)						Natural Water Enhancement $k_{natural}/k_{DI}$
	Direct Photolysis				Hydroxyl Radical		
	30° latitude		45° latitude		$[^{\bullet}OH]_{SS}$		
	Summer	Winter	Summer	Winter	1 fM	0.01 fM	
sulfamethazine	28	72	31	180	38	3800	2.8
sulfamerazine	51	120	55	300	51	5100	1.6
sulfadiazine	28	69	31	160	52	5200	1.4
sulfachloropyridazine	9	21	9	48	44	4400	1.3
sulfadimethoxole	35	120	45	420	32	3200	0.9

13.8.1.3 Other PPCPs

The photochemical behavior of the antiviral drug Tamiflu was studied by Goncalves *et al.* (2011). They found that the parent drug degraded much faster than its major metabolite ($t_{1/2}$ of about 15 days versus 150 days, respectively). Photoproducts from Tamiflu and its major metabolite tended to be more stable than the parent chemicals, meaning that these products could build up in natural waters. The antiepileptic drug carbamazepine was found to degrade through direct photolysis and by reaction with photogenerated •OH and that the relative importance of the two pathways depended on the specific field conditions (De Laurentiis *et al.* 2012). They developed a model to predict expected reaction rates and pathways based on variations in water depth and concentrations of natural water constituents. They also identified several photoproducts, including the mutagenic compound acridine. Personal care products have also been extensively studied, with the photochemistry of sunscreens being an example (MacManus-Spencer *et al.* 2011; Vione *et al.* 2013). The Vione *et al.* (2013) study included models to predict behavior under different environmental conditions.

13.8.2 Agrochemicals

Agrochemicals (pesticides, herbicides, fungicides) are ubiquitous contaminants in surface waters and have been the subject of many photochemical studies. Fenner *et al.* (2013) recently reviewed the degradation of pesticides in the environment and determined that while biodegradation is the dominant transformation pathway, direct and indirect photochemical reactions are important for some classes of pesticides. They also caution that pesticide degradation products may retain their biological activity (or even increase in potency).

An early study demonstrated that atrazine-like triazines were degraded upon exposure to UV irradiation (Pape & Zabik, 1972). Building on that work, Minero *et al.* have shown that atrazine is degraded upon exposure to environmentally relevant light energy (Minero *et al.* 1992). Furthermore, they demonstrated that the atrazine photodegradation rate was enhanced by OM, indicating that indirect photolysis pathways may be an important environmental fate process for this herbicide. Marchetti *et al.* (2013) followed up on that prior work to show that direct photolysis and reactions with •OH and ^{3}OM may be important to the fate of atrazine. They include a model to show how reaction rates and pathways are expected to change with depth and concentrations of dissolved natural water constituents. Lam *et al.* (2003) used a similar approach to analyze the photochemical behavior of atrazine and several other pesticides and pharmaceuticals.

Several studies have demonstrated the importance of photochemistry to the fate of chlorinated pesticides (Pillai, 1977; Quistad & Mulholland, 1983; Palumbo & Garcia, 1988; Fulkerson Brekken & Brezonik, 1998). In these reports, both direct photolysis and indirect photolysis mechanisms involving $^1O_2$ and $^{\bullet}OH$ have been implicated as degradation pathways. Indirect photolysis mechanisms have also been involved in the degradation of other pesticides. Alar, a pesticide used by apple growers, has demonstrable reactivity with $^1O_2$ (Brown & Casida, 1988), as do a series of thioether pesticides, including disulfoton and fenthion (Gohre & Miller, 1986). Indirect photochemistry involving ^3OM leads to enhanced degradation rates for organophosphorous pesticides (Gohre & Miller, 1986; Kamiya & Kameyama, 2001), models of the pesticide fenuron (Richard *et al.* 1997), and phenylurea herbicides (Gerecke *et al.* 2001). Gerecke *et al.* (2001) have shown that triplet sensitization reaction rates measured in well-controlled laboratory experiments can explain seasonal and depth-dependent phenylurea concentration profiles in natural waters. Gerecke *et al.* (2001) and Canonica *et al.* (2006) have reported on the rapid reaction of phenylurea herbicides with excited state triplets, with second-order reaction rate constants approaching the diffusion-controlled limit (ca. $10^9\ M^{-1}s^{-1}$). Rate constants show a strong negative correlation to the sum of the phenylureas' σ_{meta}^+ and σ_{para}^+ Hammett constants (Gerecke *et al.* 2001 and Schwarzenbach *et al.* 2002). The proposed general mechanism for the oxidation of phenylurea herbicides by excited state triplets is shown in Figure 13.11.

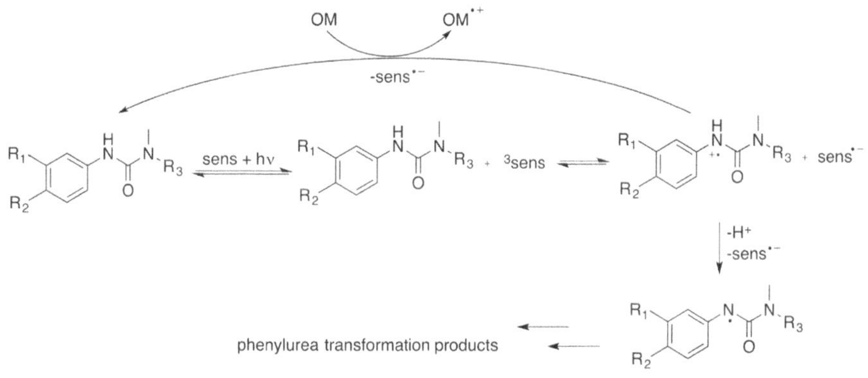

Compound	R_1	R_2	R_3	k_{sens} (h^{-1})
Metoxuron	OCH_3	Cl	CH_3	0.63
CGA 24482	3,4-tetramethylene		CH_3	0.52
GGA 16519	CH_2CH_3	H	CH_3	0.47
IPU	$CH(CH_3)_2$	H	CH_3	0.44
CGA 17767	$CH(CH_3)_2$	H	CH_2CH_3	0.32
CGA 17092	$C(CH_3)_3$	H	CH_3	0.31
Fenuron	H	H	CH_3	0.18
Chlorotoluron	CH_3	Cl	CH_3	0.22
GCA 18414	$CH(CH_3)_2$	Cl	CH_2CH_3	0.16
Fluometuron	H	CF_3	CH_3	0.051
Diuron	Cl	Cl	CH_3	0.055

Figure 13.11. Scheme depicting the oxidation of phenylurea herbicides initiated by a triplet sensitizer (sens) and OM's ability to act as an antioxidant, shown here quenching the partially oxidized phenylurea radical cation. Also included are measured rate constants for the triplet-sensitized oxidation of several phenylurea herbicides with Suwannee River fulvic acid as sensitizer. (Adapted from Gerecke *et al.* 2001; Schwarzenbach *et al.* 2002; Canonica *et al.* 2006; Canonica & Glaubscher, 2008.)

The relatively recalcitrant herbicide alachlor was found to undergo a pH-sensitive reaction with •OH produced in water from treatment wetlands (Miller & Chin, 2005). Hand and Oliver (2010) describe the indirect photodegradation of the relatively new fungicide isopyrazam in laboratory and mesocosm experiments and highlight the need to perform higher order studies that more closely mimic field conditions. The herbicide imazethapyr was also found to be photodegradable, with direct photolysis being the main reaction pathway (Espy *et al.* 2011). The authors also determined that higher energy UV light (254 nm) promoted faster degradation than solar light, due to greater photon flux and energy of the 254 nm light. They also found that NOM acted primarily as a light filter in the photolysis process, photolysis proceeded more rapidly above pH 4, and identified several photoproducts. Two interesting studies reported the role of NOM in the photolysis of mirex, a perchlorinated pesticide (Mudawbi & Hassett, 1988; Burns *et al.* 1997). Through careful experimentation, the authors of the two studies determined that photodechlorination reactions of mirex occurred within the NOM matrix (i.e. the highly hydrophobic mirex was partitioned within the NOM rather than freely dissolved in the aqueous phase during the photolysis process).

Zeng and Arnold (2013) surveyed the photochemical activity of sixteen pesticides in pothole lakes in North Dakota, USA. Prairie pothole lakes are very small wetland areas found throughout the Great Plains of the United States and Canada. They found that photochemical reactions were elevated in these pothole lakes relative to those performed in deionized water, indicating that these systems promote indirect photolysis.

Burrows *et al.* (1998) wrote a review that summarizes environmental direct and indirect photochemical degradation kinetics for a wide variety of phenolic pesticides. Another review by Burrows *et al.* (2002) reported on the photochemical reaction pathways and mechanisms undertaken by a broader set of pesticides. A recent review by Pace and Barreca (2013) describes surface water photochemical reaction mechanisms for several anthropogenic chemicals, including pesticides. Remucal (2014) recently wrote an extensive review covering the reported indirect photochemistry of a wide range of pesticides and highlighted areas for future study, including assessing the importance of specific formulations on pesticide photolysis, identification of photoproducts and analyzing their ecotoxicity, and better identification of PPRIs involved in indirect photodegradation reactions.

13.8.3 Other photochemically active pollutants detected in surface waters

In addition to transforming organic pollutants, photochemical processes can be involved in the cycling of metals and the breakdown of organometallic species, as in the case of mercury. Mercury is one of the most problematic pollutants found in surface waters. It is most commonly found in three different forms in the surface waters: its neutral form (Hg^0), as a divalent cation (Hg^{2+}), and as the organometallic complex methylmercury $(CH_3)Hg^+$, ($MeHg^+$). Photochemical processes have been implicated in the environmental cycling and transport of mercury (Nriagu, 1994; Sellers *et al.* 1996; Lalonde *et al.* 2001; Zhang & Hsu-Kim, 2010). Amyot *et al.* (1997) have shown that Hg^{2+} is readily photoreduced to Hg^0. Among others, Lalonde and coworkers (2001, 2004) demonstrate that the reverse photooxidation reaction (Hg^0 to Hg^{2+}) also occurs. In a 2005 study, Garcia *et al.* monitored diel variations in Hg^0 photooxidation and determined that Hg^{2+} reduction was the dominant redox reaction occurring during daylight hours (i.e. reduction of Hg^{2+} was faster than Hg^0 photooxidation). From a human health perspective, $MeHg^+$ is the most problematic of the common Hg forms due to its toxicity and bioaccumulative properties. Sellers *et al.* (1996) reported that $MeHg^+$ photodemethylates upon exposure to sunlight. A later study attributed the photodemethylation reaction an intra-OM reaction of $MeHg^+$ with $^1O_2$ (Zhang & Hsu-Kim, 2010). Two later studies point out the importance of binding between $MeHg^+$ and reduced sulfur sites of OM in the photodemethylation of $MeHg^+$ (Black *et al.* 2012; Fernandez-Gomez *et al.* 2013). Both studies also report that shorter irradiation wavelengths elicited faster degradation rates. Black *et al.* (2012) note that although shorter wavelengths lead to faster photodemethylation rates at the surface, longer wavelengths in the visible region may be more important in the degradation of $MeHg^+$ due to their penetration depth in natural waters. Jeremiason *et al.* (2015) show that thiol binding sites on OM are involved in the photoreduction of Hg^{2+} and the photodemethylation of $MeHg^+$. The Hg^{2+} photoreduction appears to proceed through direct photolysis of the Hg^{2+}:thiol bond while the $MeHg^+$ photodemethylation reaction involves PPRIs.

Photochemistry has been implicated in the degradation of many other compounds of environmental concern, including dichlorinated dioxins (Choudry & Webster, 1987; Kim & O'Keefe, 2000), polycyclic aromatic hydrocarbons (PAHs) (Karthikeyan & Chorover, 2000), and various phenolic pollutants (Garcia 1994; Canonica & Hoigne, 1995; Canonica *et al.* 1995; Burrows *et al.* 1998; Vialoton *et al.* 1998; Chun *et al.* 2000; Richard & Grabner, 2001; Vialoton & Richard, 2002; Javier Benitez *et al.* 2003). Nonylphenol and nonylphenol ethoxylates are potent endocrine disruptors detected in the environment that have also been shown to undergo photo-induced degradation, primarily via indirect pathways (Ahel *et al.* 1994; Hale *et al.* 2000; Neamtu & Frimmel, 2006). The potent toxin produced by cyanbacteria, microcystin-LR is subject to indirect photodegradation, though this will likely only be a significant loss process in shallow water (Welker & Steinberg, 2000).

13.9 NOTABLE EXAMPLES OF AQUATIC POLLUTANTS TRANSFORMED THROUGH PHOTOCHEMICAL REACTIONS

13.9.1 Triclosan

Triclosan is a widely-employed antibacterial compound used in many consumer products such as soaps, shampoos, mouth rinses, deodorants, and polymer mixtures to give them bacteriostatic properties (Heath *et al.* 1998; McMurry *et al.* 1998; Schweizer, 2001). Triclosan is one of the most ubiquitous personal care products and its consumption in the U.S. is on the order of 600,000 kg/yr (Anderson *et al.* 2004). As such, it has drawn significant attention from environmental scientists and engineers. Many publications regarding the occurrence and fate of triclosan and triclosan-related compounds have been published (Okomura & Nishikawa, 1996; Müller *et al.* 2000; Kolpin *et al.* 2002; Lindstrom *et al.* 2002; Tixier *et al.* 2002; Bester, 2003; Boyd *et al.* 2003; Anderson *et al.* 2004; Balmer *et al.* 2004; Boyd *et al.* 2004; Morrall *et al.* 2004; Kanda *et al.* 2003; Quintana & Reemtsma, 2004; Kronimus *et al.* 2004; Mezcua *et al.* 2004). The interest in triclosan as an aquatic pollutant stems from three main factors: its potential to induce bacterial resistance, its ubiquitousness, and its pre-dioxin structure.

A mechanism for bacterial resistance to triclosan is known (Schweizer, 2001). McBain *et al.* (2002) report that exposure to low concentrations may cause triclosan-resistant *staphylococci* to develop. Triclosan-resistant bacteria have been isolated from industrial plants, which use relatively high amounts of triclosan for sterilization purposes (Lear *et al.* 2002). While most of the strains that they isolated were naturally resistant, others had developed a low tolerance to triclosan. In a review of the relevant triclosan literature, Aiello and Larson (2003) also report that there is some evidence for triclosan-resistance development in certain bacteria. Drury *et al.* (2013) found that while triclosan did not alter bacterial abundance or respiration rates, triclosan addition to an artificial stream led to reduced benthic bacterial diversity and could lead to triclosan resistance. Due to the potential harm that triclosan and triclocarban, another antimicrobial, pose to aquatic environments and their limited effectiveness in household use, Halden (2014) has called for regulating their use.

Anderson *et al.* (2004) report that the triclosan removal efficacy at wastewater treatment facilities varies widely, from only about 32% removal at plants using primary treatment methods to 90–95% at plants using primary methods in combination with secondary and tertiary treatments. Triclosan was detected in 57% of the 139 streams tested in a USGS national sampling survey (using a report level of 50 ng/L), at a maximum concentration of 2.3 μg/L and a median value of 140 ng/L (Kolpin *et al.* 2002). Additional studies by the USGS report the detection of triclosan at levels of up to 140 ng/L or higher in 10% of streams in Iowa (USA) during low flow conditions (Kolpin *et al.* 2004). Triclosan was also detected in the influent of water entering a wastewater treatment plant in the Netherlands, with trace levels seen exiting in the discharged effluent (van Stee *et al.* 1999). In a survey of PPCPs in six wastewater treatment plants in the United Kingdom, Kanda *et al.* (2003) detected triclosan in all of the crude water samples and in the effluent.

A study from the UK found triclosan in both the influent and effluent of multiple wastewater treatment plants (Sabalunius *et al.* 2003). While the researchers determined that various combination of treatment methods led to between 95% and 99.5% removal, triclosan was still detectable in the effluent leaving the facilities. Furthermore, they tracked the degradation of triclosan as it traveled downstream from the treatment plants; in-stream processes degraded triclosan with rate constants of 0.21–0.33 hr^{-1}. Consistent with this finding, Morrall *et al.* (2004) also detected triclosan concentrations at various sites downstream from a sewage treatment facility in Texas. After accounting for sorption and settling, Morrall *et al.* determined that other in-stream processes remove or degrade triclosan with a rate constant of 0.25 hr^{-1}. In a pair of studies, Singer *et al.* (2002) and Tixier *et al.* (2002) found that photochemical processes are the major summertime degradation process for triclosan in Lake Greifensee. In a study of a Spanish wastewater treatment facility, Mezcua *et al.* (2004) detected triclosan and an unresolved mixture of the triclosan photoproducts 2,7- and 2,8-dichlorodibenzo-*p*-dioxin (2,7/2,8-DCDD) in the influent and effluent. Bester (2003) has detected triclosan in the influent (median = 1200 ng/L) and effluent (median = 51 ng/L) of wastewater treatment facilities in Germany, indicating that between 4–5% of the incoming triclosan evades removal at the treatment facilities. Triclosan has also been detected at concentrations between 1.6–29 ng/L in stormwater canals and recreational urban waterways in New Orleans, LA (Boyd *et al.* 2004). The presence of triclosan in sediments from Jamaica Bay (New York, USA) shows that it is persistent over the span of several decades (Miller *et al.* 208). In addition to these findings, others have also detected methyl triclosan, produced from the biological methylation of triclosan, as a ubiquitous contaminant in German sediments (Kronimus *et al.* 2004). Methyl triclosan has also been detected in Swiss lakes and in the tissue of fish taken from those lakes (Balmer *et al.* 2004).

Like other polychlorinated phenoxyphenols, triclosan is a pre-dioxin that forms a polychlorodibenzo-*p*-dioxin upon photolysis (Crosby & Wong, 1976; Choudry & Webster, 1987; Freeman & Srinivasa, 1986; Kanetoshi *et al.* 1988; Kanetoshi *et al.* 1992 Latch *et al.* 2003b; Latch *et al.* 2005). Dioxins are a family of compounds that have been extensively studied due to their potential toxicity to animals and ecosystems. An early set of studies in the Pawtuxet and Providence Rivers detected triclosan along with structurally related compounds, including 2,8-DCDD, which were hypothesized to be derived from the synthesis of triclosan (Jungclaus *et al.* 1978; Hites & Lopez-Avila, 1979; Lopez-Avila & Hites, 1980). It is also possible that these triclosan-related compounds have been formed from the photochemical degradation of triclosan in the natural waters. Müller and coworkers concluded that photochemical transformation of triclosan accounts for up to 80% of its loss from the epilimnion in Lake Greifensee during the summer months (Singer *et al.* 2002; Tixier *et al.* 2002). Latch *et al.* (2003b) showed that triclosan photochemically converts to 2,8-dichlorodibenzo-*p*-dioxin (2,8-DCDD) under environmental conditions. A follow up study identified the EPA priority pollutant 2,4-dichlorophenol (2,4-DCP) as an additional triclosan photolysis product (Latch *et al.* 2005). Earlier studies of 2,4-DCP report that it is also photoreactive, producing several chlorinated products (Boule *et al.* 1984). Kliegman *et al.* (2013) identified a dichlorophenoxyphenol and a chlorinated biphenyl as additional triclosan photoproducts. A summary of the photochemical behavior of triclosan is given in Figure 13.12.

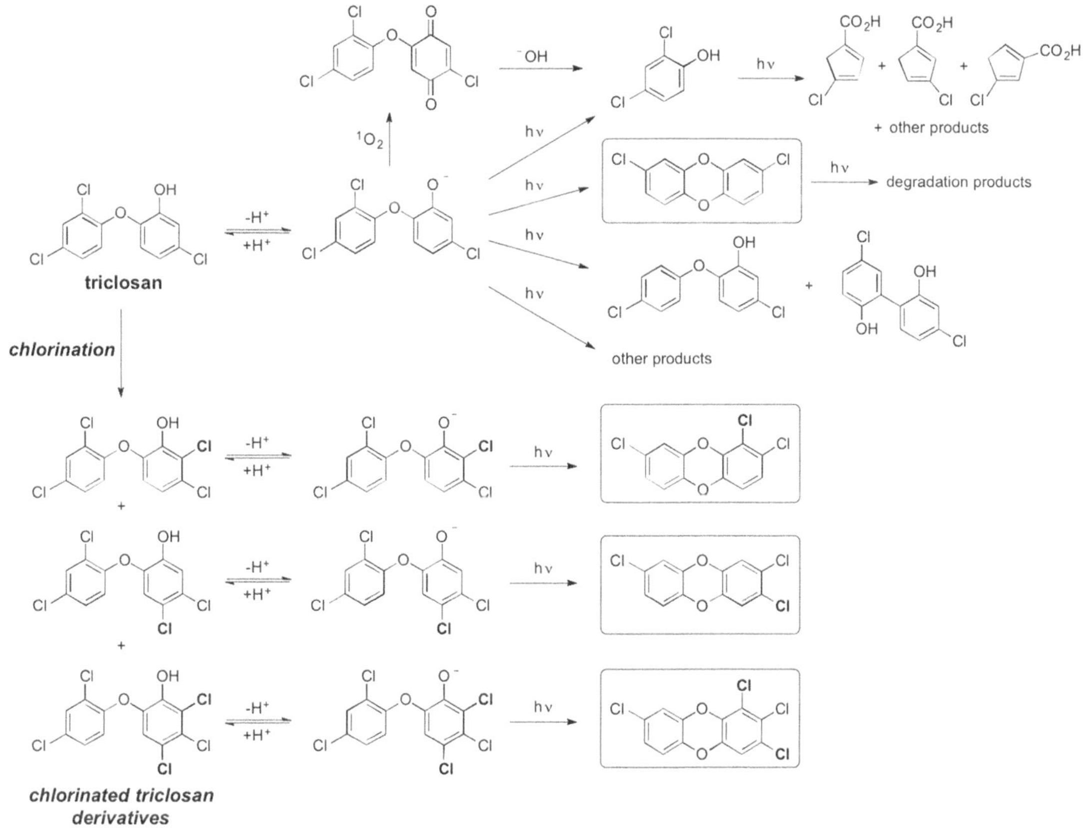

Figure 13.12. Reactivity of triclosan under photochemical and chlorination conditions. All of the polychlorinated dioxins depicted in the boxes in this figure have been detected in environmental samples. (Adapted from Latch *et al.* 2003b; Latch *et al.* 2005; Rule *et al.* 2005; Buth *et al.* 2009; Anger *et al.* 2013; Kliegman *et al.* 2013; Boule *et al.* 1984.)

In addition to being photochemically active, triclosan is susceptible to chlorination, with chlorine preferentially adding to *ortho-* and *para*-positions of phenolic ring (Rule *et al.* 2005). Indeed, these three chlorinated triclosan species with chlorination at those positions were frequently detected (from 2 to 98 ng/L) in wastewaters from a treatment plant employing chlorination, but were largely absent from wastewater from a plant that used UV treatment (Buth *et al.* 2011). Like triclosan, these chlorinated triclosan derivatives produce polychlorodibenzo-*p*-dioxins upon photolysis (Buth *et al.* 2009). Sediment cores from their study reveal an interesting temporal trend with these dioxins relative to other incineration-derived polychlorodibenzo-*p*-dioxins and polychlorodibenzofurans (PCDDs and PCDFs). While sediment concentrations of incineration-derived PCDDs and PCDFs peaked in the mid-1970s and have since decreased, the PCDDs from triclosan and chlorinated triclosan were largely absent in the 1960s and 1970s, but have increased in concentration since as the commercial use of triclosan expanded. A later study gave similar results for lakes impacted by wastewater, while no triclosan or chlorinated triclosan species were detected in a lake that had no wastewater input (Anger *et al.* 2013). Figure 13.12 shows the chlorination of triclosan to produce three chlorinated triclosan derivatives and the subsequent conversion of these species to PCDDs. A recent study by Bianco *et al.* (2015) indicates that other processes, namely reaction with •OH and ^{3}OM, may also be important to the fate of triclosan, especially in acidic waters where triclosan is expected to exist primarily in its protonated form.

13.9.2 Steroid hormones and related EDCs

Steroid hormones and other endocrine disrupting chemicals (EDCs) are among the most important pollutants found in natural waters due to their prevalence and noted biological potency. Steroid hormones and EDCs have been frequently detected in surface waters (Desbrow *et al.* 1998; Baronti *et al.* 2000; Kolpin *et al.* 2002; Phillips *et al.* 2012). The presence of these chemicals in natural waters, even at very low concentrations, is problematic due to their abilities to alter the development of aquatic organisms (Jobling *et al.* 1998; Sumpter & Johnson, 2005; Woodling *et al.* 2006). A set of studies by Vajda *et al.* (2008, 2011) report demasculinization and associated reproductive disruption of male fish exposed to wastewater effluent containing estrogens. Intersex development in male fish is linked to anthropogenic sources, as indicated by the correlation of vitellogenin, a precursor egg yolk protein, with population size upstream from WWTPs (Desforges *et al.* 2010). Al Ansari *et al.* (2010) detected the estrogen 17α-ethinylestradiol at levels of about 1.5 ng/g in about half of the shorthead redhorse suckers sampled near a WWTP outfall, while it could not be detected at a reference downstream site. In a controlled manipulation of an experimental lake, Kidd *et al.* (2007) report feminization of male fathead minnows exposed to 5 ng/L concentrations of 17α-ethinylestradiol. Over the seven-year study period, the fathead minnow population reached near extinction due to the impacts of 17α-ethinylestradiol on reproduction.

The fates of steroid hormones have been studied under photochemical conditions and using advanced oxidation processes that produce reactive intermediates that are also found in sunlit waters. Lee *et al.* (2008) reported the rapid loss of estrogenicity of hormones exposed to several oxidative treatment conditions, including •OH-induced processes. It is possible that such loss in activity could also occur in surface waters high in •OH. Bledzka *et al.* (2010) report that the loss of two steroid hormones, trenbolone and boldenone, was rapid when •OH was the oxidizing agent. The important estrogenic hormones 17β-estradiol and 17α-ethinylestradiol degraded under simulated sunlight with direct photolysis quantum yields of 0.07 and 0.08, respectively. Whidbey *et al.* (2012) tracked the photochemical loss of five estrogens and also monitored the estrogenic activity of the photosolutions using a yeast estrogen screen. They found that four of the estrogens (17β-estradiol, 17α-ethinylestradiol, equilin, and equilenin) produced inactive products, but that estrone produced lumiestrone, a mildly estrogenic estrone epimer. This result is consistent with other reports of lumiestrone photoproduction from estrone and highlights the importance of identifying photoproducts and assessing their activity (Atkinson *et al.* 2011; Trudeau *et al.* 2011). Androstenedione and testosterone are rapidly degraded by direct photolysis, with half-lives on the order of 3–12 hours in surface waters (Young *et al.* 2013). These androgens produced inactive products upon photolysis, as determined by the yeast androgen screen. Phytoestrogens are non-steroid chemicals derived from plants that can act as estrogens. Studies by Kelly and Arnold (2012) and Felcyn *et al.* (2012) show that a set of five phytoestrogens are photolabile, with direct and indirect photolysis pathways involved in their loss.

The photochemical behavior of the androgen trenbolone has recently been explored. Trenbolone is the main metabolite of trenbolone acetate, a high volume growth promoter for cattle. It and other trenbolone acetate metabolites are rapidly degraded by direct photolysis (Kolodziej *et al.* 2013). Several photoproducts were identified, including mono- and multiply-hydroxylated species. Importantly, bioassays revealed that the mixture of photoproducts retained biological activity, indicating that these species, some of unknown chemical structure, are themselves EDCs. A later study with trenbolone revealed an even more important aspect regarding the activity of photolysis solutions, namely that biological activity can rebound in the dark (Qu *et al.* 2013). In this study, the concentration of trenbolone was followed over several light and dark cycles. While trenbolone was rapidly degraded by direct photolysis, it (and its biological activity) was regenerated under dark conditions. These trenbolone studies should give caution in assuming that activity is lost (and *permanently* lost) as a pollutant degrades. With these results in mind, Cwiertny *et al.* (2014) recently wrote a feature about the importance of tracking biological activity over time rather than focusing solely on the loss of parent pollutants in photolysis studies. Figure 13.13 highlights some of the important aspects in the photochemical behavior of steroid hormones and related EDCs.

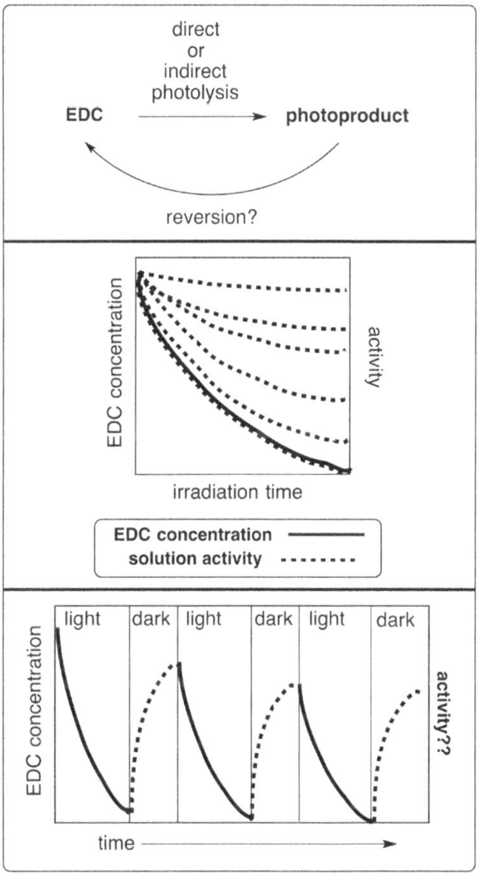

Figure 13.13. Top panel: Simplified scheme showing the photochemical transformation of an EDC. This scheme indicates that EDC photoproducts may revert back to the original EDC under certain conditions. Middle panel: Plot depicting possible changes in biological activity as an EDC is degraded (assuming no reversion). The resulting solution activity depends on the activity of the product(s) relative to the original EDC. In this example, the product is shown to have reduced activity relative to the initial EDC. Bottom panel: Plot showing regeneration of an EDC due to the dark reversion reaction of the product. In the scenario depicted, incomplete reversion occurs and a net decrease in EDC concentration is observed over time. In this case, the activity of the solution is expected to also increase/decrease according to the light/dark cycles, but will also reflect the activities of any other products formed during this cycle. (Adapted from Kolodziej *et al.* 2013; Qu *et al.* 2013; Cwiertny *et al.* 2014; Whidbey *et al.* 2012; Young *et al.* 2013.)

13.9.3 Waterborne viruses and similar model pathogens

The idea of using sunlight to disinfect water has been around for quite some time, as summarized by Wegener *et al.* (1994). Several recent studies have explored the role of direct and indirect photochemical reactions on waterborne pathogens. A primary goal of these studies is to identify whether sunlight can be used to inactivate viruses and to determine the mechanisms responsible for the inactivation, as shown in Figure 13.14.

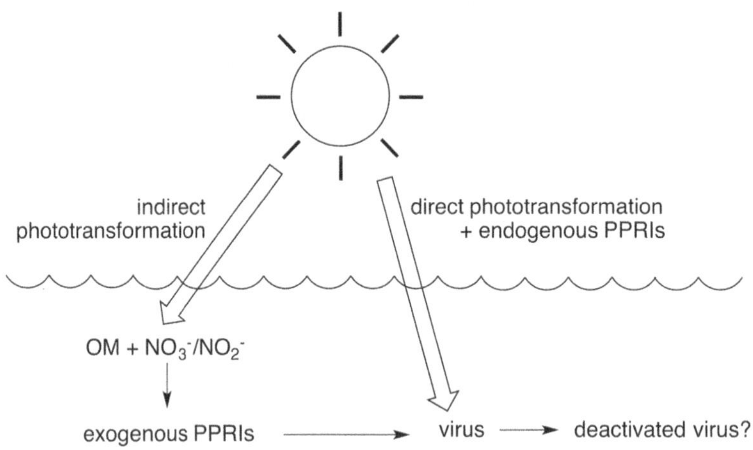

Figure 13.14. Scheme depicting the photochemical transformation of waterborne viruses. In addition to direct phototransformation viruses may react with PPRIs generated endogenously (i.e. "self-sensitized") and exogenously (i.e. from other natural water constituents such as organic matter and nitrate). (Adapted from Kohn & Nelson, 2007; Kohn *et al.* 2007; Nguyen *et al.* 2014; Mattle *et al.* 2015.)

Several studies have used the bacteriophage MS2, a surrogate for human enteric viruses, to explore photochemical disinfection pathways. The overall goal of these studies is to determine the extent to which direct UV photolysis, PPRIs produced from sensitizers contained within the virus ("endogenous"), and PPRIs generated by other water constituents such as OM and nitrate ("exogenous") inactivate the MS2. The direct UV inactivation of viruses may be limited in natural systems due to the poor depth penetration of high energy UVB light (Kohn & Nelson, 2007). Kohn and Nelson (2007) observed that indirect photoreactions efficiently inactivated MS2 under simulated solar light with various NOM sources present as sensitizers. Quenching experiments and experiments performed in D_2O implicated $^1O_2$ as the PPRI responsible for the inactivation. Measured MS2 inactivation rate constants also correlated well with $[^1O_2]_{SS}$ values determined using furfuryl alcohol as a $^1O_2$ probe. Association of the relatively large MS2 structure with OM leads to enhanced $^1O_2$-mediated inactivation rates of MS2 due to the greater concentrations of $^1O_2$ in and near OM relative to the bulk solution (Kohn *et al.* 2007; Latch & McNeill, 2006; Grandbois *et al.* 2008). Rate enhancements increased with adsorption to OM. Mass spectra analysis indicates that oxidation of capsid proteins by $^1O_2$ is an important process in the $^1O_2$-mediated inactivation of MS2 (Rule-Wigginton *et al.* 2010). A study using several bacteriophages and human viruses showed that MS2 was photochemically inactivated at slower rates than the other viruses, indicating that using MS2 as a probe for virus inactivation may underestimate inactivation rates of other viruses. Another important aspect of this study was that it showed that inactivation rates determined in the field correlated well with those measured in the laboratory using a solar simulator.

Cho *et al.* (2010) demonstrate that synthetic $^1O_2$ sensitizers based on C_{60} nanoparticles could be used to efficiently inactivate MS2 using visible light, highlighting their potential use as disinfectants. In a similar study, the potential for MS2 inactivation by Fenton and Fenton-like reactions was examined (Nieto-Juarez *et al.* 2010). They determined that these reaction systems were efficient at inactivating MS2, and that the inactivation rate could not be explained entirely by the concentration of $^{\bullet}OH$ present in solution.

The ability of MS2 to generate endogenous PPRIs in the absence of exogenous sensitizers was explored by Nguyen *et al.* (2014). They found that inactivation rates could be predicted reasonably well based on MS2 absorption, light screening, and diurnal and seasonal changes in light intensity. The higher energy region of the solar spectrum (~280–320 nm) was particularly important in the inactivation process. Mattle *et al.* (2015) have developed a model to predict photochemical inactivation rates for MS2, adenovirus, and phiX174 and to assess the relative roles of direct photolysis and various PPRIs in the inactivation process. The relative importance of different PPRIs was determined in a similar manner to predicting indirect loss rates of micropollutants (i.e. by using $k_{PPRI,pollutant}$ values and $[PPRI]_{SS}$). Their results implicated mostly direct phototransformations in the inactivation of these viruses, with some disinfection due to $^1O_2$.

13.10 FUTURE RESEARCH NEEDS

As new commercial products are continuously introduced and added to waste streams, there will always be a need to identify emerging pollutants that could potentially cause adverse ecological effects. A thorough examination of such species should include fate studies, including an analysis of their potential to undergo direct and indirect photoreactions. A current example might be nanomaterials currently added to consumer products. As has been noted by several researchers (Schwarzenbach *et al.* 2006; Oulton *et al.* 2010; Fenner *et al.* 2013; Cwiertny *et al.* 2014), we must also attempt to identify and assess the reactivity and ecotoxicology of novel byproducts produced in WWTPs, in the biodegradation of pesticides and other pollutants, and in aquatic photochemical reactions. Quantitative structure activity relationships (QSARs) have been used to predict the reactivity of pollutants and certain moieties found in pollutants with various PPRIs (Tratnyek & Hoigne, 1991; Canonica & Tratnyek, 2003). While QSARs developed with Hammett constants are useful, they are limited to relatively simple and related molecular structures. The use of computational chemistry methods may increase our abilities to predict the photochemical fates of new pollutants. Photochemical studies should focus not just on reporting degradation rates, but also on the identification of photoproducts. When possible, the activity of photoproducts or photolysis reaction mixtures should be determined. Laboratory studies should be carefully designed so that the results can be better extrapolated to field conditions (Challis *et al.* 2014), and more field and mesocosm studies should be performed to supplement laboratory results. Lastly, the mechanisms by which OM (both NOM and EfOM) produces and quenches PPRIs should continue to be studied in an effort to better understand indirect photochemistry across a wide range of sites and environmental conditions, including in effluent-impacted waters.

13.11 ACKNOWLEDGEMENTS

I gratefully acknowledge Seattle University's Center for Environmental Justice and Sustainability for support.

13.12 REFERENCES

Ackerman R. A., Pitts Jr. J. N. and Rosenthal I. (1971). Singlet oxygen in the environmental sciences. Reactions of singlet oxygen, O2(1Δg) with olefins, sulfides, and compounds of biological significance. *American Chemical Society, Division of Petroleum Chemistry*, **16**, A25–A34.

Aeschbacher M., Graf C., Schwarzenbach R. P. and Sander M. (2012). Antioxidant properties of humic substances. *Environmental Science and Technology*, **46**, 4916–4925.

Aguer J. P. and Richard C. (1996). Reactive species produced on irradiation at 365 nm of aqueous solutions of humic acids. *Journal of Photochemistry and Photobiology, A*, **93**, 193–8.

Aguer J. P., Richard C. and Andreux F. (1999). Effect of light on humic substances: production of reactive species. *Analusis*, **27**, 387–390.

Ahel M., Scully Jr. F. E., Hoigne J. and Giger W. (1994). Photochemical degradation of nonylphenol and nonylphenol ethoxylates in natural waters. *Chemosphere*, **28**, 1361–1368.

Aiello A. E. and Larson E. (2003). Antibacterial cleaning and hygiene products as an emerging risk factor for antibiotic resistance in the community. *Lancet Infectious Diseases*, **3**, 501–6.

Al-Ansari A. M., Saleem A., Kimpe L. E., Sherry J. P., McMaster M. E., Trudeau V. L. and Blais J. M. (2010). Bioaccumulation of the pharmaceutical 17alpha-ethinylestradiol in shorthead redhorse suckers (*Moxostoma macrolepidotum*) from the St. Clair River, Canada. *Environmental Pollution*, **158**(8), 2566–2571.

Al Housari F., Vione D., Chiron S. and Barbati S. (2010). Reactive photoinduced species in estuarine waters. Characterization of hydroxyl radical, singlet oxygen and dissolved organic matter triplet state in natural oxidation processes. *Photochemical and Photobiological Sciences*, **9**, 78–86.

Amyot M., Gill G. A. and Morel F. M. M. (1997). Production and loss of dissolved gaseous mercury in coastal seawater. *Environmental Science and Technology*, **31**, 3606–3611.

An AP Investigation. Pharmaceuticals Found in Drinking Water. http://hosted.ap.org/specials/interactives/pharmawater_site/ (accessed 1 April 2015).

Anderson P. D., D'Aco V. J., Shanahan P., Chapra S. C., Buzby M. E., Cunningham V. L., DuPlessie B. M., Hayes E. P., Mastrocco F. J., Parke N. J., Rader J. C., Samuelian J. H. and Schwab B. W. (2004). Screening analysis of human pharmaceutical compounds in U.S. surface waters. *Environmental Science and Technology*, **38**, 838–849.

Anger C. T., Sueper C., Blumentritt D. J., McNeill K., Engstrom D. R. and Arnold W. A. (2013). Quantification of triclosan, chlorinated triclosan derivatives, and their dioxin photoproducts in lacustrine sediment cores. *Environmental Science and Technology*, **47**, 1833–1843.

Arnold W. A. and McNeill K. (2007) Transformation of pharmaceuticals in the environment: photolysis and other abiotic processes. In: *Comprehensive Analytical Chemistry 50*, M. Petrovic and D. Barcelo (eds), Elsevier, Amsterdam, pp. 361–385.

Atkinson S. K., Marlatt V. L., Kimpe L. E., Lean D. R., Trudeau V. L. and Blais J. M. (2011). Environmental factors affecting ultraviolet photodegradation rates and estrogenicity of estrone and ethinylestradiol in natural waters. *Archives of Environmental Contamination and Toxicology*, **60**(1), 1–7.

Balmer M. E., Poiger T., Droz C., Romanin K., Bergqvist P. A., Mueller M. D. and Buser H. R. (2004). Occurrence of methyl triclosan, a transformation product of the bactericide triclosan, in fish from various lakes in Switzerland. *Environmental Science and Technology*, **38**, 390–395.

Bahnmuller S., von Gunten U. and Canonica S. (2014). Sunlight-induced transformation of sulfadiazine and sulfamethoxazole in surface waters and wastewater effluents. *Water Research*, **57**, 183–192.

Barcelo D. and Petrovic M. (2007). Conclusion and future research needs. In: Comprehensive Analytical Chemistry 50, M. Petrovic and D. Barcelo (eds.), Elsevier, Amsterdam, pp. 515–527.

Baronti C., Curini R., D'Ascenzo G., Di Corcia A., Gentili A. and Samperi R. (2000). Monitoring natural and synthetic estrogens at activated sludge sewage treatment plants and in a receiving river water. *Environmental Science and Technology*, **34**(24), 5059–5066.

Baxter R. M. and Carey J. H. (1983). Evidence for photochemical generation of superoxide ion in humic waters. *Nature*, **306**, 575–6.

Belfroid A. C., Van der Horst A., Vethaak A. D., Schafer A. J., Rijs G. B. J., Wegener J. and Cofino W. P. (1999). Analysis and occurrence of estrogenic hormones and their glucuronides in surface water and waste water in the Netherlands. *Science of the Total Environment*, **225**, 101–108.

Bester K. (2003). Triclosan in a sewage treatment process—Balances and monitoring data. *Water Research*, **37**, 3891–6.

Bianco A., Fabbri D., Minella M., Brigante M., Mailhot G., Maurino V., Minero C. and Vione D. (2015). New insights into the environmental photochemistry of 5-chloro-2-(2,4-dichlorophenoxy)phenol (triclosan): reconsidering the importance of indirect photoreactions. *Water Research*, **172**, 271–280.

Black F. J., Poulin B. A. and Flegal A. R. (2012). Factors controlling the abiotic photo-degradation of monomethylmercury in surface waters. *Geochimica et Cosmochimica Acta*, **84**, 492–507.

Bledzka D., Gmurek M., Gryglik M., Olak M., Miller J. S. and Ledakowicz S. (2010). Photodegradation and advanced oxidation of endocrine disruptors in aqueous solutions. *Catalysis Today*, **151**, 125–130.

Blough N. V. and Zepp R. G. (1995). Reactive oxygen species in natural waters. In: Active Oxygen in Chemistry, C. Foote, J. Valentine, A. Greenberg and Liebman S. (eds), Chapman and Hall, New York, pp. 280–333.

Bodhipaksha L. C., Sharpless C. M., Chin Y. P., Sander M., Langston W. K. and MacKay A. A. (2015). Triplet photochemistry of effluent and natural organic matter in whole water and isolates from effluent-receiving rivers. *Environmental Science and Technology*, **49**(6), 3453–3463.

Bodrato M. and Vione D. (2014). APEX (Aqueous Photochemistry of Environmentally occurring Xenobiotics): a free software tool to predict the kinetics of photochemical processes in surface waters. *Environmental Science: Processes & Impacts*, **16**, 732–740.

Boreen A. L., Arnold W. A. and McNeill K. (2003). Photodegradation of pharmaceuticals in the aquatic environment: a review. *Aquatic Sciences*, **65**, 320–341.

Boreen A. L., Arnold W. A. and McNeill K. (2004). Photochemical fate of sulfa drugs in the aquatic environment: sulfa drugs containing five-membered heterocycic groups. *Environmental Science and Technology*, **38**, 3933–3940.

Boreen A. L., Arnold W. A. and McNeill K. (2005). Triplet-sensitized photodegradation of sulfa drugs containing six-membered heterocyclic groups: identification of an SO_2 extrusion photoproduct. *Environmental Science and Technology*, **39**, 3630–3638.

Boule P., Guyon C. and Lemaire J. (1984). Photochemistry and the environment. VIII. Photochemical behavior of dichlorophenols in a dilute aqueous solution. *Chemosphere*, **13**, 603–612.

Boyd G. R., Reemtsma H., Grimm D. A. and Mitra S. (2003). Pharmaceuticals and personal care products (PPCPs) in surface and treated waters of Louisiana, USA and Ontario, Canada. *Science of the Total Environment*, **311**, 135–149.

Boyd G. R., Palmeri J. M., Zhang S. and Grimm D. A. (2004). Pharmaceuticals and personal care products (PPCPs) and endocrine disrupting chemicals (EDCs) in stormwater canals and Bayou St. John in New Orleans, Louisiana, USA. *Science of the Total Environment*, **333**, 137–48.

Brezonik P. L. and Arnold W. A. (2011). Water Chemistry. An Introduction to the Chemistry of Natural and Engineered Aquatic Systems. Oxford University Press, New York.

Brezonik P. L. and Fulkerson-Brekken J. (1998). Nitrate-induced photolysis in natural waters: controls on concentrations of hydroxyl radical photo-intermediates by natural scavenging agents. *Environmental Science and Technology*, **32**, 3004–3010.

Brooks B. W., Riley T. M. and Taylor R. D. (2006). Water quality of effluent-dominated ecosystems: ecotoxicological, hydrological, and management considerations. *Hydrobiologia*, **556**, 365–379.

Brown M. A. and Casida J. E. (1988). Daminozide: oxidation by photochemically generated singlet oxygen to dimethylnitrosamine and succinic anhydride. *Journal of Agricultural and Food Chemistry*, **36**, 1064–1066.

Burns J. M., Cooper W. J., Ferry J. L., King D. W., DiMento B. P., McNeill K., Miller C. J., Miller W. L., Peake B. M., Rusak S. A., Rose A. L. and Waite T. D. (2012). Methods for reactive oxygen species (ROS) detection in aqueous environments. *Aquatic Sciences*, **74**, 683–734.

Burns S. E., Hassett J. P. and Rossi M. V. (1997). Mechanistic implications of the intrahumic dechlorination of mirex. *Environmental Science and Technology*, **31**, 1365–1371.

Burrows H. D., Ernestova L., Kemp T. J., Skurlatov Y. I., Purmal A. P. and Yermakov A. N. (1998). Kinetics and mechanism of photodegradation of chlorophenols. *Progress in Reaction Kinetics and Mechanism*, **23**, 145–207.

Burrows H. D., Canle L. M., Santabella J. A. and Steenken S. (2002). Reaction pathways and mechanisms of photodegradation of pesticides. *Journal of Photochemistry and Photobiology B*, **67**, 71–108.

Buser H.-R., Poiger T. and Mueller M. D. (1998). Occurrence and fate of the pharmaceutical drug diclofenac in surface waters: rapid photodegradation in a lake. *Environmental Science and Technology*, **32**, 3449–3456.

Buth J. M., Grandbois M., Vikesland P. J., McNeill K. and Arnold W. A. (2009). Aquatic photochemistry of chlorinated triclosan derivatives: potential source of polychlorinateddibenxo-*p*-dioxins. *Environmental Toxicology and Chemistry*, **28**, 2555–2563.

Buth J. M., Steen P. O., Sueper C., Blumentritt D., Vikesland P. J., Arnold W. A. and McNeill K. (2010). Dioxin photoproducts of triclosan and its chlorinated derivatives in sediment cores. *Environmental Science and Technology*, **44**, 4545–4551.

Buth J. M., Ross M. R., McNeill K. and Arnold W. A. (2011). Removal and formation of chlorinated triclosan derivatives in wastewater treatment plants using chlorine and UV disinfection. *Chemosphere*, **84**, 1238–1243.

Buxton G. V., Greenstock C. L., Helman N. P. and Ross A. B. (1988). Critical review of rate constants for reaction of hydrated electrons and hydroxyl radicals in aqueous solution. *Journal of Physical Chemistry Reference Data*, **17**, 513–886.

Calamari D., Zuccato E., Castiglioni S., Bagnati R. and Fanelli R. (2003). Strategic survey of therapeutic drugs in the Rivers Po and Lambro in Northern Italy. *Environmental Science and Technology*, **37**, 1241–1248.

Canonica S. (2007). Oxidation of aquatic organic contaminants induced by excited triplet states. *Chimia*, **61**, 641–644.

Canonica S. and Freiburghaus M. (2001). Electron-rich phenols for probing the photochemical reactivity of freshwaters. *Environmental Science and Technology*, **35**, 690–695.

Canonica S. and Hoigne J. (1995). Enhanced oxidation of methoxy phenols at micromolar concentration photosensitized by dissolved natural organic material. *Chemosphere*, **30**, 2365–74.

Canonica S. and Laubscher H. U. (2008). Inhibitory effect of dissolved organic matter on triplet-induced oxidation of aquatic contaminants. *Photochemical and Photobiological Sciences*, **7**, 547–551.

Canonica S. and Tratnyek P. G. (2003). Quantitative structure-activity relationships for oxidation reactions of organic chemicals in water. *Environmental Toxicology and Chemistry*, **22**, 1743–1754.

Canonica S., Jans U., Stemmler K. and Hoigne J. (1995). Transformation kinetics of phenols in water: photosensitization by dissolved natural organic material and aromatic ketones. *Environmental Science and Technology*, **29**, 1822–1831.

Canonica S., Hellrung B. and Wirz J. (2000). Oxidation of phenols by triplet aromatic ketones in aqueous solutions. *Journal of Physical Chemistry A*, **104**, 1226–1232.

Canonica S., Kohn T., Mac M., Real F. J., Wirz J. and von Gunten U. (2005). Photosensitizer method to determine rate constants for the reaction of carbonate radical with organic compounds. *Environmental Science and Technology*, **39**, 9182–9188.

Canonica S., Hellrung B., Mueller P. and Wirz J. (2006). Aqueous oxidation of phenylurea herbicides by triplet aromatic ketones. *Environmental Science and Technology*, **40**, 6636–6641.

Challis J. K., Carlson J. C., Friesen K. J., Hanson M. L. and Wong C. S. (2013). Aquatic photochemistry of the sulfonamide antibiotic sulfapyridine. *Journal of Photochemistry and Photobiology A.*, **262**, 14–21.

Challis J. K., Hanson M. L., Friesen K. J. and Wong C. S. (2014). A critical assessment of the photodegradation of pharmaceuticals in aquatic environments: defining our current understanding and identifying knowledge gaps. *Environmental Science Processes & Impacts*, **16**, 672–696.

Christensen F. M. (1998). Pharmaceuticals in the environment-a human risk? *Regulatory Toxicology Pharmacology*, **28**, 212–221.

Cho M., Lee J., Mackeyev Y., Wilson L. J., Alvarez P. J. J., Hughes J. B. and Kim J. H. (2010). Visible light sensitized inactivation of MS-2 bacteriophage by a cationic amine-functionalized C_{60} derivative. *Environmental Science and Technology*, **44**, 6685–6691.

Choudhry G. G. and Webster G. R. B. (1987). Environmental photochemistry of polychlorinated dibenzofurans (PCDFs) and dibenzo-*p*-dioxins (PCDDs): a review. *Toxicological and Environmental Chemistry*, **14**, 43–61.

Chun H., Yizhong W. and Hongxiao T. (2000). Destruction of phenol aqueous solution by photocatalysis or direct photolysis. *Chemosphere*, **41**, 1205–9.

Conceicao M., Mateus D. A., Da Silva A. M. and Burrows H. D. Kinetics of the fungicide fenarimol in natural waters and in various salt solutions: salinity effects and mechanistic considerations. *Water Research*, **34**, 1119–1126.

Cooper W. J. and Zika R. G. (1983). Photochemical formation of hydrogen peroxide in surface and ground waters exposed to sunlight. *Science*, **220**, 711–712.

Cooper W. J., Zika R. G., Petasne R. G. and Fischer A. M. (1989). Sunlight-induced photochemistry of humic substances in natural waters: major reactive species. In: Aquatic Humic Substances, I. H. Suffet and P. MacCarthy (eds), American Chemical Society, Washington, D.C., pp. 333–362.

Crosby D. G. and Wong A. S. (1976). Photochemical generation of chlorinated dioxins. *Chemosphere*, **5**, 327–332.

Cwiertny D. M., Snyder S. A., Schlenk D. and Kolodziej E. P. (2014). Environmental designer drugs: when transformation may not eliminate risk. *Environmental Science and Technology*, **48**(20), 11737–11745.

Daughton C. G. and Ternes T. A. (1999). Pharmaceuticals and personal care products in the environment: agents of subtle change? *Environmental Health Perspectives Supplement*, **107**, 907–938.

De Laurentiis E., Chiron S., Kouras-Hadef S., Richard C., Minella M., Maurino V., Minero C. and Vione D. (2012). Photochemical fate of carbamazepine in surface freshwaters: laboratory measures and modeling. *Environmental Science and Technology*, **46**, 8164–8173.

De Laurentiis E., Prasse C., Ternes T. A., Minella M., Maurino V., Minero C., Sarakha M., Brigante M. and Vione D. (2014). Assessing the photochemical transformation pathways of acetaminophen relevant to surface waters: transformation kinetics, intermediates, and modelling. *Water Research*, **53**, 235–248.

Desbrow C., Routledge E. J., Brighty G. C., Sumpter J. P. and Waldock M. (1998). Identification of estrogenic chemicals in STW effluent. 1. Chemical fractionation and in vitro biological screening. *Environmental Science and Technology*, **32**(11), 1549–1558.

Desforges J. P. W., Peachey, B. D. L., Sanderson, P. M., White P. A. and Blais J. M. (2010). Plasma vitellogenin in male teleost fish from 43 rivers worldwide is correlated with upstream population size. *Environmental Pollution*, **158**, 3279–3284.

Dong M. M. and Rosario-Ortiz R. L. (2012). Photochemical formation of hydroxyl radical from effluent organic matter. *Environmental Science and Technology*, **46**, 3788–3794.

Drewes J. E. and Croue J. P. (2002). New approaches for structural characterization of organic matter in drinking water and wastewater effluents. *Water Supply*, **2**, 1–10.

Drury B., Scott J., Rosi-Marshall E. J. and Kelly J. J. (2013). Triclosan exposure increases triclosan resistance and influences taxonomic composition of benthic bacterial communities. *Environmental Science and Technology*, **47**, 8923–8930.

Dulin D. and Mill T. (1982). Development and evaluation of sunlight actinometers. *Environmental Science and Technology*, **16**, 815–20.

Espy R., Pelton E., Opseth A., Kasprisin J. and Nienow A. M. (2011). Photodegradation of the herbicide imazethapyr in aqueous solutions: effects of wavelength, pH, and natural organic matter (NOM) and analysis of photoproducts. *Journal of Agricultural and Food Chemistry*, **59**, 7277–7285.

Felcyn J. R., Davis J. C. C., Tran L. H., Berude J. C. and Latch D. E. (2012). Aquatic photochemistry of isoflavone phytoestrogens: degradation kinetics and pathways. *Environmental Science and Technology*, **46**, 6698–6704.

Fenner K., Canonica S., Wackett L. P. and Elsner M. (2013). Evaluating pesticide degradation in the environment: blind spots and emerging opportunities. *Science*, **341**, 752–758.

Fernandez-Gomez C., Drott A., Bjorn E., Diez S., Bayona J. M., Tesfalidet S., Lindfors A. and Skyllberg U. (2013). Towards universal wavelength-specific photodegradation rate constants for methylmercury in humic waters, exemplified by a boreal lake-wetland gradient. *Environmental Science and Technology*, **47**, 6279–6287.

Foreman, W. T., Gray, J. L., ReVello, R. C., Lindley, C. E., Losche, S. A. and Barber L. B. (2012). Determination of steroid hormones and related compounds in filtered and unfiltered water by solid-phase extraction, derivatization, and gas chromatography with tandem mass spectrometry, U.S. Geological Survey Techniques and Methods, book 5, chapter B9, 1–118.

Freeman P. K. and Srinivasa R. (1986). Photochemistry of polyhaloarenes. 4. Phototransformations of perchloro-*o*-phenoxyphenol in basic media. *Journal of Organic Chem*istry, **51**, 3939–3942.

Fulkerson-Brekken J. and Brezonik P. L. (1998). Indirect photolysis of acetochlor: rate constant of a nitrate-mediated hydroxyl radical reaction. *Chemosphere*, **36**, 2699–2704.

Garcia E., Poulain A. J., Amyot M. and Ariya P. A. (2005). Diel variations in photoinduced oxidation of Hg0 in freshwater. *Chemosphere*, **59**, 977–981.

Garcia N. A. (1994). New trends in photobiology: singlet-molecular-oxygen-mediated photodegradation of aquatic phenolic pollutants. A kinetic and mechanistic overview. *Journal of Photochemistry and Photobiology*, **22**, 185–196.

Gerecke A. C., Canonica S., Mueller S. R., Schaerer M. and Schwarzenbach R. P. (2001). Quantification of dissolved natural organic matter (DOM) mediated phototransformation of phenylurea herbicides in lakes. *Environmental Science and Technology*, **35**, 3915–3923.

Glover C. M. and Rosario-Ortiz F. L. (2013). Impact of halides on the photoproduction of reactive intermediates from organic matter. *Environmental Science and Technology*, **47**, 13949–13956.

Gohre K. and Miller G. C. (1986). Photooxidation of thioether pesticides on soil surfaces. *Journal of Agricultural and Food Chemistry*, **34**, 709–13.

Goldberg M. C., Cunningham K. M., Aiken G. R. and Weiner E. R. (1992). The aqueous photolysis of α-pinene in solution with humic acid. *Journal of Contaminant Hydrology*, **9**, 79–89.

Goncalves C., Perex S., Osorio V., Petrovic M., Alpendurada M. F. and Barcelo D. (2011). Photofate of Oseltamivir (Tamiflu) and oseltamivir carboxylate under natural and simulated solar irradiation: kinetics, identification of the transformation products, and environmental occurrence. *Environmental Science and Technology*, **45**, 4307–4314.

Grandbois M., Latch D. E. and McNeill K. (2008). Microheterogeneous concentrations of singlet oxygen in natural organic matter isolate solutions. *Environmental Science and Technology*, **42**, 9184–9190.

Grebel J. E., Pignatello J. J., Song W., Cooper W. J. and Mitch W. A. (2009). Impact of halides on the photobleaching of dissolved organic matter. *Marine Chemistry*, **115**, 134–144.

Grebel J. E., Pignatello J. J. and Mitch W. A. (2010). Effect of halide ions and carbonates on organic contaminant degradation by hydroxyl radical-based advanced oxidation processes in saline waters. *Environmental Science and Technology*, **44**, 6822–6828.

Grebel J. E., Pignatello J. J. and Mitch W. A. (2011). Sorbic acid as a quantitative probe for the formation, scavenging and steady-state concentrations of the triplet excited state of organic compounds. *Water Research*, **45**, 6535–6544.

Grebel J. E., Pignatello J. J. and Mitch W. A. (2012). Impact of halide ions on natural organic matter-sensitized photolysis of 17β-estradiol in saline waters. *Environmental Science and Technology*, **46**, 7128–7134.

Guerard J. J., Chin Y.-P., Mash H. and Hadad C. M. (2009a). Photochemical fate of sulfadimethoxine in aquaculture waters. *Environmental Science and Technology*, **43**, 8587–8592.

Guerard J. J., Miller P. L., Trouts T. D. and Chin Y.-P. (2009b). The role of fulvic acid composition in the photosensitized degradation of aquatic contaminants. *Aquatic Sciences*, **71**, 160–169.

Gurr C. J. and Reinhard M. (2006). Harnessing natural attenuation of pharmaceuticals and hormones in rivers. *Environmental Science and Technology*, **40**(9), 2872–2876.

Haag W. R. and Hoigné J. (1986). Singlet oxygen in surface waters. 3. Photochemical formation and steady-state concentrations in various types of waters. *Environmental Science and Technology*, **20**, 341–8.

Haag W. R. and Mill T. (1987). Rate constants for interaction of singlet oxygen ($^1\Delta g$) with azide ion in water. *Photochemistry and Photobiology*, **45**, 317–321.

Haag W. R. and Yao C. C. D. (1992). Rate constants for reaction of hydroxyl radicals with several drinking water contaminants. *Environmental Science and Technology*, **26**, 1005–1013.

Haag W. R., Hoigné J., Gassman E. and Braun A. M. (1984a). Singlet oxygen in surface waters – part I: furfuryl alcohol as a trapping agent. *Chemosphere*, **13**, 631–640.

Haag W. R., Hoigne J., Gassman E. and Braun A. M. (1984b). Singlet oxygen in surface waters – part II: quantum yields of its production by some natural humic materials as a function of wavelength. *Chemosphere*, **13**, 641–650.

Halden R. (2014). On the need and speed of regulating triclosan and triclocarban in the United States. *Environmental Science and Technology*, **48**, 3603–3611.

Hale R. C., Smith C. L., De Fur P. O., Harvey E., Bush E. O., La Guardia M. J. and Vadas G. G. (2000). Nonylphenols in sediments and effluents associated with diverse wastewater outfalls. *Environmental Toxicology and Chemistry*, **19**, 946–952.

Halladja S., ter Halle A., Aguer J.-P., Boulkamh A. and Richard C. (2007). Inhibition of humic substances mediated photooxygenation of furfuryl alcohol by 2,4,6-trimethylphenol. Evidence for reactivity of the phenol with humic triplet excited states. *Environmental Science and Technology*, **41**, 6066–6073.

Halling-Sorensen B., Nielsen S. N., Lanzky P. F., Ingerslev F., Lutzhft H. C. H. and Jorgensen S. E. (1998). Occurrence, fate and effects of pharmaceutical substances in the environment – a review. *Chemosphere*, **36**, 357–393.

Hand L. H. and Oliver R. G. (2010). The behavior of isopyrazam in aquatic ecosystems: implementation of a tiered investigation. *Environmental Toxicology and Chemistry*, **29**, 2702–2712.

Heath R. J., Yu Y. T., Shapiro M. A., Olson E. and Rock C. O. (1998). Broad spectrum antimicrobial biocides target the FabI component of fatty acid synthesis. *Journal of Biological Chemistry*, **273**, 30316–30320.

Hites R. A. and Lopez-Avila V. (1979). Identification of organic compounds in an industrial waste water. *Analytical Chemistry*, **51**, 1452A–1456A.

Hoigné J., Faust B. C., Haag W. R., Scully F. E. and Zepp R. G. (1989). Aquatic humic substances as sources and sinks of photochemically produced transient reactants. In: Aquatic Humic Substances, I. H. Suffet and P. MacCarthy (eds), American Chemical Society, Washington, D.C., pp. 363–81.

Holbrook R. D., Love N. G. and Novak J. T. (2004). Sorption of 17β-estradiol and 17α-ethinylestradiol by colloidal organic carbon derived from biological wastewater treatment systems. *Environmental Science and Technology*, **38**, 3322–3329.

Huang J. and Mabury S. A. (2000a). A new method for measuring carbonate radical reactivity toward pesticides. *Environmental Toxicology and Chemistry*, **19**, 1501–1507.

Huang J. and Mabury S. A. (2000b). Steady-state concentrations of carbonate radicals in field waters. *Environmental Toxicology and Chemistry*, **19**, 2181–2188.

Huang J. and Mabury S. A. (2000c). The role of carbonate radical in limiting the persistence of sulfur-containing chemicals in sunlit waters. *Chemosphere*, **41**, 1775–1782.

Jacobs L. E., Fimmen R. L., Chin Y. P., Mash H. E. and Weavers L. K. (2011). Fulvic acid mediated photolysis of ibuprofen in water. *Water Research*, **45**, 4449–4458.

Jarusutthiraka C. and Amy G. (2007). Understanding soluble microbial products (SMP) as a component of effluent organic matter (EfOM). *Water Research*, **41**, 2787–2793.

Javier Benitez F., Acero J. L., Real F. J. and Garcia J. (2003). Kinetics of photodegradation and ozonation of pentachlorophenol. *Chemosphere*, **51**, 651–662.

Jeremiason J. D., Portner J. C., Aiken G. R., Hiranaka A. J., Dvorak M. T., Tran K. T. and Latch D. E. (2015). Photoreduction of Hg(II) and photodemethylation of methylmercury: the key role of thiol sites on dissolved organic matter. *Environmental Science: Processes & Impacts*, **17**, 1892–1903.

Jobling S., Nolan M., Tyler C. R., Brighty G. and Sumpter J. P. (1998). Widespread sexual disruption in wild fish. *Environmental Science and Technology*, **32**(17), 2498–2506.

Jones O. A. H., Voulvoulis N. and Lester J. N. (2002). Aquatic environmental assessment of the top 25 English prescription pharmaceuticals. *Water Research*, **36**, 5013–5022.

Jungclaus G. A., Lopez-Avila V. and Hites R. A. (1978). Organic compounds in an industrial waste water: a case study of their environmental impact. *Environmental Science and Technology*, **12**, 88–96.

Kamiya M. and Kameyama K. (2001). Effects of selected metal ions on photodegradation of organophosphorus pesticides sensitized by humic acids. *Chemosphere*, **45**, 231–235.

Kanda R., Griffin P., James Huw A. and Fothergill J. (2003). Pharmaceutical and personal care products in sewage treatment works. *Journal of Environmental Monitoring*, **5**, 823–30.

Kanetoshi A., Ogawa H., Katsura E., Kaneshima H. and Miura T. (1988). Formation of polychlorinated dibenzo-p-dioxin from 2,4,4′-trichloro-2′-hydroxydiphenyl ether (Irgasan DP300) and its chlorinated derivatives by exposure to sunlight. *Journal of Chromatography*, **454**, 145–55.

Kanetoshi A., Ogawa H., Katsura E., Kaneshima H. and Miura T. (1992). Study on the environmental hygienic chemistry of chlorinated 2-hydroxydiphenyl ethers: photolytic conversion to polychlorinated dibenzo-*p*-dioxins. *Kankyo Kagaku*, **2**, 515–522.

Karthikeyan K. G. and Chorover J. (2000). Effects of Solution Chemistry on the Oxidative Transformation of 1-Naphthol and Its Complexation with Humic Acid. *Environmental Science and Technology*, **34**, 2939–2946.

Kelly M. M. and Arnold W. A. (2012). Direct and indirect photolysis of the phytoestrogens genistein and daidzein. *Environmental Science and Technology*, **46**(10), 5396–5403.

Kidd K. A., Blanchfield P. J., Mills K. H., Palace V. P., Evans R. E., Lazorchak J. M. and Flick R. W. (2007). Collapse of a fish population after exposure to a synthetic estrogen. *Proceedings of the National Academy of Sciences*, **104**(21), 8897–8901.

Kim M. and O'Keefe P. W. (2000). Photodegradation of polychlorinated dibenzo-*p*-dioxins and dibenzofurans in aqueous solutions and in organic solvents. *Chemosphere*, **41**, 793–800.

Khetan S. K. and Collins T. J. (2007). Human pharmaceuticals in the environment: a challenge to green chemistry. *Chemical Reviews*, **107**, 2319–2364.

Klan P. and Wirz J. (2010). Photochemistry of Organic Compounds. From Concepts to Practice. Wiley, West Sussox.

Kliegman S., Eustic S. N., Arnold, W. A. and McNeill K. (2013). Experimental and theoretical insights into the involvement of radicals in triclosan phototransformation. *Environmental Science and Technology*, **43**, 6756–6763.

Kohn T. and Nelson K. L. (2007). Sunlight-mediated inactivation of MS2 coliphage via exogenous singlet oxygen produced by sensitizers in natural waters. *Environmental Science and Technology*, **41**, 192–197.

Kohn T., Grandbois M., McNeill K. and Nelson K. L. (2007). Association with Natural Organic Matter Enhances the Sunlight-Mediated Inactivation of MS2 Coliphage by Singlet Oxygen. *Environmental Science and Technology*, **41**, 4626–4632.

Kolodziej E. P. and Sedlak D. L. (2007). Rangeland grazing as a source of steroid hormones to surface waters. *Environmental Science and Technology*, **41**(10), 3514–3520.

Kolodziej E. P., Gray J. L. and Sedlak D. L. (2003). Quantification of steroid hormones with pheromonal properties in municipal wastewater effluent. *Environmental Toxicology and Chemistry*, **22**(11), 2622–2629.

Kolodziej E. P., Harter T. and Sedlak D. L. (2004). Dairy wastewater, aquaculture, and spawning fish as sources of steroid hormones in the aquatic environment. *Environmental Science and Technology*, **38**(23), 6377–6384.

Kolodziej E. P., Qu S., Forsgren K. L., Long S. A., Gloer J. B., Jones G. D., Schlenk D., Baltrusaitis J. and Cwiertny D. M. (2013). Identification and environmental implications of photo-transformation products of trenbolone acetate metabolites. *Environmental Science and Technology*, **47**(10), 5031–5041.

Kolpin D. W., Furlong E. T., Meyer M. T., Thurman E. M., Zaugg S. D., Barber L. B. and Buxton H. T. (2002). Pharmaceuticals, hormones, and other organic wastewater contaminants in U.S. streams, 1999–2000: a national reconnaissance. *Environmental Science and Technology*, **36**, 1202–1211.

Kolpin D. W., Skopec M., Meyer M. T., Furlong E. T. and Zaugg S. D. (2004). Urban contribution of pharmaceuticals and other organic wastewater contaminants to streams during differing flow conditions. *Science of the Total Environment*, **328**, 119–130.

Krasnovsky A. A. (1976). Photosensitized luminescence of singlet oxygen in solution. *Biofizika*, **21**, 748–749.

Krasnovsky A. A. (1977). Photoluminescence of singlet oxygen in solutions of chlorophylls and pheophytins. *Biofizika*, **22**, 927–928.

Kronimus A., Schwarzbauer J., Dsikowitzky L., Heim S. and Littke R. (2004). Anthropogenic organic contaminants in sediments of the Lippe River, Germany. *Water Research*, **38**, 3473–3484.

Kummerer K. (2009). Antibiotics in the environment – A review – Part I. *Chemosphere*, **75**, 417–434.

Lalonde J. D., Amyot M., Kraepiel A. M. L. and Morel F. M. M. (2001). Photooxidation of Hg(0) in artificial and natural waters. *Environmental Science and Technology*, **35**, 1367–1372.

Lalonde J. D., Amyot M., Orvoine J., Morel F. M. M., Auclair J.-C. and Ariya P. A. (2004). Photoinduced oxidation of Hg^0(aq) in waters from the St. Lawrence estuary. *Environmental Science and Technology*, **38**, 508–514.

Lam M. W., Tantuco K. and Mabury S. A. (2003). PhotoFate: a new approach in accounting for the contribution of indirect photolysis of pesticides in surface waters. *Environmental Science and Technology*, **37**, 899–907.

Larson R. A. and Weber E. J. (1994). Reaction Mechanisms in Environmental Organic Chemistry. CRC Press, Inc. Boca Raton, FL.

Latch D. E. (2005). Environmental photochemistry: Studies on the degradation of pharmaceutical pollutants and the microheterogeneous distribution of singlet oxygen. Ph.D. Thesis, Department of Chemistry, University of Minnesota, Minneapolis, MN.

Latch D. E. and McNeill K. (2006). Microheterogeneity of singlet oxygen distributions in irradiated humic acid solutions. *Science*, **311**, 1743–1747.

Latch D. E., Stender B. L., Packer J. L., Arnold W. A. and McNeill K. (2003a). Photochemical fate of pharmaceuticals in the environment: cimetidine and ranitidine. *Environmental Science and Technology*, **37**, 3342–3350.

Latch D. E., Packer J. L., Arnold W. A. and McNeill K. (2003b). Photochemical conversion of triclosan to 2,8-dichlorodibenzo-*p*-dioxin in aqueous solution. *Journal of Photochemistry and Photobiology, A.*, **158**, 63–66.

Latch D. E., Packer J. L., Stender B. L., VanOverbeke J., Arnold W. A. and McNeill K. (2005). Aqueous photochemistry of triclosan: formation of 2,4-dichlorophenol, 2,8-dichlorodibenzo-*p*-dioxin and oligomerization products. *Environmental Toxicology and Chemistry*, **24**, 517–525.

Lear J. C., Maillard J. Y., Dettmar P. W., Goddard P. A. and Russell A. D. (2002). Chloroxylenol- and triclosan-tolerant bacteria from industrial sources. *Journal of Industrial Microbiology and Biotechnology*, **29**, 238–242.

Lee E., Glover C. M. and Rosario-Ortiz F. L. (2013). Photochemical formation of hydroxyl radical from effluent organic matter: role of Composition. *Environmental Science and Technology*, **47**, 12073–12080.

Lee Y., Escher B. I. and von Gunten U. (2008). Efficient removal of estrogenic activity during oxidative treatment of waters containing steroid estrogens. *Environmental Science and Technology*, **42**(17), 6333–6339.

Leifer A. (1988). The Kinetics of Environmental Aquatic Photochemistry: Theory and Practice. American Chemical Society, Washington, D.C.

Lin A. Y. and Reinhard M. (2005). Photodegradation of common environmental pharmaceuticals and estrogens in river water. *Environmental Toxicology and Chemistry*, **24**(6), 1303–1309.

Lindström A., Buerge I. J., Poiger T., Bergqvist P. A., Müller M. D. and Buser H. R. (2002). Occurrence and environmental behavior of the bactericide triclosan and its methyl derivative in surface waters and in wastewater. *Environmental Science and Technology*, **36**, 2322–2329.

Lopez-Avila V. and Hites R. A. (1980). Organic compounds in an industrial wastewater. Their transport into sediments. *Environmental Science and Technology*, **14**, 1382–90.

Love D. C., Silverman A. and Nelson K. L. (2010). Human virus and bacteriophage inactivation in clear water by simulated sunlight compared to bacteriophage inactivation at a southern California beach. *Environmental Science and Technology*, **44**, 6965–6970.

MacManus-Spencer L. A., Tse M. L., Klein J. L. and Kracunas A. E. (2011). Aqueous photolysis of the organic ultraviolet filter chemical octyl methoxycinnamate. *Environmental Science and Technology*, **45**, 3931–3937.

Mansell D. M., Bryson R. J., Harter T., Webster J. P., Kolodziej E. P. and Sedlak D. L. (2011). Fate of endogenous steroid hormones in steer feedlots under simulated rainfall-induced runoff. *Environmental Science and Technology*, **45**(20), 8811–8818.

Marchetti G., Minella M., Maurino V., Minero C. and Vione D. (2013). Photochemical transformation of atrazine and formation of photointermediates under conditions relevant to sunlit surface waters: laboratory measures and modelling. *Water Research*, **47**, 6211–6222.

Mattle M. J., Vione D. and Kohn T. (2014). Conceptual model and experimental framework to determine the contributions of direct and indirect photoreactions to the solar disinfection of MS2, phiX174, and adenovirus *Environmental Science and Technology*, **49**, 334–342.

McBain A. J., Rickard A. H. and Gilbert P. (2002). Possible implications of biocide accumulation in the environment on the prevalence of bacterial antibiotic resistance. *Journal of Industrial Microbiology and Biotechnology*, **29**, 326–330.

McMurry L. M., Oethinger M. and Levy S. B. (1998). Triclosan targets lipid synthesis. *Nature*, **394**, 531–532.

Merkel P. B., Nilsson R. and Kearns D. R. (1972). Deuterium effects on singlet oxygen lifetimes in solutions. New test of singlet oxygen reactions. *Journal of the American Chemical Society*, **94**, 1030–1031.

Mezcua M., Gomez M. J., Ferrer I., Aguera A., Hernando M. D. and Fernandez-Alba A. R. (2004). Evidence of 2,7/2,8-dibenzodichloro-p-dioxin as a photodegradation product of triclosan in water and wastewater samples. *Analytica Chimica Acta*, **524**, 241–247.

Micinski E., Ball L. A. and Zafiriou O. C. (1993). Photochemical oxygen activation: superoxide radical detection and production rates in the eastern Caribbean. *Journal of Geophysical Research*, **98**, 2299–306.

Mill T., Hendry D. G. and Richardson H. (1980). Free-radical oxidants in natural waters. *Science*, **207**, 886–887.

Miller P. L. and Chin Y. P. (2005). Indirect photolysis promoted by natural and engineered wetland water constituents: processes leading to alachlor degradation. *Environmental Science and Technology*, **39**, 4454–4462.

Minero C., Pramauro E., Pelizzetti E., Dolci M. and Marchesini A. (1992). Photosensitized transformations of atrazine under simulated sunlight in aqueous humic acid solution. *Chemosphere*, **24**, 1597–1606.

Morrall D., McAvoy D., Schatowitz B., Inauen J., Jacob M., Hauk A. and Eckhoff W. (2004). A field study of triclosan loss rates in river water (Cibolo Creek, TX). *Chemosphere*, **54**, 653–60.

Mostafa S. and Rosario-Ortiz F. L. (2013). Singlet oxygen formation from wastewater organic matter. *Environmental Science and Technology*, **47**, 8179–8186.

Mudambi A. R. and Hassett J. P. (1988). Photochemical activity of mirex associated with dissolved organic matter. *Chemosphere*, **17**, 1133–1146.

Nam S. N. and Amy G. (2002). Differentiation of wastewater effluent organic matter (EfOM) from natural organic matter (NOM) using multiple analytical techniques. *Water Science and Technology*, **57**, 1009–1015.

Neamtu M. and Frimmel F. H. (2006). Photodegradation of endocrine disrupting chemical nonylphenol by simulated solar UV-irradiation. *Science of the Total Environment*, **369**, 295–306.

Nieto-Juarez J. I., Pierzchla K., Sienkiewicz A. and Kohn T. (2010). Inactivation of MS2 coliphage in Fenton and Fenton-like systems: role of transition metals, hydrogen peroxide and sunlight. *Environmental Science and Technology*, **44**, 3351–3356.

Nguyen M. T., Silverman A. I. and Nelson K. L. (2014). Sunlight inactivation of MS2 coliphage in the absence of photosensitizers: modeling the endogenous inactivation rate using a photoaction spectrum. *Environmental Science and Technology*, **48**, 3891–3898.

Nonell S. and Braslavsky S. E. (2000). Time-resolved singlet oxygen detection. *Methods in Enzymology*, **319**, 37–49.

Nriagu J. O. (1994). Mechanistic steps in the photoreduction of mercury in natural waters. *Science of the Total Environment*, **154**, 1–8.

Okumura T. and Nishikawa Y. (1996). Gas chromatography-mass spectrometry determination of triclosans in water, sediment and fish samples via methylation with diazomethane. *Analytica Chimica Acta*, **325**, 175–184.

Oulton R. L., Kohn T. and Cwiertny D. M. (2010). Pharmaceuticals and personal care products in effluent matrices: A survey of transformation and removal during wastewater treatment and implications for wastewater management. *Journal of Environmental Monitoring*, **12**, 1956–1978.

Pace A. and Barreca S. (2013). Environmental organic photochemistry: advances and perspectives. *Current Organic Chemistry*, **17**, 3032–3041.

Packer J. L., Werner J. J., Latch D. E., McNeill K. and Arnold W.A. (2003). Photochemical fate of pharmaceuticals in the environment: naproxen, diclofenac, clofibric acid, and ibuprofen. *Aquatic Sciences*, **65**, 342–351.

Page S. E., Arnold W. A. and McNeill K. (2010). Terephthalate as a probe for photochemically produced hydroxyl radical. *Journal of Environmental Monitoring*, **12**, 1658–1665.

Page S. E., Arnold W. A. and McNeill K. (2011). Assessing the contribution of free hydroxyl radical in organic matter-sensitized photo-hydroxylation reactions. *Environmental Science and Technology*, **45**, 2818–2825.

Page S. E., Sander M., Arnold W. A. and McNeill K. (2012). Hydroxyl radical formation upon oxidation of reduced humic acids by oxygen in the dark. *Environmental Science and Technology*, **46**, 1590–1597.

Palumbo M. C. and Garcia N. A. (1988). On the mechanism of quenching of singlet oxygen by chlorinated phenolic pesticides. *Toxicological and Environmental Chemistry*, **17**, 103–116.

Pape B. E. and Zabik M. J. (1972). Photochemistry of bioactive compounds. Solution-phase photochemistry of symmetrical triazines. *Journal of Agricultural and Food Chemistry*, **20**, 316–20.

Parker K. M., Pignatello J. J. and Mitch W. A. (2013). Influence of ionic strength on triplet-state natural organic matter loss by energy transfer and electron transfer pathways. *Environmental Science and Technology*, **47**, 10987–10994.

Paul A., Hackbarth S., Vogt R. D., Roeder B., Burnison B. K. and Steinberg C. W. (2004). Photogeneration of singlet oxygen by humic substances: comparison of humic substances of aquatic and terrestrial origin. *Photochemical and Photobiological Sciences*, **3**, 273–280.

Petasne R. G. and Zika R. G. (1987). Fate of superoxide in coastal sea water. *Nature*, **325**, 516–518.

Petrovic M., Gros M. and Barcelo D. (2007). Multi-residue analysis of pharmaceuticals using LC-tandem MS and LC-hybrid MS. Comprehensive Analytical Chemistry 50, M. Petrovic and D. Barcelo (eds), Elsevier, Amsterdam, pp. 157–183.

Phillips P. J., Chalmers A. T., Gray J. L., Kolpin D. W., Foreman W. T. and Wall G. R. (2012). Combined sewer overflows: an environmental source of hormones and wastewater micropollutants. *Environmental Science and Technology*, **46**, 5336–5343.

Pillai V. N. R. (1977). Role of singlet oxygen in the environmental degradation of chlorthiamid to dichlobenil. *Chemosphere*, **6**, 777–782.

Qu S., Kolodziej E. P., Long S. A., Gloer J. B., Patterson E. V., Baltrusaitis J., Jones G. D., Benchetler P. V., Cole E. A., Kimbrough K. C., Tarnoff M. D. and Cwiertny D. M. (2013). Product-to-parent reversion of trenbolone: unrecognized risks for endocrine disruption. *Science*, **342**(6156), 347–351.

Quintana J. B. and Reemtsma T. (2004). Sensitive determination of acidic drugs and triclosan in surface and wastewater by ion-pair reverse-phase liquid chromatography/tandem mass spectrometry. *Rapid Communications in Mass Spectrometry*, **18**, 765–774.

Quistad G. B. and Mulholland K. M. (1983). Photodegradation of dienochlor [bis(pentachloro-2,4-cyclopentadien-1-yl)] by sunlight. *Journal of Agricultural and Food Chemistry*, **31**, 621–624.

Ramirez A. J., Brain R. A., Usenko S., Mottaleb M. A., O'Donnell J. G., Stahl L. L., Wathen J. B., Snyder B. D., Pitt J. L., Perez-Hurtado P., Dobbins L. L., Brooks B. W. and Chambliss C. K. (2009). Occurrence of pharmaceuticals and personal care products in fish: results of a national pilot study in the United States. *Environmental Toxicology and Chemistry*, **28**, 2587–2597.

Remucal, C. K. (2014). The role of indirect photochemical degradation in the environmental fate of pesticides: a review. *Environmental Science: Processes and Impacts*, **16**, 628–653.

Richard C. and Canonica S. (2005). Aquatic phototransformation of organic contaminants induced by coloured dissolved organic matter. *Handbook of Environmental Chemistry*, **2**, 299–323.

Richard C. and Grabner G. (1999). Mechanism of phototransformation of phenol and derivatives in aqueous solution. *Handbook of Environmental Chemistry*, **2**, 217–240.

Richard C., Vialaton D., Aguer J. P. and Andreux F. (1997). Transformation of monuron photosensitized by soil extracted humic substances: energy or hydrogen transfer mechanism? *Journal of Photochemistry and Photobiology A*, **111**, 265–271.

Richardson M. L. and Bowron J. M. (1985). The fate of pharmaceuticals in the aquatic environment. *Journal of Pharmacy and Pharmacology*, **37**, 1–12.

Routledge E. J. and Sumpter J. P. (1996). Estrogenic activity of surfactants and some of their degradation products assessed using a recombinant yeast screen. *Environmental Toxicology and Chemistry*, **15**(3), 241–248.

Ruggeri G., Ghigo G., Maurino V., Minero C. and Vione D. (2013). Photochemical transformation of ibuprofen into harmful 4-isobutylacetophenone: pathways, kinetics, and significance for surface waters. *Water Research*, **47**, 6109–6121.

Rule K. L., Ebbett V. R. and Vikesland P. J. (2005). Formation of chloroform and chlorinated organics by free-chlorine-mediated oxidation of triclosan. *Environmental Science and Technology*, **39**, 3176–3185.

Rule-Wigginton K., Menin L., Montoya J. P. and Kohn T. (2010). Oxidation of virus proteins during UV_{254} and singlet oxygen mediated inactivation. *Environmental Science and Technology*, **44**, 5437–5443.

Ryan C. C., Tan D. T. and Arnold W. A. (2011). Direct and indirect photolysis of sulfamethoxazole and trimethoprim in wastewater treatment plant effluent. *Water Research*, **45**, 1280–1286.

Sabaliunas D., Webb S. F., Hauk A., Jacob M. and Eckhoff W. S. (2003). Environmental fate of triclosan in the River Aire Basin, UK. *Water Research*, **37**, 3145–3154.

Saran M. and Summer K. H. (1999). Assaying for hydroxyl radicals: hydroxylated terephthalate is a superior fluorescence marker than hydroxylated benzoate. *Free Radical Research*, **31**, 429–436.

Schmidt R., Tanielian C., Dunsbach R. and Wolff C. J. (1994). Phenalenone, a universal reference compound for the determination of quantum yields of singlet oxygen ($O_2{}^1\Delta_g$) sensitization. *Photochemistry and Photobiology A*, **79**, 11–17.

Schwarzenbach R. P., Escher B. I., Fenner K., Hofstetter T., Johnson C. A., von Gunten U. and Wehrli B. (2006). Challenge of micropollutants in aquatic systems. *Science*, **313**, 1072–1077.

Schwarzenbach R. P., Gschwend P. M. and Imboden D. M. (2002). Environmental Organic Chemistry, 2nd ed. Wiley-Interscience, New York.

Schweizer H. P. (2001). Triclosan: a widely used biocide and its link to antibiotics. *FEMS Microbiology Letters*, **202**, 1–7.

Scully F. E. and Hoigné J. (1987). Rate constants for reactions for singlet oxygen with phenols and other compounds in water. *Chemosphere*, **16**, 681–694.

Scully N. M., Vincent W. F., Lean D. R. S. and Cooper W. J. (1997). Implications of ozone depletion for surface-water photochemistry: sensitivity of clear lakes. *Aquatic Sciences*, **59**, 260–274.

Sellers P., Kelly C. A., Rudd J. W. M. and MacHutchon A. R. (1996). Photodegradation of methylmercury in lakes. *Nature*, **380**, 694–7.

Shon H. K., Vigneswaran S. and Snyder S. A. (2006). Effluent organic matter (EfOM) in wastewater: constituents, effects, and treatment. *Critical Reviews in Environmental Science and Technology*, **36**, 327–374.

Singer H., Müller S., Tixier C. and Pillonel L. (2002). Triclosan: occurrence and fate of a widely used biocide in the aquatic environment: field measurements in wastewater treatment plants, surface waters, and lake sediments. *Environmental Science and Technology*, **36**, 4998–5004.

Snyder S. A., Westerhoff P., Yoon Y. and Sedlak D. L. (2003). Pharmaceuticals, personal care products, and endocrine disruptors in water: implications for the water industry. *Environmental Engineering Science*, **20**(5), 449–469.

Stumpf M., Ternes T. A., Wilken R. D., Rodrigues S. V. and Baumann W. (1999). Polar drug residues in sewage and natural waters in the State of Rio de Janeiro, Brazil. *Science of the Total Environment*, **225**, 135–141.

Sumpter J. P. and Johnson A. C. (2005). Lessons from endocrine disruption and their application to other issues concerning trace organics in the aquatic environment. *Environmental Science and Technology*, **39**(12), 4321–4332.

Thomas-Smith T. E. and Blough N. V. (2001). Photoproduction of hydrated electron from constituents of natural waters. *Environmental Science and Technology*, **35**, 2721–2726.

Tixier C., Singer H. P., Canonica S. and Müller S. R. (2002). Phototransformation of triclosan in surface waters: a relevant elimination process for this widely used biocide—Laboratory studies, field measurements, and modeling. *Environmental Science and Technology*, **36**, 3482–3489.

Tixier C., Singer H. P., Oellers S. and Müller S. R. (2003). Occurrence and fate of carbamazepine, clofibric acid, diclofenac, ibuprofen, ketoprofen, and naproxen in surface waters. *Environmental Science and Technology*, **37**, 1061–1068.

Tranyek P. G. and Hoigne J. (1991). Oxidation of substituted phenols in the environment: a QSAR analysis of rate constants for reaction with singlet oxygen. *Environmental Science and Technology*, **25**, 1596–1604.

Trudeau V. L., Heyne B., Blais J. M., Temussi F., Atkinson S. K., Pakdel F., Popesku J. T., Marlatt V. L., Scaiano J. C., Previtera L. and Lean D. R. S. (2011). Lumiestrone is photochemically derived from estrone and may be released to the environment without detection. *Frontiers in Endocrinolology*, **2**, 1–83.

Turro N. J. (1991). Modern Molecular Photochemistry, University Science Books, Sausalito, California.

United States Environmental Protection Agency Exposure Assessment Models, GC Solar. http://www2.epa.gov/exposure-assessment-models/gcsolar (accessed 1 April 2015).

Vajda A. M., Barber L. B., Gray J. L., Lopez E. M., Woodling J. D. and Norris D. O. (2008). Reproductive disruption in fish downstream from an estrogenic wastewater effluent. *Environmental Science and Technology*, **42**(9), 3407–3414.

Vajda A. M., Barber L. B., Gray J. L., Lopez E. M., Bolden A. M., Schoenfuss H. L. and Norris D. O. (2011). Demasculinization of male fish by wastewater treatment plant effluent. *Aquatic Toxicology*, **103**(3–4), 213–221.

van Stee L. L. P., Leonards P. E. G., Vreuls R. J. J. and Brinkman U. A. T. (1999). Identification of non-target compounds using gas chromatography with simultaneous atomic emission and mass spectrometric detection (GC-AED/MS): analysis of municipal wastewater. *Analyst*, **124**, 1547–1552.

Velagaleti R. (1997). Behavior of pharmaceutical drugs (human and animal health) in the environment. *Drug Information Journal*, **31**, 715–722.

Vialaton D., Richard C., Baglio D. and Paya-Perez A.-B. (1998). Phototransformation of 4-chloro-2-methylphenol in water: influence of humic substances on the reaction. *Journal of Photochemistry and Photobiology A*, **119**, 39–45.

Vione D., Ponzo M., Bagnus D., Maurino V., Minero C. and Carlotti M. E. (2010). Comparison of different probe molecules for the quantification of hydroxyl radicals in aqueous solutions. *Environmental Chemistry Letters*, **8**, 95–100.

Vione D., Caringella R., De Laurentiis E., Pazzi M. and Minero C. (2013). Phototranformation of sunlight filter benzopheone03 (2-hydroxy-4-methoxybenzophenone) under conditions relevant to surface waters. *Science of the Total Environment*, **463–464**, 243–251.

Webster J. P., Kover S. C., Bryson R. J., Harter T., Mansell D. S., Sedlak D. L. and Kolodziej E. P. (2012). Occurrence of trenbolone acetate metabolites in simulated confined animal feeding operation (CAFO) runoff. *Environmental Science and Technology*, **46**(7), 3803–3810.

Wegelin M., Canonica S., Mechsner K., Fleischmann T., Pesaro F. and Metzler A. (1994). Solar water disinfection: scope of the process and analysis of radiation experiments. *Journal of Water Supply and Technology-Aqua*, **43**(3), 154–169.

Welker M. and Steinberg C. (2000). Rates of humic substance shotosensitized degradation of microcystin-LR in natural waters. *Environmental Science and Technology*, **34**, 3415–3419.

Wenk J. and Canonica S. (2012). Phenolic antioxidants inhibit the triplet-induced transformation of anilines and sulfonamide antibiotics in aqueous solution. *Environmental Science and Technology*, **46**, 5455–5462.

Wenk, J., von Gunten, U. and Canonica, S. (2011). Effect of dissolved organic matter on the transformation of contaminants induced by excited triplet states and the hydroxyl radical. *Environmental Science and Technology*, **45**, 1334–1340.

Wenk J., Eustis S. N., McNeill K., Canonica S. (2013). Quenching of excited triplet states by dissolved natural organic matter. *Environmental Science and Technology*, **47**, 12802–12810.

Werner J. J., Arnold W. A. and McNeill K. (2006). Water hardness as a photochemical parameter: tetracycline photolysis as a function of calcium concentration, magnesium concentration, and pH. *Environmental Science and Technology*, **40**(23), 7236–7241.

Werner J. J., Wammer K. H., Chintapalli M., Arnold W. A. and McNeill K. (2007). Environmental photochemistry of tylosin: efficient, reversible photoisomerization to a less – active isomer, followed by photolysis. *Journal of Agricultural and Food Chemistry*, **55**, 7062–7068.

Whidbey C. M., Daumit K. E., Nguyen T. H., Ashworth D. D., Davis J. C. C. and Latch D. E. (2012). Photochemical induced changes of in vitro estrogenic activity of steroid hormones. *Water Research*, **46**, 5287–5296.

Wilkinson F., Helman W. P. and Ross A. B. (1995). Rate constants for the decay and reactions of the lowest electronically excited singlet state of molecular oxygen in solution. An expanded and revised compilation. *Journal of Physical Chemistry Reference Data*, **24**, 663–1021.

Woodling J. D., Lopez E. M., Maldonado T. A., Norris D. O. and Vadja A. M. (2006). Intersex and other reproductive disruption of fish in wastewater effluent dominated Colorado streams. *Comparative Biochemistry and Physiology – Part C: Toxicology*, **144C**(1), 10–15.

Xu H., Cooper W. J., Jung J. and Song W. (2011). Photosensitized degradation of amoxicillin in natural organic matter isolate solutions. *Water Research*, **45**, 632–638.

Young R. B., Latch D. E., Mawhinney D. B., Nguyen T. H., Davis J. C. C. and Borch T. (2013). Direct photodegradation of androstenedione and testosterone in natural sunlight and its effect on endocrine disrupting potential. *Environmental Science and Technology*, **47**, 8416–8424.

Zafiriou O. C., Blough N. V., Micinski E., Dister B., Kieber D. and Moffett J. (1990). Molecular probe systems for reactive transients in natural waters. *Marine Chemistry*, **30**, 45–70.

Zeng T. and Arnold W. A. (2013). Pesticide photolysis in prairie potholes: probing photosensitized processes. *Environmental Science and Technology*, **47**, 6735–6745.

Zepp R. and Cline D. (1977). Rates of direct photolysis in aquatic environment. *Environmental Science and Technology*, **11**(4), 359–366.

Zepp R. G. (1978). Quantum yields for reactions of pollutants in dilute aqueous solutions. *Environmental Science and Technology*, **12**, 327–329.

Zepp R. G., Wolfe N. L., Baughman G. L. and Hollis R. C. (1977). Singlet oxygen in natural waters. *Nature*, **267**, 421–423.

Zepp R. G., Baughman G. L. and Schlotzhauer P. F. (1981a). Comparison of the photochemical behavior of various humic substances in water: I. Sunlight induced reactions of aquatic pollutants photosensitized by humic substances. *Chemosphere*, **10**, 109–17.

Zepp R. G., Baughman G. L. and Schlotzhauer P. F. (1981b). Comparison of photochemical behavior of various humic substances in water: II. Photosensitized oxygenations. *Chemosphere*, **10**, 119–26.

Zepp R. G., Schlotzhauer P. F. and Sink R. M. (1985). Photosensitized transformations involving electronic energy transfer in natural waters: role of humic substances. *Environmental Science and Technology*, **19**, 48–55.

Zepp R. G., Braun A. M., Hoigne J. and Leenheer J. A. (1987). Photoproduction of hydrated electrons from natural organic solutes in aquatic environments. *Environmental Science and Technology*, **21**, 485–90.

Zhang T. and Hsu-Kim H. (2010). Photolytic degradation of methylmercury enhanced by binding to natural organic ligands. *Nature Geoscience*, **3**(7), 473–476.

Zuccato E., Calamari D., Natangelo M. and Fanelli R. (2000). Presence of therapeutic drugs in the environment. *Lancet*, **355**, 1789–1790.

Zuo Y., Zhang K. and Deng Y. (2006). Occurrence and photochemical degradation of 17α-ethinylestradiol in Acushnet River Estuary. *Chemosphere*, **63**, 1583–1590.

Chapter 14

Advanced treatment for potable water reuse

Stuart J. Khan, Troy Walker, Benjamin D Stanford and Jörg E. Drewes

14.1 PLANNED POTABLE WATER REUSE

With increasing demands on existing water supplies and limited access to new conventional water resources, some municipalities have begun to intentionally reuse highly treated municipal wastewater effluents to augment drinking water supplies. Advanced oxidation processes (AOPs) are increasingly being adopted as an important barrier to chemical and pathogenic contaminants in these types of potable water reuse projects.

In the context of potable reuse (and hence in this chapter), the term 'AOP' has been applied primarily to systems specifically designed for the enhanced generation of radical oxidants. Although some applications of ozonation to treat reclaimed water may produce radical species (particularly when operated at elevated pH), ozonation processes have generally not been classified as AOPs unless radical formation has been intentionally enhanced. As such, (non-enhanced) ozonation processes are not described as AOPs in this chapter.

Throughout the world, treated and untreated municipal effluents are discharged to waterways including streams and rivers. In many cases, towns and cities downstream draw upon such streams and rivers for municipal drinking water supplies. As such, water that was discharged as treated wastewater is unintentionally reused for drinking water supplies. This practice is commonly termed 'unplanned' or '*de facto*' potable reuse, indicating that although it is not usually seen as an intentional water supply strategy, it is nonetheless, a reality in many places (Rice & Westerhoff, 2015).

Planned potable water reuse involves the purposeful addition of highly treated wastewater (i.e., reclaimed or recycled water) to a drinking water supply. The distinction between 'unplanned' and 'planned' potable reuse is significant since the acknowledgement of intention and more holistic view of the overall urban water cycle has led to changes in implementation (Drewes & Khan, 2011). These changes have included increased attention to health risk assessment and management. In turn, these have led to the incorporation of enhanced or additional water quality treatment barriers in some cases (Drewes & Khan, 2015).

Table 14.1 Examples of planned potable reuse schemes incorporating AOP treatment.

Project Location	Project Size (ML/day)	AOP Implementation (Year)	Status (2016)	Advanced Treatment Technologies (Abbreviated)	Type of Reuse
Groundwater Replenishment System, Orange County, CA, USA	350	2008	Operational	UF → RO → UV/H_2O_2 (AOP)	IPR: Groundwater recharge via direct injection and spreading basins
Western Corridor Project, Southeast Queensland, Australia	232	2008	Not operational; largely decommissioned.	UF → RO → UV/H_2O_2 (AOP) → Cl_2	IPR: Surface water augmentation into drinking water reservoir
Arapahoe County/ Cottonwood, CO, USA	34	2009	Operational	Media filtration → RO → UV/H_2O_2 (AOP) → Cl_2	IPR: Groundwater recharge via spreading
Prairie Waters Project, Aurora, CO, USA	190	2010	Operational	Riverbank filtration → SAT → softening → UV/ H_2O_2 (AOP) → BAC → GAC → Cl_2	IPR: Groundwater recharge via riverbank filtration (note: the environmental buffer is used early in the treatment process)
Beaufort West Municipality, South Africa	2	2011	Operational	Sand filtration → UF → RO → UV/ H_2O_2 (AOP) → Cl_2	DPR: Blending with conventionally treated surface water sources
West Basin Water Recycling Plant, CA, USA (Edward C. Little Water Recycling Facility – Phase V Expansion Project)	47	2013	Operational	O_3 → MF → RO → UV/H_2O_2 (AOP) → Cl_2	IPR: Groundwater recharge via direct injection
Raw Water Production Facility, Big Spring, TX, USA	7	2013	Operational	MF → RO → UV/H_2O_2 (AOP)	DPR: Blending with raw surface water and then conventional water treatment
Terminal Island Water Reclamation Plant, Los Angeles, CA, USA	45	2016	Under construction	MF → RO → UV/Cl_2 (AOP)	IPR: Groundwater recharge via direct injection

Cl_2 = chlorine disinfection, RO = reverse osmosis, UV = ultraviolet radiation, AOP = advanced oxidation process, UF = ultrafiltration, MF = microfiltration, SAT = soil-aquifer treatment, GAC = granular activated carbon, BAC = biological activated carbon.

Practices of planned potable water reuse have been categorised as one of either 'indirect potable reuse' (IPR) or 'direct potable reuse' (DPR). The distinction is made on the inclusion or exclusion of what has been referred to as an 'environmental buffer' (Leverenz *et al.* 2011). The incorporation of an environmental buffer involves transferring the water, at some appropriate point in the treatment train, to an environmental system such as a surface water reservoir or groundwater aquifer. The environmental buffer may serve a number of functions including storage, dilution and the opportunity for further water quality improvement by natural treatment processes such as sunlight-induced photolysis, biotransformation and natural pathogen inactivation. Furthermore, passing reclaimed water through an environmental buffer has been perceived to be beneficial regarding enhancing public perception of potable water reuse projects. This is achieved, in part, by providing a 'disconnection' between sewage as the source of the water and potable use as the final application. Projects that have incorporated the use of an environmental buffer are examples of IPR, while projects that omit any significant[1] environmental buffer have been referred to as examples of DPR (Arnold *et al.* 2012).

A range of planned potable reuse schemes, employing various natural and engineered treatment processes, have been developed internationally since the early 1960s (Drewes & Khan, 2011). Some of the most prominent projects which have included AOPs in the water treatment train are summarised in Table 14.1. The majority of these projects are examples of IPR schemes. However, they include two currently operational municipal DPR projects. These include one in Beaufort West, South Africa and one in Big Spring, Texas, USA.

Advanced treatment technologies used in each case are shown in Table 14.1 to indicate the additional treatment that is applied, compared to treatment trains of the more common practice of *de facto* potable reuse. In all cases, these advanced treatment processes are preceded by variations of conventional biological secondary or tertiary wastewater unit processes. Similarly, the IPR schemes are succeeded by subsequent drinking water treatment of recovered recycled water that is blended with water from other supplies.

The potable reuse schemes included in Table 14.1 have employed a range of advanced treatment processes to achieve various water quality objectives. While there has been some evolution in treatment technologies over the decades, there remains no 'standard' treatment train for potable reuse (Gerrity *et al.* 2013). Instead, specific treatment processes have been selected according to numerous local conditions and constraints. Regulatory requirements and perceived performance of specific treatment trains have led to favouring certain combinations. However, water quality objectives suitable for potable reuse can be achieved by a range of treatment processes with a wide variety of configurations and variations. It may be anticipated that flexibility in design will increase as project objectives tend to be defined in terms of required treatment performances (fit-for-purpose), rather than required treatment processes.

14.2 TREATMENT OBJECTIVES AND DRIVERS FOR THE ADOPTION OF AOPs IN POTABLE REUSE

Given the nature of the source, public health concerns regarding potable reuse are foremost related to the presence of pathogens along with organic and inorganic chemical constituents in reclaimed water. Thus, potable reuse projects must integrate appropriate water treatment processes that are capable of providing effective, reliable and redundant barriers to those pathogens and chemicals (Drewes & Khan, 2015). However, not all wastewater treatment processes will result in the same quality of water that will feed the advanced treatment facility. Consequently, potable reuse projects are not a one-size-fits-all solution and

[1]Note that there is no clear or consistent definition of what constitutes a "significant" environmental buffer. Accordingly, the precise distinction between IPR and DPR remains somewhat vague and inconsistent.

the barriers selected must be chosen considering site-specific constraints. Fundamental to the design of potable reuse schemes is the concept of multiple barriers to address site-specific risks and to ensure that acceptable final water quality is reliably achieved.

For many potable reuse projects, the use of integrated membrane systems incorporating microfiltration (MF) or ultrafiltration (UF) followed by reverse osmosis (RO) has been perceived as the industry standard (Gerrity *et al.* 2013), though the genesis of the RO concept was for salinity control rather than for specific pathogen reduction or other soluble organic chemical constituents. Such desalting-based schemes are mostly located in coastal areas where concentrated waste streams (i.e., RO concentrates) may be conveniently discharged to the ocean or in areas where discharge of concentrated brine is allowed into sewer systems, saline surface waters or deep well injection. In the USA, Singapore, and Australia, utilities have favoured this treatment train, in some cases coupled with subsequent AOPs. For inland projects, however, membrane-based treatment is often limited by a lack of concentrate stream disposal or cost prohibitive zero-liquid discharge options. Instead, various non-RO combinations of low-pressure membranes, granular activated carbon (GAC) filtration, chemical oxidation (i.e., ozone, UV/H_2O_2), and biological treatment processes have evolved. These practices underscore that multiple options exist for the design of potable reuse schemes that consider regional conditions but are unified in the goal to lower or eliminate the risk from constituents of concern.

By far the most common application of AOP for potable water reuse is currently ultraviolet (UV) radiation with hydrogen peroxide (H_2O_2) addition (Figure 14.1). This technique is described in detail in Chapter 3. Less common techniques include the use of ozone with enhanced radical formation (Chapter 4) and UV/Cl_2 systems (Chapter 9). Techniques using solid state photo-catalysis such as titanium dioxide (Chapter 8) or other electrochemical processes like boron-doped diamond electrodes are of emerging interest for full scale water treatment applications.

Figure 14.1 UV/H_2O_2 AOP System at the Groundwater Replenishment System, Orange County, CA, USA.

Inactivation of pathogenic organisms by UV irradiation (40–180 mJ/cm^2) is an important and well established treatment process in municipal wastewater and drinking water treatment plants (Hijnen *et al.* 2006). In such applications, two types of mercury based UV lamps are commonly used, distinguished by the mercury vapour pressure inside of the operating lamp. Low pressure lamps operate at approximately

0.01 mbar (1 Pa) and medium pressure lamps at higher than 1 bar (100 kPa). The spectral radiation from low pressure mercury plasma is dominated by wavelengths of 253.7 nm and 184.9 nm. UV radiation is absorbed by DNA disrupting its structure and leading to deactivation of living cells. Disinfection depends on the UV dose and the wavelength, with 253.7 nm being well suited. In contrast, medium pressure-Hg vapour lamps produce a more continuous emission spectrum. Their greatest advantage is their very high specific UV-flux (power output) per unit arc length, but they operate at a lower electrical efficiency and therefore can be more expensive.

At sufficient energies, UV radiation can also be effective for the photolytic degradation of many trace organic contaminants of concern in potable water reuse projects (Wols & Hofman-Caris, 2012). However, the susceptibility of organic molecules to UV-photolysis is highly dependent on molecular features, leaving many important water contaminants effectively resistant or very slow to degrade photolytically (Yan & Song, 2014).

AOPs widen the range of organic chemicals that may be oxidised as well as significantly increase the reaction rates (von Gunten, 2003; Yan & Song, 2014). All AOPs involve the generation of highly reactive species, most commonly, hydroxyl radicals (·OH). With sufficient radiation energies, organic chemicals may be mineralised, that is, converted to carbon dioxide and other inorganic species such as water, chloride and nitrate ions. However, complete mineralisation is neither an objective nor likely to be achieved for full scale water treatment systems due to the energy inputs required to achieve complete mineralisation.

In AOP applications, system performance and sizing is commonly based on the electrical energy per Log_{10} order (E_{EO}) of specific target contaminants. The E_{EO} is a semi-empirical parameter defined as the electrical energy (kWh) required to reduce the concentration of a contaminant by one order of magnitude (90% removal) in one cubic meter of water (kWh/m^3/order) (Bolton & Stefan, 2002). While these are the definition and standard units adopted by the International Union of Pure and Applied Chemists (IUPAC) (Bolton *et al.* 2001), in North America E_{EO} values are commonly expressed in terms of 1000 gallons, instead of one cubic meter. The E_{EO} parameter is both contaminant and reactor specific, and also depends on the water UV transmittance (UVT).

14.2.1 Pathogen inactivation

The UV dose for pathogen inactivation depends on the UV radiation intensity distribution in the reactor (measured by UV sensors), the flow rate and flow patterns inside the UV reactor, and the water UVT.

UV/H_2O_2 AOPs have an important role in water disinfection for potable reuse projects. Compared to a typical drinking water UV disinfection dose (40–180 mJ/cm^2), the applied UV fluence in a UV/H_2O_2 AOP (usually >500 mJ/cm^2) is many times greater. The US EPA has developed UV dose requirements for potable water systems to receive credit for inactivation of Cryptosporidium, Giardia, and viruses (US EPA 2006). Since virus inactivation requires significantly higher UV doses than inactivation of the other two organisms, UV dose requirements are primarily determined by virus inactivation. These range from 39 mJ.cm^{-2} for 0.5 Log_{10} virus inactivation to 186 mJ.cm^{-2} for 4.0 Log_{10} virus inactivation (US EPA 2006).

These UV inactivation credits were developed in the context of a US drinking water regulation (The Long Term 2 Enhanced Surface Water Treatment Rule). Consequently, they have mostly been applied to drinking water applications and do not account for the much higher UV fluences used for AOPs in potable reuse projects. It is generally recognized that the much higher energy applied for UV AOPs is more than sufficient to achieve effective disinfection to the limits at which regulatory agencies commonly credit disinfection performance. For example, Californian potable reuse regulations provide a maximum credit of 6 Log_{10} reduction for each pathogen by any advanced water treatment process (California Office of Administrative Law, 2015).

Little attention is currently given to the potential role of secondary reactive species in the disinfection process. Nonetheless, research has demonstrated that disinfection of some organisms can be enhanced by AOP-produced species such as hydroxyl radicals (Mamane *et al.* 2007; Labas *et al.* 2008). In particular, enhancement of UV inactivation of adenovirus by the addition of H_2O_2 has been demonstrated (Bounty *et al.* 2012). This is significant since adenovirus has consistently been observed to be among the most resistant known pathogens to UV disinfection.

14.2.2 Trace chemical contaminants

As can be observed from Table 14.1, a common use of UV-based AOPs in potable reuse projects has been following RO treatment. This is advantageous for the AOP since the effectiveness of the process relies upon low turbidity, low concentrations of common UV-absorbing water constituents, and low concentrations of chemicals other than the targeted micropollutants which may react with (or 'scavenge') hydroxyl radicals, once produced.

While system configuration and operational conditions may lead to some variation, in general RO is a highly effective barrier to most chemical contaminants of concern in potable reuse projects. However, there are a small number of chemical contaminants, which are not well removed by RO (Bellona *et al.* 2004; Libotean *et al.* 2008; Fujioka *et al.* 2012). These tend to be small uncharged molecules with molecular weights typically less than about 100 g/mol. A small subset of these are of human health concern at concentrations in which they may occur in RO permeates. In many cases, AOPs have been included in potable reuse projects specifically to act as a barrier for the removal of these contaminants. The two most important chemical drivers for the inclusion of AOPs have been N-nitrosodimethylamine (NDMA) (Mitch *et al.* 2003b) and 1,4-dioxane (Zenker *et al.* 2003).

14.2.2.1 N-Nitrosodimethylamine (NDMA)

NDMA is a small uncharged, but polar organic chemical. This combination of physical properties is known to be among those that lead to high water solubility and poor rejection by RO membranes (Bellona *et al.* 2004). The molecular structure and properties of NDMA are presented in Figure 14.2.

Molecular structure	Molecular properties
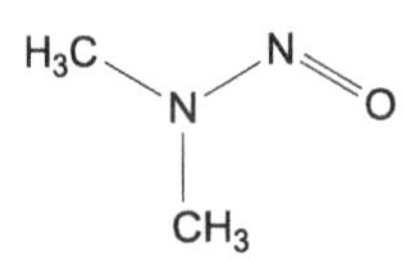	CAS Registry Number: 62-75-9
	Molecular formula: $C_2H_6N_2O$
	Molar mass: 74.08 g/mol.
	Solubility in water: 290 mg/mL at 20°C
	Log_{10} Kow: -0.50

Figure 14.2 Molecular structure and properties of N-nitrosodimethylamine (NDMA).

NDMA can be found in both domestic and industrial wastewater and concentrations up to 63 ng/L have been reported in raw domestic sewage (Sedlak *et al.* 2005; Fujioka *et al.* 2012). Much higher concentrations may be derived from industrial wastewaters. Furthermore, NDMA can readily be formed during the disinfection of biologically-treated effluent using chlorine or chloramines (Mitch *et al.* 2003a; Pehlivanoglu-Mantas *et al.* 2006). Since biofouling of water treatment membranes is commonly controlled by chlorine or chloramination, this is sometimes an important point in the treatment train for NDMA formation.

It is now well established that oxidation of NDMA precursors by strong oxidants such as during chlorination or chloramination leads to a formation of NDMA (Charrois & Hrudey, 2007). The formation of NDMA during chloramination is believed to be related to the presence of dichloramine ($NHCl_2$) (Shah & Mitch, 2012). Although monochloramine (NH_2Cl) is the target chloramine species, dichloramine coexists according to Equation 14.1:

$$2NH_2Cl + H^+ \leftrightarrow NHCl_2 + NH_4^+ \quad (14.1)$$

The role of hydrogen ions in Equation 14.1 renders this reaction highly pH-dependant. At pH > 8.5, monochloramine predominates. However, at pH < 5 dichloramine is dominant (Schreiber & Mitch, 2006). In potable reuse plants, RO feeds are typically at pH 6–7, dropping to pH 5–6 in RO permeates. Under these conditions, monochloramine and dichloramine may both be present at significant concentrations.

It is believed that NDMA formation occurs via the formation of a chlorinated unsymmetrical dimethylhydrazine intermediate (Cl-UDMH) from a nucleophilic substitution reaction between dimethylamine and $NHCl_2$ (Figure 14.3) (Schreiber & Mitch, 2006; Shah & Mitch, 2012). Oxidation of Cl-UDMH by oxygen to form NDMA competes with its oxidation by chloramines to form other products.

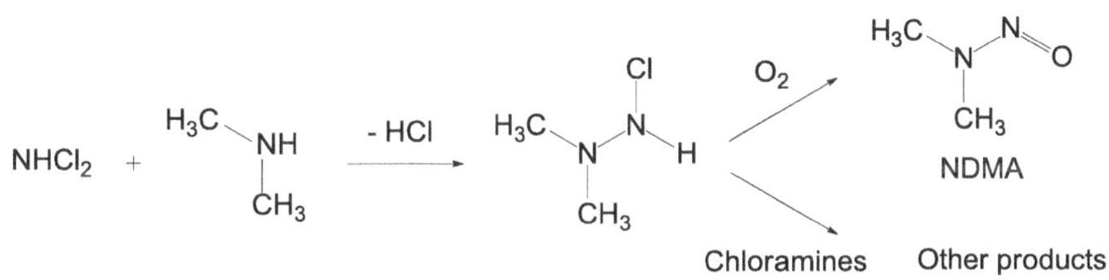

Figure 14.3 Formation of NDMA during chloramination processes (Shah & Mitch, 2012).

The formation of NDMA by chloramination can vary considerably, depending on the conditions of the chloramination process. For example, NDMA concentration may be substantially increased with increasing reaction time and chloramine (or chlorine) dose (Pehlivanoglu-Mantas *et al.* 2006). In investigations undertaken at a full scale advanced water treatment plant, it was shown that, while 1–2 h of chloramination exposure resulted in around 7 ng/L NDMA, 20–22 h contact time led to around 170 ng/L NDMA formation (Farré *et al.* 2011).

In addition to chloramination pathways, NDMA may also form during the ozonation of wastewater effluents (Pisarenko *et al.* 2012). While several mechanisms have been proposed (Andrzejewski *et al.* 2008; von Gunten *et al.* 2010), the balance between ozone and NDMA formation in wastewaters is at times paradoxical. On the one hand ozonation of the wastewater reduces the subsequent NDMA formation potential of the treated water. On the other hand, ozone directly forms NDMA in the wastewater, sometimes as concentrations as high as several hundred parts per trillion (Andrzejewski *et al.* 2008; Pisarenko *et al.* 2012).

NDMA is considered a probable human carcinogen by the U.S. Environmental Protection Agency, with a calculated life time cancer risk of 10^{-6} for a concentration of 0.7 ng/L in drinking water (US EPA, 1987). The CDPH established a notification level for NDMA of 10 ng/L and a response level of 300 ng/L. Drinking water NDMA guideline values set by the World Health Organization (2011) and Australian Drinking Water Guidelines (NHMRC & NRMMC, 2011) are higher at 100 ng/L.

For potable water reuse applications involving the use of the RO process, concentrations of NDMA in final product water can be controlled via several strategies. NDMA concentration can be minimised by reducing the formation of NDMA during the chloramination process. This can be achieved by dosing pre-formed chloramine (Farré *et al.* 2011) and reducing the contact time of chloramination (Mitch *et al.* 2005; Schreiber & Mitch, 2005). However, reducing the NDMA formation may not be sufficient if a higher NDMA concentration than the regulatory level occurs in the inflow of the wastewater treatment plant. An alternative approach is to use an additional treatment process for the removal of NDMA. The rejection of NDMA by RO membranes is highly variable, with published reports ranging at least between 10–70% rejection (Fujioka *et al.* 2012). Numerous factors, including membrane selection water temperatures and permeate flux, play a role in this observed variability (Fujioka *et al.* 2013; Fujioka *et al.* 2014).

Photolysis is the most commonly employed mitigation strategy for NDMA in potable water systems (Nawrocki & Andrzejewski, 2011). However, since a UV fluence approaching 1000 mJ/cm^2 is required for a 1 Log_{10} reduction in NDMA, UV treatment for nitrosamine destruction is more energy intensive and expensive than are UV applications strictly for disinfection (Krasner *et al.* 2013).

While not of significant health concern themselves, some NDMA formation precursors (such as dimethylamine shown in Figure 14.3) are also small, polar, uncharged organic molecules. Like NDMA, these molecules may also be poorly rejected by RO membranes. As a consequence, additional NDMA may be formed during final chloramination of RO permeates (Sgroi *et al.* 2015). In fact, there is evidence that the use of H_2O_2 dosing in the UV reactor to achieve AOP conditions may exacerbate this problem by producing oxidation byproducts with higher NDMA formation potential than the parent compounds (Farré *et al.* 2012; Sgroi *et al.* 2015).

14.2.2.2 1,4-Dioxane

1,4-Dioxane is a heterocyclic saturated organic compound, primarily used as a stabilizer for the solvent 1,1,1-trichloroethane during storage and transport in aluminium containers. Like NDMA, it has low molecular mass and is uncharged. While it is not technically a polar molecule (it has a symmetrical structure), the rich electron density of two oxygen atoms leads to miscibility with water and poor rejection by RO membranes. The molecular structure and properties of 1,4-dioxane are presented in Figure 14.4.

Molecular structure	Molecular properties
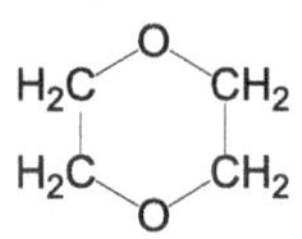	CAS Registry Number: 123-91-1 Molecular formula: $C_4H_8O_2$ Molar mass: 88.11 g/mol. Solubility in water: Miscible

Figure 14.4 Molecular structure and properties of 1,4-dioxane.

1,4-Dioxane is a significant groundwater contaminant in some parts of the USA and Europe, -most notably where groundwater has been contaminated with chlorinated solvents (Adamson *et al.* 2014; Stepien *et al.* 2014). It is relatively recalcitrant to biodegradation and its physicochemical properties preclude effective removal from contaminated groundwater by volatilization or adsorption (Stepien *et al.* 2014; Li *et al.* 2015). 1,4-Dioxane may also be present in municipal and industrial wastewaters where solvent mixtures have been discharged to sewers. Conventional wastewater treatment processes are ineffective for the degradation or removal of 1,4-dioxane (Stepien *et al.* 2014). In fact, 1,4-dioxane concentrations may increase during biological treatment as a consequence of impurities introduced during methanol dosing for post-anoxic denitrification processes (Stepien *et al.* 2014). In a survey of effluents from 40 municipal wastewater treatment plants across the USA, measured 1,4-dioxane concentrations ranged from <0.3 to 3.3 μg/L, with a mean concentration of 1.1 μg/L (Simonich *et al.* 2013).

1,4-dioxane is poorly removed by reverse osmosis and consequently has been commonly detected in permeates of reverse osmosis plants treating municipal effluents (Linge *et al.* 2012). However, removal by AOPs can be highly effective (Chitra *et al.* 2012; Antoniou & Andersen, 2015).

The US EPA considers 1,4-dioxane to be "likely to be carcinogenic to humans", with a 10^{-6} cancer risk level 0.35 μg/L (US EPA 2016). Studies in rodents show liver tumors to be consistently reported after chronic oral exposure (Dourson *et al.* 2014). 1,4-Dioxane does not cause point mutations, DNA repair, or initiation. However, it appears to promote tumors and stimulate DNA synthesis (Dourson *et al.* 2014).

14.3 VALIDATION AND PROCESS CONTROL

Most existing potable reuse facilities have not required a formal validation (i.e., a bioassay) to demonstrate disinfection performance of AOP systems. Instead, it has generally been recognized that the energy for UV-photolysis is far higher than that required for disinfection (as discussed above) and therefore disinfection occurs simultaneously to photolysis.

In circumstances where disinfection validation is required, the principles described in the US EPA UV Disinfection Guidance Manual (US EPA 2006) would normally be applicable. In that case, validation testing would be required to determine the operating conditions under which the reactor delivers the required UV dose for treatment credit. These operating conditions must include flow rate, UV intensity as measured by a UV sensor, and UV lamp status. The validated operating conditions must account for a number of factors including (US EPA 2006):

- UV absorbance of the water
- Lamp fouling and ageing
- Measurement uncertainty of online sensors
- UV dose distributions arising from the velocity profiles through the reactor
- Failure of UV lamps or other critical system components
- Inlet and outlet piping or channel configurations of the UV reactor

For a UV/H_2O_2 AOP system, validation may also be required for advanced oxidation performance. While less formal guidance is available for this process, it may involve similar protocols to those used for disinfection validation. In this case, controlled dosing of the H_2O_2 solution is a further additional performance factor and would need to be accounted for in the validation testing. Instead of relating pathogen disinfection performance to UV dose, it is appropriate to relate the oxidative transformation of key chemical groups to UV energy (fluence). The use of indicator chemicals, with diverse chemical properties and functional groups, to assess organic chemical oxidation was described in detail by Dickenson *et al.* (2009).

The California Division of Drinking Water has implemented environmental health regulations pertaining to potable reuse projects incorporating groundwater recharge, commonly referred to as the Title 22 Code of Regulations (California Office of Administrative Law, 2015). These regulations refer to a concept of "full advanced treatment", which includes the incorporation of reverse osmosis and an oxidation treatment process. To demonstrate a sufficient oxidation process has been designed for implementation, a project proponent is required to perform an occurrence study on the project's municipal wastewater to identify indicator compounds and select a total of at least nine indicator compounds, with at least one from each of the following nine functional groups:

1. Hydroxy aromatic
2. Amino/acylamino aromatic
3. Nonaromatic with carbon double bonds
4. Deprotonated amine
5. Alkoxy polyaromatic
6. Alkoxy aromatic
7. Alkyl aromatic
8. Saturated aliphatic
9. Nitro aromatic

The project proponent must then demonstrate that the oxidation process can achieve at least 0.5 Log_{10} removal for at least 5 of the indicator compounds representing the functional groups 1 to 7, and at least 0.3 Log_{10} removal for one indicator compound representing the functional groups 8 and 9. At least one surrogate or operational parameter must be selected that reflects the removal of at least five of the nine indicator compounds, and at least one surrogate or operational parameter must be capable of being monitored continuously, recorded, and have associated alarms. Project proponents must then demonstrate that these requirements can be met in full scale operation, using challenge or spiking tests to determine the removal differential under normal operating conditions.

Frequently, in lieu of using the suite of indicator chemicals described above, a project sponsor may conduct testing demonstrating that the oxidation process will achieve at least 0.5 Log_{10} reduction of 1,4-dioxane. The proponent must then establish surrogate and/or operational parameters that reflect this design criterion. At least one such surrogate or operational parameter must be capable of being monitored continuously, recorded, and have associated alarms that indicate when the process is not operating as designed. To date, this simpler validation approach has most commonly been applied for potable reuse projects in the USA.

Key operational control parameters are commonly termed 'critical control points' (CCPs). Ongoing monitoring of UV/H_2O_2 AOPs may include monitoring of identified CCPs to ensure performance. Examples of potentially useful CCPs for UV/H_2O_2 AOPs include electrical energy per order (calculated as a function of a target contaminant e.g. NDMA and compared to a minimum threshold value), UV transmittance (UVT) of incoming water, hydraulic flow rates, lamp status, UV intensity sensor readings and peroxide doses. This type of CCP monitoring is increasingly required for potable reuse systems in some countries including the USA and Australia. Consequently, the availability of suitable CCPs for a particular treatment process are a significant advantage for that process and may have major impacts on overall process selection.

14.4 PROCESS PERFORMANCE

The oxidation efficiency of any AOP depends on several factors, including equipment design (e.g., UV reactor design), process optimization (e.g., oxidant dose, type of lamps and their spectral power distribution – if UV-AOP, flowrate), yield of oxidizing species, reactivity of target compounds toward that/those radical species, and water quality parameters.

Monitoring the performance of UV/H_2O_2 AOPs for the removal of target contaminants in potable reuse projects is hindered by the very low concentrations of the contaminants present. Following RO treatment, contaminants such as NDMA and 1,4-dioxane tend to be mostly present in reclaimed waters at concentrations of up to tens of nanograms per litre only. Analytical detection limits for these chemicals are commonly no better than 0.3 ng/L for NDMA (Munch & Bassett, 2004) and 20 ng/L for 1,4-dioxane (Munch & Grimmett, 2008). Consequently, these chemicals are not always detectable in influents to AOP systems and are only rarely detectable in the effluents from these systems. Therefore, in order to assess overall performance, either long-term monitoring data are required or focused investigations using artificially elevated (spiked) influent concentrations.

Long-term monitoring results for NDMA concentration in AOP influent and effluent (along with final product water) at a potable reuse plant are presented as lognormal probability density plots in Figure 14.5. These data reveal that, over a period of six years, the AOP influent concentrations ranged from around the (laboratory-specific) detection limit of 2 ng/L to concentrations approaching 500 ng/L. The vast majority of AOP effluent concentrations were below the 2 ng/L analytical detection limit, but NDMA was occasionally detected at up to 5 ng/L.

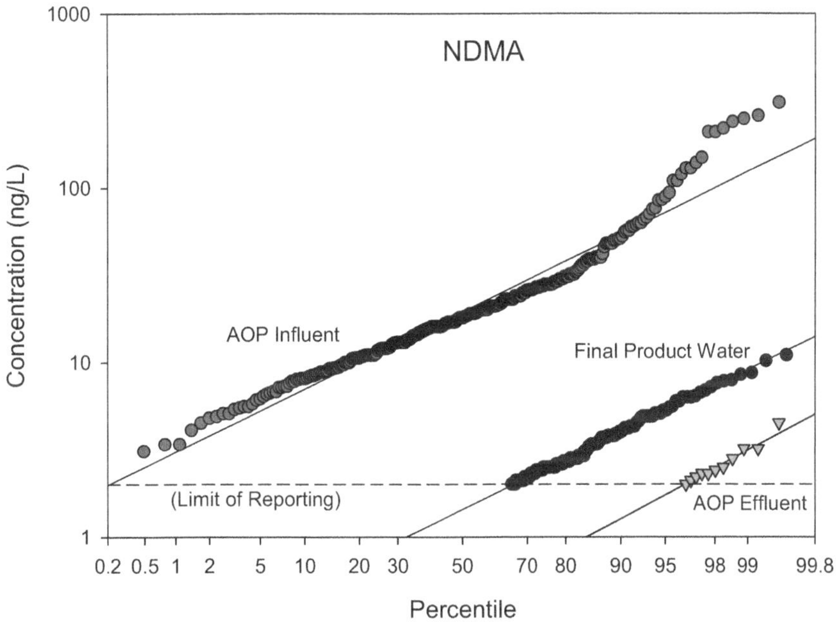

Figure 14.5 Lognormal probability plots for NDMA in AOP influent, AOP effluent, and final product water.

The lognormal probability plots reveal that NDMA was measured above the analytical detection limit in less than 5% of AOP effluent samples. Nonetheless, these limited data are sufficient to indicate that NDMA was removed from AOP influent to AOP effluent by a mean factor of around 1.6 Log_{10} units. Notably, the distribution of NDMA concentrations in final (chloraminated) effluent was higher than the distribution of concentrations in AOP effluent. This is consistent with experimental results indicating that that AOP process may produce byproducts that subsequently act as NDMA formation precursors during the chloramination process (Sgroi *et al.* 2015).

Similar results could not be achieved for 1,4-dioxane since AOP effluent concentrations were never observed to be above the analytical detection limit of 1 μg/L (Figure 14.6). The AOP 1,4-dioxane influent concentrations were only observed above this detection limit in less than 5% of samples collected throughout the 6 year monitoring campaign.

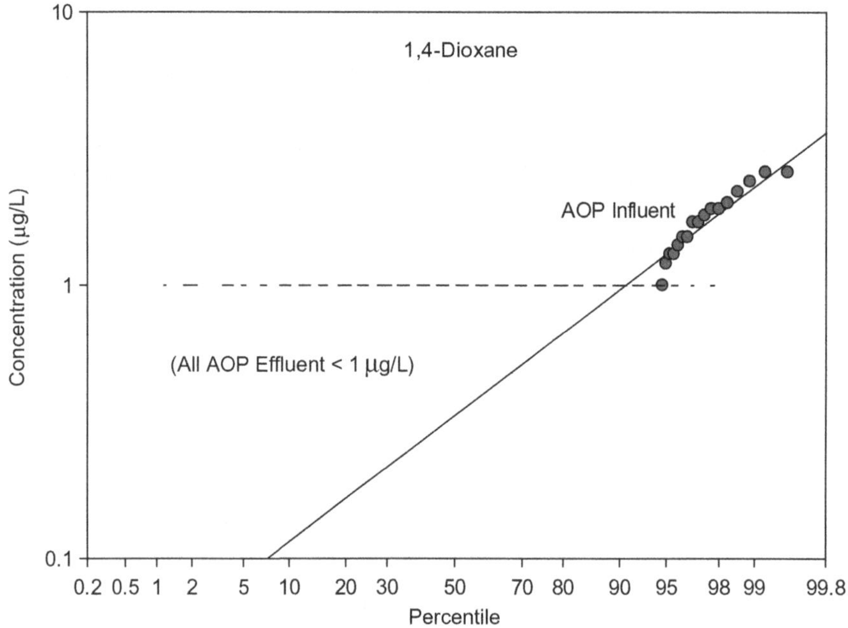

Figure 14.6 Lognormal probability plot for 1,4-dioxane in AOP influent (all AOP effluent < 1 µg/L).

14.5 INTERNATIONAL EXAMPLES OF AOP USE IN POTABLE REUSE PROJECTS

As summarised in Table 14.1, there are now a small number of potable reuse projects that have been designed with the incorporation of AOPs within the advanced treatment train. These include plants that have been constructed in the USA, South Africa and Australia. In this section, focused case studies are presented for the Groundwater Replenishment System (CA, USA), the Prairie Waters Project (CO, USA), the Western Corridor Water Recycled Water Project (QLD, Australia), and the Beaufort West Water Reclamation Plant (South Africa) since these provide examples of AOP-related challenges which were required to be overcome due to specific local considerations. A further case study is presented for the Terminal Island Water Reclamation Plant (CA, USA) since it provides an example of plant currently under construction, which incorporates new AOP technology for potable reuse.

14.5.1 Groundwater Replenishment System, Orange County, CA, USA (2008)

The Orange County Water District in California has been pioneering planned potable reuse since the establishment of the now superseded "Water Factory 21" project in 1976. The Water Factory 21 advanced water treatment plant had been treating municipal effluent by reverse osmosis since 1977 and incorporated high energy UV treatment for NDMA destruction in 2001. Since this system was not designed for the enhanced generation of radical oxidants (i.e., it did not include peroxide dosing or alternative catalysts), it does not meet all commonly used definitions of an AOP. However, it was a significant departure from regular UV disinfection practise since it was designed and sized specially to deliver high UV energies required for photolytic degradation of a chemical substance.

The Water Factory 21 plant was decommissioned in 2004 due to the need to expand capacity and to introduce updated treatment technologies. The Groundwater Replenishment System (GWRS) jointly funded and operated by the Orange County Water District and the Orange County Sanitation District was subsequently constructed and is now the world's largest water purification system for potable reuse. It takes wastewater that would have previously been discharged into the Pacific Ocean and purifies it by MF, RO and UV/H_2O_2. The purified water is then used to recharge underground drinking water supplies in Orange County, California. Operational since January 2008, the recently expanded GWRS can produce up to 350 ML/day (80 MGD) of purified water.

The California Division of Drinking Water (DDW) Title 22 Regulations require that groundwater recharge IPR projects must achieve at least 12 Log_{10} enteric virus reduction, 10 Log_{10} Giardia cyst reduction, and 10 Log_{10} Cryptosporidium oocyst reduction (California Office of Administrative Law, 2015). A precise treatment train to achieve this is not stipulated. However, at least three separate processes must be credited with at least 1 Log_{10} reduction for each pathogen. Furthermore, any engineered treatment processes may be credited with a maximum of 6 Log_{10} reduction for each pathogen. These requirements made the incorporation of either UV-disinfection or UV/H_2O_2 an attractive option for inclusion in the overall GWRS treatment train.

The AOP UV system selected for the GWRS consists of TrojanUVPhox™D72AL75 UV reactors manufactured by Trojan Technologies, and placed in full time service in January 2008. This is a closed, in-vessel type UV system based on low pressure-high output lamps (each 250 watts). The UV system is arranged with eight main trains and three redundant trains. Each of the main trains contains six reactors, with 72 lamps per reactor. Each train has a rated maximum treatment capacity of 33 ML/day (8.75 MGD) for a total of 265 ML/day (70 MGD). Hydrogen peroxide is added for advanced oxidation to transform UV-resistant chemical contaminants.

In addition to meeting virus Log_{10} removal requirements for Title 22 Regulations, the principle reasons for inclusion of the AOP were for photolysis of NDMA and advanced oxidation of 1,4 dioxane and any other remaining trace organic chemicals after RO treatment. The system was designed to achieve the required 1.2 Log_{10} NDMA and 0.5 Log_{10} 1,4 dioxane reduction at 95% $T_{254nm,1cm}$ (California Office of Administrative Law, 2015). It was given a 6 Log_{10} virus removal credit by the California DDW. Based on validation and certification testing of the AOP, conducted by Orange County Water District in 2004–2005, a minimum UV dose of 101 mJ/cm^2 was required to achieve the virus removal credit. Performance validation for this system required full scale virus challenge test using MS2 coliphage. However, the AOP system provides a significantly greater UV fluence (typically >400 mJ/cm^2) to support the photolysis and AOP effectiveness.

Two critical control limits for $\%T_{254nm,1cm}$ and average UV train power are constantly monitored to ensure that the calculated UV dose per train is achieved at all times (Patel, 2014). The minimum $\%T_{254nm,1cm}$ must be maintained at 95%. The minimum UV train power is 74 kW per train for the full scale trains and 24.6 kW for the partial trains. The partial trains are being expanded to full scale with the GWRS Initial Expansion Project. Monitoring these two factors ensures that the UV dose per train is significantly greater than 101 mJ/cm^2 at all times for disinfection (Patel, 2014). Redundant UV trains and reactors are automatically activated if on-line sensors detect lamp failures or other issues.

14.5.2 Western Corridor Recycled Water Project, Queensland, Australia (2008)

The Western Corridor Recycled Water Project in Queensland, Australia, incorporated three advanced water treatment plants, each with AOP process units (Poussade *et al.* 2009). These were the Bundamba AWTP (66 ML/day), the Luggage Point AWTP (66 ML/day) and the Gibson Island AWTP (100 ML/day) to provide a total project capacity of 232 ML/day. The AWTPs were interconnected into an overall system with secondary effluent sourced from six separate wastewater treatment plants in and around Australia's third largest city of Brisbane. They produced water for potable substitution at two nearby coal fired power stations, with a sizeable volume apportioned for IPR by augmentation of the region's largest surface water reservoir, Lake Wivenhoe. However, shortly following construction in 2009, drought breaking rains reduced the immediate water shortage and, while water continued to be produced for industrial use, the purified water has thus far not been required for potable use.

A combination of MF/RO/AOP and free chlorine was selected as the treatment train for each AWTP. UV/H_2O_2 AOP was selected for the project as it was considered a best available technology for potable reuse – based on the examples in California of both Orange County Water District and West Basin Municipal Water District groundwater replenishment systems. Contrary to the study conducted in Orange County, there was no local data available for NDMA or NDMA precursors, nor for 1,4-dioxane. As the system was being developed as a fast-tracked project, the technology was selected as an insurance policy against the possibility of their occurrence.

All three plants are equipped with multi-trains of TrojanUVPhox™ D72AL75 advanced oxidation systems that include a H_2O_2 dosing system with online mixing and sequential reactors, each equipped with 72 low-pressure high output, monochromatic radiation mercury vapour lamps. The design of each system was similar, with Luggage Point and Gibson Island having identical design configurations. The Bundamba plant was built as two separate plants on one site, and as a result, had a slightly different configuration in place. The configurations of the three AOP systems are summarised in Table 14.2 (Poussade *et al.* 2009).

Table 14.2 Configurations of the AOP systems for the Western Corridor Water Recycling Project (Poussade *et al.* 2009).

AWTP	No Trains	No Vessels/ Train	No Reactors/ Vessel	No Lamps/ Reactor	H_2O_2 Dosing Rate (mg/L)
Bundamba	4 x duty	2	2	72	4–5
Gibson Island	3 x duty, 1 x standby	3	2	72	6–10
Luggage Point	3 x duty, 1 x standby	3	2	72	6–8

The Gibson Island and Bundamba AOP systems were designed to achieve 1.0 Log_{10} NDMA removal, while the Luggage Point system was designed for 1.2 Log_{10} NDMA removal. All three systems were designed for 0.5 Log_{10} 1,4-dioxane removal.

Due to time constraints, system validation was not conducted specifically on the units at these AWTPs. Instead, as the configuration was determined to be similar to the systems in California, the design configuration was considered acceptable. These AOP systems were considered to provide a critical control point for both NDMA and 1,4-dioxane, as well as for inactivation of pathogenic microorganisms. The system was primarily designed for the destruction of the two chemicals, and microorganism removal credit was considered to be a useful additional benefit due to the high UV fluences required for the AOP.

The critical monitoring point for AOP was determined in close consultation with the system vendor and was based on the E_{EO} of NDMA removal. The E_{EO} values for the Western Corridor AWTPs were determined in consultation with system supplier engineers, based on previous validation work in the USA, supplemented with system-specific hydraulic modelling. The design basis of each of the three plants is shown in Table 14.3.

Table 14.3 AOP Design summary for the three Western Corridor project AWTPs.

AWTP	Target E_{EO} (kWh/kgal/order). New Lamps – Old Lamps	% Minimum Design UV Transmittance	End of Lamp Life Factor % New Lamp Intensity	Lamp Fouling Factor	NDMA Log_{10} Reduction	Max Inlet NDMA (ng/L)
Bundamba	0.24–0.33	95	90	0.8	1.0	50
Gibson Island	0.25–0.29	95	92	1.0	1.0	50
Luggage Point	0.29–0.37	95	80[1]	None[1]	1.2	80

[1]Lamp fouling factor was accounted for in the End of Lamp Life Factor for Luggage Point design.

The End of Lamp Life (EOLL) factor was used to anticipate lower UV light emissions by lamps at end of life (after 12,000 hours). The fouling factor was used to compensate for lamps fouling, which considering the corrosive nature of unstabilized RO permeate water, is unlikely to occur.

The critical control system used was based on the E_{EO} metric and Electrical Energy Dose (EED) parameter. The target E_{EO} for the system was calculated based on the measured $\%T_{254nm,1cm}$ after hydrogen peroxide addition. The target EED (kWh/kgal) was determined by multiplying the target E_{EO} value (kWh/kgal/order) by the target Log_{10} reduction (order). The measured flow rate across the UV train was then used to calculate a target power, in kW, for the train which was then used to determine how much UV energy is required for NDMA treatment.

The critical monitoring point for the AOP system used was the Present Power Ratio (PPR) which is the ratio of actual operating power to the target power. The UV control system was designed to operate the UV system above a power ratio of 100% to ensure sufficient NDMA treatment. The UV reactor energy input is based on volumetric throughput (flow rate). The established critical control points (critical alerts and critical limits) are presented in Table 14.4.

Table 14.4 Critical control points for the three AOP systems.

Critical Control Point	Alert Limit	Critical Limit
Present Power Ratio	<100% for more than 10 mins	<90% for more than 10 mins
Hydrogen peroxide dosing flow	+/– 20% of setpoint for more than 10 mins	+/– 50% of setpoint for more than 10 mins
Lamp failure	6 or more lamps failed on an individual reactor	30 or more lamps failed on an individual reactor (this will initiate train shutdown)

H_2O_2 dosing rate and RO permeate water $\%T_{254nm,1cm}$ are measured online. Water flow is measured independently for each UV train. Temperature, UV intensity and present power are monitored online for each reactor. The target power and PPR were calculated for each train from the E_{EO} at measured UV transmittance, and flow through the train.

As the UV fluence for advanced oxidation is very high compared with a typical disinfection dose, 4 Log_{10} inactivation of viruses, bacteria, cryptosporidium and giardia was credited, provided that the AOP system was operated to the AOP critical control point requirements.

Actual NDMA removal was measured at the Bundamba AWTP in the first year of operation. At this plant, chloramine was formed upstream of the membrane processes to minimize biological fouling particularly on the RO system. When initially designed, chlorine and ammonia were dosed separately into the secondary effluent, resulting in significant formation of NDMA upstream of the AWTP (Farré *et al.* 2011). As expected, the RO membranes were not very efficient in removing NDMA (Figure 14.7). Nonetheless, the presence of NDMA in RO permeate water madeit possible to verify the efficiency of the AOP system and validate its performance in achieving lower than 10 ng/L of NDMA in treated water at all times. A sampling campaign to assess NDMA removal across the AOP revealed consistently greater than 1.0 Log_{10} removal (Poussade *et al.* 2009). Examples of these data, collected on two days during 2008 are presented in Figure 14.7.

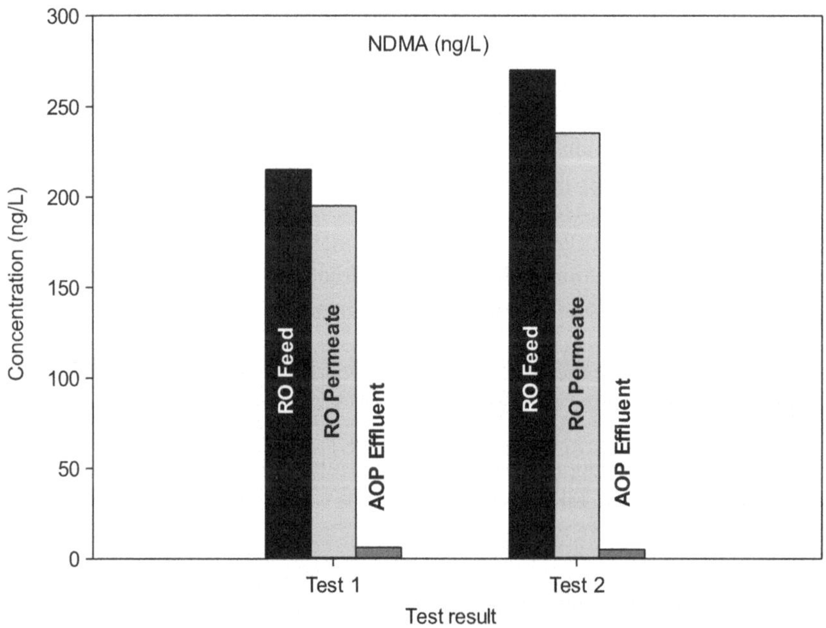

Figure 14.7 NDMA concentrations in RO feed, RO permeate and AOP-treated samples from the Bundamba AWTP.

A change to Bundamba AWTP configuration was implemented in order to pre-form chloramine prior to dosing. The effect of this was to dramatically reduce NDMA formation within the plant, and consequently reduce the NDMA concentrations to which the AOPs were exposed. Following this change, NDMA was rarely measured above the analytical detection limit of 5 ng/L in RO permeates. 1,4-Dioxane was never detected in any influent samples to the Western Corridor AWTPs.

14.5.3 Prairie Waters Project, Aurora, CO, USA (2010)

The 190 ML/day Peter Binney Water Purification Plant is operated by Aurora Water, Colorado, USA. It employs a sequence of conventional and advanced water treatment processes to augment the city's drinking water supplies with highly treated water from a stream that contains a high degree of wastewater effluent discharge (>80%). Pretreatment to the plant is provided by a natural treatment train consisting of a series of 17 vertical riverbank filtration wells followed by an artificial recharge and recovery facility located north of Brighton, Colorado. From there, water is pumped through a 36-mile long pipeline to the Peter Binney Water Purification Plant. The advanced water treatments consist of partial softening, UV/H_2O_2 AOP, biologically activated carbon filtration, and final granular activated carbon filtration. Subsequently, the water is blended in a ratio of 1:2 with Aurora's current supply (mountain run-off after conventional surface water treatment), disinfected with chlorine and delivered to the city's distribution system.

The main motivation for selecting the advanced treatment processes was the desired removal of chemicals of emerging concern and pathogens to deliver a water quality that was indistinguishable to Aurora's current supply. Based on experience in California, in particular the GWRS in Orange County, a UV/ H_2O_2 process using a TrojanUVPhox™ D72AL75 system manufactured by Trojan Technologies was selected in 2005. The principal driver for selecting UV/H_2O_2 was NDMA removal efficiency and pathogen inactivation credits using UV. After construction of the facility had started, studies at the site demonstrated that NDMA and other nitrosamines were efficiently removed during the natural treatment processes (riverbank filtration followed by artificial recharge and recovery) (Drewes *et al.* 2006) and concentrations of NDMA at the influent of the AOP process were consistently below detection levels (<10 ng/L).

The UV/H_2O_2 process was designed to achieve ≥1.2 log NDMA removal in RO permeates of ≥85% $T_{254nm,1cm}$, at maximum flowrate of 190 ML/day (50 MGD) and minimum flowrate of 19 ML/day (5 MGD). Hydrogen peroxide is dosed at 4 mg/L for 1,4-dioxane control. The UV system consists of 12 parallel trains with four TrojanUVPhox™D72AL74 reactors (i.e., 8 reactor vessels) per train, meeting California design requirements since the State of Colorado didn't specify any treatment requirements for AOP. Typical feed water conditions for the AOP are summarized in Table 14.5. Given the effective natural pre-treatment, UV transmittance is above 90% and the DOC represents a well-degraded carbon source providing favourable conditions for an AOP.

Table 14.5 Typical bulk water quality parameters for the UV/H_2O_2 train.

	UV/H_2O_2
Temperature (°C)	26
pH	7.21
Turbidity (NTU)	0.23
Discharge Flow (MGD)	2.73 per UV-unit
Influent *T*(%) at 254 nm	90.9
DOC (mg L^{-1})	3–3.1

Given the extensive biological treatment of the organic matter during river bank filtration and artificial recharge and recovery, the DOC exhibited only a small additional oxidation demand and was reduced during AOP by 2–4%. The UV absorbance, however, decreased by between 16 and 19%, suggesting a preferred transformation of humic- and fulvic-like material of the organic matter during oxidation.

The removal of trace organic chemicals during AOP provides an additional treatment barrier after the natural treatment step. Chemicals that are not amendable to oxidation, such as chlorinated flame retardants or the artificial sweetener sucralose are retained during the activated carbon filtration.

Finally, it is noteworthy that only approximately 1.5 mg/L of the hydrogen peroxide dose (4 mg/L) is being utilized in the AOP and the residual peroxide is carried into the subsequent BAC filter. Plant operators reported benefits of this residual peroxide resulting in longer filter run times.

14.5.4 Beaufort West Water Reclamation Plant (South Africa)

Beaufort West Municipality is situated in central Karoo, one of the driest areas in South Africa. It functions as the economic, political and administrative centre of central Karoo. There are roughly 40,000 inhabitants in Beaufort West Municipality spread across three towns, one of which is called Beaufort West.

In 2010, a severe drought nearly depleted the town's raw water sources, resulting in an immediate shortage of drinking water. By January 2011, the town was relying on trucks delivering additional drinking water to support its inhabitants. At that time, 5 litres of drinking water per person per day were being trucked to over 8,000 homes. Frequent droughts in combination with predicted population growth and large informal housing areas that are yet to be connected to the water supply system are expected to increase the pressure on raw water sources in future.

The situation in Beaufort West led to the construction of a DPR plant known as the Beaufort West Water Reclamation Plant (BWWRP). The plant was constructed as a 'design, build and operate' project with local contracting firm 'Water & Wastewater Engineering' (Marais & von Durckheim, 2012). It was commissioned in January 2011 and has been providing drinking water since. Subsequent to conventional tertiary treatment, the additional treatment processes used at the BWWRP include UF, RO, UV/H_2O_2 AOP and final chlorination. The plant is designed for a capacity of 2.1 ML/day.

The inclusion of the AOP was decided during the final stages of the BWWRP construction. The main objectives were to ensure improved public perception and improved removal of trace organics. The principal reason for selecting AOP for these tasks was largely to provide a 'best available technology' solution due to the sensitive nature of the project. While there were no specifically targeted contaminants, the system was seen as an effective additional barrier to trace organic chemicals including hormones and pharmaceuticals. The system operates with a UV fluence of 250 mJ/cm^2, which would be considered very high for disinfection, but relatively low for AOP. The decision to operate at a fluence exceeding disinfection requirements was justified based on the fact that the cost of UV power was very low relative to other operational costs.

The system is manufactured by Hanovia and uses medium pressure UV lamps. Peroxide is dosed, ahead of the UV reactors, at a dosing rate of 0.6 mg/L of a 50% H_2O_2 solution. No performance validation was required or implemented due to the high analytical costs involved. Online monitoring is provided through a programmable logic controller program, which initiates a plant shut down when critical limits are exceeded.

14.5.5 Terminal Island Water Reclamation Plant, Los Angeles, CA, USA (2016)

The City of Los Angeles Department of Public Works Bureau of Sanitation (LASAN) is expanding the Terminal Island Water Reclamation Plant to comply with the Division of Drinking Water's Groundwater Recharge Regulations in California (California Office of Administrative Law, 2015). The new expansion will produce 45 ML/day (12 MGD) of purified water that will be used for multiple purposes including injection at the Dominguez Gap Barrier to prevent seawater intrusion; thereby, reducing the need for importing potable water and increasing local water supply reliability.

The Terminal Island Water Reclamation Plant currently treats up to 23 ML/day (6 MGD) of tertiary effluent by MF, RO and chloramination disinfection. The expansion will increase the capacity and replace chloramination disinfection with a UV-based AOP using low pressure UV lamps with sodium hypochlorite (UV/Cl_2). This advanced oxidation process is described elsewhere in Chapter 9.

A feasibility study, including AOP bench-scale testing on treated water from the existing MF-RO treatment process, was conducted in 2013 to evaluate commercially-viable processes that could meet both the disinfection and contaminant removal requirements of the regulations. The bench-scale testing evaluated ozone with hydrogen peroxide (O_3/H_2O_2), UV/H_2O_2, and UV/Cl_2 using both medium pressure and low pressure UV lamps. UV/Cl_2 was known to be uniquely efficient in low pH waters, which is typical in RO permeate applications for potable reuse (Watts *et al.* 2012). The results of the feasibility study concluded that all the evaluated AOPs were worthy of further consideration via pilot-scale testing.

Pilot-scale testing was conducted throughout 2014 using an AOP pilot system from Xylem that had the capability to evaluate all candidate AOPs including O_3/H_2O_2 (with H_2O_2 injection before or after ozone) along with UV/Cl_2 and UV/H_2O_2 (both with either LP or MP UV irradiation). During the pilot study, all these processes were tested and benchmarked regarding the efficiency and total cost of ownership. The pilot setup at the Terminal Island Water Reclamation Plant is shown in Figure 14.8. In this installation, RO effluent was delivered to the pilot, but a spiking system for 1,4-dioxane was employed since this chemical was not typically detected at this facility.

Figure 14.8 AOP Pilot setup at the Terminal Island Water Reclamation Plant.

Initial rounds of pilot-scale testing revealed that O_3/H_2O_2 AOP was very effective at removing 1,4-dioxane but not effective at removing NDMA, which is required to be reduced to below 10 ng/L, so it was removed from further consideration for this project. Additionally, MP UV was eliminated from further consideration after results showed that it is effective, but the total cost of ownership was significantly higher than LP UV. Both UV/Cl_2 and UV/H_2O_2 performed similarly regarding the 1,4-dioxane and NDMA removal efficiency, but UV/Cl_2 was considered to have a significant advantage with respect to total cost of ownership because of the following factors:

- The plant already has sodium hypochlorite storage facilities – a commonly used chemical in water and wastewater treatment,
- A residual of chlorine is desired after the UV reactor as an additional disinfection barrier, and
- Hydrogen peroxide is poorly consumed through the AOP process and requires costly quenching.

The full-scale AOP design is based on a minimum UV fluence of 920 mJ/cm^2 and a minimum free chlorine residual of 2 mg/L at the inlet to the reactor in order to achieve greater than 0.5 Log_{10} removal of 1,4-dioxane and less than 10 ng/L NDMA. The design conditions include a minimum $\%T_{254nm,1cm}$ of 96% and pH less than 6.5 in order to ensure compliance with the regulations.

The AOP selected for TIWRP is a MiPRO photo system manufactured by Xylem/Wedeco. The equipment will consist of two Wedeco K143 closed-vessel, low pressure UV reactors with a total of 504 lamps (600 W per lamp) per reactor (Figure 14.9). The MiPRO photo system includes two sodium hypochlorite dosing and injection systems along with all instruments and controls for 100% redundancy. The K143 reactors also have the capability for high hydraulic flow and power turndown by turning on or off rows of lamps in order to reduce energy cost. It was intended that this system would be installed during 2016.

Figure 14.9 A K143 reactor of the MiPRO photo system manufactured by Xylem/Wedeco.

The K143 reactor, previously validated per the UV Design Guidance Manual (US EPA 2006) for virus inactivation in drinking water, will achieve 6 Log_{10} virus disinfection credit in addition to meeting the 1,4-dioxane and NDMA removal requirements. The system will be controlled to a UV fluence set point and ensure regulatory compliance by real-time monitoring of $\%T_{254nm,1cm}$, flow rate, and UV intensity (one sensor per row of lamps) to calculate a UV fluence based on equations that have been validated through computational fluid dynamics modelling, bioassay testing, and collimated beam testing. Chemical disinfection parameters including pH, free chlorine, and total chlorine will also be measured in real-time to ensure optimal sodium hypochlorite dosing and AOP performance.

14.6 CONCLUSIONS AND FUTURE PROJECTIONS

The incorporation of AOPs in treatment trains for potable reuse projects is a relatively recent phenomenon. However, it provides significant additional disinfection potential as well as the control of chemical constituents that are amenable to degradation by advanced oxidation and, in the case of UV/AOP, by photolysis. The Orange County Water Factory 21 project paved the way with the establishment of a post-RO UV irradiation process, implemented specifically for the photolysis of NDMA in 2001. A similar treatment design was subsequently incorporated in a number of other projects including in Singapore. However, it was not until 2008 that true AOP systems, incorporating the use of H_2O_2 for the production of secondary oxidants were commissioned at the GWRS in California and the Western Corridor Water Recycling Project in Queensland, Australia.

Since then, there have been a number of important drivers for the inclusion of AOPs in potable reuse projects. These have included the applicable pathogen Log_{10} reduction credits, photolytic transformation of NDMA, a desire to use what is widely perceived to be 'best available technology' and an expectation that AOP inclusion will enhance public acceptance for a project. Actual chemical targets for the advanced oxidation process (requiring radical oxidants) have been few, but 1,4-dioxane has been a target in some cases.

The successful development of the Prairie Waters Project at Aurora, Colorado confirms the potential flexibility in potable reuse treatment train selection and that high pressure membrane treatment is not always necessary to precede AOP treatment. In that case, natural treatment by river bank filtration and soil aquifer treatment provide effective polishing steps prior to UV/H_2O_2 AOP.

More recently, interest in UV/Cl_2 AOPs for potable reuse projects has emerged, notably in current planning for Terminal Island Water Reclamation Plant owned by the Los Angeles Department of Public Works Bureau of Sanitation. The use of ozone-based AOPs, designed with enhanced radical production, may also be developed for potable reuse projects in the future. Ozone AOPs may be coupled with biologically active carbon (BAC) and granular activated carbon (GAC) systems to provide a mix of oxidation, biological removal, and physical-chemical adsorption, especially in areas where brine disposal options are limited from RO facilities. However, ozone systems may well be limited by a lack of performance in NDMA removal, which has been a major driver for AOP selection in potable reuse projects. Indeed, recent evidence indicates that ozonation may lead to increased, rather than decreased, NDMA concentrations.

Heterogeneous catalytic processes, such as those which use TiO_2 for the production of hydroxyl radicals have not yet been seriously considered for large-scale potable reuse projects. If they were to be considered, important questions would need to be addressed such as capacity for pathogen inactivation. Importantly, techniques for validation and ongoing performance monitoring would be required, both for chemical oxidation and any pathogen inactivation. Without the advantage of high UV fluence, it is likely that these systems would struggle to provide effective removal of some contaminants such as NDMA.

There are many combinations of technologies that can be used to achieve the reliability and redundancy needed to manage risks from pathogens and chemicals in potable reuse systems, providing myriad approaches that can be tailored to each application site. That said, a proper risk assessment and understanding of the interplay between processes is critical to ensuring the proper functioning of the combined system and to ensure that it is operable and maintainable from a long-term asset management perspective.

Current methodology for AOP performance assessment and validation rely heavily on the oxidation of 1,4-dioxane. While this has been a useful approach for characterising performance, it has a number of important limitations. Notably, 1,4-dioxane is not a ubiquitous water contaminant and thus is often not present at sufficiently high concentrations to observe effective AOP removal. In some cases, this contaminant can be spiked into AOP feed waters for performance testing, but this approach limits monitoring to relatively short controlled periods. The incorporation of critical control points around measurable operational parameters such as the Present Power Ratio will enhance future validation and monitoring.

In addition to capital and operational costs, regulatory goals are a major driver for selection and implementation of AOPs for potable reuse projects. The trend toward DPR may lead to increased requirements, or more rigorous performance assessment for chemical and pathogen control. AOPs have been shown to be amenable to stringent performance validation and ongoing monitoring requirements. This has made AOPs an attractive technology choice for potable reuse projects. Even in cases where comprehensive validation has not been applied, a general perception that a treatment train incorporating an AOP process will provide optimum water treatment and quality has been a further important motivating factor for their use. As potable reuse grows internationally as an important municipal water supply strategy, so too can the adoption of AOP systems be expected to grow.

14.7 REFERENCES

Adamson D. T., Mahendra S., Walker K. L., Rauch S. R., Sengupta S. and Newell C. J. (2014). A multisite survey to identify the scale of the 1,4-Dioxane problem at contaminated groundwater sites. *Environmental Science & Technology Letters*, **1**(5), 254–258.

Andrzejewski P., Kasprzyk-Hordern B. and Nawrocki J. (2008). N-nitrosodimethylamine (NDMA) formation during ozonation of dimethylamine-containing waters. *Water Research*, **42**(4–5), 863–870.

Antoniou M. G. and Andersen H. R. (2015). Comparison of UVC/S2O82- with UVC/H2O2 in terms of efficiency and cost for the removal of micropollutants from groundwater. *Chemosphere*, **119**, S81–S88.

Arnold R. G., Sáez A. E., Snyder S., Maeng S. K., Lee C., Woods G. J., Li X. and Choi H. (2012). Direct potable reuse of reclaimed wastewater: It is time for a rational discussion. *Reviews on Environmental Health*, **27**(4), 197–206.

Bellona C., Drewes J. E., Xu P. and Amy G. (2004). Factors affecting the rejection of organic solutes during NF/RO treatment—a literature review. *Water Research*, **38**(12), 2795–2809.

Bolton J. R., Bircher K. G., Tumas W. and Chadwick A. T. (2001). Figures-of-merit for the technical development and application of advanced oxidation technologies for both electric- and solar-driven systems (IUPAC Technical Report). *Pureand Applied Chemistry*, **73**(4), 627–637.

Bolton J. R. and Stefan M. I. (2002). Fundamental photochemical approach to the concepts of fluence (UV dose) and electrical energy efficiency in photochemical degradation reactions. *Research on Chemical Intermediates*, **28**(7–9), 857–870.

Bounty S., Rodriguez R. A. and Linden K. G. (2012). Inactivation of adenovirus using low-dose UV/H2O2 advanced oxidation. *Water Research*, **46**(19), 6273–6278.

California Office of Administrative Law (2015). California Code of Regulations, Title 22: Social Security, Division 4: Environmental Health, Chapter 3: Water Recycling Criteria.

Charrois J. W. A. and Hrudey S. E. (2007). Breakpoint chlorination and free-chlorine contact time: implications for drinking water N-nitrosodimethylamine concentrations. *Water Research*, **41**(3), 674–682.

Chitra S., Paramasivan K., Cheralathan M. and Sinha P. K. (2012). Degradation of 1,4-dioxane using advanced oxidation processes. *Environmental Science and Pollution Research*, **19**(3), 871–878.

Dickenson E. R. V., Drewes J. E., Sedlak D. L., Wert E. C. and Snyder S. A. (2009). Applying Surrogates and Indicators to Assess Removal Efficiency of Trace Organic Chemicals during Chemical Oxidation of Wastewaters. *Environmental Science and Technology*, **43**(16), 6242–6247.

Dourson M., Reichard J., Nance P., Burleigh-Flayer H., Parker A., Vincent M. and McConnell E. E. (2014). Mode of action analysis for liver tumors from oral 1,4-dioxane exposures and evidence-based dose response assessment. *Regulatory Toxicology and Pharmacology*, **68**(3), 387–401.

Drewes J. E. and Khan S. J. (2011). Chapter 16: Water reuse for drinking water augmentation. In: Water Quality & Treatment: A Handbook on Drinking Water, J. K. Edzwald, (ed.), 6th edition, McGraw-Hill Professional, New York, pp. 16.1–16.48.

Drewes J. E. and Khan S. J. (2015). Contemporary design, operation, and monitoring of potable reuse systems. *Journal of Water Reuse and Desalination*, **5**(1), 1–7.

Drewes J. E., Hoppe C. and Jennings T. (2006). Fate and transport of N-nitrosamines under conditions simulating full-scale groundwater recharge operations. *Water Environment Research*, **78**(13), 2466–73.

Farré M. J., Döderer K., Hearn L., Poussade Y., Keller J. and Gernjak W. (2011). Understanding the operational parameters affecting NDMA formation at Advanced Water Treatment Plants. *Journal of Hazardous Materials*, **185**(2–3), 1575–1581.

Farré M. J., Radjenovic J. and Gernjak W. (2012). Assessment of Degradation Byproducts and NDMA Formation Potential during UV and UV/H2O2 Treatment of Doxylamine in the Presence of Monochloramine. *Environmental Science and Technology*, **46**(23), 12904–12912.

Fujioka T., Khan S. J., Poussade Y., Drewes J. E. and Nghiem L. D. (2012). N-nitrosamine removal by reverse osmosis for indirect potable water reuse – A critical review based on observations from laboratory-, pilot- and full-scale studies. *Separation and Purification Technology*, **98**(0), 503–515.

Fujioka T., Khan S. J., McDonald J. A., Roux A., Poussade Y., Drewes J. E. and Nghiem L. D. (2013). N-nitrosamine rejection by reverse osmosis membranes: a full-scale study. *Water Research*, **47**(16), 6141–6148.

Fujioka T., Khan S. J., McDonald J. A., Roux A., Poussade Y., Drewes J. E. and Nghiem L. D. (2014). Modelling the rejection of N-nitrosamines by a spiral-wound reverse osmosis system: Mathematical model development and validation. *Journal of Membrane Sciences*, **454**, 212–219.

Gerrity D., Pecson B., Trussell R. S. and Trussell R. R. (2013). Potable reuse treatment trains throughout the world. *Journal of Water Supply Research and Technology-Aqua*, **62**(6), 321–338.

Hijnen W. A. M., Beerendonk E. F. and Medema G. J. (2006). Inactivation credit of UV radiation for viruses, bacteria and protozoan (oo)cysts in water: A review. *Water Research*, **40**(1), 3–22.

Krasner S. W., Mitch W. A., McCurry D. L., Hanigan D. and Westerhoff P. (2013). Formation, precursors, control, and occurrence of nitrosamines in drinking water: A review. *Water Research*, **47**(13), 4433–4450.

Labas M. D., Zalazar C. S., Brandi R. J. and Cassano A. E. (2008). Reaction kinetics of bacteria disinfection employing hydrogen peroxide. *Biochemical Engineering Journal*, **38**(1), 78–87.

Leverenz H. L., Tchobanoglous G. and Asano T. (2011). Direct potable reuse: A future imperative. *Journal of Water Reuse and Desalination*, **1**(1), 2–10.

Li M. Y., van Orden E. T., DeVries D. J., Xiong Z., Hinchee R. and Alvarez P. J. (2015). Bench-scale biodegradation tests to assess natural attenuation potential of 1,4-dioxane at three sites in California. *Biodegradation*, **26**(1), 39–50.

Libotean D., Giralt J., Rallo R., Cohen Y., Giralt F., Ridgway H. F., Rodriguez G. and Phipps D. (2008). Organic compounds passage through RO membranes. *Journal of Membrane Sciences*, **313**(1–2), 23–43.

Linge K. L., Blair P., Busetti F., Rodriguez C. and Heitz A. (2012). Chemicals in reverse osmosis-treated wastewater: Occurrence, health risk, and contribution to residual dissolved organic carbon. *Journal of Water Supply Research and Technology-Aqua*, **61**(8), 494–505.

Mamane H., Shemer H. and Linden K. G. (2007). Inactivation of E-coli, B-subtilis spores, and MS2, T4, and T7 phage using UV/H_2O_2 advanced oxidation. *Journal of Hazardous Materials*, **146**(3), 479–486.

Marais P. and von Durckheim F. (2012). Beaufort West Water Reclamation Plant: First Direct (Toilet-to-Tap) Water Reclamation Plant in South Africa. Water & Wastewater Engineering, Stellenbosch, South Africa.

Mitch W. A., Gerecke A. C. and Sedlak D. L. (2003a). A N-Nitrosodimethylamine (NDMA) precursor analysis for chlorination of water and wastewater. *Water Research*, **37**(15), 3733–3741.

Mitch W. A., Sharp J. O., Trussell R. R., Valentine R. L., Alvarez-Cohen L. and Sedlak D. L. (2003b). N-Nitrosodimethylamine (NDMA) as a drinking water contaminant: A review. *Environmental Engineering Sciences*, **20**(5), 389–404.

Mitch W. A., Oelker G. L., Hawley E. L., Deeb R. A. and Sedlak D. L. (2005). Minimization of NDMA formation during chlorine disinfection of municipal wastewater by application of pre-formed chloramines. *Environmental Engineering Sciences*, **22**(6), 882–890.

Munch J. W. and Bassett M. V. (2004). Method 521: Determination of nitrosamines in drinking water by solid phase extraction and capillary column gas chromatography with large volume injection and chemical ionization tandem mass spectrometry (MS/MS); Version 1.0. EPA Document #: EPA/600/R-05/054., National Exposure Research Laboratory, Office of Research and Development, US EPA., Cincinnati, Ohio.

Munch J. W. and Grimmett P. E. (2008). Method 522: Determination of 1,4-dioxane in drinking water by solid phase extraction (SPE) and gas chromatography/mass spectrometry (GC/MS) with selection ion monitoring (SIM); Version 1.0. EPA Document #: EPA/600/R-08/101., National Exposure Research Laboratory, Office of Research and Development, US EPA., Cincinnati, Ohio.

National Health and Medical Research Council and Natural Resource Management Ministerial Council (2011). Australian Drinking Water Guidelines. Government of Australia, Canberra.

Nawrocki J. and Andrzejewski P. (2011). Nitrosamines and water. *Journal of Hazardous Materials*, **189**(1–2), 1–18.

Patel M. (2014). OCWD's Groundwater Replenishment System uses critical control points to monitor project performance. *Source: California Section of AWWA*, **28**(3), 10–12.

Pehlivanoglu-Mantas E., Hawley E. L., Deeb R. A. and Sedlak D. L. (2006). Formation of nitrosodimethylamine (NDMA) during chlorine disinfection of wastewater effluents prior to use in irrigation systems. *Water Research*, **40**(2), 341–347.

Pisarenko A. N., Stanford B. D., Yan D., Gerrity D. and Snyder S. A. (2012). Effects of ozone and ozone/peroxide on trace organic contaminants and NDMA in drinking water and water reuse applications. *Water Research*, **46**(2), 316–326.

Poussade Y., Roux A., Walker T. and Zavlanos V. (2009). Advanced oxidation for indirect potable reuse: A practical application in Australia. *Water Science and Technology*, **60**(9), 2419–2424.

Rice J. and Westerhoff P. (2015). Spatial and Temporal Variation in De Facto Wastewater Reuse in Drinking Water Systems across the USA. *Environmental Science and Technology*, **49**(2), 982–989.

Schreiber I. M. and Mitch W. A. (2005). Influence of the Order of Reagent Addition on NDMA Formation during Chloramination. *Environmental Science andTechnology*, **39**(10), 3811–3818.

Schreiber I. M. and Mitch W. A. (2006). Nitrosamine formation pathway revisited: The importance of chloramine speciation and dissolved oxygen. *Environmental Science and Technology*, **40**(19), 6007–6014.

Sedlak D. L., Deeb R. A., Hawley E. L., Mitch W. A., Durbin T. D., Mowbray S. and Carr S. (2005). Sources and fate of nitrosodimethylamine and its precursors in municipal wastewater treatment plants. *Water Environment Research*, **77**(1), 32–39.

Sgroi M., Roccaro P., Oelker G. L. and Snyder S. A. (2015). N-nitrosodimethyl amine (NDMA) formation at an indirect potable reuse facility. *Water Research*, **70**, 174–183.

Shah A. D. and Mitch W. A. (2012). Halonitroalkanes, halonitriles, haloamides, and N-nitrosamines: A critical review of nitrogenous disinfection byproduct formation pathways. *Environmental Science and Technology*, **46**(1), 119–131.

Simonich S. M., Sun P., Casteel K., Dyer S., Wernery D., Garber K., Carr G. and Federle T. (2013). Probabilistic Analysis of Risks to US Drinking Water Intakes from 1,4-Dioxane in Domestic Wastewater Treatment Plant Effluents. *Integrated Environmental Assessment and Management*, **9**(4), 554–559.

Stepien D. K., Diehl P., Helm J., Thorns A. and Puttmann W. (2014). Fate of 1,4-dioxane in the aquatic environment: From sewage to drinking water. *Water Research*, **48**, 406–419.

US Environmental Protection Agency (1987). Integrated Risk Information System (IRIS), N-nitrosodimethylamine. Office of Research and Development (ORD), National Center for Environmental Assessment (1987) www.epa.gov/iris/subst/0045.htm (accessed May 2, 2016).

US Environmental Protection Agency (2006). Ultraviolet Disinfection Guidance Manual for the Final Long Term 2 Enhanced Surface Water Treatment Rule. US Environmental Protection Agency, Washington, DC.

US Environmental Protection Agency (2016). Integrated Risk Information System (IRIS). http://www.epa.gov/iris/subst/index.html (accessed May 2, 2016).

von Gunten U. (2003). Ozonation of drinking water: part I. Oxidation kinetics and product formation. *Water Research*, **37**(7), 1443–1467.

von Gunten U., Salhi E., Schmidt C. K. and Arnold W. A. (2010). Kinetics and mechanisms of N-Nitrosodimethylamine formation upon ozonation of N,N-Dimethylsulfamide-Containing waters: Bromide catalysis. *Environmental Science and Technology*, **44**(15), 5762–5768.

Watts M. J., Hofmann R. and Rosenfeldt E. J. (2012). Low-pressure UV/Cl2 for advanced oxidation of taste and odor. *Journal of the American Water Works Association (JAWWA)*, **104**(1), 47–48.

Wols B. A. and Hofman-Caris C. H. M. (2012). Review of photochemical reaction constants of organic micropollutants required for UV advanced oxidation processes in water. *Water Research*, **46**(9), 2815–2827.

World Health Organization (2011). Guidelines for Drinking-Water Quality. 4th edition, WHO Press, World Health Organization, Geneva, Switzerland.

Yan S. W. and Song W. H. (2014). Photo-transformation of pharmaceutically active compounds in the aqueous environment: A review. *Environmental Science-Processes & Impacts*, **16**(4), 697–720.

Zenker M. J., Borden R. C. and Barlaz M. A. (2003). Occurrence and treatment of 1,4-dioxane in aqueous environments. *Environmental Engineering Sciences*, **20**(5), 423–432.

Chapter 15

Advanced treatment for drinking water production

Gilbert Galjaard, Bram Martijn, Erik Koreman and Holly Shorney-Darby

15.1 INTRODUCTION

This chapter focuses on pretreatment strategies and technologies which have been shown to synergistically produce a water quality for efficient and cost-effective AOP performance (Galjaard *et al.* 2011; Martijn *et al.* 2011). Tailored pretreatment is key for each source water and it must target specific compounds which impact the AOP. Because the pool of options and process combinations is too large to be covered herein, only enhanced coagulation, ion exchange and ceramic microfiltration in combination with possible AOP-processes and post-treatment with granular activated carbon (GAC)/ biologically active carbon (BAC) are discussed. Research showing the impact of dissolved organic carbon (DOC) and nitrate on the UV/H_2O_2 process performance prompted studies on pre-treatment alternatives. This chapter highlights findings on each of these processes relative to the AOP performance, mainly based on the experiences of PWN Water Supply Company North-Holland at their Andijk WTP.

The Andijk WTP was constructed in 1968 as a conventional surface water treatment plant based on breakpoint chlorination and coagulation sedimentation and filtration (CSF), servicing water from the IJssel Lake as raw water source. In 1978, the treatment process was upgraded by implementation of granular activated carbon (GAC) filtration to address customer complaints regarding taste and odour were the main cause for this modification. Strict TTHM regulation in the EC and The Netherlands in combination with required extra disinfection capacity in view of protozoa. In addition, the awareness that a barrier for organic micropollutants based on only GAC adsorption would not be sufficient to deal with emerging contaminants with a more polar character resulted in the desire to explore advanced oxidation processes (Kamp *et al.* 1997).

After exploring applicability of ozone based AOP, UV based AOP was selected by PWN for the Andijk WTP. In view of bromate formation, ozone based AOP was not pursued (Kruithof & Kamp, 1997; Kruithof, 2005). Since 2004 the Andijk WTP is operated with conventional pretreatment based on coagulation, sedimentation and rapid sand filtration, followed by MP UV/H_2O_2 treatment for disinfection and organic contaminant control (0.54 kWh/m^3; 6 mg/L H_2O_2) and biological activated carbon filtration, resulting in a process providing a very robust barrier for organic micropollutants and superior disinfection.

The UV-based AOP at the Andijk WTP was designed for 80% atrazine degradation in conventionally pretreated surface water from the IJssel Lake ($\%T_{254\ nm,\ 1\ cm}$ ~ 82%), requiring 0.54 kWh/m^3 and 6 mg/L H_2O_2. The installation consists of three UV treatment trains with four TrojanUVSwiftECT™ 16L30 reactors, designed for AOP. Medium pressure UV-lamp technology was applied, using 16 lamps of 12 kW each. All individual lamps can be controlled by electronic ballasts, allowing ballast power variation between 30% and 100%. Each UV train is equipped with an on/line %*T* monitor and flow meter, allowing accurate control both to meet the treatment objective and to operate energy efficient.

In 2008, PWN installed a similar UV AOP system in their Heemskerk WTP, prior to the artificial dune recharge system, treating conventionally pretreated surface water. This UV AOP was constructed with the same UV technology as applied at the Andijk WTP. The installed capacity was slightly higher (0.6 kWh/m^3) and 6 mg/L H_2O_2 due to a poorer water quality ($\%T_{254\ nm,\ 1\ cm}$ ~ 80%). The hydraulic capacity of the Heemskerk WTP UV AOP system is 6,000 m^3/h and consists of 4 UV trains with 5 UV reactors each. Before infiltration into the dunes, the excess of H_2O_2 is quenched over GAC (9 min EBCT, 50 m/h hydraulic loading rate).

In 2015, the conventional pretreatment, dating from 1968, was replaced by ion exchange followed by ceramic microfiltration. The nitrate ion and increased DOC removal via ion-exchange process allowed the use of ceramic microfiltration membranes on surface water treatment, and improved the water quality for the MP UV AOP, which resulted in substantial energy reduction while maintaining the AOP treatment objectives. For the Andijk WTP, the energy savings were used to allow an increase in capacity from 3,500 m^3/h to 5,000 m^3/h gross max capacity.

15.2 UV/H_2O_2 PROCESS: ANDIJK WATER TREATMENT PLANT (WTP) CASE STUDY

Nowadays thousands low molecular weight organic chemicals are used in daily life, improving the quality of life and health. Because of their abundant use, these compounds are found everywhere including in sources for drinking water (Houtman, 2010). The first emerging compounds were taste and odour compounds, algal toxins, phenolic compounds and pesticides, followed by endocrine disruptors and pharmaceuticals (Snyder *et al.* 2007), and more recently, perfluorinated compounds (Eschauzier *et al.* 2013) and nanoparticles (Van Wezel *et al.* 2011) are emerging. Thus, a wide variety of organic chemicals with different characteristics (e.g., absorbability, degradability, etc.) could be present in drinking water sources, requiring multiple barrier control strategies.

Organic contaminant control is a major issue in advanced drinking water treatment. The presence of pesticides, endocrine disruptors and pharmaceuticals has caused PWN to implement multiple barriers in their surface water treatment plants. For improved organics control, a combination of advanced oxidation by UV/H_2O_2 treatment and GAC filtration has been implemented. The design of these processes was based on bench- and pilot-scale evaluations. In bench-scale experiments, dose-response relationships have been established for these organic pollutants and degradation at specific process conditions (i.e., 0.54 kWh/m^3; 6 mg/L H_2O_2) was confirmed at the pilot-scale. Process conditions were based on 80% removal of atrazine, a pesticide identified in the drinking water source (Figure 15.1). Other commonly found pesticides, such as 2,4-D, diuron, and nicosulfuron were also removed by 80%.

Degradation of pharmaceuticals (Figure 15.2) shows a more varying profile. Some compounds, such as diclofenac and sotalol, show an almost complete removal while metformine and its metabolite guanylurea, show poor removal with the MP UV/ H_2O_2 AOP. Post-AOP treatment by GAC filtration removes the non-oxidizable compounds such as perfluorooctanoic acid (PFOA) and perfluorooctanesulfonic acid (PFOS), resulting in a very robust multi-barrier against micropollutants (Hofs *et al.* 2014). The widespread production and use of new organic chemicals cause the contamination of drinking water sources now and in the future, and climate changes influencing algae blooms and improper water discharge into the aquatic environment would impact negatively the quality of natural waters.

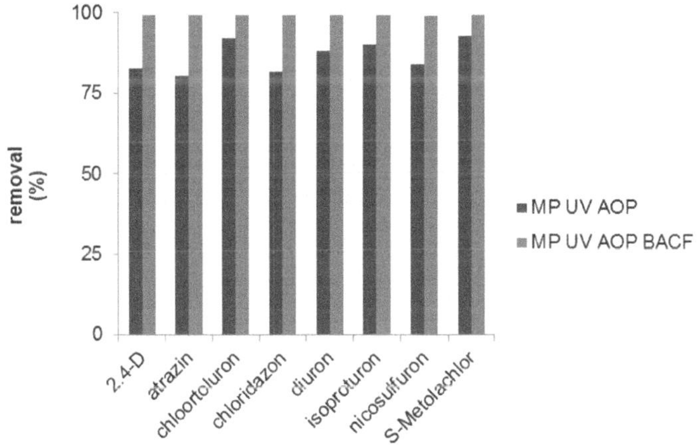

Figure 15.1 Removal of selected pesticides by medium pressure lamp (MP) UV/H_2O_2 AOP (0.54 kWh/m^3, 6 ppm H_2O_2) with and without biological activated carbon filtration (BACF) at 25 min empty bed contact time (EBCT).

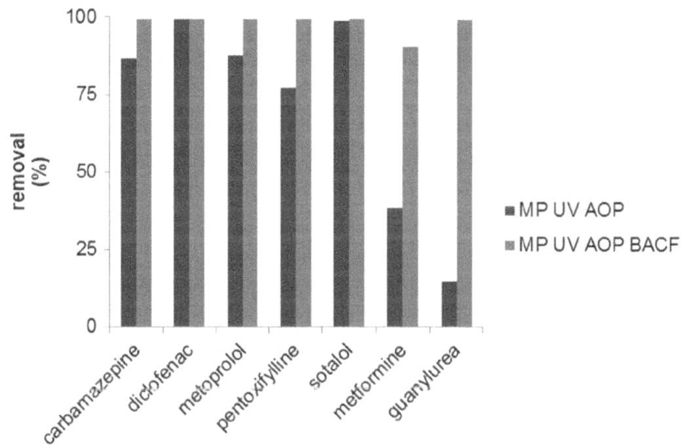

Figure 15.2 Removal of selected pharmaceuticals by MP UV/H_2O_2 AOP (0.54 kWh/m^3; 6 ppm H_2O_2) treatment with and without BACF (25 min EBCT).

At Andijk WTP, the MP UV/ H_2O_2 AOP represents one step in a multi-barrier strategy against organic micropollutants (Kruithof *et al.* 2007). Given the wide variety and diversity in micropollutants, the performance of an installation cannot be monitored or judged by the removal of all compounds; therefore, PWN has specified a monitoring compound, atrazine, and relates the atrazine removal to the removal of other contaminants. The full scale MP UV/ H_2O_2 AOP installation at Andijk WTP was designed based on modelling, bench-scale work, and small-scale piloting (Stefan *et al.* 2005). The performance of the full-scale equipment was evaluated for the removal of atrazine at the design process condition (0.56 kWh/m^3; 6 ppm H_2O_2) and at a lower electrical energy dose (0.42 kWh/m^3) (Figure 15.3). The predicted performance, based on the design models, and the experimentally observed degradation of atrazine matched satisfactorily, confirming the accuracy of the modelling and representativeness of the previous pilot-scale work.

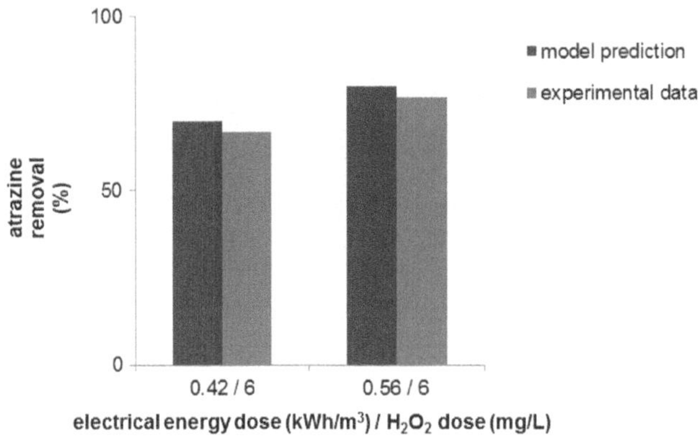

Figure 15.3 Atrazine removal in full scale MP UV/H_2O_2 AOP installation at two electrical energy doses.

Although the required energy consumption of UV/H_2O_2 treatment is still substantial, the process has proven to be economically feasible for NDMA removal (e.g., Orange County, CA), taste-and-odour causing compounds removal (e.g., Aurora, CO) and as a non-selective barrier against various organic micropollutants PWN Water Supply Company North-Holland sites in the Netherlands (Figure 15.4). The economic feasibility would increase significantly if the energy consumption could be lowered, and this would make this technology more attractive to solve a wider range of water treatment problems.

Figure 15.4 Full scale MP UV/H_2O_2 AOP installation at PWN (6,000 m^3/h; 0.54 kWh/m^3; 0–15 ppm H_2O_2).

15.3 PRETREATMENT STRATEGIES FOR AOP IN DRINKING WATER TREATMENT

Pre-treatment to an AOP is critical for sustainable and efficient operation of the AOP. As mentioned earlier, a significant reduction of energy consumption by UV/H_2O_2 can be achieved when pre-treatment has removed UV light-absorbing water constituents, which compete for UV light with the added oxidant (e.g., H_2O_2). In addition, the removal of these compounds may reduce the OH-radical water background demand. Hydroxyl radicals are produced through the photolysis of H_2O_2 and drive the efficiency of the AOP. When assessing pre-treatment needs for any AOP, it is critical to understand the composition of the water matrix. The most important water constituents influencing the efficacy of the UV/H_2O_2 process are the natural organic matter (NOM), often measured as dissolved organic carbon (DOC) or correlated with the UV transmittance (*%T* or UVT) at certain wavelengths, and nitrate ion. Nitrate ion absorbs the UV light more or less over the same wavelength range as H_2O_2, but has larger molar absorption coefficients than H_2O_2 and this introduces competition for photons particularly when a broad spectrum UV light is applied. NOM is also a strong UV absorber and is the most important OH-radical scavenger in the water. Pre-treatment technologies to achieve DOC and nitrate removal are discussed herein.

Standard coagulation, sedimentation, and sand filtration (CSF) lower the DOC content to a certain extent, reducing the water background UV absorbance and OH-radical scavenging capacity, but does not impact the nitrate concentration. Enhancing the coagulation to optimise DOC removal leads in most cases to a significant reduction in energy consumption by an AOP, however if nitrate is present, operation of the AOP can still be cost-prohibitive. Advanced pre-treatment with ion exchange and micro- and ultrafiltration (IX-MF/UF) removes both nitrate and DOC (Galjaard *et al.* 2005). The superior removal of DOC and nitrate by IX-MF/UF relative to CSF pre-treatment sets the conditions for increased OH-radical through H_2O_2 photolysis and reduced OH-radical water demand during the UV-AOP. An additional advantage of pretreatment with IX-MF/UF is that specific levels of DOC and nitrate reduction can be achieved through variable settings for process operation. IX-MF/UF pre-treatment creates favourable conditions for the UV/H_2O_2 process.

Membranes as a filtration barrier are excellent for reliably removing suspended particulate matter and some pathogens (e.g., *Giardia lamblia* and *Cryptosporidium parvum*). However, many membranes do not remove dissolved species, such as low-molecular weight DOC and nitrate ions. Membranes are, however, susceptible to fouling by DOC, so there can be a shared benefit if DOC removal is optimized during the pretreatment processes upstream to membranes and UV/H_2O_2. Polymeric membranes still dominate this sector of the water industry; however, ceramic membranes have some unique resilient properties which make them a favourable option upstream of an AOP with peroxide. Ceramic membranes can withstand a variety of chemicals and concentrations including hydrogen peroxide, which makes feasible the H_2O_2 dosing upstream the ceramic membranes when the membranes are followed by the UV/H_2O_2 treatment process. This can be regarded as a membrane fouling control strategy.

All types of membranes are susceptible to fouling. Most studies suggest that coagulation upstream of the membrane controls colloidal fouling (i.e., fouling caused by pore blockage) and removes the high molecular weight (HMW) fraction of natural organic matter (NOM) which therefore reduces NOM-driven membrane fouling. Gray *et al.* (2008) indicated that the mechanism of fouling control by coagulation consisted in the removal of low molecular weight (LMW) organics (with an adsorption peak at 220 nm) that were responsible for 'gluing' colloids to the membrane surface. Galjaard *et al.* (2002, 2005) proposed and demonstrated that coagulation removes HMW organics, but the coagulation also forms organometallic complexes. These complexes could interact with smaller organic molecules and with the membrane resulting in a film formation on the membrane surface which leads to an irreversible fouling (Galjaard *et al.* 2005). Given that each water source presents its own characteristics, adequate studies are necessary to determine the pretreatment strategy, i.e., use of coagulation to remove HMW organics or use of IX to remove LMW organics, or a combination of both. When used upstream of the membranes, IX removes LMW organics, which can often have high UV-absorbance, such that this strategy would also be beneficial for a downstream UV/H_2O_2 process.

The combination of ozone and membrane filtration has been largely unexplored due to the potential for extreme damage to polymeric membranes. The durability of ceramic membranes (the aluminium-, titanium oxide, or silicon carbide membrane materials are resistant to oxidative damage) sets the unique opportunity to combine ozone with membrane filtration. Recent data suggests that catalytic decomposition of ozone may occur within the ceramic membrane pore structure, depending on the composition of the membrane. This would lead to hydroxyl radical formation which subsequently minimizes membrane fouling while providing oxidation and removal of some emerging contaminants and also decreasing DOC by oxidation. These processes result in a higher UV transmittance of the treated water. A combination of ceramic filtration with ozonation has been demonstrated in laboratory or small bench-scale studies (Schlichter *et al.* 2004; Zhu *et al.* 2011) and more recently in pilot and demonstration plants (Galjaard *et al.* 2013). The available data suggest that ozone applied to ceramic membranes yields a synergistic effect toward reduced operational costs and improved water quality. The following subparagraphs give more detail about the different pretreatment technologies mentioned above.

15.3.1 Enhanced coagulation

In general, the main purpose of coagulation is removal of finely dispersed matter in water for turbidity reduction. In principle, coagulation consists in the "destabilization" of dispersed matter to enable aggregation and the subsequent mechanical removal of aggregated particles. Coagulation is also recognized as an attractive "conventional" method for NOM removal, particularly in developing countries because of low costs and its relative simplicity (Crozes *et al.* 1995), and it implemented for waters with TOC > 2 mg/L (Archer & Singer, 2006). When the coagulation process is designed for maximum removal of disinfection by-product (DBP) precursors – among which the most important are the NOM constituents – one speaks of 'enhanced' coagulation (Edzwald & Tobiasson, 1999).

Four main aggregation mechanisms can be distinguished in the coagulation treatment step (Amirtharajah & Mills, 1982):

i. Reduction of electrical repulsive force by charge neutralization, i.e., surface adsorption of and/or diffuse double layer compression or destabilization;
ii. Bridging across polymeric structures;
iii. Entrapment of colloids and small aggregates by large settling aggregates or "flocs", also called floc enmeshment or "sweep" floc coagulation;
iv. Complexation/precipitation.

Unlike conventional coagulation – where the third mechanism (sweep coagulation) dominates - in the enhanced coagulation, i.e., removal of negatively charged NOM (macro) molecules, all mechanisms are important. Al/Fe salts are most widely used for enhanced coagulation, since they produce intermediate compounds that are required in order for the above mechanisms to take place. The aquatic chemistry of both the aluminium and ferric ion is complex. Al/Fe cations do not exist as sole ions, but in aquatic environment, depending on water pH, they either precipitate as insoluble amorphous hydroxides or form charged complexes according to several pathways (Duan & Gregory, 2002).

Polynuclear complexes can form cross-linked gel-like constituents or adsorb onto surfaces macromolecular structures and colloidal fractions, reducing or neutralizing surface charge and zeta potential and promoting further agglomeration. Some polymeric species could adsorb onto micro agglomerates followed by hydrolysis to amorphous hydroxide agglomerates as a hydroxide-like coating. These 'coated' particles behave like pure $Me(OH)_3$ precipitates with comparable electro-chemical properties, so that further "floc" growth may occur. In addition, some micro agglomerates could be enmeshed by hydroxide-like macrostructures. The latter process is promoted by shifting pH toward neutral or slightly alkaline conditions (Amritharajah & Mills, 1982; Tambo & Kamei, 1999).

At the WTP in Andijk (Netherlands) it was also concluded that pH during coagulation and flocculation had a significant effect on NOM removal (Koreman *et al.* 2009). The pH impact was first investigated at a laboratory scale through a series of jar tests. These experiments clearly showed that adjusting pH with sodium hydroxide from 6.4 just before the end of the coagulation step to 8.0 (target level) instead of directly after rapid mixing, improved $\%T_{254\ nm}$ through higher NOM removal (Figure 15.5). Further, extending the flocculation time from 30 to 120 min – which is more representative for the full scale process – presented the advantage of significant NOM removal.

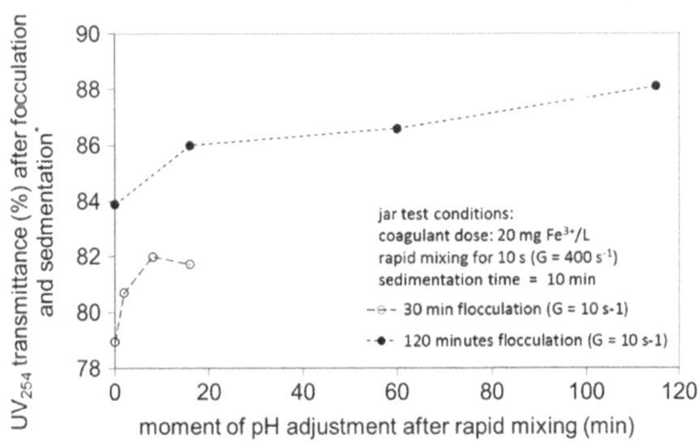

Figure 15.5 $\%T_{254\ nm,\ 1\ cm}$ as a function of pH adjustment time interval, and the flocculation time (30 and 120 min); *: UV_{254} was measured after syphoning 300 mL supernatant, 10 min sedimentation time, and 0.45 μm pore size sample filtration (Koreman *et al.* 2009).

After it was decided to operate the full scale coagulation process, June 2008, without adjusting pH with sodium hydroxide during coagulation, but just before rapid sand filtration, $\%T_{254\ cm}$ increase was even more dramatic ($5\% < \Delta\%T_{254\ cm} < 10\%$) than jar tests indicated. As a consequence, DOC removal levels increased from around 30% (from about 5.8 mg C/L to 4.0 mg C/L) to more than 50% (further reduction to 2.6 mg C/L), most likely do to both further enmeshment by the ferric hydroxide matrix and surface adsorption of NOM conglomerates and molecules, since the full scale process is performed by sludge blanket contactors, where increased floc volume concentrations can be observed, while total flocculation time is about 120 minutes.

15.3.2 Ion exchange

Ion exchange (IX) is considered to be both an adsorption and a sorption process (Wachinski, 2006). The term ion exchange describes the unit process of IX, which is widely used in water treatment to remove specific water constituents. The most common application of IX is softening, but there are many kinds of resins, including anion exchange resins which can be used for the removal of a diversity of anions such as nitrate, negatively charged-DOC species, perchlorate, arsenate. IX has been introduced by the WHO as a nitrate removal technology and was approved as the Best Available Technology (BAT) for nitrate removal by the US EPA. Cation-exchange resins are used for water softening when calcium and magnesium ions are removed from water.

Wachinski (2006) emphasized that the role of IX in water treatment is changing as a result of many environmental, economical, and health-driving factors, among which the proposed brine discharge legislation in California, Montana, and Texas, the advancements in membrane technologies and AOPs, and the stringent requirements to lower DBP formation potential that the water utilities need to follow. Ion exchange processes generate a waste stream which is usually referred to as *brine,* because salts are most commonly used to regenerate the IX resins. This wastewater contains high concentrations of the contaminant ions and high concentrations of regenerant solution, usually sodium chloride. Management and disposal of brines present a formidable challenge to engineers. Engineers must not only select the proper pre-treatment and IX process but also the proper brine recovery and/or brine disposal options. Conventional treatment alternatives like coagulation or nanofiltration which remove DOC also generate waste. In some cases, these waste streams are more difficult to discharge or dispose of than the brine of an anion IX process. This problem is very site specific and depends on local legislation and the water characteristics.

Anion-IX is a good option for enhanced removal of organic matter from source waters that contain medium to high concentrations of NOM because the majority of compounds which make up NOM are negatively charged.

The size exclusion chromatography-liquid chromatography with organic carbon detection (SEC-LC-OCD), developed by Huber *et al.* (2011), is a powerful analytical tool to characterize organic matter and to observe relative differences in DOC. The SEC broadly groups the organics into five fractions: biopolymers, humics, building blocks, LMW acids and LMW neutrals (listed herein in the order of retention time). Two detectors, organic carbon detector (OCD) and ultraviolent detector (UVD) detect the organics. The OCD spectrum is used to determine the total mass of organic carbon, whereas the UVD spectrum detects only the UV adsorbing species (i.e., compounds containing chromophores such as double bonds, aromatic rings, heterocycles, etc.). Figure 15.6 shows an example of OCD signals where the same surface water is treated with enhanced coagulation and anion IX. The enhanced coagulation removes a significant portion of the biopolymer fraction (around 30 min retention time) and a small portion of humics (around 40 min retention time). The IX removed most of the humics and LMW fractions (around 40 and 50 min retention time, respectively) but it had almost no impact on biopolymer fraction of NOM. As a result, the UV transmittance at 254 nm after enhanced coagulation was ~82%, whereas the anion-IX treated water showed ~94%; the untreated IJssel Lake water had ~72% T(254 nm, 1 cm). These studies indicated that for this particular water source, the most UV absorbing species are in the humic fraction of the DOC.

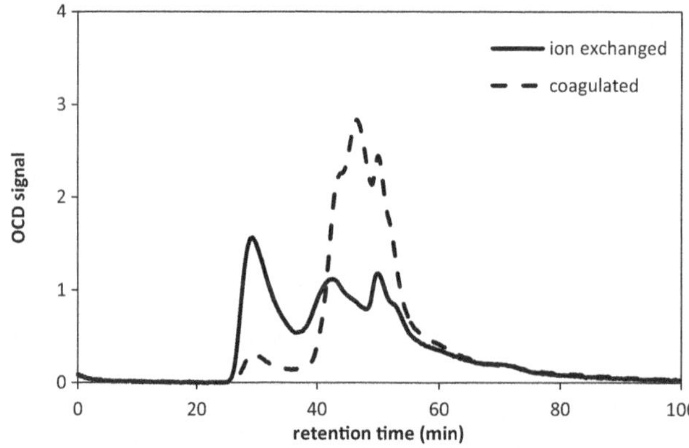

Figure 15.6 SEC-LC-OCD chromatograms of IJssel Lake water samples treated by IX or in-line coagulation processes.

15.3.2.1 Ion exchange technologies and related-challenges

Some waters can have elevated concentrations of DOC, especially sources like surface waters that are under the influence of wastewater secondary effluent, recreational waters, farming and industry. In places like the United Kingdom and Scandinavia, the DOC levels in natural waters are on the rise due to climate change. For these types of waters, IX as a pre-treatment is of interest because the removal of DOC by IX will increase the efficiency of all downstream processes, including coagulation, membrane filtration, AOP and granular activated carbon (GAC) filtration. These waters also contain significant levels of suspended and colloidal matter which makes it nearly impossible to use standard state-of-the-art, fixed bed IX columns. The fixed bed IX columns foul quickly and the head loss increases, while the resin will 'blind' over time through the formation of a biofilm. When these phenomena happen, the ion exchange bed starts to function as a filtration bed rather than as an adsorption media.

Currently available technologies to treat such waters containing suspended and colloidal matter are based on fluidized bed reactors or on totally-mixed reactors with very high concentrations of resin (>400 mL/L), like in the MIEX® process (an acronym for Magnetic Ion Exchange, manufactured and commercialized by Orica) or in suspended plug flow reactors like in the SIX®-process of PWN Technologies. The SIX® process evolved out of the disadvantages of the fluidised bed and the MIEX®-process. The MIEX® process has several different process configurations, but the main two are: the classical dual-stage MIEX® process and the high-rate MIEX® process. In the classical dual-stage MIEX® process, one or two totally mixed contactors in series are fed with raw water and resin leading to an average resin concentration of 20 to 40 mL/L of resin in the contactors. Approximately 5 to 10% of the settled resin is pumped to a regeneration station, regenerated and re-loaded to the remainder of 90 to 95% resin.

In the case of the high-rate MIEX® process, the resin concentration in the contactors is 200 to 500 mL/L. The resin stays in the contactor and the water flows through. Only a very small amount (~0.1%) of the resin is pumped to regeneration vessel while, at the same time, regenerated resin is pumped into the contactors. From the available literature on the high-rate MIEX® process, it is not exactly clear how much resin is pumped to the regeneration system over the course of time, as the 0.1% of the total resin is being regenerated without a direct correlation to time or capacity. Based on the information reported by end users and customers, one concludes that the effective resin concentration in the process is between 1 to 2.5 mL/L. This theoretically yields a treated BV of 1000 to 400 before regeneration and a resin residence time 1000 times higher than the hydraulic residence time, which is not known. This may explain why the process has been called high-rate. The hydraulic residence time and contact time of the water may be short (although not confirmed in the literature) compared to the dual-stage process, but the resin inventory and residence time until regeneration are much larger.

The fluidised bed and the MIEX®-process treat a high number of bed volumes (i.e., >1000 volumes of water per volume of resin) until the resin is regenerated [Slunski, 1999]. Under some water quality conditions, this approach could present quite a few disadvantages which would make the technology less economically attractive. For example, the anion exchange MIEX® process used to remove dissolved organic carbon from raw surface waters is sometimes not feasible due to the high levels of phosphates in the water, which would adsorb onto the resin during the long retention times (designed to minimize the number of regenerations) and, along with the existing carbon and porous characteristic of the resin beads create an ideal environment for bacteria growth. The biofilm formed on the resin "blinds" the active sites of the resin resulting in serious operational and performance challenges. In addition, the loss of adsorption capacity (Wachinski, 2006) requires higher resin concentration or shorter contact time (Verdickt *et al.* 2011; Cornelissen *et al.* 2009), factors which increase the operational costs and/or lower the plant capacity. Another real problem is that eventually the biofilm releases organic matter or adenosine triphosphate

(ATP), both known as detrimental to downstream water treatment processes, particularly to membranes (Cornelissen *et al.* 2010).

To overcome the 'resin blinding', fixed bed reactor systems are flushed regularly with high pH solution to destroy the biofilm as much as possible. In the case of MIEX® processes, which use hydrophobic resins, it is not possible to use high pH solutions to control the biofilm formation since the resin is not resistant to alkaline agents, and eventually the resin will be degraded, thus shortening the lifetime of this relatively expensive resin.

A PWN study (Galjaard *et al.* 2005) showed that the direct treatment of IJssel Lake water with MF/UF was only possible after the removal of the low molecular weight (LMW) DOC fractions with an anion resin PWN researched extensively all aspects of the MIEX® process, including kinetics, regeneration, influence of ions, biogrowth, and overall performance. The major issues related to the application of the MIEX® technology at PWN WTPs were the resin blinding and the fact that a public utility like PWN did not want to be constrained to the resin availability from only one supplier.

Consequently, PWNT pursued further research studies which resulted in the development of a new ion exchange process, namely the suspended ion exchange process (Galjaard *et al.* 2009). Unlike other ion exchange processes treating waters with high content of suspended matter, the 'single pass' or 'suspended' ion exchange process (SIX®) developed by PWNT achieves full control on the adsorption process minimising 'blinding' the resin or producing biomass (www.pwntechnologies.com). The process uses any commercially available resin, and all resins experimented to-date can be treated with alkaline solutions for biofilm control, if necessary. In the SIX® process, the resin is dosed from a dosing tank into the raw water at a low concentration of 4 to 20 mL resin/L, depending on the raw water quality, desired treated water quality, and resin type. This mixture then flows through plug flow contactors. In these contactors, the resin has the same residence time as the treated water, because the resin travels together with the water through these contactors. This is different than MIEX®, in which the resin is retained in the contactors. The number, shape, and design of the contactors play an important role in the adsorption kinetics. The aim of design is to approach the ideal contactor conditions of a plug-flow reactor (Ramaswamy *et al.* 1995) characterized by shorter residence time of the resin, and therefore shorter contact time. After the contact time in the contactors, the resin is separated from the treated water using a customized lamella settler. The resin collects in the hopper, and is then immediately regenerated and returned to the dosing tank (Figure 15.7).

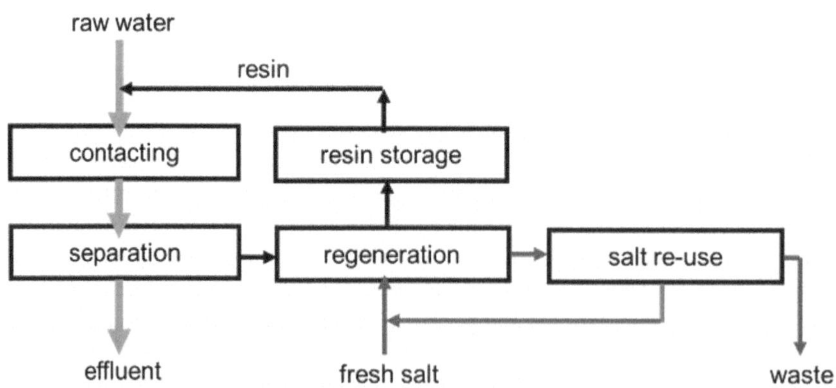

Figure 15.7 Schematic diagram of the SIX® process developed and implemented by PWN Technologies.

Knowing the exact residence time of the resin makes it possible to regenerate all of the resin equally, leading to an equally low number of regenerations. The relatively short contact time (e.g., 10 min < t < 30 min) of the treated water with the resin before the regeneration procedure overcomes the problematic issues of bacteria growth, and resin blinding, and ensures that the resin operate continuously at stable adsorption kinetics.

Another advantage of this process is that because the resin is not fully loaded prior to starting the regeneration, the regeneration, which is an equilibrium process, requires less salt and lower contact times for the regeneration procedure.

15.3.2.2 Case studies on IX in combination with AOPs

There are many case and pilot studies available in the literature, which describe the benefits of an IX process upstream of UV – or O_3-based AOPs. The most relevant example of IX-use as pre-treatment step for improving the AOP performance is represented by the WTP in Andijk, the Netherlands. This case will be discussed to a greater detail later on in this chapter. Andijk WTP is the first large scale application of the SIX® process as a first step in a surface water treatment train which also uses an AOP.

Dunea Duin en Water, the water supply company of Southwestern Holland investigated the organic micro-pollutant (OMP) removal by the existing treatment plant and the necessity to improve the pretreatment with a barrier against OMPs, such as an AOP (Scheideler *et al.* 2015). Pre- and post AOP treatment options, such as membrane filtration, IX, and activated carbon filtration, and dune filtration column experiments were investigated but not selected. Dunea selected implementation of $UV/O_3/H_2O_2$ technology, treating conventionally pretreated surface water (UVT254 approximately 75%) before infiltration in the dunes for artificial replenishing and subsequent abstraction.

Echigo *et al.* (2014) demonstrated the efficacy of combined processes, such as oxidation (ozonation or advanced oxidation processes) and IX treatment for chlorine odour control in tap water. During the oxidation step, the hydrophilic neutral fraction of NOM is converted partly to ionic species, which are effectively removed through the subsequent IX process. The combination of ozone/vacuum UV AOP with a high ozone dose and cation- and anion-exchange resins was the most effective, and the chlorine odour formation potential was reduced from ~100 TON (threshold odour number) to approx. 30 TON. The effectiveness of this process was also confirmed in continuous mode by pilot-scale experiment.

He (2014) reports results on the reverse osmosis (RO) concentrate management studies using advanced oxidation, biofiltration (BAF), and IX pretreatment for electrodialysis reversal (EDR) or advanced oxidation. Pretreatment with ozone and BAF was demonstrated to be effective for controlling organic fouling of the EDR membranes. Pretreatment using IX was demonstrated to be effective for controlling inorganic scaling on EDR membranes when used for RO concentrate management, either alone or in combination with an AOP for further micro-contaminant control. On the basis of the pilot testing findings and modelling, the costs associated with eight concentrate treatment alternatives were evaluated and compared. The project demonstrated a new potentially cost-effective approach to minimizing brine production and maximizing water recovery from the membrane treatment of high-salinity reclaimed water.

IX has also been used after the AOP as a polishing step or to remove DOC/color. Orica tested the UV/H_2O_2-AOP as a pretreatment step to IX (MIEX® process) for further removal of DOC, which was identified as the main challenge at one site (www.miexresin.com). A raw water sample from Airey's Inlet Water Treatment Plant in Victoria, Australia, was tested with MIEX®DOC Resin alone, advanced oxidation alone, and MIEX®DOC Resin in conjunction with an advanced oxidation as a pretreatment step. The advanced oxidation pretreatment with high H_2O_2 concentrations extended UV exposure (approximately 60 min) resulted in high removal rates of DOC.

A similar set-up where IX is used as a polishing step is developed by ESCO (www.escouk.com), for heavily polluted waste waters. The AOP-CATADOX process of ESCO combines treatment by ozone, UV, hydrogen peroxide, and a proprietary catalyst to enable industrial customers to meet the tightening environmental standards for wastewater and gas discharge. The treated water goes through either an IX or GAC bed for further polishing. This process is claimed to be very effective in challenging applications.

15.3.3 Ceramic membranes and hybrid combinations

Membrane filtration is a well-established technology for water treatment. Over the past decade, there has been a growing interest in the water industry sector in achieving higher efficiency through the implemented water treatment processes. This interest also encompasses environmental efficiency, as a key factor of sustainability and adaptability. Climate change and rapid urbanisation have been playing a major role in re-thinking the water treatment, consumption and conservation. One technology that has been identified has highly resilient and has been the topic of many world conferences is ceramic membranes, which are primarily considered for the removal of suspended and colloidal matter. Polymeric membranes also are effective filtration barriers; however, there are some distinct advantages of ceramic membranes relative to the polymeric membranes (Galjaard *et al.* 2011):

- No fibre breakage;
- Very long life span (generally considered > 20+ years);
- High resistance to strong chemicals used to clean the membranes and to minimize fouling;
- Allow high backwash pressure to restore permeability;
- Narrow pore size distribution for better separation.

Unlike dual media sand filtration, microfiltration is an absolute barrier for particles, nematodes, bacteria and other "carriers" of 'biological energy'. To combat the main challenge of membrane fouling by NOM, ozone oxidation upstream of or on the membranes has been explored. It is found that often ozonation reduces significantly the membrane fouling; therefore, there exists the potential for synergy in using ceramic membranes in combination with ozone alone or with the ozone/ peroxide-AOP. With these combined processes, membrane fouling is controlled, turbidity is removed, and oxidation or advanced oxidation occurs.

The benefit of ozone for controlling NOM fouling can be explained by the structural changes of NOM present in the feed water. For instance, it was observed that NOM became more hydrophilic and more negatively charged after ozonation in the absence of calcium (Kim *et al.* 2009). This is because the ozonation produces LMW-, polar, and oxygen-rich compounds with a higher content of hydroxyl, carbonyl, and carboxyl groups. The authors also observed that the oxidized NOM became less negatively charged after ozonation in the presence of calcium. This is consistent with the observation that the ozonation of aqueous NOM modifies the functional groups in the oxidized NOM, which are prone to complexation with calcium ions.

Recent information suggested that the reduction of membrane fouling by pre-ozonation is the result of microflocculation (Stanford *et al.* 2011). Although microflocculation as a result of ozonation process is well known (Edwards & Benjamin, 1992), it was not previously regarded as an advantage toward reducing membrane fouling. When ozone is in combination with a ceramic membrane, the decomposition of ozone at the membrane surface could result the formation of hydroxyl radicals and other reactive species. This is also considered to be a possible anti-fouling effect of pre-ozonation.

Lehman and Liu (2009) reported the combination of ceramic microfiltration, ozonation, and in-line coagulation to treat wastewater effluent. In the pilot study using a single Metawater membrane element of 25 m^2, stable performance with very high specific flux (>1000 LMH/bar @ 20°C) was achieved. The pre-treatment included ozonation (4 mg/L O_3) and coagulation with polyaluminum chloride (PACl, 1 mg/L as

Al), whereas the ceramic microfiltration was performed at 170 LMH with 2-h backwash intervals without chemical addition.

The national water agency of Singapore (PUB) has studied ceramic microfiltration at different sites in Singapore. In 2007, PUB worked in collaboration with Black & Veatch and METAWATER (formerly NGK) on pilot testing of ceramic membranes at the Bedok NEWater Factory (Clement, 2008). One major finding was that the application of ozone upstream of ceramic membranes improved the operation of ceramic membranes by controlling fouling as well as providing a good degree of oxidation and disinfection. An additional benefit was the enhanced coagulation in the presence of ozone. The potential benefits to PUB of applying ozone and coagulant upstream of ceramic membranes appear promising, but further studies at larger scale and for extended period of time are required in order to confirm the pilot conclusions.

The CeraMac® demonstration plant operated by PUB and PWN Technologies at Chou Chu Kang Water Works is so far the largest demonstration of the combination of ozone and ceramic microfiltration (475 m^2 membrane surface area) for drinking water purposes. When treating clarified water, dosing ozone prior to the membrane significantly improved the filtration performance. A high specific flux (>700 LMH/bar @ 25°C) was obtained in long term trial at filtration fluxes between 275 and 315 LMH, with a 3-h backwash interval and enhanced backwash after the 4th backwash. All these results are the basis for more testing of the hybrid version of an AOP with ceramic microfiltration.

Ozone in combination with ceramic membrane filtration has been evaluated for water reuse purposes. The first water reuse plant using this concept was built in 2010 by Metawater in Japan for Tokyo Metropolitan Government with a capacity of 7000 m^3/day, and consists of biological filtration followed by ozonation, coagulation, and ceramic membrane microfiltration This process is certified by Title 22 of CDPH (California Department of Public Health).

Because the ceramic membranes can sustain long-term exposure to hydrogen peroxide, such membranes are less susceptible to fouling caused by biological activity, should the UV/H_2O_2-AOP be implemented downstream of ceramic membrane and H_2O_2 be dosed upstream the membrane. During the experimental tests on water advanced treatment at Andijk WTP (see also paragraph 15.5.1) with IX followed by ceramic membrane filtration as pre-treatment steps to the UV/H_2O_2 AOP, the H_2O_2 dosage required to meet the AOP micropollutant treatment performance was moved upstream of the ceramic membrane. Since then, an even more stable ceramic membrane filtration process was observed. Figure 15.8 presents the effect of H_2O_2 dosage upstream of ceramic membrane. After 47 days of continuing H_2O_2 dosage upstream of ceramic membrane, the membrane operating pressure (measured as the transmembrane pressure, TMP) increased gradually. The anti-fouling effect of H_2O_2 can be ascribed to:

(1) partial degradation of organic compounds by oxidation, probably through reactive species formed from peroxide decomposition onto the membrane surface,
(2) significant inhibition of biological activity;

The exact underlying mechanism is not clear yet and is actively investigated.

These studies on hybrid processes led to the concept of integrating the O_3/H_2O_2-AOP in treatment trains for water reuse applications and other challenging waters. However the main challenge of ceramic membrane technology is to achieve economic feasibility for large-scale applications like potable water treatment. PWN Technologies (PWNT) has developed a system known as CeraMac (Kamp *et al.* 2009) which greatly reduces the installation cost of a ceramic membrane system to a level which is cost competitive with a polymeric membrane system. The key design feature is that rather than having ceramic membrane modules in individual stainless steel casings, 7 up to almost 200 ceramic modules can now be housed in a single stainless steel vessel (Figure 15.9). This innovative design resulted in a significant reduction in stainless steel costs and in the number of valves, and increased the productivity as all elements are backwashed at the same time which reduces the installation downtime during a backwash from 10 minutes to a few seconds.

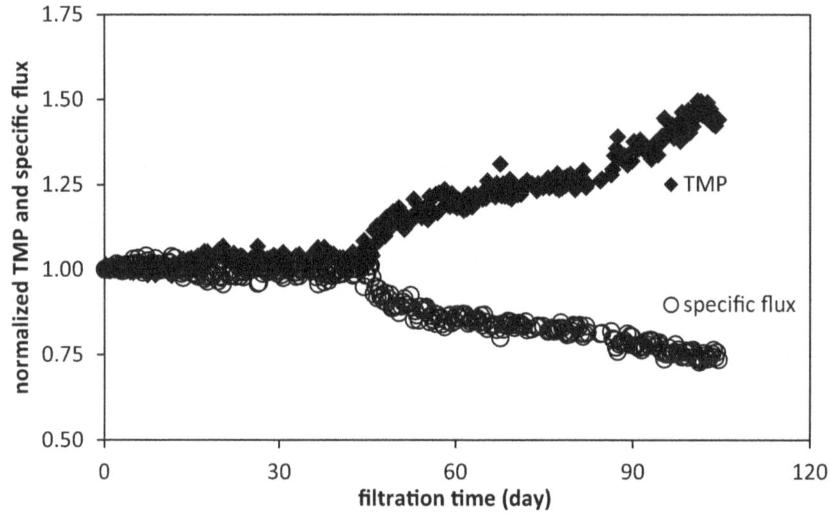

Figure 15.8 Normalized TMP and membrane permeability over time for treating ion exchanged stream with ceramic MF during the hydrogen peroxide dosing experiments.

Figure 15.9 The CeraMac vessel system with 200 ceramic elements.

15.4 THE EFFECT OF PRETREATMENT ON MP UV/H_2O_2 AOP

At PWN's surface water treatment plant in Andijk, UV/H_2O_2 treatment with medium pressure lamps is applied as a non-selective barrier against organic micropollutants. The energy consumption is strongly dependent on the water quality, particularly on the NOM content and the UV transmittance of the water over the MP lamp emission spectrum in the UV range. The most important UV-absorbing compounds in raw IJssel Lake water are NOM and nitrate. The NOM content (~6.5 mg/L as C) is rather stable over the year, while there is a strong seasonal variation of the nitrate content with an annual range typically from 1 to 14 mg/L as nitrate.

Table 15.2 presents the dynamics of DOC and nitrate in IJssel Lake water samples subjected to coagulation, sedimentation and sand filtration (CSF) or IX-UF treatment, relative to the untreated samples.

Both DOC and nitrate content can be lowered by pretreatment. In the existing water treatment facility at Andijk, conventional pretreatment with coagulation and sand filtration (CSF) is used. CSF lowers the DOC, on average, from 6.5 mg/L to 5.8 mg/L with a ferric coagulant. Nitrate removal by CSF is insignificant. Advanced pretreatment with IX followed by microfiltration, allows further removal of DOC and nitrate.

Table 15.2 Impact of pretreatment on DOC and nitrate concentrations for IJssel Lake water at the Andijk treatment facility.

	Min		Max		Average	
	mg C/L	mg NO_3/L	mg C/L	mg NO_3/L	mg C/L	mg NO_3/L
raw water	5.1	1.2	7.2	12.4	6.0	6.5
CSF treatment	2.9	1.7	5.1	9.4	4.0	5.8
IX-UF treatment	2.1	0.5	4.4	4.2	2.9	2.3

Note: CSF is coagulation followed by sand filtration, and the coagulant was ferric chloride. IX-UF is ion exchange using LanXess S5128 anion exchange resin, which is a DOC selective resin followed by ultrafiltration, and no coagulant was used in this process train.

Application of MP UV/H_2O_2 AOP allows utilization of the full UV spectral emittance of MP lamps which cover the 200–400 nm range. Table 15.3 provides the calculated fractions of light absorbed by 6 mg/L H_2O_2 at 240 and 254 nm in raw and pre-treated IJssel Lake water. The 240 nm wavelength was selected arbitrarily from the spectral emission of MP lamp in the short UV range where both nitrate ion and NOM absorb the UV light, whereas the 254 nm wavelength was selected because it is commonly used to express the NOM specific absorption of UV light (usually referred to as SUVA), and 253.7 nm is the resonance emission line of mercury atom. Hydrogen peroxide absorption coefficients at 240 and 254 nm are 44.4 and 21.2 M^{-1} cm^{-1}, respectively. The OH radical yield is directly proportional to the fraction of light absorbed by H_2O_2. Therefore, the data in Table 15.2 can be interpreted with respect to OH radical yield dynamics at a given wavelength as a function of the pre-treatment process applied to the raw water. The data shows that the IX-UF process is significantly more effective at the removal of the absorbing species at the two wavelengths than the CSF, under the experimental conditions used for the two pretreatment processes. This observation is consistent with the efficacy of IX-UF at the removal of both nitrate and ionized NOM fractions, whereas CSF is ineffective at nitrate removal. Both processes lead to substantial increase of fractions of light absorbed by H_2O_2 at both wavelengths, which translates to higher OH radical yields upon pre-treatment application.

Table 15.3 Impact of pretreatment on the fraction of absorbed photons by 6 mg/L H_2O_2.

	240 nm	**254 nm**
raw water	4.5%	2.6%
CSF treated water	8.2%	5.3%
IX-UF treatment	19.4%	14.7%

A significant reduction of the energy consumption is expected to be achieved due to the strong improvement of the UV-transmittance of the raw water upon advanced pretreatment. Two monitoring compounds which are regulated in drinking water by various legislations around the world, 1,4-dioxane and *N*-nitrosodimethylamine (NDMA), were used in pilot experiments to quantify the effect of improved pretreatment on the energy consumption of the MP UV/ H_2O_2 AOP. NDMA was selected to quantify the impact on photolysis only, whereas 1,4-dioxane was selected to monitor the effect of pretreatment on hydroxyl radical oxidation. The energy consumption is expressed as electrical energy per order (EE/O), which represents the electrical energy required for 90% conversion of a compound in a given volume of water. The water quality after pretreatment, DOC and nitrate content, are represented by the averages mentioned in Table 15.2. Figure 15.10 illustrates the substantial effect pretreatment can have on the required electrical energy consumption of MP UV/ H_2O_2 AOP. In the case of Andijk WTP, where the eutrophic surface water is pretreated with a conventional process (i.e., CSF), the implementation of advanced pretreatment with IX followed by microfiltration, results in up to 50% reduction in the required electrical energy for both photolysis and hydroxyl radical oxidation.

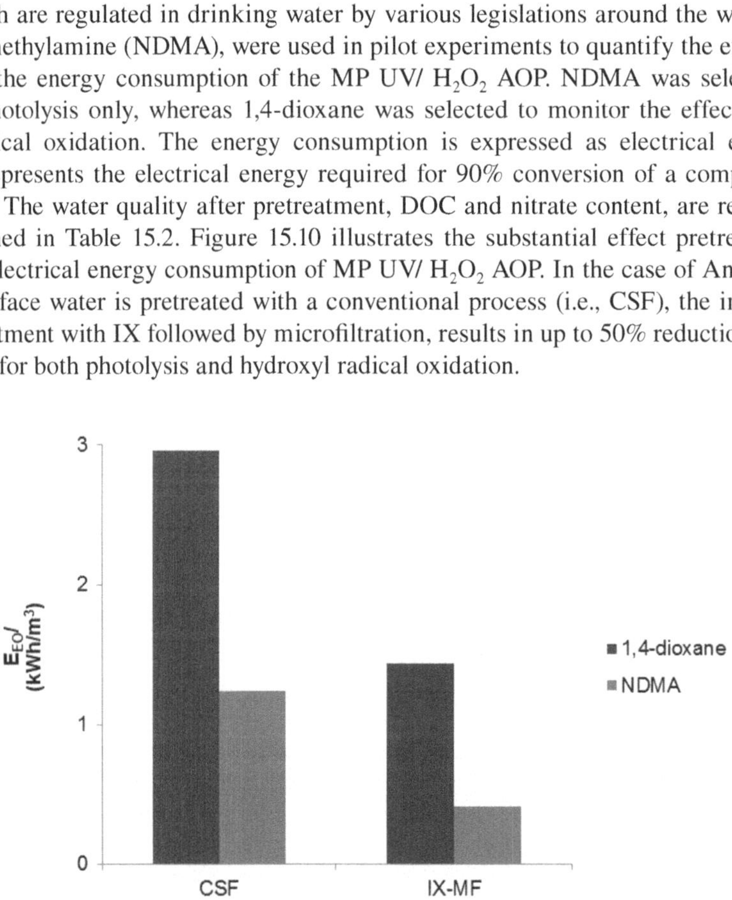

Figure 15.10 Required electrical energy dose for 90% conversion of NDMA and 1,4-dioxane by MP UV/ H_2O_2 AOP treatment in surface water pretreated with either CSF or IX-MF in the presence of 6 mg/L H_2O_2.

In a recent study, Metcalfe *et al.* (2015) investigated the DBP-formation potential in three UK surface waters of different qualities, treated by (a) a full-scale conventional process and (b) pilot-scale processes using suspended ion-exchange (SIX) process and in-line coagulation (ILCA), followed by ceramic membrane filtration (CMF). The study revealed a synergetic effect of SIX-ILCA combined processes relative to the individual SIX and coagulation processes, with respect to reductions in dissolved organic carbon, UV absorbance, and DBP-formation potentials. Furthermore, the new processes resulted in reduced concentrations in brominated-DBPs. The SIX/ILCA/CMF combined process removed 50, 62, 62, 62% and 47% more of DOC, UV absorbance, THM-formation potential, HAA-formation potential, and brominated DBPs, respectively, compared with the full-scale conventional treatment process.

15.5 SIDE EFFECTS OF MP UV/H_2O_2 AOP AND MITIGATION STRATEGIES

The objective of advanced oxidative treatment as a barrier for organic micropollutant control is to degrade the micropollutant into harmless metabolites. This is true for all commonly applied oxidative technologies in drinking water treatment, such as ozonation and advanced oxidation by the combined use of O_3/H_2O_2, UV/H_2O_2 or O_3/UV (Glaze, 1987). The effect of advanced oxidation processes is based on the formation and reaction of hydroxyl radicals. Hydroxyl radicals are rather non-selective oxidizing species, reacting with most organic compounds, including NOM, by addition to aromatic or unsaturated sites or by abstracting hydrogen atoms. In principle, a complete mineralization can be achieved by advanced oxidation; however, under economically feasible conditions this does not take place and reaction byproducts are formed. As with ozonation, intensive research has shown that the reaction byproducts are, in general, more biodegradable and less harmful than the parent compounds.

Under normal MP UV/H_2O_2 AOP conditions for trace organic contaminant control, the effect of advanced oxidation on the removal of NOM is minimal, regardless of the pretreatment. Because NOM absorbs UV light and reacts with hydroxyl radicals, the NOM will still be affected. It is important to understand the biological stability of water after an AOP and the potential formation of byproducts. Figure 15.11 shows the effect of AOP on DOC in a bench-scale study performed under conditions similar to the full scale AOP implemented at Andijk WTP (0.54 kWh/m^3; 6 mg/L H_2O_2). The water samples used in the AOP experiments were surface water from the IJssel Lake pretreated with either the conventional process or the IX-MF advanced pre-treatment process. Only a minor reduction of DOC by MP UV/H_2O_2 AOP treatment was observed as opposed to the substantial DOC removal by advanced pretreatment.

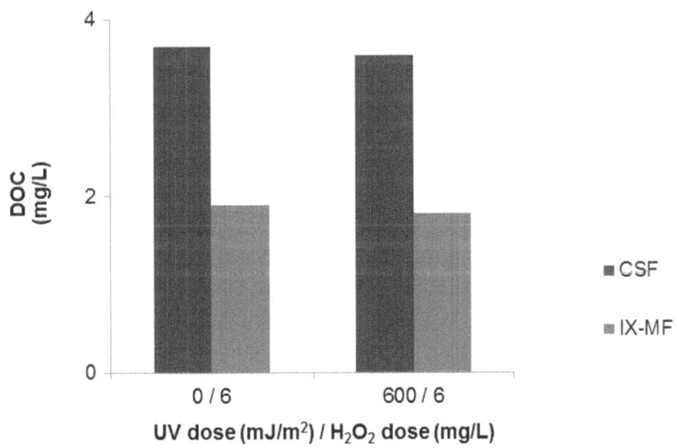

Figure 15.11 DOC content before and after MP UV/H_2O_2 AOP treatment (600 mJ/cm^2; 6 mg/L H_2O_2) in pre-treated surface water.

Due to the non-selective behaviour of the hydroxyl radicals and the absorbed photons, the NOM is altered. Under process conditions for trace chemical contaminant control, such changes are minimal and cannot be measured accurately by parameters such as DOC or SEC-LC-OCD-OND data. Instead, the effect can be measured as assimilable organic carbon (AOC)) (Van der Kooij, 1992), a bulk parameter that quantifies the biodegradable organic carbon used by microorganisms for growth. Figure 15.12 shows the formation of AOC by MP UV/H_2O_2 AOP and the removal of AOC in the biological activated carbon filters. In the distribution systems that do not keep a residual disinfectant such as chlorine, chloramine, chlorine dioxide, the reduction of biologically degradable compounds at the water treatment plant is paramount in order to minimize bacterial regrowth. Biologically active carbon filters with an EBCT of 25 min were installed at Andijk WTP to remove the AOC formed by during the MP UV/H_2O_2 AOP water treatment. Chlorine dioxide at very low levels of 0.07 mg/L is dosed to the finished water exiting the WTP in order to ensure residual disinfection in the distribution system.

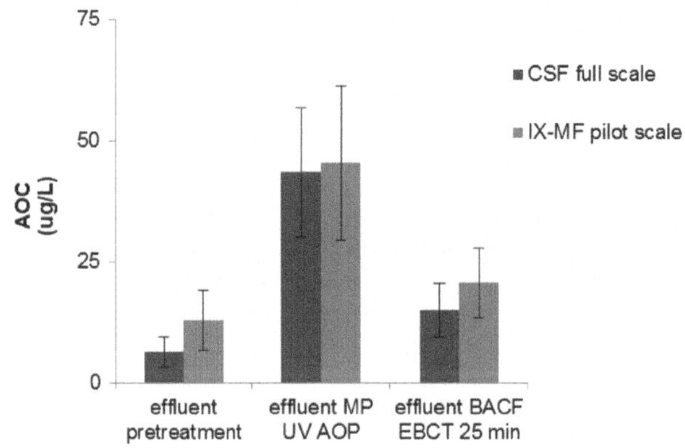

Figure 15.12 Formation of assimilable organic carbon during the MP UV/H_2O_2 treatment of CSF or IX-MF pre-treated surface water and subsequent removal through BACF.

The UV/H_2O_2 AOP relies on H_2O_2 photolysis to yield the hydroxyl radical concentration required to meet the design micropollutant treatment performance. Typically, 10–15% of the dosed H_2O_2 is consumed in the advanced oxidation process. Hydrogen peroxide is not allowed in drinking water, such that the residual from the AOP must be removed. Post-UV/H_2O_2 chlorination is a common practice to address both the residual H_2O_2 removal and the residual disinfection requirements. An alternative method to chlorination is activated carbon filtration which relies on the catalytic properties of GAC. At the WTP in Andijk, activated carbon filters were installed after the MP UV/H_2O_2 AOP, which presents multiple benefits, among which the efficient removal of H_2O_2 residual without chemical addition (Figure 15.13).

The decomposition of hydrogen peroxide by activated carbon is independent of the hydraulic loading rate as shown in figure 15.14. Typically 99% removal of hydrogen peroxide through biologically activated carbon filtration requires 6 – 9 min EBCT.

An important side effect of the application of MP UV technology is the photolysis of nitrate with nitrite formation in nitrate-rich waters. Nitrate has two absorption bands in the UV region, one in the near-UV range from 260 nm to 350 nm with a maximum at 300 nm, and a much stronger band in the short UV wavelength range (λ <~ 240 nm) with a maximum at 200 nm (Krishnan & Guha, 1934), which overlaps the high-energy spectral emittance range of MP lamps. Therefore, at λ <~ 240 nm, the yield of nitrite formation through nitrate photolysis in MP UV applications to water treatment could be significant due to a favourable combination of large molar absorption coefficients and high quantum yields of nitrate (Mack & Bolton, 1999).

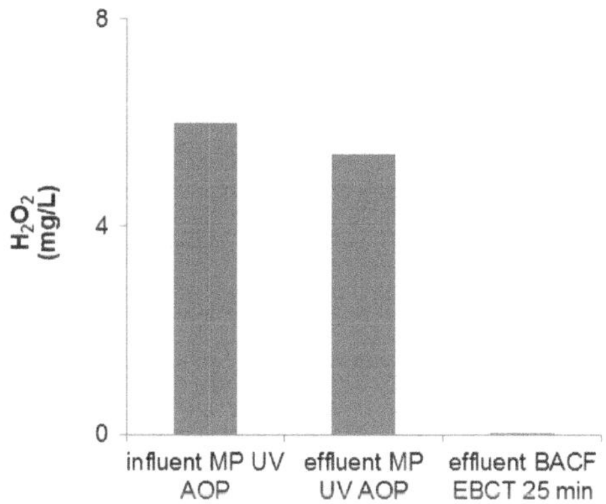

Figure 15.13 Profile of H_2O_2 concentration before and after MP UV AOP, and after BACF.

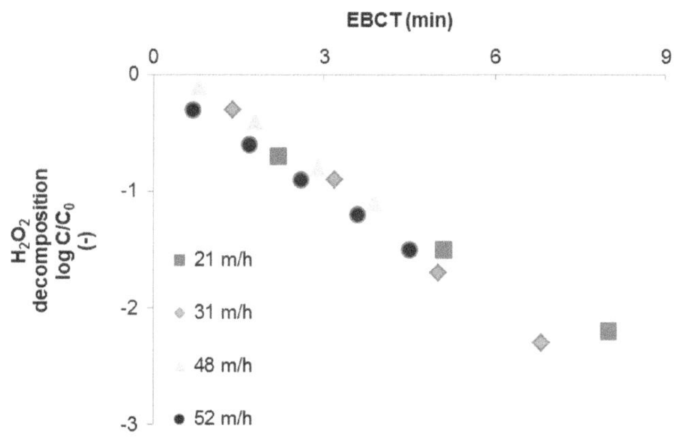

Figure 15.14 H_2O_2 removal through BAC (Norit RO 3502) filtration at four high hydraulic loading rates as a function of the empty bed contact time.

The UV lamps in the UV reactors are housed by quartz sleeves, which display different UV light transmittance properties depending on the quartz quality. The quartz solarization upon prolonged exposure to high energy UV radiation with the loss of transmittance in the short UV wavelength range (λ <~ 240 nm) is a known phenomenon. Such processes are significantly more pronounced in natural quartz than in synthetic quartz. The impact of quartz transmitting properties of MP radiation on nitrate photolysis with nitrite formation in CSF-pretreated water as observed in the full-scale MP UV/H_2O_2 AOP implemented at PWN is described by Figure 15.15. The increasing electrical energy dose (EED) corresponds to the cumulated electrical energy from the first MP reactor to the last in the MP UV AOP train. The high transmittance quartz allows the full spectrum emitted by the lamp to enter the water, favoring the nitrite formation to levels which may require post-AOP treatment. The aged natural quartz sleeves block the low wavelengths and as a result limit the nitrate photolysis. However, the unavailability of the lower UV wavelength range for the AOP may result in reduced energy efficiency of the UV system.

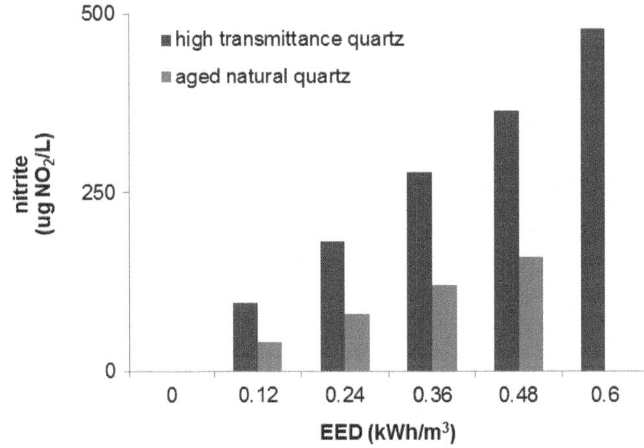

Figure 15.15 Nitrite formation in full scale MP UV/H_2O_2 AOP treatment as a function of the electrical energy dose for high transmittance quartz sleeves (full MP spectrum) and aged natural quartz sleeves (blocking the short UV wavelengths) in CSF pretreated water containing 9 mg NO_3^-/L.

The maximum nitrite concentration allowed in drinking water in the EU is 100 µg NO_2^-/L. The biological activated carbon filtration after the AOP treatment with 25 min EBCT as installed at Andijk WTP ensures nitrite removal to non-detect levels. At water temperatures <10°C, phosphate is dosed to support the biological processes in the BACF. Pilot studies showed that the oxidation of nitrite to nitrate in the carbon filters is biologically driven; a fresh (virgin) activated carbon filter does not lower the nitrite concentration (Figure 15.16).

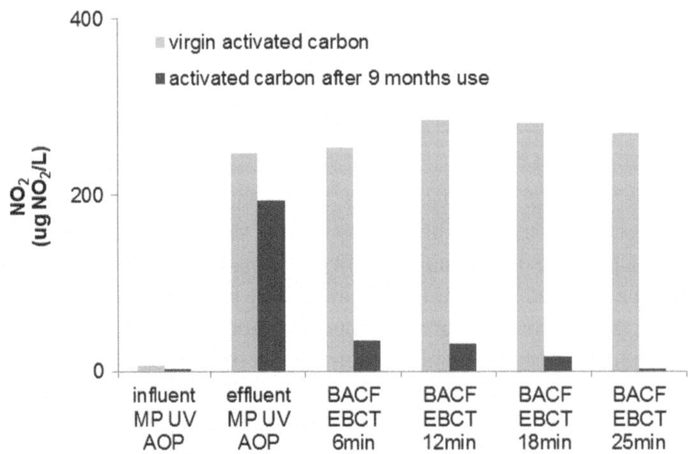

Figure 15.16 Nitrite formation during the MP UV AOP and subsequent biological re-oxidation on virgin activated carbon and biologically activated carbon.

The stable inorganic end product of nitrate photolysis is nitrite. The UV photolysis of nitrate is rather complex and nitroso-, nitro- and hydroxyl radical species are formed as reactive intermediates. These radicals may react with the organic compounds which constitute the NOM or from other sources, resulting in nitroso- and nitro-derivatives (Goldstein & Rabani, 2007). Given the high reactivity of radical species generated during the MP UV/H_2O_2 AOP, numerous reaction products, most of which are unknown, are formed as a result of radical reactions with water matrix constituents and targeted organic micropollutants. In order to assess the quality of treated water, PWN Technologies undertook extensive studies, among which the water mutagenicity using Ames II bioassay. An increase in the Ames test response, indicating the formation of genotoxic compounds, was observed in water samples collected from the full-scale MP UV/H_2O_2 installation with synthetic quartz sleeves (Martijn & Kruithof, 2012). As discussed earlier, the high quality quartz allowed significant UV treatment at wavelengths <240 nm, enabling nitrate photolysis in the site nitrate-rich surface water.

As described in a previous section, the water pre-treatment with IX and MF lowered considerably the nitrate concentration in the UV-AOP influent water. As a result, much lower nitrite was measured in the MP UV/H_2O_2 AOP effluent water. The Ames test response after MP UV/H_2O_2 treatment was substantially lower as a result of improved pre-treatment. The Ames test on samples collected after the BAC filtration indicated that BAC removed completely the mutagenic compounds.

The robust water treatment trains implemented at PWN full-scale facilities, with advanced pre-treatment and biological activated carbon filtration as a final treatment step, represent a multi-barrier approach to both micropollutants present in the surface water used as a drinking water source and potential harmful byproducts generated during the oxidative conditions created in MP UV/H_2O_2 AOP. At the same time, the continuing investment in research and innovation practiced by PWN Technologies results in technological advancements in water treatment sector which respond to the increasingly challenging goals such as energy and water conservation, and delivering safe drinking water to people.

15.6 REFERENCES

Amirtharajah A. and Mills K. M. (1982). Rapid mix design for mechanisms of alum coagulation. *Journal of the American Water Works Association*, **74**(4), 210–216.

Archer A. D. and Singer P. C. (2006). Effect of SUVA and enhanced coagulation on removal of TOX precursors. *Journal of the American Water Works Association*, **98**(8), 97–107.

Clement J. A. (2008). Evaluation of ceramic membranes with ozone for water re-use. *Proceedings of Singapore International Water Week Conference*, Singapore.

Cornelissen E. R., Beerendonk E. F., Nederlof M. N., Van der Hoek J. P. and Wessels L. P. (2009). Fluidised ion exchange (FIX) to control NOM-fouling in Ultrafiltration. *Desalination*, **236**(1–3), 334–341.

Cornelissen E. R., Chasseriaud D., Siegers W. G., Beerendonk E. F. and Van der Kooij D. (2010). Effect of anionic fluidized ion exchange (FIX) pre-treatment on nanofiltration (NF) membrane fouling. *Water Research*, **44**(10), 3283–3293.

Duan J. and Gregory J. (2002). Coagulation by hydrolysing metal salts. *Advances in Colloid Interface Science*, **100**(102), 475–502.

Echigo S., Itoh S., Ishihara S., Aoki Y. and Hisamoto Y. (2014). Reduction of chlorinous odor by the combination of oxidation and ion-exchange treatments. *Journal of Water Supply: Research and Technology – AQUA*, **63**(2), 106–113.

Edwards M. and Benjamin M. M. (1992). Transformation of NOM by ozone and its effect on iron and aluminum solubility. *Journal of the American Water Works Association*, **84**(6), 56–66.

Edzwald J. K. and Tobiasson J. E. (1999). Enhanced coagulation: US requirements and a broader view. *Water Science and Technology*, **40**(9), 63–70.

Eschauzier C., Hoppe M., Schlummer M. and de Voogt P. (2013). Presence and sources of anthropogenic perfluoroalkyl acids in high-consumption tap-water based beverages. *Chemosphere*, **90**, 36–41.

Galjaard G. and Kruithof J. C. (2002). Enhanced Pre-coat Engineering (EPCE) for MF and UF: steps to full-scale application. *Proceedings of IWA Leading Edge Technology Conference*, Mülheim, Germany, ISSN 0941-0961.

Galjaard G., Kruithof J. C. and Kamp, P. C. (2005). Influence of NOM and membrane surface charge on UF-membrane fouling. *Proceedings of the AWWA Membrane Technology Conference*, Phoenix, USA.

Galjaard G., Kamp P. C. and Koreman E. (2009). SIX®: A new resin treatment technology for drinking water. *Proceedings of Singapore International Water Week Conference*, Singapore.

Galjaard G., Martijn B., Koreman E., Bogosh M. and Malley J. (2011). Performance evaluation of SIX® – Ceramac® in comparison with conventional pre-treatment techniques. *Water Practice and Technology*, **6**(4), doi: 10.2166/wpt.2011.0066 (online).

Galjaard G., Clement J., Ang W. S. and Lim M. H. (2013). CeraMac®-19 Demonstration Plant Ceramic Microfiltration at Choa Chu Kang Waterworks. *Proceedings of the AMTA/AWWA Membrane Technology Conference*, San Antonio, Texas.

Goldstein S. and Rabani J. (2007). Mechanism of nitrite formation by nitrate photolysis in aqueous solutions: the role of peroxynitrite, nitrogendioxide and hydroxyl radical. *Journal of the American Chemical Society*, **129**, 10597–10601.

Gray S. R., Ritchie C. B., Tran T., Bolto B. A., Greenwood P., Busetti F. and Allpike B. (2008). Effect of membrane character and solution chemistry on microfiltration performance. *Water Research*, **42**(3), 743–753.

He C., Carpentier G. and Westerhoff P. (2014). Minimizing Concentrate using Advanced Oxidation. Biofiltration and Ion-Exchange Pretreatment for Electrodialysis Reversal. WRRF-12-01 Report, WateReuse Research Foundation.

Hofs B., Baggelaar P., Harmsen D. and Siegers W. (2014). Robuustheid zuiveringen DPW 2012–2013; zomer en winter. KWR Report 2014.022, KWR Watercycle Research Institute, The Netherlands.

Houtman C. J. (2010). Emerging contaminants in surface waters and their relevance for the production of drinking water in Europe. *Journal of Integrative Environmental Sciences*, **7**, 271–295.

Huber S. A., Balz A., Albert M. and Pronk W. (2011). Characterisation of aquatic humic and non humic matter with size-exclusion chromatography-organic carbon detection – organic nitrogen detection (LC-OCD-OND). *Water Research*, **45**, 879–885.

Kamp P. C., Willemsen-Zwaagstra J., Kruithof J. C. and Schippers J. C. (1997). Treatment strategy PWN Water Supply Company North Holland (in Dutch). *H_2O*, **30**, 386–390.

Kim J., Shan W., Davies S. H. R., Baumann M. J., Masten S. J. and Tarabara V. V. (2009). Interactions of aqueous NOM with nanoscale TiO_2: implications for ceramic membrane filtration-ozonation hybrid process. *Environmental Science and Technology*, **43**(14), 5488–5494.

Koreman E. A., Visser M. and Welling M. (2009). Enhanced coagulation at SWTP Andijk (in Dutch: Klassiek werkpaard presteert boven verwachting). *H_2O*, **16**(17), 48–51.

Kooij van der D. (1992). Assimilable organic carbon as an indicator of bacterial growth. *Journal of the American Water Works Association*, **84**(2), 57–65.

Kruithof J. C. and Kamp P. C. (1997). *The Dilemma of Pesticide Control and By-Product Formation in the Heemskerk Water Treatment Plant Design: Selected Topics on New Developments in Physico-Chemical Water Treatment*, Leuven University, Leuven.

Kruithof J. C. (2005). State of the art of the use of ozone and related oxidants in Dutch drinking water treatment. *Proceedings 17th IOA World Congress*, Strasbourg.

Kruithof J. C., Kamp P. C. and Martijn A. J. (2007). UV/H_2O_2 treatment: a practical solution for organic contaminant control and primary disinfection. *Ozone Science and Engineering*, **29**, 273–280.

Lehman S. G. and Liu L. (2009). Application of ceramic membranes with pre-ozonation for treatment of secondary wastewater effluent. *Water Research*, **43**(7), 2020–2028.

Mack J. and Bolton J. R. (1999). Photochemistry of nitrate and nitrite in aqueous solution: a review. *Journal of Photochemistry and Photobiology A: Chemistry*, **128**, 1–13.

Martijn A. J., Fuller A. L., Malley J. P. and Kruithof J. C. (2010). Impact of IX-UF pretreatment on the feasibility of UV/H_2O_2 treatment for degradation of NDMA and 1,4 dioxane. *Ozone Science and Engineering*, **30**(6), 383–390.

Martijn A. J. and Kruithof J. C. (2012). UV and UV/H_2O_2 treatment: the silver bullet for by product and genotoxicity formation in water production. *Ozone Science and Engineering*, **34**, 92–99.

Metcalfe D., Rockey C., Jefferson B., Judd S. and Jarvis P. (2015). Removal of disinfection by-product precursors by coagulation and an innovative suspended ion exchange process. *Water Research*, **87**, 20–28.

Ramaswamy H. S., Abdelrahim K. A., Simpson B. K. and Smith J. P. (1995). Residence time distribution (RTD) in aseptic processing of particulate foods: a review. *Food Research International*, **28**(3), 291–310.

Scheideler J., Lekkerkerker-Teunissen K., Knol T., Ried A., Verberk J. and Van Dijk J. (2011). Combination of O_3/H_2O_2 and UV for multiple barrier micropollutant treatment and bromate formation control – an economic attractive option. *Water Practice and Technology*, **6**(4), doi: 10.2166/wpt.2011.063 (online).

Schlichter B., Mavrov V. and Chmiel H. (2004). Study of a hybrid process combining ozonation and microfiltration/ultrafiltration for drinking water production from surface water. *Desalination*, **168**, 307–317.

Snyder S. A., Wert E. C., Lei H. X., Westerhoff P. and Yoon Y. (2007). *Removal of EDC's and Pharmaceuticals in Drinking and Reuse Treatment Processes*, AWWA Research Foundation, Denver CO, 331.

Stanford B. D., Pisarenko A. N., Holbrook R. D. and Snyder S. A. (2011). Preozonation effects on the reduction of Reverse Osmosis membrane fouling in water reuse. *Ozone Science and Engineering*, **33**(5), 379–388.

Stefan M. I., Kruithof J. C. and Kamp P. C. (2005). Advanced oxidation treatment of herbicides: from bench scale studies to full scale installation. *Proceedings of the 3rd IUVA World Congress*, Whistler, BC, Canada.

Verdickt L., Closset W., D'Haeseleer V. and Cromphout J. (2011). Applicability of ion exchange for NOM removal from a sulphate-rich surface water incorporating full reuse of the brine. *Proceedings of the IWA NOM-Specialty Conference*, Los Angeles, CA, USA.

Wachinski A. M. (2006). *Ion Exchange Treatment for Water*. American Water Works Association, Denver, CO, USA.

Van Wezel A. P., Morinière V., Emke E., ter Laak, T. and Hogenboom A. C. (2011). Quantifying summed fullerene nC60 and related transformation products in water using LC LTQ Orbitrap MS and application to environmental samples. *Environ. Int.*, **37**(6), 1063–1067.

Zhu H., Wen X. and Huang, X. (2011). Characterization of membrane fouling in a microfiltration ceramic membrane system treating secondary effluent. *Desalination*, **284**, 324–331.

Chapter 16

AOPs for municipal and industrial wastewater treatment

Jianlong Wang and Lejin Xu

16.1 INTRODUCTION

Over the past decades, the influx of significant quantities of organic, inorganic, and mineral substances into the aquatic environment has been substantially increasing mainly by anthropogenic activities such as excessive population growth, fast urban encroachment, rapid industrialization, and improved agricultural operations. Approximately 2 million tonnes of sewage, agricultural, and industrial wastes are being disposed of every day into the water courses worldwide (Kansal & Kumari, 2014). It is of utmost importance to dispose of these residues in a proper manner as well as to develop the sustainable water treatment strategies.

Due to the high chemical stability and/or low biodegradability of most of water contaminants, one feasible option for removing organic pollutants from wastewater is the use of Advanced Oxidation Processes (AOPs), such as direct UV photolysis, UV/H_2O_2, ozone, Fenton- and photo-Fenton, photocatalysis, Ultrasound, and their selective combinations. These processes have been widely recognized as highly efficient treatments for recalcitrant wastewater or used as pretreatment to convert pollutants into shorter-chain compounds that can then be treated by conventional biological methods. The efficacy of AOPs depends on the formation of reactive free radicals, especially hydroxyl radicals ($^{\bullet}OH$) that are highly reactive and non-selective. One or more processes must be selected appropriately for remediation of a specific industrial wastewater considering various factors including wastewater characteristics, technical applicability and potential constraints, effluent discharge standards, regulatory requirements, economical aspects and long-term environmental impacts. Oneby *et al.* (2010) have summarized the ozone treatment of secondary effluent at U.S. municipal wastewater treatment plants (MWWTPs). Brillas *et al.* (2009) have described and thoroughly discussed the fundamentals, experimental setups, and lab and pilot-plant experiments related to the major applications of electrochemical advanced oxidation processes for treating synthetic solutions and real wastewaters containing pollutants such as pesticides, dyestuffs, pharmaceuticals and personal care products (PPCPs), and industrial compounds. Oller *et al.* (2011) have reviewed recent research combining AOPs (as a pre-treatment or post-treatment stage) and bioremediation technologies for the decontamination of a wide range of synthetic and real industrial wastewater. Wang and Xu (2012) have offered a genuine critical and complete insight into the formation mechanisms of •OH in AOPs and their application for wastewater treatment.

Municipal wastewater is usually qualitatively and quantitatively similar in its composition, while industrial wastewater characteristics vary not only with the different industries but also within the same industry. A large number of pollutants have been found in wastewater, mainly characterized by chemical oxygen demand (COD), 5-day biochemical oxygen demand (BOD_5), suspended solids (SS), toxicity and color. Various organic compounds contained in industrial wastewater have been degraded by AOPs, which include dyestuffs (e.g., acid, basic, direct, disperse, etc.), pharmaceuticals (e.g., antibiotics, blood lipid regulators derivatives, neuroprotectors, etc.), pesticides (e.g., herbicides, insecticides, fungicides, rodenticides, etc.), aromatics (e.g., benzenes, phenols, anilines, etc.), aliphatics, and so on. This chapter covers a wide spectrum of municipal wastewater and industrial pollutants (i.e., textile wastewater, pharmaceutical wastewater, pesticide wastewater, papermaking wastewater, and others) treated by AOPs both in bench-scale and pilot-plant studies. Some of the representative applications of AOPs to the wastewater treatment are summarized, and relevant examples of AOPs applied to wastewater treatment especially used in China are given. Economic evaluation, ongoing research and perspectives for the future work required to facilitate efficient large-scale operation are presented.

16.2 MUNICIPAL WASTEWATER TREATMENT

Several potential applications of AOPs at MWWTPs include: 1. pre-oxidation of primary effluent, 2. sludge conditioning, 3. post-treatment of secondary effluent, 4. treatment of reverse osmosis concentrate, and 5. disinfection of effluent. These five application points are shown in a generalized process flow diagram in Figure 16.1. Municipal wastewater is collected by pipes and goes through preliminary treatment including screening, degritting and primary clarification. Primary effluent goes to a secondary treatment including conventional activated sludge system, and then treated by filtration or reverse osmosis. Finally, after disinfection the effluent is discharged.

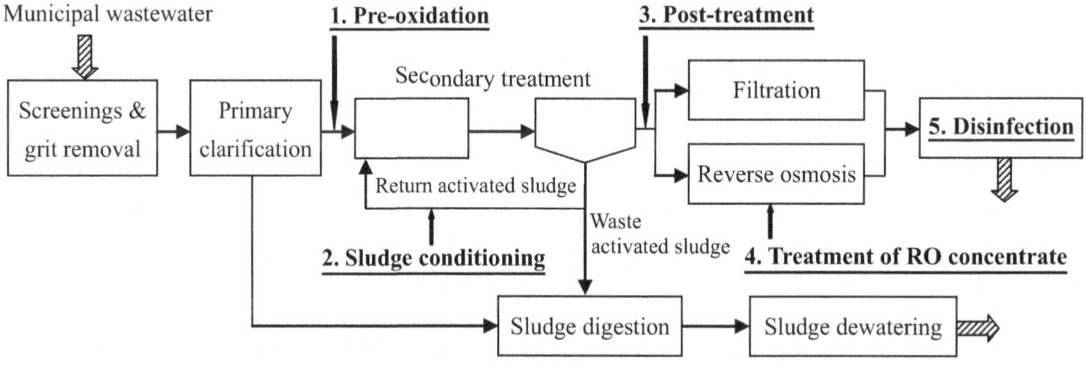

Figure 16.1 Potential AOPs applications for municipal wastewater treatment.

Preoxidation of primary effluent by AOPs can remove refractory or non-biodegradable organic compounds improving biological removal efficiency of the secondary treatment process. The combination of ozonation and ion exchange processes has been applied for removal of primary effluent from a MWWTP in Taipei, Taiwan (Chou *et al.* 2011). After 30 min pre-ozonation at initial pH 7 and 20 min pre-ozonation at initial pH 11 followed by ion exchange treatment, about 94% TOC removal could be achieved. Nielson and Smith (2005) have investigated the ability of ozone-enhanced electroflocculation to treat municipal effluent from the Gold Bar wastewater treatment plant, Edmonton, Alberta, Canada, which even was of interest as a potential option for the replacement of conventional secondary treatment. The results showed that total suspended solids (TSS) and total phosphorus (TP) were decreased from 82 and 6 mg/L to 15–45 and 0.3–0.6 mg/L, respectively, which met the regulatory guidelines for these parameters; COD and BOD_5 were reduced from 359 and 158 mg/L to 130–200 and 50–80 mg/L, respectively, which were inadequate to meet the regulatory guidelines. Tahri *et al.* (2010) used gamma irradiation to reuse domestic wastewater from the urban wastewater treatment station of Tétouan (Morocco). The results indicated that gamma irradiation is effective for wastewater treatment, not only for the elimination of total bacterial flora, faecal coliforms, total coliforms, and reduction of BOD_5 and COD, which improve the quality of wastewater, but also in keeping the nutrients elements (N, P, and K) for reuse as a natural fertilizer.

Researches indicate that return activated sludge (RAS) treated by AOPs can reduce sludge production and improve settling and dewatering. Electrochemical oxidation technique has been used to enhance dewaterability and to stabilize wastewater sludge from MWWTPs (Bureau *et al.* 2012). The best performance was obtained when the electrolytic cell was operated for 60 min at 8.0 A, with energy consumption of 856 kWh/tDM, in the presence of 177 kg NaCl/tDM and 23.3 kg H_2SO_4/tDM. The electrochemical process was efficient in increasing the dryness of sludge as well as removing indicator pathogens and unpleasant odors, which simultaneously preserved the fertilizing properties of the sludge by maintaining inorganic nutrients and organic matter in dewatered sludge.

Advanced oxidation processes (e.g., ozonation, photocatalysis, and photo-Fenton) have been proposed as tertiary treatments for MWWTP effluents due to their versatility and ability to detoxify effluent streams containing emerging contaminants (e.g., pharmaceuticals, personal-care products, steroid sex hormones, illicit drugs, etc.). Giannakis *et al.* (2015) have adopted various oxidation processes to treat wastewater from the output of three different secondary treatment facilities (Activated Sludge, Moving Bed Bioreactor and Coagulation-Flocculation) present in the MWWTP of Vidy, Lausanne (Switzerland). Six organic micropollutants (Carbamazepine, Clarithromycin, Diclofenac, Metoprolol, Benzotriazole, Mecoprop) in agreement with the new environmental legislation requirements in Switzerland were monitored throughout the treatment at laboratory scale. After secondary treatment, the efficiency of the AOPs increased in the following order: Moving Bed Bioreactor > Activated Sludge > Coagulation-Flocculation. The order of organics degradation rates was: UV/H_2O_2 > UV-C radiation and photo-Fenton > solar radiation > Fenton. Hollender *et al.* (2009) have assessed ozonation as a post-treatment step for the removal of organic micropollutants (220 compounds, ng/L–µg/L concentration range) from a secondary effluent at the scale of the MWWTP Wüeri in Regensdorf, Switzerland. The plant consists of a grit chamber, a primary clarifier, conventional activated sludge treatment with a denitrifying and nitrifying zone, a full-scale ozonation reactor, and sand filtration (Figure 16.2). During post-ozonation, most micropollutants were eliminated to a large extent, indicating that ozonation is a promising technology to significantly reduce the micropollutant load of full-scale wastewater treatment plants. The removal efficiency by ozonation can be predicted with acceptable accuracy by combination of reactor hydraulics, reaction kinetics, and measurement of ozone and hydroxyl radical exposures in laboratory-scale experiments, which provides the basis for the design and operation of new treatment steps for wastewater ozonation, and allows a direct upscaling to full scale without costly and labor intensive piloting.

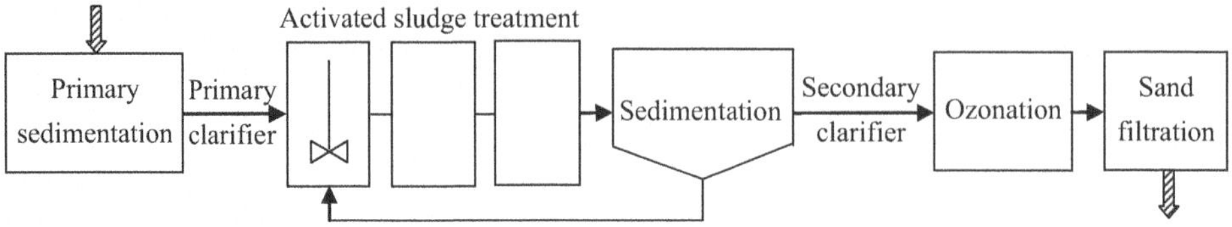

Figure 16.2 Scheme of the wastewater treatment plant Wüeri in Regensdorf, Switzerland. Adapted from Hollender *et al.* 2009.

Several AOPs, including UV/H_2O_2, UV/TiO_2, ozone and electrochemical oxidation, have been demonstrated to be effective for treating the reverse osmosis concentrate (ROC) produced from the reclamation of municipal secondary effluent. Roddick's group (Liu *et al.* 2012; Umar *et al.* 2014; Umar *et al.* 2015) has used UV-C/H_2O_2 to treat reverse osmosis concentrate collected from a wastewater reclamation facility at a local MWWTP in Victoria, Australia. The results showed that >54% chemical oxygen demand, 27–38% dissolved organic carbon, >90% A_{254} and almost complete color were reduced, and the biodegradability of the ROC was at least doubled after UV-C/H_2O_2 treatment. Thus, the UV-C/H_2O_2 process was efficient for the degradation of high levels of effluent organic pollutants as well as the increase of their biodegradability over a wide range of salinity and initial concentrations of dissolved organic carbon, demonstrating the wide applicability of this technique for treating municipal ROC or potentially ROCs from brackish sources. The combination of UV photolysis with electrochemical process led to non-selective and enhanced degradation of dissolved organic matter in RO concentrate produced from municipal wastewater (Hurwitz *et al.* 2014). After 5 h treatment, more than 80% of dissolved organic carbon was degraded, showing the effective treatment of a wide array of organics found in effluent at a rate that exceeded treatment by UV and anodic oxidation individually or additively.

AOPs mainly including ultraviolet radiation and ozonation are widely implemented at full-scale in municipal wastewater treatment for the disinfection of secondary effluent. TiO_2 photocatalysis applied to wastewater effluent disinfection is studied for its application. The disinfection by AOPs can enhance the bacteria destruction, eliminate antibiotic resistance in effluent, and thus ensure public health security. Municipal sewage, coming from the Zaohua sewage plant of Shenyang in China, is treated by an A^2/O + Advanced treatment process (Ya *et al.* 2011). The A^2/O process is a nitrogen and phosphorus removal process consisting of anaerobic, anoxic and aerobic stages. Current treatment volume of this sewage plant is 1×10^4 m^3/d with a future capacity of 6×10^4 m^3/d, and the whole investment of this process is about \$2.4 million dollars. The influent characteristics are 420 mg/L COD, 180 mg/L BOD_5, 200 mg/L SS, 30 mg/L NH_3-N, 40 mg/L TN and 4 mg/L TP, and the treated water is designed to be discharged to Liao River nearby. The treatment process flow diagram is shown in Figure 16.3. In this process, ultraviolet is used in disinfection room with more than 65% penetration rate and 15 kW running power. The average influent flow is 1.1×10^4 m^3/d, and the TSS value is lower than 20 mg/L. After treatment, the effluent qualities are: COD $\leq$ 50 mg/L, BOD_5 and SS $\leq$ 10 mg/L, NH_3-N $\leq$ 5 mg/L, TN $\leq$ 15 mg/L, and TP $\leq$ 0.5 mg/L, which can meet the requirement of first-A grand of discharge standard of pollutants for municipal wastewater treatment plant in China (GB189118–2002).

16.3 INDUSTRIAL WASTEWATER TREATMENT

Advanced oxidation processes are efficient for wastewater treatment; however, common drawbacks are the high electrical energy demand and the consumption of chemical reagents. The advantages and disadvantages of various oxidation processes are summarized in Table 16.1. AOPs can be used as pre-treatment to improve biological treatability of wastewater, and/or post-treatment to directly remove the contaminants that not completely degraded by the biological treatment as shown in Figure 16.4 (Cesaro *et al.* 2013).

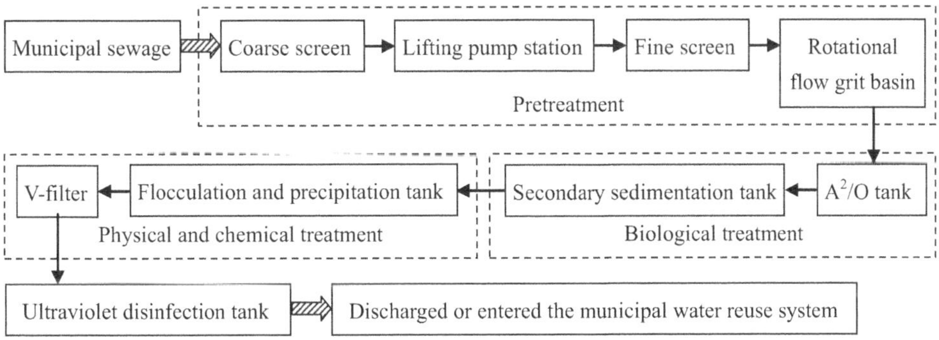

Figure 16.3 Process flow diagram of Zaohua municipal sewage plant in Shenyang, China. Adapted from Ya *et al.* 2011.

16.3.1 Textile wastewater

The release of wastewaters containing textile dyes and industrial dyestuffs in the environment is a considerable source of non-aesthetic pollution, eutrophication, and long-term bioaccumulation. These dyes can present toxic effects to microorganisms, aquatic life and human beings, and reduce light penetration in contaminated waters. Most pollution in textile wastewater comes from dyeing and finishing processes including bleaching, dyeing, printing and stiffening. Major concerns particularly regarding the textile wastewater are suspended solids, highly recalcitrant chemical oxygen demand, dyes giving intense color and other soluble substances, and typical characteristics are summarized as follows: 150–12000 mg/L COD, 2900–3100 mg/L total suspended solids, 70–80 mg/L total Kjeldahl nitrogen, and 80–6000 mg/L BOD.

Various AOPs have been widely shown to have the greatest promise for treating textile wastewater both in lab-scale and pilot-plant scale, as summarized in Table 16.2. From all of these studies, it can be concluded that effects of operating parameters, decolorization and degradation efficiencies, COD or TOC removal, toxicity reduction, intermediates analysis, reaction kinetics and mechanisms have been investigated. For example, Zhao *et al.* (2005) have studied the degradation of Diacryl Red X-GRL by photolysis and investigated the effects of temperature, pH, the incident photon flux of UV lamp, flow rate of purging gas (O_2 or N_2 or the mixture of both), initial dye concentration, concentration of *t*-BuOH (scavenger of superoxide radical anion $O_2^{\bullet -}$ and singlet oxygen $^1O_2$) and the concentration of dissolved oxygen, and proposed the photodegradation kinetics and mechanism. Song's group (Song *et al.* 2007; Song *et al.* 2008) have optimized the operational conditions for photocatalytic degradation of azo dyes (C.I. Reactive Black 5 and C.I. Direct Red 23) under UV irradiation using $SrTiO_3/CeO_2$ composite as the catalyst, and deduced the tentative degradation pathways and possible mechanisms. These experimental studies with model wastewaters provide useful information for treating more complex real effluents. Punzi *et al.* (2015) have applied a novel setup composed of an anaerobic biofilm reactor followed by ozonation to treat artificial and real textile effluents containing azo dyes, and the results suggested that the use of ozonation as short post-treatment after a biological process can be beneficial for the degradation of recalcitrant compounds and the removal of toxicity of textile wastewater.

Table 16.1 Advantages and limitations of AOPs for wastewater treatment.

AOP	Advantages	Limitations
Ionizing radiation	Relatively pH insensitive Not necessary to add chemical compounds Simultaneously kill bacteria and inactive viruses and remove pollutants	No extensive use
US	Versatile technology Suitable for small volumes Interesting upgrade applications	Energy consuming technology Sonotrode erosion issues
O_3	Strong oxidative power Effective for a wide spectrum of pollutants Existing full-scale applications	Energy consuming High operating costs Risks associated to ozone generation
UV/O_3 UV/H_2O_2 $UV/H_2O_2/O_3$	More effective than single processes UV promote •OH formation	More energy intensive than single processes Turbidity can interfere with UV radiation
Photocatalysis	Can be performed at higher wavelengths than other UV-based processes Carried out at ambient conditions Nontoxic and inexpensive inputs	Developing and expensive technology Need for pretreatment
Fenton-based reactions	Not as energy intensive as other AOPs Nontoxic reagents and easy reactor design	Developing technology Need for acidic conditions Sludge generation
Electrochemical oxidation	No additional chemicals required End products are non-hazardous High efficiency	Excessive energy requirements
Electro-Fenton	On-site production of H_2O_2 Higher removal efficiency than single processes	Requirement of acidic pH
Electrocoagulation	More effective and rapid than coagulation Not necessary for pH control Small amount of chemicals required and sludge produced Low operating costs	Anode passivation and sludge deposition on the electrodes High concentrations of iron and aluminium ions in the effluent

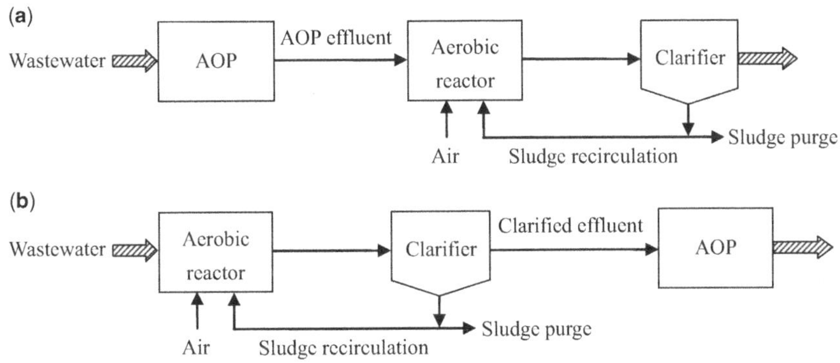

Figure 16.4 AOP as pre-treatment (a) and post-treatment (b) for wastewater treatment. Adapted from Cesaro *et al.* 2013.

At a printing and dyeing wastewater treatment plant in Changzhou, China, effluent underwent five major processes as seen in Figure 16.5: adjusting tank, push flow aeration tank, secondary sedimentation tank, modified magnetic powder/ClO_2 oxidation process, and tertiary sedimentation tank (Zhang *et al.* 2011). When the operational conditions of modified magnetic powder/ClO_2 oxidation process were pH 4, modified magnetic powder 10 g/L, and ClO_2 dosage 88.2 mg/L, about 95% COD removal and 60% decolorization were achieved, indicating that the process is reasonable and economic.

Guo *et al.* (2011) used a combination of biological process followed by ozonation for textile wastewater treatment in a 500 m^3/d pilot plant in Guangdong, China. The COD concentration decreased from 600–1200 mg/L to about 40 mg/L, and color decreased from 200–800 (multiple) to below 28 (determined with standard dilution multiple method by Wei, 2002), which means that about 95% COD reduction and 94% color removal were obtained in the pilot plant.

16.3.2 Pharmaceutical wastewater

A large amount of pharmaceutical compounds including antipyretics, analgesics, blood lipid regulators, antibiotics, antidepressants, chemotherapy agents and contraceptive drugs are continuously introduced in the aquatic environment. Environmental emerging contaminants, such as personal care products, illicit drugs, flame retardants and perfluorinated compounds, are causing particular concerns because of their endocrine-disrupting properties. Available studies indicate that these contaminants occurring in the aquatic environment as well as in finished drinking water may exert toxic effects on algae and invertebrates, favor the development of multiresistant strains of microorganisms, and cause the genotoxicity and endocrine system disruption. More sophisticated and laborious analytical tools for their accurate determination are required, because they are present in trace amounts. More efficient techniques need to be developed to increase the biodegradability and to detoxify effluent streams containing pharmaceuticals and emerging contaminants.

Recent work on AOPs has shown an efficient oxidation of pharmaceuticals from wastewater (Table 16.3), even allowing the complete mineralization of their oxidation intermediates. Effects of operating parameters, presence of radical promoters and scavengers, influence of chemical composition of waters, reaction kinetics, degradation mechanisms, as well as biodegradability improvement have been studied. Sánchez-Polo *et al.* (2009) have investigated the decomposition of Metronidazole in waste and drinking

Table 16.2 Selected results reported for the treatment of textile or dyeing wastewater by AOPs.

Advanced Oxidation Processes	Effluent	Wastewater Characteristics	Parameters Assessed	Concluding Remarks	References
Photolysis	Synthetic solution containing Diacryl Red X-GRL	$(1.33–2.80) \times 10^5$ M Diacryl Red X-GRL; pH 3.15–9.24; 3.5 L solution	Effects of temperature, pH, radiation flow rate of UV lamp, carrier gas flow rate, initial dye concentration, concentration of *t*-BuOH and the concentration of dissolved oxygen; kinetics; mechanism	About 70% dye degraded after 300 min	Zhao *et al.* 2005
γ-irradiation	Synthetic solution containing methyl orange	0.5–2.0 mM methyl orange; neutral pH; 100 mL solution	Effects of γ-ray dose rate, initial dye concentration, and saturated with gases; kinetics; intermediates analysis; degradation mechanism	At a dose rate of 92 Gy/min, more than 90% dye degradation after 120 min	Chen *et al.* 2008
O_3 UV/O_3 UV/H_2O_2 $UV/H_2O_2/O_3$ Fe^{2+}/H_2O_2	Effluent from a polyester and acetate fiber dyeing process in Turkey	930 mg/L COD; 375 mg/L BOD_5; 95 mg/L SS; 23 mg/L TN; 1.13 mg/L TP; pH 9.2	Compared to conventional chemical treatment methods: $Al_2(SO_4)_3 \cdot 18H_2O$, $FeCl_3$ and $FeSO_4$	Superior performance of $UV/H_2O_2/O_3$ system: 99% COD removal and 96% color removal from a fibre dyeing effluent	Azbar *et al.* 2004
Ozonation	Effluent from dyeing processes of woolen textile finishing industries	1700 mg/L COD; toxicity GL-value = 24; conductivity 41.4 mS/cm; pH 8.30	Effect of time intervals; Oxidative treatment carried out before and after biological treatment	Around 98–99% decolorization efficiency in 40 min	Baban *et al.* 2003

Anaerobic-ozonation	Artificial and real textile effluents containing azo dyes	100–1000 mg/L Remazol Red concentration and 60–600 mg/L COD for artificial textile effluent; 1190 mg/L COD and pH 7.5–8 for real textile effluent	Use of ozonation as short post-treatment after a biological process for artificial and real textile effluents; toxicity tests	More than 99% of decolorization, 85–90% of COD reduction and toxicity reduction for the artificial textile wastewater; 70% COD reduction and efficient toxicity removal for the real textile effluent	Punzi *et al.* 2015
Photocatalysis	Effluent from a textile industry in Brazil	569 mg/L TOC; 7.3 mg/L total phenols; pH 11.5	Comparison of processes using free TiO_2, free ZnO and ZnO-silica; evaluation of decolorization, total phenols reduction, TOC reduction, and toxicity reduction	Efficient in the decolorization process; about 40% total phenols reduction; 48% TOC reduction; about 56% toxicity reduction	Peralta-Zamora *et al.* 1998
Fenton Photo-Fenton	Wastewater from a textile manufacturer in Spanish	605 ± 9 mg/L TOC; 1669 ± 4 mg O_2/L COD; 40 ± 8 mg Pt/L color	Effects of light intensity, temperature, pH, Fe(II) and H_2O_2 initial concentrations, and different light sources	Highly effective for textile wastewater treatment	Pérez *et al.* 2002a
Electrochemical peroxidation	Textile wastewater from a plant in Brazil	1136 mg/L COD; 80 mg/L TSS; conductivity 3.7 mS/cm; pH 8.1; 500 mL solution	Effects of concentration of pollutants, applied current, temperature, and electrode material	55% COD removal and 95% removal of nonylphenol polyethoxylate in 30 min	Martins *et al.* 2006
Electro-Fenton	Textile wastewater from dyeing plant in Taiwan	2942 mg/L COD; 238 mg/L Cl^-; ADMI color = 1094	Effects of oxygen contact mode, oxygen sparging rate, applied current density, Fe^{2+} concentration, solution temperature and pH	71% color removal in 150 min	Wang *et al.* 2008a

water using gamma irradiation, and the results indicated that the decomposition yield was higher under acidic conditions with the appropriate concentrations of radical promoter (H_2O_2). The presence of *t*-BuOH ($^{\bullet}OH$ scavenger) and thiourea ($^{\bullet}OH$, $H^{\bullet}$ and e_{aq}^{-} scavenger) decreased the degradation of Metronidazole, suggesting that the mineralization of this pollutant took place *via* two pathways: oxidation by $^{\bullet}OH$ radicals and reduction by $H^{\bullet}$ and e_{aq}^{-}. Rosal *et al.* (2010) have done a systematic survey of over seventy individual pollutants including mainly pharmaceuticals, personal care products and metabolites from a sewage treatment plant located in Alcalá de Henares (Madrid), and wastewater samples were treated through biological treatment followed by ozonation. The results showed that the ozonation treatment yielded high removal efficiencies of most individual pollutants detected in treated wastewater. The kinetic analysis allowed the determination of second order kinetic constants for the ozonation of bezafibrate, cotinine, diuron, ketoprofen and metronidazole, and the hydroxyl radical reaction played a significant role in the oxidative transformation of these compounds. It should be noted that improvement of biodegradability is perhaps more important than complete mineralization in wastewater treatment, especially for the pharmaceuticals that are resistant to biodegradation.

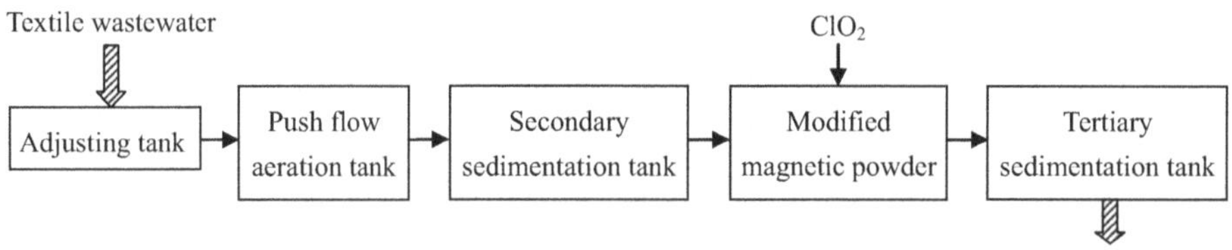

Figure 16.5 Process flow diagram of textile wastewater treatment plant in Changzhou, China. Adapted from Zhang *et al.* 2011.

Pharmaceutical wastewater containing heterocyclic compounds from a company in China was treated by cooling crystallization to deposit salt, followed by Fenton oxidation and activated sludge treatment (Huang *et al.* 2001). The COD concentration decreased from 906.5 mg/L to 93.4 mg/L, BOD decreased from 550.1 mg/L to 28.6 mg/L, and color decreased from 150 times to 40 times, with 90%, 95% and 73% of COD, BOD and color removal rates, respectively. Three combined processes, including the multi-level biological treatment system and two-stage ozonation and biological activated carbon adsorption, were conducted at pilot scale to treat effluent water from a pharmaceutical plant in China, and were proven highly effective for COD, NH_3-N, TN and TP removal (Geng *et al.* 2014).

16.3.3 Pesticide wastewater

Pesticides, divided into several groups depending on their usage including herbicides, insecticides, fungicides, rodenticides, nematicides and microbiocides, have been extensively used in agricultural activities and non-agricultural sectors. These compounds often migrate from the application site into the environment resulting in unacceptable levels of contamination of surface waters and groundwater aquifers. The wide variety of these chemical compounds makes their disposal difficult due to their persistence, bio-refractory nature, and high solubility in aquatic environment. The potential adverse health effects are carcinogenesis, neurotoxicity, and the possible chronic effects on reproduction and cell development, particularly in the early stages of life.

AOPs can be an alternative for treating pesticide-contaminated water streams; such processes have been proven to be very efficient in destroying a large variety of pesticides (Table 16.4). Many researches

Table 16.3 Selected results reported for the treatment of pharmaceutical wastewater by AOPs.

Advanced Oxidation Processes	Effluent	Wastewater Characteristics	Parameters Assessed	Concluding Remarks	References
Photolysis	Pharmaceutical effluents containing ibuprofen, naproxen, ketoprofen, carbamazepine and diclofenac	0.5–1 μg/L pharmaceuticals; pH 2–3	The use of biological treatment, cavitation/hydrogen peroxide process and UV treatment; optimization of reaction conditions	Higher than 98% removal for carbamazepine and diclofenac; higher than 90% removal for clofibric acid	Zupanc *et al.* 2013
Photolysis Photocatalysis	Synthetic solution containing tetracycline, lincomycin and ranitidine	10–50 mg/L drugs; pH 5.7–6.0	Photochemical degradation and photocatalytic degradation of the drugs; mechanistic aspects; kinetics aspects	Almost complete mineralization for tetracycline and 60% TOC reduction for lincomycin and ranitidine by UV/TiO_2	Addamo *et al.* 2005
γ-irradiation	Synthetic solution containing nitroimidazoles (Metronidazole, Dimetridazole, Tinidazole) and water samples from Aguas y Servicios de la Costa Tropical company (Motril, Spain)	140 μM nitroimidazole; pH 6; 5 mL solution	Effects of absorbed dose, contaminant concentration and pH; decomposition kinetics; effect of radical promoters and scavengers; influence of chemical composition of waters	62–68% removal after 700 Gy of treatment for nitroimidazoles; around 70% TOC reduction for Metronidazole	Sánchez-Polo *et al.* 2009
γ-irradiation/H_2O_2 γ-irradiation/Fe^{2+}	Synthetic solution containing sulfamethazine	20 mg/L sulfamethazine concentration; pH 6.0–7.5; 20 mL solution	Effects of H_2O_2 and Fe^{2+} concentration; intermediates analysis; degradation kinetics and mechanism	Highly effective for sulfamethazine removal, and almost complete removal	Liu and Wang, 2013 Liu *et al.* 2014
Ozonation	Emerging contaminants (25 compounds detected in μg/L range)	269 mg/L COD; 42 mg/L BOD_5; 67.8 mg/L TSS; 4.8 mg/L TP; pH 7.54	Biological treatment followed ozonation; kinetic analysis; oxidation mechanism	High removal efficiencies for most individual pollutants	Rosal *et al.* 2010

(Continued)

Table 16.3 Selected results reported for the treatment of pharmaceutical wastewater by AOPs (*Continued*).

Advanced Oxidation Processes	Effluent	Wastewater Characteristics	Parameters Assessed	Concluding Remarks	References
O_3 O_3/H_2O_2	Penicillin formulation effluent from a pharmaceutical company in Istanbul, Turkey	830 mg/L COD; 450 mg/L DOC; 150 mg/L SS; pH 6.9	Effects of pH and initial H_2O_2 concentrations; comparison of optimized O_3/OH^- and O_3/H_2O_2 processes; biological treatment after ozonation	10–56% COD removal for ozonation; 30–83% for the O_3/H_2O_2 process	Alaton *et al.* 2004
UV/H_2O_2 $UV/Fe^{2+}/H_2O_2$ $UV/Fe^{3+}/H_2O_2$ Fe^{2+}/H_2O_2 Fe^{3+}/H_2O_2 O_3/OH^-	Real penicillin formulation effluent from a pharmaceutical company in Istanbul, Turkey	1395 mg/L average COD; 920 mg/L TOC; 145 mg/L TSS; pH 6.95	Various AOPs for the oxidative pre-treatment of effluent; changes in biodegradability; active oxidant and oxidant requirements; advanced oxidation of penicillin active substance	49–66% COD reduction and 42–52% TOC removal by alkaline ozonation and the photo-Fenton's reagents; slight improvements of the biodegradability	Arslan-Alaton and Dogruel, 2004
Fenton	Pharmaceutical wastewater	362000 mg O_2/L COD; 2900 mg O_2/L BOD; 45.95 g/L SS; 89.7 mS/cm conductivity; pH 5.32	Effects of temperature, ferrous ion and hydrogen peroxide concentrations	56.4% COD global reduction	San Sebastián Martínez *et al.* 2003
Anodic oxidation Electro-Fenton Photoelectro-Fenton	Synthetic solution containing drug paracetamol	157 mg/L paracetamol; 100 mg/L TOC; pH 3.0	Effects of current, pH and drug concentration; intermediates analysis; degradation mechanism	Total mineralization of paracetamol by $UVA/Fe^{2+}/Cu^{2+}$	Sirés *et al.* 2006

Table 16.4 Selected results reported for the treatment of pesticide wastewater by AOPs.

Advanced Oxidation Processes	Effluent	Wastewater Characteristics	Parameters Assessed	Concluding Remarks	References
γ-irradiation	Synthetic solution containing diazinon	0.329–3.286 μmol/ dm^3; pH 6.8; 40 mL solution	Effects of initial concentration of contaminant and irradiation doses; intermediates detection; degradation mechanism	90% destruction for diazinon; complete degradation at the lowest targeted concentration	Basfar *et al.* 2007
Sonolysis	Synthetic solution containing 2,4-dichloro-phenoxyacetic acid	0.2 mM 2,4-dichloro-phenoxyacetic acid; 600 mL solution	Effect of pH; purging various gases; intermediates generation; effect of •OH scavenger	Nearly 90% degradation in less than 100 min	Peller *et al.* 2001
UV photolysis Visible light photocatalysis Ozonation	Effluent containing herbicides, insecticides, antiseptics and polyaromatic hycrocarbons from the secondary clarifier of a 3000 m^3/h wastewater treatment plant placed in Alcalá de Henares, Madrid	28 mg/L COD; 20 mg/L TSS; 875 μS/cm conductivity; pH 7.8	Removal of pollutants by various processes; energy efficiencies; Evaluation of toxicity	Over 95% removal by ozone treatment; an average of 63% removal with 254 nm UV irradiation; about 70% overall removal by visible light Ce–TiO_2 photocatalysis	Santiago-Morales *et al.* 2013
O_3 UV radiation Fenton UV/O_3 UV/H_2O_2 Photo-Fenton	Synthetic solution containing carbofuran	0.452 mM carbofuran; 350 mL solution	Comparison of various methods; effects of temperature, pH, H_2O_2 and Fe^{2+} concentrations; reaction kinetics	The removal rates: photo-Fenton > UV/ H_2O_2 > UV/ O_3 > O_3 > UV > Fenton	Benitez *et al.* 2002

(*Continued*)

Table 16.4 Selected results reported for the treatment of pesticide wastewater by AOPs (*Continued*).

Advanced Oxidation Processes	Effluent	Wastewater Characteristics	Parameters Assessed	Concluding Remarks	References
UV photolysis UV/H_2O_2 UV/Fe^{2+} Photo-Fenton	Synthetic solution containing pesticide (Vydine)	25 mg/L pesticide; pH 6.4–7; 1.5 L solution	Pesticide removal, TOC mineralization, COD removal and BOD by various processes; toxicity analysis	88% pesticide elimination and 65% toxicity reduction by photo-Fenton process after 40 min reaction	Al Momani *et al.* 2007
Photocatalysis	Synthetic solution containing pesticides chlorpyrifos, cypermethrin and chlorothalonil	400 mg/L pesticides (100 mg/L of chlorpyrifos, 50 mg/L of cypermethrin and 250 mg/L of chlorothalonil); 1130 mg/L COD; 274 mg/L TOC; biodegradability index ~0; pH 6	Effects of UV irradiation, TiO_2 concentration, and H_2O_2 addition; degradation kinetics	54% COD removal and 22% TOC removal by $UV/TiO_2/H_2O_2$ photocatalysis; biodegradability index 0.26	Affam and Chaudhuri, 2013
Solar photo-Fenton	Pesticides (alaclor, atrazine, chlorfenvinphos, diuron, isoproturon)	50 mg/L alaclor, chlorfenvinphos and isoproturon; 25 mg/L atrazine; 30 mg/L diuron	Contaminant disappearance and mineralization; toxicity reduction; enhancement of biodegradability	Complete mineralization of all of the pesticides tested; the enhancement of biodegradability (70% considered biodegradable) after 12–25 min	Lapertot *et al.* 2006
Electrochemical oxidation Electro-Fenton Photoelectro-Fenton	Synthetic solution containing herbicide 2-(2,4-dichlorophenoxy)-propionic acid	217 mg/L herbicide; 100 mg/L TOC; pH 3.0; 100 mL solution	Comparison of various methods; intermediates analysis; proposed reaction sequence	Complete mineralization of herbicide	Brillas *et al.* 2007

on pesticide removal from wastewater by AOPs have assessed the effects of operating parameters to obtain the optimum operating conditions. The toxicity of pesticide wastewaters can be reduced, and the biodegradability can be enhanced through the treatment. Benitez *et al.* (2002) have applied Fenton reaction, UV radiation, ozonation, UV/O_3, UV/H_2O_2 and photo-Fenton processes to degrade carbofuran, and the apparent pseudo-first-order rate constants were reported as (2.2–7.1) × 10^{-4} s^{-1}, 3.3 × 10^{-4} s^{-1}, (5.1–19.5) × 10^{-4} s^{-1}, 22.8 × 10^{-4} s^{-1}, (7.1–43.5) × 10^{-4} s^{-1}, (17.2–200) × 10^{-4} s^{-1}, respectively, which were further evaluated to compare the efficiencies of the investigated processes. Badawy *et al.* (2006) have used photo-Fenton process in combination with biological filtration to treat wastewater from the Pesticides and Chemical Company located at New Dammata in the north of Egypt. The effect of pH, irradiation time, and initial concentrations of both H_2O_2 and Fe^{2+} on the photo-Fenton process have been studied to obtain the optimal operating conditions, and TOC and COD removal were 82% and 85.6%, respectively. The biodegradability of industrial wastewater containing pesticides was improved; after combined with biological treatment, nearly complete mineralization could be achieved. Xia and Sun (2005) have applied ozonation to pretreat pesticide wastewater from a company in Hefei, China, using two ozone generators with a capacity of 800 g/h. After ozonation, COD concentration decreased from 38,341 mg/L to 18,787 mg/L with a COD removal of 51%. BOD concentration increased from 5,751 mg/L to 7,703 mg/L, while pH value changed from 12 to 8. The increase of BOD/COD ratio from 0.15 to 0.41 indicated the enhancement of biodegradability. Santiago-Morales *et al.* (2013) have assessed the removal of a set of non-polar pollutants in biologically treated wastewater containing UV filters, synthetic musks, herbicides, insecticides, antiseptics and polyaromatic hydrocarbons using ozonation, ultraviolet (UV 254 nm) radiation, visible light (Xe-arc lamp) radiation as well as visible light photocatalysis (using Ce-doped TiO_2). Wastewater was sampled from the secondary clarifier of a 3000 m^3/h wastewater treatment plant placed in Alcalá de Henares, Madrid. In terms of pollutant removal and power usage efficiency, ozonation was by far the most efficient process. The toxicity of treated wastewater decreased upon water ozonation, whereas UV irradiation and photocatalysis processes led to a toxicity increase because of the accumulation of toxic transformation products. Thus, special attention should be paid to monitoring the evolution of toxic intermediates as well as toxicity during wastewater treatment.

16.3.4 Paper mill wastewater

The characteristics of effluent primarily from the pulping, bleaching and paper-making production stages in pulp and paper industry vary over a wide range, and depend upon the type of pulp process and the unique characteristics of each factory. For example, the organic material content of the effluent from mechanical pulp process varies between 1000 and 5600 mg/L COD, while it increases to 2500–13000 mg/L COD in chemical pulp process. Paper mill wastewater mainly contains color, phenolic compounds, non-biodegradable organic materials, adsorbable organic halogens (AOX), etc., among which high organic material and suspended solid contents are considered major pollutants. The effluent discharge into the environment could cause coloration, thermal impact, slime growth, scum formation, as well as the increase in the amount of toxic substances in the water, severely affecting the terrestrial ecosystem.

The current state of the art concerning advanced oxidation treatment of the organic load of effluents from the paper industry has been discussed and reviewed in terms of the degree of reaction, effects of operating parameters, identification of compounds, reaction kinetics and cost estimation. Selected results reported for the application of various AOPs to effluents of the pulp and paper industry, such as ionizing radiation, ultrasonic, photocatalysis, ozonation, Fenton, photo-Fenton and electro-oxidation, are listed in Table 16.5. Generally, all AOPs have been reported to be efficient for the treatment of paper mill wastewater with over 40% COD removal and about an extra 20% COD removal if AOPs are combined with biological

Table 16.5 Selected results reported for the treatment of paper mill wastewater by AOPs.

Advanced Oxidation Processes	Effluent	Wastewater Characteristics	Parameters Assessed	Concluding Remarks	References
Ionizing radiation Coagulation Coagulation/ionizing radiation	Paper mill effluent from a factory in Wuhan, China	722 mg/L COD; 427 mg/L BOD_5; 173.6 mg/L DOC; pH 6.9	Comparison of various processes; effects of irradiation dose, pH and coagulant dosage	Efficient for the removal of organic compounds in paper mill effluent; the increase of the biodegradability	Wan *et al.* 2011
Ultrasonic irradiation Ultrasonic/Fe^{3+}/H_2O_2 Photo-Fenton (UV/Fe^{3+}/H_2O_2) Electro-oxidation Chemical precipitation	Bleaching effluents from local pulp and paper mills in Eastern Finland	1170–1510 mg/L COD; 142–221 mg/L BOD_5; 234–543 mg/L TOC; 354–563 mg/L suspended solids; 660–1230 Pt-Co unit color; pH 6.4–7.3	Comparison of various treatments; effects of operating parameters such as pH and oxidant's dose	The AOP could convert toxic materials in bleaching effluents into relatively less harmful compounds, enhancing the treatability of the wastewater	Eskelinen *et al.* 2010
Ozonation	Effluents from the pulp and paper mill in Kanchanaburi, Thailand	2000 ± 100 mg/L COD; 550 ± 50 mg/L BOD; 650 ± 50 mg/L TOC; 1100 ± 100 ADMI value; pH 7.5 ± 0.5	Identification of compounds in effluents; the ozone mass transfer efficiency; effects of pH on decolorization and TOC removal; effects of ozonation process on biodegradability	Over 90% decolorization efficiency; the BOD/COD ratio increased from 0.10 to 0.32	Kreetachat *et al.* 2007
Ozone/UVA or visible light/Fe(II)/Fe(III)	Industrial bleaching Kraft mill effluent supplied by a Spanish paper manufacturer	441 ± 8 mg/L TOC; 1384 ± 24 mg O_2/L COD; 197 ± 25 mg Pt/L color	Comparison of various processes; cost estimation; application in the synthetic sample	High efficiency (more than 90%) removal of TOC and COD	Pérez *et al.* 2002c

Photocatalysis	Effluent from a pulp and paper industry in Brazil	57.1–445.2 mg/L TOC; 10.2–29.7 mg/L total phenols; pH 2.5–9.4	Comparison of processes using free TiO_2, free ZnO and ZnO-silica; evaluation of decolorization, total phenols reduction, TOC reduction, and toxicity reduction	Efficient in the total phenols reduction; 51% TOC reduction; about 50% toxicity reduction	Peralta-Zamora *et al.* 1998
Solar photocatalysis Photo-Fenton	Synthetic solution containing *p*-toluenesulfonic acid, eugenol and guaiacol; real cardboard effluents from a factory in Alcoy, Spain	0.001 M eugenol and guaiacol; 0.005 M *p*-toluenesulfonic acid; 3500–11000 mg/L COD for real effluents	Comparison of various processes; pseudo-first-order kinetics; intermediates detection	About 55% COD removal; 30–50% BOD decrease	Amat *et al.* 2005
Fenton Photo-Fenton	Bleaching Kraft mill effluent and cellulose bleaching effluent supplied by paper manufacturers in Spain	1250–1384 mg O_2/L COD; 441–537 mg/L TOC; 197–649 mg Pt/L color; pH 1.74	Effects of initial concentrations of Fenton reagents, temperature, pH, light intensity and O_2 presence; identification of compounds	Highly effective for the treatment of the effluents; 90% TOC removal	Pérez *et al.* 2002d Torrades *et al.* 2003
Electrocoagulation Electro-oxidation	Kraft bagasse pulp bleach effluent from a large-scale pulping industry in India	1230 mg/L COD; 175 mg/L BOD; 1952 mg/L total solids; 56 mg/L suspended solids; 1228 Pt-Co unit color; pH 9.1	Comparison of various electrochemical processes; optimization of operating variables; cost estimation	68% COD removal and 52% TOC removal by electrocoagulation–electro-oxidation; 80% COD removal by electrocoagulation and biological system	Antony and Natesan, 2012

stages. It comes out that about 40%, 50% and 70% COD removal is achieved by ozonation, photocatalysis, and Fenton processes, respectively. In particular, ozonation has been the most frequently studied and successfully applied process at an industrial scale for effluent treatment or reuse within pulp and paper mills. Although Fenton processes (particularly photo-Fenton) have shown the best oxidative results at laboratory scale, these processes still need further development for large scale applications (Hermosilla *et al.* 2015).

A schematic representation of the Fenton oxidation treatment of industrial wastewaters (including pulp and paper, pharmaceutical, textile, etc.) is shown in Figure 16.6 (Bautista *et al.* 2008). Wastewater goes through a stirred batch reactor where Fenton reagents (i.e., Fe^{2+} and H_2O_2) are added and the solution pH is controlled commonly within 3–3.5 range. The discharge from the Fenton oxidation tank passes to a neutralization tank with the addition of alkaline agent for pH control, and then after flocculant addition, the $Fe(OH)_3$ and other solids are separated by settling. A large-scale papermaking enterprise in China has applied the combination of hydrolytic acidification, aerobic biological treatment, and Fenton process to treat the regenerated papermaking and wood pulping middle-stage wastewater (Shi *et al.* 2012). The results showed that when the influent flux was (0.77–2.91) × 10^4 m^3/d, the effluent COD decreased from 2150–4430 mg/L to 67–98 mg/L, and SS decreased from 1316–2414 mg/L to 21–29 mg/L, respectively, after bio-chemical treatment, Fenton oxidation, and flocculation sedimentation process. The treatment cost was about \$2.4/m^3, indicating good economic and environmental benefits.

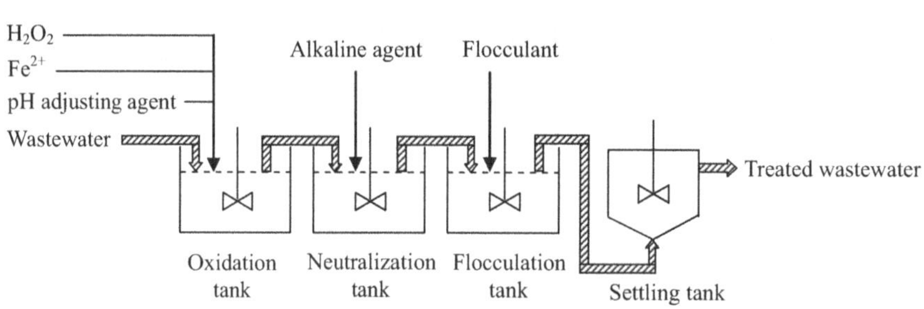

Figure 16.6 Typical scheme for Fenton treatment of industrial wastewaters. Adapted from Bautista *et al.* 2008.

16.3.5 Petrochemical wastewater

The wastewater generated from petrochemical industry contains large quantities of organic and inorganic pollutants, including petroleum hydrocarbons, phenols, aniline, nitrobenzene, naphthalenic acids, organochlorines, surface active substances, sulfides, metal derivatives, and other refractory and toxic compounds. These substances are characterized by high COD, complex chemical compositions, and low biodegradability, which pose a health risk both to humans and to other living organisms, including the aquatic species.

AOPs, such as ultraviolet radiation, UV/H_2O_2, photocatalysis, sonication, ozonation, and Fenton-based processes, have shown significant advantages in petrochemical wastewater treatment (Table 16.6). The treatment train shown in Figure 16.7 is employed at a major wastewater reclamation petrochemical plant in USA (Wong, 2000). In this system, ultraviolet is applied for the disinfection of the reverse osmosis feed water. Effluents (110 m^3/d) from cooling tower blowdown and organic wastewater treatment system are blended, and treated by chemical oxidation to precipitate heavy metals that are then removed by a dual

Table 16.6 Selected results reported for the treatment of petrochemical wastewater by AOPs.

Advanced Oxidation Processes	Effluent	Wastewater Characteristics	Parameters Assessed	Concluding Remarks	References
Ultraviolet (UV)	Wastewater from a major petrochemical plant in USA	51–74 mg/L COD; 10–24 mg/L TSS; pH 7.32–8.45	Ultraviolet disinfection; the treatment process train including chemical oxidation, filtration, adsorption, ultrafiltration, disinfection, reverse osmosis and degasification; economic analysis	A total overall water recovery of 73.6% was achieved consistently in the pilot plant by the treatment process train	Wong, 2000
Sonication Sonication/TiO_2/Fe^{2+}	Petrochemical industry wastewater treatment plant in Izmir, Turkey	1465 mg/L COD; 37.4 mg/L BOD_5; 20 mg/L TN; 11 mg/L TP; pH 7.2	Effects of ambient conditions, sonication time and temperature, TiO_2 and Fe^{2+} concentrations;	Efficient removal of polyaromatic hydrocarbons (PAHs) and the acute toxicity	Sponza and Oztekin, 2010
Ozonation	Secondary effluent from a petrochemical wastewater treatment plant in China	70–120 mg/L COD; 1–5 mg/L BOD_5; 16–30 mg/L DOC; 10–15 mg/L TN; 0.5–2.0 mg/L TP; pH 6–8	Biological aerated filter followed ozonation; evaluation of DOC, UV_{254} and genotoxicity removal; the changes of biodegradability and organics	Ozonation can decrease the organics molecular size, increase the biodegradability and obviously reduce the genotoxicity	Wu *et al.* 2015
Catalytic ozonation	Organic contaminants in petrochemical wastewater	200 mg/L aniline concentration; 491.4 mg/L COD; pH 7.2; 100 mL solution	Effects of pH, catalyst dose, temperature and •OH scavenger; reaction kinetics; degradation pathways; characterization and reusability of catalysts	77.5% aniline removal and 67.6% COD removal at 10 min	Chen *et al.* 2015

(Continued)

Table 16.6 Selected results reported for the treatment of petrochemical wastewater by AOPs (*Continued*).

Advanced Oxidation Processes	Effluent	Wastewater Characteristics	Parameters Assessed	Concluding Remarks	References
Photo-Fenton (Fe^{3+}/H_2O_2/Solar-UV)	Petrochemical refinery wastewater in Izmir, Turkey	1200–2800 mg/L COD; 800–1400 mg/L BOD; 820–1385 mg/L TOC; pH 8.2–8.75	Effects of H_2O_2 dosage, Fe(III) concentrations and flow rate; evaluation of color, TOC, BOD and COD removal	49% TOC removal; 53% BOD removal; 58% COD removal	Parilti, 2010
H_2O_2 UV UV/H_2O_2 Photocatalysis Ozonation Fenton Photo-Fenton	Sourwater from a petroleum refinery in Brazil	850–1020 mg/L COD; 570 mg/L BOD_5; 300–440 mg/L DOC; pH 8.0–8.2; 553 mg/L COD; 147 mg/L BOD; 160.5 mg/L DOC; pH 9.5	Comparison of various processes; the evaluation of process conditions; DOC removal kinetics	Over 55% and 83% DOC removal by Fenton and photo-Fenton, respectively; nearly 95% DOC reduction by UV/H_2O_2; 8–35% DOC removal by others	Coelho *et al.* 2006 Guimarães *et al.* 2012
Electro-Fenton	Wastewater from a petrochemical plant in Taiwan	226–435 mg/L COD; 103–128 mg/L Cl^-; pH 7.32–8.20	Effects of H_2O_2 dosage and reaction time; application of batch-scale and pilot-scale processes	More than 80% COD removal	Ting *et al.* 2007
Electro-oxidation Electrocoagulation Electro-Fenton	Simulated oil tanker truck washing wastewater	456 mg/L COD; pH 6.5	Effects of pH, applied current density, initial oil concentration and reaction time	Exceeding 80%, 90% and 98% COD removal by electrocoagulation, electro-Fenton and electro-oxidation processes, respectively	Dermentzis *et al.* 2014

media filter (0.91 m diameter). Two granular activated carbon filters (0.91 m diameter) remove residual organics, while the cartridge filter (10 μm) and ultrafiltration remove suspended and colloidal particles larger than 10 and 0.01 μm, respectively. A three-stage (3:2:1) reverse osmosis system is the major part of the treatment system, which removes dissolved inorganic and organic compounds, silica and residual colloidal materials. After removing carbon dioxide by degasification, the reclaimed water is suitable for reuse as deionization feed water. Long-term pilot testing is evaluated, and the results verify both the technical and economic feasibility of the project.

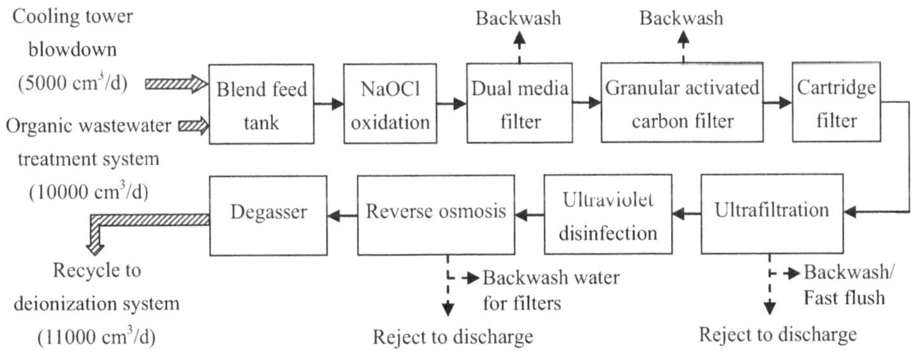

Figure 16.7 Block flow diagram of wastewater recovery treatment system in a major petrochemical plant in USA. Adapted from Wong, 2000.

The combination of high efficiency nitrifying tank (HENT) and catalytic ozone oxidation has been used for the treatment of acrylonitrile wastewater from a petrochemical company in China (Lu *et al.* 2014). The influent NH_3-N decreased from 88–286 mg/L to 0.53 mg/L with the removal efficiency of 99.7% in HENT system; COD was reduced from 259 mg/L to 57 mg/L with the removal efficiency of 75.6% by coupling the catalytic ozonation and biological aerated filtration. Furthermore, total cyanide, suspended solids, sulfide and total phosphorus were removed to a certain extent in the pilot-scale testing, e.g., 49.3%, 50.2%, 33.5% and 11.1%, respectively.

16.3.6 Landfill leachate

Landfill leachate is a type of extremely polluted wastewater produced from municipal sanitary landfill, which contains large amounts of refractory organics, xenobiotic compounds, ammonia-nitrogen, inorganic salts, and heavy metals. The quantity and quality of landfill leachate vary with a number of factors including waste composition, climate, seasonal precipitation, hydrology and geology sites, landfill design and operation, and mainly, the age of the landfill. Generally, leachate from young landfills (<2 years old) contains large amount of relatively low-molecular organics (such as volatile organic acids), resulting in high COD (3000–60000 mg/L), TOC, BOD_5 and BOD_5/COD ratio (>0.6). In contrast, mature leachate from old landfills (>10 years old) has a high fraction of relatively high-molecular organics (such as humic and fulvic substances), low COD (100–500 mg/L), TOC, BOD_5 and BOD_5/COD ratio (<0.3).

Examples of the AOPs used for stabilized leachate treatment include UV, ozonation, Fenton reaction, electro-oxidation, or combinations of two processes or more. Table 16.7 provides literature data, which indicates that AOPs are highly efficient processes for color removal and treatment of refractory organic

Table 16.7 Selected results reported for the treatment of landfill leachate by AOPs.

Advanced Oxidation Processes	Effluent	Wastewater Characteristics	Parameters Assessed	Concluding Remarks	References
γ-irradiation/ H_2O_2	Leachate from a landfill at Nanjing, China	5000 mg/L COD; turbidity 241; pH 8.24	Effects of absorbed dose, the concentration of H_2O_2 and initial pH; evaluation of COD and turbidity removal	44.3% COD removal; 98.3% turbidity removal	Jia *et al.* 2013
Microwave/ $S_2O_8^{2-}$	Leachate from the Sanjuku Landfill in Taipei, Taiwan	80–600 mg/L COD; 50–400 mg/L BOD_5; 20–200 mg/L TOC; pH 7.1–7.6	Effects of pH, temperature, oxidant doses, microwave power setting, and irradiation time	Removal of color, UV254, and TOC were 81.0%, 69.9% and 77.9%, respectively; COD reduced from 268 mg/L to 108 mg/L with 60% removal	Chou *et al.* 2013
Ozonation	Leachates from the Ljubljana landfill in the south of Ljubljana city, Slovenia	Young leachate: 2790 mg/L COD; 230 mg/L BOD_5; 700 mg/L DOC; 463 mg/L N-NH_4^+; pH 8.2; Stabilized leachate: 490 mg/L COD; 28 mg/L BOD_5; 93 mg/L DOC; 316 mg/L N-NH_4^+; pH 8.1	Effect of ozonation time; evaluation of COD, BOD_5, DOC and nitrogen compounds removal; acute toxicity test; kinetics	42% removal of organics for ozonation of young leachate, and 65% removal of organics for treatment of stabilized leachate after 120 min	Gotvajn *et al.* 2009
Photocatalysis (UV/TiO_2)	Oued Smar landfill leachate in Algeria	1233–16500 mg/L COD; 220–750 mg/L BOD_5; 0.045–0.178 BOD_5/COD; 166–392 mg/L NH_4^+; pH 7.3–7.8	The evaluation of COD removal, recalcitrant material reduction, BOD_5/COD ratio and toxicity; DOC biodegradation kinetics	50–84% COD reduction and 0.2–0.6 BOD_5/COD ratio by UV/TiO_2; 90% BOD_5 removal and 87% COD reduction by the AOP–bioreactor treatment	Chemlal *et al.* 2014

Fenton Photo-Fenton	Leachates from the Colmenar Viejo Landfill in Madrid, Spain	836.5–6118.75 mg/L COD; 42.5–175 mg/L BOD_5; 223–1481 mg/L TOC; 199.5–1965 mg/L $N–NH_3$; pH 8.34–8.47	Effects of temperature, pH, chemical reagents, and photo-irradiation	Over 80% COD removal for a "young" leachate, and 70% COD removal for "old" and "mixed" leachates by Fenton; About 70% COD and TOC removals by photo-Fenton processes	Hermosilla *et al.* 2009
O_3 O_3/OH^- O_3/H_2O_2 Fenton	Leachate from a municipal landfill in the North of Portugal	743 ± 14 mg/L COD; 10 ± 1 mg/L BOD_5; 284 ± 6 mg/L TOC; 1824 ± 103 mg/L $N\text{-}NO_3^-$; 714 ± 23 mg/L $N\text{-}NH_4^+$; pH 3.5 ± 0.1	Comparison of various processes; optimization of operational conditions of each process; evaluation of biodegradability; cost estimation	About 46% COD was removed, and BOD_5/COD ratio increased from 0.01 to 0.15 by Fenton; 72% COD was reduced, and BOD_5/COD increased from 0.01 to 0.24 by ozone-based process	Cortez *et al.* 2011
Microwave-Fenton	Leachate from the landfill leachate process plant of Jiangcungou landfill site in Xi'an, China	20000–40000 mg/L COD; 5000 mg/L BOD_5; 1000 mg/L $N–NH_3$; pH 8.5	Effects of the amount of loading Fe^{2+}, the GAC dosage the microwave irradiation time, the microwave power, the H_2O_2 dosage and pH; microanalysis; kinetics	93% COD removal; 86% NH_3-N removal	Ding and Guan, 2013
Electro-Fenton	Leachate from a landfill at Wuhan, China	2720 mg/L COD; 2850 mg/L $N\text{-}NH_4^+$; 11000 mg $CaCO_3$/L alkalinity; pH 8.03	Effects of initial pH, inter-electrode gap, H_2O_2 to Fe^{2+} molar ratio, H_2O_2 dosage and hydraulic retention time; kinetics; organics detection	62% COD removal; about 73 pollutants were detected, of which 50 organics were completely removed	Zhang *et al.* 2012
Photoelectro-Fenton	Landfill leachates from solid waste disposal area of Sivas city in Turkey	2350 ± 310 mg/L COD; 915 ± 110 mg/L BOD5; 310 ± 56 mg/L NH^4-N; 10.25 ± 2.0 mg/L PO_4-P; color 1.143 ± 0.105; pH 8.36 ± 0.08	Effects of pH, H_2O_2 concentration and applied current; comparison of different treatment processes	94% COD removal; 97% color removal; 96% phosphate removal	Altin, 2008

compounds as well as for oxidizing ammonia from raw and pretreated landfill leachate. Among these processes, ozonation and Fenton oxidation are the most extensively reported and widely implemented methods for leachate treatment, both of which can achieve 40–89% of COD removal with the COD concentrations ranging from 560 to 8894 mg/L. By integrating the Fenton oxidation and coagulation–flocculation process, 69–90% of COD removal is obtained with its concentrations ranging from 417 to 7400 mg/L. By combing the activated sludge and the Fenton oxidation or wet air oxidation, almost complete COD removal is attained (Kurniawan *et al.* 2006).

The integrated system of a Fenton unit followed by an electro-oxidation unit (Figure 16.8) has been applied at pilot scale to treat the leachate from the municipal landfill site of Meruelo in Cantabria, Spain (Urtiaga *et al.* 2009). Biologically treated leachate went through the Fenton unit formed by a feed tank, a continuous stirred oxidation reactor, a neutralization tank, and a solids separation step by ultrafiltration, and then the leachate was treated by the electro-oxidation pilot plant using stainless steel for the cathode and boron-doped diamond (BDD) on silicon for the anode. Almost complete removal of the organic matter and ammonium nitrogen was achieved, and the landfill leachate characterized by an average value of 1750 mg/L COD and 750 mg/L ammonium was transformed into the effluent with COD (< 40 mg/L) and ammonium contents (< 15 mg/L) below the discharge limits into natural watercourses. The specific energy consumption necessary to electro-oxidize the organic load below the disposal limit (COD < 160 mg/L) at pilot plant scale was 35 kWh/m^3 (120 kWh/kg COD).

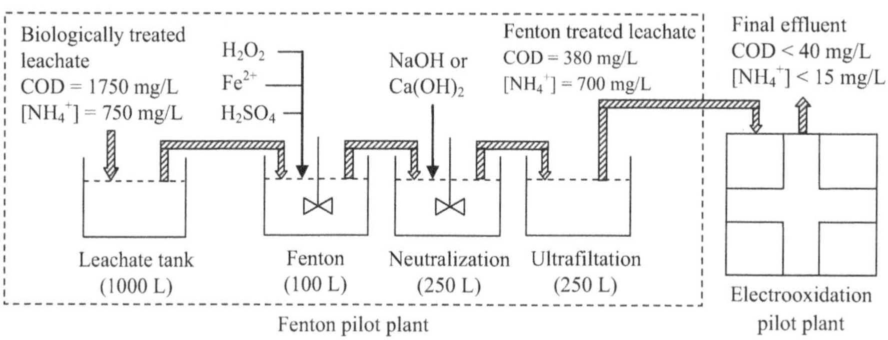

Figure 16.8 Schematic diagram of the integrated Fenton–electrooxidation pilot plant. Adapted from Urtiaga *et al.* 2009.

16.3.7 Other pollutants

The AOPs have also been used to address industrial wastewater decontamination from other kinds of pollutants, such as olive mill wastewater, winery and distillery wastewater, tannery wastewater, miscellaneous wastewater, pathogens, and heavy metals, and some examples are provided in Table 16.8. The operating conditions using various AOPs have been optimized, reaction kinetics have been investigated, and degradation mechanisms have been deduced by determining the influence of scavengers and identifying intermediates. Because industrial wastewater treatment is a complex issue with no single optimum solution, process integration of AOPs with conventional biological treatment may be an attractive option. For instance, in order to improve the biodegradability of table olive processing wastewater, several studies have dealt with the application of AOP as a suitable pre-treatment for reducing the COD and phenolic content. The majority of research on specific AOPs/biological systems for the

Table 16.8 Selected results reported for the treatment of other pollutants by AOPs.

Pollutant	Advanced Oxidation Processes	Wastewater Characteristics	Parameters Assessed	Concluding Remarks	References
Olive mill wastewater	Heterogeneous Fenton process (zero valent iron/ H_2O_2)	4000–85000 mg/L COD; 672 mg/L total polyphenol; pH 4.8–5.2	Effects of H_2O_2 dose, Fe^0 dose, pH and initial COD concentration; phenolic compounds analyses	Complete decolorization; 50% phenolic compound removal; 78–92% COD removal	Kallel *et al.* 2009a, b
Table olive processing wastewater	Electrochemical oxidation	60000 mg/L COD; 5200 mg/L total phenols; 111.5 mS/cm conductivity; pH 4.5	Effects of initial organic loading, reaction time, current intensity, initial pH and the addition of H_2O_2	73% COD removal; zero-order kinetic constant of 8.5 mg/L/min; energy consumption efficiency of 16.3 g COD/(m^3 A h)	Deligiorgis *et al.* 2008
Cork boiling wastewater	O_3 O_3/H_2O_2 Solar photo-Fenton	1240 mg/L COD; 586 mg/L DOC; 290 mg/L TSS; 163 NTU turbidity; 1.1 mS/cm conductivity; pH 7.2	Effect of different flocculants and pH; comparison of various processes; toxicity and biodegradability assays	52–59% COD removal and 16–39% DOC removal by solar photo-Fenton after 377–435 min; 62% COD removal and 47% DOC removal by ozonation after 15 h; 82% COD removal and 60% DOC removal by O_3/H_2O_2 after 15 h	De Torres-Socías *et al.* 2013
Tannery wastewater	UV/TiO_2 O_3 UV/O_3	2365 mg/L COD; 820 mg/L TOC; 720 mg/L DOC; 1010 mg/L BOD_5; 172 mg/L NH_4^+	Effect of pH; substance chemical and specific analysis; biotoxicity testing	6–21% COD removal; 11–13% TOC removal; 7% DOC removal; 15–36% BOD_5 removal	Schrank *et al.* 2004

(*Continued*)

Table 16.8 Selected results reported for the treatment of other pollutants by AOPs (*Continued*).

Pollutant	Advanced Oxidation Processes	Wastewater Characteristics	Parameters Assessed	Concluding Remarks	References
Surfactant wastewater containing abundant sulfate	Fenton	1500 ± 47 mg/L COD; 332 ± 13 mg/L BOD_5; 213 ± 8 mg/L SS; 490 ± 11 mg/L linear alkylbenzene sulfonate; pH 8.0 ± 0.1	Effects of pH, Fe^{2+} and H_2O_2 dosage; biodegradability enhancement; combined with aerobic biological process	85% COD removal; 95% removal of linear alkylbenzene sulfonate after 40 min	Wang *et al.* 2008b
Phenolic wastewater	Heterogeneous Fenton	0.10–1.05 mM chlorophenols; 7.3–59 mg/L TOC; pH 2.0–6.1	Characterization of catalysts; effects of pH, catalyst addition, H_2O_2 concentration, temperature, and radical scavengers; reaction kinetics; degradation mechanism	Highly efficient for chlorophenols removal and TOC reduction	Xu and Wang, 2011; 2012a, b; 2015
Reverse osmosis brine from water reclamation facilities	Ozonation	15.6 mg/L TOC; <2.0 mg/L BOD_5; 89.0 Pt-Co unit color; pH 6.9	Effects of ozone dosage and contact time; biodegradability enhancement; combined with biological activated carbon	5.3–24.5% TOC removal and complete decolorization were achieved, and the BOD_5/TOC ratios were increased by 1.8–3.5 times	Lee *et al.* 2009
Precursors of halogenated nitrogenous disinfection by-products	Ozonation	3.24–6.20 mg/L DOC; 0.34–1.00 mg/L NH_4^+-N; 0.039–0.112 cm^{-1} UV_{254}; 33.7–95.7 NTU turbidity; pH 7.4–7.8	Evaluation of turbidity, DOC, UV_{254}, inorganic and organic nitrogen removal; measurement of chloroform, haloacetamides, haloacetonitriles and trichloronitromethane	The removal of turbidity, dissolved organic carbon, UV_{254}, NH_4^+ and dissolved organic nitrogen were 98–99%, 58–72%, 31–53%, 16–93% and 35–74%, respectively	Chu *et al.* 2012

Real industrial wastewater	Sono-Fenton	42000 mg/L COD; 14000 mg/L TOC; pH 1.7	Effects of inlet pressure, temperature and the presence of copper windings; latent remediation	About 70% TOC removal after 150 min	Chakinala *et al.* 2009
Coliphage MS2	Photocatalysis (visible light + TiON/PdO catalyst)	$(3 \pm 2) \times 10^8$ plaque forming units (pfu)/mL; pH 8.1–8.2	Catalyst characterization; photochemical removal of coliphage MS2; •OH formation	99.75–99.94% removal of virus by combining adsorption and photocatalysis	Li *et al.* 2008
Arsenic (III)	Photocatalysis (UV/TiO_2)	40–333 μM As(III); pH 3–9	Effects of As(III) concentration, pH, catalyst loading, light intensity, dissolved oxygen concentration, type of TiO_2 surfaces and ferric ions; kinetics	Complete removal of arsenic by photocatalytic oxidation of As(III) to As(V) followed by adsorption	Dutta *et al.* 2005
As(III)	Sonolysis	1.3–26.8 μM As(III); pH 7	Effects of benzoic acid, radical scavengers, dissolved nitrogen and oxygen, arsenic concentration, pH, acoustic amplitude, solution volume of the reactor, and humic acid	Complete oxidation of As(III) after 30 min	Neppolian *et al.* 2009

treatment of raw tannery effluents is focused on using the AOP as a post-treatment step after a previous biological treatment. For effluents containing a rather large biodegradable COD fraction, a primary biological treatment is applied to remove biodegradable compounds, and an AOP as a post-oxidation step converts the non-biodegradable portion into biodegradable compounds with less chemical consumption.

A pilot plant composed of an integrated ozonation and biological activated carbon (O_3–BAC) treatment process following conventional water treatment processes (coagulation–sedimentation–sand filtration) as shown in Figure 16.9 has been used to reduce the formation of several nitrogenous and carbonaceous disinfection by-products (Chu *et al.* 2012). This pilot is located at a wastewater treatment plant of Wuxi City near Lake Taihu, China, operating at a design flow rate of 1 m^3/h. The results indicated that addition of the O_3–BAC treatment to an existing conventional water treatment indeed improved the quality of the effluent, particularly improved the removal of turbidity (from 98% to 99%), dissolved organic carbon (from 57% to 72%), UV_{254} (from 31% to 53%), NH_4^+ (from 16% to 93%), and dissolved organic nitrogen (from 35% to 74%). The findings also demonstrated that the combined O_3–BAC enhanced the removal efficiency of the precursors for the measured disinfection by-products, chemically altered the molecular structures of the precursors, increased the biodegradability of N-containing organic compounds, and effectively controlled the formation of the total organic halogen.

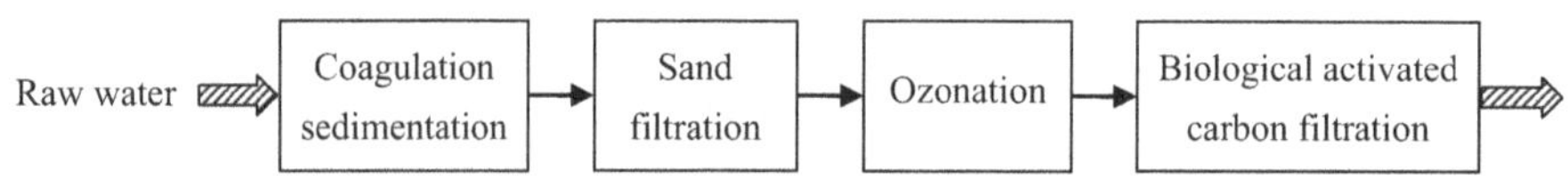

Figure 16.9 Flow chart of the pilot water treatment plant located in Wuxi city near Lake Taihu, China. Adapted from Chu *et al.* 2012.

16.4 ECONOMIC ANALYSIS

The overall costs of AOPs, usually high, drive the application of these processes to wastewater treatment. The major problems among AOPs are the high electrical energy demand and the consumption of chemical reagents, which are directly associated with the removal efficiency of pollutants. Many papers have estimated the cost of a wastewater treatment process through various aspects, including investment and operational cost. Because the cost analyses are often based on different assumptions, it is very difficult to compare the treatment costs from different papers.

Oxygen-equivalent Chemical-oxidation Capacity (OCC) has been proposed to compare the cost of the oxidant generated for the conductive-diamond electrochemical oxidation, Fenton oxidation and ozonation processes (Cañizares *et al.* 2009). Within the discharge limits that all of the technologies can reach, the average estimated operation costs were 0.7–3.0 EUR/kg equivalent O_2 for Fenton, 2.4–4.0 EUR/kg equivalent O_2 for conductive-diamond electrochemical oxidation, and 8.5–10.0 EUR/kg equivalent O_2 for ozonation. Regardless of the pollutant treated, investment for the ozonation process seemed to be the highest, and the Fenton process usually had the lowest investment costs. The investment required for conductive-diamond electrochemical oxidation was consistently lower than ozonation, which could be competitive with the Fenton approach in the treatment of some wastes.

The costs for the waste water treatment using AOPs involving ultrasound have been calculated based upon the rate constants of pollutant degradation for 1000 L/min capacity treatment plant (Mahamuni & Adewuyi, 2010). The treatment cost was estimated by considering capital cost of equipments and operating cost calculated based on annual energy usage. The costs of AOPs for phenol degradation followed the order of US > US/photocatalysis > US/UV > US/H_2O_2/CuO > US/O_3 > US/Fenton > US/UV/O_3 with the range of \$89–15536 per 1000 gallons. The cost sequences for trichloroethylene (TCE) removal were US > US/UV > photocatalysis > UV > UV/H_2O_2 > O_3 in the range of \$25–91 per 1000 gallons. The cost of waste water treatment for reactive azo dyes gave the following sequence: US > US/UV > US/O_3 > US/UV/O_3 > US/H_2O_2 > sonophotocatalysis > US/UV/H_2O_2, in the range of \$65–14203 per 1000 gallons.

The costs of wastewater treatment vary greatly with several factors, including the AOP, wastewater type and characteristics, contaminant removal efficiencies and target pollutant concentrations or COD/TOC concentrations. Pérez *et al.* (2002b) have used the AOP or combination of AOPs for the degradation of a cellulose bleaching effluent, and compared the costs of these systems per unit of TOC reduction, which included direct UV photolysis, TiO_2 assisted-photocatalysis (TiO_2/UV), Fenton, Fenton-like, and photo-Fenton reactions (Fe^{2+}/H_2O_2/UV), UV-assisted ozonation (O_3/UV) and addition of Fe^{2+} and/or H_2O_2 to the TiO_2/UV and O_3/UV systems. Fenton, Fenton-like and photo-Fenton reactions achieved better levels of TOC degradation with lower costs than photocatalysis, while ozonation was an effective but rather expensive process. The sequence of TOC reduction after 3 h treatment was ozone/UVA > ozone > photocatalysis/ozone > photocatalysis, whereas the estimated costs per kg TOC reduction followed the order of ozone > 1 h photocatalysis + 2 h ozone > ozone/UVA > 1 h ozone + 2 h photocatalysis > photocatalysis. Krichevskaya *et al.* (2011) have presented the energy cost calculations made for the pollutants removal by ozonation and Fenton oxidation processes. The costs of treatment of compounds by AOPs could be rated as follows: unsaturated aliphatic compounds < phenols < lignin < humic substances < oxygenated compounds. The ozonation treatment costs of various compounds were: 0.3–0.61 EUR/kg for unsaturated aliphatic compounds, 0.9–3.2 EUR/kg for lignin, 0.4–17 EUR/kg for phenol, 2.2–23 EUR/kg for oxygenated compounds, and 11.1 EUR/kg for humic substances. The oxidation costs of compounds by Fenton reagent and photo-Fenton system were: 5.0–23 EUR/kg for phenols, 4.9–84 EUR/kg for oxygenated compounds, 3.5–11.2 EUR/kg TOC for humic substances.

16.5 CONCLUDING REMARKS AND PROSPECTS

Various AOPs have been demonstrated in full-scale applications as feasible technologies for the treatment of municipal and industrial wastewaters, which cover a wide diversity of wastewaters from model solutions with individual substances to real effluents containing a mixture of various persistent pollutants. These processes have been applied to remove contaminant from wastewater or used as pre-treatment/post-treatment stage for wastewater remediation. As compared to individual process, combination of AOPs appears to be much more beneficial for wastewater treatment, resulting from the enhancement in the generation of oxidative radicals and the better utilization of the oxidants and catalytic activity. Combined AOPs with biological systems may become applicable and cost-competitive on an industrial scale, which eliminate some of the drawbacks of the individual techniques for the final goal to achieve maximum efficiency with minimum energy input.

Many efforts have been made to resolve the main problem of AOPs restricting their industrial applications, which are the high costs of energy source (such as γ-ray, ultraviolet light) and reagents (such as ozone, hydrogen peroxide). Effects of various operating parameters in AOPs have been investigated to maximize the degradation efficiency and to reduce the overall cost of treatment. Under optimum operating conditions, contaminant can be completely removed or converted into less harmful compounds which are amenable to biodegradation. In photocatalysis, novel photocatalysts have been developed to use solar radiation instead of UV as an energy source, which may make the treatment far more cost affordable. For electrochemical technologies, the energy demand can be considerably decreased by in situ electrochemical reagent generation, and electrode materials play an important role. Solid iron-containing catalysts can react with dissolved oxygen instead of H_2O_2 in Fenton processes, which can also reduce costs. Furthermore, many catalysts have been used to improve the degradation efficiency of ozonation. Thus, the development of novel catalytic materials may lead to improved and expanded application of various types of AOPs, which is also the future research direction.

16.6 REFERENCES

Addamo M., Augugliaro V., Di Paola A., García-López E., Loddo V., Marcì G. and Palmisano L. (2005). Removal of drugs in aqueous systems by photoassisted degradation. *Journal of Applied Electrochemistry*, **35**(7–8), 765–774.

Affam A. C. and Chaudhuri M. (2013). Degradation of pesticides chlorpyrifos, cypermethrin and chlorothalonil in aqueous solution by TiO2 photocatalysis. *Journal of Environmental Management*, **130**, 160–165.

Alaton I. A., Dogruel S., Baykal E. and Gerone G. (2004). Combined chemical and biological oxidation of penicillin formulation effluent. *Journal of Environmental Management*, **73**(2), 155–163.

Al Momani F. A., Shawaqfeh A. T. and Shawaqfeh M. S. (2007). Solar wastewater treatment plant for aqueous solution of pesticide. *Solar Energy*, **81**(10), 1213–1218.

Altin A. (2008). An alternative type of photoelectro-Fenton process for the treatment of landfill leachate. *Separation and Purification Technology*, **61**(3), 391–397.

Amat A. M., Arques A., López F. and Miranda M. A. (2005). Solar photo-catalysis to remove paper mill wastewater pollutants. *Solar Energy*, **79**(4), 393–401.

Antony S. P. and Natesan B. (2012). Optimization of integrated electro-bio process for bleaching effluent treatment. *Industrial & Engineering Chemistry Research*, **51**(24), 8211–8221.

Arslan-Alaton I. and Dogruel S. (2004). Pre-treatment of penicillin formulation effluent by advanced oxidation processes. *Journal of Hazardous Materials*, **112**(1–2), 105–113.

Azbar N., Yonar T. and Kestioglu K. (2004). Comparison of various advanced oxidation processes and chemical treatment methods for COD and color removal from a polyester and acetate fiber dyeing effluent. *Chemosphere*, **55**(1), 35–43.

Baban A., Yediler A., Lienert D., Kemerdere N. and Kettrup A. (2003). Ozonation of high strength segregated effluents from a woollen textile dyeing and finishing plant. *Dyes and Pigments*, **58**(2), 93–98.

Badawy M. I., Gad-Allah T. A., Ghaly M. Y. and Lopez A. (2006). Combination of photocatalytic and biological processes as an integrated system for treatment and recovery of industrial wastewater containing pesticides. *Afinidad*, **63**(526), 478–487.

Basfar A. A., Mohamed K. A., Al-Abduly A. J., Al-Kuraiji T. S. and Al-Shahrani A. A. (2007). Degradation of diazinon contaminated waters by ionizing radiation. *Radiation Physics and Chemistry*, **76**(8–9), 1474–1479.

Bautista P., Mohedano A. F., Casas J. A., Zazo J. A. and Rodriguez J. J. (2008). An overview of the application of Fenton oxidation to industrial wastewaters treatment. *Journal of Chemical Technology and Biotechnology*, **83**(10), 1323–1338.

Benitez F. J., Acero J. L. and Real F. J. (2002). Degradation of carbofuran by using ozone, UV radiation and advanced oxidation processes. *Journal of Hazardous Materials*, **89**(1), 51–65.

Brillas E., Banos M. A., Skoumal M., Cabot P. L., Garrido J. A. and Rodriguez R. M. (2007). Degradation of the herbicide 2,4-DP by anodic oxidation, electro-Fenton and photoelectro-Fenton using platinum and boron-doped diamond anodes. *Chemosphere*, **68**(2), 199–209.

Brillas E., Sirés I. and Oturan M. A. (2009). Electro-Fenton process and related electrochemical technologies based on Fenton's reaction chemistry. *Chemical Reviews*, **109**(12), 6570–6631.

Bureau M. A., Drogui P., Sellamuthu B., Blais J. F. and Mercier G. (2012). Municipal wastewater sludge stabilization and treatment using electrochemical oxidation technique. *Journal of Environmental Engineering-Asce*, **138**(7), 743–751.

Cañizares P., Paz R., Sáez C. and Rodrigo M. A. (2009). Costs of the electrochemical oxidation of wastewaters: a comparison with ozonation and Fenton oxidation processes. *Journal of Environmental Management*, **90**(1), 410–420.

Cesaro A., Naddeo V. and Belgiorno V. (2013). Wastewater treatment by combination of advanced oxidation processes and conventional biological systems. *Journal of Bioremediation & Biodegradation*, **4**(8), 1–8.

Chakinala A. G., Gogate P. R., Burgess A. E. and Bremner D. H. (2009). Industrial wastewater treatment using hydrodynamic cavitation and heterogeneous advanced Fenton processing. *Chemical Engineering Journal*, **152**(2–3), 498–502.

Chemlal R., Azzouz L., Kernani R., Abdi N., Lounici H., Grib H., Mameri N. and Drouiche N. (2014). Combination of advanced oxidation and biological processes for the landfill leachate treatment. *Ecological Engineering*, **73**, 281–289.

Chen C. M., Yoza B. A., Chen H. S., Li Q. X. and Guo S. H. (2015). Manganese sand ore is an economical and effective catalyst for ozonation of organic contaminants in petrochemical wastewater. *Water Air and Soil Pollution*, **226**(6), 1–11.

Chen Y. P., Liu S. Y., Yu H. Q., Yin H. and Li Q. R. (2008). Radiation-induced degradation of methyl orange in aqueous solutions. *Chemosphere*, **72**(4), 532–536.

Chou C. Y., Yu Y. H., Chang C. Y., Lin C. F. and Shang N. C. (2011). The combination of ozonation and ion exchange processes for treatment of a municipal wastewater plant effluent. *2011 International Conference on Electric Technology and Civil Engineering (ICETCE))*, 22–24 April, Lushan in Shanxi Province, China, 6961–6964.

Chou Y. C., Lo S. L., Kuo J. and Yeh C. J. (2013). A study on microwave oxidation of landfill leachate-Contributions of microwave-specific effects. *Journal of Hazardous Materials*, **246–247**, 79–86.

Chu W. H., Gao N. Y., Yin D. Q., Deng Y. and Templeton M. R. (2012). Ozone-biological activated carbon integrated treatment for removal of precursors of halogenated nitrogenous disinfection by-products. *Chemosphere*, **86**(11), 1087–1091.

Coelho A., Castro A. V., Dezotti M. and Sant'Anna G. L. (2006). Treatment of petroleum refinery sourwater by advanced oxidation processes. *Journal of Hazardous Materials*, **137**(1), 178–184.

Cortez S., Teixeira P., Oliveira R. and Mota M. (2011). Evaluation of Fenton and ozone-based advanced oxidation processes as mature landfill leachate pre-treatments. *Journal of Environmental Management*, **92**(3), 749–755.

De Torres-Socías E., Fernández-Calderero I., Oller I., Trinidad-Lozano M. J., Yuste F. J. and Malato S. (2013). Cork boiling wastewater treatment at pilot plant scale: comparison of solar photo-Fenton and ozone (O_3,O_3/H_2O_2). Toxicity and biodegradability assessment. *Chemical Engineering Journal*, **234**, 232–239.

Deligiorgis A., Xekoukoulotakis N. P., Diamadopoulos E. and Mantzavinos D. (2008). Electrochemical oxidation of table olive processing wastewater over boron-doped diamond electrodes: treatment optimization by factorial design. *Water Research*, **42**(4–5), 1229–1237.

Dermentzis K., Marmanis D., Christoforidis A. and Ouzounis K. (2014). Electrochemical reclamation of wastewater resulted from petroleum tanker truck cleaning. *Environmental Engineering and Management Journal*, **13**(9), 2395–2399.

Ding Z. and Guan W. S. (2013). Treatment of landfill leachate by microwave-Fenton oxidation process catalyzed by Fe^{2+} loaded on GAC. *Journal of Testing and Evaluation*, **41**(5), 693–700.

Dutta P. K., Pehkonen S. O., Sharma V. K. and Ray A. K. (2005). Photocatalytic oxidation of arsenic(III): evidence of hydroxyl radicals. *Environmental Science & Technology*, **39**(6), 1827–1834.

Eskelinen K., Särkkä H., Kurniawan T. A. and Sillanpää M. E. T. (2010). Removal of recalcitrant contaminants from bleaching effluents in pulp and paper mills using ultrasonic irradiation and Fenton-like oxidation, electrochemical treatment, and/or chemical precipitation: a comparative study. *Desalination*, **255**(1–3), 179–187.

Geng A. F., Xu Z. and Wang D. (2014). Selection of advanced treatment process of pharmaceutical wastewater (in Chinese). *China Water & Wastewater*, **30**(13), 73–76.

Giannakis S., Gamarra Vives F. A., Grandjean D., Magnet A., De Alencastro L. F. and Pulgarın C. (2015). Effect of advanced oxidation processes on the micropollutants and the effluent organic matter contained in municipal wastewater previously treated by three different secondary methods. *Water Research*, **84**, 295–306.

Gotvajn A. Z., Derco J., Tišler T., Cotman M. and Zagorc-Končan J. (2009). Removal of organics from different types of landfill leachate by ozonation. *Water Science and Technology*, **60**(3), 597–603.

Guimarães J. R., Gasparini M. C., Maniero M. G. and Mendes C. G. N. (2012). Stripped sour water treatment by advanced oxidation processes. *Journal of the Brazilian Chemical Society*, **23**(9), 1680–1687.

Guo H. F., Yin H., Xiong Z. H. and Ou Z. H. (2011). Combined biological treatment and post-ozonation of dyeing and printing wastewater: a case study (in Chinese). *Journal of Chongqing University*, **34**, 41–45.

Hermosilla D., Cortijo M. and Huang C. P. (2009). Optimizing the treatment of landfill leachate by conventional Fenton and photo-Fenton processes. *Science of the Total Environment*, **407**(11), 3473–3481.

Hermosilla D., Merayo N., Gascó A. and Blanco A. (2015). The application of advanced oxidation technologies to the treatment of effluents from the pulp and paper industry: a review. *Environmental Science and Pollution Research*, **22**(1), 168–191.

Hollender J., Zimmermann S. G., Koepke S., Krauss M., McArdell C. S., Ort C., Singer H., von Gunten U. and Siegrist H. (2009). Elimination of organic micropollutants in a municipal wastewater treatment plant upgraded with a full-scale post-ozonation followed by sand filtration. *Environmental Science & Technology*, **43**(20), 7862–7869.

Huang Y. H., Lu J. X. and Wang C. (2001). The studies on the treatment process of pharmaceutical wastewater (in Chinese). *Industrial Water Treatment*, **21**(1), 29–30.

Hurwitz G., Hoek E. M. V., Liu K., Fan L. H. and Roddick F. A. (2014). Photo-assisted electrochemical treatment of municipal wastewater reverse osmosis concentrate. *Chemical Engineering Journal*, **249**, 180–188.

Jia W. B., Wei Y. H., Liu J. G., Ling Y. S., Hei D. Q., Shan Q. and Zeng J. (2013). Studying the treatment effect of γ-rays combined with H_2O_2 on landfill leachate. *Journal of Radiation Research and Radiation Processing*, **31**(1), 104021–104025.

Kallel M., Belaid C., Boussahel R., Ksibi M., Montiel A. and Elleuch B. (2009a). Olive mill wastewater degradation by Fenton oxidation with zero-valent iron and hydrogen peroxide. *Journal of Hazardous Materials*, **163**, 550–554.

Kallel M., Belaid C., Mechichi T., Ksibi M. and Elleuch B. (2009b). Removal of organic load and phenolic compounds from olive mill wastewater by Fenton oxidation with zero-valent iron. *Chemical Engineering Journal*, **150**(2–3), 391–395.

Kansal S. K. and Kumari A. (2014). Potential of *M. oleifera* for the treatment of water and wastewater. *Chemical Reviews*, **114**(9), 4993–5010.

Kreetachat T., Damrongsri M., Punsuwon V., Vaithanomsat P., Chiemchaisri C. and Chomsurin C. (2007). Effects of ozonation process on lignin-derived compounds in pulp and paper mill effluents. *Journal of Hazardous Materials*, **142**(1–2), 250–257.

Krichevskaya M., Klauson D., Portjanskaja E. and Preis S. (2011). The cost evaluation of advanced oxidation processes in laboratory and pilot-scale experiments. *Ozone-Science & Engineering*, **33**(3), 211–223.

Kurniawan T. A., Lo W. H. and Chan G. Y. S. (2006). Radicals-catalyzed oxidation reactions for degradation of recalcitrant compounds from landfill leachate. *Chemical Engineering Journal*, **125**(1), 35–57.

Lapertot M., Pulgarín C., Fernández-Ibáñez P., Maldonado M. I., Pérez-Estrada L., Oller I., Gernjak W. and Malato S. (2006). Enhancing biodegradability of priority substances (pesticides) by solar photo-Fenton. *Water Research*, **40**(5), 1086–1094.

Lee L. Y., Ng H. Y., Ong S. L., Hu J. Y., Tao G. H., Kekre K., Viswanath B., Lay W. and Seah H. (2009). Ozone-biological activated carbon as a pretreatment process for reverse osmosis brine treatment and recovery. *Water Research*, **43**(16), 3948–3955.

Li Q., Page M. A., Mariñas B. J. and Shang J. K. (2008). Treatment of coliphage MS2 with palladium-modified nitrogen-doped titanium oxide photocatalyst illuminated by visible light. *Environmental Science & Technology*, **42**(16), 6148–6153.

Liu K., Roddick F. A. and Fan L. H. (2012). Impact of salinity and pH on the UVC/H_2O_2 treatment of reverse osmosis concentrate produced from municipal wastewater reclamation. *Water Research*, **46**(10), 3229–3239.

Liu Y. K. and Wang J. L. (2013). Degradation of sulfamethazine by gamma irradiation in the presence of hydrogen peroxide. *Journal of Hazardous Materials*, **250**, 99–105.

Liu Y. K., Hu J. and Wang J. L. (2014). Fe^{2+} enhancing sulfamethazine degradation in aqueous solution by gamma irradiation. *Radiation Physics and Chemistry*, **96**, 81–87.

Lu C. X., Yang Q. H., Bai Y. B., Chen A. M., Qian G. L., Shao Q. and Zhang Y. F. (2014). Pilot-scale test on advanced treatment of petrochemical wastewater by coupling process of high efficiency nitrifying tank and catalytic ozone oxidation. *China Water & Wastewater*, **30**(15), 129–131.

Mahamuni N. N. and Adewuyi Y. G. (2010). Advanced oxidation processes (AOPs) involving ultrasound for waste water treatment: a review with emphasis on cost estimation. *Ultrasonics Sonochemistry*, **17**(6), 990–1003.

Martins A. F., Wilde M. L., Vasconcelos T. G. and Henriques D. M. (2006). Nonylphenol polyethoxylate degradation by means of electrocoagulation and electrochemical Fenton. *Separation and Purification Technology*, **50**(2), 249–255.

Neppolian B., Doronila A., Grieser F. and Ashokkumar M. (2009). Simple and efficient sonochemical method for the oxidation of arsenic(III) to arsenic(V). *Environmental Science & Technology*, **43**(17), 6793–6798.

Nielson K. and Smith D. W. (2005). Ozone-enhanced electroflocculation in municipal wastewater treatment. *Journal of Environmental Engineering and Science*, **4**(1), 65–76.

Oller I., Malato S. and Sánchez-Pérez J. A. (2011). Combination of Advanced Oxidation Processes and biological treatments for wastewater decontamination-A review. *Science of the Total Environment*, **409**(20), 4141–4166.

Oneby M. A., Bromley C. O., Borchardt J. H. and Harrison D. S. (2010). Ozone treatment of secondary effluent at US municipal wastewater treatment plants. *Ozone-Science & Engineering*, **32**(1), 43–55.

Parilti N. B. (2010). Treatment of a petrochemical industry wastewater by a solar oxidation process using the Box-Wilson experimental design method. *Ekoloji*, **19**(77), 9–15.

Peller J., Wiest O. and Kamat P. V. (2001). Sonolysis of 2,4-dichlorophenoxyacetic acid in aqueous solutions. Evidence for OH-radical-mediated degradation. *Journal of Physical Chemistry A*, **105**(13), 3176–3181.

Peralta-Zamora P., de Moraes S. G., Pelegrini R., Freire M., Reyes J., Mansilla H. and Durán N. (1998). Evaluation of ZnO, TiO_2 and supported ZnO on the photoassisted remediation of black liquor, cellulose and textile mill effluents. *Chemosphere*, **36**(9), 2119–2133.

Pérez M., Torrades F., Domènech X. and Peral J. (2002a). Fenton and photo-Fenton oxidation of textile effluents. *Water Research*, **36**(11), 2703–2710.

Pérez M., Torrades F., Domènech X. and Peral J. (2002b). Removal of organic contaminants in paper pulp effluents by AOPs: an economic study. *Journal of Chemical Technology and Biotechnology*, **77**(5), 525–532.

Pérez M., Torrades F., Domènech X. and Peral J. (2002c). Treatment of bleaching Kraft mill effluents and polychlorinated phenolic compounds with ozonation. *Journal of Chemical Technology and Biotechnology*, **77**(8), 891–897.

Pérez M., Torrades F., García-Hortal J. A., Domènech X. and Peral J. (2002d). Removal of organic contaminants in paper pulp treatment effluents under Fenton and photo-Fenton conditions. *Applied Catalysis B-Environmental*, **36**(1), 63–74.

Punzi M., Nilsson F., Anbalagan A., Svensson B. M., Jönsson M., Mattiasson B. and Jonstrup M. (2015). Combined anaerobic-ozonation process for treatment of textile wastewater: removal of acute toxicity and mutagenicity. *Journal of Hazardous Materials*, **292**, 52–60.

Rosal R., Rodríguez A., Perdigón-Melón J. A., Petre A., García-Calvo E., Gómez M. J., Agüera A. and Fernández-Alba A. R. (2010). Occurrence of emerging pollutants in urban wastewater and their removal through biological treatment followed by ozonation. *Water Research*, **44**(2), 578–588.

San Sebastián Martínez N., Fíguls Fernández J., Font Segura X. and Sánchez Ferrer A. (2003). Pre-oxidation of an extremely polluted industrial wastewater by the Fenton's reagent. *Journal of Hazardous Materials*, **101**(3), 315–322.

Sánchez-Polo M., López-Peñalver J., Prados-Joya G., Ferro-García M. A. and Rivera-Utrilla J. (2009). Gamma irradiation of pharmaceutical compounds, nitroimidazoles, as a new alternative for water treatment. *Water Research*, **43**(16), 4028–4036.

Santiago-Morales J., Gómez M. J., Herrera-López S., Fernández-Alba A. R., García-Calvo E. and Rosal R. (2013). Energy efficiency for the removal of non-polar pollutants during ultraviolet irradiation, visible light photocatalysis and ozonation of a wastewater effluent. *Water Research*, **47**(15), 5546–5556.

Schrank S. G., José H. J., Moreira R. F. P. M. and Schröder H. F. (2004). Elucidation of the behavior of tannery wastewater under advanced oxidation conditions. *Chemosphere*, **56**(5), 411–423.

Shi X. L., Li F. M. and Hu H. Y. (2012). Application of hydrolytic acidification-aerobic biological treatment-Fenton process in pulping and papermaking wastewater treatment. *Water & Wastewater Engineering*, **38**(7), 47–51.

Sirés I., Garrido J. A., Rodríguez R. M., Cabot P. I., Centellas F., Arias C. and Brillas E. (2006). Electrochemical degradation of paracetamol from water by catalytic action of Fe^{2+}, Cu^{2+}, and UVA light on electrogenerated hydrogen peroxide. *Journal of the Electrochemical Society*, **153**(1), D1–D9.

Song S., Xu L. J., He Z. Q., Chen J. M., Xiao X. Z. and Yan B. (2007). Mechanism of the photocatalytic degradation of C.I. Reactive Black 5 at pH 12.0 using $SrTiO_3/CeO_2$ as the catalyst. *Environmental Science & Technology*, **41**, 5846–5853.

Song S., Xu L. J., He Z. Q., Ying H. P., Chen J. M., Xiao X. Z. and Yan B. (2008). Photocatalytic degradation of C.I. Direct Red 23 in aqueous solutions under UV irradiation using $SrTiO_3/CeO_2$ composite as the catalyst. *Journal of Hazardous Materials*, **152**(3), 1301–1308.

Sponza D. T. and Oztekin R. (2010). Effect of sonication assisted by titanium dioxide and ferrous ions on polyaromatic hydrocarbons (PAHs) and toxicity removals from a petrochemical industry wastewater in Turkey. *Journal of Chemical Technology and Biotechnology*, **85**(7), 913–925.

Tahri L., Elgarrouj D., Zantar S., Mouhib M., Azmani A. and Sayah F. (2010). Wastewater treatment using gamma irradiation: tétouan pilot station, Morocco. *Radiation Physics and Chemistry*, **79**(4), 424–428.

Ting W. P., Huang Y. H. and Lu M. C. (2007). Catalytic treatment of petrochemical wastewater by electroassisted Fenton technologies. *Reaction Kinetics and Catalysis Letters*, **92**(1), 41–48.

Torrades F., Pérez M., Mansilla H. D. and Peral J. (2003). Experimental design of Fenton and photo-Fenton reactions for the treatment of cellulose bleaching effluents. *Chemosphere*, **53**(10), 1211–1220.

Umar M., Roddick F. and Fan L. H. (2014). Effect of coagulation on treatment of municipal wastewater reverse osmosis concentrate by UVC/H_2O_2. *Journal of Hazardous Materials*, **266**, 10–18.

Umar M., Roddick F. A., Fan L., Autin Q. and Jefferson B. (2015). Treatment of municipal wastewater reverse osmosis concentrate using UVC-LED/H_2O_2 with and without coagulation pre-treatment. *Chemical Engineering Journal*, **260**, 649–656.

Urtiaga A., Rueda A., Anglada A. and Ortiz I. (2009). Integrated treatment of landfill leachates including electrooxidation at pilot plant scale. *Journal of Hazardous Materials*, **166**(2–3), 1530–1534.

Wan J. X., He S. J., Sun W. H., Fang J. M. and Wang J. L. (2011). Pretreatment of paper mill effluent by a combined process of coagulation and ionizing radiation. *Environmental Science*, **32**(6), 1638–1643.

Wang C. T., Hu J. L., Chou W. L. and Kuo Y. M. (2008a). Removal of color from real dyeing wastewater by Electro-Fenton technology using a three-dimensional graphite cathode. *Journal of Hazardous Materials*, **152**(2), 601–606.

Wang J. L. and Xu L. J. (2012). Advanced oxidation processes for wastewater treatment: formation of hydroxyl radical and application. *Critical Reviews in Environmental Science and Technology*, **42**(3), 251–325.

Wang X. J., Song Y. and Mai J. S. (2008b). Combined Fenton oxidation and aerobic biological processes for treating a surfactant wastewater containing abundant sulfate. *Journal of Hazardous Materials*, **160**(2–3), 344–348.

Wei F. (2002). Analysis of Water and Wastewater. Chinese Environmental Science Press, Beijing.

Wong J. M. (2000). Testing and implementation of an advanced wastewater reclamation and recycling system in a major petrochemical plant. *Water Science and Technology*, **42**(5–6), 23–27.

Wu C. Y., Gao Z., Zhou Y. X., Liu M. G., Song J. M. and Yu Y. (2015). Treatment of secondary effluent from a petrochemical wastewater treatment plant by ozonation-biological aerated filter. *Journal of Chemical Technology and Biotechnology*, **90**(3), 543–549.

Xia X. W. and Sun S. Q. (2005). Study of pretreating farm chemicals wastewater by ozone (in Chinese). *Journal of Hefei University of Technology*, **28**(3), 270–273.

Xu L. J. and Wang J. L. (2011). A heterogeneous Fenton-like system with nanoparticulate zero-valent iron for oxidation of 4-chloro-3-methyl phenol. *Journal of Hazardous Materials*, **186**(1), 256–264.

Xu L. J. and Wang J. L. (2012a). Fenton-like degradation of 2,4-dichlorophenol using Fe_3O_4 magnetic nanoparticles. *Applied Catalysis B: Environmental*, **123**, 117–126.

Xu L. J. and Wang J. L. (2012b). Magnetic nanoscaled Fe_3O_4/CeO_2 composite as an efficient Fenton-like heterogeneous catalyst for degradation of 4-chlorophenol. *Environmental Science & Technology*, **46**(18), 10145–10153.

Xu L. J. and Wang J. L. (2015). Degradation of 2,4,6-trichlorophenol using magnetic nanoscaled Fe_3O_4/CeO_2 composite as a heterogeneous Fenton-like catalyst. *Separation and Purification Technology*, **149**, 255–264.

Ya F. L., Zhang X. N. and Zheng L. L. (2011). Engineering case for municipal wastewater treatment by an oxidation ditch-type A2/O + Advanced treatment process. *2011 International Conference on Consumer Electronics, Communications and Networks (CECNet))*, 16–18 April, XianNing in Hubei Province, China, 1216–1219.

Zhang F. E., Li F., Dong L. F., Song W., Tu B. H., Li W. and Zhao L. P. (2011). A study of dyeing wastewater treatment technology in Changzhou textile park (in Chinese). *Chinese Journal of Environmental Engineering*, **5**(3), 589–592.

Zhang H., Ran X. N. and Wu X. G. (2012). Electro-Fenton treatment of mature landfill leachate in a continuous flow reactor. *Journal of Hazardous Materials*, **241–242**, 259–266.

Zhao W. R., Wu Z. B., Shi H. X. and Wang D. H. (2005). UV photodegradation of azo dye Diacryl red X-GRL. *Journal of Photochemistry and Photobiology a-Chemistry*, **171**(2), 97–106.

Zupanc M., Kosjek T., Petkovšek M., Dular M., Kompare B., Širok B., Blažeka Z. and Heath E. (2013). Removal of pharmaceuticals from wastewater by biological processes, hydrodynamic cavitation and UV treatment. *Ultrasonics Sonochemistry*, **20**(4), 1104–1112.

Chapter 17

Iron-based green technologies for water remediation

Virender K. Sharma and Radek Zboril

17.1 INTRODUCTION

Metallic iron (Fe(0)), ferrous (Fe(II)), and ferric (Fe(III)) ions are commonly found in the natural environment. Among these forms of iron, Fe(0) and Fe(III) oxides have shown a wide range of applications such as catalysis, pigments, magnetic recording media, and magnetic fluids (Bauer & Knölker, 2015; Gutiérrez *et al.* 2015; Ling *et al.* 2015; Machala *et al.* 2011; Pereira *et al.* 2012; Reddy *et al.* 2012). In recent years, nanomaterials of Fe(0) (e.g. nano zero valent iron, nZVI) and Fe(III) oxides (e.g. magnetite, Fe_3O_4 and maghemite, γ-Fe_2O_3) have gain attention due to their potential in industrial and water purification applications (Guan *et al.* 2015; Noubactep, 2015; Tang & Lo, 2013; Yan *et al.* 2013). Nanoparticles have high surface-area-to-volume-ratio to result in a large removal capacity of contaminants. Additionally, fast kinetics and high reactivity due to small particle size cause efficient removal of contaminants in water. Another advantage of such nanoparticles is their magnetic properties which allow easy separation by applying magnetic fields.

In the last two decades, high-valent iron species such as ferrate (e.g. FeO_4^{2-}, Fe(VI)) has emerged as a green compound because of its role in the "super iron" battery, green chemistry synthesis, and sustainable treatment technology (e.g. non-chlorine oxidation for pollutant remediation (Carr, 2008; Delaude & Laszlo, 1996; Farmand *et al.* 2011; Licht *et al.* 1999; Sharma *et al.* 2015; Wang *et al.* 2016). The Fe(VI) ion is a powerful oxidizing agent in aqueous media and oxidation reactions carried out by Fe(VI) can be accomplished in shorter time periods than oxidations carried out by permanganate and chromate (Delaude & Laszlo, 1996; Kim *et al.* 2015). Iron(III) oxide, generated from Fe(VI) ion, is considered to be practically non-toxic; therefore, Fe(VI) can make a treatment process environmentally benign. The generated Fe(III) oxide is an excellent coagulant for removal of metals and radionuclides from contaminated water (Joshi *et al.* 2008; Potts & Churchwell, 1994; Prucek *et al.* 2013; Prucek *et al.* 2015).

Application of magnetic iron nanoparticles and ferrate in decontamination of water can thus be considered as a green treatment technology. This chapter summarizes various applications of nZVI, iron(III) oxides nanoparticles, and ferrate ion in treating inorganic and organic contaminants in water.

17.2 ZEROVALENT IRON NANOPARTICLES

Zerovalent iron (ZVI) has attracted attention for the past several years in removing various contaminants (Bae & Hanna, 2015). Basically, ZVI has high reducing capacity to remove environmental contaminants from wastewater and groundwater. In the past decade, emphasis is on the nano ZVI nanoparticles (nZVI) which have shown remarkable reduction properties, thus capable of removal of numerous inorganic and organic contaminants (Table 17.1) (Crane & Scott, 2012; Jarošová *et al.* 2015; Yan *et al.* 2013). Small size and high surface area of nZVI were largely responsible for the efficiency in contaminant removal (Baikousi *et al.* 2015; Jarošová *et al.* 2015; Mueller *et al.* 2012; Raychoudhury & Scheytt, 2013; Soukupova *et al.* 2015; Yan *et al.* 2013). Several methods have been used to synthesize nZVI, which include the aqueous reduction of iron salts, thermal reduction of oxide compounds, sputtering gas-aggregation, hydrogenation of metallic complexes, chemical vapor deposition, and pulsed laser ablation (Crane & Scott, 2012).

Table 17.1 Types of contaminants treated with nZVI.

Category	Examples
Alkaline earth metals	Barium, beryllium
Metals	Cadmium, lead, copper, arsenic, selenium, nickel, silver, vanadium, technetium
Non-metal inorganic species	Bromate, perchlorate, nitrate, nitrite
Halogenated aliphatics	Chlorinated methane, brominated methane, chlorinated ethenes
Halogenated aromatics	Chlorinated benzenes, chlorinated phenols, Chlorinated biphenyls, polybrominated diphenyl ethers
Other organic compounds	Dyes, explosives, pesticides
Pharmaceuticals	Nitroimidazole, β-lactams

In the aquatic environment, dissolution of nZVI takes place yielding ferrous (Fe^{2+}) ions (Equations 17.1 and 17.2). The spherical shaped nZVI particles are usually covered by iron (hydr)oxide layer with nanometer thickness layer. Therefore, Fe(0) and Fe(II) co-exist to act in the first stage as adsorbents to remove different environmental contaminants. In the removal mechanism, redox reaction, adsorption, precipitation, and dissolution occur individually or simultaneously on the surface of nZVI.

$$2Fe^0_{(s)} + 4H^+_{(aq)} + O_{2(aq)} \rightarrow 2Fe^{2+} + 2H_2O_{(l)} \quad E^0 = +1.67\,V \tag{17.1}$$

$$2Fe^0_{(s)} + 2H_2O_{(l)} \rightarrow 2Fe^{2+} + H_{2(g)} + 2OH^-_{(aq)} \quad E^0 = -0.39\,V \tag{17.2}$$

The Fe^{2+} ions go through oxidation to Fe^{3+} species (Equations 17.3 and 17.4).

$$2Fe^{2+}_{(s)} + 2H^+_{(aq)} + 1/2O_{2(aq)} \rightarrow 2Fe^{3+} + H_2O_{(l)} \quad E^0 = +0.46\,V \tag{17.3}$$

$$2Fe^{2+}_{(s)} + 2H_2O_{(l)} \rightarrow 2Fe^{3+} + H_{2(g)} + 2OH^-_{(aq)} \quad E^0 = -1.60\,V \tag{17.4}$$

In recent years, an enhancement in removal of contaminants by nZVI was sought by developing composite material containing nZVI. Examples include the composite nZVI/ carbonaceous material hybrid material (Baikousi *et al.* 2015). This material had homogeneous distribution of nZVI (10–20 nm) within the carbon matrix. The surface area was 141 m^2/g while a total iron loading was 1 mmol/g of the composite (Baikousi *et al.* 2015). The nZVI/ carbonaceous material had maximum removal of As(III). The hybrid material of activated carbon and nZVI also had large adsorption of arsenic (Baikousi *et al.* 2015). Both of these hybrid materials had about five to seven times more adsorption capacity than that of bare nZVI. Most of iron containing materials and other coagulants had much lower removal of As(III) than ZVI composites (Baikousi *et al.* 2015).

The presence of cations and anions as well as natural organic matter (NOM) greatly influenced decontamination by nZVI (Doong & Lai, 2005; Liu *et al.* 2013; Liu *et al.* 2014; Liu & Lowry, 2006; Tratnyek *et al.* 2001; Wang *et al.* 2011; Xie & Shang, 2007). For example, the cations such as Cu^{2+}, Co^{2+}, Ni^{2+}, Mn^{2+}, and Pb^{2+} enhanced the degradation of contaminants whereas anions (Cl^-, HCO_3^-, HPO_4^{2-}, and SO_4^{2-}) inhibited the dechlorination of trichloroethylene. Cations dissolved the passive oxide layer coating on the aged Fe(0) and thus enhanced the degradation of contaminants. In the case of anions, the formation of complexes with iron oxide coating was mainly responsible for inhibition of decontamination by nZVI. The solution pH was other water quality parameter that controlled the reactivity of nZVI, therefore greatly influenced the removal of contaminants by nZVI (Bae & Hanna, 2015).

The role of pH on the reduction of contaminants by nZVI has been studied. Enhanced degradation of 4-nitrophenol was found with increase in pH (Bae & Hanna, 2015). However, other reports on Cr(VI), 1,1,-trichloroethane, trichloroethylene, and nitrobenzene demonstrated decrease in reduction with increasing pH (Bae & Lee, 2010; Chen *et al.* 2001; Dong *et al.* 2010; Liu & Lowry, 2006; Xie & Cwiertny, 2012). The discrepancy in the results of pH dependence may be understood by evaluating Fe(II) generation and its complexation with oxide coating, formed on nZVI during reduction processes (Bae & Hanna, 2015). A thorough literature review on the application of nZVI for remediating groundwater in Europe was conducted (Mueller *et al.* 2012). Results on reduction of target contaminants were evaluated as promising. Significantly, no adverse effects on the aquatic environment were observed (Mueller *et al.* 2012).

nZVI particles were also demonstrated to inactivate *Escherichia coli* (Lee *et al.* 2008). There was a significant difference in inactivation between deaerated and air-saturated solutions. In the deaerated conditions, the relationship between log inactivation and nZVI was linear (0.82 log inactivation/(mg/L nZVI – h). Comparatively, much higher dose of nZVI was required to achieve similar inactivation of *E. coli* under air-saturated conditions Mechanism of inactivation involved (i) oxidation of nZVI by intracellular oxygen to from Fe(II), (ii) oxidative stress from the generated reactive oxygen species from the reaction of Fe(II) with oxygen or hydrogen peroxide., leading to oxidative damages via the Fenton reaction, and (iii) disruption of cell membrane of *E. coli* The role of nZVI in destruction and prevention the formation of cyanobacterial water blooms has been explored (Marsalek *et al.* 2012). The multiple modes of action of nZVI in the destruction of cyanobacterial cells and the immobilization of microcystins as well as removal of bioavailable phosphorous were demonstrated (Marsalek *et al.* 2012).

17.3 IRON(III) OXIDE NANOPARTICLES

Magnetic iron(III) oxide nanoparticles like magnetite (Fe_3O_4) and maghemite (γ-Fe_2O_3) have demonstrated applications in removing heavy metals from contaminated water (Tang & Lo, 2013). Removal of Cr(VI) by γ-Fe_2O_3 nanoparticles (multidispersed with an average diameter of around 10 nm) at pH 2.5 showed that the uptake rate of Cr(VI) was quite high for up to 14 minutes and ~90% of the Cr(VI) was removed (Hu *et al.* 2005). After an initial uptake, subsequent removal attained gradually the equilibrium. The rapid Cr(VI) adsorption suggests external adsorption by magnetite nanoparticles. Removal of Cr(VI) varied with the initial concentration of Cr(VI). Removal of Cr(VI) were 97.3%, 74.6%, and 58.9% for 50, 100, and 150 mg/L initial Cr(VI) concentrations, respectively (Hu *et al.* 2005). The fixed number of adsorption sites at constant maghemite dosages were responsible for decreasing Cr(VI) removal with increased initial Cr(VI) concentration.

The mechanisms involved in removing of metals by iron(III) oxide nanoparticles are physical and chemical adsorption. The study of Cr(VI) removal by γ-Fe_2O_3 nanoparticles using X-ray diffraction (XRD) and X-ray photoelectron spectroscopy (XPS) showed a physical adsorption removal mechanism (Hu *et al.* 2005). No change in crystal structure of γ-Fe_2O_3 was found suggesting removal of Cr(VI) without any chemical reaction. Comparatively, nanoparticles of Fe_3O_4 showed both physical and chemical adsorption of Cr(VI) (Hu *et al.* 2004).

Significantly, magnetic iron(III) oxide nanoparticles were successfully reused after desorbing contaminants (Hao *et al.* 2010). However, several environmental conditions may affect the use of magnetic iron(III) oxide nanoparticles in order to meet the criteria of the sustainable treatment process (Tang & Lo, 2013). For example, natural organic matter (NOM), pH, and competing ions present in water greatly influenced the removal of heavy metals (Tang & Lo, 2013). NOM in water lowers the adsorption capacity and kinetics (Chen & Huang, 2004). Other example is the enhancement of As(V) removal in presence of Zn(II) ions (Yang *et al.* 2010). Significantly low adsorption of metals onto iron(III) oxide nanoparticles at high pH was usually observed (Tang & Lo, 2013). Attempts are being made to minimize the pH influence by coating the magnetic nanoparticles with polymers and surfactants (Ge *et al.* 2012).

17.4 FERRATES

Fe(VI) ion in aqueous solution possesses strong oxidation capability, which is explained by its redox potentials in acidic and basic solutions (Table 17.2). Under acidic environment, the redox potential of Fe(VI) is the highest among other conventional disinfectants and oxidants used in water and wastewater treatment. However, ozone has the highest redox potential in the basic environment.

Table 17.2 Redox potentials of different disinfectant/oxidants in water.

Oxidant	Acidic	*E*° (V/SHE)	Basic	*E*° (V/SHE)
Ferrate(VI)	$Fe^{VI}O_4^{2-}/Fe^{3+}$	2.20	$Fe^{VI}O_4^{2-}/Fe(OH)_3$	0.70
Ozone	O_3/O_2	2.08	O_3/O_2	1.24
Hydrogen peroxide	H_2O_2/H_2O	1.78	H_2O_2/OH^-	0.88
Permanganate	MnO_4^-/MnO_2	1.68	MnO_4^-/MnO_2	0.59
	MnO_4^-/Mn^{2+}	1.51		
Hypochlorite	$HClO/Cl^-$	1.48	ClO^-/Cl^-	0.84

Several alkali and alkaline earth salts of ferrate(VI) have been synthesized (Sharma *et al.* 2013), among which sodium and potassium salts of Fe(VI) (Na_2FeO_4 and K_2FeO_4) are the common ones. These salts were synthesized by thermal, wet chemical, and electrochemical approaches (Sharma *et al.* 2015). The thermal synthesis of Fe(VI) entails heating a mixture of iron(III) oxides and KNO_3 above 1100°C producing K_2FeO_4 with low purity (~30%). However, temperatures below 600°C are required when using a mixture of iron(III) oxide and Na_2O_2, which generated high yield of Na_2FeO_4 (>90%) (Sharma *et al.* 2013). In the wet chemical technique, Fe(III) oxides (e.g. Fe_2O_3) or Fe(III) salts (e.g.

$FeCl_3$, $Fe(NO_3)_3$) were oxidized by strongly alkaline solution of hypochlorite (OCl^-), which yielded a highly soluble Na_2FeO_4. A much less soluble K_2FeO_4 salt, with high purity (98%) of Fe(VI), was obtained from the solution of Na_2FeO_4 by adding KOH (Luo *et al.* 2011). Ozone has also been applied to produce Na_2FeO_4(Perfiliev *et al.* 2007). In the electrochemical method, electrons and iron electrode (Fe(0)), Fe(II) salts, and oxides and salts of Fe(III) were used (Macova *et al.*, 2009; Mácová & Bouzek, 2012). The Fe(VI) synthesis efficiency varied with temperature, the strength of the alkaline solution, and the composition of the iron precursors. The optimum yield of Fe(VI) was generally obtained in 14 M NaOH electrolyte solution.

Fe(VI) ion has a characteristic violet color much like that of permanganate in solutions; the spectra of Fe(VI) solutions show one maximum at 510 nm ($\varepsilon = 1150 \pm 25\ M^{-1}cm^{-1}$) and a shoulder between 275 and 320 nm. The quantification of Fe(VI) ion in solution can also be carried out with colorimetric methods based on 2,2-azino-bis(3-ethylbenzothiazoline-6-sulfonate) (ABTS) ($Fe(VI) + ABTS \rightarrow Fe(III) + ABTS^{+\bullet}$; $\varepsilon_{415\ nm} = 3.40 \times 10^4\ M^{-1}cm^{-1}$) and iodide ($Fe(VI) + 3I^- \rightarrow Fe(III) + I_3^-$; $\varepsilon_{351\ nm} = 2.97 \times 10^4\ M^{-1}cm^{-1}$) (Lee *et al.* 2005; Luo *et al.* 2011). The solid state characterization of Fe(VI) has been done using Mössbauer spectroscopy, neutron and X-ray diffractions, X-ray absorption near edge structure (XANES) and Fourier transform infrared (FTIR) spectroscopy (Sharma *et al.* 2013). Mössbauer spectroscopy can easily distinguish different valence states of iron as the values of the isomer shift (δ) are highly sensitive to the oxidation state (OS) of iron in ferrates; δ values decrease with an increase in OS (Poleshchuk *et al.* 2010).

The self-decomposition of Fe(VI) in water forms molecular oxygen (Equation 17.5) (Goff & Murmann, 1971; Lee *et al.* 2014). The decomposition rate is strongly dependent on the initial Fe(VI) concentration, pH, temperature, and to some extent on the surface character of the hydrous iron oxide formed upon decomposition (Carr, 2008). Dilute solutions of Fe(VI) are more stable than concentrated solutions.

$$2FeO_4^{2-} + 5H_2O \rightarrow 2Fe^{3+} + 3/2O_2 + 10OH^- \qquad (17.5)$$

In presence of contaminants (X), several reactions occur in addition to reaction (17.5). These reactions may be summarized as follows (Sharma *et al.* 2015): (i) the reaction of X to generate Fe(V) and Fe(IV) through 1-e$^-$ and 2-e$^-$ transfer processes (e.g. $Fe^{VI} + X \rightarrow Fe^{V} + X^{\bullet}$; $Fe^{VI} + X \rightarrow Fe^{IV} + X(O)$), (ii) production of radical species which can also generate Fe(V) and Fe(IV) species, (iii) further reactions of Fe(V) and Fe(IV) with contaminants (e.g. $Fe^{V} + X \rightarrow Fe^{IV} + X^{\bullet}$, $Fe^{V} + X \rightarrow Fe^{III} + X(O)$, and $Fe^{IV} + X \rightarrow Fe^{III}/Fe^{II}$ + Product(s), (iv) self-decompositions of Fe(V) and Fe(IV) species (e.g. $Fe^{V} + H_2O \rightarrow Fe^{III} + O_2/H_2O_2$), (v) reactions of ferrates with reactive oxygen species, $O_2^{\bullet-}$ and H_2O_2, produced during their self-decompositions. Fe(VI) provides different oxidation capacities to oxidize contaminants while Fe(VI) reduced to Fe(III) or Fe(II). Interestingly, the rates of each of the reactions mentioned above are pH dependent, hence the oxidation capacity of Fe(VI) may vary with pH (Sharma *et al.* 2015).

A recent research on methods to synthesize stable $Fe^{VI}O_4^{2-}$ in liquid phase has shown that a hybrid process combining thermal and wet processes resulted in Fe(VI) solutions which are stable for two weeks. This hybrid approach is superior to the typical synthesis methods from which the aqueous stability of Fe(VI) is only for a few hours (Sharma *et al.* 2015).

Fe(VI) displayed simultaneously disinfection, oxidation, and coagulation properties, hence inactivation of microorganisms, oxidative transformation of industrial contaminants, micropollutants and toxins, and removal of toxic metals and phosphate can be achieved in a single dose Fe(VI) treatment (Jiang, 2014, 2015; Sharma *et al.* 2015; Sharma, 2010b; Yates *et al.* 2014). The following sections describe briefly a few examples of decontamination of water by Fe(VI).

17.4.1 Disinfection

Fe(VI) as a disinfectant to inactivate a wide range of microorganism has been examined for more than three decades (Jiang, 2014; Sharma, 2007b). Fe(VI) could achieve more than 99.9% inactivation of total coliforms, collected from a different regions of the world (Sharma *et al.* 2005). Fe(VI) has shown high effectiveness of inactivation of *Escherichia coli* over a wide pH range.

The linear plots between log inactivation level of *E. coli* and the integrated exposure (i.e., ∫[Fe(VI)] dt) of *E. coli* to Fe(VI) at various pH values (25 mM phosphate buffer) at 25°C gave the rate constants for inactivation of *E. coli*, which decreased with the increase of pH (Cho *et al.* 2006). The value of rate constant for *E. coli* inactivation by Fe(VI) was evaluated as 0.625 $(mg/L \times min)^{-1}$ at pH 7.2 and 25 °C (Cho *et al.* 2006). Other microbial pathogens susceptible to Fe(VI) include *Streptococcus bovis, Canadida albican Staphylococcus aureus, Streptococci faecalis, Bacillus cereus, Bacillus subtils, Shigella flexneri,* and *Salmonella typhimurium* (Sharma, 2007a).

Inactivation of viruses by Fe(VI) could also been accomplished (Jiang *et al.* 2006; Sharma, 2007a). Fe(VI) inactivated virus *f2* and virus *Q*β. The inactivation rate decreased with increasing pH, similarly to the inactivation of *E. coli.* A recent study on the inactivation of bacteriophage MS2 by Fe(VI) showed the damage to the capsid protein and to the virus genome (Hu *et al.* 2012).

Fe(VI) had no reactivity toward bromide ion (Sharma, 2010a). The Br^- is a common constituent of water sources. Comparatively, chlorine, chlorine dioxide, and ozone react with bromide ion to from potentially carcinogenic brominated products. The potential use of Fe(VI) as a preoxidant before chlorination has also been explored as a means to minimize the formation of disinfection byproducts (DBPs) (Gan *et al.* 2015; Yang *et al.* 2013). A study on byproduct formation from natural organic matter showed the decrease in the levels of trihalomethanes (THMs), chloral hydrate (CH), haloacetonitriles (HANs), and trichloronitromethane (TCNM) when Fe(VI) was used a preoxidation step, followed by chlorination (Gan *et al.* 2015). Decreases in DBPs were related to the ferrate dose.

17.4.2 Oxidation

The kinetics of Fe(VI) reactions with a broad range of contaminants with different molecular structures have been studied in order to assess their removal by ferrates (Lee *et al.* 2014; Lee *et al.* 2009; Lee & von Gunten, 2010; Sharma, 2011, 2013; Zimmermann *et al.* 2012). Early studies were on sulfur- and nitrogen-containing compounds (e.g. sulfide, bisulfite, cyanides, ammonia, azide, thiols, amines, amino acids). The kinetic measurements have been performed mostly in the neutral to basic pH range. The rate law were first-order with respect of each reactant for almost all contaminants, and the Fe(VI) reaction second-order rate constants k_2 were determined within the $10^{-2} - 10^5$ $M^{-1}s^{-1}$ range.

Recently, the focus of the kinetics studies turned to emerging contaminants such as endocrine disrupting chemicals (EDCs), pharmaceuticals and personal care products (PPCPs) (e.g. alkylphenols, sulfonamides, β-lactams, β-blockers, and benzophenone-3), and cyanotoxins (e.g. microcystin-LR) (Sharma *et al.* 2015, 2016). The reactivity of Fe(VI) with environmentally relevant compounds such as carbamazepine, enrofloxacin, ciprofloxacin, diclofenac, 17 α-ethinylestradiol, and sulfamethoxazole, as a function of pH is shown in Figure 17.1 (Lee *et al.* 2009). These compounds have electron-rich moieties (ERM) like olefin-, amine-, and aniline and have therefore significant reactivity with Fe(VI) (Figure 17.1). In the alkaline medium, the values of rate constants decrease with increase in pH. Consequently, the calculated half-lives increase with the increase in pH (see Figure 17.1). The pH dependence of rate constants and half-lives is related to higher reactivity of protonated species of Fe(VI) ($HFeO_4^-$) than that of de-protonated species (FeO_4^{2-}). The fraction of $HFeO_4^-$ decreases with increase in pH

($HFeO_4^- \rightleftharpoons H^+ + FeO_4^{2-}$, $pK_{a3} = 7.23$ (Sharma *et al.* 2001)) and the oxidation of compounds slowed down with increasing pH.

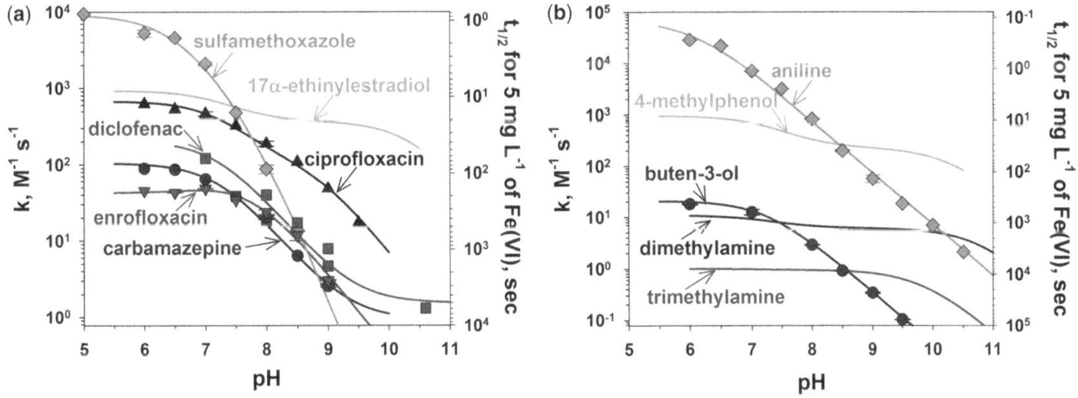

Figure 17.1 Second-order rate constants and half-lives ($t_{1/2}$) for the reactions of Fe(VI) with (a) selected micropollutants and (b) organic model compounds as a function of pH (5 – 11) and at $T = 23 \pm 2$°C. The symbols represent the measured data, and the lines represent the model fits from the present study. The half-lives are calculated for a Fe(VI) concentration of 5 mg Fe L^{-1} (90 μM) (adapted from (Lee *et al.* 2009) with the permission of the American Chemical Society).

The second-order rate constants for the reactions of Fe(VI) with various emerging contaminants at pH 7.0 are summarized in Table 17.3. The values of k_2 for oxidation of EDCs at pH 7.0 were determined in the range of 6.5×10^2–7.9×10^3 $M^{-1}s^{-1}$ with corresponding half-lives varying from 1.7 s – 21.2 s at Fe(VI) dosage of 10 mg/L (Table 17.3). In the oxidation of PPCPs by Fe(VI), a large variation in the values of of k_2 was found ($1.8 \times 10^1 - 1.5 \times 10^3$ $M^{-1}s^{-1}$ at pH 7.0). The calculated half-lives for most of the PPCPs were in the order of seconds (Table 17.3). The data in Table 17.3 indicates that a large number of compounds can be oxidized efficiently by Fe(VI). The second-order rate constants for oxidation of micropollutants by O_3 are also reported in Table 17.3, which are 3–4 orders of magnitude higher than those of Fe(VI). Most of the contaminants could be removed by O_3 in less than a second (Table 17.3). Elimination of emerging contaminants by O_3 is therefore more efficient than Fe(VI). It should be pointed out that the stability of an oxidant in the treated water also plays a significant role in determining the overall removal efficiency. Fe(VI) was more stable (more than 30 minutes) than O_3 (less than 5 minutes) in a secondary effluent from a watewater treatment plant (Lee *et al.* 2009).

Degradation of EDCs and PPCPs in secondary effluents from two wastewater treatment plants (WWTPs) by Fe(VI) was conducted by Yang *et al.* (2012). Thirty-one selected EDCs and PPCPs were detected in the effluents of the two WWTPs at concentration levels in the range from 0.2 ± 0.1 ng L^{-1} to 1156 ± 182 ng L^{-1}. Fe(VI) could oxidize most of the target micropollutants (Yang *et al.* 2012). Basically, Fe(VI) easily oxidized electron-rich organic moieties of the micropollutants such as amine-, aniline-, olefin-, and phenolic- moieties. The removal yield of the detected EDCs and PPCPs increased with the Fe(VI) dose. The results suggest the effectiveness of Fe(VI)-based treatment technology for a wide range of EDCs and PPCPs (Yang *et al.* 2012).

The oxidation kinetics of microcystin-LR (MC-LR), a cyanotoxin found in water sources, by Fe(VI) ($k_2 = 1.3 \pm 0.1 \times 10^2$ M^{-1} s^{-1} at pH 7.5 to 8.1 ± 0.08 M^{-1} s^{-1} at pH 10.0) revealed a rapid degradation of

Table 17.3 Second-order rate constants for selected EDCs and PPCPs in reaction with Fe(VI) (pH 7.0 and 25°C).

Compound	k_2 ($M^{-1}s^{-1}$) ($t_{1/2}$ (s)*) Fe(VI)	Reference	k_2 ($M^{-1}s^{-1}$) ($t_{1/2}$ (s)**) O_3	Reference
EDCs				
Bisphenol A (BPA)	6.5×10^2 (21.2 s)	(Li *et al.* 2008)	2.7×10^6 (0.01 s)	(Lee *et al.* 2009)
Tetrabromobisphenol A	7.9×10^3 (1.7 s)	(Yang *et al.* 2014)	–	–
17α-ethinylestradiol (EE2)	8.1×10^2 (17.0)	(Li *et al.* 2008)	1.6×10^6 (0.02 s)	(Lee *et al.* 2009)
Estrone (E1)	1.0×10^3 (13.7 s)	(Li *et al.* 2008)	1.0×10^6 (0.03 s)	(Jiang *et al.* 2012)
β-estradiol (E2)	1.1×10^3 (12.6 s)	(Li *et al.* 2008)	1.7×10^6 (0.02s)	(Lee *et al.* 2009)
Estriol (E3)	1.2×10^3 (10.9 s)	(Li *et al.* 2008)	1.0×10^6 (0.03 s)	(Jiang *et al.* 2012)
PPCPs				
Sulfisoxazole	1.5×10^3 (9.2 s)	(Sharma *et al.* 2006)	–	–
Sulfamethazine	1.0×10^3 (13.2 s)	(Sharma *et al.* 2006)	–	–
Sulfamethizole	4.1×10^2 (33.9 s)	(Sharma *et al.* 2006)	$2 - 3 \times 10^5$ (0.14 s)	(Garoma *et al.* 2010)
Sulfadimethoxine	0.8×0^2 (175 s)	(Sharma *et al.* 2006)	–	–
Sulfamethoxazole	1.3×10^3 (10.4 s)	(Sharma *et al.* 2006)	2.5×10^6 (0.01 s)	(Lee *et al.* 2009)
Triclosan	1.1×10^3 (12.3 s)	(Lee *et al.* 2009)	3.8×10^7 (0.01s)	(Lee *et al.* 2009)
Carbamazepine	6.7×10^1 (202 s)	(Lee *et al.* 2009)	3.0×10^5 (0.18 s)	(Lee *et al.* 2009)
Ciprofloxacin	4.7×10^2 (28.8 s)	(Lee *et al.* 2009)	1.9×10^4 (2.8 s)	(Lee *et al.* 2009)
Enrofloxacin	4.6×10^1 (294 s	(Lee *et al.* 2009)	1.5×10^5 (0.36 s)	(Lee *et al.* 2009)
Amoxicillin	2.8×10^3 (4.8 s)	(Sharma *et al.* 2013)	6.0×10^6 (0.005 s)	(Andreozzi *et al.* 2005)
Ampicillin	1.1×10^3 (12.3 s)	(Sharma *et al.* 2013)	4.1×10^5 (1.0 s)	(Jung *et al.* 2012)
Penicillin G	1.1×10^2 (123 s)	(Karlesa *et al.* 2014)	4.8×10^3 (1.2 s)	(Dodd *et al.* 2006)
Cephalexin	6.9×10^2 (19.6 s)	(Karlesa *et al.* 2014)	8.7×10^4 (0.07 s)	(Dodd *et al.* 2006)
Diclofenac	1.3×10^2 (104 s)	(Lee *et al.* 2009)	1.0×10^6 (0.03 s)	(Lee *et al.* 2009)
Tradamol	1.4×10^1 (16 min)	(Zimmermann *et al.* 2012)	2.2×10^3 (15.0 s)	(Zimmermann *et al.* 2012)
Trimethoprim	4.0×10^1 (338 s)	(Anquandah *et al.* 2011)	2.7×10^5 (0.12 s)	(Dodd *et al.* 2006)
Atenolol	1.3×10^3 (10.4 s)	(Lee & von Gunten, 2010)	1.7×10^3 (11.6 s)	(Benner *et al.* 2008)
Propranolol	1.8×10^1 (12 min)	(Anquandah *et al.* 2013)	1.0×10^5 (0.32 s)	(Benner *et al.* 2008)
Benzophenone-3	2.2×10^2 (62.6 s)	(Yang & Ying, 2013)	–	–

*calculated for 10 mg/L K_2FeO_4 dose under pseudo-order conditions with Fe(VI) in excess. **half-life at $[O_3] = 1.0$ mg L^{-1}

MC-LR (Jiang *et al.* 2014). Analysis of degradation products by liquid chromatography-mass spectrometry/mass spectrometry (LC-MS/MS) technique suggested that the oxidation products (OPs) resulted primarily from the hydroxylation of the double bond of the methyldehydroalanine (Mdha) aminoacid residue, diene, and aromatic functional groups. A fragmentation of the cyclic MC-LR structure was also seen during the Fe(VI) treatment. Significantly, the toxicity test performed using protein phosphatase (PP1) activity, demonstrated that the degradation byproducts of MC-LR were not biologically toxic. Moreover, Fe(VI) could also degrade MC-LR in water containing carbonate ions and fulvic acid (FA) and in lake water samples. However, higher amount of Fe(VI) was required to completely degrade MC-LR in lake water in comparison with deionized water (Jiang *et al.* 2014).

17.4.3 Coagulation

The ferric oxide generated from Fe(VI) acts as a powerful coagulant that is appropriate for the removal of humic acid, radionuclides, metals, and non-metals (Horst *et al.* 2013; Joshi *et al.* 2008; Potts & Churchwell, 1994; Prucek *et al.* 2015; Qu *et al.* 2003). The pre-oxidation step aids in the removal of contaminants. Fe(VI) can destruct the organic coating on the particles present in the water matrix, thus enhances the coagulation. For example, the decrease in the level of fulvic acid was larger for the combination of coagulant with Fe(VI) than for coagulant alone (Qu *et al.* 2003). Enhanced coagulation of algae was observed when water sample was pretreated with Fe(VI) (Liu & Liang, 2008).

Studies have shown that Fe(VI) can remove a wide range of toxic metals and non-metals in a laboratory synthetic water (Bartzatt *et al.* 1992). During the removal process, turbidity was also reduced. Removal of these elements was accomplished by co-precipitation process in the ferric oxide gel formed by the action of Fe(VI) ion. The mechanistic studies on removing arsenite, arsenate, copper(II), cadmium(II), nickel(II), and metals in metal-cyanide complexes showed that the main iron species responsible for the coagulation/co-precipitation were nanocrystalline Fe(III) oxides/hydroxides, produced via the reduction of Fe(VI) (Filip *et al.* 2011; Jain *et al.* 2009; Prucek *et al.* 2013; Prucek *et al.* 2015). In the case of arsenite, Fe(VI) in combination with Al(III) and Fe(III) ions could completely remove arsenic from water (Jain *et al.* 2009). However, the matrix components like phosphate, nitrate, silicate, and natural organic matter (NOM) increased the amounts of Fe(VI) required for complete removal of arsenic from the de-ionized water.

17.5 CONCLUSIONS AND FUTURE OUTLOOK

Magnetic ZVI and iron(III) oxide nanoparticles can be applied sustainably in removing a wide range of inorganic and organic contaminants from water. Importantly, magnetic separation of such nanoparticles using a low-gradient magnetic field makes it cost-effective compared to membrane filtration. It should be pointed out that aggregation of nanoparticles may reduce the removal capacity, aspect which needs to be further examined in order to recover and reuse magnetic nanoparticles. In the use of nZVI, there are technical issues including mixing and injection of the suspensions. The treatment of contaminants using nZVI in water sources is rather limited, as several difficulties are still to overcome. The surface chemistry of both nZVI and Fe(III) oxides is complex. But, there is a large scope for improving surface properties of iron nanoparticles, which would enhance practical applicability to environmental treatment.

Ferrate ion is highly promising environmentally benign compound that has shown multiple modes of action such as disinfection, high oxidation power, and coagulation effects. In the past few years, the possibility of its *in situ* application in treating water contaminated with many emerging pollutants with superior efficiency to conventional processes has been proven. However, there are still difficulties associated with ferrate solution stability control and process design for large-scale applications. Standard unit operations for treatment (e.g., rapid mix, flocculation, settling, filtration, adsorption) need to be addressed in a large scale application. Numerous examples in the water treatment in the laboratory setup have shown the main advantages of ferrate technology including process performance on oxidation/ disinfection and coagulation for a wide range of pollutants in a single-step treatment. Future work must evaluate ferrate dosages under practical conditions required in the treatment of water and wastewater from different sources.

17.6 ACKNOWLEDGMENT

V.K. Sharma acknowledges the support of the United States National Science Foundation for ferrate research (CBET-1439314). R. Zboril acknowledges the financial support from the Ministry of Education, Youth and Sports of the Czech Republic (LO1305), and by Technology Agency of the Czech Republic "Competence Centres" (project TE01020218).

17.7 REFERENCES

Andreozzi R., Canterino M., Marotta R. and Paxeus N. (2005). Antibiotic removal from wastewaters: The ozonation of amoxicillin. *Journal of Hazardours Materials*, **122**, 243–250.

Anquandah G. A. K., Sharma V. K., Knight D. A., Batchu S. R. and Gardinali P. R. (2011). Oxidation of Trimethoprim by Ferrate(VI): Kinetics, Products, and Antibacterial Activity. *Environmental Science & Technology*, **45**, 10575–10581.

Anquandah G. A. K., Sharma V. K., Panditi V. R., Gardinali P. R., Kim H. and Oturan M. A. (2013). Ferrate(VI) oxidation of propranolol: kinetics and products. *Chemosphere*, **91**, 105–109.

Bae S. and Hanna K. (2015). Reactivity of nanoscale zero-valent iron in unbuffered systems: effect of pH and Fe(II) dissolution. *Environmental Science Technology*, **49**, 1036–1045.

Bae S. and Lee W. (2010). Inhibition of nZVI reactivity by magnetite during the reductive degradation of 1,1,1-TCA in nZVI/magnetite suspension. *Application Catal. B Environmental*, **96**, 10–17.

Baikousi M., Georgiou Y., Daikopoulos C., Bourlinos A. B., Filip J., Zboril R., Deligiannakis Y. and Karakassides M. A. (2015). Synthesis and characterization of robust zero valent iron/mesoporous carbon composites and their applications in arsenic removal. *Carbon*, **93**, 636–647.

Bartzatt R., Cano M., Johnson L. and Nagel D. (1992). Removal of toxic metals and nonmetals from contaminated water. *Journal of Toxicology and Environmental Health*, **35**, 205–210.

Bauer I. and Knölker H. (2015). Iron catalysis in organic synthesis. *Chemical Reviews*, **115**, 3170–3387.

Benner J., Salhi E., Ternes T. and von Gunten U. (2008). Ozonation of reverse osmosis concentrate: Kinetics and efficiency of beta blocker oxidation. *Water Research*, **42**, 3003–3012.

Carr J. D. (2008). Kinetics and product identification of oxidation by ferrate(VI) of water and aqueous nitrogen containing solutes. *ACS Symposium Series*, **985**(Ferrates), 189–196.

Chen D. and Huang S. (2004). Fast separation of bromelain by polyacrylic acid-bound iron oxide magnetic nanoparticles. *Process Biochem*, **39**, 2207–2211.

Chen J., Al-Abed S. R., Ryan J. A. and Li Z. (2001). Effects of pH on dechlorination of trichloroethylene by zero-valent iron. *Journal of Hazardous Materials*, **83**, 243–254.

Cho M., Lee Y., Choi W., Chung H. and Yoon J. (2006). Study on Fe(VI) species as a disinfectant: Quantitative evaluation and modeling for inactivating *Escherichia coli*. *Water Research*, **40**, 3580–3586.

Crane R. A. and Scott T. B. (2012). Nanoscale zero-valent iron: Future prospects for an emerging water treatment technology. *Journal of Hazardous Materials*, 211–212, 112–125.

Delaude L. and Laszlo P. (1996). A novel oxidizing reagent based on potassium ferrate(VI). *The Journal of Organic Chemistry*, **61**, 6360–6370.

Dodd M. C., Buffle M. and Von Gunten U. (2006). Oxidation of antibacterial molecules by aqueous ozone: Moiety-specific reaction kinetics and application to ozone-based wastewater treatment. *Environmental Science & Technology*, **40**, 1969–1977.

Dong J., Zhao Y., Zhao R. and Zhou R. (2010). Effects of pH and particle size on kinetics of nitrobenzene reduction by zero-valent iron. *Journal of Environmental Sciences*, **22**, 1741–1747.

Doong R. and Lai Y. (2005). Dechlorination of tetrachloroethylene by palladized iron in the presence of humic acid. *Water Research*, **39**, 2309–2318.

Farmand M., Jiang D., Wang B., Ghosh S., Ramaker D. E. and Licht S. (2011). Super-iron nanoparticles with facile cathodic charge transfer. *Electrochemistry Communications*, **13**, 909–912.

Filip J., Yngard R. A., Siskova K., Marusak Z., Ettler V., Sajdl P., Sharma V. K. and Zboril R. (2011). Mechanisms and efficiency of the simultaneous removal of metals and cyanides by using ferrate(VI): Crucial roles of nanocrystalline iron(III) oxyhydroxides and metal carbonates. *Chemistry European Journal*, **17**, 10097–10105.

Gan W., Sharma V. K., Zhang X., Yang L. and Yang X. (2015). Investigation of disinfection byproducts formation in ferrate(VI) pre-oxidation of NOM and its model comounds followed by chlorination. *Journal of Hazardours Materials*, **292**, 197–204.

Garoma T., Umamaheshwar S. K. and Mumper A. (2010). Removal of sulfadiazine, sulfamethizole, sulfamethoxazole, and sulfathiazole from aqueous solution by ozonation. *Chemosphere*, **79**, 814–820.

Ge F., Li M., Ye H. and Zhao B. (2012). Effective removal of heavy metal ions Cd^{2+}, Zn^{2+}, Pb^{2+}, Cu^{2+} from aqueous solution by polymer-modified magnetic nanoparticles. *Journal of Hazardours Materials*, **211–212**, 366–372.

Goff H. and Murmann R. K. (1971). Studies on the mechanism of isotopic oxygen exchange and reduction of ferrate(VI) ion (FeO_4^{2-}). *Journal of the American Chemical Society*, **93**, 6058–6065.

Guan X., Sun Y., Qin H., Li J., Lo I. M. C., He D. and Dong H. (2015). The limitations of applying zero-valent iron technology in contaminants sequestration and the corresponding countermeasures: The development in zero-valent iron technology in the last two decades (1994–2014). *Water Research*, **75**, 224–248.

Gutiérrez L., Costo R., Grüttner C., Westphal F., Gehrke N., Heinke D., Fornara A., Pankhurst Q. A., Johansson C., Veintemillas-Verdaguer S. and Morales M. P. (2015). Synthesis methods to prepare single- and multi-core iron oxide nanoparticles for biomedical applications. *Dalton Transactions*, **44**, 2943–2952.

Hao Y., Man C. and Hu Z. (2010). Effective removal of Cu (II) ions from aqueous solution by amino-functionalized magnetic nanoparticles. *Journal of Hazardours Materials*, **184**, 392–399.

Horst C., Sharma V. K., Clayton Baum J. and Sohn M. (2013). Organic matter source discrimination by humic acid characterization: Synchronous scan fluorescence spectroscopy and Ferrate(VI). *Chemosphere*, **90**, 2013–2019.

Hu J., Lo I. M. C. and Chen G. (2004). Removal of Cr(VI) by magnetite nanoparticle. *Water Science and Technology*, **50**, 139–146.

Hu J., Chen G. and Lo I. M. C. (2005). Removal and recovery of Cr(VI) from wastewater by maghemite nanoparticles. *Water Research*, **39**, 4528–4536.

Hu L., Page M. A., Sigstam T., Kohn T., Mariñas B. J. and Strathmann T. J. (2012). Inactivation of bacteriophage MS2 with potassium ferrate(VI). *Environmental Science & Technology*, **46**, 12079–12087.

Jain A., Sharma V. K. and Mbuya M. S. (2009). Removal of arsenite by Fe(VI), Fe(VI)/Fe(III), and Fe(VI)/Al(III) salts: Effect of pH and anions. *Journal of Hazardours Materials*, **169**, 339–344.

Jarošová B., Filip J., Hilscherová K., Tucek J., Šimek Z., Giesy J. P., Zboril R. and Bláha L. (2015). Can zero-valent iron nanoparticles remove waterborne estrogens? *Journal of Environmental Management*, **150**, 387–392.

Jiang J., Pang S., Ma J. and Liu H. (2012). Oxidation of phenolic endocrine disrupting chemicals by potassium permanganate in synthetic and real waters. *Environmental Science & Technology*, **46**, 1774–1781.

Jiang W., Chen L., Batchu S. R., Gardinali P. R., Jasa L., Marsalek B., Zboril R., Dionysiou D. D., O'Shea K. E. and Sharma V. K. (2014). Oxidation of microcystin-LR by ferrate(VI): Kinetics, degradation pathways, and toxicity assessment. *Environmental Science & Technology*, **48**, 12164–12172 and

Jiang J. Q. (2014). Advances in the development and application of ferrate(VI) for water and wastewater treatment. *Journal Chemical Technology Biotechnology*, **89**, 165–177 and

Jiang J. Q. (2015). The role of ferrate(VI) in the remediation of emerging micropollutants: a review. *Desalination and Water Treatment*, **55**, 828–835.

Jiang J. Q., Wang S. and Panagoulopoulos A. (2006). The exploration of potassium ferrate(VI) as a disinfectant/coagulant in water and wastewater treatment. *Chemosphere*, **63**, 212–219.

Joshi U. M., Balasubramanian R. and Sharma V. K. (2008). Potential of ferrate(VI) in enhancing urban runoff water quality. *ACS Symposium Series*, **985**, 466–476.

Jung Y. J., Kim W. G., Yoon Y., Hwang T. and Kang J. (2012). pH effect on ozonation of ampicillin: Kinetic study and toxicity assessment. *Ozone Science and Engineering*, **34**, 156–162.

Karlesa A., De Vera G. A. D., Dodd M. C., Park J., Espino M. P. B. and Lee Y. (2014). Ferrate(VI) oxidation of ß-lactam antibiotics: Reaction kinetics, antibacterial activity changes, and transformation products. *Environmental Science & Technology*, **48**, 10380–10389.

Kim C., Panditi V. R., Gardinali P. R., Varma R. S., Kim H. and Sharma V. K. (2015). Ferrate promoted oxidative cleavage of sulfonamides: kinetics and product formation under acidic conditions. *Chemical Engineering Journal*, **279**, 307–316.

Lee C., Jee Y. K., Won I. L., Nelson K. L., Yoon J. and Sedlak D. L. (2008). Bactericidal effect of zero-valent iron nanoparticles on *Escherichia coli*. *Environmental Science & Technology*, **42**, 4927–4933.

Lee Y. and von Gunten U. (2010). Oxidative transformation of micropollutants during municipal wastewater treatment: Comparison of kinetic aspects of selective (chlorine, chlorine dioxide, ferrate^{VI}, and ozone) and non-selective oxidants (hydroxyl radical). *Water Research*, **44**, 555–566.

Lee Y., Yoon J. and von Gunten U. (2005). Spectrophotometeric determination of ferrate (Fe(VI)) in water by ABTS. *Water Research*, **25**, 1946–1953.

Lee Y., Zimmermann S. G., Kieu A. T. and von Gunten U. (2009). Ferrate (Fe(VI)) application for municipal wastewater treatment: A novel process for simultaneous micropollutant oxidation and phosphate removal. *Environmental Science & Technology*, **43**, 3831–3838.

Lee Y., Kissner Y. and von Gunten U. (2014). Reaction of ferrate(VI) with ABTS and self-decay of ferrate(VI): Kinetics and mechanism. *Environmental Science & Technology*, **48**, 5154–5162.

Li C., Li X. Z., Graham N. and Gao N. Y. (2008). The aqueous degradation of bisphenol A and steroid estrogens by ferrate. *Water Research*, **42**, 109–120.

Licht S., Wang B. and Ghosh S. (1999). Energetic iron(VI) chemistry: The super-iron battery. *Science*, **285**, 1039–1042.

Ling D., Lee N. and Hyeon T. (2015). Chemical synthesis and assembly of uniformly sized iron oxide nanoparticles for medical applications. *Accounts of Chemical Research*, **48**, 1276–1285.

Liu T., Li X. and Waite T. D. (2013). Depassivation of aged Fe (0) by ferrous ions: Implications to contaminant degradation. *Environmental Science & Technology*, **47**, 13712–13720.

Liu T., Li X. and Waite T. D. (2014). Depassivation of aged Fe(0) by divalent cations: Correlation between contaminant degradation and surface complexation constants. *Environmental Science & Technology*, **48**, 14564–14571.

Liu W. and Liang Y. (2008). Use of ferrate(VI) in enhancing the coagulation of algae-bearing water: effect and mechanism study. *ACS Symposium Series*, **985**(Ferrates), 434–445.

Liu Y. and Lowry G. V. (2006). Effect of particle age (Fe(0) content) and solution pH on NZVI reactivity: H2 evolution and TCE dechlorination. *Environmental Science & Technology*, **40**, 6085–6090.

Luo Z., Strouse M., Jiang J. Q. and Sharma V. K. (2011). Methodologies for the analytical determination of ferrate(VI): A Review. *Journal of Environmental Science and Health: Part A Toxic/Hazardous Substances Environmental Engineering*, **46**, 453–460.

Machala L., Tuček J. and Zbořil R. (2011). Polymorphous transformations of nanometric iron(III) oxide: A review. *Chemical Materials*, **23**, 3255–3272.

Mácová Z. and Bouzek K. (2012). The influence of electrolyte composition on electrochemical ferrate(VI) synthesis. Part III: Anodic dissolution kinetics of a white cast iron anode rich in iron carbide. *Journal of Applied Electrochemistry*, **42**, 615–626.

Macova Z., Bouzek K., Hives J., Sharma V. K., Terryn R. J. and Baum J. C. (2009). Research progress in the electrochemical synthesis of ferrate(VI). *Electrochimica Acta*, **54**, 2673–2683.

Marsalek B., Jancula D., Marsalkova E., Mashlan M., Safarova K., Tucek J. and Zboril R. (2012). Multimodal action and selective toxicity of zerovalent iron nanoparticles against cyanobacteria. *Environmental Science & Technology*, **46**, 2316–2323.

Mueller N. C., Braun J., Bruns J., Cerník M., Rissing P., Rickerby D. and Nowack B. (2012). Application of nanoscale zero valent iron (NZVI) for groundwater remediation in Europe. *Environmental Science & Pollution Research*, **19**, 550–558.

Noubactep C. (2015). Metallic iron for environmental remediation: A review of reviews. *Water Research*, **85**, 114–123.

Pereira M. C., Oliveira L. C. A. and Murad E. (2012). Iron oxide catalysts: Fenton and Fenton-like reactions – A review. *Clay Mineral*, **47**, 285–302.

Perfiliev Y. D., Benko E. M., Pankratov D. A., Sharma V. K. and Dedushenko S. D. (2007). Formation of iron(VI) in ozonolysis of iron(III) in alkaline solution. *Inorganic Chimistry Acta*, **360**, 2789–2791.

Poleshchuk O., Kruchkova N., Perfiliev Y. and Dedushenko S. (2010). Estimations of the isomer shifts for tetraoxoferrates. *Journal Physical: Conference Series*, **217**, 012041–012044.

Potts M. E. and Churchwell D. R. (1994). Removal of radionuclides in wastewaters utilizing potassium ferrate(VI). *Water Environmental Research*, **66**, 107–109.

Prucek R., Tuček J., Kolařík J., Filip J., Marušák Z., Sharma V. K. and Zbořil R. (2013). Ferrate(VI)-induced arsenite and arsenate removal by in situ structural incorporation into magnetic iron(III) oxide nanoparticles. *Environmental. Science & Technology*, **47**, 3283–3292.

Prucek R., Tucek J., Kolarik J., Huskova I., Filip J., Varma R. S., Sharma V. K. and Zboril R. (2015). Ferrate(VI)-prompted removal of metals in aqueous media: mechanistic delineation of enhanced efficiency via metal entrenchment in magnetic oxides. *Environmental Science & Technology*, **49**, 2319–2327.

Qu J., Liu H., Liu S.-X. and Lei P. J. (2003). Reduction of fulvic acid in drinking water by ferrate. *Journal of Environmental Engineering*, **129**, 17–24.

Raychoudhury T. and Scheytt T. (2013). Potential of Zerovalent iron nanoparticles for remediation of environmental organic contaminants in water: A review. *Water Science & Technology*, **68**, 1425–1439.

Reddy L. H., Arias J. L., Nicolas J. and Couvreur P. (2012). Magnetic nanoparticles: Design and characterization, toxicity and biocompatibility, pharmaceutical and biomedical applications. *Chemical Reviews*, **112**, 5818–5878.

Sharma V. K. (2007a). Disinfection performance of Fe(VI) in water and wastewater: a review. *Water Science & Technology*, **55**, 225–232.

Sharma V. K. (2007b). Ferrate Studies for Disinfection and Treatment of Drinking Water. In: Nikolaou, A., Rizzo, L., Selcuk, H. (Eds.). *Advances in Control of Disinfection by-Products in Drinking Water Systems.* Nova Science Publishers, New York, USA, pp. 373–380.

Sharma V. K. (2010a). Oxidation of inorganic compounds by Ferrate (VI) and Ferrate(V): One-electron and two-electron transfer steps. *Environmental Science & Technology*, **44**, 5148–5152.

Sharma V. K. (2010b). Oxidation of nitrogen containing pollutants by novel ferrate(VI) technology: A review. *Journal of Environmental Science and Health: Part A Toxic/Hazardous Substances Environmental Engineering*, **45**, 645–667.

Sharma V. K. (2011). Oxidation of inorganic contaminants by ferrates(Fe(VI), Fe(V), and Fe(IV))- Kinetics and Mechanisms - A Review. *Journal Environmental Manage*, **92**, 1051–1073.

Sharma V. K. (2013). Ferrate(VI) and ferrate(V) oxidation of organic compounds: Kinetics and mechanism. *Coordination Chemistry Reviews*, **257**, 495–510.

Sharma V. K., Burnett C. R. and Millero F. J. (2001). Dissociation constants of monoprotic ferrate(VI) ions in NaCl media. *Physical Chemistry Chemical Physics*, **3**, 2059–2062.

Sharma V. K., Kazama F., Jiangyong H. and Ray A. K. (2005). Ferrates as environmentally-friendly oxidants and disinfectants. *Journal of Water and Health*, **3**, 45–58.

Sharma V. K., Mishra S. K. and Nesnas N. (2006). Oxidation of sulfonamide antimicrobials by ferrate(VI) [$Fe^{VI}O_4^{2-}$]. *Environmental Science & Technology*, **40**, 7222–7227.

Sharma V. K., Liu F., Tolan S., Sohn M., Kim H. and Oturan M. A. (2013). Oxidation of β—lactam antibiotics by ferrate(VI). *Chemical Engineering Journal*, **221**, 446–451.

Sharma V. K., Perfiliev Y., Zboril R., Machala L. and Wynter C. (2013). Ferrates (IV, V, and VI): Mössbauer Spectroscopy Characterization. In: Virender K. Sharma, Gostar Klingelhofer, Tetsuaki Nishida. (Ed.). *Mossbauer Spectroscopy: Applications in Chemistry*, Biology, Industry, and Nanotechnology, Hoboken, New Jersey, USA.

Sharma V. K., Zboril R. and Varma R. S. (2015). Ferrates: Greener oxidants with multimodal action in water treatment technologies. *Accounts of Chemical Research*, **48**, 182–191.

Sharma V. K., Chen L. and Zboril R. (2016). A Review on high valent Fe^{VI} (ferrate): A sustainable green oxidant in organc chemistry and transformation of pharmaceuticals. *ACS Sustainable Chemical Eng.*, In press.

Soukupova J., Zboril R., Medrik I., Filip J., Safarova K., Ledl R., Mashlan M., Nosek J. and Cernik M. (2015). Highly concentrated, reactive and stable dispersion of zero-valent iron nanoparticles: Direct surface and site application. *Chemical Engineering Journal*, **262**, 813–822.

Tang S. C. N. and Lo I. M. C. (2013). Magnetic nanoparticles: Essential factors for sustainable environmental applications. *Water Research*, **47**, 2613–2632.

Tratnyek P. G., Scherer M. M., Deng B. and Hu S. (2001). Effects of natural organic matter, anthropogenic surfactants, and model quinones on the reduction of contaminants by zero-valent iron. *Water Research*, **35**, 4435–4443.

Wang Y., Zhou D., Wang Y., Zhu X. and Jin S. (2011). Humic acid and metal ions accelerating the dechlorination of 4-chlorobiphenyl by nanoscale zero-valent iron. *Journal Environmental Science*, **23**, 1286–1292.

Wang C., Klamerth N., Huang R., Elnakar H. and Gamal El-Din M. (2016). Oxidation of oil sands process-affected water by potassium ferrate(VI). *Environmental Science & Technology*, **50**, 4238–4247.

Xie L. and Shang C. (2007). The effects of operational parameters and common anions on the reactivity of zero-valent iron in bromate reduction. *Chemosphere*, **66**, 1652–1659.

Xie Y. and Cwiertny D. M. (2012). Influence of anionic cosolutes and ph on nanoscale zerovalent iron longevity: Time scales and mechanisms of reactivity loss toward 1,1,1,2-tetrachloroethane and Cr(VI). *Environmental Science & Technology*, **46**, 8365–8373.

Yan W., Lien H., Koel B. E. and Zhang W. (2013). Iron nanoparticles for environmental clean-up: Recent developments and future outlook. *Environmental Science: Processes & Impacts*, **15**, 63–77.

Yang B. and Ying G. (2013). Oxidation of benzophenone-3 during water treatment with ferrate(VI). *Water Research*, **47**, 2458–2466.

Yang B., Ying G. G., Zhao J., Liu S., Zhou L. J. and Chen F. (2012). Removal of selected endocrine disrupting chemicals (EDCs) and pharmaceuticals and personal care products (PPCPs) during ferrate(VI) treatment of secondary wastewater effluents. *Water Research*, **46**, 2194–2204.

Yang B., Ying G., Chen Z., Zhao J., Peng F. and Chen X. (2014). Ferrate(VI) oxidation of tetrabromobisphenol A in comparison with bisphenol A. *Water Research*, **62**, 211–219.

Yang W., Kan A. T., Chen W. and Tomson M. B. (2010). PH-dependent effect of zinc on arsenic adsorption to magnetite nanoparticles. *Water Research*, **44**, 5693–5701.

Yang X., Guo W., Zhang X., Chen F., Ye T. and Liu W. (2013). Formation of disinfection by-products after pre-oxidation with chlorine dioxide or ferrate. *Water Research*, **47**, 5856–5864.

Yates B. J., Zboril R. and Sharma V. K. (2014). Engineering aspects of ferrate in water and wastewater treatment – A Review. *Journal of Environmental Science and Health: Part A Toxic/Hazardous Substances Environmental Engineering*, **49**, 1603–1604.

Zimmermann S. G., Schmukat A., Schulz M., Benner J., von Gunten U. and Ternes T. A. (2012). Kinetic and mechanistic investigations of the oxidation of tramadol by ferrate and ozone. *Environmental Science Technology*, **46**, 876–884.

Index

IWA Publishing's authorised EU representative for General Product Safety Regulations is Diane D'Arras, 15 rue Duret, 75116 Paris, France, e-mail: safety@iwap.co.uk.

Printed and bound by CPI Group (UK) Ltd, Croydon, CR0 4YY
12/05/2026
02108863-0015